ADVANCES IN COMPUTATIONAL DYNAMICS OF PARTICLES, MATERIALS AND STRUCTURES

ADVANCES IN COMPUTATIONAL DYNAMICS OF PARTICLES, MATERIALS AND STRUCTURES

JASON HAR and KUMAR K. TAMMA
Department of Mechanical Engineering, University of Minnesota, Minneapolis, USA

WILEY
A John Wiley & Sons, Ltd., Publication

Registered office
John Wiley & Sons Ltd, The Atrium, Southern Gate, Chichester, West Sussex, PO19 8SQ, United Kingdom

For details of our global editorial offices, for customer services and for information about how to apply for permission to reuse the copyright material in this book please see our website at www.wiley.com.

Library of Congress Cataloguing-in-Publication Data

Har, Jason.
Advances in computational dynamics of particles, materials and structures/
Jason Har, Kumar K. Tamma.
p. cm.
Includes bibliographical references and index.
ISBN 978-0-470-74980-7 (hardback)
1. Dynamics. 2. Dynamics – Data processing. I. Tamma, Kumar K. II. Title.
TA352.H365 2012
531′.163 – dc23

2011044208

A catalogue record for this book is available from the British Library.

ISBN: 978-0-470-74980-7

Set in 10/12 Times by Laserwords Private Limited, Chennai, India

To our families, friends and students

CONTENTS

PREFACE xv

ACKNOWLEDGMENTS xxi

ABOUT THE AUTHORS xxiii

1 INTRODUCTION 1

1.1 Overview / 1

1.1.1 The Mechanics Underlying Computational Dynamics / 2

1.1.2 The Numerics Underlying Computational Dynamics in Space and Time / 8

1.2 Applications / 13

2 MATHEMATICAL PRELIMINARIES 15

2.1 Sets and Functions / 15

2.1.1 Sets / 15

2.1.2 Functions / 17

2.2 Vector Spaces / 18

2.2.1 Real Vector Spaces / 18

2.2.2 Linear Dependence and Independence of Vectors / 19

2.2.3 Euclidean n-Space / 20

2.2.4 Inner Product Space / 21

2.2.5 Metric Spaces / 22

2.2.6 Normed Space / 23

2.3 Matrix Algebra / 24

2.3.1 Determinant of a Coefficient Matrix / 25

2.3.2 Matrix Multiplication / 27

2.4 Vector Differential Calculus / 28
2.4.1 Scalar-Valued Functions of Multivariables / 28
2.4.2 Vector-Valued Functions of Multivariables / 30
2.5 Vector Integral Calculus / 32
2.5.1 Green's Theorem in the Plane / 32
2.5.2 Gauss's Theorem / 33
2.6 Mean Value Theorem / 33
2.6.1 Scalar Function of a Real Variable / 33
2.6.2 Scalar Function of Multivariables / 34
2.6.3 Vector Function of Multivariables / 34
2.7 Function Spaces / 34
2.7.1 Inner Product Space / 35
2.7.2 Normed Space / 35
2.7.3 Metric Space / 36
2.7.4 Lebesgue Space / 36
2.7.5 Banach Space / 36
2.7.6 Sobolev Space / 37
2.7.7 Hilbert Space / 38
2.8 Tensor Analysis / 38
2.8.1 Tensor Algebra / 39
2.8.2 Tensor Differential Calculus / 49
2.8.3 Tensor Integral Calculus / 52
Exercises / 53

PART I N-BODY DYNAMICAL SYSTEMS

3 CLASSICAL MECHANICS **57**

3.1 Newtonian Mechanics / 57
3.1.1 Newton's Laws of Motion / 57
3.1.2 Newton's Equations of Motion / 58
3.2 Lagrangian Mechanics / 60
3.2.1 Constraints / 61
3.2.2 Lagrangian Form of D'Alembert's Principle / 68
3.2.3 Configuration Space / 72
3.2.4 Generalized Coordinates / 72
3.2.5 Tangent Bundle / 74
3.2.6 Lagrange's Equations of Motion / 76
3.2.7 Kinetic Energy in Generalized Coordinates / 84
3.2.8 Lagrange Multiplier Method / 86
3.2.9 Autonomous Lagrangian Systems / 87
3.3 Hamiltonian Mechanics / 91
3.3.1 Phase Space / 91
3.3.2 Canonical Coordinates / 92

3.3.3 Cotangent Bundle / 92
3.3.4 Legendre Transformation / 93
3.3.5 Hamilton's Equations of Motion / 96
3.3.6 Autonomous Hamiltonian Systems / 99
3.3.7 Symplectic Manifold / 100
Exercises / 103

4 PRINCIPLE OF VIRTUAL WORK **108**

4.1 Virtual Work in N-Body Dynamical Systems / 108
4.2 Vector Formalism: Newtonian Mechanics in N-Body Dynamical Systems / 114
4.3 Scalar Formalisms: Lagrangian and Hamiltonian Mechanics in N-Body Dynamical Systems / 116
Exercises / 120

5 HAMILTON'S PRINCIPLE AND HAMILTON'S LAW OF VARYING ACTION **121**

5.1 Introduction / 121
5.2 Variation of the Principal Function / 122
5.3 Calculus of Variations / 125
5.4 Hamilton's Principle / 129
5.5 Hamilton's Law of Varying Action / 133
5.5.1 Newtonian Mechanics / 135
5.5.2 Lagrangian Mechanics / 136
5.5.3 Hamiltonian Mechanics / 137
Exercises / 138

6 PRINCIPLE OF BALANCE OF MECHANICAL ENERGY **141**

6.1 Introduction / 142
6.2 Principle of Balance of Mechanical Energy / 142
6.3 Total Energy Representations and Framework in the Differential Calculus Setting / 144
6.3.1 Principle of Balance of Mechanical Energy: Conservative System / 145
6.3.2 Principle of Balance of Mechanical Energy: Nonconservative System / 149
6.3.3 Newtonian Dynamical System: With/Without Constraints / 151
6.3.4 Lagrangian Dynamical System: Nonconservative/Conservative Systems - Descriptive Scalar Function, the Lagrangian / 152
6.3.5 Hamiltonian Dynamical System: Nonconservative Systems - Descriptive Scalar Function, the Hamiltonian / 154
6.4 Appendix: Total Energy Representations and Framework in the Variational Calculus Setting / 156
6.4.1 Total Energy Representation of the Equation of Motion via the Lagrangian Form of D'Alembert's Principle/Principle of Virtual Work / 156
6.4.2 Total Energy Representation of Equation of Motion via Hamilton's Principle/Hamilton's Law of Varying Action / 158
Exercises / 162

7 EQUIVALENCE OF EQUATIONS **163**

7.1 Equivalence in the Lagrangian Form of D'Alembert's Principle/Principle of Virtual Work / 163
7.2 Equivalence in Hamilton's Principle or Hamilton's Law of Varying Action / 165
7.3 Equivalence in the Principle of Balance of Mechanical Energy / 166
7.4 Equivalence Relations Between Governing Equations / 167
7.5 Conservation Laws / 171
7.6 Noether's Theorem / 171
Exercises / 172

PART II CONTINUOUS-BODY DYNAMICAL SYSTEMS

8 CONTINUUM MECHANICS **175**

8.1 Displacements, Strains and Stresses / 175
8.1.1 Configuration Space / 176
8.1.2 Riemannian Metrics / 177
8.1.3 Infinitesimal Differential Volume / 178
8.1.4 Displacements and Strains / 179
8.1.5 Stresses / 194
8.2 General Principles / 197
8.2.1 Gauss's Theorem / 198
8.2.2 Reynolds Transport Theorem / 198
8.2.3 Principle of Conservation of Mass / 200
8.2.4 Principle of Balance of Linear Momentum / 200
8.2.5 Principle of Balance of Angular Momentum / 202
8.2.6 Principle of Balance of Energy / 203
8.2.7 Principle of Entropy Inequality / 205
8.3 Constitutive Equations in Elasticity / 206
8.3.1 Cauchy Elastic Material / 206
8.3.2 Hyperelastic Material / 207
8.3.3 Hypoelastic Material / 213
8.3.4 Material Frame-Indifference: Objectivity / 214
8.3.5 Objective Stress Rates / 218
8.4 Virtual Work and Variational Principles / 220
8.4.1 Virtual Work and Potential Energy / 220
8.4.2 Principle of Virtual Work / 223
8.4.3 Principle of Virtual Power / 228
8.4.4 Principle of Stationary Potential Energy / 228
8.4.5 Principle of Complementary Virtual Work / 230
8.4.6 Principle of Stationary Complementary Energy / 232
8.4.7 Hu-Washizu Variational Principle / 233

8.4.8 Hellinger-Reissner Variational Principle / 235
8.5 Direct Variational Methods for Two-Point Boundary-Value Problems / 237
8.5.1 Rayleigh-Ritz Method / 237
8.5.2 Bubnov-Galerkin Weighted Residual Method / 244
8.5.3 Modified Bubnov-Galerkin Weighted-Residual Method / 247
8.5.4 Equivalence of the Ritz and the Galerkin Methods / 252
Exercises / 253

9 PRINCIPLE OF VIRTUAL WORK: FINITE ELEMENTS AND SOLID/STRUCTURAL MECHANICS 267

9.1 Introduction / 267
9.1.1 Vector Formalism: Cauchy's Equations of Motion, Principle of Virtual Work, and Finite Element Formulations in Continuous-Body Dynamical Systems / 267
9.2 Finite Element Library / 301
9.2.1 One-Dimensional Continuum: Axial Bar Element / 301
9.2.2 Two-Dimensional Continuum: Triangular Element / 307
9.2.3 Two-Dimensional Continuum: Quadrilateral Element / 313
9.2.4 Three-Dimensional Continuum: Tetrahedral Element / 320
9.2.5 Three-Dimensional Continuum: Hexahedral Element / 325
9.2.6 Structural Member: Euler-Bernoulli Beam Element / 329
9.2.7 Structural Member: Timoshenko Beam Element / 333
9.2.8 Structural Member: Kirchhoff-Love Plate Element / 336
9.2.9 Structural Member: Reissner-Mindlin Plate Element / 339
9.3 Nonlinear Finite Element Formulations / 343
9.3.1 Total Lagrangian Formulation / 344
9.3.2 Updated Lagrangian Formulation / 347
9.4 Scalar Formalisms: Lagrangian and Hamiltonian Mechanics and Finite Element Formulations in Continuous-Body Dynamical Systems / 350
Exercises / 353

10 HAMILTON'S PRINCIPLE AND HAMILTON'S LAW OF VARYING ACTION: FINITE ELEMENTS AND SOLID/STRUCTURAL MECHANICS 364

10.1 Introduction / 364
10.2 Hamilton's Principle and Hamilton's Law of Varying Action in Elastodynamics / 365
10.3 Lagrangian Mechanics Framework and Finite Element Formulations / 370
10.3.1 Lagrangian Density Equations of Motion / 371
10.3.2 Space-Discrete Lagrangian Finite Element Formulation / 374
10.4 Hamiltonian Mechanics Framework and Finite Element Formulations / 400
10.4.1 Hamiltonian Density Equations of Motion / 400
10.4.2 Space-Discrete Hamiltonian Finite Element Formulation / 403
Exercises / 425

11 PRINCIPLE OF BALANCE OF MECHANICAL ENERGY: FINITE ELEMENTS AND SOLID/STRUCTURAL MECHANICS **426**

11.1 Introduction / 427

11.2 Total Energy Representations and Framework in the Differential Calculus Setting and Finite Element Formulations / 429

11.2.1 Principle of Balance of Mechanical Energy/Theorem of Power Expended: Nonconservative System / 430

11.2.2 Principle of Balance of Mechanical Energy: Conservative System and Total Energy Density Equations of Motion / 433

11.2.3 Space-Discrete Total Energy Finite Element Formulation / 436

11.3 Lagrangian Mechanics Framework in the Differential Calculus Setting and Finite Element Formulations / 449

11.3.1 Lagrangian Density Equations of Motion / 449

11.3.2 Space-Discrete Lagrangian Finite Element Formulation / 451

11.4 Hamiltonian Mechanics Framework in the Differential Calculus Setting and Finite Element Formulations / 454

11.4.1 Hamiltonian Density Equations of Motion / 454

11.4.2 Space-Discrete Hamiltonian Finite Element Formulation / 456

11.5 Appendix: Total Energy Representations and Framework in the Variational Calculus Setting and Finite Element Formulations / 458

11.5.1 Infinite Dimensional Total Energy Structure / 459

11.5.2 Total Energy Density Representation of the Equation of Motion / 459

11.5.3 Space-Discrete Total Energy Finite Element Formulation / 462

Exercises / 474

12 EQUIVALENCE OF EQUATIONS **475**

12.1 Equivalence in the Principle of Virtual Work in Dynamics / 475

12.2 Equivalence in Hamilton's Principle or Hamilton's Law of Varying Action / 478

12.3 Equivalence in the Principle of Balance of Mechanical Energy / 482

12.4 Equivalence of Strong and Weak Forms for Initial Boundary-Value Problems / 483

12.5 Equivalence of the Semi-Discrete Finite Element Equations of Motion / 487

12.6 Equivalence of Finite Element Formulations / 488

12.7 Conservation Laws / 490

Exercises / 490

PART III THE TIME DIMENSION

13 TIME DISCRETIZATION OF EQUATIONS OF MOTION: OVERVIEW AND CONVENTIONAL PRACTICES **495**

13.1 Introduction / 495

13.2 Single-Step Methods for First-Order Ordinary Differential Equations / 500

13.2.1 Forward Euler Method / 501

13.2.2 Backward Euler Method / 501
13.2.3 Generalized Trapezoidal Method / 502
13.2.4 Midpoint Method / 502
13.2.5 Runge-Kutta Method / 503
13.2.6 Generalized Trapezoidal Family for a Vector-Valued Function / 504
13.3 Linear Multistep Methods / 505
13.3.1 Central Difference Method / 505
13.3.2 Linear Multistep Methods for First-Order Ordinary Differential Equations / 506
13.3.3 Linear Multistep Methods for Second-Order Ordinary Differential Equations / 507
13.4 Second-Order Systems and Single Step and/or Equivalent LMS Methods: Brief Overview of Classical Methods from Historical Perspectives and Chronological Developments / 507
13.4.1 The Houbolt Method [1950] / 508
13.4.2 Classical Midpoint Rule Method / 509
13.4.3 The Newmark Family of Algorithms [1959] / 511
13.4.4 The Wilson-θ Method [1968] / 513
13.4.5 The Park Method [1975] / 514
13.4.6 The Hilber-Hughes-Taylor-α Method [1977] / 515
13.4.7 The SSpj Family of Algorithms [1977] / 516
13.4.8 The Wood-Bosak-Zienkiewicz Method [1981] / 520
13.4.9 Velocity Based Scheme [1988] / 521
13.4.10 The Three Parameters Optimal Schemes (χ-Schemes) [1988] (and the identical Generalized-α Method) / 522
13.4.11 Optimal U0-V0 Algorithm: The Optimal Algorithm With Controllable Numerical Dissipation Within the Class of LMS Methods in the Sense of the Single-Field Form [2004] / 525
13.5 Symplectic-Momentum Conservation and Variational Time Integrators / 527
13.5.1 Discrete Euler-Lagrange Equations / 527
13.5.2 Discrete Legendre Transformation / 530
13.5.3 Symplecticness of Variational Time Integrators / 532
13.5.4 Discrete Noether's Theorem / 534
13.6 Energy-Momentum Conservation and Time Integration Algorithms / 536
13.6.1 Energy-Momentum Conserving Scheme for Discrete Systems / 536
13.6.2 Energy-Momentum Conserving Algorithms for N-Body Systems / 537
13.6.3 Energy-Momentum Conserving Algorithms for Continuum Elastodynamics / 543
13.6.4 Extension of Energy-Momentum Conserving Algorithms for a General Hyperelastic Material Model / 550

14 TIME DISCRETIZATION OF EQUATIONS OF MOTION: RECENT ADVANCES 553
14.1 Introduction / 553

14.2 Time Discretization and the Total Energy Framework: Linear Dynamic Algorithms and Designs - Generalized Single Step Single Solve [GSSSS] Unified Framework Encompassing LMS Methods / 555

14.2.1 GSSSS Framework Encompassing LMS Methods: Nonconservative Systems and Linear Dynamics Algorithms and Designs in Two-field/ Single-field Form Via the Semi-Discretized Equations of Motion / 558

14.2.2 GSSSS Framework Encompassing LMS Methods: Conservative Systems and Linear Dynamics Algorithms and Designs in Two-field/Single-field Form Via the Discrete Total Energy Framework / 566

14.2.3 Total Energy Framework and Semi-Discretized Equation of Motion for Conservative Dynamic Systems / 569

14.3 Time Discretization and the Total Energy Framework: Nonlinear Dynamics Algorithms and Designs - Generalized Single Step Single Solve [GSSSS] Framework Encompassing LMS Methods / 578

14.3.1 Classical/Normalized Time Weighted Residual Methodology / 579

14.3.2 Time Discretization and Total Energy Framework: Brief Highlights of the Two-/Single-Field Form of LMS Methods, and Conserving Algorithms and Designs - General Hyperelastic Material Models / 582

14.3.3 Numerical Implementation Aspects of Classical Framework in Single-field Form: Internal Force Based Numerically Non-Dissipative and Dissipative Algorithms and Designs / 606

14.3.4 Numerical Implementation Aspects of Symplectic-Momentum Framework in Single-field Form: Symplectic-Momentum Based Numerically Non-Dissipative and Dissipative Algorithms and Designs / 614

14.3.5 Numerical Implementation Aspects of Energy-Momentum Framework in Single-field Form: Energy-Momentum Based Numerically Non-Dissipative and Dissipative Algorithms and Designs / 622

14.4 Time Discretization and Total Energy Framework: N-Body Systems / 632

14.4.1 GSSSS Framework Encompassing LMS Methods: Conservative Systems and Linear Dynamics Algorithms and Designs in Two-field form and Single-field Form Via the Total Energy Framework / 632

14.4.2 GSSSS Framework Encompassing LMS Methods: Conservative Systems and Nonlinear Dynamics Algorithms and Designs in Two-field Form and Single-field Form Via the Total Energy Framework / 635

14.5 Time Discretization and Total Energy Framework: Nonconservative/Conservative Mechanical Systems with Holonomic-Scleronomic Constraints / 649

14.5.1 General Formulations / 650

Exercises / 662

REFERENCES **669**

INDEX **681**

PREFACE

This book treats the subject matter dealing with *advances in computational dynamics* from a unified viewpoint and approach, and thereby provides a rigorous treatment and a unique blend of the various underlying mechanics and the numerical aspects to effectively foster modeling and simulation on modern computing environments. In the broader sense, the subject matter under the umbrella of *computational dynamics* covers the necessary fundamentals associated with particle dynamics; dynamics of materials, structures, deformable continuum media and related applications to include structural/elasto-dynamics; multi-body dynamics dealing with rigid and flexible bodies; contact-impact dynamics; and so on. In particular, this book covers the classical (or traditional) practices to more contemporary aspects which include recent advances dealing with the mathematical, physical, geometrical, as well as computational aspects associated with modeling and simulation as related to numerical discretization in space and/or time. It is designed for engineers, mathematicians, physicists, and students/researchers in allied fields who wish to understand the subject matter with rigor and in a contemporary setting. We intend this book to serve as a multi-semester course at the graduate-level and/or for upper-level undergraduate students (on selected topics), advanced researchers and scientists, and engineers who are keenly interested in the fundamental aspects critical to the computational aspects of the dynamics of particles and rigid bodies, and the computational aspects dealing with structural/elasto-dynamics, continuum mechanics, the finite element method, and time integration schemes for both N-body and continuous-body dynamical systems. This book explores both classical practices as well as new avenues with differing and alternative viewpoints which additionally provide improved physical insight and new computational perspectives. With these considerations in mind, we closely embrace the underlying theme and excerpt due to Gauss as highlighted in Degas (1955): *"It is always interesting and instructive to regard the laws of nature from a new and advantageous point of view, so as to solve this or that problem more simply, or to obtain a more precise presentation"*.

We start with the premise, that in the beginning there were these landmark contributions due to Aristotle (384 BC-322 BC), Archimedes (287 BC-212 BC), Galileo (1564-1642), Kepler (1571-1630), Huygens (1629-1695), Decartes (1596-1650), and the like, and, *then there was this thing of beauty, namely, that due to Newton (1643-1727) - the famous Newton's laws of motion*. And now there are all these various fields or branches of mechanics and physics with various underlying theoretical pinning's dealing with particle dynamics; dynamics of materials, structures, and deformable continuum media and related applications to include structural/elasto-dynamics; multi-body dynamics dealing with rigid and flexible bodies; contact-impact dynamics; and the like. It is worth noting that the fundamental principles of dynamics have also been abstracted to various other fields and applications to include the theory of relativity, quantum mechanics, economics, robotics, biology, medical and allied applications such as biomechanics, virtual surgery physics based simulations for training medical residents/physicians, and the like.

Keeping the above considerations in perspective, we present an overview of not only the classical developments and the current state-of-the-art, but we also provide new and recent advances dealing with computational aspects related to the dynamics of particles, materials, and structures. In this book, we first highlight the *big picture* with consistent developments from differing viewpoints not only to derive the governing equations of motion for N-body or continuous-body dynamical systems for a wide class of engineering applications, but also to subsequently enable the discretization in space/time for numerical computations. In particular, we present our viewpoint of the evolution of a variety of numerical developments in the fields encompassing *computational dynamics* ranging from classical practices to more new and recent advances. Under the umbrella of *computational dynamics*, at the outset it should be clearly noted that this book is intended to provide a sound and fundamental background on the various theoretical and computational aspects; and we classify the evolution of the various related developments via two principal themes, namely, the *mechanics underlying computational dynamics* and the associated *numerics underlying computational dynamics*. Only in selective instances, certain theoretical bases and related considerations dealing with various aspects of classical mechanics have been carefully excerpted and interpreted from several renowned books such as Mach (1907), Pars (1965), Greenwood (1977), Rosenberg (1977), Arnold (1989), Goldstein (2002), and the like which have been some of the primary sources.

Mechanics Underlying Computational Dynamics: The terminology, namely, the *mechanics underlying computational dynamics*, implies the approach and starting point that is employed as the fundamental axiom via which one can independently derive the governing equations, and the associated strong and/or weak forms that can be readily employed for the associated numerical discretizations. Starting with the premise that in the beginning the well known Newton's law of motion for the dynamics of N-body systems is given, which reflects the statement of the principle of balance of linear momentum, subsequently, using this as a landmark, firstly, the principal relations to various other distinctly different fundamental principles which are of primary interest here are established. This is worth noting. Likewise, for the dynamics of materials, structures, and deformable continuum medium and related applications, under the premise that the governing equations such as the well known Cauchy equations of motion which also reflect the statement of the principle of balance of linear momentum are given, analogous relations as in N-body systems are also established. After first establishing the principal relations to the various fundamental principles, any of the respective principles thenceforth can serve as the standalone starting point for the subsequent theoretical and computational developments for modeling and simulation. In this book, we confine attention primarily to three distinctly different fundamental principles which comprise the *pyramid of computational dynamics*. Of particular interest are the three distinctly different fundamental principles represented as faces or planes which comprise the *pyramid of computational dynamics* (see Figure 1), namely: 1) *the Principle of Virtual work*, 2) *Hamilton's Principle, or alternatively, Hamilton's Law of Varying Action (which is not a variational principle)*, and 3) *the Principle of Balance of Mechanical Energy*. Each fundamental principle is particularly selected such that it can independently enable the theoretical and computational developments associated with and leading to the strong and/or weak forms and the corresponding numerical discretizations in space/time for applications to computational dynamics. That is, each of the above fundamental principles does not necessarily rely upon the others. However, the pros and cons, limitations of each fundamental principle, and the conditions under which equivalences of the respective formulations amongst the three fundamental principles can be drawn need to be carefully understood to avoid misinterpretation. By no means, we claim that these are the only representations for the classification as various other explanations are

Figure 1. Pyramid of computational dynamics

also plausible and could be included. Consequently, the present pyramidal structure classification could entertain other faces or planes. However, we confine attention only to the present three fundamental principles with the clear understanding of the restrictions inherent within each fundamental principle.

Numerics Underlying Computational Dynamics: Subsequently, we also describe the *numerics underlying computational dynamics* which deals with both classical (or traditional) practices and new avenues for conducting space/time discretizations to find numerical solutions useful for modeling and simulation. The terminology, *numerics underlying computational dynamics*, refers to the approach and the starting point that is employed by which we address the numerical treatments as related to spatial discretizations in the space domain and temporal discretizations in the time domain. It deals with the numerical aspects and discretization approaches in space and/or time which are necessary ingredients for modeling and simulation. Stemming independently from each of the respective fundamental principles comprising the pyramid of computational dynamics, we describe the various computational developments for the dynamics of N-body systems, and the dynamics of materials, structures, deformable continuum media and related applications. Both classical practices that are customarily followed, as well as other alternative avenues which provide new and different perspectives and/or improved physical insight for the modeling and simulation of computational dynamics applications are described. A unified viewpoint is the end result regardless of which fundamental principle serves as the starting point; and the restrictions and/or limitations associated with each of the respective fundamental principles need to be carefully understood Tamma (2012); Tamma et al. (2011) (DOI10.1007/s11831-011 9060-y).

Outline of this Book: The outline of this book is as follows. Chapter 1 presents an introduction to and an overview of the big picture and our viewpoint of the various theoretical and numerical aspects dealing with *computational dynamics*. Along the themes, namely, the *mechanics underlying computational dynamics* and under the umbrella of the *pyramid of computational dynamics*, and the associated *numerics underlying computational dynamics*, in this book, we focus attention upon the three fundamental principles comprising the pyramid structure classification, namely: 1) *the Principle of Virtual Work*, 2) *Hamilton's Principle, or alternatively, Hamilton's Law of Varying Action (which is not a variational principle)*, and 3) *the Principle of Balance of Mechanical Energy*. Each of the above fundamental principles has a wide range of applicability, and can independently describe the theoretical and numerical developments associated with and leading to the strong and/or weak forms and the corresponding numerical discretizations in space/time for applications to computational dynamics. Chapter 2 provides the basic mathematical background materials necessary for studying classical mechanics, continuum mechanics, finite element theories, and time integration schemes for integrating the equations of motion. Throughout the book, it is very important to have a fundamental grasp of the concepts of sets and functions, and the meaning of the related notations. Vector spaces with numeric entries as well as functions are addressed in this chapter. In the discussion on tensor analysis, we use not only Cartesian tensors but also general tensors, which are crucial for understanding nonlinear continuum mechanics and finite deformation theories for deformable bodies. The book is divided into three parts. Consequently, under the umbrella of the *pyramid of computational dynamics*, we devote a separate chapter in both N-Body Systems (Part 1) and Continuous-Body Systems (Part 2) to each of these respective principles which independently serve as a starting point for conducting the theoretical and numerical developments associated with and leading to the strong and/or weak forms and the corresponding numerical discretizations in space and/or time. An overview of conventional practices, and in addition, recent advances dealing with a wide variety of Time Discretization (Part 3) approaches and related time integration aspects necessary for appropriately integrating the dynamic equations of motion are finally highlighted.

Part 1: N-Body Systems With the above considerations in mind, in Part 1 which deals with N-body Systems, Chapter 3 covers classical mechanics including Newtonian, Lagrangian and Hamiltonian mechanics. In Chapter 4, after first establishing the relation between Newton's second law and the principle of virtual work (which is a restatement of the Lagrangian form of D'Alembert's principle), we directly show the subsequent theoretical and numerical developments starting from this principle. The Lagrangian form of D'Alembert's principle (or equivalently, the principle of virtual work in dynamics) is the key principle leading to analytical mechanics and descriptive scalar function formalism, in contrast to the Newtonian mechanics framework and vector formalism. Alternatively, Chapter 5 describes both Hamilton's principle and Hamilton's law of varying action for N-body dynamical systems. We draw attention in the book to the fact that Hamilton's law of varying action is equivalent to the integral form of the principle of virtual work. Consequently, it is a descriptive scalar function representation of the principle of virtual work, which naturally contains the weighted residual form in time for N-body dynamical systems. In contrast, Chapter 6 describes the principle of balance of mechanical energy as the starting point, and the corresponding formulations associated with the Total Energy representation of the equation of motion and framework in the differential calculus setting which is valid for holonomic-scleronomic systems with a new, measurable, and built-in descriptive scalar function, namely, the Total Energy (and in addition, the variational calculus setting which is valid for holonomic systems is also highlighted in the Appendix). As a descriptive scalar function analogous to the Lagrangian and the Hamiltonian, the Total Energy defined on the velocity phase space is yet another alternative and it offers good physical insight and computationally attractive features. *There exist various subject areas in mechanics and physics where it is desirable to have a direct measurable descriptive scalar function such as the Total Energy*. The related developments also readily enable the theoretical and numerical formulations for computational dynamics just as those obtained from the other two fundamental principles. Next, Chapter 7 describes equivalence relations between governing equations for N-body dynamical systems subject to holonomic constraints within the three frameworks, namely, the Lagrangian, Hamiltonian and Total Energy frameworks. Noether's Theorem for N-body dynamical systems, and the invariant properties, namely, the conservation of Linear Momentum, Angular Momentum and Total Energy of the descriptive scalar functions, such as the Lagrangian, Hamiltonian and Total Energy are also highlighted.

Part 2: Continuous-Body Systems Part 2 focuses upon Continuous-body Systems, and the continuum mechanics aspects associated with deformations, strains, and stresses in solid/structural applications. In Chapter 8, we start with basic continuum mechanics materials necessary for developing finite element formulations. Chapter 8 describes displacements, strains, and stresses with general tensors. We then discuss five fundamental principles dealing with thermo-mechanical motion which continuous bodies must obey; these include, the principle of conservation of mass, the principle of balance of linear momentum, the principle of balance of angular momentum, the principle of balance of energy, and the principle of entropy inequality. Chapter 8 also includes constitutive equations in elasticity, fundamentals of virtual work and variational principles, and direct variational methods for two-point boundary-value problems such as the Rayleigh-Ritz method, the Bubnov-Galerkin weighted residual method, and the modified Bubnov-Galerkin weighted residual method. As in N-body systems described in Part 1, we next devote a separate chapter dealing with continuous-body systems to each of the three fundamental principles comprising the *pyramid of dynamics* which independently serve as the starting point for developing the related theoretical and numerical formulations. In this regard, Chapter 9 comprehensively deals with the first of the three principles outlined earlier, namely, the principle of virtual work in dynamics; and consequently, describes conventional finite element formulations and vector formalism for continuous-body dynamical systems. We additionally describe a variety of structural members including axial bar, rotating circular bar, Euler-Bernoulli beam, Timoshenko beam, Kirchhoff-Love thin plate, and Reissner-Mindlin plate. The weak forms for continuum and structural members are derived from the weighted residual form. With regards to structural members, we first set up a free-body diagram and count on D'Alembert's principle to obtain the governing equations of motion. Then, we establish the weighted residual statement to derive the weak form by performing integration by parts and imposing natural boundary conditions. Finally, the resulting weak form is spatially discretized by using appropriate trial and test functions. In addition to the finite element formulations, we additionally describe a variety of finite elements including axial bar element, plane stress/strain two-dimensional triangular and quadrilateral elements, three-dimensional tetrahedral and hexahedral brick elements, Euler-Bernoulli beam element, Timoshenko beam element, Kirchhoff-Love plate element and Reissner-Mindlin plate element. Lastly, we highlight nonlinear finite element formulations including total and updated Lagrangian formulations. Scalar formalisms with respect to the Lagrangian and the Hamiltonian are briefly referenced. In Chapter 10, in contrast to the traditional practices described in Chapter 9, we present finite element formulations using descriptive scalar functions via Hamilton's Principle or Hamilton's Law of Varying Action as the starting point which also yield the same and/or equivalent finite element representations from another viewpoint. As yet another alternative with several computationally attractive features and good physical insight, in Chapter 11 we describe other related developments via the Total Energy representations and framework for developing the finite element formulations using the various descriptive scalar functions. This is via the theorem of power expended, and consequently the principle of balance of mechanical energy with differential calculus setting valid for holonomic-scleronomic systems. In the Appendix, we briefly also highlight the variational calculus setting in the context of the Total Energy representations and framework which is valid for holonomic systems. Chapter 12 discusses the equivalences between the strong forms and also between the weak forms which are respectively obtained via each of the three distinctly different fundamental principles, namely, the principle of virtual work, Hamilton's principle or equivalently, Hamilton's law of varying action, and the theorem of power expended and consequently the principle of balance of mechanical energy. We also present a brief discussion on Noether's Theorem for continuous-body dynamical systems, wherein after the spatial discretization they lead to finite dimensional systems analogous to the discussion highlighted in Chapter 7 for N-body systems.

Part 3: Time Discretization Finally, Part 3 is devoted to the Time Dimension and the numerical aspects that are necessary for properly dealing with the time integration of the equations of motion in both single-field and two-field forms of representation Tamma (2012); Tamma et al. (2011 (DOI 10.1007/s11831-011-9060-y). For the time discretization, an overview of the big picture and specific guidelines for developing algorithms by design that meet targeted objectives are provided and discussed. In Chapter 13 we present the following: (i) We first show starting from the standard representation of the linear semi-discretized equations of motion, the various classical and chronological developments in time integration of linear dynamical systems from historical perspectives that appear in the open literature over the past fifty years or so, (ii) Next, we highlight variational integrators stemming from the so-called Discrete Euler-Lagrange representations that inherit features which are symplectic-momentum conserving, and (iii) Following this, we highlight the so-called energy-momentum conserving/dissipating algorithm designs for finite dimensional systems following the original methods of development (classical practices) through enforcing energy constraints. Lastly, in Chapter 14, in contrast to all the previously mentioned classical and/or traditional practices described in Chapter 13, we focus special attention upon and highlight the more recent developments directly emanating from the new Total Energy framework and representations as a starting point (unlike traditional practices) in conjunction with a generalized time weighted residual approach. In particular, we provide new perspectives, a unified viewpoint, and in addition, the underlying theoretical basis on how to properly provide appropriate extensions of the parent linear dynamics algorithm designs to nonlinear dynamics applications for developing practical algorithms by design useful for integrating the equations of motion; and the associated computationally attractive features are that the developments are based upon symplectic-momentum conservation or energy-momentum conservation aspects, respectively. These latter developments via the Total Energy representations and framework, and the generalized time weighted residual approach also cover most of the developments that have been previously derived from various other classical viewpoints as mentioned in (i), (ii), and (iii) previously in Chapter 13. In summary, for both single-field and two-field forms of representation, we first describe linear dynamics algorithms by design for integrating the equations of motion, and we then provide the necessary theoretical basis for proper extensions to nonlinear dynamics algorithms by design. The overall developments are generally applicable to a wide variety of applications encompassing linear and nonlinear structural/elasto-dynamics applications in continuous-body dynamics, N-body systems, and conservative/nonconservative mechanical systems with holonomic-scleronomic constraints such as those encountered in multi-body dynamics applications.

Jason Har and Kumar K. Tamma

ACKNOWLEDGMENTS

Professor Kumar K. Tamma is particularly grateful to Dr. Jason Har for his steadfast commitment to embrace the original ideas and concepts that are put forth in this book, learn, and contribute during the five and a half year period he served as a post-doctoral associate and research assistant professor under Professor Tamma's supervision in the Department of Mechanical Engineering at the University of Minnesota. Special thanks are due to Mr. Masao Shimada, graduate Ph.D research student in the Department of Mechanical Engineering at the University of Minnesota and working under the supervision of Professor Kumar K. Tamma, for his valuable technical comments and contributions; in particular on the time integration aspects.

ABOUT THE AUTHORS

Professor Kumar K. Tamma is a highly recognized researcher and distinguished scholar, and is Professor in the Department of Mechanical Engineering, College of Science and Engineering, at the University of Minnesota, Minneapolis, Minnesota. Professor Tamma has published over 200 research papers in archival journals and book chapters, and over 300 research papers in refereed conference proceedings/abstracts. Professor Tamma's primary areas of research encompass: Computational mechanics with emphasis on multi-scale and multi-physics aspects in space and time, and on design and development of novel numerical methods and computational algorithms by design for the modeling and simulation of time dependent problems and High Performance Computing applications; multi-disciplinary computational fluid-thermal-structural interactions; structural dynamics and large deformation and large strain contact-impact-penetration-damage; multi-body dynamics of rigid and flexible bodies; computational aspects of macroscale/microscale/nanoscale heat transfer; advanced and lightweight composites and multifunctional materials manufacturing processes, and solidification. Professor Tamma serves on the editorial boards for over 15 national and international journals and is the co-editor-in-chief of an online journal. Professor Tamma is the recipient of numerous research awards including the *George Taylor Research Award* for significant and exceptional contributions to research at the University of Minnesota. Professor Tamma is also the recipient of numerous Outstanding Teacher of the Year and other national and university related awards. Professor Tamma has presented several Plenary/Semi-Plenary/Keynote lectures and various invited lectures in national/international conferences, and across various government and industrial agencies, and academic institutions. Professor Tamma is a fellow of various related societies in his field, and is also listed in various *Who's Who* of organizations and professionals.

Dr. Jason Har is a Senior Software Developer with ANSYS, Inc., in Canonsburg, Pennsylvania. Dr. Har worked under the supervision of Professor Kumar K. Tamma at the University of Minnesota in Minneapolis, Minnesota, where Dr. Har learned and embraced the original ideas and concepts being pursued by Professor Tamma that are put forth in this textbook; and contributed to the various developments for a period of five and a half years as a post doctoral associate and as a research assistant professor in the Department of Mechanical Engineering at the University of Minnesota. Dr. Jason Har has extensive industrial experience in finite element technology of structures and structural components, contact-impact, and parallel computations for over 15 years, and worked at the Korea Institute of Aerospace Technology where Dr. Har also served as Managing Research Director. Dr. Har has presented various invited and special lectures at various organizations and national/international conferences.

CHAPTER ONE

INTRODUCTION

The present book encompasses *classical (or traditional practices)* as well as *advances in computational dynamics* for computer modeling and simulation of applications in science and engineering. The highlights of this book are outlined in Chapter 1. The targeted objectives are towards a wide variety of science and engineering problems in particle dynamics; dynamics of materials, structures and deformable continuum media; and related applications which fall under this class of applications. We first introduce in this book the big picture and a unified viewpoint, and the various approaches which follow for the modeling and simulation in the broad field encompassing *computational dynamics*. In the broader sense, in this book the subject matter under the umbrella of *computational dynamics* covers the necessary fundamentals associated with particle dynamics; dynamics of materials and deformable continuum media and related applications to include structural/elasto-dynamics; multi-body dynamics dealing with rigid and flexible bodies; contact-impact dynamics; and so on. We classify the evolution of the various related developments under the umbrella of *computational dynamics* via two principal themes, namely, the *mechanics underlying computational dynamics* and the *numerics underlying computational dynamics*.

1.1 OVERVIEW

An overview of the "big picture" follows next. With regards to the *mechanics underlying computational dynamics*, we start with the premise that in the beginning, the well known Newton's law of motion for N-body systems is given, which reflects the statement of the principle of balance of linear momentum. Subsequently using this as a landmark, firstly, the principal relations to three distinctly different fundamental principles, that comprise the *pyramid of computational dynamics*, and are of primary interest here are established. Likewise, for the dynamics of materials, structures and deformable continuum media and related applications, under the premise that the governing equations such as the Cauchy's equations of motion which reflect the statement of the principle of balance of linear momentum are given, analogous developments are also established. Once the principal relations to the three fundamental principles are established, any of the respective principles can thenceforth serve as the standalone starting point for the subsequent theoretical and numerical developments because of their wide range of applicability. The overall developments provide a fundamental understanding and improved insight into the mathematical equations governing the dynamic motion for N-body and continuous body systems, and the consequent numerical discretization in space and/or time. Stemming from the three distinctly different

Advances in Computational Dynamics of Particles, Materials and Structures, First Edition. Jason Har and Kumar K. Tamma.

fundamental principles, we present recent advances in both vector and scalar formalisms for N-body dynamical systems and also continuous-body dynamical systems with focus upon the numerical aspects related to space/time discretization. The three distinctly different fundamental principles which comprise the *pyramid of computational dynamics* are the following: *the Principle of Virtual Work in Dynamics*, *Hamilton's Principle* and as an alternative (due to inconsistencies associated with *Hamilton's principle*), *Hamilton's Law of Varying Action*, and *the Principle of Balance of Mechanical Energy*. Essentially, the aforementioned three fundamental principles have been particularly highlighted and selected as each of these principles can be independently employed to derive the governing equations of motion for N-body dynamical systems, and the strong and weak forms for continuous-body dynamical systems. However, of importance and noteworthy are the various formalisms and the different ways by which one can describe the theoretical and computational developments; and there exist fundamental differences in the three distinctly different fundamental principles and their underlying axioms.

Customarily in the literature, the equations of motion, which govern the mechanical behavior of N-body or continuous-body dynamical systems for a wide class of engineering applications, have been represented by vectorial quantities in the Newtonian mechanics framework (which is referred to in this book as the vector formalism). Alternatively, in the Lagrangian or Hamiltonian mechanics framework (which is referred to in this book as the scalar formalism), they have been described by generalized or canonical coordinates with descriptive scalar functions such as the Lagrangian or the Hamiltonian; this is mostly in the sense of applications to N-body dynamical systems (Greenwood 1977; Pars 1965). This has been the traditional paradigm. It is a matter of convenience and preferred choice of the analyst in the particular selection of either vector or scalar formalism, and the corresponding framework. Although it is not customary in the classical mechanics setting, other alternative descriptive scalar functions exist and can also be employed. *The significance and importance of one such descriptive scalar function which is built-in and directly measurable, namely, the Total Energy, is additionally described in this book under the umbrella of the Total Energy representation of the equation of motion and the associated framework*. There exist various subject areas in mechanics and physics where it is desirable to have a direct measurable descriptive scalar function such as the Total Energy. It provides a new and different perspective with good physical insight and computationally attractive and convenient features in contrast to the classical mechanics setting. The end result is that any of the three previously mentioned fundamental principles can independently be employed to derive the governing equations of motion for N-body dynamical systems, and the strong and weak forms for continuous-body dynamical systems. Also, both the vector and scalar formalisms indeed can be shown to be identical and/or equivalences can be drawn. Furthermore, each respective framework has its own pros and cons which need to be carefully understood in developing the numerical discretizations in space and/or time. In summary, we describe both classical practices that are customarily followed and new avenues for conducting space/time discretizations to find numerical solutions.

Under the umbrella of *computational dynamics*, this book provides a fundamentally sound background on the various theoretical and computational aspects. The two principal themes, namely, the *mechanics underlying computational dynamics* and the associated *numerics underlying computational dynamics* are highlighted next. Although in the following the context is in the sense of N-body systems (and is described in detail in Part 1), the corresponding analogy can be equally drawn for Continuous-body dynamical systems (and is described in detail in Part 2); while the Time Discretization of the equations of motion is covered in Part 3 of this book.

1.1.1 The Mechanics Underlying Computational Dynamics

In general, classical mechanics is classified into three branches: Newtonian, Lagrangian, and Hamiltonian mechanics. It is believed that the distinction between Newtonian, Lagrangian, and Hamiltonian mechanics emanates from the notion of space (Arnold 1989). Alternative descriptive scalar functions exist, but are not the tradition. However, in this book, the significance and importance of one such descriptive scalar function which is built-in and directly measurable, namely, the Total Energy, is additionally described under the umbrella of the Total Energy representation of the equation of motion and the corresponding framework. Some brief highlights of the various frameworks follow next.

Newtonian Mechanics and Framework Newton's *Philosphiae Naturalis Principia Mathematica* (Newton, 1687) is based upon Euclid's axiomatic Elements of Geometry, which comprises of a multitude of definitions, axioms, theorems, and geometrical constructions in the course of the developments. The underlying theory as related to the so called particle mechanics which is often referred to as particle dynamics in the classical sense of the Newtonian mechanics setting has been, and is related to a fundamentally sound premise which is the main starting point, namely, Newton's second law. It reflects the statement of the principle of balance of linear momentum and is quite widely employed in dynamics. This law relates force to mass and acceleration wherein the velocities are continuous (unlike those special situations which are not Newtonian in the strict sense, such as when the velocities have isolated finite discontinuities resulting in the relation impulse equals change of linear momentum). Accepting the restrictions of Newton's law (for example, it is not applicable to a broad range of physical phenomena), and within these confines the underlying principles are, however, indeed those associated with the concepts involving the setting of vectorial dynamics; the basic formalism is with vector representations.

In three-dimensional Euclidean space, Newtonian mechanics requires the existence of an inertial frame of reference, where Newton's laws of motion hold. The inertial frame of reference is a non-rotating and non-accelerating frame of reference (Gron and Hervik 2007; McComb 1999). Then, as a direct consequence of the statement of the principle of balance of linear momentum, the Newtonian dynamical system (Newton's equation of motion) is described by physical quantities, which are represented by vectors such as position, velocity, acceleration and force in three-dimensional Euclidean space. This is referred to as vector formalism in this book. For cases when the Newtonian dynamical system is subjected to constraints in terms of Cartesian coordinate variables (which are usually used but not required to describe the motion of the dynamical system), these variables are frequently employed to impose constraints which limit the motion of the system in three-dimensional Euclidean space. The Newtonian dynamical system requires the description of field vectors in Cartesian coordinates, suffers from the presence of k-number of constraints leading an N-body dynamical system with $3N - k$ degrees of freedom (Goldstein 2002), and are given in an inertial reference frame. Note that the Newtonian dynamical system involves $3N$-number of Cartesian variables. Consequently, in Newtonian mechanics the governing equation of motion is of second-order in time (single-field form with the position as the dependent variable), represented in terms of Cartesian coordinates, and is subjected to constraint functions with Cartesian coordinate variables. Strictly speaking, this is referred to as the Newtonian mechanics framework. However, loosely speaking, we shall use the term Newtonian mechanics framework in this book for the procedure which treats the mathematical developments with vectors such as position, velocity, acceleration and force in the Cartesian coordinate system.

Subsequent major milestones were then followed by Bernoulli (1700–1782), D'Alembert (1717–1782), Euler (1707–1783), Lagrange (1736–1813), and Hamilton (1805–1865), and so on including Riemann (1822–1866), Lie (1842–1899), Poincare (1854–1912), Einstein (1879–1955), and Noether (1882–1935).

Lagrangian Mechanics and Framework In contrast to the Newtonian mechanics setting and vector formalism, and under the umbrella of analytical dynamics, this field, whose foundations were laid down by Euler as early as 1783, was put into play by Lagrange (Mechanique Analytique (Lagrange 1788), building upon the work of D'Alembert in 1743). Lagrange makes the following claim which is our interpretation of his original work: *I have set forth a theory in mechanics and the science of solution to such problems via general formulations which are simple and yield all necessary equations for their* *solution. No figures will be found in this theory, and the approaches I outline do not require geometrical constructions or discussions based on mechanics, but only simple algebraic principles. Those who enjoy (love) analysis will have great pleasure and see this science of mechanics become a new branch, and will be grateful to me for having extended this field of mechanics*. Unlike the previous vector based formalism associated with Newtonian mechanics, the developments dealing with analytical dynamics invoke instead the formalism of scalar representations and are quite often termed as *symbolic representations* that are frame invariant. They enable a generalized and unified viewpoint for dealing with the equations of motion (and although they are fairly popular for particle dynamics, they have not enjoyed much attention for deformable continuous bodies; this is especially in the sense of conducting numerical discretizations in space and time via methods such as the finite element method, and tradition has a lot to do with this issue). Furthermore, although there is no new physics that is brought forth in contrast to Newtonian mechanics, one cannot trivialize Lagrangian mechanics (it is an alternative route to the same results). Indeed, it also has certain inherent subtleties and restrictions, but it is important to note that in this book we are primarily interested in the common, but wide class of problems and applications wherein both Newtonian and Lagrangian mechanics hold.

Lagrangian mechanics does not need vector quantities requiring the inertial frame of reference. Instead of vector quantities having both magnitude and direction, a descriptive scalar function having only magnitude, called the Lagrangian, is required to be defined to describe the motion of the dynamical system (Greenwood 1977; Pars 1965). In addition, the salient feature of Lagrangian mechanics is the notion and introduction of the concept of generalized coordinates; this makes it possible to eliminate constraint equations that arise in Cartesian coordinates. Throughout this book, we refer to this procedure as a scalar formalism. The space where the constraint equations associated with the inertial frame of reference disappear is called *configuration space*. It should be noted that the Newtonian dynamical system is usually defined in the inertial frame of reference, whereas the autonomous Lagrangian is defined on the velocity phase space (tangent bundle) and the Lagrangian dynamical system (Lagrange's equation of motion) is given in the configuration space, the size of which is n_{dof}. Note that the number of degrees of freedom is $n_{\text{dof}} = 3N - k$ and the configuration space belongs to Euclidean n_{dof}-space. Consequently, in Lagrangian mechanics, the governing equations of motion (Lagrange's equations of motion), which are represented in terms of generalized coordinates and generalized velocities, are not subjected to constraint functions with Cartesian coordinate variables. The representation of the equations of motion is also of second-order in time (single-field form with position as the dependent variable) but involves a descriptive scalar function, namely, the Lagrangian. Strictly speaking, this is referred to as the Lagrangian mechanics framework in this book. However, loosely speaking, we shall use the term, the Lagrangian mechanics framework, for the procedure which treats the mathematical developments with the descriptive scalar function, namely, the Lagrangian.

Hamiltonian Mechanics and Framework Alternatively, Hamiltonian mechanics (Hamilton 1834a) has introduced the concept of the so-called phase space (co-tangent bundle configuration space) with $2n_{\text{dof}}$-number of canonical variables. Although it does not have any constraints, it inherits instead the representation of the equations of motion as a system of first-order in time via a scalar function, namely, the Hamiltonian. By introducing the notion and concepts of canonical coordinates and by means of the Legendre transformation, the descriptive scalar function, called the Hamiltonian, is defined in Hamiltonian mechanics. These canonical coordinates, as an ordered pair, belong to the domain of the Hamiltonian, namely, *phase space* (cotangent bundle), the size of which is $2n_{\text{dof}}$. This procedure is also referred to as a scalar formalism in this book. One of the most important aspects in Hamilton's mechanics is the fact that the Hamiltonian dynamical system (Hamilton's equations of motion) is a system of first-order differential equations in time unlike the Newtonian or the Lagrangian system which involves a system of second-order differential equations in time. Again, although it also has not brought forth any new physics just as in Lagrangian mechanics in contrast to that of Newtonian mechanics, this does trivialize the importance and practical utility of Hamilton's equations. This is because Hamiltonian mechanics indeed provides certain physical and geometrical interpretations significantly different from that of Lagrangian mechanics. Admittedly, it also has certain inherent restrictions; however, it has provided the necessary foundation for theoretical extensions to both within as well as outside the realm of classical mechanics (in particular, a vital route to statistical mechanics). As described subsequently in the various chapters of this book, these respective fields of classical mechanics are all equivalent to each other (to within the limitations and imposed restrictions that are inherent in each of the respective classical mechanics fields; however, of interest in this book is primarily the common ground which serves a wide class of applications in mathematical and physical sciences and engineering), and each of the three branches of classical mechanics inherits certain pros and cons. In contrast to Newtonian mechanics, the well known scalar functions, namely, the Lagrangian, $\mathcal{L}$, and the Hamiltonian, $\mathcal{H}$, are often called as *descriptive functions*. Once the descriptive functions are known for a given dynamical system, the equations of motion for the dynamical system can be readily obtained and/or constructed. The Hamiltonian dynamical system in the phase space has the property of symplecticness, thus, possessing conservation of volume and canonical transformation in the Hamiltonian flow. Strictly speaking, this is referred to as the Hamiltonian mechanics framework in this book. However, loosely speaking, we shall use the term, the Hamiltonian mechanics framework, in this book for the procedure which treats the mathematical developments with the descriptive scalar function, namely, the Hamiltonian.

Total Energy Representation of the Equation of Motion and Framework Furthermore, it is to be noted that by no means the previously mentioned descriptive scalar functions such as the Lagrangian and the Hamiltonian are the only available descriptive functions, as there also exist various others. Indeed, these descriptive functions are traditionally the most popular. Although other descriptive scalar functions exist and are customarily not the tradition in classical mechanics setting, we particularly draw attention to and place emphasis upon yet another practically useful alternative descriptive scalar function. This is the so-called Total Energy for the dynamical system which can also be employed. *There exist various subject areas in mechanics and physics where it is desirable to have a direct measurable descriptive scalar function such as the Total Energy*. For example, from a computational mechanics/dynamics perspective, the Total Energy, $\mathcal{E}$, and the corresponding Total Energy representation of the equation of motion and framework is a viable alternative which provides computationally attractive features and easy to grasp physical insight with relative simplicity and ease. It primarily fosters numerical discretization in space and/or time for modeling and simulation of dynamical applications with particular focus upon aspects dealing with computational dynamics. This is via alternative perspectives for developing the governing equations and the corresponding numerical discretizations that are applicable to a wide class of computational dynamics problems (with the clear understanding of the imposed limitations inherent in the manner in which the framework is mathematically formulated such as differential and variational calculus settings, but within the same range of common applicability to a broad class of problems as in the previous frameworks). Obviously, as in the Lagrangian and Hamiltonian mechanics frameworks which can be independently derived from the first and second fundamental principles described earlier, but with the notion of variation and variational calculus, via these fundamental principles one can also derive and/or show the equivalence respectively, to the Total Energy representation of the equation of motion through an appropriate transformation and the relation to Jacobi's integral (although this is not the formal way we present the formalism as outlined in the selected chapters of this book). In contrast to the traditional paradigm of the classical mechanics frameworks, we find that the alternative Total Energy representation of the equation of motion and the associated framework is very natural for computational dynamics developments. It is also of practical utility for fostering a unified viewpoint for conducting numerical discretization in space and time, and we sincerely hope that the reader also embraces the same. Analogous to the Lagrangian mechanics framework, the autonomous Total Energy is defined on the velocity phase space (tangent bundle). Note that the number of degrees of freedom is $n_{\text{dof}} = 3N - k$

and the configuration space belongs to Euclidean n_{dof}-space. Hence, the configuration space in the Total Energy framework involves n_{dof}-number of generalized coordinates without constraints, and the representation of the equations of motion are naturally second-order (single-field form with position as the dependent variable) in time. The Total Energy representation and computational framework specifically capitalizes upon certain inherent built-in advantages in contrast to classical mechanics frameworks. This is primarily from the viewpoint of modeling and simulation associated with contemporary computational mechanics; however, it is to be pointed out that again there is no new physics that is brought forth just as with the Lagrangian and the Hamiltonian mechanics in contrast to Newtonian mechanics. Loosely speaking, we shall use the term, the Total Energy framework, in this book for the procedure which treats the mathematical developments with the descriptive scalar function, the Total Energy. The formalism of this framework is addressed subsequently in this book.

Finally, we establish the relevant equivalences and relations amongst all the three descriptive scalar functions described above and the associated frameworks, including their relationship to the Newtonian mechanics setting. More importantly, and for obvious reasons as detailed in this book, one cannot trivialize the importance and practical utility of all three distinctly different scalar frameworks. An advantage of the scalar frameworks is that one can readily capitalize on Theorem's such as Noether to establish symmetries. This is unlike that of the Newtonian mechanics setting where establishing symmetries is not readily applicable, and it is impractical in certain cases or at the least it is very cumbersome and definitely not trivial. However, the Newtonian mechanics setting is historically very popular, and this has to do with tradition.

Noether's Theorem and Symmetries In physics, the recognition of the presence of symmetry appears to have originated from Noether (1882–1935). She discovered that the Lagrangian has invariance properties for dynamical systems because symmetries exist in physics. Noether's theorem states that the spatial translational invariance of the Lagrangian in the configuration space associated with the tangent bundle yields conservation of linear momentum, the rotational invariance yields conservation of angular momentum, and the autonomous Lagrangian has invariance in time. Likewise, it is also well known that the Hamiltonian in the phase space which is associated with the cotangent bundle has also the same invariant properties. This is the so-called Hamiltonian version of Noether's theorem. In addition, the autonomous Total Energy in the configuration space associated with the tangent bundle also possesses the same three invariant properties. Therefore, it can be regarded as a third perspective (the first and second being the Lagrangian and the Hamiltonian), and from the viewpoint of the so-called Total Energy version of Noether's theorem. Consequently, the autonomous Total Energy has time/translational/rotational symmetries for both N-body and continuous-body dynamical systems. Conservation laws indeed play an important role in physics and mechanics. Owing to Noether's celebrated works, the conservation laws are the results of symmetries in three important physical quantities: linear momentum, angular momentum, and total energy. The conservation laws can be extended to the balance laws for mechanical systems such as the linear momentum balance law, the angular momentum balance law, and the energy balance law in continuum mechanics.

Pyramid of Computational Dynamics Of particular interest and focus in this book are the three distinctly different fundamental principles whose relations to the equation of motion which reflect the principle of balance of linear momentum in N-body and/or continuous-body systems can be established; and they comprise the *pyramid of computational dynamics* (see Preface; Figure 1). The principles are: 1) *the Principle of Virtual work*, 2) *Hamilton's principle, or alternatively, Hamilton's Law of Varying Action (which is not a variational principle)*, and 3) *the Principle of Balance of Mechanical Energy*. Each respective plane (or face) of the pyramid structure classification represents a fundamental principle that can be readily employed as a starting point for the subsequent developments in computational dynamics.

Each fundamental principle that is selected can independently enable both the theoretical and computational developments associated with and leading to the strong and/or weak forms. Consequently, it readily enables the corresponding numerical discretizations in space/time for a wide variety of applications that are commonly encountered in mathematical and physical sciences, and engineering. Since each of the above fundamental principles does not necessarily rely upon the others, it is associated with an underlying axiom that serves as the basis, and is essentially the starting point of the development of the formulations. However, the pros and cons, limitations of each fundamental principle, and the conditions under which equivalences amongst the three fundamental principles can be drawn need to be carefully understood to avoid misinterpretation. Various other classifications are also plausible, and consequently, the present pyramidal structure classification could entertain other faces or planes. *Furthermore, we are primarily interested in the overlapping application areas amongst the three fundamental principles which have a common ground, and which serve a wide class of problems of interest here.*

The Three Fundamental Principles We place significant emphasis upon the applied space whose coordinates are strongly associated with the definition of the scalar functions such as the Lagrangian and the Hamiltonian. The solution of a Lagrangian dynamical system is a point $\mathbf{q}$ in configuration space Q, namely, $\mathbf{q} \in Q$ (Arnold 1989; Jose and Saletan 1998). The autonomous Lagrangian is defined on the velocity phase space TQ, to which an ordered pair belongs as a member of the domain, namely, $(\mathbf{q}, \dot{\mathbf{q}}) \in TQ$. The solution of a Hamiltonian dynamical system is an ordered pair $(\mathbf{p}, \mathbf{q})$ which defines a point in phase space T^*Q, to which canonical coordinates belong, namely, $(\mathbf{p}, \mathbf{q}) \in T^*Q$. The autonomous Hamiltonian is defined on the phase space T^*Q. The solution of a Total Energy dynamical system is also a point in configuration space Q, to which generalized coordinates belong, namely, $\mathbf{q} \in Q$. The autonomous Total Energy is defined on the velocity phase space TQ, to which an ordered pair belongs as a member of the domain, namely, $(\mathbf{q}, \dot{\mathbf{q}}) \in TQ$. We highlight both scalar and vector formalisms under the respective umbrella of each fundamental principle as a point of departure, namely, the following:

1. The fundamental principle, namely, *the Principle of Virtual Work* for developing the governing equations of motion and consequently to enable numerical discretization. The principle of balance of linear momentum naturally gives rise to Newtonian mechanics setting and vector formalism of the equation of motion (original basis utilizes Cartesian description of field variables for position and velocity, and linear momentum construction, and execution of the principle of balance of linear momentum). For N-body dynamical systems, it leads to the governing equations of motion for the Newtonian dynamical system. Next, taking the inner product with the virtual displacement, it leads to the Lagrangian form of D'Alembert's principle. Alternatively, invoking variational calculus and transformation to generalized coordinates, it first leads to the traditional development of Lagrange's equations of motion in configuration space. And then, by means of the Legendre transformation, Hamilton's equation of motion in phase space can be also established. For continuous-body dynamical systems, the principle of virtual work leads to Cauchy's equations of motion as the local form of momentum equations and Cauchy's law as natural boundary conditions. Taking the inner product between Cauchy's equations of motion and the virtual displacement leads to Bubnov-Galerkin weighted residual form. And then, carrying out integration by parts and imposing Cauchy's law, the residual form leads to the principle of virtual work, which is the weak statement for the initial-boundary value problems.

2. The fundamental principle, namely, *Hamilton's Principle* which has been regarded as the cornerstone for variational principles in dynamics (and is viewed by some to be more fundamental than Newton's laws in the sense that it does not have as severe a restriction in its broad range of the fields of applicability). Hamilton's principle can be alternatively employed for developing the governing equations of motion in scalar or vector formalism, and the associated strong and/or weak forms for N-body and continuous-body systems. Historically, variational principles have played an important role and enable formulating the governing equations of motion. It is worth recalling that Hamilton's principle naturally

first gives rise to Lagrangian formulations in configuration space. The resulting differential equations are called the Euler-Lagrange equations of motion (original basis employs description of field variables, generalized coordinates, and construction of the descriptive scalar function, namely, the Lagrangian, and utilizing Hamilton's principle). Invoking Legendre transformation, the Hamilton's equations in phase space are then established. As one of the double properties of the principal function (Hamilton 1834a) (which is often called the action), *Hamilton's law of varying action* (this is not a variational principle) is an alternative for circumventing the inconsistencies associated with Hamilton's principle as described subsequently in this book. It readily enables the theoretical and computational developments associated with and leading to the strong and/or weak forms, and the corresponding numerical discretizations.

3. The fundamental principle, namely, *the Principle of Balance of Mechanical Energy* (this is yet another viable alternative), which brings forth new and different perspectives and improved physical interpretation for developing the governing equations of motion in scalar or vector formalism, and readily enables the development of the associated strong and/or weak forms. We favor and advocate the developments emanating from the principle of balance of mechanical energy and the new representation via the Total Energy framework in contrast to the various classical frameworks from the viewpoint of providing improved physical insight, concepts that are easy to grasp, and since it also naturally inherits computationally convenient and attractive features.

For particular cases of interest in this book, the equivalences amongst the three scalar formalisms are highlighted in Figures 7.1 and 7.2 for N-body dynamical systems, and in Figure 12.2 for Continuous-body dynamical systems with dead loads and vector formalism of Cauchy equations of motion. The limitations for establishing these equivalences from amongst the various frameworks need to be clearly understood to avoid misinterpretation. In contrast to the Lagrangian and Hamiltonian frameworks, note that unlike the developments starting from variational calculus setting for the Total energy framework which is applicable to holonomic systems (see Appendices in Chapters 6 and 11), via the developments emanating from the differential calculus setting and the principle of balance of mechanical energy, the limitations are that the kinetic energy is quadratic in generalized velocities and the potential energy is velocity independent for N-body systems; and as necessary and frequently encountered in continuum dynamical applications, we additionally restrict attention to the case of holonomic-scleronomic systems.

In summary, a wide class of engineering applications can be basically represented by Newton's second law and the Newtonian framework for the equation of motion with vector based formalism, or its equivalent scalar formalisms with different descriptive functions such as the Lagrangian, the Hamiltonian, and the Total Energy. However, the selection of a particular framework is an individual choice, and it is obvious that indeed they all should ultimately lead to and yield, the "identical/equivalent final result". That is, the same or equivalent strong form of the governing equations of motion that is continuous in time for N-body systems and/or continuous in space and time for continuous-body systems, followed by the same or equivalent weak forms; and also the resulting numerically discretized equations in space and/or time must be the same and/or equivalent.

1.1.2 The Numerics Underlying Computational Dynamics in Space and Time

Besides the developments that readily enable the governing equation of motion to be formulated for dynamical systems, of interest in this book are also those ingredients additionally necessary for fostering a variety of numerical modeling and simulation aspects of N-body and deformable continuous-body dynamics. The focus is towards a wide class of applications in mathematical and physical sciences and engineering. In particular, emanating independently from each of the three distinctly different fundamental principles represented by the *Pyramid of Computational Dynamics*, the focus is upon the following:

(a) **Problem formulations**, that is, for the physical problem of interest, the objective firstly is in the development of mathematical model equations via the classical Newtonian mechanics and vector formalisms, and the consequent developments leading to the scalar formalisms via the Lagrangian mechanics framework, the Hamiltonian mechanics framework, and an alternative via the Total Energy framework. Consequently, the interest is in fostering the associated *numerics underlying computational dynamics*. In essence, firstly, the interest is in the development of the strong form of the governing equations of motion that are continuous in time for N-body systems together with the required initial conditions; and also for deformable continuous-body systems the interest is in the development of the governing equations of motion in solid/structural applications which are continuous in both space and time together with the appropriate constitutive models, boundary and initial conditions, and

(b) **Problem solutions**, that is, the *numerics underlying computational dynamics* dealing with time discretization (integration) for N-body dynamic systems, and for deformable solid/structural continuous-body dynamics the interest is in the space discretization and time discretization (integration) aspects as well for both the vector and the various scalar formalisms of the strong form of the equations of motion. Subsequently, the development of the equivalent weak forms following traditional practices or by new and alternative avenues leading to the semi-discretized and/or fully discretized equations of motion are described from different viewpoints in this book which essentially all lead to identical/equivalent results.

The terminology, namely, *numerics underlying computational dynamics* refers to the manner in which the numerical discretization approaches in space and/or time are described and developed stemming independently from each of the respective fundamental principles. A unified viewpoint is the end result regardless of which fundamental principle serves as the starting point. There indeed exist pros/cons, restrictions and limitations associated with each of the respective fundamental principles which need to be carefully understood. To date, routine finite element formulations for the space/time discretization for example, mostly describe the various developments for numerical discretization starting from the strong form of the governing equations of motion with vector formalism. This is a fairly standard procedure, and has been described extensively in the literature and in most books dealing with the subject matter. Besides traditional practices, alternative avenues which provide new and different perspectives and physical insight can also be employed. These are particularly highlighted and described in detail in this book via scalar formalisms, and employing the corresponding discretization techniques such as the space-discrete finite element formulations.

Space and Time Discretization For modeling and simulation of dynamical problems, this book offers various approaches for the numerical discretization in space and/or time for the underlying mathematical models with choices of both vector and scalar formalisms.

Space Discretization Firstly, with regards to conducting the space discretization for the physical problem of interest, various forms of representation of the mathematical model can be employed as the starting point for the numerical discretization. One of the predominant and common forms of the mathematical model or equivalently the field equations, is the mathematical representation by ordinary/partial differential equations. In general, for N-body systems the formulations are a result stemming from the satisfaction of dynamic equilibrium and can be in the form of vector or scalar formalisms. On the other hand, for deformable continua they can be a result stemming from the classical balance principles such as conservation of mass, momentum (linear and angular) balance principles, and balance of energy in either vector or scalar formalisms (although the latter is seldom discussed in the literature). For dynamics of deformable continua, historically, the traditional setting typically emanates from Cauchy's equations of motion and vector formalism in the sense of continuum mechanics for materials, solids, and structures. This is in conjunction with strain-displacement relations, constitutive equations, boundary conditions, and initial conditions, as appropriate. Therein, integral or weak statements with the vector formalism that are equivalent to the original differential equations are constructed in conjunction with a set of arbitrary functions equal to the total number of equations or components which are then set equal to zero. The assertion is that if this statement is satisfied for the set of arbitrary functions, it then follows that the original differential equation system must also be satisfied at all points in the domain. In the course of this process, the necessity to compute second derivatives can be reduced or weakened by techniques such as integration by parts. Unlike vector formalisms, historically, the scalar formalisms for continuous-body dynamic systems are neither popular nor are they the norm. However, this book describes the numerical discretization aspects for both vector and scalar formalisms of the equations of motion.

An alternative to employing field equations which are ordinary/partial differential equations as in the previous approach, is that which employs variational principles by constructing appropriate functionals that are equivalent to the differential equations. For static problems and vectorial setting, various representations that are equivalent to the vector forms of the differential equations exist and/or can be constructed such as in terms of a scalar functional (and, in the spirit of employing variational principles, the more well known and common principles are the principle of minimum or stationary potential energy, complimentary energy, Hellinger-Reissner, Hu-Washizu, and the like). Consequently, the scalar functional is extremized with respect to a set of unknown variables such as the degrees of freedom in order to derive the discretized equations of static equilibrium, and the field variables are then solved. However, for dynamical problems which are the primary focus in this book, such practices are in general, not feasible. This is because typical variational principles usually do not exist nor can they be readily constructed for most transient/dynamic situations. For example, the classical Ritz method requires a functional to be given. The functional in the Ritz method can be readily obtained, in general, for static/steady type problems. However, variational principles do not generally exist in initial-boundary-value problems (transient/dynamic problems). Therefore, the limitation of the classical Ritz method is that it is not readily applicable to transient problems such as transient heat transfer, continuum-dynamics, and so on. It is a major bottleneck via the Ritz type approach. Hence as an alternative, and from a different viewpoint and perspective, unlike traditional Newton based representations, the scalar forms of representation of the equation of motion employing the so-called space-discrete finite element formulations highlighted in this book with descriptive functions, namely, the Lagrangian, the Hamiltonian, and the Total Energy readily enable one to circumvent some of the drawbacks with existing practices, and are also additionally described which lead to the identical semi-discretized equations of motion as in traditional practices.

Time Discretization Finally, the discretization in time is also subsequently described for properly integrating the equations of motion and specific guidelines for developing algorithms by design that meet targeted objectives are also provided. In contrast to traditional practices, we additionally highlight alternative avenues which not only recover most of the traditional developments to-date, but offer new and computationally attractive alternatives with improved physical insight as well. We particularly highlight and present new avenues and optimal algorithms by design. In particular, we focus upon the class of LMS methods in both single-field and two-field forms of representation. For the time discretization, we first present the following: (i) First, we show starting from the standard representation of the semi-discretized equations of motion, the various classical and chronological developments in time integration from historical perspectives that appear in the open literature over the past fifty years or so,

(ii) Next, we highlight variational integrators that are symplectic-momentum conserving stemming from the so-called Discrete Euler-Lagrange representations, and (iii) Subsequently, we describe the so-called energy-momentum conserving/dissipating algorithm designs for finite dimensional systems following the original methods of development (classical practices) through enforcing energy constraints. Finally, in a separate chapter, in contrast to all the aforementioned classical and/or traditional practices described in the previous chapter, we focus special attention upon and highlight the more recent advances and developments directly emanating from the new Total Energy representation of the equations of motion and the associated framework as a starting point (unlike traditional practices) in conjunction with a generalized time weighted residual approach. In particular, we provide new perspectives, a unified viewpoint, and the appropriate theoretical basis on how to properly provide proper extensions of the parent linear dynamics algorithms to nonlinear dynamics applications for developing practically useful time integration algorithms by design (whose bases are based upon symplectic-momentum or energy-momentum based representations; and in addition, we describe how to embed controllable numerical dissipation features in such algorithm designs such that when the controllable numerical dissipative features are turned off, they readily recover the parent symplectic-momentum or energy-momentum representations). The presentation on recent advances in time integration in the last chapter also covers most of the time integration developments that have been previously derived from various other classical viewpoints as described in (i), (ii), and (iii) above in the previous chapter; thereby providing the impetus towards a unified approach. In addition, we particularly describe from a unified viewpoint, new avenues for the discretization of the equations of motion with improved physical insight which lead to new and optimal algorithm designs, while recovering most of the developments that currently exist from traditional and/or classical practices. Practically useful linear dynamics algorithms by design, nonlinear dynamics algorithms by design in the sense of applicability to continuum elasto-dynamics applications, algorithms by design for N-body systems, and for conservative/nonconservative mechanical systems with holonomic-scleronomic constraints such as those encountered in multi-body dynamics applications are illustrated.

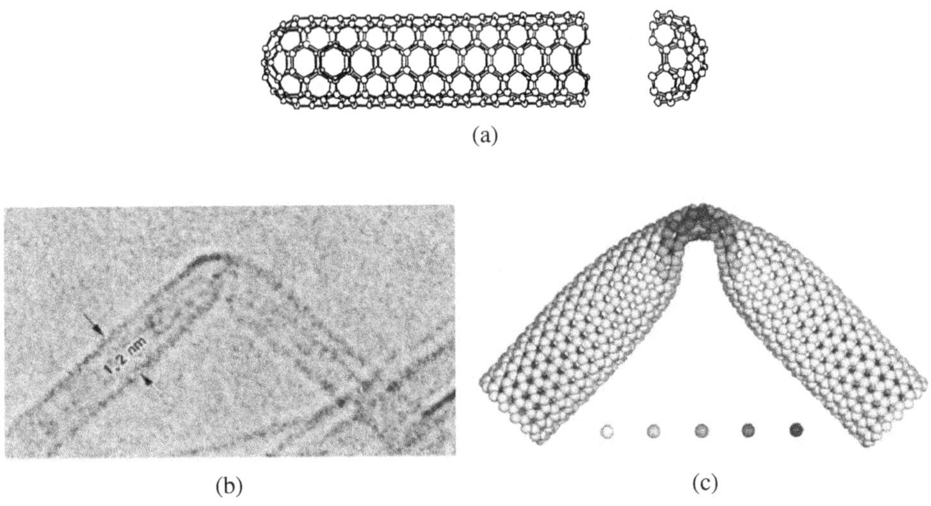

Figure 1.1. (a) Nanotube structure of armchair. (b) HREM images of kink structures formed in nanotubes under mechanical duress: Single kinks in the middle of single-walled nanotubes with diameters of 1.2 nm. Reproduced with permission from Iijima et al. 1996 © American Institute of Physics; (c) Atomic structure of a single kink obtained in the molecular dynamics simulation of bending of the single-walled tube. Reproduced with permission from Iijima et al. 1996 © American Institute of Physics

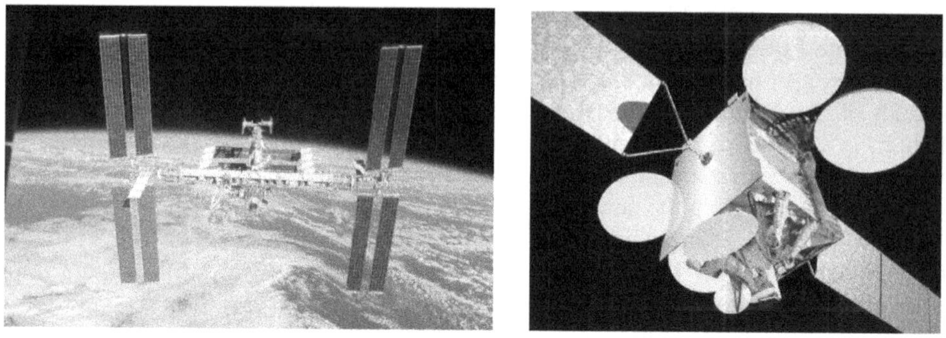

(a) (b)

Figure 1.2. (a) International space station. Courtesy of NASA (www.nasa.gov). (b) Communication satellite, Arabsat 5A. Courtesy of EADS Astrium Satellites (www.astrium.eads.net/)

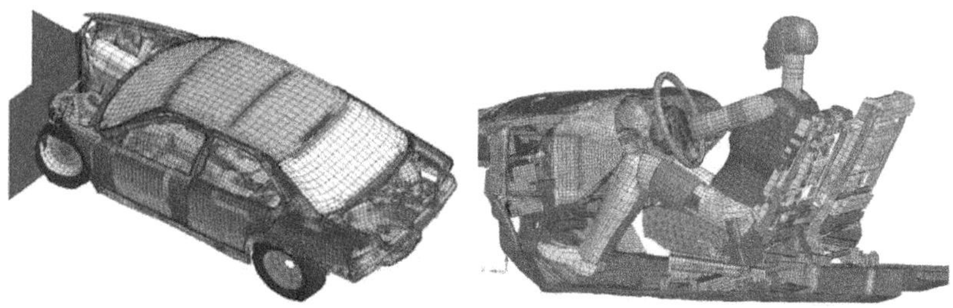

(a) (b)

Figure 1.3. (a) Car crash simulation. Reproduced from http://commons.wikimedia.org/wiki/File:FEM_car_crash1.jpg. (b) Finite element mesh with a dummy. Courtesy of http://cid-5e2d18a3fd0aa6e8.spaces.live.com/

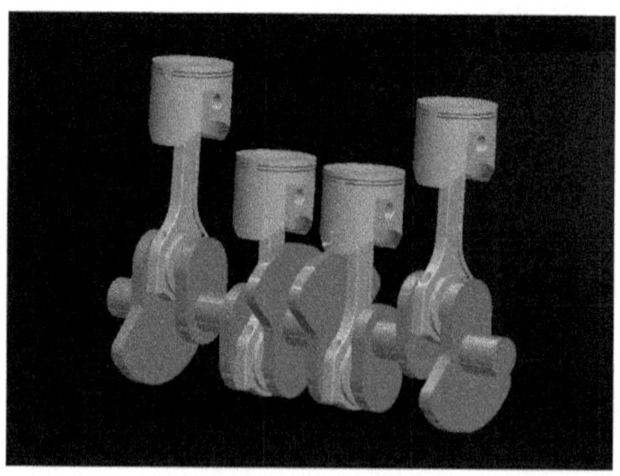

Figure 1.4. Computer graphic of crank and pistons. Courtesy of http://jordanalleycat.wordpress.com/

Contemporary computational mechanics disciplines are of focus here with particular attention towards the general areas of computational dynamics associated with N-body and continuous-body systems. In particular, we are primarily interested in applications dealing with particle dynamics, structural dynamics or elastodynamics associated with continuum mechanics, rigid/flexible structures and multi-body systems, and the like towards which the present developments are targeted.

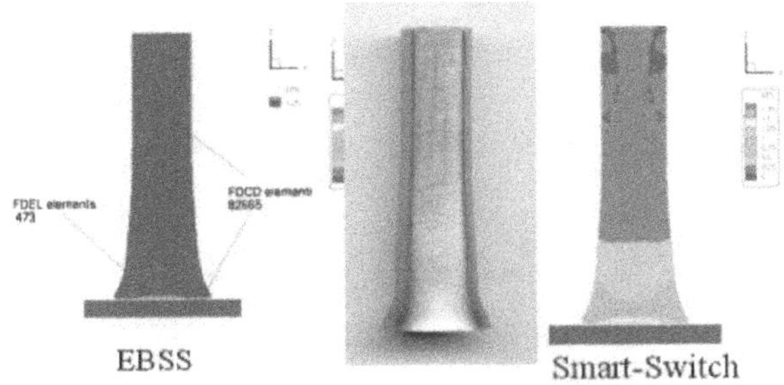

Figure 1.5. Taylor bar dynamic simulation (Zhou et al. 2008)

Figure 1.6. Bill Emerson Memorial cable-stayed bridge in Cape Girardeau, Missouri. Courtesy of UChicago Argonne LLC (www.anl.gov/TRACC/Applications/bridge_conditions.html)

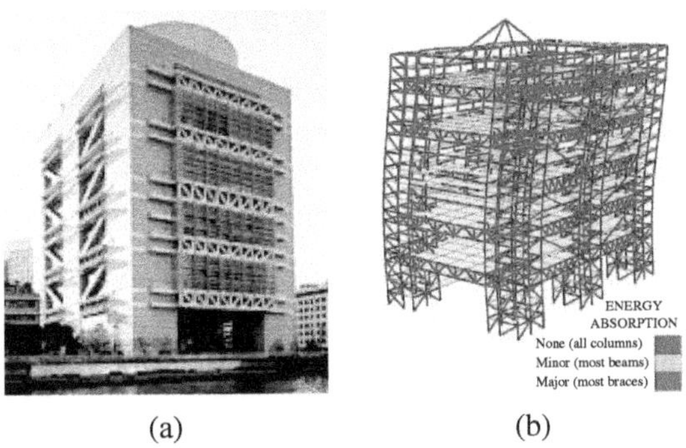

Figure 1.7. (a) Osaka international convention center; (b) During Earthquake; Full 3-D non-linear seismic response analysis for the Osaka international convention center. Reproduced with permission from ARUP (www.arup.com)

1.2 APPLICATIONS

This book deals with comprehensive topics including classical dynamics, continuum mechanics, the finite element method, and time integration schemes for both N-body and continuous-body dynamical systems. Hence, the application areas of the topics, which have been accounted for throughout the book,

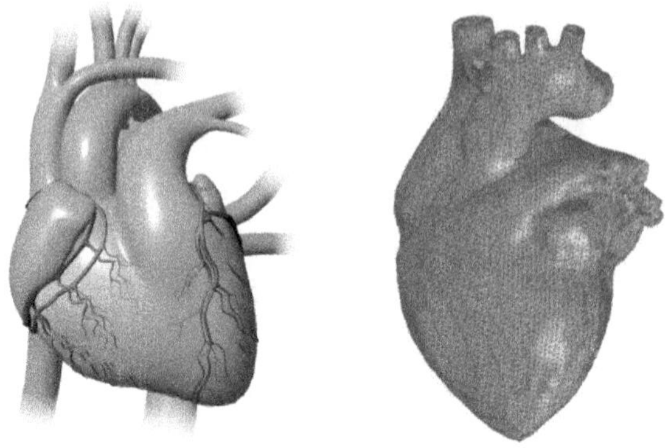

(a) (b)

Figure 1.8. (a) 3D Heart model; (b) Finite element mesh; 3D Heart model and mesh

are too broad to mention. Nonetheless, we wish to briefly present some representative application areas which are illustrated in Figures 1.1–1.8.

CHAPTER TWO

MATHEMATICAL PRELIMINARIES

The present book encompasses a wide range of topics such as classical mechanics dealing with N-body dynamical systems, continuum mechanics underlying Continuous-body dynamical systems, various numerical aspects related to finite element formulations for space discretization with both vector and scalar formalisms, and also the design and development of time discretization approaches of a variety of time integration schemes for integrating the dynamic equations of motion. The reader is encouraged to become familiar with the mathematical preliminaries highlighted in Chapter 2 as they will be helpful in making the various concepts easy to understand in the later chapters of this book. We first start with sets and functions, and then we discuss vector spaces, matrix algebra, function spaces and tensor analysis.

2.1 SETS AND FUNCTIONS

It is not at all an exaggerated expression that mathematics is a discipline dealing with sets and functions. It is important to understand both sets and functions for devel oping various mathematical formulations.

2.1.1 Sets

To begin with, we define a set. The theory of sets was first developed by the German mathematician Cantor (1845–1918) (Maddox 1988). A set in mathematics is a well defined collection of objects or entities of some kind. These objects are called members or elements of the set. Upper-case letters are used to denote the sets and lower-case letters for the elements. Calligraphy fonts are applied to the sets. For instance, $x \in \mathcal{A}$ indicates that x is a member of the set $\mathcal{A}$. Further, $x \notin \mathcal{A}$ indicates that x is not a member of the set $\mathcal{A}$.

Suppose that the set $\mathcal{A}$ has integers bigger than 3 and smaller than 9. Then, we have the set

$$\mathcal{A} = \{4, 5, 6, 7, 8\} \tag{2.1}$$

A set of all positive integers is defined as

$$\mathbb{N} = \{1, 2, 3, \ldots\} \tag{2.2}$$

Advances in Computational Dynamics of Particles, Materials and Structures, First Edition. Jason Har and Kumar K. Tamma.

Then, the set $\mathcal{A}$ is a subset of $\mathbb{N}$, denoting $\mathcal{A} \subset \mathbb{N}$, and Equation (2.1) can be rewritten as

$$\mathcal{A} = \{x \in \mathbb{N} | \, 3 < x < 9\} \tag{2.3}$$

which means that the set $\mathcal{A}$, is a set of all elements x of $\mathbb{N}$ such that x is bigger than 3 and smaller than 9. The set of real numbers $\mathbb{R}$ consists of the set of rational numbers $\mathbb{Q}$ and the set of irrational numbers $\mathbb{J}$. The set of rational numbers is defined as

$$\mathbb{Q} = \left\{ \left(\frac{m}{n}\right) | m, n \in \mathbb{Z}, n \neq 0 \right\} \tag{2.4}$$

where $\mathbb{Z}$ denotes the set of integers, i.e. $\mathbb{Z} = \{\ldots, -2, -1, 0, 1, 2, \ldots\}$.

Suppose that we have another set, denoted by $\mathcal{B}$. The set $\mathcal{B}$ is called a proper subset of $\mathcal{A}$, if the set is given as shown in Figure 2.1, i.e. $\mathcal{B} \subset \mathcal{A}$. Next we have a set $\mathcal{C}$, called the union of the sets $\mathcal{A}$ and $\mathcal{B}$. $\mathcal{A}$ union $\mathcal{B}$ is the set of all members that are in $\mathcal{A}$ or $\mathcal{B}$, or both. The union is defined by $\mathcal{A} \cup \mathcal{B} = \mathcal{C}$, and can be written formally as

$$\mathcal{A} \cup \mathcal{B} = \{x \in \mathcal{C} | \, x \in \mathcal{A} \text{ or } x \in \mathcal{B}\} \tag{2.5}$$

Hence we say that $\mathbb{R} = \mathbb{Q} \cup \mathbb{J}$. As an application of a union, the time duration with $N + 1$ time steps may be considered to be a union as follows:

$$\mathcal{T} = \bigcup_{n=0}^{N} [t_n, t_{n+1}] \tag{2.6}$$

where the step size is given by $(\Delta t)_n = t_{n+1} - t_n$.

Next, we consider a new set called intersection $\mathcal{D}$. The intersection of two sets $\mathcal{A}$ and $\mathcal{B}$ is a set of all members that are in both $\mathcal{A}$ and $\mathcal{B}$. The following mathematical expression for the intersection is represented as

$$\mathcal{A} \cap \mathcal{B} = \{x \in \mathcal{D} | \, x \in \mathcal{A} \text{ and } x \in \mathcal{B}\} \tag{2.7}$$

Consider the two sets, $a \in \mathcal{A}$ and $b \in \mathcal{B}$. Then (a, b) is called the ordered pair in the two-dimensional Cartesian space, and we define the Cartesian space or Cartesian product $\mathcal{A} \times \mathcal{B}$ as

$$\mathcal{A} \times \mathcal{B} = \{(a, b) | \, a \in \mathcal{A} \text{ and } b \in \mathcal{B}\} \tag{2.8}$$

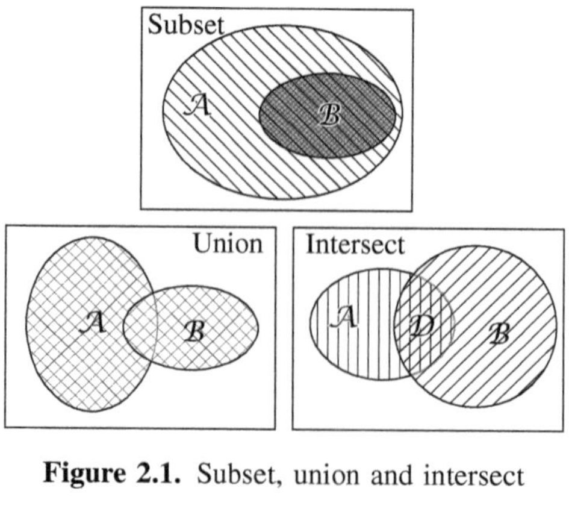

Figure 2.1. Subset, union and intersect

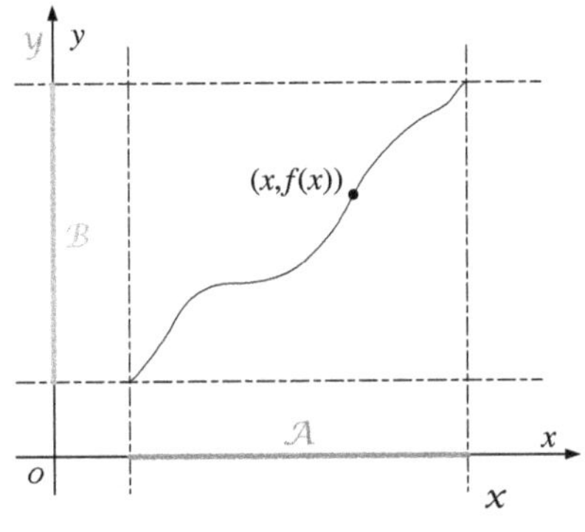

Figure 2.2. Graph of the function

2.1.2 Functions

Suppose that we are given an elementary function such as

$$y = x^3 \tag{2.9}$$

Then it is said that y is a function of x. Instead of Equation (2.9), often we use the following notation

$$f(x) = x^3 \tag{2.10}$$

We may express the above function as $x \longmapsto f(x)$, which implies that the function maps from x to $f(x)$. Alternatively the following notation $x \longmapsto x^3$ is often employed. However, it is customary to employ the notation $f(x)$: $\mathcal{A} \longmapsto \mathcal{B}$, which means that the function $f(x)$ maps $\mathcal{A}$ into $\mathcal{B}$ and the function $f(x)$ is defined on $\mathcal{A}$, as shown in Figure 2.2. The notation also implies that x is a member of $\mathcal{A}$, i.e. $x \in \mathcal{A}$, and $\mathcal{y}$ (or $f(x)$) is a member of $\mathcal{B}$, i.e. $f(x) \in B$.

The non-empty set $\mathcal{A}$ is often called the domain (or source) of the function f, whereas the non-empty set $\mathcal{B}$ is called target (or codomain) of the function $f \cdot \mathcal{y}$ (or$f(x)$) is referred to as the image (or value) of x under the mapping function f. The collection of the images of x is called the range of the function f, and is denoted by $f(\mathcal{A}) = \{f(x) \in \mathcal{B} \mid x \in \mathcal{A}\}$. In the case that $f(\mathcal{A}) \equiv \mathcal{B}$, the function $f(x) : \mathcal{A} \longmapsto \mathcal{B}$ is called a surjection (or a surjective map); in other words, it is said that the surjective function f maps $\mathcal{A}$ onto $\mathcal{B}$. If two different members in $\mathcal{A}$ correspond two different images respectively in $\mathcal{B}$ under the mapping function f, the function $f(x) : \mathcal{A} \longmapsto \mathcal{B}$ is called a injection (or a injective map) : in other words, one says that the injective function f is a one-to-one mapping. In the special case that f is a one-to-one and onto mapping, the mapping function $f(x) : \mathcal{A} \longmapsto \mathcal{B}$ is called a bijection (or a bijective map). For the bijective map, the inverse mapping function of f may be defined as a mapping $f^{-1}(\mathcal{y}) : \mathcal{B} \longmapsto \mathcal{A}$. The image of y under the inverse mapping function f^{-1} is x (or $f^{-1}(\mathcal{y})$). The range of the inverse mapping function f^{-1} can be written as $f^{-1}(\mathcal{B}) = \{x \in \mathcal{A} \mid f(x) \in \mathcal{B}\}$, and $f^{-1}(\mathcal{B}) \equiv \mathcal{A}$ for the bijective map.

Note that $\mathcal{A}$ is the subset of $\mathcal{X}$ and $\mathcal{B}$ is the subset of $\mathcal{Y}$. Therefore, the function f that maps $\mathcal{A}$ into $\mathcal{B}$ can be denoted by $f(x) : \mathcal{A} \subset \mathcal{X} \longmapsto \mathcal{B}$. The set of ordered pairs $(x, f(x)) \in \mathcal{A} \times \mathcal{B} \subseteq \mathcal{X} \times \mathcal{Y}$ is called the graph of the function $f(x)$. Furthermore, let us introduce another mapping $g(\mathcal{y}) : \mathcal{B} \longmapsto \mathcal{C}$. Then, we may define the composition or multiplication mapping such that $g \circ f(x) : \mathcal{A} \longmapsto \mathcal{C}$ or $(gf)(x) : \mathcal{A} \longmapsto \mathcal{C}$ as shown in Figure 2.3.

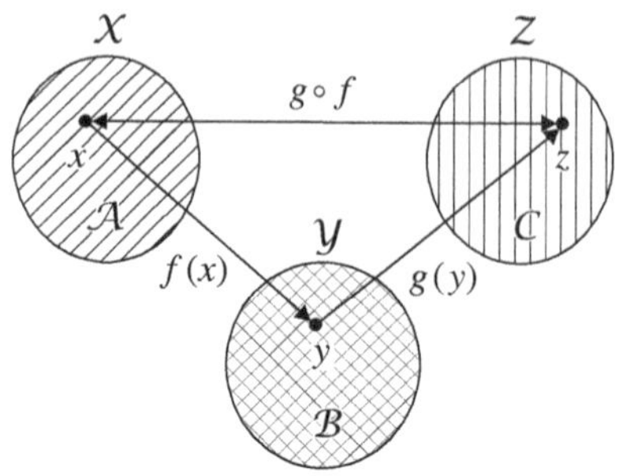

Figure 2.3. Multiplication or composition of mappings

2.2 VECTOR SPACES

A vector is defined as an object with magnitude and direction in a vector space. The magnitude of a vector is often called the length of the vector and is a scalar physical quantity. The magnitude and direction of a vector are the properties of a vector. The properties of a vector are independent of a coordinate system, but its components are dependent upon the coordinate system.

In this section, we discuss the details of a vector. In particular, we focus upon the concept of vector spaces such as the metric space, Euclidean n-space, the inner product space and normed spaces.

2.2.1 Real Vector Spaces

In the Cartesian coordinate system with which we are familiar, a point on the x-axis is a member or element of the real number line set, denoted by $\mathbb{R}^1$ or simply $\mathbb{R}$. Then the point on the axis is denoted by $x \in \mathbb{R}^1$. Note that the x-axis is regarded as a set of real numbers. An arbitrary point in a plane can be defined as an ordered pair of 2-tuple real numbers (x, y). Therefore, the point is said to be a member of the set $\mathbb{R}^1 \times \mathbb{R}^1 = \mathbb{R}^2$, which is called the Cartesian product (Bowen and Wang 1976; Maddox 1988). Consequently, in the three-dimensional space, a point can be an ordered triple of 3-tuple real numbers $(x, y, z) \in \mathbb{R}^3$. Hence, in the three-dimensional space, a point can be located by three coordinates. It should be noted that a 3-tuple means an order list of three elements in a set. Then, the real numbers x, y, z are called Cartesian coordinates. Each coordinate variable is a member of the real number line set, $\mathbb{R}$. Consequently, the ordered triple is a member of the Cartesian product, denoted by $\mathbb{R}^1 \times \mathbb{R}^1 \times \mathbb{R}^1$ or $\mathbb{R}^3$. For the case when Cartesian coordinates are defined in the three-dimensional space, the space is called Cartesian 3-space or Euclidean 3-space, $\mathbb{E}^3$ or simply 3-space as the set of real 3-tuples.

In an n-dimensional real vector space $\mathbb{V}$, a point $\mathbf{r}$ is defined as an ordered list of n-tuple real numbers, i.e.,

$$\mathbb{V} = \{\mathbf{r} = (\alpha_1, \ldots, \alpha_n) | \alpha_i \in \mathbb{R}, \ \text{for } i = 1, \ldots, n\} \tag{2.11}$$

where the entries α_i are called coordinates in the vector space $\mathbb{V}$ (Bowen and Wang 1976; Halmos 1958; Reddy 1986) and n is a positive integer. The real vector space $\mathbb{V}$ is a set of elements called real vectors. As an element contained in the vector space, every point $\mathbf{r}$ can be expressed as a linear combination of base vectors

$$\mathbf{r} = \sum_{i=1}^{n} \alpha_i \mathbf{e}_i \tag{2.12}$$

where $\mathbf{e}_i$ are called base vectors which span the real vector space $\mathbb{V}$. A set of all base vectors $\{\mathbf{e}_i\}_{i=1}^n$ or $\{\mathbf{e}_1, \ldots, \mathbf{e}_n\}$ is called a basis of the vector space. Hence, we say that $\textbf{span}\{\mathbf{e}_i\}_{i=1}^n = \mathbb{V}$. It is very important to note that every vector contained in the real vector space $\mathbb{V}$ can be expressed as a linear combination of all base vectors as follows:

$$\mathbf{v} = \sum_{i=1}^{n} v_i \mathbf{e}_i, \ \mathbf{v} \in \mathbb{V} \tag{2.13}$$

where the entries v_i are called components of the vector. The vector defined as a linear combination of base vectors is invariant like a scalar; in other words the vector is independent of a coordinate system in a vector space. Its components, however, are dependent upon the coordinate system relative to a set of base vectors. More details are presented in the sections to follow.

As defined previously, a real vector space is a collection of real vectors. The vector space is required to satisfy the following eight conditions. Let three n-vectors $\mathbf{u}, \mathbf{v}, \mathbf{w}$ be given as members of the vector space $\mathbb{V}$, i.e., $\mathbf{u}, \mathbf{v}, \mathbf{w} \in \mathbb{V}$. Then, it is said to be a linear vector space or simply a vector space if the following conditions are satisfied (Bowen and Wang 1976; Greenberg 1998; Grossman 1986; Halmos 1958; Hurley 1981; Kreyszig 2006; Marsden and Tromba 2003; Strang 1988; Zill and Cullen 2006).

(a) Commutativity: $\mathbf{u} + \mathbf{v} = \mathbf{v} + \mathbf{u}$.
(b) Associativity of vector addition: $(\mathbf{u} + \mathbf{v}) + \mathbf{w} = \mathbf{u} + (\mathbf{v} + \mathbf{w})$.
(c) Additive identity: $\mathbf{0} + \mathbf{u} = \mathbf{u} + \mathbf{0} = \mathbf{u}$.
(d) Existence of additive inverse: $\mathbf{u} + (-\mathbf{u}) = \mathbf{0}$.
(e) Associativity of scalar multiplication: $a\,(b\mathbf{u}) = (ab\mathbf{u})\,, \quad \forall a, b \in \mathbb{R}$.
(f) Distributivity of scalar sums: $(a + b)\,\mathbf{u} = a\mathbf{u} + b\mathbf{u}, \quad \forall a, b \in \mathbb{R}$.
(g) Distributivity of vector sums: $a\,(\mathbf{u} + \mathbf{v}) = a\mathbf{u} + a\mathbf{v} \quad \forall a \in \mathbb{R}$.
(h) Scalar multiplication identity: $1\mathbf{u} = \mathbf{u}$.

Therefore, as a collection of vectors, $\mathbb{V}$ is called a vector space. Consequently, Euclidean n-space $\mathbb{E}^n$ or n-space $\mathbb{R}^n$ is also a vector space.

2.2.2 Linear Dependence and Independence of Vectors

Let us consider a set of nonzero vectors, $\mathcal{A} = \{\mathbf{u}_1, \mathbf{u}_2, \ldots, \mathbf{u}_i, \ldots, \mathbf{u}_n\}$. Each vector in the set is a member of a linear vector space denoted by $\mathbb{V}$. Then, the set $\mathcal{A}$ is said to be linearly dependent if at least one vector in the set $\mathcal{A}$, $\mathbf{u}_i$, which belongs to the linear vector space $\mathbb{V}$, can be expressed as a linear combination of the other vectors (Strang 1988, page 80) (Greenberg 2001, page 140), namely,

$$\mathbf{u}_i = \alpha_1 \mathbf{u}_1 + \alpha_2 \mathbf{u}_2 + \ldots + \alpha_{i-1} \mathbf{u}_{i-1} + \ldots + \alpha_{i+1} \mathbf{u}_{i+1} + \alpha_n \mathbf{u}_n \tag{2.14}$$

where $\alpha_1, \alpha_2, \ldots, \alpha_n$ are real constants, i.e., $\alpha_i \in \mathbb{R},\ i = 1, \ldots, n$. On the contrary, a set of nonzero vectors $\mathcal{A} = \{(\mathbf{u}_1, \ldots, \mathbf{u}_i, \ldots, \mathbf{u}_n) |\, \mathbf{u}_i \in \mathbb{V},\ i = 1, \ldots, n\}$ is said to be linearly independent if all constants are zero, i.e., $\alpha_i = \alpha_2 = \ldots = \alpha_n = 0$ in the following

$$\alpha_1 \mathbf{u}_1 + \alpha_2 \mathbf{u}_2 + \ldots + \alpha_i \mathbf{u}_i + \ldots + \alpha_n \mathbf{u}_n = \mathbf{0} \tag{2.15}$$

The basis of all base vectors in the vector space $\mathbb{V}$, $\mathcal{A} = \{\mathbf{e}_1, \ldots, \mathbf{e}_i, \ldots, \mathbf{e}_n\}$, is linearly independent. Every vector $\mathbf{u} \in \mathbb{V}$ can be expressed as a linear combination of the base vectors in the set $\mathcal{A}$. Therefore $\mathcal{A}$ is called a standard basis of the linear vector space $\mathbb{V}$.

2.2.3 Euclidean *n*-Space

We next consider the space of all ordered n-tuples of real numbers $(x_1, x_2, \ldots, x_n)$ with a basis $\{\mathbf{e}_i\}_{i=1}^n$. We have the formal notation for $\mathbb{R}^n$ (Bowen and Wang 1976; Greenberg 1998; Grossman 1986; Halmos 1958; Hurley 1981; Kreyszig 2006; Marsden and Tromba 2003; Saxe 2002; Strang 1988; Zill and Cullen 2006),

$$\mathbb{R}^n = \left\{ (x_1, x_2, \ldots, x_n) | \, x_1, x_2, \ldots, x_n \in \mathbb{R}^1 \right\} \tag{2.16}$$

which stands for the set of real n-tuples.

If all the base vectors are of unit length, then the basis is said to be orthonormal. The ith base vector in the standard orthonormal basis, $\{\mathbf{e}_i\}_{i=1}^n$, is given by

$$\mathbf{e}_i = (0, \ldots, \alpha_i, \ldots, 0), \ \ \alpha_i = 1 \tag{2.17}$$

The vector space which the standard orthonormal basis spans is often called Euclidean n-space, and it is denoted by $\mathbb{E}^n$. Note that in general tensor analysis, $\mathbb{R}^n$ is not equal to $\mathbb{E}^n$ (Calin and Chang 2005; Sokolnikoff 1951). In $\mathbb{E}^n$, Christoffel symbols vanish (Sokolnikoff 1951, page 161). $\mathbb{R}^n$ is a smooth manifold (that is, infinitely differentiable) endowed with a Riemannian metric g, whereas $\mathbb{E}^n$ is a smooth manifold $\mathbb{R}^n$ equipped with a Euclidean metric ρ_2. Detailed discussions follow in the subsequent sections. In the introductory section to the real vector space, most authors do not distinguish between $\mathbb{E}^n$ and $\mathbb{R}^n$ (Marsden and Hughes 1983), i.e., $\mathbb{E}^n = \mathbb{R}^n$.

In the n-dimensional Euclidean space $\mathbb{E}^n$, a row vector $\mathbf{x}$ with n components is expressed as

$$\mathbf{x} = (x_1, x_2, \ldots, x_n) = \sum_{i=1}^{n} x_i \mathbf{e}_i, \ \ \mathbf{x} \in \mathbb{R}^n = \mathbb{E}^n \tag{2.18}$$

which is an ordered set of n numbers. The same notation for Euclidean n-space holds for column vectors. For instance, a column vector, $\mathbf{y}$, with n components is written as

$$\mathbf{y} = \begin{Bmatrix} y_1 \\ y_2 \\ \cdots \\ y_n \end{Bmatrix}, \ \ \mathbf{y} \in \mathbb{R}^n = \mathbb{E}^n \tag{2.19}$$

which is also an ordered set of n numbers. Note that both the n-component row vector $\mathbf{x}$ and the n-component column vector $\mathbf{y}$ are often simply called an n-vector.

Note that both the 3-component row vector $\mathbf{x}$ and the 3-component column vector $\mathbf{y}$ are often simply called a 3-vector. The 3-vector $\mathbf{x}$ in Equation (2.18) is often written as

$$\mathbf{x} = x_1\mathbf{i} + x_2\mathbf{j} + x_3\mathbf{k}, \ \ \mathbf{x} \in \mathbb{R}^3 \tag{2.20}$$

where an ordered set $\{\mathbf{i}, \mathbf{j}, \mathbf{k}\}$ is an orthonormal basis for the set $\mathbb{R}^3$, and base vectors $\mathbf{i}, \mathbf{j}, \mathbf{k}$ can be defined respectively as $\mathbf{i} = (1, 0, 0)$, $\mathbf{j} = (0, 1, 0)$, $\mathbf{k} = (0, 0, 1)$. It is known that Hamilton (1805–1865) first used the basis notation for a vector in $\mathbb{R}^3$ (Grossman 1986; Hurley 1981; Marsden and Tromba 2003). Therefore, in view of Equation (2.12) a position vector in $\mathbb{R}^3$ with an orthonormal basis $\{\mathbf{e}_i\}_{i=1}^3$ can be written as

$$\mathbf{r} = \alpha_i \mathbf{e}_i \tag{2.21}$$

which is called the Einstein summation convection. The summation is carried out over the repeated indices.

2.2.4 Inner Product Space

An inner product is a real-valued function defined on the domain, the Cartesian product of a vector space, i.e., $\mathbb{V} \times \mathbb{V}$. The inner product function is also a mapping function, i.e., $\langle \cdot, \cdot \rangle : \mathbb{V} \times \mathbb{V} \rightarrow \mathbb{R}^+$. Note that $\mathbb{R}^+ = (0, \infty)$ (Maddox 1988). Then, the inner product space is defined as also a vector space of an ordered pair, i.e., $(\mathbb{V}, \langle \cdot, \cdot \rangle)$.The inner product function in the inner product space satisfies the following conditions:

(a) $\langle \mathbf{x}, \mathbf{x} \rangle = 0$ if and only if $\mathbf{x} = 0,\ \forall \mathbf{x} \in \mathbb{V}$ (nondegeneracy).
(b) $\langle \mathbf{x}, \mathbf{y} + \mathbf{z} \rangle = \langle \mathbf{x}, \mathbf{y} \rangle + \langle \mathbf{x}, \mathbf{z} \rangle,\ \forall \mathbf{x}, \mathbf{y}, \mathbf{z} \in \mathbb{V}$ (distributivity).
(c) $\langle \mathbf{x}, \mathbf{y} \rangle = \langle \mathbf{y}, \mathbf{x} \rangle,\ \forall \mathbf{x}, \mathbf{y} \in \mathbb{V}$ (symmetry).
(d) $\langle \mathbf{x}, \mathbf{x} \rangle \geq 0,\ \forall \mathbf{x} \in \mathbb{V}$ (nonnegative).
(e) $\langle \alpha \mathbf{x}, \mathbf{y} \rangle = \alpha \langle \mathbf{x}, \mathbf{y} \rangle,\ \forall \mathbf{x}, \mathbf{y} \in \mathbb{V}$ and $\alpha \in \mathbb{R}$ (multiplicativity).

The inner product space is often called a pre-Hilbert space. The inner product of two vectors results in a scalar; hence it is called a scalar product.

Next, consider a row vector $\mathbf{x}$ with three components in three-dimensional Euclidean space. Therefore, $\mathbf{x}$ is a 3-component row vector, and is expressed as

$$\mathbf{x} = (x_1, x_2, x_3), \quad \mathbf{x} \in \mathbb{R}^3 \tag{2.22}$$

which is an ordered set of three numbers. Then, the inner product of the two 3-component row vectors $\mathbf{x}$, $\mathbf{z}$ is defined as

$$\langle \mathbf{x}, \mathbf{z} \rangle = \mathbf{x} \cdot \mathbf{z} = (x_1, x_2, x_3) \cdot (z_1, z_2, z_3) = x_1 z_1 + x_2 z_2 + x_3 z_3 \tag{2.23}$$

where the 3-component row vector $\mathbf{z} = (z_1, z_2, z_3)$, $\mathbf{z} \in \mathbb{R}^3$. The inner product is also called the dot product or the scalar product. Note that Equation (2.23) defines the inner product between a row vector and a row vector. And, we consider a column vector $\mathbf{y}$ with three components, namely, a 3-component column vector written as

$$\mathbf{y} = \begin{Bmatrix} y_1 \\ y_2 \\ y_3 \end{Bmatrix}, \quad \mathbf{y} \in \mathbb{R}^3 \tag{2.24}$$

which is also an ordered set of three numbers. Then, the inner product of a row vector $\mathbf{x}$ and a column vector $\mathbf{y}$ is given by

$$\mathbf{x} \cdot \mathbf{y} = (x_1, x_2, x_3) \cdot \begin{Bmatrix} y_1 \\ y_2 \\ y_3 \end{Bmatrix} = x_1 y_1 + x_2 y_2 + x_3 y_3 \tag{2.25}$$

The inner product can be also performed between a column vector and a column vector. The number of components of one vector should be in agreement with that of another vector to compute the inner product between two vectors.

There are two classes of product between vectors, namely, the scalar product and the vector product. The vector product of the two 3-component row vectors in $\mathbb{R}^3$ is defined as

$$\begin{aligned} \mathbf{x} \times \mathbf{z} &= (x_1, x_2, x_3) \times (z_1, z_2, z_2) \\ &= ((x_2 z_3 - x_3 z_2), (x_3 z_1 - x_1 z_3), (x_1 z_2 - x_2 z_1)) \end{aligned}$$

$$= (x_2z_3 - x_3z_2)\,\mathbf{i} + (x_3z_1 - x_1z_3)\,\mathbf{j} + (x_1z_2 - x_2z_1)\,\mathbf{k}$$
$$= -\mathbf{z} \times \mathbf{x} \tag{2.26}$$

The vector product is also called the outer or cross product, and can be rewritten as

$$\mathbf{x} \times \mathbf{z} = \mathbf{i}\begin{vmatrix} x_2 & x_3 \\ z_2 & z_3 \end{vmatrix} - \mathbf{j}\begin{vmatrix} x_1 & x_3 \\ z_1 & z_3 \end{vmatrix} + \mathbf{j}\begin{vmatrix} x_1 & x_2 \\ z_1 & z_2 \end{vmatrix}$$
$$= \begin{vmatrix} \mathbf{i} & \mathbf{j} & \mathbf{k} \\ x_1 & x_2 & x_3 \\ z_1 & z_2 & z_3 \end{vmatrix} \tag{2.27}$$

where we use the determinant of a square matrix, which will be discussed in detail later. It should be noted that the cross product is defined only in $\mathbb{R}^3$.

Hence, the inner product of the two vectors gives

$$\mathbf{x} \cdot \mathbf{x} = (x_1, x_2, \ldots, x_n) \cdot (x_1, x_2, \ldots, x_n) = \sum_{i=1}^{n} x_i^2 \tag{2.28}$$

Additionally, the inner product of a row vector $\mathbf{x}$ and a column vector $\mathbf{y}$ is given by

$$\mathbf{x} \cdot \mathbf{y} = (x_1, x_2, \ldots, x_n) \cdot \begin{Bmatrix} y_1 \\ y_2 \\ \cdots \\ y_n \end{Bmatrix} = \sum_{i=1}^{n} x_i y_i \tag{2.29}$$

As stated in $\mathbb{R}^3$, if the inner product of two n-vectors is zero, then one n-vector is orthogonal to another n-vector. Further, in view of Equations (2.21) and (2.13) the vector components which are scalar quantities and related to base vectors can be obtained as

$$v_i = \mathbf{v} \cdot \mathbf{e}_i = \mathbf{e}_i \cdot \mathbf{v} \tag{2.30}$$

2.2.5 Metric Spaces

A metric is a real-valued function defined on the domain, the Cartesian product of a vector space, i.e., $\mathbb{V} \times \mathbb{V}$. The metric function $\rho : \mathbb{V} \times \mathbb{V} \to \mathbb{R}$ provides a measure of the distance between elements contained in the vector space. Then, the metric space is also defined as a vector space of an ordered pair, i.e., $(\mathbb{V}, \rho)$. The metric function in the metric space satisfies the following conditions:

(a) $\rho(\mathbf{x}, \mathbf{y}) = 0$ if and only if $\mathbf{x} = \mathbf{y}$, $\forall \mathbf{x}, \mathbf{y} \in \mathbb{V}$ (nondegeneracy).
(b) $\rho(\mathbf{x}, \mathbf{z}) \le \rho(\mathbf{x}, \mathbf{y}) + \rho(\mathbf{y}, \mathbf{z})$, $\forall \mathbf{x}, \mathbf{y}, \mathbf{z} \in \mathbb{V}$ (triangle inequality).
(c) $\rho(\mathbf{x}, \mathbf{y}) = \rho(\mathbf{y}, \mathbf{x})$, $\forall \mathbf{x}, \mathbf{y} \in \mathbb{V}$ (symmetry).
(d) $\rho(\mathbf{x}, \mathbf{y}) \ge 0$, $\forall \mathbf{x}, \mathbf{y} \in \mathbb{V}$ (nonnegative).

In $\mathbb{R}^n$, there are several metrics in mathematics. A Manhattan metric $\rho_1(x, y)$ is defined as

$$\rho_1(\mathbf{x}, \mathbf{y}) = \sum_{i=1}^{n} |x_i - y_i|, \ \forall \mathbf{x}, \mathbf{y} \in \mathbb{R} \tag{2.31}$$

A Euclidean metric $\rho_2(x, y)$ is defined as

$$\rho_2(\mathbf{x}, \mathbf{y}) = \sqrt{\sum_{i=1}^{n} |x_i - y_i|^2}, \ \forall \mathbf{x}, \mathbf{y} \in \mathbb{R}^n \tag{2.32}$$

The Cartesian n-space with the Euclidean metric $(\mathbb{R}^n, \rho_2)$ is called the n-dimensional Euclidean space. A Minkowsky metric $\rho_p(\mathbf{x}, \mathbf{y})$ is defined as

$$\rho_p(\mathbf{x}, \mathbf{y}) = \left[\sum_{i=1}^{n} |x_i - y_i|^p\right]^{1/p}, \ \forall \mathbf{x}, \mathbf{y} \in \mathbb{R}^n \text{ and } p \geq 1 \tag{2.33}$$

A metric in the l^p space is defined as

$$\rho_{l^p}(\mathbf{x}, \mathbf{y}) = \left[\sum_{i=1}^{\infty} |x_i - y_i|^p\right]^{1/p}, \ \forall \mathbf{x}, \mathbf{y} \in \mathbb{R}^\infty \text{ and } p \geq 1 \tag{2.34}$$

The l^p space where the metric is defined as above is a set of all elements $\mathbf{x} = (x_1, x_2, \ldots)$

$$\sum_{i=1}^{\infty} |x_i|^p, \ \forall \mathbf{x} \in \mathbb{R}^\infty \text{ and } p \geq 1 \tag{2.35}$$

Cauchy Sequence and Completeness Consider a sequence $\{\mathbf{x}\}_{i=1}^n$ in a vector space $\mathbb{V}$ equipped with a metric ρ, i.e. a metric space $(\mathbb{V}, \rho)$. The sequence is said to be Cauchy if there is a positive integer N such that $\rho(\mathbf{x}_m, \mathbf{x}_n) < \epsilon$, $\forall m, n > N$ and $\epsilon > 0$. Cauchy sequences do not always converge, but convergent sequences are always Cauchy. The space $\mathbb{V}$ equipped with a metric ρ, i.e. a metric space $(\mathbb{V}, \rho)$, is said to be complete if every Cauchy sequence is convergent to a limit, which is a member of the vector space $\mathbb{V}$. For example, the set of rational numbers $\mathbb{Q}$ is not complete. Euclid found that a sequence in $\mathbb{Q}$ can be convergent to an irrational number $\sqrt{2}$, which is not a member of the set $\mathbb{Q}$ (Marsden and Hoffman 1993). As a subspace of $\mathbb{R}$, consider a set of real numbers $(0, 1)$. Then, the infinite sequence $\left\{\frac{1}{n}\right\}_{n=1}^{\infty}$ converges to 0, which is not a member of the set $(0, 1)$. Consequently, the set $(0, 1)$ equipped with a metric is not complete.

It is well known that the set of real numbers $\mathbb{R}^n$ is complete (Kreyszig 1978). Note that the complete inner product space is called a Hilbert space. Hence, $\mathbb{R}^n$ equipped with an inner product is a Hilbert space.

2.2.6 Normed Space

A norm is a real-valued function defined on the domain, a vector space, i.e. $\mathbb{V}$. The norm function $\rho : \mathbb{V} \to \mathbb{R}$ provides a measure of the magnitude of an element contained in the vector space. Then, the real normed linear space is also defined as a vector space of an ordered pair, i.e. $(\mathbb{V}, \|\cdot\|)$. The norm function in the real normed linear space satisfies the following conditions:

(a) $\|\mathbf{x}\| = 0$, if and only if $\|\mathbf{x}\| = 0$, $\forall \mathbf{x} \in \mathbb{V}$ (nondegeneracy).

(b) $\|\mathbf{x} + \mathbf{y}\| \leq \|\mathbf{x}\| + \|\mathbf{y}\|$, $\forall \mathbf{x}, \mathbf{y}, \mathbf{z} \in \mathbb{V}$ (triangle inequality).

(c) $\|\alpha\mathbf{x}\| = |\alpha|\,\|\mathbf{x}\|$, $\forall\mathbf{x} \in \mathbb{V}$ and $\alpha \in \mathbb{R}$ (multiplicativity).
(d) $\rho(\mathbf{x}, \mathbf{y}) \geq 0$, $\forall\mathbf{x}, \mathbf{y} \in \mathbb{V}$ (nonnegative).

In $\mathbb{R}^n$, there are several norms in mathematics. A Manhattan norm $\|\mathbf{x}\|_1$ is defined as

$$\|\mathbf{x}\|_1 = \sum_{i=1}^{n} |x_i|, \ \forall\mathbf{x} \in \mathbb{R} \tag{2.36}$$

A Minkowsky norm $\|\mathbf{x}\|_p$ is defined as

$$\|\mathbf{x}\|_p = \left[\sum_{i=1}^{n} |x_i|^p\right]^{1/p}, \ \forall\mathbf{x} \in \mathbb{R}^n \text{ and } p \geq 1 \tag{2.37}$$

A Euclidean norm $\|\mathbf{x}\|_2$ is defined as

$$\|\mathbf{x}\|_2 = \sqrt{\mathbf{x}\cdot\mathbf{x}} = \sqrt{\sum_{i=1}^{n} |x_i|^2}, \ \forall\mathbf{x} \in \mathbb{R}^n \tag{2.38}$$

which gives the magnitude or length of the vector. Further, the difference between the two row vectors is given by

$$\|\mathbf{x} - \mathbf{z}\|_2 = \sqrt{\sum_{i=1}^{n} (x_i - z_i)^2}, \ \forall\mathbf{x}, \mathbf{z} \in \mathbb{R}^n \tag{2.39}$$

which is equivalent to the Euclidean metric shown in Equation (2.32) as the distance between the two nonzero vectors. The angle between two vectors in $\mathbb{R}^3$ is defined as

$$\theta = \cos^{-1}\left(\frac{\mathbf{x}\cdot\mathbf{z}}{\|\mathbf{x}\|_2\,\|\mathbf{z}\|_2}\right) \tag{2.40}$$

As mentioned previously, the set of real numbers $\mathbb{R}^n$ is complete (Kreyszig 1978). And, the complete normed space is called a Banach space. Hence, $\mathbb{R}^n$ equipped with a norm is a Banach space.

2.3 MATRIX ALGEBRA

A matrix is defined as an array consisting of $m \times n$ entries as follows:

$$\mathbf{A} = \begin{bmatrix} a_{11} & a_{12} & \dots & a_{1n} \\ a_{21} & a_{22} & \dots & a_{2n} \\ \dots & \dots & \dots & \dots \\ a_{m1} & a_{m2} & \dots & a_{mn} \end{bmatrix}_{m\times n} \tag{2.41}$$

where the first subscript m represents the row number, and the second subscript n the column number. If m is not equal to n, $\mathbf{A}$ is called a rectangular matrix. If $m = n$, then $\mathbf{A}$ is called an $m\times$ m square matrix. If $m = 1$, then $\mathbf{A}$ becomes a n-component row vector, which is of the form

$$\mathbf{a} = \begin{bmatrix} a_{11} & a_{12} & \dots & a_{1n} \end{bmatrix} \tag{2.42}$$

If $n = 1$, then $\mathbf{A}$ becomes a m-component column vector which is of the form

$$\mathbf{a} = \begin{Bmatrix} a_{11} \\ a_{21} \\ \dots \\ a_{m1} \end{Bmatrix} \tag{2.43}$$

We frequently encounter a set of linear algebraic equations as follows:

$$\begin{bmatrix} a_{11} & a_{12} & \dots & a_{1n} \\ a_{21} & a_{22} & \dots & a_{2n} \\ \dots & \dots & \dots & \dots \\ a_{m1} & a_{m2} & \dots & a_{mn} \end{bmatrix} \begin{Bmatrix} x_1 \\ x_2 \\ \dots \\ x_n \end{Bmatrix} = \begin{Bmatrix} b_1 \\ b_2 \\ \dots \\ b_m \end{Bmatrix} \tag{2.44}$$

In matrix notation, Equation (2.44) can be written in a succinct form as

$$\mathbf{Ax} = \mathbf{b} \tag{2.45}$$

where the $m \times n$ matrix $\mathbf{A}$ is called the coefficient matrix, the unknown vector $\mathbf{x}$ is a column vector with n components, and the known vector $\mathbf{b}$ is also a column vector with n components. The above notation is called matrix and vector notation (Simo and Hughes 1998).

2.3.1 Determinant of a Coefficient Matrix

Let us consider the coefficient matrix $\mathbf{A}$ is a 2×2 matrix as follows:

$$\begin{bmatrix} a_{11} & a_{12} \\ a_{21} & a_{22} \end{bmatrix} \begin{Bmatrix} x_1 \\ x_2 \end{Bmatrix} = \begin{Bmatrix} b_1 \\ b_2 \end{Bmatrix} \tag{2.46}$$

We need to define a determinant of second-order as

$$D = \det \mathbf{A} = \begin{vmatrix} a_{11} & a_{12} \\ a_{21} & a_{22} \end{vmatrix} = a_{11}a_{22} - a_{12}a_{21} \tag{2.47}$$

For the linear system given in Equation (2.46), Cramer's rule gives the solution of the unknown variables as follows:

$$\begin{aligned} x_1 &= \frac{\begin{vmatrix} b_1 & a_{12} \\ b_2 & a_{22} \end{vmatrix}}{D} = \frac{b_1 a_{22} - a_{12} b_2}{D}, \\ x_2 &= \frac{\begin{vmatrix} a_{1a} & b_1 \\ a_{21} & b_2 \end{vmatrix}}{D} = \frac{b_2 a_{11} - a_{21} b_1}{D} \end{aligned} \tag{2.48}$$

It is easy to show that the length of the vector product of two vectors $\mathbf{a} = (a_1, a_2)$, $\mathbf{b} = (b_1, b_2)$ in $\mathbb{R}^2$ is identical to the area of the parallelogram by the two vectors as follows:

$$A = \|\mathbf{a} \times \mathbf{b}\| = \begin{vmatrix} a_1 & a_2 \\ b_1 & b_2 \end{vmatrix} \tag{2.49}$$

where A denotes the area of the parallelogram and the vector product obeys the right-hand rule. Note that in contrast to the dot product given in Equation (2.40), the right oriented area of a parallelogram can be also given by

$$A = \|\mathbf{a} \times \mathbf{b}\| = \|\mathbf{a}\| \, \|\mathbf{b}\| \sin\theta, \ 0 \leq \theta \leq \pi \tag{2.50}$$

Suppose that the linear system is a set of three linear equations as follows:

$$\begin{bmatrix} a_{11} & a_{12} & a_{13} \\ a_{21} & a_{22} & a_{23} \\ a_{31} & a_{32} & a_{33} \end{bmatrix} \begin{Bmatrix} x_1 \\ x_2 \\ x_3 \end{Bmatrix} = \begin{Bmatrix} b_1 \\ b_2 \\ b_3 \end{Bmatrix} \tag{2.51}$$

Then, a determinant of third order is defined by

$$\begin{aligned} D = \det \mathbf{A} &= \begin{vmatrix} a_{11} & a_{12} & a_{13} \\ a_{21} & a_{22} & a_{23} \\ a_{31} & a_{32} & a_{33} \end{vmatrix} \\ &= a_{11} \begin{vmatrix} a_{22} & a_{23} \\ a_{32} & a_{33} \end{vmatrix} - a_{12} \begin{vmatrix} a_{21} & a_{23} \\ a_{31} & a_{33} \end{vmatrix} + a_{13} \begin{vmatrix} a_{21} & a_{22} \\ a_{31} & a_{32} \end{vmatrix} \end{aligned} \tag{2.52}$$

The determinant of third order can be rewritten as

$$D = \det \mathbf{A} = a_{11}C_{11} + a_{12}C_{12} + a_{13}C_{13} \tag{2.53}$$

where C_{11}, C_{12}, C_{13} are called the cofactor of the matrix $\mathbf{A}$. Note that the triple scalar product of three vectors $\mathbf{a} = (a_1, a_2, a_3)$, $\mathbf{b} = (b_1, b_2, b_3)$, $\mathbf{c} = (c_1, c_2, c_3)$ in $\mathbb{R}^3$ is identical to the volume of the parallelepiped by the three vectors as follows:

$$V = (\mathbf{a} \times \mathbf{b}) \cdot \mathbf{c} = \mathbf{a} \cdot (\mathbf{b} \times \mathbf{c}) = \begin{vmatrix} a_1 & a_2 & a_3 \\ b_1 & b_2 & b_3 \\ c_1 & c_2 & c_3 \end{vmatrix} \tag{2.54}$$

where V denotes the volume of the parallelepiped and the triple scalar product obeys the right-hand rule.

We wish to define the ijth cofactor of the coefficient matrix $\mathbf{A}$ in D as

$$C_{ij} = (-1)^{i+j} M_{ij} \tag{2.55}$$

where M_{ij} is called the ijth minor of the coefficient matrix $\mathbf{A}$ in D. Then, the determinant of the $n \times n$ matrix $\mathbf{A}$ can be given by

$$\begin{aligned} \det \mathbf{A} &= \sum_{j=1}^{n} a_{ij} C_{ij} \\ &= \sum_{j=1}^{n} a_{ji} C_{ji} \end{aligned} \tag{2.56}$$

Then, Cramer's rule asserts that the solution of the unknown variables is

$$x_i = \sum_{j=1}^{n} \left(\frac{C_{ji}}{\det \mathbf{A}} \right) b_j \tag{2.57}$$

where b_j denotes components of the known vector $\mathbf{b}$.

2.3.2 Matrix Multiplication

Let us define the transpose of an $m \times n$ matrix $\mathbf{A}$ as

$$\mathbf{A}^T = \begin{bmatrix} a_{11} & a_{21} & \dots & a_{m1} \\ a_{12} & a_{22} & \dots & a_{m2} \\ \dots & \dots & \dots & \dots \\ a_{1n} & a_{2n} & \dots & a_{mn} \end{bmatrix}_{n \times m} \tag{2.58}$$

Note that for an $n \times n$ square matrix, it is said to be symmetric when $\mathbf{A} = \mathbf{A}^T$. In the preceding section, we found that the matrix multiplication of an $m \times n$ matrix $\mathbf{A}$ and an n-component vector $\mathbf{x}$ produces an m-component column vector $\mathbf{b}$. Hence, we obtain the relation

$$\mathbf{c} \cdot \mathbf{Ax} = \mathbf{c}^T \mathbf{Ax} = \mathbf{c} \cdot \mathbf{b} = \mathbf{c}^T \mathbf{b} \tag{2.59}$$

where $\mathbf{c}$ is an m-component column vector. Note that Equation (2.59) yields a certain scalar quantity. Let $\mathbf{A}$ be an $m \times p$ matrix, and let $\mathbf{B}$ be a $p \times n$ matrix. Then the product of the two matrices yields an $m \times n$ matrix as follows

$$\mathbf{A}_{m \times p} \mathbf{B}_{p \times n} = \mathbf{C}_{m \times n} \tag{2.60}$$

The Inverse of a Square Matrix For the linear system given in Equation (2.45), if $\mathbf{A}$ is invertible, then $\det \mathbf{A} \neq 0$ and the linear system has the unique solution as follows:

$$\mathbf{x} = \mathbf{Bb} \tag{2.61}$$

where $\mathbf{B}$ is called the inverse of an $n \times n$ matrix $\mathbf{A}$ and $\mathbf{B} = \mathbf{A}^{-1}$. Let $\mathbf{C}$ be the matrix of cofactors of $\mathbf{A}$, i.e.,

$$\mathbf{C} = \begin{bmatrix} C_{11} & C_{12} & \dots & C_{1n} \\ C_{21} & C_{22} & \dots & C_{2n} \\ \dots & \dots & \dots & \dots \\ C_{n1} & C_{n2} & \dots & C_{nn} \end{bmatrix} \tag{2.62}$$

Then the adjoint of $\mathbf{A}$ is defined as

$$\text{adj}\,\mathbf{A} = \mathbf{C}^T = \begin{bmatrix} C_{11} & C_{21} & \dots & C_{n1} \\ C_{12} & C_{22} & \dots & C_{n2} \\ \dots & \dots & \dots & \dots \\ C_{1n} & C_{2n} & \dots & C_{nn} \end{bmatrix} \tag{2.63}$$

The inverse of the square matrix $\mathbf{A}$ is defined as

$$\mathbf{A}^{-1} = \frac{1}{\det \mathbf{A}} \text{adj}\,\mathbf{A} \tag{2.64}$$

Therefore, it is easy to show that

$$\mathbf{A}(\text{adj}\,\mathbf{A}) = (\det \mathbf{A})\mathbf{I} \tag{2.65}$$

where $\mathbf{I}$ is called the identity matrix. Note that $\mathbf{A}(\mathbf{A}^{-1}) = (\mathbf{A}^{-1})\mathbf{A} = \mathbf{I}$.

2.4 VECTOR DIFFERENTIAL CALCULUS

Thus far, we have discussed vectors and matrices containing numeric entries. In this section, we shall deal with functions, vectors and matrices containing functions of one variable or several variables as their entries.

2.4.1 Scalar-Valued Functions of Multivariables

In analytical mechanics, the functions such as the Lagrangian and the Hamiltonian are the scalar-valued function of multivariables. In general, a scalar-valued function of multivariables is defined as $f(\mathbf{x}) : \mathbb{X} \subset \mathbb{R}^n \to \mathbb{R}$. The domain of the scalar-valued function f is $\mathbb{X}$, which is a Banach space as a subset of $\mathbb{R}^n$. The range of the scalar-valued function f is contained in $\mathbb{R}$. Then, as a set, the graph of the scalar-valued function f is given by

$$\text{graph}\, f = \left\{(\mathbf{x}, f(\mathbf{x})) \in \mathbb{R}^{n+1} | \mathbf{x} \in \mathbb{X}\right\} \tag{2.66}$$

In $\mathbb{R}^3$, for the scalar-valued function $f(x, y, z) : \mathbb{X} \subset \mathbb{R}^3 \to \mathbb{R}$, the level set $\mathcal{L}_s$ is defined as

$$\mathcal{L}_s = \{(x, y, z) | f(x, y, z) = c, \ c \in \mathbb{R}\} \subset \mathbb{R}^3 \tag{2.67}$$

where c represents a real constant. In $\mathbb{R}^2$, the above equation becomes a level contour.

Let a scalar-valued function of n-variables $f(\mathbf{x}) : \mathbb{X} \subset \mathbb{R}^n \to \mathbb{R}$ be differentiable at $\mathbf{x} \in \mathbb{X}$. Then, the first-order Taylor's formula can be expressed as

$$f(\mathbf{x} + \mathbf{h}) = f(\mathbf{x}) + \nabla f(\mathbf{x}) \cdot \mathbf{h} + R_1(\mathbf{x}, \mathbf{h}) \tag{2.68}$$

where R_1 is the remainder and $\mathbf{h} \in \mathbb{X} \subset \mathbb{R}^n$ is an increment as a real vector whose length is small. Note that the above equation can be rewritten as

$$\frac{f(\mathbf{x} + \mathbf{h}) - f(\mathbf{x}) - l(\mathbf{h})}{\|\mathbf{h}\|_{\mathbb{X}}} = \frac{1}{\|\mathbf{h}\|_{\mathbb{X}}} R_1(\mathbf{x}, \mathbf{h}) \tag{2.69}$$

where $l(\mathbf{h}) : \mathbb{X} \subset \mathbb{R}^n \to \mathbb{R}$ is a linear mapping and $l(\mathbf{h}) = \nabla f(\mathbf{x}) \cdot \mathbf{h}$. If and only if there exists the linear function such that

$$\lim_{\|\mathbf{h}\|_{\mathbb{X}} \to 0} \left\{ \frac{f\,(\mathbf{x} + \mathbf{h}) - f\,(\mathbf{x}) - l(\mathbf{h})}{\|\mathbf{h}\|_{\mathbb{X}}} \right\} = \mathbf{0} \tag{2.70}$$

then, the scalar-valued function is called Fréchet differentiable at $\mathbf{x} \in \mathbb{X} \subset \mathbb{R}^n$. Consequently, it is said that the scalar-valued function $f\,(\mathbf{x})$ is continuous at $\mathbf{x}$. The linear mapping $l(\mathbf{h}) : \mathbb{X} \subset \mathbb{R}^n \to \mathbb{R}$ is defined as

$$l(\mathbf{h}) = \nabla f\,(\mathbf{x}) \cdot \mathbf{h} \tag{2.71}$$

where the Fréchet derivative of the scalar-valued function $f(\mathbf{x})$ (Courant 1974; Marsden and Hughes 1983; Simo and Hughes 1998) is defined as

$$\nabla f\,(\mathbf{x}) = \lim_{\|\mathbf{h}\|_{\mathbb{X}} \to 0} \left\{ \frac{f\,(\mathbf{x} + \mathbf{h}) - f\,(\mathbf{x})}{\|\mathbf{h}\|_{\mathbb{X}}} \right\} \tag{2.72}$$

In addition, the linear mapping $l(\mathbf{h}) : \mathbb{X} \subset \mathbb{R}^n \to \mathbb{R}$ is called the differential of the scalar-valued function (Edwards 1994). The Fréchet derivative of the scalar-valued function $f(\mathbf{x}) : \mathbb{X} \subset \mathbb{R}^n \to \mathbb{R}$ can be expressed as an n-component row vector

$$\nabla f(\mathbf{x}) = \mathbf{grad}\, f = \frac{\partial f(\mathbf{x})}{\partial \mathbf{x}} = \begin{bmatrix} \frac{\partial f(\mathbf{x})}{\partial x_1} & \frac{\partial f(\mathbf{x})}{\partial x_2} & \cdots & \frac{\partial f(\mathbf{x})}{\partial x_n} \end{bmatrix}_{1\times n} \tag{2.73}$$

where the partial derivative $\frac{\partial f}{\partial x_i}$ is defined as

$$\frac{\partial f(\mathbf{x})}{\partial x_i} = \lim_{\epsilon \to 0} \left\{ \frac{f(\mathbf{x} + \epsilon \mathbf{e}_i) - f(\mathbf{x})}{\epsilon} \right\}, \quad i = 1, \ldots, n \tag{2.74}$$

where $\mathbf{e}_i$ denotes the ith base vector in the standard basis of the linear vector space $\mathbb{X}$. Equation (2.73) is often called the gradient of the scalar-valued function $f(\mathbf{x})$. Recall that we defined the graph of $f(\mathbf{x})$ as Equation (2.66). At the point $\mathbf{x} = \mathbf{x}_0$, the linear approximation $L_{\mathbf{x}_0} f(\mathbf{x})$ can be defined as

$$L_{\mathbf{x}_0} f(\mathbf{x}) = f(\mathbf{x}_0) + \nabla f(\mathbf{x}_0) \cdot (\mathbf{x} - \mathbf{x}_0) \tag{2.75}$$

which represents the equation of the plane tangent to the graph of $f(\mathbf{x})$ at the point $\mathbf{x}_0$ (Bonet and Wood 2008; Edwards 1994; Holzapfel 2000; Marsden and Tromba 2003; Simo and Hughes 1998; Troutman 1983).

Without any difficulty, the increment $\mathbf{h}$ can be given by

$$\mathbf{h} = \epsilon \eta \tag{2.76}$$

where η and ϵ denote an arbitrary vector and an infinitesimal parameter. Then, Equation (2.69) can be rewritten as

$$\frac{f(\mathbf{x} + \epsilon\eta) - f(\mathbf{x}) - l(\epsilon\eta)}{\epsilon \, \|\eta\|_{\mathbb{X}}} = \frac{1}{\epsilon \, \|\eta\|_{\mathbb{X}}} R_1(\mathbf{x}, \epsilon\eta) \tag{2.77}$$

Since the scalar-valued function $f(\mathbf{x})$ is Fréchet differentiable at $\mathbf{x} \in \mathbb{X} \subset \mathbb{R}^n$, there exists a linear function such that

$$\frac{1}{\|\eta\|_{\mathbb{X}}} \lim_{\epsilon \to 0} \left\{ \frac{f(\mathbf{x} + \epsilon\eta) - f(\mathbf{x}) - l(\epsilon\eta)}{\epsilon} \right\} = 0 \tag{2.78}$$

Without any difficulty, we can rewrite the above equation in the following manner

$$\lim_{\epsilon \to 0} \left\{ \frac{f(\mathbf{x} + \epsilon\eta) - f(\mathbf{x})}{\epsilon} - \nabla f(\mathbf{x}) \cdot \eta \right\} = 0 \tag{2.79}$$

Then, the Gâteaux derivative as a linear mapping is defined as

$$\mathcal{G} f(\mathbf{x}; \eta) = \lim_{\epsilon \to 0} \frac{f(\mathbf{x} + \epsilon\eta) - f(\mathbf{x})}{\epsilon} = \nabla f(\mathbf{x}) \cdot \eta \tag{2.80}$$

And then, the scalar-valued function $f(\mathbf{x})$ is said to be Gâteaux differentiable at $\mathbf{x} \in \mathbb{X}$. Let us introduce new notations as follows:

$$D_\eta f(\mathbf{x}) = \nabla f(\mathbf{x}) \cdot \eta = \left. \frac{d}{d\epsilon} f(\mathbf{x} + \epsilon\eta) \right|_{\epsilon = 0} \tag{2.81}$$

where the Gâteaux derivative $\nabla f(\mathbf{x}) \cdot \boldsymbol{\eta}$ is often referred to as the directional derivative of $f(\mathbf{x})$ in the direction $\boldsymbol{\eta}$ (Bonet and Wood 2008; Edwards 1994; Grossman 1986; Holzapfel 2000; Hurley 1981; Marsden and Tromba 2003; Simo and Hughes 1998).

Let the third derivatives of the scalar-valued function of n-variables $f(\mathbf{x}) : \mathbb{X} \subset \mathbb{R}^n \rightarrow \mathbb{R}$ be differentiable at $\mathbf{x} \in \mathbb{X}$. Then, the second-order Taylor's formula gives

$$f(\mathbf{x}+\mathbf{h}) = f(\mathbf{x}) + \nabla f(\mathbf{x}) \cdot \mathbf{h} + Hf(\mathbf{x})(\mathbf{h}) + R_2(\mathbf{x}, \mathbf{h}) \tag{2.82}$$

where $Hf(\mathbf{x})(\mathbf{h})$ is called the Hessian of $f(\mathbf{x})$ at $\mathbf{x}$. The Hessian is a quadratic function, which can be expressed as

$$Hf(\mathbf{x})(\mathbf{h}) = \frac{1}{2}\mathbf{h} \cdot \nabla(\nabla f(\mathbf{x}))\mathbf{h} \tag{2.83}$$

where $\nabla(\nabla f(\mathbf{x}))$ is called the Hessian matrix, which is such that

$$\nabla(\nabla f(\mathbf{x})) = \begin{Bmatrix} \nabla\dfrac{\partial f(\mathbf{x})}{\partial x_1} \\ \nabla\dfrac{\partial f(\mathbf{x})}{\partial x_2} \\ \cdots \\ \nabla\dfrac{\partial f(\mathbf{x})}{\partial x_n} \end{Bmatrix} = \begin{bmatrix} \dfrac{\partial^2 f(\mathbf{x})}{\partial x_1 \partial x_1} & \dfrac{\partial^2 f(\mathbf{x})}{\partial x_1 \partial x_2} & \cdots & \dfrac{\partial^2 f(\mathbf{x})}{\partial x_1 \partial x_n} \\ \dfrac{\partial^2 f(\mathbf{x})}{\partial x_2 \partial x_1} & \dfrac{\partial^2 f(\mathbf{x})}{\partial x_2 \partial x_2} & \cdots & \dfrac{\partial^2 f(\mathbf{x})}{\partial x_2 \partial x_n} \\ \cdots & \cdots & \cdots & \cdots \\ \dfrac{\partial^2 f(\mathbf{x})}{\partial x_n \partial x_1} & \dfrac{\partial^2 f(\mathbf{x})}{\partial x_n \partial x_2} & \cdots & \dfrac{\partial^2 f(\mathbf{x})}{\partial x_n \partial x_n} \end{bmatrix}_{n\times n} \tag{2.84}$$

The Hessian is called positive-definite if $Hf(\mathbf{x})(\mathbf{h}) \geq \mathbf{0}$, $\forall \mathbf{h} \in \mathbf{R}^n$, and $Hf(\mathbf{x})(\mathbf{h}) = \mathbf{0}$ only for $\mathbf{h} = \mathbf{0}$. For example, the total potential energy in the theory of elasticity is positive-definite as the determinants of all diagonal sub matrices of the constitutive matrix are greater than zero. On the contrary, the Hessian is called positive-negative if $Hf(\mathbf{x})(\mathbf{h}) \leq \mathbf{0}$, $\forall \mathbf{h} \in \mathbf{R}^n$, and $Hf(\mathbf{x})(\mathbf{h}) = \mathbf{0}$ only for $\mathbf{h} = \mathbf{0}$.

2.4.2 Vector-Valued Functions of Multivariables

In the Cartesian coordinate system, i.e. $\mathbb{R}^3$, the position of a moving particle is given by

$$\mathbf{x}(t) = (x(t), y(t), z(t)), \ t \in \mathcal{T} \subset \mathbb{R}^+ \tag{2.85}$$

which is a representation of a vector-valued function (or simply vector function) of one real variable, i.e. $\mathbf{x}(t) : \mathbb{R} \rightarrow \mathbb{R}^3$. In Euclidean n-space, the position as a vector function can be expressed as

$$\mathbf{x}(t) = (x_1(t), x_2(t), \ldots, x_n(t)) : \mathcal{T} \rightarrow \mathbb{R}^n \tag{2.86}$$

which is often called a path or trajectory in $\mathbb{R}^n$. Note that the image of the path is called a curve of $\mathbf{x}(t)$ (Marsden and Tromba 2003). The functions $x_1(t), x_2(t), \ldots, x_n(t)$ are called the component or coordinate functions of the vector-valued function $\mathbf{x}(t)$. In the above equation, the real variable t is called a parameter and the components of the vector are called the coordinate function of $\mathbf{x}(t)$. Hence, the vector $\mathbf{x}(t) : \mathcal{T} \subset \mathbb{R} \rightarrow \mathbb{R}^n$ can be called a vector-valued function of one parameter.

Consider two vector-valued functions $\mathbf{x}(t) : \mathcal{T} \subset \mathbb{R}^+ \rightarrow \mathbb{R}^n$ and $\mathbf{y}(t) : \mathcal{T} \subset \mathbb{R}^+ \rightarrow \mathbb{R}^n$ which are differentiable on the open interval $\mathcal{T} =]t_1, t_2[$. And let α be a scalar and let $f(t)$ be a real-valued function of the real variable t, which is also defined on the open interval $\mathcal{T} =]t_1, t_2[$. Then the two vector-valued functions satisfy the following properties

(a) Summation

$$\frac{d}{dt}(\mathbf{x}(t) + \mathbf{y}(t)) = \frac{d\mathbf{x}(t)}{dt} + \frac{d\mathbf{y}(t)}{dt} \tag{2.87}$$

(b) Multiplication by a scalar

$$\frac{d}{dt}(\alpha\mathbf{x}(t)) = \alpha\frac{d\mathbf{x}(t)}{dt} \tag{2.88}$$

(c) Multiplication by a real-valued function

$$\frac{d}{dt}(f(t)\mathbf{x}(t)) = \frac{df(t)}{dt}\mathbf{x}(t) + f(t)\frac{d\mathbf{x}(t)}{dt} \tag{2.89}$$

(d) Inner product

$$\frac{d}{dt}(\mathbf{x}(t) \cdot \mathbf{y}(t)) = \frac{d\mathbf{x}(t)}{dt} \cdot \mathbf{y}(t) + \mathbf{x}(t) \cdot \frac{d\mathbf{y}(t)}{dt} \tag{2.90}$$

(e) Vector product

$$\frac{d}{dt}(\mathbf{x}(t) \times \mathbf{y}(t)) = \frac{d\mathbf{x}(t)}{dt} \times \mathbf{y}(t)) + \mathbf{x}(t) \times \frac{d\mathbf{y}(t)}{dt} \tag{2.91}$$

(f) Chain rule

$$\frac{d}{dt}\mathbf{x}(f(t)) = \frac{d\mathbf{x}(f(t))}{df}\frac{df(t)}{dt} \tag{2.92}$$

As a consequence of the above property given in Equation (2.90), we have the relation

$$\frac{d}{dt}(\mathbf{x}(t) \cdot \mathbf{x}(t)) = 0 \Rightarrow \frac{d\mathbf{x}(t)}{dt} \cdot \mathbf{x}(t) = \mathbf{x}(t) \cdot \frac{d\mathbf{x}(t)}{dt} = 0 \tag{2.93}$$

Thus, the vector-valued function is orthogonal to the derivative of the vector function with respect to the real variable.

Thus far, we have discussed the derivative of a scalar-valued functions of multivariables or the derivative of a vector-valued function of one real variable. As a more complicated mathematical operation, we shall discuss the derivative of vector-valued functions of multivariables. For instance, the gradient of a scalar-valued function of multivariables shown in Equation (2.73) represents a vector-valued function of multivariables $\mathbf{F}(\mathbf{x}) : \mathbb{R}^n \rightarrow \mathbb{R}^n$.

In general, a vector-valued functions of multivariables $\mathbf{F}(\mathbf{x}) : \mathbb{R}^n \rightarrow \mathbb{R}^m$ is defined as

$$\mathbf{F}(\mathbf{x}) = [F_1(\mathbf{x}) \quad F_2(\mathbf{x}) \quad \ldots \quad F_m(\mathbf{x})]_{1\times m}, \quad \mathbf{x} \in \mathbb{R}^n \tag{2.94}$$

where the coordinate functions $F_i(\mathbf{x}) \in \mathbb{R}$, $, i = 1, \ldots, n$ are the scalar-valued functions of multivariables. The linear approximation for the vector-valued function of multivariables can be defined as

$$L_{\mathbf{x}_0}\mathbf{F}(\mathbf{x}) = \mathbf{F}(\mathbf{x}_0) + \nabla\mathbf{F}(\mathbf{x}_0) \cdot (\mathbf{x} - \mathbf{x}_0) \tag{2.95}$$

where $\nabla\mathbf{F}(\mathbf{x}_0)$ is a linear transformation, denoted by $J_{\mathbf{F}}(\mathbf{x}_0) : \mathbb{R}^n \to \mathbb{R}^m$. The linear transformation is often called the total derivative of the vector-valued function of multivariables, and it can be expressed as

$$J_{\mathbf{F}}(\mathbf{x}_0) = \begin{bmatrix} \dfrac{\partial F_1(\mathbf{x}_0)}{\partial x_1} & \dfrac{\partial F_1(\mathbf{x}_0)}{\partial x_2} & \cdots & \dfrac{\partial F_1(\mathbf{x}_0)}{\partial x_n} \\ \dfrac{\partial F_2(\mathbf{x}_0)}{\partial x_1} & \dfrac{\partial F_2(\mathbf{x}_0)}{\partial x_2} & \cdots & \dfrac{\partial F_2(\mathbf{x}_0)}{\partial x_n} \\ \cdots & \cdots & \cdots & \cdots \\ \dfrac{\partial F_m(\mathbf{x}_0)}{\partial x_1} & \dfrac{\partial F_m(\mathbf{x}_0)}{\partial x_2} & \cdots & \dfrac{\partial F_m(\mathbf{x}_0)}{\partial x_n} \end{bmatrix}_{m \times n} \tag{2.96}$$

where $J_{\mathbf{F}}(\mathbf{x}_0)$ is often called the Jacobian matrix of $\mathbf{F}(\mathbf{x})$ at $\mathbf{x}_0$ in honor of Jacobi (1804–1851).

2.5 VECTOR INTEGRAL CALCULUS

2.5.1 Green's Theorem in the Plane

Green's theorem plays an important role in two-dimensional problems such as plate problems in computational dynamics. The relation between the line integral on the boundary and the surface integral on the two-dimensional region can be obtained by Green's theorem. Consider a vector-valued function of two variables $\mathbf{F}(\mathbf{x}) : D \subset \mathbb{R}^2 \to \mathbb{R}^2$, which is given by

$$\mathbf{F}(\mathbf{x}) = P(\mathbf{x})\mathbf{i} + Q(\mathbf{x})\mathbf{j}, \ \mathbf{x} = (x, y) \tag{2.97}$$

where $P(\mathbf{x})$ *and* $Q(\mathbf{x})$ are components of $\mathbf{F}(\mathbf{x})$ and they are scalar functions of two real variables. By use of Fubini's theorem, we have the relations (Greenberg 1998; Hurley 1981; Marsden and Tromba 2003)

$$\iint_D \frac{\partial P(\mathbf{x})}{\partial y} dxdy = -\oint_{\partial D^+} P(\mathbf{x})dx, \ \iint_D \frac{\partial Q(\mathbf{x})}{\partial x} dxdy = \oint_{\partial D^+} Q(\mathbf{x})dy \tag{2.98}$$

where ∂D^+ indicates the counterclockwise orientation of the boundary curve. Then, Green's theorem asserts that

$$\iint_D \left(\frac{\partial Q(\mathbf{x})}{\partial x} - \frac{\partial P(\mathbf{x})}{\partial y} \right) dA = \oint_{\partial D^+} P(\mathbf{x})dx + Q(\mathbf{x})\,dy \tag{2.99}$$

Note that Green's theorem can be written in vectorial form as

$$\iint_D (\text{curl}\,\mathbf{F} \cdot \mathbf{k})\,dA = \oint_{\partial D^+} \mathbf{F} \cdot d\mathbf{x} \tag{2.100}$$

where $d\mathbf{x} = dx\mathbf{i} + dy\mathbf{j}$, i.e., $d\mathbf{x} = (dx, dy)$.

2.5.2 Gauss's Theorem

Green's theorem is often called the divergence theorem, because the volume integral or the surface integral involves the divergence of the vector-valued function (vector field function). The divergence theorem relating the line integral to the surface integral can be derived via Green's theorem. Consider the line integral

$$\oint_{\partial D^+} \mathbf{F}(\mathbf{x}) \cdot \mathbf{n}\, ds = \oint_{\partial D^+} \left(P(\mathbf{x})n_x + Q(\mathbf{x})n_y\right) ds \tag{2.101}$$

where ds denotes the line segment of the boundary curve and $\mathbf{n}$ is the unit normal vector to the boundary curve ∂D^+. Note that the unit vector $\mathbf{n}$ can be given by

$$\mathbf{n} = n_x\mathbf{i} + n_y\mathbf{j} = \frac{dy}{ds}\mathbf{i} - \frac{dx}{ds}\mathbf{j} \tag{2.102}$$

Thus, the term on the right-hand side of Equation (2.101) can be rewritten as

$$\oint_{\partial D^+} \left(P(\mathbf{x})n_x + Q(\mathbf{x})n_y\right) ds = \oint_{\partial D^+} \left(P(\mathbf{x})\, dy - Q(\mathbf{x})\right) dx \tag{2.103}$$

By virtue of Green's theorem shown in Equation (2.99), the above equation can be expressed as

$$\begin{aligned}\oint_{\partial D^+} \left(P(\mathbf{x})n_x + Q(\mathbf{x})n_y\right) ds &= \iint_D \left(\frac{\partial P(\mathbf{x})}{\partial x} + \frac{\partial Q(\mathbf{x})}{\partial y}\right) dA \\ &= \iint_D \nabla \cdot \mathbf{F}(\mathbf{x})\, dA\end{aligned} \tag{2.104}$$

As a result, in view of Equations (2.101) and (2.102) we obtain

$$\oint_{\partial D^+} \mathbf{F}(\mathbf{x}) \cdot \mathbf{n}\, ds = \iint_D \nabla \cdot \mathbf{F}(\mathbf{x})\, dA, \quad \mathbf{x} = (x, y) \tag{2.105}$$

which is the divergence theorem of Gauss between the line integral and the surface integral. Likewise, we have also

$$\iint_{\partial\Omega} \mathbf{F}(\mathbf{x}) \cdot \mathbf{n}\, dS = \iiint_{\Omega} \nabla \cdot \mathbf{F}(\mathbf{x})\, dV, \quad \mathbf{x} = (x, y, z) \tag{2.106}$$

which is the divergence theorem of Gauss between the surface integral and the volume integral.

2.6 MEAN VALUE THEOREM

2.6.1 Scalar Function of a Real Variable

Let $f(x)$ denote a scalar function of a real variable x as shown in Figure 2.4. The scalar function is defined on a closure $\bar{\Gamma}$, i.e., $x \in \bar{\Gamma} = [a, b] \subset \mathbb{R}$. If $f(x) : \bar{\Gamma} \to \mathbb{R}$ is continuous on the $\bar{\Gamma} = [a, b]$ and differentiable on the open set $\Gamma =]a, b[$, then there must exist a point $c \in \Gamma =]a, b[$ such that

$$f(b) - f(a) = f'(c)(b - a), \quad f'(c) = \left.\frac{df(x)}{dx}\right|_{x=c} \tag{2.107}$$

which is often called the mean value theorem for a scalar function of a real variable (Edwards 1994; Grossman 1986; Marsden and Hoffman 1993).

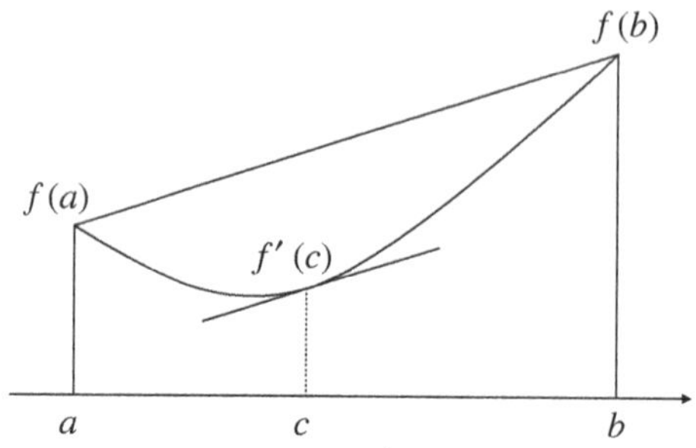

Figure 2.4. Mean value theorem

2.6.2 Scalar Function of Multivariables

The mean value theorem described above can be extended to the scalar-valued function of multivariables. A scalar function $f(\mathbf{x})$ is defined on a closure $\bar{\Omega}$, i.e, $f(\mathbf{x}) : \bar{\Omega} \subset \mathbb{R}^n \to \mathbb{R},\ n \in \mathbb{R}^+$. If $f(\mathbf{x})$ is continuous on the closure $\bar{\Omega}$ and differentiable on the open set Ω, then for every $\mathbf{a}, \mathbf{b} \in \Omega$,

$$f(\mathbf{b}) - f(\mathbf{a}) = f'(\mathbf{c}) \cdot (\mathbf{b} - \mathbf{a}), \ f'(\mathbf{c}) = \nabla f(\mathbf{x})|_{\mathbf{x}=\mathbf{c}} \tag{2.108}$$

where $\mathbf{c} \in \mathcal{L} \subset \Omega$. Note that $\mathcal{L}$ denotes a set of the line segment from $\mathbf{a}$ to $\mathbf{b}$. The above theorem can be proved in the following way. Consider a scalar function $h(t)$ defined on a closure [0,1], i.e. $h(t) : [0, 1] \to \mathbb{R}$. Then the scalar function is defined as

$$h(t) = f(\mathbf{a} + t(\mathbf{b} - \mathbf{a})), \ t \in [0, 1] \subset \mathbb{R} \tag{2.109}$$

Owing to the mean value theorem for a scalar function of a real variable, which is described previously, we can write the following

$$h(1) - h(0) = h'(t_0)(1 - 0), \ t_0 \in]0, 1[\tag{2.110}$$

Taking $\mathbf{c} = \mathbf{a} + t_0(\mathbf{b} - \mathbf{a})$, Equation (2.109) can be identified as Equation (2.108).

2.6.3 Vector Function of Multivariables

Further, we can consider the vector-valued function of multivariables. A vector function $\mathbf{f}(\mathbf{x}) = f_1(\mathbf{x}), \ldots, f_m(\mathbf{x})$ is defined on an open set Ω, i.e, $\mathbf{f}(\mathbf{x}) : \Omega \subset \mathbb{R}^n \to \mathbb{R}^m,\ n, m \in \mathbb{R}^+$. If $\mathbf{f}(\mathbf{x})$ is continuous and differentiable on Ω, then for every $\mathbf{a}, \mathbf{b} \in \Omega$,

$$f_i(\mathbf{b}) - f_i(\mathbf{a}) = f_i'(\mathbf{c}_i) \cdot (\mathbf{b} - \mathbf{a}), \ f_i'(\mathbf{c}) = \nabla f_i(\mathbf{x})|_{\mathbf{x}=\mathbf{c}}, \ i = 1, \ldots, m \tag{2.111}$$

where $\forall \mathbf{c}_i \in \mathcal{L} \subset \Omega,\ i = 1, \ldots, m$. Note that $\mathcal{L}$ denotes a set of the line segment from $\mathbf{a}$ to $\mathbf{b}$.

2.7 FUNCTION SPACES

Most of the finite element formulations have been developed based upon variational calculus. As functional analysis is an important branch of mathematics in the calculus of variations, the reader who wishes to study the finite element method is expected to understand the basics of functional analysis. We shall discuss briefly elements of functional analysis including several function spaces such as Lebesgue space, Hilbert space, Banach space, Sobolev space, and so on.

In general, a set of all scalar functions is called a function space (Kreyszig 1978), whose elements (entries) are functions, namely, zero vectors. Hence, a function space is a vector space.

2.7.1 Inner Product Space

Consider a function space $\mathcal{L}$, which is a linear set $\{f(\mathbf{x}), g(\mathbf{x})\}$. The elements are scalar-valued functions of multivariables, i.e. $f(\mathbf{x}) : \Omega \subset \mathbb{R}^n \to \mathbb{R}$ and $g(\mathbf{x}) : \Omega \subset \mathbb{R}^n \to \mathbb{R}$. The inner product of functions $\langle f, g\rangle$ in the domain Ω is defined as

$$\langle f, g\rangle = \int_\Omega f(\mathbf{x})g(\mathbf{x})\, d\mathbf{x}_n \tag{2.112}$$

where $d\mathbf{x}_n = dx_1 dx_2 \ldots dx_n$. Therefore, in $\mathbb{R}^2$, $d\mathbf{x}_2 = dxdy = dA$, and in $\mathbb{R}^3$, $d\mathbf{x}_3 = dxdydz = dV$. Then, the inner product space $(\mathcal{L}, \langle ,\rangle)$ is defined to be a set $\mathcal{L}$ together with the function $\langle ,\rangle$ called an inner product satisfying the following four conditions for all functions $f, g, h \in \mathcal{L}$ and $\alpha \in \mathbb{R}$:

(a) $\langle f, f\rangle \geq 0$, $\langle f, f\rangle = 0$ if and only if $f = 0$
(b) $\langle f, g\rangle = \langle g, f\rangle$
(c) $\langle \alpha f, g\rangle = \alpha\langle f, g\rangle$
(d) $\langle f + g, h\rangle = \langle f, h\rangle + \langle g, h\rangle$

It is known that the inner product space $(\mathcal{L}, \langle ,\rangle)$ is a vector space often called a pre-Hilbert space (Debnath 2005; Kreyszig 1978; Maddox 1988; Reddy 1986; Rektorys 1980). Therefore, the inner product space is a vector space on which an inner product is defined.

2.7.2 Normed Space

Analogous to the norm of vectors, we define the norm of a function $f(\mathbf{x}) : \Omega \subset \mathbb{R}^n \to \mathbb{R}$ in the domain Ω as

$$\|f(\mathbf{x})\| = \sqrt{\langle f, f\rangle} \tag{2.113}$$

The normed space $(\mathcal{L}, \|\cdot\|)$ is defined to be the set $\mathcal{L}$ together with the function $\|\cdot\|$ called a normed linear space, which satisfying the following three conditions for all functions $f, g \in \mathcal{L}$ and $\alpha \in \mathbb{R}$:

(a) $\|f\| \geq 0$, $\|f\| = 0$ if and only if (iff) $f = 0$
(b) $\|\alpha f\| = |\alpha|\,\|f\|$
(c) $\|f + g\| \leq \|f\| + \|g\|$

Then, the Cauchy-Schwarz inequality is given by

$$|\langle f(\mathbf{x}), g(\mathbf{x})\rangle| \leq \sqrt{\langle f, f\rangle\langle g, g\rangle} = \|f(\mathbf{x})\|\, \|g(\mathbf{x})\| \tag{2.114}$$

where the equality holds only if the set $\{f(x), g(x)\}$ is linearly dependent. As mentioned previously, the inner product space is a vector space on which an inner product is defined. Note that every inner product space is also a normed space, which is a vector space on which a norm is defined.

2.7.3 Metric Space

In addition, the distance of two functions $f(\mathbf{x}), g(\mathbf{x}) \in \mathcal{L}$ is defined as

$$d(f, g) = \|f(\mathbf{x}) - g(\mathbf{x})\| \tag{2.115}$$

where a metric space is defined as $(\mathcal{L}, d)$. Hence, the metric space $(\mathcal{L}, d)$ is a set $\mathcal{L}$ together with the distance function $d(x, y)$. Therefore, the norm of a function provides a magnitude, similar to the norm of a vector.

2.7.4 Lebesgue Space

Let $u(\mathbf{x}) : \Omega \subset \mathbb{R}^n \to \mathbb{R}$ be a finite function. The finite function $u(\mathbf{x})$ is said to be measurable when

$$\int_\Omega |u(\mathbf{x})|^p \, d\mathbf{x}_n < \infty \tag{2.116}$$

where p is a positive real number. The above integral is called the Lebesgue integral (Debnath 2005; Oden and Reddy 1976) and $d\mathbf{x}_n$ is the Lebesgue measure. The class of all measurable functions $u(\mathbf{x})$ is called the Lebesgue space, denoted by $L^p(\Omega)$. In addition, the L_p norm on $L^p(\Omega)$ is defined as

$$\|u\|_{L^p(\Omega)} = \left(\int_\Omega |u(\mathbf{x})|^p \, d\mathbf{x}_n\right)^{1/p} < \infty \tag{2.117}$$

which is also called the Lebesgue norm. For the special case of $p = 2$, it is said that the function is square integrable in the domain Ω, i.e.,

$$\int_\Omega |u(\mathbf{x})|^2 \, d\mathbf{x}_n < \infty \tag{2.118}$$

Consequently, the L_2 norm on $L^2(\Omega)$ is defined as

$$\|u\|_{L^2(\Omega)} = \left(\int_\Omega |u(\mathbf{x})|^2 \, d\mathbf{x}_n\right)^{1/2} < \infty \tag{2.119}$$

2.7.5 Banach Space

Let $C^m(\Omega)$ denote the set of all real-valued, m-times differentiable functions, which are defined on the domain Ω. For a function $u_n(\mathbf{x}) \in \mathcal{L} = C^0(\Omega)$, let us consider a normed linear space, which is given by $(\mathcal{L}, \|\cdot\|)$.

Then, as a function, a sequence $\{u_n(\mathbf{x})\}$ in metric space $(\mathcal{L}, d)$ is called Cauchy or a Cauchy sequence if for every $\varepsilon > 0$ there exists a number N such that $d(u_n(\mathbf{x}), u_m(\mathbf{x})) = \|u_n(\mathbf{x}) - u_m(\mathbf{x})\| < \varepsilon$ for all $m, n > N$. Further, a sequence $\{u_n(\mathbf{x})\}$ in metric space $(\mathcal{L}, d)$ is said to be convergent or to converge if there exists $u(\mathbf{x}) \in (\mathcal{L}, d)$ such that

$$\lim_{n\to\infty} d(u_n(\mathbf{x}), u(\mathbf{x})) = 0 \tag{2.120}$$

where $u(\mathbf{x})$ is called the limit of the sequence $\{u(\mathbf{x})_n\}$. Therefore, we have the following

$$\lim_{n\to\infty} u_n(\mathbf{x}) = u(\mathbf{x}) \tag{2.121}$$

A normed space $(\mathcal{L}, \|\cdot\|)$ is called complete if every Cauchy sequence in $\mathcal{L}$ converges to an element of $\mathcal{L}$; in other words, the Cauchy sequence has the limit. Note that a Cauchy sequence may not be convergent, but every convergent sequence should be Cauchy. A complete normed space is called a Banach space. For example, Euclidean space $\mathbb{R}^n$ is a Banach space (Kreyszig 1978).

2.7.6 Sobolev Space

To begin with, we shall learn about the multi-index notation, which leads to a succinct expression in a Sobolev space. Consider a set of ordered n-tuples of integers

$$\boldsymbol{\alpha} = (\alpha_1, \alpha_2, \ldots, \alpha_n), \quad \alpha_i \geq 0 \tag{2.122}$$

Then, the following convention is applied

$$|\boldsymbol{\alpha}| \equiv \alpha_1 + \alpha_2 + \ldots + \alpha_n = \sum_{i=1}^{n} \alpha_i \tag{2.123}$$

Assume that a function $u(\mathbf{x}) : \Omega \to \mathbb{R}$ is of class $\mathcal{C}^{|\boldsymbol{\alpha}|}(\Omega)$, where Ω is an open bounded set in $\mathbb{R}^n$. We have the following convention

$$D^{\boldsymbol{\alpha}} u(\mathbf{x}) = \frac{\partial^{|\boldsymbol{\alpha}|} u(\mathbf{x})}{\partial x_1^{\alpha_1} \partial x_2^{\alpha_2} \cdots \partial x_n^{\alpha_n}} = \left(\frac{\partial^{\alpha_1}}{\partial x_1^{\alpha_1}}\right)\left(\frac{\partial^{\alpha_2}}{\partial x_2^{\alpha_2}}\right) \cdots \left(\frac{\partial^{\alpha_n}}{\partial x_n^{\alpha_n}}\right) u(\mathbf{x}) \tag{2.124}$$

which is called the weak $\boldsymbol{\alpha}$th derivative of $u(\mathbf{x})$. Then, the Sobolev space of order m, p on Ω is given by

$$W^{m,p}(\Omega) = \left\{u(\mathbf{x}) | D^{\boldsymbol{\alpha}} u(\mathbf{x}) \in L_p;\ \forall \boldsymbol{\alpha} \text{ such that } |\boldsymbol{\alpha}| \leq m \text{ and } m \geq 1\right\} \tag{2.125}$$

For one-dimensional problems, i.e., $\Omega \subset \mathbb{R}$, the Sobolev space of order m, p on Ω means that for $1 \leq p < \infty$

$$W^{m,p}(\Omega) = \left\{u(x) | u(x) \in L_p;\ \frac{\partial u}{\partial x} \in L_p;\ \ldots;\ \frac{\partial^m u}{\partial x^m} \in L_p\right\} \tag{2.126}$$

where

$$L_p = L_p(\Omega) = \left\{u(x) | \int_\Omega |u(x)|^p dx < \infty\right\} \tag{2.127}$$

Therefore, on $C^m(\Omega)$ a Sobolev norm is defined as

$$\|u(\mathbf{x})\|_{W^{m,p}(\Omega)} = \left(\int_\Omega \sum_{|\boldsymbol{\alpha}| \leq m} \left|D^{\boldsymbol{\alpha}} u(\mathbf{x})\right|^p\right)^{1/p} \tag{2.128}$$

which is finite on Ω. Therefore, as the Sobolev space is a complete normed space, it is a Banach space. For the special case of $m = 0$, the Sobolev norm $\|u(\mathbf{x})\|_{0,p}$ becomes the Lebesgue norm $\|u(\mathbf{x})\|_p$. For two-dimensional problems, i.e., $\mathbf{x} \in \Omega \subset \mathbb{R}^2$, the Sobolev norm can be written as

$$\|u(\mathbf{x})\|_{W^{1,p}(\Omega)} = \left\{\int_\Omega \left(|u|^p + \left|\frac{\partial u}{\partial x}\right|^p + \left|\frac{\partial u}{\partial y}\right|^p\right) dx dy\right\}^{1/p} \tag{2.129}$$

Note that $W^{0,p}(\Omega) = L_p(\Omega)$.

2.7.7 Hilbert Space

A complete inner product space is called a Hilbert space, which is a special case of the Sobolev space defined in Equation (2.125), for $p = 2$. Suppose that Ω is a bounded open set in $\mathbb{R}^n$. A function $u(\mathbf{x}) \in C^m(\Omega)$ is defined on the set Ω. Then, the Hilbert space of order m is given by

$$H^m(\Omega) = \left\{ u(\mathbf{x}) : \Omega \in \mathbb{R}^n \rightarrow \mathbb{R} | D^\alpha u(\mathbf{x}) \in L_2;\ \forall\alpha\ |\alpha| \leq m \text{ and } m \geq 1 \right\} \tag{2.130}$$

where

$$L_2 = L_2(\Omega) = \left\{ u(x) | \int_\Omega |u(x)|^2 dx < \infty \right\} \tag{2.131}$$

As a special case of class $H^m(\Omega)$, the space $H_0^m(\Omega)$ is defined as

$$H_0^m(\Omega) = \left\{ \begin{array}{ll} u(\mathbf{x}) | u(\mathbf{x}) \in H^m(\Omega),\ u(\mathbf{x}) = 0,\ \dfrac{\partial u(\mathbf{x})}{\partial n} = 0, & \\ \dfrac{\partial^2 u(\mathbf{x})}{\partial n^2} = 0, \ldots, \dfrac{\partial^{m-2} u(\mathbf{x})}{\partial n^{m-2}} = 0,\ \dfrac{\partial^{m-1} u(\mathbf{x})}{\partial n^{m-1}} = 0, & \forall u(\mathbf{x}) \in \partial\Omega \end{array} \right\} \tag{2.132}$$

where n denotes the direction of the unit outward normal to the boundary $\partial\Omega$ of the open set Ω. Note that the class $H_0^m(\Omega)$ is a subspace of the class $H^m(\Omega)$.

A norm in the Hilbert space is defined as

$$\|u(\mathbf{x})\|_{H^m(\Omega)} = \left(\int_\Omega \sum_{|\alpha| \leq m} \left| D^\alpha u(\mathbf{x}) \right|^2 \right)^{1/2} \tag{2.133}$$

which is finite on Ω. Therefore, as the Hilbert space is a complete normed space, it is also a Banach space. Note that $H^m(\Omega) = W^{m,2}(\Omega)$. Also note that $H^0(\Omega) = W^{0,2}(\Omega) = L_2(\Omega)$.

2.8 TENSOR ANALYSIS

We have learned that a scalar and a vector are invariant; thus they are independent of a coordinate system. In this section, we introduce a tensor which is also invariant. Usually, a tensor means a second-order tensor which is a linear operator transforming one vector to the other vector. One vector in a vector space is given by a linear combination of base vectors spanning the vector space, while the second-order tensor is given by a linear combination of dyads. The dyad results from the tensor product of two base vectors in a vector space.

A basis in a three dimensional real vector space consists of three covariant base vectors. The components of the vector with covariant base vectors are contravariant, and they transform contravariantly. Similarly there exists the dual vector space or simply dual space which is spanned by a reciprocal basis. The reciprocal basis is composed of three contravariant base vectors, and a vector in the dual space has covariant components. Therefore, the components transform covariantly.

2.8.1 Tensor Algebra

Contravariant and covariant vectors An n-vector $\mathbf{v}$ in an n-dimensional vector space $\mathbb{V}$ can be expressed as a linear combination of a covariant basis, which is an n-tuple $\{\mathbf{g}_i\}_{i=1}^n$

$$\mathbf{v} = \sum_{i=1}^{n} v^i \mathbf{g}_i, \quad \mathbf{v} \in \mathbb{V} \tag{2.134}$$

Note that the covariant basis $\{\mathbf{g}_i\}_{i=1}^n$ spans the n-dimensional vector space $\mathbb{V}$. In the above equation, v^i are called contravariant vector components or simply contravariant components of the vector. The vector represented in terms of contravariant components with respect to covariant base vectors is referred to as a contravariant vector.

The covector $\mathbf{v}^*$ (Bowen and Wang 1976) in the n-dimensional dual vector space $\mathbb{V}^*$, which is often called the dual algebraic space (Fung 1965; Malvern 1969; Oden 1979), can also be expressed as a linear combination of a cotravariant basis $\{\mathbf{g}^i\}_{i=1}^n$, which is often called an n-tuple reciprocal basis

$$\mathbf{v}^* = \sum_{i=1}^{n} v_i \mathbf{g}^i, \quad \mathbf{v}^* \in \mathbb{V}^* \tag{2.135}$$

where the covector belongs to the dual vector space with the cotravariant basis. The covector expressed in terms of covariant components is referred to as a covariant vector. The relation between contravariant and covariant base vectors is given by

$$\mathbf{g}^i \cdot \mathbf{g}_j = \delta^i_j = \left\{ \begin{array}{l} 0, \ i \neq j \\ 1, \ i = j \end{array} \right\} \tag{2.136}$$

where δ^i_j denotes the Kronecker delta. Note that $\mathbf{g}^i \cdot \mathbf{g}_j = \mathbf{g}_i \cdot \mathbf{g}^j$ as shown in Figure 2.5.

As the duality pairing of the vector space over $\mathbb{R}$ (a bilinear map) (Oden 1979), the inner product function $(\cdot, \cdot) : \mathbb{V}^* \times \mathbb{V} \to \mathbb{R}$ of the contravariant and the covariant vectors yields

$$(\mathbf{v}^*, \mathbf{v}) = \mathbf{v}^* \cdot \mathbf{v} = \sum_{i=1}^{n} v_i v^i \tag{2.137}$$

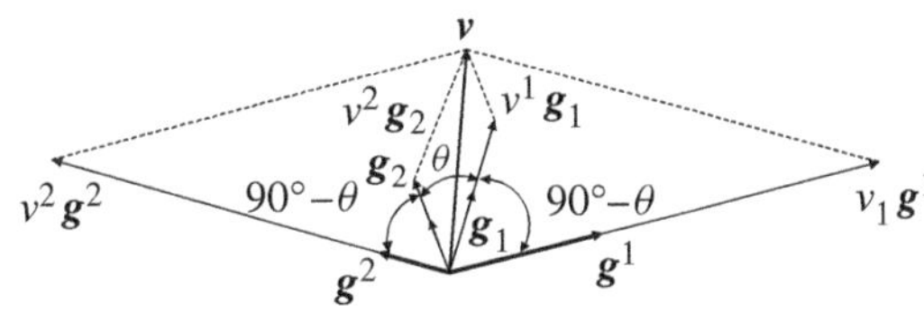

Covariant vector	$v = v^1 g_1 + v^2 g_2$	$g_i \cdot g^j = \delta_i^{\cdot j}$
Cotravariant vector	$v = v_1 g^1 + v_2 g^2$	$g^i \cdot g_j = \delta^i_{\cdot j}$
Metric tensor	$g^{ij} = g^i \cdot g^j$	$g^i = g^{ij} g_j, \quad g_i = g_{ij} g^j$

$$v^i = g^{ij} v_j, \quad v_i = g_i^{\cdot j} v_j \qquad v_i = g_{ij} v^j, \quad v^i = g^i_{\cdot j} v^j$$

Figure 2.5. Dual base vectors in two dimensions (Fung 1965; Malvern 1969)

Note that the inner product function $(\cdot,\cdot) : \mathbb{V} \times \mathbb{V} \to \mathbb{R}$ of the contravariant and the contravariant vectors gives

$$(\mathbf{v},\mathbf{v}) = \mathbf{v}\cdot\mathbf{v} = \sum_{i=1}^{n}\sum_{j=1}^{n} v^i v^j \mathbf{g}_i \cdot \mathbf{g}_j \tag{2.138}$$

while the inner product function $(\cdot,\cdot) : \mathbb{V}^* \times \mathbb{V}^* \to \mathbb{R}$ of the contravariant and the covariant vectors results in

$$(\mathbf{v}^*,\mathbf{v}^*) = \mathbf{v}^*\cdot\mathbf{v}^* = \sum_{i=1}^{n}\sum_{j=1}^{n} v_i v_j \mathbf{g}^i \cdot \mathbf{g}^j \tag{2.139}$$

In Equations (2.138) and (2.139), the inner product of two covariant or contravariant base vectors can be respectively written as

$$g^{ij} = \mathbf{g}^i \cdot \mathbf{g}^j = g^{ji}, \quad g_{ij} = \mathbf{g}_i \cdot \mathbf{g}_j = g_{ji} \tag{2.140}$$

where g^{ij} and g_{ij} are components of the metric tensor and they are often called metric coefficients. Consequently, Equations (2.138) and (2.139) can be respectively rewritten as

$$\mathbf{v}\cdot\mathbf{v} = \sum_{i=1}^{n}\sum_{j=1}^{n} v^i v^j g_{ij}, \quad \mathbf{v}^*\cdot\mathbf{v}^* = \sum_{i=1}^{n}\sum_{j=1}^{n} v_i v_j g^{ij} \tag{2.141}$$

It is known that Euclidean n-space, $\mathbb{R}^n$ is self-dual. Hence, the covariant basis $\{\mathbf{g}_i\}_{i=1}^n$ for the vector space $\mathbb{R}^n$ plays a role of a basis for the dual vector space $(\mathbb{R}^n)^*$. Conversely, the cotravariant basis $\{\mathbf{g}^i\}_{i=1}^n$ for the dual vector space $(\mathbb{R}^n)^*$ can play a role of a basis for the vector space $\mathbb{R}^n$. In other words, $\mathbb{R}^n = (\mathbb{R}^n)^*$ (Oden 1979, page 108). As a consequence, any vector in $\mathbb{R}^n$ can be represented in terms of contravariant or covariant components (Malvern 1969, page 575) (Wempner and Talaslidis 2003, page 28) (Chung 2007, page 197) (Papastavridis 1998) as

$$\mathbf{v} = \sum_{i=1}^{n} v^i \mathbf{g}_i = \sum_{i=1}^{n} v_i \mathbf{g}^i \tag{2.142}$$

From now on, we use the Einstein summation convention; hence the repeated indices imply the summation up to the size of dimension for the vector space or the dual vector space. In the Einstein summation convention, the summation is performed over the repeated indices, which are called dummy indices. The index, which is not repeated, is called the free index.

Metric Tensor By virtue of the Einstein summation convention, the above equation can be rewritten as

$$\mathbf{v} = v^i \mathbf{g}_i = v_i \mathbf{g}^i, \; i = 1, 2, \ldots, n \tag{2.143}$$

From the above equations, we can derive the following relations

$$\begin{aligned} v^i &= \mathbf{v}\cdot\mathbf{g}^i, & v_i &= \mathbf{v}\cdot\mathbf{g}_i, \\ v^i &= g^{ij} v_j, & v_i &= g_{ij} v^j \end{aligned} \tag{2.144}$$

In view of Equations (2.143) and (2.144), we have

$$\mathbf{v} = v^i \mathbf{g}_i = g^{ij} v_j \mathbf{g}_i = v_j \mathbf{g}^j \tag{2.145}$$

Therefore, we have

$$\mathbf{g}_i = g_{ij}\mathbf{g}^j, \ \mathbf{g}^i = g^{ij}\mathbf{g}_j \tag{2.146}$$

Further, we obtain the following

$$\delta_i^k = \mathbf{g}_i \cdot \mathbf{g}^k = g_{ij}\mathbf{g}^j \cdot \mathbf{g}^k = g_{ij} g^{jk} \tag{2.147}$$

Thus, it is easy to notice that the matrix $[g_{ij}]$ is the inverse of the matrix $[g^{ij}]$.

In $\mathbb{R}^3$, the contravariant base vectors can be obtained as

$$\mathbf{g}^1 = \frac{\mathbf{g}_2 \times \mathbf{g}_3}{\mathbf{g}_1 \cdot \mathbf{g}_2 \times \mathbf{g}_3}, \quad \mathbf{g}^2 = \frac{\mathbf{g}_3 \times \mathbf{g}_1}{\mathbf{g}_2 \cdot \mathbf{g}_3 \times \mathbf{g}_1}, \quad \mathbf{g}^3 = \frac{\mathbf{g}_1 \times \mathbf{g}_2}{\mathbf{g}_3 \cdot \mathbf{g}_1 \times \mathbf{g}_2} \tag{2.148}$$

Second-Order Tensor We have learned that any vector in a vector space is represented as a linear combination of base vectors. Similarly, we shall introduce a second-order tensor which is also expressed as a linear combination. The open or tensor product of two vectors, $\mathbf{u}, \mathbf{v} \in \mathbb{V} \subset \mathbb{R}^3$, is defined as

$$\begin{aligned}
\mathbf{u} \otimes \mathbf{v} &= (u^i \mathbf{g}_i) \otimes (v^j \mathbf{g}_j) \\
&= u^i v^j \mathbf{g}_i \otimes \mathbf{g}_j \\
&= u^1 v^1 \mathbf{g}_1 \otimes \mathbf{g}_1 + u^1 v^2 \mathbf{g}_1 \otimes \mathbf{g}_2 + u^1 v^3 \mathbf{g}_1 \otimes \mathbf{g}_3 \\
&\quad + u^2 v^1 \mathbf{g}_2 \otimes \mathbf{g}_1 + u^2 v^2 \mathbf{g}_2 \otimes \mathbf{g}_2 + u^2 v^3 \mathbf{g}_2 \otimes \mathbf{g}_3 \\
&= u^3 v^1 \mathbf{g}_3 \otimes \mathbf{g}_1 + u^3 v^2 \mathbf{g}_3 \otimes \mathbf{g}_2 + u^3 v^3 \mathbf{g}_3 \otimes \mathbf{g}_3
\end{aligned} \tag{2.149}$$

where the tensor product is called a dyad and the real vector space $\mathbb{V}$ is the subspace of $\mathbb{R}^3$. Hence, a second-order tensor is a linear combination of dyads with scalar coefficients. The linear combination of dyads is called a dyadic. The components of the second-order tensor can be represented in a matrix form as

$$(\mathbf{u} \otimes \mathbf{v})^{ij} = \begin{bmatrix} u^1 v^1 & u^1 v^2 & u^1 v^3 \\ u^2 v^1 & u^2 v^2 & u^2 v^3 \\ u^3 v^1 & u^3 v^2 & u^3 v^3 \end{bmatrix} \tag{2.150}$$

where the entries in the matrix are scalar coefficients.

Note that the second-order tensor in Equation (2.149) is a linear transformation from a vector space to a vector space, or from a dual vector space to a vector space, or from a dual vector space to a dual vector space, or from a vector space to a dual vector space. For example,

$$(\mathbf{u} \otimes \mathbf{v}) \cdot \mathbf{w}^* = u^i v^j w_k \mathbf{g}_i \otimes \mathbf{g}_j \cdot \mathbf{g}^k = u^i v^j w_k \mathbf{g}_i \delta_j^k = u^i v^j w_j \mathbf{g}_i \tag{2.151}$$

where $\mathbf{u}, \mathbf{v} \in \mathbb{V}$, $\mathbf{w}^* \in \mathbb{V}^*$. Thus, a second-order tensor $\mathbf{T}$ as a linear combination of dyads can be written as

$$\mathbf{T} = T_{ij}\mathbf{g}^i \otimes \mathbf{g}^j = T^{ij}\mathbf{g}_i \otimes \mathbf{g}_j = T_i^{\cdot j}\mathbf{g}^i \otimes \mathbf{g}_j = T^i_{\cdot j}\mathbf{g}_i \otimes \mathbf{g}^j \tag{2.152}$$

where T_{ij} are the covariant components, $T_{i}^{\cdot j}$ and $T_{\cdot j}^{i}$ are the mixed components, and T^{ij} are the contravariant components of $\mathbf{T}$. It should be noted that like a vector, a tensor has the tensorial property which is independent of coordinate systems, but the components of the tensor are dependent upon the coordinate systems relative to dyads. The components of $\mathbf{T}$ can be obtained as

$$\begin{aligned} T_{ij} &= \mathbf{g}_i \cdot \mathbf{T} \cdot \mathbf{g}_j, \qquad & T^{ij} &= \mathbf{g}^i \cdot \mathbf{T} \cdot \mathbf{g}^j, \\ T_{i}^{\cdot j} &= \mathbf{g}_i \cdot \mathbf{T} \cdot \mathbf{g}^j, \qquad & T_{\cdot j}^{i} &= \mathbf{g}^i \cdot \mathbf{T} \cdot \mathbf{g}_j \end{aligned} \tag{2.153}$$

A second-order tensor $\mathbf{T}$ is a member of a linear vector space, i.e. $\mathbf{T} \in \mathcal{L}(\mathbb{V}, \mathbb{V})$ or $\mathcal{L}(\mathbb{V}, \mathbb{V}^*)$ or $\mathcal{L}(\mathbb{V}^*, \mathbb{V})$ or $\mathcal{L}(\mathbb{V}^*, \mathbb{V}^*)$. For instance, a tensor satisfies the following:

$$\mathbf{T} \cdot (\alpha \mathbf{u} + \beta \mathbf{v}) = \alpha \mathbf{T} \cdot \mathbf{u} + \beta \mathbf{T} \cdot \mathbf{v}, \ \forall \alpha, \beta \in \mathbb{R} \text{ and } \mathbf{u}, \mathbf{v} \in \mathbb{V} \tag{2.154}$$

Thus, all second-order tensors, $\mathbf{U}, \mathbf{V}, \mathbf{W}, \mathbf{O}, \mathbf{I} \in L(\mathbb{V}, \mathbb{V})$ or $\mathcal{L}(\mathbb{V}, \mathbb{V}^*)$ or $\mathcal{L}(\mathbb{V}^*, \mathbb{V})$ or $\mathcal{L}(\mathbb{V}^*, \mathbb{V}^*)$, contained in the linear vector space $\mathcal{L}$, should satisfy the following

(a) Commutativity: $\mathbf{U} + \mathbf{V} = \mathbf{V} + \mathbf{U}$.
(b) Associativity of Tensor addition: $(\mathbf{U} + \mathbf{V}) + \mathbf{W} = \mathbf{U} + (\mathbf{V} + \mathbf{W})$.
(c) Additive identity: $\mathbf{O} + \mathbf{U} = \mathbf{U} + \mathbf{O} = \mathbf{U}$.
(d) Existence of additive inverse: $\mathbf{U} + (-\mathbf{U}) = \mathbf{O}$.
(e) Associativity of scalar multiplication: $\alpha\,(\beta \mathbf{U}) = (\alpha\beta \mathbf{U})\,,\ \ \forall \alpha, \beta \in \mathbb{R}$.
(f) Distributivity of scalar sums: $(\alpha + \beta)\,\mathbf{U} = \alpha \mathbf{U} + \beta \mathbf{U},\ \ \forall \alpha, \beta \in \mathbb{R}$.
(g) Distributivity of vector sums: $\alpha\,(\mathbf{U} + \mathbf{V}) = \alpha \mathbf{U} + \alpha \mathbf{V}\ \ \forall \alpha \in \mathbb{R}$.
(h) Scalar multiplication identity: $\mathbf{I} \cdot \mathbf{U} = \mathbf{U}$.

where $\mathbf{I}$ is the second-order identity tensor. These are requirements for elements in a linear vector space as shown in Section 2.2.1.

Note that the identity tensor $\mathbf{I}$ is the metric tensor, which is given by

$$\mathbf{I} = g_{ij} \mathbf{g}^i \otimes \mathbf{g}^j = g^{ij} \mathbf{g}_i \otimes \mathbf{g}_j = \mathbf{g}^i \otimes \mathbf{g}_j = \mathbf{g}_i \otimes \mathbf{g}^j \tag{2.155}$$

which is also called the unit tensor or the fundamental tensor (Malvern 1969). The components of the unit tensor are the metric coefficients. Therefore, the unit tensor satisfies the following

$$\mathbf{I} \cdot \mathbf{v} = \mathbf{v} \cdot \mathbf{I}, \ \mathbf{I} \cdot \mathbf{T} = \mathbf{T} \cdot \mathbf{I} \tag{2.156}$$

Furthermore, the second-order tensors contained in a linear vector space obey the following dot product rules

(a) Distributivity: $\mathbf{U} \cdot (\mathbf{V} + \mathbf{W}) = \mathbf{U} \cdot \mathbf{V} + \mathbf{U} \cdot \mathbf{W}$.
(b) Distributivity: $(\mathbf{U} + \mathbf{V}) \cdot \mathbf{W} = \mathbf{U} \cdot \mathbf{W} + \mathbf{V} \cdot \mathbf{W}$.
(c) Associativity: $(\mathbf{U} \cdot \mathbf{V}) \cdot \mathbf{W} = \mathbf{U} \cdot (\mathbf{V} \cdot \mathbf{W})$.
(d) Associativity: $\alpha\,(\mathbf{U} \cdot \mathbf{V}) = \alpha \mathbf{U} \cdot (\mathbf{V}) = \mathbf{U} \cdot (\alpha \mathbf{V})$.

where the dot product (also called single-dot product) between the second-order tensors is denoted by $\mathbf{U} \cdot \mathbf{U}$.

The dot product between dyads $\mathbf{p} \otimes \mathbf{q}$ or triads $\mathbf{p} \otimes \mathbf{q} \otimes \mathbf{s}$ or tetrads $\mathbf{p} \otimes \mathbf{q} \otimes \mathbf{s} \otimes \mathbf{t}$ (Drew 1961; Malvern 1969) can be given by

$$\begin{aligned} \mathbf{p} \cdot \mathbf{q} \otimes \mathbf{s} \otimes \mathbf{t} &= (\mathbf{p} \cdot \mathbf{q})\mathbf{s} \otimes \mathbf{t}, \\ \mathbf{p} \otimes \mathbf{q} \cdot \mathbf{s} \otimes \mathbf{t} &= (\mathbf{q} \cdot \mathbf{s})\mathbf{p} \otimes \mathbf{t}, \\ \mathbf{p} \otimes \mathbf{q} \otimes \mathbf{s} \cdot \mathbf{t} &= (\mathbf{s} \cdot \mathbf{t})\mathbf{p} \otimes \mathbf{q} \end{aligned} \tag{2.157}$$

It is important to note that the dot product between dyads is not commutative, i.e., $\mathbf{p} \otimes \mathbf{q} \cdot \mathbf{s} \otimes \mathbf{t} \neq \mathbf{s} \otimes \mathbf{t} \cdot \mathbf{p} \otimes \mathbf{q}$. Thus, the dot product of a contravariant tensor $\mathbf{T}$ and a covariant tensor $\mathbf{S}$ can be written as

$$\mathbf{T} \cdot \mathbf{S} = T^{ij}\mathbf{g}_i \otimes \mathbf{g}_j \cdot S_{kl}\mathbf{g}^k \otimes \mathbf{g}^l = T^{ij}S_{jk}\mathbf{g}_i \otimes \mathbf{g}^k \tag{2.158}$$

In the above, it is said that contraction is performed; in other words, the single-dot product of two second-order tensors produces a second-order tensor. Also we have the following

$$\mathbf{T} \cdot \mathbf{T} \cdot \mathbf{T} = \mathbf{T}^3 \tag{2.159}$$

which represents a tensor power. Note that $(\mathbf{T}^3)_{ij} = T_{ij}T_{jk}T_{kl}$. Therefore, the inverse of a second-order tensor can be defined as

$$\mathbf{T}^{-1} \cdot \mathbf{T} = \mathbf{T} \cdot \mathbf{T}^{-1} = \mathbf{I} \tag{2.160}$$

Hence, we have

$$\mathbf{T}^{-1} \cdot \mathbf{T}^{-1} \cdot \mathbf{T}^{-1} = \mathbf{T}^{-3} \tag{2.161}$$

Note that $(\mathbf{T} \cdot \mathbf{S})^{-1} = \mathbf{S}^{-1} \cdot \mathbf{T}^{-1}$.

Recall that the dot product of two vectors yields a scalar. Similarly, we introduce the scalar product, which is often called the double-dot product or the double contraction. The scalar product of two dyads can be defined as

$$\begin{aligned} \mathbf{p} \otimes \mathbf{q} : \mathbf{s} \otimes \mathbf{t} &= (\mathbf{p} \cdot \mathbf{s})(\mathbf{q} \cdot \mathbf{t}) \\ &= (\mathbf{s} \cdot \mathbf{p})(\mathbf{t} \cdot \mathbf{q}) \\ &= \mathbf{s} \otimes \mathbf{t} : \mathbf{p} \otimes \mathbf{q} \end{aligned} \tag{2.162}$$

where the scalar product is commutative. Consequently, the scalar product of a contravariant tensor $\mathbf{T}$ and a covariant tensor $\mathbf{S}$ can be written as

$$\mathbf{T} : \mathbf{S} = T^{ij}\mathbf{g}_i \otimes \mathbf{g}_j : S_{kl}\mathbf{g}^k \otimes \mathbf{g}^l = T^{ij}S_{ij} = \mathbf{S} : \mathbf{T} \tag{2.163}$$

which means that the scalar product of a contravariant tensor $\mathbf{T}$ and a covariant tensor $\mathbf{S}$ produces an invariant not depending on chosen coordinate systems. In general, the scalar product of two second-order tensors $\mathbf{U}, \mathbf{V}$ can be written as

$$\mathbf{U} : \mathbf{V} = U^{ij}V_{ij} = U_{ij}V^{ij} = U^{i}_{\cdot j}V_{i}^{\cdot j} = U_{i}^{\cdot j}V^{i}_{\cdot j} \tag{2.164}$$

In contrast to Equation (2.162), we have the other scalar product as follows:

$$\begin{aligned}\mathbf{p}\otimes\mathbf{q}\cdot\cdot\mathbf{s}\otimes\mathbf{t} &= (\mathbf{p}\cdot\mathbf{t})(\mathbf{q}\cdot\mathbf{s})\\ &= (\mathbf{t}\cdot\mathbf{p})(\mathbf{s}\cdot\mathbf{q})\\ &= \mathbf{t}\otimes\mathbf{q}\cdot\cdot\mathbf{p}\otimes\mathbf{s}\\ &= (\mathbf{s}\otimes\mathbf{t})^T\cdot\cdot(\mathbf{p}\otimes\mathbf{q})^T\end{aligned} \tag{2.165}$$

Therefore, the scalar product of a contravariant tensor $\mathbf{T}$ and a covariant tensor $\mathbf{S}$ can be written as

$$\mathbf{T}\cdot\cdot\mathbf{S} = T^{ij}\mathbf{g}_i\otimes\mathbf{g}_j\cdot\cdot S_{kl}\mathbf{g}^k\otimes\mathbf{g}^l = T^{ij}S_{ji} = \mathbf{S}^T\cdot\cdot\mathbf{T}^T \tag{2.166}$$

The transpose of a tensor, $\mathbf{T}^T$, is defined as

$$\mathbf{u} = \mathbf{T}\cdot\mathbf{v} = \mathbf{v}\cdot\mathbf{S},\ \ \mathbf{S} = \mathbf{T}^T,\ \ \mathbf{S}^T = \mathbf{T} \tag{2.167}$$

where the tensor $\mathbf{S}$ is the transpose of the tensor $\mathbf{T}$. Hence, we have

$$\mathbf{u}\cdot\mathbf{T}\cdot\mathbf{v} = \mathbf{v}\cdot\mathbf{T}^T\cdot\mathbf{u},\ \ \forall\mathbf{u},\mathbf{v} \tag{2.168}$$

The second-order tensor $\mathbf{T}$ is said to be symmetric if $\mathbf{T} = \mathbf{T}^T$. Then, the symmetric second-order tensor has the following properties

$$\begin{aligned}T^{ij} &= \mathbf{g}_i\cdot\mathbf{T}\cdot\mathbf{g}_j = \mathbf{g}_j\cdot\mathbf{T}^T\cdot\mathbf{g}_i = \mathbf{g}_j\cdot\mathbf{T}\cdot\mathbf{g}_i = T_{ji},\\ T_{ij} &= \mathbf{g}^i\cdot\mathbf{T}\cdot\mathbf{g}^j = \mathbf{g}^j\cdot\mathbf{T}^T\cdot\mathbf{g}^i = \mathbf{g}^j\cdot\mathbf{T}\cdot\mathbf{g}^i = T^{ji},\\ T^i_{\cdot j} &= \mathbf{g}^i\cdot\mathbf{T}\cdot\mathbf{g}_j = \mathbf{g}_j\cdot\mathbf{T}^T\cdot\mathbf{g}^i = \mathbf{g}_j\cdot\mathbf{T}\cdot\mathbf{g}^i = T_j^{\cdot i}\end{aligned} \tag{2.169}$$

Note that the second-order tensors contained in a linear vector space obey the following transpose rules

(a) $(\mathbf{p}\otimes\mathbf{q})^T = \mathbf{q}\otimes\mathbf{p}$.
(b) $(\mathbf{U}\cdot\mathbf{V})^T = \mathbf{V}^T\cdot\mathbf{U}^T$.
(c) $(\alpha\mathbf{U}+\beta\mathbf{V})^T = \alpha\mathbf{U}^T+\beta\mathbf{V}^T$.

Every second-order tensor can be divided into a symmetric tensor and a skew-symmetric tensor as follows:

$$\begin{aligned}\mathbf{T} &= \mathbf{T}_{\text{sym}} + \mathbf{T}_{\text{skew}},\\ \mathbf{T}_{\text{sym}} &= \frac{1}{2}\left(\mathbf{T}+\mathbf{T}^T\right),\ \ \mathbf{T}_{\text{skew}} = \frac{1}{2}\left(\mathbf{T}-\mathbf{T}^T\right)\end{aligned} \tag{2.170}$$

In addition, every second-order tensor can be divided into a spherical and a deviatoric tensor as below:

$$\begin{aligned}\mathbf{T} &= \mathbf{T}_{\text{dev}} + \mathbf{T}_{\text{sph}},\\ \mathbf{T}_{\text{sph}} &= \frac{1}{3}(\text{tr}\,\mathbf{T})\,\mathbf{I},\\ \mathbf{T}_{\text{dev}} &= \mathbf{T} - \mathbf{T}_{\text{sph}}\end{aligned} \tag{2.171}$$

The trace of a second-order tensor, denoted by $\mathrm{tr}\,\mathbf{T}$, produces a scalar. In view of Equation (2.152), the trace of the second-order tensor gives

$$\begin{aligned}\mathrm{tr}\,\mathbf{T} &= T_{ij}\mathbf{g}^i \cdot \mathbf{g}^j = T^{ij}\mathbf{g}_i \cdot \mathbf{g}_j = T_i^{\cdot j}\mathbf{g}^i \cdot \mathbf{g}_j = T^i_{\cdot j}\mathbf{g}_i \cdot \mathbf{g}^j \\ &= T_{ij}g^{ij} = T^{ij}g_{ij} = T_i^{\cdot i} = T^i_{\cdot i}\end{aligned} \tag{2.172}$$

where the Einstein summation convention is applied over the repeated indices. The trace of second-order tensors obey the following important rules (Chadwick 1976; Gurtin 1981; Hjelmstad 2005; Holzapfel 2000; Malvern 1969; Spencer 2004):

(a) $\mathrm{tr}\,\mathbf{U}^T = \mathrm{tr}\,\mathbf{U}$.
(b) $\mathrm{tr}\,(\mathbf{U} + \mathbf{V}) = \mathrm{tr}\,\mathbf{U} + \mathrm{tr}\,\mathbf{U}$.
(c) $\mathrm{tr}\,(\mathbf{U} \cdot \mathbf{V}) = \mathrm{tr}\,(\mathbf{V} \cdot \mathbf{U})$.
(d) $\mathbf{U} : \mathbf{V} = \mathrm{tr}\left(\mathbf{U}^T \cdot \mathbf{V}\right) = \mathrm{tr}\left(\mathbf{V}^T \cdot \mathbf{U}\right) = \mathbf{V} : \mathbf{U}$.
(e) $\mathbf{U} \cdot\cdot \mathbf{V} = \mathrm{tr}\,(\mathbf{U} \cdot \mathbf{V}) = \mathrm{tr}\left(\mathbf{V}^T \cdot \mathbf{U}^T\right) = \mathbf{V}^T \cdot\cdot \mathbf{U}^T$.
(f) $\mathrm{tr}\,(\mathbf{U} \cdot \mathbf{V} \cdot \mathbf{W}) = \mathrm{tr}\,(\mathbf{V} \cdot \mathbf{W} \cdot \mathbf{U}) = \mathrm{tr}\,(\mathbf{W} \cdot \mathbf{U} \cdot \mathbf{V})$.
(g) $\mathbf{I} : \mathbf{U} = \mathrm{tr}\,\mathbf{U}$.

Hence, the trace is a scalar invariant function producing a scalar. By use of the above rules, we can derive the following important rules

$$\begin{aligned}\mathbf{U} : (\mathbf{V} \cdot \mathbf{W}) &= \mathrm{tr}\left(\mathbf{U}^T \cdot \mathbf{V} \cdot \mathbf{W}\right) \\ &= \mathrm{tr}\left(\mathbf{W}^T \cdot \mathbf{V}^T \cdot \mathbf{U}\right) \\ &= \mathrm{tr}\left(\mathbf{U}^T \cdot \mathbf{V} \cdot \mathbf{W}\right) \\ &= \left(\mathbf{V}^T \cdot \mathbf{U}\right) : \mathbf{W} \\ &= \left(\mathbf{U} \cdot \mathbf{W}^T\right) : \mathbf{V}\end{aligned} \tag{2.173}$$

In addition, we have

$$\begin{aligned}\mathbf{U} \cdot\cdot (\mathbf{V} \cdot \mathbf{W}) &= \mathrm{tr}\,(\mathbf{U} \cdot \mathbf{V} \cdot \mathbf{W}) \\ &= \mathrm{tr}\left(\mathbf{W}^T \cdot \mathbf{V}^T \cdot \mathbf{U}^T\right) \\ &= \mathrm{tr}\,(\mathbf{U} \cdot \mathbf{V} \cdot \mathbf{W}) \\ &= (\mathbf{V} \cdot \mathbf{U}) : \mathbf{W} \\ &= (\mathbf{W} \cdot \mathbf{U}) : \mathbf{V}\end{aligned} \tag{2.174}$$

and

$$\begin{aligned}\mathbf{U} \cdot\cdot \mathbf{V} &= \mathbf{U}^T : \mathbf{V} = \mathbf{U} : \mathbf{V}^T, \\ \mathbf{U} : \mathbf{V} &= \mathbf{U}^T \cdot\cdot \mathbf{V} = \mathbf{U} \cdot\cdot \mathbf{V}^T\end{aligned} \tag{2.175}$$

The determinant of a second-order tensor, $\det \mathbf{T}$, can be defined as $\det\,[T_{ij}]$ or $\det\,[T^i_{\cdot j}]$ or $\det\,[T_i^{\cdot j}]$ or $\det\,[T^{ij}]$. Note that $\det\,(\mathbf{U} \cdot \mathbf{V}) = (\det \mathbf{U})(\det \mathbf{V})$ and $\det \mathbf{T}^T = \det \mathbf{T}$. The norm of a second-order tensor, $\|\mathbf{T}\|$, is defined as $\sqrt{\mathbf{T} \cdot \mathbf{T}}$.

Orthogonal Tensor A second-order tensor $\mathbf{Q}$ is called orthogonal if the following relation is satisfied

$$(\mathbf{Q} \cdot \mathbf{u}) \cdot (\mathbf{Q} \cdot \mathbf{v}) = \mathbf{u} \cdot \mathbf{v} \tag{2.176}$$

which means that the angle between $\mathbf{u}$ and $\mathbf{v}$ is identical to the angle between $\mathbf{Q} \cdot \mathbf{u}$ and $\mathbf{Q} \cdot \mathbf{v}$ and the inner products are preserved. The above leads to the following

$$\mathbf{u} \cdot \left(\mathbf{Q}^T \cdot \mathbf{Q} - \mathbf{I}\right) \cdot \mathbf{v} = 0 \tag{2.177}$$

In order for the second-order tensor to preserve the above dot product property, the tensor should satisfy

$$\mathbf{Q} \cdot \mathbf{Q}^T = \mathbf{Q}^T \cdot \mathbf{Q} = \mathbf{I}, \ \mathbf{Q}^T = \mathbf{Q}^{-1} \tag{2.178}$$

Hence, the second-order tensor $\mathbf{Q}$ must be orthogonal. As discussed previously, the orthogonal tensor $\mathbf{Q}$ is a linear transformation assigning a vector $\mathbf{u} \in \mathbb{U}$ to a vector $\mathbf{v} \in \mathbb{V}$,

$$\mathbf{v} = \mathbf{Q} \cdot \mathbf{u} \tag{2.179}$$

where the vector $\mathbf{v}$ is a consequence of the rotation of the vector $\mathbf{u}$, when the second-order tensor $\mathbf{Q}$ is a rotation tensor. Let $\{\mathbf{g}_i\}_{i=1}^n$ denote a covariant basis spanning an n-dimensional vector space $\mathbb{V}$. Then, the covariant basis is rotated by an orthogonal rotation tensor $\mathbf{Q}$, producing a new covariant basis $\{\hat{\mathbf{g}}_i\}_{i=1}^n$ spanning an n-dimensional vector space $\hat{\mathbb{V}}$ as follows:

$$\hat{\mathbf{g}}_j = \mathbf{Q} \cdot \mathbf{g}_j, \ \hat{\mathbf{g}}^j = \mathbf{Q} \cdot \mathbf{g}^j, \ j = 1, \ldots, n \tag{2.180}$$

The components of the orthogonal tensor can be found as

$$\begin{aligned} Q^i_{.j} &= \mathbf{g}^i \cdot \mathbf{Q} \cdot \mathbf{g}_j = \mathbf{g}^i \cdot \hat{\mathbf{g}}_j, \\ Q_j^{\cdot i} &= \mathbf{g}_j \cdot \mathbf{Q} \cdot \mathbf{g}^i = \mathbf{g}_j \cdot \hat{\mathbf{g}}^i \end{aligned} \tag{2.181}$$

A vector $\mathbf{v}$ can be represented as.

$$\mathbf{v} = v^j \mathbf{g}_j = \hat{v}^j \hat{\mathbf{g}}_j \tag{2.182}$$

Therefore, we can derive the relation

$$\begin{aligned} v^i &= \mathbf{g}^i \cdot \mathbf{v} = \mathbf{g}^i \cdot \hat{\mathbf{g}}_j \hat{v}^j = Q^i_{.j} \hat{v}^j, \\ \hat{v}^i &= \hat{\mathbf{g}}^i \cdot \mathbf{v} = \hat{\mathbf{g}}^i \cdot \mathbf{g}_j v^j = Q_j^{\cdot i} v^j \end{aligned} \tag{2.183}$$

This process is called a change of basis (Malvern 1969; Simmons 1994). Note that Equation (2.180)

$$\hat{\mathbf{g}}_k = \mathbf{Q} \cdot \mathbf{g}_k = Q^i_{.j} \mathbf{g}_i \otimes \mathbf{g}^j \cdot \mathbf{g}_k, \ \hat{\mathbf{g}}^k = \mathbf{Q} \cdot \mathbf{g}^k = Q_j^{\cdot i} \mathbf{g}^j \otimes \mathbf{g}_i \cdot \mathbf{g}^k \tag{2.184}$$

The above equations can reduce to the following

$$\hat{\mathbf{g}}_i = Q^j_{.i} \mathbf{g}_j, \ \hat{\mathbf{g}}^i = Q_j^{\cdot i} \mathbf{g}^j \ i, j = 1, \ldots, n \tag{2.185}$$

Consider a contravariant second-order tensor **T**, which can be transformed upon a change of basis as follows:

$$\mathbf{T} = \hat{T}^{ij}\hat{\mathbf{g}}_i \otimes \hat{\mathbf{g}}_j = Q^{k}_{.i} Q^{l}_{.j} \hat{T}^{ij} \mathbf{g}_k \otimes \mathbf{g}_l = T^{ij}\mathbf{g}_i \otimes \mathbf{g}_j \tag{2.186}$$

Consequently, upon a change of basis, the contravariant components of second-order tensor can be transformed as follows

$$T^{ij} = Q^{i}_{.k} Q^{j}_{.l} \hat{T}^{kl}, \; i, j = 1, \ldots, n \tag{2.187}$$

In view of Equation (2.183), we can also derive the following

$$\hat{T}^{ij} = Q^{.i}_{k} Q^{.j}_{l} T^{kl}, \; i, j = 1, \ldots, n \tag{2.188}$$

On the other hand, a covariant second-order tensor **T**, which can be transformed upon a change of basis as follows:

$$\mathbf{T} = \hat{T}_{ij}\hat{\mathbf{g}}^i \otimes \hat{\mathbf{g}}^j = Q^{.i}_{k} Q^{.j}_{l} \hat{T}_{ij} \mathbf{g}^k \otimes \mathbf{g}^l = T_{ij}\mathbf{g}^i \otimes \mathbf{g}^j \tag{2.189}$$

Consequently, upon a change of basis, the covariant components of second-order tensor can be transformed as follows

$$T_{ij} = Q^{.k}_{i} Q^{.l}_{j} \hat{T}_{kl}, \; i, j = 1, \ldots, n \tag{2.190}$$

In view of Equation (2.183), we can also derive the following

$$\hat{T}_{ij} = Q^{k}_{.i} Q^{l}_{.j} T_{kl}, \; i, j = 1, \ldots, n \tag{2.191}$$

As stated previously, a scalar, a vector and a tensor are geometric invariants, but the components of a vector and a tensor are dependent on a coordinate system. Further, every second-order tensor possesses an important invariance property, that is, the so-called principal invariants not depending upon a coordinate system. For every second-order tensor **T**, we have the equation

$$\mathbf{T} \cdot \mathbf{n}_i = \lambda_i \mathbf{n}_i, \; i = 1, 2, 3 \tag{2.192}$$

where $\mathbf{n}_i$ denotes eigenvectors and λ_i eigenvalues. The eigenvalues for corresponding eigenvectors can be uniquely determined by the characteristic equation

$$\det(\mathbf{T} - \lambda_i \mathbf{I}) = -\lambda_i^3 + \mathrm{I}_{\mathbf{T}}\lambda_i^2 - \mathrm{II}_{\mathbf{T}}\lambda_i + \mathrm{III}_{\mathbf{T}} = 0, \;\; i = 1, 2, 3 \tag{2.193}$$

where the principal invariants $\mathrm{I}_{\mathbf{T}}, \mathrm{II}_{\mathbf{T}}, \mathrm{III}_{\mathbf{T}}$ are given by

$$\begin{aligned} \mathrm{I}_{\mathbf{T}} &= \mathrm{tr}\,\mathbf{T} = \lambda_1 + \lambda_2 + \lambda_3, \\ \mathrm{II}_{\mathbf{T}} &= \frac{1}{2}\left[(\mathrm{tr}\,\mathbf{T})^2 - \mathrm{tr}\,(\mathbf{T}^2)\right] = \lambda_1\lambda_2 + \lambda_2\lambda_3 + \lambda_3\lambda_1, \\ \mathrm{III}_{\mathbf{T}} &= \det\mathbf{T} = \lambda_1\lambda_2\lambda_3 \end{aligned} \tag{2.194}$$

Multiplying both sides of Equation (2.192) by $\mathbf{T}$, we obtain

$$\begin{aligned} \mathbf{T}^2 \cdot \mathbf{n}_i &= \lambda_i \mathbf{T} \cdot \mathbf{n}_i = \lambda_i^2 \mathbf{n}_i, \\ \mathbf{T}^3 \cdot \mathbf{n}_i &= \lambda_i^2 \mathbf{T} \cdot \mathbf{n}_i = \lambda_i^3 \mathbf{n}_i \end{aligned} \tag{2.195}$$

As a consequence, the characteristic equation in Equation (2.193) yields

$$\mathbf{T}^3 - \mathrm{I}_{\mathbf{T}} \mathbf{T}^2 + \mathrm{II}_{\mathbf{T}} \mathbf{T} - \mathrm{III}_{\mathbf{T}} \mathbf{I} = \mathbf{0} \tag{2.196}$$

which is called the Cayley-Hamilton equation.

Higher-order Tensors A tensor, the order number of which is more than two, is called a higher-order tensor. For instance, a third-order tensor is a consequence of the triadic product of three vectors $\mathbf{u}, \mathbf{v}, \mathbf{w} \in \mathbb{V} \subset \mathbb{R}^n$ as follows:

$$\begin{aligned} \mathbf{u} \otimes \mathbf{v} \otimes \mathbf{w} &= (u^i \mathbf{g}_i) \otimes (v^j \mathbf{g}_j) \otimes (w^k \mathbf{g}_k) \\ &= u^i v^j w^k \mathbf{g}_i \otimes \mathbf{g}_j \otimes \mathbf{g}_k \end{aligned} \tag{2.197}$$

Single-dot product: The third-order tensor is a linear transformation that maps a vector $\mathbf{s} \in \mathbb{V} \subset \mathbb{R}^n$ to a second-order tensor

$$(\mathbf{u} \otimes \mathbf{v} \otimes \mathbf{w})\mathbf{s} = (\mathbf{w} \cdot \mathbf{s})\mathbf{u} \otimes \mathbf{v} \tag{2.198}$$

Single-dot product: The third-order tensor is a linear transformation that maps a second-order tensor $\mathbf{s} \otimes \mathbf{t}$ to a second-order tensor

$$(\mathbf{u} \otimes \mathbf{v} \otimes \mathbf{w}) \cdot (\mathbf{s} \otimes \mathbf{t}) = (\mathbf{w} \cdot \mathbf{s})(\mathbf{u} \otimes \mathbf{v} \otimes \mathbf{t}) \tag{2.199}$$

Double contraction: The third-order tensor is a linear transformation that maps a second-order tensor $\mathbf{s} \otimes \mathbf{t}$ to a vector

$$(\mathbf{u} \otimes \mathbf{v} \otimes \mathbf{w}) : (\mathbf{s} \otimes \mathbf{t}) = (\mathbf{v} \cdot \mathbf{s})(\mathbf{w} \cdot \mathbf{t})\mathbf{u} \tag{2.200}$$

Double contraction: The scalar product of a third-order tensor and a second-order identity tensor produces a vector

$$(\mathbf{u} \otimes \mathbf{v} \otimes \mathbf{w}) : \mathbf{I} = \mathrm{tr}\,(\mathbf{v} \otimes \mathbf{w})\mathbf{u} = (\mathbf{v} \cdot \mathbf{w})\mathbf{u} \tag{2.201}$$

The tensor product of two metric tensors gives a fourth-order identity tensor, i.e.

$$\begin{aligned} \mathbf{I} \otimes \mathbf{I} &= \mathbf{g}_i \otimes \mathbf{g}^j \otimes \mathbf{g}_k \otimes \mathbf{g}^l = \mathbf{g}^i \otimes \mathbf{g}_j \otimes \mathbf{g}^k \otimes \mathbf{g}_l \\ &= \mathbf{g}_i \otimes \mathbf{g}^j \otimes \mathbf{g}^k \otimes \mathbf{g}_l = \mathbf{g}^i \otimes \mathbf{g}_j \otimes \mathbf{g}_k \otimes \mathbf{g}^l \end{aligned} \tag{2.202}$$

Cartesian Tensors in the Three-Dimensional Euclidean Space Suppose that the coordinate axes, which are not coplanar lines, intersect at right angles and the basis is the orthonormal unit set, i.e. the standard orthonormal basis for Euclidean space. The coordinate system is called the orthogonal Cartesian coordinate system, or simply the Cartesian coordinate system. A tensor represented in the

Cartesian coordinate system is called a Cartesian tensor. In the oblique Cartesian coordinate system (often called the curvilinear coordinate system), the coordinate system may be oblique (Sokolnikoff 1951).

In $\mathbb{R}^3$, the contravariant base vectors can be obtained as

$$\mathbf{e}^1 = \frac{\mathbf{j} \times \mathbf{k}}{\mathbf{i} \cdot \mathbf{j} \times \mathbf{k}} = \mathbf{i}, \quad \mathbf{e}^2 = \frac{\mathbf{k} \times \mathbf{i}}{\mathbf{j} \cdot \mathbf{k} \times \mathbf{i}} = \mathbf{j}, \quad \mathbf{e}^3 = \frac{\mathbf{i} \times \mathbf{j}}{\mathbf{k} \cdot \mathbf{i} \times \mathbf{j}} = \mathbf{k} \tag{2.203}$$

Hence, the basis for the dual space $\{\mathbf{e}^i\}_{i=1}^3$ is the standard orthonormal basis for the vector space $\{\mathbf{i}, \mathbf{j}, \mathbf{k}\}$ in the three-dimensional Euclidean space; that is, span $\{\mathbf{e}^i\}_{i=1}^3 = \text{span}\{\mathbf{i}, \mathbf{j}, \mathbf{k}\} = \mathbb{R}^3$. A tensor represented in the orthonormal basis in $\mathbb{R}^3$ is called the Cartesian tensor. Note that the components of the Cartesian metric tensor are such that

$$g^{ij} = g_{ij} = g^{i}_{.j} = g_{j}^{.i} = \mathbf{e}_i \cdot \mathbf{e}_j = \delta_{ij} \tag{2.204}$$

Therefore, the identity tensor should be defined as $\mathbf{I} = \mathbf{e}_i \otimes \mathbf{e}_i$. The tensor product of two vectors, $\mathbf{u}, \mathbf{v} \in \mathbb{R}^3$, can be given by

$$\mathbf{u} \otimes \mathbf{v} = u_i v_j \mathbf{e}_i \otimes \mathbf{e}_j \tag{2.205}$$

Thus, a second-order tensor $\mathbf{T}$ as a linear combination of dyads can be also written as

$$\mathbf{T} = T_{ij} \mathbf{e}_i \otimes \mathbf{e}_j, \quad T_{ij} = \mathbf{e}_i \cdot \mathbf{T} \cdot \mathbf{e}_j, \quad \operatorname{tr} \mathbf{T} = T_{ii} \tag{2.206}$$

2.8.2 Tensor Differential Calculus

We shall begin with the derivative of a second-order tensor with respect to a real variable. Let $\mathbf{T}(t)$ denote a tensor-valued function of a real variable $t \in \mathbb{R}$. Suppose that the tensor-valued function is differentiable at the time $t \in \mathcal{T} =]t_0, t_1[$. Then, the derivative of $\mathbf{T}(t)$ with respect to t is defined as

$$\frac{d\mathbf{T}(t)}{dt} = \lim_{\Delta t \to 0} \frac{\mathbf{T}(t + \Delta t) - \mathbf{T}(t)}{\Delta t} \tag{2.207}$$

Analogous to the case of the vector-valued functions in Section 2.4.2, the two tensor-valued functions, $\mathbf{T}, \mathbf{S}$, which are differentiable at t, satisfy the following properties

(a) Summation

$$\frac{d}{dt}(\mathbf{T}(t) + \mathbf{S}(t)) = \frac{d\mathbf{T}(t)}{dt} + \frac{d\mathbf{S}(t)}{dt} \tag{2.208}$$

(b) Multiplication by a scalar

$$\frac{d}{dt}(\alpha \mathbf{T}(t)) = \alpha \frac{d\mathbf{T}(t)}{dt} \tag{2.209}$$

(c) Multiplication by a scalar-valued function

$$\frac{d}{dt}(f(t)\mathbf{T}(t)) = \frac{df(t)}{dt}\mathbf{T}(t) + f(t)\frac{d\mathbf{T}(t)}{dt} \tag{2.210}$$

(d) Single-dot contraction

$$\frac{d}{dt}(\mathbf{T}(t) \cdot \mathbf{S}(t)) = \frac{d\mathbf{T}(t)}{dt} \cdot \mathbf{S}(t) + \mathbf{T}(t) \cdot \frac{d\mathbf{S}(t)}{dt} \tag{2.211}$$

(e) Linear operator mapping from a vector to a vector

$$\frac{d}{dt}(\mathbf{T}(t) \cdot \mathbf{v}(t)) = \frac{d\mathbf{T}(t)}{dt} \cdot \mathbf{v}(t) + \mathbf{T}(t) \cdot \frac{d\mathbf{v}(t)}{dt} \tag{2.212}$$

Let us consider the derivatives of a scalar-valued function $\Psi(\mathbf{x})$. Let $\mathbf{x}$ denote a point, which is an element of the n-dimensional vector space $\mathbb{V}$ with a covariant basis $\{\mathbf{g}_i\}_{i=1}^n$, i.e.,

$$\operatorname{span}\{\mathbf{g}_i\}_{i=1}^n = \mathbb{V} = \left\{\mathbf{x} = (x^1, \ldots, x^n) | x^i \in \mathbb{R},\ \text{for } i = 1, \ldots, n\right\} \tag{2.213}$$

where the entries x^i denotes contravariant coordinates in the vector space $\mathbb{V}$. Then, the differential of the scalar-valued function is given by

$$d\Psi(\mathbf{x}) = \operatorname{grad}\Psi \cdot d\mathbf{x} = \frac{\partial \Psi(\mathbf{x})}{\partial x^i}\mathbf{g}^i \cdot dx^j \mathbf{g}_j = \frac{\partial \Psi(\mathbf{x})}{\partial x^i} dx^i \tag{2.214}$$

where the gradient of the scalar-valued function can be written as

$$\operatorname{grad}\Psi(\mathbf{x}) = \frac{\partial \Psi(\mathbf{x})}{\partial x^i}\mathbf{g}^i = \Psi_{,i}(\mathbf{x})\mathbf{g}^i = \nabla\Psi(\mathbf{x}),\ \nabla = \frac{\partial}{\partial x^k}\mathbf{g}^k \tag{2.215}$$

where ∇ denotes the gradient operator. Therefore, the gradient of the scalar-valued function $\Psi(\mathbf{x}) : \mathbb{R}^n \to \mathbb{R}$ reads as

$$\operatorname{grad}\Psi(\mathbf{x}) = \frac{\partial \Psi(\mathbf{x})}{\partial x^1}\mathbf{g}^1 + \frac{\partial \Psi(\mathbf{x})}{\partial x^2}\mathbf{g}^2 + \ldots + \frac{\partial \Psi(\mathbf{x})}{\partial x^n}\mathbf{g}^n \tag{2.216}$$

Similarly, the differential of the vector-valued function $\mathbf{v}(\mathbf{x})$ can be given by

$$d\mathbf{v}(\mathbf{x}) = \operatorname{grad}\mathbf{v}(\mathbf{x}) \cdot d\mathbf{x} = \frac{\partial \mathbf{v}(\mathbf{x})}{\partial x^i} \otimes \mathbf{g}^i \cdot dx^j \mathbf{g}_j = \frac{\partial \mathbf{v}(\mathbf{x})}{\partial x^i} dx^i \tag{2.217}$$

Hence, the gradient of the vector-valued function can be written (Irgens 2008; Malvern 1969; Simmons 1994) as

$$\operatorname{grad}\mathbf{v} = \frac{\partial \mathbf{v}(\mathbf{x})}{\partial \mathbf{x}} = \frac{\partial \mathbf{v}(\mathbf{x})}{\partial x^i} \otimes \mathbf{g}^i = \mathbf{v}_{,i}(\mathbf{x}) \otimes \mathbf{g}^i = \mathbf{v} \otimes \nabla \tag{2.218}$$

where $\mathbf{v}_{,i}$ is obtained as

$$\begin{aligned}
\mathbf{v}_{,i} &= (v^j \mathbf{g}_j)_{,i} = v^j_{,i}\mathbf{g}_j + v^j \mathbf{g}_{j,i} \\
&= v^j_{,i}\mathbf{g}_j + v^j \Gamma^k_{ij}\mathbf{g}_k \\
&= \nabla_i v^k \mathbf{g}_k = v^k|_{,i}\mathbf{g}_k,\ \nabla_i v^k = v^k_{,i} + \Gamma^k_{ij} v^j
\end{aligned} \tag{2.219}$$

where $\nabla_i v^k$ or $v^k|_{,i}$ is often called the covariant derivative of v^k with respect to x^i and Γ_{ij}^k is the Christoffel symbol of the second kind. The Christoffel symbols of the first and second kind are defined as (Haupt 2000; Malvern 1969; Narasimhan 1993; Wempner and Talaslidis 2003)

$$\Gamma_{ijk} = \mathbf{g}_k \cdot \mathbf{g}_{i,j}, \ \Gamma_{ij}^k = \mathbf{g}^k \cdot \mathbf{g}_{i,j} \tag{2.220}$$

Note that all Christoffel symbols vanish in the three-dimensional Euclidean space (Sokolnikoff 1951). Since $\mathbf{g}_{i,j} = \mathbf{g}_{j,i}$, the Christoffel symbols are symmetric such that

$$\Gamma_{ijk} = \Gamma_{jik}, \ \Gamma_{ij}^k = \Gamma_{ji}^k \tag{2.221}$$

Note that for the vector-valued function $\mathbf{v}(\mathbf{x}) : \mathbb{R}^n \to \mathbb{R}^m$, the matrix representation for $\text{grad}\,\mathbf{v}(\mathbf{x})$ is written as

$$\left[\text{grad}\,\mathbf{v}(\mathbf{x})\right] = \left[\frac{\partial \mathbf{v}(\mathbf{x})}{\partial \mathbf{x}}\right]_{m\times n} = \begin{bmatrix} \dfrac{\partial v^1(\mathbf{x})}{\partial x_1} & \dfrac{\partial v^1(\mathbf{x})}{\partial x_2} & \cdots & \dfrac{\partial v^1(\mathbf{x})}{\partial x_n} \\ \dfrac{\partial v^2(\mathbf{x})}{\partial x_1} & \dfrac{\partial v^2(\mathbf{x})}{\partial x_2} & \cdots & \dfrac{\partial v^2(\mathbf{x})}{\partial x_n} \\ \cdots & \cdots & \cdots & \cdots \\ \dfrac{\partial v^m(\mathbf{x})}{\partial x_1} & \dfrac{\partial v^m(\mathbf{x})}{\partial x_2} & \cdots & \dfrac{\partial v^m(\mathbf{x})}{\partial x_n} \end{bmatrix}_{m\times n} \tag{2.222}$$

On the other hand, the differential of the tensor-valued function is given by (Papastavridis 1998, page 93)

$$d\mathbf{T}(\mathbf{x}) = \text{grad}\,\mathbf{T}(\mathbf{x}) \cdot d\mathbf{x} = \frac{\partial \mathbf{T}(\mathbf{x})}{\partial x^i} \otimes \mathbf{g}^i \cdot dx^j \mathbf{g}_j = \frac{\partial \mathbf{T}(\mathbf{x})}{\partial x^i} dx^i \tag{2.223}$$

where the gradient of the second-order tensor gives

$$\text{grad}\,\mathbf{T}(\mathbf{x}) = \frac{\partial \mathbf{T}(\mathbf{x})}{\partial \mathbf{x}} = \frac{\partial \mathbf{T}(\mathbf{x})}{\partial x^i} \otimes \mathbf{g}^i = \mathbf{T}_{,i}(\mathbf{x}) \otimes \mathbf{g}^i = \mathbf{T} \otimes \nabla \tag{2.224}$$

where $\mathbf{T}_{,i}$ represents the covariant derivative of the second-order tensor with respect to the coordinate x^i. Note that the covariant derivative of the covariant component T_{ij} with respect to the coordinate x^k can be given by

$$\mathbf{T}_{,k} = T_{ij}|_k \mathbf{g}^i \otimes \mathbf{g}^j, \ T_{ij}|_k = \frac{\partial T_{ij}}{\partial x^k} - \Gamma_{ik}^m T_{mj} - \Gamma_{jk}^m T_{im} \tag{2.225}$$

Hence, Equation (2.224) can be rewritten as

$$\text{grad}\,\mathbf{T}(\mathbf{x}) = T_{ij}|_k \mathbf{g}^i \otimes \mathbf{g}^j \otimes \mathbf{g}^k \tag{2.226}$$

In addition, the differential of the vector-valued function multiplied by a scalar-valued function is written as follows:

$$\begin{aligned} d(\Psi(\mathbf{x})\mathbf{v}(\mathbf{x})) &= \text{grad}\,(\Psi(\mathbf{x})\mathbf{v}(\mathbf{x}))\cdot d\mathbf{x} \\ &= \frac{\partial(\Psi(\mathbf{x})\mathbf{v}(\mathbf{x}))}{\partial x^i}\otimes \mathbf{g}^i \cdot dx^j\mathbf{g}_j \\ &= \frac{\partial(\Psi(\mathbf{x})\mathbf{v}(\mathbf{x}))}{\partial x^i}dx^i \\ &= (\Psi_{,i}(\mathbf{x})\mathbf{v}(\mathbf{x}) + \Psi(\mathbf{x})\mathbf{v}_{,i}(\mathbf{x}))dx^i \end{aligned} \tag{2.227}$$

where the gradient of the function can be given by

$$\begin{aligned} \text{grad}\,(\Psi(\mathbf{x})\mathbf{v}(\mathbf{x})) &= \frac{\partial(\Psi(\mathbf{x})\mathbf{v}(\mathbf{x}))}{\partial x^i}\otimes \mathbf{g}^i \\ &= (\Psi_{,i}(\mathbf{x})\mathbf{v}(\mathbf{x}) + \Psi(\mathbf{x})\mathbf{v}_{,i}(\mathbf{x}))\otimes \mathbf{g}^i \\ &= \mathbf{v}(\mathbf{x})\otimes \text{grad}\,\Psi(\mathbf{x}) + \Psi(\mathbf{x})\text{grad}\,\mathbf{v} \end{aligned} \tag{2.228}$$

Next, consider the derivative of a scalar-valued function of a second-order tensor with respect to the second-order tensor. The differential of the scalar-valued function is given by

$$d\Phi(\mathbf{T}) = \frac{\partial \Phi}{\partial \mathbf{T}} \cdot\cdot^{\sim} d\mathbf{T} \tag{2.229}$$

where $\mathbf{T}$ is assumed to be a contravariant tensor, $T^{ij}\mathbf{g}_i \otimes \mathbf{g}_j$. Then we have

$$\frac{\partial \Phi}{\partial \mathbf{T}} = \frac{\partial \Phi}{\partial T^{ij}}\mathbf{g}^i \otimes \mathbf{g}^j \tag{2.230}$$

This is a covariant second-order tensor. Lastly, we consider the divergence of a vector $\mathbf{v}$, which is defined as

$$\text{div}\,\mathbf{v} = \mathbf{v}\cdot\nabla = \nabla\cdot\mathbf{v} = \text{tr}\,(\text{grad}\,\mathbf{v}) = \text{grad}\,\mathbf{v} : \mathbf{I} \tag{2.231}$$

The divergence of a second-order tensor $\mathbf{T}$ is defined as

$$\begin{aligned} \text{div}\,\mathbf{T} &= \mathbf{T}\cdot\nabla = \text{grad}\,\mathbf{T} : \mathbf{I} = \mathbf{T}_{,i}\cdot \mathbf{g}^i \\ &= \frac{\partial T^{ij}}{\partial x^k}\mathbf{g}_i \otimes \mathbf{g}_j \cdot \mathbf{g}^k = \frac{\partial T^{ij}}{\partial x^j}\mathbf{g}_i \end{aligned} \tag{2.232}$$

which produces a vector.

2.8.3 Tensor Integral Calculus

In integral calculus, there are important integral invariant forms providing integral identities between surface integrals and volume integrals or between line integrals and surface integrals. We first consider the divergence theorem, which is often called Gauss's theorem or Ostrogradsky's theorem or Gauss-Ostrogradsky's theorem. Let $\mathbf{v}(\mathbf{x})$ denote a vector defined on the open, bounded set Ω, which is embedded in $\mathbb{R}^3$ as a region, i.e. $\mathbf{x} \in \Omega \subset \mathbb{R}^3$. Then the Gauss's theorem is defined as

$$\int_{\Omega} \mathbf{v} \cdot \nabla \, dV = \int_{\Omega} \nabla \cdot \mathbf{v} \, dV = \int_{\partial\Omega} \mathbf{v} \cdot \mathbf{n} \, dA = \int_{\partial\Omega} \mathbf{n} \cdot \mathbf{v} \, dA \tag{2.233}$$

where $\partial\Omega$ denotes the open, bounded set as the surface of the region. In the above equation, dV and dA denote the Lebesgue measures for the volume and the surface. Instead of a vector, a second-order tensor can be applied to the equation. Then the divergence theorem reads as

$$\int_{\Omega} \nabla \cdot \mathbf{T} \, dV = \int_{\partial\Omega} \mathbf{n} \cdot \mathbf{T} \, dA \tag{2.234}$$

and

$$\int_{\Omega} \mathbf{T} \cdot \nabla dV = \int_{\partial\Omega} \mathbf{T} \cdot \mathbf{n} \, dA \tag{2.235}$$

Replacing the second-order tensor $\mathbf{T}$ by $\Psi(\mathbf{x})\mathbf{I}$, the above equation leads to

$$\int_{\Omega} \mathrm{grad}\, \Psi(\mathbf{x}) \, dV = \int_{\partial\Omega} \Psi(\mathbf{x}) \mathbf{n} \, dA \tag{2.236}$$

Note that upon a change of basis, the above forms are invariant.

EXERCISES

2.1. In mathematics, we frequently deal with the sets of numbers such as $\mathbb{N}$ for natural numbers, $\mathbb{Z}$ for integers, $\mathbb{Q}$ for rational numbers, $\mathbb{J}$ for irrational numbers, $\mathbb{R}$ for real numbers, and $\mathbb{C}$ for complex numbers. Discuss the relations.

2.2. Discuss why $\mathbb{R}^n$ can be identical to $\mathbb{E}^n$, and why $\mathbb{R}^n$ can be different from $\mathbb{E}^n$.

2.3. Let $\mathbf{u}$, $\mathbf{v}$ and $\mathbf{w}$ denote vectors. The scalar triple product is given by $\mathbf{u} \cdot (\mathbf{v} \times \mathbf{w})$. Show that the scalar triple product gives the volume of a parallelepiped as shown in Figure 2.6, i.e. $\mathrm{VOL}(\mathbf{u}, \mathbf{v}, \mathbf{w}) = \mathbf{u} \cdot (\mathbf{v} \times \mathbf{w})$. Note that the magnitude of $\mathbf{v} \cdot \mathbf{w}$ represents the area.

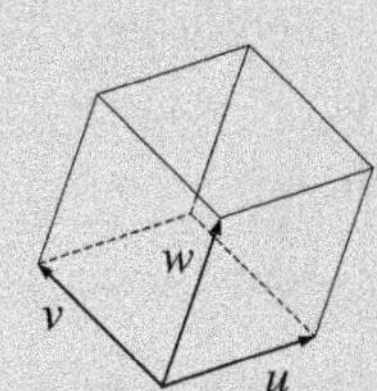

Figure 2.6. Volume of a parallelepiped

2.4. Let $\mathbf{u} = (2, 1, 5)$ and $\mathbf{v} = (1, 2, 7)$. Compute the angle between $\mathbf{u}$ and $\mathbf{v}$.

2.5. For three vectors $\mathbf{u}$, $\mathbf{v}$ and $\mathbf{w}$, the vector triple product is given by $\mathbf{u} \times (\mathbf{v} \times \mathbf{w})$. Prove that $\mathbf{u} \times (\mathbf{v} \times \mathbf{w}) = (\mathbf{u} \cdot \mathbf{w})\mathbf{v} - (\mathbf{u} \cdot \mathbf{v})\mathbf{w}$ and $(\mathbf{u} \times \mathbf{v}) \times \mathbf{w} = (\mathbf{u} \cdot \mathbf{w})\mathbf{v} - (\mathbf{v} \cdot \mathbf{w})\mathbf{u}$.

2.6. Suppose that scalar-valued functions $P(\mathbf{x}) : \Omega \in \mathbb{R}^3 \rightarrow \mathbb{R}$ and $Q(\mathbf{x}) : \mathcal{B} \in \mathbb{R}^3 \rightarrow \mathbb{R}$ are twice continuously differentiable in the domain $\mathcal{B}$. Verify Green's first identity

$$\int_{\Omega} \left(P(\mathbf{x})\nabla^2 Q(\mathbf{x}) + \nabla P(\mathbf{x}) \cdot \nabla Q(\mathbf{x})\right) dV = \int_{\partial\Omega} P(\mathbf{x}) \left(\nabla Q(\mathbf{x}) \cdot \mathbf{n}\right) dA$$

and Green's second identity

$$\int_{\Omega} \left(P(\mathbf{x})\nabla^2 Q(\mathbf{x}) - Q(\mathbf{x})\nabla^2 P(\mathbf{x})\right) dV = \int_{\Omega} \left(P(\mathbf{x})\nabla Q(\mathbf{x}) - Q(\mathbf{x})\nabla P(\mathbf{x})\right) \cdot \mathbf{n}\, dA$$

where dV and dA are volume and area segments, and $\mathbf{n}$ is the unit normal vector to the surface.

2.7. Write out

$$\text{a) } \delta^i_i, \quad \text{b) } \delta^i_j \delta^i_j, \quad \text{c) } \delta^i_j \delta^j_k \delta^k_l \delta^l_i, \quad \text{d) } \delta^i_j u_i, \quad \text{e) } \delta^i_j v^j, \quad \text{f) } \delta^i_j u_i v^j$$

2.8. Derive Equation (2.148).

2.9. Let $\mathbf{U}$ and $\mathbf{V}$ denote second-order tensors. Prove that $(\mathbf{U} \cdot \mathbf{V})^T = \mathbf{V}^T \cdot \mathbf{U}^T$.

2.10. Prove that $\operatorname{tr} \mathbf{U}^T = \operatorname{tr} \mathbf{U}$.

2.11. Prove that $\operatorname{tr}(\mathbf{U} \cdot \mathbf{V}) = \operatorname{tr}(\mathbf{V} \cdot \mathbf{U})$.

2.12. Prove that $\operatorname{tr}(\mathbf{u} \otimes \mathbf{v}) = \mathbf{u} \cdot \mathbf{v}$.

2.13. Suppose that vector-valued functions $\mathbf{u}(\mathbf{x}) : \Omega \rightarrow \mathbb{R}^n$ and $\mathbf{v}(\mathbf{x}) : \Omega \rightarrow \mathbb{R}^n$ are of class $C^1(\Omega)$, where Ω is an open bounded set in $\mathbb{R}^n$. Let $\mathbf{T}(\mathbf{x})$ denote a tensor-valued function, and $\Psi(\mathbf{x}) : \Omega \rightarrow \mathbb{R}^n$ a scalar-valued function. Prove the following properties:

- $\operatorname{div}(\mathbf{u} \otimes \mathbf{v}) = \mathbf{v} \operatorname{div} \mathbf{u} + \operatorname{grad} \mathbf{u} \cdot \mathbf{v}$.
- $\operatorname{div}(\mathbf{u} \times \mathbf{v}) = \mathbf{v} \cdot \operatorname{curl} \mathbf{u} - \mathbf{u} \cdot \operatorname{curl} \mathbf{u}$.
- $\operatorname{div}(\Psi \mathbf{u}) = \Psi \operatorname{div} \mathbf{u} + \mathbf{u} \cdot \operatorname{grad} \Psi$.
- $\operatorname{div}(\Psi \mathbf{T}) = \Psi \operatorname{div} \mathbf{T} + \mathbf{T} \cdot \operatorname{grad} \Psi$.
- $\operatorname{div}(\mathbf{u}\mathbf{T}) = \mathbf{T} : \operatorname{grad} \mathbf{u} + \mathbf{u} \cdot \operatorname{div} \mathbf{T}$.
  Note that this relation is frequently used in Part Two.
- $\operatorname{grad}(\Psi \mathbf{u}) = \Psi \operatorname{grad} \mathbf{u} + \mathbf{u} \otimes \operatorname{grad} \Psi$.
- $\operatorname{grad}(\mathbf{u} \cdot \mathbf{v}) = \mathbf{u} \cdot \operatorname{grad} \mathbf{v} + \mathbf{v} \cdot \operatorname{grad} \mathbf{w}$.

2.14. Spherical and deviatoric tensors of a second-order tensor $\mathbf{U}$ are defined in Equation (2.171). Verify that $\operatorname{tr} \mathbf{U}_{\text{dev}} = 0$.

2.15. Prove that $(\mathbf{U}_{\text{dev}})_{\text{sph}} = 0$.

2.16. Prove that $\mathrm{III}_{\mathbf{T}} = \frac{1}{3} \operatorname{tr} \mathbf{T}^3 - \frac{1}{2} \operatorname{tr} \mathbf{T}^2 \operatorname{tr} \mathbf{T} + \frac{1}{6} (\operatorname{tr} \mathbf{T})^3$, where $\mathbf{T}$ denotes a second-order tensor and $\mathrm{III}_{\mathbf{T}}$ is the third invariant scalar of $\mathbf{T}$.

PART I

N-BODY DYNAMICAL SYSTEMS

CHAPTER THREE

CLASSICAL MECHANICS

Classical mechanics as described in Chapter 3, deals with the statical or dynamical motion of material objects. The objects are mathematically regarded as point masses or rigid bodies which are dynamical systems in space and time. Here, the concept of space and time is confined to the Newtonian sense. In general, classical mechanics may be categorized as vectorial mechanics and analytical mechanics. The representative of vectorial mechanics is Newtonian mechanics, whereas analytical mechanics can be divided into Lagrangian and Hamiltonian mechanics with descriptive scalar functions, namely, the Lagrangian and the Hamiltonian. It is believed that the distinction between Newtonian, Lagrangian and Hamiltonian mechanics emanates from the notion of space.

3.1 NEWTONIAN MECHANICS

In three dimensional Euclidean space, a frame of reference, actually a set of coordinates or a coordinate system, is a mathematical abstract needed to describe a motion of a particle. In Newtonian mechanics, the frame of reference, where Newton's laws of motion hold, is called an inertial frame of reference. The inertial frame of reference is often called Galilean frame of reference (Arnold 1989; Gron and Hervik 2007; McComb 1999), in which the principle of Galilean relativity is true. The inertial frame of reference is a non-rotating and non-accelerating frame of reference (Gron and Hervik 2007; McComb 1999).

3.1.1 Newton's Laws of Motion

Then, Newton's laws of motion are stated as follows:

1. Newton's first law of motion: A particle resists to the influence of an external force causing its initial state to be changed. This resistance is brought about by the characteristic of the particle, called inertia. The inertia of the particle is quantitatively measured by an inertial mass. Therefore, the inertial mass is a quantitative measure in Newtonian mechanics. The larger the mass of a particle, then, the larger the inertia of the particle. In order to have a particle, that is at rest, to be moved, some external force should be exerted upon the particle due to the inertia of the particle. Also, in order to have a particle moving at a constant velocity to be at rest, an external force should be exerted upon the particle.

Advances in Computational Dynamics of Particles, Materials and Structures, First Edition. Jason Har and Kumar K. Tamma.

2. Newton's second law of motion: The rate of change of the linear momentum of a particle or an N-body system equals the resultant acting on the system in the rectilinear motion. For the dynamical system of the body with constant mass, the acceleration of the body is proportional to the total force exerted upon the body; thus, Newton's second law of motion can be represented by a mathematical description, which is Galilean invariant in the Galilean frame of reference (Arnold 1989; Gron and Hervik 2007; McComb 1999). The constant of proportionality is mass. Consequently, Newton's second law is often called the principle of balance of the linear momentum. Both acceleration and force are given by vector quantities defined in the inertial frame of reference to describe mathematically the motion of the system of either a particle or an N-body. Since the linear momentum of either a particle or an N-body is given by the inertial mass, Newton's first law of motion may be regarded as a consequence of Newton's second law of motion.
3. Newton's third law of motion: Interacting bodies exert equal forces and their directions are opposite each other on a line.

For convenience, it is quite natural to choose the Cartesian coordinate system $(x(t), y(t), z(t))$ (Gron and Hervik 2007) as an inertial frame of reference in three dimensional Euclidean space to describe the three-dimensional motion of either a particle or an N-body dynamical system in Newtonian mechanics as in the subsequent sections. Note that the inertial frame of reference in Newtonian mechanics can also be curvilinear coordinate systems such as the cylindrical $(r(t), \theta(t), z(t))$ or spherical polar coordinate $(r(t), \theta(t), \phi(t))$ systems.

3.1.2 Newton's Equations of Motion

Consider a system of N particles moving in the Cartesian coordinate system. Such a system is often referred to as an N-body dynamical system. The position vector of the i-th particle within the closed interval of time $t \in \bar{\mathcal{T}} = [t_0, t_1]$ is defined as a vector-valued function of one real variable t as shown below:

$$\mathbf{x}^i(t) = \left(x^i(t), y^i(t), z^i(t)\right) : \bar{\mathcal{T}} \to \mathbb{R}^3, \ \forall t \in \bar{\mathcal{T}} \subset \mathbb{R}^+ \tag{3.1}$$

where $x^i(t)$ denotes a scalar-valued function of the real variable t as a coordinate function of the i-th particle in the x-direction, $y^i(t)$ in the y-direction, and $z^i(t)$ in the z-direction in the Cartesian coordinate system. It should be noted that in the above equation the position vector is a column vector; however, for convenience the transpose notation is dropped (Marsden and Tromba 2003, page 79). The coordinate functions are referred to as Cartesian coordinate variables which are dependent variables. Hence, the range of the position vector of the i-th particle is contained in Euclidean 3-space. Note that the parameter, namely, the time variable, is considered as an independent variable.

As shown in Figure 3.1, a trajectory (Newtonian path) of the system in the three-dimensional motion is a curve of a vector-valued function of the 3N-number of multivariables as a set, i.e.,

$$\mathbf{x}(t) = \left(\mathbf{x}^1(t), \ldots, \mathbf{x}^N(t)\right) : \bar{\mathcal{T}} \to \mathbb{R}^{3N}, \ \forall t \in \bar{\mathcal{T}} \tag{3.2}$$

where the range of the vector-valued function of $3N$ variables is contained in Euclidean $3N$-space. Newton's equation of motion (Newton's second law) for the i-th particle in the open interval of time $t \in \mathcal{T} =]t_0, t_1[$, whose constant mass which is denoted by m_i, is given by

$$\frac{d}{dt}\left(m_i \dot{\mathbf{x}}^i\right) = m_i \ddot{\mathbf{x}}^i = \mathbf{F}^i_{\text{ext}} + \mathbf{F}^i_{\text{int}}, \ i = 1, \ldots, N, \ \forall t \in \mathcal{T} =]t_0, t_1[\tag{3.3}$$

where $\mathbf{F}^i_{\text{ext}}(t) : \mathcal{T} \subset \mathbb{R}^+ \to \mathbb{R}^3$ denotes the total external force acting upon the i-th particle and $\mathbf{F}^i_{\text{int}}(t) : \mathcal{T} \to \mathbb{R}^3$ denotes the total internal force as the sum of the interactions acting on the i-th particle owing

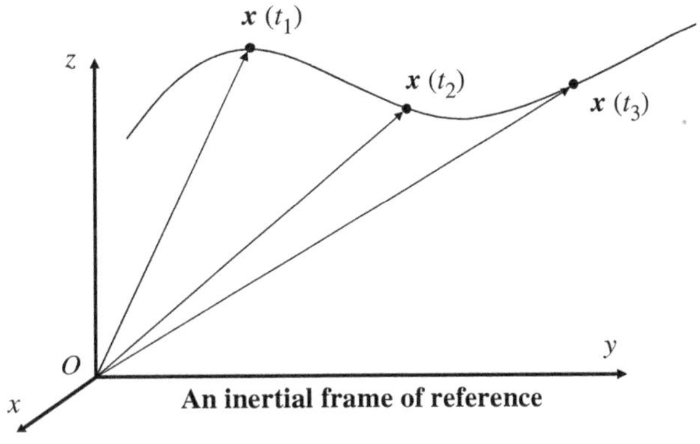

Figure 3.1. A trajectory of N particles in the three-dimensional motion

to the j-particle, i.e. $\mathbf{F}^i_{\text{int}} = \sum_{j=1}^{N} \mathbf{F}^{ij}$ and $\mathbf{F}^{ii} = \mathbf{0}$. It is important to note that Equation (3.3) represents the 3N-number of simultaneous differential equations of motion for the N-body dynamical system with 3N degrees of freedom.

Newton's third law results in the following

$$\sum_{i=1}^{N} \mathbf{F}^i_{\text{int}}(t) = \sum_{i=1}^{N}\sum_{j=1}^{N} \mathbf{F}^{ij} = \mathbf{0} \tag{3.4}$$

because $\mathbf{F}^{12} + \mathbf{F}^{21} = \mathbf{0}$ and $\mathbf{F}^{ii} = \mathbf{0}$. Hence the equation of the three-dimensional motion of the N-body system yields

$$\frac{d}{dt}\left(\sum_{i=1}^{N} m_i \dot{\mathbf{x}}^i\right) = \sum_{i=1}^{N} m_i \ddot{\mathbf{x}}^i = \sum_{i=1}^{N} \mathbf{F}^i_{\text{ext}} = \mathbf{F}_{\text{R}}(t) \tag{3.5}$$

where $\mathbf{F}_{\text{R}}(t) : \mathcal{T} \subset \mathbb{R}^+ \to \mathbb{R}^3$ is the total external force of the system is a vector-valued function of one real variable t. The total external force, which is often called the resultant force, may be dependent on the position, the velocity, and time, but it can not be a function of the acceleration.

Suppose that the N-body dynamical system is now subject to some constraints which limit the motion of the system in the Cartesian coordinate system. Then, the external force exerted on the i-th particle is divided into the applied force and the constraint force. Hence, Newton's equations of motion for the i-th particle can be rewritten as

$$m_i \ddot{\mathbf{x}}^i = \mathbf{F}^i_{\text{appl}} + \mathbf{C}^i + \sum_{j=1}^{N} \mathbf{F}^{ij} \tag{3.6}$$

where $\mathbf{C}^i$ is the constraint force acting upon the i-th particle and $\mathbf{F}^i_{\text{appl}}(t) : \mathcal{T} \subset \mathbb{R}^+ \to \mathbb{R}^3$ the applied force vector acting upon the i-th particle is a vector-valued function of one variable t. Accordingly, in favor of vector and matrix notation, Newton's second law of motion of the N-body dynamical system, which is illustrated in Figure 3.2, can be written as (noting that the total internal force is canceled (Greenwood 1977))

$$\sum_{i=1}^{N} m_i \ddot{\mathbf{x}}^i = \sum_{i=1}^{N} \mathbf{F}^i_{\text{appl}}(t) + \sum_{i=1}^{N} \mathbf{C}^i \tag{3.7}$$

Newtonian Mechanics

$$\sum_{i=1}^{N} m_i \ddot{\mathbf{x}}^i = \sum_{i=1}^{N} \mathbf{F}^i_{\text{appl}} + \sum_{i=1}^{N} \mathbf{C}^i$$

Euclidean space : $\mathbb{R}^3$

$$\mathbf{x}^i : \mathcal{T} \subset \mathbb{R}^+ \to \mathbb{R}^3,\ t \in \mathcal{T} = [t_0, t_1]$$

$$\mathbf{x} : \mathcal{T} \subset \mathbb{R} \to \mathbb{R}^{3N},\ t \in \mathcal{T}$$

$$\mathbf{x} = \left(\mathbf{x}^1, \ldots, \mathbf{x}^N\right)$$

$$f_i\left(\mathbf{x}; t\right) = 0,\ i = 1, \ldots, k$$

Newton, Isaac (1642-1727)

Figure 3.2. Newton's equations of motion in a conservative dynamical system. Reproduced with permission from O'Hanian, Boyer and Merzbach, 1987. © Dover Publications Inc.

Equation (3.6) represents 3N-number of simultaneous differential equations of motion, which are redundant because of the presence of the constraints. In other words, the total number of degrees of freedom is smaller than the number of dependent variables, i.e. the Cartesian coordinate variables. In general, the constraints are given by algebraic equations with coordinate variables in the inertial frame of reference. As a consequence, the Newtonian dynamical system is subject to the constraint equations. Suppose that the k-number of constraints ($k <$ N) are imposed in the system and these constraint equations are associated with Cartesian coordinate variables. Then, the number of degrees of freedom n_{dof} for the N-body dynamical system with k-constraints becomes $3N - k$, i.e. $n_{\text{dof}} = 3N - k$.

A conservative force can be represented by the gradient of the potential energy with respect to the position. Therefore, the conservative force $\mathbf{F}^i_{\text{c}}(\mathbf{x}) : \mathbb{R}^{3N} \to \mathbb{R}^3$ acting upon the i-th particle is given by the negative gradient of the potential energy, i.e.,

$$\mathbf{F}^i_{\text{c}}(\mathbf{x}) = -\frac{d\mathcal{U}}{d\mathbf{x}^i} = \left(f^i_{cx}(\mathbf{x}), f^i_{cy}(\mathbf{x}), f^i_{cz}(\mathbf{x})\right) : \mathbb{R}^{3N} \to \mathbb{R}^3 \tag{3.8}$$

where $\mathcal{U}(\mathbf{x}) : \mathbb{R}^{3N} \to \mathbb{R}$ denotes the potential energy of the system as a scalar-valued function of multivariables defined on Euclidean $3N$-space $\mathbb{R}^{3N}$. In addition, the applied force is divided into the conservative force and the nonconservative force as follows

$$\mathbf{F}^i_{\text{appl}} = \mathbf{F}^i_{\text{c}} + \mathbf{F}^i_{\text{nc}} \tag{3.9}$$

where $\mathbf{F}^i_{\text{nc}}$ denotes the nonconservative force acting upon the i-th particle, which cannot be expressed in terms of any potential energy.

3.2 LAGRANGIAN MECHANICS

Within the context of Newtonian mechanics, the vector quantities such as forces and accelerations are often represented in the Cartesian coordinate system. When the complexity of a problem is increased to some extent, it becomes rather difficult to solve the Newtonian differential system due to the inclusion of a number of Cartesian coordinate variables.

At the same era of Newton, it was Leibniz (1646–1716) who appears to be the first to investigate the use of the energy or work of a system to describe the motion of dynamical systems. Even though

Lagrangian Mechanics

$$\frac{\partial \mathcal{L}(\mathbf{q},\dot{\mathbf{q}},t)}{\partial \mathbf{q}} - \frac{d}{dt}\left(\frac{\partial \mathcal{L}(\mathbf{q},\dot{\mathbf{q}},t)}{\partial \dot{\mathbf{q}}}\right) = \mathbf{0}$$

$$\mathbf{q} = \left(q^1,\ldots,q^{n_{\text{dof}}}\right), \quad n_{\text{dof}} = 3N - k$$

Configuration space $Q = \mathbb{R}^{n_{\text{dof}}}$

$$\mathcal{L}(\mathbf{q},\dot{\mathbf{q}},t) = \mathcal{K}(\mathbf{q},\dot{\mathbf{q}},t) - \mathcal{U}(\mathbf{q},t)$$

$$\mathcal{L}(\mathbf{q},\dot{\mathbf{q}},t): TQ \times \bar{T} \to \mathbb{R},$$

$$TQ = Q \times T_q Q$$

Lagrange, Joseph (1736-1813)

Figure 3.3. Lagrange's equations of motion in a Lagrangian system. Courtesy of http://content.answers.com/main/content/img/scitech/HSjosepl.jpg

it was Bernoulli (1667–1748) who first embodied the principle of virtual work, his focus was confined to the case of static problems. Lagrange extended the principle of virtual work to the dynamics of particle systems via D'Alembert's principle. The resulting equation is often called the Lagrangian form of D'Alembert's principle, which still includes the Cartesian coordinate variables.

Following the above efforts and carefully noting the difficulties with Newton's laws of motion, Lagrange (see Figure 3.3) introduced the important concept of generalized coordinates which enables one to transit from Euclidean $3N$-space with Cartesian coordinate variables to the configuration space with the n_{dof}-number of generalized coordinates, where n_{dof} stands for the number of degrees of freedom. The Cartesian coordinate variables can be transformed into generalized coordinates via the use of the chain rule. It is called coordinate transformation. Further, Lagrange realized that the descriptive scalar function (Greenwood 1977; Pars 1965), called the Lagrangian, is needed to describe the dynamical motion of a system. The resulting Lagrangian dynamical system, which is described by the Lagrangian, takes a second-order differential form in the configuration space with generalized coordinates. The dimensional size of the configuration space is equivalent to the number of degrees of freedom. As a consequence, the number of differential equations becomes coincident with the number of degrees of freedom. By introducing the generalized coordinates, the constraints associated with the geometry-related Cartesian coordinate variables can be eliminated. Also, ironically Lagrangian mechanics paved the way to geometric mechanics by introducing the notion of that termed as the configuration space, which is a smooth manifold.

3.2.1 Constraints

The motion of an N-body dynamical system is often subject to constraints which are usually expressed by algebraic constraint equations as mentioned previously. Then, the constraints are naturally imposed in terms of the position or the velocity in Euclidean space. Therefore, the constraint equations are given in terms of Cartesian coordinate variables. The total external force acting on the body consists of three distinct kinds of forces, namely, the conservative force, the nonconservative force, and the constraint force. The constraint force is the one which makes the body stay on the constraint surface over which the constraint equations are defined. The direction of the constraint force is perpendicular to the surface as illustrated in Figure 3.4. Therefore, the constraint force is proportional to the gradient of the constraint surface.

In general, constraints are categorized in several ways, for example, holonomic (meaning integrable in Greek) and nonholonomic (meaning non-integrable in Greek) or, scleronomic and rheonomic in the

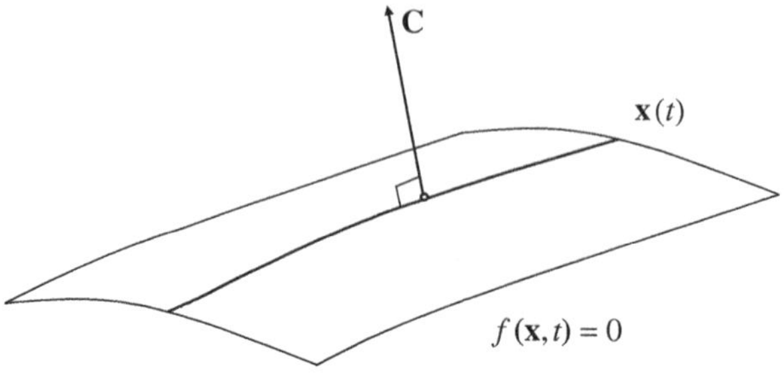

Figure 3.4. Direction of a constraint force

sense of explicit time dependence. It is usual to classify holonomic constraints into two categories, that is, holonomic-scleronomic and holonomic-rheonomic constraints. All constraints not pertaining to holonomic constraints are called nonholonomic.

Holonomic-Rheonomic Constraints Consider an N-body dynamical system that moves in three-dimensional Euclidean space. In the absence of any constraints, the dynamical system can move in any direction with $3N$ degrees of freedom in Euclidean space. Thus, the trajectory path of the system is a curve in the $3N$-dimensional configuration space. The $3N$ coordinate variables constitute the configuration space of $\left(\mathbf{x}^1, \ldots, \mathbf{x}^N\right) \in \mathbb{R}^{3N}$. Note that $\mathbb{R}^{3N}$denotes the set of $3N$-tuples of real numbers. The position vector of the i-th particle, denoted by $\mathbf{x}^i$, is a member of Euclidean 3-space, that is, $\mathbf{x}^i = (x_i, y_i, z_i) \in \mathbb{R}^3$.

Let us assume that constraints are imposed on the N-body dynamical system. Then, the system is no longer free from the constraints. Suppose that the k-number of constraint equations are endowed with the system. The total number of degrees of freedom for the system reduces to $3N - k$ due to the presence of constraints. The constraints are usually given in terms of Cartesian coordinate variables in three-dimensional Euclidean space. The holonomic constraints are given by

$$\begin{aligned} &f_i\left(\mathbf{x}(t), t\right) = 0, \quad i = 1, \ldots, k, \\ &\mathbf{x}(t) = \left(\mathbf{x}^1(t), \ldots, \mathbf{x}^N(t)\right) : \bar{\mathcal{T}} \rightarrow \mathbb{R}^{3N}, \ \forall t \in \bar{\mathcal{T}} = [t_0, t_1] \end{aligned} \tag{3.10}$$

In view of Equation (3.10), the exact differentials of the constraint equations should be zero as follows:

$$df_i = \sum_{j=1}^{N} \left(\frac{\partial f_i}{\partial \mathbf{x}^j} \cdot d\mathbf{x}^j\right) + \frac{\partial f_i}{\partial t} dt = 0, \quad i = 1, \ldots, k \tag{3.11}$$

Thus, we have the following form (referred to as holonomic constraints in Pfaffian form)

$$\sum_{j=1}^{N} \left(\frac{\partial f_i}{\partial \mathbf{x}^j} \cdot d\mathbf{x}^j\right) + \frac{\partial f_i}{\partial t} dt = 0, \quad i = 1, \ldots, k \tag{3.12}$$

In vector notation, the above can be rewritten as

$$\nabla f_i \cdot d\mathbf{x} = -\frac{\partial f_i}{\partial t} dt, \quad i = 1, \ldots, k \tag{3.13}$$

We also observe that in holonomic-rheonomic constraints the vector normal to the constraint surface is not orthogonal to the differential of the position vector. Further, the total time derivatives of the constraints are given by

$$\frac{df_i}{dt} = \sum_{j=1}^{N} \left(\frac{\partial f_i}{\partial \mathbf{x}^j} \cdot \dot{\mathbf{x}}^j \right) + \frac{\partial f_i}{\partial t} = 0, \quad i = 1, \ldots, k \tag{3.14}$$

Consequently, the above equation can be rewritten as

$$\nabla f_i \cdot \dot{\mathbf{x}} = -\frac{\partial f_i}{\partial t}, \quad i = 1, \ldots, k \tag{3.15}$$

Therefore, we note that in the case of holonomic-rheonomic constraints, the vector normal to the constraint surface is not orthogonal to the velocity vector.

Holonomic-Scleronomic Constraints We next introduce a special case in holonomic constraints, that is, holonomic-scleronomic constraints. The holonomic-scleronomic constraints do not have any explicit time dependence. Such systems are called autonomous. Hence, the holonomic-scleronomic constraints are given as constants in the motion of the system. The surface of constraints is fixed in Euclidean space, and does not depend upon time. These types of constraints are routinely dealt in this book under the terminology of an autonomous system. The holonomic-scleronomic constraints are given by

$$f_i \left(\mathbf{x}^1, \ldots, \mathbf{x}^N \right) = 0, \quad i = 1, \ldots, k, \quad \forall \mathbf{x}^i \in \mathbb{R}^3 \tag{3.16}$$

In view of Equation (3.16), the exact differentials of the holonomic-scleronomic constraint equations should equal zero as follows:

$$df_i = \sum_{j=1}^{N} \left(\frac{\partial f_i}{\partial \mathbf{x}^j} \cdot d\mathbf{x}^j \right) = 0, \quad i = 1, \ldots, k \tag{3.17}$$

Thus, the holonomic-scleronomic constraints in Pfaffian form can be given by

$$\sum_{j=1}^{N} \left(\frac{\partial f_i}{\partial \mathbf{x}^j} \cdot d\mathbf{x}^j \right) = 0, \quad i = 1, \ldots, k \tag{3.18}$$

In vector notation, the holonomic-scleronomic constraints in Pfaffian form can be rewritten as

$$\nabla f_i \cdot d\mathbf{x} = 0, \quad i = 1, \ldots, k \tag{3.19}$$

We observe that in the case of holonomic-scleronomic constraints, the vector normal to the constraint surface is obviously orthogonal to the differential of the position vector. Furthermore, the total time derivative of the constraints is given by

$$\frac{df_i}{dt} = \sum_{j=1}^{N} \left(\frac{\partial f_i}{\partial \mathbf{x}^j} \cdot \dot{\mathbf{x}}^j \right) = 0, \quad i = 1, \ldots, k \tag{3.20}$$

As a consequence, the above equation can be rewritten as

$$\nabla f_i \cdot \dot{\mathbf{x}} = 0, \quad i = 1, \ldots, k \tag{3.21}$$

Therefore, it should be noted that in holonomic-rheonomic constraints the vector normal to the constraint surface is orthogonal to the velocity vector.

Nonholonomic Constraints Constraints which cannot be expressed in the form of Equation (3.10) are called nonholonomic, and are not integrable. One example of this type is expressed by the inequality as shown below

$$f_i\left(\mathbf{x}^1, \ldots, \mathbf{x}^N; t\right) \leq 0, \quad i = 1, \ldots, k. \tag{3.22}$$

An example of nonholonomic constraints is a vertical penny rolling on a plane and the case that velocity-dependent forces are imposed (see (Greenwood 1977; Hand and Finch 1998) for details).

Illustrative Examples We demonstrate how to obtain Newton's equations of motion for single-body dynamical systems such as simple pendulums and a sliding block on an inclined plane for illustration.

Example 3.1 Planar Pendulum

Consider a planar pendulum whose origin is fixed in the x-y plane and its hinge is frictionless as shown in Figure 3.5. Furthermore, suppose that the motion is subject to a constraint such that the length $l \in \mathbb{R}$ of the pendulum bar is constant during the motion. The weight of the bar is negligible, and the mass of the pendulum is concentrated at the center of the bob. The pendulum is assumed to swing due to the gravitational force.

In the Cartesian coordinate system, the center of the bob is located at $\mathbf{x}(t) = (x_1(t), y_1(t)) \in \mathbb{R}^2$. Due to the assumption of a rigid bar, the Newtonian dynamical system is subject to a constraint equation. That is, Equation (3.16) can be written in the Cartesian coordinate system as

$$f(x_1(t), y_1(t)) = (x_1(t))^2 + (y_1(t))^2 - l^2 = 0 \tag{3.23}$$

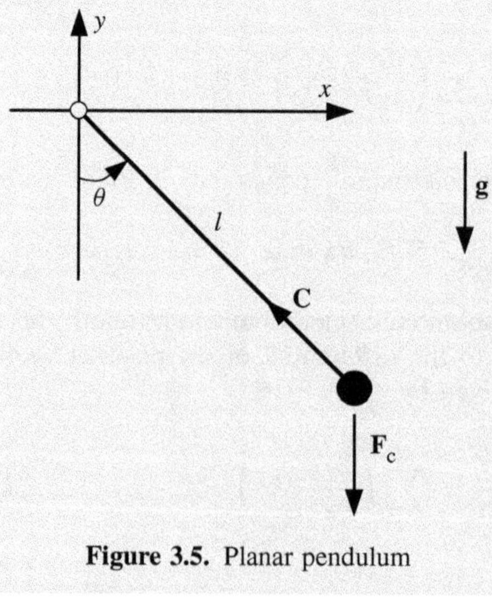

Figure 3.5. Planar pendulum

For convenience, the subscript will be dropped, and the above can be simply rewritten as

$$f(x, y) = x^2 + y^2 - l^2 = 0 \tag{3.24}$$

This example is typical of a scleronomic constraint system. We note that in the Cartesian coordinate system, the motion is described by two coordinate variables and one constraint equation. Hence, this is a dynamical system with a single degree of freedom. Notice that the trajectory on which the mass of the planar pendulum is traveling is fixed. Therefore, the configuration space is fixed in time. The trajectory is a circle in Euclidean 2-space.

Newton's second law asserts that

$$\mathbf{M}\ddot{\mathbf{x}} = \mathbf{F}_c + \mathbf{C}, \quad \mathbf{M} = \begin{bmatrix} m & 0 \\ 0 & m \end{bmatrix} \tag{3.25}$$

where $\mathbf{F}_c = (F_x, F_y) \in \mathbb{R}^2$ denotes the conservative force, m the mass of the bob, and $\mathbf{C} = (C_x, C_y) \in \mathbb{R}^2$ the constraint force. By use of the notation $C \in \mathbb{R}_+$ denoting the tension of the bar, the components of the constraint force may be calculated as $C_x = -C\sin\theta = -(Cx)/l$, $C_y = C\cos\theta - (Cy)/l$ and the components of the conservative force can be represented by $F_x = 0$, $F_y = -mg$. Therefore, Equation (3.25) can be rewritten as

$$\begin{aligned} m\ddot{x} &= -C\frac{x}{l} \\ m\ddot{y} &= -C\frac{y}{l} - mg \end{aligned} \tag{3.26}$$

where g denotes the gravitational acceleration. In Equation (3.26), we have three unknowns, namely, x,y, and C. And we have two differential equations in the above and the constraint equation in Equation (3.23). Therefore, we can solve Equation (3.26) for the three unknowns. The tension in the rigid bar can be obtained from the above equation as follows

$$C = -\frac{m}{l}\left[(x\ddot{x} + y\ddot{y}) + gy\right], \tag{3.27}$$

Taking the derivative of Equation (3.23) with respect to time twice, we have

$$\frac{d^2}{dt^2}(x^2 + y^2) = x\ddot{x} + y\ddot{y} + (\dot{x}^2 + \dot{y}^2) = 0, \tag{3.28}$$

Substituting Equation (3.28) into Equation (3.27) results in

$$C = \frac{m}{l}(\dot{x}^2 + \dot{y}^2) - mg\frac{y}{l}, \tag{3.29}$$

Therefore, Equation (3.26) can be written as

$$\begin{aligned} \ddot{x} &= -\frac{x}{l^2}(\dot{x}^2 + \dot{y}^2) + g\frac{xy}{l^2}, \\ \ddot{y} &= -g - \frac{y}{l^2}(\dot{x}^2 + \dot{y}^2) + g\frac{y^2}{l^2} \end{aligned} \tag{3.30}$$

Equation (3.30) represents the Newtonian dynamical system (Newton's equations of motion) for the planar pendulum problem. The differential equations above are the governing equations of motion in the Cartesian coordinate system.

Note that we have the two unknown variables x, y and the two differential equations. In order to solve the Newtonian dynamical system, we wish to have a single differential equation in terms of one unknown variable by eliminating the other unknown variable, because the planar pendulum is a dynamical system with one-degree of freedom. Equation (3.30) may be cast as a single equation

$$\ddot{x}y = x\ddot{y} + xg \tag{3.31}$$

Note that the above equation is still subject to the constraint equation given by Equation (3.24), because we have the two unknown dependent variables in the above equation. First, let us remove the coordinate variable y to obtain a single differential equation in terms of the coordinate variable x. Substituting the second equation in Equation (3.30) into Equation (3.31), we obtain

$$\ddot{x}y = -\frac{xy}{l^2}\left(\dot{x}^2 + \dot{y}^2\right) + xg\frac{y^2}{l^2} \tag{3.32}$$

In view of Equation (3.23), we have

$$y = -\sqrt{l^2 - x^2}, \quad x = \sqrt{l^2 - y^2} \tag{3.33}$$

which are valid for the case when $y \leq 0$ and $x \geq 0$. As a consequence, Equation (3.32) can be rewritten as

$$\ddot{x} + \frac{x}{l^2 - x^2}\dot{x}^2 + \frac{gx}{l^2}\sqrt{l^2 - x^2} = 0 \tag{3.34}$$

Further, Equation (3.31) can be represented in terms of y as shown below,

$$\ddot{y} + \frac{y}{l^2 - y^2}\dot{y}^2 + \frac{l^2 - y^2}{l^2}g = 0 \tag{3.35}$$

It is important that both Equation (3.34) and Equation (3.35) are valid for $y \leq 0$ and $x \geq 0$.

Example 3.2 Spherical Pendulum

For the spherical pendulum problem, the bob swings in the three dimensional space as shown in Figure 3.6. Then, the constraint equation is given in terms of Cartesian coordinate variables as

$$f(x, y, z) = x^2 + y^2 + z^2 - l^2 = 0 \tag{3.36}$$

This is also a holonomic-scleronomic constraint system. For this case, the surface on which the mass of the spherical pendulum is moving is also fixed in space and the surface is embedded in Euclidean 3-space. We need three Cartesian coordinates with one constraint equation to describe the motion. Hence, this is a dynamical system with two degrees of freedom.

Alternatively, we consider the case that the holonomic constraints are explicitly dependent upon time; in other words, the system is holonomic-rheonomic. The constraint surface is not fixed, but moving in Euclidean space, and depends upon time. For the case when the bar of the

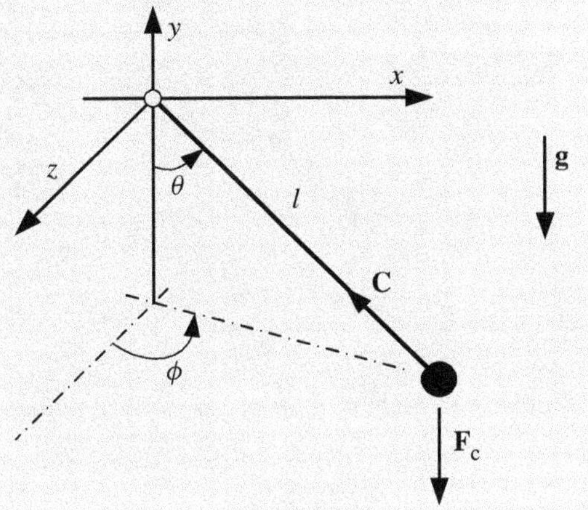

Figure 3.6. Spherical pendulum

spherical pendulum is flexible and dependent upon time, the system is holonomic-rheonomic, and the constraint equation is now given by

$$f(x, y, z, t) = x^2 + y^2 + z^2 - \{l(t)\}^2 = 0 \tag{3.37}$$

If the support point of the planar pendulum is, however, moving while the bar length does not change, then this constraint is also said to be rheonomic.

The Newtonian dynamical system for the spherical pendulum can be found by following the same procedure as discussed previously in the planar pendulum. Its derivation is skipped as it is cumbersome and repetitive to show the details.

Example 3.3 Sliding Block on an Inclined Plane

Consider a sliding block on a frictionless inclined plane under the influence of gravity as shown in Figure 3.7. The inclined surface is considered to be a constraint function given by

$$f(x, y) = y - ax - b = 0, \quad a = -\tan\theta < 0, \ b > 0 \tag{3.38}$$

Then, the constraint force $\mathbf{C} = (C_x, C_y) \in \mathbb{R}^2$, which is directed in the direction that is perpendicular to the surface, can be calculated as

$$C_x = C\sin\theta = mg\cos\theta\sin\theta, \quad C_y = C\cos\theta = mg\cos^2\theta \tag{3.39}$$

where C denotes the magnitude of the constraint force, which can be simply obtained by the relation, $C = mg\cos\theta$. Further, the conservative force $\mathbf{F}_c = (F_x, F_y) \in \mathbb{R}^2$ acting on the block due to gravity is given by

$$F_x = 0, \quad F_y = -mg \tag{3.40}$$

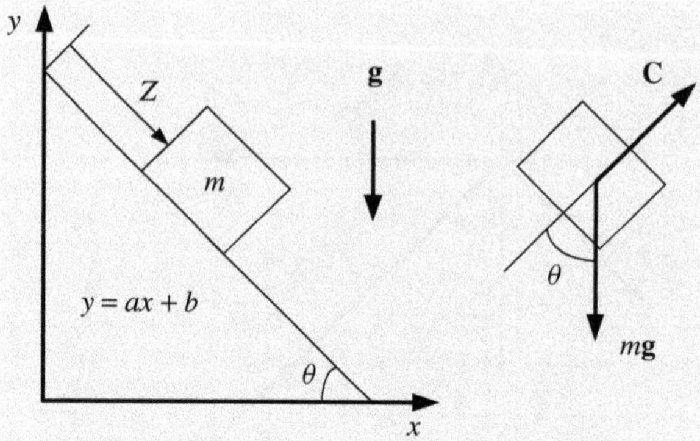

Figure 3.7. Sliding block on an inclined plane

Therefore, Newton's equations can be rewritten as

$$\begin{aligned} m\ddot{x} &= mg\cos\theta\sin\theta \\ m\ddot{y} &= -mg + mg\cos^2\theta \end{aligned} \tag{3.41}$$

In order to solve the Newtonian dynamical system for the sliding block problem, one unknown variable should be eliminated. From a geometric point of view, we have the relations

$$\cos\theta = \frac{1}{\sqrt{1+a^2}}, \quad \sin\theta = \frac{-a}{\sqrt{1+a^2}} \tag{3.42}$$

Thus, the Newtonian dynamical system for the sliding block problem reduces to a single differential equation in terms of x as

$$\ddot{x} = -\frac{ag}{1+a^2} = g\cos\theta\sin\theta \tag{3.43}$$

or in terms of y as

$$\ddot{y} = -\frac{a^2 g}{1+a^2} = -g\sin^2\theta \tag{3.44}$$

It is not necessary to use the Cartesian coordinate variable x or y in the Newtonian dynamical system to describe the differential equation. Let us next make an attempt to introduce a new coordinate z as shown in Figure 3.7. Then, in view of the following relations $x = z\cos\theta$ or $y = -z\sin\theta$, we easily obtain the differential equation

$$\ddot{z} = -\frac{ag}{\sqrt{1+a^2}} \tag{3.45}$$

which represents a single governing equation for a dynamical system with a single degree of freedom. The coordinate z selected is called the generalized coordinate, which will be later discussed in the section on Lagrangian mechanics.

3.2.2 Lagrangian Form of D'Alembert's Principle

With the notion that the constraint force makes no contribution to the virtual work, Lagrange (1736–1813) extended the principle of virtual work in static problems to the principle for the motion of a dynamical system, which is called as the Lagrangian form of D'Alembert's principle (Capecchi and Drago 2005; Greenwood 1977; Rosenberg 1977). Newton's second law states that the motion of an N-body dynamical system subject to the external and constraint forces is given by

$$\sum_{i=1}^{N} m_i \ddot{\mathbf{x}}^i = \sum_{i=1}^{N} \mathbf{F}^i_{\text{appl}} + \sum_{i=1}^{N} \mathbf{C}^i \tag{3.46}$$

where $\mathbf{F}^i_{\text{appl}} \in \mathbb{R}^3$ denotes the applied force acting on the i-th particle, $\mathbf{C}^i \in \mathbb{R}^3$ is the constraint force limiting the motion of the dynamical system and $\ddot{\mathbf{x}}^i(t) : \mathcal{T} \to \mathbb{R}^3$, $t \in \mathcal{T} \subset \mathbb{R}^+$ the acceleration vector of the i-th particle. The above equation can be rewritten as

$$\sum_{i=1}^{N} \mathbf{F}^i_{\text{appl}} + \sum_{i=1}^{N} \mathbf{C}^i + \left(- \sum_{i=1}^{N} m_i \ddot{\mathbf{x}}^i \right) = \mathbf{0} \tag{3.47}$$

where the last term above is called an inertial force or a reversed effective force. Equation (3.47) was named after D'Alembert's (1717–1783); hence, it is called D'Alembert's principle in the literature (D'Alembert 1758; Rosenberg 1977) (Greenwood 1977, page 22) (Meirovitch 2004, page 65) (Meirovitch 1967, page 42) (Moore 1983, page 37) (Baruh 1999) (Teodorescu 2007, page 388) (Chow 1995, page 107) (Fasano and Marmi 2006, page 227) (Mach 1907, page 336) (Lindsay and Margenau 1936, page 109) (Greenwood 1977) (Goldstein 2002, page 18) (Meirovitch 2004) (Capecchi and Drago 2005). Equation (3.47) is also called the Euler-D'Alembert's principle (Truesdell and Toupin 1960, page 532). Hence the above equation looks like a static equation in that the resultant consisting of the external and constraint forces plus the inertial force results in a null force.

Next, we wish to invoke the notion of the virtual displacement and the virtual work, which we shall discuss in detail in Chapter 4 including the principle of virtual work in dynamics. By virtue of Equation (3.6), taking the inner product between Newton's second law and the virtual displacement leads to

$$\sum_{i=1}^{N} \left(\mathbf{F}^i_{\text{appl}} - m_i \ddot{\mathbf{x}}^i \right) \cdot \delta \mathbf{x}^i + \sum_{i=1}^{N} \sum_{j=1}^{N} \mathbf{F}^{ij} \cdot \delta \mathbf{x}^i + \sum_{i=1}^{N} \mathbf{C}^i \cdot \delta \mathbf{x}^i = \mathbf{0} \tag{3.48}$$

The virtual work done by the constraint force should be zero, because the constraint force is workless, that is,

$$\sum_{i=1}^{N} \left(\mathbf{C}^i \cdot \delta \mathbf{x}^i \right) = 0 \tag{3.49}$$

Owing to Newton's law of action and reaction, the second term in Equation (3.48) can be rewritten as

$$\sum_{i=1}^{N} \sum_{j=1}^{N} \mathbf{F}^{ij} \cdot \delta \mathbf{x}^i = \sum_{i=1}^{N} \sum_{j=i+1}^{N} \mathbf{F}^{ij} \cdot \left(\delta \mathbf{x}^i - \delta \mathbf{x}^j \right) \tag{3.50}$$

In a rigid body motion where the distance between particles is fixed, the internal forces make no contribution to the virtual work. Hence, the above term vanishes. Consequently, Equation (3.47) reduces to

$$\sum_{i=1}^{N} \left(\mathbf{F}^i_{\text{appl}} - m_i \ddot{\mathbf{x}}^i \right) \cdot \delta \mathbf{x}^i = 0 \tag{3.51}$$

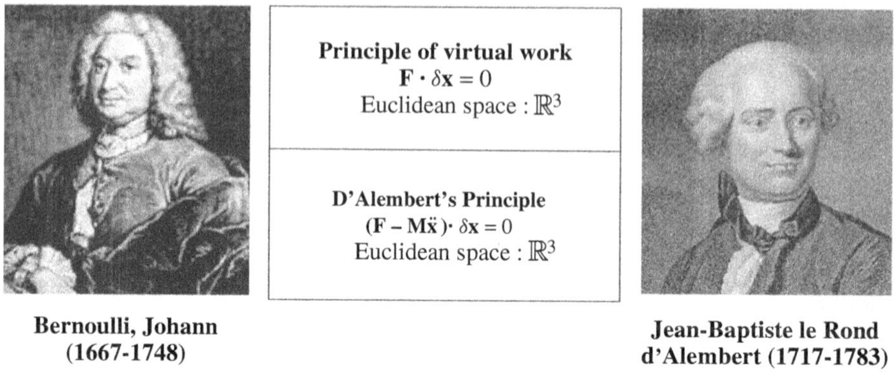

Figure 3.8. Principle of virtual work and Lagrangian form of D'Alembert's principle. Reproduced from JOC/EFR © 1998 and courtesy of ScienceWorld (http://scienceworld.wolfram.com/biography/dAlembert.html)

In favor of vector and matrix notation, Equation (3.51) can be rewritten as

$$\left(\mathbf{F}_{\mathrm{appl}} - \mathbf{M}\ddot{\mathbf{x}}\right) \cdot \delta\mathbf{x} = 0 \tag{3.52}$$

where $\mathbf{F}_{\mathrm{appl}} = (\mathbf{F}^1_{\mathrm{appl}}, \ldots, \mathbf{F}^N_{\mathrm{appl}}) \in \mathbb{R}^{3N}$ stands for the applied force applying to the system, and $\mathbf{M}$ is the global mass matrix whose off-diagonal terms are zero. The trajectory of motion of the system appears as a curve of a set, $\mathbf{x}\,(t) = \left(\mathbf{x}^1, \ldots, \mathbf{x}^N\right) : \bar{\mathcal{T}} \rightarrow \mathbb{R}^{3N},\ t \in \bar{\mathcal{T}} = [t_0, t_1]$. In Equation (3.52), $\delta\mathbf{x}\,(t) = \left(\delta\mathbf{x}^1, \ldots, \delta\mathbf{x}^N\right) : \bar{\mathcal{T}} \rightarrow \mathbb{R}^{3N}$ stands for the virtual displacements of the Cartesian coordinates. In some literature, Equation (3.51) is also referred to as D'Alembert's principle (Goldstein 2002; Hand and Finch 1998; Lanczos 1970; Synge 1960). But it is called by different names such as the application of the principle of D'Alembert (Mach 1907, page 336), the generalized principle of D'Alembert (Meirovitch 2004) or the Lagrangian form of D'Alembert's principle (Greenwood 1977; Rosenberg 1977) or the first form of the fundamental equation (Pars 1965) or the Lagrange-D'Alembert principle (Truesdell and Toupin 1960, page 595), the D'Alembert-Lagrange principle (Park 2009) and so on. Equation (3.51) implies the principle of virtual work in dynamics as illustrated in Figure 3.8. Throughout the present book, Equation (3.47) is referred to as D'Alembert's principle, whereas Equation (3.51) is the Lagrangian form of D'Alembert's principle. The dynamical equation played an important role in enabling Lagrange to open the world of analytic mechanics. Furthermore, note that Equation (3.51) is very important in connection with Hamilton's law of varying action, which will be discussed later in further detail.

Illustrative Examples We shall next show that the differential, governing equations of motion can be derived from the Lagrangian form of D'Alembert's principle given in Equation (3.52). Note that the governing equation one obtains by means of the Lagrangian form of D'Alembert's principle does not take the form of the Newtonian dynamical system, because constraint forces are not included. Furthermore, the resulting differential equations are represented in terms of Cartesian coordinate variables. Hence, the differential equations are still subject to constraint equations, as the Cartesian coordinate variables are not linearly independent in the presence of constraints.

Example 3.4 Planar Pendulum

Consider the same planar pendulum as we discussed previously. Then, the Lagrangian form of D'Alembert's principle, which is given in Equation (3.51), can be written as

$$(F_x - m\ddot{x})\,\delta x + \left(F_y - m\ddot{y}\right)\delta y = 0 \tag{3.53}$$

By use of the constraint equation given in Equation (3.24), we can obtain the following

$$x\delta x + y\delta y = 0 \tag{3.54}$$

Substituting Equation (3.54) and Equation (3.40) into Equation (3.53), we obtain

$$\ddot{x}y = x\ddot{y} + xg \tag{3.55}$$

Notice that Equation (3.55) is equivalent to Equation (3.31). This is a consequence of the Lagrangian form of D'Alembert's principle. Note that Equation (3.55) is still subject to the constraint equation given in Equation (3.24) in the Cartesian coordinate system.

Let us now introduce a new variable θ as a non-Cartesian coordinate to express the position vector $x = (x, y) \in \mathbb{R}^2$ as follows:

$$x = l\sin\theta,\quad y = -l\cos\theta \tag{3.56}$$

Substituting Equation (3.56) into Equation (3.55), we have

$$\ddot{\theta} + \frac{g}{l}\sin\theta = 0 \tag{3.57}$$

Furthermore, we can have an alternative generalized coordinate, namely the circumferential displacement $s = \theta/l$ as shown in Figure 3.9. Then, Equation (3.57) can be rewritten as

$$\ddot{s} + g\sin\frac{s}{l} = 0 \tag{3.58}$$

We realize that the introduction of the new coordinate readily enables us to have a desirable equation of motion. The coordinate is called a generalized coordinate in Lagrangian mechanics. Of course, one of the Cartesian coordinate variables x and y may be a candidate for the generalized coordinate. However, both of them can not be chosen at the same time as the generalized coordinate, because the number of generalized coordinates corresponds to the number of degrees of freedom for the system. Furthermore, the Cartesian coordinate variable x or y is not suitable as a generalized coordinate, because the choice of one of them is restricted to the condition as illustrated in Equation (3.34) or Equation (3.35). Hence, the angle for the planar pendulum is more suitable for the selection as the generalized coordinate. Consequently, the notion of generalized coordinates historically motivated Lagrange to open the world of Lagrangian mechanics.

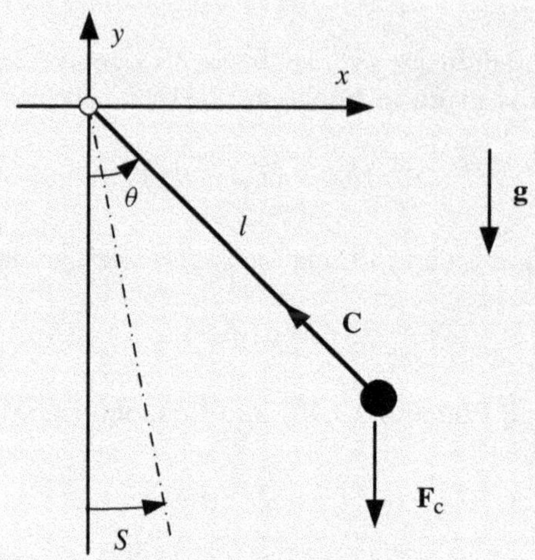

Figure 3.9. Planar pendulum

3.2.3 Configuration Space

The space in which the motion of a system can be described by generalized coordinates is called the configuration space Q. The notion of generalized coordinates was first proposed by Lagrange. The configuration space Q is an n_{dof}-dimensional smooth manifold without constraints, and it is differentiable (Abraham et al. 1988). The dimension of the configuration corresponds to the number of degrees of freedom; hence the number of generalized coordinates equals to the number of degrees of freedom. It should be noted that the number of degrees of freedom characterizes the intrinsic property of the system. A system consisting of N particles should be moving in the configuration space Q embedded as a sub-manifold in Euclidean 3-space. It is possible that the number of generalized coordinates may be larger than the number of degrees of freedom. In such a case, the space is no longer a configuration space and the generalized coordinates are not linearly independent in the space. To remove redundant coordinates, the Lagrange multiplier method should be employed to establish a new Lagrangian.

3.2.4 Generalized Coordinates

We next focus attention upon holonomic systems. In the general case of holonomic-rheonomic systems, the position vectors in the Cartesian coordinate system can be rewritten in terms of generalized coordinates as follows:

$$\mathbf{x}^i = \mathbf{x}^i\left(q^1, \ldots, q^{n_{\text{dof}}}; t\right) : Q \times \bar{\mathcal{T}} \rightarrow \mathbb{R}^3, \quad n_{\text{dof}} = 3N - k, \tag{3.59}$$

where the configuration space (Fasano and Marmi 2006) is defined as

$$Q = \left\{ (\mathbf{q}) \mid \mathbf{q} = \left(q^1, \ldots, q^{n_{\text{dof}}}\right) \in \mathbb{R}^{n_{\text{dof}}} \right\} \tag{3.60}$$

The transformation of Equation (3.59) is not a one-to-one mapping, in that the Cartesian coordinate variables in Euclidean $3N$-space are linearly dependent, whereas the generalized coordinates are linearly independent in the configuration space. Invoking the notation given in Equation (3.2), the increment of the position vector during Δt can be written as

$$\Delta \mathbf{x} = \mathbf{x}(\mathbf{q}\,(t + \Delta t)\,;\, t + \Delta t) - \mathbf{x}(\mathbf{q}\,(t)\,;\, t) \tag{3.61}$$

Note that due to Taylor's series, we have

$$\mathbf{q}\,(t + \Delta t) = \mathbf{q}\,(t) + \Delta t \frac{d\mathbf{q}}{dt} + R_1(t, \Delta t) \tag{3.62}$$

The above equation can be written as

$$\mathbf{q}\,(t + \Delta t) = \mathbf{q}\,(t) + \mathbf{h}\,(t)\,, \quad \mathbf{h}\,(t) = \Delta t \frac{d\mathbf{q}}{dt} + R_1(\mathbf{q}(t), \Delta t) \tag{3.63}$$

Therefore, the new position vector can be rewritten as

$$\begin{aligned} \mathbf{x}(\mathbf{q}(t) + \mathbf{h}(t);\, t + \Delta t) = {} & \mathbf{x}(\mathbf{q}(t);\, t) \\ & + \mathbf{h}(t) \cdot \left(\frac{\partial \mathbf{x}}{\partial \mathbf{q}}\right)_{t=constant} + \Delta t \left(\frac{\partial \mathbf{x}}{\partial t}\right)_{\mathbf{q}=constant} \\ & + R_1^{\mathbf{h}(t)}(\mathbf{q}(t), \mathbf{h}(t)) + R_1^{\Delta t}(t, \Delta t) \end{aligned} \tag{3.64}$$

Then, the differential of the position vector in conjunction with the generalized coordinates can be written as

$$d\mathbf{x} = dt\left(\lim_{\Delta t \to 0} \frac{\Delta \mathbf{x}}{\Delta t}\right) = \frac{\partial \mathbf{x}}{\partial \mathbf{q}} \cdot d\mathbf{q} + \frac{\partial \mathbf{x}}{\partial t} dt \tag{3.65}$$

On the other hand, in the particular case of holonomic-scleronomic systems, the position vectors in the Cartesian coordinate system can be given in terms of generalized coordinates by

$$\begin{aligned} & \mathbf{x}^i = \mathbf{x}^i\left(q^1, \ldots, q^{n_{\text{dof}}}\right) : Q \to \mathbb{R}^3, \quad n_{\text{dof}} = 3N - k, \\ & Q = \left\{(\mathbf{q}) \,|\, \mathbf{q} = \left(q^1, \ldots, q^{n_{\text{dof}}}\right) \in \mathbb{R}^{n_{\text{dof}}}\right\} \end{aligned} \tag{3.66}$$

Therefore, instead of Equation (3.61) the increment of the position vector during Δt can be written as

$$\Delta \mathbf{x} = \mathbf{x}(\mathbf{q}\,(t + \Delta t)) - \mathbf{x}(\mathbf{q}\,(t)) \tag{3.67}$$

Therefore, the new position vector can be rewritten as

$$\mathbf{x}(\mathbf{q}(t) + \mathbf{h}(t)) = \mathbf{x}(\mathbf{q}(t)) + \mathbf{h}(t) \cdot \frac{\partial \mathbf{x}}{\partial \mathbf{q}} + R_1(\mathbf{q}(t), \mathbf{h}(t)) \tag{3.68}$$

Then, the differential of the position vector in conjunction with the generalized coordinates can be written as

$$d\mathbf{x} = dt\left(\lim_{\Delta t \to 0}\left(\frac{\Delta \mathbf{x}}{\Delta t}\right)\right) = \frac{\partial \mathbf{x}}{\partial \mathbf{q}} \cdot d\mathbf{q} \tag{3.69}$$

The differential of the position vector can be written as

$$d\mathbf{x} = \frac{\partial \mathbf{x}}{\partial \mathbf{q}} \cdot d\mathbf{q} \tag{3.70}$$

Note that the partial derivative of the position vector with respect to time is absent in the above equation unlike Equation (3.65). Consequently, the differential of the position vectors in the holonomic-scleronomic and holonomic-rheonomic systems is different.

Next, we investigate the expression of the virtual displacement in the Cartesian coordinate system along with the virtual displacement in generalized coordinates. Previously, we defined the admissible configuration at the fixed time t in Equation (4.11). In consideration of generalized coordinates, the admissible configuration on the varied path can be rewritten by Taylor's series as

$$\begin{aligned}\tilde{\mathbf{x}}(\mathbf{q}(t)) &= \mathbf{x}(\mathbf{q}(t)+\epsilon\boldsymbol{\zeta}(t),t) = \mathbf{x}(\mathbf{q}(t),t) + \frac{\partial \mathbf{x}}{\partial \mathbf{q}}\cdot\epsilon\boldsymbol{\zeta}(t) + R_1(\mathbf{q}(t),\epsilon\boldsymbol{\zeta}(t)),\\ R_1(\mathbf{q}(t),\epsilon\boldsymbol{\zeta}(t)) &= \frac{1}{2}\epsilon\boldsymbol{\zeta}(t)\cdot\frac{\partial^2\mathbf{x}}{\partial \mathbf{q}^2}\cdot\epsilon\boldsymbol{\zeta}(t)+\ldots\end{aligned} \tag{3.71}$$

where the the variation of the generalized coordinates is given by $\delta\mathbf{q}(t)=\epsilon\boldsymbol{\zeta}(t)$. Therefore, the Gâteaux variation $\mathcal{G}\mathbf{x}(\mathbf{q}(t);\boldsymbol{\zeta}(t))$ can be defined as

$$\mathcal{G}\mathbf{x}(\mathbf{q}(t);\boldsymbol{\zeta}(t)) = \lim_{\epsilon\to 0}\frac{\mathbf{x}(\mathbf{q}(t)+\epsilon\boldsymbol{\zeta}(t)) - \mathbf{x}(\mathbf{q}(t),t)}{\epsilon} = D\mathbf{x}(\mathbf{q}(t),t)\cdot\boldsymbol{\zeta}(t) \tag{3.72}$$

As defined in the preceding section, the virtual displacement in Euclidean $3N$-space can be given in terms of the variations of the generalized coordinate parameters as

$$\delta\mathbf{x} = \frac{\partial \mathbf{x}}{\partial \mathbf{q}}\cdot\delta\mathbf{q} \tag{3.73}$$

Note that in holonomic-rheonomic constraint systems the virtual displacement is given as above, whereas the differential is given by Equation (3.65). Consequently, we observe that in holonomic-scleronomic system, the virtual displacement given in Equation (3.73) and the infinitesimal actual change given in Equation (3.70) do not have the partial derivative of the position vector with respect to time.

3.2.5 Tangent Bundle

Let $\mathbf{q}=\left(q^1,\ldots,q^{n_{\text{dof}}}\right)\in Q\subset\mathbb{R}^{n_{\text{dof}}=3N-k}$ be a set of generalized coordinates in the n_{dof}-dimensional configuration space Q (Fasano and Marmi 2006). At every point $\mathbf{q}$ on the differentiable manifold Q, a set of generalized velocities $\dot{\mathbf{q}}=\left(\dot{q}^1,\ldots,\dot{q}^{n_{\text{dof}}}\right)\in T_qQ\subset\mathbb{R}^{3N-k}$ defines an n_{dof}-dimensional tangent space or manifold T_qQ. Note that $\dim Q=\dim T_qQ$. Then the domain of the autonomous Lagrangian is an ordered pair $(\mathbf{q},\dot{\mathbf{q}})\in TQ$, i.e. the pair is a member of the tangent bundle TQ, which is also a differentiable manifold. The tangent bundle, known sometimes as the phase velocity space, is a disjoint union of the tangent spaces to each point of the configuration space Q as

$$TQ = \left\{(\mathbf{q},\dot{\mathbf{q}})\,|\,\mathbf{q}\in Q,\dot{\mathbf{q}}\in T_qQ\right\} = \bigcup_{q\in Q} T_qQ \tag{3.74}$$

Hence, the Lagrangian is defined as a scalar-valued function of multivariables, i.e.,

$$\mathcal{L}(\mathbf{q},\dot{\mathbf{q}},t): TQ\times\bar{\mathcal{T}}\to\mathbb{R},\quad \forall t\in\bar{\mathcal{T}}\subset\mathbb{R}^+ \tag{3.75}$$

We observe that $2(\dim Q)=\dim TQ$. It is important to note that the autonomous Lagrangian, a scalar function of generalized coordinates and generalized velocities, is defined on the tangent bundle, whereas

Lagrange's equations of motion are defined in the configuration space. The Lagrangian flow, as a curve of the solution of the equations of motion, is the one-parameter group of transformation in the configuration space so that $l^t : \mathbf{q}(0) \mapsto \mathbf{q}(t)$.

Illustrative Examples of Configuration Spaces For a dynamical system with constraints, the corresponding configuration space may be considered to be a smooth manifold; that is to say, it is infinitely differentiable. For certain simple cases, the configuration spaces, tangent spaces and tangent bundles can be easily visualized. In what follows, we shall demonstrate the physical interpretation of configuration spaces which are visual manifolds.

Example 3.5 Planar Pendulum

The bob of a simple planar pendulum sways on a circle as long as the bar length of the pendulum is fixed. The length of the bar may be regarded as a constraint. Due to the constraint, the bob draws a circle which is a differentiable manifold. The circle is the configuration space of the pendulum, denoted by $Q = \mathbb{S}^1 \subset \mathbb{R}^1$. Since the length of the bar remains constant, the system is holonomic-scleronomic. In other words, for a holonomic-scleronomic constraint system, the configuration space is fixed. For a simple planar pendulum with a holonomic-scleronomic constraint, the tangent bundle of the pendulum is known to be equivalent to a cylindrical shape, and the tangent bundle is a manifold denoted by $TQ = \mathbb{R}^1 \times \mathbb{S}^1 \subset \mathbb{R}^2$. Next suppose that the length of the bar changes with time. Then, the configuration space changes as shown in Figure 3.10, and the system is holonomic-rheonomic.

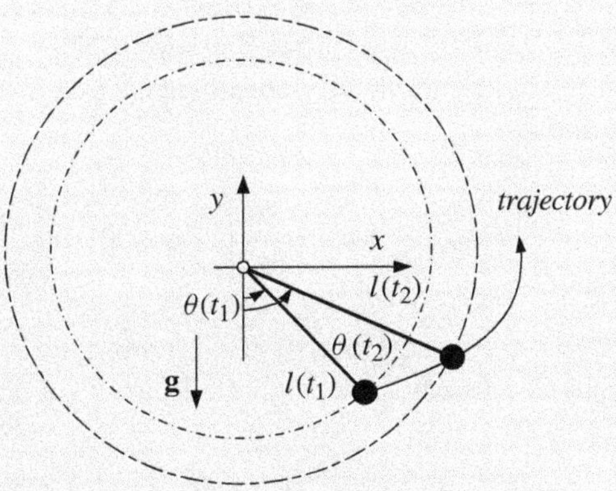

Figure 3.10. Configuration space changing in time

Example 3.6 Double Planar Pendulum

Given a fixed hinge upon which one end of the first pendulum is suspended, the bob of the other end of the pendulum is connected to one of the end's of the second pendulum with the bob of the second pendulum being free. The configuration space of the double planar pendulum now becomes a two-torus, denoted by $Q = \mathbb{T}^2 = \mathbb{S}^1 \times \mathbb{S}^1 \subset \mathbb{R}^2$. The configuration is two dimensional.

Example 3.7 Spherical Pendulum

The configuration space of a spherical pendulum is a two-dimensional manifold, denoted by $Q = \mathbb{S}^2 \subset \mathbb{R}^2$. Therefore, the center of the sphere is the origin of the spherical pendulum. The coordinates of the manifold consist of the azimuth and polar angles.

Example 3.8 Double Spherical Pendulum

The configuration space of a double spherical pendulum is represented by two spheres, denoted by $Q = \mathbb{S}^2 \times \mathbb{S}^2 \subset \mathbb{R}^4$. One spherical pendulum is attached to the center of another bob. The coordinates consist of two azimuth angles and two co-latitudes.

3.2.6 Lagrange's Equations of Motion

We readily see that Equation (3.51) is an extension of Newton's second law to the principle of virtual work in dynamics in terms of the virtual displacement. However, note that the equations of motion are defined in the Cartesian coordinate system as an inertial frame of reference. In static problems, Newton's mechanics provides an easy way to describe the equilibrium equations in Cartesian coordinates. Once motion is involved, it becomes difficult to describe the motion of each particle in terms of Cartesian coordinates. Further, the Cartesian coordinate variables, with which Equation (3.51) is defined, are linearly dependent. In other words, the number of the Cartesian coordinate variables are greater than the number of degrees of freedom. We shall next express Equation (3.51) in the configuration space with independent coordinates, that is to say, generalized coordinates. Thus, Equation (3.51) can be rewritten in terms of generalized coordinates as

$$\left[\left(\mathbf{F}_{\text{appl}} - \mathbf{M}\ddot{\mathbf{x}}\right) \cdot \frac{\partial \mathbf{x}}{\partial \mathbf{q}}\right] \cdot \delta \mathbf{q} = 0 \tag{3.76}$$

We observe that Equation (3.76) is a consequence of the Lagrangian form of D'Alembert's principle given in Equation (3.51), and $\delta \mathbf{q}$ is linearly independent. In view of Equation (3.76), we define the following

$$\mathbf{F}_{\text{appl}} \cdot \frac{\partial \mathbf{x}}{\partial \mathbf{q}} = \mathbf{Q}_{\text{appl}}, \tag{3.77}$$

where the quantity $\mathbf{Q}_{\text{appl}}$ in Equation (3.77) is referred to as the applied generalized force (Greenwood 1977). For static problems, Equation (4.19) (see Chapter 4) may take the form

$$\delta \mathcal{W} = \mathbf{Q}_{\text{appl}} \cdot \delta \mathbf{q} = 0 \tag{3.78}$$

provided that our attention is focused upon holonomic systems. Further, it is assumed that the generalized force is derivable from the potential energy

$$\mathbf{Q}_{\text{appl}} = -\frac{\partial \mathcal{U}(\mathbf{q}, t)}{\partial \mathbf{q}} = -\nabla \mathcal{U}(\mathbf{q}, t) \tag{3.79}$$

where $\mathcal{U}(\mathbf{q}, t)$ is often called the force function first introduced into dynamics by Lagrange (Hamilton 1835). It is very important to note that the special case of the velocity-dependent potential energy is not taken into account here.

Suppose the applied force is conservative. Then, the relation between the generalized force $\mathbf{Q}_c$ for the conservative force and the potential energy is given by

$$\mathbf{Q}_c = \mathbf{F}_c \cdot \frac{\partial \mathbf{x}}{\partial \mathbf{q}} = -\nabla \mathcal{U}(\mathbf{x}) \cdot \frac{\partial \mathbf{x}}{\partial \mathbf{q}} = -\frac{\partial \mathcal{U}(\mathbf{q})}{\partial \mathbf{q}} = -\nabla \mathcal{U}(\mathbf{q}) \tag{3.80}$$

where $\mathbf{F}_c$ is the negative gradient of the potential energy, $-\nabla \mathcal{U}(\mathbf{x})$ in Cartesian coordinates. We readily see that the velocity by the chain rule is given by

$$\dot{\mathbf{x}} = \frac{\partial \mathbf{x}}{\partial t} + \frac{\partial \mathbf{x}}{\partial \mathbf{q}} \cdot \dot{\mathbf{q}} \tag{3.81}$$

In view of Equation (3.81), taking the partial derivative of the velocity with respect to generalized velocities $\dot{\mathbf{q}}$, we obtain the relation

$$\frac{\partial \dot{\mathbf{x}}}{\partial \dot{\mathbf{q}}} = \frac{\partial \mathbf{x}}{\partial \mathbf{q}} \tag{3.82}$$

The above equation is the so-called dot cancellation. It is important to note that Equation (3.82) holds for holonomic constraints. By use of Equation (3.82), the left hand side of Equation (3.77) can be rewritten as

$$\mathbf{M}\ddot{\mathbf{x}} \cdot \frac{\partial \mathbf{x}}{\partial \mathbf{q}} = \frac{d}{dt}\left(\mathbf{M}\dot{\mathbf{x}} \cdot \frac{\partial \mathbf{x}}{\partial \mathbf{q}}\right) - \mathbf{M}\dot{\mathbf{x}} \cdot \frac{\partial \dot{\mathbf{x}}}{\partial \mathbf{q}} = \frac{d}{dt}\left(\mathbf{M}\dot{\mathbf{x}} \cdot \frac{\partial \dot{\mathbf{x}}}{\partial \dot{\mathbf{q}}}\right) - \mathbf{M}\dot{\mathbf{x}} \cdot \frac{\partial \dot{\mathbf{x}}}{\partial \mathbf{q}} \tag{3.83}$$

because the following is true

$$\frac{d}{dt}\left(\frac{\partial \mathbf{x}}{\partial \mathbf{q}}\right) = \frac{\partial}{\partial \mathbf{q}}\left(\frac{\partial \mathbf{x}}{\partial \mathbf{q}}\right) \cdot \dot{\mathbf{q}} + \frac{\partial}{\partial \mathbf{q}}\left(\frac{\partial \mathbf{x}}{\partial t}\right) = \frac{\partial \dot{\mathbf{x}}}{\partial \mathbf{q}} \tag{3.84}$$

For the N-body dynamical system undergoing a rectilinear motion, the kinetic energy $\mathcal{K}(\dot{\mathbf{x}}) : \mathbb{R}^{3N} \to \mathbb{R}$ in Cartesian coordinates is identical to the kinetic energy $\mathcal{K}(\mathbf{q}, \dot{\mathbf{q}}, t) : TQ \times \mathcal{T} \to \mathbb{R}$ in generalized coordinates, i.e.,

$$\mathcal{K}(\dot{\mathbf{x}}) = \frac{1}{2}\dot{\mathbf{x}} \cdot \mathbf{M}\dot{\mathbf{x}} = \mathcal{K}(\mathbf{q}, \dot{\mathbf{q}}, t) \tag{3.85}$$

since the kinetic energy is a scalar function. Note that the kinetic energy is the translational kinetic energy rather than the rotational kinetic energy. Our attention is confined to the translational kinetic energy only for the dynamical system, because the rotation effects are often neglected even in a general motion. Then, the variation of the kinetic energy can be written as

$$\delta \mathcal{K}(\dot{\mathbf{x}}) = \mathbf{M}\dot{\mathbf{x}} \cdot \delta \dot{\mathbf{x}} = \mathbf{M}\dot{\mathbf{x}} \cdot \left(\frac{\partial \dot{\mathbf{x}}}{\partial \mathbf{q}} \cdot \delta \mathbf{q} + \frac{\partial \dot{\mathbf{x}}}{\partial \dot{\mathbf{q}}} \cdot \delta \dot{\mathbf{q}}\right) \tag{3.86}$$

Also, the variation of the kinetic energy can be given by

$$\delta \mathcal{K}(\mathbf{q}, \dot{\mathbf{q}}, t) = \frac{\partial \mathcal{K}(\mathbf{q}, \dot{\mathbf{q}}, t)}{\partial \mathbf{q}} \cdot \delta \mathbf{q} + \frac{\partial \mathcal{K}(\mathbf{q}, \dot{\mathbf{q}}, t)}{\partial \dot{\mathbf{q}}} \cdot \delta \dot{\mathbf{q}} \tag{3.87}$$

In view of Equations (3.86) and (3.87), we have the relations

$$\mathbf{M}\dot{\mathbf{x}} \cdot \frac{\partial \dot{\mathbf{x}}}{\partial \dot{\mathbf{q}}} = \frac{\partial \mathcal{K}(\mathbf{q}, \dot{\mathbf{q}}, t)}{\partial \dot{\mathbf{q}}}, \quad \mathbf{M}\dot{\mathbf{x}} \cdot \frac{\partial \dot{\mathbf{x}}}{\partial \mathbf{q}} = \frac{\partial \mathcal{K}(\mathbf{q}, \dot{\mathbf{q}}, t)}{\partial \mathbf{q}} \tag{3.88}$$

Thus, in terms of the kinetic energy, Equation (3.83) may be expressed as

$$\mathbf{M}\ddot{\mathbf{x}} \cdot \frac{\partial \mathbf{x}}{\partial \mathbf{q}} = \frac{d}{dt}\left(\frac{\partial \mathcal{K}(\mathbf{q}, \dot{\mathbf{q}}, t)}{\partial \dot{\mathbf{q}}}\right) - \frac{\partial \mathcal{K}(\mathbf{q}, \dot{\mathbf{q}}, t)}{\partial \mathbf{q}} \tag{3.89}$$

Substitution of both Equation (3.89) and Equation (3.79) into Equation (3.76) leads to

$$\left[\frac{\partial \{\mathcal{K}(\mathbf{q}, \dot{\mathbf{q}}, t) - \mathcal{U}(\mathbf{q}, t)\}}{\partial \mathbf{q}} - \frac{d}{dt}\left(\frac{\partial \mathcal{K}(\mathbf{q}, \dot{\mathbf{q}}, t)}{\partial \dot{\mathbf{q}}}\right)\right] \cdot \delta \mathbf{q} = 0 \tag{3.90}$$

Since the virtual displacement $\delta \mathbf{q}$ is arbitrary in holonomic systems, we have

$$\frac{\partial \{\mathcal{K}(\mathbf{q}, \dot{\mathbf{q}}, t) - \mathcal{U}(\mathbf{q}, t)\}}{\partial \mathbf{q}} - \frac{d}{dt}\left(\frac{\partial \mathcal{K}(\mathbf{q}, \dot{\mathbf{q}}, t)}{\partial \dot{\mathbf{q}}}\right) = \mathbf{0} \tag{3.91}$$

We now introduce a new scalar function called the Lagrangian $\mathcal{L}(\mathbf{q}, \dot{\mathbf{q}}, t) : TQ \times \mathcal{T} \rightarrow \mathbb{R}$, that is,

$$\mathcal{L}(\mathbf{q}, \dot{\mathbf{q}}, t) = \mathcal{K}(\mathbf{q}, \dot{\mathbf{q}}, t) - \mathcal{U}(\mathbf{q}, t) \tag{3.92}$$

In the literature, the Lagrangian is often called the descriptive scalar function (Greenwood 1977; Pars 1965), which is needed to describe the dynamical motion of a system. Then, Equation (3.91) yields Lagrange's equations of motion as follows:

$$\frac{\partial \mathcal{L}(\mathbf{q}, \dot{\mathbf{q}}, t)}{\partial \mathbf{q}} - \frac{d}{dt}\left(\frac{\partial \mathcal{L}(\mathbf{q}, \dot{\mathbf{q}}, t)}{\partial \dot{\mathbf{q}}}\right) = \mathbf{0} \tag{3.93}$$

By virtue of Equation (3.93), Equation (3.76) can be rewritten as

$$\left[\frac{\partial \mathcal{L}(\mathbf{q}, \dot{\mathbf{q}}, t)}{\partial \mathbf{q}} - \frac{d}{dt}\left(\frac{\partial \mathcal{L}(\mathbf{q}, \dot{\mathbf{q}}, t)}{\partial \dot{\mathbf{q}}}\right)\right] \cdot \delta \mathbf{q} = 0 \tag{3.94}$$

Note that Equation (3.93) can be expressed in indicial notation as

$$\frac{\partial \mathcal{L}(\mathbf{q}, \dot{\mathbf{q}}, t)}{\partial q^i} - \frac{d}{dt}\left(\frac{\partial \mathcal{L}(\mathbf{q}, \dot{\mathbf{q}}, t)}{\partial \dot{q}^i}\right) = 0, \quad i = 1, \ldots, n_{\text{dof}} \tag{3.95}$$

which implies that Equation (3.95) indicates n_{dof}-simultaneous differential equations of the system. Note that Equation (3.94) can be expressed as

$$\delta \mathcal{L}(\mathbf{q}, \dot{\mathbf{q}}, t) - \frac{d}{dt}\left(\frac{\partial \mathcal{L}(\mathbf{q}, \dot{\mathbf{q}}, t)}{\partial \dot{\mathbf{q}}} \cdot \delta \mathbf{q}\right) = 0 \tag{3.96}$$

which is equivalent to the Lagrangian form of D'Alembert's principle given in Equation (3.52).

Further, the generalized momentum p_i is defined as

$$p_i(\mathbf{q}, \dot{\mathbf{q}}, t) = \frac{\partial \mathcal{L}(\mathbf{q}, \dot{\mathbf{q}}, t)}{\partial \dot{q}^i}, \quad i = 1, \ldots, n_{\text{dof}} \tag{3.97}$$

Therefore, Lagrange's equations of motion can be expressed as

$$\frac{d\mathbf{p}(\mathbf{q},\dot{\mathbf{q}},t)}{dt} = \frac{\partial\mathcal{L}(\mathbf{q},\dot{\mathbf{q}},t)}{\partial\mathbf{q}}, \quad \mathbf{p}(\mathbf{q},\dot{\mathbf{q}},t) = \frac{\partial\mathcal{L}(\mathbf{q},\dot{\mathbf{q}},t)}{\partial\dot{\mathbf{q}}} \tag{3.98}$$

We have defined the generalized force Q_i in Equation (3.79). Hence, from Lagrange's equations of motion given in Equation (3.95) the generalized force Q_i can be written in indicial notation as

$$Q_i = -\frac{\partial\mathcal{U}(\mathbf{q},t)}{\partial q^i} = \frac{dp_i(\mathbf{q},\dot{\mathbf{q}},t)}{dt} - \frac{\partial\mathcal{K}(\mathbf{q},\dot{\mathbf{q}},t)}{\partial q^i}, \quad i = 1,\ldots,n_{\text{dof}} \tag{3.99}$$

In the presence of the nonconservative force, which cannot be derivable from a potential energy, Lagrange's equation can be given by

$$\begin{aligned} &\frac{d}{dt}\left(\frac{\partial\mathcal{L}(\mathbf{q},\dot{\mathbf{q}},t)}{\partial\dot{\mathbf{q}}}\right) - \frac{\partial\mathcal{L}(\mathbf{q},\dot{\mathbf{q}},t)}{\partial\mathbf{q}} = \mathbf{Q}_{\text{nc}}, \\ &\mathbf{Q}_{\text{nc}} = \mathbf{F}_{\text{nc}} \cdot \frac{\partial\mathbf{x}}{\partial\mathbf{q}} \end{aligned} \tag{3.100}$$

where $\mathbf{Q}_{\text{nc}}$ denotes the nonconservative generalized force.

Illustrative Examples A variety of simple cases such as the pendulum problems are described to derive Lagrange's equations of motion in what follows.

Example 3.9 Planar Pendulum

Again, consider the same planar pendulum as we discussed previously. It is most convenient to choose θ as a general coordinate for the problem. The kinetic energy $\mathcal{K}$ can be represented by

$$\mathcal{K} = \frac{1}{2}m\left(\dot{x}^2 + \dot{y}^2\right) = \frac{1}{2}ml^2\dot{\theta}^2 = \frac{1}{2}\dot{\theta}\left(ml^2\right)\dot{\theta} \tag{3.101}$$

where $x = l\sin\theta,\ y = l(1-\cos\theta)$. Note that the generalized inertia matrix (Greenwood 1977) in the configuration space θ corresponds to ml^2. And, the potential energy of the conservative system is given by

$$\mathcal{U} = mgy = mgl(1-\cos\theta) \tag{3.102}$$

Hence, we have the Lagrangian

$$\mathcal{L}(\theta,\dot{\theta}) = \mathcal{K}(\dot{\theta}) - \mathcal{U}(\theta) = \frac{1}{2}ml^2\dot{\theta}^2 - mgl(1-\cos\theta) \tag{3.103}$$

As a consequence, Lagrange's equation of motion results in

$$\frac{\partial\mathcal{L}(\theta,\dot{\theta})}{\partial\theta} - \frac{d}{dt}\left(\frac{\partial\mathcal{L}(\theta,\dot{\theta})}{\partial\dot{\theta}}\right) = 0 \tag{3.104}$$

which results in the following differential equation

$$\ddot{\theta} + \frac{g}{l}\sin\theta = 0 \tag{3.105}$$

which represents a nonlinear ordinary differential equation.

Let us derive the generalized momentum from Lagrange's equation for the planar pendulum problem. In view of Equation (3.97), the generalized momentum can be given by

$$p_\theta = \frac{\partial \mathcal{L}(\theta, \dot{\theta})}{\partial \dot{\theta}} = ml^2\dot{\theta} = \left(ml^2\right)\dot{\theta} = l(mv_\theta) \tag{3.106}$$

where $v_\theta = l\dot{\theta}$. Therefore, as the generalized coordinate is a curvilinear coordinate, the generalized momentum becomes the angular momentum.

Next, let us examine the case that either x or y in the Cartesian coordinates is chosen as a generalized coordinate instead of θ. In view of Equation (3.10), we have

$$x = \sqrt{l^2 - y^2}, \quad y = -\sqrt{l^2 - x^2} \tag{3.107}$$

under the condition that $y \le 0, \ \ x \ge 0$. Therefore, we may have the following relations

$$\cos\theta = \frac{\sqrt{l^2 - x^2}}{l}, \quad \sin\theta = \frac{\sqrt{l^2 - y^2}}{l} \tag{3.108}$$

Taking the derivative of the first term in Equation (3.108) twice with respective to time leads to

$$\dot{\theta} = -\frac{\dot{x}}{\sqrt{l^2 - x^2}} = -\frac{\dot{y}}{\sqrt{l^2 - y^2}} \tag{3.109}$$

Thus, Equation (3.103) can be rewritten with the generalized coordinate x as

$$\mathcal{L}(x, \dot{x}) = \mathcal{K}(x, \dot{x}) - \mathcal{U}(x) = \frac{1}{2}\dot{x}\left[\frac{ml^2}{\left(l^2 - x^2\right)}\right]\dot{x} + mg\sqrt{l^2 - x^2} \tag{3.110}$$

Notice that with the generalized coordinate x, the constraint equation as given in Equation (3.10) may vanish. The generalized momentum can be given by

$$p_x = \frac{\partial \mathcal{L}(x, \dot{x})}{\partial \dot{x}} = \left[\frac{ml^2}{\left(l^2 - x^2\right)}\right]\dot{x} \tag{3.111}$$

In addition, it is significant to note that in the x-generalized coordinate, the corresponding generalized inertial matrix is $ml^2 / \left(l^2 - x^2\right)$. As a consequence, from Lagrange's equation given by

$$\frac{d}{dt}\left(\frac{\partial \mathcal{L}(x, \dot{x})}{\partial \dot{x}}\right) - \frac{\partial \mathcal{L}(x, \dot{x})}{\partial x} = 0 \tag{3.112}$$

we have

$$\ddot{x} + \frac{x}{l^2 - x^2}\dot{x}^2 + \frac{gx}{l^2}\sqrt{l^2 - x^2} = 0 \tag{3.113}$$

Note that the above equation is identical to Equation (3.34).

Suppose y is chosen as a generalized coordinate. Likewise, we can have the Lagrangian as

$$\mathcal{L}(y, \dot{y}) = \mathcal{K}(\dot{y}, y) - \mathcal{U}(y) = \frac{1}{2}\dot{y}\frac{ml^2}{(l^2 - y^2)}\dot{y} - mgy \tag{3.114}$$

Then, Lagrange's equation may be given by

$$\frac{d}{dt}\left(\frac{\partial \mathcal{L}(y, \dot{y})}{\partial \dot{y}}\right) - \frac{\partial \mathcal{L}(y, \dot{y})}{\partial y} = 0 \tag{3.115}$$

which leads to

$$\ddot{y} + \frac{y}{l^2 - y^2}\dot{y}^2 + \frac{l^2 - y^2}{l^2}g = 0 \tag{3.116}$$

Note that the above equation is identical to Equation (3.35). When we choose one of the Cartesian coordinates as a generalized coordinate, we need to put some conditions as in Equation (3.33). Therefore, the choice of Cartesian coordinates is not generally recommended as a proper candidate for the generalized coordinate.

Example 3.10 Spherical Pendulum

Consider the spherical pendulum whose bob can swing in the three-dimensional Cartesian space under the action of gravity as illustrated in Figure 3.6. First, to calculate the kinetic energy, we have the following

$$\begin{aligned} x &= l\sin\theta\sin\phi \\ y &= -l\cos\theta \\ z &= l\sin\theta\cos\phi \end{aligned} \tag{3.117}$$

We observe from Equation (3.117) that the proper generalized coordinates may be θ, ϕ. Then, we obtain the kinetic energy of the bob as

$$\mathcal{K} = \frac{1}{2}\dot{\mathbf{x}}\cdot\mathbf{M}\dot{\mathbf{x}} = \frac{1}{2}m\left(\dot{x}^2 + \dot{y}^2 + \dot{z}^2\right) = \frac{1}{2}ml^2\left(\dot{\theta}^2 + \dot{\phi}^2\sin^2\theta\right) = \frac{1}{2}\dot{\mathbf{q}}\cdot\tilde{\mathbf{M}}\dot{\mathbf{q}} \tag{3.118}$$

where $\tilde{\mathbf{M}}$ is the generalized inertia matrix which is a linear transformation mapping generalized velocities into generalized momentum (Bowen and Wang 1976). Then, we can write the kinetic energy in an alternative way as follows:

$$\mathcal{K} = \frac{1}{2}\dot{\mathbf{q}}\cdot\tilde{\mathbf{M}}\dot{\mathbf{q}},\quad \tilde{\mathbf{M}} = \begin{bmatrix} ml^2 & 0 \\ 0 & ml^2\sin^2\theta \end{bmatrix},\quad \dot{\mathbf{q}} = (\dot{\theta}, \dot{\phi}) \tag{3.119}$$

And the potential energy of the bob is given by

$$\mathcal{U} = mgy = -mgl\cos\theta \tag{3.120}$$

Therefore, the Lagrangian can be written as

$$\mathcal{L} = \frac{1}{2}ml^2\left(\dot{\theta}^2 + \dot{\phi}^2\sin^2\theta\right) + mgl\cos\theta \tag{3.121}$$

Thus, the generalized momenta p_θ, p_ϕ are

$$p_\theta = ml^2\dot{\theta}, \quad p_\phi = ml^2\dot{\phi}\sin^2\theta \tag{3.122}$$

Since the spherical pendulum has two degrees of freedom, we have two simultaneous differential equations such that

$$\begin{aligned} &\ddot{\theta} - \dot{\phi}^2\sin\theta\cos\theta + \frac{g}{l}\sin\theta = 0 \\ &\frac{d}{dt}\left(\dot{\phi}\sin^2\theta\right) = \ddot{\phi}\sin^2\theta + 2\dot{\phi}\dot{\theta}\sin\theta\cos\theta = 0 \end{aligned} \tag{3.123}$$

It is important in this problem to note that the Lagrangian given in Equation (3.121) does not depend upon the generalized coordinate ϕ. Such a generalized coordinate is called cyclic or ignorable so that the corresponding generalized momentum conjugate to the cyclic coordinate is conserved in the dynamical system, namely,

$$\frac{d}{dt}\left(p_\phi\right) = 0, \quad p_\phi = ml^2\dot{\phi}\sin^2\theta = constant \tag{3.124}$$

Hence, the generalized momentum p_ϕ is a constant of the motion.

It should be noted that the Lagrangian has a rotational symmetry in terms of ϕ. Accordingly, the generalized momentum p_ϕ can be rewritten as

$$p_\phi = l_\phi \times mv_\phi = (l\sin\theta)\left[m\,(l\sin\theta)\,\dot{\phi}\right] \tag{3.125}$$

where $l_\phi = l\sin\theta$ represents the arm of the momentum mv_ϕ. The generalized momentum p_i is conjugate with the generalized coordinate q^i. In conclusion, the angular momentum about the vertical axis is conserved as a constant of the motion, and the direction of the angular momentum vector becomes the positive y-axis.

Example 3.11 Double Planar Pendulum

Consider the double planar pendulum consisting of two bobs which have the identical mass quantity as shown in Figure 3.11. The weights of both of the bars are negligible and the hinge is frictionless. The two bobs are oscillating due to gravity. The coordinates of the lower bob can be given as

$$\begin{aligned} x &= l\,(\sin\theta + \sin\phi) \\ y &= -l\,(\cos\theta + \cos\phi) \end{aligned} \tag{3.126}$$

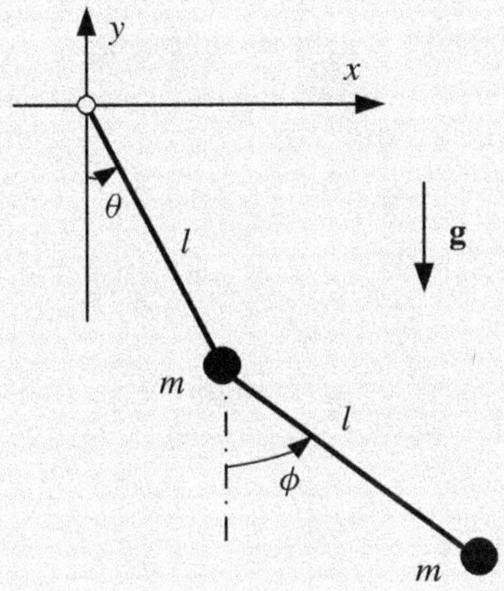

Figure 3.11. Double planar pendulum

Then, the speed of the lower bob $v \in \mathbb{R}_+$ can be calculated as

$$v = \sqrt{\dot{x}^2 + \dot{y}^2} = l\sqrt{\left[\dot{\theta}^2 + \dot{\phi}^2 + 2\dot{\theta}\dot{\phi}\cos(\theta - \phi)\right]} \tag{3.127}$$

In view of Equation (3.101), the kinetic energy of the double planar pendulum, therefore, is given by

$$\mathcal{K} = \frac{1}{2}ml^2\dot{\theta}^2 + \frac{1}{2}m\left(\dot{x}^2 + \dot{y}^2\right) = \frac{1}{2}ml^2\left[2\dot{\theta}^2 + \dot{\phi}^2 + 2\dot{\theta}\dot{\phi}\cos(\theta - \phi)\right] \tag{3.128}$$

which can be rewritten as

$$\mathcal{K} = \frac{1}{2}\dot{\mathbf{q}} \cdot \tilde{\mathbf{M}}\dot{\mathbf{q}}, \quad \tilde{\mathbf{M}} = \begin{bmatrix} 2ml^2 & ml^2\cos(\phi - \theta) \\ ml^2\cos(\phi - \theta) & ml^2 \end{bmatrix}, \quad \dot{\mathbf{q}} = (\dot{\theta}, \dot{\phi}) \tag{3.129}$$

The potential energy is

$$\mathcal{U} = -mgl\,(2\cos\theta + \cos\phi) \tag{3.130}$$

Then, we obtain the Lagrangian

$$\mathcal{L} = \frac{1}{2}ml^2\left[2\dot{\theta}^2 + \dot{\phi}^2 + 2\dot{\theta}\dot{\phi}\cos(\theta - \phi)\right] + mgl\,(2\cos\theta + \cos\phi) \tag{3.131}$$

Lagrange's equations of motion results in the following two differential equations

$$\begin{aligned} &\left[2\ddot{\theta} + \ddot{\phi}\cos(\theta - \phi) + \dot{\phi}^2\sin(\theta - \phi)\right] + 2\frac{g}{l}\sin\theta = 0 \\ &\left[\ddot{\phi} + \ddot{\theta}\cos(\theta - \phi) - \dot{\theta}^2\sin(\theta - \phi)\right] + \frac{g}{l}\sin\phi = 0 \end{aligned} \tag{3.132}$$

Example 3.12 Free Particle Motion

Consider a particle moving rectilinearly in three-dimensional Euclidean space in the absence of conservative forces. The three Cartesian coordinates x, y, z become generalized coordinates, because there are no constraint equations. Therefore, the dynamical system has three degrees of freedom, and the kinetic energy of the particle is given by

$$\mathcal{K} = \frac{1}{2}m\left(\dot{x}^2 + \dot{y}^2 + \dot{z}^2\right) = \mathcal{L} \tag{3.133}$$

Since there are no conservative forces, the Lagrangian is equal to the kinetic energy. In addition, the generalized coordinates are cyclic, as the Lagrangian does not explicitly contain these generalized coordinates. Hence, the conjugates of the generalized momenta are all conserved such that

$$\frac{\partial \mathcal{L}}{\partial x} = 0, \quad \frac{\partial \mathcal{L}}{\partial y} = 0, \quad \frac{\partial \mathcal{L}}{\partial z} = 0 \tag{3.134}$$

As a consequence, we have

$$\begin{aligned} \dot{p}_x &= m\ddot{x} = 0 \\ \dot{p}_x &= m\ddot{y} = 0 \\ \dot{p}_x &= m\ddot{z} = 0 \end{aligned} \tag{3.135}$$

All the generalized momenta are constants of motion. This is called the principle of conservation of linear momentum.

3.2.7 Kinetic Energy in Generalized Coordinates

In this section, we shall discuss the transformation of the kinetic energy from Cartesian coordinates to generalized coordinates. Consider the three-dimensional motion of an N-body dynamical system whose positions are defined as a vector-valued function of one real variable in Euclidean $3N$-space, $\mathbb{R}^{3N}$. The N-body dynamical system is assumed to be subject to the k-number of holonomic constraints. The actual change of the kinetic energy due to the inertia force can be given by

$$d\mathcal{K} = \frac{d\left(\mathbf{M}\dot{\mathbf{x}}\right)}{dt} \cdot d\mathbf{x} \tag{3.136}$$

where the rotational kinetic energy is assumed to be negligible. Therefore, the above equation can be rewritten as

$$d\mathcal{K} = \frac{d\left(\mathbf{M}\dot{\mathbf{x}}\right)}{dt} \cdot \dot{\mathbf{x}}dt = \frac{d}{dt}\left(\frac{1}{2}\dot{\mathbf{x}} \cdot \mathbf{M}\dot{\mathbf{x}}\right)dt \tag{3.137}$$

As a consequence, the energy due to the inertial force may by written in Cartesian coordinates as

$$\mathcal{K} = \frac{1}{2}\dot{\mathbf{x}} \cdot \mathbf{M}\dot{\mathbf{x}} = \mathcal{K}\left(\dot{\mathbf{x}}\right) \tag{3.138}$$

which is called the kinetic energy in Cartesian coordinates. Note that the off-diagonal terms in the $3N \times 3N$ mass matrix of the above equation are zero and diagonal terms are nonzero.

Next, we are interested in the kinetic energy in the generalized coordinate system. We have $\mathbf{q} = (q^1, \ldots, q^{n_{\text{dof}}}) \in \mathbb{R}^{n_{\text{dof}}=3N-k}$ in view of Equation (3.59), and the velocities can be written as

$$\dot{\mathbf{x}} = \frac{\partial \mathbf{x}}{\partial \mathbf{q}} \cdot \dot{\mathbf{q}} + \frac{\partial \mathbf{x}}{\partial t} \tag{3.139}$$

Substituting Equation (3.139) into Equation (3.138), we have the following

$$\mathcal{K}(\dot{\mathbf{x}}) = \frac{1}{2}\left(\frac{\partial \mathbf{x}}{\partial \mathbf{q}} \cdot \dot{\mathbf{q}} + \frac{\partial \mathbf{x}}{\partial t}\right) \cdot \mathbf{M}\left(\frac{\partial \mathbf{x}}{\partial \mathbf{q}} \cdot \dot{\mathbf{q}} + \frac{\partial \mathbf{x}}{\partial t}\right) \tag{3.140}$$

The above equation can be rewritten as

$$\mathcal{K}(\dot{\mathbf{x}}) = \frac{1}{2}\left[\dot{\mathbf{q}} \cdot \left(\frac{\partial \mathbf{x}}{\partial \mathbf{q}}\right)^T \cdot \mathbf{M}\frac{\partial \mathbf{x}}{\partial \mathbf{q}} \cdot \dot{\mathbf{q}} + 2\frac{\partial \mathbf{x}}{\partial t} \cdot \mathbf{M}\frac{\partial \mathbf{x}}{\partial \mathbf{q}} \cdot \dot{\mathbf{q}} + \left(\frac{\partial \mathbf{x}}{\partial t} \cdot \mathbf{M}\frac{\partial \mathbf{x}}{\partial t}\right)\right] \tag{3.141}$$

Therefore, the kinetic energy can be divided into the three parts (Meirovitch 2004) as follows

$$\mathcal{K}(\dot{\mathbf{x}}) = \mathcal{K}_2 + \mathcal{K}_1 + \mathcal{K}_0 \tag{3.142}$$

where $\mathcal{K}_2$ is a quadratic function of generalized velocities, $\mathcal{K}_1$ is a linear function of generalized velocities, and $\mathcal{K}_0$ is a function of generalized coordinates and time. For the holonomic-scleronomic constraint system, $\frac{\partial \mathbf{x}}{\partial t} = 0$. Hence, the kinetic energy is a quadratic function of $\dot{\mathbf{q}}$, because $\mathcal{K}_1 = \mathcal{K}_0 = 0$. Thus, the kinetic energy in the holonomic-scleronomic constraint system can be given by

$$\mathcal{K}(\dot{\mathbf{x}}) = \mathcal{K}_2(\mathbf{q}, \dot{\mathbf{q}}) = \frac{1}{2}\left(\dot{\mathbf{q}} \cdot \left(\frac{\partial \mathbf{x}}{\partial \mathbf{q}}\right)^T \cdot \mathbf{M}\frac{\partial \mathbf{x}}{\partial \mathbf{q}} \cdot \dot{\mathbf{q}}\right) = \frac{1}{2}\dot{\mathbf{q}} \cdot \tilde{\mathbf{M}}(\mathbf{q})\,\dot{\mathbf{q}} = \mathcal{K}(\mathbf{q}, \dot{\mathbf{q}}) \tag{3.143}$$

Note that the generalized inertia matrix $\tilde{\mathbf{M}}(\mathbf{q})$ is symmetric and a function of generalized coordinates. However, in the holonomic-rheonomic constraint system $\mathcal{K}_1 \neq 0, \mathcal{K}_0 \neq 0$. Consequently, we have

$$\mathcal{K}(\dot{\mathbf{x}}) = \frac{1}{2}\dot{\mathbf{q}} \cdot \tilde{\mathbf{M}}(\mathbf{q}, t)\,\dot{\mathbf{q}} + \mathcal{K}_1(\mathbf{q}, \dot{\mathbf{q}}, t) + \mathcal{K}_0(\mathbf{q}, t) = \mathcal{K}(\mathbf{q}, \dot{\mathbf{q}}, t) \tag{3.144}$$

Since the kinetic energy is positive, $\mathcal{K}_1(\mathbf{q}, \dot{\mathbf{q}}, t) + \mathcal{K}_0(\mathbf{q}, t) > 0$.

Let us first consider the case when a planar pendulum is subject to scleronomic constraints. If θ is selected as the generalized coordinate, the kinetic energy becomes a quadratic function of $\dot{\theta}$, and it does not contain the generalized coordinate as a variable. However, if x is chosen as the generalized coordinate, the kinetic energy is a quadratic function of $\dot{x}$, and it does contain the generalized coordinate. Hence, the kinetic energy is a function of both $\dot{x}$ and x. As shown in Equation (3.111), the generalized inertia matrix is a function of the generalized coordinate x. Furthermore, we previously investigated the case of the spherical pendulum. As shown in Equation (3.119), the kinetic energy is also a quadratic function of generalized velocities because the pendulum is subject to a holonomic-scleronomic constraint, and one of the inertia coefficients is a function of one of the generalized coordinates. Consequently, the kinetic energy may depend on generalized coordinates.

3.2.8 Lagrange Multiplier Method

In the previous subsection, we have shown that Lagrange's equations of motion can be derived from the Lagrangian form of D'Alembert's principle. As highlighted previously, in the Lagrangian form of D'Alembert's principle, the differential equations are still given in terms of Cartesian coordinate variables. By introducing generalized coordinates, the constraints given in terms of Cartesian coordinate variables may be canceled in configuration space. Recall that the Newtonian differential system with constraints is redundant, but the Lagrangian differential system in configuration space is not.

We often consider the case when the descriptive scalar functions need to be given in terms of Cartesian coordinate variables rather than generalized coordinates. This is because we often encounter the case when it is difficult to find the proper generalized coordinates of a system. Consider a motion of the dynamical system, which is subjected to holonomic constraints. Before we obtain the Lagrangian form of D'Alembert's principle for holonomic constraints systems, we have

$$\mathbf{M}\ddot{\mathbf{x}} \cdot \delta\mathbf{x} = \mathbf{F}_{\text{appl}} \cdot \delta\mathbf{x} + \mathbf{C}(\mathbf{x}, t) \cdot \delta\mathbf{x} \tag{3.145}$$

where $\mathbf{F}_{\text{appl}}$ and $\mathbf{C}(\mathbf{x}, t)$ denote the applied and constraint forces. As addressed previously, the constraint force $\mathbf{C}$ is orthogonal to the virtual displacement. Note that the constraint force is proportional to the gradient vector of the constraint surface and perpendicular to the surface. Therefore, the constraint forces are given by

$$\begin{aligned} \mathbf{C}(\mathbf{x}, t) &= \sum_{i=1}^{k} \lambda_i \nabla f_i \left(\mathbf{x}^1, \ldots, \mathbf{x}^N; t\right) \\ &= \nabla (\boldsymbol{\lambda} \cdot \mathbf{f}(\mathbf{x}, t)) \\ &= \frac{\partial (\boldsymbol{\lambda} \cdot \mathbf{f}(\mathbf{x}, t))}{\partial \mathbf{x}}, \quad \boldsymbol{\lambda} = (\lambda_1, \ldots, \lambda_k), \quad \mathbf{f} = (f_1, \ldots, f_k) \end{aligned} \tag{3.146}$$

where $\boldsymbol{\lambda}$ is called the Lagrange multiplier and $\mathbf{f}$ denotes the constraint function limiting the motion of the system. By use of the Lagrange multiplier and the constraint function, Equation (3.145) can be written as

$$\left(\mathbf{F}_{\text{appl}} - \mathbf{M}\ddot{\mathbf{x}}\right) \cdot \delta\mathbf{x} = -\delta (\boldsymbol{\lambda} \cdot \mathbf{f}(\mathbf{x}, t)) \tag{3.147}$$

The term on the left-hand side of Equation (3.145) can be written as

$$\mathbf{M}\ddot{\mathbf{x}} \cdot \delta\mathbf{x} = \frac{d}{dt} (\mathbf{M}\dot{\mathbf{x}} \cdot \delta\mathbf{x}) - \delta\mathcal{K}(\dot{\mathbf{x}}) \tag{3.148}$$

Next, suppose that the applied force $\mathbf{F}_{\text{appl}}$ is derivable from the potential energy $\mathcal{U}(\mathbf{x}, t)$. Then, the first term on the right-hand side in Equation (3.145) can be given by

$$\mathbf{F}_{\text{appl}} \cdot \delta\mathbf{x} = -\frac{\partial \mathcal{U}(\mathbf{x}, t)}{\partial \mathbf{x}} \cdot \delta\mathbf{x} = -\delta\mathcal{U}(\mathbf{x}, t) \tag{3.149}$$

Hence, Equation (3.147) can be written as

$$\delta\mathcal{L}(\mathbf{x}, \dot{\mathbf{x}}, t) + \delta (\boldsymbol{\lambda} \cdot \mathbf{f}(\mathbf{x}, t)) - \frac{d}{dt} (\mathbf{M}\dot{\mathbf{x}} \cdot \delta\mathbf{x}) = 0 \tag{3.150}$$

where $\mathcal{L}(\mathbf{x}, \dot{\mathbf{x}}, t) = \mathcal{K}(\dot{\mathbf{x}}) - \mathcal{U}(\mathbf{x}, t)$. Further, the above equation can be expressed as

$$\delta\tilde{\mathcal{L}}(\mathbf{x}, \dot{\mathbf{x}}, t) - \frac{d}{dt} (\mathbf{M}\dot{\mathbf{x}} \cdot \delta\mathbf{x}) = 0 \tag{3.151}$$

where $\tilde{\mathcal{L}}(\mathbf{x}, \dot{\mathbf{x}}, t) = \mathcal{K}(\dot{\mathbf{x}}) - \mathcal{U}(\mathbf{x}, t) + \boldsymbol{\lambda} \cdot \mathbf{f}(\mathbf{x}, t)$, and it is referred to as the augmented Lagrangian. Therefore, it is easy to show that the above equation leads to the following

$$\left[\frac{\partial \tilde{\mathcal{L}}(\mathbf{x}, \dot{\mathbf{x}}, t)}{\partial \mathbf{x}} - \frac{d}{dt}\left(\frac{\partial \tilde{\mathcal{L}}(\mathbf{x}, \dot{\mathbf{x}}, t)}{\partial \dot{\mathbf{x}}}\right)\right] \cdot \delta \mathbf{x} = 0 \tag{3.152}$$

As a consequence, the above equation can be rewritten as

$$\left[\frac{\partial \mathcal{L}(\mathbf{x}, \dot{\mathbf{x}}, t)}{\partial \mathbf{x}} - \frac{d}{dt}\left(\frac{\partial \mathcal{L}(\mathbf{x}, \dot{\mathbf{x}}, t)}{\partial \dot{\mathbf{x}}}\right) + \boldsymbol{\lambda} \cdot \Delta \mathbf{f}(\mathbf{x}, t)\right] \cdot \delta \mathbf{x} = 0 \tag{3.153}$$

where the term enclosed in the square bracket is identical to Newton's equations of motion in the Cartesian coordinate system. The Lagrange multiplier should be chosen to satisfy Equation (3.146). Note that the applied force is conservative, then the Lagrangian becomes autonomous so that it does not depend on time explicitly as well as the potential energy.

Illustrative Example

Example 3.13 Planar Pendulum Problem with Lagrange's Equation

Let us apply the Lagrange multiplier method to the planar pendulum problem. In the Cartesian coordinate system, the autonomous augmented Lagrangian for the planar pendulum may be expressed by

$$\tilde{\mathcal{L}}(\mathbf{x}, \dot{\mathbf{x}}) = \frac{1}{2} m \left(\dot{x}^2 + \dot{y}^2\right) - mgy + \lambda\left(l - \sqrt{x^2 + y^2}\right) \tag{3.154}$$

where λ denotes the Lagrange multiplier as an arbitrary parameter. By virtue of Equation (3.142), we obtain

$$\begin{aligned} m\ddot{x} &= -\lambda \frac{x}{l} \\ m\ddot{y} &= -\lambda \frac{y}{l} - mg \end{aligned} \tag{3.155}$$

Note that the above equation is equivalent to Equation (3.26) where the Lagrange multiplier is equal to the tension of the bar. Therefore, the Lagrange multiplier represents the magnitude of the constraint force, and the sign of the constraint force can be given by the gradient of the constraint equation.

3.2.9 Autonomous Lagrangian Systems

For the case when the Lagrangian of a system does not depend explicitly on time, we refer to the system as the autonomous Lagrangian system. A number of important problems in physics and engineering disciplines fall into this framework. However, it is very important to note that in holonomic-scleronomic systems the Lagrangian is always autonomous, but in holonomic-rheonomic systems the Lagrangian is autonomous or non autonomous. In autonomous Lagrangian systems, the autonomous Lagrangian has symmetry in time, resulting in the existence of a conserved quantity.

For a general case, we consider a holonomic system. By use of the chain rule, the total time derivative of the Lagrangian may be written as

$$\frac{d\mathcal{L}(\mathbf{q},\dot{\mathbf{q}},t)}{dt}=\frac{\partial\mathcal{L}(\mathbf{q},\dot{\mathbf{q}},t)}{\partial\mathbf{q}}\cdot\dot{\mathbf{q}}+\frac{\partial\mathcal{L}(\mathbf{q},\dot{\mathbf{q}},t)}{\partial\dot{\mathbf{q}}}\cdot\ddot{\mathbf{q}}+\frac{\partial\mathcal{L}(\mathbf{q},\dot{\mathbf{q}},t)}{\partial t} \tag{3.156}$$

After employing the product rule, the second term on the right-hand side of Equation (3.156) can be expressed as

$$\frac{\partial\mathcal{L}(\mathbf{q},\dot{\mathbf{q}},t)}{\partial\dot{\mathbf{q}}}\cdot\ddot{\mathbf{q}}=\frac{d}{dt}\left(\frac{\partial\mathcal{L}(\mathbf{q},\dot{\mathbf{q}},t)}{\partial\dot{\mathbf{q}}}\cdot\dot{\mathbf{q}}\right)-\frac{d}{dt}\left(\frac{\partial\mathcal{L}(\mathbf{q},\dot{\mathbf{q}},t)}{\partial\dot{\mathbf{q}}}\right)\cdot\dot{\mathbf{q}} \tag{3.157}$$

Substituting Equation (3.157) into Equation (3.156), yields

$$\frac{d}{dt}\left(\frac{\partial\mathcal{L}}{\partial\dot{\mathbf{q}}}\cdot\dot{\mathbf{q}}-\mathcal{L}\right)=\left[\frac{d}{dt}\left(\frac{\partial\mathcal{L}(\mathbf{q},\dot{\mathbf{q}},t)}{\partial\dot{\mathbf{q}}}\right)-\frac{\partial\mathcal{L}(\mathbf{q},\dot{\mathbf{q}},t)}{\partial\dot{\mathbf{q}}}\right]\cdot\dot{\mathbf{q}}-\frac{\partial\mathcal{L}(\mathbf{q},\dot{\mathbf{q}},t)}{\partial t} \tag{3.158}$$

In the case of a holonomic constraint system, the first term on the right-hand side of Equation (3.158) becomes zero due to Lagrange's equation of motion. Accordingly, Equation (3.158) can be recast as the following

$$\frac{d}{dt}\left(\frac{\partial\mathcal{L}(\mathbf{q},\dot{\mathbf{q}},t)}{\partial\dot{\mathbf{q}}}\cdot\dot{\mathbf{q}}-\mathcal{L}(\mathbf{q},\dot{\mathbf{q}},t)\right)=-\frac{\partial\mathcal{L}(\mathbf{q},\dot{\mathbf{q}},t)}{\partial t} \tag{3.159}$$

The term enclosed in the big brackets on the left-hand side above can be defined as

$$h(\mathbf{q},\dot{\mathbf{q}},t)=\frac{\partial\mathcal{L}(\mathbf{q},\dot{\mathbf{q}},t)}{\partial\dot{\mathbf{q}}}\cdot\dot{\mathbf{q}}-\mathcal{L}(\mathbf{q},\dot{\mathbf{q}},t) \tag{3.160}$$

where the integral function $h(\mathbf{q},\dot{\mathbf{q}},t):TQ\times\bar{\mathcal{T}}\rightarrow\mathbb{R}$ is called the Jacobi integral. If the Lagrangian does not have explicit time dependence, the term on the right-hand side of Equation (3.159) vanishes. In such a case, it is said that the Lagrangian is autonomous. In an autonomous Lagrangian system, the Jacobi integral remains as a constant of motion. In holonomic-scleronomic systems, the kinetic energy is quadratic in generalized velocities, and then the Jacobi integral equals the total energy. As a consequence, in holonomic-scleronomic systems we have the relation

$$\frac{\partial\mathcal{L}(\mathbf{q},\dot{\mathbf{q}})}{\partial\dot{\mathbf{q}}}\cdot\dot{\mathbf{q}}=2\mathcal{K}(\mathbf{q},\dot{\mathbf{q}}) \tag{3.161}$$

Thus, Equation (3.160) reduces to the total energy given by

$$\mathcal{E}(\mathbf{q},\dot{\mathbf{q}})=\mathcal{K}(\mathbf{q},\dot{\mathbf{q}})+\mathcal{U}(\mathbf{q}) \tag{3.162}$$

where $\mathcal{E}(\mathbf{q},\dot{\mathbf{q}}):TQ\rightarrow\mathbb{R}$ is called the autonomous total energy of the dynamical system. Upon introducing the canonical momentum as a new variable as shown in Equation (3.97), we have

$$\mathbf{p}=\frac{\partial\mathcal{L}(\mathbf{q},\dot{\mathbf{q}})}{\partial\dot{\mathbf{q}}} \tag{3.163}$$

By means of Equation (3.163), Equation (3.160) in the autonomous Lagrangian system can be rewritten as,

$$\mathcal{H}(\mathbf{p},\mathbf{q})=\mathbf{p}\cdot\dot{\mathbf{q}}-\mathcal{L}(\mathbf{q},\dot{\mathbf{q}}) \tag{3.164}$$

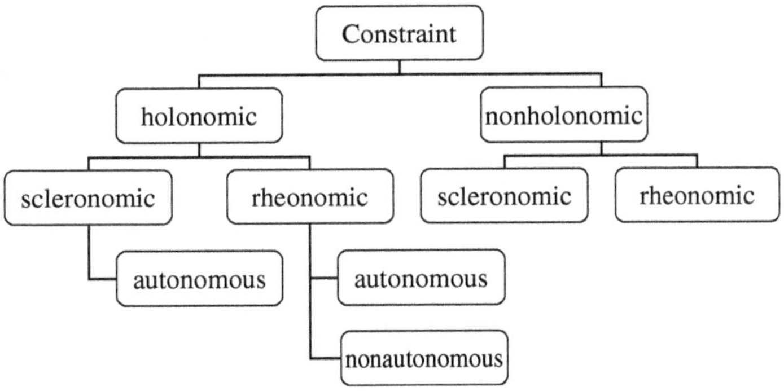

Figure 3.12. Autonomous and non autonomous systems

and the Hamiltonian $\mathcal{H}(\mathbf{p}, \mathbf{q})$ is autonomous. The above equation is called the Legendre transformation which will be discussed in detail in due course. Note that the above equation holds for the autonomous Lagrangian system, in which Lagrange's equations of motion can be given by

$$\frac{d}{dt}\left(\frac{\partial \mathcal{L}(\mathbf{q}, \dot{\mathbf{q}})}{\partial \dot{\mathbf{q}}}\right) - \frac{\partial \mathcal{L}(\mathbf{q}, \dot{\mathbf{q}})}{\partial \dot{\mathbf{q}}} = \mathbf{0} \tag{3.165}$$

instead of Equation (3.93).

The autonomous Lagrangian or Hamiltonian system can be classified as in Figure 3.12. Note that the holonomic-scleronomic constraint system may be autonomous always, but the holonomic-rheonomic constraint system can be autonomous or non autonomous.

Illustrative Example

Example 3.14 Rotating Bead Moving on a Vertical Circular Hoop

Consider a beam with mass m moving on a vertical circular hoop which rotates about the vertical axis with constant angular velocity ω. The beam is sliding without friction on the hoop whose radius is r as shown in Figure 3.13. This well known example illustrates a typical case of the autonomous Lagrangian system subject to a rheonomic constraint. The circumferential angle ϕ is constrained as

$$\phi = \omega t \tag{3.166}$$

and the radius is constrained as $x^2 + y^2 + z^2 = r^2$. Then, the Cartesian coordinates are represented as

$$x = r \sin\theta \cos\omega t, \quad y = r \sin\theta \sin\omega t, \quad z = -r\cos\theta \tag{3.167}$$

It is easy to grasp the fact that this is a dynamical system with one degree of freedom. As a consequence, the kinetic energy can be represented as

$$\mathcal{K}(\dot{x}) = \frac{1}{2}m\left(\dot{x}^2 + \dot{y}^2 + \dot{z}^2\right) = \frac{1}{2}\dot{\theta}\left(mr^2\right)\dot{\theta} + \frac{1}{2}mr^2\omega^2\sin^2\theta = \mathcal{K}(\theta, \dot{\theta}) \tag{3.168}$$

It is very important to realize that the above kinetic energy is a scalar-valued function of two variables, namely, the generalized coordinate and the generalized velocity. Furthermore, also note that the kinetic energy does not depend upon time explicitly. However, since the kinetic energy is not quadratic in the generalized velocity, this system is not scleronomic but rheonomic. The potential energy is given by

$$\mathcal{U}(z) = mgz = -mgr\cos\theta = \mathcal{U}(\theta) \tag{3.169}$$

Thus, we can establish the Lagrangian as

$$\begin{aligned}\mathcal{L}(\theta,\dot{\theta}) &= \mathcal{K}(\theta,\dot{\theta}) - \mathcal{U}(\theta)\\ &= \frac{1}{2}mr^2\left(\dot{\theta}^2 + \omega^2\sin^2\theta\right) + mgr\cos\theta\end{aligned} \tag{3.170}$$

Note that the Lagrangian does not depend upon time explicitly. In other words, the Lagrangian is autonomous. Since the kinetic energy is not a quadratic function in generalized velocity, this system is rheonomic. The corresponding momentum can be obtained as

$$p_\theta = \frac{\partial\mathcal{L}(\theta,\dot{\theta})}{\partial\dot{\theta}} = mr^2\dot{\theta} \tag{3.171}$$

Then, in view of Equation (3.168) we find

$$p_\theta\dot{\theta} = mr^2\dot{\theta}^2 \neq 2\mathcal{K}(\theta,\dot{\theta}) \tag{3.172}$$

Substituting Equation (3.170) into Equation (3.93), we have Lagrange's equation of motion as follows:

$$mr^2\ddot{\theta} - mr^2\omega^2\sin\theta\cos\theta + mgr\sin\theta = 0 \tag{3.173}$$

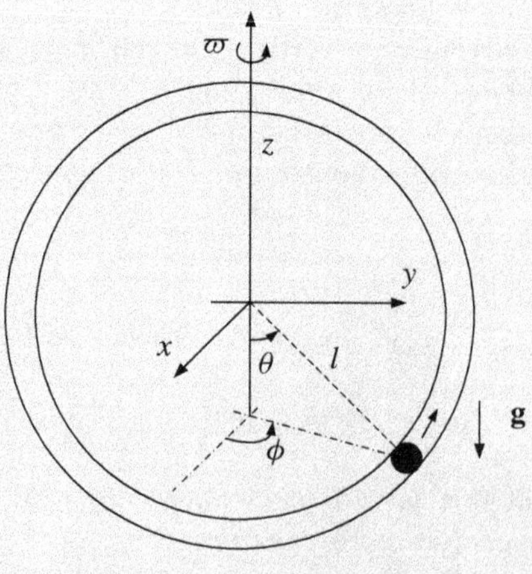

Figure 3.13. Bead circulating on a vertical circular hoop rotating

Hamiltonian Mechanics

$$\dot{\mathbf{p}} = -\frac{\partial \mathcal{H}}{\partial \mathbf{q}}, \quad \dot{\mathbf{q}} = \frac{\partial \mathcal{H}}{\partial \mathbf{p}}$$

$$\mathbf{p} = \left(p_1, \ldots, p_{n_{\text{dof}}}\right), \quad n_{\text{dof}} = 3N - k$$

$$\mathbf{q} = \left(q^1, \ldots, q^{n_{\text{dof}}}\right), \quad n_{\text{dof}} = 3N - k$$

$$\text{Phase space} \quad P \equiv T^*Q = \mathbb{R}^{2n_{\text{dof}}}$$

$$\mathcal{H}(\mathbf{p}, \mathbf{q}, t) : T^*Q \times \bar{T} \to \mathbb{R},$$

$$T^*Q = Q \times T_q^*Q$$

Hamilton, William (1805-1865)

Figure 3.14. Hamilton's equations of motion in a conservative system. Reproduced from The Word (www.theword.ie/cms/uploads/hamilton_portrait-web.jpg)

3.3 HAMILTONIAN MECHANICS

The most significant distinction between Lagrangian and Hamiltonian mechanics (see Figure 3.14) comes from the transition from configuration space to phase space via the use of Legendre transformation. Whereas the configuration space is an n_{dof}-dimensional manifold, the phase space is a $2n_{\text{dof}}$-dimensional manifold. In the Lagrangian system, the trajectory of the motion of a dynamical system is the curve of a point in the configuration space. However, in the Hamiltonian system, it is the curve of an ordered pair on the phase space. The autonomous Lagrangian is defined on the velocity phase space, whereas the autonomous Hamiltonian is defined on the phase space.

In addition, the Lagrangian differential system is of second-order in time, but the Hamiltonian system is of first-order in time. Above all things, the phase space has the property of a symplectic structure which the Hamiltonian phase flow preserves. However the velocity phase space, tangent bundle, is known to also have the symplectic structure in Lagrangian mechanics (Marsden and West 2001). In figuring out the Hamiltonian structure, it is very important to define the Hamiltonian. As in the case of the Lagrangian system, the autonomous Hamiltonian system has symmetry in time, resulting in the conservation of the total energy.

3.3.1 Phase Space

In Lagrangian mechanics, Lagrange's equations of motion for an N-body dynamical system in the n_{dof}-dimensional configuration space can be obtained via the Lagrangian form of D'Alembert's principle and the introduction of generalized coordinates. For the system with n_{dof}-degrees of freedom, the solutions for the n_{dof}-simultaneous equations of motion are determined as points on the path in the n_{dof}-dimensional configuration space. As mentioned previously, it is to be noted that the trajectory of the motion of the N-body dynamical system is a curve for a set of generalized coordinates in the configuration space. In other words, Lagrange's equations of motion represent the motion of the system at a point by the n_{dof}-generalized coordinates. Lagrange's equations of motion are second-order differential equations in the n_{dof}-generalized coordinate variables $\mathbf{q}$. In Newtonian mechanics, Newton's equations of motion are also second-order differential equations in the Cartesian coordinate system.

Legendre Transformation

$$\mathcal{H}(\mathbf{p},\mathbf{q},t)=\mathbf{p}\cdot\dot{\mathbf{q}}-\mathcal{L}(\mathbf{q},\dot{\mathbf{q}},t)$$

$$\left|\frac{\partial^2\mathcal{L}(\mathbf{q},\dot{\mathbf{q}},t)}{\partial\dot{\mathbf{q}}\partial\dot{\mathbf{q}}}\right|=\mathbf{M}\neq 0,\ TQ\rightarrow T^*Q$$

$$\mathcal{L}(\mathbf{q},\dot{\mathbf{q}},t)=\mathbf{p}\cdot\dot{\mathbf{q}}-\mathcal{H}(\mathbf{p},\mathbf{q},t)$$

$$\left|\frac{\partial^2\mathcal{H}(\mathbf{p},\mathbf{q},t)}{\partial\mathbf{p}\partial\mathbf{p}}\right|=\mathbf{M}^{-1}\neq 0,\ T^*Q\rightarrow TQ$$

Legendre, Adrien-Marie
(1752-1833)

Figure 3.15. Legendre transformation. Courtesy of Biographicon (www.biographicon.com/view/tilg0/Adrien_Marie_Legendre)

In contrast, Hamilton introduced a new function, called the Hamiltonian of an ordered pair $(\mathbf{q}, \mathbf{p})$ via employing the Legendre transformation of the Lagrangian (see Figure 3.15). He used the relation in Equation (3.164) to define the Hamiltonian. Via the Legendre transformation, transforming a function of a set of points $(\mathbf{q}, \dot{\mathbf{q}})$ to a function of a set of points $(\mathbf{q}, \mathbf{p})$, we shall derive a set of the $2n$-number of simultaneous differential equations of motion, called Hamilton's equations of motion. The ordered pair $(\mathbf{q}, \mathbf{p}) \in P$ defines a point on the phase space P or the so-called cotangent bundle T^*Q, which we subsequently discuss in further detail in what follows.

3.3.2 Canonical Coordinates

The coordinate variables specifying a point on the phase space are called canonical coordinates. The generalized momenta conjugate to the generalized coordinates are defined by Equation (3.163). Then, an ordered pair $(\mathbf{q}, \mathbf{p}) \in P$ gives the canonical coordinate system. The phase space with the canonical coordinates possesses a special form, called the canonical symplectic form that will be discussed later.

3.3.3 Cotangent Bundle

The generalized velocity $\dot{\mathbf{q}} = (\dot{q}^1, \ldots, \dot{q}^{n_{\text{dof}}}) \in T_q Q$ is a vector-valued function and is a tangent vector on the tangent space to the configuration space at every point $\dot{\mathbf{q}}$. In view of Equation (3.179), it is implied that the space having the canonical momenta $\mathbf{p} = \left(p_1, \ldots, p_{n_{\text{dof}}}\right) \in T_q^* Q$ as local coordinates is dual to the tangent space. The dual space to the tangent space $T_q Q$ is called the cotangent space $T_q^* Q$. Hence, the canonical momenta are dual vectors, covectors, covariant vectors, or one-forms on the cotangent space. The disjoint union of the cotangent and configuration spaces is called the cotangent bundle T^*Q. The cotangent bundle is the so-called phase space. Then, an ordered pair $(\mathbf{q}, \mathbf{p}) \in T^*Q$ is a point in the cotangent bundle. Note that $\dim T_q Q = \dim T_q^* Q = n_{\text{dof}}$ and $2\left(\dim T_q^* Q\right) = \dim TQ$. The phase space is mathematically defined as

$$P \equiv T^*Q = \bigcup_{q \in Q} T_q^* Q = \left\{ (\mathbf{q}, \mathbf{p}) | \, \mathbf{q} \in Q, \mathbf{p} \in T_q^* Q \right\} \tag{3.174}$$

The phase space with the canonical coordinates has special structures in the sense of mathematics, such as the canonical one-form and the canonical two-form. The canonical one-form is defined as

$$\theta = \sum_{i=1}^{n_{\text{dof}}} \left(p_i \cdot dq^i\right) = \mathbf{p} \cdot d\mathbf{q} \tag{3.175}$$

The canonical two-form is defined as

$$\omega = -d\theta = -\sum_{i=1}^{n_{\text{dof}}} \left(dp_i \wedge dq^i\right) = -d\mathbf{p} \wedge d\mathbf{q} = d\mathbf{q} \wedge d\mathbf{p} \tag{3.176}$$

The canonical two-form is skew-symmetric and constant in the phase space. This special property is called symplectic. Not only in Hamiltonian mechanics, but also in Lagrangian mechanics, there exist one-form and two-form representations (Marsden and West 2001). The Lagrangian one-form is given as

$$\theta_L = \sum_{i=1}^{n_{\text{dof}}} \left(\frac{\partial \mathcal{L}}{\partial \dot{q}^i} \cdot dq^i\right) = \frac{\partial \mathcal{L}}{\partial \dot{\mathbf{q}}} \cdot d\mathbf{q} \tag{3.177}$$

whereas the Lagrangian two-form is defined as

$$\omega_L = \sum_{i=1}^{n_{\text{dof}}} \left(d\frac{\partial \mathcal{L}}{\partial \dot{q}^i} \wedge dq^i\right) = d\frac{\partial \mathcal{L}}{\partial \dot{\mathbf{q}}} \wedge d\mathbf{q} \tag{3.178}$$

3.3.4 Legendre Transformation

For an N-body dynamical system with the holonomic constraints, canonical momenta are defined by the partial derivative of the Lagrangian with respect to generalized velocities as

$$\mathbf{p} = \frac{\partial \mathcal{L}(\mathbf{q}, \dot{\mathbf{q}}, t)}{\partial \dot{\mathbf{q}}} \tag{3.179}$$

Note that the generalized momentum in Lagrangian mechanics is referred to as the canonical momentum in Hamiltonian mechanics. Since the determinant of the Hessian matrix becomes the mass matrix of the system which is positive definite, we have

$$\left|\frac{\partial \mathcal{L}(\mathbf{q}, \dot{\mathbf{q}}, t)}{\partial \dot{q}^i \partial \dot{q}^j}\right| = \left|\frac{\partial p_i}{\partial \dot{q}^j}\right| = |\mathbf{M}| \neq 0, \tag{3.180}$$

and, we can obtain a new scalar-valued function via the Legendre transformation of the Lagrangian. Among the local coordinates on the tangent bundle, the generalized velocities may be interpreted as active variables and the generalized coordinates as passive variables. When the canonical momenta are chosen as the active variables of the new scalar-valued function $\mathcal{H}(\mathbf{q}, \mathbf{p}, t) : T_q^* Q \times \bar{\mathcal{T}} \rightarrow \mathbb{R}$, the Legendre transformation results in

$$\mathcal{H}(\mathbf{q}, \mathbf{p}, t) = \mathbf{p} \cdot \dot{\mathbf{q}} - \mathcal{L}(\mathbf{q}, \dot{\mathbf{q}}, t), \tag{3.181}$$

where $\mathcal{H}(\mathbf{q}, \mathbf{p}, t)$ is called the Hamiltonian. Hence, there should exist a new equation conjugate to Equation (3.97) given as

$$\dot{\mathbf{q}} = \frac{\partial \mathcal{H}(\mathbf{q}, \mathbf{p}, t)}{\partial \mathbf{p}} \tag{3.182}$$

Note that Equation (3.182) is one of Hamilton's equations of motion. Equation (3.181) is invertible, since

$$\left|\frac{\partial^2 \mathcal{H}(\mathbf{q}, \mathbf{p}, t)}{\partial p_i \partial p_j}\right| = \left|\frac{\partial \dot{q}^i}{\partial p_j}\right| = \left|\mathbf{M}^{-1}\right| \neq 0 \tag{3.183}$$

Therefore, we have

$$\mathcal{L}(\mathbf{q}, \dot{\mathbf{q}}, t) = \mathbf{p} \cdot \dot{\mathbf{q}} - \mathcal{H}(\mathbf{q}, \mathbf{p}, t) \tag{3.184}$$

This transformation makes it possible to open the way to the world of Hamiltonian mechanics.

Illustrative Examples Let $Q \subset \mathbb{R}^{n_{\text{dof}}}$ denote the configuration space with generalized coordinates $\mathbf{q} = \left(q^1, \ldots, q^{n_{\text{dof}}}\right) \in Q$ for a finite dimensional dynamical system. Furthermore, the kinetic energy of the dynamical system takes a quadratic form in terms of generalized velocities $\dot{\mathbf{q}} = \left(\dot{q}^1, \ldots, \dot{q}^{n_{\text{dof}}}\right) \in T_q Q \subset \mathbb{R}^{n_{\text{dof}}}$. Then, the corresponding generalized momenta can be denoted by $\mathbf{p} = \left(p_1, \ldots, p_{n_{\text{dof}}}\right) \in T_q^* Q \subset \mathbb{R}^{n_{\text{dof}}}$.

Example 3.15 Generalized Momentum and Velocity in Dynamics

Let $Q \subset \mathbb{R}^{n_{\text{dof}}}$ denote the configuration space with a set of generalized coordinates $\mathbf{q} = \left(q^1, \ldots, q^{n_{\text{dof}}}\right) \in Q$ for a finite dimensional dynamical system. Further, the kinetic energy of the dynamical system is supposed to be quadratic in terms of generalized momentum $\mathbf{p} = \left(p_1, \ldots, p_{n_{\text{dof}}}\right) \in T_{\mathbf{q}}^* Q \subset \mathbb{R}^{n_{\text{dof}}}$. Then, the corresponding generalized velocities can be denoted by $\dot{\mathbf{q}} = \left(\dot{q}^1, \ldots, \dot{q}^{n_{\text{dof}}}\right) \in T_q Q \subset \mathbb{R}^{n_{\text{dof}}}$. Thus, the autonomous kinetic energy $\mathcal{K}(\mathbf{p}, \mathbf{q}) : T^*Q \to \mathbb{R}$ can be expressed as

$$\mathcal{K}(\mathbf{p}, \mathbf{q}) = \frac{1}{2}\mathbf{p} \cdot \mathbf{M}(\mathbf{q})^{-1}\,\mathbf{p} \tag{3.185}$$

where $\mathbf{M}$ denotes the proper mass of the system. Then, the complementary kinetic energy $\mathcal{K}^*(\mathbf{q}, \dot{\mathbf{q}}) : TQ \to \mathbb{R}$ can be given by

$$\mathcal{K}^*(\mathbf{q}, \dot{\mathbf{q}}) = \frac{1}{2}\dot{\mathbf{q}} \cdot \mathbf{M}(\mathbf{q})\,\dot{\mathbf{q}} \tag{3.186}$$

because the relationship shown in Figure 3.16 is valid in motion. From both Equation (3.185) and Equation (3.186), we obtain the following equations

$$\mathbf{p} = \frac{\partial \mathcal{K}^*(\mathbf{q}, \dot{\mathbf{q}})}{\partial \dot{\mathbf{q}}}, \quad \dot{\mathbf{q}} = \frac{\partial \mathcal{K}(\mathbf{q}, \mathbf{p})}{\partial \mathbf{p}} \tag{3.187}$$

It is said that both equations in Equation (3.187) are conjugate with each other. Consequently, the Legendre transformation yields

$$\mathcal{K}^*(\mathbf{q}, \dot{\mathbf{q}}) = \mathbf{p} \cdot \dot{\mathbf{q}} - \mathcal{K}(\mathbf{q}, \mathbf{p}) \tag{3.188}$$

We observe that the generalized coordinates in the transformation are inactive. Furthermore, note that the determinants of the following Hessian matrix do not vanish.

$$|\mathbf{M}| = \left|\frac{\partial^2 \mathcal{K}^*}{\partial \dot{q}^i \partial \dot{q}^j}\right| \neq 0, \left|\mathbf{M}^{-1}\right| = \left|\frac{\partial^2 \mathcal{K}}{\partial p_i \partial p_j}\right| \neq 0, \quad for\ i \neq j \tag{3.189}$$

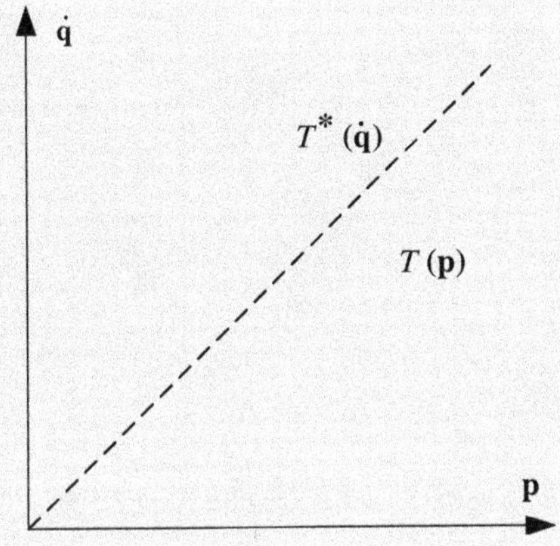

Figure 3.16. Relationship between momentum and velocity

Example 3.16 Stress and Strain in Linear Elasticity

The Legendre transformation is also used in the theory of elasticity. Using the Voigt notation, the stress is denoted by $\boldsymbol{\sigma} = (\sigma_1, \sigma_2, \sigma_3, \sigma_4, \sigma_5, \sigma_6) \in \mathbb{V}^* \subset \mathbb{R}^6$, and (see Figure 3.17) the corresponding strain is represented by $\boldsymbol{\varepsilon} = \left(\varepsilon^1, \varepsilon^2, \varepsilon^3, \varepsilon^4, \varepsilon^5, \varepsilon^6\right) \in \mathbb{V} \subset \mathbb{R}^6$ in a linear elastic mechanical body. Then, the strain energy density function $\mathcal{U}(\boldsymbol{\varepsilon}) : \mathbb{V} \to \mathbb{R}$ is given by

$$\mathcal{U}(\boldsymbol{\varepsilon}) = \frac{1}{2}\boldsymbol{\varepsilon} \cdot \mathbf{E} \cdot \boldsymbol{\varepsilon} \tag{3.190}$$

where $\mathbf{E}$ denotes the proper elastic modulus. Likewise, we have the complementary energy density function $\mathcal{U}^*(\boldsymbol{\sigma}) : \mathbb{V}^* \to \mathbb{R}$

$$\mathcal{U}^*(\boldsymbol{\sigma}) = \frac{1}{2}\boldsymbol{\sigma} \cdot \mathbf{E}^{-1} \cdot \boldsymbol{\sigma} \tag{3.191}$$

It is well known that the following conjugate relations are valid

$$\boldsymbol{\varepsilon} = \frac{\partial \mathcal{U}^*}{\partial \boldsymbol{\sigma}}, \quad \boldsymbol{\sigma} = \frac{\partial \mathcal{U}}{\partial \boldsymbol{\varepsilon}} \tag{3.192}$$

due to the Legendre transformation

$$\mathcal{U}(\boldsymbol{\varepsilon}) = \boldsymbol{\varepsilon} \cdot \boldsymbol{\sigma} - \mathcal{U}^*(\boldsymbol{\sigma}) \tag{3.193}$$

In addition, we have the following determinants for the Hessian matrix,

$$|\mathbf{E}| = \left|\frac{\partial^2 \mathcal{U}}{\partial \varepsilon^i \partial \varepsilon^j}\right| \neq 0, \quad \left|\mathbf{E}^{-1}\right| = \left|\frac{\partial^2 \mathcal{U}}{\partial \sigma^i \partial \sigma^j}\right| \neq 0, \ for\ i \neq j \tag{3.194}$$

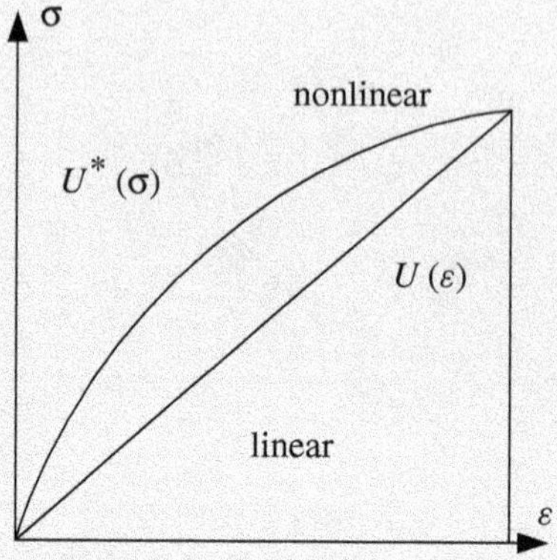

Figure 3.17. Relationship between stress and strain

3.3.5 Hamilton's Equations of Motion

In Hamiltonian mechanics, the motion of the system is described in the $2n_{\text{dof}}$-dimensional phase space. As mentioned previously, the symplectic form is preserved in the phase space. Hamilton's equations of motion are of first-order in time. In holonomic-scleronomic systems, the autonomous Hamiltonian equals the total energy of the dynamical system. In holonomic-rheonomic systems, the Hamiltonian is not identical to the total energy of the dynamical system, regardless of whether the dynamical system is holonomic-scleronomic or holonomic-rheonomic.

By virtue of Equation (3.181), the infinitesimal actual change of the Hamiltonian can be written as

$$d\mathcal{H}(\mathbf{q}, \mathbf{p}, t) = \mathbf{p} \cdot d\dot{\mathbf{q}} + \dot{\mathbf{q}} \cdot d\mathbf{p} - d\mathcal{L}(\mathbf{q}, \dot{\mathbf{q}}, t) \tag{3.195}$$

By use of the chain rule, the infinitesimal actual change of the Lagrangian may be written,

$$d\mathcal{L}(\mathbf{q}, \dot{\mathbf{q}}, t) = \frac{\partial \mathcal{L}(\mathbf{q}, \dot{\mathbf{q}}, t)}{\partial \mathbf{q}} d\mathbf{q} + \frac{\partial \mathcal{L}(\mathbf{q}, \dot{\mathbf{q}}, t)}{\partial \dot{\mathbf{q}}} d\dot{\mathbf{q}} + \frac{\partial \mathcal{L}}{\partial t} dt \tag{3.196}$$

Equation (3.196) can be rewritten in consideration of Equation (3.163) and Equation (3.165) as the form

$$d\mathcal{L}(\mathbf{q}, \dot{\mathbf{q}}, t) = \dot{\mathbf{p}} \cdot d\mathbf{q} + \mathbf{p} \cdot d\dot{\mathbf{q}} + \frac{\partial \mathcal{L}}{\partial t} dt \tag{3.197}$$

Upon substituting Equation (3.197) into Equation (3.195), we obtain

$$d\mathcal{H}(\mathbf{q}, \mathbf{p}, t) = \dot{\mathbf{q}} \cdot d\mathbf{p} - \dot{\mathbf{p}} \cdot d\mathbf{q} - \frac{\partial \mathcal{L}}{\partial t} dt \tag{3.198}$$

The infinitesimal actual change of the Hamiltonian leads to,

$$d\mathcal{H}(\mathbf{q}, \mathbf{p}, t) = \frac{\partial \mathcal{H}(\mathbf{q}, \mathbf{p}, t)}{\partial \mathbf{q}} d\mathbf{q} + \frac{\partial \mathcal{H}(\mathbf{q}, \mathbf{p}, t)}{\partial \mathbf{p}} d\mathbf{p} + \frac{\partial \mathcal{H}(\mathbf{q}, \mathbf{p}, t)}{\partial t} dt \tag{3.199}$$

Due to the equivalence of Equation (3.198) and Equation (3.199), we have

$$\dot{\mathbf{q}} = \frac{\partial \mathcal{H}(\mathbf{q}, \mathbf{p}, t)}{\partial \mathbf{p}}, \quad \dot{\mathbf{p}} = -\frac{\partial \mathcal{H}(\mathbf{q}, \mathbf{p}, t)}{\partial \mathbf{q}}, \quad \frac{\partial \mathcal{L}}{\partial t} = -\frac{\partial \mathcal{H}}{\partial t} \tag{3.200}$$

It is natural to have the third equation in Equation (3.200) in view of the following

$$\frac{\partial \mathcal{H}(\mathbf{q}, \mathbf{p}, t)}{\partial t} = \frac{\partial}{\partial t}(\mathbf{p} \cdot \dot{\mathbf{q}} - \mathcal{L}(\mathbf{q}, \dot{\mathbf{q}}, t)) = -\frac{\partial \mathcal{L}(\mathbf{q}, \dot{\mathbf{q}}, t)}{\partial t} \tag{3.201}$$

The first and second equations in Equation (3.200), which are of first-order in time, are called Hamilton's equations of motion.

Alternatively, Hamilton's equations of motion can be obtained from the Lagrangian form of D'Alembert's principle. In Lagrangian mechanics, we have shown that the Lagrangian form of D'Alembert's principle given in Equation (3.52) can be transformed into Equation (3.96) in terms of the descriptive scalar function, the Lagrangian. Then, we rewrite Equation (3.96) as

$$\delta \mathcal{L}(\mathbf{q}, \dot{\mathbf{q}}, t) - \frac{\partial \mathcal{L}(\mathbf{q}, \dot{\mathbf{q}}, t)}{\partial \dot{\mathbf{q}}} \cdot \delta \dot{\mathbf{q}} - \frac{d}{dt}\left(\frac{\partial \mathcal{L}(\mathbf{q}, \dot{\mathbf{q}}, t)}{\partial \dot{\mathbf{q}}}\right) \cdot \delta \mathbf{q} = 0 \tag{3.202}$$

By introducing the generalized momentum in Equation (3.179), the above equation can be rewritten as

$$\delta \mathcal{L}(\mathbf{q}, \dot{\mathbf{q}}, t) - \mathbf{p} \cdot \delta \dot{\mathbf{q}} - \dot{\mathbf{p}} \cdot \delta \mathbf{q} = 0 \tag{3.203}$$

By virtue of the Legendre transformation given in Equation (3.181), we obtain the expression

$$\delta \mathcal{L}(\mathbf{q}, \dot{\mathbf{q}}, t) = \mathbf{p} \cdot \delta \dot{\mathbf{q}} + \dot{\mathbf{q}} \cdot \delta \mathbf{p} - \delta \mathcal{H}(\mathbf{q}, \mathbf{p}, t) \tag{3.204}$$

Upon substitution of the above equation into Equation (3.203), we have

$$\left(\dot{\mathbf{q}} - \frac{\partial \mathcal{H}(\mathbf{q}, \mathbf{p}, t)}{\partial \mathbf{p}}\right) \cdot \delta \mathbf{p} - \left(\dot{\mathbf{p}} + \frac{\partial \mathcal{H}(\mathbf{q}, \mathbf{p}, t)}{\partial \mathbf{q}}\right) \cdot \delta \mathbf{q} = 0 \tag{3.205}$$

which is a consequence of the Lagrangian form of D'Alembert's principle in the Hamiltonian mechanics framework. Since the virtual displacement $\delta \mathbf{q}$ and the virtual momentum $\delta \mathbf{p}$ are arbitrary in holonomic systems, we have

$$\dot{\mathbf{q}} - \frac{\partial \mathcal{H}(\mathbf{q}, \mathbf{p}, t)}{\partial \mathbf{p}} = \mathbf{0}, \quad \dot{\mathbf{p}} + \frac{\partial \mathcal{H}(\mathbf{q}, \mathbf{p}, t)}{\partial \mathbf{q}} = \mathbf{0} \tag{3.206}$$

which represent Hamilton's equations of motion.

Let us consider the Hamiltonian in the Cartesian coordinate system as

$$\mathcal{H}(\mathbf{x}, \mathbf{p}, t) = \mathcal{K}(\mathbf{p}) - \mathcal{U}(\mathbf{x}, t) \tag{3.207}$$

Then, the Legendre transformation in the Cartesian coordinate system can be given by

$$\mathcal{L}(\mathbf{x}, \dot{\mathbf{x}}, t) = \mathbf{p} \cdot \dot{\mathbf{x}} - \mathcal{H}(\mathbf{x}, \mathbf{p}, t) \tag{3.208}$$

Hence, Equation (3.150) can be written as

$$\delta(\mathbf{p} \cdot \dot{\mathbf{x}} - \mathcal{H}(\mathbf{x}, \mathbf{p}, t)) + \delta(\boldsymbol{\lambda} \cdot \mathbf{f}(\mathbf{x}, t)) - \frac{d}{dt}(\mathbf{p} \cdot \delta \mathbf{x}) = 0 \tag{3.209}$$

which can be rewritten as

$$\delta\left(\mathbf{p}\cdot\dot{\mathbf{x}}-\tilde{\mathcal{H}}(\mathbf{x},\mathbf{p},t)\right)-\frac{d}{dt}(\mathbf{p}\cdot\delta\mathbf{x})=0 \tag{3.210}$$

where $\tilde{\mathcal{H}}(\mathbf{x},\mathbf{p},t)=\mathcal{K}(\mathbf{p})+\mathcal{U}(\mathbf{x},t)-\boldsymbol{\lambda}\cdot\mathbf{f}(\mathbf{x},t)$, and it is referred to as the augmented Hamiltonian. Therefore, it is easy to show that the above equation leads to the following

$$\left(\dot{\mathbf{x}}-\frac{\partial\tilde{\mathcal{H}}(\mathbf{x},\mathbf{p},t)}{\partial\mathbf{p}}\right)\cdot\delta\mathbf{p}-\left(\dot{\mathbf{p}}+\frac{\partial\tilde{\mathcal{H}}(\mathbf{x},\mathbf{p},t)}{\partial\mathbf{x}}\right)\cdot\delta\mathbf{x}=0 \tag{3.211}$$

which leads to

$$\left(\dot{\mathbf{x}}-\frac{\partial\mathcal{H}(\mathbf{x},\mathbf{p},t)}{\partial\mathbf{p}}\right)\cdot\delta\mathbf{p}-\left(\dot{\mathbf{p}}+\frac{\partial\mathcal{H}(\mathbf{x},\mathbf{p},t)}{\partial\mathbf{x}}-\boldsymbol{\lambda}\cdot\Delta\mathbf{f}(\mathbf{x},t)\right)\cdot\delta\mathbf{x}=0 \tag{3.212}$$

Illustrative Examples Next, we discuss typical examples of the pendulum problems we described previously to derive Hamilton's equations of motion.

Example 3.17 Planar Pendulum

We again consider the same simple planar pendulum as we discussed previously. The canonical momentum can be given by

$$p_\theta=\frac{\partial\mathcal{L}(\theta,\dot{\theta})}{\partial\dot{\theta}}=ml^2\dot{\theta} \tag{3.213}$$

Therefore, the kinetic energy $\mathcal{K}(p_\theta):T_q^*Q\to\mathbb{R}$ can be rewritten as

$$\mathcal{K}(p_\theta)=\frac{1}{2}p_\theta\frac{1}{ml^2}p_\theta=\frac{p_\theta^2}{2ml^2} \tag{3.214}$$

Note that the kinetic energy takes the quadratic form in terms of the canonical momentum, because the system is scleronomic. Then, the corresponding Hamiltonian can be written as

$$\mathcal{H}(\theta,p_\theta)=\frac{p_\theta^2}{2ml^2}-mgl\cos\theta \tag{3.215}$$

Hamilton's equations result in

$$\dot{\theta}=\frac{\partial H(\theta,p_\theta)}{\partial p_\theta}=\frac{p_\theta}{ml^2},\quad \dot{p}_\theta=-\frac{\partial\mathcal{H}(\theta,p_\theta)}{\partial\theta}=-mgl\sin\theta \tag{3.216}$$

Example 3.18 Spherical Pendulum

Likewise, next consider the spherical pendulum as we discussed previously. The canonical momenta were given in Equation (3.122). Then, the kinetic energy $\mathcal{K}(\mathbf{p},\mathbf{q}):T^*Q\to\mathbb{R}$ can be written as

$$\mathcal{K}=\frac{1}{2}\mathbf{p}\cdot\tilde{\mathbf{M}}^{-1}\mathbf{p},\quad \tilde{\mathbf{M}}^{-1}=\frac{1}{ml^2\sin^2\theta}\begin{bmatrix}\sin^2\theta & 0\\ 0 & 1\end{bmatrix},\quad \mathbf{p}=\left(p_\theta,p_\phi\right) \tag{3.217}$$

Note that the kinetic energy takes the quadratic form in terms of the canonical momenta, because the system is also scleronomic. Then the corresponding Hamiltonian can be written as

$$\mathcal{H}(\mathbf{p}, \mathbf{q}) = \frac{1}{2ml^2 \sin^2\theta}\left(p_\theta^2 \sin^2\theta + p_\phi^2\right) - mgl\cos\theta \tag{3.218}$$

Hamilton's equations result in

$$\begin{aligned} \dot{\theta} &= \frac{\partial \mathcal{H}(\mathbf{q}, \mathbf{p})}{\partial p_\theta} = \frac{p_\theta}{ml^2}, \ \dot{p}_\theta = -\frac{\partial \mathcal{H}(\mathbf{q}, \mathbf{p})}{\partial \theta} = \frac{p_\phi^2 \cos\theta}{ml^2 \sin^3\theta} + mgl\sin\theta \\ \dot{\phi} &= \frac{\partial \mathcal{H}(\mathbf{q}, \mathbf{p})}{\partial p_\phi} = \frac{p_\phi}{ml^2 \sin^2\theta}, \ \dot{p}_\phi = -\frac{\partial \mathcal{H}(\mathbf{q}, \mathbf{p})}{\partial \phi} = 0 \end{aligned} \tag{3.219}$$

Substituting Equation (3.122) into Equation (3.219) results in Equation (3.123).

3.3.6 Autonomous Hamiltonian Systems

As addressed previously in the section on Lagrangian mechanics, here in Hamiltonian mechanics, if the Hamiltonian of a system does not depend upon time explicitly, it is said that the Hamiltonian is autonomous. As in the case of the Lagrangian, a number of important problems in physics also fall into the category of autonomous Hamiltonian systems. Furthermore, the autonomous Hamiltonian system has symmetry in time, resulting in the conservation of the total energy if the kinetic energy is a quadratic function in generalized velocities or generalized momenta. The total derivative of the Hamiltonian with respect to time is given as

$$\frac{d\mathcal{H}(\mathbf{q}, \mathbf{p}, t)}{dt} = \frac{\partial \mathcal{H}(\mathbf{q}, \mathbf{p}, t)}{\partial \mathbf{q}}\dot{\mathbf{q}} + \frac{\partial \mathcal{H}(\mathbf{q}, \mathbf{p}, t)}{\partial \mathbf{p}}\dot{\mathbf{p}} + \frac{\partial \mathcal{H}}{\partial t} \tag{3.220}$$

By virtue of Hamilton's equations of motion, Equation (3.220) takes the form

$$\frac{d\mathcal{H}(\mathbf{q}, \mathbf{p}, t)}{dt} = \frac{\partial \mathcal{H}(\mathbf{q}, \mathbf{p}, t)}{\partial \mathbf{q}}\frac{\partial \mathcal{H}(\mathbf{q}, \mathbf{p}, t)}{\partial \mathbf{p}} - \frac{\partial \mathcal{H}(\mathbf{q}, \mathbf{p}, t)}{\partial \mathbf{p}}\frac{\partial \mathcal{H}(\mathbf{q}, \mathbf{p}, t)}{\partial \mathbf{q}} + \frac{\partial \mathcal{H}}{\partial t} = \frac{\partial \mathcal{H}}{\partial t} \tag{3.221}$$

In view of Equation (3.201), the above equation can be given by

$$\frac{d\mathcal{H}(\mathbf{q}, \mathbf{p}, t)}{dt} = \frac{\partial \mathcal{H}}{\partial t} = -\frac{\partial \mathcal{L}}{\partial t} \tag{3.222}$$

Therefore, the Hamiltonian is autonomous, as the Lagrangian is autonomous. Furthermore, in the case when the Hamiltonian is autonomous, the Hamiltonian is conserved and is a constant of motion. The autonomous Hamiltonian can be given by

$$\mathcal{H}(\mathbf{q}, \mathbf{p}) = \mathbf{p} \cdot \dot{\mathbf{q}} - \mathcal{L}(\mathbf{q}, \dot{\mathbf{q}}) \tag{3.223}$$

If the kinetic energy is a quadratic function in generalized velocities or generalized momenta, we have the following

$$\mathbf{p} \cdot \dot{\mathbf{q}} = \mathbf{p} \cdot \mathbf{M}(\mathbf{q})^{-1}\mathbf{p} = 2\mathcal{K}(\mathbf{q}, \mathbf{p}) \tag{3.224}$$

From Equations (3.223) and (3.224), we find that the autonomous Hamiltonian equals the total energy as follows:

$$\mathcal{H}(\mathbf{q},\mathbf{p}) = \mathcal{K}(\mathbf{q},\mathbf{p}) + \mathcal{U}(\mathbf{q}) \tag{3.225}$$

Therefore, Equation (3.222) implies that if the Hamiltonian does not possess explicit time dependence, then the Hamiltonian remains constant. In particular, in the holonomic-scleronomic systems where the kinetic energy is quadratic in generalized velocities or generalized momenta, the Hamiltonian equals the total energy.

3.3.7 Symplectic Manifold

In this section, we shall discuss the symplectic structures of the phase space in Hamiltonian mechanics and of the configuration space in Lagrangian mechanics. In Lagrangian mechanics, the configuration space Q is defined as n_{dof}-dimensional Euclidean space, where a point is determined by $\mathbf{q} \in Q$. On the other hand, in Hamiltonian mechanics the phase space TQ is considered to be $2n_{\text{dof}}$-dimensional Euclidean space, where a point is determined by $(\mathbf{q},\mathbf{p}) \in TQ \equiv P$. The phase space is sometimes called a symplectic manifold (P,ω), because the manifold retains a special property, called symplecticness, on the trajectory of the solution of Hamilton's equations of motion. The trajectory is called a Hamiltonian flow, which is a symplectic mapping. Symplecticness of the symplectic manifold (P,ω) implies that the volume of the Hamiltonian flow is preserved.

As a consequence, symplecticness characterizes the Hamiltonian flow. As a conserved quantity of the Hamiltonian flow, we have the volume preservation property, but this is just a result of symplecticness in the symplectic manifold (P,ω). It should be noted that the property of symplecticness is more important in the symplectic manifold than the volume conservation one. The symplectic structure of the phase space is induced by the symplectic two-form. The exact solution of Hamilton's equations requires the preservation of the symplectic two-form, and consequently the volume of the Hamiltonian flow can be conserved due to Liouville's theorem.

Symplectic Two-Form We know that the generalized velocities are members of the tangent space $\dot{\mathbf{q}} \in T_q Q$, which is an n_{dof}-dimensional Euclidean vector space. Consider a dual vector space $T_q^* Q$ which is of the same dimension as the tangent space. The dual vector space is called the cotangent space whose members are the canonical momenta, i.e. $\mathbf{p} \in T_q^* Q$. The action of the dual vector of the canonical momentum on the differential vector of the generalized coordinates is, by definition, given by

$$\theta = \mathbf{p}\cdot d\mathbf{q} = p_k dq^k, \quad k = 1,\ldots,n_{\text{dof}} \quad on\ T^*Q \tag{3.226}$$

which is called the canonical differential one-form on the phase space. Note that a Lagrangian differential one-form on the tangent bundle is defined as

$$\theta_{\mathcal{L}} = \frac{\partial \mathcal{L}(\mathbf{q},\dot{\mathbf{q}})}{\partial \dot{\mathbf{q}}}\cdot d\mathbf{q} = \frac{\partial \mathcal{L}(\mathbf{q},\dot{\mathbf{q}})}{\partial \dot{q}^k}\cdot dq^k, \quad k = 1,\ldots,n_{\text{dof}} \quad on\ TQ \tag{3.227}$$

Hence, the canonical momentum is the local component of the one-form on the phase space, and in addition, the one-form may be interpreted as a mapping from the vectors to functions.

We now define the differential two-form as a bilinear and skew-symmetric map of the vectors on the symplectic manifold (P,ω). The exterior derivative of the canonical differential one-form is given by

$$\hat{d}\theta = d\mathbf{p}\wedge d\mathbf{q} \quad on\ T^*Q \tag{3.228}$$

where $\hat{d}$ denotes the exterior derivative. Then, the symplectic two-form on the symplectic manifold (P, ω) is by definition given as

$$\omega = d\mathbf{q} \wedge d\mathbf{p} = -\hat{d}\theta \quad on\ T^*Q \tag{3.229}$$

where $d\mathbf{q}$ and $d\mathbf{p}$ are dual vectors or covectors. Note that differential canonical momenta and differential generalized coordinates are vectors on the symplectic manifold. Then, it is called symplectic that the three-form, $\hat{d}\omega$, is equal to zero, i.e.,

$$\hat{d}\omega = -\hat{d}^2\theta = 0 \tag{3.230}$$

In other words, the property, that the symplectic two-form is constant, means symplecticness.

Consider the case when $n = 1$. The dimension size of the phase space is 2. Then, we use the notation $\omega = dq \wedge dp$ in stead of $\omega = dq^1 \wedge dp_1$. The action of the two-form on two vectors $(\mathbf{v}, \mathbf{w}) \in \mathbb{R}^2 \times \mathbb{R}^2$ on the $(\mathbf{q}, \mathbf{p})$ tangent plane is given by

$$\omega(\mathbf{v}, \mathbf{w}) = d\mathbf{q} \wedge d\mathbf{p}(\mathbf{v}, \mathbf{w}) = (d\mathbf{q} \cdot \mathbf{v})(d\mathbf{p} \cdot \mathbf{w}) - (d\mathbf{q} \cdot \mathbf{w})(d\mathbf{p} \cdot \mathbf{v}) \tag{3.231}$$

The above results in the area

$$\omega(\mathbf{v}, \mathbf{w}) = v_1 w_2 - v_2 w_1 = A(\mathbf{v}, \mathbf{w}) \tag{3.232}$$

where a bilinear form $A(\mathbf{v}, \mathbf{w}) : \mathbb{R}^2 \times \mathbb{R}^2 \to \mathbb{R}$ denotes the area form.

It is said to be symplectic when the area remains constant in the Hamiltonian flow. Hence, the symplectic two-form can be $d\mathbf{q} \wedge d\mathbf{p}(\mathbf{v}, \mathbf{w}) : \mathbb{R}^2 \times \mathbb{R}^2 \to \mathbb{R}_+$ or $d\mathbf{p} \wedge d\mathbf{q}(\mathbf{v}, \mathbf{w}) : \mathbb{R}^2 \times \mathbb{R}^2 \to \mathbb{R}_-$. On the $2n_{\text{dof}}$-dimensional phase space, the sum of all areas in each of the (q^k, p_k) tangent planes corresponds to the volume of the Hamiltonian flow due to Liouville's theorem. The symplectic two-form is skew-symmetric, i.e.,

$$\omega(\mathbf{v}, \mathbf{w}) = -(v_2 w_1 - v_1 w_2) = -\omega(\mathbf{w}, \mathbf{v}) \tag{3.233}$$

The above can be written as

$$\omega(\mathbf{v}, \mathbf{w}) = \mathbf{v}^T \begin{bmatrix} 0 & 1 \\ -1 & 0 \end{bmatrix} \mathbf{w} = \mathbf{v}^T \mathbb{J}_{2\times 2}\mathbf{w} \tag{3.234}$$

As a consequence, the $2n_{\text{dof}}$-dimensional symplectic manifold (P, ω) has the following symplectic two-form for $2n_{\text{dof}}$-dimensional two vectors $(\mathbf{v}, \mathbf{w}) \in \mathbb{R}^{2n_{\text{dof}}} \times \mathbb{R}^{2n_{\text{dof}}}$

$$\omega(\mathbf{v}, \mathbf{w}) = \mathbf{v}^T \begin{bmatrix} \mathbf{0}_{n_{\text{dof}}\times n_{\text{dof}}} & \mathbf{1}_{n_{\text{dof}}\times n_{\text{dof}}} \\ -\mathbf{1}_{n_{\text{dof}}\times n_{\text{dof}}} & \mathbf{0}_{n_{\text{dof}}\times n_{\text{dof}}} \end{bmatrix} \mathbf{w} = \mathbf{v}^T \mathbb{J}_{2n_{\text{dof}}\times 2n_{\text{dof}}}\mathbf{w} \tag{3.235}$$

Consider a vector $\mathbf{z}^T := \left(q^1, \ldots, q^{n_{\text{dof}}}; p_1, \ldots, p_{n_{\text{dof}}}\right) \in \mathbb{R}^{2n_{\text{dof}}}$ using the canonical coordinates $(\mathbf{q}, \mathbf{p}) \in \mathbb{R}^n \times \mathbb{R}^{n_{\text{dof}}}$ and a gradient

$$\nabla_{qp} H(\mathbf{z}) := \left\{ \begin{array}{c} \dfrac{\partial H(\mathbf{z})}{\partial \mathbf{q}} \\ \dfrac{\partial H(\mathbf{z})}{\partial \mathbf{p}} \end{array} \right\} \tag{3.236}$$

Then Hamilton's equations of motion can be given

$$\frac{d\mathbf{z}}{dt} = \mathbb{J}_{2n_{\text{dof}}\times 2n_{\text{dof}}} \nabla_{qp} H(\mathbf{z}) \tag{3.237}$$

The trajectories of the solution of the above equations define the Hamiltonian flow in the phase space. The skew-orthogonal matrix holds for the following relations

$$\begin{aligned} \mathbb{J} + \mathbb{J}^T &= \mathbf{0}_{2n_{\text{dof}} \times 2n_{\text{dof}}}, \quad |\mathbb{J}| = 1 \\ \mathbb{J}\mathbb{J}^T &= \mathbf{I}_{2n_{\text{dof}} \times 2n_{\text{dof}}}, \quad \mathbb{J}^2 = -\mathbf{I}_{2n_{\text{dof}} \times 2n_{\text{dof}}} \end{aligned} \tag{3.238}$$

Symplectic Transformation The Hamiltonian phase flow is the one-parameter group of transformation of the phase space. Consider the following one-parameter-group

$$\mathfrak{g}^t : \mathbf{z}_0 \rightarrow \mathbf{z}_t \tag{3.239}$$

where $\mathbf{z}_0 = \mathbf{z}(t)|_{t=0} = (\mathbf{q}(0), \mathbf{p}(0))^T$ and $\mathbf{z}_t = (\mathbf{q}(t), \mathbf{p}(t))^T$. For convenience, we use the notation for the transformation as

$$\mathfrak{g}^t : \mathbf{z}_{QP} \rightarrow \mathbf{z}_{QP} \tag{3.240}$$

where $\mathbf{z}_{QP} = (\mathbf{q}, \mathbf{p})^T$ and $\mathbf{z}_{QP} = (\mathbf{Q}, \mathbf{P})^T$. Then, the transformed differential form $d\mathbf{z}_{QP}$ can be given as

$$d\mathbf{z}_{QP} = \mathbf{J}_{QP \rightarrow QP} d\mathbf{z}_{QP} \tag{3.241}$$

where the Jacobian is denoted by

$$\mathbf{J}_{QP \rightarrow QP} = \frac{\partial (\mathbf{Q}, \mathbf{P})}{\partial (\mathbf{q}, \mathbf{p})} \tag{3.242}$$

Since the differential form of the Hamiltonian can be written as

$$dH = \left\{\nabla_{QP} H(\mathbf{z})\right\}^T d\mathbf{z}_{QP} = \left\{\nabla_{QP} H(\mathbf{z})\right\}^T d\mathbf{z}_{QP} \tag{3.243}$$

we obtain the relation

$$\nabla_{QP} H(\mathbf{z}) = \left[\mathbf{J}_{QP \rightarrow QP}\right]^T \nabla_{QP} H(\mathbf{z}) \tag{3.244}$$

From Equation (3.237) and (3.241), we can derive the following relation

$$\mathbb{J}_{2n_{\text{dof}} \times 2n_{\text{dof}}} \nabla_{QP} H(\mathbf{z}) = \mathbf{J}_{QP \rightarrow QP} \mathbb{J}_{2n_{\text{dof}} \times 2n_{\text{dof}}} \nabla_{QP} H(\mathbf{z}) \tag{3.245}$$

Upon substitution of Equation (3.244) into Equation (3.245), we have

$$\mathbb{J}_{2n_{\text{dof}} \times 2n_{\text{dof}}} = \mathbf{J}_{QP \rightarrow QP} \mathbb{J}_{2n_{\text{dof}} \times 2n_{\text{dof}}} \left[\mathbf{J}_{QP \rightarrow QP}\right]^T \tag{3.246}$$

The Jacobian is called a symplectic matrix if the equation of the transformation is satisfied in the Hamiltonian flow.

Consider the exterior product in the transformed coordinates. Then the wedge product can be given by

$$d\mathbf{Q} \wedge d\mathbf{P} = \left|\mathbf{J}_{QP \rightarrow QP}\right| d\mathbf{q} \wedge d\mathbf{p} \tag{3.247}$$

Conservation of the volume requires that the determinant of the Jacobian should be 1. Such a transformation is called the symplectic or canonical transformation.

Remark 3.3.1 On the other hand, if we introduce a different perspective, namely, consider the descriptive scalar function, the Total Energy $\mathcal{E}(\mathbf{q}, \dot{\mathbf{q}}, t) : TQ \times \mathcal{T} \rightarrow \mathbb{R}$ for the holonomic constraint N-body dynamical system, the Lagrangian form of D'Alembert's Principle finally yields,

$$\frac{d}{dt}\left(\frac{\partial \mathcal{E}(\mathbf{q},\dot{\mathbf{q}},t)}{\partial \dot{\mathbf{q}}}\right)+\frac{\partial \mathcal{E}(\mathbf{q},\dot{\mathbf{q}},t)}{\partial \mathbf{q}}=\mathbf{0} \tag{3.248}$$

where the kinetic energy is independent of the generalized coordinates. The related issues dealing with the Total Energy representation of the equation of motion and the corresponding framework are described in Chapter 6.

EXERCISES

3.1. Newton's equation of motion is discussed for a sliding block on a fixed, inclined plane in Example 3.3. Find Newton's equations for a sliding block on a moving, inclined plane in Example 4.2.

3.2. Use the Lagrangian form of D'Alembert's principle given in Equation (3.51) to find the governing, differential equation for a sliding block on a moving, inclined plane in Example 4.2.

3.3. Consider a spring-mass system where an object of mass m is suspended by the spring due to the gravitational force $m\mathbf{g}$ as shown in Figure 3.18. Let x, u and k denote the position coordinate, the displacement and the spring constant. Find the displacement δ_{st} at the static equilibrium position, when $A = 0$ and $\mathbf{g} \neq \mathbf{0}$.

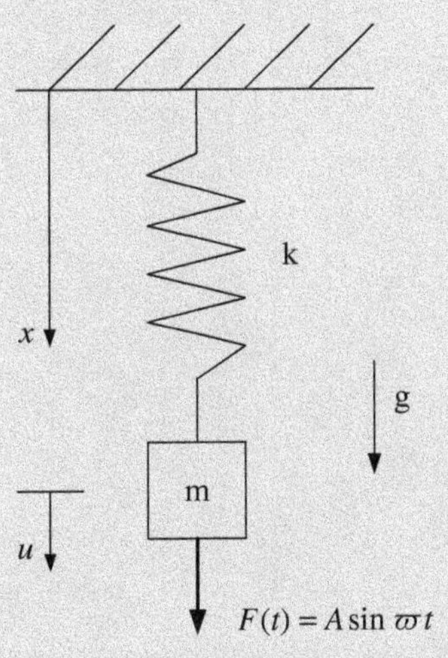

Figure 3.18. Undamped spring-mass system

3.4. In a rigid body motion, the Lagrangian form of D'Alembert's principle given in Equation (3.51) indicates that the contribution made by the internal forces vanishes. Show that $\sum_{i=1}^{N} \mathbf{F}_{\text{int}}^{i}(t) \cdot \delta \mathbf{x}^{i} = 0$.

3.5. Consider the simple planar pendulum with an oscillating hinge point as shown in Figure 3.19. The constraint equation is dependent upon time; hence the system is holonomic-rheonomic.

Examine that the constraint force is not perpendicular to the virtual displacement, and the configuration space is not fixed. Use Equation (3.48) to find the governing equations of motion.

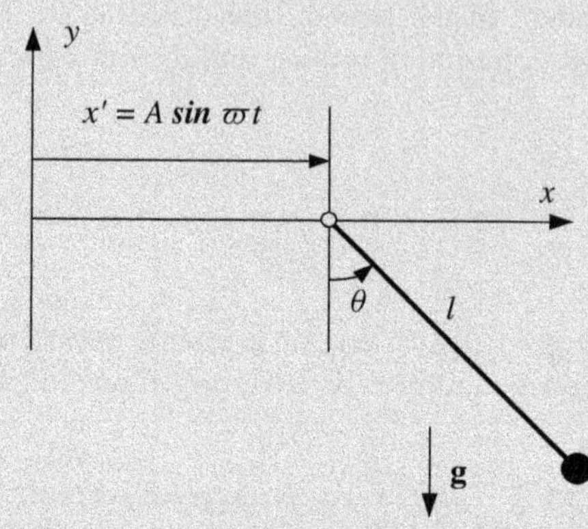

Figure 3.19. Simple pendulum with an oscillating hinge point

3.6. Three particles, whose masses are m_1, m_2 and m_3, are connected by three nonlinear springs as shown in Figure 3.20. They are assumed to be in planar motion. Each spring has the same constant k as the spring constant, and the undeformed length l. The potential energy due to internal forces is given by $U\left(\left\|\mathbf{r}_i - \mathbf{r}_j\right\|\right) = \sum_{i=1}^{3}\sum_{j=i+1}^{3}\frac{1}{2}k\left(\left\|\mathbf{r}_i - \mathbf{r}_j\right\| - l\right)^2$, where $\mathbf{r}_i$ denotes the position vector of the ith particle. Discuss why Equation (3.51) cannot be employed to derive the governing equations of motion. Show that the internal force can be given by $\mathbf{F}_{ij} = k\left(\mathbf{r}_i - \mathbf{r}_j\right)\left(1 - \frac{1}{\left\|\mathbf{r}_i - \mathbf{r}_j\right\|}\right)$. Notice that $\mathbf{F}_{ji} = -\frac{\partial U}{\partial\left\|\mathbf{r}_i - \mathbf{r}_j\right\|} = -\mathbf{F}_{ij}$ (Goldstein 2002, page 10). Use Equation (3.48) to find the governing equations of motion.

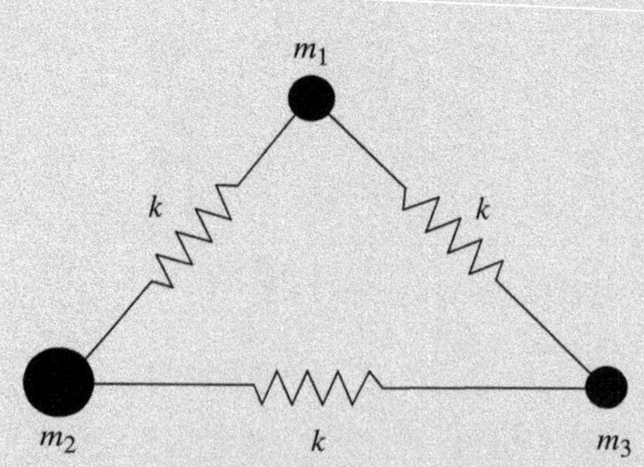

Figure 3.20. Three particle dynamical system connected by three nonlinear springs

3.7. Consider the augmented Lagrangian given in Example 3.9. Derive Newton's equations of motion given in Equation (3.30) after finding the Lagrange multiplier.

3.8. Suppose the object suspended by the spring in Figure 3.18 undergoes an harmonic oscillatory motion, i.e. $A = 0$ and $\mathbf{g} = \mathbf{0}$. Use the Lagrangian form of D'Alembert's principle given in Equation (3.51) to find the governing, differential equation.

3.9. In Exercise 3.8, find Lagrange's equation of motion in the form of a second-order differential equation of motion.

3.10. Consider a forced dynamic response of the undamped spring-mass system, i.e. $A \neq 0$ and $\mathbf{g} \neq \mathbf{0}$. in Figure 3.18. Use the Lagrangian form of D'Alembert's principle to derive the differential equation of motion.

3.11. In Exercise 3.10, establish Lagrange's equation of motion to derive the differential equation of motion in the form of a second-order differential equation of motion.

3.12. In Exercise 3.10, establish Hamilton's equations of motion in the form of first-order differential equations of motion.

3.13. Equation (3.105) represents a nonlinear ordinary differential equation for a simple planar pendulum. Suppose the oscillatory angle is very small so that $\sin\theta \approx \theta$. Show that the free response of the planar pendulum is $\theta(t) = \theta_0 \cos(\omega_n t) + \dot{\theta}_0 \sin(\omega_n t)$, where the initial displacement is $\theta(0) = \theta_0$, the initial velocity $\dot{\theta}(0) = \dot{\theta}_0$ and the circular natural frequency $\omega_n = \sqrt{\frac{g}{l}}$ (Ginsberg and Genin 1984). Find the period T of oscillation and the cyclical natural frequency f_n. Note that $T = \frac{2\pi}{\omega_n}$.

3.14. In Example 3.7, establish Hamilton's equations of motion in the form of first-order differential equations of motion.

3.15. In Exercise 3.8, show that the free response of the undamped spring-mass system is $u(t) = u_0 \cos(\omega_n t) + \dot{u}_0 \sin(\omega_n t)$, where the initial displacement is $u(0) = u_0$, the initial velocity $\dot{u}(0) = \dot{u}_0$, the circular natural frequency $\omega_n = \sqrt{\frac{k}{m}}$ and the natural frequency $f_n = \frac{\omega_n}{2\pi}$.

3.16. In Example 3.10, find the Jacobi integral and discuss why the Jacobi integral is not identical to the total energy.

3.17. In Example 3.10, establish the Hamiltonian and discuss why the Hamiltonian cannot be identical to the total energy.

3.18. In Example 3.10, establish Hamilton's equations of motion in the form of first-order differential equations of motion.

3.19. Consider the double planar pendulum as shown in Figure 3.21. One bob of mass m_1 and the other bob of m_2 are connected and suspended by rigid bars. The weights of both of the rigid bars are negligible and the hinge is frictionless. Let $\mathbf{x} = (x_1, x_2, y_1, y_2)$ and $\mathbf{q} = (\theta, \phi)$. Find the Jacobian matrix $\left[\frac{\partial \mathbf{x}}{\partial \mathbf{q}}\right]$.

3.20. Two particles are in planar motion as shown in Figure 3.22. No gravitational force is involved in the motion and the two particles are connected by a massless, rigid link. Let (x, y) denotes the point of the mass center in the 2-body dynamical system. Let $\mathbf{x} = (x_1, x_2, y_1, y_2)$ and $\mathbf{q} = (x, y, \theta)$. Find the Jacobian matrix $\left[\frac{\partial \mathbf{x}}{\partial \mathbf{q}}\right]$.

3.21. By making use of the Jacobian matrix obtained from the above exercise, establish Equation (3.76) and obtain the governing equations of motion in terms of generalized coordinates $\mathbf{q}$.

3.22. For the planar motion as shown in Figure 3.22, obtain the governing equations of motion by use of Lagrange's equations of motion.

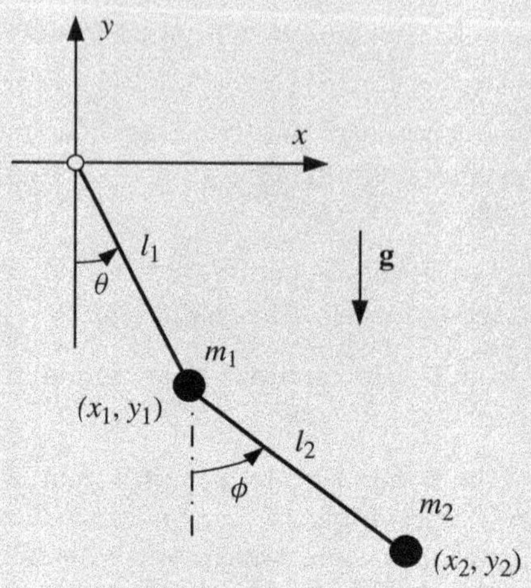

Figure 3.21. Double planar pendulum

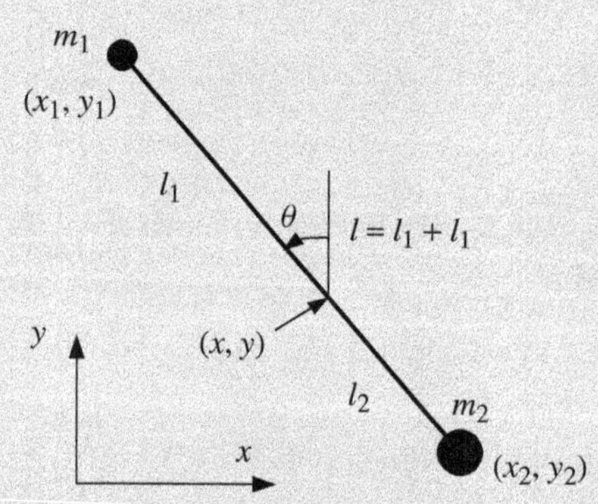

Figure 3.22. Planar motion of two particles

3.23. For the planar motion as shown in Figure 3.22, establish Hamilton's equations of motion in the form of first-order differential equations of motion.

3.24. Three objects are linked by two rigid rods as shown in Figure 3.23. The massless rigid rods are in equilibrium. Use the principle of virtual work given in Equation (4.21) to find F at the static equilibrium position in terms of θ and m, when $m_1 = 2m$ and $m_2 = m_3 = m$.

3.25. Suppose the external force F is suddenly released in Figure 3.23. Hence the 3-body dynamical system is in motion. What is the number of degrees of freedom for the dynamical system? Describe the constraint equations.

3.26. In Exercise 3.25, use the Lagrangian form of D'Alembert's principle to obtain the governing, differential equation.

3.27. In Exercise 3.25, obtain Lagrange's equation of motion in the form of a second-order differential equation of motion.

3.28. In Exercise 3.25, obtain Hamilton's equations of motion in the form of first-order differential equations of motion.

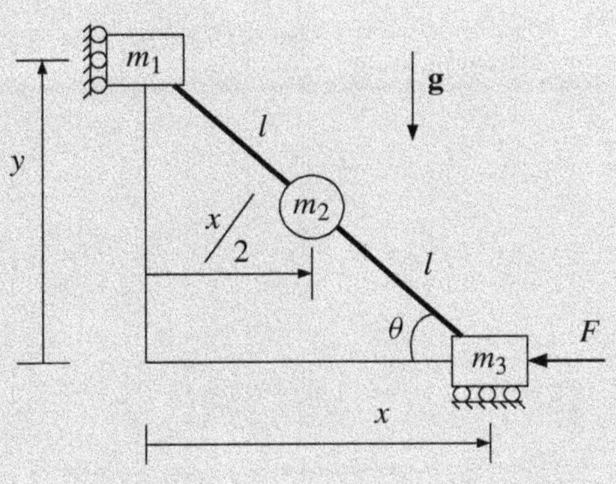

Figure 3.23. Rigid rod system

3.29. Let us consider a damped spring-mass dynamical system shown in Figure 3.24. The viscous damping force is given by $F_d = -c\dot{u}$, where c denotes the linear viscous damping coefficient. Establish Newton's equation of motion.

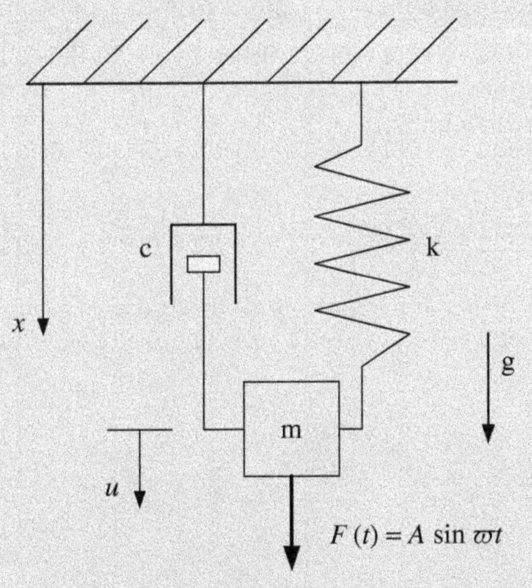

Figure 3.24. Damped spring-mass system

3.30. For the damped spring-mass dynamical system shown in Figure 3.24, establish Lagrange's equation of motion in the form of a second-order differential equation of motion.

3.31. For the damped spring-mass dynamical system shown in Figure 3.24, obtain Hamilton's equations of motion in the form of first-order differential equations of motion.

CHAPTER FOUR

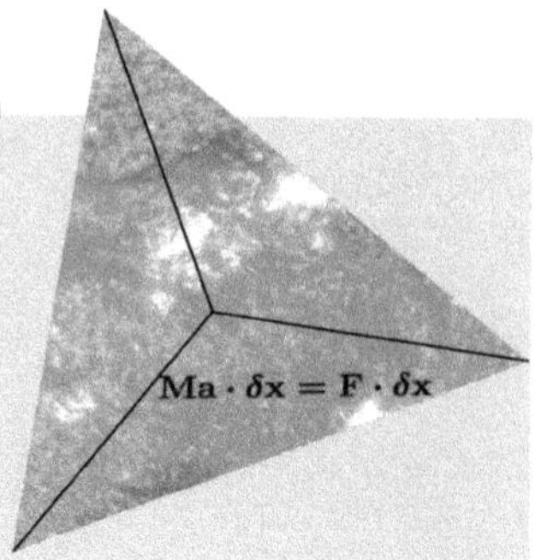

PRINCIPLE OF VIRTUAL WORK

The fundamental concepts underlying the principle of virtual work for dynamical systems are covered in Chapter 4. We first establish the relations stemming from the principle of balance of linear momentum and leading to the principle of virtual work in dynamics. Therein, directly starting from the principle of virtual work in dynamics, we then describe the development of the governing equations of motion for N-body systems using this as the standalone principle. Both vector and scalar formalisms in the sense of the Newtonian mechanics setting, and scalar formalisms as well, in the sense of the Lagrangian and the Hamiltonian descriptive functions, and an alternative via the descriptive scalar function, namely, the Total Energy are highlighted in this chapter.

4.1 VIRTUAL WORK IN N-BODY DYNAMICAL SYSTEMS

Consider a dynamical system of N particles moving in three-dimensional Euclidean space. An infinitesimal work $d\mathcal{W}$ done by an actual external force $\mathbf{F}_{\text{ext}}$ acting on the N particles on the Newtonian path (the true path) is defined as

$$d\mathcal{W} = \sum_{i=1}^{N} \mathbf{F}^{i}_{\text{ext}} \cdot d\mathbf{x}^{i} = \mathbf{F}_{\text{ext}} \cdot d\mathbf{x} \tag{4.1}$$

where the external force $\mathbf{F}_{\text{ext}}(t) : \bar{\mathcal{T}} \subset \mathbb{R}^{+} \rightarrow \mathbb{R}^{3N}$ is a vector-valued function of $3N$ variables; hence $\mathbf{F}_{\text{ext}}(t) = \left(\mathbf{F}^{1}_{\text{ext}}(t), \ldots, \mathbf{F}^{N}_{\text{ext}}(t)\right)$.

Suppose that the external force is conservative. In other words, $\mathbf{F}_{\text{ext}} = \mathbf{F}_{\text{c}}$. Note that $d\mathbf{x}$ in Equation (4.1) indicates an infinitesimal actual change or infinitesimal displacement in the inertial frame of reference. In view of Equation (3.8), the above equation can be rewritten as

$$d\mathcal{W} = \mathbf{F}_{\text{c}} \cdot d\mathbf{x} = -\frac{d\mathcal{U}(\mathbf{x})}{d\mathbf{x}} \cdot d\mathbf{x} = -d\mathcal{U} \tag{4.2}$$

Thus, the infinitesimal work done by the conservative force equals to the minus of the infinitesimal potential energy.

Advances in Computational Dynamics of Particles, Materials and Structures, First Edition. Jason Har and Kumar K. Tamma.

Consider an N-body statical system which is conservative. The statical system is assumed to be subject to the k-number of constraints. Then, the equilibrium equation can be written as

$$\sum_{i=1}^{N} \mathbf{F}_c^i + \sum_{i=1}^{N} \mathbf{C}^i = \mathbf{0} \tag{4.3}$$

In view of Equations (4.1), (4.2) and (4.3), we obtain the following

$$d\mathcal{W} = -d\mathcal{U} = -\sum_{i=1}^{N} \mathbf{C}^i \cdot d\mathbf{x}^i = \mathbf{C} \cdot d\mathbf{x} \tag{4.4}$$

where the constraint force $\mathbf{C}$ is also a vector-valued function of $3N$ variables; hence $\mathbf{C} = \left(\mathbf{C}^1, \ldots, \mathbf{C}^N\right)$. Note that in static problems, time does not appear as an independent variable in the constraint equations. Consequently, the constraint force is always orthogonal to the constraint surface; hence, it can be given by

$$\mathbf{C}(\mathbf{x}) \cdot d\mathbf{x} = \left(\sum_{i=1}^{k} \lambda_i \nabla f_i \left(\mathbf{x}^1, \ldots, \mathbf{x}^N\right) \right) \cdot d\mathbf{x} = 0 \tag{4.5}$$

where λ_i are arbitrary parameters. Note that in static problems, the constraints should be holonomic-scleronomic. Recalling Equation (3.19), then, Equation (4.5) is true in static problems.

Next, consider an N-body dynamical system subject to the constraint force $\mathbf{C}$. In holonomic-scleronomic systems, the constraint force is also independent of the time variable. Thus, the constraint force does not do an actual work; hence we have

$$\mathbf{C}(\mathbf{x}) \cdot d\mathbf{x} = \left(\sum_{i=1}^{k} \lambda_i \nabla f_i \left(\mathbf{x}^1, \ldots, \mathbf{x}^N\right) \right) \cdot d\mathbf{x} = 0 \tag{4.6}$$

which is equal to the same equation as shown in Equation (4.5). However, in holonomic-rheonomic systems the constraint force does do actual work as shown below:

$$\mathbf{C}(\mathbf{x}, t) \cdot d\mathbf{x} = \left(\sum_{i=1}^{k} \lambda_i \nabla f_i \left(\mathbf{x}^1, \ldots, \mathbf{x}^N; t\right) \right) \cdot d\mathbf{x} \neq 0 \tag{4.7}$$

The infinitesimal change $d\mathbf{x}$ represents the real change between time t and $t + dt$ (see the illustrative example to follow). Note that dt is different from Δt, which represents an increment of time. Thus, during the time increment, Taylor's series gives a representation for a new position vector as follows:

$$\mathbf{x}(t + \Delta t) = \mathbf{x}(t) + \Delta t \frac{d\mathbf{x}(t)}{dt} + R_1(t, \Delta t) \tag{4.8}$$

Then, the derivative of the position vector with respect to time is defined as

$$\frac{d\mathbf{x}(t)}{dt} = \lim_{\Delta t \to 0} \left\{ \frac{\mathbf{x}(t + \Delta t) - \mathbf{x}(t)}{\Delta t} \right\} = \dot{\mathbf{x}}(t), \quad \lim_{\Delta t \to 0} \left\{ \frac{R_1(t, \Delta t)}{\Delta t} \right\} = 0 \tag{4.9}$$

The differential can be written as

$$d\mathbf{x}(t) = \dot{\mathbf{x}}(t)\, dt \tag{4.10}$$

which represents an infinitesimal actual change of the position vector.

In contrast to the infinitesimal actual change, the infinitesimal virtual change, which we consider as a variation of the position vector, is an imaginary quantity. It is called the virtual displacement. To differentiate the terminology, namely, the variation from the actual change, the operator δ, the use of which Lagrange first proposed, has been historically employed. The virtual displacement $\delta\mathbf{x}$, which is a variation of the displacement, means an infinitesimal imaginary displacement intentionally defined in any kinematically admissible way. The difference between the infinitesimal displacement and the virtual displacement becomes distinctly clear for the dynamical system in noting that the virtual displacement $\delta\mathbf{x}$ is given as the time is held fixed during the virtual displacement, i.e., $\delta t = 0$. Note that the variable t is the independent variable. Then, the admissible configuration on the so-called varied path $\tilde{\mathbf{x}}(t)$ at the fixed time t can be given by

$$\tilde{\mathbf{x}}(t) = \mathbf{x}(t) + \epsilon\boldsymbol{\eta}(t) \tag{4.11}$$

where $\varepsilon \in \mathbb{R}$ is sufficiently a small parameter and $\boldsymbol{\eta}(t)$ is an admissible function. The variation of the position vector is defined as an infinitesimal virtual displacement as follows:

$$\delta\mathbf{x}(t) = \epsilon\boldsymbol{\eta}(t) \tag{4.12}$$

In classical mechanics, the variation is referred to as the virtual displacement, which is a vector function of time. It is well known that Galileo (1564–1642) perceived that the variation is an infinitesimal virtual change (Mach 1906). The terminology "virtual" in English means "existing in effect, but not actual" in Latin (Mach 1906).

Next, let us consider the variation of the holonomic constraints, regardless of whether the holonomic constraint is rheonomic or scleronomic. By virtue of Taylor's series, we have the expression

$$\frac{f_i(\mathbf{x}(t) + \mathbf{h}(t); t) - f_i(\mathbf{x}(t); t)}{\mathbf{h}(t)} = \nabla f_i(\mathbf{x}(t)) + \frac{1}{\mathbf{h}(t)} R_1(\mathbf{x}(t), \mathbf{h}(t)) \tag{4.13}$$

Then, the Fréchet derivative of the constraint equation is defined as

$$\begin{aligned} &\nabla f_i(\mathbf{x}(t); t) = \lim_{\|\mathbf{h}(t)\| \to 0} \left\{ \frac{f_i(\mathbf{x}(t) + \mathbf{h}(t); t) - f_i(\mathbf{x}(t); t)}{\|\mathbf{h}(t)\|} \right\}, \\ &\lim_{\|\mathbf{h}(t)\| \to 0} \frac{1}{\|\mathbf{h}(t)\|} R_1(\mathbf{x}(t), \mathbf{h}(t)) = 0 \end{aligned} \tag{4.14}$$

where the gradient of the function is called Fréchet differentiable at $\mathbf{x}(t)$, only if the derivative exists. Next, in view of Equation (4.12), Equation (4.13) can be rewritten as

$$\frac{f_i(\mathbf{x}(t) + \epsilon\boldsymbol{\eta}(t); t) - f_i(\mathbf{x}(t); t)}{\epsilon\boldsymbol{\eta}(t)} = \nabla f_i(\mathbf{x}(t); t) + \frac{1}{\epsilon\boldsymbol{\eta}(t)} R_1(\mathbf{x}(t), \epsilon\boldsymbol{\eta}(t)) \tag{4.15}$$

If and only if the constraint equation is Fréchet differentiable at $\mathbf{x}(t)$, then the Gâteaux variation $\mathcal{G}f_i(\mathbf{x}(t); \boldsymbol{\eta}(t))$ (Marsden and Hughes 1983; Simo and Hughes 1998; Troutman 1996) as a linear mapping can be defined as

$$\mathcal{G}f_i(\mathbf{x}(t); \boldsymbol{\eta}(t)) = \lim_{\epsilon \to 0} \frac{f_i(\mathbf{x}(t) + \epsilon\boldsymbol{\eta}(t); t) - f_i(\mathbf{x}(t); t)}{\epsilon} = D_{\mathbf{x}} f_i(\mathbf{x}(t); t) \cdot \boldsymbol{\eta}(t) \tag{4.16}$$

where $D_{\mathbf{x}} f_i(\mathbf{x}(t); t)$, i.e. $\nabla f_i(\mathbf{x}(t); t)$ is the Fréchet derivative (Edwards 1994). Hence, the Gâteaux variation of the function can be considered as a directional derivative of the function $f_i(\mathbf{x}(t))$ at $\mathbf{x}(t)$

in the direction $\eta(t)$ (Edwards 1994). As it is conventional to use the notation δ, the variation of the constraints can be written as

$$\begin{aligned}\delta f_i(\mathbf{x};t) &= \epsilon \mathcal{G} f_i(\mathbf{x}(t);\eta(t)) \\ &= \nabla f_i(\mathbf{x};t)\cdot\delta\mathbf{x}(t) \\ &= \varepsilon\frac{d}{d\varepsilon}f_i(\mathbf{x}+\varepsilon\eta;t)|_{\varepsilon=0}, \quad i=1,\ldots,k \\ &= \varepsilon D_{\mathbf{x}} f_i(\mathbf{x})\cdot\eta \\ &= D_{\mathbf{x}} f_i(\mathbf{x})\cdot\delta\mathbf{x}=0\end{aligned} \tag{4.17}$$

where $\varepsilon\in\mathbb{R}$ is sufficiently a small parameter and $\delta\mathbf{x}(t)=\left(\delta\mathbf{x}^1,\ldots,\delta\mathbf{x}^N\right):\bar{\mathcal{T}}\to\mathbb{R}^{3N}$ stands for the virtual displacement in Cartesian coordinates. Subsequently, the constraint force is always orthogonal to the virtual displacement as shown below:

$$\mathbf{C}(\mathbf{x},t)\cdot\delta\mathbf{x}=\left(\sum_{i=1}^{k}\lambda_i\nabla f_i\left(\mathbf{x}^1,\ldots,\mathbf{x}^N;t\right)\right)\cdot\delta\mathbf{x}=0 \tag{4.18}$$

We now introduce the concept of virtual work in conjunction with the virtual displacement and the actual force. The virtual work is an imaginary quantity due to the imaginary virtual displacement. Once the virtual displacement is defined, then we can consider the infinitesimal virtual work associated with the virtual displacement as follows:

$$\delta\mathcal{W}=\mathbf{F}_{\text{ext}}\cdot\delta\mathbf{x}=\mathbf{F}_{\text{appl}}\cdot\delta\mathbf{x} \tag{4.19}$$

where the applied force $\mathbf{F}_{\text{appl}}(t):\bar{\mathcal{T}}\subset\mathbb{R}^+\to\mathbb{R}^{3N}$ is a vector-valued function of $3N$ variables; hence $\mathbf{F}_{\text{appl}}(t)=(\mathbf{F}^1_{\text{appl}}(t),\ldots,\mathbf{F}^N_{\text{appl}}(t))$. At this stage, we would like to place a special emphasis on the notion that the virtual displacement is a tangent vector to the configuration space at the point $\mathbf{x}$, whereas the infinitesimal change is a tangent vector to the trajectory of the system as illustrated in the example to follow. In a static equilibrium state with constraints, considering Equation (3.6), we have the virtual work $\delta\mathcal{W}$ defined as

$$\delta W=\mathbf{F}_{\text{appl}}\cdot\delta\mathbf{x}=-\mathbf{C}\cdot\delta\mathbf{x}=0 \tag{4.20}$$

Note that the above equation can be rewritten as

$$\delta W=\sum_{i=1}^{N}\mathbf{F}^i_{\text{appl}}\cdot\delta\mathbf{x}^i=0 \tag{4.21}$$

Equation (4.21) is called the principle of virtual work in statics, which was proposed first in 1717 by Bernoulli (1667–1748). Note that the principle of virtual work in the nonconservative field can be rewritten as

$$\delta W=\mathbf{F}_{\text{c}}\cdot\delta\mathbf{x}+\mathbf{F}_{\text{nc}}\cdot\delta\mathbf{x}=0 \tag{4.22}$$

where $\mathbf{F}_{\text{c}}$ and $\mathbf{F}_{\text{nc}}$ denote the conservative and the nonconservative forces, respectively.

Illustrative Examples To understand the virtual work concept, in the following we shall discuss the difference between the actual infinitesimal change and the imaginary virtual displacement.

Example 4.1 Variation Versus Infinitesimal Actual Change

Consider a dynamical system of N particles moving in three-dimensional Euclidean space. An infinitesimal work $d\mathcal{W}$ done by the external force $\mathbf{F}_{ext}$ acting on the N particles on the Newtonian path (the true path) is defined as

$$d\mathcal{W} = \mathbf{F}_{ext} \cdot d\mathbf{x} \tag{4.23}$$

In Equation (4.23), $d\mathbf{x}$ indicates an infinitesimal actual change or infinitesimal displacement in the current configuration. The infinitesimal work associated with the virtual displacement is called the virtual work and is defined as

$$\delta\mathcal{W} = \mathbf{F}_{ext} \cdot \delta\mathbf{x} \tag{4.24}$$

where $\mathbf{F}_{ext}$ is the actual external force acting on the body at the equilibrium state. Recall that the virtual displacement is a tangent vector to the configuration space at the point $\mathbf{x}$ and normal to the constraints, whereas the infinitesimal change is a tangent vector to the trajectory of the system. The total constraint force $\mathbf{C}$ is orthogonal to the virtual displacement, that is, workless constraints. This means that the constraint force is perpendicular to the tangent plane where the virtual displacement lies. This statement is valid for both holonomic-scleronomic and holonomic-rheonomic cases. However, the infinitesimal real work done by the constraint forces does not vanish due to the non-orthogonal relationship between the constraint force and the infinitesimal change. Topologically, the virtual displacement $\delta\mathbf{x}$ is a tangent vector which is parallel to the tangent plane of the configuration space Q at $\mathbf{x}$. It is pointed in the direction normal to the displacement of the system. Due to the constraint force, the body manages to move on the constraint surface.

In the case of a dynamical system for a planar pendulum whose length remains constant during the motion, the infinitesimal displacement $d\mathbf{x} = (dx, dy) \in \mathbb{R}^2$ may be considered as a vector that is dependent upon time as follows

$$d\mathbf{x}(t) = (l\cos\theta, l\sin\theta)\, d\theta(t) \tag{4.25}$$

Then, the virtual displacement may be calculated as

$$\delta\mathbf{x}(t) = (l\cos\theta, l\sin\theta)\, \delta\theta(t) \tag{4.26}$$

Since the trajectory is fixed in time, both the infinitesimal and virtual displacement vectors remain on the tangent line of the curve as shown in Figure 6.2. Therefore, we note that the infinitesimal displacement shares the same direction with the virtual displacement for the simple planar pendulum problem. Further, taking the change or variation of the constraint function, namely,

$$df = 2x dx + 2y dy = 0, \ \ \delta f = 2x\delta x + 2y\delta y = 0 \tag{4.27}$$

we have,

$$\frac{dy}{dx} = \frac{\delta y}{\delta x} = \tan\theta \tag{4.28}$$

For the case of a rheonomic constraint system, the above relations are not true anymore. As a typical example, the length of the pendulum bar depends upon time. Then, Equation (4.25) should be written in a different manner as follows:

$$d\mathbf{x}(t) = (dl(t)\sin\theta + l(t)\cos\theta d\theta(t), -dl(t)\cos\theta + l(t)\sin\theta d\theta(t)) \tag{4.29}$$

However, the virtual displacement is given by

$$\delta\mathbf{x}(t) = (l\cos\theta, l\sin\theta)\,\delta\theta(t) \tag{4.30}$$

Thus, the virtual displacement in the rheonomic constraint system is the same as the virtual displacement in the scleronomic constraint system, and shares the same direction. However, the infinitesimal displacement is different from the virtual displacement in the rheonomic constraint system as shown in Figure 6.1. Furthermore, the infinitesimal change of the constraint function is also different from that of the variation as shown below,

$$\begin{aligned} df &= 2xdx + 2ydy - 2ldl = 0 \\ \delta f &= 2x\delta x + 2y\delta y - 2l\frac{\partial l}{\partial t}\delta t = 2x\delta x + 2y\delta y = 0 \end{aligned} \tag{4.31}$$

As a consequence, the virtual work done by the constraint force should vanish as follows,

$$\delta\mathcal{W} = \mathbf{C}\cdot\delta\mathbf{x} = C_x\delta x + C_y\delta y = -\frac{C}{l}(x\delta x + y\delta y) = 0 \tag{4.32}$$

which is valid for both holonomic-scleronomic and holonomic-rheonomic cases. However, the infinitesimal work done by the constraint force in the rheonomic case does not vanish as follows,

$$d\mathcal{W} = \mathbf{C}\cdot d\mathbf{x} = C_x dx + C_y dy = -\frac{C}{l}(xdx + ydy) \neq 0 \tag{4.33}$$

Example 4.2 Sliding Block on a Moving, Inclined Plane

Let us consider another example to clarify the difference between the virtual displacement and the infinitesimal actual change. This example is analogous to the example we dealt with previously, but the inclined plane is oscillating so that we have a holonomic-rheonomic constraint as shown in Figure 4.1.

The constraint function is expressed as

$$f(x, y, t) = y - ax + ac\sin\omega t - b = 0 \tag{4.34}$$

Thus, the constraint function contains the time variable explicitly so that the dynamical system is holonomic-rheonomic. The first variation of the constraint function can be given as

$$\delta f = \delta y - a\delta x = 0 \tag{4.35}$$

whereas the exact differential of the constraint function can be obtained as

$$df = dy - adx + ac\omega\cos\omega t dt = 0 \tag{4.36}$$

As a consequence, the virtual work done by the constraint force is

$$\delta\mathcal{W} = \mathbf{C}\cdot\delta\mathbf{x} = C_x\delta x + C_y\delta y = mg\cos^2\theta\,(\tan\theta + a)\,\delta x = 0 \tag{4.37}$$

On the other hand, the infinitesimal work done by the constraint force is

$$d\mathcal{W} = \mathbf{C} \cdot d\mathbf{x} = C_x dx + C_y dy = mg\cos^2\theta\,(\dot{x}\tan\theta + a\dot{x} - ac\omega\cos\omega t)\,dt \neq 0 \tag{4.38}$$

Via these illustrative examples, we can understand the underlying principal differences between the infinitesimal actual displacement and the imaginary virtual displacement.

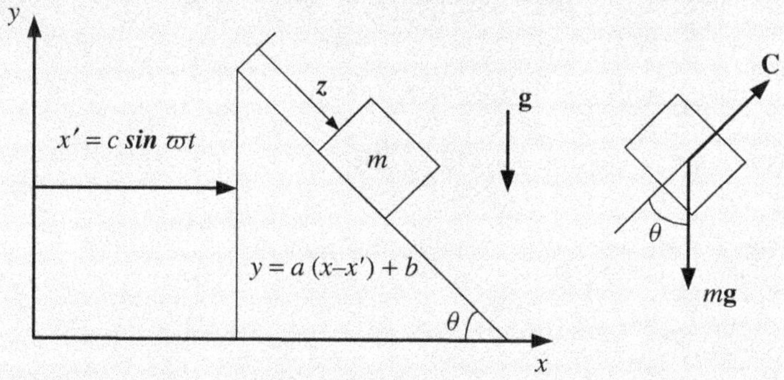

Figure 4.1. Sliding bock on a moving, inclined plane

4.2 VECTOR FORMALISM: NEWTONIAN MECHANICS IN N-BODY DYNAMICAL SYSTEMS

In the context of the Newtonian mechanics setting and the description in the Cartesian coordinate system with the vector formalism via the principle of balance of linear momentum, the governing equations of motion can be developed via different ways for the dynamic system. Next, invoking the Lagrange form of D'Alembert's principle and variational calculus, the equivalent scalar formalisms can also be readily introduced via transformation to generalized coordinates and introducing the descriptive scalar functions such as Lagrangian, Hamiltonian, Total Energy, and the like. The various developments are briefly highlighted next.

Principle of Balance of Linear Momentum Consider a system of N particles undergoing motion. The trajectory of the dynamic motion of the system is a curve of a vector-valued function of 3N-number of multivariables as a set, i.e.,

$$\mathbf{x}(t) = \left(\mathbf{x}^1(t), \ldots, \mathbf{x}^N(t)\right) : \bar{\mathcal{T}} \to \mathbb{R}^{3N},\ t \in \bar{\mathcal{T}} \tag{4.39}$$

Note that $\mathbf{x}(t) \in \mathcal{X}$ can be regarded as a point which belongs to a function space $\mathcal{X} \subset \mathbb{R}^{3N}$, a linear vector space, which is a set in Euclidean $3N$-space. For the case when there is no constraint limiting the motion of the N-body dynamical system, the function space $\mathcal{X}$ becomes the configuration space.

Newton's equations of motion for the i-th particle can be written as

$$\frac{d}{dt}\left(m_i\dot{\mathbf{x}}^i\right) = \mathbf{F}^i_{\text{appl}} + \mathbf{C}^i + \mathbf{F}^i_{\text{int}} \tag{4.40}$$

where $\mathbf{C}^i$ is the constraint force acting upon the i-th particle and $\mathbf{F}^i_{\text{appl}}(t) : \mathcal{T} =]t_0, t_1[\subset \mathbb{R} \to \mathbb{R}^3$ the applied force vector is a vector-valued function of one variable t and $\mathbf{F}^i_{\text{int}}$ denotes the internal force. Hence, Newton's second law of motion of the N-body dynamical system can be written as

$$\sum_{i=1}^{N} m_i \ddot{\mathbf{x}}^i = \sum_{i=1}^{N} \mathbf{F}^i_{\text{appl}}(t) + \sum_{i=1}^{N} \mathbf{C}^i \tag{4.41}$$

The above equation can be rewritten as

$$\sum_{i=1}^{N} \mathbf{F}^i_{\text{appl}} + \sum_{i=1}^{N} \mathbf{C}^i + \left(- \sum_{i=1}^{N} m_i \ddot{\mathbf{x}}^i \right) = \mathbf{0} \tag{4.42}$$

which is D'Alembert's principle. The constraint force $\mathbf{C}^i$ is a vector-valued function of multivariables, and is not dependent upon the velocity.

Principle of Virtual Work Taking the inner product between Equation (4.42) and the virtual displacement, we have the Lagrangian form of D'Alembert's principle as follows:

$$\sum_{i=1}^{N} \left(\mathbf{F}^i_{\text{appl}} - m_i \ddot{\mathbf{x}}^i \right) \cdot \delta \mathbf{x}^i = 0 \tag{4.43}$$

Note that the constraint force $\mathbf{C}$ is orthogonal to the virtual displacement. In favor of vector and matrix notation, Equation (4.43) can be rewritten as

$$\left(\mathbf{F}_{\text{appl}} - \mathbf{M}\ddot{\mathbf{x}} \right) \cdot \delta \mathbf{x} = 0 \tag{4.44}$$

Equation (4.44) can be rearranged as

$$\left(\mathbf{M}\ddot{\mathbf{x}}(t) \right) \cdot \delta \mathbf{x}(t) = \mathbf{F}_{\text{appl}} \cdot \delta \mathbf{x}(t) \tag{4.45}$$

Equation (4.45) is called the principle of virtual work in dynamics. It is to be noted that the term on the left-hand side in the above equation is the virtual work done by the inertia force and the term on the right is the virtual work done by the applied force. Unlike Equation (4.41), Equation (4.45) yields scalar values, in other words, work. However, the equation is still represented by vector quantities such as the position vector, the acceleration and the applied force in the Cartesian coordinate system. Note that due to the presence of constraints, the dependent variable $\mathbf{x}$ is not linearly independent. Hence, the virtual displacement $\delta \mathbf{x}$ is not arbitrary.

Remark 4.2.1 The differential, governing equations of motion can be derived from the Lagrangian form of D'Alembert's principle. Note that the governing equations one obtains by means of the Lagrangian form of D'Alembert's principle do not take the form of the Newtonian dynamical system, because constraint forces are not included. Furthermore, the resulting differential equations are represented in terms of Cartesian coordinate variables. Hence, the differential equations are still subject to constraint equations, as the Cartesian coordinate variables are not linearly independent in the presence of constraints.

Time Weighted Residual Form Taking the integration over the given closed interval of time $t \in \bar{\mathcal{T}} = [t_0, t_1]$ for Equation (4.44) leads to

$$\int_{t_0}^{t_1} \left(\mathbf{F}_{\text{appl}} - \mathbf{M}\ddot{\mathbf{x}}(t) \right) \cdot \delta \mathbf{x}(t)\, dt = 0 \tag{4.46}$$

which represents the Bubnov-Galerkin weighted residual form in time. Equation (4.46) is most frequently utilized in designing time integration approaches such as the time weighted-residual method which includes time finite element formulations (Belytschko and Hughes 1983; Sha et al. 2003; Tamma et al. 2000; Tamma et al. 2003, 2002, 2000; Wood 1990; Zhou and Tamma 2004a, c, 2006; Zhou et al. 2001, 2000; Zienkiewicz and Taylor 2006; Zienkiewicz et al. 2005, 1984).

4.3 SCALAR FORMALISMS: LAGRANGIAN AND HAMILTONIAN MECHANICS IN N-BODY DYNAMICAL SYSTEMS

We now turn attention to other alternative descriptions to the classical Newtonian mechanics setting and vector formalism described previously. In particular, we instead draw attention to scalar formalisms for N-body dynamical systems from the Lagrange form of D'Alembert's principle. Obviously, the scalar formalisms all yield the governing equations of motion which are equivalent to the classical Newtonian mechanics setting, and bring forth alternative perspectives and an improved physical insight in certain cases. It is very important to note that the various descriptive scalar functions have symmetries in the sense of Noether (Noether 1918).

Descriptive Scalar Function, the Lagrangian The virtual displacement in Euclidean $3N$-space can be given in terms of the variations of the generalized coordinate parameters as

$$\delta\mathbf{x} = \frac{\partial\mathbf{x}}{\partial\mathbf{q}} \cdot \delta\mathbf{q} \tag{4.47}$$

Equation (4.44) representing the principle of virtual work in dynamics can be recast into the equation

$$\left\{\left(\mathbf{F}_{\text{appl}} - \mathbf{M}\ddot{\mathbf{x}}(t)\right) \cdot \frac{\partial\mathbf{x}(t)}{\partial\mathbf{q}}\right\} \cdot \delta\mathbf{q} = 0 \tag{4.48}$$

As discussed in Chapter 3, the above equation leads to

$$\left[\frac{\partial\left\{\mathcal{K}(\mathbf{q},\dot{\mathbf{q}},t) - \mathcal{U}(\mathbf{q},t)\right\}}{\partial\mathbf{q}} - \frac{d}{dt}\left(\frac{\partial\mathcal{K}(\mathbf{q},\dot{\mathbf{q}},t)}{\partial\dot{\mathbf{q}}}\right)\right] \cdot \delta\mathbf{q} = 0 \tag{4.49}$$

Then, by using the Lagrangian $\mathcal{L}(\mathbf{q},\dot{\mathbf{q}},t) : TQ \times \mathcal{T} \rightarrow \mathbb{R}$, the above equation can be rewritten as

$$\left[\frac{\partial\mathcal{L}(\mathbf{q},\dot{\mathbf{q}},t)}{\partial\mathbf{q}} - \frac{d}{dt}\left(\frac{\partial\mathcal{L}(\mathbf{q},\dot{\mathbf{q}},t)}{\partial\dot{\mathbf{q}}}\right)\right] \cdot \delta\mathbf{q} = 0 \tag{4.50}$$

Since the virtual displacement $\delta\mathbf{q}$ is arbitrary and linearly independent in holonomic systems, we have

$$\frac{\partial\mathcal{L}(\mathbf{q},\dot{\mathbf{q}},t)}{\partial\mathbf{q}} - \frac{d}{dt}\left(\frac{\partial\mathcal{L}(\mathbf{q},\dot{\mathbf{q}},t)}{\partial\dot{\mathbf{q}}}\right) = \mathbf{0} \tag{4.51}$$

which is Lagrange's equation of motion. Note that Equation (4.50) can be expressed as

$$\delta\mathcal{L}(\mathbf{q},\dot{\mathbf{q}},t) - \frac{d}{dt}\left(\frac{\partial\mathcal{L}(\mathbf{q},\dot{\mathbf{q}},t)}{\partial\dot{\mathbf{q}}} \cdot \delta\mathbf{q}\right) = 0 \tag{4.52}$$

which is equivalent to the Lagrangian form of D'Alembert's principle given in Equation (4.44).

Descriptive Scalar Function, the Hamiltonian For the holonomic constraint N-body dynamical system, Equation (4.52) can be rewritten as

$$\delta \mathcal{L}(\mathbf{q}, \dot{\mathbf{q}}, t) - \frac{\partial \mathcal{L}(\mathbf{q}, \dot{\mathbf{q}}, t)}{\partial \dot{\mathbf{q}}} \cdot \delta \dot{\mathbf{q}} - \frac{d}{dt}\left(\frac{\partial \mathcal{L}(\mathbf{q}, \dot{\mathbf{q}}, t)}{\partial \dot{\mathbf{q}}}\right) \cdot \delta \mathbf{q} = 0 \tag{4.53}$$

By introducing the generalized momentum, the above equation can be rewritten as

$$\delta \mathcal{L}(\mathbf{q}, \dot{\mathbf{q}}, t) - \mathbf{p} \cdot \delta \dot{\mathbf{q}} - \dot{\mathbf{p}} \cdot \delta \mathbf{q} = 0 \tag{4.54}$$

By virtue of the Legendre transformation, we obtain the expression

$$\delta \mathcal{L}(\mathbf{q}, \dot{\mathbf{q}}, t) = \mathbf{p} \cdot \delta \dot{\mathbf{q}} + \dot{\mathbf{q}} \cdot \delta \mathbf{p} - \delta \mathcal{H}(\mathbf{q}, \mathbf{p}, t) \tag{4.55}$$

Upon substitution of the above equation into Equation (4.54), we obtain

$$\left(\dot{\mathbf{q}} - \frac{\partial \mathcal{H}(\mathbf{q}, \mathbf{p}, t)}{\partial \mathbf{p}}\right) \cdot \delta \mathbf{p} - \left(\dot{\mathbf{p}} + \frac{\partial \mathcal{H}(\mathbf{q}, \mathbf{p}, t)}{\partial \mathbf{q}}\right) \cdot \delta \mathbf{q} = 0 \tag{4.56}$$

which is a consequence of the Lagrangian form of D'Alembert's principle in the Hamiltonian mechanics framework. Since the virtual displacement $\delta \mathbf{q}$ and the virtual momentum $\delta \mathbf{p}$ are arbitrary in holonomic systems, we have

$$\dot{\mathbf{q}} - \frac{\partial \mathcal{H}(\mathbf{q}, \mathbf{p}, t)}{\partial \mathbf{p}} = \mathbf{0}, \quad \dot{\mathbf{p}} + \frac{\partial \mathcal{H}(\mathbf{q}, \mathbf{p}, t)}{\partial \mathbf{q}} = \mathbf{0} \tag{4.57}$$

which represent Hamilton's equations of motion.

Remark 4.3.1 On the other hand, if we introduce a different perspective, namely, the descriptive scalar function, the Total Energy $\mathcal{E}(\mathbf{q}, \dot{\mathbf{q}}, t) : TQ \times \mathcal{T} \rightarrow \mathbb{R}$ for the holonomic constraint N-body dynamical system, the Lagrangian form of D'Alembert's Principle can be transformed into

$$\left[2\frac{\partial \mathcal{K}(\mathbf{q}, \dot{\mathbf{q}}, t)}{\partial \mathbf{q}} - \frac{\partial \mathcal{E}(\mathbf{q}, \dot{\mathbf{q}}, t)}{\partial \mathbf{q}} - \frac{d}{dt}\left(\frac{\partial \mathcal{E}(\mathbf{q}, \dot{\mathbf{q}}, t)}{\partial \dot{\mathbf{q}}}\right)\right] \cdot \delta \mathbf{q} = 0 \tag{4.58}$$

As $\delta \mathbf{q}$ is arbitrary, Equation (4.58) gives the Total Energy representation of the equation of motion as below.

$$\frac{d}{dt}\left(\frac{\partial \mathcal{E}(\mathbf{q}, \dot{\mathbf{q}}, t)}{\partial \dot{\mathbf{q}}}\right) + \frac{\partial \mathcal{E}(\mathbf{q}, \dot{\mathbf{q}}, t)}{\partial \mathbf{q}} - 2\frac{\partial \mathcal{K}(\mathbf{q}, \dot{\mathbf{q}}, t)}{\partial \mathbf{q}} = \mathbf{0} \tag{4.59}$$

which is valid for the holonomic constraint N-body dynamical system. If the kinetic energy is independent of generalized coordinates, Equation (4.58) reduces to

$$\left[\frac{d}{dt}\left(\frac{\partial \mathcal{E}(\mathbf{q}, \dot{\mathbf{q}}, t)}{\partial \dot{\mathbf{q}}}\right) + \frac{\partial \mathcal{E}(\mathbf{q}, \dot{\mathbf{q}}, t)}{\partial \mathbf{q}}\right] \cdot \delta \mathbf{q} = 0 \tag{4.60}$$

As $\delta \mathbf{q}$ is arbitrary and linearly independent, we have

$$\frac{d}{dt}\left(\frac{\partial \mathcal{E}(\mathbf{q}, \dot{\mathbf{q}}, t)}{\partial \dot{\mathbf{q}}}\right) + \frac{\partial \mathcal{E}(\mathbf{q}, \dot{\mathbf{q}}, t)}{\partial \mathbf{q}} = \mathbf{0} \tag{4.61}$$

which is valid for the holonomic constraint N-body dynamical system whose kinetic energy is independent of generalized coordinates (see Chapter 6 for details).

Scalar Formalisms in the Cartesian Coordinate System Consider the motion of the dynamical system, which is subjected to holonomic constraints. Before we obtain the Lagrangian form of D'Alembert's principle for holonomic constraints systems, we have

$$\mathbf{M}\ddot{\mathbf{x}} \cdot \delta\mathbf{x} = \mathbf{F}_{\mathrm{appl}} \cdot \delta\mathbf{x} + \mathbf{C}(\mathbf{x}, t) \cdot \delta\mathbf{x} \tag{4.62}$$

where $\mathbf{F}_{\mathrm{appl}}$ and $\mathbf{C}(\mathbf{x}, t)$ denote the applied force and the constraint force. As addressed previously, by use of the Lagrange multiplier and the constraint function, Equation (4.62) can be written as

$$\left(\mathbf{F}_{\mathrm{appl}} - \mathbf{M}\ddot{\mathbf{x}}\right) \cdot \delta\mathbf{x} = -\delta\left(\boldsymbol{\lambda} \cdot \mathbf{f}(\mathbf{x}, t)\right) \tag{4.63}$$

The Lagrangian:

The term on the left-hand side of Equation (4.62) can be written as

$$\mathbf{M}\ddot{\mathbf{x}} \cdot \delta\mathbf{x} = \frac{d}{dt}(\mathbf{M}\dot{\mathbf{x}} \cdot \delta\mathbf{x}) - \delta\mathcal{K}(\dot{\mathbf{x}}) \tag{4.64}$$

where $\boldsymbol{\lambda}$ is the Lagrange multiplier. By introducing the Lagrangian, Equation (4.63) can be written as

$$\delta\mathcal{L}(\mathbf{x}, \dot{\mathbf{x}}, t) + \delta\left(\boldsymbol{\lambda} \cdot \mathbf{f}(\mathbf{x}, t)\right) - \frac{d}{dt}(\mathbf{M}\dot{\mathbf{x}} \cdot \delta\mathbf{x}) = 0 \tag{4.65}$$

where $\mathcal{L}(\mathbf{x}, \dot{\mathbf{x}}, t) = \mathcal{K}(\dot{\mathbf{x}}) - \mathcal{U}(\mathbf{x}, t)$. Further, the above equation can be expressed as

$$\delta\tilde{\mathcal{L}}(\mathbf{x}, \dot{\mathbf{x}}, t) - \frac{d}{dt}(\mathbf{M}\dot{\mathbf{x}} \cdot \delta\mathbf{x}) = 0 \tag{4.66}$$

where $\tilde{\mathcal{L}}(\mathbf{x}, \dot{\mathbf{x}}, t) = \mathcal{K}(\dot{\mathbf{x}}) - \mathcal{U}(\mathbf{x}, t) + \boldsymbol{\lambda} \cdot \mathbf{f}(\mathbf{x}, t)$, and it is referred to as the augmented Lagrangian. Therefore, it is easy to show that the above equation leads to the following

$$\left\{\frac{\partial\tilde{\mathcal{L}}(\mathbf{x}, \dot{\mathbf{x}}, t)}{\partial\mathbf{x}} - \frac{d}{dt}\left(\frac{\partial\tilde{\mathcal{L}}(\mathbf{x}, \dot{\mathbf{x}}, t)}{\partial\dot{\mathbf{x}}}\right)\right\} \cdot \delta\mathbf{x} = 0 \tag{4.67}$$

As a consequence, the above equation can be rewritten as

$$\left\{\frac{\partial\mathcal{L}(\mathbf{x}, \dot{\mathbf{x}}, t)}{\partial\mathbf{x}} - \frac{d}{dt}\left(\frac{\partial\mathcal{L}(\mathbf{x}, \dot{\mathbf{x}}, t)}{\partial\dot{\mathbf{x}}}\right) + \boldsymbol{\lambda} \cdot \Delta\mathbf{f}(\mathbf{x}, t)\right\} \cdot \delta\mathbf{x} = 0 \tag{4.68}$$

where the term inside the curly bracket is identical to Newton's equations of motion in the Cartesian coordinate system. The Lagrange multiplier should be chosen so as to satisfy Equation (4.68).

The Hamiltonian:

Let us define the Hamiltonian in the Cartesian coordinate system as

$$\mathcal{H}(\mathbf{x}, \mathbf{p}, t) = \mathcal{K}(\mathbf{p}) - \mathcal{U}(\mathbf{x}, t) \tag{4.69}$$

Then, the Legendre transformation in the Cartesian coordinate system can be given by

$$\mathcal{L}(\mathbf{x}, \dot{\mathbf{x}}, t) = \mathbf{p} \cdot \dot{\mathbf{x}} - \mathcal{H}(\mathbf{x}, \mathbf{p}, t) \tag{4.70}$$

Hence, Equation (4.65) can be written as

$$\delta\left(\mathbf{p}\cdot\dot{\mathbf{x}}-\mathcal{H}\left(\mathbf{x},\mathbf{p},t\right)\right)+\delta\left(\boldsymbol{\lambda}\cdot\mathbf{f}\left(\mathbf{x},t\right)\right)-\frac{d}{dt}\left(\mathbf{p}\cdot\delta\mathbf{x}\right)=0 \tag{4.71}$$

which can be rewritten as

$$\delta\left(\mathbf{p}\cdot\dot{\mathbf{x}}-\tilde{\mathcal{H}}\left(\mathbf{x},\mathbf{p},t\right)\right)-\frac{d}{dt}\left(\mathbf{p}\cdot\delta\mathbf{x}\right)=0 \tag{4.72}$$

where $\tilde{\mathcal{H}}\left(\mathbf{x},\mathbf{p},t\right)=\mathcal{K}\left(\mathbf{p}\right)+\mathcal{U}\left(\mathbf{x},t\right)-\boldsymbol{\lambda}\cdot\mathbf{f}\left(\mathbf{x},t\right)$, and it is referred to as the augmented Hamiltonian. Therefore, it is easy to show that the above equation leads to the following

$$\left(\dot{\mathbf{x}}-\frac{\partial\tilde{\mathcal{H}}\left(\mathbf{x},\mathbf{p},t\right)}{\partial\mathbf{p}}\right)\cdot\delta\mathbf{p}-\left(\dot{\mathbf{p}}+\frac{\partial\tilde{\mathcal{H}}\left(\mathbf{x},\mathbf{p},t\right)}{\partial\mathbf{x}}\right)\cdot\delta\mathbf{x}=0 \tag{4.73}$$

which leads to

$$\left(\dot{\mathbf{x}}-\frac{\partial\mathcal{H}\left(\mathbf{x},\mathbf{p},t\right)}{\partial\mathbf{p}}\right)\cdot\delta\mathbf{p}-\left(\dot{\mathbf{p}}+\frac{\partial\mathcal{H}\left(\mathbf{x},\mathbf{p},t\right)}{\partial\mathbf{x}}-\boldsymbol{\lambda}\cdot\Delta\mathbf{f}\left(\mathbf{x},t\right)\right)\cdot\delta\mathbf{x}=0 \tag{4.74}$$

The Total Energy:
In addition, we define the Total Energy in the Cartesian coordinate system as

$$\mathcal{E}\left(\mathbf{x},\dot{\mathbf{x}},t\right)=\mathcal{K}\left(\dot{\mathbf{x}}\right)+\mathcal{U}\left(\mathbf{x},t\right) \tag{4.75}$$

Then, the relation between the Lagrangian and the Total Energy in the Cartesian coordinate system can be given by

$$\mathcal{L}\left(\mathbf{x},\dot{\mathbf{x}},t\right)=2\mathcal{K}\left(\dot{\mathbf{x}}\right)-\mathcal{E}\left(\mathbf{x},\dot{\mathbf{x}},t\right) \tag{4.76}$$

Hence, Equation (4.63) can be written as

$$\delta\left(2\mathcal{K}\left(\dot{\mathbf{x}}\right)-\mathcal{E}\left(\mathbf{x},\dot{\mathbf{x}},t\right)\right)+\delta\left(\boldsymbol{\lambda}\cdot\mathbf{f}\left(\mathbf{x},t\right)\right)-\frac{d}{dt}\left(\mathbf{M}\dot{\mathbf{x}}\cdot\delta\mathbf{x}\right)=0 \tag{4.77}$$

which can be rewritten as

$$\delta\left(2\mathcal{K}\left(\dot{\mathbf{x}}\right)-\tilde{\mathcal{E}}\left(\mathbf{x},\dot{\mathbf{x}},t\right)\right)-\frac{d}{dt}\left(\mathbf{M}\dot{\mathbf{x}}\cdot\delta\mathbf{x}\right)=0 \tag{4.78}$$

where $\tilde{\mathcal{E}}\left(\mathbf{x},\dot{\mathbf{x}},t\right)=\mathcal{K}\left(\dot{\mathbf{x}}\right)+\mathcal{U}\left(\mathbf{x},t\right)-\boldsymbol{\lambda}\cdot\mathbf{f}\left(\mathbf{x},t\right)$, and it is referred to as the augmented Total Energy. Therefore, it is easy to show that the above equation leads to the following

$$\left\{-\frac{\partial\tilde{\mathcal{E}}(\mathbf{x},\dot{\mathbf{x}},t)}{\partial\mathbf{x}}-\frac{d}{dt}\left(\frac{\partial\tilde{\mathcal{E}}(\mathbf{x},\dot{\mathbf{x}},t)}{\partial\dot{\mathbf{x}}}\right)\right\}\cdot\delta\mathbf{x}=0 \tag{4.79}$$

which leads to

$$\left\{-\frac{\partial\mathcal{E}(\mathbf{x},\dot{\mathbf{x}},t)}{\partial\mathbf{x}}-\frac{d}{dt}\left(\frac{\partial\mathcal{E}(\mathbf{x},\dot{\mathbf{x}},t)}{\partial\dot{\mathbf{x}}}\right)+\boldsymbol{\lambda}\cdot\Delta\mathbf{f}\left(\mathbf{x},t\right)\right\}\cdot\delta\mathbf{x}=0 \tag{4.80}$$

where the term inside the curly bracket is identical to Newton's equations of motion in the Cartesian coordinate system. The Lagrange multiplier should be chosen to satisfy Equation (3.146).

Remark 4.3.2 Upon taking the integration in the time domain for a specified time interval, Equations (4.50), (4.56), and (4.60) respectively, lead to the weighted residual form in time for the different frameworks with scalar formalisms. Hence, one can utilize the time weighted residual formulations for designing time integration approaches such as the time weighted-residual method which includes time finite element formulations (Belytschko and Hughes 1983; Sha et al. 2003; Tamma et al. 2003, 2002, 2000; Wood 1990; Zhou and Tamma 2004a,c, 2006; Zhou et al. 2001, 2000; Zienkiewicz and Taylor 2006; Zienkiewicz et al. 2005, 1984).

EXERCISES

4.1. Consider the spring-mass system shown in Figure 3.18. Use the principle of virtual work given in Equation (4.21) to find the displacement at the static equilibrium position.

4.2. Consider the simple planar pendulum with an oscillating hinge point as shown in Figure 3.19. Use the principle of virtual work given in Equation (4.45) to find the governing, differential equation of motion.

4.3. Use the principle of virtual work given in Equation (4.45) to find the governing, differential equation for a sliding block on a moving, inclined plane in Example 4.2.

4.4. Consider a spring-mass system discussed in Exercise 3.3 of Chapter 3. Use the principle of virtual work given in Equation (4.21) to find the displacement δ_{st} at the static equilibrium position, when $A = 0$ and $\mathbf{g} \neq \mathbf{0}$.

4.5. Reconsider the spring-mass system in the above exercise. Use the principle of virtual work given in Equation (4.45) to find the governing equation of motion for the undamped spring-mass system, i.e. $A \neq 0$ and $\mathbf{g} \neq \mathbf{0}$.

4.6. For the planar motion as shown in Figure 3.22, derive the governing, differential equation of motion by use of the principle of virtual work given in Equation (4.45).

4.7. For the damped spring-mass dynamical system shown in Figure 3.24, derive the governing, differential equation of motion by use of the principle of virtual work given in Equation (4.45).

CHAPTER FIVE

HAMILTON'S PRINCIPLE AND HAMILTON'S LAW OF VARYING ACTION

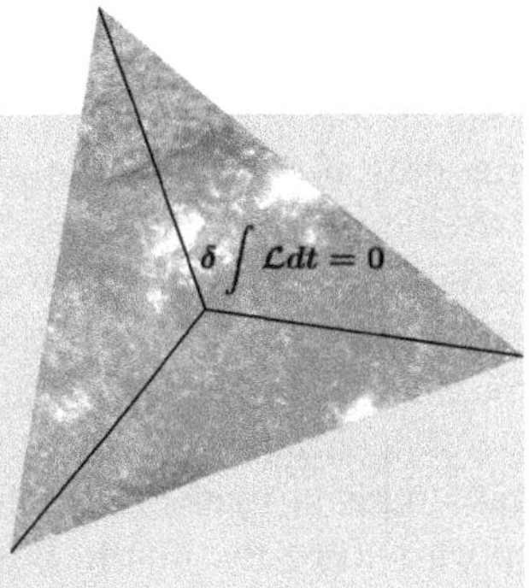

Previously in Chapter 4, after first establishing the relations stemming from the balance of linear momentum and leading to the principle of virtual work in dynamics, directly starting from this principle we then described the development of the governing equations for N-body dynamical systems suitable for the subsequent time discretization which is presented in the later chapters of this book. Here in Chapter 5, in contrast, we provide an alternative formulation and formalism directly starting from Hamilton's principle, and alternatively, also via Hamilton's law of varying action (due to certain concerns as related to logic inconsistencies that exist for transient situations with Hamiltons principle as highlighted subsequently). Nonetheless, the present developments also lead to the same and/or equivalent representations as in the previous chapter, but are formulated from different perspectives.

5.1 INTRODUCTION

It is well known that Newton (1642–1727) first described the motion of a particle by its vector quantities within the vectorial mechanics framework. On the other hand, Leibniz (1646–1716) first considered taking advantage of scalar quantities for the particle, that is, the energy quantities within the analytical mechanics framework. Leibniz's calculus leading to the principle of virtual work to describe the static equilibrium state of the body appears to have been first established by Bernoulli (1667–1748). Extensions of the concept, namely, the principle of virtual work appears to be first applied to dynamic problems by D'Alembert (1717–1783), and is the well known D'Alembert's principle. Euler (1707–1783), whom Bernoulli appears to have tutored in mathematics, gave birth to the modern calculus of variations, and Lagrange (1736–1813) first applied the calculus of variations to mechanics. Subsequently, Hamilton (1805–1865) used D'Alembert's principle and integral principle to drive Hamilton's principle and Hamilton's law of varying action. Hamilton's principle was later applied to optics by Hamilton. Historically it has been known that Hamilton's principle was reformulated and the name seems to have been coined later by Jacobi (1804–1851), who appears to be the first to apply Hamilton's principle to mechanics (Nakane and Fraser 2002).

Hamilton first defined the principal function (Hamilton 1834a) and pronounced that its first variation has the double property (Hamilton 1835, page 99), of giving the differential equations of motion for any transformed coordinates when the extreme positions are regarded as fixed, and of giving the integrals of those differential equations when the extreme positions are treated as varying.

Advances in Computational Dynamics of Particles, Materials and Structures, First Edition. Jason Har and Kumar K. Tamma.

The first property leads to Hamilton's principle (Hamilton 1835) which is variational; however, it possesses a logic inconsistency as pointed out in (Simkins 1981) regarding forcing the second end point in time to be also prescribed (Hamilton's principle is true only if the configurations at the two given instants are known). Langhaar (Langhaar 1962, page 235) also explained that Hamilton's principle is true only if the configurations at the two given instants are known. However, the configurations at both extreme positions can not be assumed to be simultaneously known (Bailey 1975b, 1982), since only initial conditions are known. The one extreme position at $t = t_0$ can be prescribed, whilst the other extreme position at $t = t_1$ cannot be prescribed. The principal quantities at the other instant can be sought by appropriate time marching schemes. Hamilton's principle is a variational principle which yields a differential equation at the extremum of the functional. Calculating either a minimum or a maximum in mathematics has for a long time attracted many scientists and mathematicians. Aristotle appears to be the first to postulate that nature requires minimum energy (Bailey 2004). This philosophy influenced many mathematicians in the 17th through the 19th century. Based upon the philosophy, which is somewhat metaphysical rather than physical, Fermat proposed the principle of least time (Yourgrau and Mandelstam 1968) in optics. Maupertuis applied the philosophy to dynamics, claiming the principle of least action (Yourgrau and Mandelstam 1968). It has been well known that his principle was never proved. Furthermore, it appears that many scholars such as Jacobi, Euler, and Lagrange investigated the principle of least action (Osgood 1937) for dynamics. These principles are restricted to the condition that the two extreme points are not too far apart in time (Lanczos 1970) (Gelfand and Fomin 1963, page 159). If the time interval is over one-half of the period, the resulting real action does not extremize the action (Gelfand and Fomin 1963; Lanczos 1970; Smith and Smith 1974). Despite such controversy, one readily comes to the popular statement, as in the modern calculus of variations, that the real solution in a dynamical system comes out from minimizing a physical quantity or finding a stationary point of the quantity, and the underlying principle is not based on any physical concept, but a metaphysical.

The second property appears to have motivated Hamilton to derive the so-called Hamilton's law of varying action (Hamilton 1834a, page 307); (Bailey 1975a) that is not variational, and it does not suffer from requiring one to take such an assumption as in the first property. Hence, Hamilton's law of varying action seems more natural to enabling a consistent numerical discretization process to be formulated. We briefly discuss both in this chapter.

In addition to the descriptive functions, namely, the Lagrangian and the Hamiltonian, we also briefly comment upon other possible descriptive functions such as the Total Energy as a viable alternative to the classical setting. We defer to more detailed discussions in a later chapter.

5.2 VARIATION OF THE PRINCIPAL FUNCTION

In the literature (Ardema 2005; Greenwood 1977; Hamel 1948; Meirovitch 2004; Pars 1965; Rosenberg 1977), it is asserted that Hamilton's principle can be deduced from the Lagrangian form of D'Alembert's principle. In this section, we show how to obtain not only Hamilton's principle but also Hamilton's law of varying action from the variation of the principal function (Hamilton 1834a,b, 1835) or the Lagrangian form of D'Alembert's principle.

With the notion that constraint forces in holonomic systems make no contribution to the virtual work, the consequence of the Lagrangian form of D'Alembert's principle (Greenwood 1977; Rosenberg 1977) given in Equation (3.76) can be rewritten as

$$\mathbf{M}\ddot{\mathbf{x}} \cdot \frac{\partial \mathbf{x}}{\partial \mathbf{q}} \cdot \delta \mathbf{q} = \mathbf{F}_{\mathrm{appl}} \cdot \frac{\partial \mathbf{x}}{\partial \mathbf{q}} \cdot \delta \mathbf{q} \tag{5.1}$$

In view of Equation (3.89), the term on the left-hand side in the above can be rewritten as

$$\mathbf{M}\ddot{\mathbf{x}} \cdot \frac{\partial \mathbf{x}}{\partial \mathbf{q}} \cdot \delta \mathbf{q} = \left[\frac{d}{dt} \left(\frac{\partial \mathcal{K}(\mathbf{q}, \dot{\mathbf{q}}, t)}{\partial \dot{\mathbf{q}}} \right) - \frac{\partial \mathcal{K}(\mathbf{q}, \dot{\mathbf{q}}, t)}{\partial \mathbf{q}} \right] \cdot \delta \mathbf{q} \tag{5.2}$$

By virtue of the product rule, the above equation can be written as

$$\mathbf{M}\ddot{\mathbf{x}} \cdot \frac{\partial \mathbf{x}}{\partial \mathbf{q}} \cdot \delta \mathbf{q} = \frac{d}{dt} \left(\frac{\partial \mathcal{K}(\mathbf{q}, \dot{\mathbf{q}}, t)}{\partial \dot{\mathbf{q}}} \cdot \delta \mathbf{q} \right) - \frac{\partial \mathcal{K}(\mathbf{q}, \dot{\mathbf{q}}, t)}{\partial \dot{\mathbf{q}}} \cdot \delta \dot{\mathbf{q}} - \frac{\partial \mathcal{K}(\mathbf{q}, \dot{\mathbf{q}}, t)}{\partial \mathbf{q}} \cdot \delta \mathbf{q} \tag{5.3}$$

Note that Equation (5.3) can be rewritten as

$$\mathbf{M}\ddot{\mathbf{x}} \cdot \frac{\partial \mathbf{x}}{\partial \mathbf{q}} \cdot \delta \mathbf{q} = \frac{d}{dt} \left(\frac{\partial \mathcal{K}(\mathbf{q}, \dot{\mathbf{q}}, t)}{\partial \dot{\mathbf{q}}} \cdot \delta \mathbf{q} \right) - \delta \mathcal{K}(\mathbf{q}, \dot{\mathbf{q}}, t) \tag{5.4}$$

The above equation is called Hamel's central principle (Hamel 1948; Rosenberg 1977). According to Hamel, the above equation states that the virtual work done by the inertial forces is equal to the time rate of change of the work done by the momentum, diminished by the virtual change in kinetic energy.

The term on the right-hand side of Equation (5.1) corresponding to the virtual work done by the applied force can be defined as

$$\mathbf{F}_{\text{appl}} \cdot \frac{\partial \mathbf{x}}{\partial \mathbf{q}} \cdot \delta \mathbf{q} = \mathbf{Q}_{\text{appl}} \cdot \delta \mathbf{q} = \delta \mathcal{W} \tag{5.5}$$

where $\mathbf{Q}_{\text{appl}}$ represents the applied force, and $\delta \mathcal{W}$ is the virtual work done by the applied force. In addition, in terms of the force function given in Equation (3.79), we can also write

$$\mathbf{F}_{\text{appl}} \cdot \frac{\partial \mathbf{x}}{\partial \mathbf{q}} \cdot \delta \mathbf{q} = -\nabla \mathcal{U}(\mathbf{q}, t) \cdot \delta \mathbf{q} = -\delta \mathcal{U}(\mathbf{q}, t) \tag{5.6}$$

Consequently, upon substitution of Equation (5.6) and (5.4) into Equation (5.1), we obtain

$$\delta \left(\mathcal{K}(\mathbf{q}, \dot{\mathbf{q}}, t) - \mathcal{U}(\mathbf{q}, t) \right) - \frac{d}{dt} \left(\frac{\partial \mathcal{K}(\mathbf{q}, \dot{\mathbf{q}}, t)}{\partial \dot{\mathbf{q}}} \cdot \delta \mathbf{q} \right) = 0 \tag{5.7}$$

Integrating the above equation between two terminal instants leads to

$$\int_{t_0}^{t_1} \delta \left(\mathcal{K}(\mathbf{q}, \dot{\mathbf{q}}, t) - \mathcal{U}(\mathbf{q}, t) \right) dt - \left. \frac{\partial \mathcal{K}(\mathbf{q}, \dot{\mathbf{q}}, t)}{\partial \dot{\mathbf{q}}} \cdot \delta \mathbf{q} \right|_{t_0}^{t_1} = 0 \tag{5.8}$$

For the case of holonomic systems, operations of variation and integration can be interchanged (Greenwood 1977; Osgood 1937; Rosenberg 1977). Therefore, the first term in the above equation can be written as (see (Osgood 1937))

$$\begin{aligned} \int_{t_0}^{t_1} \delta \left(\mathcal{K}(\mathbf{q}, \dot{\mathbf{q}}, t) - \mathcal{U}(\mathbf{q}, t) \right) dt &= \delta \int_{t_0}^{t_1} \left(\mathcal{K}(\mathbf{q}, \dot{\mathbf{q}}, t) - \mathcal{U}(\mathbf{q}, t) \right) dt \\ &\quad - \int_{t_0}^{t_1} \left(\mathcal{K}(\mathbf{q}, \dot{\mathbf{q}}, t) - \mathcal{U}(\mathbf{q}, t) \right) d\delta t \end{aligned} \tag{5.9}$$

Since the time variation should be zero, the last term in the right-hand side of Equation (5.9) may vanish. As a consequence, the above equation yields the variation of the principle action (Hamilton 1834a). In the literature, the principal function $S(\mathbf{q})$ of motion of a dynamical system, which is defined below, is often called the action (Feynman 2006) or Hamilton's integral (Osgood 1937).

$$S(\mathbf{q}) = \int_{t_0}^{t_1} \mathcal{L}(\mathbf{q}, \dot{\mathbf{q}}, t)\, dt \tag{5.10}$$

Therefore, Equation (5.8) results in the following form

$$\delta \int_{t_0}^{t_1} \mathcal{L}(\mathbf{q}, \dot{\mathbf{q}}, t)\, dt - \frac{\partial \mathcal{L}(\mathbf{q}, \dot{\mathbf{q}}, t)}{\partial \dot{\mathbf{q}}} \cdot \delta \mathbf{q} \Bigg|_{t_0}^{t_1} = 0 \tag{5.11}$$

Double Property: Hamilton's Principle and Hamilton's Law of Varying Action Hamilton first presented the above integral form with time as an independent variable in his celebrated paper (Hamilton 1834a) on the general method in particle dynamics as shown in Equation (5.11). Hamilton coined the equation given in Equation (5.11) the law of varying action. The above equation is often called Hamilton's law of varying action (Bailey 1975a, b). As the second property of the variation of the principal function, the above equation reduces to Hamilton's principle in the case that the configuration at the end points is held fixed (prescribed) (Hamilton 1835, page 99), because the virtual displacement at the end points vanishes.

In summary, for holonomic constraint systems the variation of the principal function yields

$$\begin{aligned} \delta \int_{t_0}^{t_1} \mathcal{L}(\mathbf{q}, \dot{\mathbf{q}}, t)\, dt &= \int_{t_0}^{t_1} \left[\frac{\partial \mathcal{L}(\mathbf{q}, \dot{\mathbf{q}}, t)}{\partial \mathbf{q}} - \frac{d}{dt}\left(\frac{\partial \mathcal{L}(\mathbf{q}, \dot{\mathbf{q}}, t)}{\partial \dot{\mathbf{q}}} \right) \right] \cdot \delta \mathbf{q}\, dt \\ &\quad + \frac{\partial \mathcal{L}(\mathbf{q}, \dot{\mathbf{q}}, t)}{\partial \dot{\mathbf{q}}} \cdot \delta \mathbf{q} \Bigg|_{t_0}^{t_1} \end{aligned} \tag{5.12}$$

Hamilton's law of varying action given in Equation (5.11) reads as

$$\begin{aligned} \delta \int_{t_0}^{t_1} \mathcal{L}(\mathbf{q}, \dot{\mathbf{q}}, t)\, dt - \frac{\partial \mathcal{L}(\mathbf{q}, \dot{\mathbf{q}}, t)}{\partial \dot{\mathbf{q}}} \cdot \delta \mathbf{q} \Bigg|_{t_0}^{t_1} \\ = \int_{t_0}^{t_1} \left[\frac{\partial \mathcal{L}(\mathbf{q}, \dot{\mathbf{q}}, t)}{\partial \mathbf{q}} - \frac{d}{dt}\left(\frac{\partial \mathcal{L}(\mathbf{q}, \dot{\mathbf{q}}, t)}{\partial \dot{\mathbf{q}}} \right) \right] \cdot \delta \mathbf{q}\, dt = 0 \end{aligned} \tag{5.13}$$

We observe that Hamilton's law of varying action in Equation (5.13) results from taking the time integral of the consequence of the Lagrangian form of D'Alembert's principle given in Equation (3.94) over the time interval of interest. On the other hand, Hamilton's principle states that

$$\delta \int_{t_0}^{t_1} \mathcal{L}(\mathbf{q}, \dot{\mathbf{q}}, t)\, dt = \int_{t_0}^{t_1} \left[\frac{\partial \mathcal{L}(\mathbf{q}, \dot{\mathbf{q}}, t)}{\partial \mathbf{q}} - \frac{d}{dt}\left(\frac{\partial \mathcal{L}(\mathbf{q}, \dot{\mathbf{q}}, t)}{\partial \dot{\mathbf{q}}} \right) \right] \cdot \delta \mathbf{q}\, dt = 0 \tag{5.14}$$

under the restriction that $\delta \mathbf{q}(t_0) = \delta \mathbf{q}(t_1) = \mathbf{0}$. This is a variational principle applicable to field problems including non-mechanical problems as long as the Lagrangian is defined in the sense of classical mechanics. Therefore, we conclude that both Hamilton's principle and Hamilton's law of varying action provide an alternative formulation to derive a governing equation, as they follow directly starting from the Lagrangian form of D'Alembert's principle.

It should be noted that Hamilton's law of varying action (Hamilton 1834a) is not a variational principle, whereas Hamilton's principle is a variational principle and is the cornerstone of variational principles in dynamics.

5.3 CALCULUS OF VARIATIONS

The calculus of variations is a formal mathematical tool to find an optimal differential equation, leading to the optimal solution for the problem of interest. Hamilton's principle is based upon the calculus of variations. Hence, Hamilton's principle enables one to obtain the differential equation of an N-body dynamical system, which is called the Euler-Lagrange equation of motion. Before examining Hamilton's principle in more detail, we briefly highlight the calculus of variations and some historical examples.

Illustrative Examples First, we consider the problem to find a path given by a function which satisfies the condition that the first variation of the given functional becomes zero at the point at which the extremum of the functional exists. The variation, denoted by the operator δ, varies between the two points.

Example 5.1 Euler-Lagrange Equations of Motion

We consider a typical variational problem with a single variable. Consider a path function, often called a trial function, $y = y(x)$ with the prescribed boundary conditions, that is, (x_1, y_1) and (x_2, y_2). Assume that between the two points the varied path y^* is given as

$$y^*(x, \varepsilon) = y(x) + \varepsilon\eta(x) \tag{5.15}$$

where ε is an arbitrary parameter and $\eta(x)$ is an admissible function satisfying the homogeneous conditions at the two points as in

$$\eta(x_1) = \eta(x_2) = 0 \tag{5.16}$$

Note that $\varepsilon\eta(x)$ is the first variation of the function $y = y(x)$, denoted by δy. Due to the prescribed boundary conditions, we therefore have

$$\delta y(x_1) = \delta y(x_2) = 0 \tag{5.17}$$

Consider the function

$$F = F(y, y', x), \quad y' = \frac{dy}{dx} \tag{5.18}$$

Next, consider the following functional $I(y)$ with an argument y,

$$I(y) = \int_{x_1}^{x_2} F(y, y', x)\, dx, \quad y' = \frac{dy}{dx} \tag{5.19}$$

The challenge is to find the true path between the two points, satisfying the functional at the extremum point. In order for the functional to be an extremum at the point, the first variation of the functional at the point should be zero. In addition, if the second variation of the functional is positive, the extremum becomes the minimum. On the contrary, if the second variation is negative, the extremum becomes the maximum.

A varied functional, denoted by I_ε, may be expanded by the Taylor's series expansion theorem

$$I_\varepsilon = I(y + \varepsilon\eta) = I_\varepsilon|_{\varepsilon=0} + \varepsilon \left.\frac{dI_\varepsilon}{d\varepsilon}\right|_{\varepsilon=0} + \frac{\varepsilon^2}{2!} \left.\frac{d^2 I_\varepsilon}{d\varepsilon^2}\right|_{\varepsilon=0} + O\left(\varepsilon^3\right) \tag{5.20}$$

The above equation can be recast in the form

$$I(y+\varepsilon\eta) = I(y) + \varepsilon\frac{\partial I}{\partial y}\eta + \frac{\varepsilon^2}{2!}\frac{\partial^2 I}{\partial y^2}\eta^2 + O\left(\eta^3\right) \tag{5.21}$$

Since the varied functional can be given by

$$I(y+\delta y) = I(y) + \delta I(y) + \frac{1}{2!}\delta^2 I(y) + \dots \tag{5.22}$$

the first variation of the functional is given by

$$\delta I(y) = \varepsilon\left(\left.\frac{dI(y+\varepsilon\eta)}{d\varepsilon}\right|_{\varepsilon=0}\right) = \frac{\partial I}{\partial y}\varepsilon\eta = \frac{\partial I}{\partial y}\delta y \tag{5.23}$$

In order for the varied functional I_ε to have an extremal in the vicinity of $\varepsilon = 0$, the following should be satisfied

$$\left.\frac{dI(y+\varepsilon\eta)}{d\varepsilon}\right|_{\varepsilon=0} = 0 \tag{5.24}$$

which is called the Gâteaux variation. Since the admissible function $\eta(x)$ is arbitrary within the interval, Equation (5.24) yields

$$\frac{\partial I(y)}{\partial y} = 0 \tag{5.25}$$

which is known to be the necessary condition for the extremal of the varied functional. In view of the relation

$$I(y+\varepsilon\eta) - I(y) = \int_{x_1}^{x_2} F\left(y+\varepsilon\eta, y'+\varepsilon\eta', x\right)dx - \int_{x_1}^{x_2} F\left(y, y', x\right)dx \tag{5.26}$$

and the expansion of the integrand of the varied functional

$$F\left(y+\delta y, y'+\delta y', x\right) = F\left(y, y', x\right) + \frac{\partial F}{\partial y}\delta y + \frac{\partial F}{\partial y'}\delta y' + O\left(\varepsilon^2\right), \tag{5.27}$$

where $O\left(\varepsilon^2\right)$ denotes the second order terms of the parameter. The first variation of the functional is defined as

$$\begin{aligned}\delta I(y) &= \left.\int_{x_1}^{x_2}\left(\frac{\partial F}{\partial(y+\varepsilon\eta)}\frac{d(y+\varepsilon\eta)}{d\varepsilon} + \frac{\partial F}{\partial(y+\varepsilon\eta)'}\frac{d(y+\varepsilon\eta)'}{d\varepsilon}\right)dx\right|_{\varepsilon=0} \\ &= \int_{x_1}^{x_2}\left(\frac{\partial F}{\partial y}\delta y + \frac{\partial F}{\partial y'}\delta y'\right)dx\end{aligned} \tag{5.28}$$

The product rule leads to

$$\int_{x_1}^{x_2}\frac{\partial F}{\partial y'}\delta y' dx = \int_{x_1}^{x_2}\frac{d}{dx}\left(\frac{\partial F}{\partial y'}\delta y\right)dx - \int_{x_1}^{x_2}\frac{d}{dx}\left(\frac{\partial F}{\partial y'}\right)\delta y dx \tag{5.29}$$

Substituting Equation (5.29) into Equation (5.28) yields

$$\delta I(y) = \int_{x_1}^{x_2} \left[\frac{\partial F}{\partial y} - \frac{d}{dx}\left(\frac{\partial F}{\partial y'}\right) \right] \delta y \, dx + \left. \frac{\partial F}{\partial y'} \delta y \right|_{x_1}^{x_2} = 0 \tag{5.30}$$

Finally, we have the differential equation, which is a necessary condition that the functional is stationary, i.e.,

$$\frac{\partial F}{\partial y} - \frac{d}{dx}\left(\frac{\partial F}{\partial y'}\right) = 0 \tag{5.31}$$

The above differential equation is called the Euler-Lagrange equation.

Example 5.2 Shortest Distance Problem

As a typical example in the calculus of variations, we introduce the shortest distance problem. The Pythagorean theorem states that the length of the hypotenuse of a infinitesimal right-angled triangle in the $x - y$ plane may be written as

$$ds = \sqrt{1 + y'^2} dx \tag{5.32}$$

Now we are given a problem to find the shortest length s between the two points in the plane. The total length can be considered as a function of a function y, i.e. a functional as below

$$s(y) = \int_{x_1}^{x_2} \sqrt{1 + y'^2} dx = \int_{x_1}^{x_2} F(y, y', x) \, dx \tag{5.33}$$

Since the integrand $F(y, y', x)$ is not a function of y, in view of Equation (5.31) we obtain

$$\frac{\partial F(y, y', x)}{\partial y} = 0, \quad \frac{\partial F(y, y', x)}{\partial y'} = \frac{y'}{\sqrt{1 + y'^2}} \tag{5.34}$$

Substituting the above equations into the Euler-Lagrange equation given in Equation (5.31) yields

$$\frac{d}{dx}\left(\frac{y'}{\sqrt{1 + y'^2}}\right) = 0 \tag{5.35}$$

which implies that the function y minimizing the total length s between the two points is a straight line with the form

$$y = ax + b \tag{5.36}$$

where a and b are arbitrary constants.

Example 5.3 Brachistochrone Problem

The problem is to find the curve on which a particle, which is subject to gravity, travels from a point P_1 to a point P_2 during the shortest time (see Figure 5.1). It was Euler (1707–1783) who appears to have formulated the modern calculus of variations. However, before Euler, Bernoulli (1667–1748) appears to have first posted the brachistochrone problem to the European mathematician community. This problem seems to have motivated the community, and provided the necessary challenges to develop the calculus of variations. Galileo (1564-1642) in 1638 appears to have first considered the brachistochrone problem, and not Bernoulli.

Assume that gravity is directed to the negative y-axis. Then the principle of conservation of the total energy yields

$$\frac{1}{2}mv^2 = -mgy \tag{5.37}$$

where v denotes the speed of the particle traveling from the point P_1 to the point P_2. Thus the speed may be written in terms of x

$$dt = \frac{ds}{v} = \frac{ds}{\sqrt{-2gy}} \tag{5.38}$$

Since the infinitesimal distance can be expressed by the Pythagorean theorem

$$ds = \sqrt{1 + x'^2}dy, \quad x' = \frac{dx}{dy} \tag{5.39}$$

The time for the particle, subject to the gravity, to travel from a point P_1 to a point P_2 is a functional, $T\left(x\right)$ as shown below:

$$T\left(x\right) = t_2 - t_1 = \int_1^2 \frac{ds}{\sqrt{-2gy}} = \int_1^2 \frac{\sqrt{1+x'^2}}{\sqrt{-2gy}}dy = \int_1^2 F\left(x, x', y\right) dy \tag{5.40}$$

Since the integrand does not possess an explicit x term, we have

$$\frac{\partial F\left(x, x', y\right)}{\partial x} = 0, \; \frac{\partial F\left(x, x', y\right)}{\partial x'} = \frac{x'}{\sqrt{-2gy\left(1 + x'^2\right)}} = c \tag{5.41}$$

where c is a constant. From Equation (5.41), we have

$$x' = \sqrt{\frac{-2gyc^2}{1 + 2gyc^2}} \tag{5.42}$$

Taking $y = -\frac{1}{4gc^2}\left(1 - \cos\theta\right)$, then we have $dy = -\frac{1}{4gc^2}\sin\theta d\theta$. Accordingly, we shall have

$$dx = \sqrt{\frac{-2gyc^2}{1 + 2gyc^2}}dy = \frac{1}{4gc^2}\left(1 - \cos\theta\right) d\theta \tag{5.43}$$

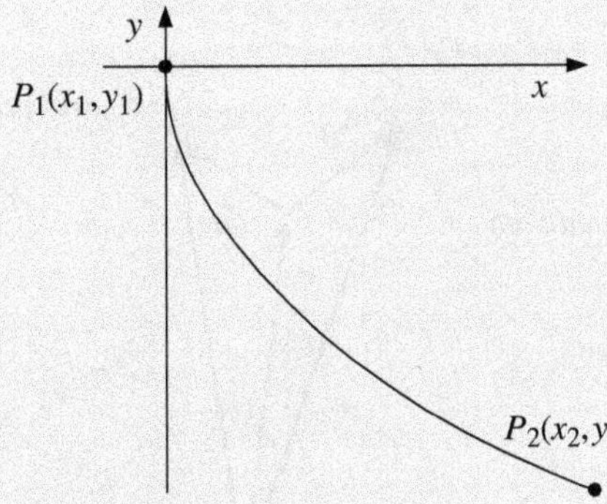

Figure 5.1. Brachistochrone problem

The function of the curve on which the particle travels is given by

$$x = \frac{1}{4gc^2}(\theta - \sin\theta) \tag{5.44}$$

because at $\theta = 0$, $x = y = 0$. We note that the curve, which is defined by Equation (5.44), extremizes the time that the particle takes to travel between the two points.

5.4 HAMILTON'S PRINCIPLE

In the previous section, we discussed a purely mathematical derivation of the Euler-Lagrange equations from the functional by use of the calculus of variations. Note that any physical meaning in the principle can not be deduced in the sense that the functional, which is called the action, does not stand for any physical state variable. Hamilton's Principle is a variation principle, which can be applied to N-body dynamical systems.

Within a closed interval of time $\bar{\mathcal{T}} = [t_0, t_1]$, the action $I(\mathbf{q})$ in generalized coordinates $\mathbf{q}$ is defined as

$$I(\mathbf{q}) = \int_{t_0}^{t_1} \mathcal{L}(\mathbf{q}, \dot{\mathbf{q}}, t)\, dt \tag{5.45}$$

The action defined in the above is subject to the restriction such that

$$\begin{aligned} \mathbf{q}(t_0) &= \bar{\mathbf{q}}_0, \ \text{at } t = t_0, \\ \mathbf{q}(t_1) &= \bar{\mathbf{q}}_1, \ \text{at } t = t_1 \end{aligned} \tag{5.46}$$

where $\bar{\mathbf{q}}_0$ and $\bar{\mathbf{q}}_1$ are assumed to be prescribed values, even though $\bar{\mathbf{q}}_1$ is not known. Therefore, the generalized coordinate $\mathbf{q}$, which is interpreted as a point in configuration space (Papastavridis 1998), is prescribed at both of the terminal points; the configuration at $t = t_0$ and $t = t_1$ is held fixed (Langhaar 1962).

As the variation of the admissible displacement of N-particles vanishes at both of the terminal points, the varied path $\mathbf{q}^*(t)$ and the real path $\mathbf{q}(t) : \bar{\mathcal{T}} = [t_0, t_1] \to \mathbb{R}^{n_{\text{dof}}}$ are contemporaneous at the terminal points as shown in Figure 5.2. Note that $\mathbf{q}^*(t) = \mathbf{q}(t) + \delta\mathbf{q}(t)$. Hamilton's Principle claims that the Euler-Lagrange's equations, the governing equations, make the action stationary under the restriction that the virtual displacement at the terminal points vanishes. It is very important to note that Hamilton's

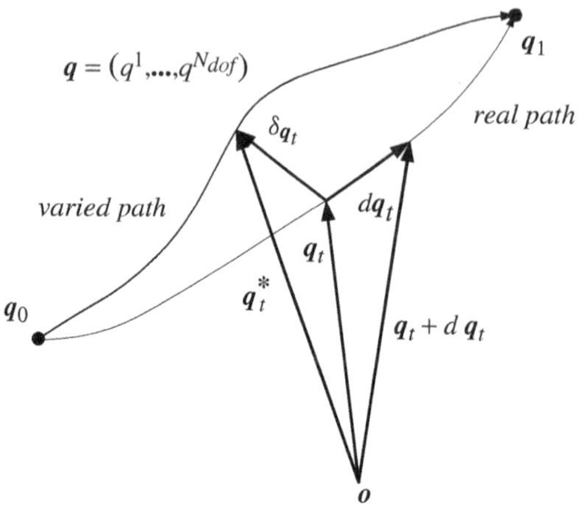

Figure 5.2. Hamilton's principle

Principle requires the virtual displacement at the terminal points to vanish as follows:

$$\begin{aligned}&\delta\int_{t_0}^{t_1}\mathcal{L}\left(\mathbf{q},\dot{\mathbf{q}},t\right)dt=0,\\&\delta\mathbf{q}\left(t_0\right)=\delta\mathbf{q}\left(t_1\right)=\mathbf{0}\end{aligned}\tag{5.47}$$

which is called Hamilton's principle.

Lagrange's Equations of Motion and Lagrangian Mechanics Framework Consider a Lagrangian flow with generalized coordinates $\mathbf{q}(t)$. In other words, it is assumed that the solution of an N-body dynamical system is a point in the configuration space, i.e. $\mathbf{q}=\left(q^1,\ldots,q^{n_{\text{dof}}}\right)\in Q$ and the Lagrangian is a mapping $\mathcal{L}(\mathbf{q},\dot{\mathbf{q}},t):TQ\times\bar{\mathcal{T}}\to\mathbb{R},\quad t\in\bar{\mathcal{T}}=[t_0,t_1]$. Then, the varied action can be represented as

$$I_\varepsilon=I\left(\mathbf{q}+\varepsilon\boldsymbol{\eta}\right)=\int_{t_0}^{t_1}\mathcal{L}\left(\mathbf{q}+\varepsilon\boldsymbol{\eta},\dot{\mathbf{q}}+\varepsilon\dot{\boldsymbol{\eta}},t\right)dt\tag{5.48}$$

where $\boldsymbol{\eta}$ denotes the admissible function satisfying the boundary conditions, that is,

$$\boldsymbol{\eta}\left(t_0\right)=\boldsymbol{\eta}\left(t_1\right)=\mathbf{0}\tag{5.49}$$

It is important to note that $\boldsymbol{\eta}(t)$ represents the direction of the virtual displacement. Accordingly, in order for the varied action to be stationary at $\varepsilon=0$, the Gateaux variation of the action as given in Equation (5.24) should vanish as follows:

$$\left.\frac{dI\left(\mathbf{q}+\varepsilon\boldsymbol{\eta}\right)}{d\varepsilon}\right|_{\varepsilon=0}=0\tag{5.50}$$

In other words, the action on the real path is stationary when the above equation is satisfied. Thus, the above equation can be recast in the form

$$\int_{t_0}^{t_1}\left.\left[\frac{\partial\mathcal{L}\left(\mathbf{q}+\varepsilon\boldsymbol{\eta},\dot{\mathbf{q}}+\varepsilon\dot{\boldsymbol{\eta}},t\right)}{\partial\left(\mathbf{q}+\varepsilon\boldsymbol{\eta}\right)}\cdot\boldsymbol{\eta}+\frac{\partial\mathcal{L}\left(\mathbf{q}+\varepsilon\boldsymbol{\eta},\dot{\mathbf{q}},t\right)}{\partial\left(\dot{\mathbf{q}}+\varepsilon\dot{\boldsymbol{\eta}}\right)}\cdot\dot{\boldsymbol{\eta}}\right]\right|_{\varepsilon=0}dt=0\tag{5.51}$$

Therefore, we find

$$\int_{t_0}^{t_1} \left[\frac{\partial \mathcal{L}(\mathbf{q}, \dot{\mathbf{q}}, t)}{\partial \mathbf{q}} - \frac{d}{dt}\left(\frac{\partial \mathcal{L}(\mathbf{q}, \dot{\mathbf{q}}, t)}{\partial \dot{\mathbf{q}}} \right) \right] \cdot \eta \, dt + \left. \frac{\partial \mathcal{L}(\mathbf{q}, \dot{\mathbf{q}}, t)}{\partial \dot{\mathbf{q}}} \cdot \eta \right|_{t_0}^{t_1} = 0 \tag{5.52}$$

Since the admissible function satisfies the boundary conditions, the second term in the above equation should vanish. Eventually, we have

$$\int_{t_0}^{t_1} \left[\frac{\partial \mathcal{L}(\mathbf{q}, \dot{\mathbf{q}}, t)}{\partial \mathbf{q}} - \frac{d}{dt}\left(\frac{\partial \mathcal{L}(\mathbf{q}, \dot{\mathbf{q}}, t)}{\partial \dot{\mathbf{q}}} \right) \right] \cdot \eta \, dt = 0 \tag{5.53}$$

The above equation results in the Euler-Lagrange equations of motion, since the admissible function is arbitrary within the time interval. Consequently, in view of Equations (5.47), (5.49), and (5.47), Hamilton's principle can be rewritten as

$$\begin{gathered} \delta \int_{t_0}^{t_1} \mathcal{L}(\mathbf{q}, \dot{\mathbf{q}}, t) \, dt = \int_{t_0}^{t_1} \left[\frac{\partial \mathcal{L}(\mathbf{q}, \dot{\mathbf{q}}, t)}{\partial \mathbf{q}} - \frac{d}{dt}\left(\frac{\partial \mathcal{L}(\mathbf{q}, \dot{\mathbf{q}}, t)}{\partial \dot{\mathbf{q}}} \right) \right] \cdot \delta \mathbf{q} \, dt = 0, \\ \delta \mathbf{q}(t_0) = \delta \mathbf{q}(t_1) = 0 \end{gathered} \tag{5.54}$$

where $\delta \mathbf{q} = \varepsilon \eta$. As the admissible function is arbitrary within the time interval, the virtual displacement $\delta \mathbf{q}$ is arbitrary within the same time interval. Note that the differential equation derived from Hamilton's principle, i.e. the Euler-Lagrange equation in dynamics, is in the equivalent form to Lagrange's equation of motion which we derived from the Lagrangian form of D'Alembert's principle in the previous chapter.

Modified Hamilton's Principle in the Hamiltonian Mechanics Framework Applying the Legendre transformation to Hamilton's Principle leads to the modified Hamilton's Principle as follows:

$$\begin{gathered} \delta \int_{t_0}^{t_1} (\mathbf{p} \cdot \dot{\mathbf{q}} - \mathcal{H}(\mathbf{q}, \mathbf{p}, t)) \, dt = 0, \\ \delta \mathbf{q}(t_0) = \delta \mathbf{q}(t_1) = \mathbf{0} \end{gathered} \tag{5.55}$$

This is a two-field variational principle, and the action given in Equation (5.45) is modified in the manner that the action has two dependent variables as follows:

$$I(\mathbf{q}, \mathbf{p}) = \int_{t_0}^{t_1} (\mathbf{p} \cdot \dot{\mathbf{q}} - \mathcal{H}(\mathbf{q}, \mathbf{p}, t)) \, dt \tag{5.56}$$

The modified Hamilton's Principle states that Hamilton's canonical equations make the action defined in Equation (5.45) stationary under the restriction that the virtual displacement at the terminal points vanishes. Therefore, the modified Hamilton's Principle gives

$$\delta I(\mathbf{q}, \mathbf{p}) = \int_{t_0}^{t_1} \left[\left(\dot{\mathbf{q}} - \frac{\partial \mathcal{H}}{\partial \mathbf{p}} \right) \cdot \delta \mathbf{p} - \left(\dot{\mathbf{p}} + \frac{\partial \mathcal{H}}{\partial \mathbf{q}} \right) \cdot \delta \mathbf{q} \right] dt + \mathbf{p} \cdot \delta \mathbf{q}|_{t_0}^{t_1} = 0 \tag{5.57}$$

As the modified Hamilton's Principle requires the virtual displacement to vanish at the terminal points, the last term in the above should vanish. As the variations of canonical coordinates are arbitrary with in the time interval, we obtain Hamilton's canonical equations of motion as follows:

$$\dot{\mathbf{q}} - \frac{\partial \mathcal{H}}{\partial \mathbf{p}} = \mathbf{0}, \; \dot{\mathbf{p}} + \frac{\partial \mathcal{H}}{\partial \mathbf{q}} = \mathbf{0} \tag{5.58}$$

Remark 5.4.1 Hamilton's principle can be also represented in terms of the Total Energy, which is yet another alternative descriptive scalar function that is measurable and directly built-in into the framework. We defer to and address the modified Hamilton's principle in the context of the Total Energy framework in Chapter 6.

Hamilton's Principle for the Applied Force not Derivable from the Potential Energy For holonomic constraint systems, Hamilton's Principle defined in Equation (5.47) can be rewritten as

$$\begin{aligned} &\int_{t_0}^{t_1} (\delta\mathcal{K}(\mathbf{q}, \dot{\mathbf{q}}, t) - \delta\mathcal{U}(\mathbf{q}, t))\, dt = 0, \\ &\delta\mathbf{q}(t_0) = \delta\mathbf{q}(t_1) = \mathbf{0} \end{aligned} \tag{5.59}$$

where the applied force given in Equation (3.51) is assumed to be derivable from the potential energy $\mathcal{U}(\mathbf{q}, t)$. Thus, Hamilton's Principle can be rewritten as

$$\begin{aligned} &\int_{t_0}^{t_1} (\delta\mathcal{K}(\mathbf{q}, \dot{\mathbf{q}}, t) + \delta\mathcal{W})\, dt = 0, \\ &\delta\mathbf{q}(t_0) = \delta\mathbf{q}(t_1) = \mathbf{0} \end{aligned} \tag{5.60}$$

where $\delta\mathcal{W}$ denotes the virtual work by the applied force and the above equation is referred to as the extended Hamilton's Principle (Meirovitch 2004) or the first form of Hamilton's Principle (Ardema 2005). Note that the virtual work by the applied generalized force $\delta\mathcal{W}$ can be written as

$$\delta\mathcal{W} = \mathbf{Q}_{\text{appl}} \cdot \delta\mathbf{q} \tag{5.61}$$

where $\mathbf{Q}_{\text{appl}}$ is the applied generalized force defined in Equation (3.77).

For the case of the applied force not derivable from the potential energy, the applied generalized force is divided into two parts: the applied force $\mathbf{Q}_{\text{appl}_a}$, which is derivable from the potential energy, and the applied force $\mathbf{Q}_{\text{appl}_b}$, which is not derivable from the potential energy. As a consequence, the Hamilton's principle for the case of the applied force not derivable from the potential energy can be rewritten as

$$\begin{aligned} &\int_{t_0}^{t_1} \left(\delta\mathcal{L}(\mathbf{q}, \dot{\mathbf{q}}, t) + \delta\mathcal{W}_{\text{appl}_b}\right) dt = 0, \\ &\delta\mathbf{q}(t_0) = \delta\mathbf{q}(t_1) = \mathbf{0} \end{aligned} \tag{5.62}$$

where $\delta\mathcal{W}_{\text{appl}_b} = \mathbf{Q}_{\text{appl}_b} \cdot \mathbf{q}$.

Illustrative Example

Example 5.4 Planar Pendulum

We shall discuss the planar pendulum problem to derive Lagrange's equation of motion using Hamilton's principle. Again consider the same planar pendulum as we discussed previously. The autonomous Lagrangian can be given as

$$\mathcal{L}(\theta, \dot{\theta}) = \mathcal{K}(\dot{\theta}) - \mathcal{U}(\theta) = \frac{1}{2} m l^2 \dot{\theta}^2 + mgl\cos\theta \tag{5.63}$$

Then, Hamilton's principle reads as

$$\delta \int_{t_0}^{t_1} \mathcal{L}(\theta, \dot{\theta}) dt = 0, \qquad \delta\theta\,(t_0) = \delta\theta\,(t_1) = 0 \tag{5.64}$$

And, we have

$$\delta \int_{t_0}^{t_1} \mathcal{L}(\theta, \dot{\theta}) dt = \int_{t_0}^{t_1} \left[\frac{\partial \mathcal{L}}{\partial \theta} - \frac{d}{dt}\left(\frac{\partial \mathcal{L}}{\partial \dot{\theta}} \right) \right] \delta\theta dt + \left. \frac{\partial \mathcal{L}}{\partial \dot{\theta}} \delta\theta \right|_{t_0}^{t_1} = 0 \tag{5.65}$$

With the virtual displacements at the end points vanishing, we obtain

$$\int_{t_0}^{t_1} \left[\frac{\partial \mathcal{L}(\theta, \dot{\theta})}{\partial \theta} - \frac{d}{dt}\left(\frac{\partial \mathcal{L}(\theta, \dot{\theta})}{\partial \dot{\theta}} \right) \right] \delta\theta dt = 0 \tag{5.66}$$

As $\delta\theta$ is arbitrary within the time interval, the above equation yields

$$\frac{\partial \mathcal{L}(\theta, \dot{\theta})}{\partial \theta} - \frac{d}{dt}\left(\frac{\partial \mathcal{L}(\theta, \dot{\theta})}{\partial \dot{\theta}} \right) = 0 \tag{5.67}$$

which is equivalent to Equation (3.104).

5.5 HAMILTON'S LAW OF VARYING ACTION

Bailey (Bailey 1975a,b, 2004) appears to be the first to use Hamilton's law of varying action to obtain the direct analytical solution by means of a power series, and discusses the inconsistency (Bailey 1975a,b; Simkins 1981) with Hamilton's principle. He applied an appropriate power series satisfying the initial condition to Equation (5.11). We observe that Equation (5.11) can be expressed as

$$\int_{t_0}^{t_1} \left[\frac{d}{dt}\left(\frac{\partial \mathcal{L}(\mathbf{q}, \dot{\mathbf{q}}, t)}{\partial \dot{\mathbf{q}}} \right) - \frac{\partial \mathcal{L}(\mathbf{q}, \dot{\mathbf{q}}, t)}{\partial \mathbf{q}} \right] \cdot \delta\mathbf{q}\, dt = 0 \tag{5.68}$$

In fact, we can readily derive Lagrange's equations from the Lagrangian form of D'Alembert's principle as discussed in the preceding chapter. Accordingly, Equation (5.11) can also yield

$$\left[\frac{d}{dt}\left(\frac{\partial \mathcal{L}(\mathbf{q}, \dot{\mathbf{q}}, t)}{\partial \dot{\mathbf{q}}} \right) - \frac{\partial \mathcal{L}(\mathbf{q}, \dot{\mathbf{q}}, t)}{\partial \mathbf{q}} \right] \cdot \delta\mathbf{q} = 0 \tag{5.69}$$

Note that the Lagrangian in Equation (5.69) may not be autonomous. From Equation (5.69), we have Lagrange's equations of motion governing the motion of the dynamical system in configuration space where the constraint force does not exist. Note that Equation (5.69) holds for holonomic constraint systems. Also note that there is a distinct difference between Equation (5.69) and Equation (5.68), in the sense that Equation (5.68) is an integral equation over the time domain. Bailey (Bailey 1975a,b) applied a simple truncated power series to Equation (5.11), whilst we will regard Equation (5.68) rather than

Equation (5.11) as an important equation, because it takes a weighted residual form to which the time weighted-residual method can be applied. The variable, time, in Equation (5.68) is the only independent variable, whereas the generalized coordinates are dependent variables.

Note that for the case of the applied force not derivable from the potential energy, Hamilton's law of varying action given in Equation (5.11) can be rewritten as

$$\delta \int_{t_0}^{t_1} \mathcal{L}(\mathbf{q}, \dot{\mathbf{q}}, t)\, dt - \frac{\partial \mathcal{L}(\mathbf{q}, \dot{\mathbf{q}}, t)}{\partial \dot{\mathbf{q}}} \cdot \delta \mathbf{q}\bigg|_{t_0}^{t_1} + \int_{t_0}^{t_1} \delta W_{\mathrm{appl_b}}\, dt = 0 \tag{5.70}$$

where $\delta \mathcal{W}_{\mathrm{appl_b}} = \mathbf{Q}_{\mathrm{appl_b}} \cdot \mathbf{q}$.

Hamilton's Law of Varying Action in the Cartesian Coordinate System Next, we examine Hamilton's law of varying action for N-body dynamical systems subject to constraints in the Cartesian coordinate system. Suppose that the domain of the Lagrangian contains Cartesian coordinates rather than generalized coordinates and velocities. Then, the Newtonian dynamical system of N particles, subject to constraints, is redundant. The solution of the system is a point in the Euclidean $3N$-space, i.e. $\mathbf{x} = \left(\mathbf{x}^1, \ldots, \mathbf{x}^N\right) \in \mathbb{R}^{3N}$. An emphasis should be placed on the fact that the domain of the Lagrangian is not the configuration space but the Euclidean $6N$-space, i.e., $(\mathbf{x}, \dot{\mathbf{x}}) \in \mathbb{R}^{6N}$. The Lagrangian form of D'Alembert's principle for conservative systems states that

$$\mathbf{M}\ddot{\mathbf{x}} \cdot \delta \mathbf{x} = \mathbf{F}_c \cdot \delta \mathbf{x} + \mathbf{C} \cdot \delta \mathbf{x} \tag{5.71}$$

The term on the left-hand side can be written as

$$\mathbf{M}\ddot{\mathbf{x}} \cdot \delta \mathbf{x} = \frac{d}{dt}(\mathbf{M}\dot{\mathbf{x}} \cdot \delta \mathbf{x}) - \delta \mathcal{K}(\dot{\mathbf{x}}) \tag{5.72}$$

The first term on the right-hand side in Equation (5.71) is given as

$$\mathbf{F}_c \cdot \delta \mathbf{x} = -\frac{\partial \mathcal{U}(\mathbf{x})}{\partial \mathbf{x}} \cdot \delta \mathbf{x} = -\delta \mathcal{U}(\mathbf{x}) \tag{5.73}$$

And as stated previously in Section 3.2.6, the second term on the right-hand side can be written as

$$\mathbf{C}(\mathbf{x}, t) \cdot \delta \mathbf{x} = \nabla (\boldsymbol{\lambda} \cdot \mathbf{f}) \cdot \delta \mathbf{x} \tag{5.74}$$

where holonomic constraints are assumed. Then, Equation (5.71) can be given as

$$\frac{d}{dt}(\mathbf{M}\dot{\mathbf{x}} \cdot \delta \mathbf{x}) = \delta(\mathcal{K}(\dot{\mathbf{x}}) - \mathcal{U}(\mathbf{x})) = \delta \mathcal{L}(\dot{\mathbf{x}}, \mathbf{x}) + \nabla(\boldsymbol{\lambda} \cdot \mathbf{f}) \cdot \delta \mathbf{x} = \delta \tilde{\mathcal{L}}(\dot{\mathbf{x}}, \mathbf{x}) \tag{5.75}$$

where the augmented Lagrangian $\tilde{\mathcal{L}}(\dot{\mathbf{x}}, \mathbf{x}, t) : \mathbb{R}^{6N} \times \bar{\mathcal{T}} \rightarrow \mathbb{R}$ is denoted by

$$\tilde{\mathcal{L}}(\dot{\mathbf{x}}, \mathbf{x}, t) = \mathcal{L}(\dot{\mathbf{x}}, \mathbf{x}) + \boldsymbol{\lambda} \cdot \mathbf{f} \tag{5.76}$$

By performing the line integral within the time interval, we obtain

$$\delta \int_{t_0}^{t_1} \tilde{\mathcal{L}}(\dot{\mathbf{x}}, \mathbf{x}, t)\, dt - \frac{\partial \mathcal{L}(\dot{\mathbf{x}}, \mathbf{x}, t)}{\partial \dot{\mathbf{x}}} \cdot \delta \mathbf{x}\bigg|_{t_0}^{t_1} = 0 \tag{5.77}$$

which is Hamilton's law of varying action in the Cartesian coordinate system. Note that if there exists a nonconservative force $\mathbf{F}_{\mathrm{nc}}$, which is not derivable from the potential energy, the above equation should be modified into the following

$$\delta \int_{t_0}^{t_1} \left(\tilde{\mathcal{L}} (\dot{\mathbf{x}}, \mathbf{x}, t) + \delta W_{\mathrm{nc}} \right) dt - \frac{\partial \mathcal{L} (\dot{\mathbf{x}}, \mathbf{x}, t)}{\partial \dot{\mathbf{x}}} \cdot \delta \mathbf{x} \bigg|_{t_0}^{t_1} = 0 \tag{5.78}$$

where δW_{nc} denotes the virtual work done by the nonconservative force $\mathbf{F}_{\mathrm{nc}}$, i.e., $\delta W_{\mathrm{nc}} = \mathbf{F}_{\mathrm{nc}} \cdot \delta \mathbf{x}$. Because of the inconsistencies associated with and cited in the literature regarding Hamilton's principle, henceforth, we mostly place emphasis and advocate the Hamilton's law of varying action as an alternative.

5.5.1 Newtonian Mechanics

In this section, we shall derive a Newtonian dynamical system from Hamilton's law of varying action in the Cartesian coordinate system. It should be noted that the Newtonian dynamical system in the Cartesian coordinate system is redundant due to the presence of constraint forces; hence the Lagrange multiplier is employed as discussed in Section 3.2.6. We emphasize the fact that Hamilton's law of varying action seems to emanate from Newton's second law. In what follows, we shall demonstrate the converse derivation only.

Newton's Equations of Motion Consider the motion of a dynamical system with N particles. The system is assumed to be subject to holonomic constraints. Our objective here is to derive the Newtonian dynamical system in the Cartesian coordinate system from Hamilton's law of varying action. Accordingly, we first start with the augmented Lagrangian defined in Equation (5.76) to re-derive Newton's equations from Hamilton's law of varying action. The variation of the action having the augmented Lagrangian as an integrand can be written as

$$\delta \int_{t_0}^{t_1} \tilde{\mathcal{L}} (\dot{\mathbf{x}}, \mathbf{x}, t) \, dt = \delta \int_{t_0}^{t_1} \left(\mathcal{K} (\dot{\mathbf{x}}) - \mathcal{U} (\mathbf{x}) + \boldsymbol{\lambda} \cdot \mathbf{f} \right) dt \tag{5.79}$$

The kinetic energy $\mathcal{K} (\dot{\mathbf{x}}) : \mathbb{R}^{3N} \rightarrow \mathbb{R}$ is given by

$$\mathcal{K} = \frac{1}{2} \dot{\mathbf{x}} \cdot \mathbf{M} \dot{\mathbf{x}} \tag{5.80}$$

Then, the variation of the kinetic energy can be expressed as

$$\delta \mathcal{K} = \mathbf{M} \dot{\mathbf{x}} \cdot \delta \dot{\mathbf{x}} \tag{5.81}$$

Further, the variation of the potential energy $\mathcal{U} = \mathcal{U} (\mathbf{x}) : \mathbb{R}^{3N} \rightarrow \mathbb{R}$ gives

$$\delta \mathcal{U} (\mathbf{x}) = \frac{\partial \mathcal{U} (\mathbf{x})}{\partial \mathbf{x}} \cdot \delta \mathbf{x} \tag{5.82}$$

Upon substitution of Equation (5.82) and (5.81) into Equation (5.79), we have

$$\delta \int_{t_0}^{t_1} \tilde{\mathcal{L}} (\dot{\mathbf{x}}, \mathbf{x}, t) \, dt = \int_{t_0}^{t_1} \left(\mathbf{M} \dot{\mathbf{x}} \cdot \delta \dot{\mathbf{x}} - \frac{\partial \mathcal{U} (\mathbf{x})}{\partial \mathbf{x}} \cdot \delta \mathbf{x} + \frac{\partial (\boldsymbol{\lambda} \cdot \mathbf{f})}{\partial \mathbf{x}} \cdot \delta \mathbf{x} \right) dt \tag{5.83}$$

By means of the product rule, we find

$$\mathbf{M}\dot{\mathbf{x}} \cdot \delta\dot{\mathbf{x}} = \frac{d}{dt}\left(\mathbf{M}\dot{\mathbf{x}} \cdot \delta\mathbf{x}\right) - \mathbf{M}\ddot{\mathbf{x}} \cdot \delta\mathbf{x} \tag{5.84}$$

Substituting Equation (5.84) into Equation (5.83), we have

$$\begin{aligned} \delta \int_{t_0}^{t_1} \tilde{\mathcal{L}}\left(\dot{\mathbf{x}}, \mathbf{x}, t\right) dt = \int_{t_0}^{t_1} \left[\left(-\mathbf{M}\ddot{\mathbf{x}} - \frac{\partial \mathcal{U}(\mathbf{x})}{\partial \mathbf{x}} + \boldsymbol{\lambda} \cdot \nabla \mathbf{f}\right) \cdot \delta\mathbf{x}\right] dt \\ + \mathbf{M}\dot{\mathbf{x}} \cdot \delta\mathbf{x}\Big|_{t_0}^{t_1} = 0 \end{aligned} \tag{5.85}$$

Consequently, in view of Equation (5.77), Hamilton's law of varying action for nonconservative systems reduces to the following:

$$\begin{aligned} \delta \int_{t_0}^{t_1} \left(\tilde{\mathcal{L}}\left(\dot{\mathbf{x}}, \mathbf{x}, t\right) + \delta W_{\mathrm{nc}}\right) dt - \frac{\partial \mathcal{L}\left(\dot{\mathbf{x}}, \mathbf{x}, t\right)}{\partial \dot{\mathbf{x}}} \cdot \delta\mathbf{x}\Bigg|_{t_0}^{t_1} \\ = \int_{t_0}^{t_1} \left[\left(-\mathbf{M}\ddot{\mathbf{x}} - \frac{\partial \mathcal{U}(\mathbf{x})}{\partial \mathbf{x}} + \mathbf{F}_{\mathrm{nc}} + \boldsymbol{\lambda} \cdot \nabla \mathbf{f}\right) \cdot \delta\mathbf{x}\right] dt = 0 \end{aligned} \tag{5.86}$$

Note that the virtual displacement $\delta\mathbf{x}$ in the Cartesian coordinate system is not linearly independent. However, the Lagrange multiplier $\boldsymbol{\lambda}$ is chosen to make the term enclosed in the bracket vanish. Then we obtain the Newtonian dynamical system as follows

$$\mathbf{M}\ddot{\mathbf{x}} = \mathbf{F}_{\mathrm{appl}} + \mathbf{C}\left(\mathbf{x}, t\right) \tag{5.87}$$

where the applied force $\mathbf{F}_{\mathrm{appl}}$ is the summation of the conservative force and the nonconservative force, i.e., $\mathbf{F}_{\mathrm{appl}} = \mathbf{F}_{\mathrm{c}} + \mathbf{F}_{\mathrm{nc}}$. And the constraint force is denoted by $\mathbf{C}\left(\mathbf{x}, t\right) = \boldsymbol{\lambda} \cdot \nabla \mathbf{f}$. Furthermore, since $\nabla \mathbf{f} \cdot \delta\mathbf{x} = 0$, from Equation (5.86), we obtain the equation

$$\left(\mathbf{F}_{\mathrm{appl}} - \mathbf{M}\ddot{\mathbf{x}}\right) \cdot \delta\mathbf{x} = 0 \tag{5.88}$$

which is coincident with the Lagrangian form of D'Alembert's principle.

5.5.2 Lagrangian Mechanics

In Section 3.2.6, we showed that Lagrange's equations of motion are derived from the Lagrangian form of D'Alembert's principle. In the preceding section, we also demonstrated that Newton's equations of motion in the Cartesian coordinate system can be obtained from Hamilton's law of varying action. In this section, we shall show that Lagrange's equations can also be derived from Hamilton's law of varying action.

Lagrange's Equations of Motion For the holonomic systems (Osgood 1937), the variation of the action can be given by

$$\delta \int_{t_0}^{t_1} \mathcal{L}\left(\mathbf{q}, \dot{\mathbf{q}}, t\right) dt = \int_{t_0}^{t_1} \delta\mathcal{L}\left(\mathbf{q}, \dot{\mathbf{q}}, t\right) dt \tag{5.89}$$

With time fixed, the variation of the Lagrangian can be written as

$$\delta\mathcal{L}\left(\mathbf{q}, \dot{\mathbf{q}}, t\right) = \frac{\partial \mathcal{L}}{\partial \mathbf{q}} \cdot \delta\mathbf{q} + \frac{\partial \mathcal{L}}{\partial \dot{\mathbf{q}}} \cdot \delta\dot{\mathbf{q}} \tag{5.90}$$

Integration by parts yields

$$\int_{t_0}^{t_1} \left(\frac{\partial \mathcal{L}(\mathbf{q}, \dot{\mathbf{q}}, t)}{\partial \dot{\mathbf{q}}} \cdot \delta \dot{\mathbf{q}} \right) dt = \frac{\partial \mathcal{L}(\mathbf{q}, \dot{\mathbf{q}}, t)}{\partial \dot{\mathbf{q}}} \cdot \delta \mathbf{q} \bigg|_{t_0}^{t_1} - \int_{t_0}^{t_1} \left[\frac{d}{dt} \left(\frac{\partial \mathcal{L}}{\partial \dot{\mathbf{q}}} \right) \cdot \delta \mathbf{q} \right] dt \tag{5.91}$$

In view of Equations (5.11), (5.90) and (5.91), Hamilton's law of varying action yields

$$\delta \int_{t_0}^{t_1} \mathcal{L}(\mathbf{q}, \dot{\mathbf{q}}, t)\, dt - \frac{\partial \mathcal{L}(\mathbf{q}, \dot{\mathbf{q}}, t)}{\partial \dot{\mathbf{q}}} \cdot \delta \mathbf{q} \bigg|_{t_0}^{t_1} = \int_{t_0}^{t_1} \left[\frac{\partial \mathcal{L}}{\partial \mathbf{q}} - \frac{d}{dt} \left(\frac{\partial \mathcal{L}}{\partial \dot{\mathbf{q}}} \right) \right] \cdot \delta \mathbf{q}\, dt = 0 \tag{5.92}$$

Note that Hamilton's law of varying action does not require the action to be stationary when Lagrange's equations are true on the Lagrangian flow or the true path; hence, it is not a variational principle, which finds an optimal solution. Besides, Hamilton's law of varying action does not require the virtual displacement to be contemporaneous at the terminal points. Within the terminal instants, the virtual displacement is arbitrary. Therefore, we have Lagrange's equations of motion as

$$\frac{\partial \mathcal{L}(\mathbf{q}, \dot{\mathbf{q}}, t)}{\partial \mathbf{q}} - \frac{d}{dt} \left(\frac{\partial \mathcal{L}(\mathbf{q}, \dot{\mathbf{q}}, t)}{\partial \dot{\mathbf{q}}} \right) = \mathbf{0} \tag{5.93}$$

Note that Equation (5.93) implies n_{dof}-number of simultaneous differential equations of motion. We also observe that Hamilton's law of varying action takes the Bubnov-Galerkin weighted residual form in time.

In contrast to Hamilton's law of varying action in Equation (5.92), Hamilton's principle states that

$$\begin{aligned} \delta \int_{t_0}^{t_1} \mathcal{L}(\mathbf{q}, \dot{\mathbf{q}}, t)\, dt = \int_{t_0}^{t_1} & \left[\frac{\partial \mathcal{L}(\mathbf{q}, \dot{\mathbf{q}}, t)}{\partial \mathbf{q}} - \frac{d}{dt} \left(\frac{\partial \mathcal{L}(\mathbf{q}, \dot{\mathbf{q}}, t)}{\partial \dot{\mathbf{q}}} \right) \right] \cdot \delta \mathbf{q}\, dt \\ & + \frac{\partial \mathcal{L}(\mathbf{q}, \dot{\mathbf{q}}, t)}{\partial \dot{\mathbf{q}}} \cdot \delta \mathbf{q} \bigg|_{t_0}^{t_1} = 0 \end{aligned} \tag{5.94}$$

In order for the action to be stationary when the Euler-Lagrange's equations of motion are true on the Lagrangian flow, the virtual displacement should be contemporaneous at the terminal points, i.e.,

$$\delta \int_{t_0}^{t_1} \mathcal{L}(\mathbf{q}, \dot{\mathbf{q}}, t)\, dt = \int_{t_0}^{t_1} \left[\frac{\partial \mathcal{L}(\mathbf{q}, \dot{\mathbf{q}}, t)}{\partial \mathbf{q}} - \frac{d}{dt} \left(\frac{\partial \mathcal{L}(\mathbf{q}, \dot{\mathbf{q}}, t)}{\partial \dot{\mathbf{q}}} \right) \right] \cdot \delta \mathbf{q}\, dt = 0 \tag{5.95}$$

As the virtual displacement is arbitrary within the time interval, Lagrange's equations are true on the Lagrangian flow. Note that the virtual displacement at the initial and future points should be prescribed. However, the virtual displacement at the future point is obviously unknown. It can be predicted by use of an appropriate time integration scheme. As a consequence, Hamilton's principle possesses a logic inconsistency as discussed in the literature (Bailey 1975b; Simkins 1981).

5.5.3 Hamiltonian Mechanics

Thus far, we have shown that Newton's equations and Lagrange's equations can be derived from Hamilton's law of varying action. Likewise, we shall show that Hamilton's equations of motion can also be derived from Hamilton's law of varying action.

Hamilton's Equations of Motion Recall the Legendre transformation given in Equation (3.181) and the canonical momentum given in Equation (3.179); then Hamilton's law of varying action is of the form,

$$\delta \int_{t_0}^{t_1} (\mathbf{p} \cdot \dot{\mathbf{q}} - \mathcal{H}(\mathbf{q}, \mathbf{p}, t))\, dt - \mathbf{p} \cdot \delta \mathbf{q}|_{t_0}^{t_1} = 0 \tag{5.96}$$

which is referred to here as the modified Hamilton's law of varying action. For holonomic systems, the above equation can be transformed as

$$\int_{t_0}^{t_1}\left(\mathbf{p}\cdot\delta\dot{\mathbf{q}}+\dot{\mathbf{q}}\cdot\delta\mathbf{p}-\frac{\partial\mathcal{H}(\mathbf{q},\mathbf{p},t)}{\partial\mathbf{p}}\cdot\delta\mathbf{p}-\frac{\partial\mathcal{H}(\mathbf{q},\mathbf{p},t)}{\partial\mathbf{q}}\cdot\delta\mathbf{q}\right)dt-\mathbf{p}\cdot\delta\mathbf{q}|_{t_0}^{t_1}=0 \tag{5.97}$$

Since integration by parts yields,

$$\int_{t_0}^{t_1}(\mathbf{p}\cdot\delta\dot{\mathbf{q}})\,dt=\mathbf{p}\cdot\delta\mathbf{q}|_{t_0}^{t_1}-\int_{t_0}^{t_1}\left(\frac{d\mathbf{p}}{dt}\cdot\delta\mathbf{q}\right)dt, \tag{5.98}$$

Equation (5.97) reduces to

$$\int_{t_0}^{t_1}\left[\left(\dot{\mathbf{q}}-\frac{\partial\mathcal{H}(\mathbf{q},\mathbf{p},t)}{\partial\mathbf{p}}\right)\cdot\delta\mathbf{p}-\left(\dot{\mathbf{p}}+\frac{\partial\mathcal{H}(\mathbf{q},\mathbf{p},t)}{\partial\mathbf{q}}\right)\cdot\delta\mathbf{q}\right]dt=0 \tag{5.99}$$

The variations of canonical variables are arbitrary in holonomic systems. As a result, we have a set of first order differential equations, i.e., Hamilton's equations as follows:

$$\dot{\mathbf{p}}+\frac{\partial\mathcal{H}(\mathbf{q},\mathbf{p},t)}{\partial\mathbf{q}}=\mathbf{0},\quad \dot{\mathbf{q}}-\frac{\partial\mathcal{H}(\mathbf{q},\mathbf{p},t)}{\partial\mathbf{p}}=\mathbf{0} \tag{5.100}$$

Note that the solution of the Hamiltonian dynamical system is a point indicating an ordered pair $(\mathbf{p},\mathbf{q})\in T^*Q$ in phase space.

Remark 5.5.1 Hamilton's law of varying action can be also represented in terms of the Total Energy, which is yet another descriptive scalar function that is measurable. We defer to and address Hamilton's law of varying action in the context of the Total Energy framework in Chapter 6.

Remark 5.5.2 Note that Hamilton's law of varying action naturally leads to the weighted residual form in time. Hence, one can utilize the weighted residual form in time for designing time integration approaches such as the time weighted-residual method which includes time finite element formulations (Belytschko and Hughes 1983; Sha et al. 2003; Tamma et al. 2003, 2002, 2000; Wood 1990; Zhou and Tamma 2004a,c, 2006; Zhou et al. 2001, 2000; Zienkiewicz and Taylor 2006; Zienkiewicz et al. 2005, 1984).

EXERCISES

5.1. Use Hamilton's principle to obtain the Euler-Lagrange equation of motion for a sliding block on a moving, inclined plane in Example 4.2.

5.2. Use the modified Hamilton's principle to obtain Hamilton's equation of motion for a sliding block on a moving, inclined plane in Example 4.2.

5.3. Consider a forced dynamic response of the undamped spring-mass system, i.e. $A\neq 0$ and $\mathbf{g}\neq\mathbf{0}$ in Figure 3.18. Use Hamilton's law of varying action to obtain Lagrange's equation of motion.

5.4. A particle with mass m as shown in Figure 5.3, is connected by a rigid bar of length l with a frictionless hinge, and the mass is subject to the spring force and a sinusoidal external force $A\cos\varpi t$, where A is a positive parameter. Obtain Lagrange's equations of motion via Hamilton's law of varying action.

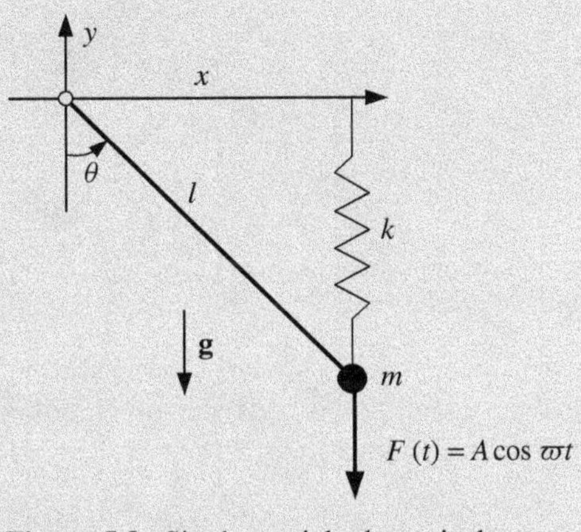

Figure 5.3. Single particle dynamical system

5.5. Consider the motion of a bead rotating about the z-axis, and sliding on a wire as shown in Figure 5.4. The wire rotates about the z-axis with a constant angular velocity α, and the angle θ is fixed. Use Hamilton's law of varying action to obtain Lagrange's equation of motion.

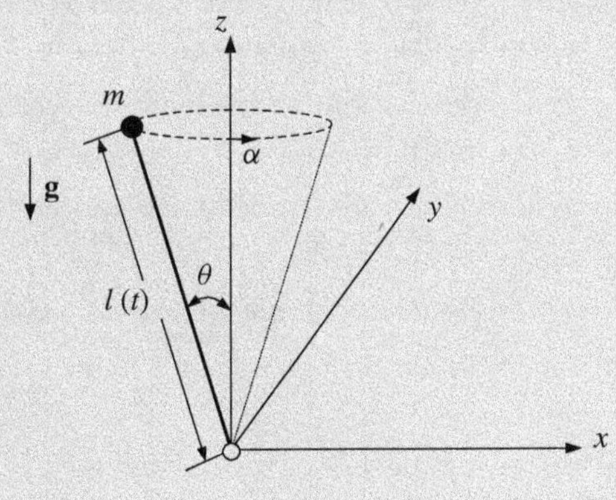

Figure 5.4. Bead rotating about the z-axis, and sliding on a wire

5.6. In Example 3.11, use Hamilton's law of varying action to obtain Hamilton's equations of motion in the form of first-order differential equations of motion.

5.7. Consider the simple planar pendulum with an oscillating, restoring hinge point as shown in Figure 5.5. At the hinge, the restoring moment is given by $M_o = k\theta$. The constraint equation is dependent on time; hence the system is holonomic-rheonomic. Use Hamilton's law of varying action to obtain Lagrange's equation of motion.

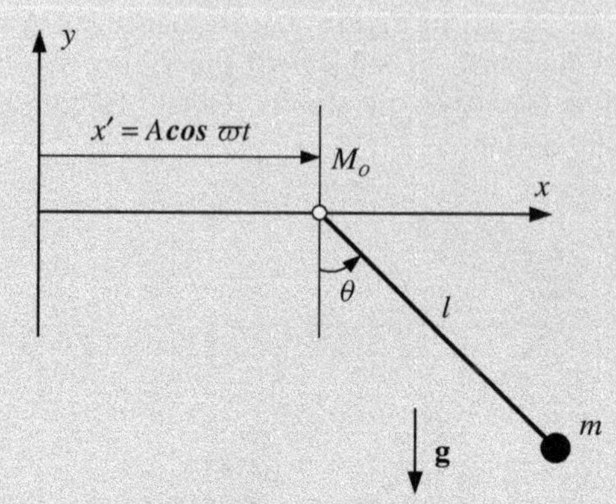

Figure 5.5. Simple pendulum with a oscillating, restoring hinge point

5.8. In Exercise 5.4, establish the Hamiltonian and derive Hamilton's equations of motion via Hamilton's law of varying action.

5.9. For the planar motion as shown in Figure 3.22, use Hamilton's principle to obtain the Euler-Lagrange equation of motion.

5.10. For the planar motion as shown in Figure 3.22, use Hamilton's law of varying action to obtain Hamilton's equations of motion.

5.11. In Exercise 3.25 of Chapter 3, use Hamilton's law of varying action to obtain Lagrange's equation of motion in the form of a second-order differential equation of motion.

5.12. Let us consider a damped spring-mass dynamical system shown in Figure 3.24. use Hamilton's law of varying action to obtain Lagrange's equation of motion.

5.13. In Exercise 3.6 of Chapter 3, establish the Lagrangian and obtain Lagrange's equations of motion via Hamilton's law of varying action.

5.14. In Exercise 3.6 of Chapter 3, establish the Hamiltonian and derive Hamilton's equations of motion via Hamilton's law of varying action.

5.15. In Exercise 5.4, use the modified Hamilton's principle to obtain Hamilton's equation of motion.

CHAPTER SIX

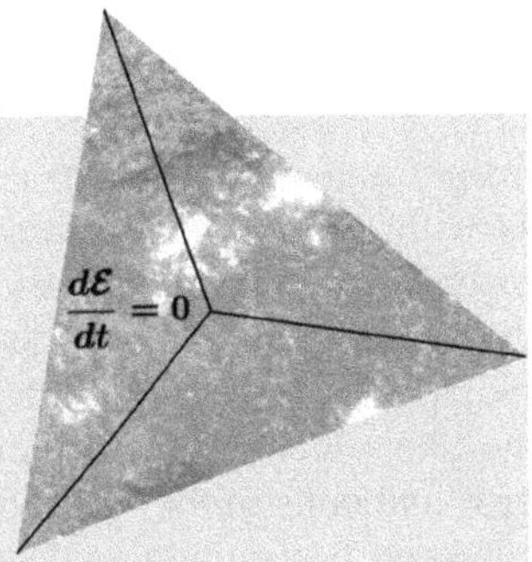

PRINCIPLE OF BALANCE OF MECHANICAL ENERGY

Thus far, we have described the fundamental aspects of classical mechanics as evident from much of the open literature, namely, the three branches, Newtonian, Lagrangian, and Hamiltonian mechanics. Summarizing, the principal conclusions are that Newtonian mechanics reflects the statement of the principle of balance of linear momentum, and it required the description of field variables as vectors with the Newtonian dynamical system represented as second-order in time, the presence of the k-number of constraints leading an N-body system to a dynamical system with $3N - k$ degrees of freedom, and an inertial reference frame. The Newtonian dynamical system involves $3N$-number of Cartesian variables. In contrast, although there is no new physics that has been brought about in comparison to Newtonian mechanics, Lagrangian mechanics introduced the concept of generalized coordinates in that termed as the configuration space, eliminated the constraints, and provided a dynamical system with the n_{dof} number of generalized coordinates (note that the number of degrees of freedom is $n_{\text{dof}} = 3N - k$). The representation of the equation of motion is also of second-order in time, but involving the scalar quantity, namely, the Lagrangian. On the other hand, Hamiltonian mechanics introduced the concept of phase space with $2n_{\text{dof}}$-number of canonical variables, also did not have any constraints, but inherits the representation of the equations of motion as a system of first-order in time via a scalar function, namely, the Hamiltonian. It also has not brought forth any new physics in contrast to the two previous branches of mechanics. As described subsequently in later chapters of this book, they are all equivalent to each other and each of the three branches of mechanics inherits certain pros and cons.

Various other descriptive scalar functions also exist in the literature, however, they have not been predominant. There exist various subject areas in mechanics and physics where it is desirable to have a direct measurable descriptive scalar function such as the Total Energy. As a viable alternative, and in contrast to the previously mentioned train of thought stemming from traditional practices, we describe in this chapter the introduction of and the notion of the so-called Total Energy representation of the equation of motion and the associated framework which is in configuration space. It involves n_{dof}-number of generalized coordinates without constraints, and with the representation of the equation of motion as a second-order in time system via a direct and measurable, built-in descriptive scalar function, namely, the Total Energy (in contrast to the scalar function, namely, the Hamiltonian in phase space and first-order system representation; and in contrast to the abstract scalar function, namely, the Lagrangian in configuration space).

Advances in Computational Dynamics of Particles, Materials and Structures, First Edition. Jason Har and Kumar K. Tamma.

6.1 INTRODUCTION

From the viewpoint of contemporary computational mechanics, the Total Energy framework is specifically targeted to capitalize upon certain computationally attractive features and it provides good physical insight and easy to grasp concepts. However, again there is no new physics that is brought forth just as with that of the Lagrangian and the Hamiltonian mechanics frameworks in contrast to Newtonian mechanics. In contemporary computational mechanics disciplines, for the general areas of computational dynamics, in particular, those applications dealing with particle dynamics, structural dynamics or elastodynamics associated with continuum mechanics, rigid/flexible structures and multi-body systems, and the like which are the principal areas of interest towards which the present book is targeted, there exist pros and cons in the use and selection of the respective frameworks. Historically, Newtonian mechanics has been fairly traditional and has been the starting point, and is popular.

Not much appears in the literature via the notion of scalar representations, and as an alternative to the scalar quantities such as the Lagrangian and the Hamiltonian, here in Chapter 6, we focus upon the notion of a measurable, built-in descriptive scalar function, namely, the Total Energy in the configuration space. The Total Energy we shall define herein is the kinetic energy plus the potential energy for N-body systems. We have previously shown that the Hamiltonian becomes equivalent to the total energy only as far as the system is conservative. Note that the autonomous Total Energy we shall employ here is defined on the velocity phase space (a tangent bundle) just as with the autonomous Lagrangian, whereas the autonomous Hamiltonian is defined on the phase space. As with the Lagrangian and the Hamiltonian scalar functions, the descriptive scalar function, namely, the Total Energy we employ in the configuration space possesses translational and rotational invariance properties, and the autonomous Total Energy has symmetry in time. Hence, for the present framework which is indeed associated with a built-in scalar function, Noether's theorem is readily applicable in that the spatial translation symmetry of the scalar quantity such as the Total Energy described in the configuration space, has symmetries in translation/rotation/time. Yet, it can be viewed as the third perspective (the first and second being the Lagrangian and the Hamiltonian, respectively) wherein, the built-in scalar function that is introduced, namely, the Total Energy [note that $\mathcal{E}(\mathbf{q}, \dot{\mathbf{q}}, t) : TQ \times \bar{\mathcal{T}} \rightarrow \mathbb{R}$] readily provides a different perspective and a new avenue. For example, it provides a meaningful and sound physical interpretation and practical utility (in contrast to a somewhat abstract scalar quantity such as the Lagrangian in the configuration space [$\mathcal{L}(\mathbf{q}, \dot{\mathbf{q}}, t) : TQ \times \bar{\mathcal{T}} \rightarrow \mathbb{R}$]) which yields a second-order system of equations that historically employ particular form of time discretization approaches just as in the Newtonian mechanics description for continuum mechanics. Alternatively, the Hamiltonian [note that $\mathcal{H}(\mathbf{p}, \mathbf{q}, t) : T^*Q \times \bar{\mathcal{T}} \rightarrow \mathbb{R}$, and it does not readily represent energy for the general cases] is not as popular for numerical discretization in the computational mechanics community due to first-order system representations.

6.2 PRINCIPLE OF BALANCE OF MECHANICAL ENERGY

In the preceding chapters, we independently described the various developments leading to the governing equations of motion for N-body systems from totally different viewpoints with both vector and scalar formalisms. This was via two fundamentally different starting points after having first established the relations stemming from Newton's second law which reflects the principle of balance of linear momentum to, namely, (i) the principle of virtual work in dynamics, and (ii) Hamilton's principle (and, equivalently, the alternative, namely, Hamilton's Law of Varying Action). Once the relations were firmly established, we instead directly employed these two fundamental principles thenceforth as the principal starting point for the subsequent theoretical and numerical developments. In both the preceding chapters, the notion of variation and variational concepts were invoked in developing the various formulations applicable to holonomic systems starting from two different fundamental principles. In contrast, in this chapter, yet as another alternative, we instead invoke the notion of differential calculus with the formalism via the principle of balance of mechanical energy as the fundamental axiom and starting point. However, it is to be noted that the developments via differential calculus setting are strictly restricted to holonomic-scleronomic systems. In particular, the limitations for establishing the equivalences amongst the various frameworks via the differential calculus setting are that the kinetic energy is quadratic in generalized velocities and the potential energy is not velocity dependent for N-body systems. In contrast to the differential calculus setting, we briefly also highlight in the Appendix, the variational calculus setting which is applicable to holonomic systems.

Differential versus Variational Calculus in Holonomic-Scleronomic and Holonomic-Rheonomic Systems Some brief highlights of the holonomic and holonomic-scleronomic systems are overviewed including the notion of variational and differential calculus. This is followed next by detailed developments for N-body dynamics starting from the principle of balance of mechanical energy which provides new and different viewpoints. The developments described in this chapter are shown to also lead to the same representations of the governing equations of motion as described in the preceding chapters, but via different viewpoints and new perspectives.

Boltzmann classified holonomic systems into scleronomic (meaning rigid) and rheonomic (meaning flow) (Lanczos 1970). In rheonomic systems, the introduction of the concept of variation enables one to eliminate the time partial derivative term. Therefore, the direction of the differential in the rheonomic systems is different from that of the variation due to the presence of the partial derivative of any physical quantity depending upon time explicitly as shown in Figure 6.1. Literature cites certain advantages of the variational calculus versus the differential calculus in rheonomic systems; but this needs to be further explored. Any advantages disappear in scleronomic systems (Greenwood 1977; Meirovitch 2004). Note that the concept of variation is imaginary, and it is an abstract mathematical quantity. Most of the science and engineering applications of focus here with the imposed restrictions for holonomic-scleronomic systems, primarily fall into the class of the scleronomic systems (de Castro 2000). In scleronomic systems, the differential shares the same direction with the variation (Lanczos 1970, page 95) (Greenwood 1977, page 14). Therefore, for science and engineering applications that are scleronomic, the variational approach is not particularly indispensable for obtaining governing equations or for finding approximate solutions; and alternatives such as the differential formulations can also be employed.

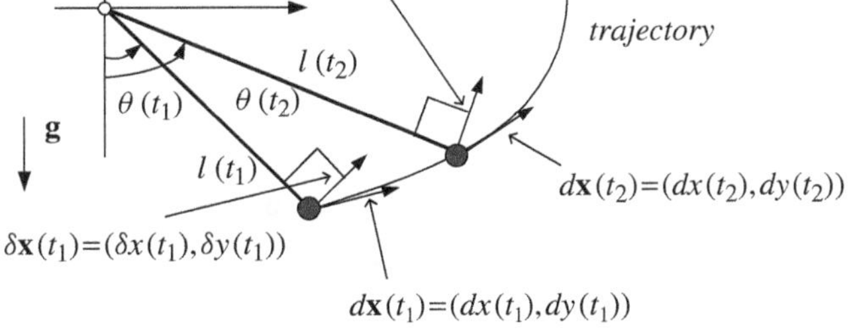

Figure 6.1. Virtual and actual infinitesimal displacements in holonomic-rheonomic constraints

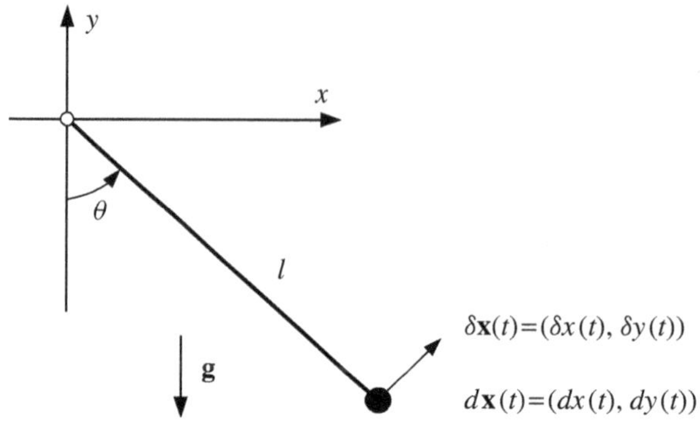

Figure 6.2. Virtual and actual infinitesimal displacements in holonomic-scleronomic constraints

As an illustrative example, recall the dynamical system of a planar pendulum whose length remains constant during the motion. In a holonomic-scleronomic constraint system where the configuration space is fixed in time, both the infinitesimal and virtual displacement vectors remain on the tangent line of the curve as shown in Figure 6.2. Therefore, the infinitesimal displacement shares the same direction with the virtual displacement for the simple planar pendulum problem. On the other hand, for the case of a holonomic-rheonomic constraint system where the configuration space is moving in time, the length of the pendulum bar depends upon time and the infinitesimal displacement does not share the same direction with the virtual displacement as shown in Figure 6.1.

In what follows, we first describe the developments with strict restrictions to mechanical systems only, and emanating from the principle of balance of mechanical energy which is also equivalent to the well known theorem of power expended for mechanical systems in the sense of continuum mechanics (and the differential calculus setting). In the literature (Truesdell and Toupin 1960, page 569) (Gurtin 1972, page 65) (Simo and Tarnow 1992), in the context of continuum mechanics, the theorem of power expended is originally shown to be able to be derived from the weighted residual form involving Cauchy's first law of motion (Truesdell and Toupin 1960, page 545). The weighting function is the velocity. In rational continuum mechanics, the theorem of power expended states that the power input, the rate of the work done by the external forces, equals the rate of change of kinetic energy and the stress power (Truesdell and Toupin 1960, page 569). In the context of N-body systems, the present formalism via the principle of balance of mechanical energy and differential calculus can readily enable the developments leading to the same or equivalent vector and scalar strong forms of representation of the equations of motion (keeping in mind the imposed restrictions for holonomic-scleronomic systems) just as those obtained from the other two fundamental principles. That is, the principle of virtual work in dynamics stemming from its relation to the balance of linear momentum and Hamilton's principle (or Hamilton's law of varying action as an alternative), respectively. Furthermore, it can subsequently enable the numerical discretizations to be conducted in a fairly natural setting with good physical insight. In addition, in the Appendix, we also address the setting of variational calculus and the Total Energy representation and framework which is applicable to holonomic systems.

6.3 TOTAL ENERGY REPRESENTATIONS AND FRAMEWORK IN THE DIFFERENTIAL CALCULUS SETTING

We first establish the relations starting with Newton's second law to the principle of balance of mechanical energy in the differential calculus setting. Thenceforth, we directly employ this principle as the starting point for all subsequent theoretical and numerical developments keeping in mind the imposed restrictions for holonomic-scleronomic systems.

6.3.1 Principle of Balance of Mechanical Energy: Conservative System

Consider a conservative dynamical system of N particles moving in three-dimensional Euclidean space. Then, Newton's second law of motion can be rewritten as

$$\mathbf{M}\ddot{\mathbf{x}} = \mathbf{F}_c(\mathbf{x}) + \mathbf{C}(\mathbf{x}) \tag{6.1}$$

where $\mathbf{F}_c(\mathbf{x}) : \mathbb{R}^{3N} \to \mathbb{R}^{3N}$ denotes the conservative force, $\mathbf{C}(\mathbf{x})$ the constraint force, $\ddot{\mathbf{x}}$ a set of acceleration vectors $\ddot{\mathbf{x}}^i$, and $\mathbf{M}$ the total diagonal mass matrix. Since the conservative system is subject to holonomic-scleronomic constraints, the above equation must be subject to the k-number of holonomic-scleronomic constraint equations, i.e.,

$$\mathbf{f}(\mathbf{x}) = \left(f_1, \ldots, f_i\left(\mathbf{x}^1, \ldots, \mathbf{x}^N\right), \ldots, f_k\right) = \mathbf{0} \tag{6.2}$$

Hence, the constraint force $\mathbf{C}(\mathbf{x})$ is given by

$$\mathbf{C}(\mathbf{x}) = \boldsymbol{\lambda} \cdot \nabla \mathbf{f}(\mathbf{x}), \quad \mathbf{f}(\mathbf{x}) = \left(f_1, \ldots, f_i\left(\mathbf{x}^1, \ldots, \mathbf{x}^N\right), \ldots, f_k\right) = \mathbf{0} \tag{6.3}$$

where $\boldsymbol{\lambda}$ denotes a set of the Lagrange multipliers and $\mathbf{C}$ is the total constraint force. Recall that in Chapter 3, we have discussed that in holonomic-scleronomic systems the constraint force is orthogonal to the velocity, i.e.,

$$\mathbf{C} \cdot \dot{\mathbf{x}} = 0 \tag{6.4}$$

In other words, the constraint force does not create power as it does not make any contribution to the virtual work. It is important to note that for the case of holonomic-rheonomic systems, the total energy is not conserved, and the constraint force does create power.

Owing to Equation (6.4), taking the inner product of Equation (6.1) and velocity leads to

$$(\mathbf{M}\ddot{\mathbf{x}} - \mathbf{F}_c) \cdot \dot{\mathbf{x}}(t) = 0 \tag{6.5}$$

Invoking the kinetic energy $\mathcal{K}(\dot{\mathbf{x}}) : \mathbb{R}^{3N} \to \mathbb{R}$ and the potential energy $\mathcal{U}(\mathbf{x}) : \mathbb{R}^{3N} \to \mathbb{R}$ for the conservative N-body dynamical system, i.e.,

$$\mathcal{K}(\dot{\mathbf{x}}) = \frac{1}{2}\dot{\mathbf{x}}(t) \cdot \mathbf{M}\dot{\mathbf{x}}(t), \quad \mathbf{F}_c = -\frac{\partial \mathcal{U}(\mathbf{x})}{\partial \mathbf{x}} \tag{6.6}$$

Equation (6.5) yields the following

$$\frac{d\mathcal{K}(\dot{\mathbf{x}})}{dt} + \frac{d\mathcal{U}(\mathbf{x})}{dt} = \frac{d\mathcal{E}(\mathbf{x}, \dot{\mathbf{x}})}{dt} = 0 \tag{6.7}$$

where the autonomous Total Energy $\mathcal{E}(\mathbf{x}, \dot{\mathbf{x}})$ is defined as

$$\mathcal{E}(\mathbf{x}, \dot{\mathbf{x}}) = \mathcal{K}(\dot{\mathbf{x}}) + \mathcal{U}(\mathbf{x}) \tag{6.8}$$

Note that $\ddot{\mathbf{x}} \cdot \mathbf{M}\dot{\mathbf{x}} = \dot{\mathbf{x}} \cdot \mathbf{M}\ddot{\mathbf{x}}$, because the mass matrix $\mathbf{M}$ is symmetric. Further, our attention is confined to the case that the potential energy is not dependent upon the velocity, since the system is conservative. The summation of the kinetic and potential energies for the conservative dynamical system is the total mechanical energy, or simply Total Energy, which remains constant. Equation (6.7) is referred to as the principle of conservation of Total Energy for N-body dynamical mechanical systems. As a consequence, we observe that for a conservative N-body dynamical system, Equation (6.5) is similar to the Lagrangian form of D'Alembert's principle shown in Equation (3.51), but the virtual displacement is replaced by the velocity.

From Equation (6.7), we cannot obtain the governing equations of motion, because $\dot{\mathbf{x}}(t)$ is not linearly independent due to the presence of constraints. Consequently, we need to rely upon the Lagrange multiplier method. Next, we take advantage of the generalized coordinates.

Generalized Coordinates and Descriptive Scalar Function: The Total Energy Just as the autonomous Lagrangian is defined on the velocity phase space, the autonomous Total Energy is also defined on the same velocity phase space. The generalized coordinates on the trajectory curve belong to the configuration space Q. In the configuration space, the dynamical system framework for the Total Energy is not subject to any constraints as in the Lagrangian dynamical system. Note that the configuration space is an n_{dof}-dimensional smooth manifold, and it is also differentiable. The dimension of the configuration corresponds to the number of degrees of freedom; hence the number of generalized coordinates equals to the number of degrees of freedom.

In the following Total Energy framework, we focus attention upon holonomic-scleronomic systems. Suppose that the k-number of scleronomic constraint equations are endowed with the system; the constraint equations are algebraic. The number of degrees of freedom for the system would be reduced to $n_{\mathrm{dof}} = 3N - k$, due to constraints; these constraints can be written in terms of the position vectors of N particles in the three-dimensional Euclidean space. In general, the scleronomic constraint equation in Equation (6.2) is given by

$$f_i\left(\mathbf{x}^1, \ldots, \mathbf{x}^N\right) = 0, \quad i = 1, \ldots, k \tag{6.9}$$

Invoking generalized coordinates $\mathbf{q}$ which belong to the configuration space Q, namely, $\mathbf{q}(t) \in Q \subset \mathbb{R}^{n_{\mathrm{dof}}}$ in the closed interval of time $t \in \bar{\mathcal{T}} = [t_0, t_1]$, the position vectors in the Cartesian coordinate system can be rewritten in terms of generalized coordinates as below:

$$\mathbf{x}^i = \mathbf{x}^i\left(q^1, \ldots, q^{n_{\mathrm{dof}}}; t\right) : Q \times \bar{\mathcal{T}} \to \mathbb{R}^3, \quad n_{\mathrm{dof}} = 3N - k, \tag{6.10}$$

where the configuration space is defined as

$$Q = \left\{ (\mathbf{q}(t)) \mid \mathbf{q} = \left(q^1(t), \ldots, q^{n_{\mathrm{dof}}}(t)\right) \in \mathbb{R}^{n_{\mathrm{dof}}} \right\} \tag{6.11}$$

Subsequently, the transformation equation in the holonomic-rheonomic constraint systems can be given by

$$d\mathbf{x} = \frac{\partial \mathbf{x}}{\partial \mathbf{q}} \cdot d\mathbf{q} + \frac{\partial \mathbf{x}}{\partial t} \tag{6.12}$$

whereas in the holonomic-scleronomic constraint system, the transformation equation yields

$$d\mathbf{x} = \frac{\partial \mathbf{x}}{\partial \mathbf{q}} \cdot d\mathbf{q} \tag{6.13}$$

Consequently, the transformation equation in the Total Energy framework is identical to that in the Lagrangian mechanics framework; in other words, the velocity in the Cartesian coordinate system can be expressed as

$$\dot{\mathbf{x}}(\mathbf{q}(\mathbf{x})) = \frac{\partial \mathbf{x}}{\partial \mathbf{q}} \cdot \dot{\mathbf{q}} \tag{6.14}$$

Then, by use of the above equation, Equation (6.5) can be rewritten in terms of generalized velocities,

$$(\mathbf{M}\ddot{\mathbf{x}} - \mathbf{F}_c) \cdot \frac{\partial \mathbf{x}}{\partial \mathbf{q}} \cdot \dot{\mathbf{q}} = 0 \tag{6.15}$$

The above equation in the differential calculus setting can be regarded as an analogy to the consequence of the Lagrangian form of D'Alembert's principle for conservative systems in the variational calculus setting. Hence, we have the relation

$$\mathbf{M}\ddot{\mathbf{x}} \cdot \frac{\partial \mathbf{x}}{\partial \mathbf{q}} = \frac{d}{dt}\left(\mathbf{M}\dot{\mathbf{x}} \cdot \frac{\partial \mathbf{x}}{\partial \mathbf{q}}\right) - \mathbf{M}\dot{\mathbf{x}} \cdot \frac{\partial \dot{\mathbf{x}}}{\partial \mathbf{q}} = \frac{d}{dt}\left(\mathbf{M}\dot{\mathbf{x}} \cdot \frac{\partial \dot{\mathbf{x}}}{\partial \dot{\mathbf{q}}}\right) - \mathbf{M}\dot{\mathbf{x}} \cdot \frac{\partial \dot{\mathbf{x}}}{\partial \mathbf{q}} \tag{6.16}$$

Since the system is holonomic-scleronomic, the kinetic energy must be quadratic in generalized velocities (Greenwood 1977) and it has no explicit time dependence. The above equation can be rewritten as

$$\mathbf{M}\ddot{\mathbf{x}} \cdot \frac{\partial \mathbf{x}}{\partial \mathbf{q}} = \frac{d}{dt}\left(\frac{\partial \mathcal{K}(\mathbf{q}, \dot{\mathbf{q}})}{\partial \dot{\mathbf{q}}}\right) - \frac{\partial \mathcal{K}(\mathbf{q}, \dot{\mathbf{q}})}{\partial \mathbf{q}} \tag{6.17}$$

In view of Equation (6.6), we have the generalized force $\mathbf{Q}_c$,

$$\mathbf{Q}_c = \mathbf{F}_c \cdot \frac{\partial \mathbf{x}}{\partial \mathbf{q}} = -\frac{\partial \mathcal{U}(\mathbf{q})}{\partial \mathbf{q}} \tag{6.18}$$

As a consequence, in terms of the kinetic energy $\mathcal{K}(\mathbf{q}, \dot{\mathbf{q}})$ and potential energy $\mathcal{U}(\mathbf{q})$, Equation (6.15) can be rewritten as

$$\frac{d\mathcal{E}(\mathbf{q}, \dot{\mathbf{q}})}{dt} = \left[\frac{d}{dt}\left(\frac{\partial \mathcal{K}(\mathbf{q}, \dot{\mathbf{q}})}{\partial \dot{\mathbf{q}}}\right) - \frac{\partial \mathcal{K}(\mathbf{q}, \dot{\mathbf{q}})}{\partial \mathbf{q}} + \frac{\partial \mathcal{U}(\mathbf{q})}{\partial \mathbf{q}}\right] \cdot \dot{\mathbf{q}} = 0 \tag{6.19}$$

Furthermore, in terms of the autonomous Total Energy, Equation (6.19) yields

$$\left[\frac{d}{dt}\left(\frac{\partial \mathcal{E}(\mathbf{q}, \dot{\mathbf{q}})}{\partial \dot{\mathbf{q}}}\right) + \frac{\partial \mathcal{E}(\mathbf{q}, \dot{\mathbf{q}})}{\partial \mathbf{q}} - 2\frac{\partial \mathcal{K}(\mathbf{q}, \dot{\mathbf{q}})}{\partial \mathbf{q}}\right] \cdot \dot{\mathbf{q}} = 0 \tag{6.20}$$

As $\dot{\mathbf{q}}$ is linearly independent, the coefficient of $\dot{\mathbf{q}}$ should vanish. Therefore, we obtain the governing equation of motion

$$\frac{d}{dt}\left(\frac{\partial \mathcal{E}(\mathbf{q}, \dot{\mathbf{q}})}{\partial \dot{\mathbf{q}}}\right) + \frac{\partial \mathcal{E}(\mathbf{q}, \dot{\mathbf{q}})}{\partial \mathbf{q}} - 2\frac{\partial \mathcal{K}(\mathbf{q}, \dot{\mathbf{q}})}{\partial \mathbf{q}} = \mathbf{0} \tag{6.21}$$

For the special case when the kinetic energy is independent of the generalized coordinates, the above equation reduces to

$$\frac{d}{dt}\left(\frac{\partial \mathcal{E}(\mathbf{q}, \dot{\mathbf{q}})}{\partial \dot{\mathbf{q}}}\right) + \frac{\partial \mathcal{E}(\mathbf{q}, \dot{\mathbf{q}})}{\partial \mathbf{q}} = \mathbf{0} \tag{6.22}$$

The above equation is referred to as the Total Energy representation of the equation of motion.

The domain of the autonomous Total Energy is also an ordered pair $(\mathbf{q}, \dot{\mathbf{q}}) \in TQ$, that is, the pair is a member of the tangent bundle TQ, which is also a differentiable manifold. The tangent bundle, namely the phase velocity space, is a disjoint union of the tangent spaces to each point of the configuration space Q as

$$TQ = \left\{(\mathbf{q}, \dot{\mathbf{q}}) | \, \mathbf{q} \in Q, \dot{\mathbf{q}} \in T_q Q\right\} = \bigcup_{q \in Q} T_q Q \tag{6.23}$$

Then, the Total Energy is defined as a scalar-valued function of multivariables, that is,

$$\mathcal{E}(\mathbf{q}, \dot{\mathbf{q}}, t) : TQ \times \bar{\mathcal{T}} \rightarrow \mathbb{R}, \quad t \in \mathcal{T} \subset \mathbb{R} \tag{6.24}$$

Note that the autonomous Total Energy, a scalar function of generalized coordinates and generalized velocities, is defined on the tangent bundle.

Illustrative Example

Example 6.1 Spring-Mass Conservative Discrete System

Consider the spring-mass discrete system under a simple-harmonic motion, which may readily help one to find an analogy between the spring-mass system and continuous-body systems in dynamical motion. In addition, this analogy makes it clear to define the total energy in conservative or nonconservative systems. Suppose that the body with mass m in Figure 6.3 is subject to a conservative force and a constant external force $\bar{f}$. Since the external force does not depend upon time, the relation between the external work $\mathcal{W}_{\text{ext}}(x)$ and the external potential energy $\mathcal{U}_{\text{ext}}(x)$ is given by

$$\mathcal{W}_{\text{ext}}(x) = \bar{f} \cdot x(t) = -\mathcal{U}_{\text{ext}}(x) \tag{6.25}$$

As there exists the external potential energy, the external work is said to be conservative and the external force is derivable from the external potential energy as follows:

$$\bar{f} = -\frac{d\mathcal{U}_{\text{ext}}(x)}{dx} \tag{6.26}$$

Then, the kinetic energy $\mathcal{K}(\dot{x}) : \mathbb{R} \rightarrow \mathbb{R}$ of the discrete system is given by

$$\mathcal{K}(\dot{x}) = \frac{1}{2} m \dot{x}^2 \tag{6.27}$$

The conservative force $f_{\text{con}}(x) : \mathbb{R} \rightarrow \mathbb{R}$ assures that there exists the potential energy $\mathcal{U}_{\text{int}}(x) : \mathbb{R} \rightarrow \mathbb{R}$ in the spring, and is given as

$$\mathcal{U}_{\text{int}}(x) = \frac{1}{2} k x^2, \quad f_{\text{con}}(x) = \frac{d\mathcal{U}_{\text{int}}(x)}{dx} = kx \tag{6.28}$$

Since the spring-mass system is conservative, the Total Energy is conserved, i.e.,

$$\begin{aligned} \frac{d\mathcal{E}(\dot{x}, x)}{dt} &= \frac{d}{dt}(\mathcal{K}(\dot{x}) + \mathcal{U}(x)) \\ &= \frac{d\mathcal{K}(\dot{x})}{dt} + \frac{d\mathcal{U}_{\text{int}}(x)}{dt} + \frac{d\mathcal{U}_{\text{ext}}(x)}{dt} \\ &= (m\ddot{x} + kx - \bar{f})\dot{x} = 0 \end{aligned} \tag{6.29}$$

where $\mathcal{U}(x)$ denotes the potential energy for the spring-mass system, i.e.,

$$\mathcal{U}(x) = \mathcal{U}_{\text{int}}(x) + \mathcal{U}_{\text{ext}}(x) \tag{6.30}$$

Hence, for conservative systems, we define the Total Energy as

$$\mathcal{E}(x, \dot{x}) = \mathcal{K}(\dot{x}) + \mathcal{U}_{\text{int}}(x) + \mathcal{U}_{\text{ext}}(x) \tag{6.31}$$

The internal potential energy in the spring-mass system corresponds to the internal potential energy in a continuous-body dynamical system.

The spring-mass conservative discrete system we have discussed so far is a single-degree-of-freedom system, and is not subject to any constraints. It is to be noted that the variable x in Figure 6.3 is a generalized coordinate. Consequently, we obtain the governing equation from Equation (6.29) as follows:

$$m\ddot{x} + kx = \bar{f} \tag{6.32}$$

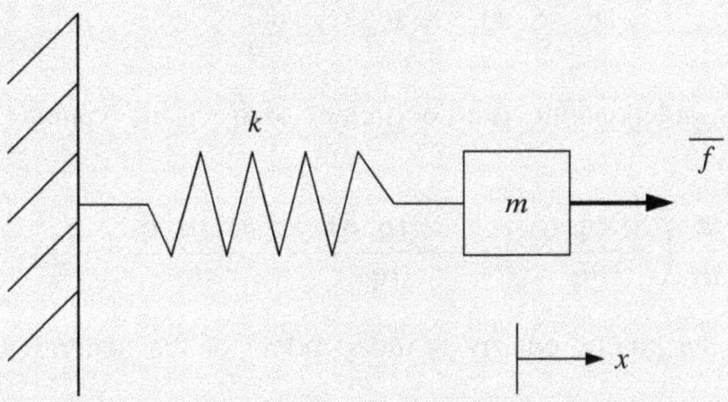

Figure 6.3. Spring-mass conservative discrete system

6.3.2 Principle of Balance of Mechanical Energy: Nonconservative System

A conservative N-body dynamical system subject to holonomic constraints is scleronomic, however, a holonomic-scleronomic constraint system is not always a conservative system. In this section, we consider a nonconservative N-body dynamical system, which is still holonomic-scleronomic. Hence, the Total Energy for the nonconservative N-body dynamical system is no longer a constant of motion. However, as the dynamical system is holonomic-scleronomic, the constraint force does not create any power, that is

$$\mathbf{C} \cdot \dot{\mathbf{x}} = (\mathbf{M}\ddot{\mathbf{x}}(t) - \mathbf{F}_c - \mathbf{F}_{nc}(t)) \cdot \dot{\mathbf{x}}(t) = 0 \tag{6.33}$$

where $\mathbf{F}_{nc}(t) : \bar{\mathcal{T}} \rightarrow \mathbb{R}^{3N}$, $t \in \bar{\mathcal{T}} = [t_0, t_1]$ denotes the total nonconservative force. Thus, Equation (6.33) can be recast into

$$\mathbf{M}\ddot{\mathbf{x}} \cdot \dot{\mathbf{x}}(t) - \mathbf{F}_c \cdot \dot{\mathbf{x}}(t) = \mathbf{F}_{nc}(t) \cdot \dot{\mathbf{x}}(t) \tag{6.34}$$

Invoking the kinetic energy $\mathcal{K}(\dot{\mathbf{x}}) : \mathbb{R}^3 \rightarrow \mathbb{R}$ and the potential energy $\mathcal{U}(\mathbf{x}) : \mathbb{R}^3 \rightarrow \mathbb{R}$ for the nonconservative N-body dynamical system, we have

$$\frac{d\mathcal{E}(\mathbf{x}, \dot{\mathbf{x}})}{dt} = \mathbf{F}_{nc}(t) \cdot \dot{\mathbf{x}}(t), \quad \mathcal{E}(\mathbf{x}, \dot{\mathbf{x}}) = \mathcal{K}(\dot{\mathbf{x}}) + \mathcal{U}(\mathbf{x}) \tag{6.35}$$

We observe that the above equation implies that the mechanical energy is balanced in the Cartesian coordinate system. Equation (6.35) is referred to as the principle of balance of mechanical energy for N-body dynamical systems. In addition, we have an alternative time differential equation in terms of the Total Energy as Newton's vectorial differential equations are transformed into a scalar time differential equation. Note that the scalar time differential equation in Equation (6.35) is subject to holonomic-scleronomic constraint equations in the Cartesian coordinate system.

Next, we shall eliminate the constraints by introducing generalized coordinates in configuration space. Since the dynamical system is holonomic-scleronomic, the transformation equation given in Equation (6.13) is still valid. In view of equation (6.20), Equation (6.35) can be rewritten as

$$\left[\frac{d}{dt}\left(\frac{\partial \mathcal{E}(\mathbf{q},\dot{\mathbf{q}})}{\partial \dot{\mathbf{q}}}\right)+\frac{\partial \mathcal{E}(\mathbf{q},\dot{\mathbf{q}})}{\partial \mathbf{q}}-2\frac{\partial \mathcal{K}(\mathbf{q},\dot{\mathbf{q}})}{\partial \mathbf{q}}\right]\cdot\dot{\mathbf{q}}=\mathbf{Q}_{\text{nc}}\cdot\dot{\mathbf{q}} \tag{6.36}$$

where the nonconservative generalized force $\mathbf{Q}_{\text{nc}}$, which is not derivable from the potential energy, is given by

$$\mathbf{Q}_{\text{nc}}=\mathbf{F}_{\text{nc}}(t)\cdot\frac{\partial \mathbf{x}}{\partial \mathbf{q}} \tag{6.37}$$

In the end, as $\dot{\mathbf{q}}$ is linearly independent, the coefficient of $\dot{\mathbf{q}}$ should vanish. Therefore, we obtain the governing equation of motion

$$\frac{d}{dt}\left(\frac{\partial \mathcal{E}(\mathbf{q},\dot{\mathbf{q}})}{\partial \dot{\mathbf{q}}}\right)+\frac{\partial \mathcal{E}(\mathbf{q},\dot{\mathbf{q}})}{\partial \mathbf{q}}-2\frac{\partial \mathcal{K}(\mathbf{q},\dot{\mathbf{q}})}{\partial \mathbf{q}}=\mathbf{Q}_{\text{nc}} \tag{6.38}$$

For the special case when the kinetic energy is independent of the generalized coordinates, the above equation reduces to

$$\frac{d}{dt}\left(\frac{\partial \mathcal{E}(\mathbf{q},\dot{\mathbf{q}})}{\partial \dot{\mathbf{q}}}\right)+\frac{\partial \mathcal{E}(\mathbf{q},\dot{\mathbf{q}})}{\partial \mathbf{q}}=\mathbf{Q}_{\text{nc}} \tag{6.39}$$

which is the Total Energy representation of the equation of motion for a nonconservative N-body dynamical system subject to holonomic-scleronomic constraints.

Illustrative Example

Example 6.2 Spring-Mass Nonconservative Discrete System

Let us again consider the spring-mass discrete system where the external force is dependent upon time. Therefore, the spring-mass discrete system is nonconservative as shown in Figure 6.4. The external work $\mathcal{W}_{\text{nc}}$ is nonconservative due to the time-dependent external force, namely,

$$\mathcal{W}_{\text{nc}}(x,t)=f_{\text{nc}}(t)\cdot x(t) \tag{6.40}$$

The rate of the external work is equivalent to the power input as follows,

$$\mathcal{P}_{\text{input}}=f_{\text{nc}}(t)\,\dot{x}(t) \tag{6.41}$$

Then, the rate of the Total Energy is given as

$$\frac{d\mathcal{E}(x,\dot{x})}{dt}=\frac{d\mathcal{K}(\dot{x})}{dt}+\frac{d\mathcal{U}_{\text{int}}(x)}{dt}=\mathcal{P}_{\text{input}} \tag{6.42}$$

Note that there does not exist the external potential energy. The total potential energy equals the internal potential energy in the case of nonconservative systems. Therefore, it is easy to observe that the total energy increases due to the nonconservative force creating the power input. The total energy is no longer constant in the dynamical system. In short, note that the definition of the Total Energy is different between conservative systems and nonconservative systems.

The spring-mass nonconservative discrete system is also a single-degree-of-freedom system, and is not subject to any constraints. It is to be noted that the variable x in Figure 6.4 is a generalized coordinate. From Equation (6.42), we obtain the governing equation

$$m\ddot{x} + kx = f_{\mathrm{nc}}(t) \tag{6.43}$$

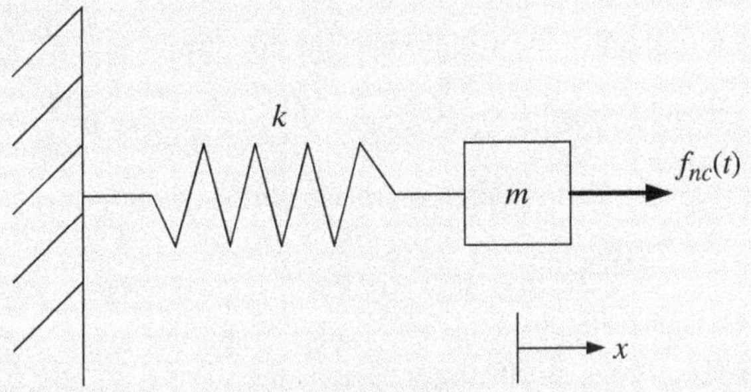

Figure 6.4. Spring-mass nonconservative discrete system

6.3.3 Newtonian Dynamical System: With/Without Constraints

In what follows, motivated by the rate of the total energy given in Equation (6.35) in the Cartesian coordinate system, we shall show that Newtonian dynamical system in the Cartesian coordinate system can be obtained from the rate of the total energy on the condition that the dynamical system is not subject to any constraints (Arnold 1989, page 22).

In three-dimensional Euclidean space where an N-body dynamical system is in motion without constraints, the rate of the total energy can be given by

$$\begin{aligned}
&\frac{d\mathcal{E}(\mathbf{x}, \dot{\mathbf{x}})}{dt} = \mathbf{F}_{\mathrm{nc}}(t) \cdot \dot{\mathbf{x}}(t), \quad \mathcal{E}(\mathbf{x}, \dot{\mathbf{x}}) = \mathcal{K}(\dot{\mathbf{x}}) + \mathcal{U}(\mathbf{x}), \\
&\mathcal{K}(\dot{\mathbf{x}}) = \frac{1}{2}\dot{\mathbf{x}}(t) \cdot \mathbf{M}\dot{\mathbf{x}}(t), \quad \mathcal{U}(\mathbf{x}) = -\mathbf{F}_{\mathrm{c}} \cdot \mathbf{x},
\end{aligned} \tag{6.44}$$

Since the mass matrix is symmetric, we have the relations

$$\frac{d\mathcal{K}(\dot{\mathbf{x}})}{dt} = \frac{d\mathcal{K}}{d\dot{\mathbf{x}}} \cdot \ddot{\mathbf{x}} = \ddot{\mathbf{x}} \cdot \mathbf{M}\dot{\mathbf{x}} = \dot{\mathbf{x}} \cdot \mathbf{M}\ddot{\mathbf{x}} \tag{6.45}$$

Therefore, the rate of the total energy yields

$$(\mathbf{M}\ddot{\mathbf{x}}(t) - \mathbf{F}_{\mathrm{c}} - \mathbf{F}_{\mathrm{nc}}(t)) \cdot \dot{\mathbf{x}}(t) = 0 \tag{6.46}$$

Since there is no constraint, the term $\dot{\mathbf{x}}$ is linearly independent for this system. Therefore, from the above equation, we obtain the equation

$$\mathbf{M}\ddot{\mathbf{x}}(t) = \mathbf{F}_{\mathrm{c}} + \mathbf{F}_{\mathrm{nc}}(t) \tag{6.47}$$

which represents the Newtonian dynamical system as a set of second-order differential equations.

Consider the case that an N-body dynamical system is subject to holonomic constraints, and is in motion in three-dimensional Euclidean space. Note that the Newtonian dynamical system is redundant due to the presence of constraints; in other words, the number of differential equations is less than the number of Cartesian coordinate variables. For an N-body dynamical system which is in motion under scleronomic constraints, the principle of balance of mechanical energy yields

$$\frac{d\mathcal{E}(\mathbf{x},\dot{\mathbf{x}})}{dt} = \mathbf{F}_{\mathrm{nc}}(t)\cdot\dot{\mathbf{x}}(t), \; \mathcal{E}(\mathbf{x},\dot{\mathbf{x}}) = \mathcal{K}(\dot{\mathbf{x}}) + \mathcal{U}(\mathbf{x}) \tag{6.48}$$

which is subject to the holonomic-scleronomic constraint equations

$$\mathbf{f} = \left(f_1,\ldots,f_i\left(\mathbf{x}^1,\ldots,\mathbf{x}^N\right),\ldots,f_k\right) = \mathbf{0} \tag{6.49}$$

By taking advantage of the Lagrange multiplier method, we have

$$\frac{d\mathcal{E}(\mathbf{x},\dot{\mathbf{x}})}{dt} = \mathbf{F}_{\mathrm{nc}}(t)\cdot\dot{\mathbf{x}}(t) + \boldsymbol{\lambda}\cdot\nabla\mathbf{f}(\mathbf{x}) \tag{6.50}$$

where $\boldsymbol{\lambda}$ denotes a set of the Lagrange multipliers. Then, the above equation leads to

$$\left(\mathbf{M}\ddot{\mathbf{x}}(t) - \mathbf{F}_{\mathrm{c}} - \mathbf{F}_{\mathrm{nc}}(t) - \boldsymbol{\lambda}\cdot\nabla\mathbf{f}(\mathbf{x})\right)\cdot\dot{\mathbf{x}}(t) = 0 \tag{6.51}$$

Since the N-body dynamical system is subject to holonomic constraints, the term $\dot{\mathbf{x}}$ is no longer linearly independent. However, if $\boldsymbol{\lambda}$ is chosen to make the coefficient of $\dot{\mathbf{x}}(t)$ in Equation (6.51) vanish, we obtain

$$\mathbf{M}\ddot{\mathbf{x}}(t) = \mathbf{F}_{\mathrm{c}} + \mathbf{F}_{\mathrm{nc}}(t) + \boldsymbol{\lambda}\cdot\nabla\mathbf{f}(\mathbf{x}) \tag{6.52}$$

which is the Newtonian dynamical system.

6.3.4 Lagrangian Dynamical System: Nonconservative/Conservative Systems - Descriptive Scalar Function, the Lagrangian

In this section, we are motivated to derive Lagrange's equations from the rate of the total energy in configuration space. Sometimes, we have Cartesian coordinate variables as dependent variables in the Cartesian coordinate system. In such a case, we introduce the Lagrange multiplier to obtain Lagrange's equations of motion in the Cartesian coordinate system.

Suppose that a conservative N-body dynamical system is subject to scleronomic constraints, where the Jacobi's integral is constant. Then, the rate of the Jacobi's integral is given by

$$\frac{d\mathcal{E}(\mathbf{q},\dot{\mathbf{q}})}{dt} = \frac{d}{dt}\left(\frac{\partial\mathcal{L}(\mathbf{q},\dot{\mathbf{q}})}{\partial\dot{\mathbf{q}}}\cdot\dot{\mathbf{q}} - \mathcal{L}(\mathbf{q},\dot{\mathbf{q}})\right) = 0 \tag{6.53}$$

Then due to the chain and product rules, the above equation reduces to

$$\left[\frac{d}{dt}\left(\frac{\partial\mathcal{L}(\mathbf{q},\dot{\mathbf{q}})}{\partial\dot{\mathbf{q}}}\right) - \frac{\partial\mathcal{L}(\mathbf{q},\dot{\mathbf{q}})}{\partial\mathbf{q}}\right]\cdot\dot{\mathbf{q}} = 0 \tag{6.54}$$

Since $\dot{\mathbf{q}}$ is linearly independent, we have Lagrange's equations of motion for a conservative system as follows,

$$\frac{d}{dt}\left(\frac{\partial \mathcal{L}(\mathbf{q},\dot{\mathbf{q}})}{\partial \dot{\mathbf{q}}}\right) - \frac{\partial \mathcal{L}(\mathbf{q},\dot{\mathbf{q}})}{\partial \mathbf{q}} = \mathbf{0} \tag{6.55}$$

Next, we consider a nonconservative N-body dynamical system, which is subject to scleronomic constraints. In this case, the Jacobi's integral is no longer constant. Then, the rate of the Jacobi's integral can be given by

$$\frac{d\mathcal{E}(\mathbf{q},\dot{\mathbf{q}})}{dt} = \frac{d}{dt}\left(\frac{\partial \mathcal{L}(\mathbf{q},\dot{\mathbf{q}})}{\partial \dot{\mathbf{q}}}\cdot\dot{\mathbf{q}} - \mathcal{L}(\mathbf{q},\dot{\mathbf{q}})\right) = \mathbf{Q}_{\mathrm{nc}}\cdot\dot{\mathbf{q}} \tag{6.56}$$

where $\mathbf{Q}_{\mathrm{nc}}$ denotes the nonconservative generalized force, which is not derivable from the potential energy. Then due to the chain and product rules, the above equation reduces to

$$\left[\frac{d}{dt}\left(\frac{\partial \mathcal{L}(\mathbf{q},\dot{\mathbf{q}})}{\partial \dot{\mathbf{q}}}\right) - \frac{\partial \mathcal{L}(\mathbf{q},\dot{\mathbf{q}})}{\partial \mathbf{q}}\right]\cdot\dot{\mathbf{q}} = \mathbf{Q}_{\mathrm{nc}}\cdot\dot{\mathbf{q}} \tag{6.57}$$

Since $\dot{\mathbf{q}}$ is linearly independent, we have Lagrange's equations of motion for nonconservative systems as follows,

$$\frac{d}{dt}\left(\frac{\partial \mathcal{L}(\mathbf{q},\dot{\mathbf{q}})}{\partial \dot{\mathbf{q}}}\right) - \frac{\partial \mathcal{L}(\mathbf{q},\dot{\mathbf{q}})}{\partial \mathbf{q}} = \mathbf{Q}_{\mathrm{nc}} \tag{6.58}$$

Illustrative Examples

Example 6.3 Spherical Pendulum

Consider the spherical pendulum problem which we discussed previously. It should be noted that the system of interest here is holonomic-scleronomic. The Jacobi's integral with $\mathbf{q} = (\theta, \phi)$ is given by

$$\frac{\partial \mathcal{L}(\mathbf{q},\dot{\mathbf{q}})}{\partial \dot{\mathbf{q}}}\cdot\dot{\mathbf{q}} - \mathcal{L}(\mathbf{q},\dot{\mathbf{q}}) = \frac{1}{2}ml^2\left(\dot{\theta}^2 + \dot{\phi}^2\sin^2\theta\right) - mgl\cos\theta \tag{6.59}$$

Taking the total derivative of the Jacobi's integral leads to the following

$$\begin{aligned}\frac{d}{dt}\left(\frac{\partial \mathcal{L}(\mathbf{q},\dot{\mathbf{q}})}{\partial \dot{\mathbf{q}}}\cdot\dot{\mathbf{q}} - \mathcal{L}(\mathbf{q},\dot{\mathbf{q}})\right) &= \left(ml^2\ddot{\theta} - ml^2\dot{\phi}^2\sin\theta\cos\theta + mgl\sin\theta\right)\dot{\theta} \\ &\quad + \left(ml^2\ddot{\phi}\sin^2\theta + 2ml^2\dot{\phi}\dot{\theta}\sin\theta\cos\theta\right)\dot{\phi} = 0\end{aligned} \tag{6.60}$$

which is equivalent to the following

$$\begin{aligned}\frac{d}{dt}\left(\frac{\partial \mathcal{L}(\mathbf{q},\dot{\mathbf{q}})}{\partial \dot{\mathbf{q}}}\cdot\dot{\mathbf{q}} - \mathcal{L}(\mathbf{q},\dot{\mathbf{q}})\right) &= \left[\frac{d}{dt}\left(\frac{\partial \mathcal{L}(\mathbf{q},\dot{\mathbf{q}})}{\partial \dot{\theta}}\right) - \frac{\partial \mathcal{L}(\mathbf{q},\dot{\mathbf{q}})}{\partial \theta}\right]\dot{\theta} \\ &\quad + \left[\frac{d}{dt}\left(\frac{\partial \mathcal{L}(\mathbf{q},\dot{\mathbf{q}})}{\partial \dot{\phi}}\right) - \frac{\partial \mathcal{L}(\mathbf{q},\dot{\mathbf{q}})}{\partial \phi}\right]\dot{\phi} = 0\end{aligned} \tag{6.61}$$

Since $\dot{\mathbf{q}}$ are linearly independent in configuration space, we have two differential equations as follows:

$$\begin{aligned}\frac{d}{dt}\left(\frac{\partial\mathcal{L}(\mathbf{q},\dot{\mathbf{q}})}{\partial\dot{\theta}}\right)-\frac{\partial\mathcal{L}(\mathbf{q},\dot{\mathbf{q}})}{\partial\theta}&=0,\\ \frac{d}{dt}\left(\frac{\partial\mathcal{L}(\mathbf{q},\dot{\mathbf{q}})}{\partial\dot{\phi}}\right)-\frac{\partial\mathcal{L}(\mathbf{q},\dot{\mathbf{q}})}{\partial\phi}&=0\end{aligned}\tag{6.62}$$

Example 6.4 Bead on a Rotating Circular Hoop

Consider the bead problem which we discussed previously. It should be noted that this system is holonomic-rheonomic. The Lagrangian of the system with a θ-generalized coordinate is given by

$$\mathcal{L}(\theta,\dot{\theta})=\frac{1}{2}mr^2\left(\dot{\theta}^2+\omega^2\sin^2\theta\right)+mgr\cos\theta\tag{6.63}$$

Taking the total derivative of the Jacobi's integral leads to

$$\frac{d}{dt}\left(\frac{\partial\mathcal{L}(\theta,\dot{\theta})}{\partial\dot{\theta}}\cdot\dot{\theta}-\mathcal{L}(\theta,\dot{\theta})\right)=\left(mr^2\ddot{\theta}+mr^2\omega^2\sin\theta\cos\theta+mgr\sin\theta\right)\dot{\theta}\tag{6.64}$$

However, the above equation can be rewritten as

$$\begin{aligned}\frac{d}{dt}\left(\frac{\partial\mathcal{L}(\theta,\dot{\theta})}{\partial\dot{\theta}}\cdot\dot{\theta}-\mathcal{L}(\theta,\dot{\theta})\right)&=\left(mr^2\ddot{\theta}-mr^2\omega^2\sin\theta\cos\theta+mgr\sin\theta\right)\dot{\theta}\\ &\quad+2mr^2\omega^2\sin\theta\cos\theta\dot{\theta}\end{aligned}\tag{6.65}$$

In view of Equation (3.173), Equation (6.65) results in

$$\frac{d}{dt}\left(\frac{\partial\mathcal{L}(\theta,\dot{\theta})}{\partial\dot{\theta}}\cdot\dot{\theta}-\mathcal{L}(\theta,\dot{\theta})\right)=2mr^2\omega^2\sin\theta\cos\theta\dot{\theta}\neq0\tag{6.66}$$

Thus, this system is not a conservative system. Note that the system is holonomic-rheonomic. This is a very important example in noting that even though the Lagrangian is autonomous, the total energy of the system is not conserved.

6.3.5 Hamiltonian Dynamical System: Nonconservative Systems - Descriptive Scalar Function, the Hamiltonian

We have shown that Lagrange's equations can be derived from the total time derivative of the Jacobi's integral for scleronomic constraint systems. Next, we shall show the derivation of Hamilton's equations of motion. Consider the total time derivative of the autonomous Hamiltonian as

$$\frac{d\mathcal{H}(\mathbf{q},\mathbf{p})}{dt}=\frac{\partial\mathcal{H}(\mathbf{q},\mathbf{p})}{\partial\mathbf{p}}\cdot\frac{d\mathbf{p}}{dt}+\frac{\partial\mathcal{H}(\mathbf{q},\mathbf{p})}{\partial\mathbf{q}}\cdot\frac{d\mathbf{q}}{dt}=\mathbf{Q}_{\mathrm{nc}}\cdot\dot{\mathbf{q}}\tag{6.67}$$

From the Legendre transformation, we also have

$$\frac{d\mathcal{H}(\mathbf{q},\mathbf{p})}{dt}=\frac{d}{dt}\left(\mathbf{p}\cdot\dot{\mathbf{q}}-\mathcal{L}(\mathbf{q},\dot{\mathbf{q}})\right)=\frac{d\mathbf{p}}{dt}\cdot\dot{\mathbf{q}}+\mathbf{p}\cdot\frac{d\dot{\mathbf{q}}}{dt}-\frac{d\mathcal{L}(\mathbf{q},\dot{\mathbf{q}})}{dt} \tag{6.68}$$

By virtue of Lagrange's equations of motion shown in Equation (6.55), the total time derivative of the Lagrangian can be given by

$$\frac{d\mathcal{L}(\mathbf{q},\dot{\mathbf{q}})}{dt}=\frac{\partial\mathcal{L}(\mathbf{q},\dot{\mathbf{q}})}{\partial\mathbf{q}}\cdot\frac{d\mathbf{q}}{dt}+\frac{\partial\mathcal{L}(\mathbf{q},\dot{\mathbf{q}})}{\partial\dot{\mathbf{q}}}\cdot\frac{d\dot{\mathbf{q}}}{dt}=(\dot{\mathbf{p}}-\mathbf{Q}_{\mathrm{nc}})\cdot\frac{d\mathbf{q}}{dt}+\mathbf{p}\cdot\frac{d\dot{\mathbf{q}}}{dt} \tag{6.69}$$

Substituting Equation (6.69) into Equation (6.68), Equation (6.67) leads to

$$\begin{aligned}\frac{d\mathcal{H}(\mathbf{q},\mathbf{p})}{dt}&=\frac{d\mathbf{p}}{dt}\cdot\dot{\mathbf{q}}+(\mathbf{Q}_{\mathrm{nc}}-\dot{\mathbf{p}})\cdot\frac{d\mathbf{q}}{dt}\\&=\mathbf{Q}_{\mathrm{nc}}\cdot\dot{\mathbf{q}}\end{aligned} \tag{6.70}$$

In view of Equation (6.67) and (6.70), we obtain the Hamilton's equations of motion as

$$\dot{\mathbf{p}}=-\frac{\partial\mathcal{H}(\mathbf{q},\mathbf{p})}{\partial\mathbf{q}}+\mathbf{Q}_{\mathrm{nc}},\quad \dot{\mathbf{q}}=\frac{\partial\mathcal{H}(\mathbf{q},\mathbf{p})}{\partial\mathbf{p}} \tag{6.71}$$

which are Hamilton's equations of motion for nonconservative scleronomic constraint systems.

Alternatively, Hamilton's equations of motion for nonconservative scleronomic constraint systems can be obtained from Equation (6.57). We can rewrite Equation (6.57) as

$$\frac{d}{dt}\left(\frac{\partial\mathcal{L}(\mathbf{q},\dot{\mathbf{q}})}{\partial\dot{\mathbf{q}}}\right)\cdot\dot{\mathbf{q}}-\frac{d\mathcal{L}(\mathbf{q},\dot{\mathbf{q}})}{dt}+\frac{\partial\mathcal{L}(\mathbf{q},\dot{\mathbf{q}})}{\partial\dot{\mathbf{q}}}\cdot\ddot{\mathbf{q}}=\mathbf{Q}_{\mathrm{nc}}\cdot\dot{\mathbf{q}} \tag{6.72}$$

By introducing the generalized momentum in Equation (3.179), the above equation can be rewritten as

$$\dot{\mathbf{p}}\cdot\dot{\mathbf{q}}+\mathbf{p}\cdot\ddot{\mathbf{q}}-\frac{d\mathcal{L}(\mathbf{q},\dot{\mathbf{q}})}{dt}=\mathbf{Q}_{\mathrm{nc}}\cdot\dot{\mathbf{q}} \tag{6.73}$$

By virtue of the Legendre transformation given in Equation (3.181), we obtain the expression

$$\frac{d\mathcal{L}(\mathbf{q},\dot{\mathbf{q}})}{dt}=\mathbf{p}\cdot\ddot{\mathbf{q}}+\dot{\mathbf{q}}\cdot\dot{\mathbf{p}}-\frac{d\mathcal{H}(\mathbf{q},\mathbf{p})}{dt} \tag{6.74}$$

Upon substitution of the above equation into Equation (6.73), we have

$$\left(\dot{\mathbf{p}}+\frac{\partial\mathcal{H}(\mathbf{q},\mathbf{p})}{\partial\mathbf{q}}-\mathbf{Q}_{\mathrm{nc}}\right)\cdot\dot{\mathbf{q}}-\left(\dot{\mathbf{q}}-\frac{\partial\mathcal{H}(\mathbf{q},\mathbf{p})}{\partial\mathbf{p}}\right)\cdot\dot{\mathbf{p}}=0 \tag{6.75}$$

Since $\dot{\mathbf{q}}$ and $\dot{\mathbf{p}}$ are linearly independent, we can obtain Equation (6.71).

Note that in the absence of nonconservative forces, we have the Hamiltonian dynamical conservative system.

Illustrative Example

Example 6.5 Bead on a Rotating Circular Hoop

We again consider the bead problem for which the dynamical system is holonomic-rheonomic. The total time derivative of the Hamiltonian gives

$$\frac{d\mathcal{H}(\theta, p_\theta)}{dt} = \dot{p}_\theta \dot{\theta} + p_\theta \ddot{\theta} - \left(mr^2\ddot{\theta} + mr^2\omega^2 \sin\theta \cos\theta - mgr \sin\theta\right)\dot{\theta} \tag{6.76}$$

By virtue of Equation (3.171), the above equation can be written as

$$\frac{d\mathcal{H}(\theta, p_\theta)}{dt} = \left(mr^2\ddot{\theta} - mr^2\omega^2 \sin\theta \cos\theta + mgr \sin\theta\right)\dot{\theta} = 0 \tag{6.77}$$

Thus, the Hamiltonian of the system is conserved, but the Hamiltonian is not equal to the total energy.

Time Weighted Residual Form Taking the integration over the given closed interval of time $t \in \bar{\mathcal{T}} = [t_0, t_1]$ for Equation (6.20), (6.54) and (6.75) leads to the weighted residual form in time. Hence, one can utilize the weighted residual form for designing time integration approaches such as the time weighted-residual method which includes time finite element formulations (Belytschko and Hughes 1983; Sha et al. 2003; Tamma et al. 2003, 2002, 2000; Wood 1990; Zhou and Tamma 2004a,c, 2006; Zhou et al. 2001, 2000; Zienkiewicz and Taylor 2006; Zienkiewicz et al. 2005, 1984).

6.4 APPENDIX: TOTAL ENERGY REPRESENTATIONS AND FRAMEWORK IN THE VARIATIONAL CALCULUS SETTING

So far, we have discussed the Total Energy representations and framework in the differential calculus setting, and starting from the principle of balance of mechanical energy where the focus and limitations were restricted to holonomic-scleronomic constraint systems. From a different perspective, next, we also discuss the Total Energy representations and framework in the setting of variational calculus. In contrast, our attention is now confined to holonomic constraint systems, where in general, the total energy for the dynamical system might not be conserved. Consequently in the variational calculus setting, we describe a more general representation, that is, we relax the condition of holonomic-scleronomic system arising in the differential calculus setting. The objective here is to describe the development of the governing equation of motion for the Total Energy framework from a different perspective: (i) D'Alembert's principle/Principle of virtual work, and (ii) Hamilton's principle/Hamilton's law of varying action. Obviously all of the above lead to the same representation of the equation of motion in the Total Energy framework. In the special case of holonomic-scleronomic systems, we obviously obtain the same representations as that from the principle of balance of mechanical energy which was described in the previous section.

6.4.1 Total Energy Representation of the Equation of Motion via the Lagrangian Form of D'Alembert's Principle/Principle of Virtual Work

Total Energy Representation of Equation of Motion in the Configuration Space In Section 3.2.6, we have shown that Lagrange's equation of motion is derived from the Lagrangian form of

D'Alembert's principle for a holonomic constraint system. In this section, similarly we would like to derive the Total Energy representation of the equation of motion from the Lagrangian form of D'Alembert's principle. The Lagrangian form of D'Alembert's principle in the Cartesian coordinate system is given by

$$\left(\mathbf{F}_{\text{appl}} - \mathbf{M}\ddot{\mathbf{x}}\right) \cdot \delta\mathbf{x} = 0 \tag{6.78}$$

By introducing generalized coordinates in holonomic systems, the above equation can be rewritten as

$$\left[\left(\mathbf{F}_{\text{appl}} - \mathbf{M}\ddot{\mathbf{x}}\right) \cdot \frac{\partial \mathbf{x}}{\partial \mathbf{q}}\right] \cdot \delta\mathbf{q} = 0 \tag{6.79}$$

Invoking the kinetic and potential energies in generalized coordinates, the above equation leads to

$$\left[\frac{\partial \{\mathcal{K}(\mathbf{q}, \dot{\mathbf{q}}, t) - \mathcal{U}(\mathbf{q}, t)\}}{\partial \mathbf{q}} - \frac{d}{dt}\left(\frac{\partial \mathcal{K}(\mathbf{q}, \dot{\mathbf{q}}, t)}{\partial \dot{\mathbf{q}}}\right)\right] \cdot \delta\mathbf{q} = 0 \tag{6.80}$$

Next, we consider the Total Energy, which is defined as the summation of the kinetic energy and the potential energy in generalized coordinates

$$\mathcal{E}(\mathbf{q}, \dot{\mathbf{q}}, t) = \mathcal{K}(\mathbf{q}, \dot{\mathbf{q}}, t) + \mathcal{U}(\mathbf{q}, t) \tag{6.81}$$

Note that our attention is confined to the case that the potential energy is not dependent upon generalized velocities. Then, by making use of the above relation, Equation (6.79) can be expressed as

$$\left[\frac{d}{dt}\left(\frac{\partial \mathcal{E}(\mathbf{q}, \dot{\mathbf{q}}, t)}{\partial \dot{\mathbf{q}}}\right) + \frac{\partial \mathcal{E}(\mathbf{q}, \dot{\mathbf{q}}, t)}{\partial \mathbf{q}} - 2\frac{\partial \mathcal{K}(\mathbf{q}, \dot{\mathbf{q}}, t)}{\partial \mathbf{q}}\right] \cdot \delta\mathbf{q} = 0 \tag{6.82}$$

Since $\delta\mathbf{q}$ is linearly independent and arbitrary, the above equation yields

$$\frac{d}{dt}\left(\frac{\partial \mathcal{E}(\mathbf{q}, \dot{\mathbf{q}}, t)}{\partial \dot{\mathbf{q}}}\right) + \frac{\partial \mathcal{E}(\mathbf{q}, \dot{\mathbf{q}}, t)}{\partial \mathbf{q}} - 2\frac{\partial \mathcal{K}(\mathbf{q}, \dot{\mathbf{q}}, t)}{\partial \mathbf{q}} = \mathbf{0} \tag{6.83}$$

If the kinetic energy is independent of generalized coordinates, the above equation reduces to

$$\frac{d}{dt}\left(\frac{\partial \mathcal{E}(\mathbf{q}, \dot{\mathbf{q}}, t)}{\partial \dot{\mathbf{q}}}\right) + \frac{\partial \mathcal{E}(\mathbf{q}, \dot{\mathbf{q}}, t)}{\partial \mathbf{q}} = \mathbf{0} \tag{6.84}$$

This is the Total Energy representation of the equation of motion for a holonomic constraint system for the case when the kinetic energy is independent of generalized coordinates.

Total Energy Representation of Equation of Motion in the Cartesian Coordinate System

In the case that the dynamical system is subject to holonomic constraints, we have previously used the Lagrange multiplier method to derive differential equations in the Cartesian coordinate system under the Lagrangian mechanics framework. Likewise, under the Total Energy framework we can also derive the Total Energy version of the equations in the Cartesian coordinate system. Note that the term enclosed in the curly bracket of Equation (3.153) can be written as

$$\frac{d}{dt}\left[\frac{\partial}{\partial \dot{\mathbf{x}}}\left(\mathcal{K}(\dot{\mathbf{x}}) + \mathcal{U}(\mathbf{x})\right)\right] + \frac{\partial}{\partial \mathbf{x}}\left(\mathcal{K}(\dot{\mathbf{x}}) + \mathcal{U}(\mathbf{x})\right) = \frac{\partial\left(\boldsymbol{\lambda} \cdot \mathbf{f}(\mathbf{x}, t)\right)}{\partial \mathbf{x}}$$

Then we obtain the expression in terms of the Total Energy as

$$\frac{d}{dt}\left(\frac{\partial \mathcal{E}(\mathbf{x},\dot{\mathbf{x}})}{\partial \dot{\mathbf{x}}}\right)+\frac{\partial \mathcal{E}(\mathbf{x},\dot{\mathbf{x}})}{\partial \mathbf{x}}=\frac{\partial(\boldsymbol{\lambda}\cdot\mathbf{f}(\mathbf{x},t))}{\partial \mathbf{x}} \tag{6.85}$$

where the Total Energy in the Cartesian coordinate system is defined as

$$\mathcal{E}(\mathbf{x},\dot{\mathbf{x}})=\mathcal{K}(\dot{\mathbf{x}})+\mathcal{U}(\mathbf{x}) \tag{6.86}$$

Equation (6.85) can be also rewritten as

$$\frac{d}{dt}\left(\frac{\partial \tilde{\mathcal{E}}(\mathbf{x},\dot{\mathbf{x}},t)}{\partial \dot{\mathbf{x}}}\right)+\frac{\partial \tilde{\mathcal{E}}(\mathbf{x},\dot{\mathbf{x}},t)}{\partial \mathbf{x}}=\mathbf{0} \tag{6.87}$$

where $\tilde{\mathcal{E}}(\mathbf{x},\dot{\mathbf{x}},t)$ will be referred to as the augmented Total Energy, that is,

$$\tilde{\mathcal{E}}(\mathbf{x},\dot{\mathbf{x}},t)=\mathcal{E}(\mathbf{x},\dot{\mathbf{x}})-\boldsymbol{\lambda}\cdot\mathbf{f}(\mathbf{x},t) \tag{6.88}$$

Just as in N-body systems, for continuous-body systems, the Total Energy representation of the equation of motion, in contrast to the Lagrangian equation of motion is particularly discussed in further detail in a subsequent chapter.

Illustrative Example

Example 6.6 Planar Pendulum Problem and Total Energy Representation of the Equation of Motion

In the $x=(x,y)\in\mathbb{R}^2$ coordinate system, the autonomous augmented total energy for the planar pendulum may be expressed by

$$\tilde{\mathcal{E}}(x,\dot{x})=\frac{1}{2}m\left(\dot{x}^2+\dot{y}^2\right)+mgy-\lambda\left(l-\sqrt{x^2+y^2}\right) \tag{6.89}$$

Substituting Equation (6.89) into Equation (6.85) yields Equation (3.155). Hence we have the identical equations of motion as illustrated previously via the Lagrange equations of motion.

6.4.2 Total Energy Representation of Equation of Motion via Hamilton's Principle/Hamilton's Law of Varying Action

Hamilton's Principle

Total Energy Representation of the Equation of Motion in the Configuration Space It is assumed that the potential energy is not velocity-dependent in this section. In the Total Energy framework, Hamilton's Principle defined in Equation (5.47) can be rewritten as

$$\delta\int_{t_0}^{t_1}\left(2\mathcal{K}(\mathbf{q},\dot{\mathbf{q}},t)-\mathcal{E}(\mathbf{q},\dot{\mathbf{q}},t)\right)dt=0,\qquad \delta\mathbf{q}(t_0)=\delta\mathbf{q}(t_1)=\mathbf{0} \tag{6.90}$$

which is the modified Hamilton's principle in the Total Energy framework. Since the dynamical system is holonomic (Osgood 1937), we have the relation

$$\delta \int_{t_0}^{t_1} (2\mathcal{K}(\mathbf{q}, \dot{\mathbf{q}}, t) - \mathcal{E}(\mathbf{q}, \dot{q}, t))\, dt = \int_{t_0}^{t_1} (2\delta\mathcal{K}(\mathbf{q}, \dot{\mathbf{q}}, t) - \delta\mathcal{E}(\mathbf{q}, \dot{\mathbf{q}}, t))\, dt \tag{6.91}$$

By means of the chain rule and integration by parts, Equation (6.91) yields

$$\int_{t_0}^{t_1} \left[2\frac{\partial \mathcal{K}(\mathbf{q}, \dot{\mathbf{q}}, t)}{\partial \mathbf{q}} - \frac{\partial \mathcal{E}(\mathbf{q}, \dot{\mathbf{q}}, t)}{\partial \mathbf{q}} - \frac{d}{dt}\left(\frac{\partial \mathcal{E}(\mathbf{q}, \dot{\mathbf{q}}, t)}{\partial \dot{\mathbf{q}}} \right) \right] \cdot \delta\mathbf{q}\, dt = 0 \tag{6.92}$$

in which the prescribed boundary conditions are applied. Since the virtual displacement is arbitrary within the time interval, we have the governing equations, that is, the Total Energy equations of motion as follows:

$$\frac{\partial \mathcal{E}(\mathbf{q}, \dot{\mathbf{q}}, t)}{\partial \mathbf{q}} + \frac{d}{dt}\left(\frac{\partial \mathcal{E}(\mathbf{q}, \dot{\mathbf{q}}, t)}{\partial \dot{\mathbf{q}}} \right) - 2\frac{\partial \mathcal{K}(\mathbf{q}, \dot{\mathbf{q}}, t)}{\partial \mathbf{q}} = \mathbf{0} \tag{6.93}$$

Suppose that the kinetic energy is not a function of generalized coordinates. Then, Equation (6.93) can be given as

$$\frac{\partial \mathcal{E}(\mathbf{q}, \dot{\mathbf{q}}, t)}{\partial \mathbf{q}} + \frac{d}{dt}\left(\frac{\partial \mathcal{E}(\mathbf{q}, \dot{\mathbf{q}}, t)}{\partial \dot{\mathbf{q}}} \right) = \mathbf{0} \tag{6.94}$$

It is important to note that the Total Energy for the above system is not conserved.

Total Energy Representation of the Equation of Motion in the Cartesian Coordinate System Next, we shall derive the Total Energy representation of the equation of motion from the modified Hamilton's principle in the Cartesian coordinate system. The dynamical system is assumed to be subject to some holonomic constraints. First, let us recall Equations (5.76) and (5.77). In terms of the so-called augmented Total Energy, $\tilde{\mathcal{E}}(\dot{\mathbf{x}}, \mathbf{x}, t) : \mathbb{R}^{6N} \times \bar{\mathcal{T}} \rightarrow \mathbb{R}$, Equation (5.76) can be rewritten as

$$\tilde{\mathcal{L}}(\dot{\mathbf{x}}, \mathbf{x}, t) = 2\mathcal{K}(\dot{\mathbf{x}}) - \tilde{\mathcal{E}}(\dot{\mathbf{x}}, \mathbf{x}, t) \tag{6.95}$$

where the augmented Total Energy is defined as

$$\tilde{\mathcal{E}}(\dot{\mathbf{x}}, \mathbf{x}, t) = \mathcal{E}(\dot{\mathbf{x}}, \mathbf{x}, t) - \boldsymbol{\lambda} \cdot \boldsymbol{f} \tag{6.96}$$

In view of Equation (5.77), the modified Hamilton's principle in terms of the Total Energy in the Cartesian coordinate system can be given by

$$\delta \int_{t_0}^{t_1} \left(2\mathcal{K}(\dot{\mathbf{x}}) - \tilde{\mathcal{E}}(\dot{\mathbf{x}}, \mathbf{x}, t) \right) dt = 0 \tag{6.97}$$

For holonomic systems, the above equation yields

$$\int_{t_0}^{t_1} \left[\frac{\partial \mathcal{K}}{\partial \dot{\mathbf{x}}} \cdot \delta\dot{\mathbf{x}} - \frac{\partial \tilde{\mathcal{E}}(\dot{\mathbf{x}}, \mathbf{x}, t)}{\partial \mathbf{x}} \cdot \delta\mathbf{x} \right] dt = 0 \tag{6.98}$$

Upon integration by parts, we obtain in terms of the augmented Total Energy

$$\int_{t_0}^{t_1} \left[\frac{d}{dt}\left(\frac{\partial \tilde{\mathcal{E}}}{\partial \dot{\mathbf{x}}}\right) + \frac{\partial \tilde{\mathcal{E}}}{\partial \mathbf{x}} \right] \cdot \delta\mathbf{x}\, dt = 0 \tag{6.99}$$

The above equation yields

$$\int_{t_0}^{t_1} \left[\frac{d}{dt}\left(\frac{\partial \mathcal{E}}{\partial \dot{\mathbf{x}}}\right) + \frac{\partial \mathcal{E}}{\partial \mathbf{x}} - \boldsymbol{\lambda} \cdot \frac{\partial \mathbf{f}}{\partial \mathbf{x}} \right] \cdot \delta\mathbf{x}\, dt = 0 \tag{6.100}$$

Note the virtual displacement $\delta\mathbf{x}$ is not arbitrary within the time interval, since it is not linearly independent. However, the Lagrange multiplier can be chosen to make the term enclosed in the bracket to vanish. Note that the term enclosed in the bracket is identical to Equation (6.87). Therefore, the differential equations in the bracket must vanish. Hence, we have the Total Energy representation of the equation of motion in the Cartesian coordinate system as follows,

$$\frac{d}{dt}\left(\frac{\partial \mathcal{E}}{\partial \dot{\mathbf{x}}}\right) + \frac{\partial \mathcal{E}}{\partial \mathbf{x}} - \boldsymbol{\lambda} \cdot \nabla \mathbf{f} = \mathbf{0} \tag{6.101}$$

Note that the above equation is identical to Equation (6.87).

Hamilton's Law of Varying Action Previously, we have shown that governing equations can be derived from Hamilton's law of varying action. For example, in Newtonian mechanics framework we obtained Newton's equations of motion in the Cartesian coordinate system, in Lagrangian mechanics framework we obtained Lagrange's equations of motion, and in Hamiltonian mechanics we obtained Hamilton's equations of motion. In this section, other than the descriptive scalar functions such as the Lagrangian and the Hamiltonian, the Total Energy which is defined on the velocity phase space will be employed to represent the governing equations, and are of a system of second-order differential equations.

Total Energy Representation of the Equation of Motion in the Configuration Space Suppose that our attention is not confined to the system whose kinetic energy is a quadratic form in terms of generalized velocities. In other words, the system can be scleronomic or rheonomic. Then the relation between the Lagrangian and the Total Energy is given by

$$\mathcal{L}(\mathbf{q}, \dot{\mathbf{q}}, t) = 2\mathcal{K}(\mathbf{q}, \dot{\mathbf{q}}, t) - \mathcal{E}(\mathbf{q}, \dot{\mathbf{q}}, t) \tag{6.102}$$

where the non-autonomous Total Energy is denoted by $\mathcal{E}(\mathbf{q}, \dot{\mathbf{q}}, t)$. Provided that the potential energy is not a function of generalized velocities, the following relation is true

$$\frac{\partial \mathcal{K}(\mathbf{q}, \dot{\mathbf{q}}, t)}{\partial \dot{\mathbf{q}}} = \frac{\partial \mathcal{E}(\mathbf{q}, \dot{\mathbf{q}}, t)}{\partial \dot{\mathbf{q}}} \tag{6.103}$$

which means that the potential energy is not velocity-dependent. In the Total Energy framework, in view of Equation (6.102) Hamilton's law of varying action can be given by

$$\delta \int_{t_0}^{t_1} \left(2\mathcal{K}(\mathbf{q}, \dot{\mathbf{q}}, t) - \mathcal{E}(\mathbf{q}, \dot{\mathbf{q}}, t)\right) dt - \left. \frac{\partial \mathcal{E}(\mathbf{q}, \dot{\mathbf{q}}, t)}{\partial \dot{\mathbf{q}}} \cdot \delta\mathbf{q} \right|_{t_0}^{t_1} = 0 \tag{6.104}$$

Since the dynamical system is holonomic, we have the relation

$$\delta \int_{t_0}^{t_1} (2\mathcal{K}(\mathbf{q}, \dot{\mathbf{q}}, t) - \mathcal{E}(\mathbf{q}, \dot{\mathbf{q}}, t))\, dt = \int_{t_0}^{t_1} (2\delta\mathcal{K}(\mathbf{q}, \dot{\mathbf{q}}, t) - \delta\mathcal{E}(\mathbf{q}, \dot{\mathbf{q}}, t))\, dt \tag{6.105}$$

By means of the chain rule and integration by part, Equation (6.102) can be transformed into

$$\int_{t_0}^{t_1} \left[2\frac{\partial \mathcal{K}(\mathbf{q}, \dot{\mathbf{q}}, t)}{\partial \mathbf{q}} - \frac{\partial \mathcal{E}(\mathbf{q}, \dot{\mathbf{q}}, t)}{\partial \mathbf{q}} - \frac{d}{dt}\left(\frac{\partial \mathcal{E}(\mathbf{q}, \dot{\mathbf{q}}, t)}{\partial \dot{\mathbf{q}}} \right) \right] \cdot \delta\mathbf{q}\, dt = 0 \tag{6.106}$$

Since the virtual displacement is arbitrary within the time interval, we have the governing equations, that is, the Total Energy representation of the equations of motion as follows,

$$\frac{\partial \mathcal{E}(\mathbf{q}, \dot{\mathbf{q}}, t)}{\partial \mathbf{q}} + \frac{d}{dt}\left(\frac{\partial \mathcal{E}(\mathbf{q}, \dot{\mathbf{q}}, t)}{\partial \dot{\mathbf{q}}} \right) - 2\frac{\partial \mathcal{K}(\mathbf{q}, \dot{\mathbf{q}}, t)}{\partial \mathbf{q}} = \mathbf{0} \tag{6.107}$$

Suppose that the kinetic energy is not a function of generalized coordinates as a special case as in the planar pendulum problem. Then, Equation (6.107) reduces to

$$\frac{\partial \mathcal{E}(\mathbf{q}, \dot{\mathbf{q}}, t)}{\partial \mathbf{q}} + \frac{d}{dt}\left(\frac{\partial \mathcal{E}(\mathbf{q}, \dot{\mathbf{q}}, t)}{\partial \dot{\mathbf{q}}} \right) = \mathbf{0} \tag{6.108}$$

Total Energy Representation of the Equation of Motion in the Cartesian Coordinate System We shall also derive the Total Energy representation of the equation of motion from Hamilton's law of varying action in the Cartesian coordinate system. The dynamical system is assumed to be subject to some holonomic constraints. First, let us recall Equations (5.76) and (5.77). In terms of the so-called augmented Total Energy, $\tilde{\mathcal{E}}(\dot{\mathbf{x}}, \mathbf{x}, t) : \mathbb{R}^{6N} \times \bar{\mathcal{T}} \rightarrow \mathbb{R}$, Equation (5.76) can be rewritten as

$$\tilde{\mathcal{L}}(\dot{\mathbf{x}}, \mathbf{x}, t) = 2\mathcal{K}(\dot{\mathbf{x}}, \mathbf{x}) - \tilde{\mathcal{E}}(\dot{\mathbf{x}}, \mathbf{x}, t) \tag{6.109}$$

where the augmented Total Energy is defined as

$$\tilde{\mathcal{E}}(\dot{\mathbf{x}}, \mathbf{x}, t) = \mathcal{E}(\dot{\mathbf{x}}, \mathbf{x}, t) - \boldsymbol{\lambda} \cdot f \tag{6.110}$$

In view of Equation (5.77), Hamilton's law of varying action in terms of the Total Energy in the Cartesian coordinate system can be given by

$$\delta \int_{t_0}^{t_1} \left(2\mathcal{K}(\dot{\mathbf{x}}) - \tilde{\mathcal{E}}(\dot{\mathbf{x}}, \mathbf{x}, t) \right) dt - \left. \frac{\partial \mathcal{E}(\dot{\mathbf{x}}, \mathbf{x}, t)}{\partial \dot{\mathbf{x}}} \cdot \delta\mathbf{x} \right|_{t_0}^{t_1} = 0 \tag{6.111}$$

For holonomic systems, the above equation yields

$$\int_{t_0}^{t_1} \left[\frac{\partial \mathcal{K}}{\partial \dot{\mathbf{x}}} \cdot \delta\dot{\mathbf{x}} - \left(\frac{\partial \mathcal{E}(\mathbf{x})}{\partial \mathbf{x}} - \boldsymbol{\lambda} \cdot \frac{\partial f}{\partial \mathbf{x}} \right) \cdot \delta\mathbf{x} \right] dt - \left. \frac{\partial \mathcal{E}(\dot{\mathbf{x}}, \mathbf{x}, t)}{\partial \dot{\mathbf{x}}} \cdot \delta\mathbf{x} \right|_{t_0}^{t_1} = 0 \tag{6.112}$$

Upon integration by parts, we obtain

$$\int_{t_0}^{t_1} \left[\frac{d}{dt}\left(\frac{\partial \mathcal{E}}{\partial \dot{\mathbf{x}}} \right) + \frac{\partial \mathcal{E}}{\partial \mathbf{x}} - \boldsymbol{\lambda} \cdot \frac{\partial \mathbf{f}}{\partial \mathbf{x}} \right] \cdot \delta\mathbf{x}\, dt = 0 \tag{6.113}$$

The virtual displacement $\delta\mathbf{x}$ is not linearly independent within the time interval. However, $\boldsymbol{\lambda}$ is chosen so that the coefficient of $\delta\mathbf{x}$ can vanish. Hence, we have the Total Energy representation of the equation of motion in the Cartesian coordinate system as follows:

$$\frac{d}{dt}\left(\frac{\partial \mathcal{E}}{\partial \dot{\mathbf{x}}}\right) + \frac{\partial \mathcal{E}}{\partial \mathbf{x}} - \boldsymbol{\lambda} \cdot \nabla \mathbf{f} = \mathbf{0} \tag{6.114}$$

EXERCISES

6.1. Derive the governing equation by use of the Total Energy representation of the equation of motion given in Equation (6.21) for the double planar pendulum problem in Example 3.11. By use of Equation (6.87), find out that the Lagrange multiplier is identical to C in Equation (3.29) and derive Newton's equation given in Equation (3.30).

6.2. Consider a sliding block on a moving, inclined plane in Example 4.2. Establish the kinetic and potential energies. Give the total energy for the holonomic-scleronomic constraint system. Obtain the governing equation by use of the Total Energy representation of the equation of motion given in Equation (6.83).

6.3. For the planar motion as shown in Figure 3.22, use the Total Energy representation of the equation of motion given in Equation (6.21) to obtain the governing equations of motion.

6.4. Reconsider the rigid rod system discussed in Exercise 3.25 of Chapter 3. Use Hamilton's law of varying action given in Equation (6.104) to obtain the Total Energy representation of the equation of motion in the form of the second-order differential equation of motion.

6.5. Consider the rotating bead moving on a vertical circular hoop in Example 3.14. Notice that this system is holonomic-rheonomic. Obtain the governing equation of motion given in Equation (3.173) by use of the Total Energy representation of the equation of motion given in Equation (6.93) via Hamilton's principle or in Equation (6.107) via Hamilton's law of varying action.

6.6. Reconsider the three particle dynamical system connected by three nonlinear springs discussed in Exercise 3.6 of Chapter 3. Establish the Total Energy, and derive the Total Energy representation of the equation of motion from the principle of conservation of total mechanical energy.

6.7. Consider the motion of a bead in Exercise 5.5 of Chapter 5. Establish the kinetic and potential energies. Give the total energy for the holonomic-rheonomic constraint system. Obtain the governing equation by use of the Total Energy representation of the equation of motion given in Equation (6.21).

6.8. Consider the motion of the simple planar pendulum with a oscillating, restoring hinge point in Exercise 5.7 of Chapter 5, establish the kinetic and potential energies. Give the total energy for the holonomic-rheonomic constraint system. Derive the governing equation by use of the Total Energy representation of the equation of motion given in Equation (6.83).

6.9. Consider a forced dynamic response of the undamped spring-mass system in Exercise 5.3 of Chapter 5, and establish the kinetic and potential energies. Give the total energy for the holonomic-rheonomic constraint system. Derive the governing equation by use of the Total Energy representation of the equation of motion given in Equation (6.83).

CHAPTER SEVEN

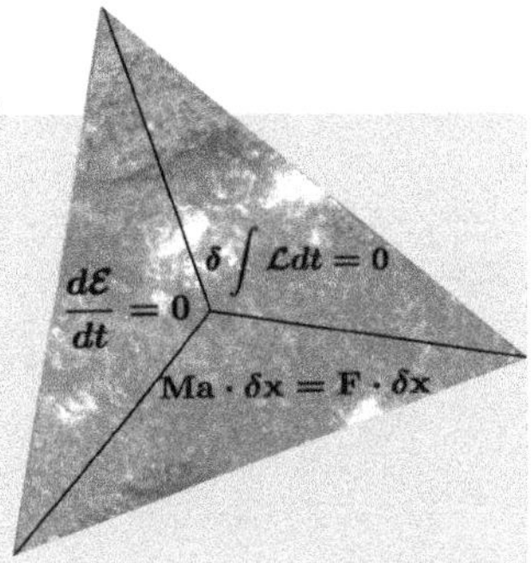

EQUIVALENCE OF EQUATIONS

In Chapter 7, we shall investigate the equivalence relations between the various governing equations for N-body dynamical systems subject to holonomic constraints. Although we have highlighted various forms of representation in the preceding chapters, here we intend to shed light upon the relationships between governing equations as differential equations, integral equations in the context of Hamilton's Principle and Hamilton's law of varying action, and time differential equations in the context of the Principle of balance of mechanical energy.

7.1 EQUIVALENCE IN THE LAGRANGIAN FORM OF D'ALEMBERT'S PRINCIPLE/PRINCIPLE OF VIRTUAL WORK

We assume that the motion of an N-body dynamical system subject to the k-number of holonomic constraint equations within the closed interval of time $t \in \bar{\mathcal{T}} = [t_0, t_1]$ is governed by Newton's law of motion as defined in Equation (3.46), which leads to D'Alembert's principle given in Equation (3.52). Multiplying Equation (3.46) by the virtual displacement $\delta \mathbf{x}(t) = \left(\delta \mathbf{x}^1, \ldots, \delta \mathbf{x}^N\right) : \bar{\mathcal{T}} \to \mathbb{R}^{3N}$ yields the Lagrangian Form of D'Alembert's Principle/Principle of virtual work, which is given by

$$\left(\mathbf{F}_{\text{appl}} - \mathbf{M}\ddot{\mathbf{x}}\right) \cdot \delta \mathbf{x} = 0, \quad \text{or equivalently,}$$

$$\mathbf{F}_{\text{appl}} \cdot \delta \mathbf{x} = \mathbf{M}\ddot{\mathbf{x}} \cdot \delta \mathbf{x} \tag{7.1}$$

Introducing generalized coordinates $\mathbf{q} = \left(q^1, \ldots, q^{n_{\text{dof}}}\right) \in Q \subset \mathbb{R}^{n_{\text{dof}} = 3N-k}$, the above equation can be rewritten as

$$\left[\left(\mathbf{F}_{\text{appl}} - \mathbf{M}\ddot{\mathbf{x}}\right) \cdot \frac{\partial \mathbf{x}}{\partial \mathbf{q}}\right] \cdot \delta \mathbf{q} = 0 \tag{7.2}$$

Advances in Computational Dynamics of Particles, Materials and Structures, First Edition. Jason Har and Kumar K. Tamma.

As addressed in detail in Chapter 3, in terms of the kinetic energy $\mathcal{K}(\mathbf{q}, \dot{\mathbf{q}}, t) : TQ \times \bar{\mathcal{T}} \to \mathbb{R}$ and the potential energy $\mathcal{U}(\mathbf{q}, t) : Q \times \mathcal{T} \to \mathbb{R}$, the above equation can be expressed as

$$\left[\frac{\partial\left\{\mathcal{K}(\mathbf{q}, \dot{\mathbf{q}}, t) - \mathcal{U}(\mathbf{q}, t)\right\}}{\partial \mathbf{q}} - \frac{d}{dt}\left(\frac{\partial \mathcal{K}(\mathbf{q}, \dot{\mathbf{q}}, t)}{\partial \dot{\mathbf{q}}}\right)\right] \cdot \delta\mathbf{q} = 0 \tag{7.3}$$

where the potential energy is assumed not to be velocity-dependent. By introducing the Lagrangian $\mathcal{L}(\mathbf{q}, \dot{\mathbf{q}}, t) : TQ \times \mathcal{T} \to \mathbb{R}$, the Lagrangian Form of D'Alembert's Principle leads to

$$\left[\frac{\partial \mathcal{L}(\mathbf{q}, \dot{\mathbf{q}}, t)}{\partial \mathbf{q}} - \frac{d}{dt}\left(\frac{\partial \mathcal{L}(\mathbf{q}, \dot{\mathbf{q}}, t)}{\partial \dot{\mathbf{q}}}\right)\right] \cdot \delta\mathbf{q} = 0 \tag{7.4}$$

On the other hand, if we introduce a different perspective, namely, the descriptive scalar function, the Total Energy $\mathcal{E}(\mathbf{q}, \dot{\mathbf{q}}, t) : TQ \times \bar{\mathcal{T}} \to \mathbb{R}$ for the holonomic constraint N-body dynamical system, the Lagrangian Form of D'Alembert's Principle/Principle of virtual work can be transformed into

$$\left[\frac{d}{dt}\left(\frac{\partial \mathcal{E}(\mathbf{q}, \dot{\mathbf{q}}, t)}{\partial \dot{\mathbf{q}}}\right) + \frac{\partial \mathcal{E}(\mathbf{q}, \dot{\mathbf{q}}, t)}{\partial \mathbf{q}} - 2\frac{\partial \mathcal{K}(\mathbf{q}, \dot{\mathbf{q}}, t)}{\partial \mathbf{q}}\right] \cdot \delta\mathbf{q} = 0 \tag{7.5}$$

As $\delta\mathbf{q}$ is arbitrary, Equations (7.4) and (7.5) yield Lagrange's equation of motion and the Total Energy representation of the equation of motion, respectively. Hence, they are equivalent to each other.

For the special case when the dynamical system is holonomic-scleronomic and conservative, both the Lagrangian and the Total Energy become autonomous. As a consequence, the Lagrangian form of D'Alembert's Principle/Principle of virtual work leads to, in terms of the Lagrangian,

$$\left[\frac{\partial \mathcal{L}(\mathbf{q}, \dot{\mathbf{q}})}{\partial \mathbf{q}} - \frac{d}{dt}\left(\frac{\partial \mathcal{L}(\mathbf{q}, \dot{\mathbf{q}})}{\partial \dot{\mathbf{q}}}\right)\right] \cdot \delta\mathbf{q} = 0 \tag{7.6}$$

In addition, if the kinetic energy is independent of generalized coordinates, Equation (7.5) reduces to

$$\left[\frac{d}{dt}\left(\frac{\partial \mathcal{E}(\mathbf{q}, \dot{\mathbf{q}})}{\partial \dot{\mathbf{q}}}\right) + \frac{\partial \mathcal{E}(\mathbf{q}, \dot{\mathbf{q}})}{\partial \mathbf{q}}\right] \cdot \delta\mathbf{q} = 0 \tag{7.7}$$

Lastly, we consider the Hamiltonian mechanics framework. For the holonomic constraint N-body dynamical system, Equation (7.4) can be rewritten as

$$\delta\mathcal{L}(\mathbf{q}, \dot{\mathbf{q}}, t) - \frac{\partial \mathcal{L}(\mathbf{q}, \dot{\mathbf{q}}, t)}{\partial \dot{\mathbf{q}}} \cdot \delta\dot{\mathbf{q}} - \frac{d}{dt}\left(\frac{\partial \mathcal{L}(\mathbf{q}, \dot{\mathbf{q}}, t)}{\partial \dot{\mathbf{q}}}\right) \cdot \delta\mathbf{q} = 0 \tag{7.8}$$

By virtue of the Legendre transformation, the Hamiltonian $\mathcal{H}(\mathbf{p}, \mathbf{q}, t) : T^*Q \times \bar{\mathcal{T}} \to \mathbb{R}$ is defined as

$$\mathcal{H}(\mathbf{q}, \mathbf{p}, t) = \mathbf{p} \cdot \dot{\mathbf{q}} - \mathcal{L}(\mathbf{q}, \dot{\mathbf{q}}, t) \tag{7.9}$$

By use of Equation (7.9), Equation (7.8) can be expressed as

$$\left(\dot{\mathbf{q}} - \frac{\partial \mathcal{H}(\mathbf{q}, \mathbf{p}, t)}{\partial \mathbf{p}}\right) \cdot \delta\mathbf{p} - \left(\dot{\mathbf{p}} + \frac{\partial \mathcal{H}(\mathbf{q}, \mathbf{p}, t)}{\partial \mathbf{q}}\right) \cdot \delta\mathbf{q} = 0 \tag{7.10}$$

which is a consequence of the Lagrangian form of D'Alembert's Principle/Principle of virtual work in the Hamiltonian mechanics framework.

7.2 EQUIVALENCE IN HAMILTON'S PRINCIPLE OR HAMILTON'S LAW OF VARYING ACTION

Let us again consider an N-body dynamical system which is subject to holonomic constraints. Then, in the Lagrangian mechanics framework Hamilton's principle is given as

$$\begin{aligned} &\delta \int_{t_0}^{t_1} \mathcal{L}(\mathbf{q}, \dot{\mathbf{q}}, t)\, dt = 0, \\ &\delta \mathbf{q}(t_0) = \delta \mathbf{q}(t_1) = \mathbf{0} \end{aligned} \tag{7.11}$$

In the Total Energy framework, Hamilton's principle is given by

$$\begin{aligned} &\delta \int_{t_0}^{t_1} (2\mathcal{K}(\mathbf{q}, \dot{\mathbf{q}}, t) - \mathcal{E}(\mathbf{q}, \dot{\mathbf{q}}, t))\, dt = 0, \\ &\delta \mathbf{q}(t_0) = \delta \mathbf{q}(t_1) = \mathbf{0} \end{aligned} \tag{7.12}$$

And, in the Hamiltonian mechanics framework, Hamilton's principle is given as

$$\begin{aligned} &\delta \int_{t_0}^{t_1} (\mathbf{p} \cdot \dot{\mathbf{q}} - \mathcal{H}(\mathbf{q}, \mathbf{p}, t))\, dt = 0, \\ &\delta \mathbf{q}(t_0) = \delta \mathbf{q}(t_1) = \mathbf{0} \end{aligned} \tag{7.13}$$

On the other hand, in the Lagrangian mechanics framework, Hamilton's law of varying action reads as

$$\delta \int_{t_0}^{t_1} \mathcal{L}(\mathbf{q}, \dot{\mathbf{q}}, t)\, dt - \left. \frac{\partial \mathcal{L}(\mathbf{q}, \dot{\mathbf{q}}, t)}{\partial \dot{\mathbf{q}}} \cdot \delta \mathbf{q} \right|_{t_0}^{t_1} = 0 \tag{7.14}$$

In the Total Energy framework, Hamilton's law of varying action is given by

$$\delta \int_{t_0}^{t_1} (2\mathcal{K}(\mathbf{q}, \dot{\mathbf{q}}, t) - \mathcal{E}(\mathbf{q}, \dot{\mathbf{q}}, t))\, dt - \left. \frac{\partial \mathcal{E}(\mathbf{q}, \dot{\mathbf{q}}, t)}{\partial \dot{\mathbf{q}}} \cdot \delta \mathbf{q} \right|_{t_0}^{t_1} = 0 \tag{7.15}$$

Lastly, in the Hamiltonian mechanics framework, Hamilton's law of varying action is given as

$$\delta \int_{t_0}^{t_1} (\mathbf{p} \cdot \dot{\mathbf{q}} - \mathcal{H}(\mathbf{q}, \mathbf{p}, t))\, dt - \mathbf{p} \cdot \delta \mathbf{q}|_{t_0}^{t_1} = 0 \tag{7.16}$$

As discussed in the preceding chapters, Hamilton's principle requires the virtual displacement to vanish at the two terminal points, but Hamilton's law of varying action does not require this condition. In the end, in the Lagrangian mechanics framework, Hamilton's law of varying action or Hamilton's principle leads to

$$\int_{t_0}^{t_1} \left[\frac{\partial \mathcal{L}(\mathbf{q}, \dot{\mathbf{q}}, t)}{\partial \mathbf{q}} - \frac{d}{dt}\left(\frac{\partial \mathcal{L}(\mathbf{q}, \dot{\mathbf{q}}, t)}{\partial \dot{\mathbf{q}}} \right) \right] \cdot \delta \mathbf{q}\, dt = 0 \tag{7.17}$$

In the Hamiltonian mechanics framework, we have

$$\int_{t_0}^{t_1} \left[\left(\dot{\mathbf{q}} - \frac{\partial \mathcal{H}(\mathbf{q}, \mathbf{p}, t)}{\partial \mathbf{p}} \right) \cdot \delta \mathbf{p} - \left(\dot{\mathbf{p}} + \frac{\partial \mathcal{H}(\mathbf{q}, \mathbf{p}, t)}{\partial \mathbf{q}} \right) \cdot \delta \mathbf{q} \right] dt = 0 \tag{7.18}$$

and, in the Total Energy framework we have,

$$\int_{t_0}^{t_1} \left[2\frac{\partial \mathcal{K}(\mathbf{q}, \dot{\mathbf{q}}, t)}{\partial \mathbf{q}} - \frac{\partial \mathcal{E}(\mathbf{q}, \dot{\mathbf{q}}, t)}{\partial \mathbf{q}} - \frac{d}{dt}\left(\frac{\partial \mathcal{E}(\mathbf{q}, \dot{\mathbf{q}}, t)}{\partial \dot{\mathbf{q}}} \right) \right] \cdot \delta \mathbf{q}\, dt = 0 \tag{7.19}$$

As the variations of dependent variables are arbitrary in the time interval of interest, the resulting governing equations are equivalent to each other. Equations (7.17) and (7.19) take the Bubnov-Galerkin weighted residual form in time as a single-field weighted residual form in time, whereas Equation (7.18) is a two-field weighted residual form in time.

7.3 EQUIVALENCE IN THE PRINCIPLE OF BALANCE OF MECHANICAL ENERGY

The principle of balance of mechanical energy for an N-body dynamical system which is subject to holonomic-scleronomic constraints can be given by

$$\frac{d\mathcal{E}(\mathbf{x}, \dot{\mathbf{x}})}{dt} = \mathbf{F}_{\text{nc}}(t) \cdot \dot{\mathbf{x}}(t) + \mathbf{C} \cdot \dot{\mathbf{x}} \tag{7.20}$$

where $\mathbf{F}_{\text{nc}}$ is the nonconservative force depending on time and $\mathbf{C}$ is the constraint force. The time-dependent nonconservative force is assumed not to be derivable from the potential energy. In holonomic-scleronomic systems, the differential of the position vector can be written as

$$d\mathbf{x} = \frac{\partial \mathbf{x}}{\partial \mathbf{q}} \cdot d\mathbf{q} \tag{7.21}$$

The partial derivative of the position vectors with respect to time vanishes. Hence, Equation (7.20) can be transformed into

$$\frac{d\mathcal{E}(\mathbf{q}, \dot{\mathbf{q}})}{dt} = \mathbf{Q}_{\text{nc}} \cdot \dot{\mathbf{q}} \tag{7.22}$$

where $\mathbf{Q}_{\text{nc}}$ is the generalized non-conservative force, which is not derivable from the potential energy. Hence, the principle of balance of mechanical energy for an N-body dynamical system includes the governing equation in terms of the Total Energy as follows

$$\left[\frac{d}{dt}\left(\frac{\partial \mathcal{E}(\mathbf{q}, \dot{\mathbf{q}})}{\partial \dot{\mathbf{q}}} \right) + \frac{\partial \mathcal{E}(\mathbf{q}, \dot{\mathbf{q}})}{\partial \mathbf{q}} - 2\frac{\partial \mathcal{K}(\mathbf{q}, \dot{\mathbf{q}})}{\partial \mathbf{q}} \right] \cdot \dot{\mathbf{q}} = \mathbf{Q}_{\text{nc}} \cdot \dot{\mathbf{q}} \tag{7.23}$$

With the assumption that there exist no nonconservative forces, we find the Bubnov-Galerkin weighted residual form in time after integration over a given time interval as follows

$$\int_{t_0}^{t_1} \left\{ \left[\frac{d}{dt}\left(\frac{\partial \mathcal{E}(\mathbf{q}, \dot{\mathbf{q}})}{\partial \dot{\mathbf{q}}} \right) + \frac{\partial \mathcal{E}(\mathbf{q}, \dot{\mathbf{q}})}{\partial \mathbf{q}} - 2\frac{\partial \mathcal{K}(\mathbf{q}, \dot{\mathbf{q}})}{\partial \mathbf{q}} \right] \cdot \dot{\mathbf{q}} \right\} dt = 0 \tag{7.24}$$

In addition, if the kinetic energy is independent of generalized coordinates, Equation (7.24) reduces to

$$\int_{t_0}^{t_1} \left\{ \left[\frac{d}{dt}\left(\frac{\partial \mathcal{E}(\mathbf{q}, \dot{\mathbf{q}})}{\partial \dot{\mathbf{q}}} \right) + \frac{\partial \mathcal{E}(\mathbf{q}, \dot{\mathbf{q}})}{\partial \mathbf{q}} \right] \cdot \dot{\mathbf{q}} \right\} dt = 0 \tag{7.25}$$

In the context of the Lagrangian mechanics framework, the principle of balance of mechanical energy for an N-body dynamical system which is subject to holonomic constraints can be written as

$$\frac{d}{dt}\left(\frac{\partial \mathcal{L}(\mathbf{q}, \dot{\mathbf{q}})}{\partial \dot{\mathbf{q}}} \cdot \dot{\mathbf{q}} - \mathcal{L}(\mathbf{q}, \dot{\mathbf{q}}) \right) = \mathbf{Q}_{\text{nc}} \cdot \dot{\mathbf{q}} \tag{7.26}$$

The above equation can be rewritten as

$$\left[\frac{d}{dt}\left(\frac{\partial \mathcal{L}(\mathbf{q},\dot{\mathbf{q}})}{\partial \dot{\mathbf{q}}}\right)-\frac{\partial \mathcal{L}(\mathbf{q},\dot{\mathbf{q}})}{\partial \mathbf{q}}\right]\cdot\dot{\mathbf{q}}=\mathbf{Q}_{\mathrm{nc}}\cdot\dot{\mathbf{q}} \tag{7.27}$$

Provided we have no nonconservative forces, we obtain the Bubnov-Galerkin weighted residual form in time after integration over a given time interval as below,

$$\int_{t_0}^{t_1}\left\{\left[\frac{d}{dt}\left(\frac{\partial \mathcal{L}(\mathbf{q},\dot{\mathbf{q}})}{\partial \dot{\mathbf{q}}}\right)-\frac{\partial \mathcal{L}(\mathbf{q},\dot{\mathbf{q}})}{\partial \mathbf{q}}\right]\cdot\dot{\mathbf{q}}\right\}dt=0 \tag{7.28}$$

Lastly, in the context of Hamiltonian mechanics framework, Hamilton's law of varying action for holonomic systems can be defined as

$$\frac{d\mathcal{H}(\mathbf{q},\mathbf{p})}{dt}=\mathbf{Q}_{\mathrm{nc}}\cdot\dot{\mathbf{q}} \tag{7.29}$$

The above equation can be rewritten as

$$\left(\dot{\mathbf{p}}+\frac{\partial \mathcal{H}(\mathbf{q},\mathbf{p})}{\partial \mathbf{q}}\right)\cdot\dot{\mathbf{q}}-\left(\dot{\mathbf{q}}-\frac{\partial \mathcal{H}(\mathbf{q},\mathbf{p})}{\partial \mathbf{p}}\right)\cdot\dot{\mathbf{p}}=\mathbf{Q}_{\mathrm{nc}}\cdot\dot{\mathbf{q}} \tag{7.30}$$

Provided that we have no nonconservative forces, then the above equation leads to the two-field weighted residual form in time after integration over a given time interval as follows

$$\int_{t_0}^{t_1}\left[\left(\dot{\mathbf{p}}+\frac{\partial \mathcal{H}(\mathbf{p},\mathbf{q})}{\partial \mathbf{q}}\right)\cdot\dot{\mathbf{q}}-\left(\dot{\mathbf{q}}-\frac{\partial \mathcal{H}(\mathbf{p},\mathbf{q})}{\partial \mathbf{p}}\right)\cdot\dot{\mathbf{p}}\right]dt=0 \tag{7.31}$$

As $\dot{\mathbf{q}}$ or $\dot{\mathbf{p}}$ is linearly independent, Equations (7.24), (7.28) and (7.31) yield the Total Energy representation of the equation of motion, Lagrange's equation of motion and Hamilton's equation of motion, respectively, and they are equivalent to each other.

7.4 EQUIVALENCE RELATIONS BETWEEN GOVERNING EQUATIONS

Holonomic Constraint Systems At first, we consider a dynamical system of N particles, which is subject to the k number holonomic constraint equations. In Newtonian mechanics, the Newtonian dynamical system in Cartesian coordinates is defined as

$$\mathbf{M}\ddot{\mathbf{x}}=\mathbf{F}_{\mathrm{appl}}+\mathbf{C} \tag{7.32}$$

where $\ddot{\mathbf{x}}$ denotes the acceleration vector, i.e. $\ddot{\mathbf{x}}(t)=\left(\ddot{\mathbf{x}}^1,\ldots,\ddot{\mathbf{x}}^N\right):\mathcal{T}\rightarrow\mathbb{R}^{3N}$. In Lagrangian mechanics, the Lagrangian $\mathcal{L}(\mathbf{q},\dot{\mathbf{q}},t):TQ\times\mathcal{T}\rightarrow\mathbb{R}$ is defined as

$$\mathcal{L}(\mathbf{q},\dot{\mathbf{q}},t)=\mathcal{K}(\mathbf{q},\dot{\mathbf{q}},t)-\mathcal{U}(\mathbf{q},t),\quad \mathbf{q}=\left(q^1,\ldots,q^{n_{\mathrm{dof}}}\right)\in Q\subset\mathbb{R}^{n_{\mathrm{dof}}=3N-k} \tag{7.33}$$

Then, the Lagrangian dynamical system in generalized coordinates is given by

$$\frac{d}{dt}\left(\frac{\partial \mathcal{L}(\mathbf{q},\dot{\mathbf{q}},t)}{\partial \dot{\mathbf{q}}}\right)-\frac{\partial \mathcal{L}(\mathbf{q},\dot{\mathbf{q}},t)}{\partial \mathbf{q}}=\mathbf{0} \tag{7.34}$$

which represents n_{dof}-number of second-order differential equations. In Hamiltonian mechanics, the Hamiltonian $\mathcal{H}(\mathbf{p}, \mathbf{q}, t) : T^*Q \times \mathcal{T} \rightarrow \mathbb{R}$ is defined as

$$\mathcal{H}(\mathbf{q}, \mathbf{p}, t) = \mathbf{p} \cdot \dot{\mathbf{q}} - \mathcal{L}(\mathbf{q}, \dot{\mathbf{q}}, t) \tag{7.35}$$

and the Hamiltonian dynamical system in canonical coordinates is given by

$$\dot{\mathbf{q}} = \frac{\partial \mathcal{H}(\mathbf{q}, \mathbf{p}, t)}{\partial \mathbf{p}}, \quad \dot{\mathbf{p}} = -\frac{\partial \mathcal{H}(\mathbf{q}, \mathbf{p}, t)}{\partial \mathbf{q}} \tag{7.36}$$

which represent $2n_{\text{dof}}$-number of first-order differential equations.

Additionally, in the Total Energy framework, the Total Energy $\mathcal{E}(\mathbf{q}, \dot{\mathbf{q}}, t) : TQ \rightarrow \mathbb{R}$ is defined as

$$\mathcal{E}(\mathbf{q}, \dot{\mathbf{q}}, t) = \mathcal{K}(\mathbf{q}, \dot{\mathbf{q}}, t) + \mathcal{U}(\mathbf{q}, t) \tag{7.37}$$

Note that the potential energy here is assumed to be velocity-independent. Hence, in view of Equations (7.33) and (7.37), we obtain the relation between the Lagrangian and the Total Energy as below:

$$\mathcal{L}(\mathbf{q}, \dot{\mathbf{q}}, t) = 2\mathcal{K}(\mathbf{q}, \dot{\mathbf{q}}, t) - \mathcal{E}(\mathbf{q}, \dot{\mathbf{q}}, t) \tag{7.38}$$

In configuration space, we have the Total Energy representation of the equation of motion in generalized coordinates as follows

$$\frac{d}{dt}\left(\frac{\partial \mathcal{E}(\mathbf{q}, \dot{\mathbf{q}}, t)}{\partial \dot{\mathbf{q}}}\right) + \frac{\partial \mathcal{E}(\mathbf{q}, \dot{\mathbf{q}}, t)}{\partial \mathbf{q}} - 2\frac{\partial \mathcal{K}(\mathbf{q}, \dot{\mathbf{q}}, t)}{\partial \mathbf{q}} = \mathbf{0} \tag{7.39}$$

Then, the generalized momentum is given by the relation

$$\mathbf{p} = \frac{\partial \mathcal{L}(\mathbf{q}, \dot{\mathbf{q}}, t)}{\partial \dot{\mathbf{q}}} = \frac{\partial \mathcal{E}(\mathbf{q}, \dot{\mathbf{q}}, t)}{\partial \dot{\mathbf{q}}} = \frac{\partial \mathcal{K}(\mathbf{q}, \dot{\mathbf{q}}, t)}{\partial \dot{\mathbf{q}}} \tag{7.40}$$

In addition to the relation given in Equation (7.40), we can have also the following relation

$$\frac{\partial \mathcal{L}(\mathbf{q}, \dot{\mathbf{q}}, t)}{\partial \mathbf{q}} = 2\frac{\partial \mathcal{K}(\mathbf{q}, \dot{\mathbf{q}}, t)}{\partial \mathbf{q}} - \frac{\partial \mathcal{E}(\mathbf{q}, \dot{\mathbf{q}}, t)}{\partial \mathbf{q}} = -\frac{\partial \mathcal{H}(\mathbf{q}, \mathbf{p}, t)}{\partial \mathbf{q}} \tag{7.41}$$

The governing equations given in Equations (7.34), (7.36) and (7.39) are equivalent to each other as shown in Figure 7.1. Note that the above governing equations are valid in holonomic constraint systems.

Holonomic-Scleronomic Constraint Systems In this section, we confine our attention to the case that an N-body dynamical system is holonomic-scleronomic and conservative.

In Lagrangian mechanics, the Lagrangian $\mathcal{L}(\mathbf{q}, \dot{\mathbf{q}}) : TQ \rightarrow \mathbb{R}$ is defined as

$$\mathcal{L}(\mathbf{q}, \dot{\mathbf{q}}) = \mathcal{K}(\mathbf{q}, \dot{\mathbf{q}}) - \mathcal{U}(\mathbf{q}), \quad \mathbf{q} = \left(q^1, \ldots, q^{n_{\text{dof}}}\right) \in Q \subset \mathbb{R}^{n_{\text{dof}} = 3N-k} \tag{7.42}$$

which is autonomous. In such a holonomic-scleronomic and conservative system, the kinetic energy is quadratic in generalized velocities. Therefore, the Lagrangian dynamical system in generalized coordinates reduces to

$$\frac{d}{dt}\left(\frac{\partial \mathcal{L}(\mathbf{q}, \dot{\mathbf{q}})}{\partial \dot{\mathbf{q}}}\right) - \frac{\partial \mathcal{L}(\mathbf{q}, \dot{\mathbf{q}})}{\partial \mathbf{q}} = \mathbf{0} \tag{7.43}$$

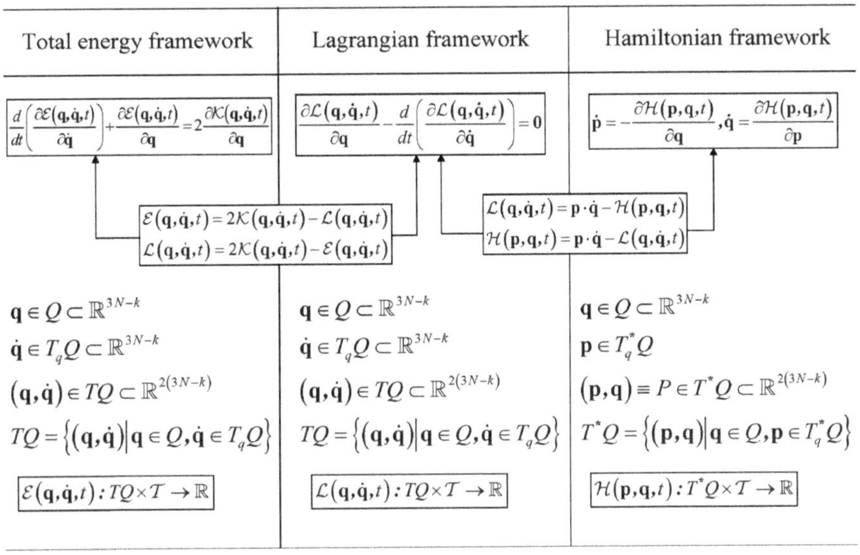

Figure 7.1. Equivalence of governing equations in holonomic constraint systems

In Hamiltonian mechanics, the Hamiltonian $\mathcal{H}(\mathbf{p},\mathbf{q}):T^*Q\rightarrow\mathbb{R}$ is defined as

$$\mathcal{H}(\mathbf{q},\mathbf{p})=\mathbf{p}\cdot\dot{\mathbf{q}}-\mathcal{L}(\mathbf{q},\dot{\mathbf{q}}) \tag{7.44}$$

which is also autonomous. And the Hamiltonian dynamical system in canonical coordinates leads to

$$\dot{\mathbf{q}}=\frac{\partial\mathcal{H}(\mathbf{q},\mathbf{p})}{\partial\mathbf{p}},\quad \dot{\mathbf{p}}=-\frac{\partial\mathcal{H}(\mathbf{q},\mathbf{p})}{\partial\mathbf{q}} \tag{7.45}$$

Additionally, in the Total Energy framework, the autonomous Total Energy $\mathcal{E}(\mathbf{q},\dot{\mathbf{q}}):TQ\times\mathcal{T}\rightarrow\mathbb{R}$ is defined as

$$\mathcal{E}(\mathbf{q},\dot{\mathbf{q}})=\mathcal{K}(\mathbf{q},\dot{\mathbf{q}})+\mathcal{U}(\mathbf{q}) \tag{7.46}$$

Note that the potential energy here is assumed to be velocity-independent. Since the kinetic energy is quadratic in generalized velocities, we have the relationship between the autonomous Lagrangian and the autonomous Total Energy as below:

$$\begin{aligned}\mathcal{L}(\mathbf{q},\dot{\mathbf{q}})&=\frac{\partial\mathcal{E}(\mathbf{q},\dot{\mathbf{q}})}{\partial\dot{\mathbf{q}}}\cdot\dot{\mathbf{q}}-\mathcal{E}(\mathbf{q},\dot{\mathbf{q}}),\\ \mathcal{E}(\mathbf{q},\dot{\mathbf{q}})&=\frac{\partial\mathcal{L}(\mathbf{q},\dot{\mathbf{q}})}{\partial\dot{\mathbf{q}}}\cdot\dot{\mathbf{q}}-\mathcal{L}(\mathbf{q},\dot{\mathbf{q}})\end{aligned} \tag{7.47}$$

Hence, the Total Energy representation of the equation of motion in generalized coordinates can be written as

$$\frac{d}{dt}\left(\frac{\partial\mathcal{E}(\mathbf{q},\dot{\mathbf{q}})}{\partial\dot{\mathbf{q}}}\right)+\frac{\partial\mathcal{E}(\mathbf{q},\dot{\mathbf{q}})}{\partial\mathbf{q}}-2\frac{\partial\mathcal{K}(\mathbf{q},\dot{\mathbf{q}})}{\partial\mathbf{q}}=\mathbf{0} \tag{7.48}$$

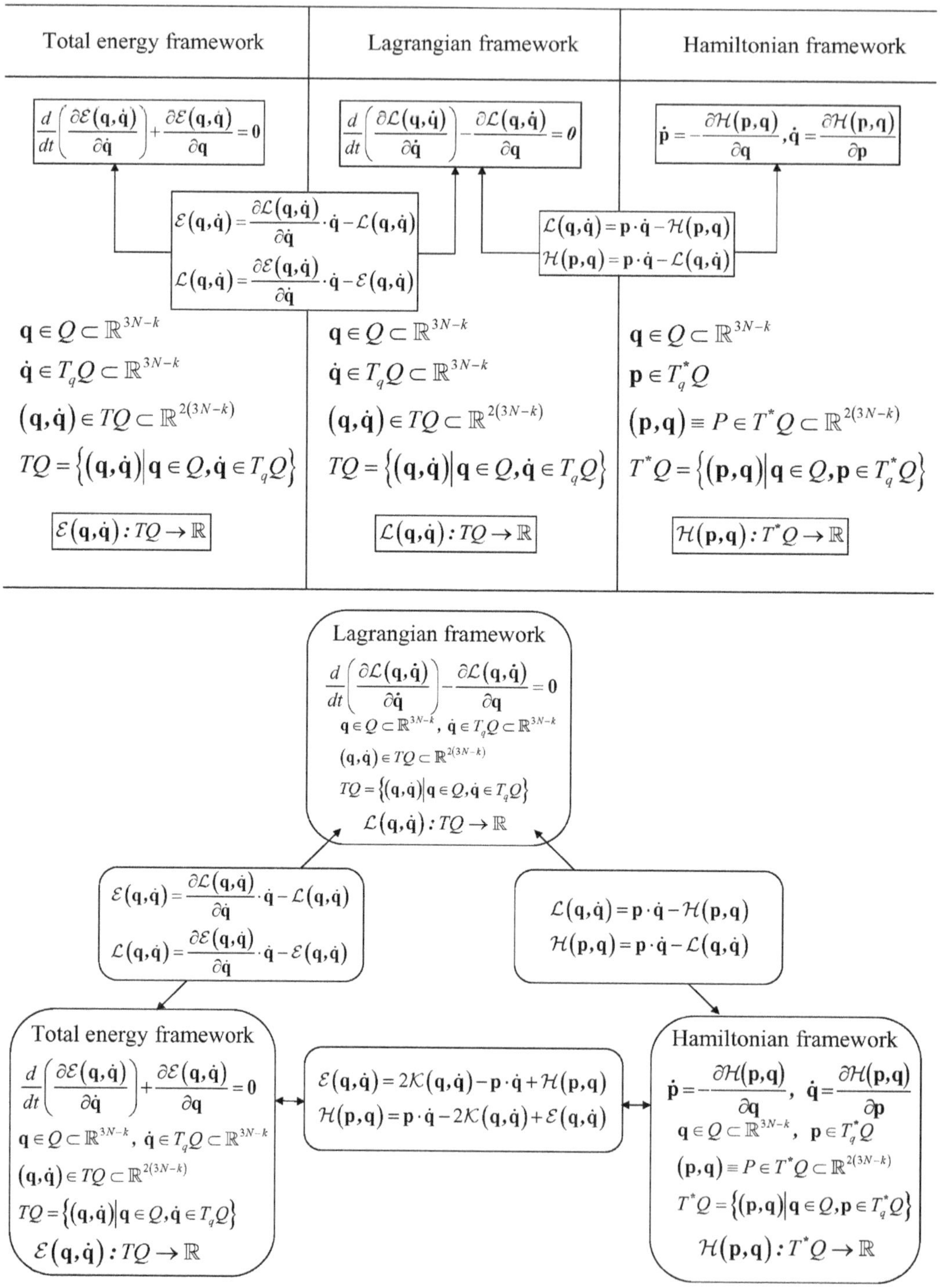

Figure 7.2. Equivalence of governing equations in holonomic-scleronomic constraint systems where the mass matrix in the kinetic energy is independent of generalized coordinates

For the special case when the kinetic energy is independent of generalized coordinates, the Total Energy representation of the equation of motion in generalized coordinates reduces to

$$\frac{d}{dt}\left(\frac{\partial \mathcal{E}(\mathbf{q}, \dot{\mathbf{q}})}{\partial \dot{\mathbf{q}}}\right) + \frac{\partial \mathcal{E}(\mathbf{q}, \dot{\mathbf{q}})}{\partial \mathbf{q}} = \mathbf{0} \tag{7.49}$$

Finally, the governing equations given in Equations (7.43), (7.45) and (7.49) are equivalent to each other as shown in Figure 7.2.

7.5 CONSERVATION LAWS

During the motion, some physical functions remain constant. Such functions are called constants of motion and it is said to be conserved in motion. The conserved quantities, namely, constants of motion, imply that the descriptive scalar functions such as the Lagrangian, the Hamiltonian, and the Total Energy, have corresponding symmetries in motion. In what follows, we simply highlight the invariance and conservation in physics via Noether's theorem.

7.6 NOETHER'S THEOREM

Symmetry in physics is very important in that it implies that there exists a corresponding conserved quantity which is called a constant of the motion, when a descriptive scalar function is invariant under a symmetry transformation. Here, symmetry means invariance under a symmetry transformation. Noether (Noether 1918; see Figure 7.3) first revealed the symmetry of the Lagrangian in field problems using the calculus of variation and the Lie group theory. This theorem is further known to be valid for even continuous-body dynamical systems. She proved that conserved quantities, i.e. linear and angular

Figure 7.3. Amalie Emmy Noether (1882-1935)

momenta, are preserved due to the symmetries of the Lagrangian. This is called Noether's theorem. Noether's theorem can be extended to not only the Hamiltonian in phase space but also the Total Energy in configuration space, as these scalar functions also possess symmetries (proofs are left as an exercise and hence are not discussed here).

EXERCISES

7.1. Discuss the equivalence relations of governing equations for the simple planar pendulum problem, we have considered in Chapter 3.

7.2. Discuss the equivalence in the principle of balance of mechanical energy for the same planar pendulum problem.

7.3. Discuss the equivalence relations of governing equations for the double planar pendulum problem, we have considered in Chapter 3.

7.4. Discuss the equivalence in the principle of balance of mechanical energy for the same double planar pendulum problem.

7.5. Discuss the equivalence relations of governing equations for the spherical pendulum problem, we have considered in Chapter 3.

7.6. Discuss the equivalence in the principle of balance of mechanical energy for the same spherical pendulum problem.

7.7. Discuss the equivalence relations of governing equations for the problem of the sliding block on an inclined plane, we have considered in Chapter 3.

7.8. Discuss the equivalence in the principle of balance of mechanical energy for the same problem of the sliding block on an inclined plane.

7.9. If the Lagrangian is translationally invariant under Euclidean group, then the linear momentum is conserved. If the Lagrangian is rotationally invariant, then the angular momentum is conserved. Lastly, the total energy is conserved, if the Lagrangian is invariant in time for holonomic-sceleronomic systems. Use Lagrange's equation of motion to examine the invariance properties (symmetries) of the descriptive scalar function, namely, the Lagrangian. (Hint: Use an infinitesimal continuous parameter in the Lie group (Bluman and Kumei 1989))

7.10. Repeat Exercise 7.9 for the descriptive scalar function, namely, the Hamiltonian.

7.11. Repeat Exercise 7.9 for the descriptive scalar function, namely, the Total Energy.

PART II

CONTINUOUS-BODY DYNAMICAL SYSTEMS

CHAPTER EIGHT

CONTINUUM MECHANICS

Some basic preliminaries dealing with three-dimensional elastodynamics are covered in Chapter 8. In general, physical quantities in elastodynamics are represented in terms of tensor-valued functions in the Euclidean three-dimensional space. These tensors may be classified as covariant, contravariant and mixed tensors. The covariant tensors are the physical quantities whose components are covariant on the contravariant base vectors, whereas the contravariant tensors are the ones whose components are contravariant on the covariant base vectors. The mixed tensors possess both components on both base vectors. The covariant and the contravariant base vectors are dual in regard to the counterpart.

In this chapter, contravariant vectors are employed to describe the kinematics of the body, as shown in Figure 8.1. This approach for continuous bodies is usually referred to as three-dimensional elastodynamics, in which strain tensors are covariant, whereas stress tensors are contravariant. The strain tensors, for illustration, are the Green strain tensor, the Almansi strain tensor and the Green deformation tensor, while the stress tensors are the Cauchy stress tensor, the Kirchhoff stress tensor and the Second Piola-Kirchhoff stress tensor. It should be noted that strain and stress tensors are dual as defined on the dual base vectors respectively.

8.1 DISPLACEMENTS, STRAINS AND STRESSES

In continuum mechanics, it is often postulated that infinite particles constitute a continuous body, but the particles are not considered to have concentrated mass other than the postulation in particle dynamics. At any instant, the simultaneous position of the particles is termed configuration or placement. The space occupied by a set of all configurations in a dynamical motion of the body is called the configuration space. In mechanics, the configuration is taken into account as a point in the configuration space (Papastavridis 1998).

The initial configuration in continuum mechanics is called the reference or material configuration, whereas the current position of all particles is the current or spatial configuration. The particles are defined as mathematical quantities. These quantities are in general called material points in the reference configuration and spatial points in the current configuration. The configuration is usually known to be embedded in the Euclidean three-dimensional space. The current configuration at any instant is a motion of the body, and the trajectory of the motion is a curve in the configuration space.

Advances in Computational Dynamics of Particles, Materials and Structures, First Edition. Jason Har and Kumar K. Tamma.

The introduction of the concept of particles in continuum mechanics leads naturally to the concept of a set in the mathematical description. The set concept plays an important role in the Lagrangian and the Hamiltonian structures for the continuous body.

8.1.1 Configuration Space

A configuration represents the current position of a body, and it is usually called a placement of the body. The configuration space is the collection of all orientation-preserving embeddings (configurations) as the body undergoes deformation in the time domain. To have a general description, let us define the flat Riemannian smooth manifolds with a reference configuration $(\mathcal{B}, \mathbf{G})$ and a spatial configuration $(\mathcal{S}, \mathbf{g})$. The manifolds are assumed to be embedded in the Euclidean three-dimensional space (Abraham et al. 1988). Every configuration is a smooth infinite-dimensional manifold. In other words, the referential (initial) manifold with the Riemannian material metric $\mathbf{G}$ is defined as an open set, denoted by $\mathcal{B}$, occupying a solid continuum, simple body in the Euclidean $\mathbb{R}^3$ space, while the spatial (current) manifold with the Riemannian spatial metric $\mathbf{g}$ is defined as the other open set, denoted by $\mathcal{S}$ as the ambient space as illustrated in Figure 8.1. The reference body $\mathcal{B}$ is bounded with the smooth reference boundary, denoted by $\partial\mathcal{B}$, which is also an open set. Then, we have a union as a closure $\bar{\mathcal{B}} := \mathcal{B} \cup \partial\mathcal{B}$. Further, the reference boundary may be partitioned into disjoint subsets including the prescribed traction boundary $\partial\mathcal{B}_\sigma$ and the prescribed displacement boundary $\partial\mathcal{B}_\varphi$ such that $\partial\mathcal{B} = \partial\mathcal{B}_\sigma \cup \partial\mathcal{B}_\varphi$. We note that the disjoint subsets are also open sets. The reference (initial) configuration is often called the undeformed configuration, whereas the spatial (current) configuration is called the deformed configuration. The spatial configuration of the body is a mapping $\boldsymbol{\varphi}_t : \mathcal{B} \to \mathbb{R}^3$ that is sufficiently smooth, orientation preserving and invertible for fixed time t. Let a contravariant set $\{X\} = \left\{X^I \mid I = 1, 2, 3\right\} \in \mathcal{B} \subset \mathbb{R}^3$ denote a Cartesian coordinate

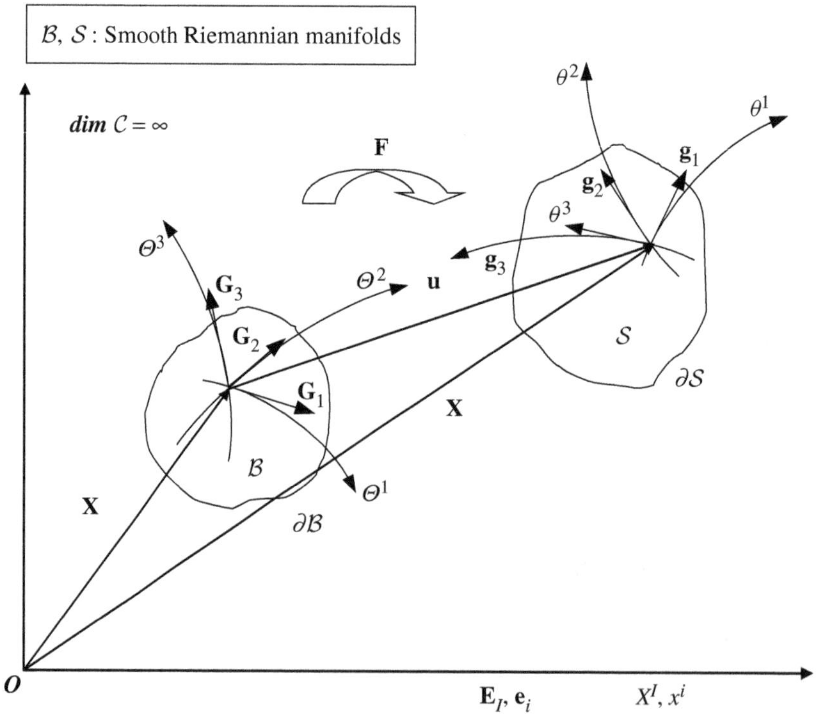

Figure 8.1. Deformation of a continuous body in configuration space

chart on the undeformed configuration $\mathcal{B}$. Therefore, the position of a continuum particle constituting the undeformed body is denoted by $\mathbf{X} = \left(X^1, X^2, X^3\right) \in \mathcal{B} \subset \mathbb{R}^3$, which is often called the material point. The corresponding covariant basis is denoted as a set $\{\mathbf{E}_I \mid I = 1, 2, 3\}$, and the covariant base vectors are unit Cartesian vectors. The coordinates indicating components of the position in the reference configuration are often called the material coordinates.

As the body moves, a series of configuration changes occurs in the configuration space. Within a closed interval of time $\bar{\mathcal{T}} = [0, T]$, a smooth motion of the body may be described as a one-to-one mapping: $\boldsymbol{\varphi}(\mathbf{X}, t) : \mathcal{B} \times \bar{\mathcal{T}} \to \mathcal{S} \subset \mathbb{R}^3, \quad \forall t \in \bar{\mathcal{T}}$, where $\mathcal{S}$ denotes a set of the infinite particles of the body in the spatial or current configuration. Note that the boundary of the closure $\bar{\mathcal{T}}$ is denoted by $\partial\mathcal{T}$, and consists of its two end points, i.e. $\partial\mathcal{T} = \{t_0, t_1\}$ and $\mathcal{T} =]t_0, t_1[$, where $t_0 = 0$ and $t_1 = T$. Hence, $\boldsymbol{\varphi}(\mathbf{X}, t)$ represents the position in the Lagrangian description. We have now a new configuration representing a deformed state of the body, being described by the other contravariant set $\{x\} = \left\{x^i \mid i = 1, 2, 3\right\} \in \mathcal{S} \subset \mathbb{R}^3$. In the same manner, we shall denote the position of a continuum particle constituting the deformed body by $\mathbf{x} = \left(x^1, x^2, x^3\right) \in \mathcal{S} \subset \mathbb{R}^3$, which is often called the spatial point. The current body $\mathcal{S}$ is bounded with the smooth current boundary, denoted by $\partial\mathcal{S}$. The current boundary $\partial\mathcal{S}$ is also an open set. Then, we have a union as a closure $\bar{\mathcal{S}} := \mathcal{S} \cup \partial\mathcal{S}$. Further, the current boundary may also be partitioned into disjoint subsets including the prescribed traction boundary $\partial\mathcal{S}_\sigma$ and the prescribed displacement boundary $\partial\mathcal{S}_\varphi$ such that $\partial\mathcal{S} = \partial\mathcal{S}_\sigma \cup \partial\mathcal{S}_\varphi$. It is noted that the disjoint subsets are also open sets. The covariant basis of the current body in the Cartesian coordinate system, then, is denoted as $\{\mathbf{e}_i \mid i = 1, 2, 3\}$, and their covariant base vectors are unit Cartesian vectors. Similarly, the coordinates indicating components of the position in the spatial configuration are often called the spatial coordinates.

Next, let us consider a set $\{\Theta\} = \left\{\Theta^A \mid A = 1, 2, 3\right\} \in \mathcal{B} \subset \mathbb{R}^3$ as a general coordinate chart embedded in the undeformed body, which is the so-called convected or intrinsic coordinate system. The corresponding covariant basis in the general or curvilinear coordinate system, then, is denoted as a set $\{\mathbf{G}_A \mid A = 1, 2, 3\}$, and the covariant base vectors are not required to be unit Cartesian vectors. Likewise, we may have the other set $\{\theta\} = \{\theta^a \mid a = 1, 2, 3\} \in \mathcal{S} \subset \mathbb{R}^3$, which is also a general coordinate chart embedded in the deformed body. The corresponding covariant basis in the general or curvilinear coordinate system is denoted as a set $\{\mathbf{g}_a \mid a = 1, 2, 3\}$, and the covariant base vectors are not required to be unit Cartesian vectors.

As mentioned previously, the space occupied by a set of all configurations in the dynamical motion is called the configuration space, denoted by $\mathcal{C}$, that is,

$$\mathcal{C} = \left\{ \begin{aligned} &\boldsymbol{\varphi}(\mathbf{X}, t) : \mathcal{B} \times \bar{\mathcal{T}} \to \mathcal{S} \subset \mathbb{R}^3 \mid \boldsymbol{\varphi} \in W_2^m(\mathcal{B}) = H^m(\mathcal{B}), \\ &\boldsymbol{\varphi}_{\mathbf{X}}(t) \in C^2\left(\bar{\mathcal{T}}\right), m > 2, \; |\mathrm{GRAD}\,\boldsymbol{\varphi}| > 0, \text{ and } \boldsymbol{\varphi}|_{\partial\mathcal{B}_\varphi} = \bar{\boldsymbol{\varphi}} \end{aligned} \right\} \tag{8.1}$$

where $\bar{\boldsymbol{\varphi}}$ stands for the prescribed displacement on the prescribed-displacement boundary $\boldsymbol{\varphi}|_{\partial\mathcal{B}_\varphi}$ and $\mathrm{GRAD}\,\boldsymbol{\varphi} = \frac{\partial\boldsymbol{\varphi}}{\partial\mathbf{X}}$. As every configuration is a smooth infinite dimensional manifold, the second derivative of the motion with respect to the independent state variable $\mathbf{X} \in \bar{\mathcal{B}}$ is required to be square-integrable in the Lebesgue sense up to the m-th order derivative with respect to the space independent variables; i.e. $\boldsymbol{\varphi} \in H^m(\mathcal{B})$. The configuration is embedded in the Euclidean three-dimensional space $\mathbb{E}^3$; for example, $\mathcal{C} = Emb(\mathcal{B}, \mathcal{S})$.

In the subsequent subsections, we shall introduce the fundamental tensors as strain and stress-related tensors to discuss the general principles in nonlinear continuum mechanics.

8.1.2 Riemannian Metrics

From now on, it is implied that the Einstein summation convention is applied when repeated indices as in $\eta^a\mathbf{g}_a = \eta^1\mathbf{g}_1 + \eta^2\mathbf{g}_2 + \eta^3\mathbf{g}_3$ occur. A vectorial line segment of the undeformed body in the reference configuration can be defined in terms of the Cartesian or convective coordinate system

$$
\begin{aligned}
d\mathbf{X} &:= dX^I \mathbf{E}_I \quad \text{or} \\
d\mathbf{X} &:= \frac{\partial \mathbf{X}}{\partial \Theta^A} d\Theta^A = \frac{\partial X^I}{\partial \Theta^A} \mathbf{E}_I d\Theta^A = d\Theta^A \mathbf{G}_A
\end{aligned} \tag{8.2}
$$

which is a contravariant vector with covariant base vectors. Therefore, the covariant basis vector in the convective coordinate system is expressed as

$$
\mathbf{G}_A = \frac{\partial \mathbf{X}}{\partial \Theta^A} = \frac{\partial X^I}{\partial \Theta^A} \mathbf{E}_I, \quad A = 1, 2, 3 \tag{8.3}
$$

where A denotes a free index, but I denotes a dummy index. Note that the contravariant basis vector in the reference configuration can be given by

$$
\mathbf{G}^A = \frac{\partial \Theta^A}{\partial \mathbf{X}} = \frac{\partial \Theta^A}{\partial X^I} \mathbf{E}^I, \quad A = 1, 2, 3 \tag{8.4}
$$

In addition, a vectorial line segment of the deformed body in the spatial configuration is defined as

$$
d\mathbf{x} := dx^i \mathbf{e}_i = \frac{\partial \mathbf{x}}{\partial \theta^a} d\theta^a = \frac{dx^i}{\partial \theta^a} \mathbf{e}_i d\theta^a = d\theta^a \mathbf{g}_a \tag{8.5}
$$

which is also a contravariant vector with covariant base vectors. The covariant basis vector in the spatial configuration can be expressed as

$$
\mathbf{g}_a = \frac{\partial \mathbf{x}}{\partial \theta^a} = \frac{\partial x^i}{\partial \theta^a} \mathbf{e}_i, \quad a = 1, 2, 3 \tag{8.6}
$$

Likewise, the contravariant basis vector in the spatial configuration can be given by

$$
\mathbf{g}^a = \frac{\partial \theta^a}{\partial \mathbf{x}} = \frac{\partial \theta^a}{\partial x^i} \mathbf{e}^i, \quad a = 1, 2, 3 \tag{8.7}
$$

Notice that the following is true due to the dual relationship

$$
\begin{aligned}
\mathbf{G}_A \cdot \mathbf{G}^B &= \mathbf{G}^A \cdot \mathbf{G}_B = \delta_A^{\cdot B} = \delta^A_{\cdot B} \\
\mathbf{g}_a \cdot \mathbf{g}^b &= \mathbf{g}^a \cdot \mathbf{g}_b = \delta_a^{\cdot b} = \delta^a_{\cdot b}
\end{aligned} \tag{8.8}
$$

where $\delta_A^{\cdot B} = \delta^A_{\cdot B} = 1$ if $A = B$ and $\delta_a^{\cdot b} = \delta^a_{\cdot b} = 1$ if $a = b$; otherwise, $\delta_A^{\cdot B} = \delta^A_{\cdot B} = 0$ if $A \neq B$ and $\delta_a^{\cdot b} = \delta^a_{\cdot b} = 0$ if $a \neq b$ as the Kronecker delta.

8.1.3 Infinitesimal Differential Volume

The definition of an infinitesimal differential volume element is important in that most crucial equations of interest are given in an integral form as described in the subsequent sections. Based on the geometrical interpretation as shown in Figure 8.1, we have an infinitesimal differential volume $dV \in \mathbb{R}^+$ in the undeformed body as follows:

$$dV = (\mathbf{E}_1 \cdot \mathbf{E}_2 \times \mathbf{E}_3)\, dX^1 dX^2 dX^3 = \sqrt{E} dX^1 dX^2 dX^3 = dX^1 dX^2 dX^3 \tag{8.9}$$

where the infinitesimal differential volume is obtained by the scalar triple product as discussed in Exercise 2.5 of Chapter 2. Alternatively, by virtue of Equation (8.2), the infinitesimal differential volume in Equation (8.9) can be expressed in the convective coordinate system as follows:

$$dV = (\mathbf{G}_1 \cdot \mathbf{G}_2 \times \mathbf{G}_3)\, d\Theta^1 d\Theta^2 d\Theta^3 = \sqrt{G} d\Theta^1 d\Theta^2 d\Theta^3 \tag{8.10}$$

Further, by virtue of Equation (8.5), the infinitesimal differential volume $dv \in \mathbb{R}^+$ in the deformed body can be written as

$$dv = (\mathbf{e}_1 \cdot \mathbf{e}_2 \times \mathbf{e}_3)\, dx^1 dx^2 dx^3 = \sqrt{e} dx^1 dx^2 dx^3 = dx^1 dx^2 dx^3 \tag{8.11}$$

whereas the infinitesimal differential volume in Equation (8.11) can be given in the convective coordinate system as follows:

$$dv = (\mathbf{g}_1 \cdot \mathbf{g}_2 \times \mathbf{g}_3)\, d\theta^1 d\theta^2 d\theta^3 = \sqrt{g} d\theta^1 d\theta^2 d\theta^3 \tag{8.12}$$

Note that Equation (8.11) can be recast into the form

$$dv = dx^1 dx^2 dx^3 = \left| \frac{\partial \mathbf{x}}{\partial \mathbf{X}} \right| dX^1 dX^2 dX^3 = \left| \frac{\partial \mathbf{x}}{\partial \mathbf{X}} \right| dV \tag{8.13}$$

Then, the infinitesimal differential volume ratio is obtained as

$$J(\mathbf{X}, t) = \frac{dv}{dV} = \frac{\sqrt{g} d\theta^1 d\theta^2 d\theta^3}{\sqrt{G} d\Theta^1 d\Theta^2 d\Theta^3} = \frac{dx^1 dx^2 dx^3}{dX^1 dX^2 dX^3} = \left| \frac{\partial \mathbf{x}}{\partial \mathbf{X}} \right| = |\mathrm{GRAD}\, \mathbf{x}| > 0 \tag{8.14}$$

which is called the Jacobian or the determinant of the Jacobian matrix. It is important to note that the infinitesimal differential volume can be defined in terms of convective coordinates or Cartesian coordinates in the spatial configuration. Therefore, there are two ways to have an integral equation in convective coordinates and in Cartesian coordinates. In the literature, one takes the convective coordinate system (see (Marsden and Hughes 1983; Oden 1972; Wempner and Talaslidis 2003), on the other hand, one may take the Cartesian coordinate system (see (Belytschko et al. 2000; Holzapfel 2000; Malvern 1969)). In this chapter, we will follow the latter for convenience.

8.1.4 Displacements and Strains

In the literature, the mathematical description of motion of a body is found in two ways. Firstly, mechanical state variables described by the material coordinates are called Lagrangian, and the way by which the mathematical differential equations are given is called the Lagrangian description. Secondly, mechanical state variables described by the spatial coordinates are called Eulerian, and the way by which the mathematical differential equations are given is called the Eulerian description.

In what follows, we shall define mechanical state variables in the two descriptions.

Displacements in the Lagrangian Description In the Lagrangian description, we have the current position of a material point describing the motion of the body as

$$\mathbf{x} = \boldsymbol{\varphi}(\mathbf{X}, t) : \mathcal{B} \times \bar{\mathcal{T}} \rightarrow \mathcal{S} \subset \mathbb{R}^3, \quad \forall t \in \bar{\mathcal{T}} \tag{8.15}$$

Sometimes, we may also adopt the following notation

$$\mathbf{x} = \varphi_t(\mathbf{X}) : \mathcal{B} \to \mathcal{S} \subset \mathbb{R}^3, \quad \forall t \in \bar{\mathcal{T}} \tag{8.16}$$

where $\varphi_t(\mathbf{X})$ represents the motion when one holds $t \in \bar{\mathcal{T}} \subset \mathbb{R}^+$ fixed. In addition, we also have

$$\mathbf{x} = \varphi_{\mathbf{X}}(t) : \bar{\mathcal{T}} \to \mathcal{S} \subset \mathbb{R}^3, \quad \forall \mathbf{X} \in \mathcal{B} \tag{8.17}$$

where $\varphi_{\mathbf{X}}(t)$ denotes the motion when one holds material points $\mathbf{X} \in \mathcal{B}$ fixed. As a consequence, the initial position of the material point in the undeformed configuration can be written as

$$\mathbf{X} = \varphi^{-1}(\mathbf{x}, t) : \mathcal{S} \times \bar{\mathcal{T}} \to \mathcal{B} \subset \mathbb{R}^3, \quad \forall t \in \bar{\mathcal{T}} \tag{8.18}$$

when φ is invertible. Likewise, we have

$$\begin{aligned} \mathbf{X} &= \varphi_t^{-1}(\mathbf{x}) : \mathcal{S} \to \mathcal{B} \subset \mathbb{R}^3, \quad \forall t \in \bar{\mathcal{T}} \\ \mathbf{X} &= \varphi_{\mathbf{x}}^{-1}(t) : \bar{\mathcal{T}} \to \mathcal{B} \subset \mathbb{R}^3, \quad \forall \mathbf{x} \in \mathcal{S} \end{aligned} \tag{8.19}$$

where φ_t and $\varphi_{\mathbf{x}}$ are assumed to be invertible. Then, the displacement field may be defined as the difference between the position in the current configuration and the initial position as follows:

$$\mathbf{U}(\mathbf{X}, t) = \mathbf{x}(X, t) - \mathbf{X} = \varphi(\mathbf{X}, t) - \mathbf{X} = \varphi_t(\mathbf{X}) - \mathbf{X} = \mathbf{u}(\mathbf{x}, t) \tag{8.20}$$

Let us study in depth the displacement given in Equation (8.20). The displacement vector can be alternatively recast as

$$\mathbf{U}(\mathbf{X}, t) = x^i \mathbf{e}_i - X^I \mathbf{E}_I = \left(x^i - X^I\right) \mathbf{E}_I = \left(x^i - X^I\right) \mathbf{e}_i = \bar{u}^i \mathbf{e}_i = \bar{U}^I \mathbf{E}_I \tag{8.21}$$

In addition, the gradient of the displacement can be represented as

$$\mathrm{GRAD}\,\mathbf{U}(\mathbf{X}, t) = \frac{\partial \mathbf{x}}{\partial \mathbf{X}} - \mathbf{E}_I \otimes \mathbf{E}^I = \frac{\partial \mathbf{x}}{\partial \mathbf{X}} - \mathbf{I}_{\mathbf{E}\otimes\mathbf{E}} \tag{8.22}$$

where $\mathbf{I}_{\mathbf{E}\otimes\mathbf{E}}$ denotes a second-order identity tensor. Suppose that the motion of the body is C^1-continuous with respect to time, namely a C^1-regular or invertible motion. The velocity field of the motion, then, may be given by

$$\mathbf{V}(\mathbf{X}, t) = \frac{\partial \varphi(\mathbf{X}, t)}{\partial t} = \frac{d\varphi_{\mathbf{X}}(t)}{dt} = \mathbf{V}_t(\mathbf{X}) = \mathbf{V}_t \circ \varphi_t^{-1}(\mathbf{x}) \tag{8.23}$$

which is sometimes referred to as the material velocity of the motion. Note that $\frac{d\varphi_{\mathbf{X}}(t)}{dt}$ is the derivative of $\varphi(\mathbf{X}, t)$ with respect to time, holding material points $\mathbf{X}$ fixed. Likewise, suppose that the motion of the body is C^2-continuous with respect to time, namely a C^2-regular or invertible motion. The acceleration field of the body motion is given by

$$\mathbf{A}(\mathbf{X}, t) = \frac{\partial \mathbf{V}(\mathbf{X}, t)}{\partial t} = \mathbf{A}_t(\mathbf{X}) = \mathbf{A}_t \circ \varphi_t^{-1}(\mathbf{x}) \tag{8.24}$$

which is often called the material acceleration of the motion. The gradient of a scalar function is defined to be a vector field given by

$$\mathrm{GRAD}\, f(\mathbf{X}, t) = \frac{\partial f(\mathbf{X}, t)}{\partial \mathbf{X}} = \nabla_{\mathbf{X}} f = \frac{\partial f(\mathbf{X}, t)}{\partial X^I} \mathbf{E}^I = \frac{\partial f(\mathbf{X}, t)}{\partial \Theta^A} \mathbf{G}^A \tag{8.25}$$

where ∇ is called the nabla operator or the gradient of the function. The gradient of a vector function is defined to be a second-order tensor field as

$$\mathrm{GRAD}\,\boldsymbol{\varphi}(\mathbf{X},t)=\frac{\partial\boldsymbol{\varphi}(\mathbf{X},t)}{\partial\mathbf{X}}=\frac{\partial\boldsymbol{\varphi}(\mathbf{X},t)}{\partial X^I}\otimes\mathbf{E}^I=\frac{\partial\boldsymbol{\varphi}(\mathbf{X},t)}{\partial\Theta^A}\otimes\mathbf{G}^A=\boldsymbol{\varphi}\otimes\nabla_{\mathbf{X}} \tag{8.26}$$

Displacements in the Eulerian Description We constructed displacements, velocities and accelerations in the material coordinate system. Likewise, we shall show those quantities in the spatial coordinate system. The displacement field in the Eulerian description $\mathbf{u}(\mathbf{x},t):\mathcal{S}\times\bar{\mathcal{T}}\rightarrow\mathbb{R}^3$ can be recast as

$$\mathbf{u}(\mathbf{x},t)=\mathbf{x}-\mathbf{X}(\mathbf{x},t)=\mathbf{x}-\boldsymbol{\varphi}^{-1}(\mathbf{x},t) \tag{8.27}$$

Accordingly, the above equation can also be rewritten as

$$\mathbf{u}(\mathbf{x},t)=x^i\mathbf{e}_i-X^I\mathbf{E}_I=\left(x^i-X^I\right)\mathbf{E}_I=\left(x^i-X^I\right)\mathbf{e}_i=\bar{u}^i\mathbf{e}_i=\bar{U}^I\mathbf{E}_I \tag{8.28}$$

In view of Equation (8.21) and (8.28), we find that

$$\mathbf{u}(\mathbf{x},t)=\bar{u}^i\mathbf{e}_i=\bar{U}^I\mathbf{E}_I=U^A\mathbf{G}_A=u^a\mathbf{g}_a=\mathbf{U}(\mathbf{X},t) \tag{8.29}$$

Likewise, the gradient of the displacement can be given as

$$\mathrm{grad}\,\mathbf{u}(\mathbf{x},t)=\mathbf{g}_a\otimes\mathbf{g}^a-\frac{\partial\mathbf{X}}{\partial\mathbf{x}}=\mathbf{I}_{\mathbf{g}\otimes\mathbf{g}}-\frac{\partial\mathbf{X}}{\partial\mathbf{x}} \tag{8.30}$$

where $\mathbf{I}_{\mathbf{g}\otimes\mathbf{g}}$ denotes a second-order identity tensor. The gradient of a scalar function is defined to be a vector field as

$$\mathrm{grad}\,f(\mathbf{x},t)=\frac{\partial f(\mathbf{x},t)}{\partial\mathbf{x}}=\nabla_{\mathbf{x}}f=\frac{\partial f(\mathbf{x},t)}{\partial x^i}\mathbf{e}^i=\frac{\partial f(\mathbf{x},t)}{\partial\theta^a}\mathbf{g}^a \tag{8.31}$$

The gradient of a vector function $\boldsymbol{\phi}(\mathbf{x},t)$ is defined to be a second-order tensor field as

$$\mathrm{grad}\,\boldsymbol{\phi}(\mathbf{x},t)=\frac{\partial\boldsymbol{\phi}(\mathbf{x},t)}{\partial\mathbf{x}}=\frac{\partial\boldsymbol{\phi}(\mathbf{x},t)}{\partial x^i}\otimes\mathbf{e}^i=\frac{\partial\boldsymbol{\phi}(\mathbf{x},t)}{\partial\theta^a}\otimes\mathbf{g}^a=\boldsymbol{\phi}\otimes\nabla_{\mathbf{x}} \tag{8.32}$$

The spatial velocity field of a motion is given by

$$\mathbf{v}(\mathbf{x},t)=\frac{D\mathbf{x}(\mathbf{X},t)}{Dt}=\frac{D\boldsymbol{\varphi}(\mathbf{X},t)}{Dt}=\frac{\partial\boldsymbol{\varphi}(\mathbf{X},t)}{\partial t}=\mathbf{V}(\mathbf{X},t)=\dot{\boldsymbol{\varphi}}(\mathbf{X},t) \tag{8.33}$$

where the motion is assumed to be C^1-continuous with respect to time, and $\frac{D}{Dt}$ denotes the material time derivative or the convective derivative which is used when the derivatives with respect to spatial coordinates are involved. The acceleration field of the motion is given by

$$\mathbf{a}(\mathbf{x},t)=\frac{D\mathbf{v}(\mathbf{x},t)}{Dt}=\frac{\partial\mathbf{V}(\mathbf{X},t)}{\partial t}=\mathbf{A}(\mathbf{X},t)=\ddot{\boldsymbol{\varphi}}(\mathbf{X},t) \tag{8.34}$$

where the motion is assumed to be C^2-continuous with respect to time.

Deformation Gradient Tensor We shall introduce the deformation gradient tensor, which is a fundamental physical quantity in continuum mechanics in noting that most tensors in continuum mechanics are defined by means of the deformation gradient tensor. Further, it can be taken as a linear transformation mapping the vectorial line segment in the reference configuration into one in the spatial configuration as follows:

$$d\mathbf{x} = \mathbf{F}(\mathbf{X}, t) \cdot d\mathbf{X} = d\mathbf{X} \cdot \mathbf{F}^T \tag{8.35}$$

where $\mathbf{F}(\mathbf{X}, t)$ denotes the deformation gradient tensor. Equation (8.35) can also be recast as follows:

$$d\mathbf{X} = \mathbf{F}^{-1} \cdot d\mathbf{x} = d\mathbf{x} \cdot \mathbf{F}^{-T} \tag{8.36}$$

when $J(\mathbf{X}, t) = |\mathrm{GRAD}\,\boldsymbol{\varphi}| > 0$. The deformation gradient tensor can be expressed in terms of generalized base vectors or unit Cartesian base vectors as

$$\begin{aligned} \mathbf{F} &= \frac{\partial \mathbf{x}}{\partial \mathbf{X}} = \frac{\partial \mathbf{x}}{\partial \Theta^A} \otimes \mathbf{G}^A = \frac{\partial \theta^a}{\partial \Theta^A} \mathbf{g}_a \otimes \mathbf{G}^A = \frac{\partial x^i}{\partial X^I} \mathbf{e}_i \otimes \mathbf{E}^I \\ &= \bar{F}^i_{\ .I} \mathbf{e}_i \otimes E^I = F^a_{\ .A} \mathbf{g}_a \otimes \mathbf{G}^A = \boldsymbol{\varphi} \otimes \nabla_{\mathbf{X}} \end{aligned} \tag{8.37}$$

which turns out to be the Jacobian matrix as shown in Equation (8.14). Accordingly, the Jacobian is equivalent to the determinant of the deformation gradient tensor. In view of Equation (8.35) and (8.37), we find that

$$d\mathbf{x} = \frac{\partial x^i}{\partial X^I} \mathbf{e}_i \otimes \mathbf{E}^I \cdot dX^K \mathbf{E}_K = \frac{\partial x^i}{\partial X^I} dX^I \mathbf{e}_i = dx^i \mathbf{e}_i \tag{8.38}$$

The deformation gradient tensor is neither a Lagrangian nor an Eulerian. As shown in Equation (8.37), it is represented by two different base vectors. Such a tensor is called the two-point tensor in the sense of Truesdell and Toupin (Truesdell and Toupin 1960). Notice that the determinant of the deformation gradient tensor cannot be negative, because the Jacobian represents the volume ratio. The deformation gradient tensor is represented in a variety of ways in the literature,

$$\mathbf{F} = \frac{\partial \mathbf{x}}{\partial \mathbf{X}} = \mathrm{GRAD}\,\mathbf{x} = \mathrm{GRAD}\,\boldsymbol{\varphi} = \mathbf{x} \otimes \nabla_{\mathbf{X}} = \mathrm{GRAD}\,\mathbf{U}(\mathbf{X}, t) + \mathbf{I}_{\mathbf{G}\otimes\mathbf{G}} \tag{8.39}$$

where $\mathbf{I}_{\mathbf{G}\otimes\mathbf{G}}$ denotes a second-order identity tensor. In a rigid body motion, the first term on the right hand side, called the displacement gradient tensor in the Lagrangian description, vanishes; then the deformation gradient tensor equals the identity tensor. The inverse of the deformation gradient tensor can be expressed as

$$\mathbf{F}^{-1} = \frac{\partial \mathbf{X}}{\partial \mathbf{x}} = \mathrm{grad}\,\mathbf{X} = \mathbf{X} \otimes \nabla_{\mathbf{x}} = \mathbf{I}_{\mathbf{g}\otimes\mathbf{g}} - \mathrm{grad}\,\mathbf{u}(\mathbf{x}, t) \tag{8.40}$$

Likewise, in a rigid body motion, the second term on the right hand side, called the displacement gradient tensor in the Eulerian description, vanishes; then the inverse of the deformation gradient tensor equals the identity tensor. Equation (8.40) can also be recast in terms of generalized base vectors or unit Cartesian base vectors as

$$\mathbf{F}^{-1} = \frac{\partial \mathbf{X}}{\partial \mathbf{x}} = \frac{\partial \mathbf{X}}{\partial x^i} \otimes \mathbf{e}^i = \frac{\partial \mathbf{X}}{\partial \theta^a} \otimes \frac{\partial \theta^a}{\partial x^i} \mathbf{e}^i = \frac{\partial \Theta^A}{\partial \theta^a} \mathbf{G}_A \otimes \mathbf{g}^a = \frac{\partial X^I}{\partial x^i} \mathbf{E}_I \otimes \mathbf{e}^i \tag{8.41}$$

Also, we have the transpose of the deformation gradient tensor given by

$$\mathbf{F}^T = \frac{\partial \theta^a}{\partial \Theta^A} \mathbf{G}^A \otimes \mathbf{g}_a = \nabla_{\mathbf{X}} \otimes \boldsymbol{\varphi} \tag{8.42}$$

The transpose of the inverse of the deformation gradient tensor is given by

$$\mathbf{F}^{-T} = \frac{\partial \Theta^A}{\partial \theta^a}\mathbf{g}^a \otimes \mathbf{G}_A = \nabla_{\mathbf{x}} \otimes \mathbf{X} \tag{8.43}$$

The deformation gradient tensor can be multiplicatively decomposed as

$$\mathbf{F} = \mathbf{R} \cdot \bar{\mathbf{U}} = \bar{\mathbf{v}} \cdot \mathbf{R} \tag{8.44}$$

where $\mathbf{R}$ is called the rotation tensor as a proper orthogonal right-handed rotation tensor, $\bar{\mathbf{U}}$ the right (material) stretch tensor and $\bar{\mathbf{v}}$ the left (spatial) stretch tensor. The stretch tensors are positive definite and symmetric tensors. The above equation is often referred to as the polar decomposition theorem, which states that any nonsingular second-order tensor can be deformed into an orthogonal rotation tensor and a positive definite symmetric tensor. Via the polar decomposition theorem, we find that

$$d\mathbf{x} = \mathbf{F} \cdot d\mathbf{X} = \mathbf{R} \cdot \bar{\mathbf{U}} \cdot d\mathbf{X} = \bar{\mathbf{v}} \cdot \mathbf{R} \cdot d\mathbf{X} \tag{8.45}$$

The above implies that the spatial line segment consists of the first stretch and second rotation of the material line segment or the first rotation and second stretch deformation. In a rigid body motion, the stretch tensors equal the identity tensor, and then the deformation gradient tensor is identical to the rotation tensor. In other words, if $\mathbf{R} = \mathbf{I}$, then $\mathbf{F} = \bar{\mathbf{U}} = \bar{\mathbf{v}}$, and if $\bar{\mathbf{U}} = \bar{\mathbf{v}} = \mathbf{I}$, then $\mathbf{F} = \mathbf{R}$.

Spatial Velocity Gradient Tensor The total time derivative of the deformation gradient tensor is called the material velocity gradient tensor, which can be represented as

$$\dot{\mathbf{F}} = \frac{d\mathbf{F}}{dt} = \frac{d}{dt}\left(\frac{\partial \mathbf{U}(\mathbf{X}, t)}{\partial \mathbf{X}} + \mathbf{I}\right) = \frac{\partial}{\partial \mathbf{X}}\left(\frac{\partial \mathbf{U}(\mathbf{X}, t)}{\partial t}\right) = \frac{\partial \mathbf{V}(\mathbf{X}, t)}{\partial \mathbf{X}} = \mathrm{GRAD}\,\mathbf{V} \tag{8.46}$$

where the superposed dot represents the total time derivative. In the above, $\mathbf{I}$ and $\mathbf{U}$ are an identity tensor and the Lagrangian displacement vector. It is important to note that the total time derivative of the deformation gradient tensor becomes equivalent to the directional derivative of the deformation gradient in the direction of the spatial velocity as follows:

$$D_{\mathbf{v}}\mathbf{F} = \frac{d}{d\varepsilon}\left(\frac{\partial(\mathbf{x} + \varepsilon \mathbf{v})}{\partial \mathbf{X}}\right)\bigg|_{\varepsilon=0} = \frac{\partial \mathbf{v}}{\partial \mathbf{X}} = \frac{\partial \mathbf{V}(\mathbf{X}, t)}{\partial \mathbf{X}} = \mathrm{GRAD}\,\mathbf{V} = \dot{\mathbf{F}} \tag{8.47}$$

where $\mathbf{v}$ denotes the spatial velocity vector defined in Equation (8.33) rather than the left stretch tensor in Equation (8.44). The partial derivative of the spatial velocity with respect to the spatial coordinates is called the spatial velocity gradient tensor $\mathbf{l}$. It can be derived as

$$\mathbf{l} = \frac{\partial \mathbf{v}}{\partial \mathbf{x}} = \mathrm{grad}\,\mathbf{v} = \frac{\partial \mathbf{v}}{\partial \mathbf{X}} \cdot \frac{\partial \mathbf{X}}{\partial \mathbf{x}} = \frac{\partial \mathbf{V}}{\partial \mathbf{X}} \cdot \frac{\partial \mathbf{X}}{\partial \mathbf{x}} = \dot{\mathbf{F}} \cdot \mathbf{F}^{-1} \tag{8.48}$$

Accordingly, the acceleration field can be recast as

$$\boldsymbol{a}(\mathbf{x}, t) = \frac{D\mathbf{v}(\mathbf{x}, t)}{Dt} = \frac{\partial \mathbf{v}(\boldsymbol{x}, t)}{\partial t} + \mathbf{l}(\mathbf{x}, t) \cdot \mathbf{v}(\mathbf{x}, t) \tag{8.49}$$

Note that the spatial velocity gradient tensor can be expressed using Equation (8.44) as

$$\mathbf{l} = \left\{(\dot{\mathbf{R}} \cdot \bar{\mathbf{U}}) + \left(\mathbf{R} \cdot \dot{\bar{\mathbf{U}}}\right)\right\} \cdot \bar{\mathbf{U}}^{-1} \cdot \mathbf{R}^T = \dot{\mathbf{R}} \cdot \mathbf{R}^T + \mathbf{R} \cdot \dot{\bar{\mathbf{U}}} \cdot \bar{\mathbf{U}}^{-1} \cdot \mathbf{R}^T \tag{8.50}$$

Lagrangian Strain Tensor The square of the infinitesimal line segment, denoted by dS, in the reference (undeformed) configuration is given by

$$dS^2 = d\mathbf{X} \cdot d\mathbf{X} = d\mathbf{x} \cdot \mathbf{F}^{-T} \cdot \mathbf{F}^{-1} \cdot d\mathbf{x} = \mathbf{G}_A \cdot \mathbf{G}_B d\Theta^A d\Theta^B = G_{AB} d\Theta^A d\Theta^B \tag{8.51}$$

whereas the square of the infinitesimal line segment, denoted by ds, in the spatial (deformed) configuration is given by

$$ds^2 = d\mathbf{x} \cdot d\mathbf{x} = d\mathbf{X} \cdot \mathbf{F}^T \cdot \mathbf{F} \cdot d\mathbf{X} = \mathbf{g}_a \cdot \mathbf{g}_b d\theta^a d\theta^b = g_{ab} d\theta^a d\theta^b \tag{8.52}$$

The Lagrangian strain tensor, denoted by $\mathbf{E}$, is defined by

$$ds^2 - dS^2 = d\mathbf{X} \cdot \left(\mathbf{F}^T \cdot \mathbf{F} - \mathbf{G}\right) \cdot d\mathbf{X} = d\mathbf{X} \cdot 2\mathbf{E} \cdot d\mathbf{X} \tag{8.53}$$

where $\mathbf{G}$ is an identity tensor which takes the form

$$\mathbf{G} = \mathbf{G}_A \otimes \mathbf{G}^A = \mathbf{G}^A \otimes \mathbf{G}_A = G_{AB} \mathbf{G}^A \otimes \mathbf{G}^B = \mathbf{I}_{\mathbf{G} \otimes \mathbf{G}} \tag{8.54}$$

It is often called the metric tensor on the reference configuration. Thus, we see that

$$\mathbf{E} = \frac{1}{2}\left(\mathbf{F}^T \cdot \mathbf{F} - \mathbf{G}\right) = \frac{1}{2}\left\{\left(\frac{\partial \mathbf{U}}{\partial \mathbf{X}}\right) + \left(\frac{\partial \mathbf{U}}{\partial \mathbf{X}}\right)^T\right\} + \frac{1}{2}\left(\frac{\partial \mathbf{U}}{\partial \mathbf{X}}\right) \cdot \left(\frac{\partial \mathbf{U}}{\partial \mathbf{X}}\right)^T \tag{8.55}$$

which is also called the Green-Lagrange strain tensor or the material strain tensor. It should be noticed that the Lagrangian strain tensor is a Lagrangian measure. Further, in a rigid body motion in which the deformation gradient tensor is identical to the rotation tensor, the Lagrangian strain tensor becomes a zero second-order tensor such that

$$\mathbf{E} = \frac{1}{2}\left(\mathbf{R}^T \cdot \mathbf{R} - \mathbf{I}\right) = \mathbf{0} \tag{8.56}$$

The Lagrangian strain tensor can be alternatively written as

$$\mathbf{E} = \frac{1}{2}(\mathbf{C} - \mathbf{G}) \tag{8.57}$$

where $\mathbf{C}$ is referred to as the Green deformation tensor or the right Cauchy-Green tensor. The Green deformation tensor can also be found from the following

$$ds^2 = d\mathbf{X} \cdot \mathbf{F}^T \cdot \mathbf{F} \cdot d\mathbf{X} = d\mathbf{X} \cdot \mathbf{C} \cdot d\mathbf{X} \tag{8.58}$$

We see that the Green deformation tensor is positive definite, that is,

$$d\mathbf{X} \cdot \mathbf{C} \cdot d\mathbf{X} = (ds)^2 > 0, \quad \forall\, d\mathbf{X} \neq 0 \tag{8.59}$$

and it is symmetric, that is,

$$\mathbf{C} = \mathbf{F}^T \cdot \mathbf{F} = \mathbf{C}^T \tag{8.60}$$

Using Equation (8.37) and (8.42), we find

$$\mathbf{C} = \frac{\partial \theta^a}{\partial \Theta^A} \mathbf{G}^A \otimes \mathbf{g}_a \cdot \frac{\partial \theta^b}{\partial \Theta^B} \mathbf{g}_b \otimes \mathbf{G}^B = \frac{\partial \theta^a}{\partial \Theta^A} \frac{\partial \theta^b}{\partial \Theta^B} g_{ab} \mathbf{G}^A \otimes \mathbf{G}^B = C_{AB} \mathbf{G}^A \otimes \mathbf{G}^B \tag{8.61}$$

In addition, the Green deformation tensor can be expressed as

$$\mathbf{C} = \frac{\partial x^i}{\partial X^I}\mathbf{E}^I \otimes \mathbf{e}_i \cdot \frac{\partial x^j}{\partial X^J}\mathbf{e}_j \otimes \mathbf{E}^J = \frac{\partial x^i}{\partial X^I}\frac{\partial x^i}{\partial X^J}\mathbf{E}^I \otimes \mathbf{E}^J = \bar{C}_{IJ}\mathbf{E}^I \otimes \mathbf{E}^J \tag{8.62}$$

We observe that from Equations (8.61) and (8.62), the Green deformation tensor is a Lagrangian measure as a second-order covariant tensor with contravariant base vectors. Hence, the Lagrangian strain tensor as a second-order covariant tensor with contravariant base vectors like the Green deformation tensor, namely,

$$\mathbf{E} = \frac{1}{2}(\mathbf{C} - \mathbf{G}) = \bar{E}_{IJ}\mathbf{E}^I \otimes \mathbf{E}^J = E_{AB}\mathbf{G}^A \otimes \mathbf{G}^B \tag{8.63}$$

The inverse of the Green deformation tensor is called the Piola deformation tensor, denoted by **B**,

$$\mathbf{B} = \mathbf{C}^{-1} = \mathbf{F}^{-1} \cdot \mathbf{F}^{-T} = \frac{\partial \Theta^A}{\partial \theta^a}\frac{\partial \Theta^B}{\partial \theta^b} g^{ab}\mathbf{G}_A \otimes \mathbf{G}_B = B^{AB}\mathbf{G}_A \otimes \mathbf{G}_B \tag{8.64}$$

which is a second-order contravariant tensor with covariant base vectors. The Green deformation tensor can be rewritten using the polar decomposition theorem as

$$\mathbf{C} = \bar{\mathbf{U}}^T \cdot \mathbf{R}^T \cdot \mathbf{R} \cdot \bar{\mathbf{U}} = \bar{\mathbf{U}}^T \cdot \bar{\mathbf{U}} = \bar{\mathbf{U}}^2 \tag{8.65}$$

Note that the right stretch tensor is also a material object or Lagrangian measure as a second-order contravariant-covariant tensor with covariant-contravariant base vectors. Thus, it can be written as

$$\begin{aligned} \bar{\mathbf{U}} &= \bar{U}^A_{.B}\mathbf{G}_A \otimes \mathbf{G}^B = \bar{U}_B^{.A}\mathbf{G}^B \otimes \mathbf{G}_A = \bar{U}_A^{.B}\mathbf{G}^A \otimes \mathbf{G}_B \\ \bar{\mathbf{U}}^T &= \bar{U}^A_{.B}\mathbf{G}^B \otimes \mathbf{G}_A = \bar{U}_B^{.A}\mathbf{G}_A \otimes \mathbf{G}^B \end{aligned} \tag{8.66}$$

Since the right stretch tensor is symmetric, we find that $\bar{U}^I_{.J} = \bar{U}_J^{.I}$. In view of Equation (8.44), the rotation tensor can be written as

$$\mathbf{R} = \mathbf{F} \cdot \bar{\mathbf{U}}^{-1} = F^a_{.A}\left(\bar{U}^B_{.C}\right)^{-1}\mathbf{g}_a \otimes \mathbf{G}^A \cdot \mathbf{G}_B \otimes \mathbf{G}^C = R^a_{.A}\mathbf{g}_a \otimes \mathbf{G}^A \tag{8.67}$$

As a consequence, the rotation tensor is a two-point tensor or mixed tensor.

Eulerian Strain Tensor As an Eulerian measure, we have the Eulerian strain tensor, denoted by **e**,

$$(ds)^2 - (dS)^2 = d\mathbf{x} \cdot \left(\mathbf{g} - \mathbf{F}^{-T} \cdot \mathbf{F}^{-1}\right) \cdot d\mathbf{x} = d\mathbf{x} \cdot 2\mathbf{e} \cdot d\mathbf{x} \tag{8.68}$$

where **g** is an identity tensor which takes the form

$$\mathbf{g} = \mathbf{g}_a \otimes \mathbf{g}^a = \mathbf{g}^a \otimes \mathbf{g}_a = g_{ab}\mathbf{g}^a \otimes \mathbf{g}^b = g^{ab}\mathbf{g}_a \otimes \mathbf{g}_b = \mathbf{I}_{\mathbf{g}\otimes\mathbf{g}} \tag{8.69}$$

It is also called the metric tensor on the spatial configuration. Thus, we see that

$$\mathbf{e} = \frac{1}{2}\left(\mathbf{g} - \mathbf{F}^{-T} \cdot \mathbf{F}^{-1}\right) = \frac{1}{2}\left\{\left(\frac{\partial \mathbf{u}}{\partial \mathbf{x}}\right) + \left(\frac{\partial \mathbf{u}}{\partial \mathbf{x}}\right)^T\right\} - \frac{1}{2}\left(\frac{\partial \mathbf{u}}{\partial \mathbf{x}}\right) \cdot \left(\frac{\partial \mathbf{u}}{\partial \mathbf{x}}\right)^T \tag{8.70}$$

which is also called the Almansi strain tensor. It should be noted that the Eulerian strain tensor is an Eulerian measure.

Further we note that in a rigid body motion, the Eulerian strain tensor becomes a zero tensor such that

$$\mathbf{e} = \frac{1}{2}\left(\mathbf{g} - \mathbf{R}^{-T} \cdot \mathbf{R}^{-1}\right) = \frac{1}{2}\left(\mathbf{g} - \mathbf{R} \cdot \mathbf{R}^{T}\right) = \mathbf{0} \tag{8.71}$$

The Eulerian strain tensor can be written alternatively as

$$\mathbf{e} = \frac{1}{2}\left(\mathbf{g} - \mathbf{b}^{-1}\right), \tag{8.72}$$

where $\mathbf{b}$ is referred to as the Cauchy deformation tensor or the left Cauchy-Green tensor, and its inverse $\mathbf{b}^{-1}$ may be replaced by a new notation $\mathbf{c}$ which is often called the Finger deformation tensor since it was first introduced by Finger (1894). The Cauchy deformation tensor can be found also from the following

$$dS^2 = d\mathbf{x} \cdot \mathbf{F}^{-T} \cdot \mathbf{F}^{-1} \cdot d\mathbf{x} = d\mathbf{x} \cdot \mathbf{b}^{-1} \cdot d\mathbf{x} = d\mathbf{x} \cdot \mathbf{c} \cdot d\mathbf{x} \tag{8.73}$$

We see that the inverse of the Cauchy deformation tensor is positive definite, that is,

$$d\mathbf{x} \cdot \mathbf{b}^{-1} \cdot d\mathbf{x} = (dS)^2 > 0, \quad \forall\, d\mathbf{x} \neq \mathbf{0} \tag{8.74}$$

and the Cauchy deformation tensor is symmetric, that is,

$$\mathbf{b} = \mathbf{F} \cdot \mathbf{F}^{T} = \mathbf{b}^{T} \tag{8.75}$$

Using Equation (8.37) and (8.42), we find

$$\mathbf{b} = \frac{\partial \theta^a}{\partial \Theta^A}\mathbf{g}_a \otimes \mathbf{G}^A \cdot \frac{\partial \theta^b}{\partial \Theta^B}\mathbf{G}^B \otimes \mathbf{g}_b = \frac{\partial \theta^a}{\partial \Theta^A}\frac{\partial \theta^b}{\partial \Theta^B} G^{AB}\mathbf{g}_a \otimes \mathbf{g}_b = b^{ab}\mathbf{g}_a \otimes \mathbf{g}_b \tag{8.76}$$

In addition, the Cauchy deformation tensor can be expressed as

$$\mathbf{b} = \frac{\partial x^i}{\partial X^I}\mathbf{e}_i \otimes \mathbf{E}^I \cdot \frac{\partial x^j}{\partial X^J}\mathbf{E}^J \otimes \mathbf{e}_j = \frac{\partial x^i}{\partial X^I}\frac{\partial x^j}{\partial X^I}\mathbf{e}_i \otimes \mathbf{e}_j = \bar{b}^{ij}\mathbf{e}_i \otimes \mathbf{e}_j \tag{8.77}$$

We observe that from Equation (8.75) and (8.76), the Cauchy deformation tensor is an Eulerian measure as a second-order contravariant tensor with covariant base vectors. The Cauchy deformation tensor can be rewritten using the polar decomposition theorem as

$$\mathbf{b} = \bar{\mathbf{v}} \cdot \mathbf{R} \cdot \mathbf{R}^{T} \cdot \bar{\mathbf{v}}^{T} = \bar{\mathbf{v}} \cdot \bar{\mathbf{v}}^{T} = \bar{\mathbf{v}}^2 \tag{8.78}$$

Note that the left stretch tensor is also a spatial object or Eulerian measure as a second-order contravariant-covariant tensor with covariant-contravariant base vectors. Thus, it can be written as

$$\bar{\mathbf{v}} = \mathbf{F} \cdot \mathbf{R}^{-1} = F^{a}_{\ .A}\left(R^{b}_{\ .B}\right)^{-1}\mathbf{g}_a \otimes \mathbf{G}^A \cdot \mathbf{G}_B \otimes \mathbf{g}^b = \bar{v}^{a}_{.b}\mathbf{g}_a \otimes \mathbf{g}^b \tag{8.79}$$

Eulerian Rate of Deformation Tensor The Eulerian rate of deformation tensor is the symmetric part of the spatial velocity gradient tensor, that is,

$$\mathbf{d} = \frac{1}{2}\left(\mathbf{l} + \mathbf{l}^{T}\right) = \frac{1}{2}\left(\dot{\mathbf{F}} \cdot \mathbf{F}^{-1} + \mathbf{F}^{-T} \cdot \dot{\mathbf{F}}^{T}\right) = \mathbf{d}^{T} \tag{8.80}$$

which is often called the rate of deformation tensor or the velocity strain tensor. Accordingly, it is symmetric. The Eulerian rate of deformation tensor may also be found as follows:

$$\begin{aligned}\frac{D}{Dt}(ds)^2 &= 2d\mathbf{x}\cdot\frac{D(d\mathbf{x})}{Dt} = 2d\mathbf{x}\cdot d\mathbf{v} = 2d\mathbf{x}\cdot\frac{\partial\mathbf{v}}{\partial\mathbf{x}}\cdot d\mathbf{x} = 2d\mathbf{x}\cdot\mathbf{l}\cdot d\mathbf{x}\\ &= d\mathbf{x}\cdot\left(\mathbf{l}+\mathbf{l}^T+\mathbf{l}-\mathbf{l}^T\right)\cdot d\mathbf{x} = d\mathbf{x}\cdot\left(\mathbf{l}+\mathbf{l}^T\right)\cdot d\mathbf{x} = 2d\mathbf{x}\cdot\mathbf{d}\cdot d\mathbf{x} = d\mathbf{X}\cdot\frac{\partial\mathbf{C}}{\partial t}\cdot d\mathbf{X}\end{aligned} \tag{8.81}$$

Hence, we obtain

$$\frac{\partial\mathbf{C}}{\partial t} = 2\mathbf{F}^T\cdot\mathbf{d}\cdot\mathbf{F} \tag{8.82}$$

Hence, it is said that the Eulerian rate of deformation tensor is a measure of the rate of change of the square of the infinitesimal line segment in the spatial (deformed) configuration. The spatial velocity gradient tensor can be recast as

$$\mathbf{l} = \boldsymbol{\Omega} + \mathbf{R}\cdot\dot{\bar{\mathbf{U}}}\cdot\bar{\mathbf{U}}^{-1}\cdot\mathbf{R}^T \tag{8.83}$$

where $\boldsymbol{\Omega}$ represents the angular velocity of the material (Atluri and Cazzani 1995; Dienes 1979) which is skew-symmetric. Since the rotation tensor $\mathbf{R}\in SO(3)$ is orthogonal, we find

$$\dot{\mathbf{R}}\cdot\mathbf{R}^T + \mathbf{R}\cdot\dot{\mathbf{R}}^T = \boldsymbol{\Omega} + \boldsymbol{\Omega}^{\mathrm{T}} = \mathbf{0} \tag{8.84}$$

The angular velocity $\boldsymbol{\Omega}$ represents the rate of change of the rotation tensor as follows:

$$\dot{\mathbf{R}} = \boldsymbol{\Omega}\cdot\mathbf{R} \tag{8.85}$$

It is associated with the angular velocity vector $\boldsymbol{\omega}$

$$\boldsymbol{\Omega}\cdot\mathbf{v} = \boldsymbol{\omega}\times\mathbf{v}, \quad \text{for } \forall\mathbf{v}\in\mathbb{R}^3 \tag{8.86}$$

And its components can be given as

$$\Omega_{ik} = \varepsilon_{ijk}\omega_j \tag{8.87}$$

where ε_{ijk} represent the components of the permutation tensor. Therefore, Equation (8.83) can be rewritten as

$$\begin{aligned}\mathbf{l} &= \frac{1}{2}\mathbf{R}\cdot\left\{\left(\dot{\bar{\mathbf{U}}}\cdot\bar{\mathbf{U}}^{-1}\right)+\left(\dot{\bar{\mathbf{U}}}\cdot\bar{\mathbf{U}}^{-1}\right)^T\right\}\cdot\mathbf{R}^T+\boldsymbol{\Omega}\\ &+\frac{1}{2}\mathbf{R}\cdot\left\{\left(\dot{\bar{\mathbf{U}}}\cdot\bar{\mathbf{U}}^{-1}\right)-\left(\dot{\bar{\mathbf{U}}}\cdot\bar{\mathbf{U}}^{-1}\right)^T\right\}\cdot\mathbf{R}^T\end{aligned} \tag{8.88}$$

In view of Equation (8.80), we find that

$$\mathbf{d} = \frac{1}{2}\mathbf{R}\cdot\left\{\left(\dot{\bar{\mathbf{U}}}\cdot\bar{\mathbf{U}}^{-1}\right)+\left(\dot{\bar{\mathbf{U}}}\cdot\bar{\mathbf{U}}^{-1}\right)^T\right\}\cdot\mathbf{R}^T \tag{8.89}$$

It is well known that the Eulerian rate of deformation tensor plays a crucial role in the hypoelasticity theory which will be discussed in detail later. From the above, we have

$$\mathbf{R}^T\cdot\mathbf{d}\cdot\mathbf{R} = \frac{1}{2}\left\{\left(\dot{\bar{\mathbf{U}}}\cdot\bar{\mathbf{U}}^{-1}\right)+\left(\dot{\bar{\mathbf{U}}}\cdot\bar{\mathbf{U}}^{-1}\right)^T\right\} = \hat{\mathbf{d}} \tag{8.90}$$

where $\hat{\mathbf{d}}$ is called the rotated rate of deformation tensor (Dienes 1979; Simo and Hughes 1998).

Rate of Rotation Tensor The skew-symmetric part of the spatial velocity gradient tensor is called the rate of rotation tensor or the spin tensor. It is often called the vorticity tensor. The rate of deformation tensor and the spin tensor are physically interpreted as the rate of deformation and rotation (Malvern 1969). The spin tensor, thus, is given by

$$\mathbf{w} = \frac{1}{2}\left(\mathbf{l} - \mathbf{l}^T\right) = -\mathbf{w}^T \tag{8.91}$$

In view of Equation (8.88), the spin tensor becomes

$$\mathbf{w} = \boldsymbol{\Omega} + \frac{1}{2}\mathbf{R}\cdot\left\{\left(\dot{\bar{\mathbf{U}}}\cdot\bar{\mathbf{U}}^{-1}\right) - \left(\dot{\bar{\mathbf{U}}}\cdot\bar{\mathbf{U}}^{-1}\right)^T\right\}\cdot\mathbf{R}^T \tag{8.92}$$

Since the rate of rotation tensor is skew-symmetric, it is associated with the rate of rotation vector $\boldsymbol{\varpi}$

$$\mathbf{w}\cdot\mathbf{v} = \boldsymbol{\varpi}\times\mathbf{v}, \qquad \text{for } \forall\mathbf{v}\in\mathbb{R}^3 \tag{8.93}$$

And its components can be given as

$$w_{ik} = \varepsilon_{ijk}\varpi_j \tag{8.94}$$

Consider a rigid body motion, in which the rate of the right stretch tensor becomes zero. In addition, the Eulerian rate of deformation tensor should be zero in view of Equation (8.86). As a consequence, the spatial velocity gradient tensor is identical to the spin tensor as follows:

$$\mathbf{l} = \mathbf{w} = \boldsymbol{\Omega} \tag{8.95}$$

Further, we find that

$$\boldsymbol{\omega} = \boldsymbol{\varpi} = \frac{1}{2}\mathrm{curl}\mathbf{v} \tag{8.96}$$

In the literature, $\boldsymbol{\omega}$ may be considered as the rate of rotation of the material and $\boldsymbol{\varpi}$ as the rate of rotation of the principal axes of the rate of deformation tensor (Atluri and Cazzani 1995; Dienes 1979).

Right/Left Stretch Tensors As mentioned previously, the Green deformation tensor is positive definite and symmetric. Therefore, the Green deformation tensor has positive real eigenvalues (Greenberg 1998) as follows:

$$\mathbf{C}\cdot\mathbf{N}_i = \Lambda_i\mathbf{N}_i, \quad i = 1, 2, 3 \tag{8.97}$$

where Λ_i and $\mathbf{N}_i$ denote the ith-eigenvalue and the ith-eigenvector of the Green deformation tensor $\mathbf{C}$. Since the Green deformation tensor is positive definite and symmetric, Λ_i is a positive real number. Further, $\mathbf{N}_i$, $i = 1, 2, 3$ are orthogonal unit vectors, and the triad $\{\mathbf{N}_1, \mathbf{N}_2, \mathbf{N}_3\}$ is called the Lagrangian axes at $\mathbf{X}\in\mathcal{B}$.

The eigenvalues for corresponding eigenvectors can be uniquely determined by the characteristic equation

$$\det\left(\mathbf{C} - \Lambda_i\mathbf{I}\right) = -\Lambda_i^3 + \mathrm{I}_\mathbf{C}\Lambda_i^2 - \mathrm{II}_\mathbf{C}\Lambda_i + \mathrm{III}_\mathbf{C} = 0, \; i = 1, 2, 3 \tag{8.98}$$

where the principal invariants $\mathrm{I}_\mathbf{C}, \mathrm{II}_\mathbf{C}, \mathrm{III}_\mathbf{C}$ are given by

$$\begin{aligned}
\mathrm{I}_\mathbf{C} &= \mathrm{tr}\,\mathbf{C} = \Lambda_1 + \Lambda_2 + \Lambda_3, \\
\mathrm{II}_\mathbf{C} &= \frac{1}{2}\left[(\mathrm{tr}\,\mathbf{C})^2 - \mathrm{tr}\,(\mathbf{C}^2)\right] = \lambda_1\Lambda_2 + \Lambda_2\Lambda_3 + \Lambda_3\Lambda_1, \\
\mathrm{III}_\mathbf{C} &= \det\mathbf{C} = \Lambda_1\Lambda_2\Lambda_3
\end{aligned} \tag{8.99}$$

where the eigenvalues of the Green deformation tensor **C** can be calculated by (Basar and Weichert 2000)

$$\Lambda_i = \frac{1}{3}\left[\mathrm{I_C} + 2\sqrt{\mathrm{I_C^2} - 3\mathrm{II_C}}\cos\frac{1}{3}(\Theta + 2i\pi)\right], \quad i = 1, 2, 3 \tag{8.100}$$

where Θ is given by

$$\Theta = \arccos\left[\frac{2\mathrm{I_C^3} - 9\mathrm{I_C}\mathrm{II_C} + 27\mathrm{III_C}}{2\sqrt[3]{\mathrm{I_C^2} - 3\mathrm{II_C}}}\right] \tag{8.101}$$

In view of Equation (2.195), Equation (8.97) leads to

$$\begin{aligned} \mathbf{C}\cdot\mathbf{N}_i &= \mathbf{U}^2\cdot\mathbf{N}_i \\ &= \lambda_i\mathbf{U}\cdot\mathbf{N}_i \\ &= \lambda_i^2\mathbf{N}_i \\ &= \Lambda_i\mathbf{N}_i \end{aligned}, \quad i = 1, 2, 3 \tag{8.102}$$

when the right stretch tensor $\bar{\mathbf{U}}$ is given by

$$\bar{\mathbf{U}}\cdot\mathbf{N}_i = \lambda_i\mathbf{N}_i, \quad i = 1, 2, 3 \tag{8.103}$$

where the eigenvalues λ_i, $i = 1, 2, 3$, which are called the principal stretches, should be real and the eigenvectors are orthogonal, since the right stretch tensor $\bar{\mathbf{U}}$ is symmetric (Greenberg 1998). Note that Equation (8.103) implies that **C** and $\bar{\mathbf{U}}$ are coaxial. Consequently, we have the relation

$$\Lambda_i = \lambda_i^2, \quad i = 1, 2, 3 \tag{8.104}$$

Analogous to Equation (8.98), the eigenvalues of the right stretch tensor $\bar{\mathbf{U}}$ can be uniquely determined by the characteristic equation

$$\det(\bar{\mathbf{U}} - \lambda_i\mathbf{I}) = -\lambda_i^3 + \mathrm{I}_{\bar{\mathrm{U}}}\lambda_i^2 - \mathrm{II}_{\bar{\mathrm{U}}}\lambda_i + \mathrm{III}_{\bar{\mathrm{U}}} = 0, \; i = 1, 2, 3 \tag{8.105}$$

where the principal invariants $\mathrm{I}_{\bar{\mathrm{U}}}$, $\mathrm{II}_{\bar{\mathrm{U}}}$, $\mathrm{III}_{\bar{\mathrm{U}}}$ are given by

$$\begin{aligned} \mathrm{I}_{\bar{\mathrm{U}}} &= \mathrm{tr}\,\bar{\mathbf{U}} = \lambda_1 + \lambda_2 + \lambda_3, \\ \mathrm{II}_{\bar{\mathrm{U}}} &= \frac{1}{2}\left[(\mathrm{tr}\,\bar{\mathbf{U}})^2 - \mathrm{tr}\,(\bar{\mathbf{U}}^2)\right] = \lambda_1\lambda_2 + \lambda_2\lambda_3 + \lambda_3\lambda_1, \\ \mathrm{III}_{\bar{\mathrm{U}}} &= \det\bar{\mathbf{U}} = \lambda_1\lambda_2\lambda_3 \end{aligned} \tag{8.106}$$

The spectral decomposition theorem enables one to express the Green deformation tensor in terms of the eigenvalues as

$$\mathbf{C} = \sum_{i=1}^{3}\Lambda_i\mathbf{N}_i\otimes\mathbf{N}_i \tag{8.107}$$

which is called the spectral decomposition of the Green deformation tensor. Hence, the spectral decomposition of the right stretch tensor $\bar{\mathbf{U}}$ can be written as

$$\bar{\mathbf{U}} = \sum_{i=1}^{3}\lambda_i\mathbf{N}_i\otimes\mathbf{N}_i \tag{8.108}$$

where $\lambda_1, \lambda_2, \lambda_3$ are often called the principal stretches of the right stretch tensor $\bar{\mathbf{U}}$, and $\lambda_i^2 = \Lambda_i$, $i = 1, 2, 3$. It is known that the right stretch tensor $\bar{\mathbf{U}}$ is positive-definite and symmetric (Marsden and Hughes 1983). Hence, $\lambda_i = \sqrt{\Lambda_i},\ i = 1, 2, 3$.

By use of the polar decomposition theorem and Equation (8.103), we have

$$\begin{aligned} \mathbf{F} \cdot \mathbf{N}_i &= \mathbf{R} \cdot \bar{\mathbf{U}} \cdot \mathbf{N}_i \\ &= \lambda_i \mathbf{R} \cdot \mathbf{N}_i = \bar{\mathbf{v}} \cdot \mathbf{R} \cdot \mathbf{N}_i \\ &= \bar{\mathbf{v}} \cdot \mathbf{n}_i \\ &= \lambda_i \mathbf{n}_i \end{aligned} \tag{8.109}$$

where $\mathbf{n}_i$ represents the ith-eigenvector of the left stretch tensor $\bar{\mathbf{v}}$ and λ_i is the corresponding eigenvalue. $\mathbf{n}_i$, $i = 1, 2, 3$ are also orthogonal unit vectors, and the triad $\{\mathbf{n}_1, \mathbf{n}_2, \mathbf{n}_3\}$ is called the Eulerian axes at $\mathbf{x} \in \mathcal{S}$. Note that $\bar{\mathbf{U}}$ and $\bar{\mathbf{v}}$ share the same eigenvalues. Thus, the spectral decomposition of the left stretch tensor $\bar{\mathbf{v}}$ can be written as

$$\bar{\mathbf{v}} = \sum_{i=1}^{3} \lambda_i \mathbf{n}_i \otimes \mathbf{n}_i \tag{8.110}$$

And, in view of Equation (8.78), the spectral decomposition of the Cauchy deformation tensor $\mathbf{b}$ can be written as

$$\mathbf{b} = \sum_{i=1}^{3} \Lambda_i \mathbf{n}_i \otimes \mathbf{n}_i \tag{8.111}$$

From Equation (8.101), we obtain

$$\mathbf{R} \cdot \mathbf{N}_i = \mathbf{n}_i \tag{8.112}$$

We also observe that the spectral decomposition of the deformation gradient tensor can be expressed as

$$\begin{aligned} \mathbf{F} &= \mathbf{R} \cdot \bar{\mathbf{U}} \\ &= \sum_{i=1}^{3} \lambda_i \mathbf{R} \cdot \mathbf{N}_i \otimes \mathbf{N}_i \\ &= \sum_{i=1}^{3} \lambda_i \mathbf{n}_i \otimes \mathbf{N}_i \end{aligned} \tag{8.113}$$

which implies that the deformation gradient tensor can be represented in terms of the principle stretches and the eigenvectors of $\bar{\mathbf{U}}$ and $\bar{\mathbf{v}}$.

Further, $\bar{\mathbf{U}}$ and $\bar{\mathbf{U}}^{-1}$ are coaxial, as $\bar{\mathbf{U}}^{-1}$ is symmetric. In other words, we can express the inverse of $\bar{\mathbf{U}}$ as

$$\bar{\mathbf{U}}^{-1} = \sum_{i=1}^{3} \frac{1}{\lambda_i} \mathbf{N}_i \otimes \mathbf{N}_i \tag{8.114}$$

Therefore, the spectral decomposition of the rotation tensor can be expressed as

$$\mathbf{R} = \sum_{i=1}^{3} \mathbf{n}_i \otimes \mathbf{N}_i \tag{8.115}$$

and the rotation tensor is a two-point tensor.

Push-Forward and Pull-Back Operation According to the manifold theory (Abraham et al. 1988; Marsden and Hughes 1983), the push-forward of the vectorial line segment on the undeformed manifold $(\mathcal{B}, \mathbf{G})$ as a contravariant vector $()^{\sharp}$, which is referred to as a vector with contravariant components on the covariant basis, results in the vectorial line segment on the deformed manifold $(\mathcal{S}, \mathbf{g})$ as follows:

$$\varphi_* \left(d\mathbf{X}^{\sharp}\right) = \mathbf{F}(\mathbf{X}, t) \cdot d\mathbf{X}^{\sharp} = d\mathbf{x}^{\sharp} \tag{8.116}$$

In view of Equations (8.2) and (8.37), the above linear transformation can be recast as

$$\mathbf{F} \cdot d\mathbf{X}^{\sharp} = \frac{\partial \theta^a}{\partial \Theta^A} \mathbf{g}_a \otimes \mathbf{G}^A \cdot d\Theta^B \mathbf{G}_B = \frac{\partial \theta^a}{\partial \Theta^A} d\Theta^A \mathbf{g}_a = d\theta^a \mathbf{g}_a = d\mathbf{x}^{\sharp} \tag{8.117}$$

Likewise, the pull-back of the vectorial line segment on the deformed manifold $(\mathcal{S}, \mathbf{g})$ as a contravariant vector results in the vectorial line segment on the undeformed manifold $(\mathcal{B}, \mathbf{G})$ as follows:

$$d\mathbf{X}^{\sharp} = \varphi^* \left(d\mathbf{x}^{\sharp}\right) = \mathbf{F}^{-1}(\mathbf{X}, t) \cdot d\mathbf{x}^{\sharp} \tag{8.118}$$

Likewise, by means of Equations (8.2) and (8.41), the above linear transformation means the following

$$\mathbf{F}^{-1} \cdot d\mathbf{x} = \frac{\partial \Theta^A}{\partial \theta^a} \mathbf{G}_A \otimes \mathbf{g}^a \cdot d\theta^b \mathbf{g}_b = \frac{\partial \Theta^A}{\partial \theta^a} d\theta^a \mathbf{G}_A = d\Theta^A \mathbf{G}_A = d\mathbf{X}^{\sharp} \tag{8.119}$$

Thus, the push-forward operation obeys the following rule depending on the characteristics of the components of the material vector $\mathbf{V}$

$$\varphi_* \left(\mathbf{V}^{\sharp}\right) = \mathbf{F} \cdot \mathbf{V}^{\sharp} \tag{8.120}$$

On the other hand, the push-forward of the vectorial line segment as a covariant vector $()^{\flat}$, can be written as

$$\varphi_* \left(\mathbf{V}^{\flat}\right) = \mathbf{F}^{-T} \cdot \mathbf{V}^{\flat} \tag{8.121}$$

For a second-order tensor $\mathbf{S}$, the push-forward operation therefore obeys the rules

$$\varphi_* \left(\mathbf{S}^{\sharp}\right) = \mathbf{F} \cdot \mathbf{S}^{\sharp} \cdot \mathbf{F}^T, \quad \varphi_* \left(\mathbf{S}^{\flat}\right) = \mathbf{F}^{-T} \cdot \mathbf{S}^{\flat} \cdot \mathbf{F}^{-1} \tag{8.122}$$

Note that the push-forward operation of a Lagrangian measure results in an Eulerian measure. The pull-back operation obeys the following rule depending on the characteristics of the components of the spatial vector $\mathbf{v}$

$$\varphi^* \left(\mathbf{v}^{\flat}\right) = \mathbf{F}^T \cdot \mathbf{v}^{\flat}, \quad \varphi^* \left(\mathbf{v}^{\sharp}\right) = \mathbf{F}^{-1} \cdot \mathbf{v}^{\sharp} \tag{8.123}$$

With regard to a second-order tensor $\boldsymbol{\tau}$, the pull-back operation obeys the rules

$$\varphi^* \left(\boldsymbol{\tau}^{\flat}\right) = \mathbf{F}^T \cdot \boldsymbol{\tau}^{\flat} \cdot \mathbf{F}, \quad \varphi^* \left(\boldsymbol{\tau}^{\sharp}\right) = \mathbf{F}^{-1} \cdot \boldsymbol{\tau}^{\sharp} \cdot \mathbf{F}^{-T} \tag{8.124}$$

Note that the push-forward operation of an Eulerian measure results in a Lagrangian measure.

Next, we examine the push-forward/pull-back operations on second-order tensors. The pull-back of the Eulerian strain tensor, which is a covariant second-order tensor, yields the Lagrangian strain tensor, which is also a covariant second-order tensor, due to the first term in Equation (8.124)

$$\varphi^* \left(\mathbf{e}^{\flat}\right) = \mathbf{F}^T \cdot \mathbf{e}^{\flat} \cdot \mathbf{F} = \mathbf{F}^T \cdot \frac{1}{2}\left(\mathbf{g} - \mathbf{F}^{-T} \cdot \mathbf{F}^{-1}\right) \cdot \mathbf{F} = \frac{1}{2}\left(\mathbf{F}^T \cdot \mathbf{F} - \mathbf{G}\right) = \mathbf{E}^{\flat} \tag{8.125}$$

Conversely, the push-forward of the Lagrangian strain tensor yields the Eulerian strain tensor due to the second term in Equation (8.122)

$$\begin{aligned}\varphi_* \left(\mathbf{E}^\flat\right) &= \mathbf{F}^{-T} \cdot \mathbf{E}^\flat \cdot \mathbf{F}^{-1} = \mathbf{F}^{-T} \cdot \frac{1}{2} \left(\mathbf{F}^T \cdot \mathbf{F} - \mathbf{G}\right) \cdot \mathbf{F}^{-1} \\ &= \frac{1}{2} \left(\mathbf{g} - \mathbf{F}^{-T} \cdot \mathbf{F}^{-1}\right) = \mathbf{e}^\flat\end{aligned} \tag{8.126}$$

Since the Green deformation tensor shown in Equation (8.62) is a covariant first-order tensor, the push-forward operation gives the metric tensor on the spatial configuration which is a first-order tensor as follows:

$$\varphi_* \left(\mathbf{C}^\flat\right) = \mathbf{F}^{-T} \cdot \mathbf{C}^\flat \cdot \mathbf{F}^{-1} = \mathbf{F}^{-T} \cdot \mathbf{F}^T \cdot \mathbf{F} \cdot \mathbf{F}^{-1} = \mathbf{g}^\flat \tag{8.127}$$

Conversely, we obtain the following

$$\varphi^* \left(\mathbf{g}^\flat\right) = \mathbf{F}^T \cdot \mathbf{g}^\flat \cdot \mathbf{F} = \mathbf{F}^T \cdot \mathbf{F} = \mathbf{C}^\flat \tag{8.128}$$

where the Green deformation tensor is often called the convected Riemannian metric tensor (Simo et al. 1988). Likewise, we can obtain the following

$$\begin{aligned}&\varphi_* \left(\mathbf{G}^\flat\right) = \mathbf{c}^\flat, \quad \varphi_* \left(\mathbf{G}^\natural\right) = \mathbf{b}^\natural, \quad \varphi_* \left(\mathbf{B}^\natural\right) = \mathbf{g}^\natural \\ &\varphi^* \left(\mathbf{c}^\flat\right) = \mathbf{G}^\flat, \quad \varphi^* \left(\mathbf{b}^\natural\right) = \mathbf{G}^\natural, \quad \varphi^* \left(\mathbf{g}^\natural\right) = \mathbf{B}^\natural\end{aligned} \tag{8.129}$$

In Table 8.1 and Table 8.2 (strain and stress), representative covariant, contravariant and mixed tensors in continuum kinematics are summarized.

Lie Derivative The Lagrangian strain tensor is a Lagrangian measure and a covariant tensor as shown in Equation (8.67). Recall that the pull-back of the Eulerian strain tensor gives the Lagrangian strain

Table 8.1. Covariant, contravariant and mixed tensors in kinematics

Measure	Tensor	Notation
Lagrangian	Lagrangian strain tensor	$\mathbf{E}^\flat = E_{AB}\mathbf{G}^A \otimes \mathbf{G}^B$
	Green deformation tensor	$\mathbf{C}^\flat = C_{AB}\mathbf{G}^A \otimes \mathbf{G}^B$
	Piola deformation tensor	$\mathbf{B}^\natural = B^{AB}\mathbf{G}_A \otimes \mathbf{G}_B$
	Lagrangian rate of deformation tensor	$\mathbf{D}^\flat = D_{AB}\mathbf{G}^A \otimes \mathbf{G}^B$
	Referential velocity tensor	$\mathbf{L}^\flat = L_{AB}\mathbf{G}^A \otimes \mathbf{G}^B$
	Right stretch tensor	$\bar{\mathbf{U}} = \bar{U}^A_{.B}\mathbf{G}_A \otimes \mathbf{G}^B$
Eulerian	Eulerian strain tensor	$\mathbf{e}^\flat = e_{ab}\mathbf{g}^a \otimes \mathbf{g}^b$
	Finger deformation tensor	$\mathbf{c}^\flat = c_{ab}\mathbf{g}^a \otimes \mathbf{g}^b$
	Eulerian rate deformation tensor	$\mathbf{d}^\flat = d_{ab}\mathbf{g}^a \otimes \mathbf{g}^b$
	Spatial velocity gradient tensor	$\mathbf{l}^\flat = l_{ab}\mathbf{g}^a \otimes \mathbf{g}^b$
	Spin tensor	$\mathbf{w}^\flat = w_{ab}\mathbf{g}^a \otimes \mathbf{g}^b$
	Left stretch tensor	$\bar{\mathbf{v}} = \bar{v}^a_{.b}\mathbf{g}_a \otimes \mathbf{g}^b$
Two-point	Deformation gradient tensor	$\mathbf{F} = F^a_{.A}\mathbf{g}_a \otimes \mathbf{G}^A$
	Rotation tensor	$\mathbf{R} = R^a_{.A}\mathbf{g}_a \otimes \mathbf{G}^A$

tensor as shown in Equation (8.125). Therefore, the material time derivative of the tensor is still a Lagrangian measure

$$\frac{\partial \mathbf{E}^{\flat}}{\partial t} = \frac{\partial}{\partial t}\left\{\varphi^{*}\left(\mathbf{e}^{\flat}\right)\right\} = \dot{E}_{AB}\mathbf{G}^{A} \otimes \mathbf{G}^{B} \tag{8.130}$$

In view of Equation (8.47) and (8.55), the above equation can be recast as

$$\dot{\mathbf{E}} = \frac{1}{2}\left(\dot{\mathbf{F}}^{T} \cdot \mathbf{F} + \mathbf{F}^{T} \cdot \dot{\mathbf{F}}\right) = \frac{1}{2}\left(D_{\mathbf{v}}\mathbf{F}^{T} \cdot \mathbf{F} + \mathbf{F}^{T} \cdot D_{\mathbf{v}}\mathbf{F}\right) = D_{\mathbf{v}}\mathbf{E} \tag{8.131}$$

It is interpreted that the material time derivative of the Lagrangian strain tensor is the directional derivative of the Lagrangian strain tensor in the direction $\mathbf{v}$ at $\mathbf{X}$. Further, since the rate of the Lagrangian strain tensor is a covariant tensor, the push-forward operation yields an Eulerian measure, that is,

$$\varphi_{*}\left(\dot{\mathbf{E}}^{\flat}\right) = \mathbf{F}^{-T} \cdot \dot{\mathbf{E}}^{\flat} \cdot \mathbf{F}^{-1} = \varphi_{*}\left[\frac{\partial}{\partial t}\left\{\varphi^{*}\left(\mathbf{e}^{\flat}\right)\right\}\right] = \varphi_{*}\left(D_{v}\mathbf{E}^{\flat}\right) \tag{8.132}$$

Due to Equation (8.84), the above equation results in the Eulerian rate of deformation tensor as follows:

$$\begin{aligned} L_{v}\left(\mathbf{e}^{\flat}\right) &= \varphi_{*}\left[\frac{\partial}{\partial t}\left\{\varphi^{*}\left(\mathbf{e}^{\flat}\right)\right\}\right] = \varphi_{*}\left(D_{v}\mathbf{E}^{\flat}\right) = \frac{1}{2}\left(\dot{\mathbf{F}} \cdot \mathbf{F}^{-1} + \mathbf{F}^{-T} \cdot \dot{\mathbf{F}}^{T}\right) \\ &= \mathbf{d}^{\flat} \end{aligned} \tag{8.133}$$

which is called the Lie derivative, named after Lie (1842-1899). It may be interpreted as the push-forward of the directional derivative of the Lagrangian strain tensor in the direction $\mathbf{v}$ at $\mathbf{X}$. This operation is very important in that objective stress rates, which will be discussed later, result from the Lie derivative operation of Eulerian stress measures. Note that the pull-back of the Eulerian rate of deformation tensor yields the Lagrangian rate of deformation tensor, namely,

$$\varphi^{*}\left(\mathbf{d}^{\flat}\right) = \mathbf{F}^{T} \cdot \mathbf{d}^{\flat} \cdot \mathbf{F} = \frac{1}{2}\left(\mathbf{F}^{T} \cdot \dot{\mathbf{F}} + \dot{\mathbf{F}}^{T} \cdot \mathbf{F}\right) = \mathbf{D}^{\flat} \tag{8.134}$$

which is given by

$$\mathbf{D}^{\flat} = \frac{1}{2}\frac{\partial \mathbf{C}^{\flat}}{\partial t} = \frac{1}{2}\left(\mathbf{L}^{\flat} + \mathbf{L}^{\flat T}\right) = \frac{1}{2}\left(\frac{\partial \mathbf{V}}{\partial \mathbf{X}} + \left(\frac{\partial \mathbf{V}}{\partial \mathbf{X}}\right)^{T}\right) \tag{8.135}$$

which is also the symmetric part of the material velocity gradient tensor $\mathbf{L}^{\flat}$.

Table 8.2. Covariant, contravariant, and mixed tensors (stress) in kinematics

Measure	Tensor	Notation
Lagrangian	Second Piola-Kirchhoff stress tensor	$\mathbf{S}^{\natural} = S^{AB}\mathbf{G}_{A} \otimes \mathbf{G}_{B}$
Eulerian	Kirchhoff stress tensor	$\boldsymbol{\tau}^{\natural} = \tau^{ab}\mathbf{g}_{a} \otimes \mathbf{g}_{a}$
	Cauchy stress tensor	$\boldsymbol{\sigma}^{\natural} = \sigma^{ab}\mathbf{g}_{a} \otimes \mathbf{g}_{b}$
Two-point	First Piola-Kirchhoff stress tensor	$\mathbf{P} = P^{aA}\mathbf{g}_{a} \otimes \mathbf{G}_{A}$

8.1.5 Stresses

In three-dimensional elastodynamics, contravariant physical state variables conjugate to covariant strain tensors are assumed to exist. These variables are called stresses in kinematics (see Table 8.2). Strains are associated with displacements, whereas stresses are defined with tractions.

Cauchy Stress Tensor The infinitesimal real force $d\mathbf{f}$ acting on the infinitesimal deformed surface da in the spatial configuration is given using the Cauchy traction vector $\mathbf{t}$ (force per unit deformed area on the surface)

$$d\mathbf{f} = \mathbf{t}da := df^a \mathbf{g}_a \tag{8.136}$$

which is a contravariant vector. Recall Equation (8.5) in which the vectorial line segment in kinematics is defined as a contravariant vector. Likewise, the infinitesimal actual force is defined as a contravariant vector as shown in Equation (8.136). Cauchy theorem states that the Cauchy traction vector $\mathbf{t}$ (force per unit deformed area acted on the surface) on the spatial configuration, which is often called the Cauchy stress vector, can be defined as

$$\mathbf{t}^{\natural}\left(\mathbf{x}, t, \hat{\mathbf{n}}^{\flat}\right) = \boldsymbol{\sigma}^{\natural}(\mathbf{x}, t) \cdot \hat{\mathbf{n}}^{\flat}(\mathbf{x}, t) \tag{8.137}$$

where $\boldsymbol{\sigma}$ denotes the Cauchy stress tensor which is a second-order contravariant stress tensor, because the unit vector outward $\hat{\mathbf{n}}^{\flat}$ normal to the surface is parallel to the gradient of the surface. The gradient of a scalar quantity is a covariant vector with contravariant base vectors. Note that the Cauchy traction vector is also a contravariant vector. Hence, the Cauchy stress tensor can be given by

$$\boldsymbol{\sigma}^{\natural}(\mathbf{x}, t) = \sigma^{ab} \mathbf{g}_a \otimes \mathbf{g}_b \tag{8.138}$$

as an Eulerian stress measure. In three-dimensional elastodynamics, stresses are conjugate to strains in association with energy. Thus, it is natural that stresses are contravariant tensors, since strains are covariant tensors.

As the Cauchy stress tensor is symmetric, its eigenvalues are real and its eigenvectors are orthogonal (Greenberg 1998) as follows

$$\boldsymbol{\sigma} \cdot \mathbf{n}_i = \sigma_i \mathbf{n}_i, \quad i = 1, 2, 3 \tag{8.139}$$

where the eigenvalues σ_i, $i = 1, 2, 3$ are called the principal stresses and the eigenvectors $\mathbf{n}_i$, $i = 1, 2, 3$ are called the principal directions. The eigenvalues for the corresponding eigenvectors can be uniquely determined by the characteristic equation

$$\det(\boldsymbol{\sigma} - \sigma_i \mathbf{I}) = -\sigma_i^3 + \mathrm{I}_\sigma \sigma_i^2 - \mathrm{II}_\sigma \sigma_i + \mathrm{III}_\sigma = 0, \; i = 1, 2, 3 \tag{8.140}$$

where the principal invariants $\mathrm{I}_\sigma, \mathrm{II}_\sigma, \mathrm{III}_\sigma$ are given by

$$\begin{aligned} \mathrm{I}_\sigma &= \mathrm{tr}\,\boldsymbol{\sigma} = \Lambda_1 + \Lambda_2 + \Lambda_3, \\ \mathrm{II}_\sigma &= \frac{1}{2}\left[(\mathrm{tr}\,\boldsymbol{\sigma})^2 - \mathrm{tr}\left(\boldsymbol{\sigma}^2\right)\right] = \lambda_1 \Lambda_2 + \Lambda_2 \Lambda_3 + \Lambda_3 \Lambda_1, \\ \mathrm{III}_\sigma &= \det \boldsymbol{\sigma} = \Lambda_1 \Lambda_2 \Lambda_3 \end{aligned} \tag{8.141}$$

Therefore, the spectral decomposition of the Cauchy stress tensor $\boldsymbol{\sigma}$ can be written as

$$\boldsymbol{\sigma} = \sum_{i=1}^{3} \sigma_i \mathbf{n}_i \otimes \mathbf{n}_i \tag{8.142}$$

where the eigenvectors $\mathbf{n}_i$, $i = 1, 2, 3$ determine the principal axes of the Cauchy stress tensor $\boldsymbol{\sigma}$.

In view of Equation (8.137), we can write the normal force $\sigma\hat{\mathbf{n}}$

$$\sigma\hat{\mathbf{n}} = (\hat{\mathbf{n}} \cdot \mathbf{t})\hat{\mathbf{n}} = (\hat{\mathbf{n}} \cdot \boldsymbol{\sigma} \cdot \hat{\mathbf{n}})\hat{\mathbf{n}} \tag{8.143}$$

where σ is called the normal stress on the surface to which the unit vector $\hat{\mathbf{n}}$ outward is normal. Now we can define the shearing force $\tau\hat{\mathbf{m}}$

$$\tau\hat{\mathbf{m}} = \mathbf{t} - \sigma\hat{\mathbf{n}} = \boldsymbol{\sigma} \cdot \hat{\mathbf{n}} - \sigma\hat{\mathbf{n}} \tag{8.144}$$

where τ denotes the shearing stress and $\hat{\mathbf{m}}$ is unit vector perpendicular to $\hat{\mathbf{n}}$. Note that when the unit vector $\hat{\mathbf{n}}$ outward is equal to the eigenvector, then the shearing force on the surface vanishes. The shearing stress τ can be determined by

$$\tau = \pm\sqrt{\mathbf{t} \cdot \mathbf{t} - \sigma^2} \tag{8.145}$$

and the unit vector $\hat{\mathbf{m}}$ is calculated by

$$\hat{\mathbf{m}} = \frac{1}{\tau}(\boldsymbol{\sigma} \cdot \hat{\mathbf{n}} - \sigma\hat{\mathbf{n}}) \tag{8.146}$$

The Cauchy stress tensor $\boldsymbol{\sigma}$ can be divided into a spherical and a deviatoric tensor (see Equation (2.171)),

$$\begin{aligned}
\boldsymbol{\sigma} &= \boldsymbol{\sigma}_{\text{dev}} + \boldsymbol{\sigma}_{\text{sph}}, \\
\boldsymbol{\sigma}_{\text{sph}} &= \frac{1}{3}(\operatorname{tr}\boldsymbol{\sigma})\,\mathbf{I} = \frac{1}{3}\mathrm{I}_{\sigma}\mathbf{I} = -p\mathbf{I}, \\
\boldsymbol{\sigma}_{\text{dev}} &= \boldsymbol{\sigma} - \boldsymbol{\sigma}_{\text{sph}} = \boldsymbol{\sigma} - \frac{1}{3}\mathrm{I}_{\sigma}\mathbf{I} = \boldsymbol{\sigma} + p\mathbf{I}
\end{aligned} \tag{8.147}$$

where p is called the hydrostatic stress or hydrostatic pressure.

Kirchhoff Stress Tensor The Kirchhoff stress tensor is an Eulerian stress measure defined by

$$\boldsymbol{\tau}^{\natural} = J\boldsymbol{\sigma} = \tau^{ab}\mathbf{g}_a \otimes \mathbf{g}_b, \quad \tau^{ab} = J\sigma^{ab} \tag{8.148}$$

where J is the Jacobian of the deformation, i.e. the determinant of the deformation gradient tensor. For incompressible materials, the Kirchhoff stress tensor is equivalent to the Cauchy stress tensor, since the Jacobian is identical to one. Since the Cauchy stress tensor is symmetric, the Kirchhoff stress tensor is symmetric.

Nanson's Formula Nanson's formula gives the relation between the infinitesimal surface da in the spatial configuration and the infinitesimal surface dA in the reference configuration. Due to Equation (8.14), the relation between the infinitesimal volume element dv in the spatial configuration and the infinitesimal volume element dV in the reference configuration is given by

$$dv = \left|\frac{\partial \mathbf{x}}{\partial \mathbf{X}}\right| dV = J(\mathbf{X}, t)\,dV \tag{8.149}$$

The volume element in the spatial (or current) configuration can be written as

$$dv = \hat{\mathbf{n}}\,da \cdot d\mathbf{x} \tag{8.150}$$

where $\hat{\mathbf{n}}$ is the unit vector outward normal to the surface in the current configuration. The volume element in the reference configuration can also written as

$$dV = \hat{\mathbf{N}}\, dA \cdot d\mathbf{X} \tag{8.151}$$

where $\hat{\mathbf{N}}$ is the unit vector outward normal to the surface in the reference configuration. By means of Equation (8.36), we find that

$$\hat{\mathbf{n}}\, da = J(\mathbf{X}, t)\, \hat{\mathbf{N}} \cdot \mathbf{F}^{-1}\, dA \tag{8.152}$$

which is called Nanson's formula.

First Piola-Kirchhoff Stress Tensor In the spatial configuration, the physical quantities including the Cauchy stress tensor, spatial coordinates and so on, are unknown. Hence, it is very important to find the relations between Lagrangian stress measures in the reference configuration and the Eulerian stress measures such as the Cauchy stress tensor in the spatial configuration.

Suppose that the infinitesimal actual force given in Equation (8.136) is given by

$$d\mathbf{f} = \mathbf{t}(\mathbf{x}, t, \hat{\mathbf{n}})\, da = \mathbf{T}\left(\mathbf{X}, t, \hat{\mathbf{N}}\right) dA = \left(\mathbf{P} \cdot \hat{\mathbf{N}}\right) dA \tag{8.153}$$

where $\mathbf{T}\left(\mathbf{X}, t, \hat{\mathbf{N}}\right)$ denotes the first Piola-Kirchhoff traction vector and $\hat{\mathbf{N}}(\mathbf{X}, t)$ a unit vector outward normal to the undeformed surface in the reference configuration. It is very important to note that the first Piola-Kirchhoff traction vector is a contravariant vector sharing the base vectors with the infinitesimal actual force. Therefore, it is an Eulerian state variable rather than a Lagrangian state variable. Then, the first Piola-Kirchhoff traction vector can be recast as

$$\mathbf{T}\left(\mathbf{X}, t, \hat{\mathbf{N}}\right) = \mathbf{P}(\mathbf{X}, t) \cdot \hat{\mathbf{N}}(\mathbf{X}, t) \tag{8.154}$$

where $\mathbf{P}(\mathbf{X}, t)$ is called the first Piola-Kirchhoff stress tensor, which is a two-point tensor as follows:

$$\mathbf{P}(\mathbf{X}, t) = P^{aA} \mathbf{g}_a \otimes \mathbf{G}_A \tag{8.155}$$

In view of Equations (8.154) and (8.155), the first Piola-Kirchhoff traction vector can be expressed as

$$\mathbf{T}\left(\mathbf{X}, t, \hat{\mathbf{N}}\right) = P^{aA} \hat{N}_A \mathbf{g}_a = T^a \mathbf{g}_a \tag{8.156}$$

By use of Nanson's formula, we have the relation between the first Piola-Kirchhoff stress and the Cauchy stress:

$$\mathbf{P}(\mathbf{X}, t) = J(\mathbf{X}, t)\, \boldsymbol{\sigma}(\mathbf{X}, t) \cdot \mathbf{F}^{-T} \tag{8.157}$$

The Cauchy stress tensor is symmetric due to the principle balance of angular momentum, which will be discussed in detail (see Equation (8.213)). Therefore, from the above, we obtain

$$\mathbf{P} \cdot \mathbf{F}^T = \mathbf{F} \cdot \mathbf{P}^T \tag{8.158}$$

On the other hand, the first Piola-Kirchhoff stress is not symmetric as follows:

$$\begin{aligned} \mathbf{P}^T &= J\mathbf{F}^{-1} \cdot \boldsymbol{\sigma}^T = P^{aA} \mathbf{G}_A \otimes \mathbf{g}_a = P^{Aa} \mathbf{g}_a \otimes \mathbf{G}_A, \\ \mathbf{P} &= P^{aA} \mathbf{g}_a \otimes \mathbf{G}_A = P^{Aa} \mathbf{G}_A \otimes \mathbf{g}_a \end{aligned} \tag{8.159}$$

The reader should be cautious in noting that in some literature (Belytschko et al. 2000; Lubarda 2002; Malvern 1969) the first Piola-Kirchhoff stress tensor is defined as the transpose of the Piola-Kirchhoff stress tensor we defined here.

Second Piola-Kirchhoff Stress Tensor As stated in Equation (8.36), the infinitesimal vectorial line segment in the reference configuration results from the pull-back of the infinitesimal vectorial line segment in the spatial configuration. Similarly, we may define the so-called pseudo-force as

$$d\tilde{\mathbf{f}}_0 = \varphi^* (d\mathbf{f}) = \mathbf{F}^{-1} \cdot d\mathbf{f} \tag{8.160}$$

The pseudo-force may be written in terms of the pseudo-traction $\tilde{\mathbf{T}}$ as

$$d\tilde{\mathbf{f}}_0 = \tilde{\mathbf{T}}\, dA \tag{8.161}$$

Note that the pseudo-force and the pseudo-traction are contravariant vectors with covariant base vectors in the reference configuration. Then, we may define a new stress tensor as a Lagrangian stress measure as follows

$$\tilde{\mathbf{T}} = \mathbf{S} \cdot \hat{\mathbf{N}} \tag{8.162}$$

where $\mathbf{S}$ is called the second Piola-Kirchhoff stress tensor. The second-order tensor is a contravariant tensor as follows:

$$\mathbf{S}^{\sharp} = S^{IJ} \mathbf{G}_I \otimes \mathbf{G}_J \tag{8.163}$$

In addition, we find that

$$\tilde{\mathbf{T}} = \varphi^* (\mathbf{T}) = \mathbf{F}^{-1} \cdot \mathbf{T} \tag{8.164}$$

In view of Equation (8.154), (8.157) and (8.164), we obtain

$$\mathbf{S} = \mathbf{F}^{-1} \cdot \mathbf{P} = J\mathbf{F}^{-1} \cdot \boldsymbol{\sigma} \cdot \mathbf{F}^{-T} = \mathbf{F}^{-1} \cdot \boldsymbol{\tau} \cdot \mathbf{F}^{-T} = \mathbf{S}^T \tag{8.165}$$

From the above, we observe that the second Piola-Kirchhoff stress tensor is symmetric. In addition, the above implies that in view of Equation (8.124),

$$\mathbf{S}^{\sharp} = \varphi^* \left(\boldsymbol{\tau}^{\sharp}\right) = \mathbf{F}^{-1} \cdot \boldsymbol{\tau}^{\sharp} \cdot \mathbf{F}^{-T} = S^{AB} \mathbf{G}_A \otimes \mathbf{G}_B \tag{8.166}$$

On the contrary, we have in view of Equation (8.122),

$$\boldsymbol{\tau}^{\sharp} = \varphi_* \left(\mathbf{S}^{\sharp}\right) = \mathbf{F} \cdot \mathbf{S}^{\sharp} \cdot \mathbf{F}^T = \tau^{ab} \mathbf{g}_a \otimes \mathbf{g}_b \tag{8.167}$$

Notice that the Cauchy stress tensor is an Eulerian stress measure indicating the real stress in the spatial configuration, whereas the second Piola-Kirchhoff stress is a Lagrangian stress measure in the reference configuration. By virtue of the principle of balance of angular momentum, the Cauchy stress is symmetric in the body. As shown previously, the first Piola-Kirchhoff stress is not symmetric, whereas the second Piola-Kirchhoff stress tensor is symmetric.

8.2 GENERAL PRINCIPLES

In this section, we examine fundamental principles that a continuous body in motion should obey. First we shall introduce the principle of conservation of mass. Then, we shall discuss classical balance principles which include the principle of balance of linear momentum, the principle of balance of angular momentum and the principle of balance of energy. Lastly, we shall introduce the principle of entropy inequality. In continuum mechanics, it is presupposed that the dynamical motion of continuum bodies obeys such fundamental principles mentioned above. Prior to the detailed discussion of the principles, we shall present the fundamental theorems including Gauss's theorem and Reynolds theorem. These theorems are of importance in developing the classical general principles.

8.2.1 Gauss's Theorem

Consider a smooth function, denoted by $f(\mathbf{X}, t) : \mathcal{B} \times \bar{\mathcal{T}} \rightarrow \mathbb{R}$, over the referential or the spatial domain. Let the function be differentiable with respect to the material coordinates. Gauss's theorem (Gauss, 1777-1855) states that

$$\int_{\mathcal{B}} \nabla f(\mathbf{X}, t)\, dV = \int_{\partial\mathcal{B}} \hat{\mathbf{N}} f(\mathbf{X}, t)\, dA \tag{8.168}$$

where $\hat{\mathbf{N}}$ is the outward normal unit vector acting on the surface of the volume, and dV denotes the volume element on $\mathcal{B}$ and dA the surface element on $\partial\mathcal{B}$. Hence, we observe that the volume integral form of the first partial derivative of the function $f(\mathbf{X}, t) : \mathcal{B} \times \bar{\mathcal{T}} \rightarrow \mathbb{R}$ can be transformed into the surface integral form of the function $f(\mathbf{X}, t) : \partial\mathcal{B} \times \bar{\mathcal{T}} \rightarrow \mathbb{R}$. Likewise, in the spatial coordinate the scalar-valued function $g(\mathbf{x}, t) : \mathcal{S} \times \bar{\mathcal{T}} \rightarrow \mathbb{R}$ is assumed to be smooth over the spatial domain. Then, Gauss's theorem states that

$$\int_{\mathcal{S}} \nabla g(\mathbf{x}, t)\, dv = \int_{\partial\mathcal{S}} \hat{\mathbf{n}} g(\mathbf{x}, t)\, da \tag{8.169}$$

where $\hat{\mathbf{n}}$ is the outward normal unit vector acting on the surface of the volume, and dv denotes the volume element on $\mathcal{S}$ and da the surface element on $\partial\mathcal{S}$.

In the case that a vector-valued function is defined over the the referential domain, Gauss's theorem yields

$$\int_{\mathcal{B}} \nabla \cdot \mathbf{V}(\mathbf{X}, t)\; dV = \int_{\mathcal{B}} \mathrm{DIV}\, \mathbf{V}(\mathbf{X}, t)\; dV = \int_{\partial\mathcal{B}} \hat{\mathbf{N}} \cdot \mathbf{V}(\mathbf{X}, t)\, dA \tag{8.170}$$

where ∇ in the the referential domain means $\nabla_{\mathbf{X}}$, which is defined in Equation (8.25). Likewise, in the spatial domain, Gauss's theorem states that

$$\int_{\mathcal{S}} \nabla \cdot \mathbf{v}(\mathbf{x}, t)\, dv = \int_{\mathcal{S}} \mathrm{div}\, \mathbf{v}(\mathbf{x}, t)\, dv = \int_{\partial\mathcal{S}} \hat{\mathbf{n}} \cdot \mathbf{v}\, da \tag{8.171}$$

where ∇ in the the spatial domain means $\nabla_{\mathbf{x}}$, which is defined in Equation (8.31). Further, Gauss's theorem of a tensor-valued function holds for $(\mathbf{X}, t) \in \mathcal{B} \times \bar{\mathcal{T}}$, i.e. for the second-order tensor we have

$$\int_{\mathcal{B}} \mathrm{DIV}\, \mathbf{S}(\mathbf{X}, t)\; dV = \int_{\mathcal{B}} (\mathbf{S}(\mathbf{X}, t) \cdot \nabla)\, dV = \int_{\partial\mathcal{B}} \hat{\mathbf{N}} \cdot \mathbf{S}(\mathbf{X}, t)\; dA \tag{8.172}$$

Likewise, in the spatial coordinate, Gauss's theorem states that

$$\int_{\mathcal{S}} \mathrm{div}\, \boldsymbol{\sigma}(\mathbf{x}, t)\, dv = \int_{\mathcal{S}} (\boldsymbol{\sigma}(\mathbf{x}, t) \cdot \nabla)\, dv = \int_{\partial\mathcal{S}} \hat{\mathbf{n}} \cdot \boldsymbol{\sigma}(\mathbf{x}, t)\; da \tag{8.173}$$

Gauss's theorem is often called the divergence theorem in the literature.

8.2.2 Reynolds Transport Theorem

We are often involved in the total time derivative of the volume integral in the spatial coordinates. Let a smooth function be given over the domain. The total time derivative of the integral form can be described as the following in the material coordinate.

$$\frac{d}{dt} \int_{\mathcal{S}} f(\mathbf{x}, t)\, dv = \frac{d}{dt} \int_{\mathcal{B}} f(\boldsymbol{\varphi}(\mathbf{X}, t), t)\, J(\mathbf{X}, t)\, dV \tag{8.174}$$

where $J(\mathbf{X}, t)$ is the Jacobian which is the determinant of the deformation gradient tensor, i.e. $J(\mathbf{X}, t) = |\mathbf{F}|$. In the sequence, the above equation may be rewritten as

$$\frac{d}{dt}\int_S f(\mathbf{x}, t)\, dv = \int_{\mathcal{B}} \frac{d}{dt}\{f(\boldsymbol{\varphi}(\mathbf{X}, t), t)\, J(\mathbf{X}, t)\}\, dV \tag{8.175}$$

Further, the integrand of the integral form can be expressed by the chain rule as

$$\frac{d}{dt}\{f(\varphi(\mathbf{X}, t), t)\, J(\mathbf{X}, t)\} = \frac{d}{dt}\{f(x, t)\, J\} = J\frac{Df(\mathbf{x}, t)}{Dt} + f(\mathbf{x}, t)\frac{dJ}{dt} \tag{8.176}$$

where it should be noted that $\frac{D}{Dt}$ stands for the material time derivative. The total time derivative of the Jacobian is derived as

$$\frac{dJ}{dt} = \frac{\partial J}{\partial \mathbf{F}}{:}\dot{\mathbf{F}} = J\mathbf{F}^{-\mathrm{T}}{:}\dot{\mathbf{F}} = J\mathbf{F}^{-\mathrm{T}}{:}\mathbf{L}\cdot\mathbf{F} = J\mathbf{I}{:}\mathbf{L} = J(tr\mathbf{d}) = J(\nabla\cdot\mathbf{v}) \tag{8.177}$$

Substituting Equation (8.177) into Equation (8.176), we find that

$$\frac{d}{dt}\{f(\mathbf{x}, t)\, J\} = \left(\frac{Df(\mathbf{x}, t)}{Dt} + f(\mathbf{x}, t)(\nabla\cdot\mathbf{v})\right) J \tag{8.178}$$

Consequently, Equation (8.174) takes the following form in the spatial coordinates

$$\frac{d}{dt}\int_S f(\mathbf{x}, t)\, dv = \int_S \left\{\frac{Df(\mathbf{x}, t)}{Dt} + f(\mathbf{x}, t)(\nabla\cdot\mathbf{v})\right\} dv \tag{8.179}$$

Equation (8.179) may be expressed as

$$\frac{d}{dt}\int_S f(\mathbf{x}, t)\, dv = \int_S \left\{\frac{\partial f(\mathbf{x}, t)}{\partial t} + \nabla\cdot(f\mathbf{v})\right\} dv \tag{8.180}$$

where $\frac{\partial f(\mathbf{x},t)}{\partial t}$ means the derivative of $f(\mathbf{x}, t)$ with respect to time holding $\mathbf{x}$ fixed. By virtue of Gauss's theorem, Equation (8.180) can take the form

$$\frac{d}{dt}\int_S f(\mathbf{x}, t)\, dv = \int_S \frac{\partial f(\mathbf{x}, t)}{\partial t} dv + \int_{\partial S} \hat{\mathbf{n}}\cdot(f\mathbf{v})\, dv \tag{8.181}$$

Equation (8.181) is the so-called Reynolds transport theorem. Let us consider the case that the physical quantity is pre-multiplied by the density function in the integrand. In consideration of Equation (8.179), we find that

$$\frac{d}{dt}\int_S \rho(\mathbf{x}, t)\, f(\mathbf{x}, t)\, dv = \int_S \left\{\frac{d(\rho f)}{dt} + \rho f(\nabla\cdot\mathbf{v})\right\} dv \tag{8.182}$$

The use of the chain rule leads to the form

$$\frac{d}{dt}\int_S \rho(\mathbf{x}, t)\, f(\mathbf{x}, t)\, dv = \int_S \left[\rho\frac{Df(\mathbf{x}, t)}{Dt} + f\left\{\frac{D\rho}{Dt} + \rho(\nabla\cdot\mathbf{v})\right\}\right] dv \tag{8.183}$$

Due to the principle of conservation of mass, which will be discussed in the next subsection, Equation (8.183) can be rewritten as

$$\frac{d}{dt}\int_S \rho(\mathbf{x}, t)\, f(\mathbf{x}, t)\, dv = \int_S \rho\frac{df}{dt} dv \tag{8.184}$$

8.2.3 Principle of Conservation of Mass

In continuum mechanics, the energy transformation of mass is not considered in a closed system in which the mass cannot cross the boundary surface. In the following, the mass of a body is postulated to be conserved during a motion. Even though the volume of the body can change for compressible materials, the mass of the body is not assumed to change, in other words, there is no increase or decrease of the mass.

$$\mathbf{M} = \int_{\mathcal{B}} \rho_0(\mathbf{X})\, dV = \int_{\mathcal{S}} \rho(\mathbf{x}, t)\, dv = \int_{\mathcal{B}} \rho(\boldsymbol{\varphi}(\mathbf{X}, t), t)\, J(\mathbf{X}, t)\, dV \tag{8.185}$$

Therefore, the Jacobian may be given by the ratio of Euclidean volume segments,

$$J(\mathbf{X}, t) = \frac{dv(\mathbf{x}, t)}{dV(\mathbf{X})} = \frac{\rho_0(\mathbf{X})}{\rho(\mathbf{x}, t)} \tag{8.186}$$

Since the mass of the body is constant during the deformation process, the rate of the mass should be zero. The mass inside the body cannot be decreased or increased. By virtue of Reynolds transport theorem, we observe that

$$\frac{d\mathbf{M}}{dt} = \frac{D}{Dt} \int_{\mathcal{S}} \rho(\mathbf{x}, t)\, dv = \int_{\mathcal{S}} \left\{ \frac{D\rho(\mathbf{x}, t)}{Dt} + \rho(\mathbf{x}, t)(\nabla \cdot \mathbf{v}) \right\} dv = 0 \tag{8.187}$$

As the above equation holds for the arbitrary domain, as an open set, it leads to

$$\frac{D\rho(\mathbf{x}, t)}{Dt} + \rho(\mathbf{x}, t)(\nabla \cdot \mathbf{v}) = 0 \tag{8.188}$$

It is known that the above equation is the continuity equation or the equation of mass conservation. If $\rho(\mathbf{x}, t)$ is constant, i.e. isochoric, then the continuity equation becomes

$$\nabla \cdot \mathbf{v}(\mathbf{x}, t) = \mathbf{v}(\mathbf{x}, t) \cdot \nabla = \operatorname{div} \mathbf{v}(\mathbf{x}, t) = 0 \tag{8.189}$$

For isochoric materials, the Jacobian is

$$J(\mathbf{X}, t) = \frac{\rho_0(\mathbf{X})}{\rho(\mathbf{x}(\mathbf{X}, t), t)} = 1 \tag{8.190}$$

8.2.4 Principle of Balance of Linear Momentum

In the Eulerian description, the total linear momentum of a continuous body, denoted by $\mathbf{p}(\mathbf{x}, t) : \mathcal{S} \times \bar{\mathcal{T}} \rightarrow \mathbb{R}^3$, may be given by the Eulerian description,

$$\mathbf{p}(\mathbf{x}, t) = \int_{\mathcal{S}} \rho(\mathbf{x}, t)\, \dot{\mathbf{x}}(\mathbf{x}, t)\, dv = \int_{\mathcal{S}} \rho(\mathbf{x}, t)\, \dot{\mathbf{u}}(\mathbf{x}, t)\, dv \tag{8.191}$$

According to Cauchy (1789-1857), Newton's second law in particle dynamics may be applied to the motion of the continuous body. The total time derivative of the total linear momentum of the body is equivalent to the translational resultant exerted on the body. The resultant consists of the surface traction $\mathbf{t}(\mathbf{x}, t, \hat{\mathbf{n}}) : \partial\mathcal{S} \times \bar{\mathcal{T}} \times \mathbb{R}^3 \rightarrow \mathbb{R}^3$ exerted on the surface of the body $\partial\mathcal{S}$ and the body force per unit mass $\mathbf{b}(\mathbf{x}, t) : \mathcal{S} \times \bar{\mathcal{T}} \rightarrow \mathbb{R}^3$ in the current configuration

$$\frac{d}{dt} \int_{\mathcal{S}} \rho \dot{\mathbf{u}}(\mathbf{x}, t)\, dv = \int_{\partial\mathcal{S}} \mathbf{t}(\mathbf{x}, t, \hat{\mathbf{n}})\, da + \int_{\mathcal{S}} \rho \mathbf{b}(\mathbf{x}, t)\, dv \tag{8.192}$$

The surface traction (the Cauchy traction vector) is given by

$$\mathbf{t}(\mathbf{x}, t, \hat{\mathbf{n}}) = \boldsymbol{\sigma} \cdot \hat{\mathbf{n}} \quad \text{for } \mathbf{x} \in \partial\mathcal{S} \tag{8.193}$$

which is also called Cauchy's fundamental theorem (Truesdell and Toupin 1960, page 543) (Truesdell 1991, page 174) or simply Cauchy's law (Belytschko et al. 2000; Holzapfel 2000; Malvern 1969; Marsden and Hughes 1983). By virtue of Gauss's theorem, the surface internal form can be converted into the volume integral form

$$\int_{\partial\mathcal{S}} \mathbf{t}(\mathbf{x}, t, \hat{\mathbf{n}})\, da = \int_{\partial\mathcal{S}} \boldsymbol{\sigma} \cdot \hat{\mathbf{n}}\, da = \int_{\mathcal{S}} \operatorname{div} \boldsymbol{\sigma}(\mathbf{x}, t)\, dv \tag{8.194}$$

Substituting Equation (8.194) into (8.192), we find

$$\int_{\mathcal{S}} \left[\rho \frac{D\dot{\mathbf{u}}}{Dt} - \rho \mathbf{b} - \boldsymbol{\sigma} \cdot \nabla \right] dv = \int_{\mathcal{S}} \left[\rho \frac{D\dot{\mathbf{u}}}{Dt} - \rho \mathbf{b} - \nabla \cdot \boldsymbol{\sigma} \right] dv = 0 \tag{8.195}$$

Note that the Cauchy stress tensor above is symmetric. As a consequence, we have

$$\rho \frac{D\dot{\mathbf{u}}}{Dt} = \operatorname{div} \boldsymbol{\sigma} + \rho \mathbf{b}, \quad \text{in } \mathcal{S} \tag{8.196}$$

Note that the above differential equations are subject to two boundary conditions, i.e. the prescribed traction boundary condition (the so-called natural boundary condition)

$$\mathbf{t}(\mathbf{x}, t, \hat{\mathbf{n}}) = \boldsymbol{\sigma}(\mathbf{x}, t) \cdot \hat{\mathbf{n}} = \bar{\mathbf{t}}(\mathbf{x}, t, \hat{\mathbf{n}}), \quad \text{on } \partial\mathcal{S}_\sigma \tag{8.197}$$

and the prescribed displacement boundary condition (the so-called essential boundary condition)

$$\mathbf{u}(\mathbf{x}, t) = \bar{\mathbf{u}}(\mathbf{x}, t), \quad \text{on } \partial\mathcal{S}_\varphi \tag{8.198}$$

In the Lagrangian description, Equation (8.192) may be described as

$$\frac{d}{dt} \int_{\mathcal{B}} \rho_0(\mathbf{X})\, \dot{\boldsymbol{\varphi}}(\mathbf{X}, t)\, dV = \int_{\partial\mathcal{B}} \mathbf{T}\left(\mathbf{X}, t, \hat{\mathbf{N}}\right) dA + \int_{\mathcal{B}} \rho_0(\mathbf{X})\, \mathbf{B}(\mathbf{X}, t)\, dV \tag{8.199}$$

where $\mathbf{B}(\mathbf{X}, t) : \mathcal{B} \times \bar{\mathcal{T}} \rightarrow \mathbb{R}^3$ denotes the body force per unit mass and $\mathbf{T}\left(\mathbf{X}, t, \hat{\mathbf{N}}\right) : \mathcal{B} \times \bar{\mathcal{T}} \times \mathbb{R}^3 \rightarrow \mathbb{R}^3$ the surface traction in the reference configuration. Equation (8.193) may be described as

$$\mathbf{T}\left(\mathbf{X}, t, \hat{\mathbf{N}}\right) = \mathbf{P}(\mathbf{X}, t) \cdot \hat{\mathbf{N}}(\mathbf{X}, t) \quad \text{for } \mathbf{X} \in \partial\mathcal{B} \tag{8.200}$$

By virtue of Gauss's theorem, the surface internal form can be converted into the volume integral form as

$$\int_{\partial\mathcal{B}} \mathbf{T}\left(\mathbf{X}, t, \hat{\mathbf{N}}\right) dA = \int_{\partial\mathcal{B}} \mathbf{P}(\mathbf{X}, t) \cdot \hat{\mathbf{N}}(\mathbf{X}, t)\; dA = \int_{\mathcal{B}} \operatorname{DIV} \mathbf{P}(\mathbf{X}, t)\, dV \tag{8.201}$$

Substituting Equation (8.201) into (8.199), we find that

$$\int_{\mathcal{B}} \left[\rho_0(\mathbf{X})\, \ddot{\boldsymbol{\varphi}}(\mathbf{X}, t) - \rho_0(\mathbf{X})\, \mathbf{B}(\mathbf{X}, t) - \operatorname{DIV} \mathbf{P}(\mathbf{X}, t) \right] dV = 0 \tag{8.202}$$

In the end, we have

$$\rho_0(\mathbf{X})\,\ddot{\boldsymbol{\varphi}}(\mathbf{X},t) - \rho_0(\mathbf{X})\,\mathbf{B}(\mathbf{X},t) - \mathrm{DIV}\,\mathbf{P}(\mathbf{X},t) = \mathbf{0} \quad \text{in } \mathcal{B} \tag{8.203}$$

with the boundary conditions, as the prescribed traction boundary condition

$$\mathbf{T}\left(\mathbf{X},t,\hat{\mathbf{N}}\right) = \mathbf{P}(\mathbf{X},t)\cdot\hat{\mathbf{N}}(\mathbf{X},t) = \bar{\mathbf{T}}\left(\mathbf{X},t,\hat{\mathbf{N}}\right) \quad \text{on } \partial\mathcal{B}_\sigma \tag{8.204}$$

and the prescribed displacement boundary condition

$$\boldsymbol{\varphi} = \bar{\boldsymbol{\varphi}} \quad \text{on } \partial\mathcal{B}_\varphi \tag{8.205}$$

Equation (8.196) or (8.203) is called Cauchy's equation of motion (Belytschko et al. 2000; Holzapfel 2000; Malvern 1969; Marsden and Hughes 1983) or Cauchy's first law of motion (Ogden 1984) (Truesdell and Toupin 1960, page 545) (Truesdell 1991, page 182). In static problems, there is no acceleration term; hence it results in Cauchy's equation of equilibrium as shown below

$$\mathrm{div}\,\boldsymbol{\sigma}(\mathbf{x}) + \rho(\mathbf{x})\,\mathbf{b}(\mathbf{x}) = \mathbf{0} \ \text{ or } \ \mathrm{DIV}\,\mathbf{P}(\mathbf{X}) + \rho_0(\mathbf{X})\,\mathbf{B}(\mathbf{X}) = \mathbf{0} \tag{8.206}$$

where $\mathrm{div}\,\boldsymbol{\sigma} = \boldsymbol{\sigma}\cdot\nabla_{\mathbf{x}}$ and $\mathrm{DIV}\,\mathbf{P} = \mathbf{P}\cdot\nabla_{\mathbf{X}}$.

8.2.5 Principle of Balance of Angular Momentum

Let us assume that we are concerned with the non polar case in which there are no distributed assigned couples over the surface or the volume. Then, the total time derivative of the total angular momentum of the body is equal to the resultant torque consisting of the torque by the surface traction acting on the surface and the torque by the body force.

$$\frac{d}{dt}\int_{\mathcal{S}} \mathbf{x}\times\rho\dot{\mathbf{u}}(\mathbf{x},t)\,dv = \int_{\partial\mathcal{S}} \mathbf{x}\times\mathbf{t}(\mathbf{x},t,\hat{\mathbf{n}})\,da + \int_{\mathcal{S}} \mathbf{x}\times\rho\mathbf{b}(\mathbf{x},t)\,dv \tag{8.207}$$

The term on the left-hand side of Equation (8.207) can be rewritten by chain rule and Reynolds transport theorem as

$$\frac{d}{dt}\int_{\mathcal{S}} \mathbf{x}\times\rho\dot{\mathbf{u}}(\mathbf{x},t)\,dv = \int_{\mathcal{S}} \rho\dot{\mathbf{u}}\times\dot{\mathbf{u}}(\mathbf{x},t)\,dv + \int_{\mathcal{S}} \rho\mathbf{x}\times\ddot{\mathbf{u}}(\mathbf{x},t)\,dv \tag{8.208}$$

The first term on the right-hand side of Equation (8.208) should be zero, and then we have

$$\int_{\mathcal{S}} \rho\mathbf{x}\times\ddot{\mathbf{u}}(\mathbf{x},t)\,dv = \int_{\partial\mathcal{S}} \mathbf{x}\times\mathbf{t}(\mathbf{x},t,\hat{\mathbf{n}})\,da + \int_{\mathcal{S}} \mathbf{x}\times\rho\mathbf{b}(\mathbf{x},t)\,dv \tag{8.209}$$

The first term of the right-hand side of Equation (8.209) can be rewritten by means of Cauchy's law and the divergence theorem as

$$\int_{\partial\mathcal{S}} \mathbf{x}\times\mathbf{t}\,da = \int_{\partial\mathcal{S}_\sigma} \mathbf{x}\times\boldsymbol{\sigma}\cdot\hat{\mathbf{n}}\,da = \int_{\mathcal{S}} \left(\mathbf{x}\times\boldsymbol{\sigma}\cdot\nabla + \boldsymbol{\varepsilon}:\boldsymbol{\sigma}^T\right)dv \tag{8.210}$$

where $\boldsymbol{\varepsilon}$ is the third order permutation tensor. Substituting Equation (8.210) into Equation (8.208), we have

$$\int_{\mathcal{S}} \mathbf{x}\times\{\rho\dot{\mathbf{u}}(\mathbf{x},t) - \boldsymbol{\sigma}\cdot\nabla - \rho\mathbf{b}\}\,dv = \int_{\mathcal{S}} \boldsymbol{\varepsilon}:\boldsymbol{\sigma}^T dv = 0 \tag{8.211}$$

Due to Cauchy's equations of motion, the left-hand side of Equation (8.211) becomes a zero vector. Note that the following holds in the spatial configuration

$$\boldsymbol{\varepsilon} : \boldsymbol{\sigma}^T = \begin{bmatrix} \sigma^{32} - \sigma^{23} & \sigma^{13} - \sigma^{31} & \sigma^{21} - \sigma^{12} \end{bmatrix}^T = 0 \tag{8.212}$$

In the end, we come to the conclusion that the Cauchy stress tensor should be symmetric, that is,

$$\boldsymbol{\sigma} = \boldsymbol{\sigma}^T \tag{8.213}$$

The above equation is known to be Cauchy's second equation of motion (Holzapfel 2000) or Cauchy's second law of motion (Ogden 1984).

8.2.6 Principle of Balance of Energy

The principle of balance of energy plays an important role in deriving the constitutive equations for hyperelastic materials. In the nonpolar case for which distributed assigned couples over the bounding surface or the volume do not exist, the power input (Malvern 1969) which is equivalent to the rate of the work done by the external forces is given by

$$\mathcal{P}_{\text{input}} = \int_{\partial \mathcal{B}_\sigma} \bar{\mathbf{T}}(\mathbf{X}, t, \hat{\mathbf{N}}) \cdot \dot{\boldsymbol{\varphi}}\,(\mathbf{X}, t)\, dA + \int_{\mathcal{B}} \rho_0\,(\mathbf{X})\, \mathbf{B}\,(\mathbf{X}, t) \cdot \dot{\boldsymbol{\varphi}}\,(\mathbf{X}, t)\, dV \tag{8.214}$$

In the literature, the power input is often called the external power (Holzapfel 2000) or the mechanical power (Marsden and Hughes 1983). By means of Cauchy's law and the divergence theorem, the first term on the right-hand side of Equation (8.214) may be expressed as

$$\int_{\partial \mathcal{B}_\sigma} \bar{\mathbf{T}} \cdot \dot{\boldsymbol{\varphi}}\, dA = \int_{\partial \mathcal{B}_\sigma} \dot{\boldsymbol{\varphi}} \cdot (\mathbf{P} \cdot \hat{\mathbf{N}})\, dA = \int_{\mathcal{B}} \text{DIV}\ (\dot{\boldsymbol{\varphi}} \cdot \mathbf{P})\, dV \tag{8.215}$$

By virtue of the product rule (Basar and Weichert 2000; Bonet and Wood 2008; Holzapfel 2000; Marsden and Hughes 1983; Simo and Hughes 1998), we have

$$\text{DIV}\ (\dot{\boldsymbol{\varphi}} \cdot \mathbf{P}) = \text{DIV}\, \mathbf{P} \cdot \dot{\boldsymbol{\varphi}} + \mathbf{P} : \text{GRAD}\, \dot{\boldsymbol{\varphi}} \tag{8.216}$$

Substituting Equation (8.215) into Equation (8.214), the power input may be given by

$$\mathcal{P}_{\text{input}} = \int_{\mathcal{B}} (\text{DIV}\, \mathbf{P}\,(\mathbf{X}, t) + \rho_0\,(\mathbf{X})\, \mathbf{B}\,(\mathbf{X}, t)) \cdot \dot{\boldsymbol{\varphi}}\, dV + \int_{\mathcal{B}} \mathbf{P} : \text{GRAD}\, \dot{\boldsymbol{\varphi}}\, dV \tag{8.217}$$

By virtue of Cauchy's equations of motion, Equation (8.217) takes the form

$$\mathcal{P}_{\text{input}} = \int_{\mathcal{B}} \rho_0 \ddot{\boldsymbol{\varphi}} \cdot \dot{\boldsymbol{\varphi}}\ dV + \int_{\mathcal{B}} \mathbf{P} : \text{GRAD}\, \dot{\boldsymbol{\varphi}}\, dV \tag{8.218}$$

Further, the second term on the right-hand side of Equation (8.218) can be rewritten as

$$\int_{\mathcal{B}} \mathbf{P} : \text{GRAD}\, \dot{\boldsymbol{\varphi}}\, dV = \int_{\mathcal{B}} \mathbf{P} : \dot{\mathbf{F}}\, dV = \int_{\mathcal{B}} \mathbf{P} \cdot\cdot \dot{\mathbf{F}}^T\, dV \tag{8.219}$$

It is important to note that the first Piola-Kirchhoff stress tensor (two-point tensor) is conjugate to the Lagrangian rate of the deformation gradient tensor (two-point tensor) in regard to the power. The term on the right hand side in Equation (8.219) is known as the stress power which is equal to the total time derivative of the internal energy $\mathcal{U}$, that is,

$$\frac{d\mathcal{U}}{dt} = \frac{d}{dt}\int_{\mathcal{B}} \rho_0 U dV = \int_{\mathcal{B}} \rho_0 \frac{\partial U}{\partial t} dV = \int_{\mathcal{B}} \mathbf{P} : \dot{\mathbf{F}} dV \tag{8.220}$$

where U denotes the specific internal energy function in the reference configuration. The material for which Equation (8.220) holds is called the perfectly elastic material (Holzapfel 2000). The first term on the right-hand side of Equation (8.218) is the total time derivative of the kinetic energy $\mathcal{K}$. Consequently, the power input is equivalent to the total time derivative of the total energy, that is,

$$\begin{aligned}\frac{d(\mathcal{K}+\mathcal{U})}{dt} &= \frac{d}{dt}\left(\int_{\mathcal{B}} \frac{1}{2}\rho_0 \dot{\boldsymbol{\varphi}} \cdot \dot{\boldsymbol{\varphi}}\, dV + \int_{\mathcal{B}} \rho_0 U\, dV\right) \\ &= \int_{\partial\mathcal{B}_\sigma} \bar{\mathbf{T}}\left(\mathbf{X}, t, \hat{\mathbf{N}}\right) \cdot \dot{\boldsymbol{\varphi}}(\mathbf{X}, t)\, dA + \int_{\mathcal{B}} \rho_0 \mathbf{B}(\mathbf{X}, t) \cdot \dot{\boldsymbol{\varphi}}(\mathbf{X}, t)\, dV = \mathcal{P}_{\text{input}}\end{aligned} \tag{8.221}$$

We observe that in the presence of symmetry the right-hand side terms are zero and the total energy remains constant.

Up to now, we have considered only the mechanical energy. In thermo-mechanical systems, thermal energies should be considered on the surface or in the volume. Then, Equation (8.221) should be modified in the following way

$$\begin{aligned}\frac{d}{dt}\left\{\int_{\mathcal{B}} \frac{1}{2}\rho_0 \dot{\boldsymbol{\varphi}} \cdot \dot{\boldsymbol{\varphi}}\, dv + \int_{\mathcal{B}} \rho_0 U\, dV\right\} &= \int_{\partial\mathcal{B}_\sigma} \bar{\mathbf{T}}\left(\mathbf{X}, t, \hat{\mathbf{N}}\right) \cdot \dot{\boldsymbol{\varphi}}(\mathbf{X}, t)\, dA + \int_{\mathcal{B}} \rho_0 \mathbf{B}(\mathbf{X}, t) \cdot \dot{\boldsymbol{\varphi}}(\mathbf{X}, t)\, dV \\ &\quad + \int_{\mathcal{B}} \rho_0 R(\mathbf{X}, t)\, dV - \int_{\partial\mathcal{B}} \mathbf{Q}(\mathbf{X}, t) \cdot \hat{\mathbf{N}}(\mathbf{X}, t)\, dA\end{aligned} \tag{8.222}$$

where $R(\mathbf{X}, t)$ denotes heat sources and $\mathbf{Q}(\mathbf{X}, t)$ the Cauchy heat flux outward the surface. Equation (8.222) is known as the first law of thermodynamics for thermo-mechanical systems. Equation (8.223) can be rewritten as

$$\int_{\mathcal{B}} (\rho_0 \ddot{\boldsymbol{\varphi}} - \mathrm{DIV}\,\mathbf{P} - \rho_0 \mathbf{B}) \cdot \dot{\boldsymbol{\varphi}}\, dV + \int_{\mathcal{B}} \rho_0 \frac{\partial U}{\partial t} dV = \int_{\mathcal{B}} \left(\mathbf{P} : \dot{\mathbf{F}} + \rho_0 R(\mathbf{X}, t) - \mathrm{DIV}\,\mathbf{Q}(\mathbf{X}, t)\right) dV \tag{8.223}$$

Since the first term on the left-hand side of Equation (8.223) should vanish, Equation (8.223) results in

$$\int_{\mathcal{B}} \rho_0 \frac{\partial U}{\partial t} dV = \int_{\mathcal{B}} \left(\mathbf{P} : \dot{\mathbf{F}} - \mathrm{DIV}\,\mathbf{Q}(\mathbf{X}, t) + \rho_0 R(\mathbf{X}, t)\right) dV \tag{8.224}$$

The above equation results in

$$\rho_0 \frac{\partial U}{\partial t} = \mathbf{P} : \dot{\mathbf{F}} - \mathrm{DIV}\,\mathbf{Q}(\mathbf{X}, t) + \rho_0 R(\mathbf{X}, t) \tag{8.225}$$

Equation (8.225) is called the local form of balance of energy or the Kirchhoff's equation of energy.

8.2.7 Principle of Entropy Inequality

In this principle, an important physical state variable is the entropy. The terminology implies the direction of thermal process in the thermo-mechanical system (Holzapfel 2000; Malvern 1969; Marsden and Hughes 1983). According to Clausius, the rate of the total entropy of the thermo-mechanical systems is equal or greater than the entropy input rate as follows:

$$\frac{d}{dt}\int_{\mathcal{B}} \rho_0(\mathbf{X})\, S(\mathbf{X},t)\, dV \geq \int_{\mathcal{B}} \rho_0(X)\, \frac{R(\mathbf{X},t)}{\Theta(\mathbf{X},t)} dV - \int_{\partial\mathcal{B}} \frac{\mathbf{Q}(\mathbf{X},t)\cdot\hat{\mathbf{N}}}{\Theta(\mathbf{X},t)} dA \tag{8.226}$$

where $S(\mathbf{X},t)$ denotes the specific entropy and $\Theta(\mathbf{X},t)$ the absolute temperature. Equation (8.226) is called the second law of thermodynamics. Further, Equation (8.226) is often called the Clasius-Duhem inequality or the entropy production inequality. Notice that in the adiabatic process without the heat sources, the total entropy should be zero or greater than zero, namely,

$$\frac{d}{dt}\int_{\mathcal{B}} \rho_0(\mathbf{X})\, S(\mathbf{X},t)\, dV \geq 0 \tag{8.227}$$

Due to Gauss's theorem, we may have the local form of the principle of entropy inequality,

$$\rho_0(\mathbf{X})\, \frac{\partial S(\mathbf{X},t)}{\partial t} - \rho_0(\mathbf{X})\, \frac{R(\mathbf{X},t)}{\Theta(\mathbf{X},t)} + \mathrm{DIV}\left(\frac{\mathbf{Q}(\mathbf{X},t)}{\Theta(\mathbf{X},t)}\right) \geq 0 \tag{8.228}$$

The totality of all the terms on the left hand side of Equation (8.228) is called the rate of entropy production. In addition, by use of the Kirchhoff energy equation in Equation (8.225) we have

$$\mathbf{P}:\dot{\mathbf{F}} - \rho_0\frac{\partial U}{\partial t} + \rho_0(\mathbf{X})\,\Theta(\mathbf{X},t)\,\frac{\partial S(\mathbf{X},t)}{\partial t} - \frac{\mathbf{Q}(\mathbf{X},t)\cdot\mathrm{GRAD}\,\Theta(\mathbf{X},t)}{\Theta(\mathbf{X},t)} \geq 0 \tag{8.229}$$

Since the entropy production by heat flux should be positive (Coleman and Noll 1963), we have

$$-\frac{\mathbf{Q}(\mathbf{X},t)\cdot\mathrm{GRAD}\,\Theta(\mathbf{X},t)}{\Theta(\mathbf{X},t)} \geq 0 \tag{8.230}$$

Equation (8.229), hence, can be rewritten as

$$\mathcal{D}_{\mathrm{int}} = \mathbf{P}:\dot{\mathbf{F}} - \rho_0\frac{\partial U}{\partial t} + \rho_0(\mathbf{X})\,\Theta(\mathbf{X},t)\,\frac{\partial S(\mathbf{X},t)}{\partial t} \geq 0 \tag{8.231}$$

which is called the Clausius-Planck inequality and $\mathcal{D}_{\mathrm{int}}$ is referred to as the internal dissipation. In a reversible process, the internal dissipation becomes zero, whereas in an irreversible process, the internal dissipation is greater than zero.

In a thermo-mechanical system, it is known that the Helmholtz free energy Ψ has the relation (Marsden and Hughes 1983):

$$\frac{\partial\Psi}{\partial\Theta} = -S \tag{8.232}$$

Via the Legendre transformation, the above equation leads to the relation between the specific internal energy function and the Helmholtz free energy:

$$U(\Phi, S) = \Psi(\Phi, \Theta) + \Theta S \tag{8.233}$$

where Φ is assumed to denote an internal state variable. Therefore, substituting Equation (8.233) into Equation (8.231) leads to

$$\mathcal{D}_{\text{int}} = \mathbf{P} : \dot{\mathbf{F}} - \rho_0 \frac{\partial \Psi}{\partial t} - \rho_0 S(\mathbf{X}, t) \frac{\partial \Theta(\mathbf{X}, t)}{\partial t} \geq 0 \tag{8.234}$$

In case of no entropy production or in an isothermal process, the above equation reduces to

$$\mathcal{D}_{\text{int}} = \mathbf{P} : \dot{\mathbf{F}} - \rho_0 \frac{\partial \Psi}{\partial t} \geq 0 \tag{8.235}$$

In a reversible process (thermal effects are neglected and the entropy is not produced in the process),

$$\mathcal{D}_{\text{int}} = \mathbf{P} : \dot{\mathbf{F}} - \rho_0 \frac{\partial \Psi}{\partial t} = 0 \tag{8.236}$$

For the case that the above equation holds, we have

$$\mathbf{P} : \dot{\mathbf{F}} = \rho_0 \frac{\partial \Psi}{\partial t} = \rho_0 \frac{\partial U}{\partial t} = \frac{\partial W}{\partial t} \tag{8.237}$$

where W stands for the elastic potential energy or the stored function energy. And, the material whose mechanical behavior obeys Equation (8.237) is called the perfect elastic material.

8.3 CONSTITUTIVE EQUATIONS IN ELASTICITY

Cauchy's equation of motion cannot be solved unless the relation between the first Piola-Kirchhoff stress tensor and the acceleration or displacement is defined, because the first Piola-Kirchhoff stress tensor depends upon the deformation. Hence, constitutive equations are needed to establish the relation between stress measures and strain measures. We shall discuss constitutive equations for elastic materials. In the literature (Truesdell and Noll 2004), elastic materials are classified into three principal categories, namely, Cauchy elastic, hyperelastic and hypoelastic materials. They are briefly highlighted next.

8.3.1 Cauchy Elastic Material

Due to Cauchy, a material is said to be elastic or Cauchy elastic when the Cauchy stress tensor within the material depends on the current configuration. The Cauchy elastic material is one for which the Cauchy stress tensor can be written by a one-to-one mapping $\hat{\boldsymbol{\sigma}}$ as

$$\boldsymbol{\sigma}(\mathbf{X}, t) = \hat{\boldsymbol{\sigma}}(\mathbf{X}, \mathbf{F}(\mathbf{X}, t)) \tag{8.238}$$

where $\hat{\boldsymbol{\sigma}}$ is called the constitutive or response function of the elastic material. It is to be noted that the current stress, the Cauchy stress, depends on only the current configuration, and not on the past history of the motion. When the Cauchy stress tensor does not depend on $\mathbf{X}$ explicitly, the elastic material is said to be homogeneous as follows:

$$\boldsymbol{\sigma} = \hat{\boldsymbol{\sigma}}(\mathbf{F}(\mathbf{X}, t)) \tag{8.239}$$

In some literature (Marsden and Hughes 1983), an elastic material is defined using the first Piola-Kirchhoff stress tensor rather than the Cauchy stress tensor as below

$$\mathbf{P}(\mathbf{X}, t) = \hat{\mathbf{P}}(\mathbf{X}, \mathbf{F}(\mathbf{X}, t)) \tag{8.240}$$

where $\hat{\mathbf{P}}$ is called a constitutive function.

8.3.2 Hyperelastic Material

An elastic material is said to be hyperelastic, or Green-elastic under the assumption that an elastic potential function for the material exists. Previously as shown in Equation (8.220), the rate of the specific internal energy function becomes equivalent to the stress power in the mechanical process only, that is,

$$\rho_0 \frac{\partial U}{\partial t} = \mathbf{P} : \dot{\mathbf{F}} \tag{8.241}$$

The above equation gives the importance of the notion of the elastic potential function.

In view of Equation (8.233), the rate of the specific internal energy function can be written as

$$\rho_0 \frac{\partial U}{\partial t} = \rho_0 \frac{\partial \Psi}{\partial t} + \rho_0 \frac{\partial S}{\partial t} \Theta + \rho_0 \frac{\partial \Theta}{\partial t} S \tag{8.242}$$

In an isentropic process ($\frac{\partial S}{\partial t} = 0$), Equation (8.242) yields

$$\rho_0 \frac{\partial U}{\partial t} = \rho_0 \frac{\partial \Psi}{\partial t} + \rho_0 \frac{\partial \Theta}{\partial t} S \tag{8.243}$$

whereas in an isothermal process ($\frac{\partial \Theta}{\partial t} = 0$), Equation (8.242) yields

$$\rho_0 \frac{\partial U}{\partial t} = \rho_0 \frac{\partial \Psi}{\partial t} + \rho_0 \frac{\partial \Theta}{\partial t} S \tag{8.244}$$

Therefore, in an isentropic-isothermal process ($\frac{\partial S}{\partial t} = 0$, $\frac{\partial \Theta}{\partial t} = 0$), Equation (8.242) yields

$$\rho_0 \frac{\partial U}{\partial t} = \rho_0 \frac{\partial \Psi}{\partial t} \tag{8.245}$$

In addition, in an adiabatic process we have

$$\rho_0 R (\mathbf{X}, t) - \mathrm{DIV}\, \mathbf{Q} (\mathbf{X}, t) = 0 \tag{8.246}$$

As a consequence, in a mechanical process, namely isentropic-isothermal-adiabatic, we arrive at Equation (8.241).

The Helmholtz free energy is, in some literature, also regarded as an elastic potential function or the strain energy density function or the stored energy function. Therefore, in hyperelastic materials with thermal effects neglected, we have the relation:

$$\rho_0 U = \rho_0 \Psi = W \tag{8.247}$$

Hyperelastic materials, often referred to as Green elastic materials, assume that the elastic potential functions can be expressed as functions of the deformation gradient tensor $\mathbf{F}$ and the position vector $\mathbf{X}$

$$\rho_0 U = \rho_0 \Psi (\mathbf{X}, \mathbf{F}) = W (\mathbf{X}, \mathbf{F}) \tag{8.248}$$

In view of Equation (8.237), the first Piola-Kirchhoff stress tensor can be given such that

$$\mathbf{P} := \rho_0 \frac{\partial \Psi (\mathbf{X}, \mathbf{F})}{\partial \mathbf{F}} = \frac{\partial W (\mathbf{X}, \mathbf{F})}{\partial \mathbf{F}} \tag{8.249}$$

In the Lagrangian description, Equation (8.237) gives the relation:

$$\rho_0 \frac{\partial U}{\partial t} = \mathbf{S} : \dot{\mathbf{E}} = \mathbf{S} : \frac{\dot{\mathbf{C}}}{2} \tag{8.250}$$

where $\dot{\mathbf{C}}$ is the material time derivative of the right Cauchy-Green tensor. Thus, we obtain

$$\mathbf{S} := \rho_0 \frac{\partial \bar{\Psi}(\mathbf{X},\mathbf{E})}{\partial \mathbf{E}} = \frac{\partial \bar{W}(\mathbf{X},\mathbf{E})}{\partial \mathbf{E}} = 2\rho_0 \frac{\partial \bar{\bar{\Psi}}(\mathbf{X},\mathbf{C})}{\partial \mathbf{C}} = 2\frac{\partial \bar{\bar{W}}(\mathbf{X},\mathbf{C})}{\partial \mathbf{C}} \tag{8.251}$$

For the case that the elastic potential functions are dependent on the deformation gradient tensor $\mathbf{F}$, the material is also called homogeneous as follows:

$$\Psi = \hat{\Psi}(\mathbf{F}) \tag{8.252}$$

otherwise, the material is said to be heterogeneous. The material time derivative of the first Piola-Kirchhoff stress tensor may be expressed as

$$\dot{\mathbf{P}} = \rho_0 \frac{\partial^2 \hat{\Psi}(\mathbf{F})}{\partial \mathbf{F} \partial \mathbf{F}} : \dot{\mathbf{F}} = \frac{\partial^2 \hat{W}(\mathbf{F})}{\partial \mathbf{F} \partial \mathbf{F}} : \dot{\mathbf{F}} \tag{8.253}$$

where $\rho_0 \frac{\partial^2 \hat{\Psi}(\mathbf{F})}{\partial \mathbf{F} \partial \mathbf{F}}$ or $\frac{\partial^2 \hat{W}(\mathbf{F})}{\partial \mathbf{F} \partial \mathbf{F}}$ is called the referential first elasticity fourth-order mixed tensor (Simo and Pister 1984). Likewise, taking the material time derivative of the second Piola-Kirchhoff stress tensor yields

$$\dot{\mathbf{S}} = 4\rho_0 \frac{\partial^2 \bar{\bar{\Psi}}(\mathbf{C})}{\partial \mathbf{C} \partial \mathbf{C}} : \frac{\dot{\mathbf{C}}}{2} = \rho_0 \frac{\partial^2 \bar{\Psi}(\mathbf{E})}{\partial \mathbf{E} \partial \mathbf{E}} : \dot{\mathbf{E}} = \mathbb{C} : \dot{\mathbf{E}} \tag{8.254}$$

where $\mathbb{C}$ is called the referential second elasticity 4th order tensor. Consequently, we have

$$\mathbb{C} = \rho_0 \frac{\partial^2 \bar{\Psi}(\mathbf{E})}{\partial \mathbf{E} \partial \mathbf{E}} = 4\rho_0 \frac{\partial^2 \bar{\bar{\Psi}}(\mathbf{C})}{\partial \mathbf{C} \partial \mathbf{C}} = C^{AB}_{\,..CD} \mathbf{G}^A \otimes \mathbf{G}^B \otimes \mathbf{G}_C \otimes \mathbf{G}_D \tag{8.255}$$

Note that the first or second elasticity tensor is a fourth order mixed tensor. More comprehensive discussions may be found in (Holzapfel 2000; Marsden and Hughes 1983).

For hyperelastic-plastic materials, the second Piola-Kirchhoff stress for elastic-plastic material can be defined as

$$\mathbf{S} = \rho_0 \frac{\partial \bar{\Psi}(\mathbf{E})}{\partial \mathbf{E}^{el}} \tag{8.256}$$

where $\mathbf{E}^{el}$ is the elastic part of the Green strain tensor. It should be noted that $\mathbf{E}^{el}$ is different from that of the intermediate configuration as shown in the literature. Green and Naghdi (Green and Naghdi 1965) advocated that an additive decomposition of the Green strain tensor into elastic-plastic components be supported by solid thermodynamic principles. Casey (Casey 1985) also justified the additive decomposition of the Green strain tensor as an approximation that is valid when: (i) small plastic deformations are accompanied by moderate elastic stains, (ii) small elastic stains are accompanied by moderate plastic deformation, or (iii) small strains are accompanied by moderate rotations (Casey 1985).

Isotropic Elastic Material An elastic material is said to be isotropic if the elastic potential function remains constant regardless of the coordinate transformation of the reference frame. The notion of the isotropic elastic response of a material is concerned with the invariance of the elastic constitutive modulus (Marsden and Hughes 1983; Simo and Hughes 1998).

Consider a rigid body motion superimposed into the reference frame. The rigid body motion $\mathbf{X}^*$ can be given by

$$\mathbf{X}^* = \mathbf{c} + \mathbf{Q} \cdot \mathbf{X} \tag{8.257}$$

where $\mathbf{X}$ denotes the position vector in the reference body $\mathcal{B}$, and $\mathbf{X}^*$ represents the position vector in the reference body $\mathcal{B}^*$ undergoing the rigid body motion. In the above equation, $\mathbf{c}$ is a constant vector representing a superimposed rigid body translation and $\mathbf{Q}$ is the proper orthogonal tensor, i.e. $\mathbf{Q} \in SO(3)$. Note that the position $\mathbf{x}$ in the current body $\mathcal{S}$ can be written as

$$\mathbf{x} = \varphi(\mathbf{X}, t) = \varphi^*(\mathbf{X}^*, t) \tag{8.258}$$

By use of the chain rule, the relation between the deformation gradient tensors can be written as

$$\mathbf{F} = \frac{\partial \mathbf{x}}{\partial \mathbf{X}} = \frac{\partial \varphi^*(\mathbf{X}^*, t)}{\partial \mathbf{X}^*} \mathbf{Q} = \mathbf{F}^* \cdot \mathbf{Q} \tag{8.259}$$

From the above, we have

$$\mathbf{F}^* = \mathbf{F} \cdot \mathbf{Q}^T \tag{8.260}$$

A hyperelastic material is said to be isotropic only if

$$\hat{\Psi}(\mathbf{F}) = \hat{\Psi}(\mathbf{F}^*) = \hat{\Psi}(\mathbf{F} \cdot \mathbf{Q}^T) \tag{8.261}$$

In view of Equation (8.58), for the Green deformation tensor, we have the relation

$$\mathbf{C}^* = \mathbf{F}^{*T} \cdot \mathbf{F}^* = \mathbf{Q} \cdot \mathbf{C} \cdot \mathbf{Q}^T \tag{8.262}$$

Therefore, a homogeneous, isotropic hyperelastic material satisfies the following (Lubarda 2002; Murnaghan 1951)

$$\bar{\bar{\Psi}}(\mathbf{C}) = \bar{\bar{\Psi}}(\mathbf{C}^*) = \bar{\bar{\Psi}}(\mathbf{Q} \cdot \mathbf{C} \cdot \mathbf{Q}^T), \text{ for } \forall\, \mathbf{Q} \in SO(3) \tag{8.263}$$

Otherwise, the material is said to be anisotropic.

In the end, the elastic potential function is an invariant under a superposed rigid body motion. For homogeneous isotropic hyperelastic materials, the elastic potential function may be represented in terms of the principal invariants of the Green deformation tensor $\mathbf{C}$ (Gurtin 1981, page 230) as follows:

$$\bar{\bar{\Psi}}(\mathbf{C}) = \bar{\bar{\Psi}}(\mathrm{I_C}, \mathrm{II_C}, \mathrm{III_C}) = \bar{\bar{\Psi}}(\lambda_1, \lambda_2, \lambda_3) \tag{8.264}$$

where $\mathrm{I_C}$, $\mathrm{II_C}$ and $\mathrm{III_C}$ are the principal invariants of the Green deformation tensor (Holzapfel 2000; Zienkiewicz and Taylor 2006), and they are defined as

$$\begin{aligned} \mathrm{I_C} &= \mathrm{tr}\mathbf{C} = \lambda_1^2 + \lambda_2^2 + \lambda_3^2, \\ \mathrm{II_C} &= \frac{1}{2}\left(\mathrm{I_C^2} - \mathrm{tr}(\mathbf{C}^2)\right) = \mathrm{tr}(\mathbf{C}^{-1})\mathrm{det}\mathbf{C} = \lambda_1^2\lambda_2^2 + \lambda_1^2\lambda_3^2 + \lambda_2^2\lambda_3^2, \\ \mathrm{III_C} &= \mathrm{det}\mathbf{C} = \lambda_1^2\lambda_2^2\lambda_3^2 \end{aligned} \tag{8.265}$$

where $\lambda_i, i = 1, 2, 3$ are the principal stretches as the eigenvalues of the right stretch tensor **U** defined in Equation (8.44). By use of the chain rule, the partial derivative of the isotropic free energy elastic function with respect to the Green deformation tensor can be written as

$$\frac{\partial \bar{\bar{\Psi}}(\mathrm{I_C}, \mathrm{II_C}, \mathrm{III_C})}{\partial \mathbf{C}} = \frac{\partial \bar{\bar{\Psi}}}{\partial \mathrm{I_C}}\frac{\partial \mathrm{I_C}}{\partial \mathbf{C}} + \frac{\partial \bar{\bar{\Psi}}}{\partial \mathrm{II_C}}\frac{\partial \mathrm{II_C}}{\partial \mathbf{C}} + \frac{\partial \bar{\bar{\Psi}}}{\partial \mathrm{III_C}}\frac{\partial \mathrm{III_C}}{\partial \mathbf{C}} \tag{8.266}$$

The partial derivatives of the principal invariants with respect to the Green deformation tensor yields (Basar and Weichert 2000; Marsden and Hughes 1983)

$$\begin{aligned}
\frac{\partial \mathrm{I_C}}{\partial \mathbf{C}} &= \frac{\partial \mathrm{tr}\mathbf{C}}{\partial \mathbf{C}} = \frac{\partial \mathbf{G}:\mathbf{C}}{\partial \mathbf{C}} = \mathbf{G}, \\
\frac{\partial \mathrm{II_C}}{\partial \mathbf{C}} &= \frac{1}{2}\left(2(\mathrm{tr}\mathbf{C})\mathbf{G} - \frac{\partial \mathrm{tr}(\mathbf{C}^2)}{\partial \mathbf{C}}\right) = \mathrm{tr}(\mathbf{C}^{-1})\mathrm{det}\mathbf{C} = \mathrm{I_C}\mathbf{G} - \mathbf{C}, \\
\frac{\partial \mathrm{III_C}}{\partial \mathbf{C}} &= \mathrm{III_C}\mathbf{C}^{-1}
\end{aligned} \tag{8.267}$$

where **G** is the second-order identity tensor given in Equation (8.54). Because of the above equations, the second Piola-Kirchhoff stress can be expressed in terms of the invariants as

$$\begin{aligned}
\mathbf{S} &= 2\rho_0 \frac{\partial \bar{\bar{\Psi}}(\mathrm{I_C}, \mathrm{II_C}, \mathrm{III_C})}{\partial \mathbf{C}} \\
&= 2\rho_0 \left(\frac{\partial \bar{\bar{\Psi}}}{\partial \mathrm{I_C}} + \frac{\partial \bar{\bar{\Psi}}}{\partial \mathrm{II_C}}\right)\mathbf{G} - 2\rho_0 \frac{\partial \bar{\bar{\Psi}}}{\partial \mathrm{II_C}}\mathbf{C} + 2\rho_0 \frac{\partial \bar{\bar{\Psi}}}{\partial \mathrm{III_C}}\mathrm{III_C}\mathbf{C}^{-1} \\
&= 2\rho_0 \left(\frac{\partial \bar{\bar{\Psi}}}{\partial \mathrm{I_C}} + \frac{\partial \bar{\bar{\Psi}}}{\partial \mathrm{II_C}}\right)\mathbf{G} - 2\rho_0 \frac{\partial \bar{\bar{\Psi}}}{\partial \mathrm{II_C}}\mathbf{B}^{-1} + 2\rho_0 \frac{\partial \bar{\bar{\Psi}}}{\partial \mathrm{III_C}}\mathrm{III_C}\mathbf{B}
\end{aligned} \tag{8.268}$$

where **B** is the Piola deformation tensor. Note that the Green deformation tensor **C** shares the same eigenvalues with the Cauchy deformation tensor **b**. Hence, the two second-order tensors have the same invariants. By virtue of the rules of the trace of the second-order tensor (see Section 2.8.1), we have

$$\begin{aligned}
\mathrm{I_C} &= \mathrm{tr}\mathbf{C} = \mathrm{tr}\mathbf{b} = \mathrm{I_b}, \\
\mathrm{II_C} &= \frac{1}{2}\left(\mathrm{I_C^2} - \mathrm{tr}(\mathbf{C}^2)\right) = \frac{1}{2}\left(\mathrm{I_b^2} - \mathrm{tr}(\mathbf{b}^2)\right) = \mathrm{II_b}, \\
\mathrm{III_C} &= \mathrm{det}\mathbf{C} = \mathrm{det}\mathbf{b} = \mathrm{III_b} = (\mathrm{det}\mathbf{F})^2 = J^2
\end{aligned} \tag{8.269}$$

As a consequence of the above relation, we have

$$\bar{\bar{\Psi}}(\mathbf{C}) = \bar{\bar{\Psi}}(\mathrm{I_C}, \mathrm{II_C}, \mathrm{III_C}) = \bar{\bar{\Psi}}(\lambda_1, \lambda_2, \lambda_3) = \bar{\bar{\Psi}}(\mathrm{I_b}, \mathrm{II_b}, \mathrm{III_b}) = \bar{\bar{\Psi}}(\mathbf{b}) \tag{8.270}$$

Since the Kirchhoff stress tensor can be given by $\boldsymbol{\tau} = \mathbf{F}\cdot\mathbf{S}\cdot\mathbf{F}^T$, we may have the following

$$\boldsymbol{\tau} = 2\rho_0 \frac{\partial \bar{\bar{\Psi}}}{\partial \mathrm{III_b}}\mathrm{III_b}\mathbf{g} + 2\rho_0 \left(\frac{\partial \bar{\bar{\Psi}}}{\partial \mathrm{I_b}} + \frac{\partial \bar{\bar{\Psi}}}{\partial \mathrm{II_b}}I_{\mathbf{b}}\right)\mathbf{b} - 2\rho_0 \frac{\partial \bar{\bar{\Psi}}}{\partial \mathrm{II_b}}\mathbf{b}^2 \tag{8.271}$$

where $\mathbf{g}$ is the second-order identity tensor given in Equation (8.69). It is to be noted that the Cauchy stress tensor can be expressed as

$$\begin{aligned}\boldsymbol{\sigma} &= J^{-1}\boldsymbol{\tau} \\ &= 2\rho_0 J^{-1}\frac{\partial\bar{\bar{\Psi}}}{\partial \mathrm{III}_{\mathbf{b}}}\mathrm{III}_{\mathbf{b}}\mathbf{g} + 2\rho_0 J^{-1}\left(\frac{\partial\bar{\bar{\Psi}}}{\partial \mathrm{I}_{\mathbf{b}}} + \frac{\partial\bar{\bar{\Psi}}}{\partial \mathrm{II}_{\mathbf{b}}}\mathrm{I}_{\mathbf{b}}\right)\mathbf{b} - 2\rho_0 J^{-1}\frac{\partial\bar{\bar{\Psi}}}{\partial \mathrm{II}_{\mathbf{b}}}\mathbf{b}^2\end{aligned} \tag{8.272}$$

which represents the spatial description of the stress for hyperelastic materials.

St. Venant-Kirchhoff Elastic Material We shall discuss the simplest isotropic hyperelastic material retaining the linear constitutive equation for nonlinear elastic materials. Consider the elastic potential function (Ciarlet 1988, page 155)

$$\rho_0\bar{\Psi} = \frac{\lambda}{2}(\mathrm{tr}\mathbf{E})^2 + \mu\mathrm{tr}\mathbf{E}^2 \tag{8.273}$$

where λ and μ denote the Lamé constants. In view of Equation (8.251), the second Piola-Kirchhoff stress yields

$$\mathbf{S} = \lambda(\mathrm{tr}\mathbf{E})\mathbf{G} + 2\mu\mathbf{E} = \mathbb{C} : \mathbf{E} \tag{8.274}$$

which shows that the second Piola-Kirchhoff stress $\mathbf{S}$ is a linear function of the Lagrangian strain tensor $\mathbf{E}$. In the above, $\mathbb{C}$ is the fourth-order elastic tensor, that is,

$$\mathbb{C} = \lambda(\mathbf{G}\otimes\mathbf{G}) + 2\mu\mathbf{II} \tag{8.275}$$

where $\mathbf{II}$ is the fourth-order identity tensor, i.e. $\mathbf{II} = \mathbf{G}_A\otimes\mathbf{G}_B\otimes\mathbf{G}^A\otimes\mathbf{G}^B$. Consequently, the first Piola-Kirchhoff stress tensor can be written as

$$\mathbf{P} = \mathbf{F}\cdot\mathbf{S} = \lambda(\mathrm{tr}\mathbf{E})\mathbf{F} + 2\mu\mathbf{F}\cdot\mathbf{E} \tag{8.276}$$

which means that $\mathbf{P}$ is not a linear function of $\mathbf{F}$ (Fosdick and Serrin 1979). In addition, we can derive the Kirchhoff stress tensor $\boldsymbol{\tau}$ as follows:

$$\boldsymbol{\tau} = \mathbf{F}\cdot\mathbf{S}\cdot\mathbf{F}^T = \lambda(\mathrm{tr}\mathbf{E})\mathbf{F}\cdot\mathbf{F}^T + 2\mu\mathbf{F}\cdot\mathbf{E}\cdot\mathbf{F}^T \tag{8.277}$$

In view of Equations (8.276) and (8.277), even though the St. Venant-Kirchhoff material represents the linear relationship between the second Piola-Kirchhoff stress tensor and the Lagrangian strain tensor, it represents a nonlinear hyperelastic material in displacements. The nonlinear hyperelastic material is considered to be compressible, namely, $j \neq 1$; hence a metal may be often regarded as the St. Venant-Kirchhoff material.

Compressible Neo-Hookean Material An isotropic free energy constitutive function $\bar{\Psi}$ for the compressible neo-Hookean elastic material is defined as (Basar and Weichert 2000; Belytschko et al. 2000; Bonet and Wood 2008; Ogden 1984; Simo and Pister 1984)

$$\rho_0\bar{\Psi} = \frac{\mu}{2}(\mathrm{I}_{\mathbf{C}} - 3) - \mu\ln\sqrt{\mathrm{III}_{\mathbf{C}}} + \frac{\lambda}{2}(\ln\sqrt{\mathrm{III}_{\mathbf{C}}})^2 \tag{8.278}$$

In view of Equation (8.251), the second Piola-Kirchhoff stress yields

$$\mathbf{S} = \mu(\mathbf{G} - \mathbf{C}^{-1}) + \lambda(\ln\sqrt{\mathrm{III}_{\mathbf{C}}})\mathbf{C}^{-1} \tag{8.279}$$

and, we can derive the Kirchhoff stress tensor $\boldsymbol{\tau}$ as follows:

$$\boldsymbol{\tau} = \mu(\mathbf{b} - \mathbf{g}) + \lambda(\ln\sqrt{\mathrm{III}_{\mathbf{b}}})\mathbf{g} \tag{8.280}$$

In the above, $\mathbb{C}$ is the fourth-order elastic tensor of the compressible Neo-Hookean elastic material, i.e.

$$\mathbb{C} = \lambda\mathbf{C}^{-1} \otimes \mathbf{C}^{-1} + 2(\mu - \lambda\ln\sqrt{\mathrm{III}_{\mathbf{C}}})\frac{\partial\mathbf{C}^{-1}}{\partial\mathbf{C}} \tag{8.281}$$

Incompressible Neo-Hookean Material We have learned that in the infinitesimal strain theory, the linear strain tensor may be decomposed into two parts, namely, deviatoric and volumetric parts. In the hyperelasticity theory, it is postulated that the deformation gradient tensor can be multiplicatively decomposed into volumetric and isochoric (volume-preserving) parts (Flory 1961) as follows:

$$\mathbf{F} = (J^{\frac{1}{3}}\mathbf{I}) \cdot \bar{\mathbf{F}} \tag{8.282}$$

where $\bar{\mathbf{F}}$ denotes the isochoric part of the deformation gradient tensor, while $J^{\frac{1}{3}}\mathbf{I}$ corresponds to the volumetric part. It is easy to show that consequently the above equation yields

$$\mathbf{C} = (J^{\frac{2}{3}}\mathbf{I}) \cdot \bar{\mathbf{C}} \tag{8.283}$$

Consequently, we can observe that

$$\det\bar{\mathbf{F}} = \det(J^{-\frac{1}{3}}\mathbf{F}) = J^{-1}\det\mathbf{F} = \bar{\lambda}_1\bar{\lambda}_2\bar{\lambda}_3 = 1 \tag{8.284}$$

where $\det(J^{-\frac{1}{3}}\mathbf{I}) = J^{-1}$. From Equation (8.282), we may write by use of the polar decomposition theorem

$$\bar{\mathbf{F}} = (J^{-\frac{1}{3}}\mathbf{I}) \cdot \mathbf{F} = (J^{-\frac{1}{3}}\mathbf{I}) \cdot \mathbf{R} \cdot \mathbf{U} = \mathbf{R} \cdot (J^{-\frac{1}{3}}\mathbf{U}) = \mathbf{R} \cdot \bar{\mathbf{U}} \tag{8.285}$$

where $\bar{\mathbf{U}} = J^{-\frac{1}{3}}\mathbf{U}$. Therefore, the principal stretches of $\bar{\mathbf{U}}$ can be written as

$$\bar{\lambda}_i = J^{-\frac{1}{3}}\lambda_i, \ i = 1, 2, 3 \tag{8.286}$$

Finally, we have the incompressibility condition for hyperelastic materials (Ogden 1984; Simo and Taylor 1991) as follows:

$$\det\bar{\mathbf{F}} = \det\bar{\mathbf{U}} = \bar{\lambda}_1\bar{\lambda}_2\bar{\lambda}_3 = 1 \tag{8.287}$$

For incompressible hyperelastic materials (for instance, rubber), $\det\mathbf{F} = J = 1$. Therefore, in view of Equations (8.282) and (8.286) we have the following

$$\begin{aligned} &\bar{\mathbf{F}} = \mathbf{F}, \\ &\bar{\mathbf{U}} = \mathbf{U}, \\ &\bar{\lambda}_i = \lambda_i, \ i = 1, 2, 3 \end{aligned} \tag{8.288}$$

Due to Mooney (Mooney 1940) and Rivlin (Rivlin 1948), a stored energy function for isotropic, incompressible hyperelastic materials with negligible hysteresis may be written in terms of the principal stretches as

$$\begin{aligned} W(\lambda_1, \lambda_2, \lambda_3) &= \sum_{i=1}^{N} \alpha_i (\lambda_1^{a_i} + \lambda_2^{a_i} + \lambda_3^{a_i} - 3) \\ &+ \sum_{j=1}^{M} \beta_j ((\lambda_1\lambda_2)^{b_j} + (\lambda_2\lambda_3)^{b_j} + (\lambda_3\lambda_2)^{b_j} - 3) \\ &+ \gamma(\lambda_1\lambda_2\lambda_3) \end{aligned} \tag{8.289}$$

the form of which is due to Ogden (Ogden 1972; Ogden 1984). In the special case when $N = M = 1$, $a_1 = b_1 = 2$ and $\gamma = 0$, the stored energy function reduces to

$$\begin{aligned} W(\lambda_1, \lambda_2, \lambda_3) &= \alpha_1(\lambda_1^2 + \lambda_2^2 + \lambda_3^2 - 3) + \beta_1((\lambda_1\lambda_2)^2 + (\lambda_2\lambda_3)^2 + (\lambda_3\lambda_2)^2 - 3) \\ &= \alpha_1(\mathrm{I_C} - 3) + \beta_1(\mathrm{II_C} - 3) \end{aligned} \tag{8.290}$$

Such a material is called the incompressible Mooney-Rivlin material (Holzapfel 2000; Marsden and Hughes 1983; Ogden 1984).

Incompressible Mooney-Rivlin Material In Equation (8.290), if $\beta_1 = 0$, then we obtain

$$\begin{aligned} W(\lambda_1, \lambda_2, \lambda_3) &= \alpha_1(\lambda_1^2 + \lambda_2^2 + \lambda_3^2 - 3) \\ &= \alpha_1(\mathrm{I_C} - 3) \end{aligned} \tag{8.291}$$

which is a stored energy function for an incompressible neo-Hookean material. It is to be noted that substituting $J = 1$ into Equation (8.278), we obtain the stored energy function shown in Equation (8.291), i.e. $\alpha_1 = \frac{\mu}{2}$.

8.3.3 Hypoelastic Material

The terminology, hypoelasticity, is due to Truesdell (Truesdell 1955). A material is said to be hypoelastic, if any objective stress rate is a homogeneous linear function depending on the rate of deformation tensor and the Cauchy stress tensor as

$$\boldsymbol{\sigma}^{\nabla} = \bar{\mathfrak{M}}(\boldsymbol{\sigma}, \mathbf{d}) \tag{8.292}$$

where $\boldsymbol{\sigma}^{\nabla}$ denotes any objective rate of the Cauchy stress, which is discussed later. The function $\bar{\mathfrak{M}}$ should be an objective function of the Cauchy stress tensor and the rate of deformation tensor. For most hypoelastic materials, the function $\bar{\mathfrak{M}}$ may be assumed to be linear between the objective Cauchy stress rate and the rate of deformation as shown below:

$$\boldsymbol{\sigma}^{\nabla} = \mathbb{C}(\sigma) : \mathbf{d} \tag{8.293}$$

where $\mathbb{C}(\boldsymbol{\sigma})$ is the fourth-order tensor of the elastic moduli depending on stress, and $\mathbf{d}$ represents the Eulerian rate of deformation tensor.

For hypoelastic-plastic materials, the rate-independent constitutive equation may be defined as follows:

$$\boldsymbol{\sigma}^{\nabla} = \mathbb{C} : \mathbf{d}^{el} \tag{8.294}$$

where the total Eulerian rate of deformation tensor $\mathbf{d}$ can be additively decomposed as follows:

$$\mathbf{d} = \mathbf{d}^{el} + \mathbf{d}^{pl}. \tag{8.295}$$

in which $\mathbf{d}^{el}$ and $\mathbf{d}^{pl}$ are the elastic and plastic parts respectively. In phenomenological hypoelastic-plastic materials with combined isotropic and kinematic hardening effects, the hypoelastic rate constitutive equations are based on the postulation that the elastic part of the rate of deformation tensor is hypoelastic with grade zero. The postulation may be accepted on the ground that the elastic strains are small compared to plastic strains and the large volumetric strains do not occur in the elastic deformation. Further a series of objective stress rates may be deduced using the Lie derivative.

According to Prager (Prager 1961), the constitutive relation of hypoelastic material is rate-independent and incrementally linear and reversible. In other words, for small increments about a finitely deformed state, the increments in stress and strain can be linearly related and are reversible upon unloading. However, for large deformation, energy is not necessarily conserved and the work done in a closed deformation path is not necessarily zero. Prager (Prager 1961) first suggested the use of the objective rate that Jaumann (Jaumann 1911) defined for the hypoelasticity rate constitutive equation. However, Dienes (Dienes 1979) first disclosed the anomaly of the Jaumann stress rate that the shear stress shows the spurious oscillation as the shear strain over 40% increases under pure shear load. Lee (Lee and Liu 1967) proposed a modified Zaremba-Jaumann-Noll rate to remedy the oscillation anomaly. Many researchers advocated the replacement of the Jaumann stress rate by the Green-Naghdi stress rate (Har 2007).

Atluri (Atluri 1984) discussed that the anomaly stemmed from improper simple generalizations of constitutive relations of infinitesimal strain to finite strain, and showed that the anomaly disappears with the use of the hypoelasticity of grade one instead of grade zero. He exploited the generalized objective rates with the hypoelasticity with grade one. He observed that in isotropic-hardening rigid-plasticity theory, the hypoelasticity with grade zero even does not show any oscillations of the shear stress under pure shear. He pointed out that in spite of the hypoelasticity with grade one, the Jaumann stress rate shows the oscillations in elastic-plastic classical phenomenological isotropic or kinematic hardening cases due to the elastic portion of the hypoelasticity rate constitutive equations. On the other hand, Simo and Pister (Simo and Pister 1984) proved that the use of the constant isotropic elasticity tensor of the infinitesimal in the hypoelasticity rate makes the material be inelastic due to the loss of the major symmetries of the elasticity tensors. Furthermore, the hypoelasticity rate constitutive equations in which the elastic response of the rate of deformation tensor is characterized are also no longer appropriate for an elastic material due to the major symmetry failure. Nonetheless, they concluded the assumption, particularly in the context of finite deformation metal plasticity, of the constant isotropic rate constitutive equation may be valid, provided large volumetric strains do not occur. In other words, the Jacobian should be almost one in the elastic region. It should be noted that the demonstrations given by Simo and Pister (Simo and Pister 1984) are contradictory to the arguments associated with the anomaly of the Jaumann stress rate as mentioned before.

8.3.4 Material Frame-Indifference: Objectivity

The notion of objectivity is very important in constructing a constitutive equation for a material undergoing a large deformation, because constitutive equations should be invariant under a change of frame. It should be noted that a change of frame does not mean a change of coordinate system. A change of frame is obtained when a rigid body motion is superimposed into the deformed configuration.

$$\mathbf{x}^* = \mathbf{c}(t) + \mathbf{Q}(t) \cdot \mathbf{x}(t), \quad \mathbf{x} \in \mathcal{S} \tag{8.296}$$

Then, the difference vector **a**, which is an Eulerian vector, between two positions $\mathbf{x}_1(t)$ and $\mathbf{x}_2(t)$ in the deformed body $\mathcal{S}$ can be written as

$$\mathbf{a} = \mathbf{x}_2(t) - \mathbf{x}_2(t) \tag{8.297}$$

The difference vector $\mathbf{a}^*$ between two spatial points $\mathbf{x}_1^*$ and $\mathbf{x}_2^*$ in the starred frame undergoing a superimposed rigid body motion is given by

$$\mathbf{a}^* = \mathbf{x}_2^* - \mathbf{x}_1^* = \mathbf{Q}(t) \cdot \mathbf{a} \tag{8.298}$$

Note that the two vectors do not share the same direction, but the same distance. Hence, under a change of frame the distance between the two points is preserved as follows:

$$d^* = \sqrt{\mathbf{a}^* \cdot \mathbf{a}^*} = \sqrt{\mathbf{a} \cdot \mathbf{a}} = d \tag{8.299}$$

where d^* and d denote the distances in the starred frame and in the unstarred frame. Therefore, the distance is frame-indifferent (objective) under a change of frame. Note that the scalar function $d(\mathbf{a})$ is (left) rotationally invariant in the sense that

$$d(\mathbf{a}) = d(\mathbf{Q}(t) \cdot \mathbf{a}) \tag{8.300}$$

where the scalar function d is objective.

In continuum mechanics, a frame is taken into account as an observer. Therefore, under a change of observer the difference vectors $\mathbf{a}^*$ and **a** are regarded as the same vector to each observer (Malvern 1969). Consequently, under a change of frame an objective vector **a** is transformed in the following manner

$$\mathbf{a}^* = \mathbf{Q}(t) \cdot \mathbf{a} \tag{8.301}$$

Next, consider a second-order tensor $\hat{\boldsymbol{\sigma}}$ (an Eulerian second-order tensor) mapping an objective vector **a** into an objective vector **b** as follows:

$$\mathbf{b} = \hat{\boldsymbol{\sigma}} \cdot \mathbf{a} \tag{8.302}$$

As the two vectors **a**, **b** are objective, we have

$$\begin{aligned} \mathbf{a}^* &= \mathbf{Q}(t) \cdot \mathbf{a}, \\ \mathbf{b}^* &= \mathbf{Q}(t) \cdot \mathbf{b} \end{aligned} \tag{8.303}$$

Thus, in view of Equations (8.302) and (8.303), we obtain

$$\begin{aligned} \mathbf{b}^* &= \mathbf{Q}(t) \cdot \mathbf{b} \\ &= \mathbf{Q}(t) \cdot \hat{\boldsymbol{\sigma}} \cdot \mathbf{a} \\ &= \mathbf{Q}(t) \cdot \hat{\boldsymbol{\sigma}} \cdot \mathbf{Q}^T(t) \cdot \mathbf{a}^* \end{aligned} \tag{8.304}$$

As a result of the above equation, if a second-order tensor $\hat{\boldsymbol{\sigma}}$ transforms under a change of frame in the following manner

$$\hat{\boldsymbol{\sigma}}^* = \mathbf{Q}(t) \cdot \hat{\boldsymbol{\sigma}} \cdot \mathbf{Q}^T(t) \tag{8.305}$$

then the second-order tensor $\hat{\boldsymbol{\sigma}}$ is said to be frame-indifferent or objective.

Cauchy Stress Tensor As a representative form of Equation (8.302), we can recall Cauchy's first law, that is,

$$\mathbf{t} = \boldsymbol{\sigma} \cdot \hat{\mathbf{n}} \tag{8.306}$$

where $\mathbf{t}$, $\boldsymbol{\sigma}$ and $\hat{\mathbf{n}}$ denote the traction, the Cauchy stress tensor and the unit vector outward normal to the surface of the body. Hence, in view of Equation (8.305) the Cauchy stress tensor is objective, that is,

$$\boldsymbol{\sigma}^* = \mathbf{Q}(t) \cdot \boldsymbol{\sigma} \cdot \mathbf{Q}^T(t) \tag{8.307}$$

Deformation Gradient Tensor Let us consider the objectivity of the deformation gradient tensor defined in Equation (8.35). The vectorial line segment in the reference configuration is transformed into one in the spatial configuration by the deformation gradient tensor. Then, under a change of frame the vectorial line segment in the starred frame can be written as

$$\begin{aligned} d\mathbf{x}^* &= \mathbf{Q}(t) \cdot d\mathbf{x} \\ &= \mathbf{Q}(t) \cdot \mathbf{F} \cdot d\mathbf{X} \end{aligned} \tag{8.308}$$

Since $d\mathbf{x}^* = \mathbf{F}^* d\mathbf{X}$, in view of the above equation, the deformation gradient tensor is frame-indifferent (Belytschko et al. 2000; Holzapfel 2000). Note that since the deformation gradient tensor is a two-point tensor, it transforms objectively like an objective vector as follows:

$$\mathbf{F}^* = \mathbf{Q}(t) \cdot \mathbf{F} \tag{8.309}$$

Green Deformation Tensor Under spatially superimposed rigid body motions, the Green deformation tensor remains unchanged, that is,

$$\mathbf{C}^* = (\mathbf{F}^*)^T \cdot \mathbf{F}^* = \mathbf{F}^T \cdot \mathbf{Q}^T \cdot \mathbf{Q} \cdot \mathbf{F} = \mathbf{F}^T \cdot \mathbf{F} = \mathbf{C} \tag{8.310}$$

It should be noted that all Lagrangian measures such as $\mathbf{E}$, $\mathbf{S}$, $\mathbf{U}$ and $\mathbf{B}$ remain unchanged under spatially superimposed rigid body motions. All Lagrangian second-order tensors are objective (Belytschko et al. 2000; Holzapfel 2000; Simo and Hughes 1998); hence they can be used to describe material responses (Holzapfel 2000; Simo and Hughes 1998).

Rate of Deformation Tensor Consider the spatial velocity gradient tensor $\mathbf{l}$. Under superimposed rigid body motions, the Eulerian tensor can be given by

$$\mathbf{l}^* = \dot{\mathbf{F}}^* \cdot (\mathbf{F}^*)^{-1} \tag{8.311}$$

By use of Equation (8.309), the above equation can be rewritten as

$$\mathbf{l}^* = \dot{\mathbf{Q}} \cdot \mathbf{Q}^T + \mathbf{Q} \cdot \mathbf{l} \cdot \mathbf{Q}^T \tag{8.312}$$

Thus, the spatial velocity gradient tensor is not objective. From the above equation, the transpose of $\mathbf{l}^*$ can be written as

$$(\mathbf{l}^*)^T = \mathbf{Q} \cdot \dot{\mathbf{Q}}^T + \mathbf{Q} \cdot \mathbf{l}^T \cdot \mathbf{Q}^T \tag{8.313}$$

As a consequence of the above, the spin tensor in the starred frame $\mathbf{w}^*$ is given by

$$\begin{aligned} \mathbf{w}^* &= \frac{1}{2}(\mathbf{l}^* - (\mathbf{l}^*)^T) \\ &= \frac{1}{2}(\dot{\mathbf{Q}} \cdot \mathbf{Q}^T - \mathbf{Q} \cdot \dot{\mathbf{Q}}^T) + \mathbf{Q} \cdot \mathbf{w} \cdot \mathbf{Q}^T \end{aligned} \tag{8.314}$$

Hence, the spin tensor is not objective either. However, the rate of deformation tensor is objective as follows:

$$\begin{aligned}\mathbf{d}^* &= \frac{1}{2}(\mathbf{l}^* + (\mathbf{l}^*)^T) \\ &= \mathbf{Q} \cdot \mathbf{d} \cdot \mathbf{Q}^T\end{aligned} \tag{8.315}$$

since

$$\frac{d}{dt}(\mathbf{Q} \cdot \mathbf{Q}^T) = \dot{\mathbf{Q}} \cdot \mathbf{Q}^T + \mathbf{Q} \cdot \dot{\mathbf{Q}}^T = 0 \tag{8.316}$$

Co-rotational Stress Rates Consider an objective vector **a** in a spatial frame, which transforms objectively as follows:

$$\mathbf{a}^* = \mathbf{Q} \cdot \mathbf{a} \tag{8.317}$$

Then, the material time derivative of $\mathbf{a}^*$ can be written as

$$\dot{\mathbf{a}}^* = \dot{\mathbf{Q}} \cdot \mathbf{a} + \mathbf{Q} \cdot \dot{\mathbf{a}} \tag{8.318}$$

Hence, it is easy to find that $\dot{\mathbf{a}}$ is not objective. From Equations (8.314) and (8.316), we obtain the following

$$\dot{\mathbf{Q}} = \mathbf{w}^* \cdot \mathbf{Q} - \mathbf{Q} \cdot \mathbf{w} \tag{8.319}$$

since $(\mathbf{w}^*)^T = -\mathbf{w}^*$. Substituting Equation (8.319) into Equation (8.318), we obtain

$$(\dot{\mathbf{a}} - \mathbf{w} \cdot \mathbf{a})^* = \mathbf{Q}(\dot{\mathbf{a}} - \mathbf{w} \cdot \mathbf{a}) \tag{8.320}$$

which can be expressed as

$$\mathring{\mathbf{a}}^* = \mathbf{Q} \cdot \mathring{\mathbf{a}} \tag{8.321}$$

where the co-rotational rate of the objective vector **a** is defined as

$$\mathring{\mathbf{a}} = \dot{\mathbf{a}} - \mathbf{w} \cdot \mathbf{a} \tag{8.322}$$

Note that the co-rotational rate of an objective vector is objective, namely, frame-indifferent.

Next, consider the co-rotational rate of an objective second-order tensor given in Equation (8.305). Then its material time derivative of $\boldsymbol{\sigma}^*$ can be written as

$$\dot{\boldsymbol{\sigma}}^* = \dot{\mathbf{Q}} \cdot \boldsymbol{\sigma} \cdot \mathbf{Q}^T + \mathbf{Q} \cdot \dot{\boldsymbol{\sigma}} \cdot \mathbf{Q}^T + \mathbf{Q} \cdot \boldsymbol{\sigma} \cdot \dot{\mathbf{Q}}^T \tag{8.323}$$

From Equation (8.319), we have

$$\dot{\mathbf{Q}}^T = -\mathbf{Q}^T \cdot \mathbf{w}^* + \mathbf{w} \cdot \mathbf{Q}^T \tag{8.324}$$

Substituting Equations (8.319) and (8.324) into Equation (8.323), we obtain

$$(\dot{\boldsymbol{\sigma}} - \mathbf{w} \cdot \boldsymbol{\sigma} + \boldsymbol{\sigma} \cdot \mathbf{w})^* = \mathbf{Q} \cdot (\dot{\boldsymbol{\sigma}} - \mathbf{w} \cdot \boldsymbol{\sigma} + \boldsymbol{\sigma} \cdot \mathbf{w}) \cdot \mathbf{Q}^T \tag{8.325}$$

which can be expressed as

$$\mathring{\boldsymbol{\sigma}}^* = \mathbf{Q} \cdot \mathring{\boldsymbol{\sigma}} \cdot \mathbf{Q}^T \tag{8.326}$$

where the co-rotational rate of the objective vector **a** is defined as

$$\mathring{\boldsymbol{\sigma}} = \dot{\boldsymbol{\sigma}} - \mathbf{w} \cdot \boldsymbol{\sigma} + \boldsymbol{\sigma} \cdot \mathbf{w} \tag{8.327}$$

which is called the Jaumann stress rate. Note that the co-rotational rate of an objective second-order tensor is objective, namely, frame-indifferent. The notion of objectivity is specially important in discussing objective stress rates such as co-rotational stress rates and convected stress rates, because they are all objective (Holzapfel 2000; Malvern 1969; Simo and Hughes 1998).

8.3.5 Objective Stress Rates

The pull-back of the contravariant Kirchhoff stress tensor becomes the second Piola-Kirchhoff stress tensor as shown in Equation (8.167). Likewise, the pull-back of the contravariant Cauchy stress tensor leads to the contravariant convected stress tensor $\mathbf{\Sigma}^{\natural}$

$$\mathbf{\Sigma}^{\natural} = \varphi^{*}\left(\boldsymbol{\sigma}^{\natural}\right) \tag{8.328}$$

Furthermore, the push-forward of the rate of the contravariant convected stress tensor in the direction $\mathbf{v}$ leads to the Lie derivative of the contravariant Cauchy stress tensor as

$$L_{v}\left(\boldsymbol{\sigma}^{\natural}\right) = \varphi_{*}\left[\frac{\partial}{\partial t}\left\{\varphi^{*}\left(\boldsymbol{\sigma}^{\natural}\right)\right\}\right] = \varphi_{*}\left(\dot{\mathbf{\Sigma}}^{\natural}\right) = \mathbf{F}\cdot\dot{\mathbf{\Sigma}}^{\natural}\cdot\mathbf{F}^{T} \tag{8.329}$$

Hence, the Lie derivative of the contravariant Cauchy stress tensor is equivalent to the push-forward of the material time derivative of the contravariant convected stress tensor $\mathbf{\Sigma}^{\natural}$. The above equation may be rewritten as

$$\begin{aligned}\mathbf{F}\cdot\dot{\mathbf{\Sigma}}^{\natural}\cdot\mathbf{F}^{T} &= \mathbf{F}\cdot\frac{d}{dt}\left(\mathbf{F}^{-1}\cdot\boldsymbol{\sigma}^{\natural}\cdot\mathbf{F}^{-T}\right)\cdot\mathbf{F}^{T} \\ &= \mathbf{F}\cdot\left(\dot{\mathbf{F}}^{-1}\cdot\boldsymbol{\sigma}^{\natural}\cdot\mathbf{F}^{-T}+\mathbf{F}^{-1}\cdot\dot{\boldsymbol{\sigma}}^{\natural}\cdot\mathbf{F}^{-T}+\mathbf{F}^{-1}\cdot\boldsymbol{\sigma}^{\natural}\cdot\dot{\mathbf{F}}^{-T}\right)\cdot\mathbf{F}^{T}\end{aligned} \tag{8.330}$$

The above equation can be recast as

$$\mathbf{F}\cdot\dot{\mathbf{\Sigma}}^{\natural}\cdot\mathbf{F}^{T} = \dot{\boldsymbol{\sigma}}^{\natural}+\mathbf{F}\cdot\dot{\mathbf{F}}^{-1}\cdot\boldsymbol{\sigma}^{\natural}+\boldsymbol{\sigma}^{\natural}\cdot\dot{\mathbf{F}}^{-T}\cdot\mathbf{F}^{T} \tag{8.331}$$

Note that the following holds

$$\mathbf{F}\cdot\dot{\mathbf{F}}^{-1} = -\dot{\mathbf{F}}\cdot\mathbf{F}^{-1}, \quad \therefore\ \mathbf{F}\cdot\mathbf{F}^{-1} = \mathbf{I} \tag{8.332}$$

In view of Equation (8.48), we have the equation, $\dot{\mathbf{F}} = \mathbf{l}\cdot\mathbf{F}$. Consequently, we find

$$\mathbf{F}\cdot\dot{\mathbf{\Sigma}}^{\natural}\cdot\mathbf{F}^{T} = \dot{\boldsymbol{\sigma}}^{\natural}-\mathbf{l}\cdot\boldsymbol{\sigma}^{\natural}-\boldsymbol{\sigma}^{\natural}\cdot\mathbf{l}^{T} = \left(\boldsymbol{\sigma}^{\nabla O}\right)^{\natural} \tag{8.333}$$

where $\left(\boldsymbol{\sigma}^{\nabla O}\right)^{\natural}$ is called the Oldroyd stress rate of the contravariant Cauchy stress rate. Further, the push-forward of the rate of the second Piola-Kirchhoff stress tensor in the direction v leads to the Lie derivative of the Kirchhoff stress tensor as

$$L_{v}\left(\boldsymbol{\tau}^{\natural}\right) = \varphi_{*}\left[\frac{\partial}{\partial t}\left\{\varphi^{*}\left(\boldsymbol{\tau}^{\natural}\right)\right\}\right] = \varphi_{*}\left(\dot{\mathbf{S}}^{\natural}\right) = \mathbf{F}\cdot\dot{\mathbf{S}}^{\natural}\cdot\mathbf{F}^{T} \tag{8.334}$$

The above equation may be rewritten as

$$L_{v}\left(\boldsymbol{\tau}^{\natural}\right) = \dot{\boldsymbol{\tau}}-\mathbf{l}\cdot\boldsymbol{\tau}-\boldsymbol{\tau}\cdot\mathbf{l}^{T} = \left(\boldsymbol{\tau}^{\nabla T}\right)^{\natural} \tag{8.335}$$

which is called the Truesdell stress rate of the contravariant Kirchhoff stress tensor (Simo and Hughes 1998). The Truesdell stress rate can be rewritten in terms of the contravariant Cauchy stress tensor as

$$L_v\left(\boldsymbol{\tau}^\natural\right) = \left(\boldsymbol{\tau}^{\nabla T}\right)^\natural = J\left(\left(\boldsymbol{\sigma}^\nabla\right)^\natural + (tr\mathbf{l})\,\boldsymbol{\sigma}\right) = J\left(\boldsymbol{\sigma}^{\nabla T}\right)^\natural \tag{8.336}$$

where $\left(\boldsymbol{\sigma}^{\nabla T}\right)^\natural$ is called the Truesdell stress rate of the contravariant Cauchy stress tensor (Belytschko et al. 2000).

In the literature, it is found that there are several forms for the Lie derivative of the Kirchhoff stress tensor. The Cotter-Rivlin stress rate results from the Lie derivative of the covariant Kirchhoff stress tensor

$$\boldsymbol{\tau}^{\nabla CR} = L_v\left(\boldsymbol{\tau}^\flat = \tau_{ij}\mathbf{g}^i \otimes \mathbf{g}^j\right) = \varphi_*\left[\frac{\partial}{\partial t}\left\{\varphi^*\left(\boldsymbol{\tau}^\flat\right)\right\}\right] \tag{8.337}$$

In view of Equation (8.122) and (8.124), the above equation can be given as

$$\varphi_*\left[\frac{\partial}{\partial t}\left\{\varphi^*\left(\boldsymbol{\tau}^\flat\right)\right\}\right] = \mathbf{F}^{-T}\cdot\frac{d}{dt}\left(\mathbf{F}^T\cdot\boldsymbol{\tau}^\flat\cdot\mathbf{F}\right)\cdot\mathbf{F}^{-1} \tag{8.338}$$

Hence, we find that

$$\begin{aligned} &\mathbf{F}^{-T}\cdot\frac{d}{dt}\left(\mathbf{F}^T\cdot\boldsymbol{\tau}^\flat\cdot\mathbf{F}\right)\cdot\mathbf{F}^{-1} \\ &= \mathbf{F}^{-T}\cdot\left(\dot{\mathbf{F}}^T\cdot\boldsymbol{\tau}^\flat\cdot\mathbf{F} + \mathbf{F}^T\cdot\dot{\boldsymbol{\tau}}^\flat\cdot\mathbf{F} + \mathbf{F}^T\cdot\boldsymbol{\tau}^\flat\cdot\dot{\mathbf{F}}\right)\cdot\mathbf{F}^{-1} \end{aligned} \tag{8.339}$$

The above equation can be recast as

$$\mathbf{F}\cdot\dot{\boldsymbol{\Sigma}}^\natural\cdot\mathbf{F}^T = \dot{\boldsymbol{\tau}}^\flat + \mathbf{l}^T\cdot\boldsymbol{\tau}^\flat + \boldsymbol{\tau}^\flat\cdot\mathbf{l} = \left(\boldsymbol{\tau}^{\nabla CR}\right)^\flat \tag{8.340}$$

where $\left(\boldsymbol{\tau}^{\nabla CR}\right)^\flat$ is called the Cotter-Rivlin stress rate of the Kirchhoff stress tensor. The Jaumann stress rate, the most popular objective stress rate, can be obtained by the following. First we take the Lie derivative of the mixed Kirchhoff stress tensor

$$L_v\left(\boldsymbol{\tau} = \tau^i_{.j}\mathbf{g}_i \otimes \mathbf{g}^j\right) = \boldsymbol{\tau}^{\nabla m1} = \dot{\boldsymbol{\tau}} - \mathbf{l}\cdot\boldsymbol{\tau} + \boldsymbol{\tau}\cdot\mathbf{l} \tag{8.341}$$

and

$$L_v\left(\boldsymbol{\tau} = \tau_{i.}^{\ j}\mathbf{g}^i \otimes \mathbf{g}_j\right) = \boldsymbol{\tau}^{\nabla m2} = \dot{\boldsymbol{\tau}} + \mathbf{l}^T\cdot\boldsymbol{\tau} - \boldsymbol{\tau}\cdot\mathbf{l}^T \tag{8.342}$$

Then, taking the average leads to

$$\boldsymbol{\tau}^{\nabla J} = \frac{1}{2}\left(\boldsymbol{\tau}^{\nabla m1} + \boldsymbol{\tau}^{\nabla m2}\right) = \dot{\boldsymbol{\tau}} + \boldsymbol{\tau}\cdot\frac{1}{2}\left(\mathbf{l} - \mathbf{l}^T\right) - \frac{1}{2}\left(\mathbf{l} - \mathbf{l}^T\right)\cdot\boldsymbol{\tau} \tag{8.343}$$

Thus, we find that

$$\boldsymbol{\tau}^{\nabla J} = \dot{\boldsymbol{\tau}} + \boldsymbol{\tau}\cdot\mathbf{w} - \mathbf{w}\cdot\boldsymbol{\tau} \tag{8.344}$$

In terms of the Cauchy stress tensor, the Jaumann stress tensor can be written as

$$\boldsymbol{\sigma}^{\nabla J} = \dot{\boldsymbol{\sigma}} + \boldsymbol{\sigma}\cdot\mathbf{w} - \mathbf{w}\cdot\boldsymbol{\sigma} \tag{8.345}$$

Due to the polar decomposition, the rotation tensor $\mathbf{R}$, a two point tensor, can be written as

$$\mathbf{R} = \mathbf{F} \cdot \mathbf{U}^{-1} \tag{8.346}$$

In view of Equation (8.124), the push-back of the contravariant Kirchhoff tensor under the rotation part $\mathbf{R}$ of $\mathbf{F}$ leads to

$$\mathbf{R}^* \left(\boldsymbol{\tau}^\natural = \tau^{ij} \mathbf{g}_i \otimes \mathbf{g}_j \right) = \mathbf{R}^{-1} \cdot \boldsymbol{\tau}^\natural \cdot \mathbf{R}^{-T} \tag{8.347}$$

Since the rotation tensor is orthogonal, we know that $\mathbf{R}^{-1} = \mathbf{R}^T, \quad \mathbf{R} = \mathbf{R}^{-T}$. The above equation can be given by

$$\mathbf{R}^* \left(\boldsymbol{\tau}^\natural = \tau^{ij} \mathbf{g}_i \otimes \mathbf{g}_j \right) = \mathbf{R}^T \cdot \boldsymbol{\tau}^\natural \cdot \mathbf{R} = \boldsymbol{\Upsilon}^\natural = \Upsilon^{IJ} \mathbf{G}_I \otimes \mathbf{G}_J \tag{8.348}$$

where $\boldsymbol{\Upsilon}^\natural$ is referred to as the unrotated Kirchhoff stress tensor. Note that the push-back of the contravariant Kirchhoff tensor results in a Lagrangian contravariant tensor and the push-back of the contravariant Kirchhoff tensor results in a Lagrangian covariant tensor as follows:

$$\mathbf{R}^* \left(\boldsymbol{\tau}^\flat = \tau_{ij} \mathbf{g}^i \otimes \mathbf{g}^j \right) = \mathbf{R}^T \cdot \boldsymbol{\tau}^\flat \cdot \mathbf{R} = \boldsymbol{\Upsilon}^\flat = \Upsilon_{IJ} \mathbf{G}^I \otimes \mathbf{G}^J \tag{8.349}$$

The Lie derivative of the contravariant Kirchhoff stress tensor leads to

$$L_v^R \left(\boldsymbol{\tau}^\natural \right) = \mathbf{R}_* \left[\frac{\partial}{\partial t} \left\{ \mathbf{R}^* \left(\boldsymbol{\tau}^\natural \right) \right\} \right] = \dot{\boldsymbol{\tau}}^\natural + \boldsymbol{\tau}^\natural \cdot \boldsymbol{\Omega} - \boldsymbol{\Omega} \cdot \boldsymbol{\tau}^\natural = \left(\boldsymbol{\tau}^{\nabla G} \right)^\natural \tag{8.350}$$

where $\boldsymbol{\Omega}$ represents the angular velocity defined in Equation (8.85). The above equation is called the Green-McInnis-Naghdi stress rate or simply the Green-Naghdi stress rate of the Kirchhoff stress. Likewise, the Lie derivative of the contravariant Cauchy stress tensor leads to

$$L_v^R \left(\boldsymbol{\sigma}^\natural \right) = \mathbf{R}_* \left[\frac{\partial}{\partial t} \left\{ \mathbf{R}^* \left(\boldsymbol{\sigma}^\natural \right) \right\} \right] = \dot{\boldsymbol{\sigma}}^\natural + \boldsymbol{\sigma}^\natural \cdot \boldsymbol{\Omega} - \boldsymbol{\Omega} \cdot \boldsymbol{\sigma}^\natural = \left(\boldsymbol{\sigma}^{\nabla G} \right)^\natural \tag{8.351}$$

Suppose that the rate of the right stretch tensor is zero as shown in Equation (8.95). Then, the angular velocity may be replaced by the spin tensor. Consequently, the Green-Naghdi stress rate becomes equal to the Jaumann stress rate in Equation (8.345).

8.4 VIRTUAL WORK AND VARIATIONAL PRINCIPLES

8.4.1 Virtual Work and Potential Energy

Above all, it is crucial for one to grasp the notion of virtual work to understand the basics underlying the principle of virtual work. We also have to stress the importance of the notion of the potential energy in continuum mechanics. Then, we shall examine the relation between the variations of both the virtual work and the potential energy.

In the Eulerian description, we consider the definition of the work done by the external forces acting on a deformable continuous body. The infinitesimal work done by the prescribed traction $\bar{\mathbf{t}}(\mathbf{x}, t, \hat{\mathbf{n}}) : \partial\mathcal{S}_\sigma \times \bar{\mathcal{T}} \times \mathbb{R}^3 \to \mathbb{R}^3$ per unit area on the infinitesimal surface da is defined as

$$d\mathcal{W}_{\text{ext}_{\partial\mathcal{S}_\sigma}}(\mathbf{x}, t) = \int_{\partial\mathcal{S}_\sigma} \bar{\mathbf{t}}(\mathbf{x}, t, \hat{\mathbf{n}}) \cdot d\mathbf{u}(\mathbf{x}, t)\, da, \quad \text{for } \mathbf{x} \in \partial\mathcal{S}_\sigma \tag{8.352}$$

where $d\mathbf{u}(\mathbf{x},t) : \partial\mathcal{S}_\sigma \times \bar{\mathcal{T}} \rightarrow \mathbb{R}^3$ denotes the infinitesimal displacement. The infinitesimal work done by the body force $\mathbf{b}(\mathbf{x},t)$ per unit mass in the body is defined as

$$d\mathcal{W}_{\text{ext}_\mathcal{S}}(\mathbf{x},t) = \int_\mathcal{S} \rho\mathbf{b}(\mathbf{x},t)\cdot d\mathbf{u}(\mathbf{x},t)\,dv, \quad \text{for } \mathbf{x}\in\mathcal{S} \tag{8.353}$$

Note that the external forces $\bar{\mathbf{t}}(\mathbf{x},t,\hat{\mathbf{n}})$, $\mathbf{b}(\mathbf{x},t)$ in Equation (8.352) and (8.353) do depend on time explicitly. Then, it is said that the external work $\mathcal{W}_{\text{nc}}(\mathbf{u},t)$ is nonconservative, and we have

$$\mathcal{W}_{\text{nc}}(\mathbf{u},t) = \int_{\partial\mathcal{S}_\sigma} \bar{\mathbf{t}}(\mathbf{x},t,\hat{\mathbf{n}})\cdot\mathbf{u}(\mathbf{x},t)\,da + \int_\mathcal{S}\rho\mathbf{b}(\mathbf{x},t)\cdot\mathbf{u}(\mathbf{x},t)\,dv \tag{8.354}$$

In the case that the external forces are constant, namely, dead loads, the external work not depending on both time and motion can be given by

$$\mathcal{W}_{\text{ext}}(\mathbf{u}) = \int_{\partial\mathcal{S}_\sigma}\bar{\mathbf{t}}\cdot\mathbf{u}(\mathbf{x},t)\,da + \int_\mathcal{S}\rho\mathbf{b}\cdot\mathbf{u}(\mathbf{x},t)\,dv = \mathcal{W}_{\text{con}}(\mathbf{u}) \tag{8.355}$$

Note that the external forces in Equation (8.355), $\bar{\mathbf{t}}$ and $\mathbf{b}$, above do not depend on both time and motion explicitly. In other words, they remain constant all the time. In such a case, it is said that the external work is conservative. Thus, Equation (8.355) is the expression of the conservative work $\mathcal{W}_{\text{con}}(\mathbf{u})$.

Next, let us begin with the conservative work. First, consider that the admissible neighboring displacement $\tilde{\boldsymbol{u}}(\mathbf{x},t) : \mathcal{S}\times\bar{\mathcal{T}}\rightarrow\mathbb{R}^3$ satisfying the boundary conditions can be given as

$$\tilde{\boldsymbol{u}} = \mathbf{u} + \varepsilon\eta = \mathbf{u} + \delta\mathbf{u} = \mathbf{x} - \mathbf{X} + \delta\mathbf{x} \tag{8.356}$$

where $\delta\mathbf{u}$ is the virtual displacement, ε an infinitesimal parameter and $\eta(\mathbf{x},t) : \mathcal{S}\times\bar{\mathcal{T}}\rightarrow\mathbb{R}^3$ an admissible vector satisfying the boundary conditions. It is very important to note that the dependent variable is the displacement $\mathbf{u}$ and the independent variables $\mathbf{X}$, t are held fixed. The virtual displacement $\delta\mathbf{u}$ is not real. The external work at the admissible neighboring displacement may be expanded by the Taylor's series expansion about $\varepsilon = 0$ as

$$\mathcal{W}_{\text{ext}}(\tilde{\boldsymbol{u}}) = \mathcal{W}_{\text{ext}}(\tilde{\boldsymbol{u}})\big|_{\varepsilon=0} + \varepsilon\left.\frac{d\mathcal{W}_{\text{ext}}(\tilde{\boldsymbol{u}})}{d\varepsilon}\right|_{\varepsilon=0} + \frac{\varepsilon^2}{2}\left.\frac{d^2\mathcal{W}_{\text{ext}}(\tilde{\boldsymbol{u}})}{d\varepsilon^2}\right|_{\varepsilon=0} + \ldots \tag{8.357}$$

The above equation can be rewritten in the form

$$\mathcal{W}_{\text{ext}}(\tilde{\boldsymbol{u}}) = \mathcal{W}_{\text{ext}}(\mathbf{u}) + \delta\mathcal{W}_{\text{ext}}(\mathbf{u},\delta\mathbf{u}) + \frac{1}{2}\delta^2\mathcal{W}_{\text{ext}}(\mathbf{u},\delta\mathbf{u}) + \ldots \tag{8.358}$$

The variation of the external work may be described by the Gâteaux derivative (or called the directional derivative),

$$\begin{aligned}\delta\mathcal{W}_{\text{ext}}(\mathbf{u},\delta\boldsymbol{u}) &= \varepsilon\frac{d}{d\varepsilon}\mathcal{W}_{\text{ext}}(\mathbf{u}+\varepsilon\eta)|_{\varepsilon=0}\\ &= \varepsilon D_\mathbf{u}\mathcal{W}_{\text{ext}}(\mathbf{u})\cdot\eta\\ &= D_\mathbf{u}\mathcal{W}_{\text{ext}}(\mathbf{u})\cdot\delta\mathbf{u}\end{aligned} \tag{8.359}$$

Therefore, the above equation is called the Gâteaux variation of the external work at $\mathbf{u} \in \mathbb{R}^3$ in the direction $\boldsymbol{\eta} \in \mathbb{R}^3$. Simply, we have

$$\delta \mathcal{W}_{\text{ext}}(\mathbf{u}, \delta\mathbf{u}) = D_{\mathbf{u}} \mathcal{W}_{\text{ext}}(\mathbf{u}) \cdot \delta\mathbf{u} \tag{8.360}$$

where $D_u \mathcal{W}_{\text{ext}}(\mathbf{u})$ is called the Gâteaux operator (Holzapfel 2000; Marsden and Hughes 1983), and $D_u \mathcal{W}_{\text{ext}}(\mathbf{u}) \cdot \delta\boldsymbol{\eta}$ is the directional derivative or the Gâteaux derivative (Ciarlet 1978; Holzapfel 2000; Simo and Hughes 1998). The corresponding external virtual work as the first variation of the external work can be described as

$$\delta \mathcal{W}_{\text{ext}}(\mathbf{u}, \delta\mathbf{u}) = \int_{\partial S_\sigma} \bar{\mathbf{t}} \cdot \delta\mathbf{u}(\mathbf{x}, t)\, da + \int_S \rho\mathbf{b} \cdot \delta\mathbf{u}(\mathbf{x}, t)\, dv \tag{8.361}$$

which means that the actual external forces such as the traction and the body forces do work on the fictitious displacement, the so-called virtual displacement. As mentioned previously, the external forces given in Equation (8.355) are dead loads. The dead load is one case of conservative loading. In general, the external force must be derivable from the external potential energy (Washizu 1982). Let us introduce physical state variables, namely, the external potential energy density functions as follows:

$$_{\bar{\mathbf{t}}}u_{\text{ext}}(\mathbf{u}) = -\bar{\mathbf{t}} \cdot \mathbf{u}(\mathbf{x}, t), \quad _{\mathbf{b}}u_{\text{ext}}(\mathbf{u}) = -\rho\mathbf{b} \cdot \mathbf{u}(\mathbf{x}, t) \tag{8.362}$$

where the traction and the body forces are functions of time and motion. Note that the density of the material is assumed to be constant. Then, the external forces are derivable, namely,

$$\bar{\mathbf{t}} = -\frac{\partial_{\bar{\mathbf{t}}}u_{\text{ext}}(\mathbf{u})}{\partial\mathbf{u}}, \quad \rho\mathbf{b} = -\frac{\partial_{\mathbf{b}}u_{\text{ext}}(\mathbf{u})}{\partial\mathbf{u}} \tag{8.363}$$

Then, Equation (8.361) can be rewritten as

$$\begin{aligned}
\delta \mathcal{W}_{\text{ext}}(\mathbf{u}) &= \int_{\partial S_\sigma} \delta\left(\bar{\mathbf{t}} \cdot \mathbf{u}(\mathbf{x}, t)\right) da + \int_S \delta\left(\rho\mathbf{b} \cdot \mathbf{u}(\mathbf{x}, t)\right) dv \\
&= -\int_{\partial S_\sigma} \delta\left(_{\bar{\mathbf{t}}}u_{\text{ext}}(\mathbf{u})\right) da - \int_S \delta\left(_{\mathbf{b}}u_{\text{ext}}(\mathbf{u})\right) dv \\
&= -\delta \int_{\partial S_\sigma} \left(_{\bar{\mathbf{t}}}u_{\text{ext}}(\mathbf{u})\right) da - \delta \int_S \left(_{\mathbf{b}}u_{\text{ext}}(\mathbf{u})\right) dv \\
&= -\delta \mathcal{U}_{\text{ext}}(\mathbf{u})
\end{aligned} \tag{8.364}$$

Therefore, when the external forces are dead loads, we have the relation between the external potential energy and the external work as well as the one between their variations. The external potential energy and its variation are given by,

$$\mathcal{U}_{\text{ext}}(\mathbf{u}) = -\mathcal{W}_{\text{ext}}(\mathbf{u}), \quad \delta\mathcal{U}_{\text{ext}}(\mathbf{u}) = -\delta\mathcal{W}_{\text{ext}}(\mathbf{u}) \tag{8.365}$$

where the minus sign implies the loss of the potential energy. For nonconservative loading cases where the external forces do depend on time and motion, the potential energy cannot be assured to exist. Hence, we have

$$\delta \mathcal{W}_{\text{nc}}(\mathbf{u}, \delta\mathbf{u}, t) = \int_{\partial S_\sigma} \bar{\mathbf{t}}(\mathbf{x}, t, \hat{\mathbf{n}}) \cdot \delta\mathbf{u}(\mathbf{x}, t)\, da + \int_S \rho\mathbf{b}(\mathbf{x}, t) \cdot \delta\mathbf{u}(\mathbf{x}, t)\, dv \tag{8.366}$$

where $\delta\mathcal{W}_{nc}$ denotes the variation of the nonconservative work, and the external forces, $\bar{\mathbf{t}}(\mathbf{x}, t, \hat{\mathbf{n}})$ and $\mathbf{b}(\mathbf{x}, t)$, are said to be nonconservative.

The internal potential energy by the internal forces in both elastodynamics and elastostatics is given by

$$\mathcal{U}_{\text{int}}(\mathbf{u}) = \int_{\mathcal{S}} \rho\psi(\text{grad}\,\mathbf{u})\,dv = \frac{1}{2}\int_{\mathcal{S}} \boldsymbol{\sigma} : \mathbf{e}\,dv, \quad \text{in } \mathcal{S}, \tag{8.367}$$

where $\psi(\text{grad}\,\mathbf{u})$ is the strain energy function or stored energy function per unit volume in the spatial configuration and $\text{grad}\,\mathbf{u}$ represents the gradient of the displacement in the spatial configuration. In Equation (8.367), it is shown that the strain energy density function is expressed in terms of the Cauchy stress tensor and the Eulerian strain tensor. Note that the Cauchy stress tensor is conjugate to the Eulerian strain tensor. Then, the relation between the internal potential energy and the internal work is defined as

$$\mathcal{W}_{\text{int}}(\mathbf{u}) = \mathcal{U}_{\text{int}}(\mathbf{u}) \tag{8.368}$$

and its variation can be expressed as

$$\delta\mathcal{W}_{\text{int}}(\mathbf{u}, \delta\mathbf{u}) = \delta\mathcal{U}_{\text{int}}(\mathbf{u}, \delta\mathbf{u}) = \int_{\mathcal{S}} \boldsymbol{\sigma} : \delta\mathbf{e}\,dv \tag{8.369}$$

In solid mechanics, the total potential energy with conservative loadings is defined as

$$\mathcal{U}(\mathbf{u}) = \mathcal{U}_{\text{int}}(\mathbf{u}) + \mathcal{U}_{\text{ext}}(\mathbf{u}) = \mathcal{W}_{\text{int}}(\mathbf{u}) - \mathcal{W}_{\text{ext}}(\mathbf{u}) \tag{8.370}$$

Consequently, the total potential energy can be described as

$$\mathcal{U}(\mathbf{u}) = \int_{\mathcal{S}} \rho\psi(\text{grad}\,\mathbf{u})\,dv - \int_{\partial\mathcal{S}_\sigma} \bar{\mathbf{t}}\cdot\mathbf{u}(\mathbf{x}, t)\,da - \int_{\mathcal{S}} \rho\mathbf{b}\cdot\mathbf{u}(\mathbf{x}, t)\,dv \tag{8.371}$$

where the external potential energy exists when the external forces are dead loads.

8.4.2 Principle of Virtual Work

The principle of virtual work in continuum dynamics plays an important role in the conventional finite element formulation in that admissible trial and test functions are applied into the principle, the weak form for the initial boundary-value problems, to discretize the space. It can be shown that most of variational principles can be derived from the principle of virtual work. Further, the principle of virtual work can be used for linear or nonlinear or conservative or nonconservative systems.

Bubnov-Galerkin Weighted Residual Form In Section 3.2.2, we have shown that the principle of virtual work in particle dynamics is deduced from the Lagrangian form of D'Alembert's principle. Firstly, D'Alembert's principle in Equation (3.52) is established by use of Newton's second law of motion. The inner product of D'Alembert's principle and the virtual displacement leads to the Lagrangian form of D'Alembert's principle.

As an analogy of D'Alembert's principle in particle dynamics, in continuum dynamics with Cauchy's equations of motion, the local form of momentum balance equations, can be written in the Eulerian description as

$$\text{div}\,\boldsymbol{\sigma} + \rho\mathbf{b} - \rho\ddot{\mathbf{u}} = \mathbf{0}, \quad \text{in } \mathcal{S} \tag{8.372}$$

The second-order time differential equations above are subject to two boundary conditions, i.e., the natural boundary condition (the mechanical boundary condition)

$$\mathbf{t}\left(\mathbf{x}, t, \hat{\mathbf{n}}\right) = \boldsymbol{\sigma}\left(\mathbf{x}, t\right) \cdot \hat{\mathbf{n}} = \bar{\mathbf{t}}\left(\mathbf{x}, t, \hat{\mathbf{n}}\right), \quad \text{on } \partial\mathcal{S}_{\sigma} \tag{8.373}$$

and the essential boundary condition (the geometrical boundary condition)

$$\mathbf{u}\left(\mathbf{x}, t\right) = \bar{\mathbf{u}}\left(\mathbf{x}, t\right), \quad \text{on } \partial\mathcal{S}_{\varphi} \tag{8.374}$$

We say that Equation (8.372), (8.373) and (8.374) constitute the strong form in space for the initial boundary-value problems. As an analogy of the Lagrangian form of D'Alembert's principle in particle dynamics, taking the inner product between the virtual displacement and the above equation, and integrating the resulting equation over the domain of interest yields the following

$$\int_{\mathcal{S}} \left(\operatorname{div}\boldsymbol{\sigma} + \rho\mathbf{b} - \rho\ddot{\mathbf{u}}\right) \cdot \delta\mathbf{u}\, dv = 0 \tag{8.375}$$

which is called the Bubnov-Galerkin weighted residual form in continuum dynamics (Cook et al. 2002; Hughes 1987; Mikhlin 1964). It is very important to note that the virtual displacement employed to establish the Bubnov-Galerkin weighted residual form is required to be kinematically admissible, that is,

$$\delta\mathbf{u}\left(\mathbf{x}, t\right) = \mathbf{0}, \quad \text{on } \partial\mathcal{S}_{\varphi} \tag{8.376}$$

Further, the variation of the Eulerian strain tensor **e** should satisfy the compatibility (Pian and Wu 2006; Pian and Tong 1969; Richards 1977; Washizu 1982) in the body

$$\delta\mathbf{e} = \frac{1}{2}\left[\left(\operatorname{grad}\delta\mathbf{u}\right)^{T} + \operatorname{grad}\delta\mathbf{u}\right] \tag{8.377}$$

which means that the Eulerian strain tensor should be compatible with the displacement; therefore, the kinematically admissible displacement should be continuous and differentiable in the body. The equations of compatibility are satisfied when the Eulerian strain tensor is expressed in terms of the displacements. The Eulerian strain tensor is compatible when it is definable in terms of a single-valued, continuously differentiable displacement (Barber 2010). By use of the product rule, we have the relation (Basar and Weichert 2000; Bonet and Wood 2008; Holzapfel 2000; Marsden and Hughes 1983; Simo and Hughes 1998)

$$\delta\mathbf{u} \cdot \left(\operatorname{div}\boldsymbol{\sigma}\right) = \operatorname{div}\left(\delta\mathbf{u} \cdot \boldsymbol{\sigma}\right) - \boldsymbol{\sigma} : \operatorname{grad}\delta\mathbf{u} \tag{8.378}$$

By making use of the above equation, Equation (8.375) can be rewritten as

$$\int_{\mathcal{S}} \rho\ddot{\mathbf{u}} \cdot \delta\mathbf{u}\, dv + \int_{\mathcal{S}} \boldsymbol{\sigma} : \operatorname{grad}\delta\mathbf{u}\, dv = \int_{\mathcal{S}} \rho\mathbf{b} \cdot \delta\mathbf{u}\, dv + \int_{\mathcal{S}} \operatorname{div}\left(\delta\mathbf{u} \cdot \boldsymbol{\sigma}\right) dv \tag{8.379}$$

Next, we shall go through the details related to the second term on the left-hand side above.

Note that the Cauchy stress tensor is symmetric. And in view of Equation (8.377), we obtain the relation

$$\begin{aligned} \boldsymbol{\sigma} : \operatorname{grad}\delta\mathbf{u} &= \boldsymbol{\sigma} : \frac{1}{2}\left[\left(\operatorname{grad}\delta\mathbf{u}\right)^{T} + \operatorname{grad}\delta\mathbf{u}\right] \\ &= \boldsymbol{\sigma} : \delta\mathbf{e} \end{aligned} \tag{8.380}$$

As a consequence, the second term on the left-hand side in Equation (8.379) can be rewritten as

$$\delta \mathcal{W}_{\text{int}}(\mathbf{u}, \delta\mathbf{u}) = \int_{\mathcal{S}} \boldsymbol{\sigma} : \text{grad}\, \delta\mathbf{u}\, dv = \int_{\mathcal{S}} \boldsymbol{\sigma} : \delta\mathbf{e}\, dv \tag{8.381}$$

which corresponds to the internal virtual work.

Principle of Virtual Work in the Eulerian Description Next we recall Gauss's theorem and apply the natural boundary condition to Equation (8.379), then we obtain

$$\int_{\mathcal{S}} \rho\ddot{\mathbf{u}} \cdot \delta\mathbf{u}\, dv + \int_{\mathcal{S}} \boldsymbol{\sigma} : \delta\mathbf{e}\, dv = \int_{\partial\mathcal{S}_\sigma} \bar{\mathbf{t}}(\mathbf{x}, t, \hat{\mathbf{n}}) \cdot \delta\mathbf{u}\, da + \int_{\mathcal{S}} \rho\mathbf{b}(\mathbf{x}, t) \cdot \delta\mathbf{u}\, dv \tag{8.382}$$

which is called the principle of virtual work. It is also called the principle of virtual displacements. Note that the virtual displacement on the prescribed displacement boundary surface should vanish. The above equation may be interpreted in another way, namely,

$$\delta \mathcal{W}_{\text{kin}}(\mathbf{u}, \delta\mathbf{u}) + \delta \mathcal{W}_{\text{int}}(\mathbf{u}, \delta\mathbf{u}) = \delta \mathcal{W}_{\text{nc}}(\mathbf{u}, \delta\mathbf{u}) \tag{8.383}$$

where $\delta \mathcal{W}_{\text{kin}}$ is the contribution by the inertia term, i.e. $\delta \mathcal{W}_{\text{kin}}(\mathbf{u}) = \int_{\mathcal{S}} \rho\ddot{\mathbf{u}} \cdot \delta\mathbf{u}\, dv$. Accordingly, the principle of virtual work for continuous-body dynamical systems with conservative loadings can be written as

$$\delta \mathcal{W}_{\text{kin}}(\mathbf{u}, \delta\mathbf{u}) + \delta \mathcal{W}_{\text{int}}(\mathbf{u}, \delta\mathbf{u}) = \delta \mathcal{W}_{\text{con}}(\mathbf{u}, \delta\mathbf{u}) \tag{8.384}$$

which is equivalent to

$$\int_{\mathcal{S}} \rho\ddot{\mathbf{u}} \cdot \delta\mathbf{u}\, dv + \int_{\mathcal{S}} \boldsymbol{\sigma} : \delta\mathbf{e}\, dv = \int_{\partial\mathcal{S}_\sigma} \bar{\mathbf{t}} \cdot \delta\mathbf{u}(\mathbf{x}, t)\, da + \int_{\mathcal{S}} \rho\mathbf{b} \cdot \delta\mathbf{u}(\mathbf{x}, t)\, dv \tag{8.385}$$

where $\bar{\mathbf{t}}$ and $\mathbf{b}$ are conservative loadings, namely, dead loads, in the current configuration.

By virtue of Equation (8.378), the principle of virtual work in Equation (8.382) or Equation (8.383) for nonconservative systems can be recast into

$$\begin{aligned} &\int_{\mathcal{S}} (\rho\ddot{\mathbf{u}} - \rho\mathbf{b}(\mathbf{x}, t) - \text{div}\, \boldsymbol{\sigma}(\mathbf{x}, t)) \cdot \delta\mathbf{u}\, dv \\ &+ \int_{\partial\mathcal{S}_\sigma} \left(\boldsymbol{\sigma}(\mathbf{x}, t) \cdot \hat{\mathbf{n}} - \bar{\mathbf{t}}(\mathbf{x}, t, \hat{\mathbf{n}})\right) \cdot \delta\mathbf{u}\, da = 0 \end{aligned} \tag{8.386}$$

As the virtual displacement, which is kinematically admissible, is arbitrary in the domain $\mathcal{S}$ and on the domain $\partial\mathcal{S}_\sigma$, the principle of virtual work produces Cauchy's equations of motion

$$\text{div}\, \boldsymbol{\sigma} + \rho\mathbf{b} - \rho\ddot{\mathbf{u}} = \mathbf{0}, \quad \text{in } \mathcal{S} \tag{8.387}$$

and the natural boundary conditions

$$\boldsymbol{\sigma}(\mathbf{x}, t) \cdot \hat{\mathbf{n}} = \bar{\mathbf{t}}(\mathbf{x}, t, \hat{\mathbf{n}}), \quad \text{on } \partial\mathcal{S}_\sigma \tag{8.388}$$

Besides, the essential boundary conditions are reflected in Equation (8.386), since the kinematically admissible virtual displacements on the prescribed displacement boundary surface should vanish. Hence, when the principle of virtual work is employed for direct variational methods, the admissible displacement functions should satisfy the essential boundary conditions.

Principle of Virtual Work in the Lagrangian Description First, let us consider the variation of the deformation gradient tensor. In view of Equation (4.16), we have the Gâteaux variation of the deformation gradient tensor as a directional derivative of the tensor the function $\mathbf{F}(\mathbf{U})$ at $\mathbf{U}$ in the direction η (Edwards 1994) as follows:

$$\begin{aligned}
\delta\mathbf{F} &= \epsilon\mathcal{G}\mathbf{F}(\mathbf{U}(\mathbf{X},t)\,;\,\eta(\mathbf{X},t)) \\
&= \varepsilon\frac{d}{d\varepsilon}\mathbf{F}(\mathbf{U}+\varepsilon\eta)|_{\varepsilon=0} \\
&= \varepsilon\frac{d}{d\varepsilon}\left(\frac{\partial\mathbf{U}}{\partial\mathbf{X}}+\varepsilon\frac{\partial\eta}{\partial\mathbf{X}}\right)\Big|_{\varepsilon=0} \\
&= \frac{\partial\varepsilon\eta}{\partial\mathbf{X}} = \frac{\partial\delta\mathbf{U}}{\partial\mathbf{X}} \\
&= \mathrm{GRAD}\,\delta\mathbf{U} = \frac{\partial\mathbf{U}}{\partial\mathbf{x}}\cdot\frac{\partial\mathbf{x}}{\partial\mathbf{X}} \\
&= \mathrm{grad}\,\delta\mathbf{u}\cdot\mathbf{F}
\end{aligned} \tag{8.389}$$

since $\delta\mathbf{u}(\mathbf{x},t) = \delta\mathbf{u}(\varphi(\mathbf{X},t),t) = \delta\mathbf{U}(\mathbf{X},t)$. From the above, we obtain

$$\begin{aligned}
\mathrm{grad}\,\delta\mathbf{u} &= \mathrm{GRAD}\,\delta\mathbf{U}\cdot\mathbf{F}^{-1} \\
&= \delta\mathbf{F}\cdot\mathbf{F}^{-1}
\end{aligned} \tag{8.390}$$

Therefore, the internal virtual work can be rewritten as

$$\begin{aligned}
\delta\mathcal{W}_{\mathrm{int}}(\mathbf{u},\delta\mathbf{u}) &= \int_{\mathcal{S}}\boldsymbol{\sigma}:\delta\mathbf{e}\,dv = \int_{\mathcal{S}}\boldsymbol{\sigma}:\mathrm{grad}\,\delta\mathbf{u}\,dv = \int_{\mathcal{S}}\boldsymbol{\sigma}:\delta\mathbf{F}\cdot\mathbf{F}^{-1}\,dv \\
&= \int_{\mathcal{B}} J\boldsymbol{\sigma}:\delta\mathbf{F}\cdot\mathbf{F}^{-1}\,dV \\
&= \int_{\mathcal{B}} J\boldsymbol{\sigma}\mathbf{F}^{-T}:\delta\mathbf{F}\,dV \\
&= \int_{\mathcal{B}} \mathbf{P}:\delta\mathbf{F}\,dV
\end{aligned} \tag{8.391}$$

where the tensor property in Equation (2.173) and the Piola transformation (see Exercise 28) are employed.

In view of Equation (8.55), the variation of the Lagrange strain tensor can be written as

$$\delta\mathbf{E} = \frac{1}{2}\left((\delta\mathbf{F})^T\cdot\mathbf{F}+\mathbf{F}^T\cdot\delta\mathbf{F}\right) \tag{8.392}$$

Recall that the push-forward of the Lagrangian strain tensor yields the Eulerian strain tensor as shown in Equation (8.126). Therefore, we derive the variation of the Almansi strain tensor (the Eulerian strain tensor) as follows

$$
\begin{aligned}
\varphi_* (\delta \mathbf{E}) &= \mathbf{F}^{-T} \cdot \delta \mathbf{E} \cdot \mathbf{F}^{-1} \\
&= \frac{1}{2} \left(\mathbf{F}^{-T} \cdot (\delta \mathbf{F})^T + \delta \mathbf{F} \cdot \mathbf{F}^{-1} \right) \\
&= \frac{1}{2} \left[\left(\mathrm{GRAD}\, \delta \mathbf{U} \cdot \mathbf{F}^{-1} \right)^T + \mathrm{GRAD}\, \delta \mathbf{U} \cdot \mathbf{F}^{-1} \right] \\
&= \frac{1}{2} \left[(\mathrm{grad}\, \delta \mathbf{u})^T + \mathrm{grad}\, \delta \mathbf{u} \right] \\
&= \delta \mathbf{e}
\end{aligned}
\tag{8.393}
$$

Consequently, in view of Equation (8.167), the internal virtual work can be rewritten as

$$
\begin{aligned}
\delta \mathcal{W}_{\text{int}} &= \int_{\mathcal{S}} \boldsymbol{\sigma} : \delta \mathbf{e}\, dv = \int_{\mathcal{S}} \boldsymbol{\sigma} : \mathrm{grad}\, \delta \mathbf{u}\, dv \\
&= \int_{\mathcal{B}} \mathbf{P} : \delta \mathbf{F}\, dV = \int_{\mathcal{B}} \mathbf{P} : \mathrm{GRAD}\, \delta \mathbf{U}\, dV \\
&= \int_{\mathcal{B}} \mathbf{S} : \delta \mathbf{E}\, dV
\end{aligned}
\tag{8.394}
$$

where $\mathbf{S}$ and $\mathbf{E}$ denote the second Piola-Kirchhoff stress tensor and the Lagrangian strain tensor, and dV represents the Lebesgue measures in the body $\mathcal{B}$. As a consequence, the principle of virtual work in the Lagrangian description can be given by

$$
\int_{\mathcal{B}} \rho_0 \ddot{\boldsymbol{\varphi}} \cdot \delta \boldsymbol{\varphi}\; dV + \int_{\mathcal{B}} \mathbf{P} : \delta \mathbf{F}\; dV = \int_{\mathcal{B}} \rho_0 \mathbf{B} (\mathbf{X}, t) \cdot \delta \boldsymbol{\varphi}\; dV + \int_{\partial \mathcal{B}_\sigma} \bar{\mathbf{T}} \cdot \delta \boldsymbol{\varphi}\; dA \tag{8.395}
$$

where $\mathbf{B} (\mathbf{X}, t) : \mathcal{B} \times \mathcal{T} \rightarrow \mathbb{R}^3$ denotes the body force per unit mass and $\bar{\mathbf{T}}(\mathbf{X}, t, \hat{\mathbf{N}}) : \partial \mathcal{B} \times \mathcal{T} \times \mathbb{R}^3 \rightarrow \mathbb{R}^3$ the surface traction in the reference configuration. By use of Cauchy's law and Gauss' theorem, the above equation leads to

$$
\int_{\mathcal{B}} (\mathrm{DIV}\, \mathbf{P} + \rho_0 \mathbf{B} - \rho_0 \ddot{\boldsymbol{\varphi}}) \cdot \delta \boldsymbol{\varphi}\; dV + \int_{\partial \mathcal{B}_\sigma} \left(\bar{\mathbf{T}} - \mathbf{P} \cdot \hat{\mathbf{N}} \right) \cdot \delta \boldsymbol{\varphi}\; dV = 0 \tag{8.396}
$$

As the virtual displacement, which is kinematically admissible, is arbitrary in the domain $\mathcal{B}$ and on the domain $\partial \mathcal{B}_\sigma$, the above equation yields Cauchy's equations of motion

$$
\rho_0 \ddot{\boldsymbol{\varphi}} (\mathbf{X}, t) = \mathrm{DIV}\, \mathbf{P} (\mathbf{X}, t) + \rho_0 \mathbf{B} (\mathbf{X}, t), \quad \text{for } (\mathbf{X}, t) \in \mathcal{B} \times \mathcal{T} \tag{8.397}
$$

and the natural boundary conditions

$$
\bar{\mathbf{T}} \left(\mathbf{X}, t, \hat{\mathbf{N}} \right) = \mathbf{P} (\mathbf{X}, t) \cdot \hat{\mathbf{N}} (\mathbf{X}, t), \quad \text{for } (\mathbf{X}, t) \in \partial \mathcal{B}_\sigma \times \mathcal{T} \tag{8.398}
$$

which means that the body is governed by the equations of motion and the natural boundary conditions.

8.4.3 Principle of Virtual Power

If the virtual displacement $\delta \mathbf{u}$ in Equation (8.375) is replaced by the virtual velocity $\delta \mathbf{v}$, the Bubnov-Galerkin weighted residual form can be given by

$$\int_{\mathcal{S}} (\operatorname{div} \boldsymbol{\sigma} + \rho \mathbf{b} - \rho \ddot{\mathbf{u}}) \cdot \delta \mathbf{v} \, dv = 0 \tag{8.399}$$

where $\delta \mathbf{v} = \delta \dot{\mathbf{u}}$. In the case that the virtual displacement is replaced by the virtual velocity, Equation (8.399) yields the principle of virtual power, that is,

$$\langle \rho \ddot{\mathbf{u}}, \delta \dot{\mathbf{u}} \rangle + \langle \boldsymbol{\sigma}(\mathbf{x}, t), \operatorname{grad} \delta \dot{\mathbf{u}} \rangle = \langle \bar{\mathbf{t}}(\mathbf{x}, t, \hat{\mathbf{n}}), \delta \dot{\mathbf{u}} \rangle + \langle \rho \mathbf{b}(\mathbf{x}, t), \delta \dot{\mathbf{u}} \rangle \tag{8.400}$$

The virtual power is done by the actual external forces on the virtual velocities. The principle of virtual power in Equation (8.400) results from the Bubnov-Galerkin weighted residual form in Equation (8.399). The principle of virtual power is also called the principle of virtual velocities. The principle of virtual power in Equation (8.400) can be written as

$$\int_{\mathcal{S}} \rho \ddot{\mathbf{u}} \cdot \delta \dot{\mathbf{u}} \, dv + \int_{\mathcal{S}} \boldsymbol{\sigma} : \delta \boldsymbol{d} \, dv = \int_{\partial \mathcal{S}_\sigma} \bar{\mathbf{t}}(\mathbf{x}, t, \hat{\mathbf{n}}) \cdot \delta \dot{\mathbf{u}} \, da + \int_{\mathcal{S}} \rho \mathbf{b}(\mathbf{x}, t) \cdot \delta \dot{\mathbf{u}} \, dv \tag{8.401}$$

where $\delta \boldsymbol{d}$ denotes the variation of the rate of deformation tensor. By use of Gauss' theorem and Cauchy's law, the principle of virtual power can be recast into the form

$$\int_{\mathcal{S}} (\rho \ddot{\mathbf{u}} - \rho \mathbf{b}(\mathbf{x}, t) - \operatorname{div} \boldsymbol{\sigma}) \cdot \delta \dot{\mathbf{u}} \, dv + \int_{\partial \mathcal{S}_\sigma} \left(\boldsymbol{\sigma} \cdot \hat{\mathbf{n}} - \bar{\mathbf{t}}(\mathbf{x}, t, \hat{\mathbf{n}}) \right) \cdot \delta \dot{\mathbf{u}} \, da = 0 \tag{8.402}$$

Comparing the above to Equation (8.386), Equation (8.402) takes the physical insight and concept of the virtual velocity instead of the virtual displacement as a weighting function. In the Lagrangian description, the principle of virtual power is expressed as

$$\int_{\mathcal{B}} \rho_0 \ddot{\boldsymbol{\varphi}} \cdot \delta \dot{\boldsymbol{\varphi}} \, dV + \int_{\mathcal{B}} \mathbf{P} : \delta \dot{\mathbf{F}} \, dV = \int_{\mathcal{B}} \rho_0 \mathbf{B} \cdot \delta \dot{\boldsymbol{\varphi}} \, dV + \int_{\partial \mathcal{B}_\sigma} \bar{\mathbf{T}} \cdot \delta \dot{\boldsymbol{\varphi}} \, dA \tag{8.403}$$

which gives

$$\int_{\mathcal{B}} (\operatorname{DIV} \mathbf{P} + \rho_0 \mathbf{B} - \rho_0 \ddot{\boldsymbol{\varphi}}) \cdot \delta \dot{\boldsymbol{\varphi}} \, dV + \int_{\partial \mathcal{B}_\sigma} \left(\bar{\mathbf{T}} - \mathbf{P} \cdot \hat{\mathbf{N}} \right) \cdot \delta \dot{\boldsymbol{\varphi}} \, dV = 0 \tag{8.404}$$

As the virtual velocity, which is kinematically admissible, is arbitrary, Equations (8.402) and (8.404) yield Cauchy's equations of motion and the natural boundary conditions.

8.4.4 Principle of Stationary Potential Energy

Consider the principle of virtual work for static problems with conservative loadings. Then the virtual work done by the internal forces is identical to the variation of the internal potential energy as follows:

$$\delta \mathcal{U}_{\text{int}}(\mathbf{u}) = \int_{\mathcal{S}} \boldsymbol{\sigma} : \operatorname{grad} \delta \mathbf{u} \, dv = \delta \mathcal{W}_{\text{int}}(\mathbf{u}) \tag{8.405}$$

Hence, we can have an alternative expression of the principle of virtual work for static problems in terms of potential energies,

$$\delta \mathcal{U}_{\text{int}}(\mathbf{u}) = -\delta \mathcal{U}_{\text{ext}}(\mathbf{u}) \tag{8.406}$$

The total potential energy is given by

$$\begin{aligned}\mathcal{U}(\mathbf{u}) &= \mathcal{U}_{\text{int}}(\mathbf{u}) + \mathcal{U}_{\text{ext}}(\mathbf{u}) \\ &= \int_{\mathcal{S}} \rho\psi\,(\text{grad}\,\mathbf{u})\,dv - \int_{\partial\mathcal{S}_\sigma} \bar{\mathbf{t}}\cdot\mathbf{u}\,da - \int_{\mathcal{S}} \rho\mathbf{b}\cdot\mathbf{u}\,dv\end{aligned} \tag{8.407}$$

in which tractions and body forces are assumed to be dead loads not depending on the motion of the body.

Consider a set of kinematically admissible displacements satisfying the essential boundary conditions as

$$\tilde{\mathbf{u}} = \mathbf{u} + \delta\mathbf{u} \tag{8.408}$$

Then the total potential energy at the admissible displacements may be given by

$$\mathcal{U}(\tilde{\mathbf{u}}) = \mathcal{U}(\mathbf{u}) + \delta\mathcal{U}(\mathbf{u}) + \delta^2\mathcal{U}(\mathbf{u}) + \ldots. \tag{8.409}$$

In order for the total potential energy to be extremized at the stationary point, the following is considered. The first necessary condition for the extremum of the total potential energy is that the variation should vanish at the stationary point, that is,

$$\delta\mathcal{U}(\mathbf{u}) = 0 \tag{8.410}$$

which is called the principle of stationary potential energy, which states that among all kinematically admissible displacements, the actual displacements at the stationary point make the variation of the total potential energy vanish. Consequently, the resulting governing equation is represented in terms of the actual displacement. As shown in Equation (8.407), the primary dependent variable is the displacement function satisfying the essential boundary conditions. Hence, the principle of stationary potential energy is a single field variational principle.

As a special case of the principle of stationary potential energy, we shall discuss the principle of minimum potential energy in a small deformation. In the small displacement theory, the strain energy density function $w(\boldsymbol{\varepsilon})$ is quadratic in infinitesimal strain and positive definite in displacement. Hence, the internal potential energy can be written as

$$\mathcal{U}_{\text{int}}(\mathbf{u}) = \int_{\mathcal{S}} \rho\psi\,(\text{grad}\,\mathbf{u})\,dv = \int_{\mathcal{S}} w(\boldsymbol{\varepsilon})dv \tag{8.411}$$

where $\boldsymbol{\varepsilon}$ denotes the infinitesimal strain tensor. And the variation of the total potential energy can be represented as

$$\delta\mathcal{U}_{\text{int}}(\mathbf{u}) = \int_{\mathcal{S}} \frac{\partial w(\boldsymbol{\varepsilon})}{\partial\boldsymbol{\varepsilon}} : \delta\boldsymbol{\varepsilon}dv = \int_{\mathcal{S}} \boldsymbol{\sigma} : \delta\boldsymbol{\varepsilon}dv \tag{8.412}$$

At the stationary point, the variation of the total potential energy is assumed to be zero as given in Equation (8.410). And, the second sufficient condition for the minimum of the total potential energy requires that the second variation,

$$\delta^2\mathcal{U}(\mathbf{u}) > 0 \tag{8.413}$$

and, for the maximum that the second variation,

$$\delta^2\mathcal{U}(\mathbf{u}) < 0 \tag{8.414}$$

at the stationary point. In view of Equation (8.410), Equation (8.409) can be rewritten as

$$\mathcal{U}(\tilde{\mathbf{u}}) - \mathcal{U}(\mathbf{u}) = \delta^2\mathcal{U}(\mathbf{u}) + \ldots. \tag{8.415}$$

Since the strain energy density function is positive definitive, the second variation is positive as follows:

$$\delta\mathcal{U}^2(\mathbf{u}) = \frac{1}{2}\int_{\mathcal{S}} \delta\boldsymbol{\varepsilon} : \frac{\partial\boldsymbol{\sigma}}{\partial\boldsymbol{\varepsilon}} : \delta\boldsymbol{\varepsilon}\, dv + \ldots > 0 \tag{8.416}$$

Consequently, at the stationary point the total potential energy becomes a minimum. This is called the principle of minimum potential energy which is valid in a small, linear elastic deformation.

8.4.5 Principle of Complementary Virtual Work

In the principle of virtual work or the principle of stationary or minimum potential energy, the primary field variable is the displacement. Furthermore, the displacement field variable should be continuous over the domain of interest. On the prescribed displacement surface, the reaction is unknown, whilst the displacement is unknown on the prescribed traction boundary surface. Hence, this is called the displacement-based variational formulation (Pian and Tong 1969; Pian and Wu 2006). The displacement field is also assumed to satisfy the compatibility equation, namely, the strain-displacement relationship.

On the contrary, consider the stress field as the primary field variable. Then, the complementary virtual work done by the actual displacement on the virtual forces is defined as

$$\delta\mathcal{W}^*_{\text{ext}} = \int_{\partial\mathcal{S}_\varphi} \delta\mathbf{t}(\mathbf{x}) \cdot \mathbf{u}(\mathbf{x})\, da + \int_{\mathcal{S}} \rho(\mathbf{x})\,\delta\mathbf{b}(\mathbf{x}) \cdot \mathbf{u}(\mathbf{x})\, dv \tag{8.417}$$

The stress field should be statically admissible so that the virtual stresses and the virtual forces satisfy the following

$$\operatorname{div}\delta\boldsymbol{\sigma}(\mathbf{x}) + \rho(\mathbf{x})\,\delta\mathbf{b}(\mathbf{x}) = \mathbf{0} \quad \text{in } \mathcal{S} \tag{8.418}$$

and

$$\hat{\mathbf{n}} \cdot \delta\boldsymbol{\sigma} = \delta\mathbf{t}(\mathbf{x}) \quad \text{on } \partial\mathcal{S}_\varphi \tag{8.419}$$

where $\hat{\mathbf{n}}$ denotes the unit vector outward normal to the surface of the body. Note that on the prescribed traction boundary surface $\partial\mathcal{S}_\sigma$, the virtual traction $\delta\mathbf{t}$ should vanish, that is,

$$\delta\mathbf{t} = \delta\bar{\mathbf{t}} = \mathbf{0}, \quad \text{on } \partial\mathcal{S}_\sigma \tag{8.420}$$

Additionally, via the Legendre transformation the complementary energy density function is also obtained as

$$W^*_{\text{int}}(\boldsymbol{\sigma}) = \boldsymbol{\sigma} : \mathbf{e} - W_{\text{int}}(\mathbf{e}) \tag{8.421}$$

where $W_{\text{int}}(\mathbf{e})$ stands for the strain energy density function. The complementary internal virtual work is given by

$$\delta\mathcal{W}^*_{\text{int}} = \int_{\mathcal{S}} \delta W^*_{\text{int}}(\boldsymbol{\sigma})\, dv = \int_{\mathcal{S}} \delta\boldsymbol{\sigma} : \mathbf{e}\, dv \tag{8.422}$$

Substituting Equation (8.419) into Equation (8.417), the complementary virtual work can be written as

$$\begin{aligned}\delta\mathcal{W}^*_{\text{ext}} &= \int_{\partial\mathcal{S}_\varphi} \hat{\mathbf{n}} \cdot \delta\boldsymbol{\sigma} \cdot \mathbf{u}(\mathbf{x})\; da + \int_{\mathcal{S}} \rho(\mathbf{x})\,\delta\mathbf{b}(\mathbf{x}) \cdot \mathbf{u}(\mathbf{x})\, dv \\ &= \int_{\mathcal{S}} \operatorname{div}(\mathbf{u} \cdot \delta\boldsymbol{\sigma})\, dv + \int_{\mathcal{S}} \rho(\mathbf{x})\,\delta\mathbf{b}(\mathbf{x}) \cdot \mathbf{u}(\mathbf{x})\, dv\end{aligned} \tag{8.423}$$

where Gauss's theorem is applied. Next, we recall the relation

$$\operatorname{div}\,(\mathbf{u}\cdot\delta\boldsymbol{\sigma}) = \delta\boldsymbol{\sigma} : \operatorname{grad}\mathbf{u} + \mathbf{u}\cdot(\operatorname{div}\delta\boldsymbol{\sigma}) \tag{8.424}$$

As a consequence of the above relation, the complementary virtual work can be written as

$$\delta\mathcal{W}^*_{\text{ext}} = \int_{\mathcal{S}} \delta\boldsymbol{\sigma} : \operatorname{grad}\mathbf{u}\,dv + \int_{\mathcal{S}} (\operatorname{div}\delta\boldsymbol{\sigma}\,(\mathbf{x}) + \rho\,(\mathbf{x})\,\delta\mathbf{b}\,(\mathbf{x}))\cdot\mathbf{u}\,(\mathbf{x})\,dv \tag{8.425}$$

By virtue of Equation (8.418), the second term on the right of Equation (8.425) vanishes. Hence, we have

$$\delta\mathcal{W}^*_{\text{int}} = \delta\mathcal{W}^*_{\text{ext}} \tag{8.426}$$

which yields

$$\int_{\mathcal{S}} \delta\boldsymbol{\sigma} : \mathbf{e}\,dv = \int_{\partial\mathcal{S}_\varphi} \delta\mathbf{t}\,(\mathbf{x})\cdot\mathbf{u}\,(\mathbf{x})\,da + \int_{\mathcal{S}} \rho\,(\mathbf{x})\,\delta\mathbf{b}\,(\mathbf{x})\cdot\mathbf{u}\,(\mathbf{x})\,dv \tag{8.427}$$

The above equation is called the principle of complementary virtual work or the principle of virtual forces.

Alternatively, the principle of complementary virtual work given in Equation (8.427) can be derived from the weighted residual form, which is given by

$$\int_{\mathcal{S}} (\operatorname{div}\delta\boldsymbol{\sigma}\,(\mathbf{x}) + \rho\,(\mathbf{x})\,\delta\mathbf{b}\,(\mathbf{x}))\cdot\mathbf{u}\,(\mathbf{x})\,dv = 0 \tag{8.428}$$

where the stress field is subject to the boundary condition given in Equation (8.419). Thus, by means of the relation of Equation (8.424), the above equation can be rewritten as

$$\int_{\mathcal{S}} \operatorname{div}\,(\mathbf{u}\cdot\delta\boldsymbol{\sigma})\,dv - \int_{\mathcal{S}} \delta\boldsymbol{\sigma} : \operatorname{grad}\mathbf{u}\,dv + \int_{\mathcal{S}} \rho\,(\mathbf{x})\,\delta\mathbf{b}\,(\mathbf{x})\cdot\mathbf{u}\,(\mathbf{x})\,dv = 0 \tag{8.429}$$

Thanks to Gauss's theorem, the above equation leads to

$$\int_{\mathcal{S}} \delta\boldsymbol{\sigma} : \mathbf{e}\,dv = \int_{\partial\mathcal{S}_\varphi} (\hat{\mathbf{n}}\cdot\delta\boldsymbol{\sigma})\cdot\mathbf{u}\,(\mathbf{x})\,da + \int_{\mathcal{S}} \rho\,(\mathbf{x})\,\delta\mathbf{b}\,(\mathbf{x})\cdot\mathbf{u}\,(\mathbf{x})\,dv \tag{8.430}$$

Imposing the boundary condition given in Equation (8.419), the above equation becomes identical to the principle of complementary virtual work given in Equation (8.427).

By use of Equation (8.418), the principle of complementary virtual work in Equation (8.427) can be written as

$$\begin{aligned}\int_{\mathcal{S}} \delta\boldsymbol{\sigma} : \mathbf{e}\,dv &= \int_{\partial\mathcal{S}_\varphi} \bar{\mathbf{u}}\cdot\delta\mathbf{t}\,(\mathbf{x})\,da - \int_{\mathcal{S}} \mathbf{u}\,(\mathbf{x})\cdot(\operatorname{div}\delta\boldsymbol{\sigma})\,dv \\ &= \int_{\partial\mathcal{S}_\varphi} \bar{\mathbf{u}}\cdot\delta\mathbf{t}\,(\mathbf{x})\,da + \int_{\mathcal{S}} \delta\boldsymbol{\sigma} : \operatorname{grad}\mathbf{u}\,dv - \int_{\partial\mathcal{S}_\varphi} \mathbf{u}\cdot\delta\mathbf{t}\,(\mathbf{x})\,da\end{aligned} \tag{8.431}$$

In view of Equation (8.424), we have

$$\int_{\mathcal{S}} \delta\boldsymbol{\sigma} : (\mathbf{e} - \operatorname{grad}\mathbf{u})\,dv + \int_{\partial\mathcal{S}_\varphi} \delta\mathbf{t}\,(\mathbf{x})\cdot(\mathbf{u} - \bar{\mathbf{u}})\,da = 0 \tag{8.432}$$

Hence, the principle of complementary virtual work includes the strain-displacement relation and the essential boundary conditions.

8.4.6 Principle of Stationary Complementary Energy

The principle of stationary complementary energy is derived from the principle of complementary virtual work. The relation between the external complementary energy and the external complementary work is given by

$$\mathcal{U}^*_{\text{ext}}(\boldsymbol{\sigma}) = -\mathcal{W}^*_{\text{ext}}(\boldsymbol{\sigma}) \tag{8.433}$$

and the relation the internal complementary energy and the internal complementary work is expressed as

$$\mathcal{U}^*_{\text{int}}(\boldsymbol{\sigma}) = \mathcal{W}^*_{\text{int}}(\boldsymbol{\sigma}) \tag{8.434}$$

From the principle of complementary virtual work, we have

$$\delta\mathcal{W}^*_{\text{int}}(\boldsymbol{\sigma}) - \delta\mathcal{W}^*_{\text{ext}}(\boldsymbol{\sigma}) = \delta\mathcal{U}^*_{\text{int}}(\boldsymbol{\sigma}) + \delta\mathcal{U}^*_{\text{ext}}(\boldsymbol{\sigma}) = \delta\left(\mathcal{U}^*_{\text{int}} + \mathcal{U}^*_{\text{ext}}\right) = \delta\mathcal{U}^* = 0 \tag{8.435}$$

As usual, the body force is prescribed; hence the virtual body force vanishes. Consequently, instead of the above equation the total complementary energy $\mathcal{U}^*$ is defined as

$$\mathcal{U}^*(\boldsymbol{\sigma}) = \int_S w^*(\boldsymbol{\sigma})\, dv - \int_{\partial S_\varphi} \mathbf{t}(\mathbf{x}) \cdot \bar{\mathbf{u}}(\mathbf{x})\, da \tag{8.436}$$

where $w^*(\boldsymbol{\sigma})$ is the complementary energy density function per unit volume. Equation (8.435) is called the principle of stationary total complementary energy. At the minimum total complementary energy point, the governing equations hold when the variation of the total complementary energy vanishes, and we obtain the principle of complementary virtual work.

Consider a set of statically admissible stresses satisfying the natural boundary conditions as

$$\tilde{\boldsymbol{\sigma}} = \boldsymbol{\sigma} + \delta\boldsymbol{\sigma} \tag{8.437}$$

Then the total complementary energy at the statically admissible stresses may be given by

$$\mathcal{U}^*(\tilde{\boldsymbol{\sigma}}) = \mathcal{U}^*(\boldsymbol{\sigma}) + \delta\mathcal{U}^*(\boldsymbol{\sigma}) + \delta^2\mathcal{U}^*(\boldsymbol{\sigma}) + \ldots. \tag{8.438}$$

In order for the total complementary energy to be extremized at the stationary point, the following is considered. The first necessary condition for the extremum of the total complementary energy is that the variation should vanish at the stationary point, that is,

$$\delta\mathcal{U}^*(\boldsymbol{\sigma}) = 0 \tag{8.439}$$

which is called the principle of stationary complementary energy, which states that among all statically admissible stresses, the actual stresses at the stationary point make the variation of the total complementary energy vanish. As shown in Equation (8.439), the primary dependent variable is the stress function satisfying the natural boundary conditions. Hence, the principle of stationary complementary energy is also a single field variational principle.

As a special case of the principle of stationary complementary energy, we shall discuss the principle of minimum complementary energy in a small deformation. In the small displacement theory, the complementary energy density function $w^*(\boldsymbol{\sigma})$ is quadratic in stress and positive definite. Hence, the internal complementary energy can be written as

$$\mathcal{U}^*_{\text{int}}(\boldsymbol{\sigma}) = \int_S w^*(\boldsymbol{\sigma}) dv \tag{8.440}$$

where $\boldsymbol{\sigma}$ denotes the Cauchy stress tensor. And the variation of the total complementary energy can be represented as

$$\delta \mathcal{U}^*_{\text{int}}(\boldsymbol{\sigma}) = \int_{\mathcal{S}} \frac{\partial w^*(\boldsymbol{\sigma})}{\partial \boldsymbol{\sigma}} : \delta\boldsymbol{\sigma}\, dv = \int_{\mathcal{S}} \boldsymbol{\varepsilon} : \delta\boldsymbol{\sigma}\, dv \tag{8.441}$$

At the stationary point, the variation of the total complementary energy is assumed to be zero as given in Equation (8.439). And, the second sufficient condition for the minimum of the total complementary energy requires that the second variation,

$$\delta^2 \mathcal{U}^*(\boldsymbol{\sigma}) > 0 \tag{8.442}$$

and, for the maximum that the second variation,

$$\delta^2 \mathcal{U}^*(\boldsymbol{\sigma}) < 0 \tag{8.443}$$

at the stationary point. In view of Equation (8.439), Equation (8.438) can be rewritten as

$$\mathcal{U}^*(\tilde{\boldsymbol{\sigma}}) - \mathcal{U}^*(\boldsymbol{\sigma}) = \delta^2 \mathcal{U}^*(\boldsymbol{\sigma}) + \ldots. \tag{8.444}$$

Since the complementary energy density function is positive definitive, the second variation is positive as follows:

$$\delta^2 \mathcal{U}^*(\boldsymbol{\sigma}) = \frac{1}{2}\int_{\mathcal{S}} \delta\boldsymbol{\sigma} : \frac{\partial \boldsymbol{\varepsilon}}{\partial \boldsymbol{\sigma}} : \delta\boldsymbol{\sigma}\, dv + \ldots > 0 \tag{8.445}$$

Consequently, at the stationary point the total complementary energy becomes a minimum. This is called the principle of minimum complementary energy which is valid in a small, linear elastic deformation.

8.4.7 Hu-Washizu Variational Principle

Thus far, we have discussed a single field variational formulation. As a general variational principle, we shall discuss the Hu-Washizu variation principle (Holzapfel 2000; Pian and Tong 1969; Richards 1977; Washizu 1982), which is often called the three field variational formulation (Pian 1995).

Finite Deformation The total potential energy with a displacement field can be given by

$$\mathcal{U}(\boldsymbol{\varphi}) = \int_{\mathcal{B}} (\rho_0 \Psi(\mathbf{F}) - \rho_0 \mathbf{B} \cdot \boldsymbol{\varphi})\, dV - \int_{\partial\mathcal{B}_\sigma} \bar{\mathbf{T}} \cdot \boldsymbol{\varphi}\, dA \tag{8.446}$$

In the above equation, the single field state variable is $\boldsymbol{\varphi}$. As the compatibility condition in the body, the deformation gradient tensor obeys the following

$$\mathbf{F} = \text{GRAD}\,\boldsymbol{\varphi} \quad \text{for } \mathbf{X} \in \mathcal{B} \tag{8.447}$$

and the kinematically admissible displacement on the prescribed displacement boundary surface satisfies

$$\boldsymbol{\varphi} = \bar{\boldsymbol{\varphi}} \quad \text{for } \mathbf{X} \in \partial\mathcal{B}_\varphi \tag{8.448}$$

which is the essential boundary condition. Via introducing the Lagrange multipliers $\boldsymbol{\lambda}_1, \boldsymbol{\lambda}_2$, the modified total potential energy $\mathcal{U}_{\text{HW}}(\mathbf{F}, \boldsymbol{\varphi}, \boldsymbol{\lambda}_1, \boldsymbol{\lambda}_2)$ can be given by

$$\begin{aligned}\mathcal{U}_{\text{HW}} &= \int_{\mathcal{B}} (\rho_0 \Psi(\mathbf{F}) - \rho_0 \mathbf{B} \cdot \boldsymbol{\varphi} + \boldsymbol{\lambda}_1 : (\mathbf{F} - \text{GRAD}\,\boldsymbol{\varphi}))\, dV \\ &\quad - \int_{\partial\mathcal{B}_\sigma} \bar{\mathbf{T}} \cdot \boldsymbol{\varphi}\, dA + \int_{\partial\mathcal{B}_\varphi} \boldsymbol{\lambda}_2 \cdot (\boldsymbol{\varphi} - \bar{\boldsymbol{\varphi}})\, dA\end{aligned} \tag{8.449}$$

Then, the variation of the modified total potential energy can be represented as

$$\begin{aligned}\delta\mathcal{U}_{HW} &= \int_{\mathcal{B}} \begin{bmatrix} \left(\rho_0 \dfrac{\partial \Psi(\mathbf{F})}{\partial \mathbf{F}} + \boldsymbol{\lambda}_1\right) : \delta\mathbf{F} + \delta\boldsymbol{\lambda}_1 : (\mathbf{F} - \text{GRAD}\,\boldsymbol{\varphi}) \\ -\rho_0 \mathbf{B} \cdot \delta\boldsymbol{\varphi} - \boldsymbol{\lambda}_1 : \text{GRAD}\,\delta\boldsymbol{\varphi} \end{bmatrix} dV \\ &\quad - \int_{\partial\mathcal{B}_\sigma} \bar{\mathbf{T}} \cdot \delta\boldsymbol{\varphi}\, dA + \int_{\partial\mathcal{B}_\varphi} \delta\boldsymbol{\lambda}_2 \cdot (\boldsymbol{\varphi} - \bar{\boldsymbol{\varphi}})\, dA + \int_{\partial\mathcal{B}_\varphi} \boldsymbol{\lambda}_2 \cdot \delta\boldsymbol{\varphi}\, dA\end{aligned} \tag{8.450}$$

By use of the relation:

$$\boldsymbol{\lambda}_1 : \text{GRAD}\,\delta\boldsymbol{\varphi} = \text{DIV}\,(\delta\boldsymbol{\varphi} \cdot \boldsymbol{\lambda}_1) - \delta\boldsymbol{\varphi} \cdot (\text{DIV}\,\boldsymbol{\lambda}_1) \tag{8.451}$$

we have

$$\begin{aligned}\delta\mathcal{U}_{HW} &= \int_{\mathcal{B}} \begin{bmatrix} \left(\rho_0 \dfrac{\partial \Psi(\mathbf{F})}{\partial \mathbf{F}} + \boldsymbol{\lambda}_1\right) : \delta\mathbf{F} + \delta\boldsymbol{\lambda}_1 : (\mathbf{F} - \text{GRAD}\,\boldsymbol{\varphi}) \\ -(\rho_0 \mathbf{B} - (\text{DIV}\,\boldsymbol{\lambda}_1)) \cdot \delta\boldsymbol{\varphi} \end{bmatrix} dV \\ &\quad - \int_{\partial\mathcal{B}_\sigma} \left(\bar{\mathbf{T}} + \boldsymbol{\lambda}_1 \cdot \hat{\mathbf{N}}\right) \cdot \delta\boldsymbol{\varphi}\, dA + \int_{\partial\mathcal{B}_\varphi} \delta\boldsymbol{\lambda}_2 \cdot (\boldsymbol{\varphi} - \bar{\boldsymbol{\varphi}})\, dA \\ &\quad + \int_{\partial\mathcal{B}_\varphi} \left(\boldsymbol{\lambda}_2 - \boldsymbol{\lambda}_1 \cdot \hat{\mathbf{N}}\right) \cdot \delta\boldsymbol{\varphi}\, dA\end{aligned} \tag{8.452}$$

For the variation to be stationary, the following are required

$$\begin{aligned}\boldsymbol{\lambda}_1 &= -\rho_0 \frac{\partial \Psi(\mathbf{F})}{\partial \mathbf{F}} = -\mathbf{P} \\ \boldsymbol{\lambda}_2 &= \boldsymbol{\lambda}_1 \cdot \hat{\mathbf{N}} = -\mathbf{P} \cdot \hat{\mathbf{N}} = -\mathbf{T}\end{aligned} \tag{8.453}$$

Accordingly, we may reconstruct the modified total potential energy

$$\begin{aligned}\mathcal{U}_{HW}(\mathbf{F}, \boldsymbol{\varphi}, \mathbf{P}) &= \int_{\mathcal{B}} \left[\rho_0 \Psi(\mathbf{F}) - \rho_0 \mathbf{B} \cdot \boldsymbol{\varphi} - \mathbf{P} : (\mathbf{F} - \text{GRAD}\,\boldsymbol{\varphi})\right] dV \\ &\quad - \int_{\partial\mathcal{B}_\sigma} \bar{\mathbf{T}} \cdot \boldsymbol{\varphi}\, dA - \int_{\partial\mathcal{B}_\varphi} \mathbf{T} \cdot (\boldsymbol{\varphi} - \bar{\boldsymbol{\varphi}})\, dA\end{aligned} \tag{8.454}$$

The modified total potential energy in Equation (8.454) is called the Hu-Washizu functional (Marsden and Hughes 1983; Pian 1995; Washizu 1982). When the variation of the functional vanishes, the following equations can be obtained in the body as

$$
\begin{aligned}
&\mathbf{P} = \rho_0 \frac{\partial \Psi(\mathbf{F})}{\partial \mathbf{F}}, \\
&\mathbf{F} = \mathrm{GRAD}\, \boldsymbol{\varphi}, \qquad \text{for } \mathbf{X} \in \mathcal{B} \\
&\rho_0 \mathbf{B} + \mathrm{DIV}\, \mathbf{P} = \mathbf{0}
\end{aligned}
\tag{8.455}
$$

and on the surface as

$$
\begin{aligned}
&\mathbf{T} = \bar{\mathbf{T}}, \qquad \text{for } \mathbf{X} \in \partial\mathcal{B}_\sigma \\
&\mathbf{T} = \mathbf{P} \cdot \hat{\mathbf{N}}, \quad \boldsymbol{\varphi} = \bar{\boldsymbol{\varphi}}, \quad \text{for } \mathbf{X} \in \partial\mathcal{B}_\varphi
\end{aligned}
\tag{8.456}
$$

This is the Hu-Washizu variational principle, in which the variation of the functional includes the constitutive equation, the compatibility equation, and the equilibrium equation in the body, and the natural and essential boundary conditions on the surface. The three field dependent variables are the displacement, the deformation gradient tensor and the first Piola-Kirchhoff stress tensor.

Small Deformation In the infinitesimal displacement theory, the Hu-Washizu functional can be given by

$$
\begin{aligned}
\mathcal{U}_{\mathrm{HW}}(\boldsymbol{\varepsilon}, \mathbf{u}, \boldsymbol{\sigma}) = &\int_{\mathcal{S}} \left\{ w(\boldsymbol{\varepsilon}) - \rho \mathbf{b} \cdot \mathbf{u} - \boldsymbol{\sigma} : \left[\boldsymbol{\varepsilon} - \frac{1}{2}\left(\mathrm{grad}\, \mathbf{u} + (\mathrm{grad}\, \mathbf{u})^T\right) \right] \right\} dv \\
&- \int_{\partial\mathcal{S}_\sigma} \bar{\mathbf{t}} \cdot \mathbf{u}\, d\Gamma - \int_{\partial\mathcal{S}_\varphi} \mathbf{t} \cdot (\mathbf{u} - \bar{\mathbf{u}})\, da
\end{aligned}
\tag{8.457}
$$

When the variation of the function vanishes, the following equations can be obtained in the body

$$
\begin{aligned}
&\boldsymbol{\sigma} = \rho \frac{\partial \Psi(\boldsymbol{\varepsilon})}{\partial \boldsymbol{\varepsilon}}, \\
&\boldsymbol{\varepsilon} = \frac{1}{2}\left(\mathrm{grad}\, \mathbf{u} + (\mathrm{grad}\, \mathbf{u})^T\right), \qquad \text{for } \mathbf{x} \in \mathcal{S} \\
&\rho \mathbf{b} + \mathrm{div}\, \boldsymbol{\sigma} = \mathbf{0}
\end{aligned}
\tag{8.458}
$$

and on the surface

$$
\begin{aligned}
&\mathbf{t} = \bar{\mathbf{t}}, \qquad \text{for } \mathbf{x} \in \partial\mathcal{S}_\sigma \\
&\mathbf{t} = \boldsymbol{\sigma} \cdot \hat{\mathbf{n}}, \quad \mathbf{u} = \bar{\mathbf{u}}, \quad \text{for } \mathbf{x} \in \partial\mathcal{S}_\varphi
\end{aligned}
\tag{8.459}
$$

Next, from the three field variational formulation, we shall deduce the two field variational formulation.

8.4.8 Hellinger-Reissner Variational Principle

Most of mixed variational formulations can be derived from this principle (Pian and Tong 1969). The first Piola-Kirchhoff stress tensor can be given by

$$
\rho_0 \frac{\partial \Psi(\mathbf{F})}{\partial \mathbf{F}} = \frac{\partial W(\mathbf{F})}{\partial \mathbf{F}} = \mathbf{P}
\tag{8.460}
$$

The Legendre transformation gives

$$W(\mathbf{F}) = \mathbf{P} : \mathbf{F} - W^*(\mathbf{P}), \quad \because \ \rho_0 \Psi(\mathbf{F}) = W(\mathbf{F}) \tag{8.461}$$

where $W^*(\mathbf{P})$ stands for the complementary energy density function. Therefore, we have

$$\mathbf{F} = \frac{\partial W^*(\mathbf{P})}{\partial \mathbf{P}} \tag{8.462}$$

Substituting Equation (8.461) into Equation (8.454) leads to

$$\begin{aligned} \mathcal{U}_{\text{HR}}(\boldsymbol{\varphi}, \mathbf{P}) = & \int_{\mathcal{B}} \left(-W^*(\mathbf{P}) - \rho_0 \mathbf{B} \cdot \boldsymbol{\varphi} + \mathbf{P} : \text{GRAD}\, \boldsymbol{\varphi}\right) dV \\ & - \int_{\partial \mathcal{B}_\sigma} \bar{\mathbf{T}} \cdot \boldsymbol{\varphi}\, dA - \int_{\partial \mathcal{B}_\varphi} \mathbf{T} \cdot (\boldsymbol{\varphi} - \bar{\boldsymbol{\varphi}})\, dA \end{aligned} \tag{8.463}$$

This is the Hellinger-Reissner functional (Pian 1995; Reissner 1950; Washizu 1982) whose variation yields

$$\begin{aligned} \delta \mathcal{U}_{\text{HR}}(\boldsymbol{\varphi}, \mathbf{P}) = & \int_{\mathcal{B}} \left[\left(\mathbf{F} - \frac{\partial W^*(\mathbf{P})}{\partial \mathbf{P}}\right) : \delta \mathbf{P} - \rho_0 \mathbf{B} \cdot \delta \boldsymbol{\varphi} + \mathbf{P} : \text{GRAD}\, \delta \boldsymbol{\varphi}\right] dV \\ & - \int_{\partial \mathcal{B}_\sigma} \bar{\mathbf{T}} \cdot \delta \boldsymbol{\varphi}\, dA - \int_{\partial \mathcal{B}_\varphi} \delta \mathbf{T} \cdot (\boldsymbol{\varphi} - \bar{\boldsymbol{\varphi}})\, dA - \int_{\partial \mathcal{B}_\varphi} \mathbf{T} \cdot \delta \boldsymbol{\varphi}\, dA \end{aligned} \tag{8.464}$$

Recall that the relation holds in the body:

$$\mathbf{P} : \text{GRAD}\, \boldsymbol{\varphi} = \text{DIV}\, (\delta \boldsymbol{\varphi} \cdot \mathbf{P}) - \delta \boldsymbol{\varphi} \cdot \text{DIV}\, \mathbf{P} \tag{8.465}$$

Substituting Equation (8.465) into Equation (8.464) yields the following

$$\begin{aligned} \delta \mathcal{U}_{\text{HR}}(\boldsymbol{\varphi}, \mathbf{P}) = & \int_{\mathcal{B}} \left[\left(\mathbf{F} - \frac{\partial W^*(\mathbf{P})}{\partial \mathbf{P}}\right) : \delta \mathbf{P} - (\rho_0 \mathbf{B} + \text{DIV}\, \mathbf{P}) \cdot \delta \boldsymbol{\varphi}\right] dV \\ & + \int_{\partial \mathcal{B}_\sigma} (\mathbf{P} \cdot \hat{\mathbf{N}} - \bar{\mathbf{T}}) \cdot \delta \boldsymbol{\varphi} dA - \int_{\partial \mathcal{B}_\varphi} \delta \mathbf{T} \cdot (\boldsymbol{\varphi} - \bar{\boldsymbol{\varphi}})\, dA \\ & - \int_{\partial \mathcal{B}_\varphi} (\mathbf{T} - \mathbf{P} \cdot \hat{\mathbf{N}}) \cdot \delta \boldsymbol{\varphi}\, dA \end{aligned} \tag{8.466}$$

At the stationary point, we have the following equations,

$$\left.\begin{aligned} \mathbf{F} &= \frac{\partial W^*(\mathbf{P})}{\partial \mathbf{P}}, \\ \rho_0 \mathbf{B} + \text{DIV}\, \mathbf{P} &= \mathbf{0} \end{aligned}\right\} \quad \text{for } \mathbf{X} \in \mathcal{B} \tag{8.467}$$

and on the surface

$$\begin{aligned} \mathbf{P} \cdot \hat{\mathbf{N}} = \bar{\mathbf{T}}, \quad & \text{for } \mathbf{X} \in \partial \mathcal{B}_\sigma \\ \boldsymbol{\varphi} = \bar{\boldsymbol{\varphi}}, \quad & \text{for } \mathbf{X} \in \partial \mathcal{B}_\varphi \end{aligned} \tag{8.468}$$

Notice that when one applies admissible trial and test functions to Equation (8.466), the errors arising from the satisfaction of the above equations could be minimized.

8.5 DIRECT VARIATIONAL METHODS FOR TWO-POINT BOUNDARY-VALUE PROBLEMS

According to Sobolev (Mikhlin 1964), direct variational methods are those for obtaining the approximate solution of problems of the theory of differential and integral equations which reduce these problems to finite systems of algebraic equations. It is well known that the inclusion of the variation in formulating the problems provides a mechanism to facilitate the discretization of the domain. The discretization can be carried out by use of a finite number of approximate functions which are linearly independent. As representative methods of this class, we shall discuss the Rayleigh-Ritz method and the Bubnov-Galerkin weighted residual method.

Strong Form To Be Solved In three dimensional Euclidean space, let Ω denote a domain of interest as an open bounded set. The boundary of the domain as an open set $\partial\Omega$ is divided into two open sets, i.e. $\partial\Omega = \partial\Omega_u \cup \partial\Omega_\sigma$, where $\partial\Omega_u$ denotes the prescribed displacement boundaries and $\partial\Omega_\sigma$ represents the prescribed traction boundaries. Next, one wants to solve a two-point boundary value problem whose strong form can be given as follows

$$\begin{aligned} &L_{2m}u(\mathbf{x}) - f(\mathbf{x}) = 0, \quad \forall \mathbf{x} \in \Omega \subset \mathbb{R}^n \\ &E_\alpha u(\mathbf{x}) = \bar{u}_{eb\alpha}, \quad \alpha = 0, \ldots, m-1, \quad \forall \mathbf{x} \in \partial\Omega_u \\ &N_\beta u(\mathbf{x}) = \hat{u}_{nb\beta}, \quad \beta = m, \ldots, 2m-1, \quad \forall \mathbf{x} \in \partial\Omega_\sigma \end{aligned} \tag{8.469}$$

where $L_{2m} : C^{2m}(\Omega) \rightarrow L_2(\Omega)$ denotes a linear differential operator of order $2m$ (Bathe 1996; Carey and Oden 1983; Mikhlin 1964; Reddy 1986; Rektorys 1980; Strang and Fix 1973), E_α a linear differential operator of up to order $m-1$ for essential boundary conditions, and N_β a linear differential operator involving derivatives of order m to $2m-1$ for natural boundary conditions (Bathe 1996; Carey and Oden 1983; Ciarlet 1978). The dependent variable $u(\mathbf{x}) : \Omega \rightarrow \mathbb{R}$ denotes a real-valued function. And, $f(\mathbf{x}) : \Omega \rightarrow \mathbb{R}$ denotes the forcing function as a real-valued function. Note that as a continuity condition, an exact solution of the dependent variable $u(\mathbf{x})$ is of class $C^{2m}(\Omega)$, whereas the forcing function $f(\mathbf{x})$ is of class $C^0(\Omega)$. It is said to be homogeneous when all of the boundary conditions are zero, i.e. $E_\alpha u(\mathbf{x}) = N_\beta u(\mathbf{x}) = 0, \quad \forall \mathbf{x} \in \partial\Omega$.

8.5.1 Rayleigh-Ritz Method

Historically, Lord Rayleigh (1842-1919) first employed an admissible trial solution to find an approximate fundamental natural frequency in structural dynamics. Ritz (1878-1909) generalized Rayleigh's idea to find approximate solutions in elastostatics. Therefore, our attention in this subsection is confined to elastostatics, where external potential energies with dead loads are assumed to exist. Suppose that the linear differential operator is symmetric (self-adjoint) and positive definite, that is,

$$\begin{aligned} &(L_{2m}u(\mathbf{x}), v(\mathbf{x}))_\Omega = (u(\mathbf{x}), L_{2m}v(\mathbf{x}))_\Omega, \quad u(\mathbf{x}), v(\mathbf{x}) \in H_0^{2m}(\Omega), \\ &(L_{2m}u(\mathbf{x}), u(\mathbf{x}))_\Omega > 0, \quad , u(\mathbf{x}) \neq 0, v(\mathbf{x}) \neq 0 \end{aligned} \tag{8.470}$$

where $(,)_{(\Omega)}$ denotes the inner product operator (Rektorys 1980) over the open bounded set $\Omega \subset \mathbb{R}^n$. Then, we can define a functional $I(w(\mathbf{x})) : C^{m-1}(\Omega) \rightarrow \mathbb{R}$ as a functional mapping a dependent variable into a real value (Reddy 1986). Note that the functional is a reformulation which allows a weaker continuity condition for the admissible trial function (Carey and Oden 1983). Then, the functional can be defined as

$$
\begin{aligned}
& I(w(\mathbf{x})) = \frac{1}{2}B(w(\mathbf{x}), w(\mathbf{x})) - l(w(\mathbf{x})), \quad w(\mathbf{x}) \in U \subset H^m(\Omega), \\
& l(w(\mathbf{x})) = (\hat{u}_{nb\beta}, w(\mathbf{x}))_{\partial\Omega_\sigma} + (f(\mathbf{x}), w(\mathbf{x}))_\Omega, \\
& U = \left\{ w(\mathbf{x}) : \Omega \to \mathbb{R} | w(\mathbf{x}) \in H^m(\Omega) \subset C^{m-1}(\Omega),\ E_\alpha u(\mathbf{x})|_{\partial\Omega_u} = \bar{u}_{eb\alpha} \right\}
\end{aligned}
\tag{8.471}
$$

where $B(w(\mathbf{x}), w(\mathbf{x})) : U \times U \to \mathbb{R}$ denotes a bilinear quadratic functional involving derivatives of order m and $l(w(\mathbf{x}))_\Omega$ a linear functional (Ciarlet 1978; Johnson 1987). Since the bilinear quadratic functional is positive definite, the functional $I(w(\mathbf{x}))$ has a convexity (Troutman 1983). Subsequently, the functional $I(w(\mathbf{x}))$ has a minimum at a local point $u(\mathbf{x})$. Hence, when its variation vanishes at the critical point $w(\mathbf{x}) = u(\mathbf{x})$, we can obtain a weak form as

$$
\begin{aligned}
& D_{v(\mathbf{x})} I(u(\mathbf{x})) = B(u(\mathbf{x}), v(\mathbf{x})) - (\hat{u}_{nb\beta}, v(\mathbf{x}))_{\partial\Omega_\sigma} - (f(\mathbf{x}), v(\mathbf{x}))_\Omega = 0, \\
& U = \left\{ u(\mathbf{x}) : \Omega \to \mathbb{R} | u(\mathbf{x}) \in H^m(\Omega) \subset C^{m-1}(\Omega),\ E_\alpha u(\mathbf{x})|_{\partial\Omega_u} = \bar{u}_{eb\alpha} \right\}, \\
& V = \left\{ v(\mathbf{x}) : \Omega \to \mathbb{R} | v(\mathbf{x}) \in H_0^m(\Omega) \subset C_0^{m-1}(\Omega),\ E_\alpha v(\mathbf{x})|_{\partial\Omega_u} = 0 \right\}
\end{aligned}
\tag{8.472}
$$

where $D_{v(\mathbf{x})} I(u(\mathbf{x}))$ denotes the Gâteaux variation of the functional (Marsden and Hughes 1983; Simo and Hughes 1998; Suetake et al. 2003; Troutman 1996) as a directional derivative of the functional $I(u(\mathbf{x}))$ at $u(\mathbf{x})$ in the direction $v(\mathbf{x})$ (Edwards 1994). Note that $H_0^m(\Omega)$ is a subspace of $H^m(\Omega)$ and $C_0^{m-1}(\Omega)$ is a subspace of $C^{m-1}(\Omega)$, that is,

$$
\begin{aligned}
H_0^m(\Omega) &= \left\{ v(\mathbf{x}) : \Omega \to \mathbb{R} | v(\mathbf{x}) \in H^m(\Omega),\ E_\alpha v(\mathbf{x})|_{\partial\Omega_u} = 0 \right\}, \\
C_0^{m-1}(\Omega) &= \left\{ v(\mathbf{x}) : \Omega \to \mathbb{R} | v(\mathbf{x}) \in C^{m-1}(\Omega),\ E_\alpha v(\mathbf{x})|_{\partial\Omega_u} = 0 \right\}
\end{aligned}
\tag{8.473}
$$

The Gâteaux variation as a linear functional (Simo and Hughes 1998) is often written as

$$
D_{v(\mathbf{x})} I(u(\mathbf{x})) = DI(u(\mathbf{x})) \cdot v(\mathbf{x}) \tag{8.474}
$$

where $DI(u(\mathbf{x}))$ is called the Fréchet derivative (Ciarlet 1978). The Gâteaux variation can be defined if and only if the functional $I(w(\mathbf{x}))$ is Fréchet differentiable at $w(\mathbf{x}) = u(\mathbf{x})$. Note that the equation given in Equation (8.472) can be rewritten as

$$
\begin{aligned}
D_{v(\mathbf{x})} I(u(\mathbf{x})) &= (L_{2m} u(\mathbf{x}), v(\mathbf{x}))_\Omega - (f(\mathbf{x}), v(\mathbf{x}))_\Omega \\
&\quad + (N_\beta u(\mathbf{x}), v(\mathbf{x}))_{\partial\Omega_\sigma} - (\hat{u}_{nb\beta}, v(\mathbf{x}))_{\partial\Omega_\sigma} = 0
\end{aligned}
\tag{8.475}
$$

since the linear differential operator is symmetric. Consequently, we obtain

$$
\begin{aligned}
D_{v(\mathbf{x})} I(u(\mathbf{x})) &= ((L_{2m} u(\mathbf{x}) - f(\mathbf{x})), v(\mathbf{x}))_\Omega \\
&\quad + ((N_\beta u(\mathbf{x}) - \hat{u}_{nb\beta}), v(\mathbf{x}))_{\partial\Omega_\sigma} = 0
\end{aligned}
\tag{8.476}
$$

which consists of two residual forms including the natural boundary conditions.

In the Rayleigh-Ritz method (Rayleigh 1870; Ritz 1909), by means of a trial function $\tilde{u}(\mathbf{x}) \in \tilde{U} \subset H^m(\Omega) \subset C^{m-1}(\Omega)$, in view of Equation (8.471) the functional $I(\tilde{u}(\mathbf{x}))$ can be approximated as follows

$$
\begin{aligned}
I(\tilde{u}(\mathbf{x})) &= \frac{1}{2}B(\tilde{u}(\mathbf{x}), \tilde{u}(\mathbf{x})) - l(\tilde{u}(\mathbf{x})), \quad \tilde{u}(\mathbf{x}) \in \tilde{U} \subset H^m(\Omega), \\
l(\tilde{u}(\mathbf{x})) &= (\hat{u}_{nb\beta}, \tilde{u}(\mathbf{x}))_{\partial\Omega_\sigma} + (f(\mathbf{x}), \tilde{u}(\mathbf{x}))_\Omega, \\
\tilde{U} &= \left\{\tilde{u}(\mathbf{x}) : \Omega \to \mathbb{R} | \tilde{u}(\mathbf{x}) \in H^m(\Omega) \subset C^{m-1}(\Omega), E_\alpha \tilde{u}(\mathbf{x}) = \bar{u}_{eb\alpha}\right\}
\end{aligned}
\tag{8.477}
$$

where $B(\tilde{u}(\mathbf{x}), \tilde{u}(\mathbf{x})) : \tilde{U} \times \tilde{U} \to \mathbb{R}$ denotes a bilinear quadratic functional involving derivatives of order m. $\tilde{U}$ is a subspace of U. Note that the trial function should be chosen to satisfy only the essential boundary conditions. At the minimum point of the approximate functional $I(\tilde{u}(\mathbf{x}))$, its variation should vanish as follows

$$
D_{\tilde{v}(\mathbf{x})} I(\tilde{u}(\mathbf{x})) = B(\tilde{u}(\mathbf{x}), \tilde{v}(\mathbf{x})) - (\hat{u}_{nb\beta}, \tilde{v}(\mathbf{x}))_{\partial\Omega_\sigma} - (f(\mathbf{x}), \tilde{v}(\mathbf{x}))_\Omega = 0 \tag{8.478}
$$

where $\tilde{v}(\mathbf{x})$ denotes the variation of the trial function $\tilde{u}(\mathbf{x})$. As a consequence, the above equation can be rewritten as

$$
\begin{aligned}
B(\tilde{u}(\mathbf{x}), \tilde{v}(\mathbf{x})) &= l(\tilde{v}(\mathbf{x})), \quad \tilde{u}(\mathbf{x}) \in H^m(\Omega), \ \tilde{v}(\mathbf{x}) \in H_0^m(\Omega) \\
l(\tilde{v}(\mathbf{x})) &= (\hat{u}_{nb\beta}, \tilde{v}(\mathbf{x}))_{\partial\Omega_\sigma} + (f(\mathbf{x}), \tilde{v}(\mathbf{x}))_\Omega
\end{aligned}
\tag{8.479}
$$

Next, the n-parameter Ritz approximate solution $\tilde{u}(\mathbf{x}) \in H^m(\Omega) \subset C^{m-1}(\Omega)$ is expressed as a linear combination of the n number of independent, differentiable coordinate functions $\phi_i(\mathbf{x}) \in H_0^m(\Omega) \subset C_0^{m-1}(\Omega)$, $i = 1, \ldots, n$

$$
\tilde{u}(\mathbf{x}) = \phi_0(\mathbf{x}) + \sum_{i=1}^{n} b_i \phi_i(\mathbf{x}) \tag{8.480}
$$

where b_i are unknown Ritz coefficients and $\phi_0(\mathbf{x})$ are differentiable functions satisfying the essential boundary conditions. The coordinate functions $\phi_i(\mathbf{x}) \in H_0^m(\Omega) \subset C_0^{m-1}(\Omega)$, $i = 1, \ldots, n$ should vanish at the prescribed displacement boundaries such that the Ritz approximate solution is kinematically admissible. Hence, the variation $\tilde{v}(\mathbf{x})$ of the trial function $\tilde{u}(\mathbf{x})$ can be written as

$$
\tilde{v}(\mathbf{x}) = \sum_{i=1}^{n} \phi_i(\mathbf{x}) \delta b_i \tag{8.481}
$$

The Gâteaux variation can be written as

$$
\begin{aligned}
D_{\tilde{v}(\mathbf{x})} I(\tilde{u}(\mathbf{x})) &= DI(\tilde{u}(\mathbf{x})) \cdot \tilde{v}(\mathbf{x}) = DI(\tilde{u}(\mathbf{x})) \sum_{i=1}^{n} \phi_i(x) \delta b_i \\
&= DI(\tilde{u}(\mathbf{x})) \sum_{i=1}^{n} \frac{\partial \tilde{u}(x)}{\partial b_i} \delta b_i = \sum_{j=1}^{n} \frac{\partial I(\tilde{u}(\mathbf{x}))}{\partial b_i} \delta b_i = 0
\end{aligned}
\tag{8.482}
$$

Consequently, the Ritz method claims that at the minimum point of the approximate functional $I(\tilde{u}(\mathbf{x}))$, its variation should vanish if

$$
\frac{\partial I(\tilde{u}(\mathbf{x}))}{\partial b_i} = 0, \quad i = 1, \ldots, n \tag{8.483}
$$

since δb_i in Equation (8.482) is arbitrary. Therefore, Substituting Equations (8.532) into Equation (8.529), we obtain n number of simultaneous algebraic equations as follows:

$$\begin{aligned} &B(\tilde{u}(\mathbf{x}), \phi_i(\mathbf{x})) = l(\phi_i(\mathbf{x})), \quad \tilde{u}(\mathbf{x}) \in H^m(\Omega), \ \tilde{v}(\mathbf{x}) \in H_0^m(\Omega) \\ &l(\phi_i(\mathbf{x})) = (\hat{u}_{nb\beta}, \phi_i(\mathbf{x}))_{\partial\Omega_\sigma} + (f(\mathbf{x}), \phi_i(\mathbf{x}))_\Omega, \ \beta = m, \ldots, 2m-1 \end{aligned} \tag{8.484}$$

since δb_i $i = 1, \ldots, n$ are arbitrary. Substituting the test function given in Equation (8.481) into Equation (8.479) yields

$$\begin{aligned} \frac{\partial I(\tilde{u}(\mathbf{x}))}{\partial b_i} &= B(\tilde{u}(\mathbf{x}), \phi_i(\mathbf{x})) - l(\phi_i(\mathbf{x})) \\ &= \sum_{j=1}^{n} A_{ij} b_j - c_i = 0 \end{aligned} \quad , \quad i = 1, \ldots, n \tag{8.485}$$

where the coefficients A_{ij} and c_i are determined by

$$\begin{aligned} A_{ij} &= B(\phi_i(\mathbf{x}), \phi_j(\mathbf{x})), \\ c_i &= l(\phi_i(\mathbf{x})) - B(\phi_i(\mathbf{x}), \phi_0(\mathbf{x})) \end{aligned} \tag{8.486}$$

Thus, the n-number of undetermined parameters b_i, $i = 1, \ldots, n$ can be determined by the above n algebraic equations.

Note that Equation (8.479) is identical to the weak form in the modified Bubnov-Galerkin method we shall discuss in the subsequent section. Further, for homogeneous boundary conditions, i.e. $\tilde{u}(\mathbf{x}) \in H_0^m(\Omega)$ and $\hat{u}_{nb\beta} = 0$, Equation (8.476) reduces to the Bubnov-Galerkin weighted residual form, which will be discussed in the sequel.

Illustrative Example

Example 8.1 Rayleigh-Ritz Method, Axial Bar

Consider the axial bar as shown in Figure 8.2. The functional as the total potential energy for the axial bar is given by

$$\mathcal{U}(u) = \int_L \frac{1}{2} EA(x) \left(\frac{du(x)}{dx}\right)^2 dx - \int_L \bar{b}(x) u(x) dx - F_L u_L \tag{8.487}$$

The above functional is subjected to the essential boundary condition

$$u = \bar{u}_0 = 0, \quad \text{for } x = 0 \tag{8.488}$$

Hence, the bilinear quadratic functional $B(u(x), u(x)) : U \times U \to \mathbb{R}$ involving derivatives of order 1 and the linear functional $l(u(x))$ are defined as

$$\begin{aligned} B(u(x), u(x)) &= \int_L EA(x) \left(\frac{du(x)}{dx}\right)\left(\frac{du(x)}{dx}\right) dx, \\ l(u(x)) &= \int_L \bar{b}(x) u(x) dx + F_L u_L \end{aligned} \tag{8.489}$$

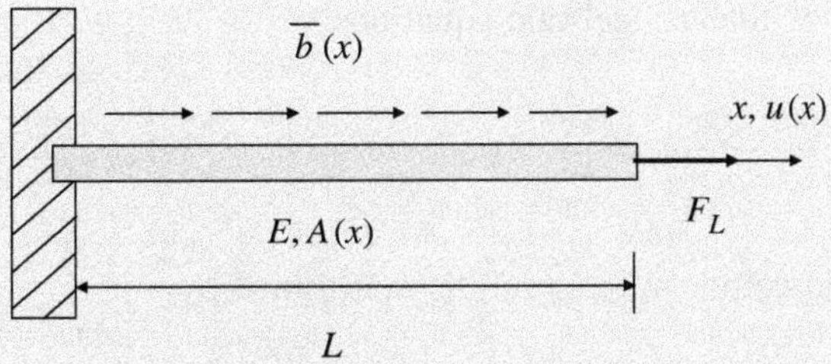

Figure 8.2. Axial bar in statics

Thus, the Gâteaux variation of the functional, which is the weak form for the boundary-value problem, gives

$$\begin{gathered} D_{v(x)}I(u(x)) = B(u(x), v(x)) - l(v(x)) = 0, \\ u(x) \in H^1(L) \subset C^0(L),\ v(x) \in H_0^1(L) \subset C_0^0(L) \end{gathered} \tag{8.490}$$

which can be written as

$$\begin{gathered} \int_L EA(x)\left(\frac{du(x)}{dx}\right)\left(\frac{dv(x)}{dx}\right)dx = \int_L \bar{b}(x)v(x)dx + F_L v_L, \\ u(x) \in H^1(L) \subset C^0(L),\ v(x) \in H_0^1(L) \subset C_0^0(L) \end{gathered} \tag{8.491}$$

The n-parameter Ritz approximate solution $\tilde{u}(x) \in H^1(L \subset C^0(L)$ is expressed as a linear combination of the n number of independent, differentiable coordinate functions $\phi_i(x) \in H_0^1(L) \subset C_0^0(L),\ i = 1, \ldots, n$

$$\tilde{u}(x) = \phi_0(x) + \sum_{i=1}^{n} b_i \phi_i(x) \tag{8.492}$$

where b_i are unknown Ritz coefficients and $\phi_0(\mathbf{x})$ are differentiable functions satisfying the essential boundary conditions. Upon the substitution of the n-parameter Ritz approximate solution into the functional, we have

$$\begin{aligned} \mathcal{U}(\tilde{u}(x)) = &\int_L \frac{1}{2}EA(x)\left(\sum_{i=j}^{n} b_j \frac{d\phi_j(x)}{dx} + \frac{d\phi_0(x)}{dx}\right)^2 dx \\ &- \int_L \bar{b}(x)\left(\sum_{j=1}^{n} b_j\phi_j(x) + \phi_0(x)\right)dx - F_L\left(\sum_{j=1}^{n} b_j\phi_i(L) + \phi_0(L)\right) \end{aligned} \tag{8.493}$$

The Ritz method requires that the variation of the functional vanish

$$\delta\mathcal{U}(\tilde{u}(x)) = \sum_{i=1}^{n} \frac{\partial\mathcal{U}(\tilde{u}(x))}{\partial b_i}\delta b_i = 0 \tag{8.494}$$

which yields n simultaneous linear algebraic equations

$$\frac{\partial \mathcal{U}(\tilde{u}(x))}{\partial b_i} = \sum_{j=1}^{n} A_{ij} b_j - c_i = 0, \quad i = 1, \ldots, n \tag{8.495}$$

where the Ritz matrix coefficients A_{ij} and c_i are determined by

$$\begin{aligned} A_{ij} &= B(\phi_i(x), \phi_j(x)) = \int_L EA\,(x) \frac{d\phi_i(x)}{dx} \frac{d\phi_j(x)}{dx} dx, \\ c_i &= l(\phi_i(x)) - B(\phi_i(x), \phi_0(x)) \\ &= \int_L \bar{b}(x)\phi_i(x)dx + F_L \phi_i(L) - \int_L EA\,(x) \frac{d\phi_i(x)}{dx} \frac{d\phi_0(x)}{dx} dx \end{aligned} \tag{8.496}$$

Thus, the n Ritz coefficients b_i, $i = 1, \ldots, n$ can be determined by the above n algebraic equations.

Example 8.2 Rayleigh-Ritz Method, Euler-Bernoulli Beam

Consider the Euler-Bernoulli beam clamped at the end as shown in Figure 8.3. The functional as the total potential energy for the Euler-Bernoulli beam is given by

$$\mathcal{U}(u) = \int_L \frac{1}{2} EI\,(x) \left(\frac{d^2 w(x)}{dx^2} \right)^2 dx - \int_L \bar{w} w(x) dx - Q_L w_L + M_L \frac{dw(L)}{dx} \tag{8.497}$$

The above functional is subjected to the essential boundary condition

$$\begin{aligned} \left. \frac{d\delta w}{dx} \right|_{x=0} &= 0, \qquad \text{for } x = 0, \\ \delta w|_{x=0} &= 0, \qquad \text{for } x = 0 \end{aligned} \tag{8.498}$$

Hence, the bilinear quadratic functional $B(w(x), w(x)) : U \times U \to \mathbb{R}$ involving derivatives of order 1 and the linear functional $l(w(\mathbf{x}))$ are defined as

$$\begin{aligned} B(w(x), w(x)) &= \int_L EI\,(x) \left(\frac{d^2 w(x)}{dx^2} \right) \left(\frac{d^2 w(x)}{dx^2} \right) dx, \\ l(w(x)) &= \int_L \bar{w} w(x) dx + Q_L w_L - M_L \frac{dw(L)}{dx} \end{aligned} \tag{8.499}$$

The Gâteaux variation of the functional yields

$$\begin{aligned} D_{v(x)} I(w(x)) &= B(w(x), v(x)) - l(v(x)) = 0, \\ w(x) \in H^2(L) \subset C^1(L), &\ \ v(x) \in H_0^2(L) \subset C_0^1(L) \end{aligned} \tag{8.500}$$

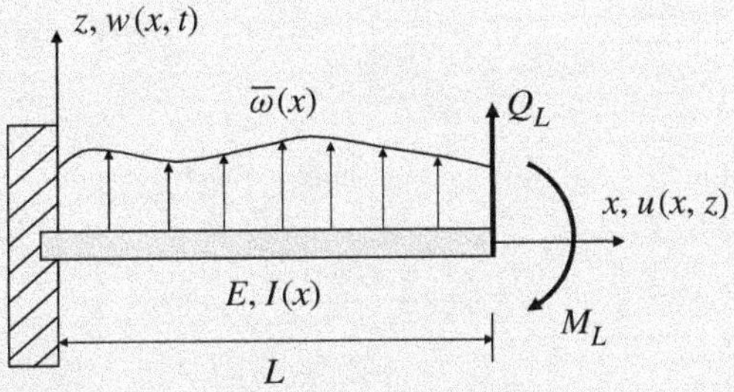

Figure 8.3. Cantilever Euler-Bernoulli beam in statics

which can be written as

$$\int_L EI(x)\left(\frac{d^2w(x)}{dx^2}\right)\left(\frac{d^2v(x)}{dx^2}\right)dx = \int_L \bar{w}v(x)dx + Q_L vL - M_L\frac{dv(L)}{dx},$$
$$w(x) \in H^2(L) \subset C^1(L),\ v(x) \in H_0^2(L) \subset C_0^1(L) \tag{8.501}$$

Then, the n-parameter Ritz approximate solution $\tilde{w}(x) \in H^2(L) \subset C^1(L)$ is expressed as a linear combination of the n number of independent, differentiable coordinate functions $\phi_i(x) \in H_0^2(L) \subset C_0^1(L),\ i = 1, \ldots, n$

$$\tilde{w}(x) = \phi_0(x) + \sum_{j=1}^{n} b_j\phi_j(x) \tag{8.502}$$

where b_j are unknown Ritz coefficients and $\phi_0(x)$ are coordinate functions satisfying the essential boundary conditions. Upon the substitution of the n-parameter Ritz approximate solution into the functional, we have

$$\begin{aligned}\mathcal{U}(\tilde{w}(x)) = &\int_L \frac{1}{2}EI(x)\left(\sum_{j=1}^{n} b_j\frac{d^2\phi_j(x)}{dx^2} + \frac{d^2\phi_0(x)}{dx^2}\right)^2 dx \\ &- \int_L \bar{w}\left(\sum_{j=1}^{n} b_j\phi_j(x) + \phi_0(x)\right)dx - Q_L\left(\sum_{j=1}^{n} b_j\phi_j(L) + \phi_0(L)\right) \\ &+ M_L\left(\sum_{j=1}^{n} b_j\frac{d\phi_j(L)}{dx} + \frac{d\phi_0(L)}{dx}\right)\end{aligned} \tag{8.503}$$

The Ritz method claims that

$$\delta\mathcal{U}(\tilde{w}(x)) = \sum_{i=1}^{n}\frac{\partial\mathcal{U}(\tilde{w}(x))}{\partial b_i}\delta b_i = 0 \tag{8.504}$$

which yields

$$\frac{\partial \mathcal{U}(\tilde{w}(x))}{\partial b_i} = \sum_{j=1}^{n} A_{ij} b_j - c_i = 0, \quad i = 1, \ldots, n \tag{8.505}$$

where the Ritz matrix coefficients A_{ij} and c_i are determined by

$$\begin{aligned} A_{ij} &= B(\phi_i(x), \phi_j(x)) = \int_L EI\,(x) \frac{d^2\phi_i(x)}{dx^2} \frac{d^2\phi_j(x)}{dx^2} dx, \\ c_i &= l(\phi_i(x)) - B(\phi_i(x), \phi_0(x)) = \int_L \bar{w}\phi_i(x)dx + Q_L\phi_i(L) - M_L \frac{d\phi_i(L)}{dx} \\ &- \int_L EI\,(x) \frac{d^2\phi_i(x)}{dx^2} \frac{d^2\phi_0(x)}{d^2x} dx \end{aligned} \tag{8.506}$$

Thus, the n Ritz coefficients b_i, $i = 1, \ldots, n$ can be determined by the above n algebraic equations.

8.5.2 Bubnov-Galerkin Weighted Residual Method

According to Finlayson (Finlayson 1972), the terminology, 'the weighted residual method' or 'the method of weighted residuals', is credited to Crandall (Crandall 1956). The Galerkin method (Galerkin 1915) is one of the methods of weighted residuals, which are also called the direct approximate methods which include the subdomain method, the least squares method, the collocation method and the Galerkin method (Crandall 1956). The classical Galerkin method is often called the Bubnov-Galerkin method (Bubnov 1913; Finlayson 1972; Galerkin 1915; Hughes 1987; Mikhlin 1964).

Suppose that the strong form in Equation (8.469) is given. Then, a residual function $R(u(\mathbf{x}))$ is defined as

$$R(u(\mathbf{x})) = L_{2m}u(\mathbf{x}) - f(\mathbf{x}) = 0, \quad u(\mathbf{x}) \in C^{2m}(\Omega) \tag{8.507}$$

To find an approximate solution of Equation (8.469), the residual function in terms of a trial function $\tilde{u}(\mathbf{x}) \in \tilde{U} \subset H^{2m}(\Omega) \subset C^{2m-1}(\Omega)$ is given by

$$R(\tilde{u}(\mathbf{x})) = L_{2m}\tilde{u}(\mathbf{x}) - f(\mathbf{x}) \neq 0, \quad \tilde{u}(\mathbf{x}) \in C^{2m-1}(\Omega) \tag{8.508}$$

Note that the trial function $\tilde{u}(\mathbf{x})$ above is differentiable 2m times over the domain Ω. The Bubnov-Galerkin weighted-residual form Crandall (1956), Finlayson (1972), Galerkin (1915), Mikhlin (1964) claims that the residual function in terms of the trial function $\tilde{u}(\mathbf{x}) \in \tilde{U} \subset H^{2m}(\Omega)$ is orthogonal to all of the basis functions (the basis functions are also called coordinate functions, which are linearly independent.) of a test function $\tilde{v}(\mathbf{x}) \in \tilde{V} \subset H_0^{2m}$.

$$(L_{2m}\tilde{u}(\mathbf{x}) - f(\mathbf{x}), \tilde{v}(\mathbf{x}))_\Omega = 0, \quad \tilde{u}(\mathbf{x}) \in H^{2m}(\Omega), \tilde{v}(\mathbf{x}) \in H_0^{2m}(\Omega) \tag{8.509}$$

which means that the weighted averages of the residual are required to vanish (Crandall 1956, page 149). The trial function $\tilde{u}(\mathbf{x})$ belongs to a Sobolev space of order $2m$, $W^{2m,2}(\Omega)$ and the forcing function $f(\mathbf{x})$ is a member of a Sobolev space (Adams 1975) of order 0; in other words $\tilde{u}(\mathbf{x}) \in H_0^{2m}(\Omega)$ and $f(\mathbf{x}) \in H^0(\Omega)$. Then, we wish to solve for the admissible approximate functions $\tilde{u}(\mathbf{x})$ such that

Equation (8.509) is true. The admissible trial function $\tilde{u}(\mathbf{x})$ belongs to a function space $\tilde{U}$, whereas the admissible test function $\tilde{v}(\mathbf{x})$ belongs to a function space $\tilde{V}$ as follows

$$\begin{aligned} \tilde{U} &= \left\{\tilde{u}(\mathbf{x}) : \Omega \to \mathbb{R} | \tilde{u} \in H^{2m}(\Omega),\ E_\alpha \tilde{u}(\mathbf{x}) = \bar{u}_{eb\alpha},\ N_\beta u(\mathbf{x}) = \hat{u}_{nb\beta}\right\}, \\ \tilde{V} &= \left\{\tilde{v}(\mathbf{x}) : \Omega \to \mathbb{R} | \tilde{v} \in H_0^{2m}(\Omega),\ E_\alpha \tilde{u}(\mathbf{x}) = 0,\ N_\beta u(\mathbf{x}) = 0\right\} \end{aligned} \tag{8.510}$$

In the case of two-point boundary-value problems with homogeneous boundary conditions, the Bubnov-Galerkin method requires that $\tilde{U}$ be equal to $\tilde{V}$.

Next, the n-parameter admissible trial function $\tilde{u}(\mathbf{x}) \in H^{2m}(\Omega) \subset C^{2m-1}(\Omega)$ is expressed as a linear combination of the n number of independent, differentiable coordinate functions $\phi_i(\mathbf{x}) \in H_0^{2m}(\Omega) \subset C_0^{2m-1}(\Omega),\ i = 1, \ldots, n$

$$\tilde{u}(\mathbf{x}) = \phi_0(\mathbf{x}) + \sum_{i=1}^{n} b_i \phi_i(\mathbf{x}) \tag{8.511}$$

where b_i are unknown coefficients and $\phi_0(\mathbf{x})$ are differentiable functions satisfying the essential boundary conditions. The coordinate functions $\phi_i(\mathbf{x}) \in H_0^{2m}(\Omega) \subset C_0^{2m-1}(\Omega),\ i = 1, \ldots, n$ should vanish at the prescribed displacement boundaries such that the trial function is kinematically admissible. Hence, the variation $\tilde{v}(\mathbf{x})$ of the trial function $\tilde{u}(\mathbf{x})$ can be written as

$$\tilde{v}(\mathbf{x}) = \sum_{i=1}^{n} \phi_i(\mathbf{x}) \delta b_i = \sum_{i=1}^{n} \frac{\partial \tilde{v}(\mathbf{x})}{\partial b_i} \delta b_i = \sum_{i=1}^{n} W_i(\mathbf{x}) \delta b_i \tag{8.512}$$

where $W_i(\mathbf{x})$ is called the weighting function, that is,

$$W_i(\mathbf{x}) = \frac{\partial \tilde{v}(\mathbf{x})}{\partial b_i} = \phi_i(\mathbf{x}) \in H_0^{2m}(\Omega) \subset C_0^{2m-1}(\Omega),\quad i = 1, \ldots, n \tag{8.513}$$

Therefore, substituting Equations (8.512) into Equation (8.509), we obtain the n number of simultaneous algebraic equations as follows:

$$(L_{2m}\tilde{u}(\mathbf{x}) - f(\mathbf{x}), \phi_i(\mathbf{x}))_\Omega = 0,\quad i = 1, \ldots, n\ \ \tilde{u}(\mathbf{x}) \in H^{2m}(\Omega) \tag{8.514}$$

since $\delta b_i\ i = 1, \ldots, n$ are arbitrary. Consequently, the Galerkin integral requires that the residual function $R(\tilde{u}(\mathbf{x}))$ be orthogonal to every weighting function $W_i(\mathbf{x}),\ i = 1, \ldots, n$ as follows

$$(R(\tilde{u}(\mathbf{x})), W_i(\mathbf{x}))_\Omega = 0,\quad i = 1, \ldots, n \tag{8.515}$$

which means that the weighted averages of the residual should vanish (Crandall 1956). The Galerkin method takes the weighting function

$$W_i(\mathbf{x}) = \frac{\partial \tilde{v}(\mathbf{x})}{\partial b_i} = \phi_i(\mathbf{x}),\quad i = 1, \ldots, n \tag{8.516}$$

which means that the weighting functions are identical to the basis functions of the test function. And then substituting Equations (8.511) into Equation (8.514), we have

$$(R(\tilde{u}(\mathbf{x})), \phi_i(\mathbf{x}))_\Omega = \sum_{j=1}^{n} A_{ij} b_j - c_i = 0,\quad i = 1, \ldots, n \tag{8.517}$$

where the Galerkin matrix coefficients A_{ij} and c_i are determined by

$$\begin{aligned} A_{ij} &= (\phi_i(\mathbf{x}), L_{2m}\phi_j(\mathbf{x}))_\Omega, \\ c_i &= (f, \phi_i(\mathbf{x}))_\Omega - (L_{2m}\phi_0(\mathbf{x}), \phi_i(\mathbf{x}))_\Omega \end{aligned} \tag{8.518}$$

Note that the matrix $[A_{ij}]$ is not symmetric in general and the coordinate functions are required to be differential $2m$ times. Thus, the n-number of undetermined parameters b_i, $i = 1, \ldots, n$ can be determined by the above n algebraic equations.

Illustrative Example

Example 8.3 Bubnov-Galerkin Method, Euler-Bernoulli Beam

Consider the simply supported Euler-Bernoulli beam as shown in Figure 8.4. The governing equation and the essential boundary condition are given by

$$-\frac{d^2}{dx^2}\left(EI(x)\frac{d^2w(x,t)}{dx^2}\right) + \bar{\omega}(x) = 0, \quad \text{for } x \in]0, L[\tag{8.519}$$

which is subject to the essential boundary conditions

$$\left.\frac{dw}{dx}\right|_{x=0} = 0, \quad w|_{x=0} = 0 \tag{8.520}$$

and the natural boundary condition

$$\left.\frac{\partial^2 w}{\partial x^2}\right|_{x=0} = 0 \tag{8.521}$$

Then, the n-parameter trial function $\tilde{w}(x)$ is expressed as a linear combination of the n number of independent test functions $\phi_i(x)$, $i = 1, \ldots, n$

$$\tilde{w}(x) = \phi_0(x) + \sum_{j=1}^{n} b_j\phi_j(x) \tag{8.522}$$

where b_j are undetermined parameters and $\phi_0(x)$ is a differentiable function satisfying the essential boundary conditions. Note that the Bubnov-Galerkin method requires the trial function $\tilde{w}(x)$ to satisfy all boundary conditions. Then, the residual $R(\tilde{w}(x))$ is defined as

$$R(\tilde{w}(x)) = \frac{d^2}{dx^2}\left(EI(x)\left(\frac{d^2\tilde{w}(x,t)}{dx^2}\right)\right) - \bar{\omega}(x) \neq 0 \tag{8.523}$$

Then, the Galerkin integral requires that the residual function be orthogonal to every weighting function $W_i(x)$, $i = 1, \ldots, n$ as follows

$$\int_L R(\tilde{w}(x))W_i(x)dx = 0, \quad i = 1, \ldots, n \tag{8.524}$$

which means that the weighted averages of the residual should vanish (Crandall 1956). The Galerkin method takes the weighting function

$$W_i(x) = \frac{\partial \tilde{w}(x)}{\partial b_i} = \phi_i(x), \quad i = 1, \ldots, n \tag{8.525}$$

which means that the weighting function is identical to the test function. Therefore, we have

$$\int_L R(\tilde{w}(x))\phi_i(x)dx = \sum_{j=1}^{n} A_{ij}b_j - c_i = 0, \quad i = 1, \ldots, n \tag{8.526}$$

where the Galerkin matrix coefficients A_{ij} and c_i are determined by

$$\begin{aligned} A_{ij} &= \int_L \phi_i(x)\frac{d^2}{dx^2}\left(EI(x)\frac{d^2\phi_j(x)}{dx^2}\right)dx, \\ c_i &= -\int_L \phi_i(x)\frac{d^2}{dx^2}\left(EI(x)\frac{d^2\phi_0(x)}{dx^2}\right)dx + \int_L \bar{\omega}\phi_i(x)dx \end{aligned} \tag{8.527}$$

Thus, the n-number of undetermined parameters b_i, $i = 1, \ldots, n$ can be determined by the above n algebraic equations.

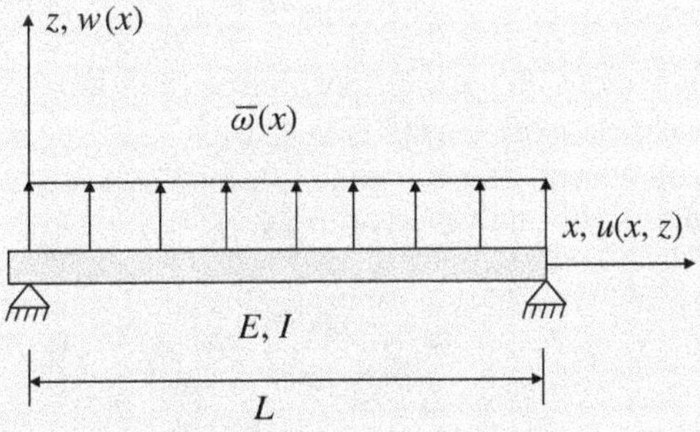

Figure 8.4. Simply supported Euler-Bernoulli beam

8.5.3 Modified Bubnov-Galerkin Weighted-Residual Method

In view of Equation (8.509), we have

$$(L_{2m}u(\mathbf{x}), v(\mathbf{x}))_\Omega = (f(\mathbf{x}), v(\mathbf{x}))_\Omega \tag{8.528}$$

Upon integration by parts, we can have a reformulation allowing a weaker continuity condition for the admissible functions. The modified Bubnov-Galerkin method gives a weak form (Mikhlin 1964, page 489)

$$\begin{aligned} B(\tilde{u}(\mathbf{x}), \tilde{v}(\mathbf{x})) &= l(\tilde{v}(\mathbf{x})), \quad \tilde{u}(\mathbf{x}) \in H^m(\Omega), \quad \tilde{v}(\mathbf{x}) \in H_0^m(\Omega) \\ l(\tilde{v}(\mathbf{x})) &= (\hat{u}_{nb\beta}, \tilde{v}(\mathbf{x}))_{\partial\Omega_\sigma} + (f(\mathbf{x}), \tilde{v}(\mathbf{x}))_\Omega, \quad \beta = m, \ldots, 2m-1 \end{aligned} \tag{8.529}$$

where $B(\tilde{u}(\mathbf{x}), \tilde{v}(\mathbf{x})) : \tilde{U} \times \tilde{V} \to \mathbb{R}$ denotes a bilinear functional involving derivatives of order m and $l(\tilde{v}(\mathbf{x})) : \tilde{V} \to \mathbb{R}$ a linear functional (Ciarlet 1978; Gelfand and Fomin 1963; Johnson 1987; Oden 1979). In the above equation, the natural boundary conditions, i.e. $N_\beta\tilde{u}(\mathbf{x}) = \hat{u}_{nb\beta}$, are imposed on the weak form. We note that in Equation (8.529), both the forcing function $f(\mathbf{x})$ and the natural boundary terms $\hat{u}_{nb\beta}$ are fixed members (Reddy 2006). Furthermore, the function spaces to which the admissible functions such as trial and test functions belong are given by

$$\begin{aligned}\tilde{U} &= \left\{\tilde{u}(\mathbf{x}) : \Omega \to \mathbb{R} | \tilde{u}(\mathbf{x}) \in H^m(\Omega) \subset C^{m-1}(\Omega),\ E_\alpha \tilde{u}(\mathbf{x}) = \bar{u}_{eb\alpha}\right\}, \\ \tilde{V} &= \left\{\tilde{v}(\mathbf{x}) : \Omega \to \mathbb{R} | \tilde{v}(\mathbf{x}) \in H_0^m(\Omega) \subset C_0^{m-1}(\Omega), E_\alpha \tilde{v}(\mathbf{x}) = 0\right\}\end{aligned} \tag{8.530}$$

The trial function should satisfy the essential boundary conditions only, and the test function should vanish on the prescribed displacement boundaries. The test functions need not satisfy the natural boundary conditions, since the conditions are automatically imposed in deriving the weak form. Furthermore, it should be noted that the continuity condition for the admissible functions above becomes weakened. Therefore, Equation (8.529) is called a weak or variational form in the literature (Hughes 1987; Reddy 2006).

Next, the n-parameter admissible trial function $\tilde{u}(\mathbf{x}) \in H^{2m}(\Omega) \subset C^{2m-1}(\Omega)$ is expressed as a linear combination of the n number of independent, differentiable coordinate functions $\phi_i(\mathbf{x}) \in H_0^{2m}(\Omega) \subset C_0^{2m-1}(\Omega),\ i = 1, \ldots, n$

$$\tilde{u}(\mathbf{x}) = \phi_0(\mathbf{x}) + \sum_{i=1}^{n} b_i \phi_i(\mathbf{x}) \tag{8.531}$$

where b_i are unknown coefficients and $\phi_0(\mathbf{x})$ are differentiable functions satisfying the essential boundary conditions. The coordinate functions $\phi_i(\mathbf{x}) \in H_0^{2m}(\Omega) \subset C_0^{2m-1}(\Omega),\ i = 1, \ldots, n$ should vanish at the prescribed displacement boundaries such that the trial function is kinematically admissible. Hence, the variation $\tilde{v}(\mathbf{x})$ of the trial function $\tilde{u}(\mathbf{x})$ can be written as

$$\tilde{v}(\mathbf{x}) = \sum_{i=1}^{n} \phi_i(\mathbf{x}) \delta b_i = \sum_{i=1}^{n} \frac{\partial \tilde{v}(\mathbf{x})}{\partial b_i} \delta b_i = \sum_{i=1}^{n} W_i(\mathbf{x}) \delta b_i \tag{8.532}$$

Therefore, substituting Equation (8.532) into Equation (8.529), we obtain the n number of simultaneous algebraic equations as follows:

$$\begin{aligned}&B(\tilde{u}(\mathbf{x}), \phi_i(\mathbf{x})) = l(\phi_i(\mathbf{x})),\ \ \tilde{u}(\mathbf{x}) \in H^m(\Omega),\ \ \tilde{v}(\mathbf{x}) \in H_0^m(\Omega) \\ &l(\phi_i(\mathbf{x})) = (\hat{u}_{nb\beta}, \phi_i(\mathbf{x}))_{\partial\Omega_\sigma} + (f(\mathbf{x}), \phi_i(\mathbf{x}))_\Omega,\ \ \beta = m, \ldots, 2m-1\end{aligned} \tag{8.533}$$

since $\delta b_i\ i = 1, \ldots, n$ are arbitrary. Substituting the trial function given in Equation (8.531) into the weak form in Equation (8.533) yields

$$\sum_{j=1}^{n} A_{ij} b_j - c_i = 0, \quad i = 1, \ldots, n \tag{8.534}$$

where the coefficients A_{ij} and c_i are determined by

$$\begin{aligned}A_{ij} &= B(\phi_i(\mathbf{x}), \phi_j(\mathbf{x})), \\ c_i &= l(\phi_i(\mathbf{x})) - B(\phi_i(\mathbf{x}), \phi_0(\mathbf{x}))\end{aligned} \tag{8.535}$$

Thus, the n-number of undetermined parameters $b_i,\ i = 1, \ldots, n$ can be determined by the above n algebraic equations.

Note that the matrix $[A_{ij}]$ is symmetric in general and the coordinate functions are required to be differential m times. We also note that Equation (8.529) means a reformulation requiring a weaker continuity condition for admissible functions. Consequently, the trial and test functions which are applied to Equation (8.529) are of class $C^{m-1}(\Omega)$. Furthermore, we notice that the weak form we obtained in the modified Bubnov-Galerkin method as shown in Equation (8.529) is equivalent to Equation (8.479) and the coefficients in Equation (8.535) is identical to the coefficients in Equation (8.486).

Illustrative Examples

Example 8.4 Modified Bubnov-Galerkin Method, Axial Bar

Consider the same axial bar discussed in Example 8.1. The governing equation is given by

$$\frac{d}{dx}\left(EA(x)\frac{du}{dx}\right)+\bar{b}(x)=0, \quad \text{for } x\in]0,L[\tag{8.536}$$

which is subject to the essential boundary condition

$$u|_{x=0}=0 \tag{8.537}$$

and the natural boundary condition

$$F_L=EA(x)\left(\frac{du(x)}{dx}\right)\bigg|_{x=L}, \quad \text{for } x=L \tag{8.538}$$

Note that the modified Bubnov-Galerkin method requires the trial function $\tilde{u}(x)$ to satisfy the essential boundary condition (Mikhlin 1964). Then, the residual $R(u(x))$ is defined as

$$R(u(x))=\frac{d}{dx}\left(EA(x)\left(\frac{du(x)}{dx}\right)\right)+\bar{b}(x)=0 \tag{8.539}$$

Then, the Galerkin integral requires that the residual be orthogonal to every test function $\phi_i(x)$, $i=1,\ldots,n$ as follows

$$\int_L R(u(x))\phi_i(x)dx=0, \quad i=1,\ldots,n \tag{8.540}$$

which can be rewritten as

$$\int_L\left(\frac{d}{dx}\left(EA(x)\frac{du(x)}{dx}\right)+\bar{b}(x)\right)\phi_i(x)dx=0, \quad i=1,\ldots,n \tag{8.541}$$

Note that every test function must satisfy the essential boundary condition. Performing integration by parts, the above equation leads to a weak form, that is

$$\begin{aligned}&-\int_L EA(x)\frac{du(x)}{dx}\frac{d\phi_i}{dx}dx+\int_L\bar{b}(x)\phi_i(x)dx\\&+EA(x)\left(\frac{du(x)}{dx}\right)\bigg|_{x=L}\phi_i(L)=0\end{aligned}, \quad i=1,\ldots,n \tag{8.542}$$

After imposing the natural boundary condition into the weak form, we have

$$-\int_L EA(x)\frac{du(x)}{dx}\frac{d\phi_i}{dx}dx + \int_L \bar{b}(x)\phi_i(x)dx + F_L\phi_i(L) = 0, \quad i = 1, \ldots, n \tag{8.543}$$

Then, the n-parameter trial function $\tilde{u}(x)$ is expressed as a linear combination of the n number of independent test functions $\phi_i(x)$, $i = 1, \ldots, n$

$$\tilde{u}(x) = \phi_0(x) + \sum_{j=1}^{n} b_j\phi_j(x) \tag{8.544}$$

where b_j is undetermined parameters and $\phi_0(x)$ is a differentiable function satisfying the essential boundary conditions. Substituting the trial function $\tilde{u}^h(x)$ into the weak form yields

$$\int_L EA(x)\frac{d\tilde{u}^h}{dx}\frac{d\phi_i}{dx}dx - \int_L \bar{b}(x)\phi_i(x)dx - F_L\phi_i(L) = 0, \quad i = 1, \ldots, n \tag{8.545}$$

which can be rewritten as

$$\sum_{j=1}^{n} A_{ij}b_j = c_i, \quad i = 1, \ldots, n \tag{8.546}$$

where the Galerkin matrix coefficients A_{ij} and c_i are determined by

$$\begin{aligned} A_{ij} &= \int_L EA(x)\frac{d\phi_i}{dx}\frac{d\phi_j}{dx}dx, \\ c_i &= \int_L \phi_i(x)\bar{b}(x)dx + F_L\phi_i(L) \end{aligned} \tag{8.547}$$

Thus, the n-number of undetermined parameters b_i, $i = 1, \ldots, n$ can be determined by the above n algebraic equations. Note that Equation (8.547) is identical to Equation (8.496).

Example 8.5 Modified Bubnov-Galerkin Method, Euler-Bernoulli Beam

Consider the Euler-Bernoulli beam as shown in Figure 8.3. The governing equation is given by

$$-\frac{d^2}{dx^2}\left(EI(x)\frac{d^2w(x,t)}{dx^2}\right) + \bar{\omega}(x) = 0, \quad \text{for } x \in]0, L[\tag{8.548}$$

which is subject to the essential boundary condition

$$\begin{aligned} \left.\frac{dw}{dx}\right|_{x=0} &= 0, \\ w|_{x=0} &= 0 \end{aligned} \tag{8.549}$$

and the natural boundary condition

$$M_L = -\,EI(x)\left.\frac{d^2w}{dx^2}\right|_{x=L}, \qquad \text{for } x = L,$$
$$Q_L = -\frac{d}{dx}\left(EI(x)\frac{d^2w}{dx^2}\right)\Bigg|_{x=L}, \qquad \text{for } x = L \tag{8.550}$$

Note that the modified Bubnov-Galerkin method requires the trial function $\tilde{w}(x)$ to satisfy the essential boundary conditions (Mikhlin 1964). Then, the residual $R(w(x))$ is defined as

$$R(u(x)) = -\frac{d^2}{dx^2}\left(EI(x)\frac{d^2w(x,t)}{dx^2}\right) + \bar{\omega}(x) = 0 \tag{8.551}$$

Then, the Galerkin integral requires that the residual be orthogonal to every test function $\phi_i(x),\ i = 1,\ldots,n$ as follows

$$\int_L R(u(x))\phi_i(x)dx = 0, \quad i = 1,\ldots,n \tag{8.552}$$

which can be rewritten as

$$\int_L \left(-\frac{d^2}{dx^2}\left(EI(x)\frac{d^2w(x,t)}{dx^2}\right) + \bar{\omega}(x)\right)\phi_i(x)dx = 0, \quad i = 1,\ldots,n \tag{8.553}$$

Note that every test function must satisfy the essential boundary conditions. Performing integration by parts, the above equation leads to a weak form, that is

$$\begin{aligned}&-\int_L EI(x)\frac{d^2w(x)}{dx^2}\frac{d^2\phi_i}{dx^2}dx + \int_L \bar{\omega}(x)\phi_i(x)dx \\ &-\frac{d}{dx}\left(EI(x)\frac{d^2w(x)}{dx^2}\right)\Bigg|_{x=L}\phi_i(L) + \left(EI(x)\frac{d^2w(x)}{dx^2}\right)\left.\frac{d\phi_i(x)}{dx}\right|_{x=L} = 0\end{aligned}, \quad i = 1,\ldots,n \tag{8.554}$$

After imposing the natural boundary condition into the weak form, we have

$$\begin{aligned}&-\int_L EI(x)\frac{d^2w(x)}{dx^2}\frac{d^2\phi_i}{dx^2}dx + \int_L \bar{\omega}(x)\phi_i(x)dx \\ &+ Q_L\phi_i(L) - M_L\frac{d\phi_i(L)}{dx}\end{aligned}, \quad i = 1,\ldots,n \tag{8.555}$$

Then, the n-parameter trial function $\tilde{u}(x)$ is expressed as a linear combination of the n number of independent test functions $\phi_i(x),\ i = 1,\ldots,n$

$$\tilde{w}(x) = \phi_0(x) + \sum_{j=1}^{n} b_j\phi_j(x) \tag{8.556}$$

where b_j is undetermined parameters and $\phi_0(x)$ is a differentiable function satisfying the essential boundary conditions. Substituting the trial function $\tilde{w}(x)$ into the weak form yields

$$\begin{aligned} &\int_L EI(x)\frac{d^2w^h(x)}{dx^2}\frac{d^2\phi_i}{dx^2}dx - \int_L \bar{\omega}(x)\phi_i(x)dx \\ &- Q_L\phi_i(L) + M_L\frac{d\phi_i(L)}{dx} = 0 \end{aligned} \quad , \quad i = 1, \ldots, n \tag{8.557}$$

which can be rewritten as

$$\sum_{j=1}^{n} A_{ij}b_j = c_i, \quad i = 1, \ldots, n \tag{8.558}$$

where the Galerkin matrix coefficients A_{ij} and c_i are determined by

$$\begin{aligned} A_{ij} &= \int_L EI(x)\frac{d^2\phi_i}{dx^2}\frac{d^2\phi_j}{dx^2}dx, \\ c_i &= \int_L \phi_i(x)\bar{\omega}(x)dx + Q_L\phi_i(L) - M_L\frac{d\phi_i(L)}{dx} \end{aligned} \tag{8.559}$$

Thus, the n-number of undetermined parameters b_i, $i = 1, \ldots, n$ can be determined by the above n algebraic equations. Note that Equation (8.559) is identical to Equation (8.506).

8.5.4 Equivalence of the Ritz and the Galerkin Methods

The Ritz method requires a functional to be given. The functional for the Ritz method can be easily obtained in steady-state problems, but variational principles do not generally exist in initial boundary-value problems (transient problems). Therefore, the classical Ritz method is not readily applicable to transient problems such as transient heat transfer, continuum-dynamics and so on. The Ritz method uses the functional which allows a weak continuity condition for trial functions satisfying essential boundary conditions. Since both the governing equations and natural boundary conditions are contained implicitly in the functional in the Ritz solution which operates on the functional corresponding to the problem, there is a mechanism by which the errors in the satisfaction of the governing equations and natural boundary conditions are minimized, when substituting the trial functions into the functional to discretize the space (Bathe 1982, page 108) (Cook et al. 2002, page 136) (Zienkiewicz et al. 2005, page 61) (Reddy 2006, page 74).

On the other hand, the classical Bubnov-Galerkin method requires a strong form to be given. The method requires two admissible functions such as trial and test functions. These functions should satisfy both essential and natural boundary conditions. Furthermore, the continuity condition for these admissible functions is stronger than the functional in the Ritz method. Unless the differential operator is symmetric, the resulting coefficient matrix is not guaranteed to be symmetric. It is very hard to find suitable admissible functions satisfying natural boundary conditions. Therefore, we consider an alternative. Once the natural boundary conditions are given, the modified Bubnov-Galerkin method gives a weak form by integration by parts and imposing natural boundary conditions on the weak form. The weak form we obtain by integration by parts allows a weaker continuity condition for both trial and test functions as found in the Ritz method.

Consequently, the Ritz method can be identical to the modified Bubnov-Galerkin method as far as the linear differential operator is positive definite and symmetric. In such a case, the functional should be bilinear-quadratic. Due to the limitations and drawbacks for transient/dynamic problems with methods such as the Ritz type where variational principles usually do not exist for these situations, in the subsequent parts of this book, we describe new and different perspectives and alternative formulations with proper context for appropriate extensions to transient/dynamic problems (see space-discrete finite element formulations).

EXERCISES

8.1. Consider a covariant basis $\{\mathbf{G}_A | A = 1, 2, 3\}$, where $\mathbf{G}_1 = 2\mathbf{E}_1 + 3\mathbf{E}_2$, $\mathbf{G}_2 = -2\mathbf{E}_1 + 3\mathbf{E}_2$ and $\mathbf{G}_A = \mathbf{E}_2 + \mathbf{E}_3$. Obtain the metric tensor $\mathbf{G}$ on the reference configuration. Use Equation (2.148) to find the corresponding contravariant basis $\left\{\mathbf{G}^A | A = 1, 2, 3\right\}$. Show that the covariant metric tensor $\mathbf{G}^\flat = G_{IJ}\mathbf{G}^I \otimes \mathbf{G}^J$ obeys the property

$$\mathbf{G}_I = G_{IJ} \cdot \mathbf{G}^J$$

Show that the contravariant metric tensor $\mathbf{G}^\natural = G^{IJ}\mathbf{G}_I \otimes \mathbf{G}_J$ obeys the property

$$\mathbf{G}^I = G^{IJ} \cdot \mathbf{G}_J$$

8.2. Suppose the material velocity vector $\mathbf{V}$ is given in the Cartesian coordinate system by $\mathbf{V} = \mathbf{E}_1 + 2\mathbf{E}_2 + 3\mathbf{E}_3$. Use the covariant/contravariant base vectors in Exercise 1 to find the covariant and contravariant components of $\mathbf{V}$.

8.3. Use the covariant basis $\{\mathbf{G}_A | A = 1, 2, 3\}$ and the contravariant basis $\left\{\mathbf{G}^A | A = 1, 2, 3\right\}$ in Exercise 8.1 and the covariant and contravariant components of $\mathbf{V}$ in Exercise 2 to show that

$$\mathbf{V}^\natural = \mathbf{G}^\natural \cdot \mathbf{V}^\flat,$$
$$\mathbf{V}^\flat = \mathbf{G}^\flat \cdot \mathbf{V}^\natural$$

8.4. As a convective coordinate chart embedded in the undeformed body, a set is given by $\{\Theta\} = \{R, \Theta, Z\}$ representing the cylindrical coordinates. Find the covariant basis $\{\mathbf{G}_A | A = 1, 2, 3\}$ and the contravariant basis $\left\{\mathbf{G}^A | A = 1, 2, 3\right\}$. Examine the components of the covariant/cotravariant metric tensors respectively.

8.5. As a convective coordinate chart embedded in the deformed body (see Figure 8.5), a set is given by $\{\theta\} = \{r, \theta, \phi\}$ representing the spherical coordinates. Find the covariant basis $\{\mathbf{g}_a | a = 1, 2, 3\}$ and the contravariant basis $\{\mathbf{g}^a | a = 1, 2, 3\}$. Examine the components of the covariant/cotravariant metric tensors respectively.

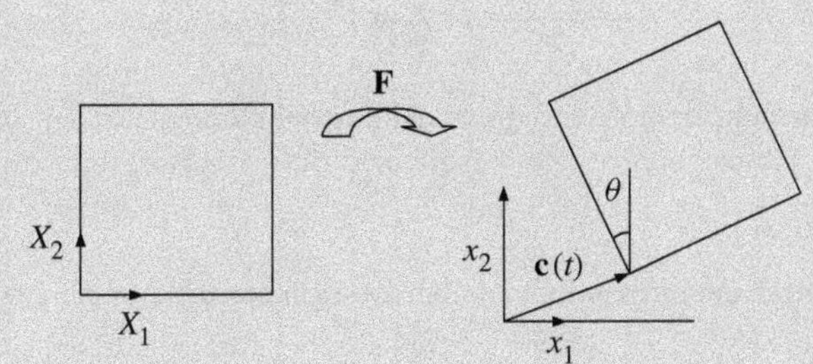

Figure 8.5. Planar rigid body motion

8.6. Consider a rigid body motion $\mathbf{x} = \mathbf{Q}(t) \cdot \mathbf{X} + \mathbf{c}(t)$, where $\mathbf{Q}(t)$ is a proper orthogonal tensor and $\mathbf{c}(t)$ is a translation vector. Find $\mathbf{F}$, $\mathbf{R}$, $\mathbf{U}$, $\mathbf{v}$, $\mathbf{E}$, $\mathbf{C}$, $\mathbf{b}$, $\mathbf{B}$, $\mathbf{l}$, $\mathbf{e}$, $\mathbf{d}$ and $\mathbf{w}$. Show that $J = 1$.

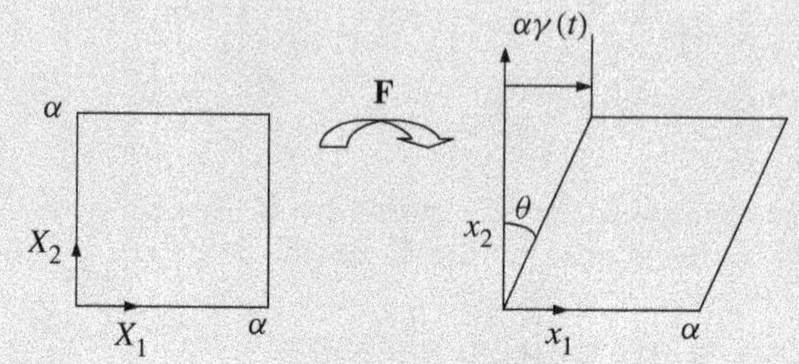

Figure 8.6. Simple shear

8.7. Prove (see Figure 8.6) that $|d\mathbf{x}_1 \cdot (d\mathbf{x}_2 \times d\mathbf{x}_3)| = J(\mathbf{X}, t)\,|d\mathbf{X}_1 \cdot (d\mathbf{X}_2 \times d\mathbf{X}_3)|$.

8.8. Consider a deformation field undergoing simple shear; namely the displacements are given by

$$U_1 = \gamma(t)X_2, \quad U_2 = U_3 = 0$$

where $\gamma(t) = \tan\theta(t)$ denotes shear strain. Find $\mathbf{F}$, $\mathbf{R}$, $\mathbf{U}$, $\mathbf{v}$, $\mathbf{E}$, $\mathbf{C}$, $\mathbf{b}$, $\mathbf{B}$, $\mathbf{l}$, $\mathbf{e}$, $\mathbf{d}$ and $\mathbf{w}$. Show that $J = 1$.

8.9. Consider a deformation field undergoing pure extension (uniform tension with lateral contraction), i.e. namely the displacements are given by

$$U_1 = \epsilon(t)X_1, \quad U_2 = -\nu\epsilon(t)X_2, \quad U_3 = -\nu\epsilon(t)X_3$$

where ν and ϵ denote the Poisson's ratio and axial strain. Find $\mathbf{F}$, $\mathbf{R}$, $\mathbf{U}$, $\mathbf{v}$, $\mathbf{E}$, $\mathbf{C}$, $\mathbf{b}$, $\mathbf{B}$, $\mathbf{l}$, $\mathbf{e}$, $\mathbf{d}$ and $\mathbf{w}$.

8.10. Consider a deformation field undergoing pure dilatation. Suppose a deformation field is given by

$$x_1 = \lambda_1(t)X_1, \quad x_2 = \lambda_2(t)X_2, \quad x_3 = \lambda_3(t)X_3$$

where λ_i, $i = 1, 2, 3$ can be scalar-valued function of a single variable t. Find $\mathbf{F}$, $\mathbf{R}$, $\mathbf{U}$, $\mathbf{v}$, $\mathbf{E}$, $\mathbf{C}$, $\mathbf{b}$, $\mathbf{B}$, $\mathbf{l}$, $\mathbf{e}$, $\mathbf{d}$ and $\mathbf{w}$. Show that the Jacobian $J = \lambda_1\lambda_2\lambda_3$. Therefore, the volumetric strain is expressed as

$$\frac{dv - dV}{dV} = J - 1 = \lambda_1\lambda_2\lambda_3 - 1$$

Note that the deformation is called pure dilatation when $\lambda_1 = \lambda_2 = \lambda_3 = \lambda$, and is called isochoric if $\lambda = 1$. Consequently, for a isochoric deformation, the volumetric strain is zero.

8.11. Prove $\det\mathbf{F} = \det\bar{\mathbf{U}} = \det\bar{\mathbf{v}}$.

8.12. Show that the spectral decomposition of the rotation tensor can be expressed as

$$\mathbf{R} = \sum_{i=1}^{3} \mathbf{n}_i \otimes \mathbf{N}_i$$

8.13. Show that the Lagrangian and Eulerian strain tensors can be represented as

$$\mathbf{E} = \frac{1}{2}\sum_{i=1}^{3}(\Lambda_i - 1)\mathbf{N}_i \otimes \mathbf{N}_i,$$

$$\mathbf{e} = \frac{1}{2}\sum_{i=1}^{3}(1 - \Lambda_i^{-1})\mathbf{n}_i \otimes \mathbf{n}_i$$

where $\Lambda_i,\ i = 1, 2, 3$ represent the principal stretches of the Green deformation tensor $\mathbf{C}$.

8.14. Show that the deformation gradient tensor $\mathbf{F}$ can be given in terms of cylindrical polar coordinates by

$$\mathbf{F} = \begin{bmatrix} \dfrac{\partial r}{\partial R} & \dfrac{1}{R}\dfrac{\partial r}{\partial \Theta} & \dfrac{\partial r}{\partial Z} \\ r\dfrac{\partial \theta}{\partial R} & \dfrac{r}{R}\dfrac{\partial \theta}{\partial \Theta} & r\dfrac{\partial \theta}{\partial Z} \\ \dfrac{\partial z}{\partial R} & \dfrac{1}{R}\dfrac{\partial z}{\partial \Theta} & \dfrac{\partial z}{\partial Z} \end{bmatrix}$$

where $\{R, \Theta, Z\}$ denotes a general coordinate chart embedded in the undeformed body and $\{r, \theta, z\}$ represents a general coordinate chart embedded in the deformed body.

8.15. Show that the deformation gradient tensor $\mathbf{F}$ can be given in terms of spherical coordinates by

$$\mathbf{F} = \begin{bmatrix} \dfrac{\partial r}{\partial R} & \dfrac{1}{R}\dfrac{\partial r}{\partial \Theta} & \dfrac{1}{R\sin\Theta}\dfrac{\partial r}{\partial \Phi} \\ r\dfrac{\partial \theta}{\partial R} & \dfrac{r}{R}\dfrac{\partial \theta}{\partial \Theta} & \dfrac{r}{R\sin\Theta}\dfrac{\partial \theta}{\partial \Phi} \\ r\sin\theta\dfrac{\partial \phi}{\partial R} & \dfrac{r\sin\theta}{R}\dfrac{\partial \phi}{\partial \Theta} & \dfrac{r\sin\theta}{R\sin\Theta}\dfrac{\partial \phi}{\partial \Phi} \end{bmatrix}$$

where $\{R, \Theta, \Phi\}$ denotes a general coordinate chart embedded in the undeformed body and $\{r, \theta, \phi\}$ represents a general coordinate chart embedded in the deformed body.

8.16. Show that the deviatoric stress tensor $\boldsymbol{\sigma}_{\text{dev}}$ and the Cauchy stress tensor $\boldsymbol{\sigma}$ have the same principal directions.

8.17. Examine that $(\boldsymbol{\sigma}_{\text{dev}})_{\text{sph}} = \mathbf{0}$.

8.18. Let $s_i,\ i = 1, 2, 3$ denote the principal deviatoric stresses. Show that $s_i = \sigma_i + p,\ ,i = 1, 2, 3$.

8.19. Show that the principal invariants $J_i,\ i = 1, 2, 3$ of the deviatoric stress tensor $\boldsymbol{\sigma}_{\text{dev}}$ can be given by

$$J_1 = 0$$

$$J_2 = \frac{1}{2}\mathrm{tr}\,\boldsymbol{\sigma}_{\text{dev}}^2 = \frac{1}{3}(\mathrm{I}_\sigma^2 - 3\mathrm{II}_\sigma)$$

$$J_3 = \frac{1}{3}\mathrm{tr}\,\boldsymbol{\sigma}_{\text{dev}}^3 = \frac{1}{27}(2I_\sigma^3 - 9\mathrm{I}_\sigma\mathrm{II}_\sigma + 27\mathrm{III}_\sigma)$$

8.20. An octahedral plane is the plane whose normal $\mathbf{n} = (n_1, n_2, n_3)$ satisfies

$$n_1^2 = n_2^2 = n_3^2 = 1,$$
$$\mathbf{n} \cdot \mathbf{n} = 1$$

which yields

$$n_1 = \pm\frac{1}{\sqrt{3}},\; n_2 = \pm\frac{1}{\sqrt{3}},\; n_3 = \pm\frac{1}{\sqrt{3}}$$

Verify that the octahedral normal stress is $\sigma_{\text{oct}} = \frac{1}{3}\mathbf{I}_\sigma$ and the octahedral shear stress is $\tau_{\text{oct}} = \sqrt{\frac{2}{3}J_2}$.

8.21. The Maxwell-Huber-von Mises yield function $\Upsilon(\boldsymbol{\sigma})$ (Khan and Huang 1995) is given by

$$\Upsilon(\boldsymbol{\sigma}) = \sqrt{J_2} - k(\sigma_y, \tau_y)$$
$$= \sqrt{\frac{1}{2}(\boldsymbol{\sigma}_{\text{dev}} : \boldsymbol{\sigma}_{\text{dev}})} - k(\sigma_y, \tau_y) = 0$$

where k is a parameter. Verify that the parameter k can be given by

$$k = \frac{1}{\sqrt{3}}\sigma_y = \tau_y$$

where σ_y and τ_y denote the yield stress in uniaxial tension and the yield stress in pure shear.

8.22. In a small deformation theory, i.e. ($\boldsymbol{\sigma} \cong \boldsymbol{\tau} \cong \mathbf{S} \cong \mathbf{P}$) and ($\boldsymbol{\epsilon} \cong \mathbf{e} \cong \mathbf{E}$), the strain energy density function $W(\boldsymbol{\epsilon})$of an isotropic, linearly elastic material is given by

$$W(\boldsymbol{\epsilon}) = \frac{\lambda}{2}(\text{tr}\,\boldsymbol{\epsilon})^2 + \mu\text{tr}\,\boldsymbol{\epsilon}^2$$

where λ, μ are Lamé constants. Note that the Lamé constants can be represented as

$$\lambda = \frac{E\nu}{(1+\nu)(1-2\nu)},\quad \mu = \frac{E}{2(1+\nu)}$$

where E and ν denote Young's modulus and Poisson's ratio. Show that the Cauchy stress tensor $\boldsymbol{\sigma}$ for a Hookean material can be rewritten as

$$\boldsymbol{\sigma} = \lambda(\text{tr}\,\boldsymbol{\epsilon})\mathbf{I} + 2\mu\boldsymbol{\epsilon}$$

where $\mathbf{I} = \mathbf{G}_A \otimes \mathbf{G}^A$ and $\boldsymbol{\epsilon}$ is the infinitesimal strain tensor. Note that the relation between the Cauchy stress tensor $\boldsymbol{\sigma}$ and the infinitesimal strain tensor $\boldsymbol{\epsilon}$ is given by

$$\boldsymbol{\sigma} = \mathbb{C} : \boldsymbol{\epsilon}$$

where $\mathbb{C} = \lambda\mathbf{I} \otimes \mathbf{I} + 2\mu\mathbf{II}$, $\mathbf{II} = \mathbf{G}_i \otimes \mathbf{G}_j \otimes \mathbf{G}^i \otimes \mathbf{G}^j$. For the Hookean elastic material, show that $\boldsymbol{\sigma}_{\text{dev}} = 2\mu\boldsymbol{\epsilon}_{\text{dev}}$.

8.23. Verify that for the linearly elastic material in Exercise 22, $\boldsymbol{\sigma}_{\text{sph}} = K(\text{tr}\,\boldsymbol{\epsilon})\mathbf{I}$, where $K = \lambda + \frac{2\mu}{3}$ as the bulk modulus.

8.24. Demonstrate that for the linear elastic material in Exercise 8.22, the infinitesimal strain tensor can be expressed as

$$\boldsymbol{\epsilon} = -\frac{\lambda}{2\mu(3\lambda + 2\mu)}(\mathrm{tr}\,\boldsymbol{\sigma})\mathbf{I} + \frac{1}{2\mu}\boldsymbol{\sigma}$$

8.25. Show that $\frac{\partial J(\mathbf{X},t)}{\partial \mathbf{C}} = \frac{1}{2}J(\mathbf{X},t)\mathbf{C}^{-1}$.

8.26. Show that $d\frac{J(\mathbf{X},t)}{dt} = \frac{1}{2}J(\mathbf{X},t)\mathbf{C}^{-1} : \frac{d\mathbf{C}}{dt}$.

8.27. Show that in indicial notation, Equation (8.203) gives

$$\rho(\mathbf{X})A^i(\mathbf{X},t) - \rho(\mathbf{X})B^i(\mathbf{X},t) - \frac{\partial P^{iJ}}{\partial X^J} = 0$$

Note that if the first Piola-Kirchhoff stress tensor is defined as $\mathbf{T}\left(\mathbf{X},t,\hat{\mathbf{N}}\right) = \hat{\mathbf{N}}(\mathbf{X},t)\cdot\mathbf{P}(\mathbf{X},t)$, Equation (8.203) in indicial notation gives

$$\rho(\mathbf{X})A^i(\mathbf{X},t) - \rho(\mathbf{X})B^i(\mathbf{X},t) - \frac{\partial P^{Ji}}{\partial X^J} = 0$$

8.28. Show that by use of Nason's formula and Equation (8.154), the relation between the first Piola-Kirchhoff stress and the Cauchy stress can be given by

$$\mathbf{P}(\mathbf{X},t) = J(\mathbf{X},t)\,\boldsymbol{\sigma}(\mathbf{X},t)\cdot\mathbf{F}^{-T}$$

which is often called the Piola transformation.

8.29. Demonstrate that the relation holds

$$\mathbf{P}\mathbf{F}^T = \mathbf{F}\mathbf{P}^T$$

8.30. Assume that the elastic potential function for a compressible Neo-Hookean material is given by (Bonet and Wood 2008; Ciarlet 1988; Simo and Hughes 1998)

$$\rho_0\bar{\Psi} = \lambda\frac{(J^2-1)}{4} + \frac{1}{2}\mu(\mathrm{tr}\,\mathbf{C} - 3) - \left(\frac{\lambda}{2} + \mu\right)\ln J$$

where λ, μ are Lamé constants. Find $\mathbf{S}$, $\boldsymbol{\tau}$ and $\boldsymbol{\sigma}$.

8.31. Assume that the elastic potential function for a compressible Neo-Hookean material is given by (Bonet and Wood 2008; Simo and Pister 1984)

$$\rho_0\bar{\Psi} = \frac{\lambda}{2}(\ln J)^2 - \mu\ln J + \frac{1}{2}\mu(\mathrm{tr}\,\mathbf{C} - 3)$$

Find $\mathbf{S}$, $\boldsymbol{\tau}$ and $\boldsymbol{\sigma}$.

8.32. Derive the Jaumann stress rate by use of the Lie derivative.

8.33. Derive the Truesdell stress rate of the contravariant Kirchhoff stress tensor.

8.34. Derive the Cotter-Rivlin stress rate from the Lie derivative of the covariant Kirchhoff stress tensor.

8.35. Derive the Green-Naghdi stress rate from the Lie derivative of the contravariant Cauchy stress tensor.

8.36. Principle of virtual work, Axial bar Consider a dynamic problem of an elastic axial bar with a varying cross-section $A(x)$ and length L as shown in Figure 8.7. The material is linearly elastic, isotropic and homogeneous. The axial bar is vibrating longitudinally in the x-direction at the time $t \in \mathcal{T}=]t_0, t_1[$. The differential, governing equation and the natural boundary condition are given by

$$\rho A(x)\ddot{u}(x,t) = \frac{\partial F_x}{\partial x} + \bar{b}(x), \qquad \text{for } (x,t) \in]0, L[\times \mathcal{T},$$

$$F_L = EA(x)\left(\frac{\partial u(x,t)}{\partial x}\right)\bigg|_{x=L}, \qquad \text{for } x = L$$

where F_x, E and $\bar{b}(x)$ denote the uniform axial force at the point x, elastic modulus and the body force per unit length. An end force F_L is applied at $x = L$. By use of the kinematically admissible virtual displacement satisfying the essential boundary condition, i.e.

$$\delta u = \delta \bar{u}_0 = 0, \quad \text{for } x = 0$$

show that the principle of virtual work in elastodynamics is equal to

$$\int_L (\rho A(x)\,\ddot{u})\,\delta u\,dx + \int_L F_x \delta\varepsilon_x dx = \int_L \bar{b}(x)\delta u dx + F_L \delta u(L,t)$$

which is subject to the compatibility equation, i.e. $\delta\varepsilon_x = \frac{\partial \delta u}{\partial x}$. By use of the principle of virtual work given above, derive the differential, governing equation and the natural boundary condition.

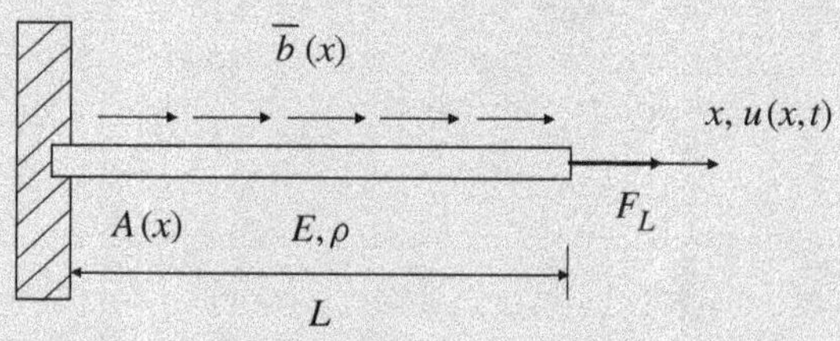

Figure 8.7. Axial bar in dynamics

8.37. Principle of minimum potential energy, Axial bar Consider the same axial bar. But the system is in statics; hence, there is no inertial force. Write down the total potential energy of the axial bar in terms of the displacement. Use the principle of minimum potential energy to derive the governing equation and the natural boundary condition

$$\frac{d}{dx}\left(EA(x)\frac{du}{dx}\right) + \bar{b}(x) = 0, \qquad \text{for } x \in]0, L[,$$

$$F_L = EA(x)\left(\frac{du}{dx}\right)\bigg|_{x=L}, \qquad \text{for } x = L$$

Note that the internal potential energy $\mathcal{U}_{int}$ can be written as

$$\mathcal{U}_{int} = \int_L \frac{1}{2} EA(x)\, \varepsilon_x^2 dx = \int_L \frac{1}{2} EA(x) \left(\frac{du}{dx}\right)^2 dx = \int_L w(\varepsilon_x) dx, \ \varepsilon_x = \frac{du}{dx}$$

8.38. Principle of complementary virtual work, Axial bar Consider the same axial bar in statics. Show that the principle of complementary virtual work in statics is equal to

$$\int_L \varepsilon_x \delta F_x \, dx = \int_L u \delta \bar{b}(x) \, dx + \bar{u}_0 \delta F_0$$

which is subject to the statically compatible equilibrium equation, that is,

$$\frac{d\delta F_x}{dx} + \delta \bar{b}(x) = 0, \ \text{for } x \in]0, L[$$

Further, show that the principle of complementary virtual work in statics yields

$$\int_L \left(\varepsilon_x - \frac{du}{dx}\right) \delta F_x \, dx = 0$$

8.39. Principle of minimum complementary energy, Axial bar Consider the same axial bar in statics. Write down the total complementary energy of the axial bar. Use the principle of minimum complementary energy to derive the compatibility equation and the essential boundary condition can be written as

$$\begin{aligned} &\frac{F_x}{EA(x)} - \frac{du}{dx} = 0, && \text{for } x \in]0, L[\\ &u = \bar{u}_0 = 0, && \text{for } x = 0 \end{aligned}$$

Note that the internal complementary energy $\mathcal{U}^*_{int}$ can be written as

$$\mathcal{U}^*_{int} = \int_L \frac{1}{2} \frac{F_x^2}{EA(x)} dx = \int_L w^*(F_x) dx$$

8.40. Hu-Washizu variational principle, Axial bar Establish the Hu-Washizu functional for the same axial bar in statics. Use the Hu-Washizu variational principle to obtain the following

$$\left.\begin{aligned} &F_x = \frac{dw(\varepsilon_x)}{d\varepsilon_x}, \\ &\varepsilon_x = \frac{du}{dx}, \\ &\frac{dF_x}{dx} + \bar{b}(x) = 0 \end{aligned}\right\} \ , \ \text{for } x \in]0, L[$$

and at both ends

$$F_L = EA(x) \left(\frac{du}{dx}\right)\bigg|_{x=L}, \quad \text{for } x = L,$$

$$F_x = EA(x)\left(\frac{du}{dx}\right)\bigg|_{x=0}, \quad \text{for } x = 0,$$

$$u = \bar{u}_0 = 0, \quad \text{for } x = 0$$

where $w(\varepsilon_x) = \frac{1}{2}EA(x)\,\varepsilon_x^2$.

8.41. Hellinger-Reissner variational principle, Axial bar Establish the Hellinger-Reissner functional for the same axial bar in statics. Use the Hellinger-Reissner variational principle to obtain the following

$$\varepsilon_x = \frac{dw^*(F_x)}{dF_x},$$

$$\frac{dF_x}{dx} + \bar{b}(x) = 0$$

and at the both ends

$$F_L = EA(x)\left(\frac{du}{dx}\right)\bigg|_{x=L}, \quad \text{for } x = L,$$

$$u = \bar{u}_0 = 0, \quad \text{for } x = 0$$

8.42. Principle of virtual work, Prismatic torsional bar Consider the Saint-Venant torsion problem (Love 1927) for a prismatic, elastic solid bar of constant cross-section subjected to equal and opposite twisting moments applied at both ends (see Figure 8.8). Let $\Theta(z)$ denote the angle of twist on the cross section $\mathcal{A}$ at the axial position z and let θ denote the rate of twist, i.e. $\theta = \frac{d\Theta}{dz}$. The prismatic torsional bar with unit length ($L = 1$) is subject to the end torque M. The equilibrium equation and the natural boundary condition are given by

$$\frac{\partial \tau_{xz}}{\partial x} + \frac{\partial \tau_{yz}}{\partial y} = 0, \quad \text{for } x \in \mathcal{A},$$

$$\tau_{xz}n_x + \tau_{yz}n_y = 0, \quad \text{for } x \in \partial\mathcal{A}$$

and

$$M = \int_{\mathcal{A}} \left(\tau_{yz}x - \tau_{xz}y\right) dA$$

where n_x and n_y denote the directional cosines of the normal outward on the boundary $\partial\mathcal{A}$. In the above equation, τ_{xz} and τ_{yz} denote the non-zero shear stresses on the cross section. Show that the principle of virtual work in statics is equal to

$$\int_{\mathcal{A}} \left(\tau_{xz}\delta\gamma_{xz} + \tau_{yz}\delta\gamma_{yz}\right) dA = M\delta\theta$$

which is subject to the compatibility equations, that is,

$$\delta\gamma_{xz} = \left(\frac{\partial \delta w(x,y)}{\partial x} - y\delta\theta\right), \quad \delta\gamma_{yz} = \frac{\partial \delta w(x,y)}{\partial y} + x\delta\theta$$

where $w(x, y)$ denotes the displacement in the z-direction.

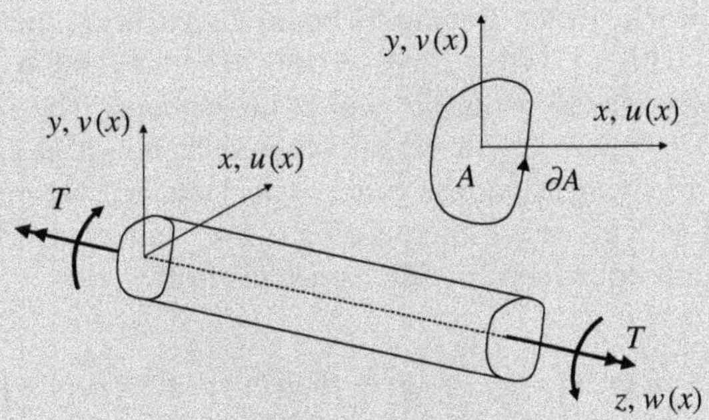

Figure 8.8. Prismatic torsional bar

8.43. Principle of minimum potential energy, Prismatic torsional bar Consider the torsion of a prismatic bar. Write down the total potential energy of the prismatic torsion bar in terms of the displacement and the rate of twist. Show that the total potential energy of the prismatic torsion bar can be written as

$$\mathcal{U} = \frac{1}{2}\int_A G\left[\left(\frac{\partial w(x,y)}{\partial x} - y\theta\right)^2 + \left(\frac{\partial w(x,y)}{\partial y} + x\theta\right)^2\right] dA - M\theta$$

where G is the shear modulus of the bar. Note that the above equation can be expressed as

$$\mathcal{U} = \frac{1}{2}\int_A G\left(\gamma_{xz}^2 + \gamma_{yz}^2\right)\, dA - M\theta$$

Use the principle of minimum potential energy to derive the governing equation (Laplace's equation) and the natural boundary conditions as follows:

$$\Delta\psi(x,y) = \frac{\partial^2\psi}{\partial x^2} + \frac{\partial^2\psi}{\partial x^2} = 0, \qquad \text{for } x \in \mathcal{A},$$

$$M = G\theta\int_A\left[\frac{\partial\psi}{\partial y}x - \frac{\partial\psi}{\partial x}y + (x^2+y^2)\right] dA,$$

$$\left(\frac{\partial\psi}{\partial x} - y\right)n_x + \left(\frac{\partial\psi}{\partial y} + x\right)n_y = 0, \qquad \text{for } x \in \partial\mathcal{A}$$

where $\psi(x,y)$ is called the warping function and $w(x,y) = \theta\psi(x,y)$.

8.44. Show that Laplace's equation above can be written as Poisson's equation

$$\frac{\partial^2\Psi}{\partial x^2} + \frac{\partial^2\Psi}{\partial x^2} = H(x,y)$$

where $H(x,y) = -2G\theta$ and $\Psi(x,y)$ is called the Prandtl's stress function, that is,

$$\tau_{xz} = \frac{\partial\Psi}{\partial y}, \quad \tau_{yz} = -\frac{\partial\Psi}{\partial x}$$

8.45. Principle of virtual work, Euler-Bernoulli beam Consider a dynamic problem of an elastic Euler-Bernoulli beam with a varying cross-section $A(x)$ and length L as shown in Figure 8.9. The material is linearly elastic, isotropic and homogeneous. The beam being subject to the distributed load $\bar{\omega}(x)$ per unit length is cantilevered; hence it is subject to an end shearing force Q_L and an end moment M_L at the end $x = L$. Details and notations regarding the beam are found in Chapter 9. The rotary inertia effects are neglected. The differential, governing equation and the natural boundary conditions are given by

$$\rho A(x)\,\ddot{w}(x,t) = \frac{\partial^2 M(x,t)}{\partial x^2} + \bar{\omega}(x), \quad \text{for } (x,t) \in]0, L[\times T,$$

subject to the natural boundary conditions

$$M_L = -\,EI(x)\left.\frac{\partial^2 w}{\partial x^2}\right|_{x=L},$$

$$Q_L = -\frac{\partial}{\partial x}\left(EI(x)\frac{\partial^2 w}{\partial x^2}\right)\bigg|_{x=L}$$

where $EI(x)$ denote the elastic modulus and the second moment of inertia about the neutral axis. By use of the kinematically admissible virtual displacement satisfying the essential boundary conditions, that is,

$$\left.\frac{\partial \delta w}{\partial x}\right|_{x=0} = 0,$$

$$\delta w|_{x=0} = 0$$

show that the principle of virtual work in elastodynamics can be written as

$$\int_L (\rho A(x)\,\ddot{w})\,\delta w\,dx + \int_L M\delta\kappa_x dx$$

$$= \int_L \bar{\omega}(x)\delta w dx + Q_L\delta w(L,t) - M_L\frac{\partial \delta w(L,t)}{\partial x}$$

which is subject to the compatibility equation, i.e. $\delta\kappa_x = -\frac{\partial^2 \delta w}{\partial x^2}$. By use of the principle of virtual work given above, derive the differential, governing equation and the natural boundary conditions.

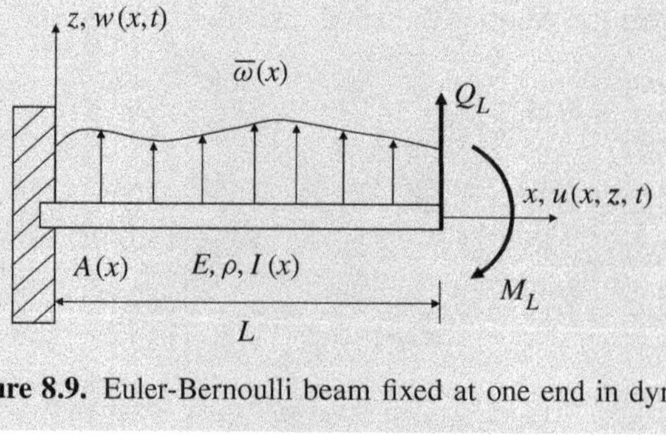

Figure 8.9. Euler-Bernoulli beam fixed at one end in dynamics

8.46. Principle of minimum potential energy, Euler-Bernoulli beam Consider the same Euler-Bernoulli beam. But the system is static; hence, there is no inertial force. Write down the total potential energy of the Euler-Bernoulli beam in terms of the deflection. Use the principle of minimum potential energy to derive the differential, governing equation

$$-\frac{d^2}{dx^2}\left(EI(x)\frac{d^2w}{dx^2}\right)+\bar{\omega}(x)=0, \quad \text{for } x\in]0,L[,$$

and the natural boundary conditions

$$M_L=-\left.EI(x)\frac{d^2w}{dx^2}\right|_{x=L},$$

$$Q_L=-\frac{d}{dx}\left.\left(EI(x)\frac{d^2w}{dx^2}\right)\right|_{x=L}$$

Note that the internal potential energy $\mathcal{U}_{int}$ can be written as

$$\mathcal{U}_{int}=\int_L\frac{1}{2}EI(x)\kappa_x^2dx=\int_L\frac{1}{2}EI(x)\left(\frac{d^2w}{dx^2}\right)^2dx=\int_L w(\kappa_x)dx, \quad \kappa_x=-\frac{d^2w}{dx^2}$$

8.47. Principle of complementary virtual work, Euler-Bernoulli beam Consider the same Euler-Bernoulli beam in statics. Show that the principle of complementary virtual work in statics is equal to

$$\int_L\kappa_x\delta M\,dx=\int_L w\delta\bar{\omega}(x)dx$$

which is subject to the statically compatible equilibrium equation, that is,

$$\frac{d^2\delta M(x,t)}{dx^2}+\delta\bar{\omega}(x)=0, \quad \text{for } x\in]0,L[$$

Further, show that the principle of complementary virtual work in statics yields

$$\int_L\left(\kappa_x+\frac{d^2w}{dx^2}\right)\delta M\,dx=0$$

8.48. Principle of minimum complementary energy, Euler-Bernoulli beam Consider the Euler-Bernoulli beam in statics. Write down the total complementary energy of the Euler-Bernoulli beam. Use the principle of minimum complementary energy to derive the compatibility equation and the essential boundary condition can be written as

$$\frac{M}{EI(x)}+\frac{d^2w}{dx^2}=0, \quad \text{for } x\in]0,L[$$

$$\left.\frac{d\delta w}{dx}\right|_{x=0}=0, \quad \text{for } x=0,$$

$$\delta w|_{x=0}=0, \quad \text{for } x=0$$

Note that the internal complementary energy $\mathcal{U}^*_{int}$ can be written as

$$\mathcal{U}^*_{int} = \int_L \frac{1}{2} \frac{M^2}{EI(x)} dx = \int_L w^*(M) dx$$

8.49. Hu-Washizu variational principle, Euler-Bernoulli beam Establish the Hu-Washizu functional for the Euler-Bernoulli beam in statics. Use the Hu-Washizu variational principle to obtain the following

$$\left.\begin{aligned} M &= \frac{dw(\kappa_x)}{d\kappa_x}, \\ \kappa_x &+ \frac{d^2 w}{dx^2} = 0, \\ &\frac{d^2 M(x,t)}{dx^2} + \bar{\omega}(x) = 0 \end{aligned}\right\} \quad \text{, for } x \in]0, L[$$

and at both ends

$$\begin{aligned} M_L &= -\, EI(x) \left.\frac{d^2 w}{dx^2}\right|_{x=L}, && \text{for } x = L, \\ Q_L &= -\frac{d}{dx}\left(EI(x) \frac{d^2 w}{dx^2}\right)\bigg|_{x=L}, && \text{for } x = L, \\ M &= -\, EI(x) \left.\frac{d^2 w}{dx^2}\right|_{x=0}, && \text{for } x = 0, \\ Q &= -\frac{d}{dx}\left(EI(x) \frac{d^2 w}{dx^2}\right)\bigg|_{x=0}, && \text{for } x = 0, \\ &\left.\frac{d\delta w}{dx}\right|_{x=0} = 0, && \text{for } x = 0, \\ &\delta w|_{x=0} = 0, && \text{for } x = 0 \end{aligned}$$

8.50. Hellinger-Reissner variational principle, Euler-Bernoulli beam Establish the Hellinger-Reissner functional for the Euler-Bernoulli beam in statics. Use the Hellinger-Reissner variational principle to obtain the following

$$\left.\begin{aligned} \kappa_x &= \frac{dw^*(M)}{dM}, \\ &\frac{d^2 M(x,t)}{dx^2} + \bar{\omega}(x) = 0 \end{aligned}\right\} \quad \text{, for } x \in]0, L[$$

and at both ends

$$\begin{aligned} M_L &= -\, EI(x) \left.\frac{d^2 w}{dx^2}\right|_{x=L}, && \text{for } x = L, \\ Q_L &= -\frac{d}{dx}\left(EI(x) \frac{d^2 w}{dx^2}\right)\bigg|_{x=L}, && \text{for } x = L, \end{aligned}$$

$$\left.\frac{d\delta w}{dx}\right|_{x=0} = 0, \quad \text{for } x = 0,$$

$$\delta w|_{x=0} = 0, \quad \text{for } x = 0$$

8.51. Rayleigh-Ritz method, Axial bar Consider the same axial bar discussed in Example 8.1.

(a) Determine the Ritz coefficients if $\bar{b}(x) = x$ and the 2-parameter Ritz approximate solution for the axial bar is given by

$$\tilde{u}(x) = \phi_0(x) + \sum_{i=1}^{2} b_i x^i$$

(b) Determine the Ritz coefficients if $\bar{b}(x) = x^2$ and the 3-parameter Ritz approximate solution for the axial bar is given by

$$\tilde{u}(x) = \phi_0(x) + \sum_{i=1}^{2} b_i \sin(\frac{2i-1}{2L}\pi x)$$

8.52. Rayleigh-Ritz method, Euler-Bernoulli beam Consider the Euler-Bernoulli beam as shown in Figure 8.3.

(a) Determine the Ritz coefficients and the maximum deflection at the center point of the beam if $\bar{\omega}(x) = -\bar{\omega}_0$ and the n-parameter Ritz approximate solution for the simply supported Euler-Bernoulli beam as shown in Figure 8.4 is given by

$$\tilde{u}(x) = \phi_0(x) + \sum_{i=1}^{n} b_i \sin\frac{i\pi x}{L}$$

(b) Determine the Ritz coefficients and the maximum deflection at the center point of the beam if $\bar{\omega}(x) = -\bar{\omega}_0$ and the 3-parameter Ritz approximate solution for the simply supported Euler-Bernoulli beam as shown in Figure 8.4 is given by

$$\tilde{u}(x) = \phi_0(x) + \sum_{i=1}^{3} b_i x^i (L-x)^i$$

(c) Determine the Ritz coefficients and the maximum deflection at the center point of the beam if $\bar{\omega}(x) = -\bar{\omega}_0$ and the n-parameter Ritz approximate solution for the simply supported Euler-Bernoulli beam supporting a concentrated load as shown in Figure 8.10 is given by

$$\tilde{u}(x) = \phi_0(x) + \sum_{i=1}^{n} b_i \sin\frac{i\pi x}{L}$$

8.53. Bubnov-Galerkin method, Euler-Bernoulli beam Consider the simply supported Euler-Bernoulli beam as shown in Figure 8.4. The trial solution for the simply supported Euler-Bernoulli beam is given by

$$\tilde{u}(x) = b_1\phi_1(x),$$

$$\phi_1(x) = a_0 + \sum_{i=1}^{4} a_i x^i$$

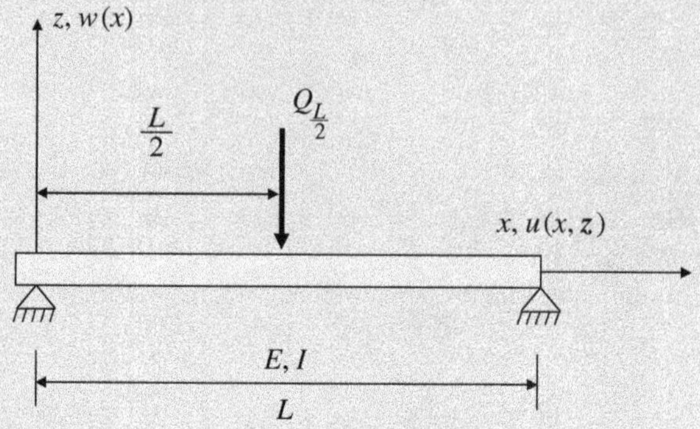

Figure 8.10. Simply supported Euler-Bernoulli beam supporting a concentrated load

where a_i, $i = 0, \ldots, 4$ are constant parameters. Note that the test function $\phi_1(x)$ should satisfy all boundary conditions. Determine the undetermined parameter b_1 and the maximum deflection at the center point of the beam when $\bar{\omega}(x) = -\bar{\omega}_0$.

8.54. Modified Bubnov-Galerkin method, Axial bar Consider the axial bar discussed in Example 8.4.

(a) Determine the undetermined parameters if $\bar{b}(x) = x$ and the 2-parameter trial solution for the axial bar is given by

$$\tilde{u}(x) = \phi_0(x) + \sum_{i=1}^{2} b_i x^i$$

(b) Determine the undetermined parameters if $\bar{b}(x) = x^2$ and the 3-parameter trial solution for the axial bar is given by

$$\tilde{u}(x) = \phi_0(x) + \sum_{i=1}^{2} b_i \sin(\frac{2i-1}{2L}\pi x)$$

8.55. Modified Bubnov-Galerkin method, Euler-Bernoulli beam Consider the cantilever Euler-Bernoulli beam supporting a concentrated load at the free end.

(a) Determine the undetermined parameters if the 2-parameter trial solution for the beam is given by

$$\tilde{w}(x) = \phi_0(x) + \sum_{i=1}^{2} b_i x^{i+1}$$

(b) Determine the undetermined parameters if the 3-parameter trial solution for the beam is given by

$$\tilde{w}(x) = \phi_0(x) + \sum_{i=1}^{3} b_i x^{i+1}$$

CHAPTER NINE

PRINCIPLE OF VIRTUAL WORK: FINITE ELEMENTS AND SOLID/STRUCTURAL MECHANICS

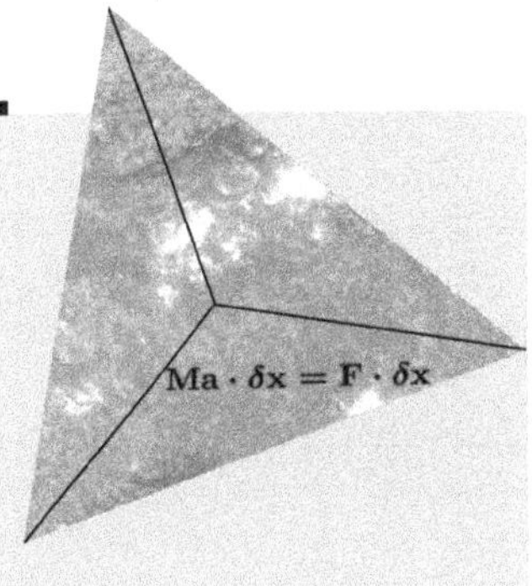

The Galerkin method (Galerkin 1915) is one of the methods of weighted residuals, which is also called a direct approximation method. The method of weighted residuals includes the subdomain method, the least squares method, the collocation method and the Galerkin method (Crandall 1956). According to Finlayson (Finlayson 1972), the terminology, namely, the method of weighted residuals, is credited to Crandall (Crandall 1956). The Galerkin method is often called the Bubnov-Galerkin method (Bubnov 1913; Finlayson 1972; Galerkin 1915; Hughes 1987; Mikhlin 1964). In Chapter 9, the Galerkin finite element method (Bathe 1996; Cook et al. 2002; Huebner 1982; Hughes 1987; Reddy 2006; Strang and Fix 1973; Zienkiewicz and Taylor 2006) in association with the modified Bubnov-Galerkin method to find approximate solutions of time-dependent continuum-elastodynamic problems is discussed; and the relation to the principle of virtual work in dynamics is also highlighted.

9.1 INTRODUCTION

9.1.1 Vector Formalism: Cauchy's Equations of Motion, Principle of Virtual Work, and Finite Element Formulations in Continuous-Body Dynamical Systems

In three dimensional Euclidean space, let $\mathcal{B}$ denote the reference configuration of a continuous body as an open, bounded, and simply connected set with a smooth boundary open set $\partial\mathcal{B}$ in a topological space. In continuum mechanics, it is often postulated that the continuous body is constructed by infinite particles which are not considered to have a concentrated mass. The position of a particle in the reference configuration, which can be defined as a mathematical quantity, is sometimes called a material point, being denoted by $\mathbf{X} \in \mathcal{B} \subset \mathbb{R}^3$. The smooth motion of the body may be described as a one-to-one mapping, $\mathbf{x} = \boldsymbol{\varphi}(\mathbf{X}, t) : \mathcal{B} \times \bar{\mathcal{T}} \rightarrow \mathcal{S} \subset \mathbb{R}^3, \quad \forall t \in \bar{\mathcal{T}} = [t_0, t_1]$, where $\mathcal{S}$ denotes the current configuration.

Principle of Balance of Linear Momentum According to Cauchy, Newton's second law of motion in particle dynamics may be extended to the principle of balance of linear momentum in continuum-elastodynamics, which can be described in the Lagrangian description as follows:

$$\frac{d}{dt}\int_{\mathcal{B}} \rho_0(\mathbf{X})\,\dot{\boldsymbol{\varphi}}(\mathbf{X}, t)\,dV = \int_{\partial\mathcal{B}} \mathbf{T}\left(\mathbf{X}, t, \hat{\mathbf{N}}\right) dA + \int_{\mathcal{B}} \rho_0(\mathbf{X})\,\mathbf{B}(\mathbf{X}, t)\,dV \tag{9.1}$$

Advances in Computational Dynamics of Particles, Materials and Structures, First Edition. Jason Har and Kumar K. Tamma.

Invoking Reynolds transport theorem and Cauchy's law $\bar{\mathbf{T}}\left(\mathbf{X}, t, \hat{\mathbf{N}}\right) = \mathbf{P}(\mathbf{X}, t) \cdot \hat{\mathbf{N}}(\mathbf{X}, t)$, Equation (9.1) yields

$$\int_{\mathcal{B}} \left[\rho_0(\mathbf{X})\ddot{\boldsymbol{\varphi}} - \mathrm{DIV}\,\mathbf{P}(\mathbf{X}, t) - \rho_0(\mathbf{X})\,\mathbf{B}(\mathbf{X}, t)\right] dV + \int_{\mathcal{B}} \dot{\boldsymbol{\varphi}} \left\{\frac{d\rho_0(\mathbf{X})}{dt}\right\} dv = 0 \tag{9.2}$$

In the above, the last term should vanish due to the law of conservation of mass, which yields the continuity equation. As shown in the above, in the Lagrangian description the continuity equation is readily satisfied due to the assumption of constant density in time, but in the Eulerian description it should be incorporated with the governing equation. Thus, the initial boundary-value problem for elasto-dynamical systems (Hughes 1987; Simo and Hughes 1998) is formally stated as follows:

$$\begin{aligned}
\rho_0\ddot{\boldsymbol{\varphi}}(\mathbf{X}, t) &= \mathrm{DIV}\,\mathbf{P}(\mathbf{X}, t) + \rho_0\mathbf{B}(\mathbf{X}, t), && \text{for } (\mathbf{X}, t) \in \mathcal{B} \times \mathcal{T}, \\
\bar{\mathbf{T}}\left(\mathbf{X}, t, \hat{\mathbf{N}}\right) &= \mathbf{P}(\mathbf{X}, t) \cdot \hat{\mathbf{N}}(\mathbf{X}, t), && \text{for } (\mathbf{X}, t) \in \partial\mathcal{B}_\sigma \times \mathcal{T}, \\
\boldsymbol{\varphi} &= \bar{\boldsymbol{\varphi}}, && \text{for } (\mathbf{X}, t) \in \partial\mathcal{B}_\varphi \times \mathcal{T}, \\
\boldsymbol{\varphi}(\mathbf{X}, t_0) &= \boldsymbol{\varphi}_0, && \text{for } (\mathbf{X}, t_0) \in \mathcal{B} \times \partial\mathcal{T}, \\
\dot{\boldsymbol{\varphi}}(\mathbf{X}, t_0) &= \dot{\boldsymbol{\varphi}}_0, && \text{for } (\mathbf{X}, t_0) \in \mathcal{B} \times \partial\mathcal{T}
\end{aligned} \tag{9.3}$$

where t_0 denotes the initial time. In the literature, Equation (9.3) is formally called the strong form of the initial boundary-value problem. In the literature, Equation $(9.3)_1$ is referred to as Cauchy's first law of motion (Truesdell and Toupin 1960, page 545) (Truesdell 1991, page 182) or simply Cauchy's equation of motion (Belytschko et al. 2000; Holzapfel 2000; Malvern 1969; Marsden and Hughes 1983). One interpretation is that it is the local form of balance of linear momentum. In addition, Equation $(9.3)_2$ stands for the prescribed traction boundary condition as the natural boundary condition, which is called Cauchy's fundamental theorem (Truesdell and Toupin 1960, page 543) (Truesdell 1991, page 174) or simply Cauchy's law (Belytschko et al. 2000; Holzapfel 2000; Malvern 1969; Marsden and Hughes 1983), and Equation $(9.3)_3$ indicates the prescribed displacement boundary condition as the essential boundary condition. In the sense of functional analysis, the dependent variable $\boldsymbol{\varphi}$ is of class $C^{2m}(\mathcal{B})$ in space and of class $C^2\left(\bar{\mathcal{T}}\right)$ in time. As described in the preceding section, the configuration space $\mathcal{C}$ can be defined as,

$$\mathcal{C} = \left\{\begin{array}{l} \boldsymbol{\varphi} : \mathcal{B} \times \bar{\mathcal{T}} \rightarrow \mathcal{S} \subset \mathbb{R}^3 | \boldsymbol{\varphi} \in C^{2m}(\mathcal{B}) \\ \boldsymbol{\varphi}_{\mathbf{X}} \in C^2\left(\bar{\mathcal{T}}\right), \; m \geq 1, \; |\mathbf{F}| > 0, \text{ and } \boldsymbol{\varphi}|_{\partial\mathcal{B}_\varphi} = \bar{\boldsymbol{\varphi}} \end{array}\right\} \tag{9.4}$$

where $\bar{\boldsymbol{\varphi}}$ stands for the prescribed quantity on the prescribed displacement boundary.

By use of the governing equation given in Equation $(9.3)_1$, we establish the following equation (Belytschko et al. 2000; Fish and Belytschko 2007; Hughes 1987; Reddy 2002, 2006; Simo and Hughes 1998)

$$\int_{\mathcal{B}} \left(\mathrm{DIV}\,\mathbf{P}(\mathbf{X}, t) + \rho_0\mathbf{B}(\mathbf{X}, t) - \rho_0\ddot{\boldsymbol{\varphi}}(\mathbf{X}, t)\right) \cdot \delta\boldsymbol{\varphi}(\mathbf{X}, t)\, dV = 0 \tag{9.5}$$

where $\delta\boldsymbol{\varphi}(\mathbf{X}, t)\,(\mathbf{X}, t)$ denotes the virtual displacement. It is important to note that the dependent variable to be solved is the position, i.e. $\boldsymbol{\varphi}(\mathbf{X}, t) : \mathcal{B} \times \bar{\mathcal{T}} \rightarrow \mathbb{R}^3$. In the literature, the above equation is often called the Bubnov-Galerkin weighted residual form (Bubnov 1913; Crandall 1956; Finlayson 1972; Galerkin 1915; Mikhlin 1964). As the subsidiary conditions, the above equation is subject to the equations of compatibility and the geometrical boundary conditions

$$\begin{aligned}
\delta\mathbf{F} &= \mathrm{GRAD}\,\delta\boldsymbol{\varphi} && \text{for } (\mathbf{X}, t) \in \mathcal{B} \times \bar{\mathcal{T}}, \\
\delta\boldsymbol{\varphi} &= \delta\bar{\boldsymbol{\varphi}} = \mathbf{0}, && \text{for } (\mathbf{X}, t) \in \partial\mathcal{B}_\varphi \times \bar{\mathcal{T}}
\end{aligned} \tag{9.6}$$

where $\delta\mathbf{F}$ denotes the variation of the deformation gradient tensor $\mathbf{F}$. It is important to note that Equation (9.5) can be regarded as a key equation to the virtual work expression for the principle of balance of linear momentum. By invoking the product rule, the following relation is obtained:

$$\mathrm{DIV}\,\mathbf{P}\cdot\delta\boldsymbol{\varphi} = \mathrm{DIV}\,(\delta\boldsymbol{\varphi}\cdot\mathbf{P}) - \mathbf{P} : \mathrm{GRAD}\,\delta\boldsymbol{\varphi} = \mathrm{DIV}\,(\delta\boldsymbol{\varphi}\cdot\mathbf{P}) - \mathbf{P} : \delta\mathbf{F} \tag{9.7}$$

Upon substitution of Equation (9.7) into Equation (9.5), one can obtain the following

$$\int_{\mathcal{B}} \rho_0\ddot{\boldsymbol{\varphi}}\cdot\delta\boldsymbol{\varphi}\,dV + \int_{\mathcal{B}} \mathbf{P} : \delta\mathbf{F}\,dV = \int_{\mathcal{B}} \rho_0\mathbf{B}\cdot\delta\boldsymbol{\varphi}\,dV + \int_{\mathcal{B}} \mathrm{DIV}\,(\delta\boldsymbol{\varphi}\cdot\mathbf{P})\,dV \tag{9.8}$$

Recall Gauss's divergence theorem, the natural boundary condition (Cauchy's law) given in Equation $(9.3)_2$, and the essential boundary condition given in Equation $(9.3)_3$. Then, the above equation yields the principle of virtual work in dynamics for continuous-body dynamical systems as follows:

$$\int_{\mathcal{B}} \rho_0\ddot{\boldsymbol{\varphi}}\cdot\delta\boldsymbol{\varphi}\,dV + \int_{\mathcal{B}} \mathbf{P} : \delta\mathbf{F}\,dV = \int_{\mathcal{B}} \rho_0\mathbf{B}\,(\mathbf{X},t)\cdot\delta\boldsymbol{\varphi}\,dV + \int_{\partial\mathcal{B}_\sigma} \bar{\mathbf{T}}\cdot\delta\boldsymbol{\varphi}\,dA \tag{9.9}$$

It should be noted that the above equation is obtained by imposing both natural and essential boundary conditions onto Equation (9.8). This is often called the modified Bubnov-Galerkin weighted residual method (Mikhlin 1964) or the generalized Galerkin method (Washizu 1982). Accordingly, the above equation implicitly contains both natural and essential boundary conditions as well as the governing equation. As such, by means of Equation (9.7) the above equation can be rewritten as

$$\int_{\mathcal{B}} (\mathrm{DIV}\,\mathbf{P} + \rho_0\mathbf{B} - \rho_0\ddot{\boldsymbol{\varphi}})\cdot\delta\boldsymbol{\varphi}\,dV + \int_{\partial\mathcal{B}_\sigma} \left(\bar{\mathbf{T}} - \mathbf{P}\cdot\hat{\mathbf{N}}\right)\cdot\delta\boldsymbol{\varphi}\,dV = 0 \tag{9.10}$$

We refer to the above equation as the modified Bubnov-Galerkin weighted residual form, which leads to the weak form given in Equation (9.9) without imposing the natural boundary conditions. Notice that Equation (9.5) is different from the above equation in terms of the second integral equation. Further, it is not suitable for the admissible functions to be readily applied to the above equation due to continuity requirements. The requirements can be relaxed when Equation (9.9) is applied (Fish and Belytschko 2007; Gosz 2006; Hughes 1987; Pian and Tong 1969, 1972; Reddy 2006). Equation (9.9) is weaker than Equation (9.5) or Equation (9.10) in the sense of continuity requirements for the displacement-based finite element method. Therefore, rather than Equation (9.10), it is quite natural for Equation (9.9) to be employed to discretize the space.

Remark 9.1.1 Note that in view of Equation (9.10), the principle of virtual work in dynamics for continuous-body dynamical systems yields Cauchy's equations of motion and Cauchy's law.

Spatial Discretization First, within the conventional framework of the finite element method the admissible trial functions $\tilde{\boldsymbol{\varphi}}$ which can be applied to the principle of virtual work in dynamics are examined. The functions should be continuous along the inter-element boundaries (Pian and Tong 1969, 1972), and the $(m-1)$th order derivatives are required to be spatially piece-wise continuous within the element. They may be represented by a linear combination of coordinate functions $\mathbf{N}_i\,(\mathbf{X})$ as follows:

$$\tilde{\boldsymbol{\varphi}}\,(\mathbf{X},t) \approx \tilde{\boldsymbol{\varphi}}^h\,(\mathbf{X},t) = \mathbf{X} + \sum_{i=1}^{n_{\mathrm{node}}} \mathbf{N}_i\,(\mathbf{X})\,\mathbf{U}_i^e\,(t) \tag{9.11}$$

where n_{node} denotes the number of nodes within the eth element and the superscript h indicates that the continuous function is discretized in space. In the above the nodal displacement vector $\mathbf{U}_i^e(t)$ and the shape function $\mathbf{N}_i(\mathbf{X})$ are given by

$$\mathbf{N}_i(\mathbf{X}) = \begin{bmatrix} N_i & 0 & 0 \\ 0 & N_i & 0 \\ 0 & 0 & N_i \end{bmatrix}, \quad \mathbf{U}_i^e(t) = \begin{Bmatrix} U_{xi}^e \\ U_{yi}^e \\ U_{zi}^e \end{Bmatrix} \tag{9.12}$$

Thus, Equation (9.11) can be rewritten as

$$\begin{aligned} &\tilde{\boldsymbol{\varphi}}^h(\mathbf{X}, t) = \mathbf{X} + \mathbf{N}_e(\mathbf{X})\,\mathbf{U}^e(t), \quad \mathbf{N}_e(\mathbf{X}) = [\mathbf{N}_1, \ldots, \mathbf{N}_{n\text{node}}], \\ &\mathbf{U}^e(t) = \left(\mathbf{U}_1^e(t), \ldots, \mathbf{U}_{n\text{node}}^e\right)^T \end{aligned} \tag{9.13}$$

The trial functions, which are of class $C^{m-1}(\mathcal{B})$ over the entire domain, belong to the subspace, denoted by $\mathcal{C}^h \subset \mathcal{C}$, i.e.,

$$\mathcal{C}^h = \left\{ \begin{array}{l} \tilde{\boldsymbol{\varphi}}^h(\mathbf{X}, t) : \mathcal{B} \times \bar{\mathcal{T}} \to \mathbb{R}^3 | \tilde{\boldsymbol{\varphi}}^h \in W_2^m(\mathcal{B}) \subset C^{m-1}(\mathcal{B}), \text{ in } \mathcal{B} \\ \tilde{\boldsymbol{\varphi}}^h = \bar{\boldsymbol{\varphi}}, \quad \text{on } \partial\mathcal{B}_\varphi \end{array} \right\} \tag{9.14}$$

In traditional practices, namely, the conventional Galerkin finite element method (Hughes 1987; Reddy 2006), Equation (9.11) is substituted into Equation (9.9), i.e. the weak form. Then, the domain of interest can be discretized as follows:

$$\sum_{e=1}^{n_{\text{elem}}} \left(\begin{array}{l} \displaystyle\int_{\mathcal{B}_e} \rho_0 \ddot{\tilde{\boldsymbol{\varphi}}}^h \cdot \delta\tilde{\boldsymbol{\varphi}}^h dV + \int_{\mathcal{B}_e} \tilde{\mathbf{P}}^h \cdot \cdot \delta\tilde{\mathbf{F}}^h dV \\ \displaystyle - \int_{\mathcal{B}_e} \rho_0 \mathbf{B} \cdot \delta\tilde{\boldsymbol{\varphi}}^h dV - \int_{\partial\mathcal{B}_{e\sigma}} \bar{\mathbf{T}} \cdot \delta\tilde{\boldsymbol{\varphi}}^h dA \end{array} \right) = 0 \tag{9.15}$$

where $\mathcal{B}_e$ stands for the eth element domain and n_{elem} denotes the number of elements. The variation of the admissible trial function is given by $\delta\tilde{\boldsymbol{\varphi}}^h(\mathbf{X}, t) = \mathbf{N}_e(\mathbf{X})\,\delta\mathbf{U}^e(t)$. Therefore, substituting Equation (9.13) into the above equation leads to

$$\sum_{e=1}^{n_{\text{elem}}} \left\{ \left(\mathbf{M}_e \ddot{\mathbf{U}}^e(t) + \mathbf{F}_{\text{int}}^e - \mathbf{F}_{\text{ext}}^e\right) \cdot \delta\mathbf{U}^e(t) \right\} = 0 \tag{9.16}$$

where

$$\begin{aligned} &\mathbf{M}_e = \int_{\mathcal{B}_e} \rho_0 \mathbf{N}_e^T \mathbf{N}_e \, dV, \quad \mathbf{F}_{\text{int}}^e = \int_{\mathcal{B}_e} \bar{\mathbf{B}}_e^T \tilde{\mathbf{P}}^h \, dV, \\ &\mathbf{F}_{\text{ext}}^e = \int_{\mathcal{B}_e} \rho_0 \mathbf{N}_e^T \mathbf{B} \, dV + \int_{\partial\mathcal{B}_{e\sigma}} \mathbf{N}_e^T \bar{\mathbf{T}} \, dA \end{aligned} \tag{9.17}$$

where the strain matrix $\bar{\mathbf{B}}_e$ is given by $\frac{\partial \mathbf{N}_e(\mathbf{X})}{\partial \mathbf{X}}$. Notice that Equation (9.15) yield a set of ordinary time differential simultaneous equations to be subsequently discretized in time. Therefore, the weak form is crucial to the finite element method, and is employed as a variational equation to which one must apply the admissible trial function, $\tilde{\boldsymbol{\varphi}}(\mathbf{X}, t)$. Thus, to discretize the entire domain of interest in the modified Bubnov-Galerkin weighted residual method, Equation (9.15) can be rewritten as

$$(\mathbf{M}\ddot{\mathbf{q}} + \mathbf{F}_{\text{int}} - \mathbf{F}_{\text{ext}}) \cdot \delta\mathbf{q} = 0 \tag{9.18}$$

where

$$\begin{aligned}
&\mathbf{q}(t) = \{U_1(t), \ldots, U_{n_{\text{dof}}}(t)\}^T, \; n_{\text{dof}} = 3n_{\text{node}}, \\
&\ddot{\mathbf{q}}(t) = \{\ddot{U}_1(t), \ldots, \ddot{U}_{n_{\text{dof}}}(t)\}^T, \delta\mathbf{q}(t) = \{\delta U_1(t), \ldots, \delta U_{n_{\text{dof}}}(t)\}^T, \\
&\mathbf{M} = \int_{\mathcal{B}} \rho_0 \mathbf{N}^T \mathbf{N}\, dV, \;\; \mathbf{F}_{\text{int}} = \int_{\mathcal{B}} \bar{\mathbf{B}}^T \tilde{\mathbf{P}}^h \, dV, \\
&\mathbf{F}_{\text{ext}} = \int_{\mathcal{B}} \rho_0 \mathbf{N}^T \mathbf{B}\, dV + \int_{\partial\mathcal{B}_\sigma} \mathbf{N}^T \bar{\mathbf{T}}\, dA
\end{aligned} \tag{9.19}$$

where $\mathbf{M}$ denotes the global mass matrix, $\mathbf{F}_{\text{int}}$ the global internal force vector, $\mathbf{F}_{\text{ext}}$ the global external force vector, $\mathbf{N}$ the interpolation matrix, $U_i(t)$ the ith nodal variable and n_{node} the total number of nodes in the entire body. Note that the jth virtual displacement δU_j is independent of the kth virtual displacement δU_k; in other words, the set $\delta \boldsymbol{q}$ is linearly independent. Therefore, the only way for Equation (9.18) to hold is for the individual coefficients to vanish. The terms enclosed in the parenthesis in Equation (9.18) can be regarded as individual coefficients (Lanczos 1970, page 39) (Greenwood 1977, page 51) (Goldstein 2002, page 17 and 20) (Ardema 2005, page 70, page 113). Therefore, Equation (9.18) stands for a linear combination of functions in the sense of functional analysis (Rektorys 1980, page 47) (Reddy 1986, page 64). Consequently, since the variation of nodal variables is arbitrary, we obtain the following

$$\mathbf{M}\ddot{\mathbf{q}} + \mathbf{F}_{\text{int}} - \mathbf{F}_{\text{ext}} = \mathbf{0} \tag{9.20}$$

which is the semi-discrete second-order differential equations in time, i.e. the finite element equations of motion. Taking the integration over the time domain for Equation (9.18) leads to

$$\int_{t_0}^{t_1} (\mathbf{F}_{\text{ext}} - \mathbf{F}_{\text{int}} - \mathbf{M}\ddot{\mathbf{q}}) \cdot \delta\mathbf{q}\, dt = 0 \tag{9.21}$$

which represents the Bubnov-Galerkin weighted residual form in time. Consequently, the generalized time weighted residual method which includes time finite element formulations (Belytschko and Hughes 1983; Sha et al. 2003; Tamma et al. 2003, 2002, 2000; Wood 1990; Zhou and Tamma 2004a,c, 2006; Zhou et al. 2001, 2000; Zienkiewicz and Taylor 2006; Zienkiewicz et al. 2005, 1984) can be readily employed for conducting the time discretization as described in the later chapters of this book.

A variety of illustrative examples commonly encountered as structural finite elements are described next for completeness.

Illustrative Examples

Example 9.1 Elastic Bar

Problem Description

Consider an elastic axial bar vibrating longitudinally in the x-direction at the time $t \in \bar{\mathcal{T}} = [t_0, t_1]$. The material is linearly elastic, isotropic and homogeneous. The cross-sectional area is denoted by $A(x)$. The axial bar with elastic modulus E has a length, denoted by L. The bar is assumed to be subject to a body force per unit length $\bar{b}$, and an end force F_L at $x = L$.

Free-Body Diagram

The axial bar consists of continuum elements having only translational degrees of freedom. As shown in Figure 9.1(b), only axial stress is defined. In Figure 9.1(a), the axial stress $\sigma(x, t)$ transforms in value to $\sigma'(x,t)$, which can be expressed as

$$\sigma'(x,t) = \sigma(x,t) + \frac{\partial \sigma(x,t)}{\partial x}\Delta x \tag{9.22}$$

as we move from the negative face to the positive face on a small element. It is assumed that the area $A(x)$ also changes in the same way. By virtue of D'Alembert's principle given in Equation (3.47), on the small element of the axial bar the continuous-body dynamical systems is dynamically in equilibrium as shown in Figure 9.1(a). Therefore, we have the following

$$\begin{aligned}&\left(\sigma(x,t) + \frac{\partial \sigma(x,t)}{\partial x}\Delta x\right)\left(A(x) + \frac{\partial A(x)}{\partial x}\Delta x\right) - \sigma(x,t)A(x)\\ &+\bar{b}\Delta x - \rho A\left(x + \frac{\Delta x}{2}\right)\ddot{u}\left(x + \frac{\Delta x}{2}, t\right)\Delta x = 0\end{aligned} \tag{9.23}$$

where $\sigma(x,t)$ is the axial stress over a cross-section of the bar, and ρ is the mass density. The axial stress is assumed to remain constant over the cross-section. On the cross-section, the uniform axial stress is calculated as follows:

$$\sigma(x,t) = E\varepsilon(x,t) = E\frac{\partial u}{\partial x} = \frac{F_x}{A(x)} \tag{9.24}$$

where $\varepsilon(x,t)$ denotes the axial elastic strain and F_x is the axial force on the cross-section (Crandall et al. 1978; Meirovitch 1967; Timoshenko et al. 1974). In the limit as Δx goes to zero, we obtain the strong form as follows:

$$\begin{aligned}&\rho A(x)\ddot{u} = \frac{\partial}{\partial x}\left(EA(x)\frac{\partial u}{\partial x}\right) + \bar{b}, && \text{for } (x,t) \in]0, L[\times T,\\ &F_L = EA(x)\left(\frac{\partial u}{\partial x}\right)\bigg|_{x=L}, && \text{for } x = L,\\ &u = 0, && \text{for } x = 0\end{aligned} \tag{9.25}$$

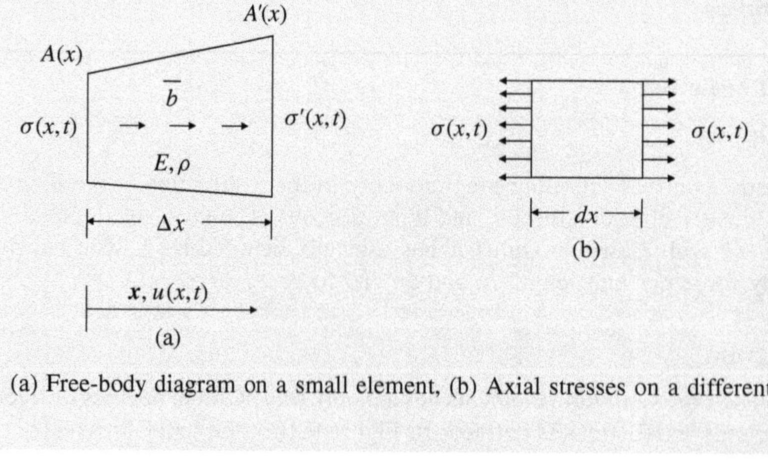

Figure 9.1. (a) Free-body diagram on a small element, (b) Axial stresses on a differential element

Principle of Virtual Work

The dependent variable $u(x,t)$ satisfying the governing equation in the above is of class $C^2\,(]0,L[)$ in space and of class $C^2\left(\bar{T}\right)$ in time. Next, the Bubnov-Galerkin weighted residual form is set up as follows:

$$\int_L \left[\rho A(x)\,\ddot{u} - \frac{\partial}{\partial x}\left(EA(x)\frac{\partial u}{\partial x}\right) + \bar{b}\right]\delta u dx = 0 \tag{9.26}$$

where dx denotes the line segment. Upon integration by parts, the Bubnov-Galerkin weighted residual form results in the weak form as follows:

$$\int_L (\rho A(x)\,\ddot{u})\,\delta u dx + \int_L \left(EA(x)\frac{\partial u}{\partial x}\right)\frac{\partial \delta u}{\partial x}dx = \int_L \bar{b}\delta u dx + F_L\delta u(L,t) \tag{9.27}$$

where the natural boundary condition given in Equation $(9.25)_2$ is applied. The above equation is the principle of virtual work in dynamics for the longitudinal motion of the axial bar.

Spatial Discretization

The admissible trial function is chosen by a linear combination of interpolation functions $\mathbf{N}_e(x)$ and element nodal variables $\tilde{\mathbf{u}}^e(t)$ as follows:

$$\begin{aligned}
&\tilde{u}^h(x,t) = \mathbf{N}_e(x)\,\mathbf{u}^e(t), \quad \ddot{\tilde{u}}^h(x,t) = \mathbf{N}_e(x)\,\ddot{\mathbf{u}}^e(t),\\
&\mathbf{u}^e(t) = \left(u_1^e,\ldots,u_{n_\text{node}}^e\right)^T, \; \ddot{\mathbf{u}}^e(t) = \left(\ddot{u}_1^e,\ldots,\ddot{u}_{n_\text{node}}^e\right)^T,\\
&\mathbf{N}_e(x) = \left[N_1,\ldots,N_{n_\text{node}}\right]
\end{aligned} \tag{9.28}$$

The trial function is of class $C^0\,(]0,L[)$ over the entire domain. As the variation of the dependent variable, the test function is given by $\delta\tilde{u}^h(x,t) = \mathbf{N}_e(x)\,\delta\mathbf{u}^e(t)$. Substituting both the trial and test functions into the weak form in Equation (9.27), the domain of interest can be discretized as follows:

$$\begin{aligned}
&\sum_{e=1}^{n_\text{elem}}\left[\int_{L_e}\rho A(x)\ddot{\tilde{u}}^h\delta\tilde{u}^h dx + \int_{L_e}\left(EA(x)\frac{\partial \tilde{u}^h}{\partial x}\right)\frac{\partial \delta\tilde{u}^h}{\partial x}dx - \int_{L_e}\bar{b}\delta\tilde{u}^h dx\right]\\
&- F_L\delta\tilde{u}^h(L,t) = 0
\end{aligned} \tag{9.29}$$

Therefore, Equation (9.29) reduces to the following

$$\sum_{e=1}^{n_\text{elem}}\left(\mathbf{M}_e\ddot{\mathbf{u}}^e(t) + \mathbf{F}_\text{int}^e - \mathbf{F}_\text{ext}^e\right)\cdot\delta\mathbf{u}^e(t) = 0 \tag{9.30}$$

where

$$\begin{aligned}
&\mathbf{M}_e = \int_{L_e}\rho A(x)\mathbf{N}_e^T\mathbf{N}_e dx, \quad \mathbf{F}_\text{int}^e = \mathbf{K}^e\mathbf{u}^e, \quad \mathbf{F}_\text{ext}^e = \int_{L_e}\bar{b}\mathbf{N}_e^T dx + \mathbf{F}_L^e,\\
&\mathbf{K}^e = \int_{L_e}EA(x)\,\bar{\mathbf{B}}_e^T\bar{\mathbf{B}}_e dx, \quad \bar{\mathbf{B}}_e = \frac{\partial \mathbf{N}_e(x)}{\partial x}
\end{aligned} \tag{9.31}$$

where for the last element only, $\mathbf{F}_L^e = (0,\ldots,F_L)^T$, and for the others, $\mathbf{F}_L^e = (0,\ldots,0)^T$.

Example 9.2 Rotating Circular Bar

Problem Description

Consider the dynamical behavior of a rotating, circular bar with $J(x)$, the polar moment of inertia of the cross-section about the center of the cross-section. The cross-section is assumed to be circular; hence warping effects are neglected. The bar is subject to an external torque T_L at the end, $x = L$. As the dependent variable, we have $\theta(x, t)$, which represents the angle of rotation about the axial axis. In addition, the bar is subject to a distributed torque $\bar{m}(x)$ per unit length over the length of the bar. The material is assumed to be linearly elastic, isotropic and homogeneous.

Free-Body Diagram

The free-body diagram for a rotating bar with circular cross-section is presented in Figure 9.2. By virtue of D'Alembert's principle given in Equation (3.47), on the small element of the rotating bar, the dynamical system is dynamically in equilibrium as follows:

$$\begin{aligned} &T(x,t) + \frac{\partial T(x,t)}{\partial x}\Delta x - T(x,t) + \bar{m}(x)\Delta x \\ &- \rho J(x + \frac{\Delta x}{2})\ddot{\theta}(x + \frac{\Delta x}{2}, t)\Delta x = 0 \end{aligned} \tag{9.32}$$

The shear stress $\tau(x, t)$ at the radius r over the cross-section is given by

$$\tau(x,r,t) = G\gamma(x,r,t) = Gr\frac{\partial\theta(x,t)}{\partial x} \tag{9.33}$$

where $\gamma(x, r, t)$ denotes the shear strain at the radius r over the cross-section, and G represents the shear modulus. The torque $T(x, t)$ causing the bar to rotate (Timoshenko et al. 1974) is given by

$$T(x,t) = \int_A r\tau(x,r,t)dA = GJ(x)\frac{\partial\theta(x,t)}{\partial x} \tag{9.34}$$

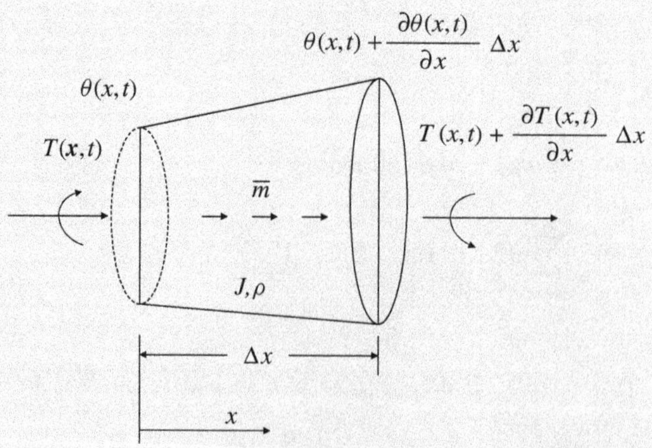

Figure 9.2. Free-body diagram of a rotating circular bar

where $GJ(x)$ is called the torsional stiffness. In the limit as the increment Δx goes to zero, we obtain the strong form as follows:

$$\begin{aligned} &\rho J(x)\ddot{\theta}(x,t) = \frac{\partial}{\partial x}\left(GJ(x)\frac{\partial \theta(x,t)}{\partial x}\right) + \bar{m}(x), \\ &GJ(x)\left.\frac{\partial \theta(x,t)}{\partial x}\right|_{x=L} = T_L, \\ &\theta(x,t)|_{x=0} = 0 \end{aligned} \tag{9.35}$$

Principle of Virtual Work

The dependent variable $\theta(x,t)$ satisfying the governing equation in the above is of class $C^2(]0, L[)$ in space and of class $C^2(\bar{T})$ in time. Next, the Bubnov-Galerkin weighted residual form is set up as follows:

$$\int_L \left[\rho J(x)\ddot{\theta} - \frac{\partial}{\partial x}\left(GJ(x)\frac{\partial \theta}{\partial x}\right) - \bar{m}\right]\delta\theta\, dx = 0 \tag{9.36}$$

Upon integration by parts, the Bubnov-Galerkin weighted residual form results in the weak form as follows:

$$\int_L \left(\rho J(x)\ddot{\theta}\right)\delta\theta\, dx + \int_L \left(GJ(x)\frac{\partial \theta}{\partial x}\right)\frac{\partial \delta\theta}{\partial x}dx = \int_L \bar{m}\delta\theta\, dx + T_L\delta\theta(L,t) \tag{9.37}$$

where the natural boundary condition given in Equation $(9.35)_2$ is applied. The above equation is the principle of virtual work in dynamics for the dynamic motion of the rotating circular bar.

Spatial Discretization

The admissible trial function is chosen by a linear combination of interpolation functions $\mathbf{N}_e(x)$ and element nodal variables $\boldsymbol{\theta}^e(t)$ as follows:

$$\begin{aligned} &\tilde{\theta}^h(x,t) = \mathbf{N}_e(x)\cdot\boldsymbol{\theta}^e(t), \quad \tilde{\boldsymbol{\theta}}^e(t) = \left(\theta_1^h(t), \theta_2^h(t), \ldots, \theta_{n_{\text{node}}}^h(t)\right)^T, \\ &\ddot{\tilde{\theta}}^h(x,t) = \mathbf{N}_e(x)\cdot\ddot{\boldsymbol{\theta}}^e(t), \quad \ddot{\boldsymbol{\theta}}^e(t) = \left(\ddot{\theta}_1^h(t), \ddot{\theta}_2^h(t), \ldots, \ddot{\theta}_{n_{\text{node}}}^h(t)\right)^T, \\ &\mathbf{N}_e(x) = \left[N_1, \ldots, N_{n_{\text{node}}}\right] \end{aligned} \tag{9.38}$$

The trial function is of class $C^0(]0, L[)$ over the entire domain. As the variation of the dependent variable, the test function is given by $\delta\tilde{\theta}^h(x,t) = \mathbf{N}_e(x)\,\delta\boldsymbol{\theta}^e(t)$. Substituting both the trial and test functions into the weak form in Equation (9.37), the domain of interest can be discretized as follows:

$$\begin{aligned} &\sum_{e=1}^{n_{\text{elem}}}\left[\int_{L_e}\rho J(x)\ddot{\tilde{\theta}}^h\delta\tilde{\theta}^h dx + \int_{L_e}\left(GJ(x)\frac{\partial\tilde{\theta}^h}{\partial x}\right)\frac{\partial\delta\tilde{\theta}^h}{\partial x}dx - \int_{L_e}\bar{m}\delta\tilde{\theta}^h dx\right] \\ &- T_L\delta\tilde{\theta}^h(L,t) = 0 \end{aligned} \tag{9.39}$$

Therefore, Equation (9.39) reduces to the following

$$\sum_{e=1}^{n_{\text{elem}}} \left(\mathbf{M}_e \ddot{\boldsymbol{\theta}}^e (t) + \mathbf{F}^e_{\text{int}} - \mathbf{F}^e_{\text{ext}} \right) \cdot \delta \boldsymbol{\theta}^e (t) = 0 \tag{9.40}$$

where

$$\begin{aligned} &\mathbf{M}_e = \int_{L_e} \rho J(x) \mathbf{N}_e^T \mathbf{N}_e dx, \quad \mathbf{F}^e_{\text{int}} = \mathbf{K}^e \boldsymbol{\theta}^e, \quad \mathbf{F}^e_{\text{ext}} = \int_{L_e} \bar{m} \mathbf{N}_e^T dx + \mathbf{T}^e_L, \\ &\mathbf{K}^e = \int_{L_e} GJ(x) \bar{\mathbf{B}}_e^T \bar{\mathbf{B}}_e dx, \quad \bar{\mathbf{B}}_e = \frac{\partial \mathbf{N}_e (x)}{\partial x} \end{aligned} \tag{9.41}$$

where for the last element only, $\mathbf{T}^e_L = (0, \ldots, T_L)^T$ and for the others, $\mathbf{T}^e_L = (0, \ldots, 0)^T$.

Example 9.3 Euler Bemoulli Beam

Problem Description

Let us consider the dynamical behavior of the Euler-Bernoulli beam, which is a prismatic beam of the rectangular cross-section as shown in Figure 9.4(c). The beam is cantilevered; hence it is subject to an end shearing force Q_L and an end moment M_L at the end, $x = L$. The material is assumed to be linearly elastic, isotropic and homogeneous. The beam bending is subjected to the distributed transverse load $\bar{\omega}(x)$ per unit length acting in the z-direction as shown in Figure 9.4. In the Euler-Bernoulli beam theory (Crandall et al. 1978; Gear 2004; Ugural 2008), the normal plane is assumed to be perpendicular to the axial axis of the beam and it remains as a plane normal to the axial axis after deformation and the beam is not subject to the axial force. The derivative of the transverse deflection is the angle of rotation $\varphi(x, t)$ due to bending moment acting on the beam. The total rotation of the Euler-Bernoulli beam is defined as

$$\varphi (x, t) \approx \frac{\partial w (x, t)}{\partial x} \tag{9.42}$$

where $w(x, t)$ denotes the transverse deflection of the neutral axis as shown in Figure 9.3. Thus, we assume that the strain energy is totally due to the bending moment $M(x, t)$ only. Then, the dynamical behavior of the Euler-Bernoulli beam is governed by one differential equation, because of the presence of one dependent variable, namely, the transverse deflection $w(x, t)$. Let $I(x)$ denote the second moment of inertia (also called the moment of inertia of the cross-sectional area) about the neutral axis (Crandall et al. 1978; Gear 2004; Reismann and Pawlik 1980; Timoshenko and Goodier 1951; Ugural 2008), $\rho(x)$ density, and $A(x)$ the cross-sectional area.

As the Euler-Bernoulli beam undergoes the deformation caused by the bending moment, the point p moves to the point p' as shown in Figure 9.3. The displacement of any point on the cross-section in the x-direction is defined as

$$u(x, z, t) = -z\varphi (x, t) = -z \frac{\partial w (x, t)}{\partial x} \tag{9.43}$$

Thus, the normal and shear strains are obtained as

$$\begin{aligned} &\varepsilon_x = \frac{\partial u}{\partial x} = -z \frac{\partial \varphi (x, t)}{\partial x} = -z \frac{\partial^2 w (x, t)}{\partial x^2}, \\ &\varepsilon_y = 0, \quad \varepsilon_z \approx 0, \quad \gamma_{xz} \approx 0, \quad \gamma_{xy} \approx 0, \quad \gamma_{yz} \approx 0 \end{aligned} \tag{9.44}$$

The elastic bending stress in the x-direction on the cross-section is given by

$$\sigma_{xx} = E\varepsilon_{xx} = -zE\frac{\partial \varphi}{\partial x} = -zE\frac{\partial^2 w}{\partial x^2} \tag{9.45}$$

where ε_{xx} denotes the elastic bending strain in the x-direction on the cross-section. Therefore, the bending moment at any point on the cross-section can be written as

$$M(x,t) = \int_A \sigma_{xx} z dA = -EI(x)\frac{\partial^2 w}{\partial x^2} \tag{9.46}$$

In view of Equations (9.45) and (9.46), we obtain the relationship between the bending moment and the bending stress

$$\sigma_{xx} = \frac{M(x,t)}{I(x)}z \tag{9.47}$$

Free-Body Diagram

A free-body diagram for the Euler-Bernoulli beam is shown in Figure 9.4. By virtue of D'Alembert's principle given in Equation (3.47), the small element of the Euler-Bernoulli beam is dynamically in equilibrium in the y-direction as follows:

$$\begin{aligned} &Q(x,t) + \frac{\partial Q(x,t)}{\partial x}\Delta x - Q(x,t) + \bar{\omega}\left(x + \frac{\Delta x}{2}\right)\Delta x \\ &- \rho A(x + \frac{\Delta x}{2})\ddot{w}(x + \frac{\Delta x}{2}, t)\Delta x = 0 \end{aligned} \tag{9.48}$$

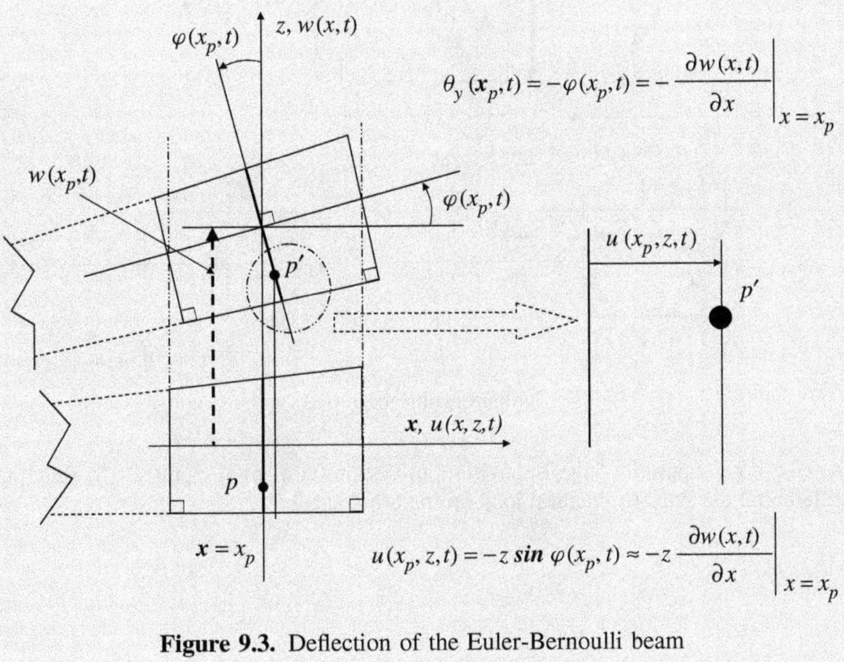

Figure 9.3. Deflection of the Euler-Bernoulli beam

Furthermore, about the y-axis, the rotational motion of the small element is dynamically in equilibrium as follows:

$$\begin{aligned} M(x,t) + \frac{\partial M(x,t)}{\partial x}\Delta x - M(x,t) - Q(x,t)\Delta x \\ - \frac{Q(x,t)}{\partial x}\Delta x\frac{\Delta x}{2} + \rho I(x + \frac{\Delta x}{2})\frac{\partial \ddot{w}}{\partial x}\Delta x = 0 \end{aligned} \tag{9.49}$$

In the limit as the increment Δx goes to zero, Equations (9.48) and (9.49) yield a single differential equation as follows:

$$\frac{\partial^2 M(x,t)}{\partial x^2} + \frac{\partial}{\partial x}\left(\rho I(x)\frac{\partial \ddot{w}(x,t)}{\partial x}\right) + \bar{\omega}(x) - \rho A(x)\ddot{w}(x,t) = 0 \tag{9.50}$$

where $I(x) = \frac{b(x)h^3(x)}{12}$. If the cross-section is not prismatic, the second moment of inertia $I(x)$ should be calculated as the area inertia about the neutral axis.

As the line increment Δx goes to zero, Equation (9.49) results in the following

$$Q(x,t) = \rho I(x)\frac{\partial \ddot{w}}{\partial x} - \frac{\partial}{\partial x}\left(EI(x)\left(\frac{\partial^2 w}{\partial x^2}\right)\right) \tag{9.51}$$

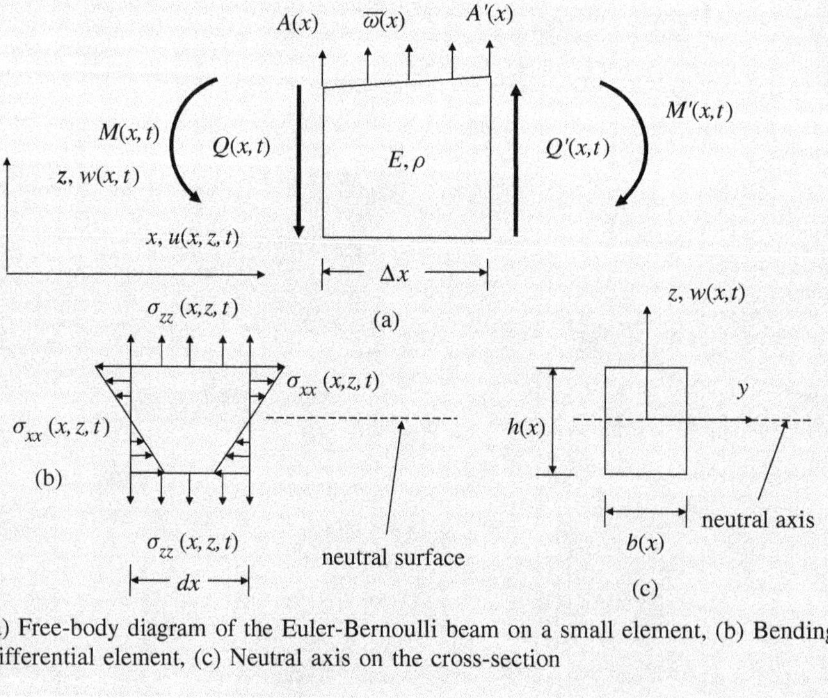

Figure 9.4. (a) Free-body diagram of the Euler-Bernoulli beam on a small element, (b) Bending and shear stresses in a differential element, (c) Neutral axis on the cross-section

As usual, the first term on the left-hand side of Equation (9.51) is ignored in the Euler-Bernoulli beam theory (Zienkiewicz and Taylor 2006). Consequently, upon substitution of Equation (9.46) into (9.50), the strong form for the Euler-Bernoulli Beam is given by

$$\begin{aligned}\rho A(x)\,\ddot{w} = {} & \frac{\partial}{\partial x}\left(\rho I(x)\,\frac{\partial \ddot{w}}{\partial x}\right)\\ & - \frac{\partial^2}{\partial x^2}\left(EI(x)\left(\frac{\partial^2 w}{\partial x^2}\right)\right) + \bar{\omega}(x), \quad \text{for } (x,t) \in]0, L[\times\mathcal{T},\end{aligned} \tag{9.52}$$

subject to the boundary conditions

$$\begin{aligned} M_L &= -\,EI(x)\left(\frac{\partial^2 w}{\partial x^2}\right)\bigg|_{x=L},\\ Q_L &= \left(\rho I(x)\left(\frac{\partial \ddot{w}}{\partial x}\right) - \frac{\partial}{\partial x}\left(EI(x)\left(\frac{\partial^2 w}{\partial x^2}\right)\right)\right)\bigg|_{x=L},\\ \frac{\partial w}{\partial x}\bigg|_{x=0} &= 0,\\ w|_{x=0} &= 0 \end{aligned} \tag{9.53}$$

Principle of Virtual Work

The dependent variable $w(x,t)$ satisfying the governing equation in the above is of class $C^4\,(]0, L[)$ in space and of class $C^2\left(\bar{\mathcal{T}}\right)$ in time. Next, the Bubnov-Galerkin weighted residual form is set up as follows:

$$\int_L \left[\begin{array}{c} \rho A(x)\,\ddot{w} - \dfrac{\partial}{\partial x}\left(\rho I(x)\,\dfrac{\partial \ddot{w}}{\partial x}\right)\\ + \dfrac{\partial^2}{\partial x^2}\left(EI(x)\left(\dfrac{\partial^2 w}{\partial x^2}\right)\right) - \bar{\omega}(x)\end{array}\right]\delta w\, dx = 0 \tag{9.54}$$

Carrying out integration by parts twice, we obtain the weak form as follows:

$$\begin{aligned}&\int_L (\rho A(x)\,\ddot{w})\,\delta w\, dx + \int_L \left(EI\frac{\partial^2 w}{\partial x^2}\right)\frac{\partial^2 \delta w}{\partial x^2}dx\\ &+ \int_L \left(\rho I\frac{\partial \ddot{w}}{\partial x}\right)\frac{\partial \delta w}{\partial x}dx = \int_L \bar{\omega}(x)\delta w dx + Q_L\delta w(L,t) + M_L\frac{\partial \delta w(L,t)}{\partial x}\end{aligned} \tag{9.55}$$

where the natural boundary conditions given in Equation (9.53) are applied. The above equation is the principle of virtual work in dynamics for the Euler-Bernoulli Beam.

Spatial Discretization

The trial displacement function may be approximated using the Hermite polynomial interpolation functions as follows:

$$\tilde{w}^h(x,t) = \mathbf{N}_e(x)\,\mathbf{u}^e(t), \qquad \tilde{\mathbf{u}}^e(t) = (w_1(t), \theta_1(t), w_2(t), \theta_2(t)),$$
$$\frac{\partial \tilde{w}^h(x,t)}{\partial x} = \bar{\mathbf{B}}_1(x)\,\mathbf{u}^e(t), \quad \bar{\mathbf{B}}_1(x) = \frac{\partial \mathbf{N}(x)}{\partial x}, \quad \theta_i = -\frac{\partial w_i(t)}{\partial x}, \ i = 1, 2 \tag{9.56}$$

The trial function is of class C^1 (]0, L[) over the entire domain. As the variation of the dependent variable, the test function is given by $\delta\tilde{u}^h(x,t) = \mathbf{N}_e(x)\,\delta\mathbf{u}^e(t)$. Substituting both the trial and test functions into the weak form in Equation (9.55), the domain of interest can be discretized as follows:

$$\sum_{e=1}^{n_{\text{elem}}} \left[\begin{array}{l} \int_{L_e} (\rho A(x)\,\ddot{w})\,\delta w\,dx + \int_{L_e} \left(EI\dfrac{\partial^2 w}{\partial x^2} \right) \dfrac{\partial^2 \delta w}{\partial x^2} dx \\ + \int_{L_e} \left(\rho I \dfrac{\partial \ddot{w}}{\partial x} \right) \dfrac{\partial \delta w}{\partial x} dx - \displaystyle\int_{L_e} \bar{\omega}(x)\,\delta w\,dx \end{array} \right] = M_L \delta w(L,t) + Q_L \frac{\partial \delta w(L,t)}{\partial x} \tag{9.57}$$

Therefore, Equation (9.57) reduces to the following:

$$\sum_{e=1}^{n_{\text{elem}}} \left(\mathbf{M}_e \ddot{\mathbf{u}}^e(t) + \mathbf{F}^e_{\text{int}} - \mathbf{F}^e_{\text{ext}} \right) \cdot \delta\mathbf{u}^e(t) = 0 \tag{9.58}$$

where

$$\mathbf{M}_e = \int_{L_e} \rho A(x) \mathbf{N}_e^T \mathbf{N}_e dx + \int_{L_e} \rho I(x) \bar{\mathbf{B}}_1^T \bar{\mathbf{B}}_1 dx,$$
$$\mathbf{F}^e_{\text{int}} = \mathbf{K}^e \mathbf{u}^e, \quad \mathbf{F}^e_{\text{ext}} = \int_{L_e} \mathbf{N}_e^T \bar{\omega}(x) dx + \mathbf{M}^e_L + \mathbf{Q}^e_L, \tag{9.59}$$
$$\mathbf{K}^e = \int_{L_e} EI(x)\, \bar{\mathbf{B}}_2^T \mathbf{B}_2 dx, \quad \bar{\mathbf{B}}_2(x) = \frac{\partial^2 \mathbf{N}_e(x)}{\partial x^2}$$

where for the last element only, $\mathbf{M}^e_L = (0, 0, 0, M_L)^T$ and $\mathbf{Q}^e_L = (0, 0, Q_L, 0)^T$.

Example 9.4 Timoshenko Beam

Problem Description

Next consider the dynamical behavior of the Timoshenko beam with the cross-sectional area, $A(x)$, and the moment of inertia of area with respect to the neutral axis of the cross-section, $I(x)$. The material is assumed to be linearly elastic, isotropic and homogeneous. The Timoshenko beam is subject to the end force Q_L and the end moment M_L at the end, $x = L$. The beam in dynamic motion is governed by two differential equations, because of the presence of two dependent variables such as the transverse deflection $w(x,t)$ and the angle of rotation $\varphi(x,t)$ due to bending moment acting on the beam subject to the transversely distributed load $\bar{\omega}(x)$. The total rotation of the Timoshenko beam is composed of two contributions as

$$\frac{\partial w(x,t)}{\partial x} = \varphi(x,t) + \chi(x,t) \tag{9.60}$$

where $\chi(x,t)$ denotes the shear angle due to shear. This means that the beam undergoes translational and rotational motions at the same time.

As the Timoshenko beam undergoes the deformation, the point p moves to the point p' as shown in Figure 9.5. The displacement of any point on the cross-section in the x-direction is defined as

$$\begin{aligned} u(x,t) &= -z\varphi(x,t), \\ \theta_y(x,t) &= -\varphi(x,t) = -\left(\frac{\partial w}{\partial x} - \gamma_{xz}\right) \end{aligned} \tag{9.61}$$

where the rotation $\theta_y(x,t)$ is defined in Figure 9.5. Thus, the normal and shear strains are obtained as

$$\begin{aligned} \varepsilon_x &= \frac{\partial u}{\partial x} = -z\frac{\partial \varphi(x,t)}{\partial x} = z\frac{\partial \theta_y}{\partial x}, \\ \varepsilon_y &= 0, \quad \varepsilon_z \approx 0, \quad \gamma_{xz} = \chi(x,t), \quad \gamma_{xy} \approx 0, \quad \gamma_{yz} \approx 0 \end{aligned} \tag{9.62}$$

The elastic stress in the x-direction on the cross-section is given by

$$\sigma_{xx} = E\varepsilon_{xx} = -zE\frac{\partial \varphi}{\partial x} = zE\frac{\partial \theta_y}{\partial x} \tag{9.63}$$

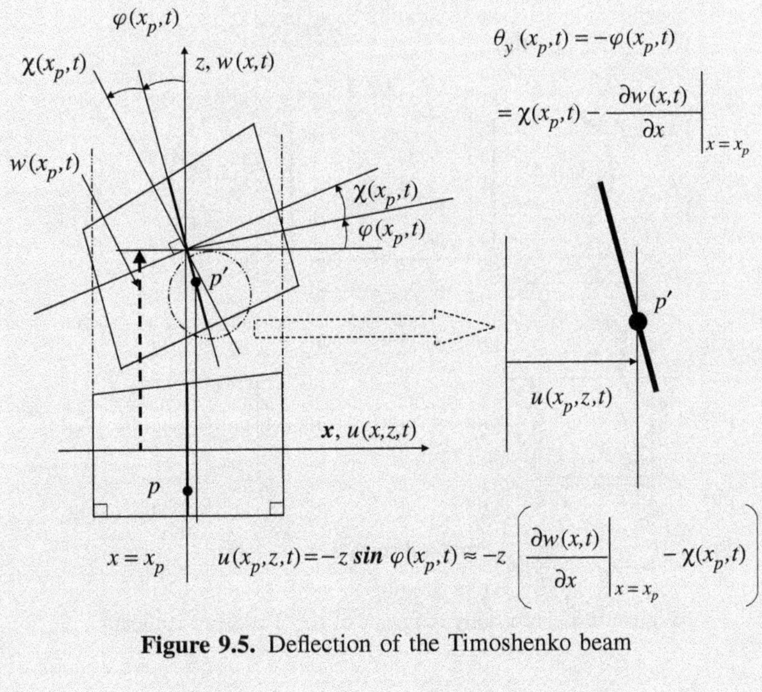

Figure 9.5. Deflection of the Timoshenko beam

where ε_{xx} denotes the elastic strain in the x-direction on the cross-section. Therefore, the bending moment can be written as

$$M(x,t) = \int_A \sigma_{xx} z dA = -EI(x)\frac{\partial \varphi}{\partial x} = EI(x)\frac{\partial \theta_y}{\partial x} \tag{9.64}$$

Unlike the Euler-Bernoulli beam theory, the shearing force $Q(x,t)$ is defined as (Timoshenko et al. 1974)

$$Q(x,t) = \kappa GA(x)\,\chi(x,t) = \kappa GA(x)\left(\frac{\partial w(x,t)}{\partial x} + \theta_y(x,t)\right) \tag{9.65}$$

where κ denotes the shear correction factor. For a rectangular cross-section beam, $\kappa = \frac{5}{6}$ (Crandall et al. 1978).

Free-Body Diagram

A free-body diagram for the Timoshenko beam is shown in Figure 9.6. By virtue of D'Alembert's principle given in Equation (3.47), the infinitesimal element of the Timoshenko beam is dynamically in equilibrium in the y-direction as follows:

$$\begin{aligned} &Q(x,t) + \frac{\partial Q(x,t)}{\partial x}\Delta x - Q(x,t) + \bar{\omega}\left(x + \frac{\Delta x}{2}\right)\Delta x \\ &- \rho A(x + \frac{\Delta x}{2})\ddot{w}(x + \frac{\Delta x}{2}, t)\Delta x = 0 \end{aligned} \tag{9.66}$$

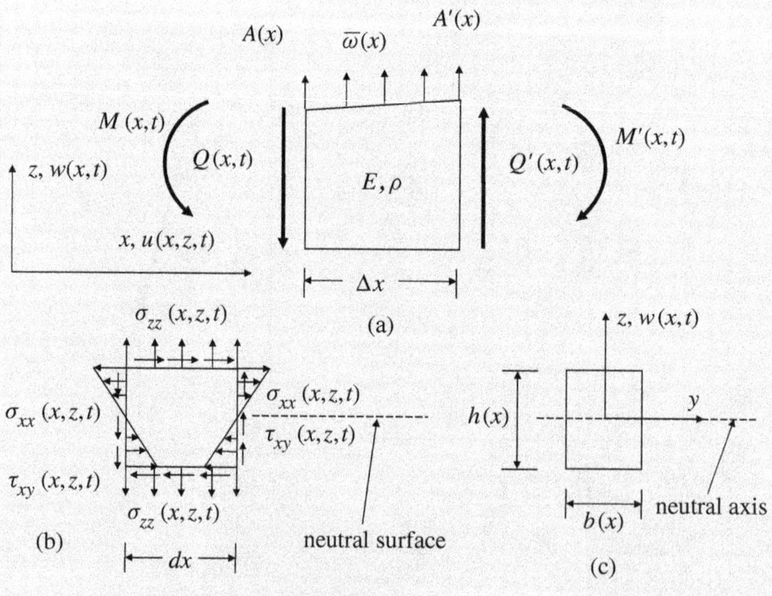

Figure 9.6. Free-body diagram of the Timoshenko beam

Furthermore, about the z-axis, the rotational motion of the infinitesimal element is dynamically in equilibrium as follows:

$$\begin{aligned} &M(x,t)+\frac{\partial M(x,t)}{\partial x}\Delta x-M(x,t)-Q(x,t)\Delta x \\ &-\frac{Q(x,t)}{\partial x}\Delta x\frac{\Delta x}{2}+\rho I(x+\frac{\Delta x}{2})\ddot{\varphi}\Delta x=0 \end{aligned} \tag{9.67}$$

In the limit as the increment Δx goes to zero, Equations (9.66) and (9.67) yield the following two differential equations

$$\begin{aligned} &\frac{\partial Q(x,t)}{\partial x}+\bar{\omega}(x)-\rho A(x)\ddot{w}(x,t)=0, \\ &\frac{\partial M(x,t)}{\partial x}-Q(x,t)-\rho I(x)\ddot{\theta}_y=0 \end{aligned} \tag{9.68}$$

where $I(x)=\frac{b(x)h^3(x)}{12}$. If the cross-section is not prismatic, the moment of inertia of the cross-sectional area $I(x)$ should be calculated as the area inertia about the neutral axis.

Then, in view of Equations (9.64), (9.65) and (9.68), the strong form can be given by

$$\begin{aligned} &\rho A(x)\ddot{w}-\frac{\partial}{\partial x}\left(\kappa GA(x)\left(\frac{\partial w}{\partial x}+\theta_y\right)\right)-\bar{\omega}(x)=0, \\ &\rho I(x)\ddot{\theta}_y+\kappa GA(x)\left(\frac{\partial w}{\partial x}+\theta_y\right)-\frac{\partial}{\partial x}\left(EI(x)\frac{\partial\theta_y}{\partial x}\right)=0 \end{aligned} \tag{9.69}$$

which are subject to the natural boundary conditions

$$M_L=EI(x)\frac{\partial\theta_y}{\partial x}\bigg|_{x=L},\quad Q_L=\kappa GA(x)\left(\frac{\partial w}{\partial x}+\theta_y\right)\bigg|_{x=L} \tag{9.70}$$

and the essential boundary conditions

$$w|_{x=0}=0,\quad \theta_y|_{x=0}=0 \tag{9.71}$$

Principle of Virtual Work

The transverse deflection and rotation field functions for the governing equation in Equation (9.69) should satisfy the continuity condition, viz., $w(x,t)\in C^2(]0,L[)$ and $\varphi(x,t)\in C^2(]0,L[)$. Next, the Bubnov-Galerkin weighted residual forms are set up as follows:

$$\begin{aligned} &\int_L\left\{\rho A(x)\ddot{w}-\frac{\partial}{\partial x}\left[\kappa GA(x)\left(\frac{\partial w}{\partial x}+\theta_y\right)\right]-\bar{\omega}(x)\right\}\delta w dx=0, \\ &\int_L\left\{\rho I(x)\ddot{\theta}_y+\kappa GA(x)\left(\frac{\partial w}{\partial x}+\theta_y\right)-\frac{\partial}{\partial x}\left(EI(x)\frac{\partial\theta_y}{\partial x}\right)\right\}\delta\theta_y dx=0 \end{aligned} \tag{9.72}$$

Carrying out integration by parts twice, we obtain the weak form as follows:

$$\begin{aligned}
&\int_L \rho A(x)\,\ddot{w}\delta w dx + \int_L \kappa GA(x)\left(\frac{\partial w}{\partial x} + \theta_y\right)\frac{\partial \delta w}{\partial x}dx - \int_L \bar{\omega}(x)\,\delta w dx = Q_L \delta w_L, \\
&\int_L \rho I(x)\,\ddot{\theta}_y \delta\theta_y dx + \int_L EI(x)\left(\frac{\partial \theta_y}{\partial x}\right)\frac{\partial \delta\theta_y}{\partial x}dx \\
&+ \int_L \kappa GA(x)\left(\frac{\partial w}{\partial x} + \theta_y\right)\delta\theta_y dx = M_L \delta\theta_{yL}
\end{aligned} \tag{9.73}$$

where the natural boundary conditions given in Equation (9.71) are applied. Equation (9.73) represents the principle of virtual work in dynamics for the Timoshenko Beam.

Note that the admissible trial functions $\tilde{w}^h(x,t)$ and $\tilde{\theta}_y^h(x,t)$ for the scalar functions are of class $H^1(]0, L[) \subset C^0(]0, L[)$.

Spatial Discretization

The admissible trial functions $w^h(x,t)$ and $\theta_y^h(x,t)$ may be approximated using a linear combination of coordinate functions and element nodal variables as follows:

$$\begin{aligned}
&\tilde{w}^h(x,t) = \mathbf{N}_w(x)\,\mathbf{w}^e(t), \quad \mathbf{w}^e(t) = \begin{bmatrix} w_1(t) & w_2(t) & \dots & w_{n_{\text{node}}}(t) \end{bmatrix}^T, \\
&\tilde{\theta}_y^h(x,t) = \mathbf{N}_\theta(x)\,\boldsymbol{\theta}_y^e(t), \quad \boldsymbol{\theta}_y^e(t) = \begin{bmatrix} \theta_{y1}(t) & \theta_{y2}(t) & \dots & \theta_{yn_{\text{node}}}(t) \end{bmatrix}^T
\end{aligned} \tag{9.74}$$

Introducing a different notation $\mathbf{u}^e(t) = \begin{bmatrix} \mathbf{w}^e(t) & \boldsymbol{\theta}_y^e(t) \end{bmatrix}^T$, the trial functions are rewritten as

$$\tilde{w}^h(x,t) = \mathbf{H}_w(x)\,\mathbf{u}^e(t), \quad \tilde{\theta}_y^h(x,t) = \mathbf{H}_\theta(x)\,\mathbf{u}^e(t) \tag{9.75}$$

where $\mathbf{H}_w(x) = \begin{bmatrix} \mathbf{N}_w(x) & \mathbf{0} \end{bmatrix}$ and $\mathbf{H}_\theta(x) = \begin{bmatrix} \mathbf{0} & \mathbf{N}_\theta(x) \end{bmatrix}$. Thus, we have

$$\tilde{\mathbf{u}}^h = \left\{ \begin{array}{c} \tilde{w}^h(x,t) \\ \tilde{\theta}_y^h(x,t) \end{array} \right\} = \mathbf{H}_e(x)\,\mathbf{u}^e(t), \quad \mathbf{H}_e(x) = \begin{bmatrix} \mathbf{H}_w(x) \\ \mathbf{H}_\theta(x) \end{bmatrix} \tag{9.76}$$

The trial function is of class $C^0(]0, L[)$ over the entire domain. As the variation of the dependent variable, the test function is given by $\delta\tilde{\mathbf{u}}^h(x,t) = \mathbf{H}_e(x)\,\delta\mathbf{u}^e(t)$. Substituting both the trial and test functions into the weak form in Equation (9.55), the domain of interest can be discretized as follows:

$$\begin{aligned}
&\sum_{e=1}^{n_{\text{elem}}} \begin{bmatrix} \int_{L_e} \rho A(x)\,\ddot{w}\delta w dx + \int_{L_e} \kappa GA(x)\left(\frac{\partial w}{\partial x} + \theta_y\right)\frac{\partial \delta w}{\partial x}dx \\ - \int_{L_e} \bar{\omega}(x)\,\delta w dx \end{bmatrix} = Q_L \delta w_L, \\
&\sum_{e=1}^{n_{\text{elem}}} \begin{bmatrix} \int_{L_e} \rho I(x)\,\ddot{\theta}_y \delta\theta_y dx + \int_{L_e} EI(x)\left(\frac{\partial \theta_y}{\partial x}\right)\frac{\partial \delta\theta_y}{\partial x}dx \\ + \int_{L_e} \kappa GA(x)\left(\frac{\partial w}{\partial x} + \theta_y\right)\delta\theta_y dx \end{bmatrix} = M_L \delta\theta_{yL},
\end{aligned} \tag{9.77}$$

Therefore, Equation (9.77) reduces to the following:

$$\sum_{e=1}^{n_{\text{elem}}} \left(\mathbf{M}^e \ddot{\mathbf{u}}^e(t) + \mathbf{F}^e_{\text{int}} - \mathbf{F}^e_{\text{ext}} \right) \cdot \delta \mathbf{u}^e(t) = 0 \tag{9.78}$$

where

$$\begin{aligned}
\mathbf{M}^e &= \int_{L_e} \mathbf{H}_e^T(x)\, \boldsymbol{\rho} \mathbf{H}_e(x)\, dx, \quad \boldsymbol{\rho} = \begin{bmatrix} \rho A(x) & 0 \\ 0 & \rho I(x) \end{bmatrix} \\
\mathbf{K}_b &= \int_{L_e} \mathbf{B}_b^T(x)\, \mathbf{D}_b \mathbf{B}_b(x)\, dx, \quad \mathbf{K}_s = \int_{L_e} \mathbf{B}_s^T(x)\, \mathbf{G}_s \mathbf{B}_s(x)\, dx, \\
\mathbf{B}_b &= \left[\frac{\partial \mathbf{H}_\theta(x)}{\partial x} \right], \quad \mathbf{B}_s = \left[\frac{\partial \mathbf{H}_w(x)}{\partial x} + \mathbf{H}_\theta(x) \right], \\
\mathbf{F}^e_{\text{int}} &= \mathbf{K}^e \mathbf{u}^e(t), \quad \mathbf{K}^e = \mathbf{K}_b + \mathbf{K}_s, \quad \mathbf{F}^e_{\text{ext}} = \int_{L_e} \bar{\omega}(x)\, \mathbf{H}_w^T(x)\, dx
\end{aligned} \tag{9.79}$$

where for the last element only, $\mathbf{M}^e_L = \begin{bmatrix} 0 & 0 & 0 & M_L \end{bmatrix}^T$ and $\mathbf{Q}^e_L = \begin{bmatrix} 0 & 0 & Q_L & 0 \end{bmatrix}^T$.

Example 9.5 Kirchhoff-Love Thin Plate

Previously, we discussed the Euler-Bernoulli beam theory, in which the transverse shearing strains are neglected. Here, we consider the Kirchhoff-Love thin plate theory (also called the classical plate theory), which can be viewed as an extension of the Euler-Bernoulli beam theory to the thin plate. The material is assumed to be linearly elastic, isotropic and homogeneous.

Problem Description

The following assumptions, known as Kirchhoff's hypotheses (Kirchhoff 1877), are applied to the thin plate.

1. The plate of uniform thickness h is very thin compared to the width of the plate; the transverse normal stress σ_z is negligible. Thickness change is neglected. The plate is uniformly subjected to a transversely distributed load $\omega(x, y)$ per unit area of the mid-surface in the z-direction. Then, the plane stress constitutive relationship is employed.
2. The transverse deflection of the Kirchhoff-Love thin plate is due to the bending strains; hence the transverse shear strains γ_{xz} and γ_{yz} are neglected. As such, it is assumed that undeformed normal planes perpendicular to the mid surface remain normal during deformation.
3. The deflection of the mid surface is very small compared to the plate thickness; elastic deformation is assumed.

Furthermore, it is assumed that there are no body forces acting in the plate and no shearing forces on the flat surfaces. Then, the dynamical behavior of the Kirchhoff-Love thin plate is governed by one governing equation, because of the presence of only one variable as in the Euler-Bernoulli beam, namely, the transverse deflection $w(x, y, t)$. Being confined to the small displacement theory, the displacements corresponding to such deformations at any point of the plate are given by

$$
\begin{aligned}
u(x, y, z, t) &= -z\varphi(x, y, t) = -z\frac{\partial w_0(x, y, t)}{\partial x}, \\
v(x, y, z, t) &= -z\phi(x, y, t) = -z\frac{\partial w_0(x, y, t)}{\partial y}, \\
w(x, y, z, t) &= w_0(x, y, t)
\end{aligned} \tag{9.80}
$$

where the rotations $\varphi(x, y, t)$ and $\phi(x, y, t)$ are defined in Figure 9.7 and $w_0(x, y, t)$ is the deflection of a point on the mid surface of the plate. Thus, the normal and shear strains are obtained as

$$
\begin{aligned}
\varepsilon_x &= \frac{\partial u}{\partial x} = -z\frac{\partial \varphi(x, y, t)}{\partial x} = -z\frac{\partial^2 w(x, y, t)}{\partial x^2}, \\
\varepsilon_y &= \frac{\partial v}{\partial y} = -z\frac{\partial \phi(x, y, t)}{\partial y} = -z\frac{\partial^2 w(x, y, t)}{\partial y^2}, \\
\gamma_{xy} &= \frac{\partial u}{\partial y} + \frac{\partial v}{\partial x} = -2z\frac{\partial^2 w(x, y, t)}{\partial x \partial y}, \\
\varepsilon_z &= \frac{\partial w}{\partial z} \approx 0, \quad \gamma_{xz} \approx 0, \quad \gamma_{yz} \approx 0
\end{aligned} \tag{9.81}
$$

The relation between the plane strains $\boldsymbol{\varepsilon}$ and the curvatures $\boldsymbol{\kappa}$ of the plate is given by

$$
\boldsymbol{\varepsilon} = z\boldsymbol{\kappa} \tag{9.82}
$$

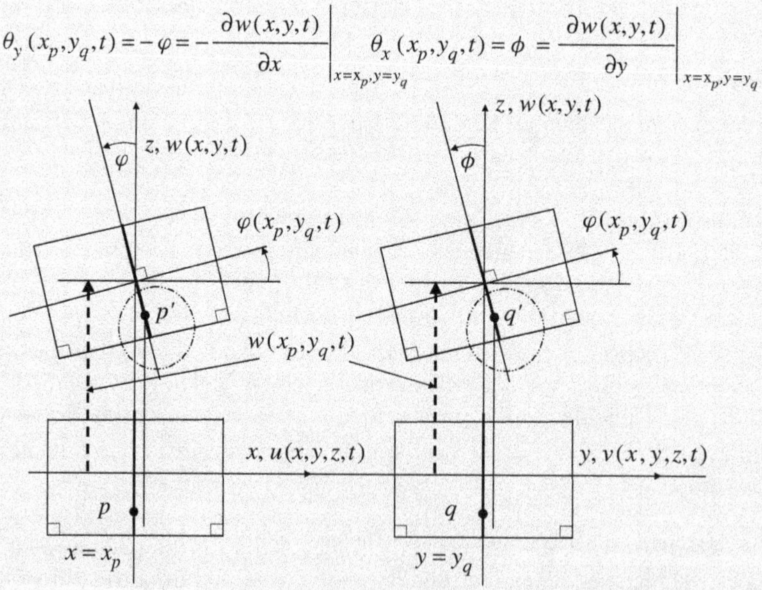

Figure 9.7. Deflection of the Kirchhoff-Love thin plate

where

$$\boldsymbol{\varepsilon} = \left\{ \begin{array}{c} \varepsilon_x \\ \varepsilon_y \\ \gamma_{xy} \end{array} \right\}, \quad \boldsymbol{\kappa} = \left\{ \begin{array}{c} \kappa_x \\ \kappa_y \\ \kappa_{xy} \end{array} \right\} = \left\{ \begin{array}{c} -\dfrac{\partial^2 w\,(x, y, t)}{\partial x^2} \\ -\dfrac{\partial^2 w\,(x, y, t)}{\partial y^2} \\ -2\dfrac{\partial^2 w\,(x, y, t)}{\partial x \partial y} \end{array} \right\} \tag{9.83}$$

And, the in-plane stress components are given by

$$\boldsymbol{\sigma} = \left\{ \begin{array}{c} \sigma_x \\ \sigma_y \\ \tau_{xy} \end{array} \right\} = \frac{zE}{1-\nu^2} \left[\begin{array}{ccc} 1 & \nu & 0 \\ \nu & 1 & 0 \\ 0 & 0 & \frac{1-\nu}{2} \end{array} \right] \left\{ \begin{array}{c} \kappa_x \\ \kappa_y \\ \kappa_{xy} \end{array} \right\} \tag{9.84}$$

In view of Equation (9.80), the above equation reduces to

$$\left\{ \begin{array}{c} \sigma_x \\ \sigma_y \\ \tau_{xy} \end{array} \right\} = -\frac{zE}{1-\nu^2} \left\{ \begin{array}{c} \dfrac{\partial^2 w}{\partial x^2} + \nu \dfrac{\partial^2 w}{\partial y^2} \\ \dfrac{\partial^2 w}{\partial y^2} + \nu \dfrac{\partial^2 w}{\partial x^2} \\ (1-\nu)\dfrac{\partial^2 w}{\partial x \partial y} \end{array} \right\} \tag{9.85}$$

Therefore, the bending and twisting moments per unit length (Timoshenko and Woinowsky-Krieger 1959; Ugural and Fenster 2003) can be given by

$$\begin{aligned} M_x &= \int_{-\frac{h}{2}}^{\frac{h}{2}} z\sigma_x dz = -D\left(\frac{\partial^2 w}{\partial x^2} + \nu \frac{\partial^2 w}{\partial y^2}\right), \\ M_y &= \int_{-\frac{h}{2}}^{\frac{h}{2}} z\sigma_y dz = -D\left(\frac{\partial^2 w}{\partial y^2} + \nu \frac{\partial^2 w}{\partial x^2}\right), \\ M_{xy} &= \int_{-\frac{h}{2}}^{\frac{h}{2}} z\tau_{xy} dz = -D\,(1-\nu)\frac{\partial^2 w}{\partial x \partial x} \end{aligned} \tag{9.86}$$

where D denotes the flexural rigidity of the plate, which is defined as

$$D = \frac{Eh^3}{12\left(1-\nu^2\right)} \tag{9.87}$$

In addition, the shearing forces per unit length are given by

$$Q_x = \int_{-\frac{h}{2}}^{\frac{h}{2}} \tau_{xz} dz \neq 0, \quad Q_y = \int_{-\frac{h}{2}}^{\frac{h}{2}} \tau_{yz} dz \neq 0 \tag{9.88}$$

The shearing forces above are considered not to be negligible, even though the transverse shearing strains in Equation (9.81) are negligible.

Free-Body Diagram

A free-body diagram for the Kirchhoff-Love plate is shown in Figure 9.8, where we have the following relations

$$\begin{aligned} Q'_x &= Q_x + \frac{\partial Q_x}{\partial x}\Delta x, \quad Q'_y = Q_y + \frac{\partial Q_y}{\partial y}\Delta y, \\ M'_{xx} &= M_{xx} + \frac{\partial M_{xx}}{\partial x}\Delta x, \quad M'_{yy} = M_{yy} + \frac{\partial M_{yy}}{\partial y}\Delta y, \\ M'_{xy} &= M_{xy} + \frac{\partial M_{xy}}{\partial x}\Delta x, \quad M'_{yx} = M_{yx} + \frac{\partial M_{yx}}{\partial y}\Delta y \end{aligned} \tag{9.89}$$

where Q_x and Q_y are shearing forces per unit length, M_{xx} and M_{xx} are bending moments per unit length, and M_{xy} and M_{xy} are twisting moments per unit length.

By virtue of D'Alembert's principle given in Equation (3.47), the infinitesimal element of the Kirchhoff-Love thin plate is dynamically in equilibrium in the z-direction as follows:

$$\begin{aligned} &\frac{\partial Q_x}{\partial x}\Delta x\Delta y + \frac{\partial Q_y}{\partial y}\Delta x\Delta y + \bar{\omega}\left(x + \frac{\Delta x}{2}, y + \frac{\Delta y}{2}\right)\Delta x\Delta y \\ &- \rho h\ddot{w}(x + \frac{\Delta x}{2}, y + \frac{\Delta y}{2}, t)\Delta x\Delta y = 0 \end{aligned} \tag{9.90}$$

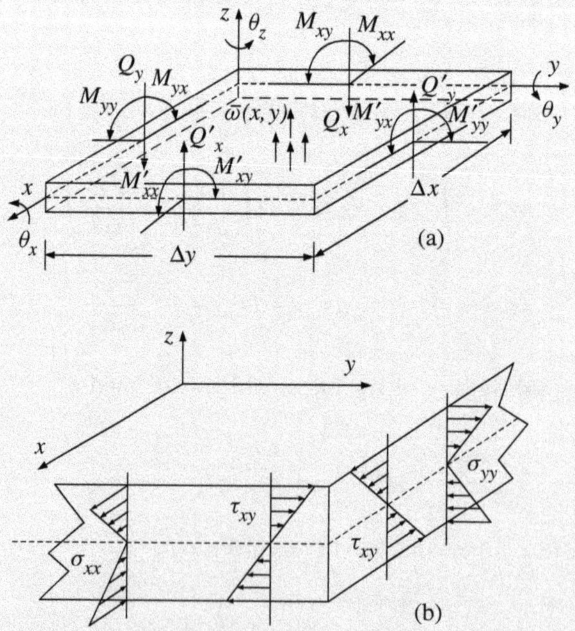

Figure 9.8. (a) Free-body diagram of the Kirchhoff-Love thin plate, (b) Stresses on the cross-section of the plate

Further, as the bending moments about the x-axis and about the y-axis are dynamically in equilibrium, we have

$$\begin{aligned}
&\frac{M_{xy}}{\partial x}\Delta x\Delta y + \frac{M_{yy}}{\partial y}\Delta x\Delta y - Q_y\Delta x\Delta y - \frac{\partial Q_y}{\partial y}\Delta x\Delta y\frac{\Delta y}{2} + \rho I\frac{\partial \ddot{w}}{\partial y}\Delta x\Delta y = 0, \\
&\frac{M_{xx}}{\partial x}\Delta x\Delta y + \frac{M_{xy}}{\partial y}\Delta x\Delta y - Q_x\Delta x\Delta y - \frac{\partial Q_x}{\partial x}\Delta x\Delta y\frac{\Delta x}{2} + \rho I\frac{\partial \ddot{w}}{\partial x}\Delta x\Delta y = 0
\end{aligned} \tag{9.91}$$

where the moment of inertia of area per unit length I is defined as

$$I = I_x = I_y = \int_{-\frac{h}{2}}^{\frac{h}{2}} z^2 dz \tag{9.92}$$

In view of Equation (9.86), the shearing forces can be written as

$$\begin{aligned}
Q_x &= \frac{M_{xx}}{\partial x} + \frac{M_{xy}}{\partial y} + \rho I\frac{\partial \ddot{w}}{\partial x} = -D\frac{\partial}{\partial x}\left(\frac{\partial^2 w}{\partial x^2} + \frac{\partial^2 w}{\partial y^2}\right) + \rho I\frac{\partial \ddot{w}}{\partial x}, \\
Q_y &= \frac{M_{xy}}{\partial x} + \frac{M_{yy}}{\partial y} + \rho I\frac{\partial \ddot{w}}{\partial y} = -D\frac{\partial}{\partial y}\left(\frac{\partial^2 w}{\partial x^2} + \frac{\partial^2 w}{\partial y^2}\right) + \rho I\frac{\partial \ddot{w}}{\partial y}
\end{aligned} \tag{9.93}$$

We first substitute the shearing forces in the above into Equation (9.90). Next, taking the limit of the element segment leads to the governing equation of the Kirchhoff-Love plate as follows:

$$\frac{\partial^2 M_{xx}}{\partial x^2} + 2\frac{\partial^2 M_{xy}}{\partial x\partial y} + \frac{\partial^2 M_{yy}}{\partial y^2} + \rho I\left(\frac{\partial^2 \ddot{w}}{\partial x^2} + \frac{\partial^2 \ddot{w}}{\partial y^2}\right) + \bar{\omega}(x, y) - \rho h\ddot{w} = 0 \tag{9.94}$$

By use of Equation (9.86), the above equation can be rewritten as

$$\rho h\ddot{w} + D\left(\frac{\partial w^4}{\partial x^4} + 2\frac{\partial^4 w}{\partial x^2\partial y^2} + \frac{\partial w^4}{\partial y^4}\right) - \rho I\left(\frac{\partial^2 \ddot{w}}{\partial x^2} + \frac{\partial^2 \ddot{w}}{\partial y^2}\right) - \bar{\omega}(x, y) = 0 \tag{9.95}$$

which can be rewritten in a simplified form as

$$\rho h\ddot{w} + D\nabla^4 w(x, y, t) - \rho I\left(\frac{\partial^2 \ddot{w}}{\partial x^2} + \frac{\partial^2 \ddot{w}}{\partial y^2}\right) - \bar{\omega}(x, y) = 0 \tag{9.96}$$

where the biharmonic operator is denoted by ∇^4.

Boundary Conditions

Essential Boundary Conditions

$$w(x, y, t) = \bar{w}, \quad \frac{\partial w}{\partial x} = \bar{\psi}_x, \quad \frac{\partial w}{\partial y} = \bar{\psi}_y, \quad \frac{\partial w}{\partial n} = \bar{\psi}_n \tag{9.97}$$

Natural Boundary Conditions

$$Q_x + \frac{\partial M_{xy}}{\partial y} = \bar{V}_x, \quad Q_y + \frac{\partial M_{xy}}{\partial x} = \bar{V}_y,$$
$$M_x = \bar{M}_x, \quad M_y = \bar{M}_y, \quad M_{nn} = \bar{M}_{nn} \tag{9.98}$$

Simply Supported Edge

$$w(x, y, t) = 0, \quad M_x = 0 \text{ or } M_y = 0 \text{ or } M_{nn} = 0 \tag{9.99}$$

Built-in Edge

$$w(x, y, t) = 0, \quad \frac{\partial w}{\partial x} = 0 \text{ or } \frac{\partial w}{\partial y} = 0 \text{ or } \frac{\partial w}{\partial n} = 0 \tag{9.100}$$

Free Edge

$$Q_x + \frac{\partial M_{xy}}{\partial y} = 0 \text{ or } Q_y + \frac{\partial M_{xy}}{\partial x} = 0,$$
$$M_x = 0 \text{ or } M_y = 0 \text{ or } M_{nn} = 0 \tag{9.101}$$

Sliding Edge

$$\frac{\partial w}{\partial x} = 0 \text{ or } \frac{\partial w}{\partial y} = 0 \text{ or } \frac{\partial w}{\partial n} = 0,$$
$$Q_x + \frac{\partial M_{xy}}{\partial y} = 0 \text{ or } Q_y + \frac{\partial M_{xy}}{\partial x} = 0 \tag{9.102}$$

Principle of Virtual Work

The transverse deflection for the governing equation in Equation (9.96) should satisfy the continuity condition, viz., $w(x, t) \in C^4(\Omega)$. Next, in view of Equation (9.94) the Bubnov-Galerkin weighted residual forms are set up as follows:

$$\int_{\Omega} \left[\begin{array}{c} \rho h\ddot{w} - \rho I \left(\dfrac{\partial^2 \ddot{w}}{\partial x^2} + \dfrac{\partial^2 \ddot{w}}{\partial y^2} \right) - \bar{\omega}(x, y) \\ -\dfrac{\partial^2 M_{xx}}{\partial x^2} - 2\dfrac{\partial^2 M_{xy}}{\partial x \partial y} - \dfrac{\partial^2 M_{yy}}{\partial y^2} \end{array} \right] \delta w \, dxdy = 0 \tag{9.103}$$

By making use of Green's theorem and Equation (9.93), the above equation can be written as

$$\begin{aligned} &\int_{\Omega} \rho h \ddot{w} \delta w dxdy + \int_{\Omega} \rho I \left(\frac{\partial \ddot{w}}{\partial x}\frac{\partial \delta w}{\partial x} + \frac{\partial \ddot{w}}{\partial y}\frac{\partial \delta w}{\partial y} \right) dxdy \\ &- \int_{\Omega} \left(M_{xx}\frac{\partial^2 \delta w}{\partial x^2} + 2M_{xy}\frac{\partial^2 \delta w}{\partial x \partial y} + M_{yy}\frac{\partial^2 \delta w}{\partial y^2} \right) dxdy \\ &= \int_{\Omega} \bar{\omega}(x, y)\, \delta w dxdy + \int_{\partial\Omega^+} (Q_x n_x + Q_y n_y)\, \delta w d\Gamma \\ &- \int_{\partial\Omega^+} (M_{xx} n_x + M_{xy} n_y) \frac{\partial \delta w}{\partial x} d\Gamma - \int_{\partial\Omega^+} (M_{yy} n_y + M_{xy} n_x) \frac{\partial \delta w}{\partial y} d\Gamma \end{aligned} \tag{9.104}$$

where $\partial\Omega^+$ represents the counterclockwise orientation on the boundary of the region Ω (Marsden and Tromba 2003) where $\partial\Omega^+$ represents the counterclockwise orientation on the boundary of the region Ω (Marsden and Tromba 2003), and both n_x and n_y are the directional cosines of the unit outward normal vector $\mathbf{n}$ as shown in Figure 9.9. The (x, y) coordinate system is transformed to the (n, s) coordinate system in the following way on the edge of the plate

$$\left\{ \begin{array}{c} \mathbf{i}_x \\ \mathbf{i}_y \end{array} \right\} = \left[\begin{array}{cc} cos\theta & -sin\theta \\ sin\theta & cos\theta \end{array} \right] \left\{ \begin{array}{c} \mathbf{i}_n \\ \mathbf{i}_s \end{array} \right\} = \left[\begin{array}{cc} n_x & -n_y \\ n_y & n_x \end{array} \right] \left\{ \begin{array}{c} \mathbf{i}_n \\ \mathbf{i}_s \end{array} \right\} \tag{9.105}$$

Note that the following relation can be obtained

$$\begin{aligned} \frac{\partial \delta w}{\partial x} &= \nabla \delta w \cdot \mathbf{i}_x = \left(\frac{\partial \delta w}{\partial n}\mathbf{i}_n + \frac{\partial \delta w}{\partial s}\mathbf{i}_s \right) \cdot \mathbf{i}_x = n_x \frac{\partial \delta w}{\partial n} - n_y \frac{\partial \delta w}{\partial s}, \\ \frac{\partial \delta w}{\partial y} &= \nabla \delta w \cdot \mathbf{i}_y = \left(\frac{\partial \delta w}{\partial n}\mathbf{i}_n + \frac{\partial \delta w}{\partial s}\mathbf{i}_s \right) \cdot \mathbf{i}_y = n_y \frac{\partial \delta w}{\partial n} + n_x \frac{\partial \delta w}{\partial s} \end{aligned} \tag{9.106}$$

Also, we have

$$\begin{aligned} \frac{\partial \delta w}{\partial n} &= n_x \frac{\partial \delta w}{\partial x} + n_y \frac{\partial \delta w}{\partial y}, \\ \frac{\partial \delta w}{\partial s} &= n_x \frac{\partial \delta w}{\partial y} - n_y \frac{\partial \delta w}{\partial s} \end{aligned} \tag{9.107}$$

the boundary terms on the right-hand side of Equation (9.104) can be expressed as

$$\begin{aligned} & - \int_{\partial\Omega^+} (M_{xx}n_x + M_{xy}n_y) \frac{\partial \delta w}{\partial x} d\Gamma - \int_{\partial\Omega^+} (M_{yy}n_y + M_{xy}n_x) \frac{\partial \delta w}{\partial y} d\Gamma \\ & = - \int_{\partial\Omega^+} M_{nn} \frac{\partial \delta w}{\partial n} d\Gamma - \int_{\partial\Omega^+} M_{ns} \frac{\partial \delta w}{\partial s} d\Gamma \end{aligned} \tag{9.108}$$

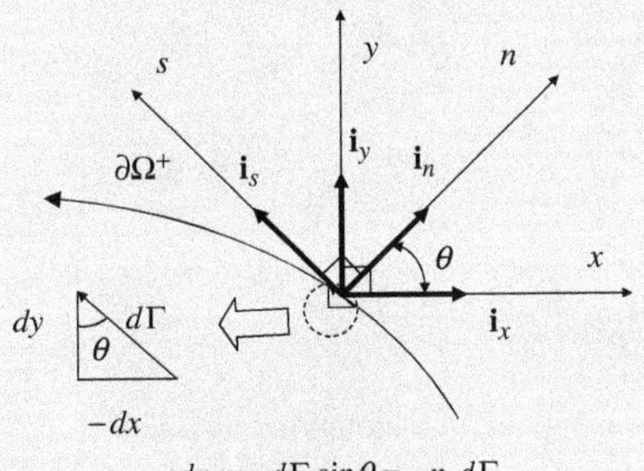

$dx = -d\Gamma\ \sin\theta = -n_y d\Gamma$

$dy = d\Gamma\ \cos\theta = n_x d\Gamma$

Figure 9.9. Normal and tangential on the counterclockwise boundary

where

$$M_{nn} = M_{xx}n_x^2 + 2M_{xy}n_xn_y + M_{yy}n_y^2,$$
$$M_{ns} = \left(M_{yy} - M_{xx}\right)n_xn_y + M_{xy}\left(n_x^2 - n_y^2\right) \tag{9.109}$$

The last term in Equation (9.108) can be integrated by parts as follows:

$$\int_{\partial\Omega+} M_{ns}\frac{\partial\delta w}{\partial s}d\Gamma = \left(M_{ns}\delta w\right)|_{\Gamma} - \int_{\partial\Omega+}\frac{\partial M_{ns}}{\partial s}\delta w d\Gamma = -\int_{\partial\Omega+}\frac{\partial M_{ns}}{\partial s}\delta w d\Gamma \tag{9.110}$$

By introducing a concise notation

$$Q_xn_x + Q_yn_y = \mathbf{Q}\cdot\mathbf{n} = Q_n \tag{9.111}$$

then, Equation (9.104) can be rewritten in a simplified form as

$$\begin{aligned}
&\int_{\Omega}\rho h\ddot{w}\delta w dxdy + \int_{\Omega}\rho I\left(\frac{\partial\ddot{w}}{\partial x}\frac{\partial\delta w}{\partial x} + \frac{\partial\ddot{w}}{\partial y}\frac{\partial\delta w}{\partial y}\right)dxdy\\
&-\int_{\Omega}\left(M_{xx}\frac{\partial^2\delta w}{\partial x^2} + 2M_{xy}\frac{\partial^2\delta w}{\partial x\partial y} + M_{yy}\frac{\partial^2\delta w}{\partial y^2}\right)dxdy = \int_{\Omega}\bar{\omega}\left(x, y\right)\delta w dxdy\\
&+\int_{\partial\Omega+}\left(Q_n + \frac{\partial M_{ns}}{\partial s}\right)\delta w d\Gamma - \int_{\partial\Omega+}M_{nn}\delta\left(\frac{\partial w}{\partial n}\right)d\Gamma
\end{aligned} \tag{9.112}$$

which is the weak form of the Kirchhoff-Love thin plate. Next, we define the externally applied loadings such as

$$\left\{\begin{array}{c} Q_n + \frac{\partial M_{ns}}{\partial s} = \bar{V}_n \\ M_{nn} = \bar{M}_{nn} \end{array}\right\}, \quad \text{on } \partial\Omega_\sigma^+ \tag{9.113}$$

where $\bar{V}_n$ is known as Kirchhoff's force on the edge of the plate (Reddy 2002; Ugural and Fenster 2003). As a consequence, the weak form of the Kirchhoff-Love thin plate can be written as

$$\begin{aligned}
&\int_{\Omega}\rho h\ddot{w}\delta w dxdy + \int_{\Omega}\rho I\left(\frac{\partial\ddot{w}}{\partial x}\frac{\partial\delta w}{\partial x} + \frac{\partial\ddot{w}}{\partial y}\frac{\partial\delta w}{\partial y}\right)dxdy\\
&-\int_{\Omega}\left(M_{xx}\frac{\partial^2\delta w}{\partial x^2} + 2M_{xy}\frac{\partial^2\delta w}{\partial x\partial y} + M_{yy}\frac{\partial^2\delta w}{\partial y^2}\right)dxdy - \int_{\Omega}\bar{\omega}\left(x, y\right)\delta w dxdy\\
&=\int_{\partial\Omega_\sigma^+}\bar{V}_n\delta w d\Gamma - \int_{\partial\Omega_\sigma^+}\bar{M}_{nn}\delta\left(\frac{\partial w}{\partial n}\right)d\Gamma
\end{aligned} \tag{9.114}$$

For convenience, we introduce some new notations as follows:

$$\mathbf{m} = \left\{\begin{array}{c} M_{xx} \\ M_{yy} \\ M_{xy} \end{array}\right\} = \mathbf{D}\boldsymbol{\kappa}, \quad \mathbf{D} = \frac{Eh^3}{12\left(1-\nu^2\right)}\left[\begin{array}{ccc} 1 & \nu & 0 \\ \nu & 1 & 0 \\ 0 & 0 & \frac{1-\nu}{2} \end{array}\right] \tag{9.115}$$

Then, Equation (9.114) can be rewritten as

$$\begin{aligned}&\int_{\Omega}\rho h\ddot{w}\delta w dxdy+\int_{\Omega}\rho I\left(\frac{\partial\ddot{w}}{\partial x}\frac{\partial\delta w}{\partial x}+\frac{\partial\ddot{w}}{\partial y}\frac{\partial\delta w}{\partial y}\right)dxdy+\int_{\Omega}\delta\boldsymbol{\kappa}^T\boldsymbol{m}dxdy\\&=\int_{\partial\Omega_\sigma^+}\bar{V}_n\delta w d\Gamma+\int_{\partial\Omega_\sigma^+}\bar{M}_{nn}\delta\theta_n d\Gamma,\ \ \delta\theta_n=-\delta\left(\frac{\partial w}{\partial n}\right)\end{aligned}\tag{9.116}$$

The above equation represents the weak form of the Kirchhoff-Love thin plate, namely, the principle of virtual work in dynamics for the Kirchhoff-Love thin plate.

Spatial Discretization

The trial displacement function may be approximated using the Hermite polynomial interpolation functions as follows:

$$\begin{aligned}&\tilde{w}^h(x,t)=\mathbf{N}(x,y)\,\mathbf{u}^e(t),\ \ \mathbf{u}_i^e(t)=\left(w_i(t),\theta_{xi}(t),\theta_{yi}(t)\right),\\&\boldsymbol{\kappa}^h(x,y,t)=\bar{\mathbf{B}}(x,y)\,\mathbf{u}^e(t),\mathbf{m}^h(x,y,t)=\mathbf{D}\boldsymbol{\kappa}^h(x,y,t)\end{aligned}\tag{9.117}$$

The trial function is of class $C^1(\Omega)$ over the entire domain. As the variation of the dependent variable, the test function is given by $\delta\tilde{w}^h(x,y,t)=\mathbf{N}(x,y)\,\delta\mathbf{u}^e(t)$. Substituting both the trial and test functions into the weak form in Equation (9.116), the domain of interest can be discretized as follows:

$$\begin{aligned}&\sum_{e=1}^{n_{\text{elem}}}\left[\begin{aligned}&\int_{\Omega_e}(\rho h\ddot{w})\,\delta w dxdy+\int_{\Omega_e}\rho I\left(\frac{\partial\ddot{w}}{\partial x}\frac{\partial\delta w}{\partial x}+\frac{\partial\ddot{w}}{\partial y}\frac{\partial\delta w}{\partial y}\right)dxdy\\&+\int_{\Omega_e}\delta\boldsymbol{\kappa}^T\boldsymbol{m}dxdy-\int_{\Omega_e}\bar{\omega}(x,y)\delta w dxdy\end{aligned}\right]\\&=\int_{\partial\Omega_\sigma^+}\bar{V}_n\delta w d\Gamma+\int_{\partial\Omega_\sigma^+}\bar{M}_{nn}\delta\theta_n d\Gamma\end{aligned}\tag{9.118}$$

Therefore, Equation (9.118) reduces to the following

$$\sum_{e=1}^{n_{\text{elem}}}\left(\mathbf{M}_e\ddot{\mathbf{u}}^e(t)+\mathbf{F}_{\text{int}}^e-\mathbf{F}_{\text{ext}}^e\right)\cdot\delta\mathbf{u}^e(t)=0\tag{9.119}$$

where

$$\begin{aligned}&\mathbf{M}_e=\int_{\Omega_e}\rho h\mathbf{N}_e^T\mathbf{N}_e dx+\int_{\Omega_e}\rho I\mathbf{B}_1^T\bar{\mathbf{B}}_1 dx+\int_{\Omega_e}\rho I\mathbf{B}_2^T\bar{\mathbf{B}}_2 dx,\\&\mathbf{F}_{\text{int}}^e=\mathbf{K}_e\mathbf{u}^e,\ \ \mathbf{K}_e=\int_{\Omega_e}\bar{\mathbf{B}}^T\mathbf{D}\mathbf{B}dx,\ \ \mathbf{B}_1=\frac{\partial\mathbf{N}_e}{\partial x}\ \ \mathbf{B}_2=\frac{\partial\mathbf{N}_e}{\partial y}\\&\mathbf{F}_{\text{ext}}^e=\int_{\Omega_e}\mathbf{N}_e^T\bar{\omega}(x)dx+\mathbf{F}_{\text{bound}}^e,\\&\mathbf{F}_{\text{bound}}^e=\int_{\partial\Omega_\sigma^+}\mathbf{N}_e^T\bar{V}_n d\Gamma+\int_{\partial\Omega_\sigma^+}\left[\frac{\partial\mathbf{N}_e^T}{\partial n}\right]^T\bar{M}_{nn}d\Gamma\end{aligned}\tag{9.120}$$

where $\mathbf{F}_{\text{bound}}^e$ denotes the boundary external force.

Example 9.6 Reissner-Mindlin plate

We next describe the Reissner-Mindlin plate of uniform thickness h. It is assumed that the transverse normal stress of the moderately thick plate is negligible in plane stress, compared to the other stresses components, and the plate is uniformly subject to a transversely distributed load $\omega(x, y)$ per unit area of the mid-surface in the z-direction.

Problem Description

Furthermore, it is assumed that there are no body forces acting in the plate and no shearing forces on the flat surfaces. The material is assumed to be linearly elastic, isotropic and homogeneous. For the sake of brevity, the membrane behavior of the plate is not considered; the deformation is due to the bending and shear. In contrast to the Kirchhoff-Love thin plate, the transverse shear strains due to the moderate thickness effect (Mindlin 1951; Reissner 1945) are taken into account as an extension of the Timoshenko beam to the plate. Consequently, the dynamical behavior of the Reissner-Mindlin plate is governed by three differential governing equations, because of the presence of three dependent variables, namely, the transverse deflection $w(x, y, t)$ and two rotations, $\theta_x(x, y, t), \theta_y(x, y, t)$. The reason for us to illustrate the Reissner-Mindlin plate is that the plate reflects the rotary inertia effect, as well as the translational kinetic motion effect and the strain energy consists of the bending strain and shear strain energies. Being confined to the small displacement theory, the displacements corresponding to such deformations at any point of the plate are given by

$$\begin{aligned}
u(x, y, z, t) &= -z\varphi(x, y, t) = -z\left(\frac{\partial w_0(x, y, t)}{\partial x} - \gamma_{xz}(x, y, t)\right), \\
v(x, y, z, t) &= -z\phi(x, y, t) = -z\left(\frac{\partial w_0(x, y, t)}{\partial y} - \gamma_{yz}(x, y, t)\right), \\
w(x, y, z, t) &= w_0(x, y, t), \\
\theta_y &= -\varphi(x, y, t) = -\left(\frac{\partial w_0}{\partial x} - \gamma_{xz}\right), \quad \theta_x = \phi(x, y, t) = \frac{\partial w_0}{\partial y} - \gamma_{yz}
\end{aligned} \tag{9.121}$$

where the rotations $\varphi(x, y, t)$ and $\phi(x, y, t)$ are defined in Figure 9.10 and $w_0(x, y, t)$ is the deflection of a point on the midsurface of the plate. which is often called the first order shear deformation theory of the plate. Thus, the normal and shear strains are obtained as

$$\begin{aligned}
\varepsilon_x &= \frac{\partial u}{\partial x} = -z\frac{\partial \varphi(x, y, t)}{\partial x} = z\frac{\partial \theta_y}{\partial x}, \\
\varepsilon_y &= \frac{\partial v}{\partial y} = -z\frac{\partial \phi(x, y, t)}{\partial y} = -z\frac{\partial \theta_x}{\partial y}, \\
\varepsilon_z &= \frac{\partial w}{\partial z} \approx 0, \\
\gamma_{xy} &= \frac{\partial u}{\partial y} + \frac{\partial v}{\partial x} = -z\left(\frac{\partial \varphi(x, y, t)}{\partial y} + \frac{\partial \phi(x, y, t)}{\partial x}\right) = z\left(\frac{\partial \theta_y}{\partial y} - \frac{\partial \theta_x}{\partial x}\right), \\
\gamma_{xz} &= \frac{\partial u}{\partial z} + \frac{\partial w}{\partial x} = \frac{\partial w}{\partial x} - \varphi = \frac{\partial w}{\partial x} + \theta_y, \\
\gamma_{yz} &= \frac{\partial v}{\partial z} + \frac{\partial w}{\partial y} = \frac{\partial w}{\partial y} - \phi = \frac{\partial w}{\partial y} - \theta_x
\end{aligned} \tag{9.122}$$

The relation between the plane strains $\boldsymbol{\varepsilon}$ and the curvatures $\boldsymbol{\kappa}$ of the plate is given by

$$\boldsymbol{\varepsilon} = z\boldsymbol{\kappa} \tag{9.123}$$

where

$$\boldsymbol{\varepsilon} = \begin{Bmatrix} \varepsilon_x \\ \varepsilon_y \\ \gamma_{xy} \end{Bmatrix}, \quad \boldsymbol{\kappa} = \begin{Bmatrix} \kappa_x \\ \kappa_y \\ \kappa_{xy} \end{Bmatrix} = \begin{Bmatrix} -\dfrac{\partial \varphi}{\partial x} \\ -\dfrac{\partial \phi}{\partial y} \\ -\dfrac{\partial \phi}{\partial x} - \dfrac{\partial \varphi}{\partial y} \end{Bmatrix} = \begin{Bmatrix} \dfrac{\partial \theta_y}{\partial x} \\ -\dfrac{\partial \theta_x}{\partial y} \\ \dfrac{\partial \theta_y}{\partial y} - \dfrac{\partial \theta_x}{\partial x} \end{Bmatrix} \tag{9.124}$$

where κ_x and κ_y are the direct curvatures, and κ_{xy} is the twisting curvature of the plate (Gallagher 1975). The in-plane stress components are given by

$$\boldsymbol{\sigma} = \begin{Bmatrix} \sigma_x \\ \sigma_y \\ \tau_{xy} \end{Bmatrix} = \frac{zE}{1-\nu^2} \begin{bmatrix} 1 & \nu & 0 \\ \nu & 1 & 0 \\ 0 & 0 & \frac{1-\nu}{2} \end{bmatrix} \begin{Bmatrix} \kappa_x \\ \kappa_y \\ \kappa_{xy} \end{Bmatrix} \tag{9.125}$$

The transverse shear stress components are given by

$$\boldsymbol{\tau} = G\boldsymbol{\gamma} \tag{9.126}$$

where

$$\boldsymbol{\tau} = \begin{Bmatrix} \tau_{xz} \\ \tau_{yz} \end{Bmatrix}, \quad \boldsymbol{\gamma} = \begin{Bmatrix} \gamma_{xz} \\ \gamma_{yz} \end{Bmatrix} = \begin{Bmatrix} \dfrac{\partial w}{\partial x} - \varphi \\ \dfrac{\partial w}{\partial y} - \phi \end{Bmatrix} = \begin{Bmatrix} \dfrac{\partial w}{\partial x} + \theta_y \\ \dfrac{\partial w}{\partial y} - \theta_x \end{Bmatrix} \tag{9.127}$$

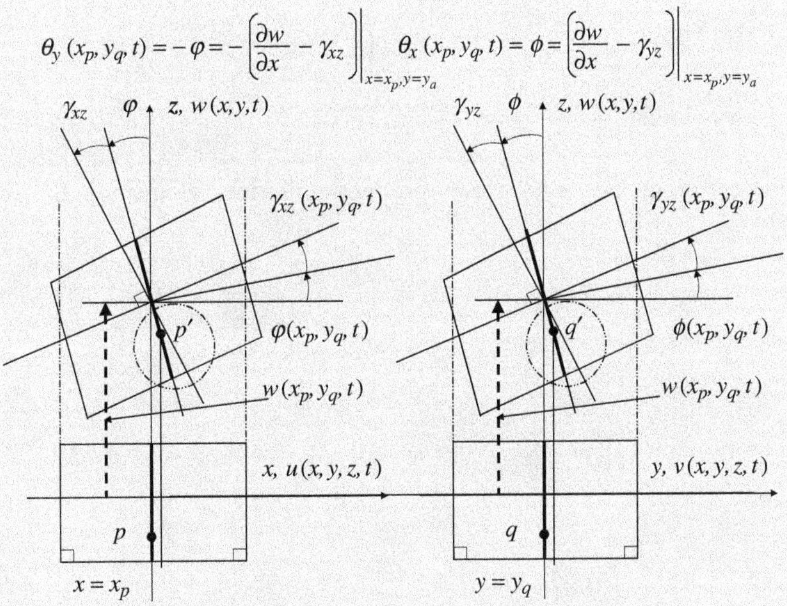

Figure 9.10. Deflection of the Reissner-Mindlin plate

Therefore, the line loadings $\mathbf{m} = \{M_x, M_y, M_{xy}\}^T$ (Gallagher 1975) can be given by

$$\mathbf{m} = \left\{ \begin{array}{c} M_{xx} \\ M_{yy} \\ M_{xy} \end{array} \right\} = \left\{ \begin{array}{c} \int_{-\frac{h}{2}}^{\frac{h}{2}} z\sigma_x dz \\ \int_{-\frac{h}{2}}^{\frac{h}{2}} z\sigma_y dz \\ \int_{-\frac{h}{2}}^{\frac{h}{2}} z\tau_{xy} dz \end{array} \right\} = \mathbf{D}\kappa, \quad \mathbf{D} = \frac{Eh^3}{12(1-\nu^2)} \begin{bmatrix} 1 & \nu & 0 \\ \nu & 1 & 0 \\ 0 & 0 & \frac{1-\nu}{2} \end{bmatrix} \tag{9.128}$$

In addition, the shearing forces per unit length are given by

$$Q_x = \int_{-\frac{h}{2}}^{\frac{h}{2}} \tau_{xz} dz = \kappa G h \gamma_{xz}, \quad Q_y = \int_{-\frac{h}{2}}^{\frac{h}{2}} \tau_{yz} dz = \kappa G h \gamma_{yz} \tag{9.129}$$

which can be rewritten as

$$\mathbf{Q} = \left\{ \begin{array}{c} Q_x \\ Q_y \end{array} \right\} = \left\{ \begin{array}{c} \kappa G h \gamma_{xz} \\ \kappa G h \gamma_{yz} \end{array} \right\} = \left\{ \begin{array}{c} \kappa G h \left(\dfrac{\partial w}{\partial x} + \theta_y \right) \\ \kappa G h \left(\dfrac{\partial w}{\partial y} - \theta_x \right) \end{array} \right\} \tag{9.130}$$

Free-Body Diagram

A free-body diagram for the Reissner-Mindlin plate is shown in Figure 9.11. By virtue of D'Alembert's principle given in Equation (3.47), the infinitesimal element of the Reissner-Mindlin plate is dynamically in equilibrium in the y-direction as follows:

$$\begin{aligned} &\frac{\partial Q_x}{\partial x}\Delta x \Delta y + \frac{\partial Q_y}{\partial y}\Delta x \Delta y + \bar{\omega}\left(x + \frac{\Delta x}{2}, y + \frac{\Delta y}{2}\right)\Delta x \Delta y \\ &- \rho h \ddot{w}(x + \frac{\Delta x}{2}, y + \frac{\Delta y}{2}, t)\Delta x \Delta y = 0 \end{aligned} \tag{9.131}$$

Further, about the z-axis, the rotational motion of the infinitesimal element is dynamically in equilibrium as follows:

$$\begin{aligned} &\frac{\partial M_{xy}}{\partial x}\Delta x \Delta y + \frac{\partial M_{yy}}{\partial y}\Delta x \Delta y - Q_y \Delta x \Delta y - \frac{\partial Q_y}{\partial y}\Delta x \Delta y \frac{\Delta y}{2} + \rho I_p \ddot{\phi} \Delta x \Delta y = 0, \\ &\frac{\partial M_{xx}}{\partial x}\Delta x \Delta y + \frac{\partial M_{xy}}{\partial y}\Delta x \Delta y - Q_x \Delta x \Delta y - \frac{\partial Q_x}{\partial x}\Delta x \Delta y \frac{\Delta x}{2} + \rho I_p \ddot{\varphi} \Delta x \Delta y = 0 \end{aligned} \tag{9.132}$$

where $I_p = \frac{h^3}{12}$. In the limit as Δx and Δy go to zero, Equations (9.131) and (9.132) yield

$$\begin{aligned} &\frac{\partial Q_x}{\partial x} + \frac{\partial Q_y}{\partial y} + \bar{\omega}(x, y) - \rho h \ddot{w} = 0, \\ &\frac{\partial M_{xy}}{\partial x} + \frac{\partial M_{yy}}{\partial y} - Q_y + \rho I_p \ddot{\theta}_x = 0, \end{aligned}$$

$$\frac{\partial M_{xx}}{\partial x} + \frac{\partial M_{xy}}{\partial y} - Q_x - \rho I_p \ddot{\theta}_y = 0 \tag{9.133}$$

The above equation can be rewritten as

$$\begin{Bmatrix} \rho \ddot{w} h \\ \rho \ddot{\theta}_x I_p \\ \rho \ddot{\theta}_y I_p \end{Bmatrix} + \begin{Bmatrix} 0 \\ -Q_x \\ Q_y \end{Bmatrix} - \frac{\partial}{\partial x} \begin{Bmatrix} Q_y \\ -M_{xy} \\ M_{xx} \end{Bmatrix} - \frac{\partial}{\partial y} \begin{Bmatrix} Q_x \\ -M_{yy} \\ M_{xy} \end{Bmatrix} - \begin{Bmatrix} \omega \\ 0 \\ 0 \end{Bmatrix} = \begin{Bmatrix} 0 \\ 0 \\ 0 \end{Bmatrix} \tag{9.134}$$

Consequently, the strong form can be given by

$$\begin{aligned}
&\rho h \ddot{w} - \kappa G h \frac{\partial}{\partial x}\left(\frac{\partial w}{\partial x} + \theta_y\right) - G h k \frac{\partial}{\partial y}\left(\frac{\partial w}{\partial y} - \theta_x\right) - \bar{\omega}(x, y) = 0, \\
&\rho I_p \ddot{\theta}_x - \kappa G h \left(\frac{\partial w}{\partial y} - \theta_x\right) - \frac{1}{2} D \left\{ \begin{array}{l} (1-\nu)\nabla^2 \theta_x \\ + (1+\nu)\left(\dfrac{\partial^2 \theta_x}{\partial y^2} - \dfrac{\partial^2 \theta_y}{\partial x \partial y}\right) \end{array} \right\} = 0, \\
&\rho I_p \ddot{\theta}_y + \kappa G h \left(\frac{\partial w}{\partial x} + \theta_y\right) - \frac{1}{2} D \left\{ \begin{array}{l} (1-\nu)\nabla^2 \theta_y \\ + (1+\nu)\left(\dfrac{\partial^2 \theta_y}{\partial x^2} - \dfrac{\partial^2 \theta_x}{\partial x \partial y}\right) \end{array} \right\} = 0
\end{aligned} \tag{9.135}$$

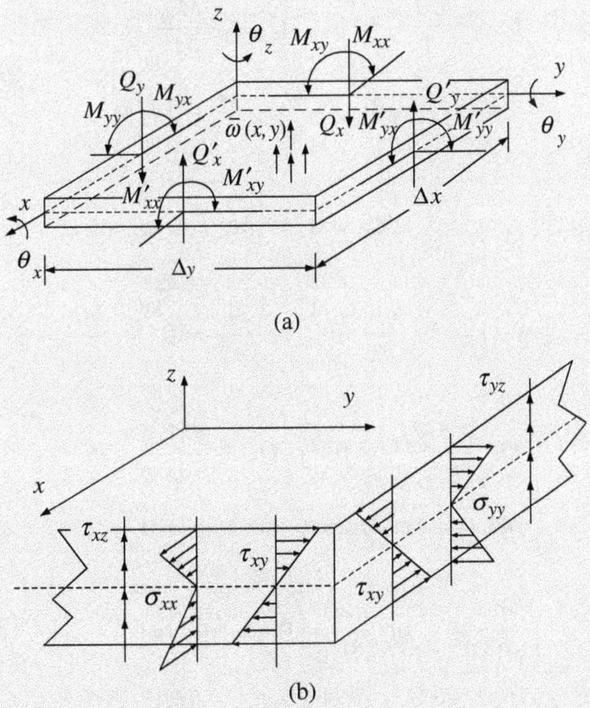

Figure 9.11. (a) Free-body diagram of the Reissner-Mindlin plate, (b) Stresses on the cross-section of the plate

which are subject to the natural boundary conditions

$$\begin{aligned} n_x Q_y + n_y Q_x &= 0, \\ n_x M_{xx} + n_y M_{xy} &= 0, \qquad \text{on } \partial\Omega_\sigma \\ n_x M_{xy} + n_y M_{yy} &= 0 \end{aligned} \tag{9.136}$$

and the essential boundary conditions

$$w = \bar{w}, \quad \theta_x = \bar{\theta}_x, \quad \theta_y = \bar{\theta}_y \qquad \text{on } \partial\Omega_u \tag{9.137}$$

where $\bar{w}$, $\bar{\theta}_x$, and $\bar{\theta}_y$ are the prescribed displacement and rotation values. Note that the exact solutions of the transverse deflection $w(x, t)$ and the two rotations $\theta_x(x, y, t), \theta_y(x, y, t)$ should satisfy the continuity condition, namely, $w(x, y, t), \theta_x(x, y, t), \theta_y(x, y, t) \in C^2(\Omega)$.

Boundary Conditions

Essential Boundary Conditions

$$w(x, y, t) = \bar{w}, \quad \frac{\partial w}{\partial x} = \bar{\psi}_x, \quad \frac{\partial w}{\partial y} = \bar{\psi}_y, \quad \frac{\partial w}{\partial n} = \bar{\psi}_n \tag{9.138}$$

Natural Boundary Conditions

$$\begin{aligned} &Q_x + \frac{\partial M_{xy}}{\partial y} = \bar{V}_x, \quad Q_y + \frac{\partial M_{xy}}{\partial x} = \bar{V}_y, \\ &M_x = \bar{M}_x, \quad M_y = \bar{M}_y, \quad M_{nn} = \bar{M}_{nn} \end{aligned} \tag{9.139}$$

Simply Supported Edge

$$w(x, y, t) = 0, \quad M_x = 0 \text{ or } M_y = 0 \text{ or } M_{nn} = 0 \tag{9.140}$$

Built-in Edge

$$w(x, y, t) = 0, \quad \frac{\partial w}{\partial x} = 0 \text{ or } \frac{\partial w}{\partial y} = 0 \text{ or } \frac{\partial w}{\partial n} = 0 \tag{9.141}$$

Free Edge

$$\begin{aligned} &Q_x + \frac{\partial M_{xy}}{\partial y} = 0 \text{ or } Q_y + \frac{\partial M_{xy}}{\partial x} = 0, \\ &M_x = 0 \text{ or } M_y = 0 \text{ or } M_{nn} = 0 \end{aligned} \tag{9.142}$$

Sliding Edge

$$\begin{aligned} &\frac{\partial w}{\partial x} = 0 \text{ or } \frac{\partial w}{\partial y} = 0 \text{ or } \frac{\partial w}{\partial n} = 0, \\ &Q_x + \frac{\partial M_{xy}}{\partial y} = 0 \text{ or } Q_y + \frac{\partial M_{xy}}{\partial x} = 0 \end{aligned} \tag{9.143}$$

Principle of Virtual Work

Next, the Bubnov-Galerkin weighted residual forms are set up as follows:

$$\begin{aligned}
&\int_{\Omega}\left(\frac{\partial Q_x}{\partial x}+\frac{\partial Q_y}{\partial y}+\bar{\omega}(x,y)-\rho h\ddot{w}\right)\delta w dxdy=0,\\
&\int_{\Omega}\left(\frac{\partial M_{xy}}{\partial x}+\frac{\partial M_{yy}}{\partial y}-Q_y+\rho I_p\ddot{\theta}_x\right)\delta\theta_x dxdy=0,\\
&\int_{\Omega}\left(\frac{\partial M_{xx}}{\partial x}+\frac{\partial M_{xy}}{\partial y}-Q_x-\rho I_p\ddot{\theta}_y\right)\delta\theta_y dxdy=0
\end{aligned} \tag{9.144}$$

Carrying out integration by parts twice, we obtain

$$\begin{aligned}
&\int_{\Omega}\rho h\ddot{w}\delta w dxdy+\int_{\Omega}\rho I_p\ddot{\theta}_x\delta\theta_x dxdy+\int_{\Omega}\rho I_p\ddot{\theta}_y\delta\theta_y dxdy\\
&+\int_{\Omega}Q_x\left(\frac{\partial\delta w}{\partial x}+\delta\theta_y\right)dxdy+\int_{\Omega}Q_y\left(\frac{\partial\delta w}{\partial y}-\delta\theta_x\right)dxdy\\
&+\int_{\Omega}M_{xx}\frac{\partial\delta\theta_y}{\partial x}dxdy+\int_{\Omega}M_{yy}\left(-\frac{\partial\delta\theta_x}{\partial y}\right)dxdy\\
&+\int_{\Omega}M_{xy}\left(\frac{\partial\delta\theta_y}{\partial y}-\frac{\partial\delta\theta_x}{\partial x}\right)dxdy-\int_{\Omega}\bar{\omega}(x,y)\delta w dxdy\\
&=\int_{\partial\Omega^+}M_{xx}\delta\theta_y dy+\int_{\partial\Omega^+}M_{yy}\delta\theta_x dx-\int_{\partial\Omega^+}M_{xy}\delta\theta_y dx-\int_{\partial\Omega^+}M_{xy}\delta\theta_x dy\\
&+\int_{\partial\Omega^+}Q_x\delta w dy-\int_{\partial\Omega^+}Q_y\delta w dx
\end{aligned} \tag{9.145}$$

and

$$\begin{aligned}
&\int_{\partial\Omega^+}M_{xx}\delta\theta_y dy+\int_{\partial\Omega^+}M_{yy}\delta\theta_x dx-\int_{\partial\Omega^+}M_{xy}\delta\theta_y dx-\int_{\partial\Omega^+}M_{xy}\delta\theta_x dy\\
&+\int_{\partial\Omega^+}Q_x\delta w dy-\int_{\partial\Omega^+}Q_y\delta w dx\\
&=\int_{\partial\Omega^+}\left(M_{xx}n_x+M_{xy}n_y\right)\delta\theta_y d\Gamma-\int_{\partial\Omega^+}\left(M_{yy}n_y+M_{xy}n_x\right)\delta\theta_x d\Gamma\\
&+\int_{\partial\Omega^+}\left(Q_xn_x+Q_yn_y\right)\delta w d\Gamma+\int_{\Omega}\bar{\omega}(x,y)\delta w dxdy
\end{aligned} \tag{9.146}$$

Consequently, Equation (9.145) reduces to the following

$$\begin{aligned}
&\int_{\Omega}\delta\mathbf{u}^T\boldsymbol{\rho}\mathbf{u}dxdy+\int_{\Omega}\delta\boldsymbol{\gamma}^T\mathbf{Q}dxdy+\int_{\Omega}\delta\boldsymbol{\kappa}^T\mathbf{m}dxdy\\
&=\int_{\partial\Omega^+}\left(M_{xx}n_x+M_{xy}n_y\right)\delta\theta_y d\Gamma-\int_{\partial\Omega^+}\left(M_{yy}n_y+M_{xy}n_x\right)\delta\theta_x d\Gamma\\
&+\int_{\partial\Omega^+}\left(Q_xn_x+Q_yn_y\right)\delta w d\Gamma
\end{aligned} \tag{9.147}$$

where the dependent variable $\mathbf{u}$ and the density matrix $\boldsymbol{\rho}$ are given by

$$\mathbf{u} = \left\{ \begin{array}{c} w \\ \theta_x \\ \theta_y \end{array} \right\}, \quad \boldsymbol{\rho} = \begin{bmatrix} \rho h & 0 & 0 \\ 0 & \rho I_p & 0 \\ 0 & 0 & \rho I_p \end{bmatrix} \tag{9.148}$$

In virtue of Equation (9.106), one obtains the relation (Rao 1954, page 98),

$$\begin{aligned} \delta\varphi &= n_x \delta\psi_n - n_y \delta\psi_s = -\delta\theta_y, \\ \delta\phi &= n_y \delta\psi_n + n_x \delta\psi_s = \delta\theta_x \end{aligned} \tag{9.149}$$

Consequently, one can obtain the following

$$\begin{aligned} &\int_{\partial\Omega^+} \left(M_{xx} n_x + M_{xy} n_y\right) \delta\theta_y d\Gamma - \int_{\partial\Omega^+} \left(M_{yy} n_y + M_{xy} n_x\right) \delta\theta_x d\Gamma \\ &+ \int_{\partial\Omega^+} \left(Q_x n_x + Q_y n_y\right) \delta w d\Gamma \\ &= -\int_{\partial\Omega^+} \left(M_{xx} n_x^2 + 2M_{xy} n_x n_y + M_{yy} n_y^2\right) \delta\psi_n d\Gamma \\ &- \int_{\partial\Omega^+} \left(M_{yy} - M_{xx}\right) n_x n_y + M_{xy} \left(n_x^2 - n_y^2\right) \delta\psi_s d\Gamma + \int_{\partial\Omega^+} Q_n \delta w d\Gamma \\ &= -\int_{\partial\Omega^+} M_{nn} \delta\psi_n d\Gamma - \int_{\partial\Omega^+} M_{ns} \delta\psi_s d\Gamma + \int_{\partial\Omega^+} Q_n \delta w d\Gamma \end{aligned} \tag{9.150}$$

Finally, Equation (9.147) can be rewritten as

$$\begin{aligned} &\int_\Omega \delta\mathbf{u}^T \boldsymbol{\rho} \mathbf{u} dxdy + \int_\Omega \delta\boldsymbol{\gamma}^T \mathbf{G} \boldsymbol{\gamma} dxdy + \int_\Omega \delta\boldsymbol{\kappa}^T \mathbf{D} \boldsymbol{\kappa} dxdy \\ &= \int_\Omega \bar{\omega}(x, y)\, \delta w dxdy - \int_{\partial\Omega^+} M_{nn} \delta\psi_n d\Gamma - \int_{\partial\Omega^+} M_{ns} \delta\psi_s d\Gamma + \int_{\partial\Omega^+} Q_n \delta w d\Gamma \end{aligned} \tag{9.151}$$

which is the weak form of the Reissner-Mindlin plate. Note that the admissible trial functions $w^h(x, y, t)$, $\theta_x^h(x, y, t)$, and $\theta_y^h(x, y, t)$ are of class $H^1(\Omega) \subset C^0(\Omega)$.

Spatial Discretization

The admissible trial functions, namely, $w^h(x, y, t), \theta_x^h(x, y, t), \theta_y^h(x, y, t) \in H^1(\Omega) \subset C^0(\Omega)$, are written as.

$$\begin{aligned} &w^h = \mathbf{H}_w(x, y)\,\mathbf{u}(t), \quad \theta_x^h = \mathbf{H}_{\theta_x}(x, y)\,\mathbf{u}(t), \quad \theta_y^h(x, t) = \mathbf{H}_{\theta_y}\mathbf{u}(t) \\ &\mathbf{H}_w = \begin{bmatrix} \mathbf{N}_w & \mathbf{0} & \mathbf{0} \end{bmatrix}, \mathbf{H}_{\theta_x} = \begin{bmatrix} \mathbf{0} & \mathbf{N}_{\theta_x} & \mathbf{0} \end{bmatrix}, \mathbf{H}_{\theta_y} = \begin{bmatrix} \mathbf{0} & \mathbf{0} & \mathbf{N}_{\theta_y} \end{bmatrix} \\ &\mathbf{u}(t) = \left(w_1^h(t), \ldots, w_{n_{node}}^h(t), \theta_{x1}^h(t), \ldots, \theta_{xn_{node}}^h(t), \theta_{y1}^h(t), \ldots, \theta_{yn_{node}}^h(t)\right), \\ &\dot{\mathbf{u}}(t) = \left(\dot{w}_1^h(t), \ldots, \dot{w}_{n_{node}}^h(t), \dot{\theta}_{x1}^h(t), \ldots, \dot{\theta}_{xn_{node}}^h(t), \dot{\theta}_{y1}^h(t), \ldots, \dot{\theta}_{yn_{node}}^h(t)\right) \end{aligned} \tag{9.152}$$

The trial functions are of class $C^0(\Omega)$ over the entire domain. As the variation of the dependent variable, the test functions are given by $\delta\tilde{\mathbf{u}}^h(x, y, t) = \mathbf{H}_e(x)\,\delta\tilde{\mathbf{u}}^e(t)$. Substituting both the trial and test functions into the weak form in Equation (9.55), the domain of interest can be discretized as follows:

$$\sum_{e=1}^{n_{\text{elem}}}\left(\begin{array}{l}\displaystyle\int_{\Omega_e}\delta\mathbf{u}^T\rho\mathbf{u}dxdy + \int_{\Omega_e}\delta\boldsymbol{\gamma}^T\mathbf{G}\boldsymbol{\gamma}dxdy\\ \displaystyle+\int_{\Omega_e}\delta\boldsymbol{\kappa}^T\mathbf{D}\boldsymbol{\kappa}dxdy - \int_{\Omega_e}\bar{\omega}(x, y)\,\delta w dxdy\end{array}\right)$$
$$= -\int_{\partial\Omega^+}M_{nn}\delta\psi_n d\Gamma - \int_{\partial\Omega^+}M_{ns}\delta\psi_s d\Gamma + \int_{\partial\Omega^+}Q_n\delta w d\Gamma \tag{9.153}$$

Therefore, the above equation reduces to the following

$$\sum_{e=1}^{n_{\text{elem}}}\left(\mathbf{M}^e\ddot{\mathbf{u}}^e(t) + \mathbf{F}^e_{\text{int}} - \mathbf{F}^e_{\text{ext}}\right)\cdot\delta\mathbf{u}^e(t) = 0 \tag{9.154}$$

where

$$\mathbf{M}^e = \int_A \mathbf{H}_e^T\rho\mathbf{H}_e dxdy,\ \mathbf{F}^e_{\text{ext}} = \int_\Omega \bar{\omega}(x, y)\,\mathbf{H}_w^T dxdy,$$

$$\mathbf{K}^e = \mathbf{K}_b + \mathbf{K}_s,\ \mathbf{K}_b = \iint_\Omega \mathbf{B}_b^T\mathbf{D}_b\mathbf{B}_b dxdy,\ \ \mathbf{K}_s = k\iint_\Omega \mathbf{B}_s^T\mathbf{G}_s\mathbf{B}_s dxdy,$$

$$\mathbf{B}_b = \begin{bmatrix}\dfrac{\partial\mathbf{H}_{\theta_y}}{\partial x}\\ -\dfrac{\partial\mathbf{H}_{\theta_x}}{\partial y}\\ \left(\dfrac{\partial\mathbf{H}_{\theta_y}}{\partial y} - \dfrac{\partial\mathbf{H}_{\theta_x}}{\partial x}\right)\end{bmatrix},\ \mathbf{B}_s = \begin{bmatrix}\dfrac{\partial\mathbf{H}_w}{\partial x} + \mathbf{H}_{\theta_y}\\ \dfrac{\partial\mathbf{H}_w}{\partial y} - \mathbf{H}_{\theta_x}\end{bmatrix} \tag{9.155}$$

9.2 FINITE ELEMENT LIBRARY

For several structural elements presented earlier, the underlying finite element developments and relevant details are described next. Typical linear continuum elements are shown in Figure 9.12.

9.2.1 One-Dimensional Continuum: Axial Bar Element

Global Coordinate System As the simplest finite element, we consider a one-dimensional continuum element with n nodes. Each node has one degree of freedom, since each node has a translational degree of freedom in the continuum. Hence a field function $u(x, t)$, which is the exact solution, may be approximated by a trial function $\tilde{u}^h(x, t)$ as a polynomial function of degree n (a generalization is shown next),

$$\tilde{u}^h(x, t) = \sum_{i=0}^{n}\alpha_i(t)x^i \tag{9.156}$$

where n denotes the number of nodes, and $\alpha_i(t)$ are the coefficients of the polynomial function which are dependent upon time in dynamics. For instance, the displacement field function $u(x, t)$ in the

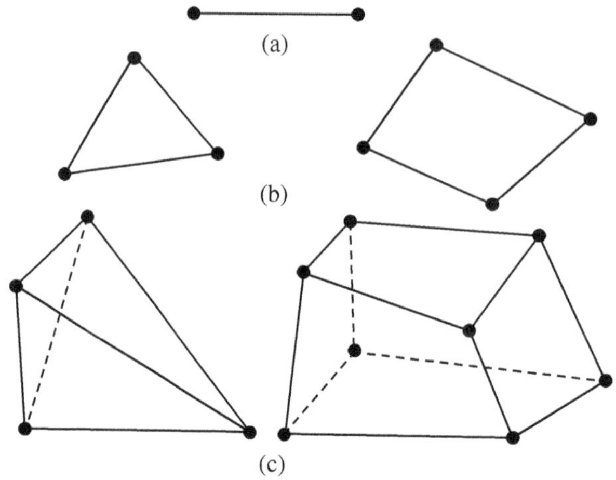

Figure 9.12. Linear continuum elements: (a) One-dimensional line element, (b) Two-dimensional triangular and quadrilateral elements, (c) Three-dimensional tetrahedral and hexahedral elements

x-direction, which is the exact solution, is approximated by a trial function $\tilde{u}^h(x,t)$ as a polynomial function of degree 1 as follows

$$\tilde{u}^h(x,t) = \alpha_0(t) + \alpha_1(t)x = \begin{bmatrix} 1 & x \end{bmatrix} \begin{Bmatrix} \alpha_0(t) \\ \alpha_1(t) \end{Bmatrix} \tag{9.157}$$

The trial function is assumed to be linear; $n = 1$. Also, the function is of class $C^0(]0, L[)$ at the inter element boundary and of class $C^1(]0, L[)$ within an element (Hughes 1987, page 110). This kind of element is often called a C^0-bar element. For an element with nodes 1 and 2, one readily obtains the expression

$$\begin{Bmatrix} \tilde{u}^h(x_1,t) \\ \tilde{u}^h(x_2,t) \end{Bmatrix} = \begin{bmatrix} 1 & x_1 \\ 1 & x_2 \end{bmatrix} \begin{Bmatrix} \alpha_0(t) \\ \alpha_1(t) \end{Bmatrix} = \begin{Bmatrix} u_1(t) \\ u_2(t) \end{Bmatrix} \tag{9.158}$$

where $u_1(t)$ and $u_2(t)$ are called nodal variables depending upon time. Consequently, the trial function can be rewritten as

$$\tilde{u}^h(x,t) = \begin{bmatrix} \dfrac{x_2 - x}{x_2 - x_1} & \dfrac{x - x_1}{x_2 - x_1} \end{bmatrix} \begin{Bmatrix} u_1(t) \\ u_2(t) \end{Bmatrix} \tag{9.159}$$

The above equation can be rewritten in a concise form as

$$\tilde{u}^h(x,t) = \mathbf{N}_e(x)\mathbf{u}^e(t) \tag{9.160}$$

where the interpolation functions $\mathbf{N}_e(x)$ within one element are defined as

$$\mathbf{N}_e(x) = \begin{bmatrix} N_1(x) & N_2(x) \end{bmatrix} \tag{9.161}$$

where

$$N_1(x) = \frac{x_2 - x}{x_2 - x_1}, \quad N_2(x) = \frac{x - x_1}{x_2 - x_1} \tag{9.162}$$

In view of the above equation, one can readily note the interpolation function property such that

$$N_i(x_j) = \delta_{ij}, \quad \sum_{i=1}^{n} N_i = 1 \tag{9.163}$$

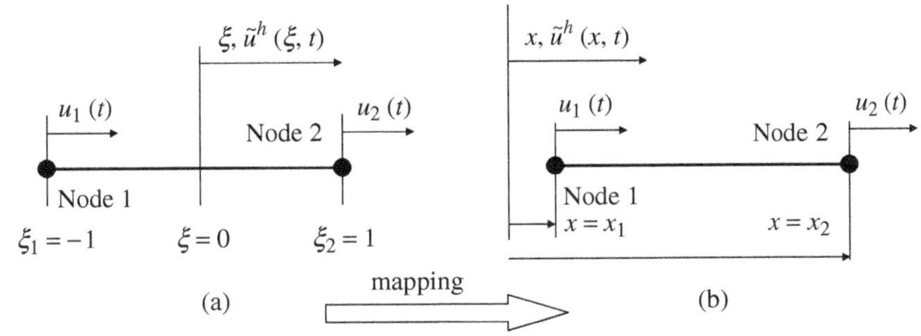

Figure 9.13. One-dimensional linear element: (a) Natural coordinate system, (b) Global coordinate system

where δ_{ij} is the Kronecker delta. Furthermore, it should be noted that the interpolation functions are of class C^{n-1}; the $(n-1)$th derivative of the interpolation functions (or the trial function) with respect to x is piece-wise continuous over the entire domain, but the function is continuous over the inter-element boundary. For a nonlinear field, the trial function may be approximated and designed with higher order polynomials. For higher order elements, the interpolation function of the jth node is computed in the following way,

$$\begin{aligned} N_j(x) &= \prod_{\substack{i=1 \\ i \neq j}}^{n} \frac{x - x_i}{x_j - x_i} \\ &= \frac{(x - x_1)(x - x_2)\ldots\left(x - x_{j-1}\right)\left(x - x_{j+1}\right)\ldots(x - x_n)}{\left(x_j - x_1\right)\left(x_j - x_2\right)\ldots\left(x_j - x_{j-1}\right)\left(x_j - x_{j+1}\right)\ldots\left(x_j - x_n\right)} \end{aligned} \tag{9.164}$$

This function is often called the Lagrange interpolation function, which does not involve any derivative of the field function. Note that the Lagrange interpolation function described here holds for the case that nodes are equally spaced.

Natural Coordinate System Thus far, we have discussed the derivation of interpolation functions in the physical coordinate system. We now introduce the natural coordinate system which is established in the isoparametric element (see Figure 9.13). The use of the isoparametric element allows one to take advantage of the Gauss-Legendre quadrature rule (Irons and Ahmad 1980), whenever some numerical integrations over the element domain are carried out in the finite element computation. In contrast to Equation (9.156), the trial function $\tilde{u}^h(\xi, t)$ as a polynomial function in the natural coordinate system is defined as

$$\tilde{u}^h(\xi, t) = \sum_{i=0}^{n} \alpha_i(t)\xi^i \tag{9.165}$$

For a linear isoparametric element with nodes 1 and 2, it is easy to approximate

$$\tilde{u}^h(x, t) = \left[\frac{1-\xi}{2} \;\; \frac{1+\xi}{2}\right]\left\{\begin{array}{c} u_1(t) \\ u_2(t) \end{array}\right\} = \left[N_1(\xi)\; N_2(\xi)\right] \tag{9.166}$$

The differential in the natural coordinate system is related to the differential in the physical coordinate system in the following manner

$$dx = \frac{dx}{d\xi}d\xi = Jd\xi \tag{9.167}$$

where J is called the Jacobian for the one-dimensional linear element. The Jacobian is a linear transformation mapping the line segment in the isoparametric domain $d\xi$ to the line segment dx in the physical domain. Thus, the isoparametric domain is considered to be a parent domain. Consequently, the natural coordinate system is often called the parent coordinate system. When the relation between the physical coordinate and the natural coordinate is given by

$$x = \sum_{i=1}^{2} N_i(\xi)x_i \tag{9.168}$$

then, the Jacobian J is given by

$$J = \frac{dx}{d\xi} = \frac{x_2 - x_1}{2} \tag{9.169}$$

Analogous to Equation (9.164), the interpolation function of the jth node for higher order isoparametric elements is given by

$$\begin{aligned} N_j(\xi) &= \prod_{\substack{i=1 \\ i \neq j}}^{n} \frac{\xi - \xi_i}{\xi_j - \xi_i} \\ &= \frac{(\xi - \xi_1)(\xi - \xi_2)\ldots(\xi - \xi_{j-1})(\xi - \xi_{j+1})\ldots(\xi - \xi_n)}{(\xi_j - \xi_1)(\xi_j - \xi_2)\ldots(\xi_j - \xi_{j-1})(\xi_j - \xi_{j+1})\ldots(\xi_j - \xi_n)} \end{aligned} \tag{9.170}$$

In the above, it is also assumed that nodes in the isoparametric element are equally spaced.

By making use of Equation (9.160), the elastic axial strain can be written as

$$\varepsilon^e(t) = \frac{\partial \tilde{u}^h(x,t)}{\partial x} = \bar{\mathbf{B}}_e \mathbf{u}^e(t) \tag{9.171}$$

where $\bar{\mathbf{B}}_e$ is given by

$$\bar{\mathbf{B}}_e = \frac{d\mathbf{N}_e(x)}{dx} = \frac{1}{x_2 - x_1}\begin{bmatrix} -1 & 1 \end{bmatrix} \tag{9.172}$$

which is often called the strain matrix. Note that the coefficients in the strain matrix are constant; this element is a constrain-strain line element. As a consequence, the element strain is constant within the element. In addition, in the isoparametric formulation, the strain matrix can be calculated as

$$\bar{\mathbf{B}}_e = \frac{d\mathbf{N}_e(\xi)}{d\xi}\frac{d\xi}{dx} = J^{-1}\frac{d\mathbf{N}_e(\xi)}{d\xi} = \frac{1}{x_2 - x_1}\begin{bmatrix} -1 & 1 \end{bmatrix} \tag{9.173}$$

It should be observed that Equations (9.172) and (9.173) share the identical result. As a consequence, the elastic axial stress in Equation (9.24) is given by

$$\sigma^e(t) = E\bar{\mathbf{B}}_e \mathbf{u}^e(t) \tag{9.174}$$

Note that the above element stress is constant within the element. For convenience, the cross-section of the element is constant, namely, $A(x) = A$. Then, the element stiffness matrix in Equation (9.31) can be expressed as

$$\mathbf{K}^e = EA\int_{-1}^{1} \bar{\mathbf{B}}_e^T \bar{\mathbf{B}}_e J\,d\xi = \frac{EA}{L_e}\begin{bmatrix} 1 & -1 \\ -1 & 1 \end{bmatrix} \tag{9.175}$$

where L_e denotes the length of the element.

Table 9.1. Number of intervals and Newton-Cotes coefficients in the Newton-Cotes quadrature

Number of Intervals, n_{nc}	H_0	H_1	H_2	H_3	H_4	H_5	H_6
1	$\frac{1}{2}$	$\frac{1}{2}$					
2	$\frac{1}{6}$	$\frac{2}{3}$	$\frac{1}{6}$				
3	$\frac{1}{8}$	$\frac{3}{8}$	$\frac{3}{8}$	$\frac{1}{8}$			
4	$\frac{7}{90}$	$\frac{16}{45}$	$\frac{2}{15}$	$\frac{16}{45}$	$\frac{7}{90}$		
5	$\frac{19}{288}$	$\frac{25}{96}$	$\frac{25}{144}$	$\frac{25}{96}$	$\frac{19}{288}$		
6	$\frac{41}{840}$	$\frac{9}{35}$	$\frac{9}{280}$	$\frac{34}{105}$	$\frac{9}{280}$	$\frac{9}{35}$	$\frac{41}{840}$

Numerical Integration Procedure In the literature, to compute an integral of a function, there are three numerical integration procedures, which are popular in the finite element analysis. These procedures include the Newton-Cotes quadrature, the Gauss-Legendre quadrature and the Lobatto quadrature (Fried and Malkus 1976; see Tables 9.1, 9.2, and 9.3 respectively). The first two quadratures are used to compute the internal and external force matrices, while the last one is employed for calculating the lumped mass matrix. The Newton-Cotes quadrature uses the sampling points which are equally spaced at the interval distances. As representative rules, the trapezoidal rule and Simpson's rule are pertinent to the Newton-Cotes quadrature. However, it is well-known that the Gauss-Legendre quadrature which allows unequally spaced sampling points is more efficient and suitable for finite elements (Bathe 1996; Irons 1966; Irons and Ahmad 1980; Zienkiewicz et al. 2005). In nonlinear analysis, the Newton-Cotes quadrature may be efficient (Bathe 1996, page 465).

In the Newton-Cotes quadrature (Bathe 1996; Whittaker 1942; Zienkiewicz et al. 2005), the integral of a function can be numerically integrated in the following manner

$$\int_{x_1}^{x_2} f(x)dx = (x_2 - x_1)\sum_{i=0}^{n_{nc}} H_i f(x_1 + i \times w), \quad w = \frac{(x_2 - x_1)}{n_{nc}} \tag{9.176}$$

where H_i are the Newton-Cotes constants (or coefficients) and $f(x_1 + i \times w)$ are the values of the integrand at predetermined x-values. Note that in the Newton-Cotes quadrature $\sum_{i=1}^{n_{nc}} H_i = 1$. In Equation (9.176), the integral domain is equally spaced with n intervals. Therefore, the number of sampling points is equal to $n_{nc} + 1$. Further, when the integrand function to be integrated is a polynomial function of degree n_{nc} or less, the numerical integration can be exactly performed by $n_{nc} + 1$ function values. The cases, namely, $n_{nc} = 1$ and $n_{nc} = 2$, correspond to the trapezoidal rule and Simpson's one-third rule respectively. For the Gauss-Legendre quadrature, the integral of a function can be numerically integrated in the following manner

$$\int_{x_1}^{x_2} f(x(\xi))dx = \int_{-1}^{1} \bar{f}(\xi)J(\xi)d\xi = \sum_{i=1}^{n_{gp}} J(\xi_i)w_i \bar{f}(\xi_i) \tag{9.177}$$

where n_{gp} and w_i denote the number of sampling points (or called Gauss points) and weight factors. When the integrand is a polynomial function of degree n, the number of Gauss points n_{gp} is given by

$$n_{gp} \geq \frac{n+1}{2} \tag{9.178}$$

Table 9.2. Gauss points and weights in the Gauss-Legendre quadrature

Number of Points, n_{gp}	Gauss Points, ξ_i	Weights, w_i
1	0.0	2.0
2	$\pm\frac{1}{\sqrt{3}}$	1.0
3	0.0	$\frac{8}{9}$
	$\pm\sqrt{\frac{3}{5}}$	$\frac{5}{9}$
4	$\pm\sqrt{\frac{3+\sqrt{4.8}}{7}}$	$\frac{1}{2}-\frac{\sqrt{30}}{36}$
	$\pm\sqrt{\frac{3-\sqrt{4.8}}{7}}$	$\frac{1}{2}+\frac{\sqrt{30}}{36}$

Consequently, it is easy to come to a conclusion that the Gauss-Legendre quadrature requires less number of sampling points than the Newton-Cotes quadrature to achieve the same degree of accuracy. Accordingly, the Gauss-Legendre quadrature is more efficient and popular than the Newton-Cotes quadrature in the finite element method. For example, to exactly integrate a polynomial function of degree 7, the Newton-Cotes quadrature requires 8 sampling points, while the Gauss-Legendre quadrature needs only 4 Gauss points (Zienkiewicz et al. 2005).

For computational dynamic problems, the explicit time integration scheme requires the mass matrix to be diagonalized. The lumped or diagonalized mass matrix can be easily obtained by the row-sum technique (Belytschko et al. 2000; Hughes 1987; Reddy 2006). For instance, for a linear axial bar element with a constant area A, the consistent mass matrix in Equation (9.31) is expressed as

$$\mathbf{M}^e = \rho A \int_{-1}^{1} \mathbf{N}_e^T(\xi)\mathbf{N}_e(\xi) J d\xi = \frac{\rho A L_e}{2} \int_{-1}^{1} \frac{1}{4} \begin{bmatrix} (1-\xi)^2 & (1-\xi^2) \\ (1-\xi^2) & (1+\xi)^2 \end{bmatrix} d\xi \tag{9.179}$$

Performing the integration above, one obtains the mass matrix

$$\mathbf{M}^e_{\text{consistent}} = \frac{\rho A L_e}{6} \begin{bmatrix} 2 & 1 \\ 1 & 2 \end{bmatrix} \tag{9.180}$$

which is the consistent matrix for the linear axial bar element with a constant cross-sectional area. Applying the row-sum technique leads to the following

$$\mathbf{M}^e_{\text{lumped}} = \frac{\rho A L_e}{2} \begin{bmatrix} 1 & 0 \\ 0 & 1 \end{bmatrix} \tag{9.181}$$

which is the lumped mass matrix of the element. The row-sum technique can be carried out in the following manner. The row-sum technique asserts that the mass matrix can be written as

$$\begin{aligned} \mathbf{M}_e &= \rho A \int_{-1}^{1} \begin{bmatrix} N_1 \sum_{i=1}^{2} N_i & 0 \\ 0 & N_2 \sum_{i=1}^{2} N_i \end{bmatrix} J d\xi \\ &= \frac{\rho A L_e}{2} \int_{-1}^{1} \begin{bmatrix} N_1(\xi) & 0 \\ 0 & N_2(\xi) \end{bmatrix} d\xi \end{aligned} \tag{9.182}$$

Table 9.3. Lobatto points and weights in the Lobatto quadrature

Number of Points, n_{lp}	Lobatto Points, ξ_i	Weights, w_i
2	± 1	1
3	± 1	$\frac{1}{3}$
	0	$\frac{4}{3}$
4	± 1	$\frac{1}{6}$
	± 0.44721360	$\frac{1}{12}$

Since the interpolation function is a polynomial of degree 1, the number of Gauss points is equal to 1, namely, $n_{gp} = 1$. Thus, the Gauss-Legendre quadrature yields the lumped mass matrix shown in Equation (9.181).

Alternatively, the lumped mass matrix is also readily obtained by Lobatto quadrature (Fried and Malkus 1976), which applies sampling points to nodes exactly. Accordingly, in the Lobatto quadrature rule, the location of a quadrature point is coincident with that of the corresponding node point.

Suppose that the distributed axial traction $\bar{b}(x)$ per unit length is constant, namely, $\bar{b}(x) = \bar{b}$. Then the external force due to the distributed axial traction is given by

$$\mathbf{F}_{\text{ext}}^e = \int_{L_e} \mathbf{N}_e^T \bar{b} dx = \frac{\bar{b} L_e}{2} \begin{bmatrix} 1 \\ 1 \end{bmatrix} \tag{9.183}$$

where the components of the external force $\mathbf{F}_{\text{ext}}^e$ are termed work-equivalent loads (Gallagher 1975).

9.2.2 Two-Dimensional Continuum: Triangular Element

Global Coordinate System We begin with a 3-node triangular element, which is often called the constant-strain triangular element. In the two-dimensional continuum, each node has two degrees of freedom, namely, the translational displacement $u(x, y, t)$ in the x-direction and the translational displacement $v(x, y, t)$ in the y-direction. Since three nodal variables are employed to approximate the displacement field function in each direction, the trial functions should be a polynomial function of degree 1 which is a linear function as shown in Figure 9.14. The polynomial function involves three unknown

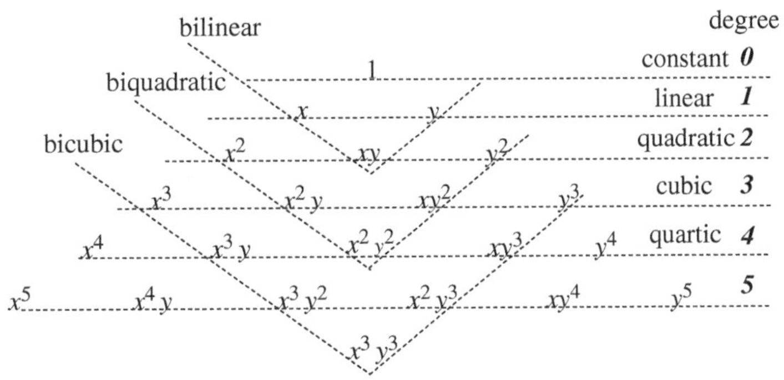

Figure 9.14. Pascal's triangle and Lagrange quadrilateral representations for complete polynomials in two-dimensions

coefficients. For instance, the exact displacement field function $u(x, y, t)$ in the x-direction, which is the exact solution, is approximated by a trial function $u^h(x, y, t)$ which is a polynomial function of degree 1

$$\tilde{u}^h(x, y, t) = \alpha_0(t) + \alpha_1(t)x + \alpha_2(t)y = \begin{bmatrix} 1 & x & y \end{bmatrix} \begin{Bmatrix} \alpha_0(t) \\ \alpha_1(t) \\ \alpha_2(t) \end{Bmatrix} \tag{9.184}$$

The trial function is assumed to be linear; the degree is one in the Pascal's triangle as shown in Figure 9.14. Thus, the function is of class $C^0(\Omega)$ at the inter element boundary and of class $C^1(\Omega_e)$ within an element (Hughes 1987, page 110) (Pian and Tong 1969). This is called a C^0-triangular element. Therefore, upon substitution of the three nodal variables into the above equation for a 3-node triangular element with nodes 1, 2 and 3, one readily obtains the expression

$$\begin{Bmatrix} \tilde{u}^h(x_1, y_1, t) \\ \tilde{u}^h(x_2, y_2, t) \\ \tilde{u}^h(x_3, y_3, t) \end{Bmatrix} = \begin{bmatrix} 1 & x_1 & y_1 \\ 1 & x_2 & y_2 \\ 1 & x_3 & y_3 \end{bmatrix} \begin{Bmatrix} \alpha_0(t) \\ \alpha_1(t) \\ \alpha_2(t) \end{Bmatrix} = \begin{Bmatrix} u_1(t) \\ u_2(t) \\ u_3(t) \end{Bmatrix} \tag{9.185}$$

where $u_1(t)$, $u_2(t)$ and $u_3(t)$ are nodal variables in the x-direction. Accordingly, the three unknown coefficients can be determined by the nodal variables. Note that the inverse matrix is given by

$$\begin{bmatrix} 1 & x_1 & y_1 \\ 1 & x_2 & y_2 \\ 1 & x_3 & y_3 \end{bmatrix}^{-1} = \frac{1}{2A} \begin{bmatrix} a_1 & a_2 & a_3 \\ b_1 & b_2 & b_3 \\ c_1 & c_2 & c_3 \end{bmatrix}, \quad 2A = \begin{vmatrix} 1 & x_1 & y_1 \\ 1 & x_2 & y_2 \\ 1 & x_3 & y_3 \end{vmatrix} \tag{9.186}$$

where A denotes the area of the triangular element, and the cofactors are given by

$$\begin{aligned} a_1 &= x_2y_3 - y_2x_3, & a_2 &= y_1x_3 - x_1y_3, & a_3 &= x_1y_2 - y_1x_2 \\ b_1 &= y_2 - y_3, & b_2 &= y_3 - y_1, & b_3 &= y_1 - y_2 \\ c_1 &= x_3 - x_2, & c_2 &= x_1 - x_3, & c_3 &= x_2 - x_1 \end{aligned} \tag{9.187}$$

which shows a cyclic permutation of subscripts (1, 2, 3), namely, $1 \to 2 \to 3$ or $2 \to 3 \to 1$ or $3 \to 1 \to 2$. Consequently, the trial function can be rewritten as

$$\tilde{u}^h(x, y, t) = \begin{bmatrix} N_1(x, y) & N_2(x, y) & N_3(x, y) \end{bmatrix} \begin{Bmatrix} u_1(t) \\ u_2(t) \\ u_3(t) \end{Bmatrix} \tag{9.188}$$

where

$$\begin{Bmatrix} N_1(x, y) \\ N_2(x, y) \\ N_3(x, y) \end{Bmatrix} = \frac{1}{2A} \begin{bmatrix} a_1 & b_1 & c_1 \\ a_2 & b_2 & c_2 \\ a_3 & b_3 & c_3 \end{bmatrix} \begin{Bmatrix} 1 \\ x \\ y \end{Bmatrix} \tag{9.189}$$

Likewise, the displacement field function $v(x, y, t)$ in the y-direction can be also written as

$$\tilde{v}^h(x, y, t) = \begin{bmatrix} N_1(x, y) & N_2(x, y) & N_3(x, y) \end{bmatrix} \begin{Bmatrix} v_1(t) \\ v_2(t) \\ v_3(t) \end{Bmatrix} \tag{9.190}$$

In vector form, the displacements are written as

$$\left\{\begin{array}{c}\tilde{u}^h(x,y,t)\\ \tilde{v}^h(x,y,t)\end{array}\right\} = \left[\begin{array}{cccccc} N_1 & 0 & N_2 & 0 & N_3 & 0\\ 0 & N_1 & 0 & N_2 & 0 & N_3\end{array}\right]\left\{\begin{array}{c}u_1(t)\\ v_1(t)\\ u_2(t)\\ v_2(t)\\ u_3(t)\\ v_3(t)\end{array}\right\} \tag{9.191}$$

The above equation can be rewritten in a more concise form as

$$\tilde{\mathbf{u}}^h(x,y,t) = \mathbf{N}_e(x,y)\mathbf{u}^e(t) \tag{9.192}$$

where the interpolation function $\mathbf{N}_e(x,y)$ within one element is defined as

$$\mathbf{N}_e = \left[\begin{array}{cccccc} N_1(x,y) & 0 & N_2(x,y) & 0 & N_3(x,y) & 0\\ 0 & N_1(x,y) & 0 & N_2(x,y) & 0 & N_3(x,y)\end{array}\right] \tag{9.193}$$

Element Strains. In plane state problems such as plane strain and plane stress states, the element strains are given by

$$\boldsymbol{\varepsilon} = \left\{\begin{array}{c}\varepsilon_{xx}\\ \varepsilon_{yy}\\ \gamma_{xy}\end{array}\right\} = \left\{\begin{array}{c}\dfrac{\partial u(x,y,t)}{\partial x}\\ \dfrac{\partial v(x,y,t)}{\partial y}\\ \dfrac{\partial u(x,y,t)}{\partial y} + \dfrac{\partial v(x,y,t)}{\partial x}\end{array}\right\} \tag{9.194}$$

In view of Equation (9.189), without difficulty one can obtain the relation

$$\left.\begin{array}{l} B_{1i} = \dfrac{\partial N_i(x,y)}{\partial x} = \dfrac{b_i}{2A}\\ B_{2i} = \dfrac{\partial N_i(x,y)}{\partial y} = \dfrac{c_i}{2A}\end{array}\right., \quad i = 1, 2, 3 \tag{9.195}$$

As a consequence, the element strains can be rewritten as

$$\boldsymbol{\varepsilon} = \left\{\begin{array}{c}\varepsilon_{xx}\\ \varepsilon_{yy}\\ \gamma_{xy}\end{array}\right\} = \sum_{i=1}^{3}\left\{\begin{array}{c}B_{1i}u_i(t)\\ B_{2i}v_i(t)\\ B_{2i}u_i(t) + B_{1i}v_i(t)\end{array}\right\} = \bar{\mathbf{B}}_e\mathbf{u}^e(t) \tag{9.196}$$

where the strain matrix $\bar{\mathbf{B}}_e$ is given by

$$\begin{aligned}\bar{\mathbf{B}}_e &= \left[\begin{array}{cccccc} B_{11} & 0 & B_{12} & 0 & B_{13} & 0\\ 0 & B_{21} & 0 & B_{22} & 0 & B_{23}\\ B_{21} & B_{12} & B_{22} & B_{12} & B_{23} & B_{13}\end{array}\right]\\ &= \frac{1}{2A}\left[\begin{array}{cccccc} b_1 & 0 & b_2 & 0 & b_3 & 0\\ 0 & c_1 & 0 & c_2 & 0 & c_3\\ c_1 & b_1 & c_2 & b_2 & c_3 & b_3\end{array}\right]\end{aligned} \tag{9.197}$$

In addition, the element strains in Equation (9.194) can also be written in a concise form as

$$\boldsymbol{\varepsilon} = \begin{Bmatrix} \varepsilon_{xx} \\ \varepsilon_{yy} \\ \gamma_{xy} \end{Bmatrix} = \sum_{i=1}^{3} \bar{\mathbf{B}}_i \mathbf{u}_i(t) \tag{9.198}$$

where

$$\bar{\mathbf{B}}_i = \begin{bmatrix} \frac{\partial N_i(x,y)}{\partial x} & 0 \\ 0 & \frac{\partial N_i(x,y)}{\partial y} \\ \frac{\partial N_i(x,y)}{\partial y} & \frac{\partial N_i(x,y)}{\partial x} \end{bmatrix}, \quad \mathbf{u}_i(t) = \begin{bmatrix} u_i(t) \\ v_i(t) \end{bmatrix} \tag{9.199}$$

Note that

$$\bar{\mathbf{B}}_e = \begin{bmatrix} \bar{\mathbf{B}}_1 & \bar{\mathbf{B}}_2 & \bar{\mathbf{B}}_3 \end{bmatrix}, \quad \mathbf{u}^e(t) = \begin{bmatrix} \mathbf{u}_1(t) \\ \mathbf{u}_2(t) \\ \mathbf{u}_3(t) \end{bmatrix} \tag{9.200}$$

Area Coordinate System. From the last term on the right-hand side of Equation (9.186), we have the relation

$$2A = 1\,(x_2 y_3 - y_2 x_3) + x_1\,(y_2 - y_3) + y_1\,(x_3 - x_2) \tag{9.201}$$

By use of the cofactors, Equation (9.201) can be written as

$$2A = \begin{bmatrix} a_1 & b_1 & c_1 \end{bmatrix} \begin{Bmatrix} 1 \\ x_1 \\ y_1 \end{Bmatrix} \tag{9.202}$$

Consequently, the areas A_1, A_2 and A_3 in Figure 9.15 are given by

$$\begin{Bmatrix} A_1(x,y) \\ A_2(x,y) \\ A_3(x,y) \end{Bmatrix} = \frac{1}{2} \begin{bmatrix} a_1 & b_1 & c_1 \\ a_2 & b_2 & c_2 \\ a_3 & b_3 & c_3 \end{bmatrix} \begin{Bmatrix} 1 \\ x \\ y \end{Bmatrix} \tag{9.203}$$

As a consequence, upon introducing new notations for convenience, without difficulty one can rewrite Equation (9.189) as

$$\begin{Bmatrix} N_1(x,y) \\ N_2(x,y) \\ N_3(x,y) \end{Bmatrix} = \frac{1}{A} \begin{Bmatrix} A_1(x,y) \\ A_2(x,y) \\ A_3(x,y) \end{Bmatrix} = \begin{Bmatrix} L_1(x,y) \\ L_2(x,y) \\ L_3(x,y) \end{Bmatrix} \tag{9.204}$$

where $L_1(x, y)$, $L_2(x, y)$ and $L_3(x, y)$ are often called the area coordinates (Taylor 1972). Therefore, the area coordinates can be linearly associated with the Cartesian coordinates as follows:

$$\begin{Bmatrix} 1 \\ x \\ y \end{Bmatrix} = \begin{bmatrix} 1 & 1 & 1 \\ x_1 & x_2 & x_3 \\ y_1 & y_2 & y_3 \end{bmatrix} \begin{Bmatrix} L_1(x,y) \\ L_2(x,y) \\ L_3(x,y) \end{Bmatrix} \tag{9.205}$$

where

$$L_i(x,y) = \frac{1}{2A}\,(a_i + b_i x + c_i y), \quad i = 1, 2, 3 \tag{9.206}$$

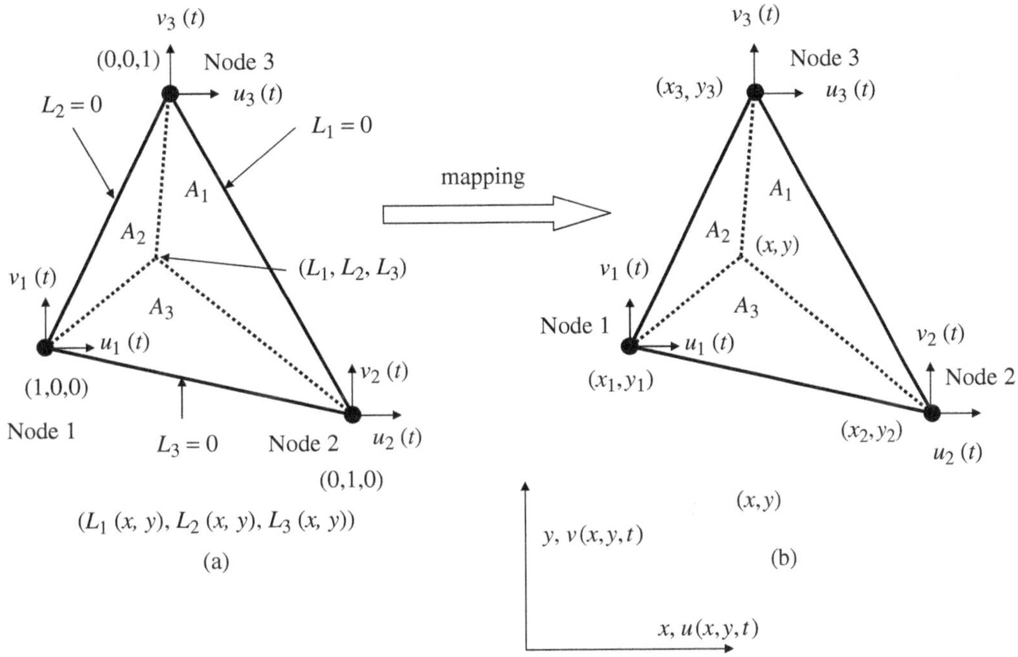

Figure 9.15. Two-dimensional linear triangular element: (a) Area coordinate system, (b) Global coordinate system

It is well known that the use of the area coordinate system is often attractive, when one takes advantage of the following formulae for the line integral,

$$\int_{\mathcal{L}} L_1^{\alpha} L_2^{\beta} d\mathcal{L} = \frac{\alpha!\beta!}{(\alpha+\beta+1)!} \times \mathcal{L},$$
$$\int_{\mathcal{L}} L_2^{\alpha} L_3^{\beta} d\mathcal{L} = \frac{\alpha!\beta!}{(\alpha+\beta+1)!} \times \mathcal{L}, \tag{9.207}$$
$$\int_{\mathcal{L}} L_3^{\alpha} L_1^{\beta} d\mathcal{L} = \frac{\alpha!\beta!}{(\alpha+\beta+1)!} \times \mathcal{L}$$

and for the area integral,

$$\int_{\Omega} L_1^{\alpha} L_2^{\beta} L_3^{\gamma} dA = \frac{\alpha!\beta!\gamma!}{(\alpha+\beta+\gamma+2)!} \times 2A \tag{9.208}$$

Element Strains. In view of Equation (9.206), the entries in the strain matrix are as follows:

$$B_{1i} = \frac{\partial L_i(x, y)}{\partial x} = \frac{b_i}{2A}, \quad B_{2i} = \frac{\partial L_i(x, y)}{\partial y} = \frac{c_i}{2A} \tag{9.209}$$

As a consequence, the element strains can be rewritten as

$$\boldsymbol{\varepsilon} = \begin{Bmatrix} \varepsilon_{xx} \\ \varepsilon_{yy} \\ \gamma_{xy} \end{Bmatrix} = \sum_{i=1}^{3} \begin{Bmatrix} B_{1i} u_i(t) \\ B_{2i} v_i(t) \\ B_{2i} u_i(t) + B_{1i} v_i(t) \end{Bmatrix} = \bar{\mathbf{B}}_e \mathbf{u}^e(t) \tag{9.210}$$

where the strain matrix $\bar{\mathbf{B}}_e$ is given by

$$\begin{aligned}\bar{\mathbf{B}}_e &= \begin{bmatrix} B_{11} & 0 & B_{12} & 0 & B_{13} & 0 \\ 0 & B_{21} & 0 & B_{22} & 0 & B_{23} \\ B_{21} & B_{12} & B_{22} & B_{12} & B_{23} & B_{13} \end{bmatrix} \\ &= \frac{1}{2A}\begin{bmatrix} b_1 & 0 & b_2 & 0 & b_3 & 0 \\ 0 & c_1 & 0 & c_2 & 0 & c_3 \\ c_1 & b_1 & c_2 & b_2 & c_3 & b_3 \end{bmatrix}\end{aligned} \tag{9.211}$$

In addition, the element strains in Equation (9.206) can also be written in a concise form as

$$\boldsymbol{\varepsilon} = \begin{Bmatrix} \varepsilon_{xx} \\ \varepsilon_{yy} \\ \gamma_{xy} \end{Bmatrix} = \sum_{i=1}^{3} \bar{\mathbf{B}}_i \mathbf{u}_i(t) \tag{9.212}$$

where

$$\bar{\mathbf{B}}_i = \begin{bmatrix} \dfrac{\partial L_i(x,y)}{\partial x} & 0 \\ 0 & \dfrac{\partial L_i(x,y)}{\partial y} \\ \dfrac{\partial L_i(x,y)}{\partial y} & \dfrac{\partial L_i(x,y)}{\partial x} \end{bmatrix}, \quad \mathbf{u}_i(t) = \begin{Bmatrix} u_i(t) \\ v_i(t) \end{Bmatrix} \tag{9.213}$$

Note that

$$\bar{\mathbf{B}}_e = \begin{bmatrix} \bar{\mathbf{B}}_1 & \bar{\mathbf{B}}_2 & \bar{\mathbf{B}}_3 \end{bmatrix}, \quad \mathbf{u}^e(t) = \begin{bmatrix} \mathbf{u}_1(t) \\ \mathbf{u}_2(t) \\ \mathbf{u}_3(t) \end{bmatrix} \tag{9.214}$$

Element Stresses. In general, many engineering applications in two-dimensional problems can be divided into two sub-classes such as the class of plane stress and the class of plane strain problems. For example, a thin plate with loads acting on the edge is pertinent to the plane stress analysis, while a long cylinder, which is subject to lateral loading and whose ends are fixed, behaves like a state of plain strain. In both classes, the element stresses can be given by

$$\begin{Bmatrix} \sigma_{xx} \\ \sigma_{yy} \\ \tau_{xy} \end{Bmatrix} = \mathbf{D} \begin{Bmatrix} \varepsilon_{xx} \\ \gamma_{yy} \\ \gamma_{xy} \end{Bmatrix} \tag{9.215}$$

where $\mathbf{D}$ denotes the constitutive modulus matrix. The above equation can be rewritten in direct notation as

$$\boldsymbol{\sigma} = \mathbf{D}\boldsymbol{\varepsilon} \tag{9.216}$$

For a state of plane stress, the constitutive modulus matrix $\mathbf{D}_{\text{plane stress}}$ is given by

$$\mathbf{D}_{\text{plane stress}} = \frac{E}{(1-\nu^2)} \begin{bmatrix} 1 & \nu & 0 \\ \nu & 1 & 0 \\ 0 & 0 & \frac{1-\nu}{2} \end{bmatrix} \tag{9.217}$$

And the stress components, i.e. σ_{zz}, τ_{xz} and τ_{yz}, are assumed to be zero. In addition, the rest of strain components are given by

$$\gamma_{xz} = \gamma_{yz} = 0, \quad \varepsilon_{zz} = -\frac{\nu}{E}\left(\sigma_{xx} + \sigma_{yy}\right) \tag{9.218}$$

On the other hand, for a plane strain state the constitutive modulus matrix $\mathbf{D}_{\text{plane strain}}$ is given by

$$\mathbf{D}_{\text{plane strain}} = \frac{E(1-\nu)}{(1+\nu)(1-2\nu)} \begin{bmatrix} 1 & \dfrac{\nu}{1-\nu} & 0 \\ \dfrac{\nu}{1-\nu} & 1 & 0 \\ 0 & 0 & \dfrac{1-2\nu}{2(1-\nu)} \end{bmatrix} \tag{9.219}$$

And the strain components, i.e. ε_{zz}, γ_{xz} and γ_{yz}, are assumed to be zero. In addition, the rest of stress components are given by

$$\tau_{xz} = \tau_{yz} = 0, \quad \sigma_{zz} = -\nu\left(\sigma_{xx} + \sigma_{yy}\right) \tag{9.220}$$

Element Stiffness and Mass Matrices. For plane stress problems, the element stiffness and consistent mass matrices are given by

$$\mathbf{K}^e = \int_\Omega \mathbf{B}_e^T \mathbf{D} \mathbf{B}_e t\, dA, \quad \mathbf{M}^e = \int_\Omega \rho \mathbf{N}_e^T \mathbf{N}_e t\, dA \tag{9.221}$$

where t and ρ denote the thickness and the density of the plate. And, for plane strain problems, the element stiffness and consistent mass matrices per unit thickness are given by

$$\mathbf{K}^e = \int_\Omega \mathbf{B}_e^T \mathbf{D} \mathbf{B}_e\, dA, \quad \mathbf{M}^e = \int_\Omega \rho \mathbf{N}_e^T \mathbf{N}_e\, dA \tag{9.222}$$

9.2.3 Two-Dimensional Continuum: Quadrilateral Element

Four-Node Rectangular Element in the Global Coordinate System Before dealing with a 4-node quadrilateral element which is not a rectangle, we shall first consider a 4-node rectangular element (see Figure 9.16) with sides parallel to the global axes, which is often called the bilinear strain element. In the two-dimensional continuum, each node has also two degrees of freedom, namely, the translational displacement $u(x, y, t)$ in the x-direction and the translational displacement $v(x, y, t)$ in the y-direction. Since four nodal variables are employed to approximate the displacement field function in each direction, the trial functions should be a bi-polynomial function which is a bilinear function. The polynomial function involves four unknown coefficients. For instance, the displacement field function $u(x, y, t)$ in the x-direction, which is the exact solution, is approximated by a bilinear function $\tilde{u}^h(x, y, t)$

$$\tilde{u}^h(x, y, t) = \alpha_0(t) + \alpha_1(t)x + \alpha_2(t)y + \alpha_3(t)xy \tag{9.223}$$

The trial function above is assumed to be bilinear in Pascal's triangle as shown in Figure 9.14. Thus, the function preserves $C^0(\Omega)$ continuity along edges and is of class $C^1(\Omega_e)$ within an element (Hughes 1987, page 110) (Pian and Tong 1969). This is also called a C^0-rectangular element. Therefore, upon substitution of the four nodal variables into the above equation for a 4-node triangular element, one readily obtains the expression

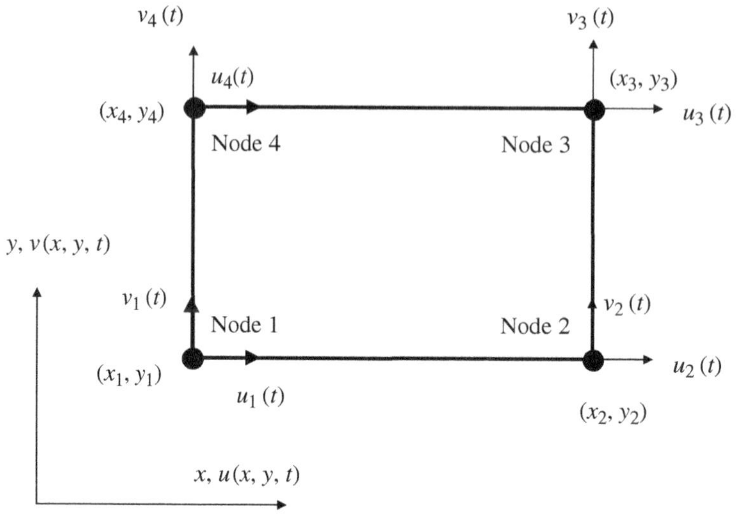

Figure 9.16. Two-dimensional linear element: Rectangular element in the global coordinate system

$$\begin{Bmatrix} \tilde{u}^h(x_1, y_1, t) \\ \tilde{u}^h(x_2, y_2, t) \\ \tilde{u}^h(x_3, y_3, t) \\ \tilde{u}^h(x_4, y_4, t) \end{Bmatrix} = \begin{bmatrix} 1 & x_1 & y_1 & x_1 y_1 \\ 1 & x_2 & y_2 & x_2 y_2 \\ 1 & x_3 & y_3 & x_3 y_3 \\ 1 & x_4 & y_4 & x_4 y_4 \end{bmatrix} \begin{Bmatrix} \alpha_0(t) \\ \alpha_1(t) \\ \alpha_2(t) \\ \alpha_3(t) \end{Bmatrix} = \begin{Bmatrix} u_1(t) \\ u_2(t) \\ u_3(t) \\ u_4(t) \end{Bmatrix} \tag{9.224}$$

where $u_1(t)$, $u_2(t)$, $u_3(t)$ and $u_4(t)$ are nodal variables in the x-direction. Accordingly, the four unknown coefficients can be determined by the nodal variables. However, instead of computing the inverse of the matrix, interpolation functions for rectangular elements can be readily obtained by use of the Lagrangian polynomials as follows:

$$\begin{aligned} N_1(x, y) &= \frac{x - x_2}{x_1 - x_2} \frac{y - y_4}{y_1 - y_4} = +\frac{1}{A}(x - x_2)(y - y_4), \\ N_2(x, y) &= \frac{x - x_1}{x_2 - x_1} \frac{y - y_4}{y_1 - y_4} = -\frac{1}{A}(x - x_1)(y - y_4), \\ N_3(x, y) &= \frac{x - x_1}{x_2 - x_1} \frac{y - y_1}{y_4 - y_1} = +\frac{1}{A}(x - x_1)(y - y_1), \\ N_4(x, y) &= \frac{x - x_2}{x_1 - x_2} \frac{y - y_4}{y_4 - y_1} = -\frac{1}{A}(x - x_2)(y - y_1) \end{aligned} \tag{9.225}$$

In vector form, the displacements are written as

$$\begin{Bmatrix} \tilde{u}^h(x, y, t) \\ \tilde{v}^h(x, y, t) \end{Bmatrix} = \begin{bmatrix} N_1 & 0 & N_2 & 0 & N_3 & 0 & N_4 & 0 \\ 0 & N_1 & 0 & N_2 & 0 & N_3 & 0 & N_4 \end{bmatrix} \begin{Bmatrix} u_1(t) \\ v_1(t) \\ u_2(t) \\ v_2(t) \\ u_3(t) \\ v_3(t) \\ u_4(t) \\ v_4(t) \end{Bmatrix} \tag{9.226}$$

The above equation can be rewritten in a more concise form as

$$\tilde{\mathbf{u}}^h(x, y, t) = \mathbf{N}_e(x, y)\mathbf{u}^e(t) \tag{9.227}$$

We introduce the notations

$$B_{1i} = \frac{\partial N_i(x, y)}{\partial x}, \quad B_{2i} = \frac{\partial N_i(x, y)}{\partial y} \tag{9.228}$$

For plane stress/strain problems, the relation between the strains and the displacements is found as

$$\boldsymbol{\varepsilon} = \begin{Bmatrix} \varepsilon_{xx} \\ \varepsilon_{yy} \\ \gamma_{xy} \end{Bmatrix} = \begin{Bmatrix} \dfrac{\partial u(x, y, t)}{\partial x} \\ \dfrac{\partial v(x, y, t)}{\partial y} \\ \dfrac{\partial u(x, y, t)}{\partial y} + \dfrac{\partial v(x, y, t)}{\partial x} \end{Bmatrix} = \sum_{i=1}^{4} \begin{Bmatrix} B_{1i}u_i(t) \\ B_{2i}v_i(t) \\ B_{2i}u_i(t) + B_{1i}v_i(t) \end{Bmatrix}$$
$$= \bar{\mathbf{B}}_e\mathbf{u}^e(t) \tag{9.229}$$

where the strain matrix $\bar{\mathbf{B}}_e$ for the four-node rectangular element can be easily obtained as

$$\bar{\mathbf{B}}_e = \begin{bmatrix} B_{11} & 0 & B_{12} & 0 & B_{13} & 0 & B_{14} & 0 \\ 0 & B_{21} & 0 & B_{22} & 0 & B_{23} & 0 & B_{24} \\ B_{21} & B_{12} & B_{22} & B_{12} & B_{23} & B_{13} & B_{24} & B_{14} \end{bmatrix} \tag{9.230}$$

Note that the strain matrix $\bar{\mathbf{B}}_e$ can be expressed in a concise form as

$$\bar{\mathbf{B}}_e = \begin{bmatrix} \bar{\mathbf{B}}_1 & \bar{\mathbf{B}}_2 & \bar{\mathbf{B}}_3 & \bar{\mathbf{B}}_4 \end{bmatrix} \tag{9.231}$$

where

$$\bar{\mathbf{B}}_1 = \frac{1}{A}\begin{bmatrix} (y - y_4) & 0 \\ 0 & (x - x_2) \\ (x - x_2) & (y - y_4) \end{bmatrix}, \bar{\mathbf{B}}_2 = \frac{1}{A}\begin{bmatrix} -(y - y_4) & 0 \\ 0 & -(x - x_1) \\ -(x - x_1) & -(y - y_4) \end{bmatrix},$$
$$\bar{\mathbf{B}}_3 = \frac{1}{A}\begin{bmatrix} (y - y_1) & 0 \\ 0 & (x - x_1) \\ (x - x_1) & (y - y_1) \end{bmatrix}, \bar{\mathbf{B}}_4 = \frac{1}{A}\begin{bmatrix} -(y - y_1) & 0 \\ 0 & -(y - y_1) \\ -(y - y_1) & -(y - y_1) \end{bmatrix} \tag{9.232}$$

Consequently, it is easy to see that the strain matrix is a linear function depending on x and y. Hence, the element strains and stresses are allowed to change linearly within an element.

Natural Coordinate System In the preceding section, we discussed the interpolation functions and the strain matrix for the rectangular element in a global coordinate system. For the finite element model with a complex geometry, we have to resort to the Taig quadrilateral (Irons and Ahmad 1980). For these cases, it is crucial to take advantage of the isoparametric finite elements. As illustrated in the triangular element section previously, we introduce the isoparametric element in the natural coordinate system (Irons and Ahmad 1980; Taig 1961), which is mapped into the physical quadrilateral element in the global coordinate system as shown in Figure 9.17. Note that the isoparametric element in the

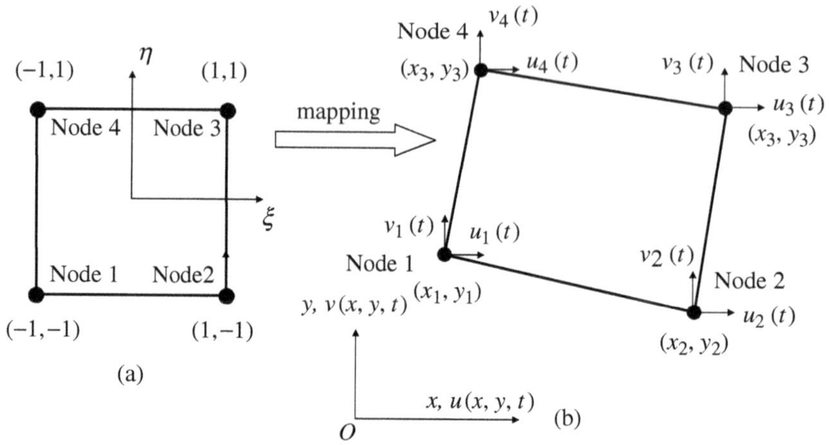

Figure 9.17. Two-dimensional quadrilateral element: (a) Natural coordinate system, (b) Global coordinate system

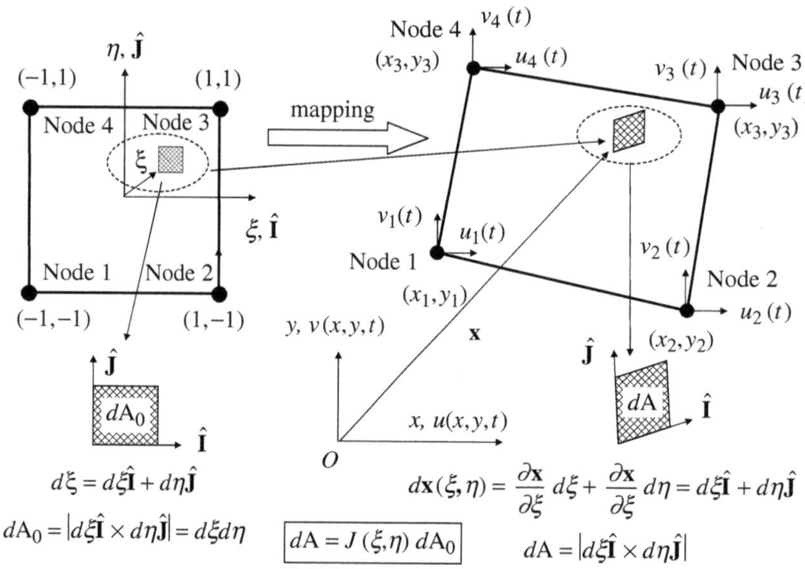

Figure 9.18. Mapping of an infinitesimal area

natural coordinate system, which is called the Taig quadrilateral (Irons and Ahmad 1980) is a rectangular element. Therefore, by use of the Lagrangian polynomials the interpolation functions are given by

$$\begin{aligned} N_1(\xi, \eta) &= \frac{1}{4}(1-\xi)(1-\eta), N_3(\xi, \eta) = \frac{1}{4}(1+\xi)(1+\eta), \\ N_2(\xi, \eta) &= \frac{1}{4}(1+\xi)(1-\eta), N_4(\xi, \eta) = \frac{1}{4}(1-\xi)(1+\eta) \end{aligned} \tag{9.233}$$

The above equations can be succinctly written as a single equation

$$N_i(\xi, \eta) = \frac{1}{4}(1+\xi\xi_i)(1+\eta\eta_i), \quad i = 1, 2, 3, 4 \tag{9.234}$$

where ξ_i and η_i indicate the nodal coordinates of the ith node in the natural coordinate system. Consequently, the displacements, in vector form, are expressed as

$$\left\{ \begin{array}{c} \tilde{u}^h(\xi,\eta,t) \\ \tilde{v}^h(\xi,\eta,t) \end{array} \right\} = \left[\begin{array}{cccccccc} N_1 & 0 & N_2 & 0 & N_3 & 0 & N_4 & 0 \\ 0 & N_1 & 0 & N_2 & 0 & N_3 & 0 & N_4 \end{array} \right] \left\{ \begin{array}{c} u_1(t) \\ v_1(t) \\ u_2(t) \\ v_2(t) \\ u_3(t) \\ v_3(t) \\ u_4(t) \\ v_4(t) \end{array} \right\} \tag{9.235}$$

The differential in the natural coordinate system is related to the differential in the physical coordinate system in the following manner

$$d\mathbf{x} = \left\{ \begin{array}{c} dx \\ dy \end{array} \right\} = \left[\begin{array}{cc} \dfrac{\partial x}{\partial \xi} & \dfrac{\partial x}{\partial \eta} \\ \dfrac{\partial y}{\partial \xi} & \dfrac{\partial y}{\partial \eta} \end{array} \right] \left\{ \begin{array}{c} d\xi \\ d\eta \end{array} \right\} = \mathbf{J}_c d\boldsymbol{\xi} \tag{9.236}$$

where the Jacobian matrix $\mathbf{J}_c$ (Chung 2007; Malvern 1969) is a transformation matrix mapping the line segment in the isoparametric domain $d\boldsymbol{\xi}$ to the line segment $d\mathbf{x}$ in the physical domain. On the other hand, to compute the strain matrix we need to evaluate the partial derivatives of the interpolation function N_i with respect to global coordinates as follows:

$$\left\{ \begin{array}{c} \dfrac{\partial N_i}{\partial \xi} \\ \dfrac{\partial N_i}{\partial \eta} \end{array} \right\} = \left[\begin{array}{cc} \dfrac{\partial x}{\partial \xi} & \dfrac{\partial y}{\partial \xi} \\ \dfrac{\partial x}{\partial \eta} & \dfrac{\partial y}{\partial \eta} \end{array} \right] \left\{ \begin{array}{c} \dfrac{\partial N_i}{\partial x} \\ \dfrac{\partial N_i}{\partial y} \end{array} \right\} = \mathbf{J}_f \left\{ \begin{array}{c} \dfrac{\partial N_i}{\partial x} \\ \dfrac{\partial N_i}{\partial y} \end{array} \right\} \tag{9.237}$$

where $\mathbf{J}_f$ is also called the Jacobian matrix (Bathe 1996; Zienkiewicz et al. 2005) in the finite element literature. Note that $\mathbf{J}_f = \mathbf{J}_c^T$. Accordingly, without difficulty one can obtain

$$\left\{ \begin{array}{c} \dfrac{\partial N_i}{\partial x} \\ \dfrac{\partial N_i}{\partial y} \end{array} \right\} = \mathbf{J}_f^{-1} \left\{ \begin{array}{c} \dfrac{\partial N_i}{\partial \xi} \\ \dfrac{\partial N_i}{\partial \eta} \end{array} \right\} = \left\{ \begin{array}{c} B_{1i} \\ B_{2i} \end{array} \right\} \tag{9.238}$$

where the inverse of the Jacobian matrix $\mathbf{J}_f^{-1}$ can be written as

$$\mathbf{J}_f^{-1} = \frac{1}{|\mathbf{J}_f|} \left[\begin{array}{cc} \dfrac{\partial y}{\partial \eta} & -\dfrac{\partial y}{\partial \xi} \\ -\dfrac{\partial x}{\partial \eta} & \dfrac{\partial x}{\partial \xi} \end{array} \right] \tag{9.239}$$

where $|\mathbf{J}_f|$ is called the Jacobian determinant or simply the Jacobian.

Element Strains. In planar problems such as plane strain and plane stress states, the element strains are given by

$$\boldsymbol{\varepsilon} = \begin{Bmatrix} \varepsilon_{xx} \\ \varepsilon_{yy} \\ \gamma_{xy} \end{Bmatrix} = \begin{Bmatrix} \dfrac{\partial u(x,y,t)}{\partial x} \\ \dfrac{\partial v(x,y,t)}{\partial y} \\ \dfrac{\partial u(x,y,t)}{\partial y} + \dfrac{\partial v(x,y,t)}{\partial x} \end{Bmatrix} \tag{9.240}$$

In view of Equation (9.228), the element strains can be rewritten as

$$\boldsymbol{\varepsilon} = \begin{Bmatrix} \varepsilon_{xx} \\ \varepsilon_{yy} \\ \gamma_{xy} \end{Bmatrix} = \sum_{i=1}^{4} \begin{Bmatrix} B_{1i} u_i(t) \\ B_{2i} v_i(t) \\ B_{2i} u_i(t) + B_{1i} v_i(t) \end{Bmatrix} = \bar{\mathbf{B}}_e \mathbf{u}^e(t) \tag{9.241}$$

where the strain matrix $\bar{\mathbf{B}}_e$ for the four-node rectangular element can be easily obtained as

$$\bar{\mathbf{B}}_e = \begin{bmatrix} B_{11} & 0 & B_{12} & 0 & B_{13} & 0 & B_{14} & 0 \\ 0 & B_{21} & 0 & B_{22} & 0 & B_{23} & 0 & B_{24} \\ B_{21} & B_{12} & B_{22} & B_{12} & B_{23} & B_{13} & B_{24} & B_{14} \end{bmatrix} \tag{9.242}$$

Note that the strain matrix $\bar{\mathbf{B}}_e$ can be also expressed in a concise form as

$$\bar{\mathbf{B}}_e = \begin{bmatrix} \bar{\mathbf{B}}_1 & \bar{\mathbf{B}}_2 & \bar{\mathbf{B}}_3 & \bar{\mathbf{B}}_4 \end{bmatrix} \tag{9.243}$$

Mapping of Infinitesimal Areas. An infinitesimal area in natural coordinates $dA_0(\xi,\eta)$ can be mapped to an infinitesimal area in global coordinates $dA(x,y)$ as shown in Figure 9.18. The diagonal vector of the infinitesimal rectangle in the natural coordinate system $d\boldsymbol{\xi}(\xi,\eta)$ can be defined as

$$d\boldsymbol{\xi}(\xi,\eta) = d\xi\hat{\mathbf{I}} + d\eta\hat{\mathbf{J}} \tag{9.244}$$

where $\hat{\mathbf{I}}$ and $\hat{\mathbf{J}}$ are unit vectors in natural coordinates. The infinitesimal area of the rectangle in natural coordinates $dA_0(\xi,\eta)$ is given by

$$dA_0 = \left| d\xi\hat{\mathbf{I}} \times d\eta\hat{\mathbf{J}} \right| = d\xi d\eta \tag{9.245}$$

The diagonal vector of the infinitesimal rectangle in the global coordinate system $d\mathbf{x}(x,y)$ can be defined as

$$d\mathbf{x}(x(\xi,\eta), y(\xi,\eta)) = \frac{\partial \mathbf{x}}{\partial \xi} d\xi + \frac{\partial \mathbf{x}}{\partial \eta} d\eta = d\xi\hat{\mathbf{i}} + d\eta\hat{\mathbf{j}} \tag{9.246}$$

where $\hat{\mathbf{i}}$ and $\hat{\mathbf{j}}$ are unit vectors in global coordinates. The infinitesimal area of the quadrilateral in global coordinates $dA(x,y)$ is given by

$$dA = \left| d\xi\hat{\mathbf{i}} \times d\eta\hat{\mathbf{j}} \right| = \left| \hat{\mathbf{i}} \times \hat{\mathbf{j}} \right| d\xi d\eta \tag{9.247}$$

where the unit vectors in global coordinates are described as

$$\hat{\mathbf{i}} = \frac{\partial x}{\partial \xi}\hat{\mathbf{I}} + \frac{\partial y}{\partial \xi}\hat{\mathbf{J}}, \qquad \hat{\mathbf{j}} = \frac{\partial x}{\partial \eta}\hat{\mathbf{I}} + \frac{\partial y}{\partial \eta}\hat{\mathbf{J}} \tag{9.248}$$

Therefore, Equation (9.247) can be rewritten as

$$dA = |\mathbf{J}_\mathrm{c}|\, d\xi d\eta = |\mathbf{J}_\mathrm{f}|\, d\xi d\eta = J dA_0 \tag{9.249}$$

where J is called the Jacobian of transformation, namely,

$$J(\xi, \eta) = |\mathbf{J}_\mathrm{c}| = |\mathbf{J}_\mathrm{f}| = \left|\frac{\partial(x, y)}{\partial(\xi, \eta)}\right| = \frac{\partial x}{\partial \xi}\frac{\partial y}{\partial \eta} - \frac{\partial y}{\partial \xi}\frac{\partial x}{\partial \eta} \tag{9.250}$$

Element Stresses. The element stresses are given by

$$\boldsymbol{\sigma} = \begin{Bmatrix} \sigma_{xx} \\ \sigma_{yy} \\ \tau_{xy} \end{Bmatrix} = \mathbf{D}\boldsymbol{\varepsilon} \tag{9.251}$$

where $\mathbf{D}$ denotes the constitutive modulus matrix. For a state of plane stress (for instance, a stress state of a thin plate), the constitutive modulus matrix $\mathbf{D}_{\text{plane stress}}$ is previously shown in Equation (9.217). The stress components, i.e. σ_{zz}, τ_{xz} and τ_{yz}, are assumed to be zero. In addition, the strain components, i.e. γ_{xz} and γ_{yz}, are zero and the transverse normal strain ε_{zz} is given by

$$\varepsilon_{zz} = -\frac{\nu}{E}\left(\sigma_{xx} + \sigma_{yy}\right) \tag{9.252}$$

For plane strain problems, the constitutive modulus matrix $\mathbf{D}_{\text{plane strain}}$ is given by Equation (9.219). The strain components, i.e. ε_{zz}, γ_{xz} and γ_{yz}, are assumed to be zero. In addition, the rest of the stress components are given by

$$\tau_{xz} = \tau_{yz} = 0, \qquad \sigma_{zz} = -\nu\left(\sigma_{xx} + \sigma_{yy}\right) \tag{9.253}$$

Element Stiffness and Mass Matrices. For plane stress problems, the element stiffness and mass matrices are given by

$$\mathbf{K}^e = \int_\Omega \mathbf{B}_e^T \mathbf{D} \mathbf{B}_e t J(\xi, \eta)\, d\xi d\eta, \qquad \mathbf{M}^e = \int_\Omega \rho \mathbf{N}_e^T \mathbf{N}_e t J(\xi, \eta)\, d\xi d\eta \tag{9.254}$$

where t and ρ denote the thickness and the density of the plate. For plane strain problems, the element stiffness and consistent mass matrices per unit thickness are given by

$$\begin{aligned} \mathbf{K}^e &= \int_\Omega \mathbf{B}_e^T \mathbf{D} \mathbf{B}_e J(\xi, \eta)\, d\xi d\eta \\ &= \sum_{i=1}^{n_{\mathrm{gp}}} \sum_{j=1}^{n_{\mathrm{gp}}} w_i w_j \mathbf{B}_e^T\left(\xi_i, \eta_j\right) \mathbf{D} \mathbf{B}_e\left(\xi_i, \eta_j\right) J\left(\xi_i, \eta_j\right), \end{aligned} \tag{9.255}$$

$$\mathbf{M}^e = \int_\Omega \rho \mathbf{N}_e^T \mathbf{N}_e J(\xi, \eta)\, d\xi d\eta$$
$$= \sum_{i=1}^{n_{gp}} \sum_{j=1}^{n_{gp}} w_i w_j \mathbf{N}_e^T(\xi_i, \eta_j) \mathbf{N}_e(\xi_i, \eta_j) J(\xi_i, \eta_j)$$

where n_{gp} denotes the number of Gauss points, w_i and w_j weights, and ξ_i and η_j are the corresponding Gauss points.

Illustrative Example

Example 9.7 Elastic Transient Response of the Cantilever Under a Distributed Load

By making use of the Newmark implicit method (see Chapter 13), the elastic transient response of the cantilever under a distributed step load of 2.85 lb/in in Figure 9.19 is investigated in the literature (Bathe et al. 1975; Yang and Saigal 1984). While Bathe et al. used five 8-node plane stress elements, Yang and Saigal used four Euler-Bernoulli beam elements. They all used the Newmark implicit method, and the time step size is taken as $\Delta t = 45$ μsecs. The dimensions of the beam: are $L = 10$in and $h = b = 1$in. The material properties are: $E = 1.2 \times 10^4$ psi, $\rho = 1.0 \times 10^{-6}$ lb − sec^2/in^4 psi, and $\nu = 0.2$. Figure 9.19 indicates that good agreement is observed between Yang and Saigal's Euler-Bernoulli beam element formulation and Bathe et al.'s plane stress element formulation.

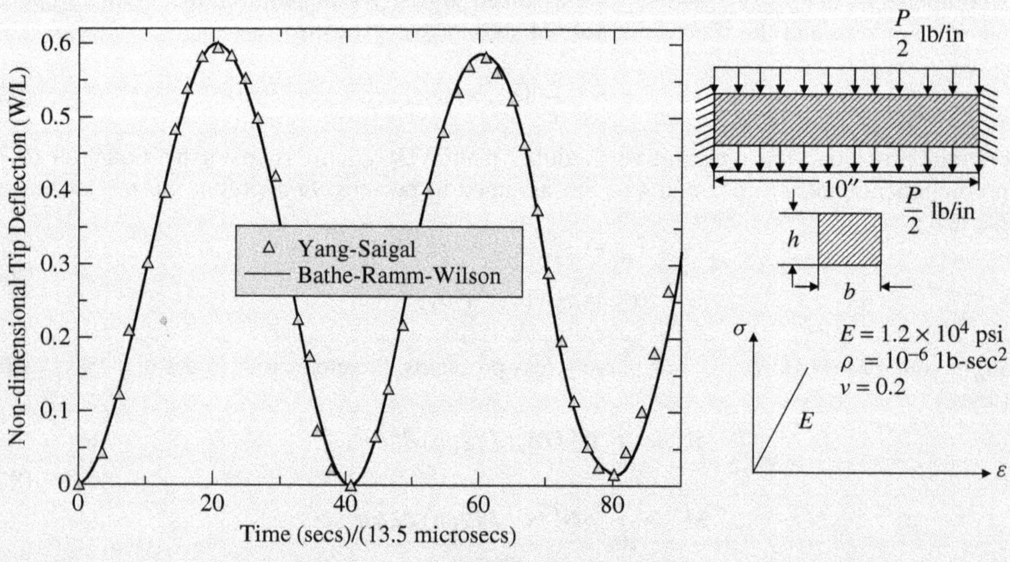

Figure 9.19. Elastic transient response for a cantilever under a distributed load using Plane stress element (Bathe et al. 1975)

9.2.4 Three-Dimensional Continuum: Tetrahedral Element

Global Coordinate System For a three-dimensional continuum element, we begin with the common 4-node tetrahedral element, which is often called the constant-strain tetrahedral element. In the three-dimensional continuum, each node has three translational degrees of freedom, namely, the translational displacement $u(x, y, z, t)$ in the x-direction, the translational displacement $v(x, y, z, t)$ in the

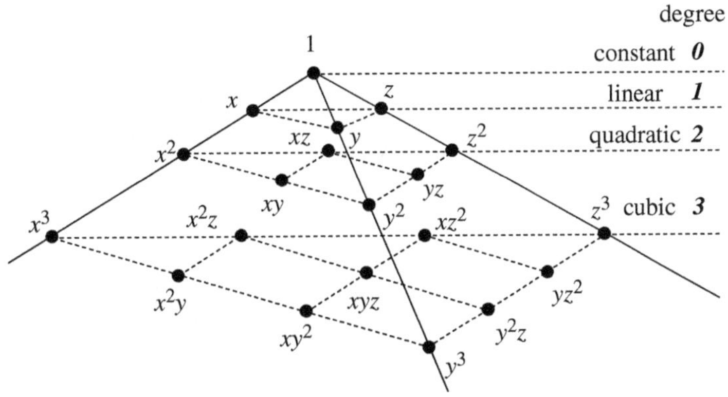

Figure 9.20. Pascal's tetrahedron (Pascal's pyramid) representations for complete polynomials in three-dimensions

y-direction, and the translational displacement $w(x, y, z, t)$ in the z-direction. Since four nodal variables are employed to approximate the displacement field function in each direction, the trial functions should be a polynomial function of degree 1 which is a linear function as shown in Figure 9.20. The polynomial function involves four unknown coefficients. For instance, the displacement field function $u(x, y, z, t)$ in the x-direction, which is the exact solution, is approximated by a trial function $\tilde{u}^h(x, y, z, t)$ as a polynomial function of degree 1 as follows:

$$\tilde{u}^h(x, y, t) = \alpha_0(t) + \alpha_1(t)x + \alpha_2(t)y + \alpha_3(t)z \tag{9.256}$$

As the trial function is assumed to be linear, the function is of class $C^0(\Omega)$ at the inter element boundary and of class $C^1(\Omega_e)$ within an element. This is called a C^0-tetrahedral element. Therefore, upon substitution of the four nodal variables into the above equation for a 4-node tetrahedral element with nodes 1, 2, 3 and 4, one readily obtains the expression

$$\begin{Bmatrix} \tilde{u}^h(x_1, y_1, z_1, t) \\ \tilde{u}^h(x_2, y_2, z_2, t) \\ \tilde{u}^h(x_3, y_3, z_3, t) \\ \tilde{u}^h(x_4, y_4, z_4, t) \end{Bmatrix} = \begin{bmatrix} 1 & x_1 & y_1 & z_1 \\ 1 & x_2 & y_2 & z_2 \\ 1 & x_3 & y_3 & z_3 \\ 1 & x_4 & y_4 & z_4 \end{bmatrix} \begin{Bmatrix} \alpha_0(t) \\ \alpha_1(t) \\ \alpha_2(t) \\ \alpha_3(t) \end{Bmatrix} = \begin{Bmatrix} u_1(t) \\ u_2(t) \\ u_3(t) \\ u_4(t) \end{Bmatrix} \tag{9.257}$$

where $u_1(t)$, $u_2(t)$, $u_3(t)$ and $u_4(t)$ are nodal variables in the x-direction. Accordingly, the four unknown coefficients can be determined by the nodal variables. Note that the inverse matrix is given by

$$\begin{bmatrix} 1 & x_1 & y_1 & z_1 \\ 1 & x_2 & y_2 & z_2 \\ 1 & x_3 & y_3 & z_3 \\ 1 & x_4 & y_4 & z_4 \end{bmatrix}^{-1} = \frac{1}{6V^e} \begin{bmatrix} a_1 & a_2 & a_3 & a_4 \\ b_1 & b_2 & b_3 & b_4 \\ c_1 & c_2 & c_3 & c_4 \\ d_1 & d_2 & d_3 & d_4 \end{bmatrix}, \quad 6V_e = \begin{vmatrix} 1 & x_1 & y_1 & z_1 \\ 1 & x_2 & y_2 & z_2 \\ 1 & x_3 & y_3 & z_3 \\ 1 & x_4 & y_4 & z_4 \end{vmatrix} \tag{9.258}$$

where V_e denotes the volume of the tetrahedral element and $a_i, b_i, c_i, d_i, i = 1, \ldots, 4$ are the cofactors of the ith row of the determinant. Consequently, the trial function can be rewritten as

$$\tilde{u}^h(x, y, z, t) = \begin{bmatrix} N_1 & N_2 & N_3 & N_4 \end{bmatrix} \begin{Bmatrix} u_1(t) \\ u_2(t) \\ u_3(t) \\ u_4(t) \end{Bmatrix} \tag{9.259}$$

where

$$\begin{Bmatrix} N_1(x, y, z) \\ N_2(x, y, z) \\ N_3(x, y, z) \\ N_4(x, y, z) \end{Bmatrix} = \frac{1}{6V_e} \begin{bmatrix} a_1 & b_1 & c_1 & d_1 \\ a_2 & b_2 & c_2 & d_2 \\ a_3 & b_3 & c_3 & d_3 \\ a_4 & b_4 & c_4 & d_4 \end{bmatrix} \begin{Bmatrix} 1 \\ x \\ y \\ z \end{Bmatrix} \tag{9.260}$$

Likewise, the other displacement field functions $\tilde{v}(x, y, z, t)$ in the y-direction and $\tilde{w}^h(x, y, z, t)$ in the z-direction can be also written in the same manner. In vector form, the displacements are written as

$$\begin{Bmatrix} \tilde{u}^h(x, y, z, t) \\ \tilde{v}^h(x, y, z, t) \\ \tilde{w}^h(x, y, z, t) \end{Bmatrix} = \begin{bmatrix} \mathbf{N}_1 & \mathbf{N}_2 & \mathbf{N}_3 & \mathbf{N}_4 \end{bmatrix} \begin{Bmatrix} \mathbf{u}_1(t) \\ \mathbf{u}_2(t) \\ \mathbf{u}_3(t) \\ \mathbf{u}_4(t) \end{Bmatrix} \tag{9.261}$$

where the interpolation function submatrix $\mathbf{N}_i(x, y, z)$ and the displacement vectors $\mathbf{u}_i(t)$ are given by

$$\mathbf{N}_i(x, y, z) = \begin{bmatrix} N_i & 0 & 0 \\ 0 & N_i & 0 \\ 0 & 0 & N_i \end{bmatrix}, \quad \mathbf{u}_i(t) = \begin{Bmatrix} u_i(t) \\ v_i(t) \\ w_i(t) \end{Bmatrix}, \quad i = 1, 2, 3 \tag{9.262}$$

It should be noted that Equation (9.261) can be rewritten in a more concise form as

$$\tilde{\mathbf{u}}^h(x, y, z, t) = \mathbf{N}_e(x, y, z)\mathbf{u}^e(t) \tag{9.263}$$

where the interpolation function $\mathbf{N}_e(x)$ within one element is defined as

$$\mathbf{N}_e = \begin{bmatrix} \mathbf{N}_1 & \mathbf{N}_2 & \mathbf{N}_3 & \mathbf{N}_4 \end{bmatrix} \tag{9.264}$$

Volume Coordinate System. Analogous to the procedure previously outlined for the triangular element, we can follow similar steps. Since $N_i\ (x_i, y_i, z_i) = 1,\ i = 1, 2, 3$, in view of Equation (9.260) we have the relation

$$6V_e = \begin{bmatrix} a_i & b_i & c_i & d_i \end{bmatrix} \begin{Bmatrix} 1 \\ x_i \\ y_i \\ z_i \end{Bmatrix}, \quad i = 1, 2, 3, 4 \tag{9.265}$$

Consequently, the volumes V_1, V_2, V_3 and V_4 in Figure 9.21 are given by

$$\begin{Bmatrix} V_1(x, y, z) \\ V_2(x, y, z) \\ V_3(x, y, z) \\ V_4(x, y, z) \end{Bmatrix} = \frac{1}{6} \begin{bmatrix} a_1 & b_1 & c_1 & d_1 \\ a_2 & b_2 & c_2 & d_2 \\ a_3 & b_3 & c_3 & d_3 \\ a_4 & b_4 & c_4 & d_4 \end{bmatrix} \begin{Bmatrix} 1 \\ x \\ y \\ z \end{Bmatrix} \tag{9.266}$$

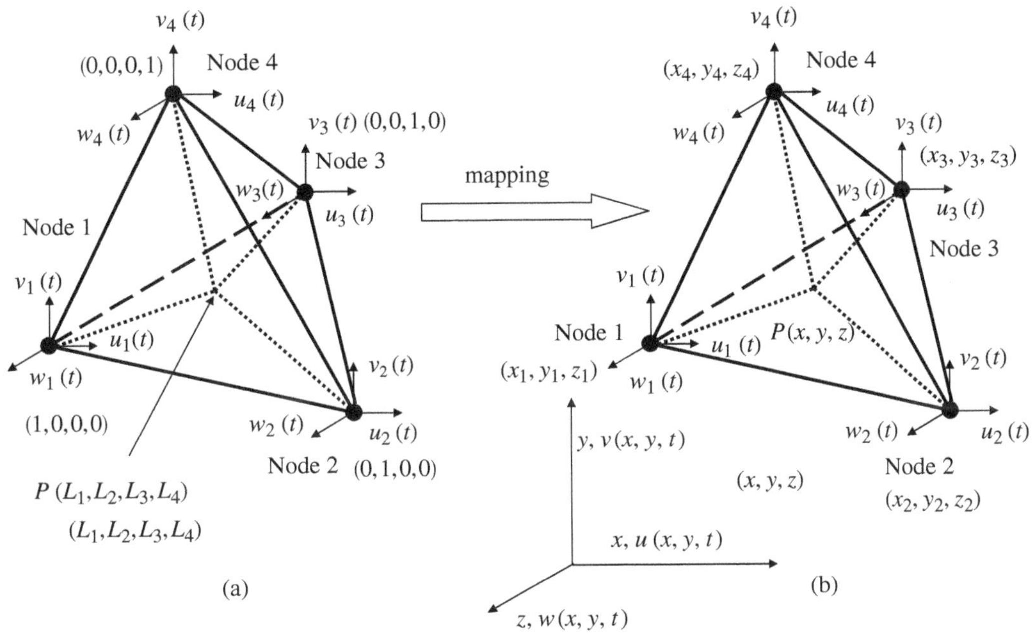

Figure 9.21. Three-dimensional linear tetrahedral element: (a) Volume coordinate system, (b) Global coordinate system

where V_1 indicates the volume of $P234$, V_2 the volume of $P134$, V_3 the volume of $P124$, and V_4 the volume of $P123$. As a consequence, upon introducing new notations for convenience, without difficulty one can rewrite Equation (9.260) as

$$\begin{Bmatrix} N_1(x,y,z) \\ N_2(x,y,z) \\ N_3(x,y,z) \\ N_4(x,y,z) \end{Bmatrix} = \frac{1}{V_e}\begin{Bmatrix} V_1(x,y,z) \\ V_2(x,y,z) \\ V_3(x,y,z) \\ V_4(x,y,z) \end{Bmatrix} = \begin{Bmatrix} L_1(x,y,z) \\ L_2(x,y,z) \\ L_3(x,y,z) \\ L_4(x,y,z) \end{Bmatrix} \tag{9.267}$$

where $L_1(x, y, z)$, $L_2(x, y, z)$, $L_3(x, y, z)$ and $L_4(x, y, z)$ are often called the volume coordinates (Taylor 1972). Therefore, the volume coordinates can be linearly associated with the Cartesian coordinates as follows:

$$\begin{Bmatrix} 1 \\ x \\ y \\ z \end{Bmatrix} = \begin{bmatrix} 1 & 1 & 1 & 1 \\ x_1 & x_2 & x_3 & x_4 \\ y_1 & y_2 & y_3 & y_4 \\ z_1 & z_2 & z_3 & z_4 \end{bmatrix} \begin{Bmatrix} L_1(x,y,z) \\ L_2(x,y,z) \\ L_3(x,y,z) \\ L_4(x,y,z) \end{Bmatrix} \tag{9.268}$$

where

$$L_i(x,y,z) = \frac{1}{6V_e}\left(a_i + b_i x + c_i y + d_i z\right), \quad i = 1, 2, 3, 4 \tag{9.269}$$

It is well known that the formula for the volume integral can be given by

$$\int_{\Omega} L_1^{\alpha} L_2^{\beta} L_3^{\gamma} L_4^{\delta} dA = \frac{\alpha!\beta!\gamma!\delta!}{(\alpha+\beta+\gamma+\delta+3)!} \times 6V_e \tag{9.270}$$

Element Strains. In the three-dimensional continuum, the element strains can be rewritten as

$$\boldsymbol{\varepsilon} = \begin{Bmatrix} \varepsilon_{xx} \\ \varepsilon_{yy} \\ \varepsilon_{zz} \\ \gamma_{xy} \\ \gamma_{yz} \\ \gamma_{zx} \end{Bmatrix} = \sum_{i=1}^{4} \begin{Bmatrix} B_{1i} u_i(t) \\ B_{2i} v_i(t) \\ B_{3i} w_i(t) \\ B_{2i} u_i(t) + B_{1i} v_i(t) \\ B_{3i} v_i(t) + B_{2i} w_i(t) \\ B_{3i} u_i(t) + B_{1i} w_i(t) \end{Bmatrix} = \bar{\mathbf{B}}_e \mathbf{u}^e(t) \tag{9.271}$$

where the strain matrix $\bar{\mathbf{B}}_e$ and the nodal variables are given by

$$\bar{\mathbf{B}}_e = \begin{bmatrix} \bar{\mathbf{B}}_1 & \bar{\mathbf{B}}_2 & \bar{\mathbf{B}}_3 & \bar{\mathbf{B}}_4 \end{bmatrix}, \quad \mathbf{u}^e(t) = \begin{bmatrix} \mathbf{u}_1(t) \\ \mathbf{u}_2(t) \\ \mathbf{u}_3(t) \end{bmatrix} \tag{9.272}$$

where

$$\bar{\mathbf{B}}_i = \begin{bmatrix} \frac{\partial L_i(x,y,z)}{\partial x} & 0 & 0 \\ 0 & \frac{\partial L_i(x,y,z)}{\partial y} & 0 \\ 0 & 0 & \frac{\partial L_i(x,y,z)}{\partial z} \\ \frac{\partial L_i(x,y,z)}{\partial y} & \frac{\partial L_i(x,y,z)}{\partial x} & 0 \\ 0 & \frac{\partial L_i(x,y,z)}{\partial z} & \frac{\partial L_i(x,y,z)}{\partial y} \\ \frac{\partial L_i(x,y,z)}{\partial z} & 0 & \frac{\partial L_i(x,y,z)}{\partial x} \end{bmatrix} = \frac{1}{6V_e} \begin{bmatrix} b_i & 0 & 0 \\ 0 & c_i & 0 \\ 0 & 0 & d_i \\ c_i & b_i & 0 \\ 0 & d_i & c_i \\ d_i & 0 & b_i \end{bmatrix} \tag{9.273}$$

and

$$\mathbf{u}^e(t) = \begin{bmatrix} \mathbf{u}_1(t) \\ \mathbf{u}_2(t) \\ \mathbf{u}_3(t) \end{bmatrix}, \quad \mathbf{u}_i(t) = \begin{bmatrix} u_i(t) \\ v_i(t) \\ w_i(t) \end{bmatrix} \tag{9.274}$$

Element Stresses. In the three-dimensional problems, the element stresses are given by

$$\begin{Bmatrix} \sigma_{xx} \\ \sigma_{yy} \\ \sigma_{zz} \\ \tau_{xy} \\ \tau_{yz} \\ \tau_{zx} \end{Bmatrix} = \mathbf{D} \begin{Bmatrix} \varepsilon_{xx} \\ \varepsilon_{yy} \\ \varepsilon_{zz} \\ \gamma_{xy} \\ \gamma_{yz} \\ \gamma_{zx} \end{Bmatrix} \tag{9.275}$$

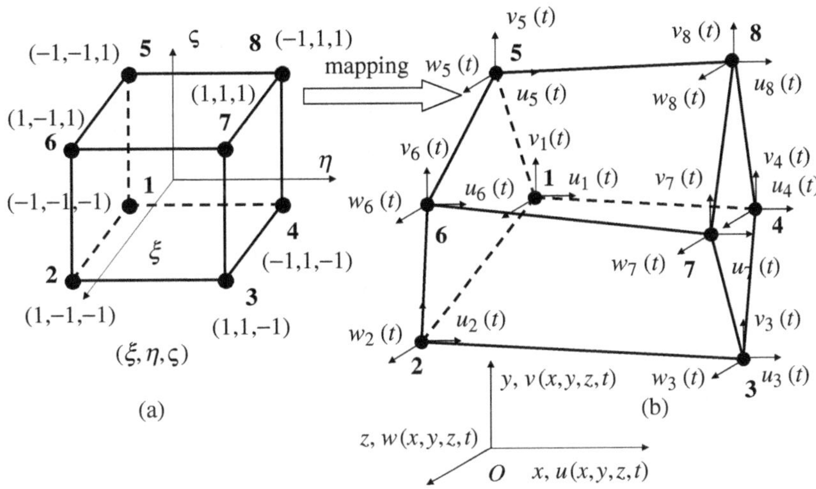

Figure 9.22. Three-dimensional linear hexahedral element: (a) Volume coordinate system, (b) Global coordinate system

where the constitutive modulus matrix **D** is given by

$$\mathbf{D} = \frac{E(1-\nu)}{(1+\nu)(1-2\nu)} \begin{bmatrix} 1 & \frac{\nu}{(1-\nu)} & \frac{\nu}{(1-\nu)} & 0 & 0 & 0 \\ \frac{\nu}{(1-\nu)} & 1 & \frac{\nu}{(1-\nu)} & 0 & 0 & 0 \\ \frac{\nu}{(1-\nu)} & \frac{\nu}{(1-\nu)} & 1 & 0 & 0 & 0 \\ 0 & 0 & 0 & \frac{(1-2\nu)}{2(1-\nu)} & 0 & 0 \\ 0 & 0 & 0 & 0 & \frac{(1-2\nu)}{2(1-\nu)} & 0 \\ 0 & 0 & 0 & 0 & 0 & \frac{(1-2\nu)}{2(1-\nu)} \end{bmatrix} \tag{9.276}$$

Element Stiffness Matrix. The element stiffness and mass matrices are given by

$$\mathbf{K}^e = \int_\Omega \mathbf{B}_e^T \mathbf{D} \mathbf{B}_e dV, \quad \mathbf{M}^e = \int_\Omega \rho \mathbf{N}_e^T \mathbf{N}_e dV \tag{9.277}$$

9.2.5 Three-Dimensional Continuum: Hexahedral Element

Natural Coordinate System. By the Lagrangian polynomials, the isoparametric hexahedral element (Figure 9.22) in the natural coordinate system has the interpolation functions

$$N_i(\xi,\eta,\varsigma) = \frac{1}{8}(1+\xi\xi_i)(1+\eta\eta_i)(1+\varsigma\varsigma_i) \tag{9.278}$$

where ξ_i, η_i and ς_i indicate the nodal coordinates of the ith node in the natural coordinate system. Consequently, the displacements, in vector form, are expressed as

$$\tilde{\mathbf{u}}^h(\xi,\eta,\varsigma,t) = \mathbf{N}_e(\xi,\eta,\varsigma)\,\mathbf{u}^e(t) \tag{9.279}$$

where

$$\tilde{\mathbf{u}}^h(\xi,\eta,\varsigma,t)=\begin{Bmatrix}\tilde{u}^h(\xi,\eta,t)\\ \tilde{v}^h(\xi,\eta,t)\\ \tilde{w}^h(\xi,\eta,t)\end{Bmatrix},\quad \mathbf{u}_e(t)=\begin{Bmatrix}\mathbf{u}_1(t)\\ \mathbf{u}_2(t)\\ \mathbf{u}_3(t)\end{Bmatrix}$$
$$\mathbf{N}_e(\xi,\eta,\varsigma)=\begin{bmatrix}\mathbf{N}_1 & \mathbf{N}_2 & \mathbf{N}_3 & \mathbf{N}_4\end{bmatrix} \tag{9.280}$$

where

$$\mathbf{N}_i=\begin{bmatrix}N_i & 0 & 0\\ 0 & N_i & 0\\ 0 & 0 & N_i\end{bmatrix},\quad \mathbf{u}_i(t)=\begin{Bmatrix}u_i(t)\\ v_i(t)\\ w_i(t)\end{Bmatrix},\quad i=1,2,3,4 \tag{9.281}$$

The differential in the natural coordinate system is related to the differential in the physical coordinate system in the following manner

$$d\mathbf{x}=\begin{Bmatrix}dx\\ dy\\ dz\end{Bmatrix}=\begin{bmatrix}\frac{\partial x}{\partial\xi} & \frac{\partial x}{\partial\eta} & \frac{\partial x}{\partial\varsigma}\\ \frac{\partial y}{\partial\xi} & \frac{\partial y}{\partial\eta} & \frac{\partial y}{\partial\varsigma}\\ \frac{\partial z}{\partial\xi} & \frac{\partial z}{\partial\eta} & \frac{\partial z}{\partial\varsigma}\end{bmatrix}\begin{Bmatrix}d\xi\\ d\eta\\ d\varsigma\end{Bmatrix}=\mathbf{J}_c d\boldsymbol{\xi} \tag{9.282}$$

where the Jacobian matrix $\mathbf{J}_c$ (Chung 2007; Malvern 1969) is a transformation matrix mapping the line segment in the isoparametric domain $d\boldsymbol{\xi}$ to the line segment $d\mathbf{x}$ in the physical domain. On the other hand, to compute the strain matrix we need to evaluate the partial derivatives of the interpolation function N_i with respect to global coordinates such as

$$\begin{Bmatrix}\frac{\partial N_i}{\partial\xi}\\ \frac{\partial N_i}{\partial\eta}\\ \frac{\partial N_i}{\partial\varsigma}\end{Bmatrix}=\begin{bmatrix}\frac{\partial x}{\partial\xi} & \frac{\partial y}{\partial\xi} & \frac{\partial z}{\partial\xi}\\ \frac{\partial x}{\partial\eta} & \frac{\partial y}{\partial\eta} & \frac{\partial z}{\partial\eta}\\ \frac{\partial x}{\partial\varsigma} & \frac{\partial y}{\partial\varsigma} & \frac{\partial z}{\partial\varsigma}\end{bmatrix}\begin{Bmatrix}\frac{\partial N_i}{\partial x}\\ \frac{\partial N_i}{\partial y}\\ \frac{\partial N_i}{\partial z}\end{Bmatrix}=\mathbf{J}_f\begin{Bmatrix}\frac{\partial N_i}{\partial x}\\ \frac{\partial N_i}{\partial y}\\ \frac{\partial N_i}{\partial z}\end{Bmatrix} \tag{9.283}$$

where $\mathbf{J}_f$ is also called the Jacobian matrix (Bathe 1996; Zienkiewicz et al. 2005) in the finite element literature. Note that $\mathbf{J}_f=\mathbf{J}_c^T$. Accordingly, without difficulty one can obtain

$$\begin{Bmatrix}\frac{\partial N_i}{\partial x}\\ \frac{\partial N_i}{\partial y}\\ \frac{\partial N_i}{\partial z}\end{Bmatrix}=\mathbf{J}_f^{-1}\begin{Bmatrix}\frac{\partial N_i}{\partial\xi}\\ \frac{\partial N_i}{\partial\eta}\\ \frac{\partial N_i}{\partial\varsigma}\end{Bmatrix}=\begin{Bmatrix}B_{1i}\\ B_{2i}\\ B_{3i}\end{Bmatrix} \tag{9.284}$$

Element Strains. In view of Equation (9.284), without difficulty one can obtain the relations

$$B_{1i}=\frac{\partial N_i(\xi,\eta,\varsigma)}{\partial x},\quad B_{2i}=\frac{\partial N_i(\xi,\eta,\varsigma)}{\partial y},\quad B_{3i}=\frac{\partial N_i(\xi,\eta,\varsigma)}{\partial z} \tag{9.285}$$

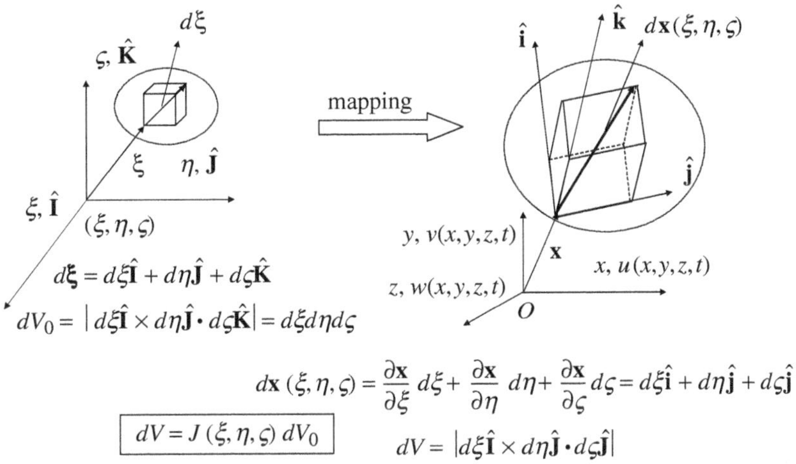

Figure 9.23. Mapping of an infinitesimal volume

In the three-dimensional continuum, the element strains can be rewritten as

$$\boldsymbol{\varepsilon} = \begin{Bmatrix} \varepsilon_{xx} \\ \varepsilon_{yy} \\ \varepsilon_{zz} \\ \gamma_{xy} \\ \gamma_{yz} \\ \gamma_{zx} \end{Bmatrix} = \sum_{i=1}^{8} \begin{Bmatrix} B_{1i} u_i(t) \\ B_{2i} v_i(t) \\ B_{3i} w_i(t) \\ B_{2i} u_i(t) + B_{1i} v_i(t) \\ B_{3i} v_i(t) + B_{2i} w_i(t) \\ B_{3i} u_i(t) + B_{1i} w_i(t) \end{Bmatrix} = \bar{\mathbf{B}}_e \mathbf{u}^e(t) \tag{9.286}$$

where the strain matrix $\bar{\mathbf{B}}_e$ for the four-node rectangular element can also be expressed in a concise form as

$$\bar{\mathbf{B}}_e = \begin{bmatrix} \bar{\mathbf{B}}_1 & \bar{\mathbf{B}}_2 & \bar{\mathbf{B}}_3 & \bar{\mathbf{B}}_4 & \bar{\mathbf{B}}_5 & \bar{\mathbf{B}}_6 & \bar{\mathbf{B}}_7 & \bar{\mathbf{B}}_8 \end{bmatrix} \tag{9.287}$$

where

$$\bar{\mathbf{B}}_i = \begin{bmatrix} B_{1i} & 0 & 0 \\ 0 & B_{2i} & 0 \\ 0 & 0 & B_{3i} \\ B_{2i} & B_{1i} & 0 \\ 0 & B_{3i} & B_{2i} \\ B_{3i} & 0 & B_{1i} \end{bmatrix}, \quad i = 1, \ldots, 8 \tag{9.288}$$

Mapping of Infinitesimal Volume. An infinitesimal volume in natural coordinates $dV_0\,(\xi, \eta, \varsigma)$ can be mapped to an infinitesimal volume in global coordinates $dV\,(x, y, z)$ as shown in Figure 9.23. The diagonal vector of the infinitesimal hexahedron in the natural coordinate system $d\boldsymbol{\xi}\,(\xi, \eta, \varsigma)$ can be defined as

$$d\boldsymbol{\xi}\,(\xi, \eta, \varsigma) = d\xi\hat{\mathbf{I}} + d\eta\hat{\mathbf{J}} + d\varsigma\hat{\mathbf{K}} \tag{9.289}$$

where $\hat{\mathbf{I}}$, $\hat{\mathbf{J}}$ and $\hat{\mathbf{K}}$ are unit vectors in natural coordinates. The infinitesimal volume of the hexahedron in natural coordinates $dV_0\,(\xi, \eta, \varsigma)$ is given by

$$dV_0 = \left| d\xi\hat{\mathbf{I}} \times d\eta\hat{\mathbf{J}} \cdot d\varsigma\hat{\mathbf{K}} \right| = d\xi d\eta d\varsigma \tag{9.290}$$

The diagonal vector of the infinitesimal volume in the global coordinate system $d\mathbf{x}(x, y, z)$ can be defined as

$$\begin{aligned} d\mathbf{x}(x(\xi,\eta,\varsigma), y(\xi,\eta,\varsigma), z(\xi,\eta,\varsigma)) &= \frac{\partial \mathbf{x}}{\partial \xi}d\xi + \frac{\partial \mathbf{x}}{\partial \eta}d\eta + \frac{\partial \mathbf{x}}{\partial \varsigma}d\varsigma \\ &= d\xi\hat{\mathbf{i}} + d\eta\hat{\mathbf{j}} + d\varsigma\hat{\mathbf{k}} \end{aligned} \tag{9.291}$$

where $\hat{\mathbf{i}}$, $\hat{\mathbf{j}}$ and $\hat{\mathbf{k}}$ are unit vectors in global coordinates. The infinitesimal volume of the hexahedron in global coordinates $dV(x, y, z)$ is given by

$$dV(x,y,z) = \left|d\xi\hat{\mathbf{i}} \times d\eta\hat{\mathbf{j}} \cdot d\varsigma\hat{\mathbf{k}}\right| = \left|\hat{\mathbf{i}} \times \hat{\mathbf{j}} \cdot \hat{\mathbf{k}}\right| d\xi d\eta d\varsigma \tag{9.292}$$

where the unit vectors in global coordinates are described as

$$\begin{aligned} \hat{\mathbf{i}} &= \frac{\partial x}{\partial \xi}\hat{\mathbf{I}} + \frac{\partial y}{\partial \xi}\hat{\mathbf{J}} + \frac{\partial z}{\partial \xi}\hat{\mathbf{K}}, \\ \hat{\mathbf{j}} &= \frac{\partial x}{\partial \eta}\hat{\mathbf{I}} + \frac{\partial y}{\partial \eta}\hat{\mathbf{J}} + \frac{\partial z}{\partial \eta}\hat{\mathbf{K}}, \\ \hat{\mathbf{k}} &= \frac{\partial x}{\partial \varsigma}\hat{\mathbf{I}} + \frac{\partial y}{\partial \varsigma}\hat{\mathbf{J}} + \frac{\partial z}{\partial \varsigma}\hat{\mathbf{K}} \end{aligned} \tag{9.293}$$

Consequently, Equation (9.292) can be rewritten as

$$dV(x,y,z) = |\mathbf{J}_{\mathrm{c}}(\xi,\eta,\varsigma)|\, d\xi d\eta d\varsigma = |\mathbf{J}_{\mathrm{f}}(\xi,\eta,\varsigma)|\, d\xi d\eta d\varsigma = J(\xi,\eta,\varsigma)\, dA_0 \tag{9.294}$$

where $J(\xi, \eta, \varsigma)$ is called the Jacobian of transformation, namely,

$$J(\xi,\eta,\varsigma) = |\mathbf{J}_{\mathrm{c}}(\xi,\eta,\varsigma)| = |\mathbf{J}_{\mathrm{f}}(\xi,\eta,\varsigma)| = \left|\frac{\partial(x,y,z)}{\partial(\xi,\eta,\varsigma)}\right| \tag{9.295}$$

Element Stiffness and Mass Matrices. The element stresses can be obtained from Equation (9.275). Then, the element stiffness and mass matrices per unit length are given by

$$\begin{aligned} \mathbf{K}^e &= \int_\Omega \mathbf{B}_e^T \mathbf{D} \mathbf{B}_e J(\xi,\eta,\varsigma)\, d\xi d\eta d\varsigma \\ &= \sum_{i=1}^{n_{\mathrm{gp}}}\sum_{j=1}^{n_{\mathrm{gp}}}\sum_{k=1}^{n_{\mathrm{gp}}} w_i w_j w_k \mathbf{B}_e^T(\xi_i,\eta_j,\varsigma_k)\, \mathbf{D}\mathbf{B}_e(\xi_i,\eta_j,\varsigma_k)\, J(\xi_i,\eta_j,\varsigma_k), \\ \mathbf{M}^e &= \int_\Omega \rho \mathbf{N}_e^T \mathbf{N}_e J(\xi,\eta,\varsigma)\, d\xi d\eta d\varsigma \\ &= \sum_{i=1}^{n_{\mathrm{gp}}}\sum_{j=1}^{n_{\mathrm{gp}}}\sum_{k=1}^{n_{\mathrm{gp}}} w_i w_j w_k \rho \mathbf{N}_e^T(\xi_i,\eta_j,\varsigma_k)\, \mathbf{N}_e(\xi_i,\eta_j,\varsigma_k)\, J(\xi_i,\eta_j,\varsigma_k) \end{aligned} \tag{9.296}$$

where n_{gp} denotes the number of Gauss points, w_i and w_j weights, and ξ_i and η_j corresponding Gauss points.

9.2.6 Structural Member: Euler-Bernoulli Beam Element

Global Coordinate System As a member of those termed as structural finite elements, we shall consider the Euler-Bernoulli beam element, which is a C^1-beam element. Note that the weak form we obtained previously in Equation (9.55) contains the second-order derivatives in both space and time. Therefore, even though the two-node Euler-Bernoulli beam element is the simplest, the interpolation function should be of class C^1 to satisfy the C^1 continuity condition at the inter element nodes. As shown in Figure 9.24, each node has two degrees of freedom, although there is only one dependent variable, i.e. $w(x)$. We shall now discuss the two-node Euler-Bernoulli beam element. Hence the field function $w(x, t)$, which is the exact solution, is approximated by a trial function $\tilde{w}^h(x, t)$ as a Hermite cubic polynomial function,

$$\tilde{w}^h(x,t) = \alpha_0(t) + \alpha_1(t)x + \alpha_2(t)x^2 + \alpha_3(t)x^3 \tag{9.297}$$

where $\alpha_i(t)$ are the coefficients of the polynomial function and dependent upon time in dynamics. It is noted that the trial function is a linear combination of $1, x, x^2, x^3$. The trial function is of class $C^1(]0, L[)$ at the inter element boundary and of class $C^2(]0, L_e[)$ within an element. This type of element is often called a C^1-beam element. For an element with nodes 1 and 2, one readily obtains the expression

$$\begin{Bmatrix} \tilde{w}^h(0,t) \\ -\dfrac{\partial \tilde{w}^h(0,t)}{\partial x} \\ \tilde{w}^h(L_e,t) \\ -\dfrac{\partial \tilde{w}^h(L_e,t)}{\partial x} \end{Bmatrix} = \begin{bmatrix} 1 & 0 & 0 & 0 \\ 0 & 1 & 0 & 0 \\ 1 & L_e & L_e^2 & L_e^3 \\ 0 & 1 & 2L_e & 3L_e \end{bmatrix} \begin{Bmatrix} \alpha_0(t) \\ \alpha_1(t) \\ \alpha_2(t) \\ \alpha_3(t) \end{Bmatrix} = \begin{Bmatrix} w_1(t) \\ \theta_{y1}(t) \\ w_2(t) \\ \theta_{y2}(t) \end{Bmatrix} \tag{9.298}$$

where $w_1(t)$, $\theta_{y1}(t)$, $u_2(t)$ and $\theta_{y2}(t)$ are the nodal variables of the Euler-Bernoulli beam element as shown in Figure 9.24. Consequently, the trial function can be rewritten as

$$\tilde{w}^h(x,t) = \begin{bmatrix} N_1(x) & N_2(x) & N_3(x) & N_4(x) \end{bmatrix} \begin{Bmatrix} w_1(t) \\ \theta_{y1}(t) \\ w_2(t) \\ \theta_{y2}(t) \end{Bmatrix} \tag{9.299}$$

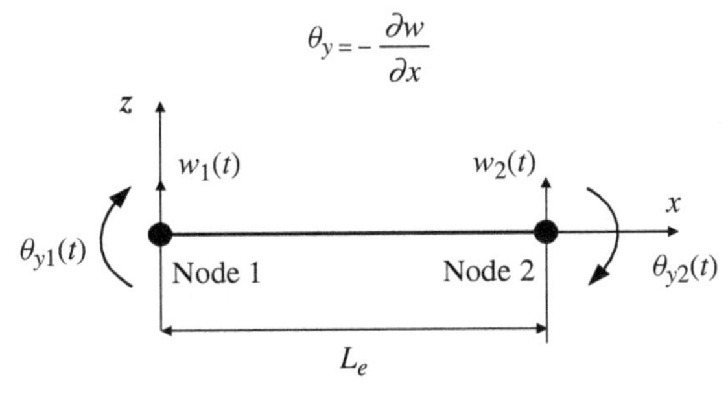

Figure 9.24. Euler-Bernoulli beam element

where the interpolation functions are given by

$$N_1(x) = 1 - 3\frac{x^2}{L_e^2} + 2\frac{x^3}{L_e^3}, N_2(x) = -x + 2\frac{x^2}{L_e} - \frac{x^3}{L_e^2},$$
$$N_3(x) = 3\frac{x^2}{L_e^2} - 2\frac{x^3}{L_e^3}, \quad N_4(x) = \frac{x^2}{L_e} - \frac{x^3}{L_e^2} \tag{9.300}$$

which are often called the Hermite cubic interpolation functions. The above equation can be rewritten in a concise form as

$$\tilde{w}^h(x,t) = \mathbf{N}_e(x)\mathbf{u}^e(t) \tag{9.301}$$

where the interpolation function matrix $\mathbf{N}_e(x)$ within one element is defined as

$$\mathbf{N}_e(x) = \begin{bmatrix} N_1(x) & N_2(x) & N_3(x) & N_4(x) \end{bmatrix} \tag{9.302}$$

and the displacement vector $\mathbf{u}^e(t)$ is defined as

$$\mathbf{u}^e(t)^T = \begin{bmatrix} w_1(t) & \theta_{y1}(t) & w_2(t) & \theta_{y2}(t) \end{bmatrix} \tag{9.303}$$

It should be noted that the interpolation functions have the following characteristics

$$\mathbf{N}_e(x)|_{x=0} = \begin{bmatrix} 1 & 0 & 0 & 0 \end{bmatrix}, \quad \mathbf{N}_e(x)|_{x=L_e} = \begin{bmatrix} 0 & 0 & 1 & 0 \end{bmatrix},$$
$$\left.\frac{d\mathbf{N}_e(x)}{dx}\right|_{x=0} = \begin{bmatrix} 0 & 1 & 0 & 0 \end{bmatrix}, \quad \left.\frac{d\mathbf{N}_e(x)}{dx}\right|_{x=L_e} = \begin{bmatrix} 0 & 0 & 0 & 1 \end{bmatrix} \tag{9.304}$$

The curvature of the neutral surface of the beam is given by

$$\kappa_x = -\frac{\partial^2 w^h(x,t)}{\partial x^2} = \begin{bmatrix} B_1(x) & B_2(x) & B_3(x) & B_4(x) \end{bmatrix} \begin{Bmatrix} w_1(t) \\ \theta_{y1}(t) \\ w_2(t) \\ \theta_{y2}(t) \end{Bmatrix} \tag{9.305}$$

where

$$B_1(x) = \frac{6}{L_e^2} - 12\frac{x}{L_e^3}, \quad B_2(x) = -\frac{4}{L_e} + 6\frac{x}{L_e^2},$$
$$B_3(x) = -\frac{6}{L_e^2} + 12\frac{x}{L_e^3}, B_4(x) = -\frac{2}{L_e} + 6\frac{x}{L_e^2} \tag{9.306}$$

The curvature can be succinctly expressed as

$$\kappa_x^h(x,t) = \bar{\mathbf{B}}_2(x)\mathbf{u}^e(t) \tag{9.307}$$

where $\bar{\mathbf{B}}_2(x)$ can be called the strain matrix of the Euler-Bernoulli beam element. The bending moment given in Equation (9.46) can be written as

$$M^h(x,t) = \mathbf{D}_{\text{bending}}\kappa_x^h(x,t) = \mathbf{D}_{\text{bending}}\bar{\mathbf{B}}_2(x)\mathbf{u}^e(t) \tag{9.308}$$

where $\mathbf{D}_{\text{bending}} = EI(x)$. Suppose that the inertial moment of the area about the neutral axis $I(x)$ is constant. Then, in view of Equation (9.59), one can have the element stiffness matrix given by

$$\mathbf{K}^e = \int_{L_e} \bar{\mathbf{B}}_2^T \mathbf{D}_{\text{bending}} \mathbf{B}_2 dx = \frac{2EI}{L_e^3} \begin{bmatrix} 6 & -3L_e & -6 & -3L_e \\ -3L_e & 2L_e^2 & 3L_e & L_e^2 \\ -6 & 3L_e & 6 & 3L_e \\ -3L_e & L_e^2 & 3L_e & 2L_e^2 \end{bmatrix} \tag{9.309}$$

The consistent mass matrix is given by

$$\begin{aligned} \mathbf{M}^e &= \int_{L_e} \rho A(x) \mathbf{N}_e^T(x) \mathbf{N}_e(x)\, dx \\ &= \int_{L_e} \rho A(x) \begin{bmatrix} N_1N_1 & N_1N_2 & N_1N_3 & N_1N_4 \\ N_2N_1 & N_2N_2 & N_2N_3 & N_2N_4 \\ N_3N_1 & N_3N_2 & N_3N_3 & N_3N_4 \\ N_4N_1 & N_4N_2 & N_4N_3 & N_4N_4 \end{bmatrix} dx \end{aligned} \tag{9.310}$$

In the case that the cross-sectional area $A(x)$ is constant, namely, $A(x) = A$, then the consistent mass matrix yields

$$\mathbf{M}^e_{\text{consistent}} = \frac{\rho A L_e}{420} \begin{bmatrix} 156 & -22L_e & 54 & 13L_e \\ -22L_e & 4L_e^2 & -13L_e & -3L_e^2 \\ 54 & -13L_e & 156 & 22L_e \\ 13L_e & -3L_e^2 & 22L_e & 4L_e \end{bmatrix} \tag{9.311}$$

In addition, the lumped mass matrix is given by

$$\mathbf{M}^e_{\text{lumped}} = \frac{\rho A L_e}{2} \begin{bmatrix} 1 & 0 & 0 & 0 \\ 0 & \alpha L_e^2 & 0 & 0 \\ 0 & 0 & 1 & 0 \\ 0 & 0 & 0 & \alpha L_e^2 \end{bmatrix} \tag{9.312}$$

where α is a parameter (Hughes 1987, page 520) (Key and Beisinger 1971). Suppose that the transversely distributed load is constant, namely, $\omega(x) = \bar{\omega}$. By using the Hermite interpolation functions, then the external force due to the distributed load is given by

$$\mathbf{F}^e_{\text{ext}} = \int_{L_e} \mathbf{N}_e^T \bar{\omega} dx = \frac{\bar{\omega} L_e}{12} \begin{bmatrix} 6 \\ L_e \\ 6 \\ -L_e \end{bmatrix} \tag{9.313}$$

where the components of the external force $\mathbf{F}^e_{\text{ext}}$ are work-equivalent loads (Gallagher 1975).

Illustrative Examples

Example 9.8 Elastic-Plastic Transient Response of the Euler-Bernoulli Beam Under a Distributed Load Over Center Portion (Wu and Witmer 1971)

A numerical comparison with experimental results (Balmer and Witmer 1964) for the transient response of the Euler-Bernoulli beam can be found in the reference (Wu and Witmer 1971). A clamped-clamped beam with the length 10 in is impulsively loaded around the midspan as shown in Figure 9.25. The T6061-T6 aluminum beam material is considered rate-independent elastic-perfectly-plastic, and the material properties are: $E = 10.4 \times 10^6$ psi,

$\rho = 2.61 \times 10^{-4}\,\text{lb} - \text{sec}^2/\text{in}^4\,\text{psi}$, $\sigma_y = 41,400\,\text{psi}$, and $\nu = 0.3$. The half-portion of the beam is modeled with 10 Euler-Bernoulli beam elements with three degrees of freedom per node, namely, two displacements and one rotation. Wu and Witmer used the central difference time integration scheme (see Chapter 13) with 1 μsec to predict the midspan deflection of the beam (Wu and Witmer 1971). Figure 9.25 indicates that the finite element prediction using the Euler-Bernoulli beam theory is in good agreement with experimental results.

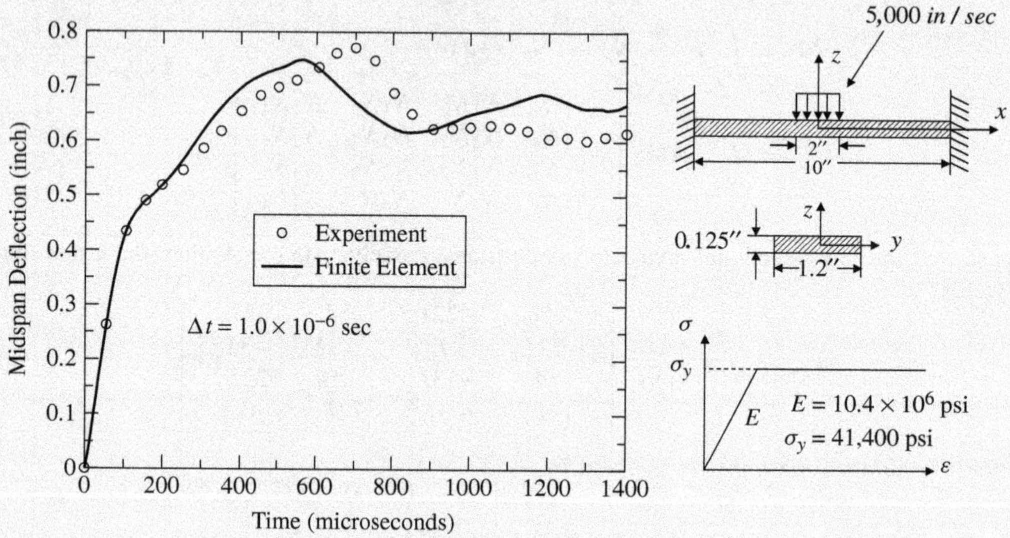

Figure 9.25. Elastic-plastic transient response of the Euler-Bernoulli beam under a distributed load (Wu and Witmer 1971)

Example 9.9 Nonlinear Dynamic Response of the Clamped-Clamped Euler-Bernoulli Beam Under a Pin Step Load

A 2-node Euler-Bernoulli beam element was investigated for the dynamical response of the beam exhibiting geometric nonlinearity in the literature (Yang and Saigal 1984). The clamped-clamped Euler-Bernoulli beam with the length 20 in is subjected to a concentrated step load 640 lb acting at the midspan as shown in Figure 9.26. The beam material is considered elastic, and the material properties are: $E = 3.0 \times 10^7\,\text{psi}$, $\rho = 26.0 \times 10^{-4}\,\text{lb} - \text{sec}^2/\text{in}^4\,\text{psi}$, and $\nu = 0.0$. This problem was previously analyzed by many researchers in the literature, (Weeks 1972), (McNamara 1974) and (Mondkar and Powell 1977). Yang and Saigal compared their Euler-Bernoulli beam results with five 8-node plane stress element results by Mondkar and Powell. Yang and Saigal modeled the half-portion of the beam with 6 Euler-Bernoulli beam elements with three degrees of freedom per node, namely, two displacements and one rotation. They used the Newmark implicit scheme (see Chapter 13) with 5 μsec to predict the midspan deflection of the beam (Yang and Saigal 1984). Mondkar and Powell also employed the Newmark implicit method with a time step size 50 μsec. Figure 9.26 indicates that Yang and Saigal's beam results are in good agreement with the numerically generated transient response by Mondkar and Powell.

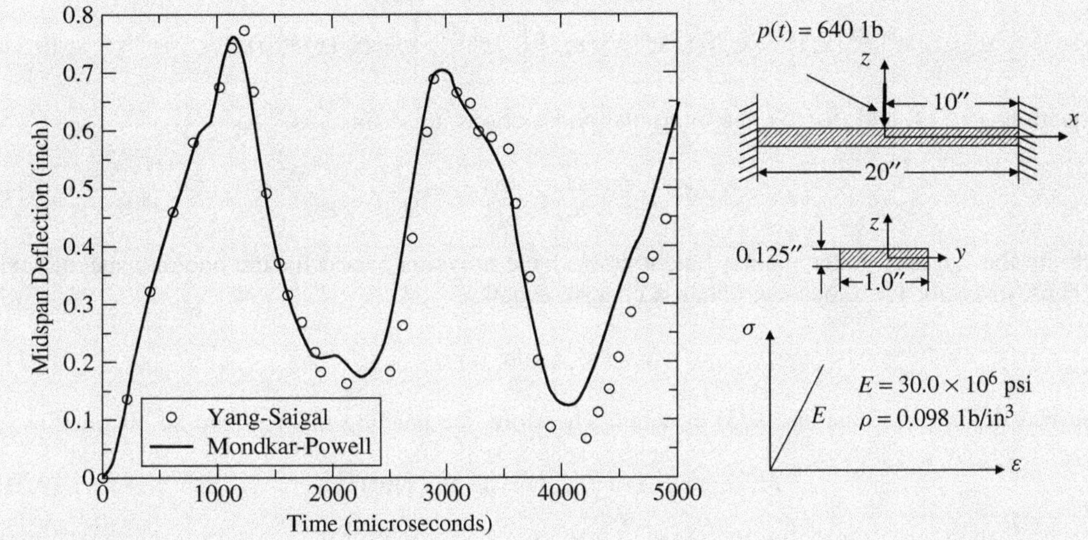

Figure 9.26. Nonlinear dynamic response of the clamped-clamped Euler-Bernoulli beam (Yang and Saigal 1984)

9.2.7 Structural Member: Timoshenko Beam Element

Unlike the Euler-Bernoulli beam element which we have so far discussed, the Timoshenko beam element has two dependent variables, namely, the transverse deflection $w(x, t)$ and the rotation $\theta_y(x, t)$ due to the bending moment. The inclusion of the additional dependent variable enables one to take advantage of the C^0 continuity for the interpolation functions at the inter element nodes. As such, we can use the Lagrange interpolation functions which we discussed in the section on the one-dimensional axial bar element. For a two-node Timoshenko beam element, which is shown in Figure 9.27, the trial function $\tilde{w}^h(x, t)$ is given by

$$\tilde{w}^h(x, t) = \begin{bmatrix} N_1(x) & N_2(x) \end{bmatrix} \begin{Bmatrix} w_1(t) \\ w_2(t) \end{Bmatrix} = \mathbf{N}_e^w(x)\mathbf{w}^e(t) \tag{9.314}$$

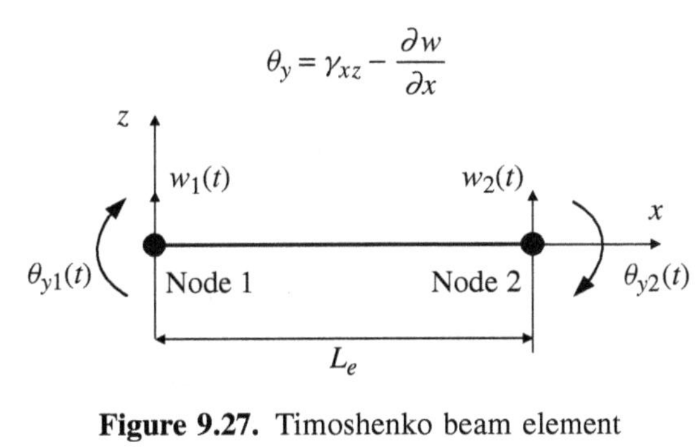

Figure 9.27. Timoshenko beam element

In addition, the other trial function $\tilde{\theta}_y^h(x,t)$ is also given by

$$\tilde{\theta}_y^h(x,t) = \begin{bmatrix} N_1(x) & N_2(x) \end{bmatrix} \begin{Bmatrix} \theta_{y1}(t) \\ \theta_{y2}(t) \end{Bmatrix} = \mathbf{N}_e^{\theta}(x)\boldsymbol{\theta}_y^e(t) \tag{9.315}$$

In Equations (9.314) and (9.315), the interpolation functions are defined as

$$N_1(x) = 1 - \frac{x}{L_e}, \quad N_2(x) = \frac{x}{L_e} \tag{9.316}$$

which are the Lagrange interpolation functions we have previously used for the one-dimensional axial bar. Thus, in vector form, one can obtain a concise equation

$$\tilde{\mathbf{u}}^h = \mathbf{H}_e\,(x)\,\mathbf{u}^e\,(t) \tag{9.317}$$

which was defined in Equation (9.75) in detail. Therefore, the bending moment can be obtained as

$$M^h(x,t) = \mathbf{D}_b\kappa_x^h(x,t) = \mathbf{D}_b\mathbf{B}_b(x)\mathbf{u}^e(t) \tag{9.318}$$

where $\mathbf{D}_b = EI(x)$, and the strain matrix $\mathbf{B}_b(x)$ can be written as

$$\mathbf{B}_b = \left[\frac{\partial \mathbf{H}_e^{\theta}\,(x)}{\partial x}\right] = \frac{EI}{L_e}\begin{bmatrix} 0 & 0 & -1 & 1 \end{bmatrix} \tag{9.319}$$

Then, the bending stiffness matrix $\mathbf{K}_b$ can be given by

$$\mathbf{K}_b = \int_{L_e} \mathbf{B}_b^T\,(x)\,\mathbf{D}_b\mathbf{B}_b\,(x)\,dx = \frac{EI}{L_e}\begin{bmatrix} 0 & 0 & 0 & 0 \\ 0 & 0 & 0 & 0 \\ 0 & 0 & 1 & -1 \\ 0 & 0 & -1 & 1 \end{bmatrix} \tag{9.320}$$

where $\mathbf{D}_b$ is assumed to be EI.

On the other hand, the other strain matrix $\mathbf{B}_s(x)$ can be written as

$$\mathbf{B}_s = \left[\frac{\partial \mathbf{H}_e^{w}\,(x)}{\partial x} + \mathbf{H}_e^{\theta}\,(x)\right] = \frac{1}{L_e}\begin{bmatrix} -1 & 1 & (L_e - x) & x \end{bmatrix} \tag{9.321}$$

Then, the shearing stiffness matrix $\mathbf{K}_s$ is given by

$$\mathbf{K}_s = \int_{L_e} \mathbf{B}_s^T\,(x)\,\mathbf{G}_s\mathbf{B}_s\,(x)\,dx = \frac{\kappa GA}{6L_e}\begin{bmatrix} 6 & -6 & -3L_e & -3L_e \\ -6 & 6 & 3L_e & 3L_e \\ -3L_e & 3L_e & 2L_e^2 & L_e^2 \\ -3L_e & 3L_e & L_e^2 & 2L_e^2 \end{bmatrix} \tag{9.322}$$

Finally, the element stiffness matrix $\mathbf{K}^e$ is written as

$$\mathbf{K}^e = \mathbf{K}_b + \mathbf{K}_s = \frac{1}{6L_e}\begin{bmatrix} 6\kappa GA & -6\kappa GA & -3\kappa GAL_e & -3\kappa GAL_e \\ -6\kappa GA & 6\kappa GA & 3\kappa GAL_e & 3\kappa GAL_e \\ -3\kappa GAL_e & 3\kappa GAL_e & 6EI + 2\kappa GAL_e^2 & -6EI + \kappa GAL_e^2 \\ -3\kappa GAL_e & 3\kappa GAL_e & -6EI + \kappa GAL_e^2 & 6EI + 2\kappa GAL_e^2 \end{bmatrix} \tag{9.323}$$

The consistent mass matrix can be given by

$$\mathbf{M}^e = \int_{L_e} \mathbf{H}_e^T(x)\,\rho \mathbf{H}_e(x)\,dx$$

$$= \int_{L_e} \begin{bmatrix} \rho A(x) N_1 N_1 & \rho A(x) N_1 N_2 & 0 & 0 \\ \rho A(x) N_2 N_1 & \rho A(x) N_2 N_2 & 0 & 0 \\ 0 & 0 & \rho I(x) N_1 N_1 & \rho I(x) N_1 N_2 \\ 0 & 0 & \rho I(x) N_2 N_1 & \rho I(x) N_2 N_2 \end{bmatrix} dx \qquad (9.324)$$

In the case that the cross-sectional area $A(x)$ and the second moment of inertia $I(x)$ are constant, the consistent mass matrix yields

$$\mathbf{M}^e_{\text{consistent}} = \frac{\rho L_e}{6} \begin{bmatrix} 2A & A & 0 & 0 \\ A & 2A & 0 & 0 \\ 0 & 0 & 2I & I \\ 0 & 0 & I & 2I \end{bmatrix} \qquad (9.325)$$

The row-sum technique yields the lumped mass matrix

$$\mathbf{M}^e_{\text{lumped}} = \frac{\rho L_e}{2} \begin{bmatrix} A & 0 & 0 & 0 \\ 0 & A & 0 & 0 \\ 0 & 0 & I & 0 \\ 0 & 0 & 0 & I \end{bmatrix} \qquad (9.326)$$

The external force due to the transversely distributed load is given by

$$\mathbf{F}^e_{\text{ext}} = \int_{L_e} \bar{\omega}(x)\,\mathbf{H}_w^T(x)\,dx = \frac{\bar{\omega} L_e}{2} \begin{Bmatrix} 1 \\ 1 \\ 0 \\ 0 \end{Bmatrix} \qquad (9.327)$$

Illustrative Example

Example 9.10 Timoshenko Pin-Ended Beam Subjected a Uniform Transverse Pulse Loading

The transient response of the Timoshenko beam can be found in the reference (Kouhia and Tuomala 1993). One-half of the beam is modeled with 10 linear Timoshenko beam elements. A pin-ended Timoshenko beam with the length 40 *cm* is subjected to a uniform transverse loading $p_0 = 19.2\,\text{N/mm}$ as shown in Figure 9.28. The material is considered elastic-perfectly-plastic, and the material properties are: $E = 80\,\text{GPa}$ and $\rho = 2700\,\text{Kg/m}^3$. Several time integration schemes are employed; the central difference method with $\Delta t = 1.5\,\mu\text{s}$, the midpoint rule with $\Delta t = 25\,\mu\text{s}$, and the trapezoidal rule with $\Delta t = 25\,\mu\text{s}$. Kouhia and Tuomala used a consistent mass matrix for the central difference method and the trapezoidal rule, and a diagonal mass matrix for the midpoint rule. The deflection histories at the midspan of the Timoshenko beam, predicted by Kouhia and Tuomala, are shown in Figure 9.28.

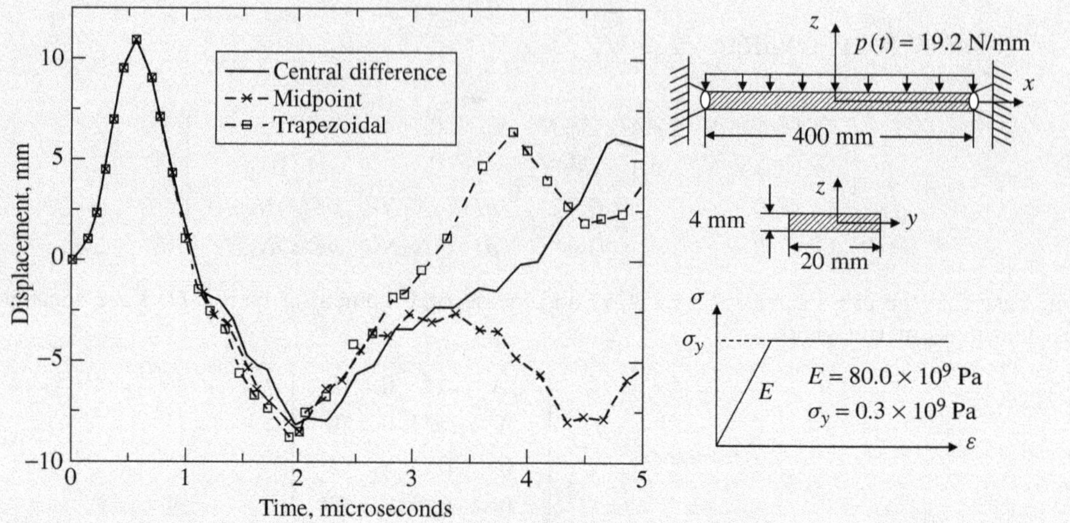

Figure 9.28. The transient response of the Timoshenko pin-ended beam subjected a uniform transverse pulse loading (Kouhia and Tuomala 1993)

9.2.8 Structural Member: Kirchhoff-Love Plate Element

Global Coordinate System As an analogy of the Euler-Bernoulli beam element, we shall consider next the Kirchhoff-Love plate element, which is also a C^1 element for thin plates. Note that the weak form we obtained previously in Equation (9.112) contains the second-order derivatives in both space and time. Therefore, the interpolation function should be of class C^1 to satisfy the C^1 continuity condition at the inter element nodes. In the literature, there are a number of finite element developments based on the Kirchhoff-Love plate theory. As one of the most popular elements, we shall discuss here the 4-node rectangular thin plate element (Bathe 1996; Cook et al. 2002; Gallagher 1975; Huebner 1982; Zienkiewicz and Taylor 2006), as shown in Figure 9.29, where each node has three degrees of freedom, namely, one translational and two rotational degrees of freedom. Hence, each element has 12 degrees of freedom. As in the Euler-Bernoulli beam element, we have only one dependent variable, the transverse deflection $w(x, y, t)$, which is the exact solution. The deflection function, which is approximated by a trial function $\tilde{w}^h(x, y, t)$, is a 12-term polynomial function,

$$\begin{aligned}\tilde{w}^h(x, y, t) &= \alpha_0(t) + \alpha_1(t)x + \alpha_2(t)y + \alpha_3(t)x^2 + \alpha_4(t)xy + \alpha_5(t)y^2 \\ &+ \alpha_6(t)x^3 + \alpha_7(t)x^2y + \alpha_8(t)xy^2 + \alpha_9(t)y^3 + \alpha_{10}(t)x^3y + \alpha_{11}(t)xy^3\end{aligned} \tag{9.328}$$

where $\alpha_i(t)$ are the coefficients of the polynomial function and dependent on time in dynamics. As shown in Figure 9.30, the above polynomial function with the 12 terms chosen is an incomplete quartic in the sense of Gallagher (Gallagher 1975, Section 8.1). Thus, the resulting issue is the satisfaction of the C^1 continuity requirement on the interface line between elements. The unknown 12-coefficients in Equation (9.328) can be uniquely determined by 12 nodal variables as demonstrated thus far for the previous elements. However, the deformed shape of the interface line cannot be uniquely specified by the 12 nodal variables. In the end, the C^1 continuity requirement is violated. Therefore, the finite element being developed using Equation (9.328) is often called the nonconforming or incompatible plate

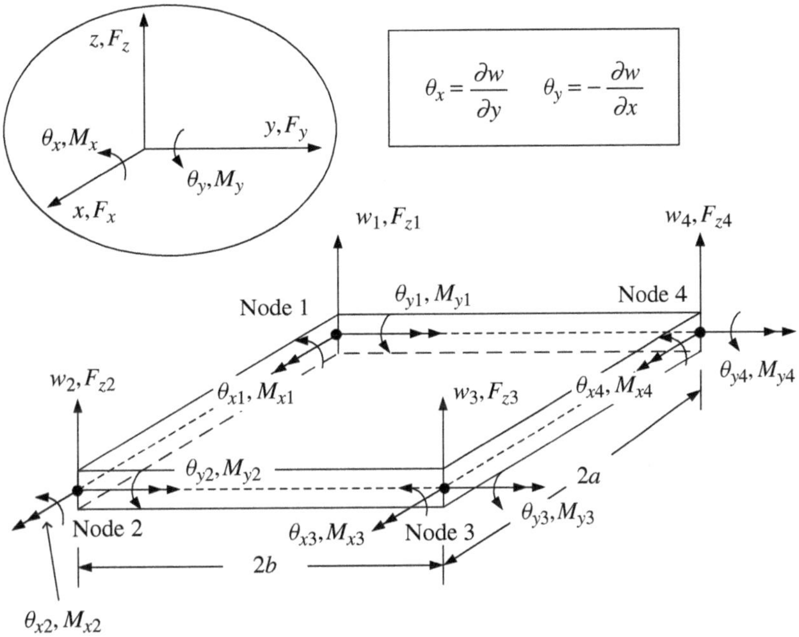

Figure 9.29. Four-node rectangular element based on the Kirchhoff-Love plate theory

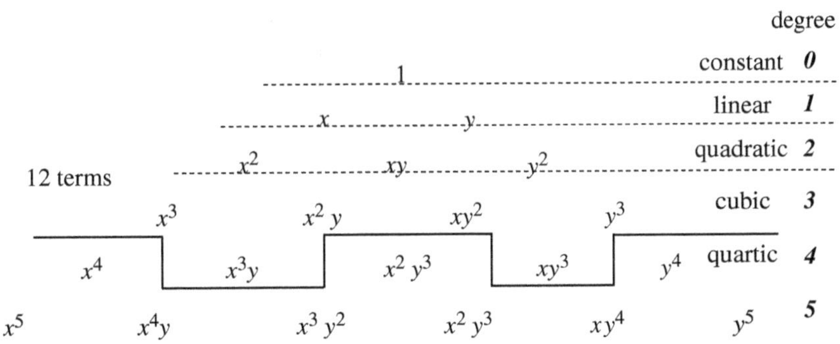

Figure 9.30. Selection of 12 terms for the Kirchhoff-Love plate element

element. It is to be noted that the trial function is a linear combination of the 12 polynomial terms. Hence, Equation (9.328) is written in direct notation as

$$\tilde{w}^h(x, y, t) = \mathbf{P}(x, y)\boldsymbol{\alpha}(t) \tag{9.329}$$

where $\mathbf{P} = \begin{bmatrix} 1 & x & y & x^2 & xyy^2 & x^3 & x^2y & xy^2 & y^3 & x^3y & xy^3 \end{bmatrix}$ and $\boldsymbol{\alpha}(t)$ denotes a coefficient column vector whose components correspond to the 12 unknown coefficients in Equation (9.328). For each node, the nodal values can be evaluated in the following fashion

$$\begin{aligned} w_i(t) &= \tilde{w}^h(x_i, y_i, t) = \alpha_0(t) + \alpha_1(t)x_i + \alpha_2(t)y_i + \alpha_3(t)x_i^2 + \alpha_4(t)xy + \ldots + \\ \theta_{xi}(t) &= \frac{\partial \tilde{w}^h(x_i, y_i, t)}{\partial y} = \alpha_2(t) + \alpha_4(t)x_i + 2\alpha_5(t)y_i + \ldots + \end{aligned} \tag{9.330}$$

$$\theta_{yi}(t) = -\frac{\partial \tilde{w}^h(x_i, y_i, t)}{\partial x} = -\alpha_1(t) - 2\alpha_3(t)x_i - \alpha_4(t)y_i + \ldots +$$

where $w_i(t)$, $\theta_{xi}(t)$ and $\theta_{yi}(t)$ are the nodal variables for the ith node as shown in Figure 9.29. Consequently, the coefficient column vector $\boldsymbol{\alpha}(t)$ can be expressed in terms of the nodal variables as follows:

$$\boldsymbol{\alpha}(t) = \mathbf{C}^{-1}\mathbf{u}^e \tag{9.331}$$

where the matrix $\mathbf{C}$ can be determined in view of Equation (9.330), and the nodal vector $\mathbf{u}^e$ is given by

$$\mathbf{u}^e = \begin{bmatrix} \mathbf{u}_1^e & \mathbf{u}_2^e & \mathbf{u}_3^e & \mathbf{u}_4^e \end{bmatrix}^T \tag{9.332}$$

where $\mathbf{u}_i^e = \begin{bmatrix} w_i & \theta_{xi} & \theta_{yi} \end{bmatrix}^T$, $i = 1, 2, 3, 4$. Hence, the trial function in Equation (9.329) can be rewritten as

$$\tilde{w}^h(x, t) = \mathbf{P}(x, y)\mathbf{C}^{-1}\mathbf{u}^e = \mathbf{N}(x, y)\mathbf{u}^e \tag{9.333}$$

where $\mathbf{N}(x, y) = \mathbf{P}(x, y)\mathbf{C}^{-1}$. Note that $\mathbf{N}(x, y)$ is the interpolation function matrix and it can be found in the literature (Zienkiewicz and Cheung 1964). In view of Equation (9.83), the admissible curvature of the thin plate is given by

$$\boldsymbol{\kappa}^h(x, y, t) = \begin{Bmatrix} -\dfrac{\partial^2 w^h(x, y, t)}{\partial x^2} \\ -\dfrac{\partial^2 w^h(x, y, t)}{\partial y^2} \\ -2\dfrac{\partial^2 w^h(x, y, t)}{\partial x \partial y} \end{Bmatrix} = \mathbf{E}(x, y)\boldsymbol{\alpha}(t) \tag{9.334}$$

where the matrix $\mathbf{E}(x, y)$ is given by

$$\mathbf{E} = \begin{bmatrix} 0 & 0 & 0 & -2 & 0 & 0 & -6x & -2y & 0 & 0 & -6xy & 0 \\ 0 & 0 & 0 & 0 & 0 & -2 & 0 & 0 & -2x & -6y & 0 & -6xy \\ 0 & 0 & 0 & 0 & -2 & 0 & 0 & -4x & -4y & 0 & -6x^2 & -6y^2 \end{bmatrix} \tag{9.335}$$

Upon substitution of Equation (9.331) into Equation (9.334), the admissible curvature of the thin plate can be rewritten as

$$\boldsymbol{\kappa}^h(x, y, t) = \bar{\mathbf{B}}(x, y)\mathbf{u}^e(t) \tag{9.336}$$

where the strain matrix of the Kirchhoff-Love plate element, $\bar{\mathbf{B}}(x, y)$ can be written as

$$\bar{\mathbf{B}}(x, y) = \mathbf{E}(x, y)\mathbf{C}^{-1} \tag{9.337}$$

Therefore, the bending moment $\mathbf{m}^h(x, y, t)$ can be written as

$$\mathbf{m}^h(x, y, t) = \mathbf{D}\boldsymbol{\kappa}^h(x, y, t) \tag{9.338}$$

where $\mathbf{D}$ is given in Equation (9.115). Consequently, one can formulate the element stiffness matrix

$$\mathbf{K}^e = \int_{\Omega_e} \bar{\mathbf{B}}^T \mathbf{D} \bar{\mathbf{B}} dV \tag{9.339}$$

9.2.9 Structural Member: Reissner-Mindlin Plate Element

Unlike the Kirchhoff-Love Plate element which we discussed thus far, the Reissner-Mindlin plate element has three dependent variables, namely, the transverse deflection $w(x,t)$ and the rotations $\theta_x(x,t)$ and $\theta_y(x,t)$ due to the bending moment. The inclusion of the additional dependent variables enables one to take advantage of the C^0 continuity for the interpolation functions at the inter element nodes. For the Mindlin-Reissner plate element, which is shown in Figure 9.31, the trial functions $\tilde{w}^h(x,t)$, $\tilde{\theta}_x(x,t)$ and $\tilde{\theta}_y(x,t)$ can be given by

$$\begin{aligned}\tilde{w}^h(\xi,\eta,t) &= \sum_{i=1}^{4} N_i(\xi,\eta)w_i(t) = \mathbf{N}_w(\xi,\eta)\mathbf{w}^e(t),\\ \tilde{\theta}_x^h(\xi,\eta,t) &= \sum_{i=1}^{4} N_i(\xi,\eta)\theta_{xi}(t) = \mathbf{N}_{\theta_x}(\xi,\eta)\boldsymbol{\theta}_x^e(t),\\ \tilde{\theta}_y^h(\xi,\eta,t) &= \sum_{i=1}^{4} N_i(\xi,\eta)\theta_{yi}(t) = \mathbf{N}_{\theta_y}(\xi,\eta)\boldsymbol{\theta}_y^e(t)\end{aligned} \tag{9.340}$$

where the interpolation functions ($\mathbf{N}_w(\xi,\eta)$, $\mathbf{N}_{\theta_x}(\xi,\eta)$, and $\mathbf{N}_{\theta_y}(\xi,\eta)$ are bilinear and they are identical to the interpolation functions $\mathbf{N}_e(\xi,\eta)$. Nodal variables ($\mathbf{w}^e(t)$, $\boldsymbol{\theta}_x^e(t)$, and $\boldsymbol{\theta}_y^e(t)$) and the bilinear interpolation function $\mathbf{N}_e(\xi,\eta)$ are defined as

$$\begin{aligned}\mathbf{N}_e(\xi,\eta) &= \begin{bmatrix} N_1(\xi,\eta) & N_2(\xi,\eta) & N_3(\xi,\eta) & N_4(\xi,\eta) \end{bmatrix},\\ \mathbf{w}^e(t) &= \begin{bmatrix} w_1(t) & w_2(t) & w_3(t) & w_4(t) \end{bmatrix}^T,\end{aligned} \tag{9.341}$$

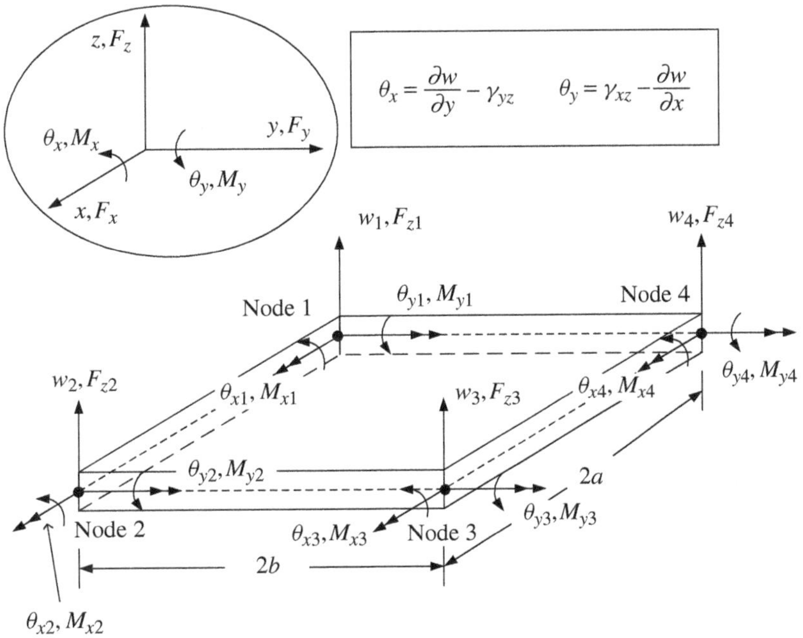

Figure 9.31. Four-node quadrilateral element based on the Reissner-Mindlin plate theory

$$\boldsymbol{\theta}_x^e(t) = \begin{bmatrix} \theta_{x1}(t) & \theta_{x2}(t) & \theta_{x3}(t) & \theta_{x4}(t) \end{bmatrix}^T,$$
$$\boldsymbol{\theta}_y^e(t) = \begin{bmatrix} \theta_{y1}(t) & \theta_{y2}(t) & \theta_{y3}(t) & \theta_{y4}(t) \end{bmatrix}^T$$

where $N_1(\xi,\eta)$, $N_2(\xi,\eta)$, $N_3(\xi,\eta)$, and $N_4(\xi,\eta)$ can be obtained from Equation (9.234). Next, we rewrite the three dependent variables in the following manner

$$\begin{aligned} \tilde{w}^h(\xi,\eta,t) &= \mathbf{H}_w(\xi,\eta)\,\mathbf{u}^e(t), \\ \tilde{\theta}_x^h(\xi,\eta,t) &= \mathbf{H}_{\theta_x}(\xi,\eta)\,\mathbf{u}^e(t), \\ \tilde{\theta}_y^h(\xi,\eta,t) &= \mathbf{H}_{\theta_y}(\xi,\eta)\,\mathbf{u}^e(t) \end{aligned} \tag{9.342}$$

where the interpolation function matrices are given by

$$\mathbf{H}_w = \begin{bmatrix} \mathbf{N}_w & \mathbf{0} & \mathbf{0} \end{bmatrix},\ \mathbf{H}_{\theta_x} = \begin{bmatrix} \mathbf{0} & \mathbf{N}_{\theta_x} & \mathbf{0} \end{bmatrix},\ \mathbf{H}_{\theta_y} = \begin{bmatrix} \mathbf{0} & \mathbf{0} & \mathbf{N}_{\theta_y} \end{bmatrix} \tag{9.343}$$

and the nodal variable vector $\mathbf{u}^e(t)$ is given by

$$\mathbf{u}^e(t) = \begin{bmatrix} \mathbf{w}^e(t) & \boldsymbol{\theta}_x^e(t) & \boldsymbol{\theta}_y^e(t) \end{bmatrix} \tag{9.344}$$

The trial functions are rewritten in vector form as

$$\begin{Bmatrix} \tilde{w}^h(\xi,\eta,t) \\ \tilde{\theta}_x^h(\xi,\eta,t) \\ \tilde{\theta}_y^h(\xi,\eta,t) \end{Bmatrix} = \begin{bmatrix} \mathbf{N}_w & \mathbf{0} & \mathbf{0} \\ \mathbf{0} & \mathbf{N}_{\theta_x} & \mathbf{0} \\ \mathbf{0} & \mathbf{0} & \mathbf{N}_{\theta_y} \end{bmatrix} \tag{9.345}$$

Thus, in vector form, one can obtain a concise equation as follows:

$$\tilde{\mathbf{u}}^h(\xi,\eta,t) = \mathbf{H}_e(\xi,\eta)\,\mathbf{u}^e(t) \tag{9.346}$$

where

$$\mathbf{H}_e(\xi,\eta) = \begin{bmatrix} \mathbf{H}_w(x,y) \\ \mathbf{H}_{\theta_x}(\xi,\eta) \\ \mathbf{H}_{\theta_y}(\xi,\eta) \end{bmatrix} \tag{9.347}$$

which was defined in Equation (9.75) in detail. Therefore, the bending moment is given by

$$\mathbf{m}^h(\xi,\eta,t) = \mathbf{D}\boldsymbol{\kappa}^h(\xi,\eta,t) = \mathbf{D}\mathbf{B}_b(\xi,\eta)\mathbf{u}^e(t) \tag{9.348}$$

where the strain matrix $\mathbf{B}_b(x,y)$ is given by

$$\mathbf{B}_b = \begin{Bmatrix} \dfrac{\partial \mathbf{H}_{\theta_y}}{\partial x} \\ -\dfrac{\partial \mathbf{H}_{\theta_x}}{\partial y} \\ \left(\dfrac{\partial \mathbf{H}_{\theta_y}}{\partial y} - \dfrac{\partial \mathbf{H}_{\theta_x}}{\partial x}\right) \end{Bmatrix} \tag{9.349}$$

Then, the bending stiffness matrix $\mathbf{K}_b$ is given by

$$\mathbf{K}_b = \int_\Omega \mathbf{B}_b^T(\xi,\eta)\,\mathbf{D}\mathbf{B}_b(\xi,\eta)\,dx \tag{9.350}$$

On the other hand, the other strain matrix $\mathbf{B}_s(x)$ is written as

$$\mathbf{B}_s = \left\{ \begin{array}{c} \dfrac{\partial \mathbf{H}_w}{\partial x} + \mathbf{H}_{\theta_y} \\ \dfrac{\partial \mathbf{H}_w}{\partial y} - \mathbf{H}_{\theta_x} \end{array} \right\} \tag{9.351}$$

Then, the shearing stiffness matrix $\mathbf{K}_s^e$ is given by

$$\mathbf{K}_s^e = \int_{\Omega_e} \mathbf{B}_s^T(\xi,\eta)\,\mathbf{G}\mathbf{B}_s(\xi,\eta)\,J(\xi,\eta)d\xi d\eta \tag{9.352}$$

Finally, the element stiffness matrix $\mathbf{K}^e$ can be written as

$$\mathbf{K}^e = \mathbf{K}_b + \mathbf{K}_s \tag{9.353}$$

The consistent mass matrix is given by

$$\begin{aligned} \mathbf{M}^e_{\text{consistent}} &= \int_{\Omega_e} \mathbf{H}_e^T(x)\,\boldsymbol{\rho}\mathbf{H}_e(x)\,J(\xi,\eta)d\xi d\eta \\ &= \begin{bmatrix} \rho h \mathbf{N}_w^T\mathbf{N}_w & \mathbf{0} & \mathbf{0} \\ \mathbf{0} & \rho I \mathbf{N}_{\theta_x}^T\mathbf{N}_{\theta_x} & \mathbf{0} \\ \mathbf{0} & \mathbf{0} & \rho I \mathbf{N}_{\theta_y}^T\mathbf{N}_{\theta_y} \end{bmatrix}_{(12\times 12)} \end{aligned} \tag{9.354}$$

In the case that the cross-sectional area $A(x)$ and the second moment of inertia $I(x)$ are constant, the consistent mass matrix yields

$$\mathbf{N}_w^T\mathbf{N}_w = \mathbf{N}_{\theta_x}^T\mathbf{N}_{\theta_x} = \mathbf{N}_{\theta_y}^T\mathbf{N}_{\theta_y} = \begin{bmatrix} N_1N_1 & N_1N_2 & N_1N_3 & N_1N_4 \\ N_2N_1 & N_2N_2 & N_2N_3 & N_2N_4 \\ N_3N_1 & N_3N_2 & N_3N_3 & N_3N_4 \\ N_4N_1 & N_4N_2 & N_4N_3 & N_4N_4 \end{bmatrix} \tag{9.355}$$

The row-sum technique yields the lumped mass matrix

$$\mathbf{M}^e_{\text{lumped}} = \begin{bmatrix} \rho h \mathbf{N}_{\text{diag}} & \mathbf{0} & \mathbf{0} \\ \mathbf{0} & \rho I \mathbf{N}_{\text{diag}} & \mathbf{0} \\ \mathbf{0} & \mathbf{0} & \rho I \mathbf{N}_{\text{diag}} \end{bmatrix}_{(12\times 12)} \tag{9.356}$$

where

$$\mathbf{N}_{\text{diag}} = \begin{bmatrix} N_1 & 0 & 0 & 0 \\ 0 & N_2 & 0 & 0 \\ 0 & 0 & N_3 & 0 \\ 0 & 0 & 0 & N_4 \end{bmatrix} \tag{9.357}$$

The external force due to the transversely distributed load is given by

$$\mathbf{F}_{\text{ext}}^{e} = \int_{\Omega_e} \bar{\omega}(\xi, \eta)\, \mathbf{H}_e^T J(\xi, \eta) d\xi d\eta \tag{9.358}$$

Illustrative Example

Example 9.11 Transient Response of Reissner-Mindlin Plate Subjected to a Distributed Step Load at the Center Portion

A finite element implementation of the 8-node isoparametric quadrilateral Reissner-Mindlin plate bending element can be found in the reference (Rock and Hinton 1974). The Reissner-Mindlin's thick plate theory was used to predict the transient response of three rectangular plates of aspect ratio $\sqrt{2}$ and length to thickness ratios of 5:1, 10:1 and 20:1. The finite element results generalized numerically by Rock and Hinton are compared with the analytical solutions for both the Reissner-Mindlin's thick plate and the Kirchhoff-Love thin plate (Reismann and Lee 1969). The transverse deflections at the center of the plates are shown in Figure 9.32. Rock and Hinton employed the Newmark implicit scheme (see Chapter 13) with $\Delta t = 0.1$ (where T is non-dimensionalized time) (Rock and Hinton 1974). The material is considered elastic, and the shear coefficient $k = 0.86$

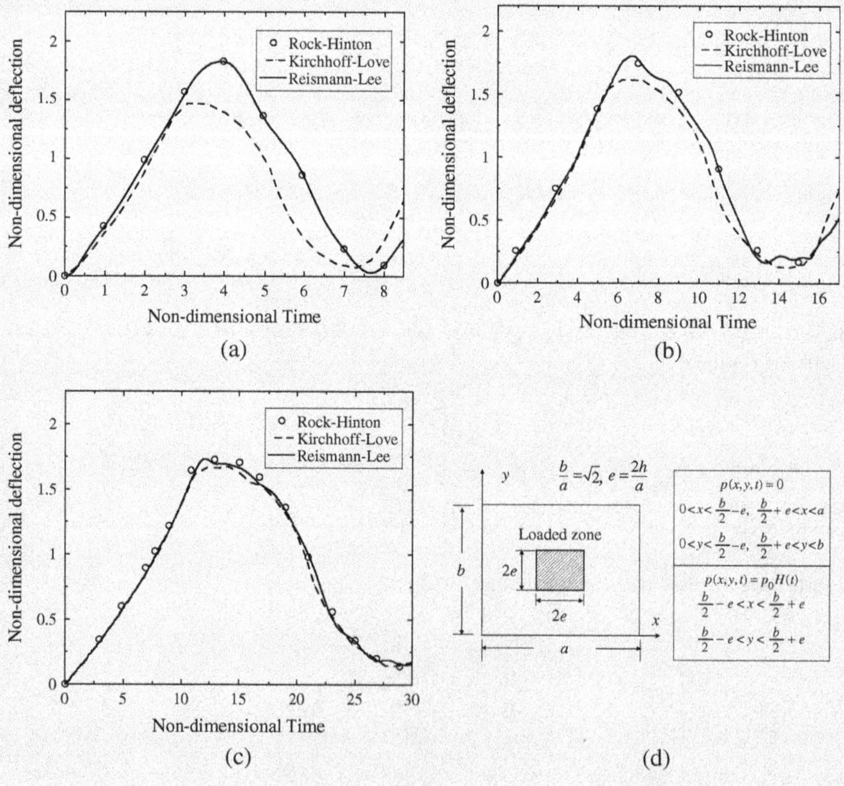

Figure 9.32. (a) Transient response of thick plate, plate thickness $h = a/5$, and aspect ratio $\beta = b/a = \sqrt{2}$, (b) Transient response of moderately thick plate, plate thickness $h = a/10$, and aspect ratio $\beta = b/a = \sqrt{2}$, (c) Transient response of moderately thick plate, plate thickness $h = a/20$, and aspect ratio $\beta = b/a = \sqrt{2}$, (d) Rectangular plate subjected to a distributed step load, $H(t)$ is a Heaviside step function

and $\nu = 0.3$. Figure 9.32 indicates that the finite element prediction using the Reissner-Mindlin's thick plate theory is in good agreement with analytical results. It should be noted that for a thin plate, the analytical solution for the Kirchhoff-Love plate is in good agreement with the finite element prediction for the Reissner-Mindlin's thick plate as well as the analytical solution for the Reissner-Mindlin's thick plate.

9.3 NONLINEAR FINITE ELEMENT FORMULATIONS

In the previous chapter, we have shown several fundamental principles including the principle of virtual work. It has been well known that the principle of virtual work is crucial for discretizing the space of interest in the sense of the finite element method, and it can be derived from the weighted residual statement.

In nonlinear solid mechanics, the principle of virtual work can be described by either the Eulerian or Lagrangian descriptions (Belytschko et al. 2000). In the literature, the solution procedure for the finite element developments can be described by the following two formulations respectively, i.e. the total Lagrangian formulation (Hibbitt et al. 1970) and the updated Lagrangian formulation (McMeeking and Rice 1975). In these formulations, meshes and nodes can move along with the spatial points of the deformed body due to the nature of Lagrangian formulations. It is natural that the Gauss quadrature points also move just as the spatial points. The distinct difference between the two formulations comes from the chosen configuration where the integral equations are computed. In the total Lagrangian formulation, the integral is carried out in the reference configuration, whereas in the updated Lagrangian formulation the integration is performed in the previous current configuration, which is known. Here, the previous current configuration means the configuration which is constructed from the time integration just carried out. The current configuration which one desires to predict is unknown so that the integration in that configuration cannot be carried out as shown in Figure 9.33.

Using the Lagrangian measures including the second Piola-Kirchhoff stress and Green strain tensors, integral expressions in the reference configuration in association with the current configuration are continuously established and computed as the material undergoes nonlinear deformation. All the Lagrangian

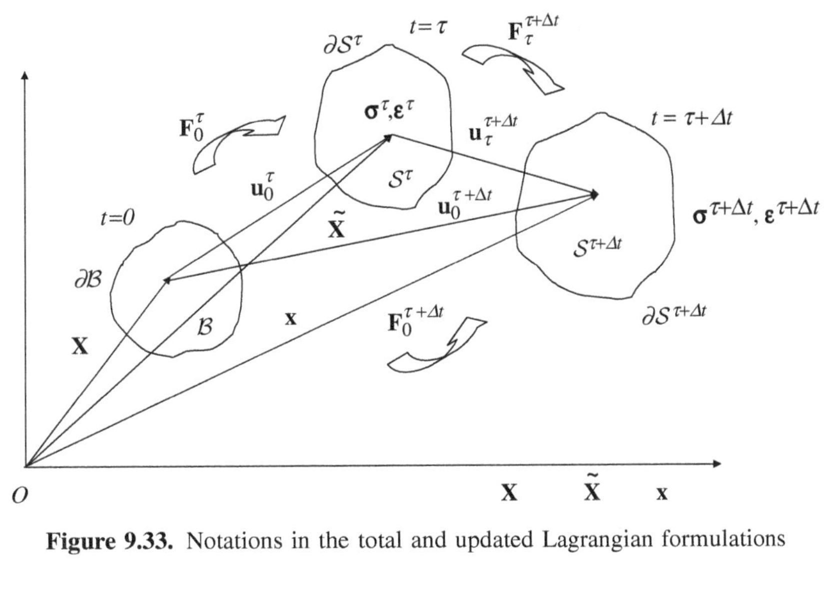

Figure 9.33. Notations in the total and updated Lagrangian formulations

measures are always referred to as the reference configuration. This formulation is known as the total Lagrangian approach (Hibbitt et al. 1970). On the other hand, the previous current configuration, which is just updated by the time integration scheme, is employed as the new reference configuration, which is not the undeformed configuration. The new reference configuration should be continuously updated as the time integration is carried out.

To start with, we shall first present the total Lagrangian formulation widely used for the hyperelastic materials. Next we shall discuss the updated Lagrangian formulation.

9.3.1 Total Lagrangian Formulation

In this approach (Hibbitt et al. 1970), the integration is always carried out in the reference configuration, and the Lagrangian measures are employed to express the principle of virtual work. To begin with, consider the internal virtual work in the current configuration as

$$\delta \mathcal{W}_{\text{int}}(\mathbf{u}) = \int_{\mathcal{S}} \boldsymbol{\sigma} : \delta \mathbf{e}\, dv \tag{9.359}$$

The infinitesimal volume element can be expressed using the Jacobian

$$dv = J dV \tag{9.360}$$

In view of Equations (8.391), (9.360) and (8.390), Equation (9.359) can be given by

$$\begin{aligned}\delta \mathcal{W}_{\text{int}}(\mathbf{u}(\mathbf{x},t)) &= \int_{\mathcal{S}} \boldsymbol{\sigma} : \delta \mathbf{e}\, dv = \int_{\mathcal{B}} J\boldsymbol{\sigma} : \text{grad}\mathbf{u}\, dV \\ &= \int_{\mathcal{B}} \boldsymbol{\tau} : \text{GRAD}\,\delta\mathbf{U}\cdot\mathbf{F}^{-1}\, dV \\ &= \delta \mathcal{W}_{\text{int}}(\mathbf{U}(\mathbf{X},t))\end{aligned} \tag{9.361}$$

where $\boldsymbol{\tau}$ is the Kirchhoff stress tensor, $\mathbf{u}(\mathbf{x},t)$ the Eulerian displacement vector and $\mathbf{U}(\mathbf{X},t)$ the Lagrangian displacement vector. Due to the double contraction property in tensor algebra, the internal virtual work can be rewritten as

$$\delta \mathcal{W}_{\text{int}}(\mathbf{U}(\mathbf{X},t)) = \int_{\mathcal{B}} \boldsymbol{\tau}\cdot\mathbf{F}^{-T} : \delta\mathbf{F}\, dV = \int_{\mathcal{B}} \mathbf{P} : \delta\mathbf{F}\, dV \tag{9.362}$$

In view of (8.165), we have the following relation

$$\boldsymbol{\tau} = \mathbf{F}\cdot\mathbf{S}\cdot\mathbf{F}^T = \mathbf{P}\cdot\mathbf{F}^T = \mathbf{F}\cdot\mathbf{P}^T \tag{9.363}$$

as the Kirchhoff stress tensor given in Equation (8.148) is symmetric. Further, we have the relation

$$\mathbf{P} : \delta\mathbf{F} = \mathbf{S} : \left(\mathbf{F}^T\cdot\delta\mathbf{F}\right) = \mathbf{S} : \frac{1}{2}\left\{\left(\mathbf{F}^T\cdot\delta\mathbf{F}\right)^T + \mathbf{F}^T\cdot\delta\mathbf{F}\right\} = \mathbf{S} : \delta\mathbf{E} \tag{9.364}$$

Consequently, we consider the following conjugate relations in the internal virtual work,

$$\boldsymbol{\tau} : \delta\mathbf{e} = \mathbf{P} : \delta\mathbf{F} = \mathbf{S} : \delta\mathbf{E} \tag{9.365}$$

Next, we consider the inertia virtual work

$$\delta\mathcal{W}_{\text{kin}} = \int_{\mathcal{S}} \rho\ddot{\mathbf{u}} \cdot \delta\mathbf{u}\, dv = \int_{\mathcal{B}} \rho J \ddot{\mathbf{U}} \cdot \delta\mathbf{U}\, dV = \int_{\mathcal{B}} \rho_0 \ddot{\boldsymbol{\varphi}} \cdot \delta\boldsymbol{\varphi}\, dV \tag{9.366}$$

where $\rho J = \rho_0$. Note that $\ddot{\boldsymbol{\varphi}} = \ddot{\mathbf{u}} = \ddot{\mathbf{U}}, \quad \delta\boldsymbol{\varphi} = \delta\mathbf{u} = \delta\mathbf{U}$.

On the other hand, the external virtual work is given by

$$\delta\mathcal{W}_{\text{ext}} = \int_{\partial\mathcal{B}_\sigma} \bar{\mathbf{T}} \cdot \delta\boldsymbol{\varphi}(\mathbf{X}, t)\, dA + \int_{\mathcal{B}} \rho_0 \mathbf{B} \cdot \delta\boldsymbol{\varphi}(\mathbf{X}, t)\, dV \tag{9.367}$$

Therefore, the principle of virtual work in the Lagrangian description can be given by,

$$\int_{\mathcal{B}} \rho_0 \ddot{\boldsymbol{\varphi}} \cdot \delta\boldsymbol{\varphi}\, dV + \int_{\mathcal{B}} \mathbf{S} : \delta\mathbf{E}\, dV = \int_{\partial\mathcal{B}_\sigma} \bar{\mathbf{T}} \cdot \delta\boldsymbol{\varphi}\, dA + \int_{\mathcal{B}} \rho_0 \mathbf{B} \cdot \delta\boldsymbol{\varphi}\, dV \tag{9.368}$$

The above is the weak form for the initial boundary-value problems in elastodynamics.

Suppose that the admissible trial functions $\tilde{\boldsymbol{\varphi}}^h$ can be given by (see Equation (9.11) for details)

$$\tilde{\boldsymbol{\varphi}}^h(\mathbf{X}, t) = \mathbf{X} + \mathbf{N}(\mathbf{X})\, \mathbf{q}(t) \tag{9.369}$$

where $\mathbf{N}(\mathbf{X})$ denotes an interpolation matrix, and $\mathbf{q}(t)$ is the displacements as nodal variables. Then, the admissible variation of the displacements can be written as

$$\delta\tilde{\boldsymbol{\varphi}}^h(\mathbf{X}, t) = \mathbf{N}(\mathbf{X})\, \delta\mathbf{q}(t) \tag{9.370}$$

where $\delta\mathbf{q}(t)$ denotes the virtual displacement of nodal variables. Notice that the displacement field should be continuous along the inter element boundary. If we use the isoparmetric coordinates $\boldsymbol{\xi}$, we see that

$$\delta\tilde{\boldsymbol{\varphi}}^h(\boldsymbol{\xi}, t) = \mathbf{N}(\boldsymbol{\xi})\, \delta\mathbf{q}(t) \tag{9.371}$$

The acceleration can be also written as

$$\ddot{\tilde{\boldsymbol{\varphi}}}^h = \mathbf{N}\ddot{\mathbf{q}} \tag{9.372}$$

where $\ddot{\mathbf{q}}$ denotes the nodal accelerations. We shall discretize the space using the above approximation. The first term on the left-hand side of Equation (9.368) can be discretized approximately in matrix form as

$$\int_{\mathcal{B}} \rho_0 \ddot{\tilde{\boldsymbol{\varphi}}}^h \cdot \delta\tilde{\boldsymbol{\varphi}}^h\, dV = \delta\mathbf{q} \cdot \mathbf{M}\ddot{\mathbf{q}} \tag{9.373}$$

where the mass matrix can be rewritten as

$$\mathbf{M} = \int_{\mathcal{B}} \rho_0 \mathbf{N}^T \mathbf{N}\, dV \tag{9.374}$$

The external virtual work can be discretized as follows:

$$\int_{\partial\mathcal{B}_\sigma} \bar{\mathbf{T}} \cdot \delta\tilde{\boldsymbol{\varphi}}^h\, dA = \delta\mathbf{q} \cdot \int_{\partial\mathcal{B}_\sigma} \mathbf{N}^T \bar{\mathbf{T}}\, dA \tag{9.375}$$

and for the body force part we have

$$\int_{\mathcal{B}} \rho_0 \mathbf{B} \cdot \delta\tilde{\boldsymbol{\varphi}}^h dV = \delta\mathbf{q} \int_{\mathcal{B}} \rho_0 \mathbf{N}^T \mathbf{B}\, dV \tag{9.376}$$

Recall that the internal virtual work is

$$\int_{\mathcal{B}} \mathbf{S}(\mathbf{X}, t) : \delta\mathbf{E}(\mathbf{X}, t)\; dV = \int_{\mathcal{B}} \mathbf{P}(\mathbf{X}, t) : \delta\mathbf{F}(\mathbf{X}, t)\; dV \tag{9.377}$$

The variation of the deformation gradient tensor can be given as

$$\delta\tilde{\mathbf{F}}^h = \frac{\partial \delta\tilde{\boldsymbol{\varphi}}^h}{\partial \mathbf{X}} = \bar{\mathbf{B}}\delta\mathbf{q}, \quad \bar{\mathbf{B}} = \frac{\partial \mathbf{N}}{\partial \mathbf{X}} \tag{9.378}$$

where $\bar{\mathbf{B}}$ is called the strain-displacement matrix. Hence, we have

$$\int_{\mathcal{B}} \tilde{\mathbf{S}}^h : \delta\tilde{\mathbf{E}}^h\; dV = \delta\mathbf{q} \cdot \int_{\mathcal{B}} \bar{\mathbf{B}}^T \tilde{\mathbf{P}}^h\; dV \tag{9.379}$$

Consequently, the discretized equations of motion yields

$$\delta\mathbf{q} \cdot \left(\mathbf{M}\ddot{\mathbf{q}} + \int_{\mathcal{B}} \bar{\mathbf{B}}^T \tilde{\mathbf{P}}^h\; dV - \int_{\partial\mathcal{B}_\sigma} \mathbf{N}^T \bar{\mathbf{T}}\, dA - \int_{\mathcal{B}} \rho_0 \mathbf{N}^T \mathbf{B}\, dV \right) = 0 \tag{9.380}$$

Since $\delta\mathbf{q}$ is arbitrary, the above yields

$$\mathbf{M}\ddot{\mathbf{q}} + \mathbf{F}_{\text{int}} = \mathbf{F}_{\text{ext}} \tag{9.381}$$

where the internal force may be given as

$$\mathbf{F}_{\text{int}} = \int_{\mathcal{B}} \bar{\mathbf{B}}^T \tilde{\mathbf{P}}^h\; dV \tag{9.382}$$

and the external force can be expressed as follows:

$$\mathbf{F}_{\text{ext}} = \int_{\partial\mathcal{B}_\sigma} \mathbf{N}^T \bar{\mathbf{T}}\, dA + \int_{\mathcal{B}} \rho_0 \mathbf{N}^T \mathbf{B}\, dV \tag{9.383}$$

Equation (9.381) is known as the semi-discrete second-order differential finite element equations of motion. Note that Equation (9.381) is discrete in space but continuous in time. Therefore, in the time domain, the finite element equations of motion should be discretized, and the displacement should be updated. By use of an appropriate implicit or explicit time integration scheme, Equation (9.381) should be integrated to update the displacement.

If the material undergoes plastic deformation, the stress update algorithm must be employed to compute the second Piola-Kirchhoff stress tensor on the yield surface. Once the displacement is updated, for the next time step, the same total Lagrangian procedure can be carried out using the updated displacement.

9.3.2 Updated Lagrangian Formulation

In the literature, two representative and different formulations appear in the updated Lagrangian approach (McMeeking and Rice 1975). The first one is to use the Cauchy stress tensor as a stress measure. It is to be noted that the Cauchy stress tensor as an Eulerian measure is conjugate to the rate of deformation tensor in power, whereas it is conjugate to the Almansi strain tensor in energy. Thus, it is preferable to use the principle of virtual power for the finite element formulation as given in Equation (8.401). The rate of deformation tensor is linear in terms of velocities, whereas the Almansi strain tensor is nonlinear in terms of displacements. The second one is to use the second Piola-Kirchhoff stress tensor as a stress measure like the total Lagrangian approach, but a new reference configuration is updated as the configuration at time $t = \tau$ in Figure 9.33. The displacement at time $t = \tau + \Delta t$ is desired to be calculated.

Updated Lagrangian Formulation I We shall begin with the first updated Lagrangian approach.

We suppose that the configuration at time $t = \tau$ is known. It is reasonable to assume that the configuration at time $t = \tau$ is known, since the configuration can be obtained from the displacements updated in the previous time step. We now discretize the principle of virtual work shown in Equation (8.382) in the configuration at $t = \tau$. As in the total Lagrangian approach, we suppose that the admissible variation of the velocity can be discretized as

$$\delta \mathbf{v}\left(\tilde{\mathbf{X}}, t\right) = \mathbf{N}\left(\tilde{\mathbf{X}}\right)\delta\dot{\mathbf{q}}\,(t) \tag{9.384}$$

where the configuration at $t = \tau$ is given by

$$\tilde{\mathbf{X}} = \varphi\left(\tilde{\mathbf{X}}, t = \tau\right) \tag{9.385}$$

Accordingly we note that $\delta\dot{\mathbf{q}}\,(t)$ indicates the nodal velocity variables for the configuration at the time $t = \tau$. When we use the isoparmetric coordinates, we have

$$\delta \mathbf{v}\,(\xi, t) = \mathbf{N}\,(\xi)\,\delta\dot{\mathbf{q}}\,(t) \tag{9.386}$$

The acceleration can be also written as follows:

$$\ddot{\mathbf{u}} = \mathbf{N}\ddot{\mathbf{q}} \tag{9.387}$$

The inertial virtual power can be discretized approximately in matrix form as

$$\int_S \rho_{\tilde{\mathbf{X}}}\ddot{\mathbf{u}} \cdot \delta\mathbf{v}\, dv_{\tilde{\mathbf{X}}} = \delta\dot{\mathbf{q}} \cdot \mathbf{M}\ddot{\mathbf{q}} \tag{9.388}$$

where the mass matrix is given by

$$\mathbf{M} = \int_S \rho\mathbf{N}^T\mathbf{N}\, dv \tag{9.389}$$

and is unchanged during the deformation process. The external virtual power by the traction can be discretized as follows:

$$\int_{\partial S_\sigma} \bar{\mathbf{t}} \cdot \delta\mathbf{v}\, da_{\tilde{\mathbf{X}}} = \delta\dot{\mathbf{q}} \cdot \int_{\partial S_\sigma} \mathbf{N}^T\bar{\mathbf{t}}da_{\tilde{\mathbf{X}}} \tag{9.390}$$

and for the body force we have

$$\int_S \rho\left(\tilde{\mathbf{X}},t\right)\mathbf{b}\left(\tilde{\mathbf{X}},t\right)\cdot\delta\mathbf{v}\,dv_{\tilde{\mathbf{X}}} = \delta\dot{\mathbf{q}}\cdot\int_S \rho\left(\tilde{\mathbf{X}},t\right)\mathbf{N}^T\mathbf{b}\left(\tilde{\mathbf{X}},t\right)dv_{\tilde{\mathbf{X}}}$$
$$= \delta\dot{\mathbf{q}}\cdot\int_S \rho_{\tilde{\mathbf{X}}}(t)\,\mathbf{N}^T\mathbf{b}_{\tilde{\mathbf{X}}}(t)\,dv_{\tilde{\mathbf{X}}} \tag{9.391}$$

Finally, the internal virtual power is given by

$$\delta\dot{W}_{\text{int}}(\mathbf{v}) = \int_S \boldsymbol{\sigma}\left(\tilde{\mathbf{X}},t\right):\delta\mathbf{d}\,dv_{\tilde{\mathbf{X}}} = \int_S \boldsymbol{\sigma}\left(\tilde{\mathbf{X}},t\right):\frac{\partial\delta v\left(\tilde{\mathbf{X}},t\right)}{\partial\tilde{X}}dv_{\tilde{\mathbf{X}}} \tag{9.392}$$

The virtual deformation gradient tensor can be given as follows:

$$\frac{\partial\delta\mathbf{v}\left(\tilde{\mathbf{X}},t\right)}{\partial\tilde{\mathbf{X}}} = \frac{\partial\mathbf{N}}{\partial\tilde{\mathbf{X}}}\delta\mathbf{q} = \bar{\mathbf{B}}_{\tilde{\mathbf{X}}}\delta\mathbf{q} \tag{9.393}$$

where $\bar{\mathbf{B}}_{\tilde{\mathbf{X}}}$ is the strain-displacement matrix in the $\tilde{\mathbf{X}}$ configuration. Hence, we have

$$\int_S \boldsymbol{\sigma}\left(\tilde{\mathbf{X}},t\right):\delta\mathbf{d}\,dv_{\tilde{\mathbf{X}}} = \delta\mathbf{q}\cdot\int_S \bar{\mathbf{B}}_{\tilde{\mathbf{X}}}^T\boldsymbol{\sigma}\left(\tilde{\mathbf{X}},t\right)dv_{\tilde{\mathbf{X}}} \tag{9.394}$$

Consequently, the discretized equations of motion yields

$$\delta\dot{\mathbf{q}}\cdot\begin{bmatrix}\mathbf{M}\ddot{\mathbf{q}} + \int_S \bar{\mathbf{B}}_{\tilde{\mathbf{X}}}^T\boldsymbol{\sigma}\left(\tilde{\mathbf{X}},t\right)dv_{\tilde{\mathbf{X}}} \\ -\int_{\partial S_\sigma}\mathbf{N}^T\bar{\mathbf{t}}_{\tilde{\mathbf{X}}}(t)\,da_{\tilde{\mathbf{X}}} - \int_S \rho_{\tilde{\mathbf{X}}}\mathbf{N}^T\mathbf{b}_{\tilde{\mathbf{X}}}(t)\,dv_{\tilde{\mathbf{X}}}\end{bmatrix} = 0 \tag{9.395}$$

Since $\delta\dot{\mathbf{q}}$ is arbitrary, the above gives

$$\mathbf{M}\ddot{\mathbf{q}} + \mathbf{F}_{\tilde{\mathbf{X}}\text{int}} = \mathbf{F}_{\tilde{\mathbf{X}}\text{ext}} \tag{9.396}$$

where the internal force may be given as follows:

$$\mathbf{F}_{\tilde{\mathbf{X}}_{\text{int}}} = \int_S \bar{\mathbf{B}}_{\tilde{\mathbf{X}}}^T\boldsymbol{\sigma}\left(\tilde{\mathbf{X}},t\right)dv_{\tilde{\mathbf{X}}} \tag{9.397}$$

and the external force can be expressed as

$$\mathbf{F}_{\tilde{\mathbf{X}}_{\text{ext}}} = \int_{\partial S_\sigma}\mathbf{N}^T\bar{\mathbf{t}}_{\tilde{\mathbf{X}}}da_{\tilde{\mathbf{X}}} + \int_S \rho_{\tilde{\mathbf{X}}}\mathbf{N}^T\mathbf{b}_{\tilde{\mathbf{X}}}(t)\,dv_{\tilde{\mathbf{X}}} \tag{9.398}$$

For hypoelastic materials, objective stress rates such as the Jaumann stress rate or the Green-Naghdi stress rate should be employed to compute the Cauchy stress in Equation (9.397). For hypoelastic-plastic materials undergoing plastic deformation, an appropriate stress update algorithm is also employed in the formulation (Har 2007). Notice that Equation (9.396) is the semi-discrete second-order differential finite element equations of motion. Equation (9.396) should be discretized in time to update the displacement. By updating the displacement, a new reference configuration can be established. The typical example for this formulation may be found in the Belytschko-Lin-Tsay shell element formulation (Belytschko et al. 1984).

Updated Lagrangian Formulation II The second approach for the updated Lagrangian formulation is similar to the total Lagrangian approach except that the $\tilde{\mathbf{X}}$ configuration as the reference configuration rather than the undeformed configuration is employed. For convenience, we will use the following notation instead of $\tilde{\mathbf{X}}$ at time $t = \tau$ as

$$\mathbf{x} = \varphi\left(\tilde{\mathbf{X}}, t = \tau\right) \tag{9.399}$$

Thus, the internal virtual work can be written as

$$\delta W_{\text{int}}\left(\tilde{\mathbf{X}}\right) = \int_{\mathcal{B}} \mathbf{S}\left(\tilde{\mathbf{X}}, t\right) : \delta\mathbf{E}\left(\tilde{\mathbf{X}}, t\right) dV_{\tilde{\mathbf{X}}} = \int_{\mathcal{S}} \boldsymbol{\sigma}\left(\mathbf{x}, t\right) : \delta\mathbf{e}\, dv = \delta W_{\text{int}}\left(\mathbf{x}\right) \tag{9.400}$$

In view of Equation (9.397), we may express the internal virtual work as

$$\int_{\mathcal{B}} \mathbf{S}\left(\tilde{\mathbf{X}}, t\right) : \delta\mathbf{E}\left(\tilde{\mathbf{X}}, t\right) dV_{\tilde{\mathbf{X}}} = \int_{\mathcal{B}} \mathbf{P}\left(\tilde{\mathbf{X}}, t\right) : \delta\mathbf{F}\left(\tilde{\mathbf{X}}, t\right) dV_{\tilde{\mathbf{X}}} \tag{9.401}$$

The variation of the deformation gradient tensor is approximated as

$$\delta\tilde{\mathbf{F}}^h = \frac{\partial\delta\tilde{\boldsymbol{\varphi}}^h}{\partial\tilde{\mathbf{X}}} = \frac{\partial\mathbf{N}}{\partial\tilde{\mathbf{X}}}\delta\mathbf{q} = \bar{\mathbf{B}}_{\tilde{\mathbf{X}}}\delta\mathbf{q} \tag{9.402}$$

Hence, we have

$$\int_{\mathcal{B}} \tilde{\mathbf{P}}^h\left(\tilde{\mathbf{X}}, t\right) \cdot\cdot\, \delta\tilde{\mathbf{F}}^h\left(\tilde{\mathbf{X}}, t\right) dV_{\tilde{\mathbf{X}}} = \delta\mathbf{q} \cdot \int_{\mathcal{B}} \bar{\mathbf{B}}_{\tilde{\mathbf{X}}}^T\left(\tilde{\mathbf{F}}^h \cdot \tilde{\mathbf{S}}^h\right) dV_{\tilde{\mathbf{X}}} \tag{9.403}$$

Consequently, the discretized equations of motion yields

$$\delta\mathbf{q} \cdot \begin{bmatrix} \mathbf{M}\ddot{\mathbf{q}} + \int_{\mathcal{B}} \bar{\mathbf{B}}_{\tilde{\mathbf{X}}}^T\left(\tilde{\mathbf{F}}^h \cdot \tilde{\mathbf{S}}^h\right) dV_{\tilde{\mathbf{X}}} - \int_{\partial\mathcal{B}_\sigma} \mathbf{N}^T\bar{\mathbf{T}}_{\tilde{\mathbf{X}}}\left(t\right) dA_{\tilde{\mathbf{X}}} \\ -\int_{\mathcal{B}} \rho_{\tilde{\mathbf{X}}}\mathbf{N}_{\tilde{\mathbf{X}}}^T\mathbf{B}_{\tilde{\mathbf{X}}}\left(t\right) dV_{\tilde{\mathbf{X}}} \end{bmatrix} = 0 \tag{9.404}$$

Since $\delta\mathbf{q}$ is arbitrary, the above yields

$$\mathbf{M}\ddot{\mathbf{q}} + \mathbf{F}_{\tilde{\mathbf{X}}\text{int}} = \mathbf{F}_{\tilde{\mathbf{X}}\text{ext}}, \tag{9.405}$$

where the internal force may be given as follows:

$$\mathbf{F}_{\tilde{\mathbf{X}}_{\text{int}}} = \int_{\mathcal{B}} \bar{\mathbf{B}}_{\tilde{\mathbf{X}}}^T\left(\tilde{\mathbf{F}}^h \cdot \tilde{\mathbf{S}}^h\right) dV_{\tilde{\mathbf{X}}} \tag{9.406}$$

and the external force can be expressed as follows:

$$\mathbf{F}_{\tilde{\mathbf{X}}_{\text{ext}}} = \int_{\partial\mathcal{B}_\sigma} \mathbf{N}^T\bar{\mathbf{T}}_{\tilde{\mathbf{X}}}\left(t\right) dA_{\tilde{\mathbf{X}}} + \int_{\mathcal{B}} \rho_{\tilde{\mathbf{X}}}\mathbf{N}^T\mathbf{B}_{\tilde{\mathbf{X}}}\left(t\right) dV_{\tilde{\mathbf{X}}} \tag{9.407}$$

In Equation (9.397), the second Piola-Kirchhoff stress can be obtained by appropriate hyperelastic constitutive equations. Notice that Equation (9.405) is the semi-discrete second-order differential finite element equations of motion. Equation (9.405) should be discretized in time to update the displacement. By updating the displacement, a new reference configuration can be established. The typical example for this formulation can be found in the shell element formulation (Bathe and Bolourchi 1980).

9.4 SCALAR FORMALISMS: LAGRANGIAN AND HAMILTONIAN MECHANICS AND FINITE ELEMENT FORMULATIONS IN CONTINUOUS-BODY DYNAMICAL SYSTEMS

Note that Cauchy's equation motion is a vector representation of the governing equation of motion, and the virtual displacement is also a vector. The inner product operation between the virtual displacement and Cauchy's equation motion, i.e., the Bubnov-Galerkin weighted residual form, readily enables a descriptive scalar function representation.

Descriptive Scalar Function, the Lagrangian The Bubnov-Galerkin weighted residual form is given by the inner product of Cauchy's equation of motion and the virtual displacement. By use of Gauss's divergence theorem and integration by parts, the Bubnov-Galerkin weighted residual form gives a weak form. We have shown that the weak form leads to the modified Bubnov-Galerkin weighted residual form. For convenience, we confine attention to the case when the external forces are dead loads. For such cases, the external potential energy can be defined. Then, the principle of virtual work in elastodynamics is written as

$$\int_{\mathcal{B}} \rho_0 \mathbf{B}(\mathbf{X}) \cdot \delta\boldsymbol{\varphi}\, dV + \int_{\partial\mathcal{B}_\sigma} \bar{\mathbf{T}}\left(\mathbf{X}, \hat{\mathbf{N}}\right) \cdot \delta\boldsymbol{\varphi}\, dA - \int_{\mathcal{B}} \mathbf{P} : \delta\mathbf{F}\, dV - \int_{\mathcal{B}} \rho_0 \ddot{\boldsymbol{\varphi}} \cdot \delta\boldsymbol{\varphi}\, dV = 0 \tag{9.408}$$

We introduce the total potential and kinetic energy for the continuous-body dynamical system. The total potential energy $\mathcal{U}(\boldsymbol{\varphi}) : \mathcal{C} \to \mathbb{R}$ is defined as,

$$\mathcal{U}(\boldsymbol{\varphi}) = \int_{\mathcal{B}} \rho_0 \Psi(\mathbf{F})\, dV - \int_{\partial\mathcal{B}_\sigma} \bar{\mathbf{T}}\left(\mathbf{X}, \hat{\mathbf{N}}\right) \cdot \boldsymbol{\varphi}(\mathbf{X}, t)\, dA - \int_{\mathcal{B}} \rho_0 \mathbf{B}(\mathbf{X}) \cdot \boldsymbol{\varphi}(\mathbf{X}, t)\, dV \tag{9.409}$$

where $\Psi(\mathbf{F})$ denotes the Helmholtz free energy per unit mass in homogeneous materials as an elastic potential function in the reference configuration, $\mathbf{F}$ the deformation gradient tensor, $\bar{\mathbf{T}}\left(\mathbf{X}, \hat{\mathbf{N}}\right) : \mathcal{B} \times \mathbb{R}^3 \to \mathbb{R}^3$ the prescribed traction on the boundary surface $\partial\mathcal{B}$, $\rho_0(\mathbf{X}) : \mathcal{B} \to \mathbb{R}$ the referential density, $\mathbf{B}(\mathbf{X}) : \mathcal{B} \to \mathbb{R}^3$ the body force, and dA, dV the Lebesgue measures on the prescribed traction boundary $\partial\mathcal{B}_\sigma$ and in the body $\mathcal{B}$. The kinetic energy $\mathcal{K}(\dot{\boldsymbol{\varphi}}) : T_{\varphi}\mathcal{C} \to \mathbb{R}$ of the dynamical system of the deformable continuous body may be written in terms of the material velocity $\dot{\boldsymbol{\varphi}} \in T_{\varphi}\mathcal{C}$ as follows:

$$\mathcal{K}(\dot{\boldsymbol{\varphi}}) = \frac{1}{2} \int_{\mathcal{B}} \rho_0 \dot{\boldsymbol{\varphi}} \cdot \dot{\boldsymbol{\varphi}}\, dV \tag{9.410}$$

By introducing the total potential and kinetic energy for the continuous-body dynamical system, Equation (9.408) can be rewritten as

$$\delta\mathcal{K}(\dot{\boldsymbol{\varphi}}) - \delta\mathcal{U}(\boldsymbol{\varphi}) - \frac{d}{dt} \int_{\mathcal{B}} \rho_0 \dot{\boldsymbol{\varphi}} \cdot \delta\boldsymbol{\varphi}\, dV = 0 \tag{9.411}$$

which is a restatement of the principle of virtual work in terms of the kinetic and total potential energies. Next, we wish to introduce a descriptive scalar function, that is, the autonomous Lagrangian $\mathcal{L}(\boldsymbol{\varphi}, \dot{\boldsymbol{\varphi}}) : T\mathcal{C} \to \mathbb{R}$ in the Lagrangian mechanics framework. Then we have

$$\delta\mathcal{L}(\boldsymbol{\varphi}, \dot{\boldsymbol{\varphi}}) - \frac{d}{dt} \int_{\mathcal{B}} \rho_0 \dot{\boldsymbol{\varphi}} \cdot \delta\boldsymbol{\varphi}\, dV = 0 \tag{9.412}$$

where the autonomous Lagrangian is defined as

$$\begin{aligned}\mathcal{L}(\boldsymbol{\varphi}, \dot{\boldsymbol{\varphi}}) &= \mathcal{K}(\dot{\boldsymbol{\varphi}}) - \mathcal{U}(\boldsymbol{\varphi}) \\ &= \frac{1}{2}\int_{\mathcal{B}} \rho_0 \dot{\boldsymbol{\varphi}} \cdot \dot{\boldsymbol{\varphi}} dV - \int_{\mathcal{B}} \rho_0 \Psi(\mathbf{F})\, dV \\ &+ \int_{\partial\mathcal{B}_\sigma} \bar{\mathbf{T}}(\mathbf{X}, \hat{\mathbf{N}}) \cdot \boldsymbol{\varphi}(\mathbf{X}, t)\, dA + \int_{\mathcal{B}} \rho_0 \mathbf{B}(\mathbf{X}) \cdot \boldsymbol{\varphi}(\mathbf{X}, t)\, dV \\ &= \int_{\mathcal{B}} \bar{\mathcal{L}}(\boldsymbol{\varphi}, \dot{\boldsymbol{\varphi}}, \mathbf{F})\, dV + \int_{\partial\mathcal{B}_\sigma} \bar{\mathbf{T}}(\mathbf{X}, \hat{\mathbf{N}}) \cdot \boldsymbol{\varphi}(\mathbf{X}, t)\, dA\end{aligned} \tag{9.413}$$

where the Lagrangian density function $\bar{\mathcal{L}}(\boldsymbol{\varphi}, \dot{\boldsymbol{\varphi}}, \mathbf{F})$ is given by

$$\bar{\mathcal{L}}(\boldsymbol{\varphi}, \dot{\boldsymbol{\varphi}}, \mathbf{F}) = \frac{1}{2}\rho_0 \dot{\boldsymbol{\varphi}}(\mathbf{X}, t) \cdot \dot{\boldsymbol{\varphi}}(\mathbf{X}, t) - \rho_0 \Psi(\mathbf{F}) + \rho_0 \mathbf{B} \cdot \boldsymbol{\varphi}(\mathbf{X}, t) \tag{9.414}$$

In the Lagrangian framework, Equation (9.412) can be expressed in terms of the Lagrangian density function as

$$\begin{aligned}&\int_{\mathcal{B}} \left[\frac{\partial \bar{\mathcal{L}}}{\partial \boldsymbol{\varphi}} - \left(\mathrm{DIV}\frac{\partial \bar{\mathcal{L}}}{\partial \mathbf{F}}\right) - \frac{d}{dt}\left(\frac{\partial \bar{\mathcal{L}}}{\partial \dot{\boldsymbol{\varphi}}}\right)\right] \cdot \delta\boldsymbol{\varphi}\, dV \\ &+ \int_{\partial\mathcal{B}_\sigma} \left(\bar{\mathbf{T}}(\mathbf{X}, \hat{\mathbf{N}}) + \frac{\partial \bar{\mathcal{L}}}{\partial \mathbf{F}} \cdot \hat{\mathbf{N}}\right) \cdot \delta\boldsymbol{\varphi}\, dA = 0\end{aligned} \tag{9.415}$$

As the virtual displacement is arbitrary, we have the governing equation in the body

$$\frac{\partial \bar{\mathcal{L}}}{\partial \boldsymbol{\varphi}} - \frac{d}{dt}\left(\frac{\partial \bar{\mathcal{L}}}{\partial \dot{\boldsymbol{\varphi}}}\right) - \left(\mathrm{DIV}\frac{\partial \bar{\mathcal{L}}}{\partial \mathbf{F}}\right) = \mathbf{0}, \text{ for } (\mathbf{X}, t) \in \mathcal{B} \times \mathcal{T} \tag{9.416}$$

and the traction boundary condition on the prescribed traction boundary surface

$$\bar{\mathbf{T}}(\mathbf{X}, \hat{\mathbf{N}}) = -\frac{\partial \bar{\mathcal{L}}}{\partial \mathbf{F}} \cdot \hat{\mathbf{N}}, \text{ for } (\mathbf{X}, t) \in \partial\mathcal{B}_\sigma \times \mathcal{T} \tag{9.417}$$

Remark 9.4.1 We defer to the various approaches for space and time discretization and related discussions to a subsequent chapter.

Descriptive Scalar Function, the Hamiltonian Meanwhile, in the Hamiltonian mechanics framework Equation (9.411) can be represented as

$$\delta\left\{\int_{\mathcal{B}} \boldsymbol{\phi} \cdot \dot{\boldsymbol{\varphi}}\, dV - \mathcal{H}(\boldsymbol{\phi}, \boldsymbol{\varphi})\right\} - \frac{d}{dt}\int_{\mathcal{B}} \boldsymbol{\phi} \cdot \delta\boldsymbol{\varphi}\, dV = 0 \tag{9.418}$$

where the descriptive scalar function, that is, the autonomous Hamiltonian $\mathcal{H}(\boldsymbol{\phi}, \boldsymbol{\varphi}) : T^*\mathcal{C} \to \mathbb{R}$, is defined as

$$\begin{aligned}\mathcal{H}(\boldsymbol{\phi}, \boldsymbol{\varphi}) &= \mathcal{K}(\boldsymbol{\phi}) + \mathcal{U}(\boldsymbol{\varphi}) \\ &= \int_{\mathcal{B}} \bar{\mathcal{H}}(\boldsymbol{\phi}, \boldsymbol{\varphi}, \mathbf{F})\, dV - \int_{\partial\mathcal{B}_\sigma} \bar{\mathbf{T}}(\mathbf{X}, \hat{\mathbf{N}}) \cdot \boldsymbol{\varphi}(\mathbf{X}, t)\, dA\end{aligned} \tag{9.419}$$

where the Hamiltonian density function is given by

$$\bar{\mathcal{H}}(\boldsymbol{\phi}, \boldsymbol{\varphi}, \mathbf{F}) = \frac{1}{2\rho_0}\boldsymbol{\phi}(\mathbf{X}, t) \cdot \boldsymbol{\phi}(\mathbf{X}, t) + \rho_0 \Psi(\mathbf{F}) - \rho_0 \mathbf{B}(\mathbf{X}) \cdot \boldsymbol{\varphi}(\mathbf{X}, t) \tag{9.420}$$

Consequently, we have

$$\begin{aligned} &\int_{\partial\mathcal{B}_\sigma} \left[\left(\bar{\mathbf{T}}(\mathbf{X}, \hat{\mathbf{N}}) - \frac{\partial\bar{\mathcal{H}}}{\partial\mathbf{F}} \cdot \hat{\mathbf{N}}\right) \cdot \delta\boldsymbol{\varphi}\right] dA \\ &+ \int_{\mathcal{B}} \left[\left(\dot{\boldsymbol{\varphi}} - \frac{\partial\bar{\mathcal{H}}}{\partial\boldsymbol{\phi}}\right) \cdot \delta\boldsymbol{\phi} - \left(\dot{\boldsymbol{\phi}} + \frac{\partial\bar{\mathcal{H}}}{\partial\boldsymbol{\varphi}} - \mathrm{DIV}\frac{\partial\bar{\mathcal{H}}}{\partial\mathbf{F}}\right) \cdot \delta\boldsymbol{\varphi}\right] dV = 0 \end{aligned} \tag{9.421}$$

Since the two variations, $\delta\boldsymbol{\varphi}$ and $\delta\boldsymbol{\varphi}$, are arbitrary, we have the Hamiltonian density equations of motion, which hold inside the body

$$\dot{\boldsymbol{\phi}} + \frac{\partial\bar{\mathcal{H}}}{\partial\boldsymbol{\varphi}} = \mathrm{DIV}\frac{\partial\bar{\mathcal{H}}}{\partial\mathbf{F}}, \quad \dot{\boldsymbol{\varphi}} = \frac{\partial\bar{\mathcal{H}}}{\partial\boldsymbol{\phi}}, \quad \text{for } (\mathbf{X}, t) \in \mathcal{B} \times \mathcal{T} \tag{9.422}$$

and the natural boundary condition on the prescribed traction boundary surface $\partial\mathcal{B}_\sigma$

$$\bar{\mathbf{T}}(\mathbf{X}, \hat{\mathbf{N}}) = \frac{\partial\bar{\mathcal{H}}}{\partial\mathbf{F}} \cdot \hat{\mathbf{N}} \quad \text{for } (\mathbf{X}, t) \in \partial\mathcal{B}_\sigma \times \mathcal{T} \tag{9.423}$$

Remark 9.4.2 We defer to the various approaches for space and time discretization and related discussions to a subsequent chapter.

Remark 9.4.3 In addition, with the Total Energy representation of the equation of motion, Equation (9.411) can be written as

$$\delta\left\{\int_{\mathcal{B}} \rho_0\dot{\boldsymbol{\varphi}} \cdot \dot{\boldsymbol{\varphi}}\, dV - \mathcal{E}(\boldsymbol{\varphi}, \dot{\boldsymbol{\varphi}})\right\} - \frac{d}{dt}\int_{\mathcal{B}} \rho_0\dot{\boldsymbol{\varphi}} \cdot \delta\boldsymbol{\varphi}\, dV = 0 \tag{9.424}$$

where the descriptive scalar function, i.e., the autonomous Total Energy $\mathcal{E}(\boldsymbol{\varphi}, \dot{\boldsymbol{\varphi}}) : TC \to \mathbb{R}$, for dead load cases is defined as

$$\begin{aligned} \mathcal{E}(\boldsymbol{\varphi}, \dot{\boldsymbol{\varphi}}) &= \mathcal{K}(\boldsymbol{\phi}) + \mathcal{U}(\boldsymbol{\varphi}) \\ &= \int_{\mathcal{B}} \bar{\mathcal{E}}(\boldsymbol{\varphi}, \dot{\boldsymbol{\varphi}}, \mathbf{F})\, dV - \int_{\partial\mathcal{B}_\sigma} \bar{\mathbf{T}}(\mathbf{X}, \hat{\mathbf{N}}) \cdot \boldsymbol{\varphi}(\mathbf{X}, t)\, dA \end{aligned} \tag{9.425}$$

where the Total Energy density function $\bar{\mathcal{E}}(\boldsymbol{\varphi}, \dot{\boldsymbol{\varphi}}, \mathbf{F})$ is defined as

$$\bar{\mathcal{E}}(\boldsymbol{\varphi}, \dot{\boldsymbol{\varphi}}, \mathbf{F}) = \frac{1}{2}\rho_0\dot{\boldsymbol{\varphi}} \cdot \dot{\boldsymbol{\varphi}} + \rho_0\Psi(\mathbf{F}) - \rho_0\mathbf{B}(\mathbf{X}) \cdot \boldsymbol{\varphi} \tag{9.426}$$

Consequently, we have

$$\begin{aligned} &\int_{\mathcal{B}} \left[\mathrm{DIV}\frac{\partial\bar{\mathcal{E}}}{\partial\mathbf{F}} - \frac{\partial\bar{\mathcal{E}}}{\partial\boldsymbol{\varphi}} - \frac{d}{dt}\left(\frac{\partial\bar{\mathcal{E}}}{\partial\dot{\boldsymbol{\varphi}}}\right)\right] \cdot \delta\boldsymbol{\varphi}\, dV \\ &+ \int_{\partial\mathcal{B}_\sigma} \left[\left(\bar{\mathbf{T}}(\mathbf{X}, \hat{\mathbf{N}}) - \frac{\partial\bar{\mathcal{E}}}{\partial\mathbf{F}} \cdot \hat{\mathbf{N}}\right) \cdot \delta\boldsymbol{\varphi}\right] dA = 0 \end{aligned} \tag{9.427}$$

Since the virtual displacement $\delta\boldsymbol{\varphi}$ is arbitrary, we have the Total Energy density representation of the equation of motion, which holds inside the body as

$$\frac{d}{dt}\left(\frac{\partial\bar{\mathcal{E}}}{\partial\dot{\boldsymbol{\varphi}}}\right)+\frac{\partial\bar{\mathcal{E}}}{\partial\boldsymbol{\varphi}}-\left(\text{DIV}\,\frac{\partial\bar{\mathcal{E}}}{\partial\mathbf{F}}\right)=\mathbf{0},\ \text{for}\ (\mathbf{X},t)\in\mathcal{B}\times\mathcal{T} \tag{9.428}$$

and the natural boundary condition on the prescribed traction boundary surface $\partial\mathcal{B}_\sigma$

$$\bar{\mathbf{T}}(\mathbf{X},\hat{\mathbf{N}})=\frac{\partial\bar{\mathcal{E}}}{\partial\mathbf{F}}\cdot\hat{\mathbf{N}}\ \ \text{for}\ \ (\mathbf{X},t)\in\partial\mathcal{B}_\sigma\times\mathcal{T} \tag{9.429}$$

Remark 9.4.4 We defer to the various approaches for space and time discretization and related discussions to a subsequent chapter.

EXERCISES

9.1. Circular-Axial Bar undergoing Rotational and Longitudinal Vibration Let us consider an elastic bar undergoing axial and rotational deformation at the same time as shown in Figure 9.34. The cross-section, denoted by $A(x)$, is circular; hence it is assumed that there is no warping of the cross-sectional plane relative to the torque. Furthermore, about the twist axis the radial displacement is assumed to be proportional to the radial distance because the bar is uniform and the cross-section is circular. Elastic modulus and density are denoted by E and ρ respectively. As boundary conditions, at the one end $x=0$ the bar is built-in, and at the other end $x=L$ it is free. As loading conditions, the bar is subjected to both the distributed axial traction $\bar{b}(x)$ per unit length and the distributed torque $\bar{m}(x)$ per unit length over the length of the bar. At the end $x=L$, both the axial force F_L and the torque T_L are applied to the bar. Thus, we consider the bar exhibiting a dynamical behavior as not only a rotating bar with $J(x)$, the polar moment of area about the center of the cross-section, but also an axial bar with sectional stiffness $EA(x)$. As dependent variables, we have $\theta(x,t)$ and $u(x,t)$, which represent the angle of rotation about the axial axis, and the axial displacement at the point x and at the time t.

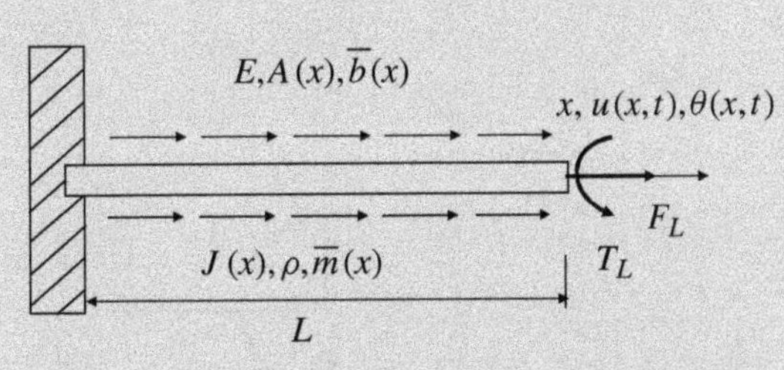

Figure 9.34. Circular-axial bar undergoing rotating and longitudinal vibration

(a) Draw the free-body diagram for the circular axial bar undergoing Rotational and Longitudinal Vibration.

(b) Establish the strong form by use of D'Alembert's principle and the free-body diagram.
(c) Derive the weak form via the Bubnov-Galerkin weighted residual form.
(d) Obtain the finite element equation for the linear axial bar problem.

9.2. Euler-Bernoulli Beam under Axial and Transverse Loads Let us consider the dynamical behavior of the Euler-Bernoulli beam under axial and transverse loads as shown in Figure 9.35. The beam is cantilevered; hence it is subject to the end shearing force Q_L, the axial load F_L and the end moment M_L at the end $x = L$. The material is assumed to be linearly elastic, isotropic and homogeneous. The beam bending is subjected to the distributed transverse load $\bar{\omega}$ per unit length in the z-direction and the distributed axial traction $\bar{b}(x)$ per unit length in the x-direction. Then, the dynamical behavior of the Euler-Bernoulli beam is governed by two differential equations in terms of two dependent variables, namely, the transverse deflection $w(x, t)$ and the axial displacement $u(x, t)$. Let $I(x)$ denote the second moment of inertia about the neutral axis, $\rho(x)$ density, and $A(x)$ the cross-sectional area.

(a) Draw the free-body diagram for the Euler-Bernoulli beam under axial and transverse loads.
(b) Establish the strong form by use of D'Alembert's principle and the free-body diagram.
(c) Derive the weak form via the Bubnov-Galerkin weighted residual form.
(d) Obtain the finite element equation for the linear Euler-Bernoulli beam problem.

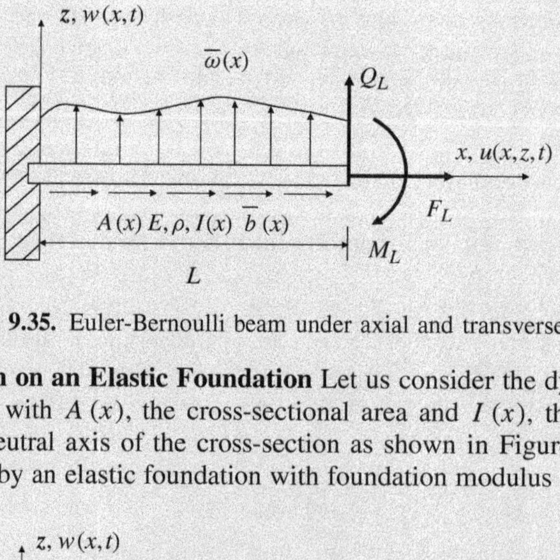

Figure 9.35. Euler-Bernoulli beam under axial and transverse loads

9.3. Timoshenko Beam on an Elastic Foundation Let us consider the dynamical behavior of the Timoshenko beam with $A(x)$, the cross-sectional area and $I(x)$, the area moment per unit length about the neutral axis of the cross-section as shown in Figure 9.36. The Timoshenko beam is supported by an elastic foundation with foundation modulus k_f over the length of the

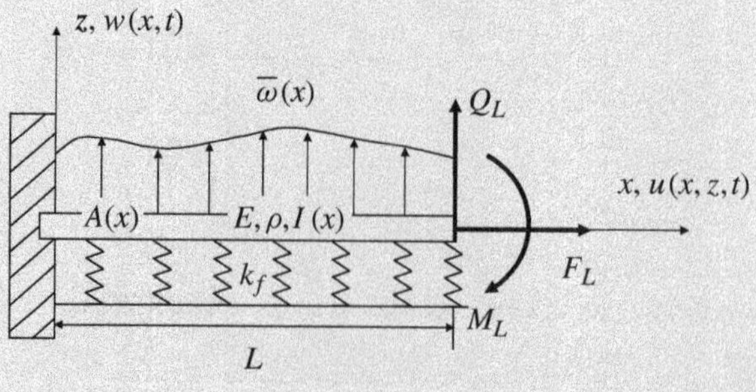

Figure 9.36. Timoshenko Beam on an Elastic Foundation

beam and subject to the end shearing force Q_L and the end moment M_L at the end $x = L$. In Timoshenko beam theory, there are two dependent variables such as the transverse deflection $y(x, t)$ and the angle of rotation $\varphi(x, t)$ due to bending moment acting on the beam supporting the transversely distributed load $\bar{\omega}(x)$.

(a) Draw the free-body diagram for the Timoshenko beam on an elastic foundation.
(b) Establish the strong form by use of D'Alembert's principle and the free-body diagram.
(c) Derive the weak form via the Bubnov-Galerkin weighted residual form.
(d) Obtain the finite element equation for the linear Timoshenko beam problem.

9.4. Kirchhoff-Love Plate under In-plane and Transverse Loads Let us consider the dynamical behavior of the Kirchhoff-Love plate subjected to in-plane loads (N_x, N_y, N_{xy}) and transverse loads. Hence, instead of Equation (9.80), the displacement field is given by

$$u(x, y, z, t) = u_0(x, y, t) - z\varphi(x, y, t) = u_0(x, y, t) - z\frac{\partial w_0(x, y, t)}{\partial x},$$
$$v(x, y, z, t) = v_0(x, y, t) - z\phi(x, y, t) = v_0(x, y, t) - z\frac{\partial w_0(x, y, t)}{\partial y},$$
$$w(x, y, t) = w_0(x, y, t)$$

where $u_0(x, y, t)$, $v_0(x, y, t)$ and $w_0(x, y, t)$ represent the displacements of a point on the midsurface of the plate.

(a) Draw the free-body diagram for the Kirchhoff-Love plate.
(b) Establish the strong form by use of D'Alembert's principle and the free-body diagram.
(c) Derive the weak form via the Bubnov-Galerkin weighted residual form.
(d) Obtain the finite element equation for the linear Kirchhoff-Love plate.

9.5. Reissner-Mindlin Plate under In-plane and Transverse Loads Let us consider the dynamical behavior of the Reissner-Mindlin plate subjected to in-plane loads (N_x, N_y, N_{xy}) and transverse loads. Hence, instead of Equation (9.121), the displacement field is given by

$$u(x, y, z, t) = u_0(x, y, t) + z\theta_y(x, y, t),$$
$$v(x, y, z, t) = v_0(x, y, t) - z\theta_x(x, y, t),$$
$$w(x, y, t) = w_0(x, y, t)$$

where $u_0(x, y, t)$, $v_0(x, y, t)$ and $w_0(x, y, t)$ represent the displacements of a point on the midsurface of the plate.

(a) Draw the free-body diagram for the Reissner-Mindlin plate.
(b) Establish the strong form by use of D'Alembert's principle and the free-body diagram.
(c) Derive the weak form via the Bubnov-Galerkin weighted residual form.
(d) Obtain the finite element equation for the linear Reissner-Mindlin plate.

9.6. One-Dimensional Truss Structure Consider one two-node linear element with length L. The distributed axial traction per unit length $\bar{b}(x) = b_0(1 + \frac{x}{L})$ is applied to the element. Determine the work-equivalent loads.

9.7. Consider one three-node quadratic element with length L. The distributed axial traction per unit length $\bar{b}(x) = b_0(1 + \frac{x}{L})$ is applied to the element. Determine the work-equivalent loads.

9.8. Consider a one-dimensional pin-jointed structure, which consists of two two-node elements. The two elements have different material properties each other. Node 1 is held fixed, whereas at Node 2 and Node 3, axial forces F_2 and F_3 are applied as shown in Figure 9.37. The distributed axial traction per unit length is given by $\bar{b}(x) = b_0(1 - x/2L)$. Determine the displacements at Node 2 and Node 3 and the axial stresses and strains of Element 1 and Element 2.

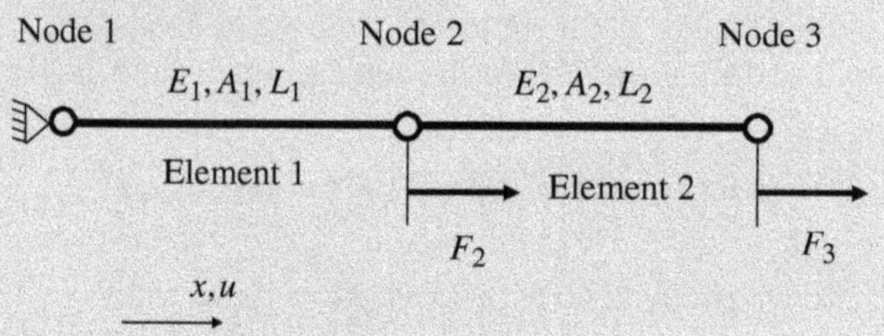

Figure 9.37. One-Dimensional truss structure

9.9. In Figure 9.37, we suppose that $E_1 = E_2 = E$, $A_1 = A$, $A_2 = A\left(1 + \frac{x}{L}\right)$, and $L_2 = L_1 = L$. Establish the global consistent mass matrix of the truss structure with the material density ρ. Construct the finite element equation of motion for the one-dimensional truss structure.

9.10. Two-Dimensional Plane Truss Structure Consider a two-dimensional planar truss structure, where F axial bars are pin-jointed and carries axial loads. It is assumed that each element has constant area and Young's modulus for convenience. The stiffness matrix of the two-node element in the local coordinate system shown in Figure 9.38 is given by

$$\hat{\mathbf{K}}^e\hat{\mathbf{u}}^e = \frac{E_eA_e}{L_e}\begin{bmatrix} 1 & 0 & -1 & 0 \\ 0 & 0 & 0 & 0 \\ -1 & 0 & 1 & 0 \\ 0 & 0 & 0 & 0 \end{bmatrix}\begin{Bmatrix} \hat{u}_1 \\ \hat{v}_1 \\ \hat{u}_2 \\ \hat{v}_2 \end{Bmatrix}$$

The displacement vector at Node 1 can be represented in global and local coordinates as

$$\hat{\mathbf{u}}_1^e = \hat{u}_1\hat{\mathbf{e}}_1 + \hat{v}_1\hat{\mathbf{e}}_2 = u_1\mathbf{e}_1 + v_1\mathbf{e}_2 = \mathbf{u}_1^e$$

Hence, in matrix form we have the relation between the nodal displacement vectors at Node 1 in global and local coordinates

$$\begin{Bmatrix} \hat{u}_1 \\ \hat{v}_1 \end{Bmatrix} = \begin{bmatrix} \hat{\mathbf{e}}_1 \cdot \mathbf{e}_1 & \hat{\mathbf{e}}_1 \cdot \mathbf{e}_2 \\ \hat{\mathbf{e}}_2 \cdot \mathbf{e}_1 & \hat{\mathbf{e}}_2 \cdot \mathbf{e}_2 \end{bmatrix}\begin{Bmatrix} u_1 \\ v_1 \end{Bmatrix}$$

where the directional cosines are determined by

$$\hat{\mathbf{e}}_1 \cdot \mathbf{e}_1 = \cos\theta_x = \frac{x_2 - x_1}{L_e}$$

$$\hat{\mathbf{e}}_1 \cdot \mathbf{e}_2 = \cos\theta_y = \frac{y_2 - y_1}{L_e}$$

$$L_e = \sqrt{(x_2 - x_1)^2 + (y_2 - y_1)^2}$$

As a consequence, we can write

$$\left\{ \begin{array}{c} \hat{u}_1 \\ \hat{v}_1 \end{array} \right\} = \left[\begin{array}{cc} \cos\theta_x & \cos\theta_y \\ -\cos\theta_y & \cos\theta_x \end{array} \right] \left\{ \begin{array}{c} u_1 \\ v_1 \end{array} \right\}$$

In matrix form, the relation between the displacement variables of the element e in the local and global coordinate systems is given by

$$\left\{ \begin{array}{c} \hat{u}_1 \\ \hat{v}_1 \\ \hat{u}_2 \\ \hat{v}_2 \end{array} \right\} = \left[\begin{array}{cccc} \cos\theta_x & \cos\theta_y & 0 & 0 \\ -\cos\theta_y & \cos\theta_x & 0 & 0 \\ 0 & 0 & \cos\theta_x & \cos\theta_y \\ 0 & 0 & -\cos\theta_y & \cos\theta_x \end{array} \right] \left\{ \begin{array}{c} u_1 \\ v_1 \\ u_2 \\ v_2 \end{array} \right\}$$

which can be rewritten as

$$\hat{\mathbf{u}}^e = \mathbf{Q}\mathbf{u}^e$$

where $\mathbf{Q}$ represents the orthogonal transformation matrix. Show that the stiffness matrix $\mathbf{K}^e$ in the global coordinate system can be written as

$$\mathbf{K}^e = \mathbf{Q}^T \hat{\mathbf{K}}^e \mathbf{Q} = \frac{E_e A_e}{L_e} \left[\begin{array}{cc} \mathbf{k}_2 & -\mathbf{k}_2 \\ -\mathbf{k}_2 & \mathbf{k}_2 \end{array} \right]$$

$$\mathbf{k}_2 = \left[\begin{array}{cc} \cos^2\theta_x & \cos\theta_x \cos\theta_y \\ \cos\theta_x \cos\theta_y & \cos^2\theta_y \end{array} \right]$$

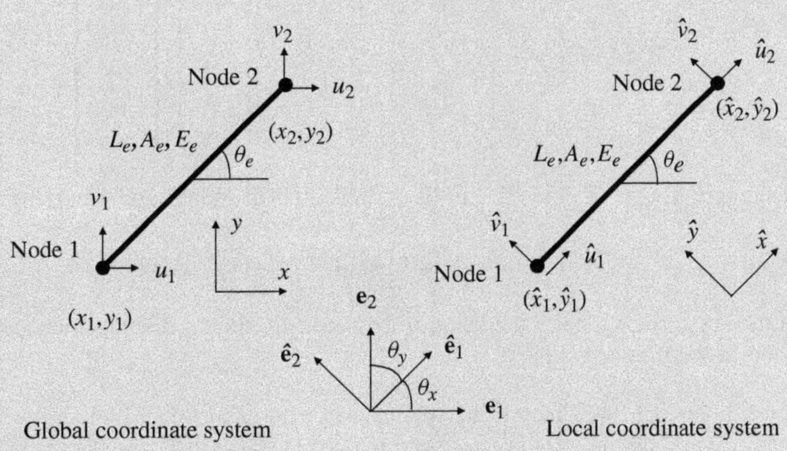

Figure 9.38. Displacements in the global and local coordinates for the two-dimensional planar truss structure

9.11. Consider a two-dimensional truss structure shown in Figure 9.39. All axial bars are pin-jointed and carries axial loads. Find the expression of the displacements at Node 2 in terms of the pin load F.

9.12. In Figure 9.39, all elements with constant area A have the material properties, i.e. EA. Establish the global consistent mass matrix of the plane truss structure with the material density ρ. Construct the finite element equation of motion for the two-dimensional truss structure.

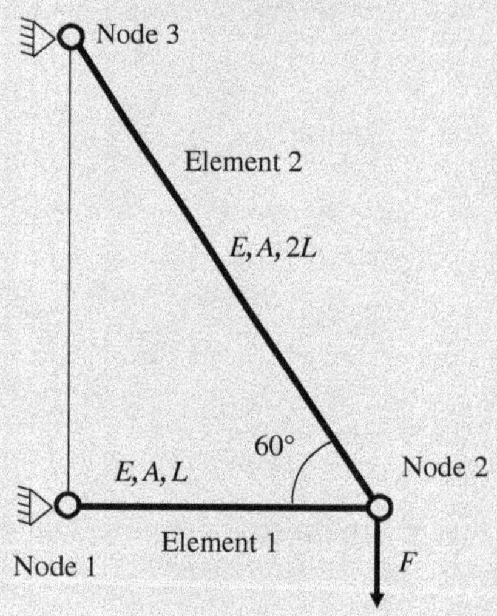

Figure 9.39. Plane truss under a pin load

9.13. Three-Dimensional Space Truss Structure Consider a three-dimensional space truss structure, where all axial bars are pin-jointed and carries axial loads; the stiffness matrix of the two-node element in the local coordinate system is given by

$$\hat{\mathbf{K}}^e \hat{\mathbf{u}}^e = \frac{E_e A_e}{L_e} \begin{bmatrix} 1 & 0 & 0 & -1 & 0 & 0 \\ 0 & 0 & 0 & 0 & 0 & 0 \\ 0 & 0 & 0 & 0 & 0 & 0 \\ -1 & 0 & 0 & 1 & 0 & 0 \\ 0 & 0 & 0 & 0 & 0 & 0 \\ 0 & 0 & 0 & 0 & 0 & 0 \end{bmatrix} \begin{Bmatrix} \hat{u}_1 \\ \hat{v}_1 \\ \hat{w}_1 \\ \hat{u}_2 \\ \hat{v}_2 \\ \hat{w}_2 \end{Bmatrix}$$

The displacement vector at Node 1 can be represented in global and local coordinates as

$$\hat{\mathbf{u}}_1^e = \hat{u}_1 \hat{\mathbf{e}}_1 + \hat{v}_1 \hat{\mathbf{e}}_2 + \hat{w}_1 \hat{\mathbf{e}}_3 = u_1 \mathbf{e}_1 + v_1 \mathbf{e}_2 + w_1 \mathbf{e}_3 = \mathbf{u}_1^e$$

Hence, in matrix form we have the relation between the nodal displacement vectors at Node 1 in global and local coordinates

$$\begin{Bmatrix} \hat{u}_1 \\ \hat{v}_1 \\ \hat{w}_1 \end{Bmatrix} = \begin{bmatrix} \hat{\mathbf{e}}_1 \cdot \mathbf{e}_1 & \hat{\mathbf{e}}_1 \cdot \mathbf{e}_2 & \hat{\mathbf{e}}_1 \cdot \mathbf{e}_3 \\ \hat{\mathbf{e}}_2 \cdot \mathbf{e}_1 & \hat{\mathbf{e}}_2 \cdot \mathbf{e}_2 & \hat{\mathbf{e}}_2 \cdot \mathbf{e}_3 \\ \hat{\mathbf{e}}_3 \cdot \mathbf{e}_1 & \hat{\mathbf{e}}_3 \cdot \mathbf{e}_2 & \hat{\mathbf{e}}_3 \cdot \mathbf{e}_3 \end{bmatrix} \begin{Bmatrix} u_1 \\ v_1 \\ w_1 \end{Bmatrix}$$

where the directional cosines are determined by

$$\hat{\mathbf{e}}_1 \cdot \mathbf{e}_1 = \cos\theta_x = \frac{x_2 - x_1}{L_e}$$
$$\hat{\mathbf{e}}_1 \cdot \mathbf{e}_2 = \cos\theta_y = \frac{y_2 - y_1}{L_e}$$
$$\hat{\mathbf{e}}_1 \cdot \mathbf{e}_3 = \cos\theta_z = \frac{z_2 - z_1}{L_e}$$
$$L_e = \sqrt{(x_2 - x_1)^2 + (y_2 - y_1)^2 + (z_2 - z_1)^2}$$

The relation between the displacement variables in the local and global coordinate systems is given by

$$\hat{\mathbf{u}}^e = \mathbf{Q}\mathbf{u}^e$$

where the transformation matrix $\mathbf{Q}$ is given by

$$\begin{Bmatrix} \hat{u}_1 \\ \hat{v}_1 \\ \hat{w}_1 \\ \hat{u}_2 \\ \hat{v}_2 \\ \hat{w}_2 \end{Bmatrix} = \begin{bmatrix} \hat{\mathbf{e}}_1 \cdot \mathbf{e}_1 & \hat{\mathbf{e}}_1 \cdot \mathbf{e}_2 & \hat{\mathbf{e}}_1 \cdot \mathbf{e}_3 & 0 & 0 & 0 \\ \hat{\mathbf{e}}_2 \cdot \mathbf{e}_1 & \hat{\mathbf{e}}_2 \cdot \mathbf{e}_2 & \hat{\mathbf{e}}_2 \cdot \mathbf{e}_3 & 0 & 0 & 0 \\ \hat{\mathbf{e}}_3 \cdot \mathbf{e}_1 & \hat{\mathbf{e}}_3 \cdot \mathbf{e}_2 & \hat{\mathbf{e}}_3 \cdot \mathbf{e}_3 & 0 & 0 & 0 \\ 0 & 0 & 0 & \hat{\mathbf{e}}_1 \cdot \mathbf{e}_1 & \hat{\mathbf{e}}_1 \cdot \mathbf{e}_2 & \hat{\mathbf{e}}_1 \cdot \mathbf{e}_3 \\ 0 & 0 & 0 & \hat{\mathbf{e}}_2 \cdot \mathbf{e}_1 & \hat{\mathbf{e}}_2 \cdot \mathbf{e}_2 & \hat{\mathbf{e}}_2 \cdot \mathbf{e}_3 \\ 0 & 0 & 0 & \hat{\mathbf{e}}_3 \cdot \mathbf{e}_1 & \hat{\mathbf{e}}_3 \cdot \mathbf{e}_2 & \hat{\mathbf{e}}_3 \cdot \mathbf{e}_3 \end{bmatrix} \begin{Bmatrix} u_1 \\ v_1 \\ w_1 \\ u_2 \\ v_2 \\ w_2 \end{Bmatrix}$$

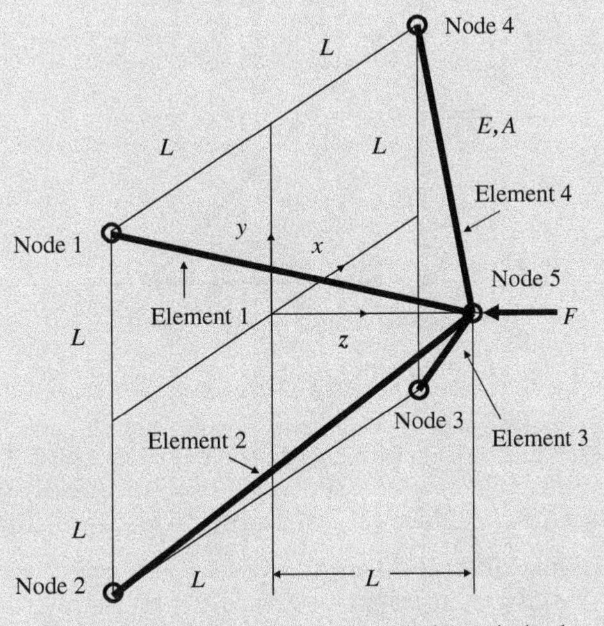

Figure 9.40. Space truss structure under a pin load

Show that the stiffness matrix $\mathbf{K}^e$ in the global coordinate system can be written as

$$\mathbf{K}^e = \mathbf{Q}^T \hat{\mathbf{K}}^e \mathbf{Q} = \frac{E_e A_e}{L_e} \begin{bmatrix} \mathbf{k}_3 & -\mathbf{k}_3 \\ -\mathbf{k}_3 & \mathbf{k}_3 \end{bmatrix}$$

$$\mathbf{k}_3 = \begin{bmatrix} \cos^2\theta_x & \cos\theta_x\cos\theta_y & \cos\theta_x\cos\theta_z \\ \cos\theta_y\cos\theta_x & \cos^2\theta_y & \cos\theta_y\cos\theta_z \\ \cos\theta_x\cos\theta_x & \cos\theta_z\cos\theta_y & \cos^2\theta_z \end{bmatrix}$$

9.14. Consider the space truss structure shown in Figure 9.40. All axial bars are pin-jointed and carry axial loads. Find the nodal displacement at Node 5 in terms of the pin load F and the stress and strain for each element. All elements have the same Young's modulus E and area A (Nodes 1,2,3, and 4 are pinned).

9.15. In Figure 9.40, all elements with constant area A have the material properties, i.e. EA. Establish the global consistent mass matrix of the space truss structure with the material density ρ. Construct the finite element equation of motion for the three-dimensional truss structure (Nodes 1,2,3, and 4 are pinned).

9.16. Two-dimensional Constant-Strain Triangular Element Consider the constant-strain triangular element shown in Figure 9.41. The material is in plane strain state. Determine the element stiffness matrix per unit length. The uniform surface traction is applied along the edge 1-3. Calculate the work-equivalent loads.

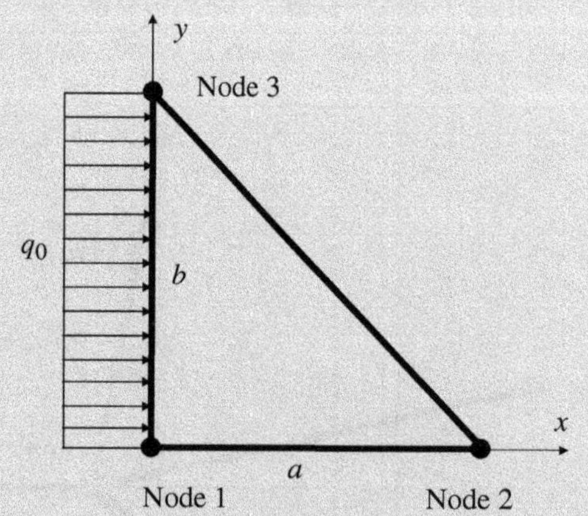

Figure 9.41. Constant-strain triangular element with the uniform edge load

9.17. Consider the constant-strain triangular element shown in Figure 9.42. The plate with thickness t is in plane stress state. Find the global stiffness matrix. The linearly varying surface traction is applied along the edge 2-3 of Element 1. Calculate the work-equivalent loads for Element 1.

9.18. Two-dimensional Quadrilateral Element Consider the quadrilateral element shown in Figure 9.43. The material is in plane strain state. Determine the element stiffness matrix per unit length. The uniform surface traction is applied along the edge 2-3. Calculate the work-equivalent loads.

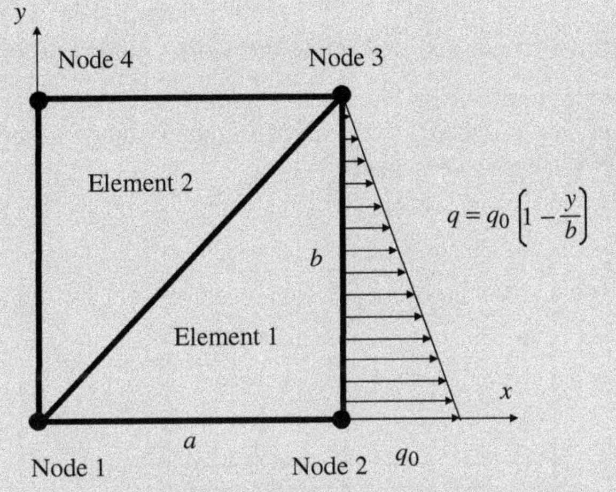

Figure 9.42. Constant-strain triangular element with the linearly varying edge load

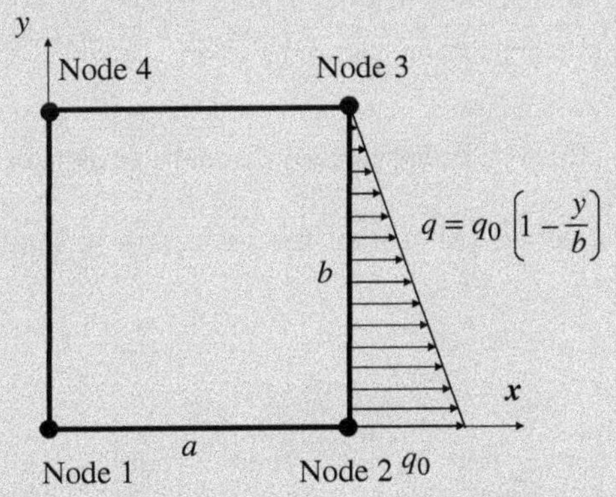

Figure 9.43. Quadrilateral element with the uniform edge load

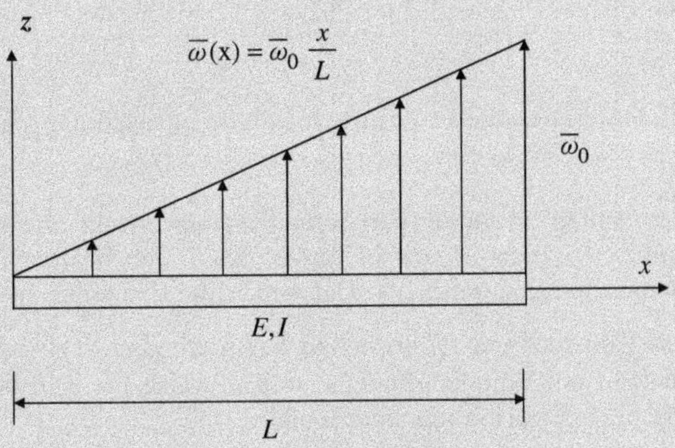

Figure 9.44. Euler-Bernoulli beam supporting a transversely distributed load

9.19. Euler-Bernoulli Beam Consider the Euler-Bernoulli beam shown in Figure 9.44. The beam supports the transversely distributed load per unit length $\overline{w}(x) = \overline{w}_0 \frac{x}{L}$. Use the two-node Euler-Bernoulli beam element and determine the work-equivalent loads.

9.20. Consider the Euler-Bernoulli beam shown in Figure 9.45. The beam supports the distributed bending moment per unit length $m(x) = m_0 \frac{x}{L}$. Use the two-node Euler-Bernoulli beam element and determine the work-equivalent loads.

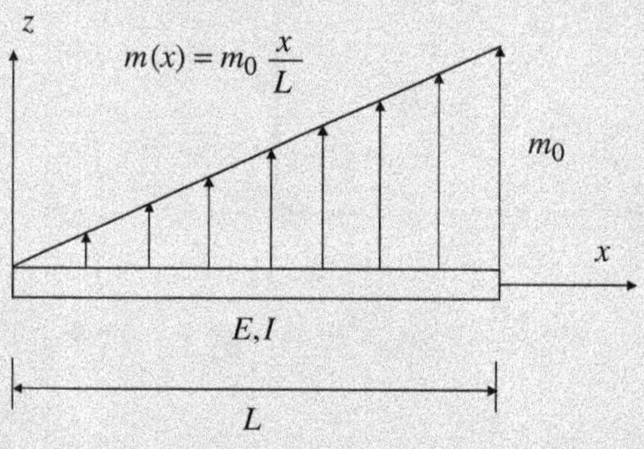

Figure 9.45. Euler-Bernoulli beam supporting a distributed bending moment

9.21. Consider the propped cantilever Euler-Bernoulli beam shown in Figure 9.46. The beam supports various loads as shown in Figure 9.46. The beam is modeled with the two two-node Euler-Bernoulli beam elements. Determine the deflection at the end of the beam.

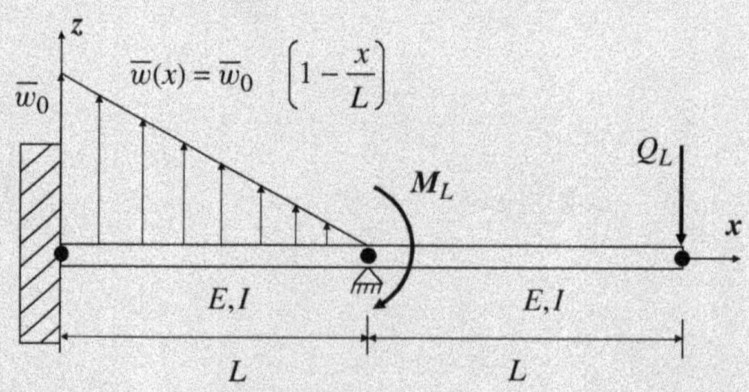

Figure 9.46. Propped cantilever Euler-Bernoulli beam supporting various loads

9.22. Consider the spring-supported cantilever Euler-Bernoulli beam shown in Figure 9.47. The beam supports various loads as shown in Figure 9.47. The beam is modeled with the two two-node Euler-Bernoulli beam elements. Determine the deflection at the end of the beam.

9.23. Timoshenko Beam Consider one Timoshenko beam element. The beam supports the transversely distributed load per unit length $\overline{w}(x) = \overline{w}_0 \frac{x}{L}$. Use the two-node Timoshenko beam element and determine the work-equivalent loads.

9.24. Consider one Timoshenko beam element supporting the distributed bending moment $m(x)$. The distributed bending moment per unit length is given by $m(x) = m_0 \frac{x}{L}$. Use the two-node Timoshenko beam element and determine the work-equivalent loads.

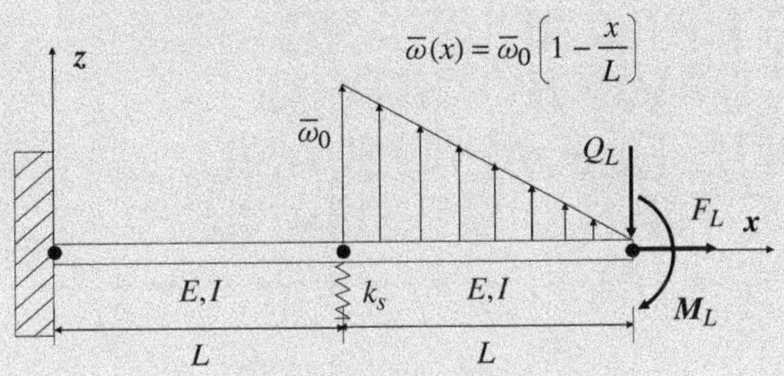

Figure 9.47. Spring-supported cantilever Euler-Bernoulli beam supporting various loads

9.25. Assume that the propped cantilever beam shown in Figure 9.46 is modeled by the two two-node Timoshenko beam elements. Determine the deflection at the end of the beam.

9.26. Assume that the spring-supported cantilever beam shown in Figure 9.47 is modeled with the two two-node Timoshenko beam elements. Determine the deflection at the end of the beam.

CHAPTER TEN

HAMILTON'S PRINCIPLE AND HAMILTON'S LAW OF VARYING ACTION: FINITE ELEMENTS AND SOLID/STRUCTURAL MECHANICS

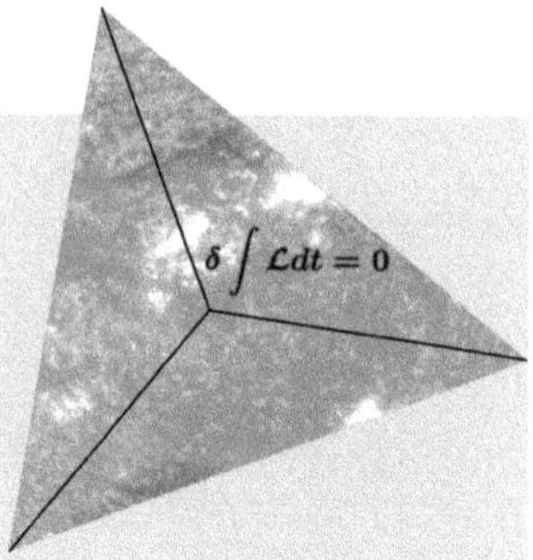

The focus of Chapter 10 is on continuum elastodynamics, and the related finite element developments are described via Hamilton's principle as well as Hamilton's law of varying action. Both vector and scalar formalisms are highlighted. The finite element formulations for a variety of structural dynamical systems such as an axial bar, rotating bar, Euler-Bernoulli beam, Timoshenko beam, and Reissner-Mindlin plate, are demonstrated for illustration. The resulting developments provide a different viewpoint and perspective in contrast to the previous chapter, but yield the same equivalent finite element representations.

10.1 INTRODUCTION

For modeling and simulation in elastodynamics applications, traditional numerical discretization methods in space and time, have been mostly developed and based upon Cauchy equations of motion which reflect the statement of the principle of the balance of linear momentum, or those governing equations derived from variational principles such as Hamilton's principle in dynamics (Simkins 1981), (Marsden and Hughes 1983, page 13), (Holzapfel 2000, page 391), and so on.

In computational mechanics, the basis of numerical methods such as finite element developments, as evidenced from the literature and historically from existing books is predominantly via the principle of balance of linear momentum (Newton's second law with vector quantities). It involves the Cauchy equations of motion in the sense of Newtonian mechanics framework to develop the principle of virtual work, which can be derived from the Bubnov-Galerkin weighted residual form. This is the traditional paradigm. One can argue that in order to ensure that the discretized equations inherit the same physics as in the continuous governing equations in space and time, it is not readily straightforward to develop such proofs via traditional practices with vector formalism since no scalar function readily exists to capitalize on theorems such as Noether's theorem (Noether 1918). In this regard, we particularly stress the significance of various scalar formalisms besides the vector formalism as in traditional practices.

Historically for finite element developments for elastodynamics applications, not much is described in the literature utilizing representations involving descriptive scalar functions. Indeed, there exist frameworks involving scalar function representations involving the Lagrangian and the Hamiltonian; however

Advances in Computational Dynamics of Particles, Materials and Structures, First Edition. Jason Har and Kumar K. Tamma.

the various efforts are towards addressing certain time discretization aspects with particular desired attributes for integrating the equations of motion. Nonetheless, for describing numerical developments in space such as the finite element method, and to therein enable numerical discretizations for applications typical of elastodynamics, firstly, the scalar formalisms are neither trivial nor popular. In contrast to Newtonian mechanics and traditional practices, for these frameworks which are associated with the concept of generalized coordinates, not much has been discussed regarding how to foster numerical discretizations. For these frameworks which are associated with descriptive scalar functions, Noether's theorem states that the spatial translation symmetry of the Lagrangian in the configuration space gives conservation of linear momentum, the rotational symmetry gives conservation of angular momentum, and the autonomous Lagrangian has symmetry in time. Likewise, it is well known that the Hamiltonian in the phase space has also the same invariance properties (Abraham et al. 1988; Arnold 1989) via the so-called Hamiltonian version of Noether's theorem (Noether 1918). As a consequence, it is now fairly straightforward to ensure that the discretized equations also inherit the same physics as in the continuous governing equations for space and time. Some background on Hamilton's principle, and consequently Hamilton's law of varying action is noteworthy.

It is widely recognized and well known that Hamilton's principle (Fung 1965; Gelfand and Fomin 1963; Goldstein 2002; Lanczos 1970; Langhaar 1962; Meirovitch 2004; Osgood 1937; Pars 1965; Washizu 1982; Yourgrau and Mandelstam 1968), one of the variational principles, is a fundamental axiom in mechanics and science. It claims that extremizing the action as a functional, first both defined and referred to as the principal function by Hamilton (Hamilton 1834a, b; Hamilton 1835), first naturally leads to Lagrange's equations of motion as the governing equations for N-body systems, as well as its ability to be able to derive both the governing equations and boundary conditions in continuum-elastodynamics that are continuous in space and time. Traditionally, these resulting governing equations and the associated boundary conditions are then used as a starting point via the weighted residual statement leading to the weak form or equivalently the principle of virtual work to conduct numerical discretization in space and time. This is quite standard and well known and appears in most books. Alternatively, Lagrange's equations of motion in generalized coordinates can also be transformed into Hamilton's equations of motion through the introduction of canonical coordinates and using the Legendre transformation, wherein, the configuration space becomes transformed into the phase space. Besides these considerations, it is also to be noted that from Lagrange's equations of motion, the associated transformations map the governing equations from a second-order equation system involving the scalar function $\mathcal{L}$ in time to a first-order form of representation involving the scalar function, $\mathcal{H}$ (introducing the Hamiltonian) or from a second-order equation system involving the scalar function, $\mathcal{L}$, to a second-order equation system involving the scalar function, $\mathcal{E}$. *All are equivalent and there is no new physics in either frameworks that arises in contrast to that of the original Newtonian mechanics framework*. It is a matter of the choice of the analyst, in selecting the appropriate framework and for convenience in conducting numerical discretization.

In this chapter, we additionally provide new perspectives for enabling numerical discretizations with particular attention to complementing and/or for providing an alternative viewpoint and physical insight in contrast to traditional practices. Consequently, in this chapter we resort to more natural, simpler and contemporary approaches for numerical discretization with focus upon space.

10.2 HAMILTON'S PRINCIPLE AND HAMILTON'S LAW OF VARYING ACTION IN ELASTODYNAMICS

In this section, we shall show that both Hamilton's principle and Hamilton's law of varying action in elastodynamics are deduced from an analogy to the Lagrangian form of D'Alembert's principle. The analogy in elastodynamics corresponds to the Bubnov-Galerkin weighted residual form, which leads to the principle of virtual work. We show subsequently that Hamilton's principle or Hamilton's law of varying action in continuous-body elastodynamics take a double weighted residual statement in space and time domain. Once discretization in space is carried out in the sense of Galerkin projection (that is, an infinite dimensional space is projected into a finite dimensional space), Hamilton's principle or Hamilton's law of varying action naturally results in a weighted residual form in time, and the governing equation in time equals the finite element equation of motion.

As shown in Chapter 9, the principle of balance of linear momentum in elastodynamics gives rise to Cauchy's equation of motion, which can be described in the Lagrangian description as follows:

$$\rho_0(\mathbf{X})\,\ddot{\boldsymbol{\varphi}}(\mathbf{X},t) - \rho_0(\mathbf{X})\,\mathbf{B}(\mathbf{X},t) - \mathrm{DIV}\,\mathbf{P}(\mathbf{X},t) = \mathbf{0} \quad \text{in } \mathcal{B} \tag{10.1}$$

We observe that the above equation in elastodynamics, Cauchy's first law of motion, is the counter part of Newton's second law of motion in particle dynamics. Note that the above governing equation is subject to the prescribed traction boundary condition, i.e., Cauchy's law

$$\mathbf{T}\left(\mathbf{X},\hat{\mathbf{N}},t\right) = \mathbf{P}(\mathbf{X},t)\cdot\hat{\mathbf{N}}(\mathbf{X},t) = \bar{\mathbf{T}}\left(\mathbf{X},\hat{\mathbf{N}},t\right) \quad \text{on } \partial\mathcal{B}_\sigma \tag{10.2}$$

and the prescribed displacement boundary condition

$$\boldsymbol{\varphi} = \bar{\boldsymbol{\varphi}} \quad \text{on } \partial\mathcal{B}_\varphi \tag{10.3}$$

As an analogy of D'Alembert's principle in particle dynamics, we have for continuum elastodynamics

$$\mathrm{DIV}\,\mathbf{P}(\mathbf{X},t) + \rho_0(\mathbf{X})\,\mathbf{B}(\mathbf{X},t) - \rho_0(\mathbf{X})\,\ddot{\boldsymbol{\varphi}}(\mathbf{X},t) = \mathbf{0} \quad \text{in } \mathcal{B} \tag{10.4}$$

We have shown that in particle dynamics, the inner product of D'Alembert's principle and the virtual displacement leads to the Lagrangian form of D'Alembert's principle. Likewise, in elastodynamics, taking the inner product between Equation (10.4) and the virtual displacement, and integrating the resulting equation over the open, bounded domain $\mathcal{B}$ yields the Bubnov-Galerkin weighted residual form as follows:

$$\int_{\mathcal{B}} \left(\mathrm{DIV}\,\mathbf{P}(\mathbf{X},t) + \rho_0\mathbf{B}(\mathbf{X},t) - \rho_0\ddot{\boldsymbol{\varphi}}(\mathbf{X},t)\right)\cdot\delta\boldsymbol{\varphi}(\mathbf{X},t)\,dV = 0 \tag{10.5}$$

It should be noted that this is the counterpart of the Lagrangian form of D'Alembert's principle given in Equation (3.51). Recall the following product rule

$$(\mathrm{DIV}\,\mathbf{P})\cdot\delta\boldsymbol{\varphi} = \mathrm{DIV}\,(\delta\boldsymbol{\varphi}\cdot\mathbf{P}) - \mathbf{P} : (\mathrm{GRAD}\,\delta\boldsymbol{\varphi}) = \mathrm{DIV}\,(\delta\boldsymbol{\varphi}\cdot\mathbf{P}) - \mathbf{P} : \delta\mathbf{F} \tag{10.6}$$

As discussed in detail in Chapter 9, by taking advantage of both Gauss' divergence theorem and Cauchy's law, Equation (10.5) leads to

$$\begin{aligned} &\int_{\mathcal{B}} \rho_0\mathbf{B}(\mathbf{X},t)\cdot\delta\boldsymbol{\varphi}\,dV + \int_{\partial\mathcal{B}_\sigma} \bar{\mathbf{T}}\left(\mathbf{X},\hat{\mathbf{N}},t\right)\cdot\delta\boldsymbol{\varphi}\,dA - \int_{\mathcal{B}} \mathbf{P} : \delta\mathbf{F}\,dV \\ &- \int_{\mathcal{B}} \rho_0\ddot{\boldsymbol{\varphi}}\cdot\delta\boldsymbol{\varphi}\,dV = 0 \end{aligned} \tag{10.7}$$

which is the principle of virtual work in elastodynamics.

Hamilton's Law of Varying Action in Elastodynamics From now on, we shall consider two cases. One case is that the external forces are dead loads; hence the external potential energy can be defined. The other case is that the external forces are not dead loads; hence the external potential energy cannot be defined. Firstly, we consider the case when the external forces are dead loads. The principle of virtual work in elastodynamics is written as

$$\begin{aligned}&\int_{\mathcal{B}} \rho_0 \mathbf{B}(\mathbf{X}) \cdot \delta\boldsymbol{\varphi}\, dV + \int_{\partial\mathcal{B}_\sigma} \bar{\mathbf{T}}\left(\mathbf{X}, \hat{\mathbf{N}}\right) \cdot \delta\boldsymbol{\varphi}\, dA - \int_{\mathcal{B}} \mathbf{P} : \delta\mathbf{F}\, dV \\ &- \int_{\mathcal{B}} \rho_0 \ddot{\boldsymbol{\varphi}} \cdot \delta\boldsymbol{\varphi}\, dV = 0\end{aligned} \tag{10.8}$$

where the body force $\mathbf{B}(\mathbf{X})$ and the Cauchy traction vector $\bar{\mathbf{T}}\left(\mathbf{X}, \hat{\mathbf{N}}\right)$ are independent of time. In the Lagrangian description, the total potential energy $\mathcal{U}(\boldsymbol{\varphi}) : \mathcal{C} \to \mathbb{R}$ is defined as,

$$\mathcal{U}(\boldsymbol{\varphi}) = \int_{\mathcal{B}} \rho_0 \Psi(\mathbf{F})\, dV - \int_{\partial\mathcal{B}_\sigma} \bar{\mathbf{T}}\left(\mathbf{X}, \hat{\mathbf{N}}\right) \cdot \boldsymbol{\varphi}(\mathbf{X}, t)\, dA - \int_{\mathcal{B}} \rho_0 \mathbf{B}(\mathbf{X}) \cdot \boldsymbol{\varphi}(\mathbf{X}, t)\, dV \tag{10.9}$$

where $\Psi(\mathbf{F})$ denotes the Helmholtz free energy per unit mass in homogeneous materials as an elastic potential function in the reference configuration. The kinetic energy $\mathcal{K}(\dot{\boldsymbol{\varphi}}) : T_{\varphi}\mathcal{C} \to \mathbb{R}$ of the dynamical system of the deformable continuous body may be written in terms of the material velocity $\dot{\boldsymbol{\varphi}} \in T_{\varphi}\mathcal{C}$ as follows:

$$\mathcal{K}(\dot{\boldsymbol{\varphi}}) = \frac{1}{2} \int_{\mathcal{B}} \rho_0 \dot{\boldsymbol{\varphi}} \cdot \dot{\boldsymbol{\varphi}}\, dV \tag{10.10}$$

Hence, in view of Equations (10.9) and (10.10), Equation (10.8) can be rewritten as

$$\delta\mathcal{K}(\dot{\boldsymbol{\varphi}}) - \delta\mathcal{U}(\boldsymbol{\varphi}) - \frac{d}{dt} \int_{\mathcal{B}} \rho_0 \dot{\boldsymbol{\varphi}} \cdot \delta\boldsymbol{\varphi}\, dV = 0 \tag{10.11}$$

which is a restatement of the principle of virtual work in terms of the kinetic and total potential energies. Next, we wish to introduce a descriptive scalar function, i.e. the autonomous Lagrangian $\mathcal{L}(\boldsymbol{\varphi}, \dot{\boldsymbol{\varphi}}) : T\mathcal{C} \to \mathbb{R}$ in the Lagrangian mechanics framework. Then we have

$$\delta\mathcal{L}(\boldsymbol{\varphi}, \dot{\boldsymbol{\varphi}}) - \frac{d}{dt} \int_{\mathcal{B}} \rho_0 \dot{\boldsymbol{\varphi}} \cdot \delta\boldsymbol{\varphi}\, dV = 0 \tag{10.12}$$

where the autonomous Lagrangian $\mathcal{L}(\boldsymbol{\varphi}, \dot{\boldsymbol{\varphi}}) : T\mathcal{C} \to \mathbb{R}$ is defined as

$$\mathcal{L}(\boldsymbol{\varphi}, \dot{\boldsymbol{\varphi}}) = \mathcal{K}(\dot{\boldsymbol{\varphi}}) - \mathcal{U}(\boldsymbol{\varphi}) \tag{10.13}$$

where the autonomous Lagrangian $\mathcal{L}(\boldsymbol{\varphi}, \dot{\boldsymbol{\varphi}}) : T\mathcal{C} \to \mathbb{R}$

$$\begin{aligned}\mathcal{L}(\boldsymbol{\varphi}, \dot{\boldsymbol{\varphi}}) &= \frac{1}{2} \int_{\mathcal{B}} \rho_0 \dot{\boldsymbol{\varphi}} \cdot \dot{\boldsymbol{\varphi}} dV - \int_{\mathcal{B}} \rho_0 \Psi(\mathbf{F})\, dV \\ &+ \int_{\partial\mathcal{B}_\sigma} \bar{\mathbf{T}}\left(\mathbf{X}, \hat{\mathbf{N}}\right) \cdot \boldsymbol{\varphi}(\mathbf{X}, t)\, dA + \int_{\mathcal{B}} \rho_0 \mathbf{B}(\mathbf{X}) \cdot \boldsymbol{\varphi}(\mathbf{X}, t)\, dV\end{aligned} \tag{10.14}$$

For holonomic systems where the prescribed displacement surfaces are given, integrating both sides of Equation (10.12) over the closed time interval of interest $\bar{\mathcal{T}} = [t_0, t_1]$ yields

$$\mathrm{HLVA}_{\mathcal{L}} : \delta \int_{t_0}^{t_1} \mathcal{L}(\boldsymbol{\varphi}, \dot{\boldsymbol{\varphi}})\, dt - \int_{\mathcal{B}} \rho_0 \dot{\boldsymbol{\varphi}} \cdot \delta \boldsymbol{\varphi}\, dV \Big|_{t_0}^{t_1} = 0 \tag{10.15}$$

Equations (10.15) is often called Hamilton's law of varying action in elastodynamics, and is the counterpart of Equation (5.11) in particle dynamics. Now we wish to recall the double property of the principal function in particle dynamics. As Hamilton claimed in particle dynamics, we may conjecture that the principal function, which is often called the action in elastodynamics, has the double property. For the special case that the configuration at the end points is assumed to be held fixed (prescribed), Equation (10.15) yields the form of equation

$$\mathrm{HP}_{\mathcal{L}} : \delta \int_{t_0}^{t_1} \mathcal{L}(\boldsymbol{\varphi}, \dot{\boldsymbol{\varphi}})\, dt = 0 \tag{10.16}$$

which is often called Hamilton's principle. Note that Hamilton's principle is a variational principle while Hamilton's law of varying action is not a variational principle.

Next, we consider the second case that the external forces are not dead loads. Since the total potential energy $\mathcal{U}(\boldsymbol{\varphi}) : \mathcal{C} \to \mathbb{R}$ is defined as,

$$\mathcal{U}(\boldsymbol{\varphi}) = \int_{\mathcal{B}} \rho_0 \Psi(\mathbf{F})\, dV \tag{10.17}$$

instead of Equation (10.15), we arrive at the following form of equation for Hamilton's law of varying action in elastodynamics

$$\begin{aligned} \mathrm{HLVA}_{\mathcal{L}} : \delta \int_{t_0}^{t_1} \mathcal{L}(\boldsymbol{\varphi}, \dot{\boldsymbol{\varphi}})\, dt - \int_{\mathcal{B}} \rho_0 \dot{\boldsymbol{\varphi}} \cdot \delta \boldsymbol{\varphi}\, dV \Big|_{t_0}^{t_1} \\ = \int_{\mathcal{B}} \rho_0 \mathbf{B}(\mathbf{X}, t) \cdot \delta \boldsymbol{\varphi}\, dV + \int_{\partial \mathcal{B}_\sigma} \bar{\mathbf{T}}(\mathbf{X}, \hat{\mathbf{N}}, t) \cdot \delta \boldsymbol{\varphi}\, dA \end{aligned} \tag{10.18}$$

where the body force $\mathbf{B}(\mathbf{X}, t)$ and the Cauchy traction vector $\bar{\mathbf{T}}(\mathbf{X}, \hat{\mathbf{N}}, t)$ are dependent on time. For the special case that the configuration at the end points is assumed to be held fixed (prescribed), Equations (10.18) instead yields the form of equation

$$\begin{aligned} \mathrm{HP}_{\mathcal{L}} : \delta \int_{t_0}^{t_1} \mathcal{L}(\boldsymbol{\varphi}, \dot{\boldsymbol{\varphi}})\, dt \\ = \int_{\mathcal{B}} \rho_0 \mathbf{B}(\mathbf{X}, t) \cdot \delta \boldsymbol{\varphi}\, dV + \int_{\partial \mathcal{B}_\sigma} \bar{\mathbf{T}}(\mathbf{X}, \hat{\mathbf{N}}, t) \cdot \delta \boldsymbol{\varphi}\, dA \end{aligned} \tag{10.19}$$

which is often called Hamilton's principle.

Hamilton's Principle in Elastodynamics As shown in particle dynamics, Hamilton's law of varying action is a consequence of integrating the Lagrangian form of D'Alembert's principle or the principle of virtual work over the given time interval. On the other hand, as one of variational principles in elastodynamics, Hamilton's principle requires the action (a functional) to exist, and it claims that both governing equations and natural boundary conditions make the action stationary under the condition that the virtual displacement at the two end points must vanish at all points of the body. As such, except for this condition, the variational formulation by use of Hamilton's principle is very similar to the one of Hamilton's law of varying action.

For convenience, we confine our attention to the dead load case. Consider a continuous body dynamical system, which we have discussed in Chapters 8 and 9. Within the closed interval of time $\bar{\mathcal{T}} = [t_0, t_1]$, a smooth motion of the body is described as a one-to-one mapping: $\boldsymbol{\varphi}(\mathbf{X}, t) : \mathcal{B} \times \bar{\mathcal{T}} \to \mathcal{S} \subset \mathbb{R}^3, \quad \forall t \in \bar{\mathcal{T}}$, where $\mathcal{S}$ denotes a set of the infinite particles of the body in the spatial or current configuration. The boundary of the closure $\bar{\mathcal{T}}$ is denoted by $\partial\mathcal{T}$, and consists of its two end points, i.e. $\partial\mathcal{T} = \{t_0, t_1\}$ and $\mathcal{T} =]t_0, t_1[$.

Then, in continuum elastodynamics Hamilton's principle for holonomic systems (the prescribed displacement surfaces are given) requires that the configuration at the end points be prescribed as follows:

$$\begin{aligned} \boldsymbol{\varphi}(\mathbf{X}, t)|_{t=t_0} &= \bar{\boldsymbol{\varphi}}_0, \\ \boldsymbol{\varphi}(\mathbf{X}, t)|_{t=t_1} &= \bar{\boldsymbol{\varphi}}_1 \end{aligned} \tag{10.20}$$

where $\bar{\boldsymbol{\varphi}}_0$ and $\bar{\boldsymbol{\varphi}}_1$ are respectively prescribed values at $t \in \partial\mathcal{T} = \{t_0, t_1\}$. It is very important to note that $\bar{\boldsymbol{\varphi}}_1$ represents the deformed configuration at the time $t = t_1$, which is in fact unknown. Under the inconsistent postulation, Hamilton's principle reads as

$$\begin{aligned} \mathrm{HP}_{\mathcal{L}} : \delta \int_{t_0}^{t_1} \mathcal{L}(\boldsymbol{\varphi}, \dot{\boldsymbol{\varphi}})\, dt &= 0, \\ \delta\boldsymbol{\varphi}(\mathbf{X}, t_0) = \delta\boldsymbol{\varphi}(\mathbf{X}, t_1) &= \mathbf{0} \end{aligned} \tag{10.21}$$

where the autonomous Lagrangian $\mathcal{L}(\boldsymbol{\varphi}, \dot{\boldsymbol{\varphi}}) : T\mathcal{C} \to \mathbb{R}$ was defined earlier.

Next, we briefly highlight the vector formalism and then discuss at length the scalar formalism for discretizing the equation of motion in space leading to the semi-discretized equations.

Vector Formalism: Strong and Weak Forms via Hamilton's Principle or Hamilton's Law of Varying Action Equation (10.21) can be rewritten as

$$\int_{t_0}^{t_1} \left\{ \begin{aligned} &\int_{\mathcal{B}} \rho_0 \dot{\boldsymbol{\varphi}} \cdot \delta\dot{\boldsymbol{\varphi}}\, dV - \int_{\mathcal{B}} \rho_0 \frac{\partial \Psi(\mathbf{F})}{\partial \mathbf{F}} : \delta\mathbf{F}\, dV \\ &+ \int_{\mathcal{B}} \rho_0 \mathbf{B}(\mathbf{X}) \cdot \delta\boldsymbol{\varphi}\, dV + \int_{\partial\mathcal{B}_\sigma} \bar{\mathbf{T}}(\mathbf{X}, \hat{\mathbf{N}}) \cdot \delta\boldsymbol{\varphi}\, dA \end{aligned} \right\} dt = 0 \tag{10.22}$$

Upon integration by parts, we obtain

$$\int_{t_0}^{t_1} \left\{ \begin{aligned} &\int_{\mathcal{B}} -\rho_0 \ddot{\boldsymbol{\varphi}} \cdot \delta\boldsymbol{\varphi}\, dV - \int_{\mathcal{B}} \rho_0 \frac{\partial \Psi(\mathbf{F})}{\partial \mathbf{F}} : \delta\mathbf{F}\, dV + \\ &\int_{\mathcal{B}} \rho_0 \mathbf{B}(\mathbf{X}) \cdot \delta\boldsymbol{\varphi}\, dV + \int_{\partial\mathcal{B}_\sigma} \bar{\mathbf{T}}(\mathbf{X}, \hat{\mathbf{N}}) \cdot \delta\boldsymbol{\varphi}\, dA \end{aligned} \right\} dt + \int_{\mathcal{B}} \rho_0 \dot{\boldsymbol{\varphi}} \cdot \delta\boldsymbol{\varphi} dV \Big|_{t_0}^{t_1} = 0 \tag{10.23}$$

In contrast to Hamilton's law of varying action, Hamilton's principle requires the virtual displacement at the two end points to vanish at all points of the body. Consequently, the above equation can be rewritten as

$$\int_{t_0}^{t_1} \left\{ \begin{aligned} &\int_{\mathcal{B}} \rho_0 \ddot{\boldsymbol{\varphi}} \cdot \delta\boldsymbol{\varphi}\, dV + \int_{\mathcal{B}} \rho_0 \frac{\partial \Psi(\mathbf{F})}{\partial \mathbf{F}} : \delta\mathbf{F}\, dV \\ &- \int_{\mathcal{B}} \rho_0 \mathbf{B}(\mathbf{X}) \cdot \delta\boldsymbol{\varphi}\, dV - \int_{\partial\mathcal{B}_\sigma} \bar{\mathbf{T}}(\mathbf{X}, \hat{\mathbf{N}}) \cdot \delta\boldsymbol{\varphi}\, dA \end{aligned} \right\} dt = 0 \tag{10.24}$$

The term enclosed in the curly bracket corresponds to the principle of virtual work in elastodynamics. In view of Equations (8.249) and (9.7), the above equation can be rewritten as

$$\int_{t_0}^{t_1} \left\{ \begin{aligned} & \int_{\mathcal{B}} (\rho_0 \ddot{\varphi} - \mathrm{DIV}\,\mathbf{P} - \rho_0 \mathbf{B}(\mathbf{X})) \cdot \delta\varphi \, dV \\ & + \int_{\mathcal{B}} \mathrm{DIV}\,(\mathbf{P} \cdot \delta\varphi) \, dV - \int_{\partial\mathcal{B}_\sigma} \bar{\mathbf{T}}(\mathbf{X}, \hat{\mathbf{N}}) \cdot \delta\varphi \, dA \end{aligned} \right\} dt = 0 \tag{10.25}$$

By use of Gauss's theorem, we obtain

$$\int_{t_0}^{t_1} \left[\int_{\mathcal{B}} (\mathrm{DIV}\,\mathbf{P} + \rho_0 \mathbf{B} - \rho_0 \ddot{\varphi}) \cdot \delta\varphi dV + \int_{\partial\mathcal{B}_\sigma} \left(\bar{\mathbf{T}} - \mathbf{P} \cdot \hat{\mathbf{N}}\right) \cdot \delta\varphi dV \right] dt = 0 \tag{10.26}$$

Note that on the prescribed displacement boundary surface, the virtual displacement vanishes. In the end, as $\delta\varphi$ is arbitrary in the reference body, the above equation yields

$$\begin{aligned} \rho_0 \ddot{\varphi}(\mathbf{X}, t) &= \mathrm{DIV}\,\mathbf{P}(\mathbf{X}, t) + \rho_0 \mathbf{B}(\mathbf{X}), \ \text{ for } (\mathbf{X}, t) \in \mathcal{B} \times \mathcal{T}, \\ \bar{\mathbf{T}}\left(\mathbf{X}, \hat{\mathbf{N}}\right) &= \mathbf{P}(\mathbf{X}) \cdot \hat{\mathbf{N}}(\mathbf{X}), \ \text{ for } (\mathbf{X}, t) \in \partial\mathcal{B}_\sigma \times \mathcal{T} \end{aligned} \tag{10.27}$$

which are Cauchy's equations of motion and the natural boundary condition.

Space and Time Discretization Using the Weak Form of Cauchy Equation of Motion: Conventional Discretization Procedure As mentioned previously, the term inside the curly bracket in Equation (10.24) represents the weak form, which is the principle of virtual work in dynamics for continuous-body dynamical systems. In traditional practices of discretization with vector formalism, the weak form is used to obtain the finite element equations of motion. Substituting the admissible trial function given in Equation (9.11) into the weak form leads to Equation (9.16). And then, the time weighted residual form can be constructed as shown in Equation (9.21).

The scalar formalisms are of interest here in this chapter and are described next. They also lead to the same semi-discretized equations.

10.3 LAGRANGIAN MECHANICS FRAMEWORK AND FINITE ELEMENT FORMULATIONS

In traditional finite element formulations, the weighted residual method has played a significant role in leading to the weak form, namely, the principle of virtual work. The approximate displacement is taken as the trial function and the virtual displacement as the test function (the weighting function) for approximating the principle of virtual work. It should be noted that the trial and test functions, as admissible functions and variations, are assumed to be smooth functions over the domain so that they can be discretized by interpolation functions and corresponding nodal variables with the finite number of degrees of freedom. This is often called the Galerkin finite element method (Cook et al. 2002; Hughes 1987; Reddy 2006), namely, the Galerkin projection of the infinite dimensional space onto the finite dimensional space.

In this section, we shall discuss an alternative finite element formulation, which is herein referred to as the space-discrete Lagrangian finite element formulation (in the subsequent sections to follow, we address other descriptive functions as well). After directly discretizing the continuous Lagrangian, we then construct the space-discrete Lagrangian which is a state variable of the Lagrangian differential operator. Because of the inconsistency associated with Hamilton's principle, henceforth, we primarily place emphasis only upon employing Hamilton's law of varying action. The discretized form of Hamilton's law of varying action contains the Lagrangian differential operator. Notice that Hamilton's law of varying action involving the continuous Lagrangian is itself the weak form in space and the weighted residual form in time. Therefore, when the function, namely the Lagrangian, is discretized in space, the admissible trial functions are $C^{m-1}(\mathcal{B})$-continuous over the entire domain. In other words, Hamilton's law of varying action for an infinite dimensional dynamical system can be directly discretized via the methodology described here, which is here referred to as the *space-discrete Lagrangian formulation*. Accordingly, we simply discretize Hamilton's law of varying action in space by only admissible trial functions, and we need not resort to traditional practices and approaches such as the Bubnov-Galerkin weighted residual form involving both admissible trial and test functions leading to constructing the weak form as in continuum-elastodynamics to enable numerical discretizations. Note that after the discretization in space, Hamilton's law of varying action involving the Lagrangian naturally takes the representation of the Bubnov-Galerkin weighted residual form in time, which then motivates one to conduct the time weighted residual method which includes time finite element formulations for the design of appropriate time marching integration schemes (Belytschko and Hughes 1983; Sha et al. 2003; Tamma et al. 2000, 2002, 2003; Wood 1990; Zhou et al. 2000; 2001; Zhou and Tamma 2004a, c, 2006; Zienkiewicz and Taylor 2006; Zienkiewicz et al. 2005, 1984).

10.3.1 Lagrangian Density Equations of Motion

In elastodynamics, a continuous body, which is a simply connected, bounded, and open set $\mathcal{B}$ with a piecewise smooth boundary open set $\partial\mathcal{B}$ in the reference configuration, is often postulated to be comprised of infinite particles constructing the body; the particles are not considered to have a concentrated mass. The particles can be defined as mathematical quantities, sometimes called material points, $\mathbf{X} \in \mathcal{B} \subset \mathbb{R}^3$. As defined in Chapter 8, the smooth motion of the body may be described as a one-to-one mapping, $\mathbf{X} = \boldsymbol{\varphi}(\mathbf{X}, t) : \mathcal{B} \times \bar{\mathcal{T}} \rightarrow \mathcal{S} \subset \mathbb{R}^3, \quad \forall t \in \bar{\mathcal{T}} = [t_0, t_1]$, where $\mathcal{B}$ stands for the reference configuration; $\mathcal{S}$ denotes the current configuration.

Recall that the configuration space, in which the deformable continuous body is in motion, is a set of all configurations as defined in Equation (8.1). With the definition of the tangent bundle,

$$T\mathcal{C} := \left\{(\boldsymbol{\varphi}, \dot{\boldsymbol{\varphi}}) : \mathcal{B} \times \bar{\mathcal{T}} \rightarrow \mathbb{R}^3 \times \mathbb{R}^3 \,|\, \boldsymbol{\varphi} \in \mathcal{C} \text{ and } \dot{\boldsymbol{\varphi}} \in T_{\boldsymbol{\varphi}}\mathcal{C}\right\} \tag{10.28}$$

for holonomic systems (since the prescribed displacement surfaces are given), we have the autonomous Lagrangian $\mathcal{L}(\boldsymbol{\varphi}, \dot{\boldsymbol{\varphi}}) : T\mathcal{C} \rightarrow \mathbb{R}$ given in Equation (10.14). Recall that the autonomous Lagrangian is defined on the tangent bundle $T\mathcal{C}$.

Next, we shall introduce the Lagrangian density function by which we can derive the strong and weak forms for the initial boundary-value problems (Marsden and Hughes 1983). The continuous Lagrangian $\mathcal{L}(\boldsymbol{\varphi}, \dot{\boldsymbol{\varphi}}) : T\mathcal{C} \rightarrow \mathbb{R}$ for a dynamical system of a continuous body with dead loads can be rewritten as

$$\mathcal{L}(\boldsymbol{\varphi}, \dot{\boldsymbol{\varphi}}) = \int_{\mathcal{B}} \bar{\mathcal{L}}(\boldsymbol{\varphi}, \dot{\boldsymbol{\varphi}}, \mathbf{F})\, dV + \int_{\partial\mathcal{B}_\sigma} \bar{\mathbf{T}}\left(\mathbf{X}, \hat{\mathbf{N}}\right) \cdot \boldsymbol{\varphi}(\mathbf{X}, t)\, dA \tag{10.29}$$

where $\bar{\mathcal{L}}(\boldsymbol{\varphi}, \dot{\boldsymbol{\varphi}}, \mathbf{F})$ denotes the Lagrangian density function given by

$$\bar{\mathcal{L}}(\boldsymbol{\varphi}, \dot{\boldsymbol{\varphi}}, \mathbf{F}) = \frac{1}{2}\rho_0 \dot{\boldsymbol{\varphi}}(\mathbf{X}, t) \cdot \dot{\boldsymbol{\varphi}}(\mathbf{X}, t) - \rho_0 \Psi(\mathbf{F}) + \rho_0 \mathbf{B} \cdot \boldsymbol{\varphi}(\mathbf{X}, t) \tag{10.30}$$

Then, we derive the strong form in terms of the Lagrangian density function from Hamilton's law of varying action. The strong form consists of two parts: governing equations and boundary conditions.

The variation of the Lagrangian can be written as

$$\delta\mathcal{L}(\boldsymbol{\varphi},\dot{\boldsymbol{\varphi}}) = \int_{\mathcal{B}} \delta\bar{\mathcal{L}}(\boldsymbol{\varphi},\dot{\boldsymbol{\varphi}},\mathbf{F})\,dV + \int_{\partial\mathcal{B}_\sigma} \bar{\mathbf{T}}(\mathbf{X},\hat{\mathbf{N}})\cdot\delta\boldsymbol{\varphi}(\mathbf{X},t)\,dA \tag{10.31}$$

By the chain and product rules for derivatives, the first variation of the Lagrangian density function can be rewritten as

$$\delta\bar{\mathcal{L}}(\boldsymbol{\varphi},\dot{\boldsymbol{\varphi}},\mathbf{F}) = \left\{\frac{\partial\bar{\mathcal{L}}}{\partial\boldsymbol{\varphi}} - \frac{d}{dt}\left(\frac{\partial\bar{\mathcal{L}}}{\partial\dot{\boldsymbol{\varphi}}}\right)\right\}\cdot\delta\boldsymbol{\varphi} + \frac{\partial\bar{\mathcal{L}}}{\partial\mathbf{F}}:\delta\mathbf{F} + \frac{d}{dt}\left(\frac{\partial\bar{\mathcal{L}}}{\partial\dot{\boldsymbol{\varphi}}}\cdot\delta\boldsymbol{\varphi}\right) \tag{10.32}$$

Note that the second term on the right-hand side of the above equation can be expressed by

$$\frac{\partial\bar{\mathcal{L}}}{\partial\mathbf{F}}:\delta\mathbf{F} = \frac{\partial\bar{\mathcal{L}}}{\partial\mathbf{F}}:\mathrm{GRAD}\,\delta\boldsymbol{\varphi} = \mathrm{DIV}\left(\delta\boldsymbol{\varphi}\cdot\frac{\partial\bar{\mathcal{L}}}{\partial\mathbf{F}}\right) - \left(\mathrm{DIV}\,\frac{\partial\bar{\mathcal{L}}}{\partial\mathbf{F}}\right)\cdot\delta\boldsymbol{\varphi} \tag{10.33}$$

Upon substitution of Equation (10.33) into (10.32), Hamilton's principle given in Equation (10.31) reads as

$$\begin{aligned}
\mathrm{HP}_{\mathcal{L}}:\ &\delta\int_{t_0}^{t_1}\mathcal{L}(\boldsymbol{\varphi},\dot{\boldsymbol{\varphi}})\,dt \\
&= \int_{t_0}^{t_1}\left\{\int_{\mathcal{B}}\left\{\frac{\partial\bar{\mathcal{L}}}{\partial\boldsymbol{\varphi}} - \frac{d}{dt}\left(\frac{\partial\bar{\mathcal{L}}}{\partial\dot{\boldsymbol{\varphi}}}\right) - \left(\mathrm{DIV}\,\frac{\partial\bar{\mathcal{L}}}{\partial\mathbf{F}}\right)\right\}\cdot\delta\boldsymbol{\varphi}\,dV\right\}dt \\
&\quad + \int_{t_0}^{t_1}\left\{\int_{\mathcal{B}}\mathrm{DIV}\left(\delta\boldsymbol{\varphi}\cdot\frac{\partial\bar{\mathcal{L}}}{\partial\mathbf{F}}\right)dV\right\}dt \\
&\quad + \int_{t_0}^{t_1}\left(\int_{\partial\mathcal{B}_\sigma}\bar{\mathbf{T}}\cdot\delta\boldsymbol{\varphi}\,dA\right)dt + \int_{\mathcal{B}}\left(\frac{\partial\bar{\mathcal{L}}}{\partial\dot{\boldsymbol{\varphi}}}\cdot\delta\boldsymbol{\varphi}\right)dV\bigg|_{t_0}^{t_1} = 0
\end{aligned} \tag{10.34}$$

It should be noted that Hamilton's principle requires the last term in the above equation to vanish. On the other hand, Hamilton's law of varying action yields

$$\begin{aligned}
\mathrm{HLVA}_{\mathcal{L}}:\ &\delta\int_{t_0}^{t_1}\mathcal{L}(\boldsymbol{\varphi},\dot{\boldsymbol{\varphi}})\,dt - \int_{\mathcal{B}}\left(\frac{\partial\bar{\mathcal{L}}}{\partial\dot{\boldsymbol{\varphi}}}\cdot\delta\boldsymbol{\varphi}\right)dV\bigg|_{t_0}^{t_1} \\
&= \int_{t_0}^{t_1}\left\{\int_{\mathcal{B}}\left\{\frac{\partial\bar{\mathcal{L}}}{\partial\boldsymbol{\varphi}} - \frac{d}{dt}\left(\frac{\partial\bar{\mathcal{L}}}{\partial\dot{\boldsymbol{\varphi}}}\right) - \left(\mathrm{DIV}\,\frac{\partial\bar{\mathcal{L}}}{\partial\mathbf{F}}\right)\right\}\cdot\delta\boldsymbol{\varphi}\,dV\right\}dt \\
&\quad + \int_{t_0}^{t_1}\left\{\int_{\mathcal{B}}\mathrm{DIV}\left(\delta\boldsymbol{\varphi}\cdot\frac{\partial\bar{\mathcal{L}}}{\partial\mathbf{F}}\right)dV\right\}dt + \int_{t_0}^{t_1}\left(\int_{\partial\mathcal{B}_\sigma}\bar{\mathbf{T}}\cdot\delta\boldsymbol{\varphi}\,dA\right)dt = 0
\end{aligned} \tag{10.35}$$

Owing to Gauss's theorem, the second integral on the right-hand side can be written as

$$\int_{t_0}^{t_1}\left\{\int_{\mathcal{B}}\mathrm{DIV}\left(\delta\boldsymbol{\varphi}\cdot\frac{\partial\bar{\mathcal{L}}}{\partial\mathbf{F}}\right)dV\right\}dt = \int_{t_0}^{t_1}\left(\int_{\partial\mathcal{B}_\sigma}\delta\boldsymbol{\varphi}\cdot\frac{\partial\bar{\mathcal{L}}}{\partial\mathbf{F}}\cdot\hat{\mathbf{N}}\,dA\right)dt \tag{10.36}$$

Note that on the prescribed displacement boundary surface $\partial\mathcal{B}_{\varphi}$, the virtual displacement $\delta\boldsymbol{\varphi} \in T_{\varphi}\mathcal{C}$ should vanish due to the Dirichlet boundary conditions. Therefore, Equation (10.35) can be written as

$$\begin{aligned}
\text{HLVA}_{\mathcal{L}} : \ & \delta \int_{t_0}^{t_1} \mathcal{L}(\boldsymbol{\varphi}, \dot{\boldsymbol{\varphi}})\, dt - \int_{\mathcal{B}} \left(\frac{\partial \bar{\mathcal{L}}}{\partial \dot{\boldsymbol{\varphi}}} \cdot \delta\boldsymbol{\varphi} \right) dV \Bigg|_{t_0}^{t_1} \\
&= \int_{t_0}^{t_1} \left\{ \int_{\mathcal{B}} \left\{ \frac{\partial \bar{\mathcal{L}}}{\partial \boldsymbol{\varphi}} - \left(\text{DIV} \frac{\partial \bar{\mathcal{L}}}{\partial \mathbf{F}} \right) - \frac{d}{dt} \left(\frac{\partial \bar{\mathcal{L}}}{\partial \dot{\boldsymbol{\varphi}}} \right) \right\} \cdot \delta\boldsymbol{\varphi}\, dV \right\} dt \\
&+ \int_{t_0}^{t_1} \left(\int_{\partial\mathcal{B}_{\sigma}} \left\{ \bar{\mathbf{T}} + \frac{\partial \bar{\mathcal{L}}}{\partial \mathbf{F}} \cdot \hat{\mathbf{N}} \right\} \cdot \delta\boldsymbol{\varphi}\, dA \right) dt = 0
\end{aligned} \tag{10.37}$$

which includes the governing equations and natural boundary condition in the Bubnov-Galerkin weighted residual form. In the end, we have the governing equations in the body

$$\frac{\partial \bar{\mathcal{L}}}{\partial \boldsymbol{\varphi}} - \frac{d}{dt} \left(\frac{\partial \bar{\mathcal{L}}}{\partial \dot{\boldsymbol{\varphi}}} \right) - \left(\text{DIV} \frac{\partial \bar{\mathcal{L}}}{\partial \mathbf{F}} \right) = \mathbf{0}, \ \text{for } (\mathbf{X}, t) \in \mathcal{B} \times \mathcal{T} \tag{10.38}$$

and the traction boundary conditions on the prescribed traction boundary surface

$$\bar{\mathbf{T}} (\mathbf{X}, \hat{\mathbf{N}}) = -\frac{\partial \bar{\mathcal{L}}}{\partial \mathbf{F}} \cdot \hat{\mathbf{N}}, \ \text{for } (\mathbf{X}, t) \in \partial\mathcal{B}_{\sigma} \times \mathcal{T} \tag{10.39}$$

We refer to Equation (10.38) as the Lagrangian density equation of motion which is the local momentum equation. Note that substituting the Lagrangian density function into Equation (10.38) reduces to Cauchy's equation of motion. Now we observe that Cauchy's equation of motion may be represented in terms of the Lagrangian density function as shown in Equation (10.38). Note that Equation (10.36) indicates the external virtual work done by the tractions on the traction boundary. We naturally have

$$-\frac{\partial \bar{\mathcal{L}}}{\partial \mathbf{F}} = \rho_0 \frac{\partial \Psi (\mathbf{F})}{\partial \mathbf{F}} = \mathbf{P} \tag{10.40}$$

where the first Piola-Kirchhoff stress is denoted by $\mathbf{P}$. Thus, Equation (10.36) may be given by

$$\int_{t_0}^{t_1} \left\{ \int_{\mathcal{B}} \text{DIV} \left(\delta\boldsymbol{\varphi} \cdot \frac{\partial \bar{\mathcal{L}}}{\partial \mathbf{F}} \right) dV \right\} dt = - \int_{t_0}^{t_1} \int_{\partial\mathcal{B}_{\sigma}} \delta\boldsymbol{\varphi} \cdot \mathbf{P} \cdot \hat{\mathbf{N}}\, dV\, dt \tag{10.41}$$

Therefore, it is natural that Equation (10.38) and (10.39) lead to the following governing equation and the natural boundary condition respectively, i.e.,

$$\begin{aligned}
\rho_0 \ddot{\boldsymbol{\varphi}} &= \text{DIV}\, \mathbf{P} + \rho_0 \mathbf{B}, && \text{in } \mathcal{B} \\
\mathbf{P} \cdot \hat{\mathbf{N}} &= \bar{\mathbf{T}}, && \text{on } \partial\mathcal{B}_{\sigma}
\end{aligned} \tag{10.42}$$

By means of Equation (10.33), Hamilton's law of varying in Equation (10.37) leads to the Lagrangian version of the weak form in the sense of the space:

$$\int_{t_0}^{t_1} \left\{ \begin{array}{l} \int_{\mathcal{B}} \frac{d}{dt}\left(\frac{\partial \bar{\mathcal{L}}}{\partial \dot{\boldsymbol{\varphi}}}\right) \cdot \delta\boldsymbol{\varphi}\, dV - \int_{\mathcal{B}} \frac{\partial \bar{\mathcal{L}}}{\partial \mathbf{F}} : \delta\mathbf{F}\, dV \\ - \int_{\mathcal{B}} \frac{\partial \bar{\mathcal{L}}}{\partial \boldsymbol{\varphi}} \cdot \delta\boldsymbol{\varphi}\, dV - \int_{\partial\mathcal{B}_\sigma} \bar{\mathbf{T}} \cdot \delta\boldsymbol{\varphi}\, dA \end{array} \right\} dt = 0 \tag{10.43}$$

where the integrand above yields the principle of virtual work in elastodynamics for dead load cases.

10.3.2 Space-Discrete Lagrangian Finite Element Formulation

In contrast to the traditional approaches for space discretization stemming from the principle of virtual work in dynamics, via the scalar formalism for the Lagrangian (Har and Tamma 2009a, 2010), two distinct approaches to spatial discretization are discussed: 1) One approach is to follow customary practices, but employing the descriptive scalar function and the associated density based strong and weak form representation shown previously which is an outcome using as a starting point Hamilton's law of varying action within the Lagrangian mechanics framework, and 2) Alternatively, a space-discrete numerical discretization approach is additionally presented. This is different from traditional practices, and it provides new avenues to enact the space discretization process. The latter space-discrete formulation emanating from Hamilton's law of varying action does not require one to derive the weak forms. It is readily shown that Hamilton's law of varying action given in Equation (10.35) can be directly discretized in space since by itself, it is already a statement of the weak form. Furthermore, the trial functions, which are employed in Hamilton's law of varying action, are required to satisfy the continuity condition of the same degree just as in the principle of virtual work in dynamics.

The following discussion is worth noting. For steady-state/static problems, approaches such as the Ritz method require a functional on which the Ritz solution operates. Natural boundary conditions are implicitly contained in the functional (Bathe 1982). However, it is well known that in general, a variational principle does not readily exist in transient/dynamic problems. Consequently, this is a drawback of the Ritz type methods for time dependent problems. To circumvent this drawback, the following discussion is useful to understand. So far we have noted that Hamilton's law of varying action in elastodynamics contains implicitly both the governing equations and natural boundary conditions in the Bubnov-Galerkin weighted residual form. In this sense, therefore, for time dependent problems, Hamilton's law of varying action naturally serves as a mechanism by which the weighted averages of the residuals are required to vanish (Crandall 1956, page 149) (Bathe 1982, page 108), when substituting the trial functions (although these satisfy only essential boundary conditions) into Hamilton's law of varying action to discretize the space. Consequently, we can naturally address time dependent problems as described next.

The admissible trial functions $\tilde{\boldsymbol{\varphi}}$ within the Lagrangian framework may be represented by a linear combination of coordinate functions $\mathbf{N}_i(\mathbf{X})$ and displacement vectors $\mathbf{U}_i^e(t)$ as follows:

$$\tilde{\boldsymbol{\varphi}}(\mathbf{X}, t) \approx \tilde{\boldsymbol{\varphi}}^h(\mathbf{X}, t) = \mathbf{X} + \sum_{i=1}^{n_{\text{node}}} \mathbf{N}_i(\mathbf{X})\, \mathbf{U}_i^e(t) \tag{10.44}$$

where $\mathbf{U}_i^e$ denotes nodal displacement vectors. Note that Equation (10.44) is identical to Equation (9.11); hence the trial functions above, which are of class $C^{m-1}(\mathcal{B})$ over the entire domain, belong to the subspace, denoted by $\mathcal{C}^h \subset \mathcal{C}$, as defined in Equation (9.14). Traditional practices starting from Cauchy's equations of motion employ the weighted residual form and lead to the weak form or principle of virtual work in dynamics for deriving the semi-discrete equations of motion. This is quite standard and was addressed previously. However, for the descriptive scalar function representations via the Lagrangian and Hamiltonian mechanics frameworks (or utilizing other potential descriptive functions) for continuous body systems, there exist two schools of thought. One may follow analogous procedures as in traditional practices, or alternatively, one may employ the space discrete finite element formulations as highlighted in this section for time dependent problems (see Approach 2 to follow).

Approach 1: Weak Form With regards to the Lagrangian density representation, substituting the approximate trial functions in Equation (10.44) into the term enclosed in the curly brackets of Equation (10.43) leads to the following:

$$\int_{t_0}^{t_1} \sum_{e=1}^{n_{elem}} \left\{ \begin{array}{l} \displaystyle\int_{\mathcal{B}_e} \frac{d}{dt}\left(\frac{\partial \bar{\mathcal{L}}^h_{\mathcal{B}_e}}{\partial \dot{\tilde{\boldsymbol{\varphi}}}^h}\right) \cdot \delta\tilde{\boldsymbol{\varphi}}^h dV - \int_{\mathcal{B}_e} \frac{\partial \bar{\mathcal{L}}^h_{\mathcal{B}_e}}{\partial \tilde{\mathbf{F}}^h} : \delta\tilde{\mathbf{F}}^h dV \\ \displaystyle - \int_{\mathcal{B}_e} \frac{\partial \bar{\mathcal{L}}^h_{\mathcal{B}_e}}{\partial \tilde{\boldsymbol{\varphi}}^h} \cdot \delta\tilde{\boldsymbol{\varphi}}^h dV - \int_{\partial\mathcal{B}_{e\sigma}} \bar{\mathbf{T}} \cdot \delta\tilde{\boldsymbol{\varphi}}^h dA \end{array} \right\} dt = 0 \tag{10.45}$$

where the space-discrete Lagrangian density function is given by

$$\bar{\mathcal{L}}^h_{\mathcal{B}_e}\left(\tilde{\boldsymbol{\varphi}}^h, \dot{\tilde{\boldsymbol{\varphi}}}^h, \tilde{\mathbf{F}}^h\right) = \frac{1}{2}\rho_0 \dot{\tilde{\boldsymbol{\varphi}}}^h \cdot \dot{\tilde{\boldsymbol{\varphi}}}^h - \rho_0 \Psi\left(\tilde{\mathbf{F}}^h\right) + \rho_0 \mathbf{B} \cdot \tilde{\boldsymbol{\varphi}}^h \tag{10.46}$$

In Equation (10.45), $\mathcal{B}_e$ stands for the eth element domain. Consequently the weak form given in Equation (10.45) simply leads to

$$\int_{t_0}^{t_1} \left\{ \sum_{e=1}^{n_{\text{elem}}} \left(\mathbf{M}^e \ddot{\mathbf{U}}^e(t) + \mathbf{F}^e_{\text{int}} - \mathbf{F}^e_{\text{ext}}\right) \cdot \delta\mathbf{U}^e(t) \right\} dt = 0 \tag{10.47}$$

where $\mathbf{M}_e$, $\mathbf{F}^e_{\text{int}}$ and $\mathbf{F}^e_{\text{ext}}$ are given by Equation (9.17).

For the entire domain of interest, we introduce a configuration space with generalized coordinates $\mathcal{Q}$, i.e.,

$$\mathcal{Q} := \left\{\mathbf{q} | \mathbf{q}(t) = \left(U_1(t), \ldots, U_{n_{\text{dof}}}(t)\right) \in \mathbb{R}^{n_{\text{dof}} = 3n_{\text{node}}}\right\} \tag{10.48}$$

Note that the subspace $\mathcal{C}^h$ is a function space whose members depend on space and time, whereas $\mathcal{Q}$ is a function space whose members depend on time only, and whose dimension is $3n_{\text{node}}$. Then, we introduce the generalized velocities which belong to a tangent space denoted by $T_q Q$, that is

$$T_q Q := \left\{\dot{\mathbf{q}} | \dot{\mathbf{q}}(t) = \left(\dot{U}_1, \ldots, \dot{U}_{n_{\text{dof}}}\right) \in \mathbb{R}^{n_{\text{dof}} = 3n_{\text{node}}}\right\} \tag{10.49}$$

Therefore, we have established the configuration and tangent spaces by use of both generalized coordinates and velocities. Further, as in particle dynamics, in the finite dimensional space we define the tangent bundle or the velocity phase space as

$$TQ := \left\{(\mathbf{q}, \dot{\mathbf{q}}) \, | \mathbf{q} \in Q \text{ and } \dot{\mathbf{q}} \in T_q Q\right\} = \bigcup_{q \in Q} T_q Q \tag{10.50}$$

Notice that the term enclosed in the curly bracket in Equation (10.47) yields the semi-discrete second-order differential equations for the dynamical system, i.e. the finite element equations of motion to be discretized subsequently in time as follows:

$$\mathbf{M}\ddot{\mathbf{q}} + \mathbf{F}_{\text{int}} - \mathbf{F}_{\text{ext}} = \mathbf{0} \tag{10.51}$$

Approach 2: Lagrangian Differential Operator In contrast to the conventional approach above but with scalar formulation which has been examined so far, a different perspective is suggested in the following. Note that the admissible trial functions for discretizing the space via Hamilton's law of varying action are of class $C^{m-1}(\mathcal{B})$ as in the trial functions applied for the principle of virtual work. In particular, an emphasis is placed on the fact that the same trial functions for the space discretization via Hamilton's law of varying action can be employed as in the principle of virtual work.

We directly start the discretization process via Hamilton's law of varying action. And then, the discrete Lagrangian differential operator is directly applied in a simple and straightforward manner, which readily yields the semi-discrete second-order differential equations. The space discrete finite element formulation follows next.

By means of Equation (10.44), Hamilton's law of varying action within the eth element domain $\mathcal{B}_e$ can be discretized in space as follows:

$$\text{HLVA}^h_{\mathcal{L}_{\mathcal{B}_e}} : \delta \int_{t_0}^{t_1} \mathcal{L}^h_{\mathcal{B}_e}\left(\tilde{\boldsymbol{\varphi}}^h, \dot{\tilde{\boldsymbol{\varphi}}}^h\right) dt - \left[\int_{\mathcal{B}_e} \frac{\partial \bar{\mathcal{L}}^h_{\mathcal{B}_e}\left(\tilde{\boldsymbol{\varphi}}^h, \dot{\tilde{\boldsymbol{\varphi}}}^h\right)}{\partial \dot{\tilde{\boldsymbol{\varphi}}}^h} \cdot \delta\tilde{\boldsymbol{\varphi}}^h dV\right]\Bigg|_{t_0}^{t_1} = 0 \tag{10.52}$$

where the autonomous space-discrete Lagrangian for the eth element domain is simply constructed as

$$\mathcal{L}^h_{\mathcal{B}_e}\left(\tilde{\boldsymbol{\varphi}}^h, \dot{\tilde{\boldsymbol{\varphi}}}^h\right) = \int_{\mathcal{B}_e} \bar{\mathcal{L}}^h_{\mathcal{B}_e}\left(\tilde{\boldsymbol{\varphi}}^h, \dot{\tilde{\boldsymbol{\varphi}}}^h, \tilde{\mathbf{F}}^h\right) dV + \int_{\partial\mathcal{B}_{e\sigma}} \bar{\mathbf{T}} \cdot \tilde{\boldsymbol{\varphi}}^h(\mathbf{X}, t)\, dA \tag{10.53}$$

Equation (10.44) can be rewritten as follows:

$$\begin{aligned} \tilde{\boldsymbol{\varphi}}^h(\mathbf{X}, t) &= \mathbf{X} + \sum_{i=1}^{n_{\text{node}}} \mathbf{N}_i(\mathbf{X})\,\mathbf{U}^e_i(t) = \mathbf{X} + \mathbf{N}_e \mathbf{q}_e, \\ \mathbf{q}_e(t) &= \left\{\mathbf{U}^e_1(t), \ldots, \mathbf{U}^e_i(t), \ldots, \mathbf{U}^e_{n_{\text{node}}}(t)\right\} \end{aligned} \tag{10.54}$$

where $\mathbf{q}_e$ denotes a set of generalized coordinates for the eth element. Hence, in terms of generalized coordinates Equation (10.52) can be rewritten as

$$\begin{aligned} \text{HLVA}^h_{\mathcal{L}_{\mathcal{B}_e}} &: \delta \int_{t_0}^{t_1} \mathcal{L}^h_{\mathcal{B}_e}(\mathbf{q}_e, \dot{\mathbf{q}}_e)\, dt - \frac{\partial \mathcal{L}^h_{\mathcal{B}_e}(\mathbf{q}_e, \dot{\mathbf{q}}_e)}{\partial \dot{\mathbf{q}}_e} \cdot \delta\mathbf{q}_e \Bigg|_{t_0}^{t_1} \\ &= \int_{t_0}^{t_1} \left\{\frac{\partial \mathcal{L}^h_{\mathcal{B}_e}(\mathbf{q}_e, \dot{\mathbf{q}}_e)}{\partial \mathbf{q}_e} - \frac{d}{dt}\left(\frac{\partial \mathcal{L}^h_{\mathcal{B}_e}(\mathbf{q}_e, \dot{\mathbf{q}}_e)}{\partial \dot{\mathbf{q}}_e}\right)\right\} \cdot \delta\mathbf{q}_e(t)\, dt = 0 \end{aligned} \tag{10.55}$$

where the space-discrete autonomous Lagrangian is constructed as follows:

$$\mathcal{L}^h_{\mathcal{B}_e}(\mathbf{q}_e, \dot{\mathbf{q}}_e) = \frac{1}{2}\dot{\mathbf{q}}_e \cdot \mathbf{M}^e \dot{\mathbf{q}}_e - \Pi^h_{\text{inte}} + \mathbf{q}_e \cdot \mathbf{F}^e_{\text{ext}}, \quad \Pi^h_{\text{inte}} = \int_{\mathcal{B}_e} \rho_0 \Psi\left(\tilde{\mathbf{F}}^h\right) dV$$

$$\mathbf{M}^e = \int_{\mathcal{B}_e} \rho_0 \mathbf{N}_e^T \mathbf{N}_e \, dV, \quad \frac{\partial \mathcal{L}_{\mathcal{B}_e}^h (\mathbf{q}_e, \dot{\mathbf{q}}_e)}{\partial \mathbf{q}_e} = \mathbf{F}_{\text{ext}}^e - \mathbf{F}_{\text{int}}^e$$
$$\mathbf{F}_{\text{int}}^e = \int_{\mathcal{B}_e} \bar{\mathbf{B}}_e^T \tilde{\mathbf{P}}^h \, dV, \quad \mathbf{F}_{\text{ext}}^e = \int_{\mathcal{B}_e} \rho_0 \mathbf{N}_e^T \mathbf{B} \, dV + \int_{\partial \mathcal{B}_{e\sigma}} \mathbf{N}_e^T \bar{\mathbf{T}} \, dA \tag{10.56}$$

Note that in the above, we use the following derivations

$$\tilde{\mathbf{F}}^h = \mathbf{I} + \bar{\mathbf{B}}_e \mathbf{q}_e, \quad \frac{\partial \tilde{\mathbf{F}}^h}{\partial \mathbf{q}_e} = \bar{\mathbf{B}}_e,$$
$$\frac{\partial \Pi_{\text{int}_e}^h}{\partial \mathbf{q}_e} = \int_{\mathcal{B}_e} \rho_0 \frac{\partial \Psi\left(\tilde{\mathbf{F}}^h\right)}{\tilde{\mathbf{F}}^h} : \frac{\partial \tilde{\mathbf{F}}^h}{\partial \mathbf{q}_e} \, dV = \int_{\mathcal{B}_e} \bar{\mathbf{B}}_e^T \tilde{\mathbf{P}}^h \, dV \tag{10.57}$$

For the entire domain, we recall the configuration space, denoted by $\mathcal{Q}$, with generalized coordinates as defined in Equation (10.48). Consequently, Hamilton's law of varying action in a finite dimensional system is given by

$$\text{HLVA}_{\mathcal{L}}^h : \delta \int_{t_0}^{t_1} \mathcal{L}^h (\mathbf{q}, \dot{\mathbf{q}}) \, dt - \left. \frac{\partial \mathcal{L}^h (\mathbf{q}, \dot{\mathbf{q}})}{\partial \dot{\mathbf{q}}} \cdot \delta \mathbf{q} \right|_{t_0}^{t_1}$$
$$= \int_{t_0}^{t_1} \left\{ \frac{\partial \mathcal{L}^h (\mathbf{q}, \dot{\mathbf{q}})}{\partial \mathbf{q}} - \frac{d}{dt} \left(\frac{\partial \mathcal{L}^h (\mathbf{q}, \dot{\mathbf{q}})}{\partial \dot{\mathbf{q}}} \right) \right\} \cdot \delta \mathbf{q} (t) \, dt = 0 \tag{10.58}$$

where the space-discrete autonomous Lagrangian is constructed as follows:

$$\mathcal{L}^h (\mathbf{q}, \dot{\mathbf{q}}) = \frac{1}{2} \dot{\mathbf{q}} \cdot \mathbf{M} \dot{\mathbf{q}} - \Pi_{\text{int}}^h + \mathbf{q} \cdot \mathbf{F}_{\text{ext}}, \quad \Pi_{\text{int}}^h = \int_{\mathcal{B}} \rho_0 \Psi \left(\tilde{\mathbf{F}}^h \right) dV$$
$$\mathbf{M} = \int_{\mathcal{B}} \rho_0 \mathbf{N}^T \mathbf{N} \, dV, \quad \frac{\partial \mathcal{L}^h (\mathbf{q}, \dot{\mathbf{q}})}{\partial \mathbf{q}} = \mathbf{F}_{\text{ext}} - \mathbf{F}_{\text{int}} \tag{10.59}$$
$$\mathbf{F}_{\text{int}} = \int_{\mathcal{B}} \bar{\mathbf{B}}^T \tilde{\mathbf{P}}^h \, dV, \quad \mathbf{F}_{\text{ext}} = \int_{\mathcal{B}} \rho_0 \mathbf{N}^T \mathbf{B} \, dV + \int_{\partial \mathcal{B}_\sigma} \mathbf{N}^T \bar{\mathbf{T}} \, dA$$

In contrast to traditional practices for discretizing in space, we simply adopt the following procedure. In view of Equation (10.58), we introduce the space-discrete Lagrangian differential operator $\mathfrak{L}_D$, by which the semi-discrete finite element equations of motion can be readily obtained after constructing the space-discrete Lagrangian which only entails the use of trial functions. Thus, we have the space-discrete Lagrangian differential operator

$$\mathfrak{L}_D = \frac{d}{dt} \left(\frac{\partial}{\partial \dot{\mathbf{q}}} \right) - \frac{\partial}{\partial \mathbf{q}} \tag{10.60}$$

The dependent state variable of the differential operator is the space-discrete Lagrangian. Upon substituting the space-discrete Lagrangian into the space-discrete Lagrangian differential operator, we obtain the following:

$$\mathfrak{L}_D \left(\mathcal{L}^h \right) = \frac{d}{dt} \left(\frac{\partial \mathcal{L}^h}{\partial \dot{\mathbf{q}}} \right) - \frac{\partial \mathcal{L}^h}{\partial \mathbf{q}} = \mathbf{M} \ddot{\mathbf{q}} + \mathbf{F}_{\text{int}} - \mathbf{F}_{\text{ext}} = 0 \tag{10.61}$$

where the dependent state variable is the space-discrete scalar function, namely, the space-discrete Lagrangian. *Note that the space-discrete Lagrangian differential operator above implies that the weighted averages of the residuals associated with governing equations and natural boundary conditions are implicitly required to vanish, when the admissible trial functions satisfying the essential boundary conditions are chosen and substituted into the principle of virtual work (Bathe* 1982, *page 108) (Cook et al.* 2002, *page 136) (Zienkiewicz et al.* 2005, *page 61) (Reddy* 2006, *page 74).* Thus, we readily have the semi-discrete finite element equations of motion governing the mechanical dynamic behavior of the continuous-body within the Lagrangian mechanics framework as follows:

$$\mathbf{M}\ddot{\mathbf{q}} + \mathbf{F}_{\text{int}} - \mathbf{F}_{\text{ext}} = \mathbf{0} \tag{10.62}$$

It should be noted that Equation (10.58) is for the Lagrangian representation with a finite number of degrees of freedom, and Hamilton's law of varying action naturally results in the following equation

$$\int_{t_0}^{t_1} \left\{ \frac{\partial \mathcal{L}^h(\mathbf{q}, \dot{\mathbf{q}})}{\partial \mathbf{q}} - \frac{d}{dt}\left(\frac{\partial \mathcal{L}^h(\mathbf{q}, \dot{\mathbf{q}})}{\partial \dot{\mathbf{q}}} \right) \right\} \cdot \delta\mathbf{q}\, dt = 0 \tag{10.63}$$

Time Weighted Residual Method We observe that Equation (10.63) can be rewritten as

$$\int_{t_0}^{t_1} \{ (\mathbf{F}_{\text{ext}} - \mathbf{F}_{\text{int}} - \mathbf{M}\ddot{\mathbf{q}}) \cdot \delta\mathbf{q} \}\, dt = 0 \tag{10.64}$$

where the integrand of Equation (10.64) is exactly identical to that obtained after discretizing the weighted residual statement or the principle of virtual work as described in the previous chapter. As a consequence, Equation (10.64) represents the weighted residual form in time. We note that Equation (10.64) leads one to the time weighted residual method which includes time finite element formulations (Belytschko and Hughes 1983; Sha et al. 2003; Tamma et al. 2000, 2002, 2003; Wood 1990; Zhou et al. 2000, 2000; Zhou and Tamma 2004a,c, 2006; Zienkiewicz and Taylor 2006; Zienkiewicz et al. 2005, 1984).

Illustrative Examples For the various illustrative examples to follow next, we first show the development of the Lagrangian density based representations of the equations of motion in the scalar formalism that are identical/equivalent to the vector formalism of the Cauchy equations of motion. One could readily employ the Lagrangian density based representations to conduct the space discretization through conventional practices described previously in Approach 1; alternatively, to illustrate the fundamental ideas, we instead focus attention upon the space discretization aspects via Approach 2 described earlier which represents the space-discrete finite element formulation and it directly leads to the semi-discretized equations of motion.

Example 10.1 Elastic Axial Bar

Let us consider the same elastic bar undergoing an axial deformation only, which we have previously discussed in Section 9.1.2.

Governing Equations: Lagrangian Density Based Representations

We shall first derive the Lagrangian density based representation of the equation of motion in scalar formalism. The total potential energy $\mathcal{U}(u)$ of the axial bar deforming longitudinally is given as

$$\mathcal{U}(u) = \frac{1}{2}\int_L EA(x)\left(\frac{\partial u(x,t)}{\partial x} \right)^2 dx - \int_L \bar{b}u(x,t)\, dx - F_L u(L,t) \tag{10.65}$$

where $\bar{b}$ and F_L denote a body force per unit length and an end force at $x = L$. For the axial bar in a longitudinal harmonic motion, the kinetic energy $\mathcal{K}(\dot{u})$ is given as

$$\mathcal{K}(\dot{u}) = \frac{1}{2}\int_L \rho A(x)\,\dot{u}^2 dx \tag{10.66}$$

Therefore, the Lagrangian can be given by

$$\begin{aligned}\mathcal{L}(u,\dot{u}) = \frac{1}{2}\int_L \rho A(x)\,\dot{u}^2 dx - \frac{1}{2}\int_l EA(x)\left(\frac{\partial u(x,t)}{\partial x}\right)^2 dx \\ + \int_L \bar{b}(x)\,u(x,t)\,dx + F_L u(L,t)\end{aligned} \tag{10.67}$$

The Lagrangian $\mathcal{L}(u(x,t), \dot{u}(x,t))$ can be rewritten as

$$\mathcal{L} = \int_L \bar{\mathcal{L}}(u,\dot{u},u')\,dx + F_L u(L,t) \tag{10.68}$$

where the Lagrangian density function $\bar{\mathcal{L}}(u,\dot{u},u')$ is defined as

$$\bar{\mathcal{L}}(u,\dot{u},u') = \frac{1}{2}\rho A(x)\,\dot{u}^2 - \frac{1}{2}EA(x)\left(u'\right)^2 + \bar{b}(x)\,u(x,t), \quad u' = \frac{\partial u}{\partial x} \tag{10.69}$$

Then, Hamilton's law of varying action reads as

$$\delta\int_{t_0}^{t_1} \mathcal{L}(u(x,t), \dot{u}(x,t))\,dt - \left.\int_L \frac{\partial\bar{\mathcal{L}}(u,\dot{u},u')}{\partial\dot{u}}\delta u(x,t)\,dx\right|_{t_0}^{t_1} = 0 \tag{10.70}$$

In the above, the Lagrangian is defined in terms of the Lagrangian density function.

Strong Form

By virtue of the product rule, the variation of the Lagrangian density function can be written as

$$\delta\bar{\mathcal{L}}(u,\dot{u},u') = \frac{\partial\bar{\mathcal{L}}}{\partial u}\delta u + \frac{\partial\bar{\mathcal{L}}}{\partial u'}\delta u' + \frac{d}{dt}\left(\frac{\partial\bar{\mathcal{L}}}{\partial\dot{u}}\delta u\right) - \frac{d}{dt}\left(\frac{\partial\bar{\mathcal{L}}}{\partial\dot{u}}\right)\delta u \tag{10.71}$$

By use of the product rule, the variation of the Lagrangian density function in Equation (10.71) can be rewritten as

$$\begin{aligned}\delta\bar{\mathcal{L}} = \frac{\partial\bar{\mathcal{L}}}{\partial u}\delta u + \frac{\partial}{\partial x}\left(\frac{\partial\bar{\mathcal{L}}}{\partial u'}\delta u\right) - \frac{\partial}{\partial x}\left(\frac{\partial\bar{\mathcal{L}}}{\partial u'}\right)\delta u \\ + \frac{d}{dt}\left(\frac{\partial\bar{\mathcal{L}}}{\partial\dot{u}}\delta u\right) - \frac{d}{dt}\left(\frac{\partial\bar{\mathcal{L}}}{\partial\dot{u}}\right)\delta u\end{aligned} \tag{10.72}$$

Substituting the above equation into Hamilton's law of varying action naturally leads to

$$\begin{aligned} &\int_{t_0}^{t_1}\left[\int_L\left\{\frac{\partial\bar{\mathcal{L}}}{\partial u}-\frac{\partial}{\partial x}\left(\frac{\partial\bar{\mathcal{L}}}{\partial u'}\right)-\frac{d}{dt}\left(\frac{\partial\bar{\mathcal{L}}}{\partial\dot{u}}\right)\right\}\delta u\,(x,t)\,dx\right]dt \\ &\qquad+\int_{t_0}^{t_1}\left(F_L\delta u(L,t)+\frac{\partial\bar{\mathcal{L}}}{\partial u'}\delta u\bigg|_0^L\right)dt=0 \end{aligned} \tag{10.73}$$

Thus, the governing equation of the axial bar dynamical motion is given by (scalar formalism)

$$\frac{d}{dt}\left(\frac{\partial\bar{\mathcal{L}}\,(u,\dot{u},u')}{\partial\dot{u}}\right)=\frac{\partial\bar{\mathcal{L}}\,(u,\dot{u},u')}{\partial u}-\frac{\partial}{\partial x}\left(\frac{\partial\bar{\mathcal{L}}\,(u,\dot{u},u')}{\partial u'}\right) \tag{10.74}$$

which is subjected to the boundary conditions

$$F_L\delta u(L,t)+\frac{\partial\bar{\mathcal{L}}\,(u,\dot{u},u')}{\partial u'}\delta u\bigg|_0^L=0 \tag{10.75}$$

Note that the governing equation of the axial bar dynamical motion of Equation (10.74) readily yields the classical form as (vector formalism)

$$\rho A\,(x)\,\ddot{u}=\frac{\partial}{\partial x}\left(EA\,(x)\,\frac{\partial u}{\partial x}\right)+\bar{b}\,(x) \tag{10.76}$$

subject to the boundary conditions

$$F_L\delta u(L,t)-EA\,(x)\left(\frac{\partial u}{\partial x}\right)\delta u\bigg|_0^L=0 \tag{10.77}$$

Note that Equation (10.77) means that there are two boundary conditions. One is called the Dirichlet or geometric or essential boundary condition, i.e. $\delta u=0$. The other is called the Neumann or dynamical or natural boundary condition, i.e.,

$$F_{\text{axial}}\,(x=L)=EA\,(x)\left(\frac{\partial u}{\partial x}\right)\bigg|_{x=L} \tag{10.78}$$

Weak Form

In addition, in view of Equation (10.71) Hamilton's law of varying action also leads to

$$\int_{t_0}^{t_1}\left\{\begin{array}{l}\displaystyle\int_L\frac{d}{dt}\left(\frac{\partial\bar{\mathcal{L}}}{\partial\dot{u}}\right)\delta u\,dx-\int_L\frac{\partial\bar{\mathcal{L}}}{\partial u'}\delta u'dx\\ \displaystyle-\int_L\frac{\partial\bar{\mathcal{L}}}{\partial u}\delta u\,dx-F_L\delta u(L,t)\end{array}\right\}dt=0 \tag{10.79}$$

Note that the integrand in the above integral equation represents the weak form corresponding to Equation (10.43).

Space-Discrete Finite Element Formulation

Of the two approaches mentioned in the previous section to discretize the space, next we focus attention upon and employ the space-discrete Lagrangian finite element formulation (Approach 2) for illustration. We shall now construct the space-discrete Lagrangian to derive the finite element equations of motion for the bar assemblage (see Figure 10.1). The linear displacement field function over one element is represented as

$$u\,(x,t) \approx a_1 + a_2 x = N_1\,(x)\,u_1^e\,(t) + N_2\,(x)\,u_2^e\,(t) = \tilde{u}^h\,(x,t) \tag{10.80}$$

where a_1, a_2 are arbitrary coefficients and the total number of the coefficients is equal to the number of degrees of freedom in the element. As a consequence, once the displacement field function, which is continuous in space and time, is discretized in space, the approximate displacement field function can be recast as

$$\tilde{u}^h\,(x,t) = \mathbf{N}_e\,(x)\,\mathbf{u}^e\,(t) = \begin{bmatrix} N_1\,(x) & N_2\,(x) \end{bmatrix} \begin{Bmatrix} u_1^e\,(t) \\ u_2^e\,(t) \end{Bmatrix} \tag{10.81}$$

We observe that the nodal displacements may be regarded as generalized coordinates. Therefore, the number of nodal displacements is equal to the number of degrees of freedom in the element. Once the displacement field function, denoted by $u\,(x,t)$ is discretized using the shape functions, we obtain the approximate displacement field function $\tilde{u}^h\,(x,t)$. Further, it is very important to note that the nodal displacements, as generalized coordinates, are time-continuous functions.

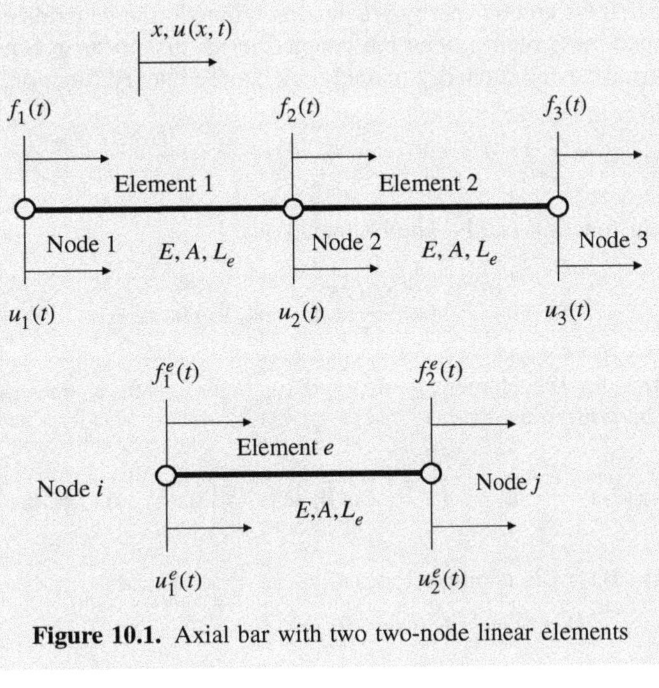

Figure 10.1. Axial bar with two two-node linear elements

The approximate velocity field function $\dot{\tilde{u}}^h(x,t)$ is given by

$$\dot{\tilde{u}}^h(x,t) = \mathbf{N}_e(x)\,\dot{\mathbf{u}}^e(t) = \begin{bmatrix} N_1(x) & N_2(x) \end{bmatrix} \begin{Bmatrix} \dot{u}_1^e(t) \\ \dot{u}_2^e(t) \end{Bmatrix} \tag{10.82}$$

The nodal velocities $\dot{u}_1^e(t), \dot{u}_2^e(t)$ may be regarded as generalized velocities. The shape function matrix is given by

$$\mathbf{N}_e(x) = \begin{bmatrix} N_1(x) & N_2(x) \end{bmatrix} = \begin{bmatrix} 1 - \dfrac{x}{L_e} & \dfrac{x}{L_e} \end{bmatrix} \tag{10.83}$$

where L_e denotes the length of the element. In addition, the consistent mass matrix can be expressed as

$$\mathbf{M}^e = \int_{L_e} \rho \mathbf{N}_e^T(x)\,\mathbf{N}_e(x)\,A\,dx = \frac{\rho A}{6} \begin{bmatrix} 2 & 1 \\ 1 & 2 \end{bmatrix} \tag{10.84}$$

Then, the linear momentum $\mathbf{p}_e$ can be written as

$$\mathbf{p}_e = \mathbf{M}^e \dot{\mathbf{q}}_e = \frac{\rho A}{6} \begin{bmatrix} 2 & 1 \\ 1 & 2 \end{bmatrix} \begin{Bmatrix} \dot{u}_1^e(t) \\ \dot{u}_2^e(t) \end{Bmatrix} = \begin{Bmatrix} p_1^e(t) \\ p_2^e(t) \end{Bmatrix} \tag{10.85}$$

where $p_1^e(t)$, $p_2^e(t)$ are called nodal linear momentum variables. Note that the nodal linear momentum variables are functions of time only.

Next, the space-discrete kinetic energy of the element reads as

$$\mathcal{K}_e^h = \frac{1}{2}\dot{\mathbf{q}}_e \cdot \mathbf{M}^e \dot{\mathbf{q}}_e = \frac{\rho A}{6}\left(\left(\dot{u}_1^e(t)\right)^2 + \dot{u}_1^e(t)\,\dot{u}_2^e(t) + \left(\dot{u}_2^e(t)\right)^2 \right) \tag{10.86}$$

It is easy to recognize that the kinetic energy is a quadratic function in generalized velocities. Further, when we have a lumped mass matrix, then the coupled terms disappear in Equation (10.86). Next, we discretize the internal potential energy, namely, the stored energy function,

$${}_e\mathcal{U}_{\text{int}} = \frac{1}{2}\int_{\Omega_e} \boldsymbol{\sigma} : \boldsymbol{\varepsilon}\,dV = \frac{A}{2}\int_{L_e} \sigma_x \varepsilon_x\,dx = \frac{EA}{2}\int_{L_e} \varepsilon_x^2\,dx \tag{10.87}$$

The infinitesimal strain function can be approximated as

$$\varepsilon_x \approx \frac{d\tilde{u}^h}{dx} = \frac{d\mathbf{N}_e(x)}{dx}\mathbf{q}_e = \bar{\mathbf{B}}_e(x)\,\mathbf{q}_e = \varepsilon_x^h \tag{10.88}$$

where the strain matrix for the element is denoted by $\bar{\mathbf{B}}_e(x)$. Hence, the space-discrete internal potential energy can be written as

$${}_e\mathcal{U}_{\text{int}}^h = \frac{EA}{2}\mathbf{q}_e \cdot \left[\int_{L_e} \bar{\mathbf{B}}_e^T(x)\,\bar{\mathbf{B}}_e(x)\,dx \right] \mathbf{q}_e = \frac{1}{2}\mathbf{q}_e \cdot \mathbf{K}^e \mathbf{q}_e \tag{10.89}$$

where the strain matrix $\bar{\mathbf{B}}_e(x)$ is computed as

$$\bar{\mathbf{B}}_e(x) = \begin{bmatrix} \bar{B}_1(x) & \bar{B}_2(x) \end{bmatrix} = \begin{bmatrix} -\dfrac{1}{L_e} & \dfrac{1}{L_e} \end{bmatrix} \tag{10.90}$$

and the elemental stiffness matrix $\mathbf{K}^e$ can be given by

$$\mathbf{K}^e = EA \int_{L_e} \bar{\mathbf{B}}_e^T(x)\,\bar{\mathbf{B}}_e(x)\;dx \tag{10.91}$$

which can be also found in Section 9.2.1. As a consequence, the space-discrete internal potential energy yields

$$\begin{aligned} {}_e\mathcal{U}^h_{\text{int}} &= \frac{EA}{2L_e}\mathbf{q}_e \cdot \begin{bmatrix} 1 & -1 \\ -1 & 1 \end{bmatrix} \mathbf{q}_e \\ &= \frac{EA}{2L_e}\left(\left(u_1^e(t)\right)^2 - 2u_1^e(t)\,u_2^e(t) + \left(u_2^e(t)\right)^2\right) \end{aligned} \tag{10.92}$$

Note that the internal potential energy is a quadratic function in generalized coordinates in the linear elastic material. Further, the internal force vector is written as

$$\mathbf{F}^e_{\text{int}} = \frac{\partial\,{}_e\mathcal{U}^h_{\text{int}}}{\partial \mathbf{q}_e} = \frac{EA}{L_e}\begin{bmatrix} 1 & -1 \\ -1 & 1 \end{bmatrix}\begin{Bmatrix} u_1^e(t) \\ u_2^e(t) \end{Bmatrix} \tag{10.93}$$

Suppose that there is no body force acting on the bar. Then, we have the following nodal forces

$$\mathbf{F}^e_{\text{ext}} = -\frac{\partial}{\partial \mathbf{q}_e}\left(-f_1^e(t)\,u_1^e(t) - f_2^e(t)\,u_2^e(t)\right) = \begin{Bmatrix} f_1^e(t) \\ f_2^e(t) \end{Bmatrix} \tag{10.94}$$

And, the space-discrete Lagrangian for the element can be constructed and written as

$$\mathcal{L}^h_e(\mathbf{q}_e, \dot{\mathbf{q}}_e) = \frac{1}{2}\dot{\mathbf{q}}_e \cdot \mathbf{M}^e\dot{\mathbf{q}}_e - \frac{1}{2}\mathbf{q}_e \cdot \mathbf{K}^e\mathbf{q}_e + \mathbf{q}_e \cdot \mathbf{F}^e_{\text{ext}} \tag{10.95}$$

Note that the first derivative of the space-discrete Lagrangian with respect to the generalized velocity results in the linear momentum, namely,

$$\mathbf{p}_e = \frac{\partial \mathcal{L}^h_e}{\partial \dot{\mathbf{q}}_e} = \begin{Bmatrix} p_1^e(t) \\ p_2^e(t) \end{Bmatrix} = \frac{\rho A}{6}\begin{bmatrix} 2 & 1 \\ 1 & 2 \end{bmatrix}\begin{Bmatrix} \dot{u}_1^e(t) \\ \dot{u}_2^e(t) \end{Bmatrix} \tag{10.96}$$

Note that the above equation is identical to Equation (10.85).

Hamilton's law of varying action for a longitudinal axial bar motion naturally yields

$$\begin{aligned} &\int_{t_1}^{t_2}\left(\frac{d}{dt}\left(\frac{\partial \mathcal{L}^h_e}{\partial \dot{\mathbf{q}}_e}\right) - \frac{\partial \mathcal{L}^h_e}{\partial \mathbf{q}_e}\right)\cdot \delta\mathbf{q}_e dt \\ &= \int_{t_1}^{t_2}\left(\mathbf{M}^e\ddot{\mathbf{q}}_e + \mathbf{K}^e\mathbf{q}_e - \mathbf{F}^e_{\text{ext}}\right)\cdot \delta\mathbf{q}_e dt = 0 \end{aligned} \tag{10.97}$$

Therefore, substituting the space-discrete Lagrangian in Equation (10.95) directly into the space-discrete Lagrangian differential operator in Equation (10.60), simply results in the semi-discrete finite element equations of motion,

$$\mathfrak{L}_D\left(\mathcal{L}_e^h\right) = \frac{d}{dt}\left(\frac{\partial \mathcal{L}_e^h}{\partial \dot{\mathbf{q}}_e}\right) - \frac{\partial \mathcal{L}_e^h}{\partial \mathbf{q}_e} = \mathbf{M}^e\ddot{\mathbf{q}}_e + \mathbf{K}^e\mathbf{q}_e - \mathbf{F}_{\text{ext}}^e = \mathbf{0} \tag{10.98}$$

Assembling the matrix equations for the two-bar assemblage, we obtain

$$\frac{\rho A}{6}\begin{bmatrix} 2 & 1 & 0 \\ 1 & 4 & 1 \\ 0 & 1 & 2 \end{bmatrix}\begin{Bmatrix} \ddot{u}_1(t) \\ \ddot{u}_2(t) \\ \ddot{u}_3(t) \end{Bmatrix} + \frac{EA}{L_e}\begin{bmatrix} 1 & -1 & 0 \\ -1 & 2 & -1 \\ 0 & -1 & 1 \end{bmatrix}\begin{Bmatrix} u_1(t) \\ u_2(t) \\ u_3(t) \end{Bmatrix} = \begin{Bmatrix} f_1(t) \\ f_2(t) \\ f_3(t) \end{Bmatrix} \tag{10.99}$$

which is the finite element equations of motion for the bar assemblage.

Example 10.2 Rotating Bar

In this example, we shall consider a rotating, circular bar. Moreover, it is assumed that the bar is not subject to the axial force. The bar is assume to rotate due to the torque.

Governing Equations: Lagrangian Density Based Representations

The torque $M_T(x,t)$ causing the bar to rotate can be given by

$$M_T(x,t) = GJ(x)\frac{\partial\theta(x,t)}{\partial x} \tag{10.100}$$

where G denotes the shear modulus, $\theta(x,t)$ is the angle of rotation about the axial axis and $J(x)$ the polar moment of inertia of the cross-section about the center of the cross-section. Therefore, the kinetic energy is only the rotational kinetic energy given by

$$\mathcal{K}\left(\dot{\theta}\right) = \frac{1}{2}\int_L \rho J(x)\dot{\theta}^2 dx \tag{10.101}$$

Further, the total potential energy of the bar, which is subject to a distributed torque $\bar{m}(x,t)$ is given by

$$\mathcal{U}(\theta) = \frac{1}{2}\int_L M_T(x,t)\frac{\partial\theta(x,t)}{\partial x}dx - \int_L \bar{m}_T(x)\theta(x,t)\,dx \tag{10.102}$$

Thus, the total potential energy may be rewritten as

$$\mathcal{U}(\theta) = \frac{1}{2}\int_L GJ(x)\left(\frac{\partial\theta(x,t)}{\partial x}\right)^2 dx - \int_L \bar{m}_T(x)\theta(x,t)\,dx \tag{10.103}$$

Eventually, the continuous Lagrangian is given by

$$\begin{aligned}\mathcal{L}(\theta,\dot{\theta}) &= \frac{1}{2}\int_L \rho J(x)\dot{\theta}^2 dx \\ &\quad - \frac{1}{2}\int_L GJ(x)\left(\frac{\partial\theta(x,t)}{\partial x}\right)^2 dx + \int_L \bar{m}_T(x)\theta(x,t)\,dx\end{aligned} \tag{10.104}$$

As a consequence, the Lagrangian density function is defined as

$$\bar{\mathcal{L}}\left(\theta,\dot{\theta},\frac{\partial\theta}{\partial x}\right)=\frac{1}{2}\rho J(x)\dot{\theta}^2-\frac{1}{2}GJ(x)\left(\frac{\partial\theta(x,t)}{\partial x}\right)^2 \tag{10.105}$$

Hamilton's law of varying action gives

$$\delta\int_{t_0}^{t_1}\mathcal{L}dt-\int_L\rho J(x)\dot{\theta}\cdot\delta\theta(x,t)\,dx\Big|_{t_0}^{t_1}=0 \tag{10.106}$$

where the Lagrangian $\mathcal{L}$ may be redefined as

$$\mathcal{L}=\int_L\bar{\mathcal{L}}\left(\theta,\dot{\theta},\frac{\partial\theta}{\partial x}\right)dx=\int_L\bar{\mathcal{L}}(\theta,\dot{\theta},\theta')dx+\int_L\bar{m}_T(x)\,\theta(x,t)\,dx \tag{10.107}$$

The variation of the Lagrangian density function can be derived as

$$\delta\bar{\mathcal{L}}(\theta,\dot{\theta},\theta')=\frac{\partial\bar{\mathcal{L}}}{\partial\theta}\delta\theta+\frac{\partial\bar{\mathcal{L}}}{\partial\dot{\theta}}\delta\dot{\theta}+\frac{\partial\bar{\mathcal{L}}}{\partial\theta'}\delta\theta' \tag{10.108}$$

Integration by parts leads to

$$\begin{aligned}\delta\bar{\mathcal{L}}(\theta,\dot{\theta},\theta')&=\left[\frac{\partial\bar{\mathcal{L}}}{\partial\theta}-\frac{d}{dt}\left(\frac{\partial\bar{\mathcal{L}}}{\partial\dot{\theta}}\right)-\frac{\partial}{\partial x}\left(\frac{\partial\bar{\mathcal{L}}}{\partial\theta'}\right)\right]\delta\theta\\&\quad+\frac{d}{dt}\left(\frac{\partial\bar{\mathcal{L}}}{\partial\dot{\theta}}\delta\theta\right)+\frac{\partial}{\partial x}\left(\frac{\partial\bar{\mathcal{L}}}{\partial\theta'}\delta\theta\right)\end{aligned} \tag{10.109}$$

Hamilton's law of varying action can be rewritten as

$$\begin{aligned}&\delta\int_{t_0}^{t_1}\mathcal{L}dt-\int_L\left(\frac{\partial\bar{\mathcal{L}}}{\partial\dot{\theta}}\delta\theta\right)dx\Big|_{t_0}^{t_1}-\int_{t_0}^{t_1}\left[\frac{\partial\bar{\mathcal{L}}}{\partial\theta'}\delta\theta\Big|_0^L\right]dt=0\\&=\int_{t_0}^{t_1}\left[\int_L\left\{\frac{\partial\bar{\mathcal{L}}}{\partial\theta}-\frac{d}{dt}\left(\frac{\partial\bar{\mathcal{L}}}{\partial\dot{\theta}}\right)-\frac{\partial}{\partial x}\left(\frac{\partial\bar{\mathcal{L}}}{\partial\theta'}\right)+\bar{m}_T(x)\right\}\delta\theta dx\right]dt\end{aligned} \tag{10.110}$$

Therefore, the governing equation of motion is given by

$$\frac{d}{dt}\left(\frac{\partial\bar{\mathcal{L}}}{\partial\dot{\theta}}\right)=\frac{\partial\bar{\mathcal{L}}}{\partial\theta}-\frac{\partial}{\partial x}\left(\frac{\partial\bar{\mathcal{L}}}{\partial\theta'}\right)+\bar{m}_T(x) \tag{10.111}$$

subject to the boundary conditions

$$\frac{\partial\bar{\mathcal{L}}}{\partial\theta'}\delta\theta\Big|_0^L=0 \tag{10.112}$$

Also note that upon substitution of Equation (10.105) into Equation (10.111), we obtain the classical form of the governing equation of motion,

$$\rho J(x)\ddot{\theta}(x,t) = \frac{\partial}{\partial x}\left(GJ(x)\frac{\partial\theta(x,t)}{\partial x}\right) + \bar{m}_T(x) \tag{10.113}$$

subject to the boundary condition

$$GJ(x)\frac{\partial\theta(x,t)}{\partial x}\delta\theta(x,t)\bigg|_0^L = 0 \tag{10.114}$$

which can be rewritten as

$$M_T(x,t)\,\delta\theta(x,t)|_0^L = 0 \tag{10.115}$$

where $M_T(x,t)$ denotes the torque as shown in Equation (10.100). Note that the boundary conditions in Equation (10.115) imply geometric and natural boundary conditions. If the end of the rotating bar is fixed, $\delta\theta(x,t)|_{x=0\ or\ x=L} = 0$. On the other hand, if the end is free, then, the torque moment vanishes, namely, $M_T(x,t)|_{x=0\ or\ x=L} = 0$.

Space-Discrete Finite Element Formulation

Henceforth, we shall discuss the development of the semi-discrete second-order finite element equation of motion via the space-discrete Lagrangian formulation. First we shall construct the space-discrete Lagrangian by constructing the space-discrete kinetic energy and the space-discrete potential energy. The trial functions are approximated as

$$\begin{aligned}&\theta^h(x,t) = \mathbf{N}_e(x)\,\mathbf{q}_e(t), \quad \dot{\theta}^h(x,t) = \mathbf{N}_e(x)\,\dot{\mathbf{q}}_e(t),\\ &\frac{\partial\theta^h(x,t)}{\partial x} = \mathbf{B}(x)\,\mathbf{q}_e(t)\end{aligned} \tag{10.116}$$

Thus, we find the space-discrete kinetic energy

$$\mathcal{K}_e^h(\dot{\mathbf{q}}_e) = \frac{1}{2}\dot{\mathbf{q}}_e\cdot\mathbf{M}^e\dot{\mathbf{q}}_e \tag{10.117}$$

where the consistent mass matrices are given as

$$\mathbf{M}^e = \frac{1}{2}\int_{L_e}\rho J(x)\,\mathbf{N}_e^T(x)\,\mathbf{N}_e(x)\,dx \tag{10.118}$$

where L_e denotes the length of one element. Likewise, the space-discrete potential energy can be obtained

$$\mathcal{U}_e^h = \frac{1}{2}\mathbf{q}_e(t)\cdot\mathbf{K}^e\mathbf{q}_e(t) - \mathbf{q}_e(t)\cdot\mathbf{F}_{\text{ext}}^e \tag{10.119}$$

where

$$\mathbf{K}^e = \int_{L_e}GJ(x)\,\mathbf{B}(x)^T\,\mathbf{B}(x)\,dx, \quad \mathbf{F}_{\text{ext}}^e = \int_{L_e}\bar{m}_T(x)\,\mathbf{N}_e^T(x)\,dx \tag{10.120}$$

In the end, the space-discrete Lagrangian can be simply constructed as

$$\mathcal{L}_e^h(\mathbf{q}_e, \dot{\mathbf{q}}_e) = \frac{1}{2}\dot{\mathbf{q}}_e \cdot \mathbf{M}^e \dot{\mathbf{q}}_e - \frac{1}{2}\mathbf{q}_e \cdot \mathbf{K}^e \mathbf{q}_e + \mathbf{q}_e \cdot \mathbf{F}^e_{\text{ext}} \tag{10.121}$$

Upon substitution of the space-discrete Lagrangian into the space-discrete Lagrangian differential operator in Equation (10.60), readily yields

$$\mathfrak{L}_D\left(\mathcal{L}_e^h\right) = \frac{d}{dt}\left(\frac{\partial \mathcal{L}_e^h}{\partial \dot{\mathbf{q}}_e}\right) - \frac{\partial \mathcal{L}_e^h}{\partial \mathbf{q}_e} = \mathbf{M}^e \ddot{\mathbf{q}}_e + \mathbf{K}^e \mathbf{q}_e - \mathbf{F}^e_{\text{ext}} = \mathbf{0} \tag{10.122}$$

Example 10.3 Euler-Bernoulli Beam

We consider the same Euler-Bernoulli beam in a transverse vibration as discussed in Section 9.1.4.

Governing Equations: Lagrangian Density Based Representations

The kinetic energy consists of the translational and rotational kinetic energies, and is given as

$$\begin{aligned} &\mathcal{K}(\dot{w}) = \tfrac{1}{2}\int_L \dot{\mathbf{u}} \cdot \boldsymbol{\rho}\dot{\mathbf{u}}\, dx, \qquad \boldsymbol{\rho} = \begin{bmatrix} \rho(x)A(x) & 0 \\ 0 & \rho(x)I(x) \end{bmatrix}, \\ &\mathbf{u}(x,t) = (w(x,t), \varphi(x,t)), \quad \dot{\mathbf{u}}(x,t) = (\dot{w}(x,t), \dot{\varphi}(x,t)) \end{aligned} \tag{10.123}$$

It should be noted that in the above equation $\mathbf{u}$ or $\dot{\mathbf{u}}$ is a column vector; but for convenience the transpose notation is deleted (Marsden and Tromba 2003, page 79). The total potential energy of the beam subject to a distributed load $\omega(x)$ is given by

$$\begin{aligned} \mathcal{U}(w) &= \frac{1}{2}\int_L M_B(x,t)\varphi' dx - \int_L \omega(x) w(x,t)\, dx \\ &= \frac{1}{2}\int_L M_B(x,t) w'' dx - \int_L \omega(x) w(x,t)\, dx, \\ w'' &= \frac{\partial^2 w(x,t)}{\partial x^2}, \quad \varphi' = \frac{\partial \varphi(x,t)}{\partial x} \end{aligned} \tag{10.124}$$

where the bending moment $M_B(x,t)$ is given as

$$M_B(x,t) = EI(x)\frac{\partial \varphi(x,t)}{\partial x} = EI(x)\frac{\partial^2 w(x,t)}{\partial x^2} \tag{10.125}$$

Thus, the total potential energy may be rewritten as

$$\begin{aligned} \mathcal{U}(w) &= \tfrac{1}{2}\int_L \left(\mathbf{u}' \cdot \mathbf{D}_b \mathbf{u}'\right) dx - \int_L \omega(x)\mathbf{a} \cdot \mathbf{u}\, dx, \ \mathbf{a} = (1,0) \\ \mathbf{D}_b &= \begin{bmatrix} 0 & 0 \\ 0 & EI(x) \end{bmatrix}, \ \mathbf{u}'(x,t) = \left(w'(x,t), \varphi'(x,t)\right) \end{aligned} \tag{10.126}$$

As a consequence, the continuous Lagrangian of the Euler-Bernoulli beam can be found as

$$\mathcal{L}(\mathbf{u}, \dot{\mathbf{u}}) = \int_L \bar{\mathcal{L}}(\dot{\mathbf{u}}, \mathbf{u}')\, dx + \int_L \omega(x)\, \mathbf{a} \cdot \mathbf{u}\, dx \tag{10.127}$$

Moreover, the Lagrangian density function can be defined as

$$\bar{\mathcal{L}}(\dot{\mathbf{u}}, \mathbf{u}') = \frac{1}{2}\dot{\mathbf{u}} \cdot \rho \dot{\mathbf{u}} - \frac{1}{2}\mathbf{u}' \cdot \mathbf{D}_b \mathbf{u}' \tag{10.128}$$

Hamilton's law of varying action yields

$$\delta \int_{t_0}^{t_1} \mathcal{L}(\mathbf{u}, \dot{\mathbf{u}})\, dt - \int_L \left(\frac{\partial \bar{\mathcal{L}}(\dot{\mathbf{u}}, \mathbf{u}')}{\partial \dot{\mathbf{u}}} \cdot \delta \mathbf{u}(x, t) \right) dx \Bigg|_{t_0}^{t_1} = 0 \tag{10.129}$$

The first term of the above equation can be derived as

$$\begin{aligned}
\delta \int_{t_0}^{t_1} \mathcal{L}(\mathbf{u}, \dot{\mathbf{u}})\, dt &= \int_{t_0}^{t_1} \left(\int_L \delta \bar{\mathcal{L}}(\dot{\mathbf{u}}, \mathbf{u}')\, dx \right) dt + \int_{t_0}^{t_1} \left(\int_L \omega(x)\, \mathbf{a} \cdot \delta \mathbf{u}\, dx \right) dt \\
&= \int_{t_0}^{t_1} \left[\int_L \left(\frac{\partial \bar{\mathcal{L}}}{\partial \dot{\mathbf{u}}} \cdot \delta \dot{\mathbf{u}} + \frac{\partial \bar{\mathcal{L}}}{\partial \mathbf{u}'} \cdot \delta \mathbf{u}' \right) dx \right] dt + \int_{t_0}^{t_1} \left(\int_L \omega(x)\, \mathbf{a} \cdot \delta \mathbf{u}\, dx \right) dt
\end{aligned} \tag{10.130}$$

Accordingly, Hamilton's law of varying action in Equation (10.129) leads to

$$\begin{aligned}
&\delta \int_{t_0}^{t_1} \mathcal{L}(\mathbf{u}, \dot{\mathbf{u}})\, dt - \int_L \frac{\partial \bar{\mathcal{L}}(\dot{\mathbf{u}}, \mathbf{u}')}{\partial \dot{\mathbf{u}}} \cdot \delta \mathbf{u}(x, t)\, dx \Bigg|_{t_0}^{t_1} \\
&= -\int_{t_0}^{t_1} \left\{ \int_L \left[\frac{d}{dt}\left(\frac{\partial \bar{\mathcal{L}}}{\partial \dot{\mathbf{u}}} \right) \cdot \delta \mathbf{u} \right] dx \right\} dt + \int_{t_0}^{t_1} \left(\int_L \omega(x)\, \mathbf{a} \cdot \delta \mathbf{u}\, dx \right) dt \\
&+ \int_{t_0}^{t_1} \left\{ \frac{\partial \bar{\mathcal{L}}}{\partial \mathbf{u}'} \cdot \delta \mathbf{u} \Bigg|_0^L - \int_L \left[\frac{\partial}{\partial x}\left(\frac{\partial \bar{\mathcal{L}}}{\partial \mathbf{u}'} \right) \cdot \delta \mathbf{u} \right] dx \right\} dt = 0
\end{aligned} \tag{10.131}$$

Note that w' is dependent on w and $\delta \mathbf{u}$ is not arbitrary. From the above equation, we have the derivations

$$\begin{aligned}
&\int_{t_0}^{t_1} \left\{ \int_L \left[\frac{d}{dt}\left(\frac{\partial \bar{\mathcal{L}}}{\partial \dot{\mathbf{u}}} \right) \cdot \delta \mathbf{u} \right] dx \right\} dt \\
&= \int_{t_0}^{t_1} \left\{ \int_L \left[\frac{d}{dt}\left(\frac{\partial \bar{\mathcal{L}}}{\partial \dot{w}} \right) - \frac{\partial}{\partial x}\left(\frac{d}{dt}\left(\frac{\partial \bar{\mathcal{L}}}{\partial \dot{w}'} \right) \right) \delta w \right] dx \right\} dt \\
&+ \int_{t_0}^{t_1} \left[\frac{d}{dt}\left(\frac{\partial \bar{\mathcal{L}}}{\partial \dot{w}'} \right) \delta w \Bigg|_0^L \right] dt
\end{aligned} \tag{10.132}$$

and

$$\begin{aligned}
&\int_{t_0}^{t_1} \left\{ \frac{\partial \bar{\mathcal{L}}}{\partial \mathbf{u}'} \cdot \delta \mathbf{u} \Big|_0^L - \int_L \left[\frac{\partial}{\partial x} \left(\frac{\partial \bar{\mathcal{L}}}{\partial \mathbf{u}'} \right) \cdot \delta \mathbf{u} \right] dx \right\} dt \\
&= \int_{t_0}^{t_1} \left\{ \begin{array}{c} \left[\dfrac{\partial \bar{\mathcal{L}}}{\partial w'} - \dfrac{\partial}{\partial x} \left(\dfrac{\partial \bar{\mathcal{L}}}{\partial \dot{w}'} \right) + \dfrac{\partial^2}{\partial x^2} \left(\dfrac{\partial \bar{\mathcal{L}}}{\partial w''} \right) \right] \delta w \\ - \left[\dfrac{\partial}{\partial x} \left(\dfrac{\partial \bar{\mathcal{L}}}{\partial w''} \right) \delta w \right] \Bigg|_0^L + \left(\dfrac{\partial \bar{\mathcal{L}}}{\partial w''} \delta w' \right) \Bigg|_0^L \end{array} \right\} dt
\end{aligned} \tag{10.133}$$

Consequently, Hamilton's law of varying action yields

$$\begin{aligned}
&\delta \int_{t_0}^{t_1} \mathcal{L}(\mathbf{u}, \dot{\mathbf{u}})\, dt - \int_L \left(\frac{\partial \bar{\mathcal{L}}(\dot{\mathbf{u}}, \mathbf{u}')}{\partial \dot{\mathbf{u}}} \cdot \delta \mathbf{u}(x, t) \right) dx \Bigg|_{t_0}^{t_1} \\
&= -\int_{t_0}^{t_1} \left\{ \int_L \left[\begin{array}{c} \dfrac{d}{dt} \left(\dfrac{\partial \bar{\mathcal{L}}}{\partial \dot{w}} \right) - \dfrac{\partial^2}{\partial x^2} \left(\dfrac{\partial \bar{\mathcal{L}}}{\partial w''} \right) \\ - \dfrac{\partial}{\partial x} \left(\dfrac{d}{dt} \left(\dfrac{\partial \bar{\mathcal{L}}}{\partial \dot{w}'} \right) - \omega(x) \right) \delta w \end{array} \right] dx \right\} dt \\
&\quad - \int_{t_0}^{t_1} \left\{ \left[\frac{d}{dt} \left(\frac{\partial \bar{\mathcal{L}}}{\partial \dot{w}'} \right) + \frac{\partial}{\partial x} \left(\frac{\partial \bar{\mathcal{L}}}{\partial w''} \right) \right] \delta w \Bigg|_0^L - \left(\frac{\partial \bar{\mathcal{L}}}{\partial w''} \delta w' \right) \Bigg|_0^L \right\} dt = 0
\end{aligned} \tag{10.134}$$

which gives the governing equation of the Euler-Bernoulli beam as follows

$$\frac{d}{dt} \left(\frac{\partial \bar{\mathcal{L}}}{\partial \dot{w}} \right) - \frac{\partial^2}{\partial x^2} \left(\frac{\partial \bar{\mathcal{L}}}{\partial w''} \right) - \frac{\partial}{\partial x} \left(\frac{d}{dt} \left(\frac{\partial \bar{\mathcal{L}}}{\partial \dot{w}'} \right) \right) - \omega(x) = 0 \tag{10.135}$$

subject to the boundary conditions

$$\begin{aligned}
&\left[\frac{d}{dt} \left(\frac{\partial \bar{\mathcal{L}}}{\partial \dot{w}'} \right) + \frac{\partial}{\partial x} \left(\frac{\partial \bar{\mathcal{L}}}{\partial w''} \right) \right] \delta w \Bigg|_0^L = 0, \\
&\left(\frac{\partial \bar{\mathcal{L}}}{\partial w''} \delta w' \right) \Bigg|_0^L = 0
\end{aligned} \tag{10.136}$$

which indicate that there are two natural boundary conditions and two essential boundary conditions. Also note that upon substitution of Equation (10.128) into Equation (10.135), the classical form of the governing equations of motion are obtained as

$$\rho(x) A(x) \ddot{w} = \frac{\partial}{\partial x} \left(\rho(x) I(x) \frac{\partial \ddot{w}}{\partial x} \right) - \frac{\partial^2}{\partial x^2} \left(EI(x) \left(\frac{\partial^2 w}{\partial x^2} \right) \right) + \omega(x) \tag{10.137}$$

subject to the boundary conditions

$$\begin{aligned}&\left[\frac{\partial}{\partial x}\left(EI(x)\left(\frac{\partial^2 w}{\partial x^2}\right)\right)-\rho I(x)\left(\frac{\partial \ddot{w}}{\partial x}\right)\right]\delta w\bigg|_0^L=0,\\&EI(x)\left(\frac{\partial^2 w}{\partial x^2}\right)\delta\left(\frac{\partial w}{\partial x}\right)\bigg|_0^L=0\end{aligned}\tag{10.138}$$

which can be rewritten as

$$Q(x,t)\,\delta w|_0^L=0,\ M_B(x,t)\,\delta\left(\frac{\partial w}{\partial x}\right)\bigg|_0^L=0\tag{10.139}$$

where $M_B(x,t)$ and $Q(x,t)$ denote the bending moment and the shearing force. Note that boundary conditions in Equation (10.139) imply both geometric and natural boundary conditions. If the end of the beam is fixed, $\delta\left(\frac{\partial w}{\partial x}\right)\big|_{x=0\ or\ x=L}=0$ and $\delta w|_{x=0\ or\ x=L}=0$. On the other hand, if the end is free, then, the bending moment and the shearing force must vanish.

Space-Discrete Finite Element Formulation

Next, we wish to construct the space-discrete Lagrangian by constructing the space-discrete kinetic and potential energies. Consider a two-node beam element. The trial displacement function may be approximated using the Hermite polynomial interpolation functions as follows:

$$\begin{aligned}&w^h(x,t)=\mathbf{N}_e(x)\,\mathbf{q}_e(t),\ \frac{\partial w^h(x,t)}{\partial x}=\mathbf{B}_1(x)\cdot\mathbf{q}_e(t),\\&\mathbf{B}_1(x)=\frac{\partial \mathbf{N}_e(x)}{\partial x}\end{aligned}\tag{10.140}$$

where

$$\begin{aligned}&\mathbf{q}_e(t)=\left(w_1^h(t),\frac{\partial w_1^h}{\partial x}(t),w_2^h(t),\frac{\partial w_2^h}{\partial x}(t)\right),\\&\dot{\mathbf{q}}_e(t)=\left(\dot{w}_1^h(t),\frac{\partial \dot{w}_1^h}{\partial x}(t),\dot{w}_2^h(t),\frac{\partial \dot{w}_2^h}{\partial x}(t)\right)\end{aligned}\tag{10.141}$$

Thus, in favor of matrix and vector notation, we construct the space-discrete kinetic energy

$$\mathcal{K}_e^h(\dot{\mathbf{q}}_e)=\frac{1}{2}\dot{\mathbf{q}}_e(t)\cdot\mathbf{M}^e\dot{\mathbf{q}}_e(t),\ \mathbf{M}^e=\mathbf{M}_A+\mathbf{M}_I\tag{10.142}$$

where the consistent mass matrices are given as

$$\mathbf{M}_A=\frac{1}{2}\int_{L_e}\rho A(x)\,\mathbf{N}_e^T\mathbf{N}_e\,dx,\quad \mathbf{M}_I=\frac{1}{2}\int_{L_e}\rho I(x)\,\mathbf{B}_1^T\mathbf{B}_1\,dx\tag{10.143}$$

where L_e denotes the length of the element. Likewise, the space-discrete potential energy can be constructed as

$$\mathcal{U}_e^h(\mathbf{q}_e)=\frac{1}{2}\mathbf{q}_e(t)\cdot\mathbf{K}^e\mathbf{q}_e(t)-\mathbf{q}_e(t)\cdot\mathbf{F}_{\text{ext}}^e\tag{10.144}$$

where

$$\mathbf{K}^e = \int_{L_e} EI(x)\,\mathbf{B}_2^T\mathbf{B}_2\,dx,\ \mathbf{B}_2(x) = \frac{\partial^2 \mathbf{N}_e(x)}{\partial x^2},\ \mathbf{F}_{\text{ext}}^e = \int_{L_e} \omega(x)\,\mathbf{N}_e^T(x)\,dx \tag{10.145}$$

whose details can be found in Section 9.2.6. Then, the space-discrete total potential energy can be written as

$$\mathcal{U}_e^h(\mathbf{q}_e) = \frac{1}{2}\mathbf{q}_e(t)\cdot\mathbf{K}^e\mathbf{q}_e(t) - \mathbf{q}_e(t)\cdot\mathbf{F}_{\text{ext}}^e \tag{10.146}$$

In the end, the space-discrete Lagrangian can be constructed as

$$\mathcal{L}_e^h(\mathbf{q}_e, \dot{\mathbf{q}}_e) = \frac{1}{2}\dot{\mathbf{q}}_e(t)\cdot\mathbf{M}^e\dot{\mathbf{q}}_e(t) - \frac{1}{2}\mathbf{q}_e(t)\cdot\mathbf{K}^e\mathbf{q}_e(t) + \mathbf{q}_e(t)\cdot\mathbf{F}_{\text{ext}}^e \tag{10.147}$$

Substituting the space-discrete Lagrangian into the space-discrete Lagrangian differential operator in Equation (10.60), yields

$$\mathfrak{L}_D\left(\mathcal{L}_e^h\right) = \frac{d}{dt}\left(\frac{\partial \mathcal{L}_e^h}{\partial \dot{\mathbf{q}}_e}\right) - \frac{\partial \mathcal{L}_e^h}{\partial \mathbf{q}_e} = \mathbf{M}^e\ddot{\mathbf{q}}_e + \mathbf{K}^e\mathbf{q}_e - \mathbf{F}_{\text{ext}}^e = \mathbf{0} \tag{10.148}$$

which is the finite element equation of motion governing the transverse dynamical motion of the beam.

Example 10.4 Timoshenko Beam

As a more complicated beam problem, we shall next consider the Timoshenko beam in transverse vibration.

Governing Equations: Lagrangian Density Based Representations

The dynamical behavior of the Timoshenko beam is governed by two differential equations, because of the presence of two dependent variables such as the transverse deflection $w(x,t)$ and the angle of rotation $\varphi(x,t)$ due to the bending moment acting on the beam subject to a transverse distributed load $\bar{\omega}(x)$. The total rotation of the Timoshenko beam is composed of two contributions as shown

$$\frac{\partial w(x,t)}{\partial x} = \varphi(x,t) + \chi(x,t) = -\theta_y(x,t) + \chi(x,t) \tag{10.149}$$

where $\chi(x,t)$ denotes the shear angle due to shear. It means that the beam undergoes transverse and rotational motions at the same time. Therefore, the kinetic energy consists of the translational and the rotational kinetic energies given as

$$\begin{gathered}\mathcal{K}(\dot{\mathbf{u}}) = \int_L \frac{1}{2}\dot{\mathbf{u}}\cdot\boldsymbol{\rho}\dot{\mathbf{u}}dx, \qquad \boldsymbol{\rho} = \begin{bmatrix} \rho A(x) & 0 \\ 0 & \rho I(x) \end{bmatrix} \\ \mathbf{u}(x,t) = \left(w(x,t), \theta_y(x,t)\right), \quad \dot{\mathbf{u}}(x,t) = \left(\dot{w}(x,t), \dot{\theta}_y(x,t)\right)\end{gathered} \tag{10.150}$$

where $I(x)$ denotes the second moment of inertia per unit length about the neutral axis of the cross-section, ρ density, and $A(x)$ the cross-sectional area. Further, the total potential energy of the beam subject to the distributed load $\bar{\omega}(x)$ is given by

$$\mathcal{U} = \frac{1}{2}\int_L M \frac{\partial \theta_y(x,t)}{\partial x} dx + \frac{1}{2}\int_L Q\chi(x,t)\,dx - \int_L \bar{\omega}(x)\, w(x,t)\,dx \tag{10.151}$$

where the bending moment $M(x,t)$ and the shearing force $Q(x,t)$ is given as

$$M(x,t) = EI(x)\frac{\partial \theta_y(x,t)}{\partial x}, \quad Q(x,t) = \kappa GA(x)\left(\frac{\partial w(x,t)}{\partial x} + \theta_y(x,t)\right) \tag{10.152}$$

where κ, E and G denote the shearing parameter, the elastic modulus, and the shear modulus. Thus, the total potential energy can be rewritten as

$$\begin{aligned}
\mathcal{U}(\mathbf{u}, \mathbf{u}') &= \frac{1}{2}\int_L \boldsymbol{\kappa} \cdot \mathbf{D}_b \boldsymbol{\kappa}\, dx + \int_L \frac{1}{2}\boldsymbol{\gamma} \cdot \mathbf{G}_s \boldsymbol{\gamma}\, dx - \int_L \bar{\omega}(x)\, \mathbf{a} \cdot \mathbf{u}\, dx, \\
\mathbf{u}' &= \left(\frac{\partial w(x,t)}{\partial x}, \frac{\partial \theta_y(x,t)}{\partial x}\right), \quad \mathbf{a} = (1, 0), \\
\boldsymbol{\kappa} &= \left[\frac{\partial \theta_y}{\partial x}\right], \; \mathbf{D}_b = [EI(x)], \; \boldsymbol{\gamma} = \left[\frac{\partial w(x,t)}{\partial x} + \theta_y(x,t)\right], \; \mathbf{G}_s = [\kappa GA(x)]
\end{aligned} \tag{10.153}$$

Eventually, the continuous Lagrangian is given as

$$\mathcal{L}(\mathbf{u}, \dot{\mathbf{u}}) = \int_L \bar{\mathcal{L}}(\mathbf{u}, \dot{u}, \mathbf{u}')\,dx + \int_L \bar{\omega}(x)\, \mathbf{a} \cdot \mathbf{u}\, dx \tag{10.154}$$

Moreover, the Lagrangian density function can be defined as

$$\bar{\mathcal{L}}(\mathbf{u}, \dot{\mathbf{u}}, \mathbf{u}') = \frac{1}{2}\dot{\mathbf{u}} \cdot \rho \dot{\mathbf{u}} - \frac{1}{2}\boldsymbol{\kappa} \cdot \mathbf{D}_b \boldsymbol{\kappa} - \frac{1}{2}\boldsymbol{\gamma} \cdot \mathbf{G}_s \boldsymbol{\gamma} \tag{10.155}$$

Thus, Hamilton's law of varying action yields

$$\delta \int_{t_0}^{t_1} \mathcal{L}(\mathbf{u}(x,t), \dot{\mathbf{u}}(x,t))\,dt - \int_L \frac{\partial \bar{\mathcal{L}}(\mathbf{u}, \dot{u}, \mathbf{u}')}{\partial \dot{\mathbf{u}}} \cdot \delta \mathbf{u}(x,t)\,dx \Bigg|_{t_0}^{t_1} = 0 \tag{10.156}$$

By use of chain rule, we have the variation of the Lagrangian density function

$$\delta \bar{\mathcal{L}}(\mathbf{u}, \dot{\mathbf{u}}, \mathbf{u}') = \frac{\partial \bar{\mathcal{L}}}{\partial \mathbf{u}} \cdot \delta \mathbf{u} + \frac{\partial \bar{\mathcal{L}}}{\partial \dot{\mathbf{u}}} \cdot \delta \dot{\mathbf{u}} + \frac{\partial \bar{\mathcal{L}}}{\partial \mathbf{u}'} \cdot \delta \mathbf{u}' \tag{10.157}$$

Consequently, Hamilton's law of varying action in Equation (10.156) can be rewritten as

$$\delta \int_{t_0}^{t_1} \mathcal{L}(\mathbf{u}(x,t), \dot{\mathbf{u}}(x,t))\,dt - \int_L \frac{\partial \bar{\mathcal{L}}(\mathbf{u}, \dot{\mathbf{u}}, \mathbf{u}')}{\partial \dot{\mathbf{u}}} \cdot \delta \mathbf{u}\, dx \Bigg|_{t_0}^{t_1}$$

$$
\begin{aligned}
&= -\int_{t_0}^{t_1} \left\{ \int_L \left(\left\{ \frac{d}{dt}\left(\frac{\partial \bar{\mathcal{L}}}{\partial \dot{\mathbf{u}}}\right) + \frac{\partial}{\partial x}\left(\frac{\partial \bar{\mathcal{L}}}{\partial \mathbf{u}'}\right) - \frac{\partial \bar{\mathcal{L}}}{\partial \mathbf{u}} - \bar{\omega}(x)\,\mathbf{a} \right\} \cdot \delta\mathbf{u} \right) dx \right\} dt \\
&+ \int_{t_0}^{t_1} \left(\frac{\partial \bar{\mathcal{L}}}{\partial \mathbf{u}'} \cdot \delta\mathbf{u} \Big|_0^L \right) dt = 0
\end{aligned} \tag{10.158}
$$

Since $\delta\mathbf{u}$ is arbitrary within the domain, we obtain the governing equation as follows:

$$
\frac{d}{dt}\left(\frac{\partial \bar{\mathcal{L}}}{\partial \dot{\mathbf{u}}}\right) + \frac{\partial}{\partial x}\left(\frac{\partial \bar{\mathcal{L}}}{\partial \mathbf{u}'}\right) - \frac{\partial \bar{\mathcal{L}}}{\partial \mathbf{u}} - \bar{\omega}(x)\,\mathbf{a} = 0 \tag{10.159}
$$

subject to the boundary conditions

$$
\frac{\partial \bar{\mathcal{L}}}{\partial \mathbf{u}'} \cdot \delta\mathbf{u} \Big|_0^L = 0 \tag{10.160}
$$

Note that Equation (10.159) can be expressed as below:

$$
\begin{aligned}
\frac{d}{dt}\left(\frac{\partial \bar{\mathcal{L}}}{\partial \dot{w}}\right) &= \frac{\partial \bar{\mathcal{L}}}{\partial w} - \frac{\partial}{\partial x}\left(\frac{\partial \bar{\mathcal{L}}}{\partial w'}\right) + \bar{\omega}(x), \\
\frac{d}{dt}\left(\frac{\partial \bar{\mathcal{L}}}{\partial \dot{\theta}_y}\right) &= \frac{\partial \bar{\mathcal{L}}}{\partial \theta_y} - \frac{\partial}{\partial x}\left(\frac{\partial \bar{\mathcal{L}}}{\partial \theta_y'}\right)
\end{aligned} \tag{10.161}
$$

while Equation (10.160) can be written as

$$
\frac{\partial \bar{\mathcal{L}}}{\partial w'}\delta w \Big|_0^L = 0, \quad \frac{\partial \bar{\mathcal{L}}}{\partial \theta_y'}\delta\theta_y \Big|_0^L = 0 \tag{10.162}
$$

Also note that, upon substitution of the Lagrangian density function in Equation (10.155) into Equation (10.161), the classical form of the resulting governing equations of motion are obtained as follows:

$$
\begin{aligned}
\rho A(x)\,\ddot{w} &= \frac{\partial}{\partial x}\left(\kappa G A(x)\left(\frac{\partial w}{\partial x} + \theta_y\right)\right) + \bar{\omega}(x), \\
\rho I(x)\,\ddot{\theta}_y &= \frac{\partial}{\partial x}\left(E I(x)\left(\frac{\partial \theta_y}{\partial x}\right)\right) - \kappa G A(x)\left(\frac{\partial y}{\partial x} + \theta_y\right)
\end{aligned} \tag{10.163}
$$

subject to the boundary condition

$$
E I(x)\left(\frac{\partial \theta_y}{\partial x}\right)\delta\theta_y \Big|_0^L = 0, \quad \kappa G A(x)\left(\frac{\partial w}{\partial x} + \theta_y\right)\delta w \Big|_0^L = 0 \tag{10.164}
$$

which can be rewritten as

$$
M(x,t)\,\delta\theta_y|_0^L = 0, \quad Q(x,t)\,\delta w|_0^L = 0 \tag{10.165}
$$

where $M(x,t)$, $Q(x,t)$ denote the bending moment and the shearing force. Note that boundary conditions in Equation (10.137) imply geometric and natural boundary conditions. If the end of the beam is fixed, $\delta\theta_y|_{x=0\ or\ x=L} = 0$ and $\delta w|_{x=0\ or\ x=L} = 0$. On the other hand, if the end is free, then, the bending moment and the shearing force must vanish, namely, $EI(x)\left(\frac{\partial\theta_y}{\partial x}\right)\Big|_{x=0\ or\ x=L} = 0$ and $\kappa GA(x)\left(\frac{\partial w}{\partial x} + \theta_y\right)\Big|_{x=0\ or\ x=L} = 0$.

We have derived the two governing equations of motion with the traction boundary condition via Hamilton's law of varying action.

Space-Discrete Finite Element Formulation

Next, we shall construct the space-discrete Lagrangian via constructing the space-discrete kinetic energy and potential energies. The trial displacement functions are approximated as

$$\begin{aligned} \tilde{w}^h(x,t) &= \mathbf{N}_w(x)\mathbf{w}^e(t), \quad \mathbf{w}^e(t) = \begin{bmatrix} w_1(t) & w_2(t) & \ldots & w_{n_{\mathrm{node}}}(t) \end{bmatrix}^T, \\ \tilde{\theta}_y^h(x,t) &= \mathbf{N}_\theta(x)\boldsymbol{\theta}_y^e(t), \quad \boldsymbol{\theta}_y^e(t) = \begin{bmatrix} \theta_{y1}(t) & \theta_{y2}(t) & \ldots & \theta_{yn_{\mathrm{node}}}(t) \end{bmatrix}^T \end{aligned} \tag{10.166}$$

Upon introducing a new notation $\mathbf{q}_e(t) = \begin{bmatrix} \mathbf{w}^e(t) & \boldsymbol{\theta}_y^e(t) \end{bmatrix}^T$, the trial functions are rewritten as

$$\tilde{w}^h(x,t) = \mathbf{H}_w(x)\mathbf{q}_e(t), \quad \tilde{\theta}_y^h(x,t) = \mathbf{H}_\theta(x)\mathbf{q}_e(t) \tag{10.167}$$

where $\mathbf{H}_w(x) = \begin{bmatrix} \mathbf{N}_w(x) & \mathbf{0} \end{bmatrix}$ and $\mathbf{H}_\theta(x) = \begin{bmatrix} \mathbf{0} & \mathbf{N}_\theta(x) \end{bmatrix}$. In vector form, we have

$$\tilde{\mathbf{u}}^h = \begin{Bmatrix} \tilde{w}^h(x,t) \\ \tilde{\theta}_y^h(x,t) \end{Bmatrix} = \mathbf{H}_e(x)\mathbf{q}_e(t), \quad \mathbf{H}_e(x) = \begin{bmatrix} \mathbf{H}_w(x) \\ \mathbf{H}_\theta(x) \end{bmatrix} \tag{10.168}$$

Thus, we obtain the space-discrete kinetic energy

$$\mathcal{K}_e^h(\dot{\mathbf{q}}_e) = \frac{1}{2}\dot{\mathbf{q}}_e(t)\cdot\mathbf{M}^e\dot{\mathbf{q}}_e(t), \quad \mathbf{M}^e = \frac{1}{2}\int_{L_e}\mathbf{H}_e^T(x)\,\rho\mathbf{H}_e(x)\,dx \tag{10.169}$$

where L_e denotes the length of the element. Likewise, the space-discrete potential energy can be obtained as

$$\begin{aligned} \mathcal{U}_e^h(\mathbf{q}_e) &= \frac{1}{2}\mathbf{q}_e(t)\cdot\mathbf{K}^e\mathbf{q}_e(t) - \mathbf{q}_e(t)\cdot\mathbf{F}_{\mathrm{ext}}^e, \\ \mathbf{F}_{\mathrm{ext}}^e &= \int_{L_e}\bar{\omega}(x)\,\mathbf{H}_w^T(x)\,dx \end{aligned} \tag{10.170}$$

where $\mathbf{K}^e = \mathbf{K}_b + \mathbf{K}_s$ and

$$\begin{aligned} \mathbf{K}_b &= \int_{L_e}\mathbf{B}_b^T(x)\,\mathbf{D}_b\mathbf{B}_b(\mathbf{x})\,dx, \quad \mathbf{K}_s = \int_{L_e}\mathbf{B}_s^T(x)\,\mathbf{G}_s\mathbf{B}_s(x)\,dx, \\ \mathbf{B}_b &= \left[\frac{\partial\mathbf{H}_\theta(x)}{\partial x}\right], \quad \mathbf{B}_s = \left[\frac{\partial\mathbf{H}_w(x)}{\partial x} + \mathbf{H}_\theta(x)\right] \end{aligned} \tag{10.171}$$

whose details can be found in Section 9.2.7. In the end, the space-discrete Lagrangian is constructed as

$$\mathcal{L}_e^h(\mathbf{q}_e,\dot{\mathbf{q}}_e) = \frac{1}{2}\dot{\mathbf{q}}_e\cdot\mathbf{M}^e\dot{\mathbf{q}}_e - \frac{1}{2}\mathbf{q}_e\cdot\mathbf{K}^e\mathbf{q}_e + \mathbf{q}_e\cdot\mathbf{F}_{\mathrm{ext}}^e \tag{10.172}$$

Substituting the space-discrete Lagrangian into the space-discrete Lagrangian differential operator in Equation (10.60), simply leads to the semi-discrete finite element equations of motion within the Lagrangian mechanics framework:

$$\mathfrak{L}_D\left(\mathcal{L}_e^h\right) = \frac{d}{dt}\left(\frac{\partial \mathcal{L}_e^h}{\partial \dot{\mathbf{q}}_e}\right) - \frac{\partial \mathcal{L}_e^h}{\partial \mathbf{q}_e} = \mathbf{M}^e \ddot{\mathbf{q}}_e + \mathbf{K}^e \mathbf{q}_e - \mathbf{F}_{\text{ext}}^e = \mathbf{0} \tag{10.173}$$

Example 10.5 Reissner-Mindlin Plate

We next consider the same Reissner-Mindlin plate of a uniform thickness h, which we previously discussed in Section 9.1.7.

Governing Equations: Lagrangian Density Based Representations

The strain energy is composed of the bending strain energy $\mathcal{U}_b(\kappa)$ and the shear strain energy $\mathcal{U}_s(\gamma)$ as below:

$$\begin{aligned}
&\mathcal{U}_b = \frac{1}{2}\iint_\Omega \boldsymbol{\kappa}\cdot\mathbf{D}_b\boldsymbol{\kappa}\,dxdy, \quad \mathcal{U}_s = \frac{1}{2}\iint_\Omega \boldsymbol{\gamma}\cdot\mathbf{G}_s\boldsymbol{\gamma}\,dxdy \\
&\boldsymbol{\kappa} = \left(\frac{\partial\theta_y}{\partial x}, -\frac{\partial\theta_x}{\partial y}, \left(\frac{\partial\theta_y}{\partial y} - \frac{\partial\theta_x}{\partial x}\right)\right), \quad \boldsymbol{\gamma} = \left(\frac{\partial w}{\partial x} + \theta_y, \frac{\partial w}{\partial y} - \theta_x\right), \\
&\mathbf{D}_b = D\begin{bmatrix} 1 & \nu & 0 \\ \nu & 1 & 0 \\ 0 & 0 & \dfrac{1-\nu}{2} \end{bmatrix}, \quad \mathbf{G}_s = \begin{bmatrix} Ghk & 0 \\ 0 & Ghk \end{bmatrix}, \quad D = \frac{Eh^3}{12\left(1-\nu^2\right)}
\end{aligned} \tag{10.174}$$

where k denotes the shear correction factor, G is the shear modulus, D is the plate flexural rigidity, and Ω is the domain of interest in the mid-surface of the flat plate. In the dynamic behavior of the plate, the deformation of the plate is associated with transverse and rotational motions at the same time. Therefore, the kinetic energy $\mathcal{K}(\dot{\mathbf{u}})$ consists of the transverse translational kinetic energy and the rotational kinetic energy such as

$$\begin{aligned}
&\mathcal{K}(\dot{\mathbf{u}}) = \tfrac{1}{2}\int_\Omega \dot{\mathbf{u}}\cdot\boldsymbol{\rho}\dot{\mathbf{u}}\,dxdy, \qquad \boldsymbol{\rho} = \begin{bmatrix} \rho h & 0 & 0 \\ 0 & \rho I_p & 0 \\ 0 & 0 & \rho I_p \end{bmatrix}, \\
&\dot{\mathbf{u}} = \left(\dot{w}, \dot{\theta}_x, \dot{\theta}_y\right), \qquad I_p = \frac{h^3}{12}
\end{aligned} \tag{10.175}$$

Consequently, Hamilton's law of varying action in the context of the Lagrangian gives

$$\begin{aligned}
&\delta\int_{t_0}^{t_1} \mathcal{L}(\mathbf{u}, \dot{\mathbf{u}})\,dt - \int_\Omega \frac{\partial\bar{\mathcal{L}}\left(\mathbf{u}, \dot{\mathbf{u}}, \mathbf{u}_{,x}, \mathbf{u}_{,y}\right)}{\partial\dot{\mathbf{u}}}\cdot\delta\mathbf{u}\,dxdy\Bigg|_{t_0}^{t_1} = 0, \\
&\mathbf{u} = \left(w, \theta_x, \theta_y\right), \quad \mathbf{u}_{,x} = \left(\frac{\partial w}{\partial x}, \frac{\partial\theta_x}{\partial x}, \frac{\partial\theta_y}{\partial x}\right), \quad \mathbf{u}_{,y} = \left(\frac{\partial w}{\partial y}, \frac{\partial\theta_x}{\partial y}, \frac{\partial\theta_y}{\partial y}\right)
\end{aligned} \tag{10.176}$$

where the continuous Lagrangian of the Reissner-Mindlin plate can be obtained as

$$\mathcal{L}(\mathbf{u}, \dot{\mathbf{u}}) = \int_{\Omega} \bar{\mathcal{L}}\left(\mathbf{u}, \dot{\mathbf{u}}, \mathbf{u}_{,x}, \mathbf{u}_{,y}\right) dxdy + \int_{\Omega} \omega(x, y)\, \mathbf{a} \cdot \mathbf{u}\, dxdy, \quad \mathbf{a} = (1, 0, 0) \tag{10.177}$$

and the Lagrangian density function is defined as

$$\begin{gathered} \bar{\mathcal{L}}\left(\mathbf{u}, \dot{\mathbf{u}}, \mathbf{u}_{,x}, \mathbf{u}_{,y}\right) = \frac{1}{2}\dot{\mathbf{u}} \cdot \rho\dot{\mathbf{u}} - \frac{1}{2}\boldsymbol{\kappa} \cdot \mathbf{D}_b\boldsymbol{\kappa} - \frac{1}{2}\boldsymbol{\gamma} \cdot \mathbf{G}_s\boldsymbol{\gamma}, \\ \mathbf{p} = \rho\dot{\mathbf{u}}, \quad \mathbf{M}_b = \mathbf{D}_b\boldsymbol{\kappa}, \quad \mathbf{q}_s = \mathbf{G}_s\boldsymbol{\gamma} \end{gathered} \tag{10.178}$$

Via the introduction of the new notations, the Lagrangian density function can be given by

$$\begin{gathered} \bar{\mathcal{L}}\left(\mathbf{u}, \dot{\mathbf{u}}, \mathbf{u}_{,x}, \mathbf{u}_{,y}\right) = \frac{1}{2}\dot{\mathbf{u}} \cdot \mathbf{p} - \frac{1}{2}\boldsymbol{\kappa} \cdot \mathbf{M}_b - \frac{1}{2}\boldsymbol{\gamma} \cdot \mathbf{q}_s, \\ \mathbf{p} = \rho\dot{\mathbf{u}}, \quad \mathbf{M}_b = \mathbf{D}_b\boldsymbol{\kappa}, \quad \mathbf{q}_s = \mathbf{G}_s\boldsymbol{\gamma} \end{gathered} \tag{10.179}$$

where the generalized momentum and the stress resultants are given by

$$\begin{aligned} \mathbf{p}^T &= \left[\begin{array}{ccc} \rho\dot{w}h & \rho\dot{\theta}_x I_p & \rho\dot{\theta}_y I_p \end{array}\right], \\ \mathbf{M}_b{}^T &= \left[\begin{array}{ccc} M_{xx} & M_{yy} & M_{xy} \end{array}\right], \\ \mathbf{q}_s{}^T &= \left[\begin{array}{cc} Q_x & Q_y \end{array}\right] \end{aligned} \tag{10.180}$$

Since the variation of the autonomous Lagrangian is given by

$$\begin{aligned} \delta\mathcal{L}(\mathbf{u}, \dot{\mathbf{u}}) &= \int_{\Omega} \delta\bar{\mathcal{L}}\left(\mathbf{u}, \dot{\mathbf{u}}, \mathbf{u}_{,x}, \mathbf{u}_{,y}\right) dxdy + \int_{\Omega} p(x, y)\, \mathbf{a} \cdot \mathbf{u}\, dxdy \\ &= \int_{\Omega} \frac{\partial\bar{\mathcal{L}}}{\partial\mathbf{u}} \cdot \delta\mathbf{u} + \frac{\partial\bar{\mathcal{L}}}{\partial\dot{\mathbf{u}}} \cdot \delta\dot{\mathbf{u}} + \frac{\partial\bar{\mathcal{L}}}{\partial\mathbf{u}_{,x}} \cdot \delta\mathbf{u}_{,x} + \frac{\partial\bar{\mathcal{L}}}{\partial\mathbf{u}_{,y}} \cdot \delta\mathbf{u}_{,y}\, dxdy \\ &\quad + \int_{\Omega} \omega(x, y)\, \mathbf{a} \cdot \delta\mathbf{u}\, dxdy \end{aligned} \tag{10.181}$$

then, integration by parts yields

$$\begin{aligned} \int_{t_0}^{t_1} \left(\int_{\Omega} \frac{\partial\bar{\mathcal{L}}}{\partial\dot{\mathbf{u}}} \cdot \delta\dot{\mathbf{u}}\, dxdy \right) dt &= \int_{t_0}^{t_1} \left(-\int_{\Omega} \frac{d}{dt}\left(\frac{\partial\bar{\mathcal{L}}}{\partial\dot{\mathbf{u}}}\right) \cdot \delta\mathbf{u}\, dxdy \right) dt \\ &\quad + \int_{\Omega} \frac{\partial\bar{\mathcal{L}}}{\partial\dot{\mathbf{u}}} \cdot \delta\mathbf{u}\, dxdy \Bigg|_{t_0}^{t_1} \end{aligned} \tag{10.182}$$

Hamilton's law of varying action leads to

$$\int_{t_0}^{t_1} \left\{ \int_{\Omega} \left[\begin{array}{l} -\dfrac{d}{dt}\left(\dfrac{\partial\bar{\mathcal{L}}}{\partial\dot{\mathbf{u}}}\right) \cdot \delta\mathbf{u} + \dfrac{\partial\bar{\mathcal{L}}}{\partial\mathbf{u}} \cdot \delta\mathbf{u} \\ +\dfrac{\partial\bar{\mathcal{L}}}{\partial\mathbf{u}_{,x}} \cdot \delta\mathbf{u}_{,x} + \dfrac{\partial\bar{\mathcal{L}}}{\partial\mathbf{u}_{,y}} \cdot \delta\mathbf{u}_{,y} + p(x, y)\, \mathbf{a} \cdot \delta\mathbf{u} \end{array} \right] dxdy \right\} dt = 0 \tag{10.183}$$

By means of Green's theorem, we have the following:

$$\int_\Omega \left(\frac{\partial \bar{\mathcal{L}}}{\partial \mathbf{u}_{,x}} \cdot \delta \mathbf{u}_{,x} + \frac{\partial \bar{\mathcal{L}}}{\partial \mathbf{u}_{,y}} \cdot \delta \mathbf{u}_{,y} \right) dxdy$$

$$= - \int_\Omega \left[\frac{\partial}{\partial x} \left(\frac{\partial \bar{\mathcal{L}}}{\partial \mathbf{u}_{,x}} \right) + \frac{\partial}{\partial y} \left(\frac{\partial \bar{\mathcal{L}}}{\partial \mathbf{u}_{,y}} \right) \right] \cdot \delta \mathbf{u} \, dxdy \tag{10.184}$$

$$+ \int_{\partial\Omega} \left(\frac{\partial \bar{\mathcal{L}}}{\partial \mathbf{u}_{,x}} \cdot \delta \mathbf{u}_x n_x + \frac{\partial \bar{\mathcal{L}}}{\partial \mathbf{u}_{,y}} \cdot \delta \mathbf{u}_y n_y \right) d\Gamma = 0, \quad n_x = \frac{\partial x}{\partial n}, \; n_y = \frac{\partial y}{\partial n}$$

where n_x, n_y denote the directional cosines respectively and n stands for the coordinate of the outward unit vector normal to the boundary surface. hence, we finally have

$$- \int_{t_0}^{t_1} \left\{ \int_\Omega \left[\begin{array}{l} \dfrac{d}{dt} \left(\dfrac{\partial \bar{\mathcal{L}}}{\partial \dot{\mathbf{u}}} \right) - \dfrac{\partial \bar{\mathcal{L}}}{\partial \mathbf{u}} + \dfrac{\partial}{\partial x} \left(\dfrac{\partial \bar{\mathcal{L}}}{\partial \mathbf{u}_{,x}} \right) \\ + \dfrac{\partial}{\partial y} \left(\dfrac{\partial \bar{\mathcal{L}}}{\partial \mathbf{u}_{,y}} \right) - \omega(x, y) \, \mathbf{a} \end{array} \right] \cdot \delta \mathbf{u} \, dxdy \right\} dt$$
$$+ \int_{\partial\Omega} \left(\frac{\partial \bar{\mathcal{L}}}{\partial \mathbf{u}_{,x}} \cdot \delta \mathbf{u}_x n_x + \frac{\partial \bar{\mathcal{L}}}{\partial \mathbf{u}_{,y}} \cdot \delta \mathbf{u}_y n_y \right) d\Gamma = 0 \tag{10.185}$$

Consequently, we have the governing equations for the Reissner-Mindlin plate as follows:

$$\frac{d}{dt} \left(\frac{\partial \bar{\mathcal{L}}}{\partial \dot{\mathbf{u}}} \right) - \frac{\partial \bar{\mathcal{L}}}{\partial \mathbf{u}} + \frac{\partial}{\partial x} \left(\frac{\partial \bar{\mathcal{L}}}{\partial \mathbf{u}_{,x}} \right) + \frac{\partial}{\partial y} \left(\frac{\partial \bar{\mathcal{L}}}{\partial \mathbf{u}_{,y}} \right) - \omega(x, y) \, \mathbf{a} = 0 \tag{10.186}$$

The above equations can be rewritten as

$$\frac{d}{dt} (\rho \dot{\mathbf{u}}) + \left[\frac{\partial \boldsymbol{\kappa}}{\partial \mathbf{u}} \right]^T \mathbf{M}_b + \left[\frac{\partial \boldsymbol{\gamma}}{\partial \mathbf{u}} \right]^T \mathbf{Q}_s + \frac{\partial \mathbf{A}_x}{\partial x} + \frac{\partial \mathbf{A}_y}{\partial y} - p(x, y) \, \mathbf{a} = 0 \tag{10.187}$$

where

$$\mathbf{A}_x = \frac{\partial \bar{\mathcal{L}}}{\partial \mathbf{u}_{,x}} = - \left[\frac{\partial \boldsymbol{\kappa}}{\partial \mathbf{u}_{,x}} \right]^T \mathbf{M}_b - \left[\frac{\partial \boldsymbol{\gamma}}{\partial \mathbf{u}_{,x}} \right]^T \mathbf{Q}_s,$$
$$\mathbf{A}_y = \frac{\partial \bar{\mathcal{L}}}{\partial \mathbf{u}_{,y}} = - \left[\frac{\partial \boldsymbol{\kappa}}{\partial \mathbf{u}_{,y}} \right]^T \mathbf{M}_b - \left[\frac{\partial \boldsymbol{\gamma}}{\partial \mathbf{u}_{,y}} \right]^T \mathbf{Q}_s \tag{10.188}$$

Note that the above matrices are given by

$$\left[\frac{\partial \boldsymbol{\kappa}}{\partial \mathbf{u}} \right] = \begin{bmatrix} 0 & 0 & 0 \\ 0 & 0 & 0 \\ 0 & 0 & 0 \end{bmatrix}, \quad \left[\frac{\partial \boldsymbol{\gamma}}{\partial \mathbf{u}} \right] = \begin{bmatrix} 0 & 0 & -1 \\ 0 & 1 & 0 \end{bmatrix},$$

$$\left[\frac{\partial \boldsymbol{\kappa}}{\partial \mathbf{u}_{,x}}\right] = \begin{bmatrix} 0 & 1 & 0 \\ 0 & 0 & 0 \\ 0 & 0 & -1 \end{bmatrix}, \quad \left[\frac{\partial \boldsymbol{\gamma}}{\partial \mathbf{u}_{,x}}\right] = \begin{bmatrix} 0 & 0 & 0 \\ 1 & 0 & 0 \end{bmatrix}, \tag{10.189}$$

$$\left[\frac{\partial \boldsymbol{\kappa}}{\partial \mathbf{u}_{,y}}\right] = \begin{bmatrix} 0 & 0 & 0 \\ 0 & 0 & -1 \\ 0 & 1 & 0 \end{bmatrix}, \quad \left[\frac{\partial \boldsymbol{\gamma}}{\partial \mathbf{u}_{,y}}\right] = \begin{bmatrix} 1 & 0 & 0 \\ 0 & 0 & 0 \end{bmatrix}$$

We observe that Equation (10.187) is identical to Equation (9.134). Notice Equation (10.186) indicates the following:

$$\begin{gathered}
\frac{d}{dt}\left(\frac{\partial \bar{\mathcal{L}}}{\partial \dot{w}}\right) - \frac{\partial \bar{\mathcal{L}}}{\partial w} + \frac{\partial}{\partial x}\left(\frac{\partial \bar{\mathcal{L}}}{\partial w_{,x}}\right) + \frac{\partial}{\partial y}\left(\frac{\partial \bar{\mathcal{L}}}{\partial w_{,y}}\right) - \omega(x, y) = 0, \\
\frac{d}{dt}\left(\frac{\partial \bar{\mathcal{L}}}{\partial \dot{\theta}_x}\right) - \frac{\partial \bar{\mathcal{L}}}{\partial \theta_x} + \frac{\partial}{\partial x}\left(\frac{\partial \bar{\mathcal{L}}}{\partial \theta_{x,x}}\right) + \frac{\partial}{\partial y}\left(\frac{\partial \bar{\mathcal{L}}}{\partial \theta_{x,y}}\right) = 0, \\
\frac{d}{dt}\left(\frac{\partial \bar{\mathcal{L}}}{\partial \dot{\theta}_y}\right) - \frac{\partial \bar{\mathcal{L}}}{\partial \theta_y} + \frac{\partial}{\partial x}\left(\frac{\partial \bar{\mathcal{L}}}{\partial \theta_{y,x}}\right) + \frac{\partial}{\partial y}\left(\frac{\partial \bar{\mathcal{L}}}{\partial \theta_{y,y}}\right) = 0
\end{gathered} \tag{10.190}$$

which leads to the same equation as Equation (9.135). The second term in the right-hand side of Equation (10.185) gives the natural boundary conditions as follows:

$$\begin{aligned}
\frac{\partial \bar{\mathcal{L}}}{\partial \mathbf{u}_{,x}} \cdot \delta \mathbf{u}_x n_x + \frac{\partial \bar{\mathcal{L}}}{\partial \mathbf{u}_{,y}} \cdot \delta \mathbf{u}_y n_y &= \mathbf{A}_x^T \delta \mathbf{u}_x n_x + \mathbf{A}_y^T \delta \mathbf{u}_y n_y \\
&= n_x\left(-Q_y \delta w - M_{xx}\delta\theta_y + M_{xy}\delta\theta_x\right) \\
&\quad + n_y\left(-Q_x \delta w - M_{xy}\delta\theta_y + M_{yy}\delta\theta_x\right) \\
&= -\delta w\left(n_x Q_y + n_y Q_x\right) \\
&\quad - \delta\theta_y\left(n_x M_{xx} + n_y M_{xy}\right) \\
&\quad + \delta\theta_x\left(n_x M_{xy} + n_y M_{yy}\right) = 0
\end{aligned} \tag{10.191}$$

In view of Equation (10.191), we can say that the governing equations in Equation (10.186) are subject to the natural boundary conditions on the mechanical boundary surface, $\partial\Omega_\sigma$ as follows,

$$\left.\begin{aligned}
n_x Q_y + n_y Q_x &= 0, \\
n_x M_{xx} + n_y M_{xy} &= 0, \\
n_x M_{xy} + n_y M_{yy} &= 0
\end{aligned}\right\} \quad \text{on } \partial\Omega_\sigma \tag{10.192}$$

Note that the essential boundary conditions on the geometric boundary surface, $\partial\Omega_u$, can be given by

$$w = \bar{w}, \quad \theta_x = \bar{\theta}_x, \quad \theta_y = \bar{\theta}_y \quad \text{on } \partial\Omega_u \tag{10.193}$$

where $\bar{w}$, $\bar{\theta}_x$, and $\bar{\theta}_y$ are the prescribed displacement and rotation values. Subsequently, we observe that Hamilton's law of varying action in Equation (10.185) includes both governing equations and natural boundary conditions as the Bubnov-Galerkin weighted residual form.

Space-Discrete Finite Element Formulation

Next, we show that the space-discrete Lagrangian formulation offers to facilitate the derivation of the semi-discrete finite element equations of motion in contrast to the principle of virtual work. Instead of substituting the admissible functions into the Bubnov-Galerkin weighted residual form in Equation (10.185), we construct first the space-discrete autonomous Lagrangian. The admissible trial functions are given by

$$\begin{aligned}
\tilde{w}^h(x,t) &= \mathbf{N}_w(x)\,\mathbf{w}^e(t), \quad \mathbf{w}^e(t) = \begin{bmatrix} w_1(t) & w_2(t) & \dots & w_{n_{\text{node}}}(t) \end{bmatrix}^T, \\
\tilde{\theta}_x^h(x,t) &= \mathbf{N}_{\theta_x}(x)\,\boldsymbol{\theta}_y^e(t), \quad \boldsymbol{\theta}_x^e(t) = \begin{bmatrix} \theta_{y1}(t) & \theta_{y2}(t) & \dots & \theta_{yn_{\text{node}}}(t) \end{bmatrix}^T, \\
\tilde{\theta}_y^h(x,t) &= \mathbf{N}_{\theta_y}(x)\,\boldsymbol{\theta}_y^e(t), \quad \boldsymbol{\theta}_y^e(t) = \begin{bmatrix} \theta_{y1}(t) & \theta_{y2}(t) & \dots & \theta_{yn_{\text{node}}}(t) \end{bmatrix}^T
\end{aligned} \tag{10.194}$$

Upon introducing a new notation $\mathbf{q}_e(t) = \begin{bmatrix} \mathbf{w}^e(t) & \boldsymbol{\theta}_x^e(t) & \boldsymbol{\theta}_y^e(t) \end{bmatrix}^T$, the trial functions are rewritten as

$$\begin{aligned}
\bar{w}^h &= \mathbf{H}_w(x,y)\,\mathbf{q}_e(t), \; \bar{\theta}_x^h = \mathbf{H}_{\theta_x}(x,y)\,\mathbf{q}_e(t), \; \bar{\theta}_y^h(x,t) = \mathbf{H}_{\theta_y}(x,y)\,\mathbf{q}_e(t), \\
\mathbf{H}_w &= \begin{bmatrix} \mathbf{N}_w & \mathbf{0} & \mathbf{0} \end{bmatrix}, \; \mathbf{H}_{\theta_x} = \begin{bmatrix} \mathbf{0} & \mathbf{N}_{\theta_x} & \mathbf{0} \end{bmatrix}, \; \mathbf{H}_{\theta_y} = \begin{bmatrix} \mathbf{0} & \mathbf{0} & \mathbf{N}_{\theta_y} \end{bmatrix}
\end{aligned} \tag{10.195}$$

The space-discrete Lagrangian for one element may be constructed as

$$\begin{aligned}
&\mathcal{L}_e^h(\mathbf{q}_e, \dot{\mathbf{q}}_e) = \frac{1}{2}\dot{\mathbf{q}}_e(t)\cdot\mathbf{M}^e\dot{\mathbf{q}}_e(t) - \frac{1}{2}\mathbf{q}_e(t)\cdot\mathbf{K}^e\mathbf{q}_e(t) + \mathbf{q}_e(t)\cdot\mathbf{F}_{\text{ext}}^e, \\
&\mathbf{M}^e = \frac{1}{2}\int_{A_e}\mathbf{H}_e^T\rho\mathbf{H}_e\,dxdy, \; \mathbf{H}_e = \begin{bmatrix} \mathbf{H}_w \\ \mathbf{H}_{\theta_x} \\ \mathbf{H}_{\theta y} \end{bmatrix}, \; \mathbf{F}_{\text{ext}}^e = \int_{\Omega_e}\omega(x,y)\,\mathbf{H}_w^T\,dxdy \\
&\mathbf{K}^e = \mathbf{K}_b + \mathbf{K}_s, \; \mathbf{K}_b^e = \frac{1}{2}\iint_{\Omega_e}\mathbf{B}_b^T\mathbf{D}_b\mathbf{B}_b\,dxdy, \; \mathbf{K}_s^e = \frac{k}{2}\iint_{\Omega_e}\mathbf{B}_s^T\mathbf{G}_s\mathbf{B}_s\,dxdy, \\
&\mathbf{B}_b = \begin{bmatrix} \dfrac{\partial\mathbf{H}_{\theta_y}}{\partial x} \\ -\dfrac{\partial\mathbf{H}_{\theta_x}}{\partial y} \\ \left(\dfrac{\partial\mathbf{H}_{\theta_y}}{\partial y} - \dfrac{\partial\mathbf{H}_{\theta_x}}{\partial x}\right) \end{bmatrix}, \; \mathbf{B}_s = \begin{bmatrix} \dfrac{\partial\mathbf{H}_w}{\partial x} + \mathbf{H}_{\theta_y} \\ \dfrac{\partial\mathbf{H}_w}{\partial y} - \mathbf{H}_{\theta_x} \end{bmatrix}
\end{aligned} \tag{10.196}$$

More details about $\mathbf{M}^e$, $\mathbf{K}^e$ and $\mathbf{F}^e_{\text{ext}}$ can be found in Section 9.2.9. Substituting the space-discrete Lagrangian in Equation (10.196) into the space-discrete Lagrangian differential operator in Equation (10.60), naturally leads to the semi-discrete finite element equations of motion:

$$\mathfrak{L}_D\left(\mathcal{L}_e^h\right) = \frac{d}{dt}\left(\frac{\partial \mathcal{L}_e^h}{\partial \dot{\mathbf{q}}_e}\right) - \frac{\partial \mathcal{L}_e^h}{\partial \mathbf{q}_e} = \mathbf{M}^e\ddot{\mathbf{q}}_e + \mathbf{K}^e\mathbf{q}_e - \mathbf{F}^e_{\text{ext}} = \mathbf{0} \tag{10.197}$$

10.4 HAMILTONIAN MECHANICS FRAMEWORK AND FINITE ELEMENT FORMULATIONS

In this section, analogous to the Lagrangian framework described previously, we embark upon starting from and selecting Hamilton's law of varying action in the context of a descriptive scalar function, namely, the Hamiltonian to illustrate the following: (1) First we explain the derivation of the continuous mathematical governing equations in space and time. Most importantly, by introducing the Hamiltonian density function, Hamilton's law of varying action can be shown to be able to be strictly recast and represent the statement of the two-field weighted residual form involving two integral equations, i.e., volume and surface integral equations. The first involves the Hamiltonian version of Cauchy's equation of motion and the second the Hamiltonian version of Cauchy's law, just as we have Cauchy's equation of motion and Cauchy's law within the Newtonian mechanics framework. Integration by parts leads to the Hamiltonian version of the weak statement or equivalently, the principle of virtual work for the two-field form in dynamics, in which the continuity requirement can be weakened in terms of the order of the highest derivative, (2) Second, whilst the two-field weighted residual form leading to the weak form or the principle of virtual work is strictly a statement of the weak form in space only, we demonstrate that Hamilton's law of varying action involving the Hamiltonian represents consequently the weak form in space and the statement of the weighted residual method in time; wherein for space, we also show that in itself it embodies the weak form that can be described by a discrete Hamiltonian Differential Operator, and (3) Third, we demonstrate a simple, space-discrete Hamiltonian formulation that directly yields the semi-discrete ordinary differential equations in time as with traditional practices but from a different viewpoint.

10.4.1 Hamiltonian Density Equations of Motion

In an autonomous Hamiltonian system, we shall introduce the generalized momentum as a dependent variable, namely,

$$\boldsymbol{\phi}(\mathbf{X}, t) := \frac{\partial \bar{\mathcal{L}}(\boldsymbol{\varphi}, \dot{\boldsymbol{\varphi}}, \mathbf{F})}{\partial \dot{\boldsymbol{\varphi}}} = \rho_0 \dot{\boldsymbol{\varphi}}(\mathbf{X}, t) \tag{10.198}$$

In the following, we define the cotangent space $T^*_{\varphi}\mathcal{C}$, dual to the tangent space $T_{\varphi}\mathcal{C}$, to which the canonical momentum belongs.

$$T^*_{\varphi}\mathcal{C} := \left\{(\boldsymbol{\phi}) : \mathcal{B} \times \bar{\mathcal{T}} \rightarrow \mathbb{R}^3\right\} \tag{10.199}$$

The autonomous Hamiltonian $\mathcal{H}(\boldsymbol{\phi}, \boldsymbol{\varphi}) : T^*\mathcal{C} \rightarrow \mathbb{R}$ takes the form

$$\mathcal{H}(\boldsymbol{\phi}, \boldsymbol{\varphi}) = \mathcal{K}(\boldsymbol{\phi}) + \mathcal{U}(\boldsymbol{\varphi}) \tag{10.200}$$

The Hamiltonian for autonomous infinite dimensional Hamiltonian structures is given by

$$\begin{aligned}\mathcal{H}(\boldsymbol{\phi}, \boldsymbol{\varphi}) = &\int_{\mathcal{B}} \frac{1}{2\rho_0} \boldsymbol{\phi}(\mathbf{X}, t) \cdot \boldsymbol{\phi}(\mathbf{X}, t)\, dV + \int_{\mathcal{B}} \rho_0 \Psi(\mathbf{F})\, dV \\ &- \int_{\mathcal{B}} \rho_0 \mathbf{B}(\mathbf{X}) \cdot \boldsymbol{\varphi}(\mathbf{X}, t)\, dV - \int_{\partial \mathcal{B}_\sigma} \bar{\mathbf{T}}(\mathbf{X}, \hat{\mathbf{N}}) \cdot \boldsymbol{\varphi}(\mathbf{X}, t)\, dA\end{aligned} \tag{10.201}$$

where the traction and the body force are assumed to be dead loads.

Via the Legendre transformation, we may define the Hamiltonian density function as

$$\bar{\mathcal{H}}(\boldsymbol{\phi}, \boldsymbol{\varphi}, \mathbf{F}) = \boldsymbol{\phi} \cdot \dot{\boldsymbol{\varphi}} - \bar{\mathcal{L}}(\boldsymbol{\varphi}, \dot{\boldsymbol{\varphi}}, \mathbf{F}) \tag{10.202}$$

where $\boldsymbol{\phi} = \rho_0 \dot{\boldsymbol{\varphi}}(\mathbf{X}, t)$. In fact, the Hamiltonian density function is given by

$$\bar{\mathcal{H}}(\boldsymbol{\phi}, \boldsymbol{\varphi}, \mathbf{F}) = \frac{1}{2\rho_0} \boldsymbol{\phi}(\mathbf{X}, t) \cdot \boldsymbol{\phi}(\mathbf{X}, t) + \rho_0 \Psi(\mathbf{F}) - \rho_0 \mathbf{B}(\mathbf{X}) \cdot \boldsymbol{\varphi}(\mathbf{X}, t) \tag{10.203}$$

Therefore, in the Hamiltonian mechanics framework, Equation (10.12) can be represented as

$$\delta \left\{ \int_{\mathcal{B}} \boldsymbol{\phi} \cdot \dot{\boldsymbol{\varphi}}\, dV - \mathcal{H}(\boldsymbol{\phi}, \boldsymbol{\varphi}) \right\} - \frac{d}{dt} \int_{\mathcal{B}} \boldsymbol{\phi} \cdot \delta \boldsymbol{\varphi}\, dV = 0 \tag{10.204}$$

Integrating both sides of Equation (10.204) over the closed time interval of interest $\bar{\mathcal{T}} = [t_0, t_1]$ yields Hamilton's law of varying action in the Hamiltonian mechanics framework for continuous-body dynamical systems as follows:

$$\text{HLVA}_{\mathcal{H}}: \ \delta \int_{t_0}^{t_1} \left\{ \int_{\mathcal{B}} \boldsymbol{\phi} \cdot \dot{\boldsymbol{\varphi}}\, dV - \mathcal{H}(\boldsymbol{\phi}, \boldsymbol{\varphi}) \right\} dt - \left[\int_{\mathcal{B}} \boldsymbol{\phi} \cdot \delta \boldsymbol{\varphi}\, dV \right] \Bigg|_{t_0}^{t_1} = 0 \tag{10.205}$$

where the Hamiltonian $\mathcal{H}(\boldsymbol{\phi}, \boldsymbol{\varphi}) : T^* \mathcal{C} \to \mathbb{R}$ is defined as

$$\mathcal{H}(\boldsymbol{\phi}, \boldsymbol{\varphi}) = \int_{\mathcal{B}} \bar{\mathcal{H}}(\boldsymbol{\phi}, \boldsymbol{\varphi}, \mathbf{F})\, dV - \int_{\partial \mathcal{B}_\sigma} \bar{\mathbf{T}}(\mathbf{X}, \hat{\mathbf{N}}) \cdot \boldsymbol{\varphi}(\mathbf{X}, t)\, dA \tag{10.206}$$

Owing to the chain rule, the variation of the Hamiltonian density function may be written as

$$\delta \bar{\mathcal{H}}(\boldsymbol{\phi}, \boldsymbol{\varphi}, \mathbf{F}) = \frac{\partial \bar{\mathcal{H}}}{\partial \boldsymbol{\phi}} \cdot \delta \boldsymbol{\phi} + \frac{\partial \bar{\mathcal{H}}}{\partial \boldsymbol{\varphi}} \cdot \delta \boldsymbol{\varphi} + \frac{\partial \bar{\mathcal{H}}}{\partial \mathbf{F}} : \delta \mathbf{F} \tag{10.207}$$

The third term on the right-hand side of the above equation can be expressed as

$$\frac{\partial \bar{\mathcal{H}}}{\partial F} : \delta \mathbf{F} = \frac{\partial \bar{\mathcal{H}}}{\partial \mathbf{F}} : \text{GRAD}\, \delta \boldsymbol{\varphi} = \text{DIV} \left(\delta \boldsymbol{\varphi} \cdot \frac{\partial \bar{\mathcal{H}}}{\partial \mathbf{F}} \right) - \left(\text{DIV} \frac{\partial \bar{\mathcal{H}}}{\partial \mathbf{F}} \right) \cdot \delta \boldsymbol{\varphi} \tag{10.208}$$

Hence, Equation (10.207) can be rewritten as

$$\delta \bar{\mathcal{H}}(\boldsymbol{\phi}, \boldsymbol{\varphi}, \mathbf{F}) = \left(\frac{\partial \bar{\mathcal{H}}}{\partial \boldsymbol{\varphi}} - \text{DIV} \frac{\partial \bar{\mathcal{H}}}{\partial \mathbf{F}} \right) \cdot \delta \boldsymbol{\varphi} + \frac{\partial \bar{\mathcal{H}}}{\partial \boldsymbol{\phi}} \cdot \delta \boldsymbol{\phi} + \text{DIV} \left(\delta \boldsymbol{\varphi} \cdot \frac{\partial \bar{\mathcal{H}}}{\partial \mathbf{F}} \right) \cdot \delta \boldsymbol{\varphi} \tag{10.209}$$

Note that for the first term in Equation (10.205), we obtain

$$\begin{aligned}\delta\int_{t_0}^{t_1}\left(\int_{\mathcal{B}}\boldsymbol{\phi}\cdot\dot{\boldsymbol{\varphi}}\,dV\right)dt &= \left[\int_{\mathcal{B}}\boldsymbol{\phi}\cdot\delta\boldsymbol{\varphi}\,dV\right]\Bigg|_{t_0}^{t_1} - \int_{t_0}^{t_1}\left(\int_{\mathcal{B}}\frac{d\boldsymbol{\phi}}{dt}\cdot\delta\boldsymbol{\varphi}\,dV\right)dt \\ &+ \int_{t_0}^{t_1}\left(\int_{\mathcal{B}}\delta\boldsymbol{\phi}\cdot\dot{\boldsymbol{\varphi}}\,dV\right)dt\end{aligned} \tag{10.210}$$

Therefore, in view of Equation (10.209) and Equation (10.210), Hamilton's law of varying action can be rewritten as

$$\begin{aligned}&\delta\int_{t_0}^{t_1}\left[\int_{\mathcal{B}}\boldsymbol{\phi}\cdot\dot{\boldsymbol{\varphi}}\,dV - \mathcal{H}(\boldsymbol{\phi},\boldsymbol{\varphi})\right]dt - \left[\int_{\mathcal{B}}\boldsymbol{\phi}\cdot\delta\boldsymbol{\varphi}\,dV\right]\Bigg|_{t_0}^{t_1} \\ &= \int_{t_0}^{t_1}\left[\int_{\partial\mathcal{B}_\sigma}\left(\bar{\mathbf{T}}(\mathbf{X},\hat{\mathbf{N}}) - \frac{\partial\bar{\mathcal{H}}}{\partial\mathbf{F}}\cdot\hat{\mathbf{N}}\right)\cdot\delta\boldsymbol{\varphi}\,dA\right]dt \\ &+ \int_{t_0}^{t_1}\left\{\int_{\mathcal{B}}\left[\left(\dot{\boldsymbol{\varphi}} - \frac{\partial\bar{\mathcal{H}}}{\partial\boldsymbol{\phi}}\right)\cdot\delta\boldsymbol{\phi} - \left(\dot{\boldsymbol{\phi}} + \frac{\partial\bar{\mathcal{H}}}{\partial\boldsymbol{\varphi}} - \mathrm{DIV}\,\frac{\partial\bar{\mathcal{H}}}{\partial\mathbf{F}}\right)\cdot\delta\boldsymbol{\varphi}\right]dV\right\}dt = 0\end{aligned} \tag{10.211}$$

Note that on the prescribed boundary, the virtual displacement $\delta\boldsymbol{\varphi} \in T_{\boldsymbol{\varphi}}\mathcal{C}$ must be null due to the Dirichlet boundary conditions. As shown in the previous section, Hamilton's law of varying action in Equation (10.211) contains governing equations and natural boundary conditions. Since the two variations, $\delta\boldsymbol{\phi}$ and $\delta\boldsymbol{\varphi}$, are arbitrary, we have Hamilton's equation of motion, as the strong form which holds inside the body

$$\dot{\boldsymbol{\phi}} + \frac{\partial\bar{\mathcal{H}}}{\partial\boldsymbol{\varphi}} = \mathrm{DIV}\,\frac{\partial\bar{\mathcal{H}}}{\partial\mathbf{F}}, \quad \dot{\boldsymbol{\varphi}} = \frac{\partial\bar{\mathcal{H}}}{\partial\boldsymbol{\phi}}, \quad \text{for } (\mathbf{X},t)\in\mathcal{B}\times\mathcal{T} \tag{10.212}$$

and the traction boundary condition on the prescribed traction boundary surface $\partial\mathcal{B}_\sigma$

$$\bar{\mathbf{T}}(\mathbf{X},\hat{\mathbf{N}}) = \frac{\partial\bar{\mathcal{H}}}{\partial\mathbf{F}}\cdot\hat{\mathbf{N}} \quad \text{for } (\mathbf{X},t)\in\partial\mathcal{B}_\sigma\times\mathcal{T} \tag{10.213}$$

Thus, we naturally have

$$\frac{\partial\bar{\mathcal{H}}}{\partial\mathbf{F}} = \rho_0\frac{\partial\Psi(\mathbf{F})}{\partial\mathbf{F}} = \mathbf{P} \tag{10.214}$$

We also have the relation,

$$\int_{t_0}^{t_1}\left[\int_{\mathcal{B}}\mathrm{DIV}\left(\delta\boldsymbol{\varphi}\cdot\frac{\partial\bar{\mathcal{H}}}{\partial\mathbf{F}}\right)dV\right]dt = \int_{t_0}^{t_1}\left(\int_{\partial\mathcal{B}_\sigma}\delta\boldsymbol{\varphi}\cdot\frac{\partial\bar{\mathcal{H}}}{\partial\mathbf{F}}\cdot\hat{\mathbf{N}}\,dA\right)dt \tag{10.215}$$

Note that Equation (10.212) can be rewritten as

$$\dot{\boldsymbol{\phi}} = \rho_0\mathbf{B} + \mathrm{DIV}\,\mathbf{P}, \quad \dot{\boldsymbol{\varphi}} = \frac{\boldsymbol{\phi}}{\rho_0}, \quad \text{for } (\mathbf{X},t)\in\mathcal{B}\times\mathcal{T} \tag{10.216}$$

which is the Hamiltonian version of Cauchy's equation of motion in continuous bodies.

Hamilton's law of varying action in Equation (10.211) can be rewritten as the weak form of the local linear momentum equations of motion. Upon integration by parts, Hamilton's law of varying action results in

$$\int_{t_0}^{t_1} \left(\int_{\mathcal{B}} \dot{\boldsymbol{\phi}} \cdot \delta \boldsymbol{\varphi} \, dV + \int_{\mathcal{B}} \frac{\partial \bar{\mathcal{H}}}{\partial \mathbf{F}} : \delta \mathbf{F} \, dV + \int_{\mathcal{B}} \frac{\partial \bar{\mathcal{H}}}{\partial \boldsymbol{\varphi}} \cdot \delta \boldsymbol{\varphi} \, dV - \int_{\partial \mathcal{B}_\sigma} \bar{\mathbf{T}} \cdot \delta \boldsymbol{\varphi} \, dA \right) dt = 0$$
$$\int_{t_0}^{t_1} \left(\int_{\mathcal{B}} \dot{\boldsymbol{\varphi}} \cdot \delta \boldsymbol{\phi} \, dV - \int_{\partial \mathcal{B}_\sigma} \frac{\partial \bar{\mathcal{H}}}{\partial \boldsymbol{\varphi}} \cdot \delta \boldsymbol{\phi} \, dA \right) dt = 0 \tag{10.217}$$

where the integrands are equivalent to the weak forms within the Hamiltonian mechanics framework.

10.4.2 Space-Discrete Hamiltonian Finite Element Formulation

In contrast to the traditional approaches for the space discretization stemming from the principle of virtual work in dynamics, via the scalar formalism for the Hamiltonian (Har and Tamma 2009b, 2010), next, we discuss two distinct approaches to spatial discretization: 1) One approach is to follow customary practices, but employing the descriptive scalar function and the associated density based strong and weak form representation shown previously which is an outcome using as a starting point, Hamilton's law of varying action within the Hamiltonian mechanics framework, and 2) Alternatively, a space-discrete finite element numerical discretization approach is additionally presented. This is different from traditional practices, and it provides new avenues to enact the space discretization process. The latter space-discrete formulation emanating from Hamilton's law of varying action does not require one to derive the weak forms. It is readily shown that Hamilton's law of varying action given in Equation (10.205) can be directly discretized in space since by itself, it is already a statement of the weak form. Furthermore, the trial functions, which are employed in Hamilton's law of varying action, are required to satisfy the continuity condition of the same degree just as in the principle of virtual work in dynamics.

In the Hamiltonian mechanics framework, the $(m-1)$th order derivatives of admissible trial functions $\tilde{\boldsymbol{\varphi}}$ are required to be piecewise continuous within the element. However, the admissible trial functions should also be continuous along the inter-element boundaries (Pian and Tong 1969). They may be represented by a combination of interpolation functions and nodal variables as follows:

$$\tilde{\boldsymbol{\varphi}}(\mathbf{X}, t) \approx \tilde{\boldsymbol{\varphi}}^h(\mathbf{X}, t) = \mathbf{X} + \sum_{i=1}^{n_{\text{node}}} \mathbf{N}_i(\mathbf{X}) \mathbf{U}_i^e(t),$$
$$\tilde{\boldsymbol{\phi}}(\mathbf{X}, t) \approx \tilde{\boldsymbol{\phi}}^h(\mathbf{X}, t) = \sum_{i=1}^{n_{\text{node}}} \mathbf{N}_i(\mathbf{X}) \tilde{\boldsymbol{\phi}}_i^e(t) \tag{10.218}$$

The above equation can be rewritten as

$$\tilde{\boldsymbol{\varphi}}^h(\mathbf{X}, t) = \mathbf{X} + \mathbf{N}_e(\mathbf{X}) \mathbf{U}^e(t), \quad \tilde{\boldsymbol{\phi}}^h(\mathbf{X}, t) = \mathbf{N}_e(\mathbf{X}) \boldsymbol{\phi}^e(t) \tag{10.219}$$

where

$$\mathbf{N}_e(\mathbf{X}) = \left[\mathbf{N}_1, \ldots, \mathbf{N}_{n_{\text{node}}}\right],$$
$$\mathbf{U}^e(t) = \left(\mathbf{U}_1^e, \ldots, \mathbf{U}_{n_{\text{node}}}^e\right)^T, \quad \tilde{\boldsymbol{\phi}}^e(t) = \left(\tilde{\boldsymbol{\phi}}_1^e, \ldots, \tilde{\boldsymbol{\phi}}_{n_{\text{node}}}^e\right)^T \tag{10.220}$$

Approach 1: Weak Form Upon substitution of Equation (10.219) into the weak form given in Equation (10.217), the domain of interest can be discretized as follows:

$$\begin{aligned} &\int_{t_0}^{t_1}\left\{\sum_{e=1}^{n_{\text{elem}}}\left(\int_{\mathcal{B}_e}\dot{\tilde{\boldsymbol{\varphi}}}^h\cdot\delta\tilde{\boldsymbol{\phi}}^h dV-\int_{\mathcal{B}_e}\left(\frac{\partial\bar{\mathcal{H}}^h_{\mathcal{B}_e}}{\partial\tilde{\boldsymbol{\phi}}^h}\right)\cdot\delta\tilde{\boldsymbol{\phi}}^h dV\right)\right\}dt,\\ &-\int_{t_0}^{t_1}\left\{\sum_{i=1}^{n_{\text{elem}}}\left(\begin{array}{c}\int_{\mathcal{B}_e}\dot{\tilde{\boldsymbol{\phi}}}^h\cdot\delta\tilde{\boldsymbol{\varphi}}^h dV+\int_{\mathcal{B}_e}\left(\frac{\partial\bar{\mathcal{H}}^h_{\mathcal{B}_e}}{\partial\tilde{\mathbf{F}}^h}\right):\delta\tilde{\mathbf{F}}^h dV\\ +\int_{\mathcal{B}_e}\left(\frac{\partial\bar{\mathcal{H}}^h_{\mathcal{B}_e}}{\partial\tilde{\boldsymbol{\varphi}}^h}\right)\cdot\delta\tilde{\boldsymbol{\varphi}}^h dV-\int_{\partial\mathcal{B}_{e\sigma}}\bar{\mathbf{T}}\cdot\delta\tilde{\boldsymbol{\varphi}}^h dA\end{array}\right)\right\}dt=0\end{aligned} \tag{10.221}$$

where $\mathcal{B}_e$ stands for the eth element domain and the space-discrete Hamiltonian density function is given by

$$\bar{\mathcal{H}}^h_{\mathcal{B}_e}\left(\tilde{\boldsymbol{\varphi}}^h,\tilde{\boldsymbol{\phi}}^h,\tilde{\mathbf{F}}^h\right)=\frac{1}{2\rho_0}\tilde{\boldsymbol{\phi}}^h\cdot\tilde{\boldsymbol{\phi}}^h+\rho_0\Psi\left(\tilde{\mathbf{F}}^h\right)-\rho_0\mathbf{B}\cdot\tilde{\boldsymbol{\varphi}}^h \tag{10.222}$$

In view of Equation (10.219), the above equation can be rewritten as

$$\begin{aligned} &\int_{t_0}^{t_1}\left\{\sum_{e=1}^{n_{\text{elem}}}\frac{1}{\rho_0}\left(\mathbf{M}^e\dot{\mathbf{U}}^e(t)-\frac{1}{\rho_0}\mathbf{M}^e\tilde{\boldsymbol{\phi}}^e(t)\right)\cdot\delta\tilde{\boldsymbol{\phi}}^e(t)\right\}dt=0,\\ &-\int_{t_0}^{t_1}\left\{\sum_{i=1}^{n_{\text{elem}}}\left(\frac{1}{\rho_0}\mathbf{M}^e\dot{\tilde{\boldsymbol{\phi}}}^e(t)+\mathbf{F}^e_{\text{int}}-\mathbf{F}^e_{\text{ext}}\right)\cdot\delta\mathbf{U}^e(t)\right\}dt=0\end{aligned} \tag{10.223}$$

where $\mathbf{M}_e$, $\mathbf{F}^e_{\text{int}}$ and $\mathbf{F}^e_{\text{ext}}$ are given by Equation (9.17).

For the entire domain, a configuration space with generalized coordinates $\mathcal{Q}$, is defined as

$$\mathcal{Q}:=\left\{\mathbf{q}|\mathbf{q}(t)=\left(U_1(t),\ldots,U_{n_{\text{dof}}}(t)\right)\in\mathbb{R}^{n_{\text{dof}}=3n_{\text{node}}}\right\} \tag{10.224}$$

In addition, we define a tangent space, denoted by $T_q Q$, i.e.,

$$T_q Q:=\left\{\dot{\mathbf{q}}|\dot{\mathbf{q}}(t)=\left(\dot{U}_1,\ldots,\dot{U}_{n_{\text{dof}}}\right)\in\mathbb{R}^{n_{\text{dof}}=3n_{\text{node}}}\right\} \tag{10.225}$$

and a cotangent space, denoted by $T_q^* Q$, i.e.,

$$T_q^* Q:=\left\{(\mathbf{p})\,|\mathbf{p}(t)=\mathbf{M}\dot{\mathbf{q}}(t)\in\mathbb{R}^{n_{\text{dof}}=3n_{\text{node}}},\ \mathbf{M}=\int_{\mathcal{B}}\rho_0\mathbf{N}^T\mathbf{N}\,dV\right\} \tag{10.226}$$

Next, we define the cotangent bundle, namely, the phase space as follows:

$$\begin{aligned} T^*Q&=\left\{\begin{array}{l}(\mathbf{q},\mathbf{p})\,|\mathbf{q}(t)=\left\{U_1(t),\ldots,U_{n_{dof}}(t)\right\}^T\in Q\subset\mathbb{R}^{n_{dof}=3n_{node}}\\ \mathbf{p}(t)=\mathbf{M}\dot{\mathbf{q}}(t)\in T_q^*Q\subset\mathbb{R}^{n_{dof}=3n_{node}},\mathbf{M}=\int_{\mathcal{B}}\rho_0\mathbf{N}^T\mathbf{N}dV\end{array}\right\}\\ &=\bigcup_{q\in Q}T_q^*Q\end{aligned} \tag{10.227}$$

Finally, Equation (10.223) yields the semi-discrete first-order differential equation system

$$\dot{\mathbf{p}}=\mathbf{F}_{\text{ext}}-\mathbf{F}_{\text{int}},\quad\dot{\mathbf{q}}=\mathbf{M}^{-1}\mathbf{p} \tag{10.228}$$

which correspond to the discretized Hamilton's equations of motion.

Approach 2: Hamiltonian Differential Operator In In contrast to the previous approach, upon substituting Equation (10.219) directly into Equation (10.205), Hamilton's law of varying action for the eth element domain $\mathcal{B}_e$ can be discretized in space as follows:

$$\text{HLVA}^h_{\mathcal{H}\mathcal{B}_e} : \delta \int_{t_0}^{t_1} \left\{ \int_{\mathcal{B}_e} \tilde{\boldsymbol{\phi}}^h \cdot \dot{\tilde{\boldsymbol{\varphi}}}^h \, dV - \mathcal{H}^h_{\mathcal{B}_e} \right\} dt - \left[\int_{\mathcal{B}_e} \tilde{\boldsymbol{\phi}}^h \cdot \delta \tilde{\boldsymbol{\varphi}}^h \, dV \right] \Bigg|_{t_0}^{t_1} = 0 \tag{10.229}$$

where the autonomous space-discrete Hamiltonian for the eth element domain is simply constructed as

$$\mathcal{H}^h_{\mathcal{B}_e}\left(\tilde{\boldsymbol{\phi}}^h, \tilde{\boldsymbol{\varphi}}^h\right) = \int_{\mathcal{B}_e} \bar{\mathcal{H}}^h_{\mathcal{B}_e}\left(\tilde{\boldsymbol{\phi}}^h, \tilde{\boldsymbol{\varphi}}^h, \tilde{\mathbf{F}}^h\right) dV - \int_{\partial \mathcal{B}_{e\sigma}} \bar{\mathbf{T}}(\mathbf{X}) \cdot \tilde{\boldsymbol{\varphi}}^h (\mathbf{X}, t) \, dA \tag{10.230}$$

In terms of generalized coordinates, Equation (10.219) can be rewritten as follows:

$$\begin{aligned} \tilde{\boldsymbol{\varphi}}^h (\mathbf{X}, t) &= \mathbf{X} + \mathbf{N}_e (\mathbf{X}) \, \mathbf{q}_e (t), \\ \tilde{\boldsymbol{\phi}}^h (\mathbf{X}, t) &= \mathbf{N}_e (\mathbf{X}) \, \boldsymbol{\phi}^e (t) \end{aligned} \tag{10.231}$$

where

$$\mathbf{q}_e (t) = \left\{ \mathbf{U}^e_1 (t), \ldots, \mathbf{U}^e_i (t), \ldots, \mathbf{U}^e_{n_{\text{node}}} (t) \right\} \tag{10.232}$$

We rewrite Equation (10.229) as

$$\begin{aligned} \text{HLVA}^h_{\mathcal{H}\mathcal{B}_e} &: \delta \int_{t_0}^{t_1} \mathbf{p}_e \cdot \dot{\mathbf{q}}_e - \mathcal{H}^h_{\mathcal{B}_e} (\mathbf{q}_e, \mathbf{p}_e) \, dt - \mathbf{p} \cdot \delta \mathbf{q}_e |_{t_0}^{t_1} \\ &= \int_{t_0}^{t_1} \left\{ \left(\dot{\mathbf{q}}_e - \frac{\partial \mathcal{H}^h_{\mathcal{B}_e}}{\partial \mathbf{p}_e} \right) \cdot \delta \mathbf{p}_e - \left(\dot{\mathbf{p}}_e + \frac{\partial \mathcal{H}^h_{\mathcal{B}_e}}{\partial \mathbf{q}_e} \right) \cdot \delta \mathbf{q}_e \right\} dt = 0 \end{aligned} \tag{10.233}$$

where the canonical momentum $\mathbf{p}_e$ is given by

$$\mathbf{p}_e (t) = \mathbf{M}^e \dot{\mathbf{q}}_e (t) = \frac{1}{\rho_0} \mathbf{M}^e \boldsymbol{\phi}^e (t), \quad \mathbf{M}^e = \int_{\mathcal{B}} \rho_0 \mathbf{N}_e^T \mathbf{N}_e \, dV \tag{10.234}$$

Note that the space-discrete Hamiltonian in Equation (10.233) is given by

$$\begin{aligned} &\mathcal{H}^h_{\mathcal{B}_e} (\mathbf{q}_e, \mathbf{p}_e) = \frac{1}{2} \mathbf{p}_e \cdot \mathbf{M}^{e-1} \mathbf{p}_e + \Pi^h_{\text{int}_e} - \mathbf{q}_e \cdot \mathbf{F}^e_{\text{ext}}, \\ &\Pi^h = \int_{\mathcal{B}_e} \rho_0 \Psi \left(\tilde{\mathbf{F}}^h \right) dV \quad \frac{\partial \mathcal{H}^h_{\mathcal{B}_e}}{\partial \mathbf{q}_e} = \mathbf{F}^e_{\text{int}} - \mathbf{F}^e_{\text{ext}}, \\ &\mathbf{F}^e_{\text{int}} = \int_{\mathcal{B}_e} \bar{\mathbf{B}}^T_e \tilde{\mathbf{P}}^h \, dV, \; \mathbf{F}^e_{\text{ext}} = \int_{\mathcal{B}_e} \rho_0 \mathbf{N}^T_e \mathbf{B} \, dV + \int_{\partial \mathcal{B}_{e\sigma}} \mathbf{N}^T_e \bar{\mathbf{T}} \, dA \end{aligned} \tag{10.235}$$

For the entire domain, a phase space with canonical coordinates is defined as shown in Equation (10.227). Consequently, Hamilton's law of varying action in a finite dimensional system is given by

$$\begin{aligned} \text{HLVA}^h_{\mathcal{H}} &: \delta \int_{t_0}^{t_1} \mathbf{p} \cdot \dot{\mathbf{q}} - \mathcal{H}^h (\mathbf{q}, \mathbf{p}) \, dt - \mathbf{p} \cdot \delta \mathbf{q} |_{t_0}^{t_1} \\ &= \int_{t_0}^{t_1} \left\{ \left(\dot{\mathbf{q}} - \frac{\partial \mathcal{H}^h}{\partial \mathbf{p}} \right) \cdot \delta \mathbf{p} - \left(\dot{\mathbf{p}} + \frac{\partial \mathcal{H}^h}{\partial \mathbf{q}} \right) \cdot \delta \mathbf{q} \right\} dt = 0 \end{aligned} \tag{10.236}$$

where the space-discrete Hamiltonian consisting of the space-discrete kinetic and potential energies is constructed as follows:

$$
\begin{aligned}
&\mathcal{H}^h(\mathbf{q},\mathbf{p}) = \mathcal{K}^h(\mathbf{p}) + \mathcal{U}^h(\mathbf{q}), \ \mathcal{K}^h(\mathbf{p}) = \frac{1}{2}\mathbf{p}\cdot\mathbf{M}^{-1}\mathbf{p}, \\
&\mathcal{U}^h(\mathbf{q}) = \Pi^h - \mathbf{q}\cdot\mathbf{F}_{\text{ext}}, \ \Pi^h = \int_{\mathcal{B}} \rho_0 \Psi\left(\tilde{\mathbf{F}}^h\right) dV \\
&\frac{\partial \mathcal{H}^h}{\partial \mathbf{q}} = \mathbf{F}_{\text{int}} - \mathbf{F}_{\text{ext}}, \ \mathbf{M} = \int_{\mathcal{B}} \rho_0 \mathbf{N}^T \mathbf{N}\, dV, \\
&\mathbf{F}_{\text{int}} = \int_{\mathcal{B}} \bar{\mathbf{B}}^T \tilde{\mathbf{P}}^h\, dV, \ \mathbf{F}_{\text{ext}} = \int_{\mathcal{B}} \rho_0 \mathbf{N}^T \mathbf{B}\, dV + \int_{\partial\mathcal{B}_\sigma} \mathbf{N}^T \bar{\mathbf{T}}\, dA
\end{aligned}
\tag{10.237}
$$

In addition, in view of Equation (10.236), the space-discrete Hamiltonian differential operators $\mathfrak{H}_{D1}$ and $\mathfrak{H}_{D2}$ can be readily defined. Hence, the semi-discrete finite-element equations of motion can be readily obtained after constructing the space-discrete Hamiltonian which only entails the use of trial functions. Thus, upon substituting the space-discrete Hamiltonian into the space-discrete Hamiltonian Differential Operator, it can be shown that

$$
\begin{aligned}
&\mathfrak{H}_{D1} : \dot{\mathbf{p}} + \frac{\partial}{\partial \mathbf{q}}, \ \mathfrak{H}_{D2} : \dot{\mathbf{q}} - \frac{\partial}{\partial \mathbf{p}}, \\
&\mathfrak{H}_{D1}\left(\mathcal{H}^h\right) = \mathbf{0} \Rightarrow \quad \dot{\mathbf{p}} + \frac{\partial \mathcal{H}^h}{\partial \mathbf{q}} = \dot{\mathbf{p}} + \mathbf{F}_{\text{int}} - \mathbf{F}_{\text{ext}} = \mathbf{0}, \\
&\mathfrak{H}_{D2}\left(\mathcal{H}^h\right) = \mathbf{0} \Rightarrow \quad \dot{\mathbf{q}} - \frac{\partial \mathcal{H}^h}{\partial \mathbf{p}} = \dot{\mathbf{q}} \cdot \mathbf{M}^{-1}\mathbf{p} = \mathbf{0}
\end{aligned}
\tag{10.238}
$$

where the dependent state variable is the space-discrete scalar function, namely, the space-discrete autonomous Hamiltonian. Thus, the semi-discrete finite-element equations of motion governing the mechanical dynamic behavior of the continuous-body within the Hamiltonian mechanics framework can be obtained as

$$
\dot{\mathbf{p}} + \mathbf{F}_{\text{int}} - \mathbf{F}_{\text{ext}} = \mathbf{0} \tag{10.239}
$$

which is subject to a subsidiary velocity condition:

$$
\dot{\mathbf{q}} - \mathbf{M}^{-1}\mathbf{p} = \mathbf{0} \tag{10.240}
$$

Time Weighted Residual Method Finally, note that in view of Equation (10.236) and Equation (10.238), the following can be also given:

$$
\text{HLVA}^h_{\mathcal{H}} : \int_{t_0}^{t_1} \left\{ \left(\dot{\mathbf{p}} + \mathbf{F}_{\text{int}} - \mathbf{F}_{\text{ext}}\right) \cdot \delta\mathbf{q}(t) - \left(\dot{\mathbf{q}} - \mathbf{M}^{-1}\mathbf{p}\right) \cdot \delta\mathbf{p}(t) \right\} dt = 0 \tag{10.241}
$$

Note that Equation (10.241), which is Hamilton's law of varying action in the Hamiltonian mechanics framework, represents the two-field weighted residual form in time. Consequently, it is observed that the space discretization via Hamilton's law of varying action naturally result in the weighted-residual form in time. Hence, the spatially discretized Hamilton's law of varying action form can be used as a variational equation to discretize the time domain. Equation (10.241) may be regarded as an unconstrained variational formulation with two dependent field variables in the time domain, namely, the two-field variational problem within the Hamiltonian mechanics framework. In general, approaches such as the time weighted-residual method which includes time finite element formulations (Belytschko and Hughes 1983; Sha et al. 2003; Tamma et al. 2000, 2002, 2003; Wood 1990; Zhou et al. 2000, 2001; Zhou and Tamma 2004a,c, 2006; Zienkiewicz and Taylor 2006; Zienkiewicz et al. 2005, 1984) can be appropriately applied to Equation (10.241) to develop time stepping schemes as highlighted in a later chapter dealing with time discretization.

Illustrative Examples For the various illustrative examples to follow next, we first show the development of the Hamiltonian density based representations of the equations of motion in the scalar formalism that are identical/equivalent to the vector formalism of the Cauchy equations of motion. One could readily employ the Hamiltonian density based representations to conduct the space discretization through conventional practices described previously in Approach 1; alternatively, to illustrate the fundamental ideas, we instead focus attention upon the space discretization aspects via Approach 2 described earlier which represents the space-discrete finite element formulation and it directly leads to the semi-discretized equations of motion.

Example 10.6 Elastic Axial Bar

For the sake of simplicity and clarity, let us again consider the same axial bar problem as discussed in Section 9.1.2.

Governing Equations: Hamiltonian Density Based Representations

Hamilton's law of varying action for the axial bar in the Lagrangian mechanics framework can be written as

$$\delta \int_{t_0}^{t_1} \mathcal{L}\left(u\left(x,t\right), \dot{u}\left(x,t\right)\right) dt - \left. \int_L \frac{\partial \bar{\mathcal{L}}}{\partial \dot{u}} \delta u\left(x,t\right) dx \right|_{t_0}^{t_1} = 0 \tag{10.242}$$

where the Lagrangian is given by Equation (10.68). The above equation can be rewritten as

$$\begin{aligned} &\int_{t_0}^{t_1} \left(\int_L \delta \bar{\mathcal{L}}\left(u, \dot{u}, u'\right) dx + F_L \delta u(L,t) \right) dt \\ &\quad - \left. \int_L \frac{\partial \bar{\mathcal{L}}}{\partial \dot{u}} \delta u\left(x,t\right) dx \right|_{t_0}^{t_1} = 0 \end{aligned} \tag{10.243}$$

The generalized momentum can be given by

$$\phi = \frac{\partial \bar{\mathcal{L}}}{\partial \dot{u}} = \rho A\left(x\right) \dot{u} \tag{10.244}$$

The Hamiltonian density function by Legendre transformation is defined as

$$\bar{\mathcal{H}}\left(\phi, u, u'\right) = \phi \dot{u} - \bar{\mathcal{L}}\left(u, \dot{u}, u'\right), \quad u' = \frac{\partial u}{\partial x} \tag{10.245}$$

which yields

$$\bar{\mathcal{H}}(\phi, u, u') = \frac{\phi^2}{2\rho A(x)} + \frac{1}{2} EA(x) \left(\frac{\partial u}{\partial x}\right)^2 - \bar{b}(x)\, u \tag{10.246}$$

The variation of the Hamiltonian density function is given by

$$\delta\bar{\mathcal{H}}(\phi, u, u') = \frac{\partial\bar{\mathcal{H}}}{\partial\phi}\delta\phi + \frac{\partial\bar{\mathcal{H}}}{\partial u}\delta u + \frac{\partial\bar{\mathcal{H}}}{\partial u'}\delta u' \tag{10.247}$$

Upon integration by parts, we have

$$\begin{aligned} \delta\bar{\mathcal{H}}(\phi, u, u') = &\left[\frac{\partial\bar{\mathcal{H}}}{\partial u} - \frac{\partial}{\partial x}\left(\frac{\partial\bar{\mathcal{H}}}{\partial u'}\right)\right]\delta u + \frac{\partial\bar{\mathcal{H}}}{\partial\phi}\delta\phi \\ &+ \frac{\partial}{\partial x}\left(\frac{\partial\bar{\mathcal{H}}}{\partial u'}\delta u\right) \end{aligned} \tag{10.248}$$

Then, the variation of the Lagrangian density function can be represented as

$$\delta\bar{\mathcal{L}} = \phi\delta\dot{u} + \dot{u}\delta\phi - \delta\bar{\mathcal{H}} \tag{10.249}$$

Upon substitution of Equation (10.248) into Equation (10.249), we obtain

$$\begin{aligned} \delta\bar{\mathcal{L}}(u, \dot{u}, u') = &-\left[\frac{\partial\bar{\mathcal{H}}}{\partial u} - \frac{\partial}{\partial x}\left(\frac{\partial\bar{\mathcal{H}}}{\partial u'}\right) + \frac{d\phi}{dt}\right]\delta u + \left(\dot{u} - \frac{\partial\bar{\mathcal{H}}}{\partial\phi}\right)\delta\phi \\ &- \frac{\partial}{\partial x}\left(\frac{\partial\bar{\mathcal{H}}}{\partial u'}\delta u\right) + \frac{d}{dt}(\phi\delta u) \end{aligned} \tag{10.250}$$

Strong Form

In view of Equation (10.244) and Equation (10.250), Hamilton's law of varying action gives

$$\begin{aligned} &\int_{t_0}^{t_1}\left[\int_L\left\{\left[\frac{\partial}{\partial x}\left(\frac{\partial\bar{\mathcal{H}}}{\partial u'}\right) - \frac{\partial\bar{\mathcal{H}}}{\partial u} - \frac{d\phi}{dt}\right]\delta u + \left(\dot{u} - \frac{\partial\bar{\mathcal{H}}}{\partial\phi}\right)\delta\phi\right\}dx\right]dt \\ &\quad + \int_{t_0}^{t_1}\left(F_L\delta u(L,t) - \left.\frac{\partial\bar{\mathcal{H}}}{\partial u'}\delta u\right|_0^L\right)dt = 0 \end{aligned} \tag{10.251}$$

From the above, the governing equations are obtained as (scalar formalism)

$$\begin{aligned} \frac{d\phi}{dt} &= \frac{\partial}{\partial x}\left(\frac{\partial\bar{\mathcal{H}}}{\partial u'}\right) - \frac{\partial\bar{\mathcal{H}}}{\partial u}, \\ \dot{u} - \frac{\partial\bar{\mathcal{H}}}{\partial\phi} &= 0 \end{aligned} \tag{10.252}$$

subject to the boundary conditions

$$F_L \delta u(L,t) - \frac{\partial \bar{\mathcal{H}}}{\partial u'}\delta u \Big|_0^L = 0 \tag{10.253}$$

Eventually, upon substitution of the Hamiltonian density function in Equation (10.246) into Equation (10.252), we have a set of the classical representation of the first-order differential equations of motion (vector formalism) as follows:

$$\begin{aligned} \dot{\phi} &= \frac{\partial}{\partial x} F_L \delta u(L,t) - \left(EA(x) \frac{\partial u}{\partial x} \right) + \bar{b} \\ \dot{u} &= \frac{\phi}{\rho A(x)} \end{aligned} \tag{10.254}$$

subject to the boundary condition

$$F_L \delta u(L,t) - EA(x) \frac{\partial u}{\partial x}\delta u \Big|_0^L = 0 \tag{10.255}$$

Weak Form

In addition, in view of Equation (10.247) Hamilton's law of varying action also leads to

$$\begin{aligned} &\int_{t_0}^{t_1} \left(\int_L \dot{\phi} \delta u \, dx + \int_L \frac{\partial \bar{\mathcal{H}}}{\partial u'} \delta u' \, dx + \int_L \frac{\partial \bar{\mathcal{H}}}{\partial u} \delta u \, dx - F_L \delta u(L,t) \right) dt = 0 \\ &\int_{t_0}^{t_1} \left(\int_L \dot{u} \delta \phi \, dx - \int_L \frac{\partial \bar{\mathcal{H}}}{\partial \phi} \delta \phi \, dx \right) dt = 0 \end{aligned} \tag{10.256}$$

Note that the above equation represents the weak form corresponding to Equation (10.217).

Space-Discrete Finite Element Formulation

Next, for illustration, we show how to derive the finite element equations of motion for the bar element assemblage via the space-discrete Hamiltonian described earlier via Approach 2. The space-discrete Hamiltonian for one element is given as

$$\mathcal{H}_e^h = \mathcal{K}_e^h + \left({}_e\mathcal{U}_{\text{int}}^h + {}_e\mathcal{U}_{\text{ext}}^h \right) \tag{10.257}$$

where the space-discrete kinetic energy is given

$$\mathcal{K}_e^h = \frac{1}{2} \mathbf{p}_e \cdot (\mathbf{M}^e)^{-1} \mathbf{p}_e \tag{10.258}$$

The above equation can be written as

$$\begin{aligned} \mathcal{K}_e^h &= \frac{1}{\rho A} \left[\, p_1^e(t) \quad p_2^e(t) \, \right] \begin{bmatrix} 2 & -1 \\ -1 & 2 \end{bmatrix} \begin{Bmatrix} p_1^e(t) \\ p_2^e(t) \end{Bmatrix} \\ &= \frac{1}{\rho A} \left(\left(p_1^e(t)\right)^2 - p_1^e(t)\, p_2^e(t) + \left(p_2^e(t)\right)^2 \right) \end{aligned} \tag{10.259}$$

Hence, the space-discrete Hamiltonian can be recast into

$$\begin{aligned}\mathcal{H}_e^h &= \frac{1}{\rho A}\left(\left(p_1^e(t)\right)^2 - p_1^e(t)\,p_2^e(t) + \left(p_2^e(t)\right)^2\right)\\ &+ \frac{EA}{2L_e}\left(\left(u_1^e(t)\right)^2 - 2u_1^e(t)\,u_2^e(t) + \left(u_2^e(t)\right)^2\right)\\ &- f_1^e(t)\,u_1^e(t) - f_2^e(t)\,u_2^e(t)\end{aligned} \tag{10.260}$$

where L_e denotes the length of one element. The first Hamilton's equation can be rewritten as

$$\dot{\mathbf{p}}_e = \begin{Bmatrix} \dot{p}_1^e(t) \\ \dot{p}_2^e(t) \end{Bmatrix} = -\frac{\partial \mathcal{H}_e^h}{\partial \mathbf{q}_e} = -\frac{EA}{L_e}\begin{bmatrix} 1 & -1 \\ -1 & 1 \end{bmatrix}\begin{Bmatrix} u_1^e(t) \\ u_2^e(t) \end{Bmatrix} + \begin{Bmatrix} f_1^e(t) \\ f_2^e(t) \end{Bmatrix} \tag{10.261}$$

The second Hamilton's equation can be rewritten as

$$\dot{\mathbf{q}}_e = \frac{\partial \mathcal{H}_e^h(\mathbf{p}_e, \mathbf{q}_e)}{\partial \mathbf{p}_e} = \begin{Bmatrix} \dot{u}_1^e(t) \\ \dot{u}_2^e(t) \end{Bmatrix} = \frac{2}{\rho A}\begin{bmatrix} 2 & -1 \\ -1 & 2 \end{bmatrix}\begin{Bmatrix} p_1^e(t) \\ p_2^e(t) \end{Bmatrix} \tag{10.262}$$

The space-discrete autonomous Hamiltonian given in Equation (10.260) can be constructed as

$$\begin{aligned}\mathcal{H}_e^h(\mathbf{p}_e, \mathbf{q}_e) &= \mathcal{K}_e^h(\mathbf{p}_e) + \mathcal{U}_e^h(\mathbf{q}_e)\\ &= \frac{1}{2}\mathbf{p}_e(t)\cdot(\mathbf{M}^e)^{-1}\mathbf{p}_e(t) + \frac{1}{2}\mathbf{q}_e(t)\cdot\mathbf{K}^e\mathbf{q}_e(t) - \mathbf{q}_e(t)\cdot\mathbf{F}_{\text{ext}}^e\end{aligned} \tag{10.263}$$

where $\mathbf{M}^e$, $\mathbf{K}^e$ and $\mathbf{F}_{\text{ext}}^e$ can be found in Section 9.2.1. Substituting the space-discrete autonomous Hamiltonian given in Equation (10.263) into the space-discrete Hamiltonian differential operators in Equation (10.238) readily yields the semi-discrete equations of motion as

$$\begin{aligned}\mathfrak{H}_{D1}\left(\mathcal{H}_e^h\right) &= \dot{\mathbf{p}}_e + \mathbf{K}^e\mathbf{q}_e - \mathbf{F}_{ext}^e = 0,\\ \mathfrak{H}_{D2}\left(\mathcal{H}_e^h\right) &= \dot{\mathbf{q}}_e - (\mathbf{M}^e)^{-1}\mathbf{p}_e = 0\end{aligned} \tag{10.264}$$

Example 10.7 Rotating Bar

We again consider the rotating circular bar which rotates due to applied torque.

Governing Equations: Hamiltonian Density Based Representations

The generalized momentum can be given by

$$\phi = \frac{\partial \bar{\mathcal{L}}}{\partial \dot{\theta}} = \rho J(x)\,\dot{\theta} \tag{10.265}$$

where the Lagrangian density function $\bar{\mathcal{L}}(\theta, \dot{\theta}, \theta')$ is given in Equation (10.105). Therefore, we find the Hamiltonian density function by the Legendre transformation

$$\bar{\mathcal{H}}(\phi, \theta, \theta') = \boldsymbol{\phi}\dot{\theta} - \bar{\mathcal{L}}(\theta, \dot{\theta}, \theta') \tag{10.266}$$

which yields

$$\bar{\mathcal{H}}(\phi, \theta, \theta') = \frac{1}{2\rho J(x)}\phi^2 + \frac{1}{2}GJ(x)\{\theta'\}^2 \tag{10.267}$$

The variation of the Hamiltonian density function is

$$\delta\bar{\mathcal{H}}(\phi, \theta, \theta') = \frac{\partial\bar{\mathcal{H}}}{\partial\theta}\delta\theta + \frac{\partial\bar{\mathcal{H}}}{\partial\phi}\delta\phi + \frac{\partial\bar{\mathcal{H}}}{\partial\theta'}\delta\theta' \tag{10.268}$$

Upon integration by parts, we find

$$\delta\bar{\mathcal{H}}(\phi, \theta, \theta') = \left[\frac{\partial\bar{\mathcal{H}}}{\partial\theta} - \frac{\partial}{\partial x}\left(\frac{\partial\bar{\mathcal{H}}}{\partial\theta'}\right)\right]\delta\theta + \frac{\partial\bar{\mathcal{H}}}{\partial\phi}\delta\phi + \frac{\partial}{\partial x}\left(\frac{\partial\bar{\mathcal{H}}}{\partial\theta'}\delta\theta\right) \tag{10.269}$$

Then, the variation of the Lagrangian density function can be represented as

$$\delta\bar{\mathcal{L}}(\theta, \dot{\theta}, \theta') = \phi\delta\dot{\theta} + \dot{\theta}\delta\phi - \delta\bar{\mathcal{H}}(\phi, \theta, \theta') \tag{10.270}$$

Upon substitution of Equation (10.269) into Equation (10.270), we obtain

$$\begin{aligned}\delta\bar{\mathcal{L}}(\theta, \dot{\theta}, \theta') = &\left\{\frac{\partial}{\partial x}\left(\frac{\partial\bar{\mathcal{H}}}{\partial\theta'}\right) - \frac{\partial\bar{\mathcal{H}}}{\partial\theta} - \frac{d\phi}{dt}\right\}\delta\theta + \left(\dot{\theta} - \frac{\partial\bar{\mathcal{H}}}{\partial\phi}\right)\delta\phi \\ &- \frac{\partial}{\partial x}\left(\frac{\partial\bar{\mathcal{H}}}{\partial\theta'}\delta\theta\right) + \frac{d}{dt}(\phi\delta\theta)\end{aligned} \tag{10.271}$$

Therefore, Hamilton's law of varying action gives

$$\begin{aligned}&\int_{t_0}^{t_1}\left[\int_L\left\{\left[\frac{\partial}{\partial x}\left(\frac{\partial\bar{\mathcal{H}}}{\partial\theta'}\right) - \frac{\partial\bar{\mathcal{H}}}{\partial\theta} - \frac{d\phi}{dt} + \bar{m}_T(x)\right]\delta\theta + \left(\dot{\theta} - \frac{\partial\bar{\mathcal{H}}}{\partial\phi}\right)\delta\phi\right\}dx\right]dt \\ &+ \int_{t_0}^{t_1}\left(\left.\frac{\partial\bar{\mathcal{H}}}{\partial\theta'}\delta\theta\right|_0^L\right)dt = 0\end{aligned} \tag{10.272}$$

Therefore, the governing equations are obtained as

$$\begin{aligned}&\frac{d\phi}{dt} = \frac{\partial}{\partial x}\left(\frac{\partial\bar{\mathcal{H}}}{\partial\theta'}\right) - \frac{\partial\bar{\mathcal{H}}}{\partial\theta} + \bar{m}_T(x), \\ &\dot{\theta} - \frac{\partial\bar{H}}{\partial\phi} = 0\end{aligned} \tag{10.273}$$

subject to the boundary condition

$$\left.\frac{\partial \bar{\mathcal{H}}}{\partial \theta'}\delta\theta\right|_0^L = 0 \tag{10.274}$$

Upon substitution of the Hamiltonian density function in Equation (10.267) into Equation (10.273), we have the classical representation of the first-order differential equations of motion as follows:

$$\rho J(x)\ddot{\theta} = \frac{\partial}{\partial x}\left(GJ(x)\frac{\partial\theta(x,t)}{\partial x}\right) + \bar{m}_T(x) \tag{10.275}$$

subject to the boundary condition

$$\left.GJ(x)\frac{\partial\theta(x,t)}{\partial x}\delta\theta(x,t)\right|_0^L = 0 \tag{10.276}$$

Space-Discrete Finite Element Formulation

Henceforth, we shall discuss the development of the semi-discrete first-order finite element equation of motion via the space-discrete Hamiltonian formulation. First we need to define the canonical momentum in the finite dimensional system as follows:

$$\mathbf{p}_e = \frac{\partial \mathcal{L}_e^h(\mathbf{q}_e,\dot{\mathbf{q}}_e)}{\partial\dot{\mathbf{q}}_e} = \mathbf{M}^e\mathbf{q}_e \tag{10.277}$$

where the consistent mass matrix is given as

$$\mathbf{M}^e = \frac{1}{2}\int_{L_e}\rho J(x)\mathbf{N}_e^T(x)\mathbf{N}_e(x)\,dx \tag{10.278}$$

where L_e denotes the length of one element. Then, we have the space-discrete kinetic energy

$$\mathcal{K}_e^h(\mathbf{p}_e) = \frac{1}{2}\mathbf{p}_e\cdot(\mathbf{M}^e)^{-1}\mathbf{p}_e \tag{10.279}$$

Likewise, the space-discrete potential energy can be obtained

$$\mathcal{U}_e^h(\mathbf{q}_e) = \frac{1}{2}\mathbf{q}_e(t)\cdot\mathbf{K}^e(x)\mathbf{q}_e(t) - \mathbf{q}_e(t)\cdot\mathbf{F}_{\text{ext}}^e \tag{10.280}$$

where the stiffness matrix and the external force are

$$\begin{aligned}\mathbf{K}^e(x) &= \int_{L_e}GJ(x)\bar{\mathbf{B}}_e^T(x)\bar{\mathbf{B}}_e(x)\,dx\\ \mathbf{F}_{\text{ext}}^e &= \int_{L_e}\bar{m}_T(x)\mathbf{N}_e^T(x)\,dx\end{aligned} \tag{10.281}$$

The space-discrete Hamiltonian is constructed as

$$\mathcal{H}_e^h(\mathbf{p}_e,\mathbf{q}_e) = \mathcal{K}^h(\mathbf{p}_e) + \mathcal{U}^h(\mathbf{q}_e) \tag{10.282}$$

Finally, Hamilton's equations of motion can be given by

$$\dot{\mathbf{p}}_e = -\frac{\partial \mathcal{H}_e^h(\mathbf{p}_e, \mathbf{q}_e)}{\partial \mathbf{q}_e}, \quad \mathbf{q}_e = \frac{\partial \mathcal{H}_e^h(\mathbf{p}_e, \mathbf{q}_e)}{\partial \mathbf{p}_e} \tag{10.283}$$

Substituting the space-discrete Hamiltonian into the space-discrete Hamiltonian differential operators in Equation (10.238) readily yields the semi-discrete first-order differential equations of motion as follows:

$$\begin{aligned} \mathfrak{H}_{D1}\left(\mathcal{H}_e^h\right) &= \dot{\mathbf{p}}_e + \mathbf{K}^e \mathbf{q}_e - \mathbf{F}_{ext}^e = 0, \\ \mathfrak{H}_{D2}\left(\mathcal{H}_e^h\right) &= \dot{\mathbf{q}}_e - (\mathbf{M}^e)^{-1}\mathbf{p}_e = 0 \end{aligned} \tag{10.284}$$

Example 10.8 Euler-Bernoulli Beam

Within the Hamiltonian framework, we shall next consider the dynamical behavior of the Euler-Bernoulli beam, which has been discussed in the Lagrangian framework in Section 10.3.3.

Governing Equations: Hamiltonian Density Based Representations

As shown in Section 10.3.3, the continuous Lagrangian of the Euler-Bernoulli beam can be found as

$$\mathcal{L}(\mathbf{u}, \dot{\mathbf{u}}) = \int_L \bar{\mathcal{L}}\left(\dot{\mathbf{u}}, u'\right) dx + \int_L \omega(x)\, \mathbf{a} \cdot \mathbf{u}\, dx \tag{10.285}$$

where the Lagrangian density function is defined as

$$\bar{\mathcal{L}}\left(\dot{\mathbf{u}}, \mathbf{u}'\right) = \frac{1}{2}\dot{\mathbf{u}} \cdot \rho \dot{\mathbf{u}} - \frac{1}{2}\mathbf{u}' \cdot \mathbf{D}_b' \mathbf{u}' \tag{10.286}$$

In view of Equation (10.286), the generalized momenta $\boldsymbol{\phi}$ can be given by

$$\boldsymbol{\phi} = \frac{\partial \bar{\mathcal{L}}\left(\dot{\mathbf{u}}, \mathbf{u}'\right)}{\partial \mathbf{R}}, \quad \boldsymbol{\phi} = \begin{Bmatrix} \dfrac{\partial \bar{\mathcal{L}}\left(\dot{\mathbf{u}}, \mathbf{u}'\right)}{\partial \dot{w}} \\ \dfrac{\partial \bar{\mathcal{L}}\left(\dot{\mathbf{u}}, \mathbf{u}'\right)}{\partial \dot{\varphi}} \end{Bmatrix} = \begin{Bmatrix} \phi \\ \phi_1 \end{Bmatrix} \tag{10.287}$$

Note that $\dot{w}'$ is dependent on $\dot{w}$. In the Euler-Bernoulli beam theory, one dependent variable is the transverse deflection, $w(x, t)$ (in other words, it is the single-field problem). Thus, the notations for the generalized momenta in Equation (10.287) are introduced for convenience.

Then, Hamilton's law of varying action gives

$$\delta \int_{t_0}^{t_1} \left\{ \int_L \boldsymbol{\phi} \cdot \dot{\mathbf{u}}(x, t)\, dx - \mathcal{H}(\boldsymbol{\phi}, \mathbf{u}) \right\} dt - \left. \int_L \boldsymbol{\phi} \cdot \delta \mathbf{u}\, dx \right|_{t_0}^{t_1} = 0, \tag{10.288}$$

where the autonomous Hamiltonian of the Euler-Bernoulli beam is given by

$$\mathcal{H}(\boldsymbol{\phi}, \mathbf{u}) = \int_L \bar{\mathcal{H}}\left(\boldsymbol{\phi}, \mathbf{u}'\right) dx - \int_L \omega(x)\, \mathbf{a} \cdot \mathbf{u} dx, \quad \mathbf{a} = (1, 0) \tag{10.289}$$

where the Hamiltonian density function is defined as

$$\bar{\mathcal{H}}\left(\boldsymbol{\phi}, \mathbf{u}'\right) = \frac{1}{2}\boldsymbol{\phi}\cdot\boldsymbol{\rho}^{-1}\boldsymbol{\phi} + \frac{1}{2}\mathbf{u}'\cdot\mathbf{D}_b\mathbf{u}' \tag{10.290}$$

Note that Equation (10.288) for holonomic systems can be rewritten as

$$\int_{t_0}^{t_1}\left\{\int_L \left(\delta\boldsymbol{\phi}\cdot\dot{\mathbf{u}} + \boldsymbol{\phi}\cdot\delta\dot{\mathbf{u}}\right)dx - \delta\mathcal{H}\left(\boldsymbol{\phi}, \mathbf{u}\right)\right\}dt - \int_L \boldsymbol{\phi}\left(x, t\right)\cdot\delta\mathbf{u}\,dx\Big|_{t_0}^{t_1} = 0 \tag{10.291}$$

Since the variation of the autonomous Hamiltonian can be given by

$$\begin{aligned}\delta\mathcal{H}\left(\boldsymbol{\phi}, \mathbf{u}\right) &= \int_L \delta\bar{\mathcal{H}}\left(\boldsymbol{\phi},\mathbf{u}'\right)dx - \int_L \omega\left(x\right)\mathbf{a}\cdot\delta\mathbf{u}\,dx \\ &= \int_L\left(\frac{\partial\bar{\mathcal{H}}}{\partial\boldsymbol{\phi}}\cdot\delta\boldsymbol{\phi} + \frac{\partial\bar{\mathcal{H}}}{\partial\mathbf{u}'}\delta\mathbf{u}'\right)dx - \int_L \omega\left(x\right)\mathbf{a}\cdot\delta\mathbf{u}\,dx\end{aligned} \tag{10.292}$$

Integration by parts gives

$$\begin{aligned}\int_{t_0}^{t_1}\left(\int_L \boldsymbol{\phi}\cdot\delta\dot{\mathbf{u}}dx\right)dt &= \int_L \boldsymbol{\phi}\left(x, t\right)\cdot\delta\mathbf{u}\,dx\Big|_{t_0}^{t_1} - \int_{t_0}^{t_1}\left(\int_L \frac{d\boldsymbol{\phi}}{dt}\cdot\delta\mathbf{u}dx\right)dt \\ &= \int_L \boldsymbol{\phi}\cdot\delta\mathbf{u}\,dx\Big|_{t_0}^{t_1} - \int_{t_0}^{t_1}\left(\int_L\left(\frac{d\phi}{dt}\delta w\right)dx\right)dt - \int_{t_0}^{t_1}\left(\frac{d\phi_1}{dt}\delta w\Big|_0^L\right)dt \\ &\quad + \int_{t_0}^{t_1}\left(\int_L\left(\frac{\partial}{\partial x}\left(\frac{d\phi_1}{dt}\right)\delta w\right)dx\right)dt,\end{aligned}$$

$$\int_{t_0}^{t_1}\left(\int_L \delta\boldsymbol{\phi}\cdot\dot{\mathbf{u}}dx\right)dt = \int_{t_0}^{t_1}\left(\int_L \dot{w}\delta\phi dx\right)dt + \int_{t_0}^{t_1}\left(\int_L \dot{w}'\delta\phi_1 dx\right)dt \tag{10.293}$$

Further, in view of Equation (10.292), we have the following:

$$\int_L\left(\frac{\partial\bar{\mathcal{H}}}{\partial\boldsymbol{\phi}}\cdot\delta\boldsymbol{\phi}\right)dx = \int_L\left(\frac{\partial\bar{\mathcal{H}}}{\partial\phi}\delta\phi + \frac{\partial\bar{\mathcal{H}}}{\partial\phi_1}\delta\phi_1\right)dx \tag{10.294}$$

and

$$\begin{aligned}\int_L \frac{\partial\bar{\mathcal{H}}}{\partial\mathbf{u}'}\cdot\delta\mathbf{u}'dx &= \frac{\partial\bar{\mathcal{H}}}{\partial\mathbf{u}'}\cdot\delta\mathbf{u}\Big|_0^L - \int_L \frac{\partial}{\partial x}\left(\frac{\partial\bar{\mathcal{H}}}{\partial\mathbf{u}'}\right)\cdot\delta\mathbf{u}\,dx \\ &= \left(\frac{\partial\bar{\mathcal{H}}}{\partial w'}\delta w + \frac{\partial\bar{\mathcal{H}}}{\partial w''}\delta w'\right)\Big|_0^L - \int_L\left(\frac{\partial}{\partial x}\left(\frac{\partial\bar{\mathcal{H}}}{\partial w'}\right)\delta w + \frac{\partial}{\partial x}\left(\frac{\partial\bar{\mathcal{H}}}{\partial w''}\right)\delta w'\right)dx\end{aligned}$$

$$= \left\{ \left(\frac{\partial \bar{\mathcal{H}}}{\partial w'} - \frac{\partial}{\partial x}\left(\frac{\partial \bar{\mathcal{H}}}{\partial w''} \right) \right) \delta w + \frac{\partial \bar{\mathcal{H}}}{\partial w''} \delta w' \right\} \Bigg|_0^L$$

$$- \int_L \left\{ \left(\frac{\partial}{\partial x}\left(\frac{\partial \bar{\mathcal{H}}}{\partial w'} \right) - \frac{\partial^2}{\partial x^2}\left(\frac{\partial \bar{\mathcal{H}}}{\partial w''} \right) \right) \delta w \right\} dx \tag{10.295}$$

Hence, in view of Equation (10.292) through Equation (10.295), Hamilton's law of varying action leads to

$$\int_{t_0}^{t_1} \left[\int_L \left\{ \begin{array}{l} \left\{ \frac{\partial}{\partial x}\left(\frac{\partial \bar{\mathcal{H}}}{\partial w'} \right) - \frac{\partial^2}{\partial x^2}\left(\frac{\partial \bar{\mathcal{H}}}{\partial w''} \right) + \frac{\partial}{\partial x}\left(\frac{d\phi}{dt} \right) - \frac{d\phi}{dt} + \omega(x) \right\} \delta w \\ \\ + \left(\dot{w} - \frac{\partial \bar{\mathcal{H}}}{\partial \phi} \right) \delta \phi + \left(\dot{w}' - \frac{\partial \bar{\mathcal{H}}}{\partial \phi_I} \right) \delta \phi_I \end{array} \right\} dx \right] dt \tag{10.296}$$

$$+ \int_{t_0}^{t_1} \left(\left\{ \frac{\partial}{\partial x}\left(\frac{\partial \bar{\mathcal{H}}}{\partial w''} \right) - \frac{d\phi_I}{dt} - \frac{\partial \bar{\mathcal{H}}}{\partial w'} \right\} \delta w \Bigg|_0^L \right) dt + \int_{t_0}^{t_1} \left(- \frac{\partial \bar{\mathcal{H}}}{\partial w''} \delta w' \Bigg|_0^L \right) dt = 0$$

where both governing equations and natural boundary conditions are included. Consequently, we have derived the governing equation of motion with the traction boundary conditions via Hamilton's law of varying action involving the Hamiltonian density function. Since the variation of the dependent variable, the transverse deflection, is arbitrary, the governing equation of the Euler-Bernoulli beam in terms of the Hamiltonian density function is obtained as

$$\begin{aligned} \frac{d\phi}{dt} &= \frac{\partial}{\partial x}\left(\frac{\partial \bar{\mathcal{H}}}{\partial w'} \right) - \frac{\partial^2}{\partial x^2}\left(\frac{\partial \bar{\mathcal{H}}}{\partial w''} \right) + \frac{\partial}{\partial x}\left(\frac{d\phi_1}{dt} \right) + \omega(x), \\ \dot{w} &= \frac{\partial \bar{\mathcal{H}}}{\partial \phi} \end{aligned} \tag{10.297}$$

subject to the boundary conditions

$$\left\{ \frac{\partial}{\partial x}\left(\frac{\partial \bar{\mathcal{H}}}{\partial w''} \right) - \frac{d\phi_1}{dt} - \frac{\partial \bar{\mathcal{H}}}{\partial w'} \right\} \delta w \Bigg|_0^L = 0, \quad \frac{\partial \bar{\mathcal{H}}}{\partial w''} \delta w' \Bigg|_0^L = 0 \tag{10.298}$$

Subsequently, upon substitution of the Hamiltonian density function in Equation (10.297) into Equation (10.298), we obtain the classical form of the differential equations of motion as follows:

$$\begin{aligned} \dot{\phi} &= \frac{\partial}{\partial x}\left(\rho(x) I(x) \ddot{w}' \right) - \frac{\partial^2}{\partial x^2}\left(EI(x) w'' \right) + \omega(x) = 0, \\ \dot{w} &= \frac{\phi}{\rho(x) A(x)} \end{aligned} \tag{10.299}$$

subject to the boundary conditions

$$EI(x)\, w'' \delta w' \big|_0^L = 0, \quad \left(\frac{\partial}{\partial x} \left(EI(x)\, w'' \right) - \rho I(x)\, \ddot{w}' \right) \delta w \bigg|_0^L = 0 \tag{10.300}$$

Space-Discrete Finite Element Formulation

Next, we discuss the development of first-order semi-discrete differential equations of motion, via the space-discrete Hamiltonian formulation. The trial displacement function may be approximated using the Hermite polynomial interpolation functions as follows

$$\begin{aligned} w^h(x,t) &= \mathbf{N}_e(x)\, \mathbf{q}_e(t), \quad \frac{\partial w^h(x,t)}{\partial x} = \mathbf{B}_1(x) \cdot \mathbf{q}_e(t), \\ \mathbf{B}_1(x) &= \frac{\partial \mathbf{N}_e(x)}{\partial x} \end{aligned} \tag{10.301}$$

where

$$\begin{aligned} \mathbf{q}_e(t) &= \left(w_1^h(t), \frac{\partial w_1^h}{\partial x}(t), w_2^h(t), \frac{\partial w_2^h}{\partial x}(t) \right), \\ \dot{\mathbf{q}}_e(t) &= \left(\dot{w}_1^h(t), \frac{\partial \dot{w}_1^h}{\partial x}(t), \dot{w}_2^h(t), \frac{\partial \dot{w}_2^h}{\partial x}(t) \right) \end{aligned} \tag{10.302}$$

Then, we have the space-discrete kinetic energy given by

$$\mathcal{K}_e^h(\mathbf{p}_e) = \frac{1}{2} \mathbf{p}_e(t) \cdot (\mathbf{M}^e)^{-1} \mathbf{p}_e(t), \quad \mathbf{p}_e(t) = \mathbf{M}^e \dot{\mathbf{q}}_e(t) \tag{10.303}$$

where the consistent mass matrix is given by

$$\begin{aligned} \mathbf{M}^e &= \mathbf{M}_A + \mathbf{M}_I, \\ \mathbf{M}_A &= \frac{1}{2} \int_{L_e} \rho A(x)\, \mathbf{N}_e^T \mathbf{N}_e\, dx, \\ \mathbf{M}_I &= \frac{1}{2} \int_{L_e} \rho I(x)\, \mathbf{B}_1^T \mathbf{B}_1\, dx \end{aligned} \tag{10.304}$$

In addition, the space-discrete potential energy can be obtained

$$\mathcal{U}_e^h(\mathbf{q}_e) = \frac{1}{2} \mathbf{q}_e(t) \cdot \mathbf{K}^e \mathbf{q}_e(t) - \mathbf{q}_e(t) \cdot \mathbf{F}_{\text{ext}}^e \tag{10.305}$$

where the stiffness matrix and the external force can be given by

$$\begin{aligned} \mathbf{K}^e &= \int_{L_e} EI(x)\, \mathbf{B}_2^T \mathbf{B}_2\, dx, \quad \mathbf{B}_2(x) = \frac{\partial^2 \mathbf{N}_e(x)}{\partial x^2}, \\ \mathbf{F}_{\text{ext}}^e &= \int_{L_e} \omega(x)\, \mathbf{N}_e^T(x)\, dx \end{aligned} \tag{10.306}$$

whose details can be found in Section 9.2.6. We now construct the space-discrete autonomous Hamiltonian as follows:

$$\begin{aligned}\mathcal{H}_e^h(\mathbf{p}_e, \mathbf{q}_e) &= \mathcal{K}_e^h(\mathbf{p}_e) + \mathcal{U}_e^h(\mathbf{q}_e) \\ &= \frac{1}{2}\mathbf{p}_e(t) \cdot (\mathbf{M}^e)^{-1}\mathbf{p}_e(t) + \frac{1}{2}\mathbf{q}_e(t) \cdot \mathbf{K}^e\mathbf{q}_e(t) - \mathbf{q}_e(t) \cdot \mathbf{F}_{\text{ext}}^e\end{aligned} \tag{10.307}$$

Substituting the space-discrete autonomous Hamiltonian given in Equation (10.307) into the space-discrete Hamiltonian differential operators in Equation (10.238) readily yields the semi-discrete equations of motion as

$$\begin{aligned}\mathfrak{H}_{D1}\left(\mathcal{H}_e^h\right) &= \dot{\mathbf{p}}_e + \mathbf{K}^e\mathbf{q}_e - \mathbf{F}_{ext}^e = 0, \\ \mathfrak{H}_{D2}\left(\mathcal{H}_e^h\right) &= \dot{\mathbf{q}}_e - (\mathbf{M}^e)^{-1}\mathbf{p}_e = 0\end{aligned} \tag{10.308}$$

Example 10.9 Timoshenko Beam

In a similar manner as previously illustrated for the Timoshenko beam problem in the context of the Lagrangian framework, we shall instead discuss the same beam problem with the space-discrete Hamiltonian formulation.

Governing Equations: Hamiltonian Density Based Representations

The Lagrangian density function is given by

$$\begin{aligned}&\bar{\mathcal{L}}(\mathbf{u}, \dot{\mathbf{u}}, \mathbf{u}') = \frac{1}{2}\dot{\mathbf{u}} \cdot \boldsymbol{\rho}\dot{\mathbf{u}} - \frac{1}{2}\boldsymbol{\kappa} \cdot \mathbf{D}_b\boldsymbol{\kappa} - \frac{1}{2}\boldsymbol{\gamma} \cdot \mathbf{G}_s\boldsymbol{\gamma}, \\ &\mathbf{u}(x,t) = \left(w(x,t), \theta_y(x,t)\right), \quad \dot{\mathbf{u}}(x,t) = \left(\dot{w}(x,t), \dot{\theta}_y(x,t)\right), \\ &\boldsymbol{\rho} = \begin{bmatrix} \rho A(x) & 0 \\ 0 & \rho I(x) \end{bmatrix}, \quad \mathbf{u}' = \left(\frac{\partial w(x,t)}{\partial x}, \frac{\partial \theta_y(x,t)}{\partial x}\right) \\ &\boldsymbol{\kappa} = \left[\frac{\partial \theta_y}{\partial x}\right], \quad \mathbf{D}_b = [EI(x)], \quad \boldsymbol{\gamma} = \left[\frac{\partial w(x,t)}{\partial x} + \theta_y(x,t)\right], \quad \mathbf{G}_s = [\kappa GA(x)]\end{aligned} \tag{10.309}$$

Let $\boldsymbol{\phi}(x,t) = (\phi_1(x,t), \phi_2(x,t))$

$$\boldsymbol{\phi} = \frac{\partial \bar{\mathcal{L}}(\mathbf{u}, \dot{\mathbf{u}}, \mathbf{u}')}{\partial \dot{\mathbf{u}}} = \boldsymbol{\rho}\dot{\mathbf{u}} \tag{10.310}$$

The Hamiltonian density function by the Legendre transformation is

$$\bar{\mathcal{H}}(\boldsymbol{\phi}, \mathbf{u}, \mathbf{u}') = \boldsymbol{\phi} \cdot \dot{\mathbf{u}} - \bar{\mathcal{L}}(\mathbf{u}, \dot{\mathbf{u}}, \mathbf{u}') \tag{10.311}$$

which yields

$$\bar{\mathcal{H}}(\boldsymbol{\phi}, \mathbf{u}, \mathbf{u}') = \frac{1}{2}\boldsymbol{\phi} \cdot \boldsymbol{\rho}^{-1}\boldsymbol{\phi} + \frac{1}{2}\boldsymbol{\kappa} \cdot \mathbf{D}_b\boldsymbol{\kappa} + \frac{1}{2}\boldsymbol{\gamma} \cdot \mathbf{G}_s\boldsymbol{\gamma} \tag{10.312}$$

Hamilton's law of varying action can be written as

$$\begin{aligned} &\delta \int_{t_0}^{t_1} \left[\int_{\mathcal{B}} \boldsymbol{\phi} \cdot \dot{\mathbf{u}} \, dV - \mathcal{H}(\boldsymbol{\phi}, \mathbf{u}) \right] dt - \int_L \boldsymbol{\phi}(x,t) \cdot \delta \mathbf{u} \, dx \Big|_{t_0}^{t_1} = \\ &= \int_{t_0}^{t_1} \left[\int_L (\delta \boldsymbol{\phi} \cdot \dot{\mathbf{u}} + \boldsymbol{\phi} \cdot \delta \dot{\mathbf{u}}) \, dx - \delta \mathcal{H}(\boldsymbol{\phi}, \mathbf{u}) \right] dt - \int_L \boldsymbol{\phi}(x,t) \cdot \delta \mathbf{u} \, dx \Big|_{t_0}^{t_1} = 0 \end{aligned} \tag{10.313}$$

where the continuous Hamiltonian of the Timoshenko beam can be found as

$$\mathcal{H}(\boldsymbol{\phi}(x,t), \mathbf{u}(x,t)) = \int_L \bar{\mathcal{H}}(\boldsymbol{\phi}(x,t), \mathbf{u}(x,t), \mathbf{u}'(x,t)) \, dx - \int_L \bar{\omega}(x) \, \mathbf{a} \cdot \mathbf{u} \, dx \tag{10.314}$$

Note that we have the relation

$$\int_{t_0}^{t_1} \left(\int_L \boldsymbol{\phi}(x,t) \cdot \delta \dot{\mathbf{u}} dx \right) dt = \int_L \boldsymbol{\phi}(x,t) \cdot \delta \mathbf{u} \, dx \Big|_{t_0}^{t_1} - \int_{t_0}^{t_1} \left(\int_L \frac{d\boldsymbol{\phi}}{dt} \cdot \delta \mathbf{u} \, dx \right) dt, \tag{10.315}$$

The variation of the Hamiltonian density function is

$$\delta \bar{\mathcal{H}}(\boldsymbol{\phi}, \mathbf{u}, \mathbf{u}') = \frac{\partial \bar{\mathcal{H}}}{\partial \mathbf{u}} \cdot \delta \mathbf{u} + \frac{\partial \bar{\mathcal{H}}}{\partial \boldsymbol{\phi}} \cdot \delta \boldsymbol{\phi} + \frac{\partial \bar{\mathcal{H}}}{\partial \mathbf{u}'} \cdot \delta \mathbf{u}' \tag{10.316}$$

Upon integration by parts, we have

$$\delta \bar{\mathcal{H}} = \left[\frac{\partial \bar{\mathcal{H}}}{\partial \mathbf{u}} - \frac{\partial}{\partial x} \left(\frac{\partial \bar{\mathcal{H}}}{\partial \mathbf{u}'} \right) \right] \cdot \delta \mathbf{u} + \frac{\partial \bar{\mathcal{H}}}{\partial \boldsymbol{\phi}} \cdot \delta \boldsymbol{\phi} + \frac{\partial}{\partial x} \left(\frac{\partial \bar{\mathcal{H}}}{\partial \mathbf{u}'} \cdot \delta \mathbf{u} \right) \tag{10.317}$$

Then, Hamilton's law of varying action can be rewritten as

$$\begin{aligned} &\delta \int_{t_0}^{t_1} \left(\int_{\mathcal{B}} \boldsymbol{\phi} \cdot \dot{\mathbf{u}} dV - \mathcal{H}(\boldsymbol{\phi}(x,t), \mathbf{u}(x,t)) \right) dt - \int_L \boldsymbol{\phi}(x,t) \cdot \delta \mathbf{u} \, dx \Big|_{t_0}^{t_1} = \\ &= \int_{t_0}^{t_1} \left\{ \int_L \left[\begin{array}{l} \left(\dot{\mathbf{u}} - \dfrac{\partial \bar{\mathcal{H}}}{\partial \boldsymbol{\phi}} \right) \cdot \delta \boldsymbol{\phi} \\ + \left\{ \dfrac{\partial}{\partial x} \left(\dfrac{\partial \bar{\mathcal{H}}}{\partial \mathbf{u}'} \right) - \dfrac{\partial \bar{\mathcal{H}}}{\partial \mathbf{u}} - \dfrac{d\boldsymbol{\phi}}{dt} + \bar{\omega}(x) \, \mathbf{a} \right\} \cdot \delta \mathbf{u} \end{array} \right] dx \right\} dt \\ &- \int_{t_0}^{t_1} \left(\frac{\partial \bar{\mathcal{H}}}{\partial \mathbf{u}'} \cdot \delta \mathbf{u} \Big|_0^L \right) dt = 0 \end{aligned} \tag{10.318}$$

in which both governing equations and natural boundary conditions are included. The governing equations are given as

$$\frac{\partial}{\partial x} \left(\frac{\partial \bar{\mathcal{H}}}{\partial \mathbf{u}'} \right) - \frac{\partial \bar{\mathcal{H}}}{\partial \mathbf{u}} - \frac{d\boldsymbol{\phi}}{dt} + \bar{\omega}(x) \, \mathbf{a} = 0, \quad \dot{\mathbf{u}} - \frac{\partial \bar{\mathcal{H}}}{\partial \boldsymbol{\phi}} = 0 \tag{10.319}$$

subject to the natural boundary conditions

$$\frac{\partial \bar{\mathcal{H}}}{\partial \mathbf{u}'} \cdot \delta \mathbf{u} \bigg|_0^L = 0 \tag{10.320}$$

Therefore, Equation (10.319) can be given by

$$\begin{aligned} \frac{d\phi_1}{dt} &= \frac{\partial}{\partial x}\left(\frac{\partial \bar{\mathcal{H}}}{\partial w'}\right) - \frac{\partial \bar{\mathcal{H}}}{\partial w} + \bar{\omega}(x), \quad \dot{w} = \frac{\partial \bar{\mathcal{H}}}{\partial \phi_1} \\ \frac{d\phi_2}{dt} &= \frac{\partial}{\partial x}\left(\frac{\partial \bar{\mathcal{H}}}{\partial \phi_1'}\right) - \frac{\partial \bar{\mathcal{H}}}{\partial \phi_1}, \quad \dot{\phi}_1 = \frac{\partial \bar{\mathcal{H}}}{\partial \phi_2} \end{aligned} \tag{10.321}$$

subject to the boundary conditions

$$\frac{\partial \bar{\mathcal{H}}}{\partial w'} \delta w \bigg|_0^L = 0, \quad \frac{\partial \bar{\mathcal{H}}}{\partial \phi_1'} \delta \phi_1 \bigg|_0^L = 0 \tag{10.322}$$

Eventually, upon substitution of the Hamiltonian density function in Equation (10.312) into Equation (10.321), we have the classical representation of the first-order differential equations of motion as follows:

$$\begin{aligned} \dot{\phi}_1 &= \frac{\partial}{\partial x}\left(\kappa G A(x)\left(\frac{\partial y}{\partial x} - \phi\right)\right) + \bar{\omega}(x), \quad \dot{y} = \frac{\partial \bar{\mathcal{H}}}{\partial \phi_1} \\ \dot{\phi}_2 &= \frac{\partial}{\partial x}\left(EI(x)\left(\frac{\partial \phi}{\partial x}\right)\right) + \kappa G A(x)\left(\frac{\partial y}{\partial x} - \phi\right), \quad \dot{\phi} = \frac{\partial \bar{\mathcal{H}}}{\partial \phi_2} \end{aligned} \tag{10.323}$$

subject to the boundary condition

$$EI(x)\left(\frac{\partial \phi}{\partial x}\right)\delta\phi \bigg|_0^L = 0, \quad \kappa G A(x)\left(\frac{\partial y}{\partial x} - \phi\right)\delta y \bigg|_0^L = 0 \tag{10.324}$$

Consequently, we have shown the derivation of the two governing equations of motion with the traction boundary condition via Hamilton's law of varying action.

Space-Discrete Finite Element Formulation

Next, we construct the space-discrete Hamiltonian. The admissible trial functions are given by

$$\begin{aligned} \tilde{w}^h(x,t) &= \mathbf{N}_w(x)\,\mathbf{w}^e(t), \quad \mathbf{w}^e(t) = \left[\begin{array}{cccc} w_1(t) & w_2(t) & \dots & w_{n_{\text{node}}}(t) \end{array}\right]^T, \\ \tilde{\theta}_y^h(x,t) &= \mathbf{N}_\theta(x)\,\boldsymbol{\theta}_y^e(t), \quad \boldsymbol{\theta}_y^e(t) = \left[\begin{array}{cccc} \theta_{y1}(t) & \theta_{y2}(t) & \dots & \theta_{yn_{\text{node}}}(t) \end{array}\right]^T \end{aligned} \tag{10.325}$$

Upon introducing a new notation $\mathbf{q}_e(t) = \left[\begin{array}{cc} \mathbf{w}^e(t) & \boldsymbol{\theta}_y^e(t) \end{array}\right]^T$, the trial functions are rewritten as

$$\tilde{w}^h(x,t) = \mathbf{H}_w(x)\,\mathbf{q}_e(t), \quad \tilde{\theta}_y^h(x,t) = \mathbf{H}_\theta(x)\,\mathbf{q}_e(t) \tag{10.326}$$

where $\mathbf{H}_w(x) = \begin{bmatrix} \mathbf{N}_w(x) & \mathbf{0} \end{bmatrix}$ and $\mathbf{H}_\theta(x) = \begin{bmatrix} \mathbf{0} & \mathbf{N}_\theta(x) \end{bmatrix}$. In vector form, we have

$$\tilde{\mathbf{u}}^h = \left\{ \begin{array}{c} \tilde{w}^h(x,t) \\ \tilde{\theta}_y^h(x,t) \end{array} \right\} = \mathbf{H}_e(x)\,\mathbf{q}_e(t), \quad \mathbf{H}_e(x) = \begin{bmatrix} \mathbf{H}_w(x) \\ \mathbf{H}_\theta(x) \end{bmatrix} \tag{10.327}$$

The space-discrete Hamiltonian may be constructed as

$$\begin{aligned}
\mathcal{H}_e^h(\mathbf{q}_e, \mathbf{p}_e) &= \left(\frac{1}{2}\mathbf{p}_e \cdot (\mathbf{M}^e)^{-1}\mathbf{p}_e\right) + \left(\frac{1}{2}\mathbf{q}_e \cdot \mathbf{K}^e\mathbf{q}_e - \mathbf{q}_e \cdot \mathbf{F}_{\text{ext}}^e\right), \\
\mathbf{M}^e &= \frac{1}{2}\int_{L_e} \mathbf{H}_e^T \rho \mathbf{H}_e(x)\,dx, \ \mathbf{p}_e(t) = \mathbf{M}^e\dot{\mathbf{q}}_e(t), \ \mathbf{K}^e = \mathbf{K}_b + \mathbf{K}_s, \\
\mathbf{K}_b &= \int_{L_e} \mathbf{B}_b^T(x)\,\mathbf{D}_b\mathbf{B}_b(x)\,dx, \ \mathbf{K}_s = \int_{L_e} \mathbf{B}_s^T(x)\,\mathbf{G}_s\mathbf{B}_s(x)\,dx, \\
\mathbf{B}_b &= \left[\frac{\partial \mathbf{H}_\theta(x)}{\partial x}\right], \ \mathbf{B}_s = \left[\frac{\partial \mathbf{H}_w(x)}{\partial x} + \mathbf{H}_\theta(x)\right], \ \mathbf{F}_{\text{ext}}^e = \int_{L_e} \bar{\omega}(x)\,\mathbf{H}_w^T(x)\,dx
\end{aligned} \tag{10.328}$$

whose details can be found in Section 9.2.7. Substituting the space-discrete Hamiltonian into the space-discrete Hamiltonian differential operators in Equation (10.238), simply and naturally leads to the straightforward derivation of the semi-discrete finite element equations of motion within the Hamiltonian mechanics framework as:

$$\begin{aligned}
\mathfrak{H}_{D1}\left(\mathcal{H}_e^h\right) &= \dot{\mathbf{p}}_e + \mathbf{K}^e\mathbf{q}_e - \mathbf{F}_{\text{ext}}^e = \mathbf{0}, \\
\mathfrak{H}_{D2}\left(\mathcal{H}_e^h\right) &= \dot{\mathbf{q}}_e - (\mathbf{M}^e)^{-1}\mathbf{p}_e = \mathbf{0}
\end{aligned} \tag{10.329}$$

Example 10.10 Reissner-Mindlin Plate

In this section we shall consider the Reissner-Mindlin plate of a uniform thickness h in the Hamiltonian mechanics framework.

Governing Equations: Hamiltonian Density Based Representations

The continuous Lagrangian of the Reissner-Mindlin plate is given by

$$\begin{aligned}
&\mathcal{L}(\mathbf{u}, \dot{\mathbf{u}}) = \int_\Omega \bar{\mathcal{L}}(\mathbf{u}, \dot{\mathbf{u}}, \mathbf{u}_{,x}, \mathbf{u}_{,y})\,dxdy + \int_\Omega \omega(x,y)\,\mathbf{a}\cdot\mathbf{u}\,dxdy, \quad \mathbf{a} = (1,0,0), \\
&\mathbf{u} = (w, \theta_x, \theta_y), \ \dot{\mathbf{u}} = (\dot{w}, \dot{\theta}_x, \dot{\theta}_y), \\
&\mathbf{u}_{,x} = \left(\frac{\partial w}{\partial x}, \frac{\partial \theta_x}{\partial x}, \frac{\partial \theta_y}{\partial x}\right), \ \mathbf{u}_{,y} = \left(\frac{\partial w}{\partial y}, \frac{\partial \theta_x}{\partial y}, \frac{\partial \theta_y}{\partial y}\right)
\end{aligned} \tag{10.330}$$

where the Lagrangian density function can be defined as

$$\bar{\mathcal{L}}(\mathbf{u}, \dot{\mathbf{u}}, \mathbf{u}_{,x}, \mathbf{u}_{,y}) = \frac{1}{2}\dot{u}\cdot\rho\dot{\mathbf{u}} - \frac{1}{2}\kappa\cdot\mathbf{D}_b\kappa - \frac{1}{2}\gamma\cdot\mathbf{G}_s\gamma \tag{10.331}$$

Therefore, the components of the canonical momenta, $\boldsymbol{\phi} = (\phi_1, \phi_2, \phi_2)$, are found as

$$\boldsymbol{\phi} = \frac{\partial \bar{\mathcal{L}}\left(\mathbf{u}, \dot{\mathbf{u}}, \mathbf{u}_{,x}, \mathbf{u}_{,y}\right)}{\partial \dot{\mathbf{u}}} \tag{10.332}$$

The Hamiltonian density function can be given via the Legendre transformation as

$$\bar{\mathcal{H}}\left(\boldsymbol{\phi}, \mathbf{u}, \mathbf{u}_{,x}, \mathbf{u}_{,y}\right) = \boldsymbol{\phi}(x, y, t) \cdot \dot{\mathbf{u}}(x, y, t) - \bar{\mathcal{L}}\left(\mathbf{u}, \dot{\mathbf{u}}, \mathbf{u}_{,x}, \mathbf{u}_{,y}\right) \tag{10.333}$$

Consequently, Hamilton's law of varying action in the context of the Hamiltonian gives

$$\delta \int_{t_0}^{t_1} \left\{ \int_\Omega \boldsymbol{\phi} \cdot \dot{\mathbf{u}} dx dy - \mathcal{H}(\boldsymbol{\phi}, \mathbf{u}) \right\} dt - \int_\Omega \boldsymbol{\phi}(x, y, t) \cdot \delta \mathbf{u}\, dx dy \bigg|_{t_0}^{t_1} = 0 \tag{10.334}$$

where the continuous Hamiltonian of the Reissner-Mindlin plate can be found as

$$\mathcal{H}(\boldsymbol{\phi}, \mathbf{u}) = \int_\Omega \bar{\mathcal{H}}\left(\boldsymbol{\phi}, \mathbf{u}, \mathbf{u}_{,x}, \mathbf{u}_{,y}\right) dx dy - \int_\Omega \omega(x, y)\, \mathbf{a} \cdot \mathbf{u}\, dx dy, \quad \mathbf{a} = (1, 0, 0) \tag{10.335}$$

We introduce new notations for convenience; then the Hamiltonian is given by

$$\begin{aligned} \bar{\mathcal{H}} &= \frac{1}{2}\boldsymbol{\phi} \cdot \boldsymbol{\rho}^{-1}\boldsymbol{\phi} + \frac{1}{2}\boldsymbol{\kappa} \cdot \mathbf{D}_b \boldsymbol{\kappa} + \frac{1}{2}\boldsymbol{\gamma} \cdot \mathbf{G}_s \boldsymbol{\gamma} \\ &= \frac{1}{2}\boldsymbol{\phi} \cdot \boldsymbol{\rho}^{-1}\boldsymbol{\phi} + \frac{1}{2}\boldsymbol{\kappa} \cdot \mathbf{M}_b + \frac{1}{2}\boldsymbol{\gamma} \cdot \mathbf{q}_s, \\ \boldsymbol{\phi} &= \boldsymbol{\rho}\dot{\mathbf{u}}, \quad \mathbf{M}_b = \mathbf{D}_b \boldsymbol{\kappa}, \quad \mathbf{q}_s = \mathbf{G}_s \boldsymbol{\gamma} \end{aligned} \tag{10.336}$$

where the generalized momentum and the stress resultants are given by

$$\begin{aligned} \mathbf{p}^T &= \left[\begin{array}{ccc} \rho \dot{w} h & \rho \dot{\theta}_x I_p & \rho \dot{\theta}_y I_p \end{array}\right], \\ \mathbf{M}_b{}^T &= \left[\begin{array}{ccc} M_{xx} & M_{yy} & M_{xy} \end{array}\right], \\ \mathbf{q}_s{}^T &= \left[\begin{array}{cc} Q_x & Q_y \end{array}\right] \end{aligned} \tag{10.337}$$

Note that Equation (10.334) for holonomic systems can be rewritten as

$$\int_{t_0}^{t_1} \left\{ \int_\Omega \left(\delta \boldsymbol{\phi} \cdot \dot{\mathbf{u}} + \boldsymbol{\phi} \cdot \delta \dot{u}\right) dx dy - \delta \mathcal{H}(\boldsymbol{\phi}, \mathbf{u}) \right\} dt - \int_\Omega \boldsymbol{\phi} \cdot \delta \mathbf{u}\, dx dy \bigg|_{t_0}^{t_1} = 0 \tag{10.338}$$

The variation of the autonomous Hamiltonian is given by

$$\begin{aligned} \delta \mathcal{H}(\boldsymbol{\phi}, \mathbf{u}) &= \int_\Omega \delta \bar{\mathcal{H}}\left(\boldsymbol{\phi}, \mathbf{u}, \mathbf{u}_{,x}, \mathbf{u}_{,y}\right) dx dy - \int_\Omega \omega(x, y)\, \mathbf{a} \cdot \mathbf{u}\, dx dy \\ &= \int_\Omega \frac{\partial \bar{\mathcal{H}}}{\partial \mathbf{u}} \cdot \delta \mathbf{u} + \frac{\partial \bar{\mathcal{H}}}{\partial \boldsymbol{\phi}} \cdot \delta \boldsymbol{\phi} + \frac{\partial \bar{\mathcal{H}}}{\partial \mathbf{u}_{,x}} \cdot \delta \mathbf{u}_{,x} + \frac{\partial \bar{\mathcal{H}}}{\partial \mathbf{u}_{,y}} \cdot \delta \mathbf{u}_{,y} dx dy \\ &\quad - \int_\Omega p(x, y)\, \mathbf{a} \cdot \delta \mathbf{u}\, dx dy \end{aligned} \tag{10.339}$$

Performing integration by parts, yields

$$\int_{t_0}^{t_1}\left(\int_{\Omega}\boldsymbol{\phi}\cdot\delta\dot{\mathbf{u}}dxdy\right)dt=\int_{\Omega}\boldsymbol{\phi}\cdot\delta\mathbf{u}\,dxdy\bigg|_{t_0}^{t_1}-\int_{t_0}^{t_1}\left(\int_{\Omega}\frac{d\boldsymbol{\phi}}{dt}\cdot\delta\mathbf{u}\,dxdy\right)dt \tag{10.340}$$

Hence, in view of Equation (10.339) and (10.340), Hamilton's law of varying action leads to

$$\int_{t_0}^{t_1}\left\{\int_{\Omega}\left(\begin{array}{c}\delta\boldsymbol{\phi}\cdot\left(\dot{\mathbf{u}}-\dfrac{\partial\bar{\mathcal{H}}}{\partial\boldsymbol{\phi}}\right)-\left(\dfrac{d\boldsymbol{\phi}}{dt}+\dfrac{\partial\bar{\mathcal{H}}}{\partial\mathbf{u}}-\omega(x,y)\,\mathbf{a}\right)\cdot\delta\mathbf{u}\\-\dfrac{\partial\bar{\mathcal{H}}}{\partial\mathbf{u}_{,x}}\cdot\delta\mathbf{u}_{,x}-\dfrac{\partial\bar{\mathcal{H}}}{\partial\mathbf{u}_{,y}}\cdot\delta\mathbf{u}_{,y}\end{array}\right)dxdy\right\}dt=0 \tag{10.341}$$

By means of Green's theorem, we have the following:

$$\begin{aligned}&\int_{\Omega}\left(\frac{\partial\bar{\mathcal{H}}}{\partial\mathbf{u}_{,x}}\cdot\delta\mathbf{u}_{,x}+\frac{\partial\bar{\mathcal{H}}}{\partial\mathbf{u}_{,y}}\cdot\delta\mathbf{u}_{,y}\right)dxdy\\&=-\int_{A}\left\{\frac{\partial}{\partial x}\left(\frac{\partial\bar{\mathcal{H}}}{\partial\mathbf{u}_{,x}}\right)+\frac{\partial}{\partial y}\left(\frac{\partial\bar{\mathcal{H}}}{\partial\mathbf{u}_{,y}}\right)\right\}\cdot\delta\mathbf{u}\,dxdy\\&+\int_{\partial\Omega}\left\{\frac{\partial\bar{\mathcal{H}}}{\partial\mathbf{u}_{,x}}\cdot\delta\mathbf{u}_{x}n_x+\frac{\partial\bar{\mathcal{H}}}{\partial\mathbf{u}_{,y}}\cdot\delta\mathbf{u}_{y}n_y\right\}d\Gamma=0,\quad n_x=\frac{\partial x}{\partial n},\ n_y=\frac{\partial y}{\partial n}\end{aligned} \tag{10.342}$$

where n_x, n_y denote the directional cosines respectively and n stands for the coordinate of the outward unit vector normal to the boundary surface.

$$\begin{aligned}&\int_{t_0}^{t_1}\left\{\int_{\Omega}\left[\begin{array}{c}\delta\boldsymbol{\phi}\cdot\left(\dot{\mathbf{u}}-\dfrac{\partial\bar{\mathcal{H}}}{\partial\boldsymbol{\phi}}\right)\\-\left(\begin{array}{c}\dfrac{d\boldsymbol{\phi}}{dt}+\dfrac{\partial\bar{\mathcal{H}}}{\partial\mathbf{u}}-\dfrac{\partial}{\partial x}\left(\dfrac{\partial\bar{\mathcal{H}}}{\partial\mathbf{u}_{,x}}\right)\\-\dfrac{\partial}{\partial y}\left(\dfrac{\partial\bar{\mathcal{H}}}{\partial\mathbf{u}_{,y}}\right)-\omega(x,y)\,\mathbf{a}\end{array}\right)\cdot\delta\mathbf{u}\end{array}\right]\cdot\delta\mathbf{u}\,dxdy\right\}dt\\&\qquad-\int_{\partial\Omega}\left(\frac{\partial\bar{\mathcal{H}}}{\partial\mathbf{u}_{,x}}\cdot\delta\mathbf{u}_{x}n_x+\frac{\partial\bar{\mathcal{H}}}{\partial\mathbf{u}_{,y}}\cdot\delta\mathbf{u}_{y}n_y\right)d\Gamma=0\end{aligned} \tag{10.343}$$

Consequently, we have the governing equations for the Reissner-Mindlin plate as follows:

$$\begin{aligned}&\frac{d\boldsymbol{\phi}}{dt}+\frac{\partial\bar{\mathcal{H}}}{\partial\mathbf{u}}-\frac{\partial}{\partial x}\left(\frac{\partial\bar{\mathcal{H}}}{\partial\mathbf{u}_{,x}}\right)-\frac{\partial}{\partial y}\left(\frac{\partial\bar{\mathcal{H}}}{\partial\mathbf{u}_{,y}}\right)-p\mathbf{a}=0,\\&\dot{\mathbf{u}}-\frac{\partial\bar{\mathcal{H}}}{\partial\boldsymbol{\phi}}=0\end{aligned} \tag{10.344}$$

The above equations can be rewritten as

$$\frac{d}{dt}(\boldsymbol{\phi}) + \left[\frac{\partial \boldsymbol{\kappa}}{\partial \mathbf{u}}\right]^T \mathbf{M}_b + \left[\frac{\partial \gamma}{\partial \mathbf{u}}\right]^T \mathbf{Q}_s - \frac{\partial \mathbf{A}_x}{\partial x} - \frac{\partial \mathbf{A}_y}{\partial y} - \mathbf{pa} = 0, \qquad (10.345)$$
$$\dot{\mathbf{u}} - \rho^{-1}\boldsymbol{\phi} = 0$$

where

$$\mathbf{A}_x = \frac{\partial \bar{\mathcal{H}}}{\partial \mathbf{u}_{,x}} = \left[\frac{\partial \boldsymbol{\kappa}}{\partial \mathbf{u}_{,x}}\right]^T \mathbf{M}_b + \left[\frac{\partial \boldsymbol{\gamma}}{\partial \mathbf{u}_{,x}}\right]^T \mathbf{Q}_s,$$
$$\mathbf{A}_y = \frac{\partial \bar{\mathcal{H}}}{\partial \mathbf{u}_{,y}} = \left[\frac{\partial \boldsymbol{\kappa}}{\partial \mathbf{u}_{,y}}\right]^T \mathbf{M}_b + \left[\frac{\partial \boldsymbol{\gamma}}{\partial \mathbf{u}_{,y}}\right]^T \mathbf{Q}_s \qquad (10.346)$$

where the matrices in the above are given in Equation (10.189). And, Equation (10.344) can be recast into

$$\begin{Bmatrix} \dot{\phi}_1(x,y,t) \\ \dot{\phi}_2(x,y,t) \\ \dot{\phi}_3(x,y,t) \end{Bmatrix} + \begin{Bmatrix} 0 \\ -Q_x \\ Q_y \end{Bmatrix} - \frac{\partial}{\partial x}\begin{Bmatrix} Q_y \\ -M_{xy} \\ M_{xx} \end{Bmatrix} - \frac{\partial}{\partial y}\begin{Bmatrix} Q_x \\ -M_{yy} \\ M_{xy} \end{Bmatrix} - \begin{Bmatrix} \omega \\ 0 \\ 0 \end{Bmatrix} = \begin{Bmatrix} 0 \\ 0 \\ 0 \end{Bmatrix},$$
$$\begin{Bmatrix} \dot{w} \\ \dot{\theta}_x \\ \ddot{\theta}_y \end{Bmatrix} - \begin{Bmatrix} \dot{\phi}_1/\rho h \\ \dot{\phi}_2/\rho I_p \\ \dot{\phi}_3/\rho I_p \end{Bmatrix} = \begin{Bmatrix} 0 \\ 0 \\ 0 \end{Bmatrix} \qquad (10.347)$$

Notice that the first of the equations in Equation (10.345) imply the following:

$$\frac{d}{dt}(\phi_1) + \frac{\partial \bar{\mathcal{H}}}{\partial w} - \frac{\partial}{\partial x}\left(\frac{\partial \bar{\mathcal{H}}}{\partial w_{,x}}\right) - \frac{\partial}{\partial y}\left(\frac{\partial \bar{\mathcal{H}}}{\partial w_{,y}}\right) - \omega(x,y) = 0,$$
$$\frac{d}{dt}(\phi_2) + \frac{\partial \bar{\mathcal{H}}}{\partial \theta_x} - \frac{\partial}{\partial x}\left(\frac{\partial \bar{\mathcal{H}}}{\partial \theta_{x,x}}\right) - \frac{\partial}{\partial y}\left(\frac{\partial \bar{\mathcal{H}}}{\partial \theta_{x,y}}\right) = 0, \qquad (10.348)$$
$$\frac{d}{dt}(\phi_3) + \frac{\partial \bar{\mathcal{H}}}{\partial \theta_y} - \frac{\partial}{\partial x}\left(\frac{\partial \bar{\mathcal{H}}}{\partial \theta_{y,x}}\right) - \frac{\partial}{\partial y}\left(\frac{\partial \bar{\mathcal{H}}}{\partial \theta_{y,y}}\right) = 0,$$

which lead to

$$\dot{\phi}_1 - Ghk\frac{\partial}{\partial x}\left(\frac{\partial w}{\partial x} + \theta_x\right) - Ghk\frac{\partial}{\partial y}\left(\frac{\partial w}{\partial y} - \theta_y\right) - \omega(x,y) = 0,$$
$$\dot{\phi}_2 - Ghk\left(\frac{\partial w}{\partial y} - \theta_x\right) - \frac{1}{2}D\left\{(1-\nu)\nabla^2\theta_x + (1+\nu)\left(\frac{\partial^2\theta_x}{\partial y^2} - \frac{\partial^2\theta_y}{\partial x \partial y}\right)\right\} = 0, \qquad (10.349)$$

$$\dot{\phi}_3 + Ghk\left(\frac{\partial w}{\partial x} + \theta_y\right) - \frac{1}{2}D\left\{(1-\nu)\nabla^2\theta_y + (1+\nu)\left(\frac{\partial^2\theta_y}{\partial x^2} - \frac{\partial^2\theta_x}{\partial y \partial x}\right)\right\} = 0$$

The second term in the right-hand side of Equation (10.343) gives the natural boundary conditions as follows:

$$\begin{aligned} &\frac{\partial \bar{\mathcal{H}}}{\partial \mathbf{u}_{,x}} \cdot \delta \mathbf{u}_x n_x + \frac{\partial \bar{\mathcal{H}}}{\partial \mathbf{u}_{,y}} \cdot \delta \mathbf{u}_y n_y = \mathbf{A}_x^T \delta \mathbf{u}_x n_x + \mathbf{A}_y^T \delta \mathbf{u}_y n_y \\ &= n_x\left(q_y \delta w - M_{xy}\delta\theta_x + M_{xx}\delta\theta_y\right) + n_y\left(q_x \delta w - M_{yy}\delta\theta_x + M_{xy}\delta\theta_y\right) \\ &= \delta w\left(n_x Q_y + n_y Q_x\right) - \delta\theta_x\left(n_x M_{xy} + n_y M_{yy}\right) + \delta\theta_y\left(n_x M_{xx} + n_y M_{xy}\right) = 0 \end{aligned} \tag{10.350}$$

In view of Equation (10.350), we say that the governing equations in Equation (10.344) are subject to the natural boundary conditions as given in Equation (10.194). Note that the essential boundary conditions on the geometric boundary surface, $\partial\Omega_u$, can be given by Equation (10.195).

Space-Discrete Finite Element Formulation

Next, we show that the space-discrete Hamiltonian formulation offers to facilitate the derivation of the semi-discrete finite element equations of motion, and is an alternative to the principle of virtual work. The admissible trial functions are given by Equation (10.197). Then, the space-discrete Hamiltonian is constructed as

$$\begin{aligned} \mathcal{H}_e^h(\mathbf{q}_e, \mathbf{p}_e) &= \frac{1}{2}\mathbf{p}_e \cdot (\mathbf{M}^e)^{-1}\mathbf{p}_e + \frac{1}{2}\mathbf{q}_e \cdot \mathbf{K}^e\mathbf{q}_e - \mathbf{q}_e \cdot \mathbf{F}_{\text{ext}}^e,\ \mathbf{p}_e = \mathbf{M}^e\dot{\mathbf{q}}_e \\ \mathbf{M}^e &= \frac{1}{2}\int_{A_e} \mathbf{H}_e^T \rho \mathbf{H}_e\, dxdy,\ \mathbf{H}_e = \begin{bmatrix} \mathbf{H}_w \\ \mathbf{H}_{\theta_x} \\ \mathbf{H}_{\theta y} \end{bmatrix},\ \mathbf{F}_{\text{ext}}^e = \int_{\Omega_e} \omega(x,y)\,\mathbf{H}_w^T\, dxdy \\ \mathbf{K}^e &= \mathbf{K}_b + \mathbf{K}_s,\ \mathbf{K}_b^e = \frac{1}{2}\iint_{\Omega_e} \mathbf{B}_b^T\mathbf{D}_b\mathbf{B}_b\, dxdy,\ \mathbf{K}_s^e = \frac{k}{2}\iint_{\Omega_e} \mathbf{B}_s^T\mathbf{G}_s\mathbf{B}_s\, dxdy, \\ \mathbf{B}_b &= \begin{bmatrix} \dfrac{\partial \mathbf{H}_{\theta_y}}{\partial x} \\ -\dfrac{\partial \mathbf{H}_{\theta_x}}{\partial y} \\ \left(\dfrac{\partial \mathbf{H}_{\theta_y}}{\partial y} - \dfrac{\partial \mathbf{H}_{\theta_x}}{\partial x}\right) \end{bmatrix},\ \mathbf{B}_s = \begin{bmatrix} \dfrac{\partial \mathbf{H}_w}{\partial x} + \mathbf{H}_{\theta_y} \\ \dfrac{\partial \mathbf{H}_w}{\partial y} - \mathbf{H}_{\theta_x} \end{bmatrix} \end{aligned} \tag{10.351}$$

More details about $\mathbf{M}^e$, $\mathbf{K}^e$ and $\mathbf{F}_{\text{ext}}^e$ can be found in Section 9.2.9. Substituting the space-discrete Hamiltonian of Equation (10.351) into the space-discrete Hamiltonian differential operator in Equation (10.238), simply and naturally, leads to the semi-discrete finite element equations of motion:

$$\begin{aligned} \mathfrak{H}_{D1}\left(\mathcal{H}_e^h\right) &= \dot{\mathbf{p}}_e + \mathbf{K}^e\mathbf{q}_e - \mathbf{F}_{\text{ext}}^e = 0, \\ \mathfrak{H}_{D2}\left(\mathcal{H}_e^h\right) &= \dot{\mathbf{q}}_e - (\mathbf{M}^e)^{-1}\mathbf{p}_e = \mathbf{0} \end{aligned} \tag{10.352}$$

Remark 10.4.1 Hamilton's law of varying action in the context of the Total Energy representation of the equation of motion can be written via the so-called single-field variational formulation. One can readily follow traditional practices of discretization in space and time and/or seek alternative avenues as described in the chapter to follow. More detailed discussion on the scalar formalism for strong and weak forms via Hamilton's law of varying action in the sense of the Total Energy framework is given in the Appendix in Section (11.5) in the context of the variational calculus setting.

EXERCISES

10.1. Circular-Axial Bar undergoing Rotating and Longitudinal Vibration Let us consider the dynamical behavior of the same axial bar in Exercise 9.1 of Chapter 9.

(a) Establish the Lagrangian density function and write down Hamilton's law of varying action in terms of the Lagrangian density function. Derive the governing equation in terms of the Lagrangian density function and the natural boundary condition from Hamilton's law of varying action.

(b) Establish the space-discrete Lagrangian for the element. Obtain the finite element equations of motion for the circular-axial bar undergoing rotating and longitudinal vibration by use of the Lagrangian differential operator.

(c) Establish the Hamiltonian density function and write down Hamilton's law of varying action in terms of the Hamiltonian density function. Derive Hamilton's equations of motion in terms of the Hamiltonian density function and the natural boundary condition from Hamilton's law of varying action.

(d) Establish the space-discrete Hamiltonian for the element. Obtain the finite element equations of motion for the circular-axial bar undergoing rotating and longitudinal vibration by use of the Hamiltonian differential operator.

10.2. Euler-Bernoulli Beam under Axial and Transverse Loads Let us consider the dynamical behavior of the same Euler-Bernoulli beam in Exercise 9.2 of Chapter 9.

(a) Repeat the same as in Exercise 10.1.

10.3. Timoshenko Beam on an Elastic Foundation Let us consider the dynamical behavior of the same Timoshenko beam in Exercise 9.3 of Chapter 9.

(a) Repeat the same as in Exercise 10.1.

10.4. Kirchhoff-Love Plate under In-plane and Transverse Loads Let us consider the dynamical behavior of the same Kirchhoff-Love plate in Exercise 9.4 of Chapter 9.

(a) Repeat the same as in Exercise 10.1.

10.5. Reissner-Mindlin Plate under In-plane and Transverse Loads Let us consider the dynamical behavior of the same Reissner-Mindlin plate in Exercise 9.5 of Chapter 9.

(a) Repeat the same as in Exercise 10.1.

CHAPTER ELEVEN

PRINCIPLE OF BALANCE OF MECHANICAL ENERGY: FINITE ELEMENTS AND SOLID/STRUCTURAL MECHANICS

In the preceding chapters, we independently described the various developments leading to the governing equations of motion for continuous-body systems from totally different viewpoints; and the representations were shown via both vector and scalar formalisms. The developments were via two fundamentally different starting points after having first established the relations stemming from Cauchy's equation of motion which reflects the principle of balance of linear momentum to, namely, (i) the principle of virtual work in dynamics, and (ii) Hamilton's principle (and, equivalently, an alternative via Hamilton's law of varying action). Once the relations were firmly established, we instead directly employed these two fundamental principles from then onwards as the principal starting point for the subsequent developments. Furthermore, both of the developments presented earlier involved the concepts invoking variational calculus.

In contrast, in Chapter 11, after having first established the relations stemming from Cauchy's equation of motion which reflects the principle of balance of linear momentum to the theorem of expended power, or equivalently, the principle of balance of mechanical energy and the descriptive scalar function, namely, the Total Energy, $\mathcal{E}$ (restricting attention to mechanical systems only), we describe yet another alternative which provides computationally convenient and attractive features and good physical insight. Here, we invoke the notion of differential calculus setting and the developments are restricted to holonomic-scleronomic systems with the formalism via the principle of balance of mechanical energy as the fundamental axiom and starting point (and alternatively, the setting of variational calculus valid for holonomic systems is additionally described in the Appendix). The present developments lead to the same and/or equivalent representations just as those obtained from the other two independent fundamental principles, namely, the principle of virtual work in dynamics and the Hamilton's principle (or equivalently, the Hamilton's law of varying action). Nonetheless, via the present developments, we describe the three distinctly different frameworks involving scalar representations, namely, the Total Energy framework, the Lagrangian and the Hamiltonian mechanics frameworks. We shall show that for mechanical systems, in itself the theorem of power expended or equivalently, the principle of balance of mechanical energy naturally embodies the weak form in space. Hence, directly emanating from the principle of balance of mechanical energy, it readily yields three distinct differential operators: the space-discrete Total Energy Differential Operator, the space-discrete Lagrangian Differential Operator, and the space-discrete Hamiltonian Differential Operator, respectively. As a consequence, unlike traditional practices for conducting the space discretization process, the corresponding straightforward and simple formulations firstly, only need to directly employ the resulting discrete scalar differential operators for enabling numerical discretization in space. Therein, they readily provide new avenues and

Advances in Computational Dynamics of Particles, Materials and Structures, First Edition. Jason Har and Kumar K. Tamma.

directly yield the semi-discrete ordinary differential equations in time which can be readily shown to preserve the same physical attributes as the continuous systems for continuous-body dynamical applications. Furthermore, the same discrete Total Energy representation of the equation of motion and the associated framework is also very natural for conducting the subsequent time discretization process as described in Chapter 14; thereby, it naturally provides a unified approach.

11.1 INTRODUCTION

Historically, besides concepts emanating from the classical Cauchy's equation of motion and vector formalism which reflects the statement of the principle of balance of linear momentum, firstly, finite element developments in elastodynamics applications that utilize formalisms involving scalar functions as a starting point are not traditionally popular nor the norm. While in the previous two chapters we introduced vector and scalar formalisms starting from the principle of virtual work in dynamics, and Hamilton's principle (and, equivalently, the Hamilton's law of varying action), here we employ the principle of balance of mechanical energy as yet another alternative. With regards to the scalar functions and formalisms, namely, the Lagrangian and the Hamiltonian, the following is worth noting. Some past efforts not particularly related to the present discussion, but which have highlighted the notion of discrete Lagrangian and Hamiltonian follow next so as to place the present developments in the proper context. In the literature, attention (Argyris and Scharpf 1969; Fried 1969) was first paid to the action, as a functional involving the scalar function, namely, the Lagrangian, to mostly develop time integration schemes. The founding theory that was employed as a starting point was Hamilton's principle (Argyris and Scharpf 1969; Fried 1969). Besides, the direct time discretization of Hamilton's principle by the mean value theorem was first proposed to obtain the so-called discrete Euler-Lagrange equations (Veselov 1988), which have been treated in detail in the context of the variational time integrators (Marsden and West 2001; Wendlandta and Marsden 1997). It appears that in developing such integrators, the Hamilton's principle was employed, although it possesses an inconsistency regarding handling the initial conditions in contrast to Hamilton's law of varying action (Har and Tamma 2009a); however this is a separate issue. In addition, in the context of discrete mechanics, many works (Greenspan 1973; LaBudde and Greenspan 1974) have been devoted to the time integration of Newton's law and the time discretization of the principle of least time (Shibberu 1992; Shibberu 1994). In addition to these efforts, as a consequence of other works such as that of Greenspan, the importance of scalar functions has been shown in the time integration efforts such as the energy-momentum conserving time integration algorithms (Simo 1992; Simo and Laursen 1992). Furthermore, all these works mentioned previously were aimed at time discretization to attain desired algorithmic attributes, and not for explaining finite element formulations in the context of space discretization. Most traditional developments in space discretization are Newton based and traditionally emanate from Cauchy's equations of motion as a starting point. In contrast to the vector formalism and traditional practices, not much is discussed in the literature on how to foster numerical spatial discretizations for frameworks which are associated with the concept of scalar representations. With these considerations in mind, the space discretzations employing descriptive scalar functions such as the Total Energy, the Lagrangian and the Hamiltonian are highlighted in this chapter via the principle of balance of mechanical energy as a starting point. They are also ideal for enabling straightforward numerical discretizations such as with the finite element type methods as in traditional practices; and in addition provide different perspectives with improved physical insight.

Some brief highlights on variational and differential calculus settings follow next. In general, firstly, variational principles (Lanczos 1970; Meirovitch 2004; Richards 1977; Washizu 1982) in physics and mechanics have played an important role for deriving the governing differential equations. The principles, namely, Hamilton's principle in dynamics, in the electromagnetic and gravitational fields, and even in quantum mechanics; the principle of stationary potential energy and Hu-Washizu's variational principle in continuum-statics; Fermat's principle of least time (Yourgrau and Mandelstam 1968) leading to

Snell's law in optics; Maupertuis' principle of least action (Yourgrau and Mandelstam 1968), Jacobi and Lagrange's principle of least action (Osgood 1937) in dynamics; Gurtin's variational principles (Gurtin 1964b; Gurtin 1964a) for initial value problems such as wave and transient heat transfer in lieu of Hamilton's principle; and the like, have been considered in engineering and science as an indispensable branch of mathematics to derive the differential equations, namely, the Euler-Lagrange type equations.

Secondly, on the other hand, variational methods (Crandall 1956; Finlayson 1972; Mikhlin 1964; Richards 1977) are critical for finding approximate solutions of the governing equations. The methods, namely, the Rayleigh-Ritz method, the principle of virtual work, the Galerkin weighted-residual method, the Kantorovitch method, the Trefftz method, and the like, have been widely used as direct methods to find approximate solutions. Regardless of whether the discipline is the variation principle or the variational method, the notion of the variation is invoked in the mathematical formulation. A question arises at this point. What about the differential instead of the variation? Note that the variation is imaginary but the differential is real! It appears that the inclusion of the variational approach has been the focus of philosophical controversies and misinterpretations due to the perception that the variational calculus seems to bring a purpose to the flow of natural events (Lanczos 1970). Newton (1642-1727) and Leibniz (1646-1716) appear to have independently developed the differential calculus and Newton appears to have first applied it to physics. Alternatively, variational calculus, which appears to be credited to Euler (1707-1783) and Lagrange (1736-1813), and often called the calculus of variations (Gelfand and Formin, 1963), has been extensively used to derive the differential equations especially for mechanical systems. The concern is with the infinitesimal changes of the physical quantities. The differential calculus uses the real quantity, whilst the variational calculus employs the imaginary quantity.

The biggest distinction of the variation over the differential seems to arise from rheonomic systems. For holonomic systems where the Lagrange's equations of motion and Hamilton's principle are true, Boltzmann (1844-1906) classified them into scleronomic and rheonomic (Lanczos 1970). In rheonomic systems, the direction of the differential is different from that of the variation (see Figure 6.1 and Figure 6.2) due to the presence of the partial derivative of any physical quantity depending on time explicitly with respect to time (see Example 4.1 in Chapter 4). The introduction of the variation enables one to eliminate the time partial derivative term. However, it is an imaginary and mathematical abstract quantity. Most of engineering and science applications do not fall into the class of the rheonomic systems (de Castro 2000). In the scleronomic systems, the direction of the differential is identical to that of the variation (Greenwood 1977). Therefore, for cases where the majority of engineering and science applications are mostly scleronomic, the variational approach is not necessarily indispensable to derive the governing equations or to find approximate solutions. Fortunately, most of engineering and science problems are scleronomic. In particular, solid mechanics or elastodynamics or continuum mechanics to which the finite element formulations are usually applied are in general, scleronomic. As a consequence, differential formulations also play an important role as described here.

In physics, the recognition of the presence of symmetry appears to have originated due to Noether (1882-1935). She discovered that the Lagrangian has invariance properties of dynamical systems because symmetries exist in physics (Byers 1996; Byers 1999; Noether 1918). Noether's theorem states that the spatial translational invariance of the Lagrangian in the configuration space gives conservation of linear momentum, the rotational invariance gives conservation of angular momentum, and the autonomous Lagrangian has invariance in time. Likewise, after Noether presented her theorem, it has been well known that the Hamiltonian in the phase space has also the same invariant properties (Abraham et al. 1988). This is also the so-called Hamiltonian version of Noether's theorem (Simo et al. 1992). In addition, we advocate that the autonomous Total Energy also possesses the three invariant properties. Therefore, it can be viewed as the so-called Total Energy version of Noether's theorem. Consequently, the autonomous Total Energy has time/translational/rotational symmetries for both continuous-body/N-body dynamical systems (Har and Tamma 2009a).

The following is worth noting regarding scalar formalisms and representation of the equations of motion. Hamilton's principle and Hamilton's law of varying action, both use the notion of variation in their formulations; and with respect to scalar representations of the governing equations of motion, the Lagrange equations of motion in configuration space are the first equations one encounters (that can be naturally derived) via these formulations, and not any other framework unlike that from differential calculus setting as highlighted in this chapter. Only from the Lagrange equations of motion, through invoking Legendre transformation one can then readily obtain the so-called Hamilton's equations in phase space. Finally, as an alternative to the Lagrangian and the Hamiltonian frameworks in the variational calculus setting, and for capitalizing on targeted benefits of a framework with a direct measurable built-in scalar quantity such as the Total Energy, as highlighted next, a more recent framework (Har and Tamma, 2008) outlined here in configuration space which we particularly advocate can also be readily employed. As mentioned previously in Chapter 6, there are various subject areas dealing with mechanics and physics where it is desirable to have a descriptive scalar function such as the Total Energy that is naturally built-in within the formulation. Note that the Total Energy has also the same symmetry properties in translation/rotation/time in the sense of Noether.

It is important to note that unlike the previous setting via variational calculus, the differential calculus setting via the theorem of power expended or consequently, the principle of balance of mechanical energy, naturally, first yields the so-called Total Energy representation of the equation of motion in configuration space. Therein, employing a particular transformation that is not a Legendre transformation, one can then derive the standard Lagrange equations of motion in configuration space. From here, through invoking Legendre transformation, we can readily obtain the standard Hamilton's equation of motion in phase space. The various developments via the principle of balance of mechanical energy with scalar fromalisms such as the Total Energy, Lagrangian, and the Hamiltonian are respectively presented in the sections to follow for continuous-body systems.

11.2 TOTAL ENERGY REPRESENTATIONS AND FRAMEWORK IN THE DIFFERENTIAL CALCULUS SETTING AND FINITE ELEMENT FORMULATIONS

The developments to follow in the differential calculus setting are restricted to holonomic-scleronomic systems in continuous-body dynamics which is of focus in this chapter (in contrast, we additionally highlight in the Appendix, the variational calculus setting which is applicable to general holonomic systems). We first establish the relation stemming from Cauchy's equation of motion which reflects the statement of the principle of balance of linear momentum to the theorem of power expended, or consequently, the principle of balance of energy for mechanical systems. With the restriction to mechanical systems, therein, we directly employ the principle of balance of mechanical energy as the starting point for developing the strong and weak forms and the consequent numerical discretization with the finite element method via both vector and scalar formalisms.

In summary, (1) Firstly, we explain how to derive the continuous mathematical governing equations in space and time in the differential calculus setting, (2) Secondly, whilst the Bubnov-Galerkin weighted residual form leads to the principle of virtual work, which is strictly a statement of the weak form in space only, we particularly also demonstrate that the total time derivative of the Total Energy which is derived from the principle of balance of mechanical energy, naturally represents the weak form in space, and integrating over a given time interval, it represents the strong form in time. For space, we also show that in itself it embodies the weak form in space that can be described by a discrete Total Energy Differential Operator, and (3) Thirdly, we demonstrate a simple, space-discrete Total Energy finite element formulation that directly yields the semi-discrete ordinary differential equations in time (when the discrete Total Energy differential operator is employed); the discretized representations can be readily shown to preserve the same physical attributes as the continuous systems for continuous-body dynamical applications unlike traditional practices. Both vector and scalar formalisms follow next.

11.2.1 Principle of Balance of Mechanical Energy/Theorem of Power Expended: Nonconservative System

Vector Formalism: Strong and Weak Forms via Principle of Balance of Mechanical Energy

In elastodynamics, a continuous body, which is a simply connected, bounded, and open set $\mathcal{B}$ with a piecewise smooth boundary open set $\partial\mathcal{B}$ in the reference configuration, is often postulated that infinite particles construct the body; the particles are not considered to have concentrated mass. The particles can be defined as mathematical quantities, sometimes called material points, $\mathbf{X} \in \mathcal{B} \subset \mathbb{R}^3$. The smooth motion of the body may be described as a one-to-one mapping, $\mathbf{x} = \boldsymbol{\varphi}(\mathbf{X}, t) : \mathcal{B} \times \bar{\mathcal{T}} \to \mathcal{S} \subset \mathbb{R}^3, \quad \forall t \in \bar{\mathcal{T}} = [t_0, t_1]$, where $\mathcal{B}$ stands for the reference configuration; $\mathcal{S}$ denotes the current configuration. The configuration space as a space occupied by a set of all configurations in the dynamical motion is defined as

$$\mathcal{C} = \left\{ \begin{array}{l} \boldsymbol{\varphi}(\mathbf{X}, t) : \mathcal{B} \times \bar{\mathcal{T}} \to \mathcal{S} \subset \mathbb{R}^3 | \boldsymbol{\varphi} \in C^{2m}(\mathcal{B}) \\ \boldsymbol{\varphi}(\mathbf{X}, t) \in C^2\left(\bar{\mathcal{T}}\right), m \geq 1, \ |\mathbf{F}| > 0, \text{ and } \boldsymbol{\varphi}|_{\partial\mathcal{B}_\varphi} = \bar{\boldsymbol{\varphi}} \end{array} \right\} \tag{11.1}$$

Let us consider nonconservative systems where the external forces are not dead loads. In contrast to Equation (10.5) for the variational calculus setting, in the differential calculus setting we have

$$\int_{\mathcal{B}} \left(\rho_0 \ddot{\boldsymbol{\varphi}} - \mathrm{DIV}\,\mathbf{P}(\mathbf{X}, t) - \rho_0 \mathbf{B}(\mathbf{X}, t)\right) \cdot \dot{\boldsymbol{\varphi}}\, dV = 0 \tag{11.2}$$

which represents the Bubnov-Galerkin weighted residual form; however, unlike Equation (10.5), the weighting function is the velocity rather than the virtual displacement. We observe that in the differential calculus, Equation (11.2) in elastodynamics is the counterpart of the Lagrangian form of D'Alembert's principle given in Equation (3.51). Recall the following product rule

$$(\mathrm{DIV}\,\mathbf{P}) \cdot \dot{\boldsymbol{\varphi}} = \mathrm{DIV}\,(\dot{\boldsymbol{\varphi}} \cdot \mathbf{P}) - \mathbf{P} : (\mathrm{GRAD}\,\dot{\boldsymbol{\varphi}}) = \mathrm{DIV}\,(\dot{\boldsymbol{\varphi}} \cdot \mathbf{P}) - \mathbf{P} : \dot{\mathbf{F}} \tag{11.3}$$

where $\dot{\mathbf{F}}$ denotes the total time derivative of the deformation gradient tensor $\mathbf{F}$. By virtue of the above equation, Equation (11.2) can be rewritten as

$$\int_{\mathcal{B}} \rho_0 \ddot{\boldsymbol{\varphi}} \cdot \dot{\boldsymbol{\varphi}}(\mathbf{X}, t)\, dV + \int_{\mathcal{B}} \mathbf{P} : \dot{\mathbf{F}}\, dV = \int_{\mathcal{B}} \rho_0 \mathbf{B}(\mathbf{X}, t) \cdot \dot{\boldsymbol{\varphi}}(\mathbf{X}, t)\, dV + \int_{\mathcal{B}} \mathrm{DIV}\,(\dot{\boldsymbol{\varphi}} \cdot \mathbf{P})\, dV \tag{11.4}$$

The first term on the right hand side in Equation (11.4) corresponds to the rate of the kinetic energy of the system. The second term is referred to as the stress power (Malvern 1969). Invoking Cauchy's law $\bar{\mathbf{T}}(\mathbf{X}, t, \hat{\mathbf{N}}) = \mathbf{P}(\mathbf{X}, t) \cdot \hat{\mathbf{N}}(\mathbf{X}, t)$ and Gauss's theorem, Equation (11.4) can be rewritten as

$$\int_{\mathcal{B}} \rho_0 \ddot{\boldsymbol{\varphi}} \cdot \dot{\boldsymbol{\varphi}} dV + \int_{\mathcal{B}} \mathbf{P} : \dot{\mathbf{F}}\, dV = \int_{\mathcal{B}} \rho_0 \mathbf{B}(\mathbf{X}, t) \cdot \dot{\boldsymbol{\varphi}} dV + \int_{\partial\mathcal{B}_\sigma} \bar{\mathbf{T}}(\mathbf{X}, t, \hat{\mathbf{N}}) \cdot \dot{\boldsymbol{\varphi}} dA \tag{11.5}$$

where $\bar{\mathbf{T}}(\mathbf{X}, t, \hat{\mathbf{N}})$ denotes the prescribed traction on the boundary $\partial\mathcal{B}$. The above equation is referred to as the theorem of power expended (Truesdell and Toupin 1960, page 569) (Gurtin 1972, page 65).

In the non polar case for which distributed assigned couples over the bounding surface or the volume do not exist, the power input $\mathcal{P}_{\text{input}}$ for the continuous-body dynamical with non dead loads is given by Equation (8.214). Notice that the body forces and tractions in Equation (11.4) are not prescribed.

Therefore, the external potential energy does not exist due to non dead loads. Then, the theorem of power expended given in Equation (11.5) can be rewritten as

$$\int_{\mathcal{B}} \rho_0 \ddot{\boldsymbol{\varphi}} \cdot \dot{\boldsymbol{\varphi}} dV + \int_{\mathcal{B}} \mathbf{P} : \dot{\mathbf{F}}\, dV = \mathcal{P}_{\text{input}} \tag{11.6}$$

Hence, the rate of change of the kinetic energy plus the stress power is identical to the power input. The theorem of power expended is often called the principle of balance of mechanical energy (Holzapfel 2000). Of critical importance is that the above equation holds regardless of whether the system is mechanical or thermo-mechanical.

In view of Equation (8.236), it is said that the first Piola-Kirchhoff stress tensor is the thermodynamics force work-conjugate to the deformation gradient tensor (Coleman and Noll 1963; Holzapfel 2000) as follows:

$$\mathbf{P} = \rho_0 \frac{\partial \Psi(\mathbf{F})}{\partial \mathbf{F}} \tag{11.7}$$

Hence, in a reversible process where thermal effects are neglected and the entropy is not produced, we have Equation (8.237). For the perfect elastic material whose mechanical behavior obeys Equation (8.237), the theorem of power expended given in Equation (11.5) can be rewritten as

$$\frac{d\mathcal{K}}{dt} + \int_{\mathcal{B}} \rho_0 \frac{\partial \Psi}{\partial t}\, dV = \int_{\mathcal{B}} \rho_0 \mathbf{B}(\mathbf{X}, t) \cdot \dot{\boldsymbol{\varphi}}\, dV + \int_{\partial \mathcal{B}_\sigma} \bar{\mathbf{T}}(\mathbf{X}, t, \hat{\mathbf{N}}) \cdot \dot{\boldsymbol{\varphi}}\, dA \tag{11.8}$$

We know that the total derivative of the internal potential energy for perfect materials is given by

$$\frac{d\Pi_{\text{int}}(\boldsymbol{\varphi})}{dt} = \frac{d}{dt} \int_{\mathcal{B}} \rho_0 \Psi dV = \int_{\mathcal{B}} \mathbf{P} : \dot{\mathbf{F}}\, dV \tag{11.9}$$

where the internal potential energy $\Pi_{\text{int}}(\boldsymbol{\varphi})$ is defined as

$$\Pi_{\text{int}}(\boldsymbol{\varphi}) = \int_{\mathcal{B}} \rho_0 \Psi(\mathbf{F})\ dV \tag{11.10}$$

Hence, Equation (11.8) means that the rate of increase in the kinetic energy and the elastic potential energy is equal to the power of the external forces in lossless motion (Ziegler 1991, page 501). In addition, Equation (11.8) also indicates that the summation of the rate of the kinetic energy and the rate of the internal potential energy equals the power input in case of perfect materials. Note that the external forces in Equation (11.8) are not dead loads and there does not exist the external potential energy. Equation (11.8) is true for nonconservative systems. Equation (11.8) can be rewritten as

$$\frac{d\mathcal{E}(\boldsymbol{\varphi}, \dot{\boldsymbol{\varphi}})}{dt} = \mathcal{P}_{\text{input}}, \quad \mathcal{E}(\boldsymbol{\varphi}, \dot{\boldsymbol{\varphi}}) = \mathcal{K}(\dot{\boldsymbol{\varphi}}) + \Pi_{\text{int}}(\boldsymbol{\varphi}) \tag{11.11}$$

which represents, in fact, the principle of balance of mechanical energy. Here, $\mathcal{E}(\boldsymbol{\varphi}, \dot{\boldsymbol{\varphi}}) : T\mathcal{C} \rightarrow \mathbb{R}$ denotes the autonomous Total Energy for nonconservative continuous-dynamical systems. The autonomous Total Energy is defined on the tangent bundle $T\mathcal{C}$

$$T\mathcal{C} := \left\{ (\boldsymbol{\varphi}, \dot{\boldsymbol{\varphi}}) \,|\, \boldsymbol{\varphi} \in \mathcal{C} \text{ and } \dot{\boldsymbol{\varphi}} \in T_{\boldsymbol{\varphi}}\mathcal{C} \right\} \tag{11.12}$$

Note that the potential energy for nonconservative continuous-dynamical systems is identical to the internal potential energy.

Cauchy's Equations of Motion Starting with the principle of balance of mechanical energy, we shall re-derive the strong form, namely, Cauchy's equations of motion and Cauchy's law (see Equation (11.11) which is employed as the starting point). We count on neither the variational principle nor the variational calculus. Besides, we do not use the principle of virtual work either. No variations are involved in the subsequent formulations. We only use the differential calculus setting including integration and differentiation aspects to obtain governing equations and natural boundary conditions.

First consider the nonconservative systems for which Equation (11.11) gives

$$\frac{d\mathcal{E}(\boldsymbol{\varphi},\dot{\boldsymbol{\varphi}})}{dt} = \frac{d\left(\mathcal{K}(\dot{\boldsymbol{\varphi}}) + \Pi_{\text{int}}(\boldsymbol{\varphi})(\boldsymbol{\varphi})\right)}{dt} = \mathcal{P}_{\text{input}} \tag{11.13}$$

which represents, in fact, the principle of balance of mechanical energy. Equation (11.13) leads to

$$\int_{\mathcal{B}} \rho_0 \ddot{\boldsymbol{\varphi}} \cdot \dot{\boldsymbol{\varphi}}\, dV + \int_{\mathcal{B}} \mathbf{P} : \dot{\mathbf{F}}\, dV = \int_{\partial\mathcal{B}_\sigma} \bar{\mathbf{T}} \cdot \dot{\boldsymbol{\varphi}}\, dA + \int_{\mathcal{B}} \rho_0 \mathbf{B}(\mathbf{X}, t) \cdot \dot{\boldsymbol{\varphi}}\, dV \tag{11.14}$$

Notice that Equation (11.14) is not the principle of balance of energy for thermo-mechanical systems, but it represents the theorem of power expended (Gurtin 1972, page 110). Thus, Equation (11.14) holds for thermo-mechanical systems of perfect materials. Next, recall the stress power which can be written as

$$\int_{\mathcal{B}} \mathbf{P} : \dot{\mathbf{F}}\, dV = \int_{\mathcal{B}} \text{DIV}\,(\dot{\boldsymbol{\varphi}} \cdot \mathbf{P}) - \dot{\boldsymbol{\varphi}} \cdot \text{DIV}\,\mathbf{P}\, dV \tag{11.15}$$

By virtue of Gauss's theorem, the above equation results in

$$\int_{\mathcal{B}} \mathbf{P} \cdot\cdot \dot{\mathbf{F}}\, dV = \int_{\partial\mathcal{B}_\sigma} \dot{\boldsymbol{\varphi}} \cdot \mathbf{P} \cdot \hat{\mathbf{N}}\, dA - \int_{\mathcal{B}} \text{DIV}\,\mathbf{P} \cdot \dot{\boldsymbol{\varphi}}\, dV \tag{11.16}$$

Substituting Equation (11.16) into Equation (11.14), we have

$$\begin{aligned}\frac{\partial\mathcal{E}(\boldsymbol{\varphi},\dot{\boldsymbol{\varphi}})}{\partial t} - \mathcal{P}_{\text{input}} &= \int_{\mathcal{B}} \left[\left(\rho_0 \ddot{\boldsymbol{\varphi}}(\mathbf{X}, t) - \text{DIV}\,\mathbf{P} - \rho_0 \mathbf{B}(\mathbf{X}, t)\right) \cdot \dot{\boldsymbol{\varphi}}\right] dV \\ &\quad - \int_{\partial\mathcal{B}_\sigma} \left[(\bar{\mathbf{T}}(\mathbf{X}, t, \hat{\mathbf{N}}) - \mathbf{P} \cdot \hat{\mathbf{N}}) \cdot \dot{\boldsymbol{\varphi}}(\mathbf{X}, t)\right] dA = 0\end{aligned} \tag{11.17}$$

On each open set (surface and volume respectively), the terms inside the brackets should vanish. As a consequence, note that for nonconservative systems, we obtain the strong form, namely, Cauchy's equations of motion and Cauchy's law, i.e.,

$$\rho_0 \ddot{\boldsymbol{\varphi}}(\mathbf{X}, t) = \text{DIV}\,\mathbf{P}(\mathbf{X}, t) + \rho_0 \mathbf{B}(\mathbf{X}, t)\,, \quad \text{for } (\mathbf{X}, t) \in \mathcal{B} \times \mathcal{T} \tag{11.18}$$

subject to Cauchy's law

$$\mathbf{P} \cdot \hat{\mathbf{N}} = \bar{\mathbf{T}}(\mathbf{X}, t, \hat{\mathbf{N}}), \quad \text{for } (\mathbf{X}, t) \in \partial\mathcal{B}_\sigma \times \mathcal{T} \tag{11.19}$$

Space and Time Discretization Employing the Principle of Balance of Mechanical Energy: Conventional Approach We note that Equation (11.14) which represents the principle of balance of energy for mechanical systems only, is in fact, the weak form in space in the differential calculus setting. Therefore, traditional practices of discretization using the principle of balance of mechanical energy lead to the finite element equations of motion.

Admissible trial functions $\tilde{\varphi}$ are required to be piece-wise continuous, but they should be continuous along the inter-element boundaries. The admissible trial functions $\tilde{\varphi}$ are represented by a linear combination of coordinate functions $\mathbf{N}_i(\mathbf{X})$ and displacement vectors $\mathbf{U}_i^e(t)$ as given by Equation (9.11); hence the trial functions, which are of class $C^{m-1}(\mathcal{B})$ over the entire domain, belong to the subspace, denoted by $\mathcal{C}^h \subset \mathcal{C}$, as shown in Equation (9.14). Then, the finite dimensional domain after performing the discretization procedure yields

$$\sum_{e=1}^{n_{elem}} \left(\int_{\mathcal{B}_e} \rho_0 \ddot{\tilde{\varphi}}^h \cdot \dot{\tilde{\varphi}}^h dV + \int_{\mathcal{B}_e} \tilde{\mathbf{P}}^h : \dot{\tilde{\mathbf{F}}}^h dV - \int_{\mathcal{B}_e} \rho_0 \mathbf{B} \cdot \dot{\tilde{\varphi}}^h dV - \int_{\partial \mathcal{B}_{e\sigma}} \bar{\mathbf{T}} \cdot \dot{\tilde{\varphi}}^h dA \right) = 0 \tag{11.20}$$

which can be rewritten as

$$\sum_{e=1}^{n_{\text{elem}}} \left\{ \left(\mathbf{M}_e \ddot{\mathbf{U}}^e(t) + \mathbf{F}_{\text{int}}^e - \mathbf{F}_{\text{ext}}^e \right) \cdot \dot{\mathbf{U}}^e(t) \right\} = 0 \tag{11.21}$$

where $\mathbf{M}_e$, $\mathbf{F}_{\text{int}}^e$ and $\mathbf{F}_{\text{ext}}^e$ are given by Equation (9.17).

For the entire domain, the above equation can also be expressed as

$$(\mathbf{M}\ddot{\mathbf{q}} + \mathbf{F}_{\text{int}} - \mathbf{F}_{\text{ext}}) \cdot \dot{\mathbf{q}} = 0 \tag{11.22}$$

where $\mathbf{q}(t)$, $\dot{\mathbf{q}}(t)$, $\ddot{\mathbf{q}}(t)$, $\mathbf{M}$, $\mathbf{F}_{\text{int}}$ and $\mathbf{F}_{\text{ext}}$ are given by Equation (9.19). Note that the j-th generalized velocity $\dot{q}_j$ is independent of the k-th generalized velocity $\dot{q}_k$; in other words, the set $\dot{\mathbf{q}}$ is linearly independent. Therefore, the only way for Equation (11.21) to hold is for the individual coefficients to vanish. The terms enclosed in the parenthesis in Equation (11.22) can be regarded as individual coefficients (Lanczos 1970, page 39) (Greenwood 1977, page 51) (Goldstein 2002, page 17 and 20) (Ardema 2005, page 70 and 113). Therefore, Equation (11.22) stands for a linear combination of functions in the sense of functional analysis (Rektorys 1980, page 47) (Oden 1979) (Reddy 1986, page 64). Subsequently, since the set $\dot{\mathbf{q}}$ is linearly independent, the following semi-discretized second-order system is obtained

$$\mathbf{M}\ddot{\mathbf{q}} + \mathbf{F}_{\text{int}} - \mathbf{F}_{\text{ext}} = \mathbf{0} \tag{11.23}$$

Time Weighted Residual Form Taking the integration over the time domain for Equation (11.22) leads to

$$\int_{t_0}^{t_1} (\mathbf{F}_{\text{ext}} - \mathbf{F}_{\text{int}} - \mathbf{M}\ddot{\mathbf{q}}) \cdot \dot{\mathbf{q}}\, dt = 0 \tag{11.24}$$

The above equation can be employed for designing time stepping methods (see later chapters of this book) via the time weighted-residual method which includes time finite element formulations (Belytschko and Hughes 1983; Sha et al. 2003; Tamma et al. 2003, 2002, 2000; Wood 1990; Zhou and Tamma 2004a,c, 2006; Zhou et a. 2001, 2000; Zienkiewicz and Taylor 2006; Zienkiewicz et al. 2005, 1984).

11.2.2 Principle of Balance of Mechanical Energy: Conservative System and Total Energy Density Equations of Motion

Scalar Formalism: Strong and Weak Forms via the Total Energy Framework In the non polar case for which distributed assigned couples over the bounding surface or the volume do not exist, the power input $\mathcal{P}_{\text{input}}$ for the continuous-body dynamical with dead loads is given by

$$\mathcal{P}_{\text{input}} = \int_{\mathcal{B}} \rho_0 \mathbf{B}(\mathbf{X}) \cdot \dot{\varphi}(\mathbf{X}, t)\, dV + \int_{\partial \mathcal{B}_\sigma} \bar{\mathbf{T}}(\mathbf{X}, \hat{\mathbf{N}}) \cdot \dot{\varphi}(\mathbf{X}, t)\, dA = -\frac{d\Pi_{\text{ext}}}{dt} \tag{11.25}$$

where the surface tractions and body forces are dead loads and Π_{ext} denotes the external potential energy. Notice that the body forces and tractions above are prescribed. Therefore, the external potential energy does exist due to dead loads.

With the definition of the tangent bundle

$$TC := \left\{ (\boldsymbol{\varphi}, \dot{\boldsymbol{\varphi}}) \,|\, \boldsymbol{\varphi} \in \mathcal{C} \text{ and } \dot{\boldsymbol{\varphi}} \in T_{\boldsymbol{\varphi}}\mathcal{C} \right\} \tag{11.26}$$

the autonomous Total Energy $\mathcal{E}(\boldsymbol{\varphi}, \dot{\boldsymbol{\varphi}}) : TC \to \mathbb{R}$ is defined as the summation of the kinetic energy and the total potential energy as follows:

$$\mathcal{E}(\boldsymbol{\varphi}, \dot{\boldsymbol{\varphi}}) = \mathcal{K}(\dot{\boldsymbol{\varphi}}) + \mathcal{U}(\boldsymbol{\varphi}), \quad \mathcal{U} = \Pi_{\text{int}} + \Pi_{\text{ext}} \tag{11.27}$$

where $\mathcal{U}$ denotes the total potential energy, and the internal potential energy Π_{int} is defined by

$$\Pi_{\text{int}} = \int_{\mathcal{B}} \rho_0 \Psi(\mathbf{F})\, dV \tag{11.28}$$

where $\Psi(\mathbf{F})$ denotes the Helmholtz free energy per unit mass in homogeneous materials. Hence, the autonomous Total Energy defined in Equation (11.27) can be written as

$$\mathcal{E}(\boldsymbol{\varphi}, \dot{\boldsymbol{\varphi}}) = \int_{\mathcal{B}} \bar{\mathcal{E}}(\boldsymbol{\varphi}, \dot{\boldsymbol{\varphi}}, \mathbf{F})\, dV - \int_{\partial \mathcal{B}_\sigma} \bar{\mathbf{T}}(\mathbf{X}, \hat{\mathbf{N}}) \cdot \boldsymbol{\varphi}(\mathbf{X}, t)\, dA \tag{11.29}$$

where the Total Energy density function $\bar{\mathcal{E}}(\boldsymbol{\varphi}, \dot{\boldsymbol{\varphi}}, \mathbf{F})$ is defined as

$$\bar{\mathcal{E}}(\boldsymbol{\varphi}, \dot{\boldsymbol{\varphi}}, \mathbf{F}) = \frac{1}{2}\rho_0 \dot{\boldsymbol{\varphi}} \cdot \dot{\boldsymbol{\varphi}} + \rho_0 \Psi(\mathbf{F}) - \rho_0 \mathbf{B}(\mathbf{X}) \cdot \boldsymbol{\varphi} \tag{11.30}$$

As the Total Energy for the conservative dynamical system is a constant of motion, the total time derivative of the Total Energy becomes zero as follows:

$$\frac{d\mathcal{E}(\boldsymbol{\varphi}, \dot{\boldsymbol{\varphi}})}{dt} = 0 \tag{11.31}$$

By use of the chain rule, the rate of the Total Energy density function gives

$$\frac{d\bar{\mathcal{E}}(\boldsymbol{\varphi}, \dot{\boldsymbol{\varphi}}, \mathbf{F})}{dt} = \frac{\partial \bar{\mathcal{E}}}{\partial \boldsymbol{\varphi}} \cdot \dot{\boldsymbol{\varphi}} + \frac{\partial \bar{\mathcal{E}}}{\partial \dot{\boldsymbol{\varphi}}} \cdot \ddot{\boldsymbol{\varphi}} + \frac{\partial \bar{\mathcal{E}}}{\partial \mathbf{F}} : \dot{\mathbf{F}} \tag{11.32}$$

The last term on the right-hand side of the above equation can be cast as

$$\frac{\partial \bar{\mathcal{E}}}{\partial \mathbf{F}} : \dot{\mathbf{F}} = \frac{\partial \bar{\mathcal{E}}}{\partial \mathbf{F}} : \text{GRAD}\, \dot{\boldsymbol{\varphi}} = \text{DIV}\left(\dot{\boldsymbol{\varphi}} \cdot \frac{\partial \bar{\mathcal{E}}}{\partial \mathbf{F}} \right) - \text{DIV}\, \frac{\partial \bar{\mathcal{E}}}{\partial \mathbf{F}} \cdot \dot{\boldsymbol{\varphi}} \tag{11.33}$$

Besides, we have the following derivation

$$\frac{\partial \bar{\mathcal{E}}}{\partial \dot{\boldsymbol{\varphi}}} \cdot \ddot{\boldsymbol{\varphi}} = \rho_0 \dot{\boldsymbol{\varphi}} \cdot \ddot{\boldsymbol{\varphi}} = \rho_0 \ddot{\boldsymbol{\varphi}} \cdot \dot{\boldsymbol{\varphi}} = \frac{d}{dt}\left(\frac{\partial \bar{\mathcal{E}}}{\partial \dot{\boldsymbol{\varphi}}} \right) \cdot \dot{\boldsymbol{\varphi}} \tag{11.34}$$

because the Total Energy density function is quadratic in $\dot{\boldsymbol{\varphi}}$. Therefore, substituting Equation (11.33) and Equation (11.34) into Equation (11.32), we find that

$$\frac{d\bar{\mathcal{E}}}{dt} = \left(\frac{\partial \bar{\mathcal{E}}(\boldsymbol{\varphi}, \dot{\boldsymbol{\varphi}}, \mathbf{F})}{\partial \boldsymbol{\varphi}} - \mathrm{DIV}\, \frac{\partial \bar{\mathcal{E}}}{\partial \mathbf{F}} + \frac{d}{dt}\left(\frac{\partial \bar{\mathcal{E}}}{\partial \dot{\boldsymbol{\varphi}}} \right) \right) \cdot \dot{\boldsymbol{\varphi}} + \mathrm{DIV}\left(\dot{\boldsymbol{\varphi}} \cdot \frac{\partial \bar{\mathcal{E}}}{\partial \mathbf{F}} \right) \tag{11.35}$$

Due to Gauss's theorem, the third term on the right hand side above gives

$$\int_{\mathcal{B}} \mathrm{DIV}\left(\dot{\boldsymbol{\varphi}} \cdot \frac{\partial \bar{\mathcal{E}}}{\partial \mathbf{F}} \right) dV = \int_{\partial \mathcal{B}_\sigma} \dot{\boldsymbol{\varphi}} \cdot \frac{\partial \bar{\mathcal{E}}}{\partial \mathbf{F}} \cdot \hat{\mathbf{N}}\, dA \tag{11.36}$$

In view of Equation (11.36) and Equation (11.35), Equation (11.27) yields

$$\begin{aligned} \frac{d\mathcal{E}(\boldsymbol{\varphi}, \dot{\boldsymbol{\varphi}})}{dt} &= \int_{\mathcal{B}} \left(\frac{\partial \bar{\mathcal{E}}(\boldsymbol{\varphi}, \dot{\boldsymbol{\varphi}}, \mathbf{F})}{\partial \boldsymbol{\varphi}} - \mathrm{DIV}\, \frac{\partial \bar{\mathcal{E}}}{\partial \mathbf{F}} + \frac{d}{dt}\left(\frac{\partial \bar{\mathcal{E}}}{\partial \dot{\boldsymbol{\varphi}}} \right) \right) \cdot \dot{\boldsymbol{\varphi}}\,(\mathbf{X}, t)\, dV \\ &- \int_{\partial \mathcal{B}_\sigma} \left(\bar{\mathbf{T}} - \frac{\partial \bar{\mathcal{E}}}{\partial \mathbf{F}} \cdot \hat{\mathbf{N}} \right) \cdot \dot{\boldsymbol{\varphi}}\,(\mathbf{X}, t)\, dA = 0 \end{aligned} \tag{11.37}$$

From the above equation it is easy to obtain governing differential equations and natural boundary conditions. In the the open set $\mathcal{B}$ and on the open set $\partial \mathcal{B}_\varphi$, the velocity $\dot{\boldsymbol{\varphi}}\,(\mathbf{X}, t)$ is linearly independent, and is vector-valued function. Hence, the terms enclosed in the parenthesis must be zero (Ginsberg and Genin 1984). It is obvious that the terms on the right-hand side of Equation (11.37) can be regarded as akin to the modified Bubnov-Galerkin weighted residual form (Mikhlin 1964) involving governing equations and natural boundary conditions with the velocity as the weighting function. Consequently, we have the Total Energy density representation of the equations of motion

$$\frac{d}{dt}\left(\frac{\partial \bar{\mathcal{E}}}{\partial \dot{\boldsymbol{\varphi}}} \right) = \mathrm{DIV}\, \frac{\partial \bar{\mathcal{E}}}{\partial \mathbf{F}} - \frac{\partial \bar{\mathcal{E}}(\boldsymbol{\varphi}, \dot{\boldsymbol{\varphi}}, \mathbf{F})}{\partial \boldsymbol{\varphi}} \quad \text{in } \mathcal{B} \tag{11.38}$$

subject to the traction boundary condition

$$\bar{\mathbf{T}}(\mathbf{X}, \hat{\mathbf{N}}) = \frac{\partial \bar{\mathcal{E}}}{\partial \mathbf{F}} \cdot \hat{\mathbf{N}} \quad \text{on } \partial \mathcal{B}_\sigma \tag{11.39}$$

We observe that the Total Energy density representation of the equations of motion are identical to Cauchy's equations of motion for a conservative system. At this juncture, we place an emphasis on the fact that the rate of the Total Energy results in the modified Bubnov-Galerkin weighted residual form.

Even though Equation (11.37) implies that both governing equations and natural boundary conditions are included, it is not readily suitable for applying the trial and test functions to Equation (11.37). This is because it is still a strong statement which does not allow a weaker continuity condition for the trial and test functions to discretize the spatial domain. Hence, from Equation (11.37) it is natural to derive a weak statement which allows a weaker continuity condition for trial and test functions. In view of Equation (11.33), Equation (11.37) can be rewritten as

$$\begin{aligned} \frac{d\mathcal{E}}{dt} &= \int_{\mathcal{B}} \frac{d}{dt}\left(\frac{\partial \bar{\mathcal{E}}}{\partial \dot{\boldsymbol{\varphi}}} \right) \cdot \dot{\boldsymbol{\varphi}}\, dV + \int_{\mathcal{B}} \frac{\partial \bar{\mathcal{E}}}{\partial \mathbf{F}} : \dot{\mathbf{F}}\, dV \\ &+ \int_{\mathcal{B}} \frac{\partial \bar{\mathcal{E}}}{\partial \boldsymbol{\varphi}} \cdot \dot{\boldsymbol{\varphi}}\, dV - \int_{\partial \mathcal{B}_\sigma} \bar{\mathbf{T}} \cdot \dot{\boldsymbol{\varphi}}\, dA = 0 \end{aligned} \tag{11.40}$$

The above equation implies that the rate of change of the Total Energy leads to the theorem of power expended for a conservative dynamical system, or equivalently, the principle of balance of mechanical energy, i.e. the weak form which is readily amenable to space discretization.

11.2.3 Space-Discrete Total Energy Finite Element Formulation

In this section, our attention is confined to conservative systems. For the scalar formalism, we highlight two distinct approaches to space discretization, namely: 1) we describe the traditional approach using as a starting point, the principle of balance of mechanical energy, or equivalently, the theorem of power expended for mechanical systems within the Total Energy framework via the scalar Total Energy density representation of the equations of motion, and 2) alternatively, we then describe a simple numerical discretization approach for deriving the semi-discretized equations of motion that is significantly different. More importantly, we provide new perspectives to enact such discretizations. We have shown that Equation (11.37) resulting from the principle of balance of mechanical energy, or equivalently, the theorem of power expended within the Total Energy framework contains implicitly both governing equations and natural boundary conditions in the Bubnov-Galerkin weighted residual form. In this sense, therefore, Equation (11.37) serves as a mechanism by which the weighted averages of the residuals are required to vanish, when substituting the trial functions satisfying only essential boundary conditions into the theorem of power expended or the principle of balance of mechanical energy to discretize the spatial domain. It is assumed that the entire domain of interest can be broken up into sub domains, called elements, and the weak form in Equation (11.40) is employed to break up the spatial domain. The admissible trial functions $\tilde{\boldsymbol{\varphi}}$ which are of class $C^{m-1}(\mathcal{B})$ over the entire domain, belong to the subspace, denoted by $\mathcal{C}^h \subset \mathcal{C}$ as given in Equation (9.14), and can be used to discretize the space.

Approach 1: Weak Form As in the traditional approach with vector formalism, the scalar formalism via the weak statement given by Equations (11.40), which is equivalent to balance of total mechanical energy or the theorem of power expended, is first employed here. It allows a weaker continuity condition for the trial and test functions for initial boundary-value problems. This corresponds to those used in traditional approaches. In the Total Energy framework, substituting the trial functions in Equation (9.11) into Equation (11.40) which is the weak statement, yields

$$\sum_{e=1}^{n_{elem}} \left\{ \begin{array}{l} \displaystyle\int_{\mathcal{B}e} \frac{d}{dt}\left(\frac{\partial \bar{\mathcal{E}}^h_{\mathcal{B}e}}{\partial \dot{\tilde{\boldsymbol{\varphi}}}^h}\right) \cdot \dot{\tilde{\boldsymbol{\varphi}}}^h \, dV + \int_{\mathcal{B}e} \frac{\partial \bar{\mathcal{E}}^h_{\mathcal{B}e}}{\partial \tilde{\mathbf{F}}^h} : \dot{\tilde{\mathbf{F}}}^h \, dV \\ \displaystyle + \int_{\mathcal{B}e} \frac{\partial \bar{\mathcal{E}}^h_{\mathcal{B}e}}{\partial \tilde{\boldsymbol{\varphi}}^h} \cdot \dot{\tilde{\boldsymbol{\varphi}}}^h \, dV - \int_{\partial \mathcal{B}e\sigma} \bar{\mathbf{T}} \cdot \dot{\tilde{\boldsymbol{\varphi}}}^h \, dA \end{array} \right\} = 0 \tag{11.41}$$

where

$$\bar{\mathcal{E}}^h_{\mathcal{B}e}\left(\tilde{\boldsymbol{\varphi}}^h, \dot{\tilde{\boldsymbol{\varphi}}}^h, \tilde{\mathbf{F}}^h\right) = \frac{1}{2}\rho_0 \dot{\tilde{\boldsymbol{\varphi}}}^h \cdot \dot{\tilde{\boldsymbol{\varphi}}}^h + \rho_0 \Psi\left(\tilde{\mathbf{F}}^h\right) - \rho_0 \mathbf{B} \cdot \tilde{\boldsymbol{\varphi}}^h \tag{11.42}$$

Consequently, Equation (11.41) results in

$$\sum_{e=1}^{n_{elem}} \left(\mathbf{M}_e \ddot{\mathbf{U}}^e(t) + \mathbf{F}^e_{\text{int}} - \mathbf{F}^e_{\text{ext}}\right) \cdot \dot{\mathbf{U}}^e(t) = 0 \tag{11.43}$$

where $\mathbf{M}_e$, $\mathbf{F}^e_{\text{int}}$ and $\mathbf{F}^e_{\text{ext}}$ are given by Equation (9.17).

For the entire domain of interest, we introduce the configuration space with generalized coordinates $\mathcal{Q}$, the tangent space $T_q Q$ and the tangent bundle TQ as shown in Equations (10.48), (10.49) and (10.50). Then, Equation (11.43) yields the semi-discrete second-order differential equations, i.e., the finite element equations of motion to be discretized subsequently in time as follows:

$$\mathbf{M}\ddot{\mathbf{q}} + \mathbf{F}_{\text{int}} - \mathbf{F}_{\text{ext}} = \mathbf{0} \tag{11.44}$$

Approach 2: Total Energy Differential Operator In contrast to using the weak statement, an alternative way is shown here to directly discretize the scalar functions such as the Total Energy with only trial functions satisfying the essential boundary conditions. Substituting the trial functions shown in Equation (9.11) directly into Equation (11.27), the rate of the total energy for the eth element domain $\mathcal{B}_e$ can be discretized in space as follows:

$$\mathrm{PBME}^h_{\mathcal{E}_{\mathcal{B}_e}}: \quad \frac{d\mathcal{E}^h_{\mathcal{B}_e}\left(\tilde{\boldsymbol{\varphi}}^h, \dot{\tilde{\boldsymbol{\varphi}}}^h\right)}{dt} = 0 \tag{11.45}$$

where the autonomous space-discrete Total Energy for the finite dimensional systems is constructed as

$$\mathcal{E}^h_{\mathcal{B}_e}\left(\tilde{\boldsymbol{\varphi}}^h, \dot{\tilde{\boldsymbol{\varphi}}}^h\right) = \int_{\mathcal{B}_e} \bar{\mathcal{E}}^h_{\mathcal{B}_e}\left(\tilde{\boldsymbol{\varphi}}^h, \dot{\tilde{\boldsymbol{\varphi}}}^h, \tilde{\mathbf{F}}^h\right) dV - \int_{\partial\mathcal{B}_{e\sigma}} \bar{\mathbf{T}} \cdot \tilde{\boldsymbol{\varphi}}^h dA \tag{11.46}$$

Equation (9.11) can be rewritten as follows:

$$\begin{aligned} &\tilde{\boldsymbol{\varphi}}^h(\mathbf{X}, t) = \mathbf{X} + \sum_{i=1}^{n_{node}} \mathbf{N}_i(\mathbf{X})\mathbf{U}_i^e(t) = \mathbf{X} + \mathbf{N}_e\mathbf{q}_e, \\ &\mathbf{q}_e(t) = \left\{\mathbf{U}_1^e(t), \ldots, \mathbf{U}_i^e(t), \ldots, \mathbf{U}_{n_{node}}^e(t)\right\} \end{aligned} \tag{11.47}$$

Thus, Equation (11.45) can be rewritten in terms of generalized coordinates as

$$\mathrm{PBME}^h_{\mathcal{E}_{\mathcal{B}_e}}: \quad \frac{d\mathcal{E}^h_{\mathcal{B}_e}(\mathbf{q}_e, \dot{\mathbf{q}}_e)}{dt} = \left\{\frac{\partial\mathcal{E}^h_{\mathcal{B}_e}(\mathbf{q}_e, \dot{\mathbf{q}}_e)}{\partial\mathbf{q}_e} + \frac{d}{dt}\left(\frac{\partial\mathcal{E}^h_{\mathcal{B}_e}(\mathbf{q}_e, \dot{\mathbf{q}}_e)}{\partial\dot{\mathbf{q}}_e}\right)\right\} \cdot \dot{\mathbf{q}}_e = 0 \tag{11.48}$$

where the space-discrete Total Energy is given by

$$\begin{aligned} &\mathcal{E}^h_{\mathcal{B}_e}(\mathbf{q}_e, \dot{\mathbf{q}}_e) = \frac{1}{2}\dot{\mathbf{q}}_e \cdot \mathbf{M}_e\dot{\mathbf{q}}_e + \Pi^h_{\mathrm{int}_e} - \mathbf{q}_e \cdot \mathbf{F}^e_{\mathrm{ext}}, \quad \Pi^h_{\mathrm{int}_e} = \int_{\mathcal{B}_e} \rho_0\Psi\left(\tilde{\mathbf{F}}^h\right) dV \\ &\mathbf{M}_e = \int_{\mathcal{B}_e} \rho_0\mathbf{N}_e^T\mathbf{N}_e dV, \quad \frac{\partial\mathcal{E}^h(\mathbf{q}_e, \dot{\mathbf{q}}_e)}{\partial\mathbf{q}_e} = \mathbf{F}^e_{\mathrm{int}} - \mathbf{F}^e_{\mathrm{ext}} \\ &\mathbf{F}_{\mathrm{int}} = \int_{\mathcal{B}_e} \bar{\mathbf{B}}_e^T\tilde{\mathbf{P}}^h dV, \quad \mathbf{F}^e_{\mathrm{ext}} = \int_{\mathcal{B}_e} \rho_0\mathbf{N}_e^T\mathbf{B} dV + \int_{\partial\mathcal{B}_{e\sigma}} \mathbf{N}_e^T\bar{\mathbf{T}} dA \end{aligned} \tag{11.49}$$

Consequently, for the entire domain we have

$$\mathrm{PBME}^h_{\mathcal{E}}: \quad \frac{d\mathcal{E}^h(\mathbf{q}, \dot{\mathbf{q}})}{dt} = \left\{\frac{\partial\mathcal{E}^h(\mathbf{q}, \dot{\mathbf{q}})}{\partial\mathbf{q}} + \frac{d}{dt}\left(\frac{\partial\mathcal{E}^h(\mathbf{q}, \dot{\mathbf{q}})}{\partial\dot{\mathbf{q}}}\right)\right\} \cdot \dot{\mathbf{q}} = 0 \tag{11.50}$$

We now introduce the space-discrete Total Energy differential operator $\mathfrak{E}_D$, by which the semi-discrete finite element equations of motion can be readily obtained after constructing the space-discrete Total Energy which only entails the use of trial functions. Thus, directly we have, upon substituting the space-discrete Total Energy into the space-discrete Total Energy differential operator, the following:

$$\mathfrak{E}_D\left(\mathcal{E}^h\right): \frac{d}{dt}\left(\frac{\partial\mathcal{E}^h}{\partial\dot{\mathbf{q}}}\right) + \frac{\partial\mathcal{E}^h}{\partial\mathbf{q}} = \mathbf{0}, \quad \mathfrak{E}_D: \frac{d}{dt}\left(\frac{\partial}{\partial\dot{\mathbf{q}}}\right) + \frac{\partial}{\partial\mathbf{q}} = \mathbf{0} \tag{11.51}$$

where the dependent state variable is the space-discrete scalar function, namely, the space-discrete Total Energy. Thus, we readily have the semi-discrete finite element equations of motion governing the mechanical dynamic behavior of the continuous-body within the total energy framework as follows:

$$\mathbf{M}\ddot{\mathbf{q}} + \mathbf{F}_{\text{int}} - \mathbf{F}_{\text{ext}} = 0 \tag{11.52}$$

Time Weighted Residual Form Taking the time integral over the given time interval $t \in \bar{\mathcal{T}} = [t_0, t_1]$ for both sides in Equation (11.50), we have

$$\begin{aligned} &\mathcal{E}^h(\mathbf{q}, \dot{\mathbf{q}})\big|_{t=t_1} - \mathcal{E}^h(\mathbf{q}, \dot{\mathbf{q}})\big|_{t=t_0} \\ &= \int_{t_0}^{t_1} \{(\mathbf{M}\ddot{\mathbf{q}} + \mathbf{F}_{\text{int}} - \mathbf{F}_{\text{ext}}) \cdot \dot{\mathbf{q}}\}\, dt = 0 \end{aligned} \tag{11.53}$$

Note that the above equation in the Total Energy framework represents a single-field weighted-residual form in time. Hence, the above representation plays an important role in facilitating the time discretization. Equation (11.53) may be regarded as an unconstrained formulation with one dependent field variable in the time domain. In general, the time weighted-residual method which includes time finite element formulations can be applied for deriving time stepping schemes (Belytschko and Hughes 1983; Sha et al. 2003; Tamma et al. 2003, 2002, 2000; Wood 1990; Zhou and Tamma 2004a,c, 2006; Zhou et al. 2001, 2000; Zienkiewicz and Taylor 2006; Zienkiewicz et al. 2005, 1984).

Illustrative Examples Various illustrative examples for discretizing the space are presented next via the scalar formalism with the descriptive scalar function, namely, the Total Energy. For the various illustrative examples to follow next, we first show the development of the Total Energy density based representations of the equations of motion in the scalar formalism that are identical/equivalent to the vector formalism of Cauchy's equations of motion. One could readily employ the Total Energy density based representations to conduct the space discretization following conventional practices that were described previously in Approach 1; alternatively, to illustrate the fundamental ideas, we instead focus attention upon the space discretization aspects via Approach 2 described earlier which represents the space-discrete finite element formulation and it directly leads to the semi-discretized equations of motion in a straigh forward manner.

Example 11.1 Elastic Axial Bar

Let us consider the same elastic bar as discussed in Section 9.1.2.

Governing Equations: Total Energy Density Based Representations

Note that the external loadings in this example are dead loads. The Total Energy can be written as

$$\mathcal{E}(u, \dot{u}) = \int_L \bar{\mathcal{E}}(u, \dot{u}, u')\, dx - F_L u(L, t) \tag{11.54}$$

where the Total Energy density function for the axial bar can be written as

$$\bar{\mathcal{E}}(u, \dot{u}, u') = \frac{1}{2}\rho A(x)\dot{u}^2 + \frac{1}{2}EA(x)\left(\frac{\partial u}{\partial x}\right)^2 - \bar{b}u \tag{11.55}$$

Then, the total time derivative of the Total Energy can be represented as

$$\frac{d\mathcal{E}(u,\dot{u})}{dt} = \int_L \frac{d\bar{\mathcal{E}}(u,\dot{u},u')}{dt} dx - F_L \dot{u}(L,t) \tag{11.56}$$

The chain rule leads to

$$\frac{d\bar{\mathcal{E}}(u,\dot{u},u')}{dt} = \frac{\partial \bar{\mathcal{E}}}{\partial u}\dot{u} + \frac{\partial \bar{\mathcal{E}}}{\partial \dot{u}}\ddot{u} + \frac{\partial \bar{\mathcal{E}}}{\partial u'}\dot{u}' \tag{11.57}$$

Note that the following derivation is true

$$\frac{\partial \bar{\mathcal{E}}}{\partial \dot{u}}\ddot{u} = \rho A(x)\dot{u}\ddot{u} = \rho A(x)\ddot{u}\dot{u} = \frac{d}{dt}\left(\frac{\partial \bar{\mathcal{E}}}{\partial \dot{u}}\right)\cdot \dot{u} \tag{11.58}$$

because $\rho A(x)\ddot{u} = \frac{d}{dt}\left(\frac{\partial \bar{\mathcal{E}}}{\partial \dot{u}}\right)$. Besides, integration by parts yields

$$\frac{\partial \bar{\mathcal{E}}}{\partial u'}\dot{u}' = \frac{\partial}{\partial x}\left(\frac{\partial \bar{\mathcal{E}}}{\partial u'}\dot{u}\right) - \frac{\partial}{\partial x}\left(\frac{\partial \bar{\mathcal{E}}}{\partial u'}\right)\dot{u} \tag{11.59}$$

Accordingly, Equation (11.57) can be rewritten as

$$\frac{d\bar{\mathcal{E}}(u,\dot{u},u')}{dt} = \left[\frac{\partial \bar{\mathcal{E}}}{\partial u} + \frac{d}{dt}\left(\frac{\partial \bar{\mathcal{E}}}{\partial \dot{u}}\right) - \frac{\partial}{\partial x}\left(\frac{\partial \bar{\mathcal{E}}}{\partial u'}\right)\right]\dot{u} + \frac{\partial}{\partial x}\left(\frac{\partial \bar{\mathcal{E}}}{\partial u'}\dot{u}\right) \tag{11.60}$$

Hence, substituting Equation (11.60) into Equation (11.56) leads to

$$\begin{aligned}\frac{d\mathcal{E}(u,\dot{u})}{dt} &= \int_L \left[\frac{\partial \bar{\mathcal{E}}}{\partial u} + \frac{d}{dt}\left(\frac{\partial \bar{\mathcal{E}}}{\partial \dot{u}}\right) - \frac{\partial}{\partial x}\left(\frac{\partial \bar{\mathcal{E}}}{\partial u'}\right)\right]\dot{u}dx \\ &\quad - F_L\dot{u}(L,t) + \left.\frac{\partial \bar{\mathcal{E}}}{\partial u'}\dot{u}\right|_0^L = 0\end{aligned} \tag{11.61}$$

For any value of the velocity in the bar, we obtain the governing equation from Equation (11.61), namely,

$$\frac{d}{dt}\left(\frac{\partial \bar{\mathcal{E}}}{\partial \dot{u}}\right) = \frac{\partial}{\partial x}\left(\frac{\partial \bar{\mathcal{E}}}{\partial u'}\right) - \frac{\partial \bar{\mathcal{E}}}{\partial u} \tag{11.62}$$

subject to the boundary conditions

$$F_L\dot{u}(L,t) - \left.\frac{\partial \bar{\mathcal{E}}}{\partial u'}\dot{u}\right|_0^L = 0 \tag{11.63}$$

Space-Discrete Finite Element Formulation

Next, we discuss the space-discrete finite element formulation via the Total Energy representation of the equation of motion.

$$\mathcal{E}_e^h(\mathbf{q}_e, \dot{\mathbf{q}}_e) = \mathcal{K}_e^h + \left({}_e\mathcal{U}_{\text{int}}^h + {}_e\mathcal{U}_{\text{ext}}^h\right) \tag{11.64}$$

where the space-discrete kinetic energy is

$$\mathcal{K}_e^h = \frac{1}{2}\dot{\mathbf{q}}_e \cdot \mathbf{M}^e \dot{\mathbf{q}}_e \tag{11.65}$$

Hence, Equation (11.64) can be recast into

$$\begin{aligned}\mathcal{E}_e^h &= \frac{\rho A}{6}\left(\left(\dot{u}_1^e(t)\right)^2 + \dot{u}_1^e(t)\,\dot{u}_2^e(t) + \left(\dot{u}_2^e(t)\right)^2\right) \\ &+ \frac{EA}{2L}\left(\left(u_1^e(t)\right)^2 - 2u_1^e(t)\,u_2^e(t) + \left(u_2^e(t)\right)^2\right) - f_1^e u_1^e(t) - f_2^e u_2^e(t)\end{aligned} \tag{11.66}$$

The space-discrete linear momentum can be rewritten as

$$\mathbf{p}_e = \left\{ \begin{array}{c} p_1^e(t) \\ p_2^e(t) \end{array} \right\} = \frac{\partial \mathcal{E}_e^h(\mathbf{q}_e, \dot{\mathbf{q}}_e)}{\partial \dot{\mathbf{q}}_e} = \frac{\rho A}{6}\begin{bmatrix} 2 & 1 \\ 1 & 2 \end{bmatrix}\left\{ \begin{array}{c} \dot{u}_1^e(t) \\ \dot{u}_2^e(t) \end{array} \right\} \tag{11.67}$$

Next, the space-discrete Total Energy representation of the equation of motion results in the finite element equations of motion,

$$\mathfrak{E}_D\left(\mathcal{E}^h\right) = \frac{d}{dt}\left(\frac{\partial \mathcal{E}_e^h}{\partial \dot{\mathbf{q}}_e}\right) + \frac{\partial \mathcal{E}_e^h}{\partial \mathbf{q}_e} = \mathbf{0} \tag{11.68}$$

which can be recast into

$$\frac{\rho A}{6}\begin{bmatrix} 2 & 1 \\ 1 & 2 \end{bmatrix}\left\{ \begin{array}{c} \ddot{u}_1^e(t) \\ \ddot{u}_2^e(t) \end{array} \right\} + \frac{EA}{L}\begin{bmatrix} 1 & -1 \\ -1 & 1 \end{bmatrix}\left\{ \begin{array}{c} u_1^e(t) \\ u_2^e(t) \end{array} \right\} = \left\{ \begin{array}{c} f_1^e \\ f_2^e \end{array} \right\} \tag{11.69}$$

Thus, substituting the space-discrete Total Energy given in Equation (11.66) into the space-discrete Total Energy differential operator in Equation (11.51) results in the semi-discrete finite element equations of motion,

$$\mathfrak{E}_D\left(\mathcal{E}_e^h\right) = \mathbf{M}^e \ddot{\mathbf{q}}_e + \mathbf{K}^e \mathbf{q}_e - \mathbf{F}_{\text{ext}}^e = \mathbf{0} \tag{11.70}$$

where $\mathbf{M}^e$, $\mathbf{K}^e$ and $\mathbf{F}_{\text{ext}}^e$ can be found in Section 9.2.1.

Example 11.2 Rotating Bar

The rotating bar example presented in previous chapters is illustrated next.

Governing Equations: Total Energy Density Based Representations

The Total Energy of the rotating bar as presented previously is given as

$$\mathcal{E}(\theta, \dot{\theta}, \theta') = \int_L \bar{\mathcal{E}}(\dot{\theta}, \theta')dx - \int_L \bar{m}_T(x)\,\theta(x,t)\,dx \tag{11.71}$$

where the Total Energy density function for the rotating bar is given by

$$\bar{\mathcal{E}}(\dot{\theta}, \theta') = \frac{1}{2}\rho J(x)\,\dot{\theta}^2 + \frac{1}{2}GJ(x)\left(\theta'\right)^2, \quad \theta' = \frac{\partial \theta(x,t)}{\partial x} \tag{11.72}$$

Then, the total time derivative of the Total Energy is

$$\frac{d\mathcal{E}(\theta, \dot{\theta}, \theta')}{dt} = \int_L \frac{d\bar{\mathcal{E}}(\theta, \dot{\theta}, \theta')}{dt} dx - \int_L \bar{m}_T(x)\,\theta(x,t)\,dx \tag{11.73}$$

The chain rule leads to

$$\frac{d\bar{\mathcal{E}}(\theta, \dot{\theta}, \theta')}{dt} = \frac{\partial \bar{\mathcal{E}}}{\partial \theta}\dot{\theta} + \frac{\partial \bar{\mathcal{E}}}{\partial \dot{\theta}}\ddot{\theta} + \frac{\partial \bar{\mathcal{E}}}{\partial \theta'}\dot{\theta}' \tag{11.74}$$

Note that the following derivation is true

$$\frac{\partial \bar{\mathcal{E}}}{\partial \dot{\theta}}\ddot{\theta} = \rho J(x)\,\dot{\theta}\ddot{\theta} = \rho J(x)\,\ddot{\theta}\dot{\theta} = \frac{d}{dt}\left(\frac{\partial \bar{\mathcal{E}}}{\partial \dot{\theta}}\right)\dot{\theta} \tag{11.75}$$

because $\rho J(x)\,\dot{\theta} = \dfrac{d}{dt}\left(\dfrac{\partial \bar{E}}{\partial \dot{\theta}}\right)$. Besides, integration by parts on the last term of Equation (11.74) yields

$$\frac{\partial \bar{\mathcal{E}}}{\partial \theta'}\dot{\theta}' = \frac{\partial}{\partial x}\left(\frac{\partial \bar{\mathcal{E}}}{\partial \theta'}\dot{\theta}\right) - \frac{\partial}{\partial x}\left(\frac{\partial \bar{\mathcal{E}}}{\partial \theta'}\right)\dot{\theta} \tag{11.76}$$

Accordingly, Equation (11.74) can be rewritten as

$$\frac{d\bar{\mathcal{E}}(\theta, \dot{\theta}, \theta')}{dt} = \left(\frac{\partial \bar{\mathcal{E}}}{\partial \theta} + \frac{d}{dt}\left(\frac{\partial \bar{\mathcal{E}}}{\partial \dot{\theta}}\right) - \frac{\partial}{\partial x}\left(\frac{\partial \bar{\mathcal{E}}}{\partial \theta'}\right)\right)\dot{\theta} + \frac{\partial}{\partial x}\left(\frac{\partial \bar{\mathcal{E}}}{\partial \theta'}\dot{\theta}\right) \tag{11.77}$$

Substituting Equation (11.77) into Equation (11.73) leads to

$$\frac{d\mathcal{E}}{dt} = \int_L \left(\frac{\partial \bar{\mathcal{E}}}{\partial \theta} + \frac{d}{dt}\left(\frac{\partial \bar{\mathcal{E}}}{\partial \dot{\theta}}\right) - \frac{\partial}{\partial x}\left(\frac{\partial \bar{\mathcal{E}}}{\partial \theta'}\right) - \bar{m}_T(x)\right)\dot{\theta}\,dx + \left.\frac{\partial \bar{\mathcal{E}}}{\partial \theta'}\dot{\theta}\right|_0^L = 0 \tag{11.78}$$

We readily obtain the governing equation from Equation (11.78), namely,

$$\frac{d}{dt}\left(\frac{\partial \bar{\mathcal{E}}}{\partial \dot{y}}\right) = \frac{\partial}{\partial x}\left(\frac{\partial \bar{\mathcal{E}}}{\partial \theta'}\right) - \frac{\partial \bar{\mathcal{E}}}{\partial \theta} + \bar{m}_T(x) \tag{11.79}$$

subject to the boundary conditions

$$\left.\left(\frac{\partial \bar{\mathcal{E}}}{\partial \theta'}\dot{\theta}\right)\right|_0^L = 0 \tag{11.80}$$

Eventually, upon substitution of the Total Energy density function in Equation (11.72) into Equation (11.79) and Equation (11.80), we obtain the same governing equation of motion as given by Equation (10.113) and the same boundary condition as given by Equation (10.114).

Space-Discrete Finite Element Formulation

Next, we discuss the space-discrete finite element formulation via the Total Energy representation of the equation of motion. The trial functions may be approximated by Equation (10.116). Thus, we obtain the space-discrete kinetic energy given by Equation (10.117) and the space-discrete potential energy given by Equation (10.119). Finally, the space-discrete Total Energy is written as

$$\mathcal{E}_e^h(\mathbf{q}_e, \dot{\mathbf{q}}_e) = \frac{1}{2}\dot{\mathbf{q}}_e \cdot \mathbf{M}^e \dot{\mathbf{q}}_e + \frac{1}{2}\mathbf{q}_e \cdot \mathbf{K}^e \mathbf{q}_e - \mathbf{q}_e \cdot \mathbf{F}^e_{\text{ext}} \tag{11.81}$$

where $\mathbf{K}^e$ and $\mathbf{F}^e_{\text{ext}}$ can be found in Equation (10.120), and $\mathbf{M}^e$ is given in Equation (10.118). Substituting the space-discrete Total Energy into the space-discrete Total Energy differential operator in Equation (11.51) simply yields

$$\mathfrak{E}_D\left(\mathcal{E}_e^h\right) = \mathbf{M}^e \ddot{\mathbf{q}}_e + \mathbf{K}^e \mathbf{q}_e - \mathbf{F}^e_{\text{ext}} = \mathbf{0} \tag{11.82}$$

which is the semi-discrete second-order finite element equation of motion.

Example 11.3 Euler-Bernoulli Beam

Next, we consider the dynamical behavior of the Euler-Bernoulli beam.

Governing Equations: Total Energy Density Based Representations

The total rotation of the Euler-Bernoulli beam is defined as

$$\frac{\partial w(x,t)}{\partial x} = w'(x,t) = \varphi(x,t) \tag{11.83}$$

The rate of the Total Energy for the Euler-Bernoulli beam yields

$$\frac{d\mathcal{E}(\mathbf{u}, \dot{\mathbf{u}})}{dt} = 0, \ \ \mathbf{u}(x,t) = (w(x,t), \varphi(x,t)), \ \ \dot{\mathbf{u}} = (\dot{w}, \dot{\varphi}) \tag{11.84}$$

and, the autonomous Total Energy is given by

$$\mathcal{E}(\mathbf{u}, \dot{\mathbf{u}}) = \int_L \bar{\mathcal{E}}(\dot{\mathbf{u}}, \mathbf{u}')\, dx - \int_L \omega(x)\, \mathbf{a} \cdot \mathbf{u}\, dx, \ \ \mathbf{a} = (0, 1) \tag{11.85}$$

where the Total Energy density function is defined as

$$\bar{\mathcal{E}}(\dot{\mathbf{u}}, \mathbf{u}') = \frac{1}{2}\dot{\mathbf{u}} \cdot \boldsymbol{\rho}\dot{\mathbf{u}} + \frac{1}{2}\mathbf{u}' \cdot \mathbf{D}_b \mathbf{u}'$$

$$\boldsymbol{\rho} = \begin{bmatrix} \rho(x) A(x) & 0 \\ 0 & \rho(x) I(x) \end{bmatrix}, \ \ \mathbf{D}_b = \begin{bmatrix} 0 & 0 \\ 0 & EI(x) \end{bmatrix} \tag{11.86}$$

Thus, the rate of the Total Energy can be rewritten as

$$\frac{d\mathcal{E}(\mathbf{u}, \dot{\mathbf{u}})}{dt} = \int_L \frac{d\bar{\mathcal{E}}(\dot{\mathbf{u}}, \mathbf{u}')}{dt} dx - \int_L \omega(x)\, \mathbf{a} \cdot \dot{\mathbf{u}}\, dx \tag{11.87}$$

The integrand of the first integral on the right hand side above gives

$$\frac{d\bar{\mathcal{E}}\left(\dot{\mathbf{u}},\mathbf{u}'\right)}{dt}=\left[\frac{d}{dt}\left(\frac{\partial\bar{\mathcal{E}}}{\partial\dot{\mathbf{u}}}\right)-\frac{\partial}{\partial x}\left(\frac{\partial\bar{\mathcal{E}}}{\partial\mathbf{u}'}\right)\right]\cdot\dot{\mathbf{u}}+\frac{\partial}{\partial x}\left(\frac{\partial\bar{\mathcal{E}}}{\partial\mathbf{u}'}\cdot\dot{\mathbf{u}}\right) \tag{11.88}$$

The above equation can be rewritten as

$$\begin{aligned}\frac{d\bar{\mathcal{E}}\left(\mathbf{u},\dot{u}'\right)}{dt}=&\left\{\begin{array}{l}\dfrac{d}{dt}\left(\dfrac{\partial\bar{\mathcal{E}}}{\partial\dot{w}}\right)-\dfrac{\partial}{\partial x}\left(\dfrac{\partial\bar{\mathcal{E}}}{\partial w'}\right)-\dfrac{\partial}{\partial x}\left(\dfrac{d}{dt}\left(\dfrac{\partial\bar{\mathcal{E}}}{\partial\dot{w}'}\right)\right)\\+\dfrac{\partial^2}{\partial x^2}\left(\dfrac{\partial\bar{\mathcal{E}}}{\partial w''}\right)\end{array}\right\}\dot{w}\\&+\frac{\partial}{\partial x}\left\{\left[\frac{d}{dt}\left(\frac{\partial\bar{\mathcal{E}}}{\partial\dot{w}'}\right)+\frac{\partial\bar{\mathcal{E}}}{\partial w'}-\frac{\partial}{\partial x}\left(\frac{\partial\bar{\mathcal{E}}}{\partial w''}\right)\right]\dot{w}+\frac{\partial\bar{\mathcal{E}}}{\partial w''}\dot{w}'\right\}\end{aligned} \tag{11.89}$$

Substituting Equation (11.89) into Equation (11.87), we have the following:

$$\begin{aligned}\frac{d\mathcal{E}\left(\mathbf{u},\dot{\mathbf{u}}\right)}{dt}=&\int_L\left\{\begin{array}{l}\dfrac{d}{dt}\left(\dfrac{\partial\bar{\mathcal{E}}}{\partial\dot{w}}\right)-\dfrac{\partial}{\partial x}\left(\dfrac{\partial\bar{\mathcal{E}}}{\partial w'}\right)-\dfrac{\partial}{\partial x}\left(\dfrac{d}{dt}\left(\dfrac{\partial\bar{\mathcal{E}}}{\partial\dot{w}'}\right)\right)\\+\dfrac{\partial^2}{\partial x^2}\left(\dfrac{\partial\bar{\mathcal{E}}}{\partial w''}\right)-\omega\left(x\right)\end{array}\right\}\dot{w}dx\\&+\left[\left\{\frac{d}{dt}\left(\frac{\partial\bar{\mathcal{E}}}{\partial\dot{w}'}\right)+\frac{\partial\bar{\mathcal{E}}}{\partial w'}-\frac{\partial}{\partial x}\left(\frac{\partial\bar{\mathcal{E}}}{\partial w''}\right)\right\}\dot{w}+\frac{\partial\bar{\mathcal{E}}}{\partial w''}\dot{w}'\right]_0^L=0\end{aligned} \tag{11.90}$$

where the governing equation of motion and the traction boundary conditions are included in terms of the Total Energy density function. The governing equation for the Euler-Bernoulli beam in terms of the Total Energy density function is obtained as

$$\frac{d}{dt}\left(\frac{\partial\bar{\mathcal{E}}}{\partial\dot{w}}\right)-\frac{\partial}{\partial x}\left(\frac{\partial\bar{\mathcal{E}}}{\partial w'}\right)-\frac{\partial}{\partial x}\left(\frac{d}{dt}\left(\frac{\partial\bar{\mathcal{E}}}{\partial\dot{w}'}\right)\right)+\frac{\partial^2}{\partial x^2}\left(\frac{\partial\bar{\mathcal{E}}}{\partial w''}\right)-\omega\left(x\right)=0 \tag{11.91}$$

subject to the boundary conditions

$$\left.\frac{\partial\bar{\mathcal{E}}}{\partial w''}\dot{w}'\right|_0^L=0,\quad\left.\left\{\frac{d}{dt}\left(\frac{\partial\bar{\mathcal{E}}}{\partial\dot{w}'}\right)+\frac{\partial\bar{\mathcal{E}}}{\partial w'}-\frac{\partial}{\partial x}\left(\frac{\partial\bar{\mathcal{E}}}{\partial w''}\right)\right\}\dot{w}\right|_0^L=0 \tag{11.92}$$

Subsequently, upon substitution of the Total Energy density function in Equation (11.86) into Equation (11.91) and Equation (11.92), we obtain the same differential equations of motion as given in Equation (10.137) and the same boundary conditions as given in Equation (10.138).

Space-Discrete Finite Element Formulation

Next, we discuss the space-discrete finite element formulation via the Total Energy representation of the equation of motion. Consider one element with two nodes. The trial displacement function may be approximated using the same Hermite polynomial interpolation functions as given in Equation (10.140). Thus, in favor of matrix and vector notation, we have the space-discrete kinetic energy given by Equation (10.142) and the space-discrete potential energy given by Equation (10.144). We now construct the space-discrete autonomous Total Energy

$$\mathcal{E}_e^h(\mathbf{q}_e, \dot{\mathbf{q}}_e) = \frac{1}{2}\dot{\mathbf{q}}_e(t) \cdot \mathbf{M}^e \dot{\mathbf{q}}_e(t) + \frac{1}{2}\mathbf{q}_e(t) \cdot \mathbf{K}^e \mathbf{q}_e(t) - \mathbf{q}_e(t) \cdot \mathbf{F}_{\text{ext}}^e \tag{11.93}$$

where $\mathbf{M}^e$, $\mathbf{K}^e$ and $\mathbf{F}_{\text{ext}}^e$ can be found in Section 9.2.6. Substituting the space-discrete autonomous Total Energy given in Equation (11.93) into the space-discrete Total Energy differential operators in Equation (11.51) readily yields the semi-discrete equations of motion as

$$\mathfrak{E}_D(\mathcal{E}_e^h) = \mathbf{M}^e \ddot{\mathbf{q}}_e + \mathbf{K}^e \mathbf{q}_e - \mathbf{F}_{\text{ext}}^e = \mathbf{0} \tag{11.94}$$

Example 11.4 Timoshenko Beam

We shall next consider the dynamical behavior of the Timoshenko beam. We wish to derive the strong form in terms of the Total Energy density representations of the equation of motion.

Governing Equations: Total Energy Density Based Representations

The Total Energy of the Timoshenko beam is given by

$$\begin{gathered}\mathcal{E}(\mathbf{u}, \dot{\mathbf{u}}) = \int_L \bar{\mathcal{E}}(\mathbf{u}, \dot{u}, \mathbf{u}')\,dx - \int_L \bar{\omega}(x)\,\mathbf{a} \cdot \mathbf{u}dx, \\ \mathbf{u}(x,t) = (w(x,t), \theta_y(x,t)), \quad \dot{\mathbf{u}}(x,t) = (\dot{w}(x,t), \dot{\theta}_y(x,t))\end{gathered} \tag{11.95}$$

where the Total Energy density function for the Timoshenko beam is given by

$$\begin{gathered}\bar{\mathcal{E}}(\mathbf{u}, \dot{\mathbf{u}}, \mathbf{u}') = \frac{1}{2}\dot{\mathbf{u}} \cdot \boldsymbol{\rho}\dot{\mathbf{u}} + \frac{1}{2}\boldsymbol{\kappa} \cdot \mathbf{D}_b\boldsymbol{\kappa} + \frac{1}{2}\boldsymbol{\gamma} \cdot \mathbf{G}_s\boldsymbol{\gamma}, \\ \boldsymbol{\kappa} = \left[\frac{\partial \theta_y}{\partial x}\right], \quad \mathbf{D}_b = [EI(x)], \quad \boldsymbol{\gamma} = \left[\frac{\partial w(x,t)}{\partial x} + \theta_y(x,t)\right], \\ \boldsymbol{\rho} = \begin{bmatrix} \rho A(x) & 0 \\ 0 & \rho I(x) \end{bmatrix}, \quad \mathbf{G}_s = [\kappa GA(x)]\end{gathered} \tag{11.96}$$

Then, the total time derivative of the Total Energy is given by

$$\frac{d\mathcal{E}}{dt} = \int_L \frac{d\bar{\mathcal{E}}(\mathbf{u}, \dot{\mathbf{u}}, \mathbf{u}')}{dt}dx - \int_L \bar{\omega}(x)\,\mathbf{a} \cdot \dot{\mathbf{u}}\,dx \tag{11.97}$$

The chain rule leads to

$$\frac{d\bar{\mathcal{E}}(\mathbf{u}, \dot{\mathbf{u}}, \mathbf{u}')}{dt} = \frac{\partial \bar{\mathcal{E}}}{\partial \mathbf{u}} \cdot \dot{\mathbf{u}} + \frac{\partial \bar{\mathcal{E}}}{\partial \dot{\mathbf{u}}} \cdot \ddot{\mathbf{u}} + \frac{\partial \bar{\mathcal{E}}}{\partial \mathbf{u}'} \cdot \dot{\mathbf{u}}' \tag{11.98}$$

Note that the following derivation is true

$$\frac{\partial \bar{\mathcal{E}}}{\partial \dot{\mathbf{u}}} \cdot \ddot{\mathbf{u}} = \ddot{\mathbf{u}} \cdot \boldsymbol{\rho} \dot{\mathbf{u}} = \dot{\mathbf{u}} \cdot \boldsymbol{\rho} \ddot{\mathbf{u}} = \frac{d}{dt}\left(\frac{\partial \bar{\mathcal{E}}}{\partial \dot{\mathbf{u}}}\right) \cdot \dot{\mathbf{u}} \tag{11.99}$$

because the density matrix $\boldsymbol{\rho}$ is not dependent upon the time. Besides, integration by parts on the last two terms of Equation (11.98) yields

$$\frac{\partial \bar{\mathcal{E}}}{\partial \mathbf{u}'} \cdot \dot{\mathbf{u}}' = \frac{\partial}{\partial x}\left(\frac{\partial \bar{\mathcal{E}}}{\partial \mathbf{u}'} \cdot \dot{\mathbf{u}}\right) - \frac{\partial}{\partial x}\left(\frac{\partial \bar{\mathcal{E}}}{\partial \mathbf{u}'}\right) \cdot \dot{\mathbf{u}} \tag{11.100}$$

Accordingly, Equation (11.98) can be rewritten as

$$\frac{d\bar{\mathcal{E}}(\mathbf{u}, \dot{\mathbf{u}}, \mathbf{u}')}{dt} = \left(\frac{\partial \bar{\mathcal{E}}}{\partial \mathbf{u}} + \frac{d}{dt}\left(\frac{\partial \bar{\mathcal{E}}}{\partial \dot{\mathbf{u}}}\right) - \frac{\partial}{\partial x}\left(\frac{\partial \bar{\mathcal{E}}}{\partial \mathbf{u}'}\right)\right) \cdot \dot{\mathbf{u}} + \frac{\partial}{\partial x}\left(\frac{\partial \bar{\mathcal{E}}}{\partial \mathbf{u}'} \cdot \dot{\mathbf{u}}\right) \tag{11.101}$$

Substituting Equation (11.101) into Equation (11.97) leads to

$$\begin{aligned} \frac{d\mathcal{E}}{dt} = \int_L \left(\frac{\partial \bar{\mathcal{E}}}{\partial \mathbf{u}} + \frac{d}{dt}\left(\frac{\partial \bar{\mathcal{E}}}{\partial \dot{\mathbf{u}}}\right) - \frac{\partial}{\partial x}\left(\frac{\partial \bar{\mathcal{E}}}{\partial \mathbf{u}'}\right) - \bar{\omega}(x)\,\mathbf{a}\right) \cdot \dot{\mathbf{u}}\, dx \\ + \left(\frac{\partial \bar{\mathcal{E}}}{\partial \mathbf{u}'} \cdot \dot{\mathbf{u}}\right)\Big|_0^L = 0 \end{aligned} \tag{11.102}$$

For any vector-valued function $\dot{\mathbf{u}}$, we obtain the governing equation from Equation (11.102), namely,

$$\frac{\partial}{\partial x}\left(\frac{\partial \bar{\mathcal{E}}}{\partial \mathbf{u}'}\right) - \frac{d}{dt}\left(\frac{\partial \bar{\mathcal{E}}}{\partial \dot{\mathbf{u}}}\right) - \frac{\partial \bar{\mathcal{E}}}{\partial \mathbf{u}} + \bar{\omega}(x)\,\mathbf{a} = 0 \tag{11.103}$$

subject to the natural boundary conditions

$$\frac{\partial \bar{\mathcal{E}}}{\partial \mathbf{u}'} \cdot \dot{\mathbf{u}}\Big|_0^L = 0 \tag{11.104}$$

Therefore, the governing equations can be written as

$$\begin{aligned} \frac{d}{dt}\left(\frac{\partial \bar{\mathcal{E}}}{\partial \dot{w}}\right) &= \frac{\partial}{\partial x}\left(\frac{\partial \bar{\mathcal{E}}}{\partial w'}\right) - \frac{\partial \bar{\mathcal{E}}}{\partial w} + \bar{\omega}(x), \\ \frac{d}{dt}\left(\frac{\partial \bar{\mathcal{E}}}{\partial \dot{\theta}_y}\right) &= \frac{\partial}{\partial x}\left(\frac{\partial \bar{\mathcal{E}}}{\partial \theta'_y}\right) - \frac{\partial \bar{\mathcal{E}}}{\partial \theta_y} \end{aligned} \tag{11.105}$$

subject to the boundary conditions

$$\frac{\partial \bar{\mathcal{E}}}{\partial w'} \dot{w}\Big|_0^L = 0, \quad \frac{\partial \bar{\mathcal{E}}}{\partial \theta'_y} \dot{\theta}_y\Big|_0^L = 0 \tag{11.106}$$

Eventually, upon substitution of the Total Energy density function in Equation (11.96) into Equation (11.105) and Equation (11.106), we obtain the same differential equations of motion as given in Equation (10.163) and the same boundary conditions as given in Equation (10.164).

Space-Discrete Finite Element Formulation

Next, we shall discuss the space-discrete finite element formulation via the Total Energy representation of the equation of motion. The admissible trial functions are given by Equation (10.166). The space-discrete Total Energy may be constructed as

$$\begin{gathered}\mathcal{E}_e^h(\mathbf{q}_e, \dot{\mathbf{q}}_e) = \frac{1}{2}\dot{\mathbf{q}}_e(t)\cdot\mathbf{M}^e\dot{\mathbf{q}}_e(t) + \frac{1}{2}\mathbf{q}_e(t)\cdot\mathbf{K}^e\mathbf{q}_e(t) - \mathbf{q}_e(t)\cdot\mathbf{F}_{\text{ext}}^e,\\ \mathbf{M}^e = \frac{1}{2}\int_{L_e}\mathbf{H}_e^T(x)\,\rho\mathbf{H}_e(x)\,dx,\\ \mathbf{K}_b = \int_{L_e}\mathbf{B}_b^T(x)\,\mathbf{D}_b\mathbf{B}_b(x)\,dx,\ \ \mathbf{K}_s = \int_L\mathbf{B}_s^T(x)\,\mathbf{G}_s\mathbf{B}_s(x)\,dx,\ \ \mathbf{K}^e = \mathbf{K}_b + \mathbf{K}_s\\ \mathbf{B}_b = \left[\frac{\partial\mathbf{H}_\theta(x)}{\partial x}\right],\ \mathbf{B}_s = \left[\frac{\partial\mathbf{H}_w(x)}{\partial x} + \mathbf{H}_\theta(x)\right],\ \mathbf{F}_{ext}^e = \int_{L_e}\bar{\omega}(x)\,\mathbf{H}_w^T(x)\,dx\end{gathered} \tag{11.107}$$

whose details can be found in Section 9.2.7. Substituting the space-discrete Total Energy into the space-discrete Total Energy differential operator in Equation (11.51) simply leads to the semi-discrete finite element equations of motion as:

$$\mathfrak{E}_D\left(\mathcal{E}_e^h\right) = \mathbf{M}^e\ddot{\mathbf{q}}_e + \mathbf{K}^e\mathbf{q}_e - \mathbf{F}_{\text{ext}}^e = \mathbf{0} \tag{11.108}$$

which is the semi-discrete second-order finite element equation of motion.

Example 11.5 Reissner-Mindlin Plate

We next consider the Reissner-Mindlin plate of uniform thickness h in the Total Energy representation and framework.

Governing Equations: Total Energy Density Based Representations

The rate of the Total Energy yields

$$\frac{d\mathcal{E}(\mathbf{u}, \dot{\mathbf{u}})}{dt} = 0,\ \ \mathbf{u} = \left(w, \theta_x, \theta_y\right),\ \ \dot{\mathbf{u}} = \left(\dot{w}, \dot{\theta}_x, \dot{\theta}_y\right) \tag{11.109}$$

where the continuous Total Energy of the Reissner-Mindlin plate can be obtained as

$$\begin{gathered}\mathcal{E}(\mathbf{u}, \dot{\mathbf{u}}) = \int_\Omega\bar{\mathcal{E}}\left(\mathbf{u}, \dot{\mathbf{u}}, \mathbf{u}_{,x}, \mathbf{u}_{,y}\right)dxdy - \int_\Omega\omega(x, y)\,\mathbf{a}\cdot\mathbf{u}dxdy,\ \ \mathbf{a} = (1, 0, 0)\\ \mathbf{u}_{,x} = \left(\frac{\partial w}{\partial x}, \frac{\partial\theta_x}{\partial x}, \frac{\partial\theta_y}{\partial x}\right),\ \ \mathbf{u}_{,y} = \left(\frac{\partial w}{\partial y}, \frac{\partial\theta_x}{\partial y}, \frac{\partial\theta_y}{\partial y}\right)\end{gathered} \tag{11.110}$$

where the Total Energy density function can be defined as

$$\bar{\mathcal{E}}\left(\mathbf{u}, \dot{\mathbf{u}}, \mathbf{u}_{,x}, \mathbf{u}_{,y}\right) = \frac{1}{2}\dot{\mathbf{u}} \cdot \rho\dot{\mathbf{u}} + \frac{1}{2}\boldsymbol{\kappa} \cdot \mathbf{D}_b\boldsymbol{\kappa} + \frac{1}{2}\boldsymbol{\gamma} \cdot \mathbf{G}_s\boldsymbol{\gamma} \tag{11.111}$$

Introducing new notations for convenience, the Total Energy density function can be given by

$$\begin{aligned} &\bar{\mathcal{E}}\left(\mathbf{u}, \dot{\mathbf{u}}, \mathbf{u}_{,x}, \mathbf{u}_{,y}\right) = \frac{1}{2}\dot{\mathbf{u}} \cdot \mathbf{p} + \frac{1}{2}\boldsymbol{\kappa} \cdot \mathbf{M}_b + \frac{1}{2}\boldsymbol{\gamma} \cdot \mathbf{q}_s, \\ &\mathbf{p} = \rho\dot{\mathbf{u}}, \quad \mathbf{M}_b = \mathbf{D}_b\boldsymbol{\kappa}, \quad \mathbf{q}_s = \mathbf{G}_s\boldsymbol{\gamma} \end{aligned} \tag{11.112}$$

where $\mathbf{p}$, $\mathbf{M}_b$ and $\mathbf{q}_s$ are given by Equation (10.180). The rate of the autonomous Total Energy is given by

$$\begin{aligned} \frac{d\mathcal{E}(\mathbf{u}, \dot{\mathbf{u}})}{dt} &= \int_\Omega \frac{d\bar{\mathcal{E}}\left(\mathbf{u}, \dot{\mathbf{u}}, \mathbf{u}_{,x}, \mathbf{u}_{,y}\right)}{dt} dxdy - \int_\Omega \omega(x, y)\, \mathbf{a} \cdot \dot{\mathbf{u}}\, dxdy \\ &= \int_\Omega \frac{\partial \bar{\mathcal{E}}}{\partial \mathbf{u}} \cdot \dot{\mathbf{u}} + \frac{\partial \bar{\mathcal{E}}}{\partial \dot{\mathbf{u}}} \cdot \ddot{\mathbf{u}} + \frac{\partial \bar{\mathcal{E}}}{\partial \mathbf{u}_{,x}} \cdot \dot{\mathbf{u}}_{,x} + \frac{\partial \bar{\mathcal{E}}}{\partial \mathbf{u}_{,y}} \cdot \dot{\mathbf{u}}_{,y}\, dxdy \\ &\quad - \int_\Omega \omega(x, y)\, \mathbf{a} \cdot \dot{\mathbf{u}}\, dxdy \end{aligned} \tag{11.113}$$

Integration by parts yields

$$\int_{t_0}^{t_1} \left(\int_\Omega \frac{\partial \bar{\mathcal{E}}}{\partial \dot{\mathbf{u}}} \cdot \ddot{\mathbf{u}}\, dxdy \right) dt = \int_{t_0}^{t_1} \left(\int_\Omega \frac{d}{dt} \left(\frac{\partial \bar{\mathcal{E}}}{\partial \dot{\mathbf{u}}} \right) \cdot \dot{\mathbf{u}}\, dxdy \right) dt \tag{11.114}$$

By means of Green's theorem, we have the following:

$$\begin{aligned} &\int_\Omega \left(\frac{\partial \bar{\mathcal{E}}}{\partial \mathbf{u}_{,x}} \cdot \dot{\mathbf{u}}_{,x} + \frac{\partial \bar{\mathcal{E}}}{\partial \mathbf{u}_{,y}} \cdot \dot{\mathbf{u}}_{,y} \right) dxdy \\ &= -\int_\Omega \left\{ \frac{\partial}{\partial x} \left(\frac{\partial \bar{\mathcal{E}}}{\partial \mathbf{u}_{,x}} \right) + \frac{\partial}{\partial y} \left(\frac{\partial \bar{\mathcal{E}}}{\partial \mathbf{u}_{,y}} \right) \right\} \cdot \dot{\mathbf{u}}\, dxdy \\ &\quad + \int_{\partial\Omega} \left\{ \frac{\partial \bar{\mathcal{E}}}{\partial \mathbf{u}_{,x}} \cdot \dot{\mathbf{u}}_x n_x + \frac{\partial \bar{\mathcal{E}}}{\partial \mathbf{u}_{,y}} \cdot \dot{\mathbf{u}}_y n_y \right\} d\Gamma = 0, \quad n_x = \frac{\partial x}{\partial n}, \; n_y = \frac{\partial y}{\partial n} \end{aligned} \tag{11.115}$$

The rate of the Total Energy leads to

$$\begin{aligned} \frac{d\mathcal{E}(\mathbf{u}, \dot{\mathbf{u}})}{dt} &= \int_\Omega \left\{ \begin{array}{l} \dfrac{d}{dt}\left(\dfrac{\partial \bar{\mathcal{E}}}{\partial \dot{\mathbf{u}}} \right) + \dfrac{\partial \bar{\mathcal{E}}}{\partial \mathbf{u}} - \dfrac{\partial}{\partial x} \left(\dfrac{\partial \bar{\mathcal{E}}}{\partial \mathbf{u}_{,x}} \right) \\ -\dfrac{\partial}{\partial y} \left(\dfrac{\partial \bar{\mathcal{E}}}{\partial \mathbf{u}_{,y}} \right) - \omega(x, y)\mathbf{a} \end{array} \right\} \cdot \dot{\mathbf{u}}\, dxdy \\ &\quad + \int_{\partial\Omega} \left\{ \frac{\partial \bar{\mathcal{E}}}{\partial \mathbf{u}_{,x}} \cdot \dot{\mathbf{u}}_x n_x + \frac{\partial \bar{\mathcal{E}}}{\partial \mathbf{u}_{,y}} \cdot \dot{\mathbf{u}}_y n_y \right\} d\Gamma \end{aligned} \tag{11.116}$$

where n_x, n_y denote the directional cosines respectively and n stands for the coordinate of the outward unit vector normal to the boundary surface.

Consequently, we have the governing equations for the Reissner-Mindlin plate as follows:

$$\frac{d}{dt}\left(\frac{\partial\bar{\mathcal{E}}}{\partial\dot{\mathbf{u}}}\right)+\frac{\partial\bar{\mathcal{E}}}{\partial\mathbf{u}}-\frac{\partial}{\partial x}\left(\frac{\partial\bar{\mathcal{E}}}{\partial\mathbf{u}_{,x}}\right)-\frac{\partial}{\partial y}\left(\frac{\partial\bar{\mathcal{E}}}{\partial\mathbf{u}_{,y}}\right)-\omega\left(x,y\right)\mathbf{a}=0 \tag{11.117}$$

The above equations can be rewritten as

$$\frac{d}{dt}\left(\rho\dot{\mathbf{u}}\right)+\left[\frac{\partial\boldsymbol{\kappa}}{\partial\mathbf{u}}\right]^T\mathbf{M}_b+\left[\frac{\partial\boldsymbol{\gamma}}{\partial\mathbf{u}}\right]^T\mathbf{Q}_s-\frac{\partial\mathbf{A}_x}{\partial x}-\frac{\partial\mathbf{A}_y}{\partial y}-\omega\left(x,y\right)\mathbf{a}=0 \tag{11.118}$$

where

$$\begin{aligned}\mathbf{A}_x&=\frac{\partial\bar{\mathcal{E}}}{\partial\mathbf{u}_{,x}}=\left[\frac{\partial\boldsymbol{\kappa}}{\partial\mathbf{u}_{,x}}\right]^T\mathbf{M}_b+\left[\frac{\partial\boldsymbol{\gamma}}{\partial\mathbf{u}_{,x}}\right]^T\mathbf{Q}_s,\\ \mathbf{A}_y&=\frac{\partial\bar{\mathcal{E}}}{\partial\mathbf{u}_{,y}}=\left[\frac{\partial\boldsymbol{\kappa}}{\partial\mathbf{u}_{,y}}\right]^T\mathbf{M}_b+\left[\frac{\partial\boldsymbol{\gamma}}{\partial\mathbf{u}_{,y}}\right]^T\mathbf{Q}_s\end{aligned} \tag{11.119}$$

Therefore, Equation (11.117) can be recast into Equation (9.134). Notice Equation (11.117) indicates the following:

$$\begin{aligned}&\frac{d}{dt}\left(\frac{\partial\bar{\mathcal{E}}}{\partial\dot{w}}\right)+\frac{\partial\bar{\mathcal{E}}}{\partial w}-\frac{\partial}{\partial x}\left(\frac{\partial\bar{\mathcal{E}}}{\partial w_{,x}}\right)-\frac{\partial}{\partial y}\left(\frac{\partial\bar{\mathcal{E}}}{\partial w_{,y}}\right)-\omega(x,y)=0,\\ &\frac{d}{dt}\left(\frac{\partial\bar{\mathcal{E}}}{\partial\dot{\theta}_x}\right)+\frac{\partial\bar{\mathcal{E}}}{\partial\theta_x}-\frac{\partial}{\partial x}\left(\frac{\partial\bar{\mathcal{E}}}{\partial\theta_{x,x}}\right)-\frac{\partial}{\partial y}\left(\frac{\partial\bar{\mathcal{E}}}{\partial\theta_{x,y}}\right)=0,\\ &\frac{d}{dt}\left(\frac{\partial\bar{\mathcal{E}}}{\partial\dot{\theta}_y}\right)+\frac{\partial\bar{\mathcal{E}}}{\partial\theta_y}-\frac{\partial}{\partial x}\left(\frac{\partial\bar{\mathcal{E}}}{\partial\theta_{y,x}}\right)-\frac{\partial}{\partial y}\left(\frac{\partial\bar{\mathcal{E}}}{\partial\theta_{y,y}}\right)=0\end{aligned} \tag{11.120}$$

which leads to the same equation as Equation (9.135). The second term in the right-hand side of Equation (11.116) gives the natural boundary conditions as follows:

$$\begin{aligned}\frac{\partial\bar{\mathcal{E}}}{\partial\mathbf{u}_{,x}}\cdot\delta\mathbf{u}_x n_x+\frac{\partial\bar{\mathcal{E}}}{\partial\mathbf{u}_{,y}}\cdot\delta\mathbf{u}_y n_y&=\mathbf{A}_x^T\delta\mathbf{u}_x n_x+\mathbf{A}_y^T\delta\mathbf{u}_y n_y\\ &=n_x\left(Q_y\delta w-M_{xy}\delta\theta_x+M_{xx}\delta\theta_y\right)\\ &\quad+n_y\left(Q_x\delta w-M_{yy}\delta\theta_x+M_{xy}\delta\theta_y\right)\\ &=\delta w\left(n_xQ_y+n_yQ_x\right)\\ &\quad-\delta\theta_x\left(n_xM_{xy}+n_yM_{yy}\right)\\ &\quad+\delta\theta_y\left(n_xM_{xx}+n_yM_{xy}\right)=0\end{aligned} \tag{11.121}$$

In view of Equation (11.121), we say that the governing equations in Equation (11.117) are subject to the natural boundary conditions as given in Equation (10.192). Note that the essential boundary conditions on the geometric boundary surface, $\partial\Omega_u$, can be given by Equation (10.193).

Space-Discrete Finite Element Formulation

Next, we shall discuss the space-discrete finite element formulation via the Total Energy representation of the equation of motion. We first construct the space-discrete autonomous Total Energy. The admissible trial functions are given by Equation (10.194). Introducing new notations for convenience, $\mathbf{q}_e(t) = \begin{bmatrix} \mathbf{w}^e(t) & \boldsymbol{\theta}_x^e(t) & \boldsymbol{\theta}_y^e(t) \end{bmatrix}^T$, the trial functions are rewritten as Equation (10.195). The space-discrete Total Energy can be constructed as

$$\mathcal{E}_e^h(\mathbf{q}_e, \dot{\mathbf{q}}_e) = \frac{1}{2}\dot{\mathbf{q}}_e(t) \cdot \mathbf{M}^e \dot{\mathbf{q}}_e(t) + \frac{1}{2}\mathbf{q}_e(t) \cdot \mathbf{K}^e \mathbf{q}_e(t) - \mathbf{q}_e(t) \cdot \mathbf{F}_{\text{ext}}^e,$$

$$\mathbf{M}^e = \frac{1}{2}\int_{A_e} \mathbf{H}_e^T \rho \mathbf{H}_e \, dxdy, \; \mathbf{H}_e = \begin{bmatrix} \mathbf{H}_w \\ \mathbf{H}_{\theta_x} \\ \mathbf{H}_{\theta y} \end{bmatrix}, \; \mathbf{F}_{\text{ext}}^e = \int_{\Omega_e} \omega(x, y)\mathbf{H}_w^T \, dxdy$$

$$\mathbf{K}^e = \mathbf{K}_b + \mathbf{K}_s, \; \mathbf{K}_b^e = \frac{1}{2}\iint_{\Omega_e} \mathbf{B}_b^T \mathbf{D}_b \mathbf{B}_b \, dxdy, \; \mathbf{K}_s^e = \frac{k}{2}\iint_{\Omega_e} \mathbf{B}_s^T \mathbf{G}_s \mathbf{B}_s \, dxdy, \tag{11.122}$$

$$\mathbf{B}_b = \begin{bmatrix} \dfrac{\partial \mathbf{H}_{\theta_y}}{\partial x} \\ -\dfrac{\partial \mathbf{H}_{\theta_x}}{\partial y} \\ \left(\dfrac{\partial \mathbf{H}_{\theta_y}}{\partial y} - \dfrac{\partial \mathbf{H}_{\theta_x}}{\partial x}\right) \end{bmatrix}, \; \mathbf{B}_s = \begin{bmatrix} \dfrac{\partial \mathbf{H}_w}{\partial x} + \mathbf{H}_{\theta_y} \\ \dfrac{\partial \mathbf{H}_w}{\partial y} - \mathbf{H}_{\theta_x} \end{bmatrix}$$

More details about $\mathbf{M}^e$, $\mathbf{K}^e$ and $\mathbf{F}_{\text{ext}}^e$ can be found in Section 9.2.9. Substituting the space-discrete Total Energy in Equation (11.122) into the space-discrete Total Energy differential operator given by Equation (11.51), simply and naturally, leads to the semi-discrete finite element equation of motion as:

$$\mathfrak{E}_D\left(\mathcal{E}_e^h\right) = \mathbf{M}^e \ddot{\mathbf{q}} + \mathbf{K}^e \mathbf{q}_e - \mathbf{F}_{\text{ext}}^e = \mathbf{0} \tag{11.123}$$

11.3 LAGRANGIAN MECHANICS FRAMEWORK IN THE DIFFERENTIAL CALCULUS SETTING AND FINITE ELEMENT FORMULATIONS

The Lagrangian mechanics framework in the context of the principle of balance of mechanical energy and scalar formalism is briefly highlighted in this section. Similar ideas as those described previously for the Total Energy framework and the differential calculus setting follow next.

11.3.1 Lagrangian Density Equations of Motion

Scalar Formalism: Strong and Weak Forms in the Lagrangian Mechanics Framework Instead of the rate of the Total Energy given in Equation (11.31), the theorem of power expended or equivalently, the principle balance of mechanical energy for continuous-body systems of perfect materials reduces to the rate of Jacobi's integral which can be rewritten as

$$\frac{d(\mathcal{E})}{dt} = \frac{d}{dt}\{2\mathcal{K}(\dot{\boldsymbol{\varphi}}) - \mathcal{L}(\boldsymbol{\varphi}, \dot{\boldsymbol{\varphi}})\} = 0, \;\; \mathcal{L} = \mathcal{K} - \mathcal{U}, \;\; \mathcal{U} = \Pi_{\text{int}} + \Pi_{\text{ext}} \tag{11.124}$$

where $\mathcal{U}$ denotes the total potential energy, and the internal potential energy is defined by

$$\Pi_{\text{int}} = \int_{\mathcal{B}} \rho_0 \Psi(\mathbf{F})\, dV \tag{11.125}$$

and $\Psi(\mathbf{F})$ denotes the Helmholtz free energy per unit mass in homogeneous materials. Note that the term enclosed in the bracket in Equation (11.124) is referred to as Jacobi's integral for a continuous-body system. In Equation (11.124), the autonomous Lagrangian $\mathcal{L}(\boldsymbol{\varphi}, \dot{\boldsymbol{\varphi}}) : T\mathcal{C} \to \mathbb{R}$, namely, the kinetic energy minus the total potential energy, with the definition of the tangent bundle $T\mathcal{C} := \{(\boldsymbol{\varphi}, \dot{\boldsymbol{\varphi}})\, | \boldsymbol{\varphi} \in \mathcal{C} \text{ and } \dot{\boldsymbol{\varphi}} \in T_{\varphi}\mathcal{C}\}$, is defined by

$$\mathcal{L}(\boldsymbol{\varphi}, \dot{\boldsymbol{\varphi}}) = \int_{\mathcal{B}} \bar{\mathcal{L}}(\boldsymbol{\varphi}, \dot{\boldsymbol{\varphi}}, \mathbf{F})\, dV + \int_{\partial\mathcal{B}_\sigma} \bar{\mathbf{T}}(\mathbf{X}, \hat{\mathbf{N}}) \cdot \boldsymbol{\varphi}(\mathbf{X}, t)\, dA \tag{11.126}$$

where the Lagrangian density function $\bar{\mathcal{L}}(\boldsymbol{\varphi}, \dot{\boldsymbol{\varphi}}, \mathbf{F})$ is defined as

$$\bar{\mathcal{L}}(\boldsymbol{\varphi}, \dot{\boldsymbol{\varphi}}, \mathbf{F}) = \frac{1}{2}\rho_0 \dot{\boldsymbol{\varphi}} \cdot \dot{\boldsymbol{\varphi}} - \rho_0 \Psi(\mathbf{F}) + \rho_0 \mathbf{B} \cdot \boldsymbol{\varphi} \tag{11.127}$$

In terms of the Lagrangian density function, Equation (11.124) can be rewritten as

$$\frac{d}{dt} \int_{\mathcal{B}} \left(\frac{\partial \bar{\mathcal{L}}}{\partial \dot{\boldsymbol{\varphi}}} \cdot \dot{\boldsymbol{\varphi}} - \bar{\mathcal{L}}(\boldsymbol{\varphi}, \dot{\boldsymbol{\varphi}}, \mathbf{F}) \right) dV - \int_{\partial\mathcal{B}_\sigma} \bar{\mathbf{T}}(\mathbf{X}, \hat{\mathbf{N}}) \cdot \dot{\boldsymbol{\varphi}}(\mathbf{X}, t)\, dA = 0 \tag{11.128}$$

Note that the above equation is true if and only if the continuous-body dynamical system is scleronomic. By use of the chain rule, the rate of the Lagrangian density function gives

$$\frac{d\bar{\mathcal{L}}(\boldsymbol{\varphi}, \dot{\boldsymbol{\varphi}}, \mathbf{F})}{dt} = \frac{\partial \bar{\mathcal{L}}(\boldsymbol{\varphi}, \dot{\boldsymbol{\varphi}}, \mathbf{F})}{\partial \boldsymbol{\varphi}} \cdot \dot{\boldsymbol{\varphi}} + \frac{\partial \bar{\mathcal{L}}(\boldsymbol{\varphi}, \dot{\boldsymbol{\varphi}}, \mathbf{F})}{\partial \dot{\boldsymbol{\varphi}}} \cdot \ddot{\boldsymbol{\varphi}} + \frac{\partial \bar{\mathcal{L}}(\boldsymbol{\varphi}, \dot{\boldsymbol{\varphi}}, \mathbf{F})}{\partial \mathbf{F}} : \dot{\mathbf{F}} \tag{11.129}$$

In view of Equation (11.129), Equation (11.128) can be recast into the form

$$\begin{aligned} &\int_{\mathcal{B}} \left(\frac{d}{dt}\left(\frac{\partial \bar{\mathcal{L}}}{\partial \dot{\boldsymbol{\varphi}}} \right) - \frac{\partial \bar{\mathcal{L}}(\boldsymbol{\varphi}, \dot{\boldsymbol{\varphi}}, \mathbf{F})}{\partial \boldsymbol{\varphi}} \right) \cdot \dot{\boldsymbol{\varphi}}\, dV - \int_{\mathcal{B}} \frac{\partial \bar{\mathcal{L}}(\boldsymbol{\varphi}, \dot{\boldsymbol{\varphi}}, \mathbf{F})}{\partial \mathbf{F}} : \dot{\mathbf{F}}\, dV \\ &\quad - \int_{\partial\mathcal{B}_\sigma} \bar{\mathbf{T}}(\mathbf{X}, \hat{\mathbf{N}}) \cdot \dot{\boldsymbol{\varphi}}(\mathbf{X}, t)\, dA = 0 \end{aligned} \tag{11.130}$$

By means of the following derivation

$$\frac{\partial \bar{\mathcal{L}}}{\partial \mathbf{F}} : \dot{\mathbf{F}} = \frac{\partial \bar{\mathcal{L}}}{\partial \mathbf{F}} : \text{GRAD}\, \dot{\boldsymbol{\varphi}} = \text{DIV}\left(\dot{\boldsymbol{\varphi}} \cdot \frac{\partial \bar{\mathcal{L}}}{\partial \mathbf{F}} \right) - \text{DIV}\, \frac{\partial \bar{\mathcal{L}}}{\partial \mathbf{F}} \cdot \dot{\boldsymbol{\varphi}} \tag{11.131}$$

and Gauss's theorem, we find that

$$\begin{aligned} &\int_{\mathcal{B}} \left(\frac{d}{dt}\left(\frac{\partial \bar{\mathcal{L}}}{\partial \dot{\boldsymbol{\varphi}}} \right) + \text{DIV}\, \frac{\partial \bar{\mathcal{L}}}{\partial \mathbf{F}} - \frac{\partial \bar{\mathcal{L}}(\boldsymbol{\varphi}, \dot{\boldsymbol{\varphi}}, \mathbf{F})}{\partial \boldsymbol{\varphi}} \right) \cdot \dot{\boldsymbol{\varphi}}\, dV \\ &\quad - \int_{\partial\mathcal{B}_\sigma} \left(\bar{\mathbf{T}}(\mathbf{X}, \hat{\mathbf{N}}) + \frac{\partial \bar{\mathcal{L}}(\boldsymbol{\varphi}, \dot{\boldsymbol{\varphi}}, \mathbf{F})}{\partial \mathbf{F}} \cdot \hat{\mathbf{N}} \right) \cdot \dot{\boldsymbol{\varphi}}(\mathbf{X}, t)\, dA = 0 \end{aligned} \tag{11.132}$$

The equation above takes the weighted residual form involving the strong form, namely, the governing equations and the natural boundary conditions. Consequently, we have the first of the Lagrangian symbolic equations referred to as the Lagrangian density equations of motion:

$$\frac{d}{dt}\left(\frac{\partial \bar{\mathcal{L}}}{\partial \dot{\boldsymbol{\varphi}}}\right) = \frac{\partial \bar{\mathcal{L}}(\boldsymbol{\varphi}, \dot{\boldsymbol{\varphi}}, \mathbf{F})}{\partial \boldsymbol{\varphi}} - \left(\nabla \cdot \frac{\partial \bar{\mathcal{L}}}{\partial \mathbf{F}}\right), \quad \text{for } (\mathbf{X}, t) \in \mathcal{B} \times \mathcal{T} \tag{11.133}$$

subject to the second of the Lagrangian symbolic equations as the traction boundary condition

$$\bar{\mathbf{T}}(\mathbf{X}, \hat{\mathbf{N}}) = -\frac{\partial \bar{\mathcal{L}}(\boldsymbol{\varphi}, \dot{\boldsymbol{\varphi}}, \mathbf{F})}{\partial \mathbf{F}} \cdot \hat{\mathbf{N}} \quad \text{for } (\mathbf{X}, t) \in \partial\mathcal{B}_\sigma \times \mathcal{T} \tag{11.134}$$

The scalar formalism of the Lagrangian density equations of motion are identical to the vector formalism of Cauchy's equations of motion, and the traction boundary condition is identical to Cauchy's law. At this juncture, we place an emphasis on the fact that the theorem of expended power, or equivalently, the principle of balance of mechanical energy, results in the Bubnov-Galerkin weighted residual form. Strictly speaking, Equation (11.132) is different from the theorem of expended power in that Equation (11.132) includes the natural boundary conditions as well as Cauchy's equations of motion. Equation (11.133) represents the scalar-termed governing equations of motion invariant to coordinate transformation within the Lagrangian mechanics framework. The scalar here is the Lagrangian density function.

In view of Equation (11.129) and Equation (11.131), Equation (11.128) leads to the weak form of the Lagrangian density equations of motion in the sense of the space:

$$\frac{d(\mathcal{E})}{dt} = \frac{d(2\mathcal{K} - \mathcal{L})}{dt} =$$
$$\int_{\mathcal{B}} \frac{d}{dt}\left(\frac{\partial \bar{\mathcal{L}}}{\partial \dot{\boldsymbol{\varphi}}}\right) \cdot \dot{\boldsymbol{\varphi}}\, dV - \int_{\mathcal{B}} \frac{\partial \bar{\mathcal{L}}}{\partial \mathbf{F}} : \dot{\mathbf{F}}\, dV - \int_{\mathcal{B}} \frac{\partial \bar{\mathcal{L}}}{\partial \boldsymbol{\varphi}} \cdot \dot{\boldsymbol{\varphi}}\, dV - \int_{\partial\mathcal{B}_\sigma} \bar{\mathbf{T}} \cdot \dot{\boldsymbol{\varphi}}\, dA = 0 \tag{11.135}$$

The above equation can be obtained when the virtual displacement in the principle of virtual work is replaced by the generalized velocity.

11.3.2 Space-Discrete Lagrangian Finite Element Formulation

Similar to that presented previously for the Total Energy framework, we again highlight two distinct approaches to space discretization, namely: (1) traditional approach using as a starting point, the principle of balance of mechanical energy, or equivalently, the theorem of power expended for mechanical systems within the Lagrangian mechanics framework via the Lagrangian density based representation of the equations of motion, and (2) alternatively, we then describe a simple numerical discretization approach that is significantly different which directly employs the Lagrangian differential operator leading to the semi-discretized equations of motion. We have demonstrated that Equation (11.132) resulting from the principle of balance of mechanical energy, or equivalently, the theorem of power expended for mechanical systems within the Lagrangian mechanics framework contains implicitly both governing equations and natural boundary conditions in the Bubnov-Galerkin weighted residual form. In this sense, therefore, Equation (11.132) serves as a mechanism by which the weighted averages of the residuals are required to vanish, when substituting the trial functions satisfying only essential boundary conditions into Equation (11.132) to discretize the space. We assume that the entire domain of interest can be broken up into sub domains, called elements, and the weak form in Equation (11.135) is employed to break up the spatial domain. Then admissible trial functions $\tilde{\varphi}$ are required to be piece-wise continuous, but they should be continuous along the inter-element boundaries.

The admissible trial functions may be represented by a linear combination of coordinate functions $\mathbf{N}_i(\mathbf{X})$ and displacement vectors $\mathbf{U}_i^e(t)$ as given by Equation (9.11); hence the trial functions, which are of class $C^{m-1}(\mathcal{B})$ over the entire domain, belong to the subspace, denoted by $\mathcal{C}^h \subset \mathcal{C}$, as shown in Equation (9.14).

Approach 1: Weak Form In the Lagrangian mechanics framework, the weak statement given in Equation (11.135) is utilized to discretize the space. Hence, substituting Equation (9.11) into Equation (11.135) yields the following

$$\sum_{e=1}^{n_{elem}} \left\{ \begin{array}{l} \displaystyle\int_{\mathcal{B}_e} \frac{d}{dt}\left(\frac{\partial \bar{\mathcal{L}}^h_{\mathcal{B}_e}}{\partial \dot{\tilde{\varphi}}^h}\right) \cdot \dot{\varphi}^h dV - \int_{\mathcal{B}_e} \frac{\partial \bar{\mathcal{L}}^h_{\mathcal{B}_e}}{\partial \tilde{\mathbf{F}}^h} : \dot{\mathbf{F}}^h dV \\ \displaystyle - \int_{\mathcal{B}_e} \frac{\partial \bar{\mathcal{L}}^h_{\mathcal{B}_e}}{\partial \tilde{\varphi}^h} \cdot \dot{\varphi}^h dV - \int_{\partial \mathcal{B}_{e\sigma}} \bar{\mathbf{T}} \cdot \dot{\varphi}^h dA \end{array} \right\} = 0 \tag{11.136}$$

where

$$\bar{\mathcal{L}}^h_{\mathcal{B}e}\left(\tilde{\varphi}^h, \dot{\tilde{\varphi}}^h, \mathbf{F}\right) = \frac{1}{2}\rho_0 \dot{\tilde{\varphi}}^h \cdot \dot{\tilde{\varphi}}^h - \rho_0 \Psi\left(\tilde{\mathbf{F}}^h\right) + \rho_0 \mathbf{B} \cdot \tilde{\varphi}^h \tag{11.137}$$

As a consequence, Equation (11.136) yields

$$\sum_{e=1}^{n_{elem}} \left(\mathbf{M}_e \ddot{\mathbf{U}}^e(t) + \mathbf{F}^e_{\text{int}} - \mathbf{F}^e_{\text{ext}}\right) \cdot \dot{\mathbf{U}}^e(t) = 0 \tag{11.138}$$

where $\mathbf{M}_e$, $\mathbf{F}^e_{\text{int}}$ and $\mathbf{F}^e_{\text{ext}}$ are given by Equation (9.17).

For the entire domain of interest, we introduce the configuration space with generalized coordinates $\mathcal{Q}$, the tangent space $T_q Q$ and the tangent bundle TQ as shown in Equations (10.48), (10.49) and (10.50). Then, Equation (11.138) yields the semi-discrete second-order differential equations, i.e. the finite element equations of motion to be discretized subsequently in time as follows:

$$\mathbf{M}\ddot{\mathbf{q}} + \mathbf{F}_{\text{int}} - \mathbf{F}_{\text{ext}} = \mathbf{0} \tag{11.139}$$

Approach 2: Lagrangian Differential Operator An alternative via the space discrete finite element formulation is highlighted next. By use of Equation (9.11), the rate of change of Jacobi's integral given in Equation (11.124) within the eth element domain $\mathcal{B}_e$ can be discretized in space as follows:

$$\text{PBME}^h_{\mathcal{L}_{\mathcal{B}_e}} : \frac{d}{dt}\left\{\int_{\mathcal{B}_e} \rho_0 \dot{\tilde{\varphi}}^h \cdot \dot{\tilde{\varphi}}^h dV - \mathcal{L}^h_{\mathcal{B}_e}\left(\tilde{\varphi}^h, \dot{\tilde{\varphi}}^h\right)\right\} = 0 \tag{11.140}$$

where the autonomous space-discrete Lagrangian for the finite dimensional systems is given by

$$\mathcal{L}^h_{\mathcal{B}_e}\left(\tilde{\varphi}^h, \dot{\tilde{\varphi}}^h\right) = \int_{\mathcal{B}_e} \bar{\mathcal{L}}^h_{\mathcal{B}_e}\left(\tilde{\varphi}^h, \dot{\tilde{\varphi}}^h, \tilde{\mathbf{F}}^h\right) dV + \int_{\partial \mathcal{B}_\sigma} \bar{\mathbf{T}} \cdot \dot{\tilde{\varphi}}^h(\mathbf{X}, t)\, dA \tag{11.141}$$

Recall that Equation (9.11) can be rewritten as Equation (11.47) Hence, in terms of generalized coordinates Equation (11.140) can be rewritten as

$$\begin{aligned}
\mathrm{PBME}^h_{\mathcal{L}_{\mathcal{B}_e}} : \; & \frac{d}{dt}\left\{\int_{\mathcal{B}_e} \rho_0 \dot{\tilde{\boldsymbol{\varphi}}}^h \cdot \dot{\tilde{\boldsymbol{\varphi}}}^h \, dV - \mathcal{L}^h_{\mathcal{B}_e}\left(\tilde{\boldsymbol{\varphi}}^h, \dot{\tilde{\boldsymbol{\varphi}}}^h\right)\right\} \\
& = \left\{\frac{d}{dt}\left(\frac{\partial \mathcal{L}^h_{\mathcal{B}_e}(\mathbf{q}_e, \dot{\mathbf{q}}_e)}{\partial \dot{\mathbf{q}}_e}\right) - \frac{\partial \mathcal{L}^h_{\mathcal{B}_e}(\mathbf{q}_e, \dot{\mathbf{q}}_e)}{\partial \mathbf{q}_e}\right\} \cdot \dot{\mathbf{q}}_e(t) = 0
\end{aligned} \tag{11.142}$$

where the space-discrete autonomous Lagrangian is constructed as follows:

$$\begin{aligned}
& \mathcal{L}^h_{\mathcal{B}_e}(\mathbf{q}_e, \dot{\mathbf{q}}_e) = \frac{1}{2}\dot{\mathbf{q}}_e \cdot \mathbf{M}^e \dot{\mathbf{q}}_e - \Pi^h_{\text{inte}} + \mathbf{q}_e \cdot \mathbf{F}^e_{\text{ext}}, \; \Pi^h_{\text{inte}} = \int_{\mathcal{B}_e} \rho_0 \Psi\left(\tilde{\mathbf{F}}^h\right) dV \\
& \mathbf{M}^e = \int_{\mathcal{B}_e} \rho_0 \mathbf{N}_e^T \mathbf{N}_e \, dV, \; \frac{\partial \mathcal{L}^h_{\mathcal{B}_e}(\mathbf{q}_e, \dot{\mathbf{q}}_e)}{\partial \mathbf{q}_e} = \mathbf{F}^e_{\text{ext}} - \mathbf{F}^e_{\text{int}} \\
& \mathbf{F}^e_{\text{int}} = \int_{\mathcal{B}_e} \bar{\mathbf{B}}_e^T \tilde{\mathbf{P}}^h \, dV, \; \mathbf{F}^e_{\text{ext}} = \int_{\mathcal{B}_e} \rho_0 \mathbf{N}_e^T \mathbf{B} \, dV + \int_{\partial \mathcal{B}_{e\sigma}} \mathbf{N}_e^T \bar{\mathbf{T}} \, dA
\end{aligned} \tag{11.143}$$

For the entire domain, a configuration space with generalized coordinates $\mathcal{Q}$, is defined in Equation (10.48). Hence, Equation (11.140) can be written as

$$\mathrm{PBME}^h_{\mathcal{L}} : \left\{\frac{d}{dt}\left(\frac{\partial \mathcal{L}^h(\mathbf{q}, \dot{\mathbf{q}})}{\partial \dot{\mathbf{q}}}\right) - \frac{\partial \mathcal{L}^h(\mathbf{q}, \dot{\mathbf{q}})}{\partial \mathbf{q}}\right\} \cdot \dot{\mathbf{q}}(t) = 0 \tag{11.144}$$

where the dependent state variable is the space-discrete scalar function, namely,

$$\begin{aligned}
& \mathcal{L}^h(\mathbf{q}, \dot{\mathbf{q}}) = \frac{1}{2}\dot{\mathbf{q}} \cdot \mathbf{M}\dot{\mathbf{q}} - \Pi^h_{\text{int}} + \mathbf{q} \cdot \mathbf{F}_{\text{ext}}, \; \Pi^h_{\text{int}} = \int_{\mathcal{B}} \rho_0 \Psi\left(\tilde{\mathbf{F}}^h\right) dV \\
& \mathbf{M} = \int_{\mathcal{B}} \rho_0 \mathbf{N}^T \mathbf{N} dV, \; \frac{\partial \mathcal{L}^h(\mathbf{q}, \dot{\mathbf{q}})}{\partial \mathbf{q}} = \mathbf{F}_{\text{ext}} - \mathbf{F}_{\text{int}} \\
& \mathbf{F}_{\text{int}} = \int_{\mathcal{B}} \bar{\mathbf{B}}^T \tilde{\mathbf{P}}^h \, dV, \; \mathbf{F}_{\text{ext}} = \int_{\mathcal{B}} \rho_0 \mathbf{N}^T \mathbf{B} \, dV + \int_{\partial \mathcal{B}_\sigma} \mathbf{N}^T \bar{\mathbf{T}} \, dA
\end{aligned} \tag{11.145}$$

In contrast to traditional practices, for discretizing in space, we simply adopt the following procedure: In view of Equation (11.144), we introduce the discrete Lagrangian differential operator $\mathfrak{L}_D$, by which the semi-discrete finite element equations of motion can be readily obtained after constructing the space-discrete Lagrangian which only entails the use of trial functions. Thus, we have, upon substituting the space-discrete Lagrangian into the space-discrete Lagrangian differential operator, the following:

$$\mathfrak{L}_D\left(\mathcal{L}^h\right) : \frac{d}{dt}\left(\frac{\partial \mathcal{L}^h}{\partial \dot{\mathbf{q}}}\right) - \frac{\partial \mathcal{L}^h}{\partial \mathbf{q}} = \mathbf{0}, \; \mathfrak{L}_D : \frac{d}{dt}\left(\frac{\partial}{\partial \dot{\mathbf{q}}}\right) - \frac{\partial}{\partial \mathbf{q}} = \mathbf{0} \tag{11.146}$$

where the dependent state variable is the space-discrete scalar function, namely, the space-discrete Lagrangian. Thus, we readily have the semi-discrete finite element equations of motion governing the mechanical dynamic behavior of the continuous-body within the Lagrangian mechanics framework as follows:

$$\mathbf{M}\ddot{\mathbf{q}} + \mathbf{F}_{\text{int}} - \mathbf{F}_{\text{ext}} = \mathbf{0} \tag{11.147}$$

Time Weighted Residual Form Taking the time integral over the given time interval $t \in \bar{\mathcal{T}} = [t_0, t_1]$ for both sides in Equation (11.144), we have

$$\int_{t_0}^{t_1} \{(\mathbf{M}\ddot{\mathbf{q}} + \mathbf{F}_{\text{int}} - \mathbf{F}_{\text{ext}}) \cdot \dot{\mathbf{q}}\}\, dt = 0 \tag{11.148}$$

Note that the above equation in the Lagrangian mechanics framework represents a single-field weighted-residual form in time. Hence, the above equation plays an important role for facilitating discretization in the time domain. Equation (11.148) may be regarded as an unconstrained formulation with one dependent field variable in the time domain. In general, the time weighted-residual method which includes time finite element formulations can be applied for deriving time stepping schemes Belytschko and Hughes (1983); Sha et al. (2003); Tamma et al. (2000, 2002, 2003); Wood (1990); Zhou and Tamma (2004a,c, 2006); Zhou et al. (2001, 2000); Zienkiewicz and Taylor (2006); Zienkiewicz et al. (2005, 1984).

Illustrative Examples

For developing the finite element formulations, similar concepts as those presented earlier via the Total Energy framework can be readily applied using instead the Lagrangian framework and the differential calculus setting to the various illustrative examples such as the Elastic Axial Bar, Bernouilli-Euler Beam, Rotating Bar, Timoshenko Beam, and the Reissner-Mindlin Plate. In this regard, the exercises at the end of the chapter will be helpful to demonstrate the various finite element formulations, and hence are not described here.

11.4 HAMILTONIAN MECHANICS FRAMEWORK IN THE DIFFERENTIAL CALCULUS SETTING AND FINITE ELEMENT FORMULATIONS

Lastly, we turn attention to the Hamiltonian mechanics framework in the context of the principle of balance of mechanical energy. The scalar formalism is briefly highlighted in this section, and also follows the same ideas as described previously for the Total Energy framework and the Lagrangian framework.

11.4.1 Hamiltonian Density Equations of Motion

Scalar Formalism: Strong and Weak Forms in the Hamiltonian Mechanics Framework In an autonomous Hamiltonian system, the canonical momentum, which belongs to the cotangent space $\boldsymbol{\phi} \in T_{\varphi}^{*}\mathcal{C}$ is represented as:

$$\boldsymbol{\phi} = \frac{\partial \bar{\mathcal{L}}(\boldsymbol{\varphi}, \dot{\boldsymbol{\varphi}}, \mathbf{F})}{\partial \dot{\boldsymbol{\varphi}}} = \rho_0 \dot{\boldsymbol{\varphi}}(\mathbf{X}, t) \tag{11.149}$$

where the Lagrangian density function is given by Equation (10.30). Via the Legendre transformation we may define the Hamiltonian density function

$$\bar{\mathcal{H}}(\boldsymbol{\phi}, \boldsymbol{\varphi}, \mathbf{F}) = \boldsymbol{\phi} \cdot \dot{\boldsymbol{\varphi}} - \bar{\mathcal{L}}(\boldsymbol{\varphi}, \dot{\boldsymbol{\varphi}}, \mathbf{F}) \tag{11.150}$$

Recall that the Hamiltonian density function is given by Equation (10.203). With the definition of the cotangent bundle (phase space), the autonomous Hamiltonian $\mathcal{H}(\boldsymbol{\phi}, \boldsymbol{\varphi}) : T^{*}\mathcal{C} \to \mathbb{R}$, whose domain is an infinite dimensional space, can be defined as

$$\mathcal{H}(\boldsymbol{\phi}, \boldsymbol{\varphi}) = \int_{\mathcal{B}} \bar{\mathcal{H}}(\boldsymbol{\phi}, \boldsymbol{\varphi}, \mathbf{F})\, dV - \int_{\partial \mathcal{B}_\sigma} \bar{\mathbf{T}}(\mathbf{X}, \hat{\mathbf{N}}) \cdot \boldsymbol{\varphi}(X, t)\, dA \tag{11.151}$$

where the external forces are dead loads. Then, the total time derivative of the autonomous Hamiltonian can be written as

$$\frac{d\mathcal{H}(\boldsymbol{\phi},\boldsymbol{\varphi})}{dt} = \int_{\mathcal{B}} \frac{d\bar{\mathcal{H}}(\boldsymbol{\phi},\boldsymbol{\varphi},\mathbf{F})}{dt} dV - \int_{\partial\mathcal{B}_\sigma} \bar{\mathbf{T}}\left(\mathbf{X},\hat{\mathbf{N}}\right)\cdot\dot{\boldsymbol{\varphi}}(\mathbf{X},t)\,dA \tag{11.152}$$

By use of the chain rule, we have

$$\frac{d\bar{\mathcal{H}}(\boldsymbol{\phi},\boldsymbol{\varphi},\mathbf{F})}{dt} = \frac{\partial\bar{\mathcal{H}}}{\partial\boldsymbol{\phi}}\cdot\frac{d\boldsymbol{\phi}}{dt} + \frac{\partial\bar{\mathcal{H}}}{\partial\boldsymbol{\varphi}}\cdot\frac{d\boldsymbol{\varphi}}{dt} + \frac{\partial\bar{\mathcal{H}}}{\partial\mathbf{F}}:\dot{\mathbf{F}} \tag{11.153}$$

The above equation can be rewritten as

$$\frac{d\bar{\mathcal{H}}(\boldsymbol{\phi},\boldsymbol{\varphi},\mathbf{F})}{dt} = \left(\frac{\partial\bar{\mathcal{H}}}{\partial\boldsymbol{\varphi}} + \dot{\boldsymbol{\phi}}\right)\cdot\frac{d\boldsymbol{\varphi}}{dt} + \left(\frac{\partial\bar{\mathcal{H}}}{\partial\boldsymbol{\phi}} - \dot{\boldsymbol{\varphi}}\right)\cdot\frac{d\boldsymbol{\phi}}{dt} + \frac{\partial\bar{\mathcal{H}}}{\partial\mathbf{F}}:\dot{\mathbf{F}} \tag{11.154}$$

Recall the following relation

$$\frac{\partial\bar{\mathcal{H}}}{\partial\mathbf{F}}:\dot{\mathbf{F}} = \frac{\partial\bar{\mathcal{H}}}{\partial\mathbf{F}}:\mathrm{GRAD}\,\dot{\boldsymbol{\varphi}} = \mathrm{DIV}\left(\dot{\boldsymbol{\varphi}}\cdot\frac{\partial\bar{\mathcal{H}}}{\partial\mathbf{F}}\right) - \left(\mathrm{DIV}\,\frac{\partial\bar{\mathcal{H}}}{\partial\mathbf{F}}\right)\cdot\dot{\boldsymbol{\varphi}} \tag{11.155}$$

Due to Gauss's theorem, the last term on the right hand side of Equation (11.155) follows the relationship:

$$\int_{\mathcal{B}} \mathrm{DIV}\left(\dot{\boldsymbol{\varphi}}\cdot\frac{\partial\bar{\mathcal{H}}}{\partial\mathbf{F}}\right) dV = \int_{\partial\mathcal{B}_\sigma} \dot{\boldsymbol{\varphi}}\cdot\frac{\partial\bar{\mathcal{H}}}{\partial\mathbf{F}}\cdot\hat{\mathbf{N}}\,dA \tag{11.156}$$

In view of Equation (11.154) through Equation (11.156), Equation (11.152) leads to

$$\frac{d\mathcal{H}}{dt} = \int_{\mathcal{B}} \left(\left(\dot{\boldsymbol{\phi}} + \frac{\partial\bar{\mathcal{H}}}{\partial\boldsymbol{\varphi}} - \mathrm{DIV}\,\frac{\partial\bar{\mathcal{H}}}{\partial\mathbf{F}}\right)\cdot\frac{d\boldsymbol{\varphi}}{dt} + \left(\frac{\partial\bar{\mathcal{H}}}{\partial\boldsymbol{\phi}} - \dot{\boldsymbol{\varphi}}\right)\cdot\frac{d\boldsymbol{\phi}}{dt}\right) dV$$

$$- \int_{\partial\mathcal{B}_\sigma} \left(\bar{\mathbf{T}}(\mathbf{X},\hat{\mathbf{N}}) - \frac{\partial\bar{\mathcal{H}}}{\partial\mathbf{F}}\cdot\hat{\mathbf{N}}\right)\cdot\dot{\boldsymbol{\varphi}}(\mathbf{X},t)\,dA = 0 \tag{11.157}$$

Firstly, we readily observe that the rate of the autonomous Hamiltonian contains the strong form; namely, the Hamiltonian version of Cauchy's equations of motion and the traction boundary conditions as follows:

$$\dot{\boldsymbol{\phi}} = -\frac{\partial\bar{\mathcal{H}}}{\partial\boldsymbol{\varphi}} + \mathrm{DIV}\,\frac{\partial\bar{\mathcal{H}}}{\partial\mathbf{F}}, \quad \dot{\boldsymbol{\varphi}} = \frac{\partial\bar{\mathcal{H}}}{\partial\boldsymbol{\phi}}, \quad \text{for } (\mathbf{X},t)\in\mathcal{B}\times\mathcal{T} \tag{11.158}$$

subject to the traction boundary condition

$$\bar{\mathbf{T}}\left(\mathbf{X},\hat{\mathbf{N}}\right) = \frac{\partial\bar{\mathcal{H}}}{\partial\mathbf{F}}\cdot\hat{\mathbf{N}}, \quad \text{for } (\mathbf{X},t)\in\partial\mathcal{B}_\sigma\times\mathcal{T} \tag{11.159}$$

It is to be noted that the rate of the autonomous Hamiltonian results in a two-field weighted residual form. Notice that Equation (11.157) is different from the theorem of power expended in that Equation (11.157) includes the natural boundary conditions as well as Cauchy's equations of motion. Equation (11.158) represents the scalar-termed governing equations of motion invariant to coordinate transformation within the Hamiltonian mechanics framework. The scalar here is the Hamiltonian density function. By use of

Equation (11.155), we obtain the weak form in the sense of the space:

$$\int_{\mathcal{B}} \dot{\boldsymbol{\phi}} \cdot \dot{\boldsymbol{\varphi}}\, dV + \int_{\mathcal{B}} \frac{\partial \bar{\mathcal{H}}}{\partial \mathbf{F}} : \dot{\mathbf{F}}\, dV + \int_{\mathcal{B}} \frac{\partial \bar{\mathcal{H}}}{\partial \boldsymbol{\varphi}} \cdot \dot{\boldsymbol{\varphi}}\, dV - \int_{\partial \mathcal{B}_\sigma} \bar{\mathbf{T}} \cdot \dot{\boldsymbol{\varphi}}\, dA = 0,$$
$$\int_{\mathcal{B}} \left(\dot{\boldsymbol{\varphi}} - \frac{\partial \bar{\mathcal{H}}}{\partial \boldsymbol{\phi}} \right) \cdot \dot{\boldsymbol{\phi}}\, dV = 0 \tag{11.160}$$

11.4.2 Space-Discrete Hamiltonian Finite Element Formulation

As presented previously, we again highlight two distinct approaches to space discretization, namely: (1) the traditional approach using as a starting point the principle of balance of mechanical energy, or equivalently, the theorem of power expended for mechanical systems within the Hamiltonian mechanics framework via the Hamiltonian density representation of the equations of motion, and (2) alternatively, we then describe a simple numerical discretization approach that is significantly different, namely, the space-discrete Hamiltonian finite element formulation which directly employs the Hamiltonian differential operator. We have discussed that Equation (11.157) resulting from the principle of balance of mechanical energy, or equivalently, the theorem of power expended contains implicitly both governing equations and natural boundary conditions in the Bubnov-Galerkin weighted residual form. In this sense, therefore, the theorem of power expended or equivalently, the principle of balance of mechanical energy serves as a mechanism by which the weighted averages of the residuals are required to vanish, when substituting the trial functions into Equation (11.157) to discretize the space. It is supposed that the entire domain having a complex geometry can be broken up into sub domains, called elements (Hughes 1987; Reddy 2006), and the weak form is employed to break up the spatial domain. Then admissible trial functions $\tilde{\boldsymbol{\phi}}$ and $\tilde{\boldsymbol{\varphi}}$ are required to be piece-wise continuous, but they should be continuous along the inter-element boundaries. They may be represented by a combination of interpolation functions and nodal variables as defined in Equations (10.218) and (10.219).

Approach 1: Weak Form The weak form given in Equation (11.160) can be discretized by using the trial functions in Equation (10.218) as follows:

$$\sum_{i=1}^{n_{\text{elem}}} \left(\begin{array}{l} \int_{\mathcal{B}e} \dot{\tilde{\boldsymbol{\phi}}}^h \cdot \dot{\tilde{\boldsymbol{\varphi}}}^h dV + \int_{\mathcal{B}e} \left(\dfrac{\partial \bar{\mathcal{H}}^h_{\mathcal{B}_e}}{\partial \tilde{\mathbf{F}}^h} \right) : \dot{\tilde{\mathbf{F}}}^h dV \\ + \int_{\mathcal{B}_e} \left(\dfrac{\partial \bar{\mathcal{H}}^h_{\mathcal{B}_e}}{\partial \tilde{\boldsymbol{\varphi}}^h} \right)^h \cdot \dot{\tilde{\boldsymbol{\varphi}}}^h dV - \int_{\partial \mathcal{B}_{e\sigma}} \bar{\mathbf{T}} \cdot \dot{\tilde{\boldsymbol{\varphi}}}^h dA \end{array} \right) = 0,$$
$$\sum_{e=1}^{n_{\text{elem}}} \left(\int_{\mathcal{B}_e} \dot{\tilde{\boldsymbol{\varphi}}}^h \cdot \dot{\tilde{\boldsymbol{\phi}}}^h dV - \int_{\mathcal{B}_e} \left(\frac{\partial \bar{\mathcal{H}}^h_{\mathcal{B}_e}}{\partial \tilde{\boldsymbol{\phi}}^h} \right) \cdot \dot{\tilde{\boldsymbol{\phi}}}^h dV \right) = 0 \tag{11.161}$$

where

$$\bar{\mathcal{H}}^h_{\mathcal{B}_e} \left(\tilde{\boldsymbol{\varphi}}^h, \tilde{\boldsymbol{\phi}}^h, \tilde{\mathbf{F}}^h \right) = \frac{1}{2\rho_0} \tilde{\boldsymbol{\phi}}^h \cdot \tilde{\boldsymbol{\phi}}^h + \rho_0 \Psi \left(\tilde{\mathbf{F}}^h \right) - \rho_0 \mathbf{B} \cdot \tilde{\boldsymbol{\varphi}}^h \tag{11.162}$$

As a consequence, the above equation in Equation (11.161) can be rewritten as

$$\sum_{i=1}^{n_{\text{elem}}} \left(\frac{1}{\rho_0} \mathbf{M}_e \cdot \dot{\tilde{\boldsymbol{\phi}}}^e (t) + \mathbf{F}^{int}_e - \mathbf{F}^{ext}_e \right) \cdot \dot{\mathbf{U}}^e (t) = 0,$$
$$\sum_{e=1}^{n_{\text{elem}}} \frac{1}{\rho_0} \left(\int_{\mathcal{B}_e} \mathbf{M}_e \cdot \dot{\mathbf{U}}^e (t) - \frac{1}{\rho_0} \mathbf{M}_e \cdot \tilde{\boldsymbol{\phi}}^e (t) \right) \cdot \dot{\tilde{\boldsymbol{\phi}}}^e (t) = 0 \tag{11.163}$$

where $\mathbf{M}_e$, $\mathbf{F}^e_{\text{int}}$ and $\mathbf{F}^e_{\text{ext}}$ are given by Equation (9.17).

From Equation (11.163), since $\dot{\tilde{\mathbf{U}}}^e(t)$ and $\dot{\tilde{\boldsymbol{\phi}}}^e(t)$ are linearly independent, we obtain the semi-discrete first-order differential equations for the Hamiltonian mechanics framework.

For the entire domain of interest, we introduce the configuration space with generalized coordinates $\mathcal{Q}$, the cotangent space T_q^*Q and the cotangent bundle, namely, the phase space. Finally, Equation (11.163) yields the semi-discrete first-order differential equations

$$\dot{\mathbf{p}} = \mathbf{F}_{\text{ext}} - \mathbf{F}_{\text{int}}, \quad \dot{\mathbf{q}} = \mathbf{M}^{-1}\mathbf{p} \tag{11.164}$$

which correspond to Hamilton's equations of motion.

Approach 2: Hamiltonian Differential Operator Upon substitution of Equation (10.218) into Equation (11.152), the rate of the Hamiltonian given in Equation (11.152) over the eth element domain $\mathcal{B}_e$ can be discretized in space as described next. Substituting Equation (10.218) into Equation (11.152), the rate of the Hamiltonian can be discretized in space as follows:

$$\text{PBME}^h_{\mathcal{H}\mathcal{B}_e} : \frac{d\mathcal{H}^h_{\mathcal{B}_e}\left(\tilde{\varphi}^h, \tilde{\varphi}^h\right)}{dt} = 0 \tag{11.165}$$

where the autonomous space-discrete Hamiltonian for the eth element domain is simply constructed as

$$\mathcal{H}^h_{\mathcal{B}_e}\left(\tilde{\boldsymbol{\phi}}^h, \tilde{\varphi}^h\right) = \int_{\mathcal{B}_e} \bar{\mathcal{H}}^h_{\mathcal{B}_e}\left(\tilde{\boldsymbol{\phi}}^h, \tilde{\varphi}^h, \mathbf{F}^h\right) dV - \int_{\partial\mathcal{B}_{e\sigma}} \bar{\mathbf{T}} \cdot \tilde{\varphi}^h(\mathbf{X}, t)\, dA \tag{11.166}$$

Next, Equation (10.219) can be rewritten as Equation (10.231). Therefore, in terms of generalized coordinates and canonical momenta, Equation (11.165) can be rewritten as

$$\text{PBME}^h_{\mathcal{H}\mathcal{B}_e} : \frac{d\mathcal{H}^h_{\mathcal{B}_e}(\mathbf{p}_e, \mathbf{q}_e)}{dt} = \left(\dot{\mathbf{p}}_e + \frac{\partial\mathcal{H}^h_{\mathcal{B}_e}}{\partial\mathbf{q}_e}\right) \cdot \dot{\mathbf{q}}_e - \left(\dot{\mathbf{q}}_e - \frac{\partial\mathcal{H}^h_{\mathcal{B}_e}}{\partial\mathbf{p}_e}\right) \cdot \dot{\mathbf{p}}_e = 0 \tag{11.167}$$

where the canonical momentum $\mathbf{p}_e$ is given by Equation (10.234). Note that the space-discrete Hamiltonian in Equation (11.167) is given by Equation (10.235).

For the entire domain, a phase space with canonical coordinates, namely the cotangent bundle, is defined as shown in Equation (10.227). Consequently, the rate of the Hamiltonian in a finite dimensional system is given by

$$\text{PBME}^h_{\mathcal{H}} : \frac{d\mathcal{H}^h(\mathbf{p}, \mathbf{q})}{dt} = \left(\dot{\mathbf{p}} + \frac{\partial\mathcal{H}^h}{\partial\mathbf{q}}\right) \cdot \dot{\mathbf{q}} - \left(\dot{\mathbf{q}} - \frac{\partial\mathcal{H}^h}{\partial\mathbf{p}}\right) \cdot \dot{\mathbf{p}} = 0 \tag{11.168}$$

where the space-discrete Hamiltonian consisting of the space-discrete kinetic and potential energies is constructed as Equation (10.237).

In contrast to traditional practices, for discretizing in space, we simply adopt the following procedure. In view of Equation (11.167), we introduce the discrete Hamiltonian differential operators $\mathfrak{H}_{D1}$ and $\mathfrak{H}_{D2}$, by which the space discrete finite element equation of motion can be readily obtained after constructing the space-discrete Hamiltonian which only entails the use of trial functions. Thus, we have, upon substituting the space-discrete Hamiltonian into the space-discrete Hamiltonian differential operators, the following:

$$\mathfrak{H}_{D1}\left(\mathcal{H}^h\right) : \dot{\mathbf{p}} + \tfrac{\partial\mathcal{H}^h}{\partial\mathbf{q}} = \mathbf{0}, \quad \mathfrak{H}_{D2}\left(\mathcal{H}^h\right) : \dot{\mathbf{q}} - \tfrac{\partial\mathcal{H}^h}{\partial\mathbf{p}} = \mathbf{0};$$

$$\mathfrak{H}_{D1} : \dot{\mathbf{p}} + \tfrac{\partial}{\partial\mathbf{q}} = \mathbf{0}, \quad \mathfrak{H}_{D2} : \dot{\mathbf{q}} - \tfrac{\partial}{\partial\mathbf{p}} = \mathbf{0} \tag{11.169}$$

where the dependent state variable is the space-discrete scalar function, namely, the space-discrete autonomous Hamiltonian. Thus, we readily have the space-discrete finite element equation of motion governing the mechanical dynamic behavior of the continuous-body within the Hamiltonian mechanics framework as follows:

$$\dot{\mathbf{p}} + \mathbf{F}_{\text{int}} - \mathbf{F}_{\text{ext}} = \mathbf{0} \tag{11.170}$$

which is subject to a subsidiary velocity condition:

$$\dot{\mathbf{q}} - \mathbf{M}^{-1}\mathbf{p} = \mathbf{0} \tag{11.171}$$

Time Weighted Residual Form Taking the time integral over the given time interval $t \in \bar{\mathcal{T}} = [t_0, t_1]$ for both sides in Equation (11.167), we have

$$\mathcal{H}^h(\mathbf{p}, \mathbf{q})\big|_{t=t_1} - \mathcal{H}^h(\mathbf{p}, \mathbf{q})\big|_{t=t_0} = \int_{t_0}^{t_1} \left\{(\dot{\mathbf{p}} + \mathbf{F}_{\text{int}} - \mathbf{F}_{\text{ext}}) \cdot \dot{\mathbf{q}} - \left(\dot{\mathbf{q}} - \mathbf{M}^{-1}\mathbf{p}\right) \cdot \dot{\mathbf{p}}\right\} dt = 0 \tag{11.172}$$

The above equation may be regarded as an unconstrained variational formulation with two dependent field variables in the time domain; thus we have the two-field problem within the Hamiltonian mechanics framework. Note that Equation (11.172) in the Hamiltonian mechanics framework represents the two-field weighted-residual form in time. Hence, Equation (11.172) plays an important role for facilitating discretization in the time domain. In general, the time weighted-residual method which includes time finite element formulations can be applied for deriving time stepping schemes (Belytschko and Hughes 1983; Sha et al. 2003; Tamma et al. 2000, 2002, 2003; Wood 1990; Zhou and Tamma 2004a,c, 2006; Zhou et al. 2001, 2000; Zienkiewicz and Taylor 2006; Zienkiewicz et al. 2005, 1984).

Illustrative Examples

> Employing the Hamiltonian framework and differential calculus setting, similar concepts as those presented previously via the Total Energy framework and the Lagrangian framework can be readily applied to the various illustrative examples such as the Elastic Axial Bar, Bernoulli-Euler Beam, Rotating Bar, Timoshenko Beam, and the Reissner-Mindlin Plate. In this regard, the exercises at the end of the chapter will be helpful to demonstrate the various finite element formulations, and hence are not described here.

11.5 APPENDIX: TOTAL ENERGY REPRESENTATIONS AND FRAMEWORK IN THE VARIATIONAL CALCULUS SETTING AND FINITE ELEMENT FORMULATIONS

Unlike the differential calculus setting which is valid for holonomic-scleronomic systems described previously, we next turn attention to the Total Energy representations and framework, and the setting of variational calculus. Note that in general, the variational setting is also valid for holonomic systems (not shown here; see Appendix in Chapter 6 for N-body systems). Next, for illustration, we primarily focus attention on the descriptive scalar function, namely, the Total Energy (and analogous formulations also hold for the other descriptive functions such as the Lagrangian and the Hamiltonian).

It is to be pointed out that due to the inconsistencies associated with Hamilton's principle, we restrict attention to the alternative formulation, namely, Hamilton's law of varying action in the variational calculus setting to describe the various developments. Consequently, starting from and selecting

Hamilton's law of varying action in the context of a descriptive scalar function, namely, the Total Energy as a starting point: (1) Firstly, we explain how to derive the scalar form of the Total Energy density representation of the continuous mathematical governing equations of motion in space and time, (2) Secondly, whilst the Bubnov-Galerkin weighted residual form leads to the principle of virtual work, which is strictly a statement of the weak form in space only, we demonstrate that Hamilton's law of varying action involving the Total Energy density represents consequently the weak form in space and the weighted residual for in time; for space, we also show that in itself it embodies the weak form in space that can be described by a discrete Total Energy Differential Operator, and (3) Thirdly, we demonstrate a simple, space-discrete Total Energy formulation that directly yields the semi-discrete ordinary differential equations in time which can be readily shown to preserve the same physical attributes as the continuous systems for continuous-body dynamical applications unlike traditional practices.

11.5.1 Infinite Dimensional Total Energy Structure

As stated previously, the Total Energy $\mathcal{E}(\boldsymbol{\varphi}, \dot{\boldsymbol{\varphi}}) : TC \rightarrow \mathbb{R}$ is defined as

$$\mathcal{E}(\boldsymbol{\varphi}, \dot{\boldsymbol{\varphi}}) = \mathcal{K}(\dot{\boldsymbol{\varphi}}) + \mathcal{U}(\boldsymbol{\varphi}) \tag{11.173}$$

which can be rewritten as

$$\begin{aligned}\mathcal{E}(\boldsymbol{\varphi}, \dot{\boldsymbol{\varphi}}) = & \frac{1}{2}\int_{\mathcal{B}} \rho_0 \dot{\boldsymbol{\varphi}}(\mathbf{X}, t) \cdot \dot{\boldsymbol{\varphi}}(\mathbf{X}, t)\, dV + \int_{\mathcal{B}} \rho_0 \Psi(\mathbf{F})\, dV \\ & - \int_{\partial \mathcal{B}_\sigma} \bar{\mathbf{T}}\left(\mathbf{X}, \hat{\mathbf{N}}\right) \cdot \boldsymbol{\varphi}(\mathbf{X}, t)\, dA - \int_{\mathcal{B}} \rho_0 \mathbf{B}(\mathbf{X}) \cdot \boldsymbol{\varphi}(\mathbf{X}, t)\, dV\end{aligned} \tag{11.174}$$

In the above, the traction and the body force are dead loads. By use of the Total Energy, we shall introduce a space-discrete Total Energy system via the Galerkin projection of the infinite dimensional Total Energy structure onto a finite dimensional discrete space.

11.5.2 Total Energy Density Representation of the Equation of Motion

Scalar Formalism: Strong and Weak Forms Via Hamilton's Law of Varying Action In this section, we describe the Total Energy framework and representations in the variational calculus setting. In an earlier chapter, we have mostly focused upon and described the Lagrangian and Hamiltonian frameworks in the variational calculus setting with Hamilton's principle and Hamilton's law of varying action, respectively. In a similar manner, here we restrict attention to Hamilton's law of varying action in conjunction with the Total Energy density to derive the scalar formalism of the governing Total Energy density based equations of motion. Naturally, these are equivalent to Cauchy's equations of motion in the vector formalism. Finally, we discuss the space-discrete Total Energy formulation in the variational calculus setting to develop the semi-discrete second-order differential finite element equations of motion.

We define the Total Energy density function by the following equation

$$\bar{\mathcal{E}}(\boldsymbol{\varphi}, \dot{\boldsymbol{\varphi}}, \mathbf{F}) = \frac{\partial \bar{\mathcal{L}}}{\partial \dot{\boldsymbol{\varphi}}} \cdot \dot{\boldsymbol{\varphi}} - \bar{\mathcal{L}}(\boldsymbol{\varphi}, \dot{\boldsymbol{\varphi}}, \mathbf{F}) \tag{11.175}$$

The above equation shows the relation between the Total Energy density function and the Lagrangian density function. Then, the Total Energy function can be rewritten as

$$\bar{\mathcal{E}}(\boldsymbol{\varphi}, \dot{\boldsymbol{\varphi}}, \mathbf{F}) = \frac{1}{2}\rho_0 \dot{\boldsymbol{\varphi}} \cdot \dot{\boldsymbol{\varphi}} + \rho_0 \Psi(\mathbf{F}) - \rho_0 \mathbf{B}(\mathbf{X}) \cdot \boldsymbol{\varphi}(\mathbf{X}, t) \tag{11.176}$$

The Total Energy $\mathcal{E}(\boldsymbol{\varphi}, \dot{\boldsymbol{\varphi}}) : T\mathcal{C} \rightarrow \mathbb{R}$ is given by

$$\mathcal{E}(\boldsymbol{\varphi}, \dot{\boldsymbol{\varphi}}) = \int_{\mathcal{B}} \bar{\mathcal{E}}(\boldsymbol{\varphi}, \dot{\boldsymbol{\varphi}}, \mathbf{F})\, dV - \int_{\partial \mathcal{B}_\sigma} \bar{\mathbf{T}}(\mathbf{X}, \hat{\mathbf{N}}) \cdot \boldsymbol{\varphi}(\mathbf{X}, t)\, dA \tag{11.177}$$

Therefore, in the Total Energy framework, Equation (10.12) can be represented as

$$\delta \left(\int_{\mathcal{B}} \frac{\partial \bar{\mathcal{E}}(\boldsymbol{\varphi}, \dot{\boldsymbol{\varphi}}, \mathbf{F})}{\partial \dot{\boldsymbol{\varphi}}} \cdot \dot{\boldsymbol{\varphi}}\, dV - \mathcal{E}(\boldsymbol{\varphi}, \dot{\boldsymbol{\varphi}}) \right) - \frac{d}{dt} \left(\int_{\mathcal{B}} \frac{\partial \bar{\mathcal{E}}}{\partial \dot{\boldsymbol{\varphi}}} \cdot \delta \boldsymbol{\varphi} dV \right) = 0 \tag{11.178}$$

Consequently, integrating both sides of Equation (11.178) over the closed time interval of interest $\bar{\mathcal{T}} = [t_0, t_1]$ yields

$$\delta \int_{t_0}^{t_1} \left\{ \int_{\mathcal{B}} \frac{\partial \bar{\mathcal{E}}(\boldsymbol{\varphi}, \dot{\boldsymbol{\varphi}}, \mathbf{F})}{\partial \dot{\boldsymbol{\varphi}}} \cdot \dot{\boldsymbol{\varphi}}\, dV - \mathcal{E}(\boldsymbol{\varphi}, \dot{\boldsymbol{\varphi}}) \right\} dt - \int_{\mathcal{B}} \frac{\partial \bar{\mathcal{E}}}{\partial \dot{\boldsymbol{\varphi}}} \cdot \delta \boldsymbol{\varphi} dV \Bigg|_{t_0}^{t_1} = 0 \tag{11.179}$$

which is here referred to as Total Energy version of Hamilton's law of varying action for continuous-body systems. The above equation leads to the following,

$$\int_{t_0}^{t_1} \left\{ \int_{\mathcal{B}} \delta \frac{\partial \bar{\mathcal{E}}}{\partial \dot{\boldsymbol{\varphi}}} \cdot \dot{\boldsymbol{\varphi}} + \frac{\partial \bar{\mathcal{E}}}{\partial \dot{\boldsymbol{\varphi}}} \cdot \delta \dot{\boldsymbol{\varphi}}\, dV - \delta \mathcal{E}(\boldsymbol{\varphi}, \dot{\boldsymbol{\varphi}}) \right\} dt - \int_{\mathcal{B}} \frac{\partial \bar{\mathcal{E}}}{\partial \dot{\boldsymbol{\varphi}}} \cdot \delta \boldsymbol{\varphi}\, dV \Bigg|_{t_0}^{t_1} = 0 \tag{11.180}$$

The first term on the left-hand side above can be written as

$$\int_{t_0}^{t_1} \left(\int_{\mathcal{B}} \delta \frac{\partial \bar{\mathcal{E}}}{\partial \dot{\boldsymbol{\varphi}}} \cdot \dot{\boldsymbol{\varphi}}\, dV \right) dt = \int_{\mathcal{B}} \frac{\partial \bar{\mathcal{E}}}{\partial \dot{\boldsymbol{\varphi}}} \cdot \delta \boldsymbol{\varphi}\, dV \Bigg|_{t_0}^{t_1} - \int_{t_0}^{t_1} \left(\int_{\mathcal{B}} \frac{d}{dt} \left(\frac{\partial \bar{\mathcal{E}}}{\partial \dot{\boldsymbol{\varphi}}} \right) \cdot \delta \boldsymbol{\varphi}\, dV \right) dt \tag{11.181}$$

because the kinetic energy is quadratic in generalized velocities, namely,

$$\begin{aligned} \int_{t_0}^{t_1} \left(\int_{\mathcal{B}} \delta \frac{\partial \bar{\mathcal{E}}}{\partial \dot{\boldsymbol{\varphi}}} \cdot \dot{\boldsymbol{\varphi}}\, dV \right) dt &= \int_{t_0}^{t_1} \left(\int_{\mathcal{B}} \rho_0 \delta \dot{\boldsymbol{\varphi}} \cdot \dot{\boldsymbol{\varphi}}\, dV \right) dt \\ &= \int_{t_0}^{t_1} \left(\int_{\mathcal{B}} \rho_0 \dot{\boldsymbol{\varphi}} \cdot \delta \dot{\boldsymbol{\varphi}}\, dV \right) dt = \int_{t_0}^{t_1} \left(\int_{\mathcal{B}} \frac{\partial \bar{\mathcal{E}}}{\partial \dot{\boldsymbol{\varphi}}} \cdot \delta \dot{\boldsymbol{\varphi}}\, dV \right) dt \end{aligned} \tag{11.182}$$

The variation of the Total Energy can be given by

$$\delta \mathcal{E}(\boldsymbol{\varphi}, \dot{\boldsymbol{\varphi}}) = \int_{\mathcal{B}} \delta \bar{\mathcal{E}}(\boldsymbol{\varphi}, \dot{\boldsymbol{\varphi}}, \mathbf{F})\, dV - \int_{\partial \mathcal{B}_\sigma} \bar{\mathbf{T}}(\mathbf{X}, \hat{\mathbf{N}}) \cdot \delta \boldsymbol{\varphi}(\mathbf{X}, t)\, dA \tag{11.183}$$

Owing to the chain rule, the variation of the Total Energy density function may be written as

$$\delta \bar{\mathcal{E}}(\boldsymbol{\varphi}, \dot{\boldsymbol{\varphi}}, \mathbf{F}) = \frac{\partial \bar{\mathcal{E}}}{\partial \boldsymbol{\varphi}} \cdot \delta \boldsymbol{\varphi} + \frac{\partial \bar{\mathcal{E}}}{\partial \dot{\boldsymbol{\varphi}}} \cdot \delta \dot{\boldsymbol{\varphi}} + \frac{\partial \bar{\mathcal{E}}}{\partial \mathbf{F}} : \delta \mathbf{F} \tag{11.184}$$

Recall that the second term on the right-hand side of the above equation can be expressed as

$$\frac{\partial \bar{\mathcal{E}}}{\partial \mathbf{F}} : \delta \mathbf{F} = \frac{\partial \bar{\mathcal{E}}}{\partial \mathbf{F}} : \mathrm{GRAD}\, \delta \boldsymbol{\varphi} = \mathrm{DIV} \left(\delta \boldsymbol{\varphi} \cdot \frac{\partial \bar{\mathcal{E}}}{\partial \mathbf{F}} \right) - \left(\mathrm{DIV} \frac{\partial \bar{\mathcal{E}}}{\partial \mathbf{F}} \right) \cdot \delta \boldsymbol{\varphi} \tag{11.185}$$

Thus, in view of Equation (11.183) and (11.184), Equation (11.182) can be rewritten as

$$\delta\mathcal{E}(\boldsymbol{\varphi},\dot{\boldsymbol{\varphi}}) = \int_{\mathcal{B}} \left[\frac{\partial\bar{\mathcal{E}}}{\partial\boldsymbol{\varphi}} - \left(\mathrm{DIV}\, \frac{\partial\bar{\mathcal{E}}}{\partial\mathbf{F}} \right) \right] \cdot \delta\boldsymbol{\varphi}\, dV + \int_{\mathcal{B}} \frac{\partial\bar{\mathcal{E}}}{\partial\dot{\boldsymbol{\varphi}}} \cdot \delta\dot{\boldsymbol{\varphi}}\, dV$$

$$+ \int_{\mathcal{B}} \mathrm{DIV} \left(\delta\boldsymbol{\varphi} \cdot \frac{\partial\bar{\mathcal{E}}}{\partial\mathbf{F}} \right) dV - \int_{\partial\mathcal{B}_\sigma} \bar{\mathbf{T}}(\mathbf{X},\hat{\mathbf{N}}) \cdot \delta\boldsymbol{\varphi}\,(\mathbf{X},t)\, dA \tag{11.186}$$

In addition, we know that the following on the prescribed boundaries is true,

$$\int_{t_0}^{t_1} \left[\int_{\mathcal{B}} \mathrm{DIV} \left(\delta\boldsymbol{\varphi} \cdot \frac{\partial\bar{\mathcal{E}}}{\partial\mathbf{F}} \right) dV \right] dt = \int_{t_0}^{t_1} \left[\int_{\partial\mathcal{B}_\sigma} \left(\delta\boldsymbol{\varphi} \cdot \frac{\partial\bar{\mathcal{E}}}{\partial\mathbf{F}} \cdot \hat{\mathbf{N}} \right) dV \right] dt \tag{11.187}$$

Note that on the prescribed boundaries, the virtual displacement $\delta\boldsymbol{\varphi} \in T_{\varphi}\mathcal{C}$ must be null due to the Dirichlet boundary conditions. Consequently, Hamilton's law of varying action can be written as

$$\int_{t_0}^{t_1} \left\{ \int_{\mathcal{B}} \left[\mathrm{DIV}\, \frac{\partial\bar{\mathcal{E}}}{\partial\mathbf{F}} - \frac{\partial\bar{\mathcal{E}}}{\partial\boldsymbol{\varphi}} - \frac{d}{dt}\left(\frac{\partial\bar{\mathcal{E}}}{\partial\dot{\boldsymbol{\varphi}}} \right) \right] \cdot \delta\boldsymbol{\varphi}\, dV \right\} dt$$

$$+ \int_{t_0}^{t_1} \left[\int_{\partial\mathcal{B}_\sigma} \left(\bar{\mathbf{T}}(\mathbf{X},\hat{\mathbf{N}}) - \frac{\partial\bar{\mathcal{E}}}{\partial\mathbf{F}} \cdot \hat{\mathbf{N}} \right) \cdot \delta\boldsymbol{\varphi}\, dA \right] dt = 0 \tag{11.188}$$

where both the governing equations of motion and natural boundary conditions are included and $\hat{\mathbf{N}}$ stands for the unit vector normal to the surface in the reference configuration. Of particular importance is the observation that the governing equations in the first integral above are Cauchy's equation of motion within the Total Energy framework as follows:

$$\frac{d}{dt}\left(\frac{\partial\bar{\mathcal{E}}}{\partial\dot{\boldsymbol{\varphi}}} \right) + \frac{\partial\bar{\mathcal{E}}}{\partial\boldsymbol{\varphi}} - \left(\mathrm{DIV}\, \frac{\partial\bar{\mathcal{E}}}{\partial\mathbf{F}} \right) = \mathbf{0} \tag{11.189}$$

subject to the traction on the traction boundary

$$\bar{\mathbf{T}}(\mathbf{X},\hat{\mathbf{N}}) = \frac{\partial\bar{\mathcal{E}}}{\partial\mathbf{F}} \cdot \hat{\mathbf{N}} \quad \text{for} \quad (\mathbf{X},t) \in \partial\mathcal{B}_\sigma \times \mathcal{T} \tag{11.190}$$

Equation (11.189) is referred to as the Total Energy density representation of the equation of motion, which results in the local momentum equation or Cauchy's equation of motion. Note that Equation (11.189) and (11.190) can be rewritten as the conventional strong form, namely,

$$\begin{aligned} \rho_0\ddot{\boldsymbol{\varphi}}\,(\mathbf{X},t) &= \rho_0\mathbf{B} + \mathrm{DIV}\,\mathbf{P}, && \text{for } (\mathbf{X},t) \in \mathcal{B}\times\mathcal{T} \\ \bar{\mathbf{T}}\left(\mathbf{X},\hat{\mathbf{N}}\right) &= \mathbf{P}\cdot\hat{\mathbf{N}} && \text{on } \mathbf{X} \in \partial\mathcal{B}_\sigma \end{aligned} \tag{11.191}$$

We observe that Cauchy's equation of motion may be represented in terms of the Total Energy density function as shown in Equation (11.189). Hence, Equation (11.189) may be referred to as the scalar formalism of the Total Energy version of Cauchy's equation of motion.

By means of Equation (11.185), Hamilton's law of varying action in Equation (11.188) leads to the weak form in space as follows:

$$\int_{t_0}^{t_1}\left\{\begin{array}{l}\displaystyle\int_{\mathcal{B}}\frac{d}{dt}\left(\frac{\partial\bar{\mathcal{E}}}{\partial\dot{\boldsymbol{\varphi}}}\right)\cdot\delta\boldsymbol{\varphi}\,dV+\int_{\mathcal{B}}\frac{\partial\bar{\mathcal{E}}}{\partial\mathbf{F}}:\delta\mathbf{F}dV\\ \displaystyle+\int_{\mathcal{B}}\frac{\partial\bar{\mathcal{E}}}{\partial\boldsymbol{\varphi}}\cdot\delta\boldsymbol{\varphi}\,dV-\int_{\partial\mathcal{B}_\sigma}\bar{\mathbf{T}}\left(\mathbf{X}\right)\cdot\delta\boldsymbol{\varphi}\,dA\end{array}\right\}dt=0 \tag{11.192}$$

The integrand in the bracket of Equation (11.192) corresponds to the principle of virtual work in the Total Energy framework, namely,

$$\int_{\mathcal{B}}\frac{d}{dt}\left(\frac{\partial\bar{\mathcal{E}}}{\partial\dot{\boldsymbol{\varphi}}}\right)\cdot\delta\boldsymbol{\varphi}\,dV+\int_{\mathcal{B}}\frac{\partial\bar{\mathcal{E}}}{\partial\mathbf{F}}:\delta\mathbf{F}dV$$

$$+\int_{\mathcal{B}}\frac{\partial\bar{\mathcal{E}}}{\partial\boldsymbol{\varphi}}\cdot\delta\boldsymbol{\varphi}\,dV-\int_{\partial\mathcal{B}_\sigma}\bar{\mathbf{T}}(\mathbf{X},\hat{\mathbf{N}})\cdot\delta\boldsymbol{\varphi}\,dA=0 \tag{11.193}$$

Accordingly, Equation (11.193) can be regarded as the so-called single-field variational formulation; the dependent variable is the generalized coordinate.

11.5.3 Space-Discrete Total Energy Finite Element Formulation

We highlight two distinct approaches to space discretization in the variational calculus setting, namely: (1) the traditional approach using as a starting point, the Total Energy density representation of the equations of motion, and (2) alternatively, we then describe a simple space-discrete finite element discretization approach that is significantly different via Hamilton's law of varying action with the descriptive scalar function, namely, the Total Energy.

The admissible trial functions $\tilde{\boldsymbol{\varphi}}$ within the Total Energy framework may be represented by a linear combination of coordinate functions $\mathbf{N}_i\,(\mathbf{X})$ and displacement vectors $\mathbf{U}_i^e\,(t)$ as given by Equation (9.11); hence the trial functions, which are of class $C^{m-1}\,(\mathcal{B})$ over the entire domain, belong to the subspace, denoted by $\mathcal{C}^h\subset\mathcal{C}$, as shown in Equation (9.14).

Approach 1: Weak Form With regards to the Total Energy density representation of the equation of motion, substituting the approximate trial functions in Equation (9.11) into the term enclosed in the curly brackets in Equation (11.192) leads to the following:

$$\int_{t_0}^{t_1}\sum_{e=1}^{n_{\text{elem}}}\left\{\begin{array}{l}\displaystyle\int_{\mathcal{B}_e}\frac{d}{dt}\left(\frac{\partial\bar{\mathcal{E}}^h_{\mathcal{B}_e}}{\partial\dot{\tilde{\boldsymbol{\varphi}}}^h}\right)\cdot\delta\tilde{\boldsymbol{\varphi}}^h\,dV+\int_{\mathcal{B}_e}\frac{\partial\bar{\mathcal{E}}^h_{\mathcal{B}_e}}{\partial\tilde{\mathbf{F}}^h}:\delta\tilde{\mathbf{F}}^h\,dV\\ \displaystyle+\int_{\mathcal{B}_e}\frac{\partial\bar{\mathcal{E}}^h_{\mathcal{B}_e}}{\partial\tilde{\boldsymbol{\varphi}}^h}\cdot\delta\tilde{\boldsymbol{\varphi}}^h\,dV-\int_{\partial\mathcal{B}_{e\sigma}}\bar{\mathbf{T}}\cdot\delta\tilde{\boldsymbol{\varphi}}^h dA\end{array}\right\}dt=0 \tag{11.194}$$

where the space-discrete Total Energy density function is given by

$$\bar{\mathcal{E}}^h_{\mathcal{B}_e}\left(\tilde{\boldsymbol{\varphi}}^h,\dot{\tilde{\boldsymbol{\varphi}}}^h,\tilde{\mathbf{F}}^h\right)=\frac{1}{2}\rho_0\dot{\tilde{\boldsymbol{\varphi}}}^h\cdot\dot{\tilde{\boldsymbol{\varphi}}}^h+\rho_0\Psi\left(\tilde{\mathbf{F}}^h\right)-\rho_0\mathbf{B}\cdot\tilde{\boldsymbol{\varphi}}^h \tag{11.195}$$

Finally, Equation (11.194) leads to

$$\int_{t_0}^{t_1}\left\{\sum_{e=1}^{n_{\text{elem}}}\left(\mathbf{M}_e\ddot{\mathbf{U}}^e\,(t)+\mathbf{F}^e_{\text{int}}-\mathbf{F}^e_{\text{ext}}\right)\cdot\delta\mathbf{U}^e\,(t)\right\}dt=0 \tag{11.196}$$

where $\mathbf{M}_e$, $\mathbf{F}^e_{\text{int}}$ and $\mathbf{F}^e_{\text{ext}}$ are given by Equation (9.17).

For the entire domain of interest, we introduce a configuration space with generalized coordinates $\mathcal{Q}$ as given by Equation (10.48).

$$\mathcal{Q} := \left\{ \mathbf{q} | \mathbf{q}(t) = \left(U_1(t), \ldots, U_{n_{\text{dof}}}(t) \right) \in \mathbb{R}^{n_{\text{dof}}=3n_{\text{node}}} \right\} \tag{11.197}$$

Finally, Equation (11.196) yields the semi-discrete second-order differential equations, i.e. the finite element equations of motion to be discretized subsequently in time as follows:

$$\mathbf{M}\ddot{\mathbf{q}} + \mathbf{F}_{\text{int}} - \mathbf{F}_{\text{ext}} = \mathbf{0} \tag{11.198}$$

Approach 2: Total Energy Differential Operator In contrast to the previous approach, Hamilton's law of varying action can be directly discretized in space as follows:

$$\begin{aligned} \text{HLVA}^h_{\mathcal{E}_{\mathcal{B}_e}} : \delta \int_{t_0}^{t_1} \left\{ \int_{\mathcal{B}_e} \rho_0 \dot{\tilde{\boldsymbol{\varphi}}}^h \cdot \dot{\tilde{\boldsymbol{\varphi}}}^h \, dV - \mathcal{E}^h_{\mathcal{B}_e}\left(\tilde{\boldsymbol{\varphi}}^h, \dot{\tilde{\boldsymbol{\varphi}}}^h \right) \right\} dt \\ - \left[\int_{\mathcal{B}_e} \rho_0 \dot{\tilde{\boldsymbol{\varphi}}}^h \cdot \delta \tilde{\boldsymbol{\varphi}}^h \, dV \right] \Bigg|_{t_0}^{t_1} = 0 \end{aligned} \tag{11.199}$$

where the autonomous space-discrete Total Energy for the finite dimensional systems is constructed as

$$\mathcal{E}^h_{\mathcal{B}_e}\left(\tilde{\boldsymbol{\varphi}}^h, \dot{\tilde{\boldsymbol{\varphi}}}^h \right) = \int_{\mathcal{B}_e} \bar{\mathcal{E}}^h_{\mathcal{B}_e}\left(\tilde{\boldsymbol{\varphi}}^h, \dot{\tilde{\boldsymbol{\varphi}}}^h, \tilde{\mathbf{F}}^h \right) dV - \int_{\partial \mathcal{B}_{e\sigma}} \bar{\mathbf{T}} \cdot \tilde{\boldsymbol{\varphi}}^h(\mathbf{X}, t) \, dA \tag{11.200}$$

In terms of generalized coordinates, Equation (11.199) can be rewritten as

$$\begin{aligned} \text{HLVA}^h_{\mathcal{E}_{\mathcal{B}_e}} : \delta \int_{t_0}^{t_1} \left\{ \frac{\partial \mathcal{E}^h_{\mathcal{B}_e}(\mathbf{q}_e, \dot{\mathbf{q}}_e)}{\partial \dot{\mathbf{q}}_e} \cdot \dot{\mathbf{q}}_e - \mathcal{E}^h_{\mathcal{B}_e}(\mathbf{q}_e, \dot{\mathbf{q}}_e) \right\} dt - \frac{\partial \mathcal{E}^h_{\mathcal{B}_e}}{\partial \dot{\mathbf{q}}_e} \cdot \delta \mathbf{q}_e \Bigg|_{t_0}^{t_1} \\ = - \int_{t_0}^{t_1} \left\{ \frac{\partial \mathcal{E}^h_{\mathcal{B}_e}}{\partial \mathbf{q}_e} + \frac{d}{dt}\left(\frac{\partial \mathcal{E}^h_{\mathcal{B}_e}}{\partial \dot{\mathbf{q}}_e} \right) \right\} \cdot \delta \mathbf{q}_e \, dt = 0 \end{aligned} \tag{11.201}$$

where the space-discrete Total Energy in the direct notation is given by Equation (11.49).

For the entire domain, a configuration space with generalized coordinates, denoted by $\mathcal{Q}$, is defined in Equation (11.197). Consequently, Hamilton's law of varying action in a finite dimensional system in terms of the Total Energy is given by

$$\begin{aligned} \text{HLVA}^h_{\mathcal{E}} : \delta \int_{t_0}^{t_1} \left\{ \frac{\partial \mathcal{E}^h(\mathbf{q}, \dot{\mathbf{q}})}{\partial \dot{\mathbf{q}}} \cdot \dot{\mathbf{q}} - \mathcal{E}^h(\mathbf{q}, \dot{\mathbf{q}}) \right\} dt - \frac{\partial \mathcal{E}^h}{\partial \dot{\mathbf{q}}} \cdot \delta \mathbf{q} \Bigg|_{t_0}^{t_1} \\ = - \int_{t_0}^{t_1} \left\{ \frac{\partial \mathcal{E}^h}{\partial \mathbf{q}} + \frac{d}{dt}\left(\frac{\partial \mathcal{E}^h}{\partial \dot{\mathbf{q}}} \right) \right\} \cdot \delta \mathbf{q} \, dt = 0 \end{aligned} \tag{11.202}$$

In view of Equation (11.201), we introduce the discrete Total Energy Differential Operator $\mathfrak{E}_D$, by which the semi-discrete finite element equations of motion can be readily obtained after constructing the space-discrete Total Energy which only entails the use of trial functions. Thus, we have, upon substituting the space-discrete Total Energy into the space-discrete Total Energy Differential Operator, the following:

$$\mathfrak{E}_D\left(\mathcal{E}^h\right) = \frac{d}{dt}\left(\frac{\partial \mathcal{E}^h}{\partial \dot{\mathbf{q}}}\right) + \frac{\partial \mathcal{E}^h}{\partial \mathbf{q}} = \mathbf{M}\ddot{\mathbf{q}} + \mathbf{F}_{\text{int}} - \mathbf{F}_{\text{ext}} = \mathbf{0},$$
$$\mathfrak{E}_D = \frac{d}{dt}\left(\frac{\partial}{\partial \dot{\mathbf{q}}}\right) + \frac{\partial}{\partial \mathbf{q}} \tag{11.203}$$

where the dependent state variable is the space-discrete scalar function, namely, the space-discrete Total Energy. Thus, we readily have the semi-discrete finite element equations of motion governing the mechanical dynamic behavior of the continuous-body within the Total Energy framework as follows:

$$\mathbf{M}\ddot{\mathbf{q}} + \mathbf{F}_{\text{int}} - \mathbf{F}_{\text{ext}} = 0 \tag{11.204}$$

In view of Equation (11.201), we have

$$\int_{t_0}^{t_1} \left(\mathbf{M}\ddot{\mathbf{q}} + \mathbf{F}_{\text{int}} - \mathbf{F}_{\text{ext}}\right) \cdot \delta\mathbf{q}\,(t)\, dt = 0 \tag{11.205}$$

Note that Equation (11.205), which results from HLVA^h in the Total Energy framework, takes strictly the Bubnov-Galerkin weighted residual form in time. Consequently, we observe that spatial discretizations of Hamilton's law of varying action results in the Bubnov-Galerkin weighted residual form in time. Hence, the spatially discretized Hamilton's law of varying action form can be used as a variational equation to discretize the time domain. Equation (11.205) may be regarded as an unconstrained variational formulation with a single dependent field variable in the time domain; thus we have the single-field variational problem within the Total Energy mechanics framework. In general, the time weighted-residual method which includes time finite element formulations can be applied for deriving time stepping schemes (Belytschko and Hughes 1983; Sha et al. 2003; Tamma et al. 2000, 2002, 2003; Wood 1990; Zhou and Tamma 2004a,c, 2006; Zhou et al. 2001, 2000; Zienkiewicz and Taylor 2006; Zienkiewicz et al. 2005, 1984).

Illustrative Examples For the various illustrative examples described next in the variational calculus setting, we first describe the development of the governing equations in the sense of the Total Energy density based representations of the equation of motion in scalar formalism and the variational calculus setting. Following this, we only present the space-discrete finite element approach via Hamilton's law of varying action and the Total Energy Differential Operator.

Example 11.6 Elastic Axial Bar

The elastic bar described in earlier sections is considered next.

Governing Equations: Total Energy Density Based Representations

The generalized momentum for the axial bar is given by

$$\frac{\partial \bar{\mathcal{L}}}{\partial \dot{u}} = \rho A\,(x)\,\dot{u} \tag{11.206}$$

where the Lagrangian density function $\bar{\mathcal{L}}\left(u, \dot{u}, u'\right)$ is defined in Equation (10.69). In view of Equation (11.175), the Total Energy density function for the axial bar can be written as

$$\bar{\mathcal{E}}\left(u, \dot{u}, u'\right) = \rho A\,(x)\,\dot{u}^2 - \bar{\mathcal{L}}\left(u, \dot{u}, u'\right), \quad u' = \frac{\partial u}{\partial x} \tag{11.207}$$

which yields

$$\bar{\mathcal{E}}\left(u, \dot{u}, u'\right) = \frac{1}{2}\rho A\left(x\right)\dot{u}^2 + \frac{1}{2}EA\left(x\right)\left(\frac{\partial u}{\partial x}\right)^2 - \bar{b}u \tag{11.208}$$

The variation of the Total Energy density function is given by

$$\delta\bar{\mathcal{E}}\left(u, \dot{u}, u'\right) = \frac{\partial\bar{\mathcal{E}}}{\partial u}\delta u + \frac{\partial\bar{\mathcal{E}}}{\partial\dot{u}}\delta\dot{u} + \frac{\partial\bar{\mathcal{E}}}{\partial u'}\delta u' \tag{11.209}$$

Then, the variation of the Lagrangian density function can be represented as

$$\delta\bar{\mathcal{L}}\left(u, \dot{u}, u'\right) = \dot{u}\delta\left(\frac{\partial\bar{\mathcal{E}}}{\partial\dot{u}}\right) + \frac{\partial\bar{\mathcal{E}}}{\partial\dot{u}}\delta\dot{u} - \delta\bar{\mathcal{E}}\left(u, \dot{u}, u'\right) \tag{11.210}$$

Upon substitution of Equation (10.248) into Equation (10.249), we obtain

$$\delta\bar{\mathcal{L}}\left(u, \dot{u}, u'\right) = \dot{u}\delta\left(\frac{\partial\bar{\mathcal{E}}}{\partial\dot{u}}\right) - \frac{\partial\bar{\mathcal{E}}}{\partial u}\delta u - \frac{\partial\bar{\mathcal{E}}}{\partial u'}\delta u' \tag{11.211}$$

The first term on the right-hand side above can be written as

$$\dot{u}\delta\left(\frac{\partial\bar{\mathcal{E}}}{\partial\dot{u}}\right) = \frac{d}{dt}\left(\frac{\partial\bar{\mathcal{E}}}{\partial\dot{u}}\cdot\delta u\right) - \frac{d}{dt}\left(\frac{\partial\bar{\mathcal{E}}}{\partial\dot{u}}\right)\delta u \tag{11.212}$$

since $\frac{\partial\bar{E}}{\partial\dot{u}} = \rho A\left(x\right)\dot{u}$. Therefore, Equation (11.211) can be given by

$$\begin{aligned}\delta\bar{\mathcal{L}}\left(u, \dot{u}, u'\right) &= \frac{d}{dt}\left(\frac{\partial\bar{\mathcal{E}}}{\partial\dot{u}}\cdot\delta u\right) - \frac{\partial}{\partial x}\left(\frac{\partial\bar{\mathcal{E}}}{\partial u'}\delta u\right)\\ &\quad + \left\{\frac{\partial}{\partial x}\left(\frac{\partial\bar{\mathcal{E}}}{\partial u'}\right) - \frac{d}{dt}\left(\frac{\partial\bar{\mathcal{E}}}{\partial\dot{u}}\right) - \frac{\partial\bar{\mathcal{E}}}{\partial u}\right\}\delta u\end{aligned} \tag{11.213}$$

Strong Form

As a consequence, Hamilton's law of varying action yields

$$\begin{aligned}&\int_{t_0}^{t_1}\left\{\int_L\left[\left(\frac{\partial}{\partial x}\left(\frac{\partial\bar{\mathcal{E}}}{\partial u'}\right) - \frac{d}{dt}\left(\frac{\partial\bar{\mathcal{E}}}{\partial\dot{u}}\right) - \frac{\partial\bar{\mathcal{E}}}{\partial u}\right)\delta u\right]dx\right\}dt\\ &\quad + \int_{t_0}^{t_1}\left(F_L\delta u(L,t) - \frac{\partial\bar{\mathcal{E}}}{\partial u'}\delta u\Big|_0^l\right)dt = 0\end{aligned} \tag{11.214}$$

Therefore, the scalar formalism of the governing equation is obtained as

$$\frac{d}{dt}\left(\frac{\partial\bar{\mathcal{E}}}{\partial\dot{u}}\right) = \frac{\partial}{\partial x}\left(\frac{\partial\bar{\mathcal{E}}}{\partial u'}\right) - \frac{\partial\bar{\mathcal{E}}}{\partial u} \tag{11.215}$$

subject to the boundary conditions

$$F_L \delta u(L,t) + \frac{\partial \bar{\mathcal{E}}}{\partial u'} \delta u \bigg|_0^L = 0 \tag{11.216}$$

Eventually, substituting the Total Energy density function in Equation (11.208) into Equation (11.215), we have the classical vector formalism and representation of the governing equation of motion as

$$\rho A(x) \ddot{u} = \frac{\partial}{\partial x}\left(EA(x) \frac{\partial u}{\partial x} \right) + \bar{b} \tag{11.217}$$

subject to the boundary condition

$$F_L \delta u(L,t) - EA(x) \frac{\partial u}{\partial x} \delta u \bigg|_0^L = 0 \tag{11.218}$$

Weak Form

Upon integration by parts, Hamilton's law of varying action also leads to

$$\int_{t_0}^{t_1} \left\{ \begin{array}{l} \displaystyle\int_L \frac{d}{dt}\left(\frac{\partial \bar{\mathcal{E}}}{\partial \dot{u}} \right) \delta u \, dx + \int_L \frac{\partial \bar{\mathcal{E}}}{\partial u'} \delta u' dx \\ \displaystyle + \int_L \frac{\partial \bar{\mathcal{E}}}{\partial u} \delta u \, dx - F_L \delta u(L,t) \end{array} \right\} dt = 0 \tag{11.219}$$

Note that the above equation is identical to the weak form given in Equation (11.193).

Space-Discrete Finite Element Formulation

Next, we discuss the space-discrete Total Energy formulation for a single axial bar element.

$$\mathcal{E}_e^h (\mathbf{q}_e, \dot{\mathbf{q}}_e) = \mathcal{K}_e^h + \left({}_e\mathcal{U}_{\text{int}}^h + {}_e\mathcal{U}_{\text{ext}}^h \right) \tag{11.220}$$

where the space-discrete kinetic energy is,

$$\mathcal{K}_e^h = \frac{1}{2} \dot{\mathbf{q}}_e \cdot \mathbf{M}^e \dot{\mathbf{q}}_e \tag{11.221}$$

Hence, Equation (11.220) can be recast into the form

$$\begin{aligned} \mathcal{E}_e^h &= \frac{\rho A}{6} \left(\left(\dot{u}_1^e(t) \right)^2 + \dot{u}_1^e(t) \dot{u}_2^e(t) + \left(\dot{u}_2^e(t) \right)^2 \right) \\ &+ \frac{EA}{2L} \left(\left(u_1^e(t) \right)^2 - 2u_1^e(t) u_2^e(t) + \left(u_2^e(t) \right)^2 \right) - f_1^e u_1^e(t) - f_2^e u_2^e(t) \end{aligned} \tag{11.222}$$

The space-discrete linear momentum can be rewritten as

$$\mathbf{p}_e = \left\{ \begin{array}{c} p_1^e(t) \\ p_2^e(t) \end{array} \right\} = \frac{\partial \mathcal{E}_e^h (\mathbf{q}_e, \dot{\mathbf{q}}_e)}{\partial \dot{\mathbf{q}}_e} = \frac{\rho A}{6} \begin{bmatrix} 2 & 1 \\ 1 & 2 \end{bmatrix} \left\{ \begin{array}{c} \dot{u}_1^e(t) \\ \dot{u}_2^e(t) \end{array} \right\} \tag{11.223}$$

Next, the space-discrete Total Energy representation of the equation of motion results in the following finite element equations of motion,

$$\mathfrak{E}_D\left(\mathcal{E}^h\right) = \frac{d}{dt}\left(\frac{\partial \mathcal{E}_e^h}{\partial \dot{\mathbf{q}}_e}\right) + \frac{\partial \mathcal{E}_e^h}{\partial \mathbf{q}_e} = \mathbf{0} \tag{11.224}$$

which can be recast into Equation (11.69). Consequently, we see that simply substituting the space-discrete Total Energy into the space-discrete Total Energy differential operator given in Equation (11.203) yields the semi-discrete finite element equations as follows:

$$\mathbf{M}^e\ddot{\mathbf{q}}_e + \mathbf{K}^e\mathbf{q}_e - \mathbf{F}^e_{\text{ext}} = 0 \tag{11.225}$$

Example 11.7 Rotating Bar

The rotating bar example is illustrated next.

Governing Equations: Total Energy Density Based Representations

The Total Energy density function for the rotating bar can be written as

$$\bar{\mathcal{E}}\left(\theta, \dot{\theta}, \theta'\right) = \frac{\partial \bar{\mathcal{L}}}{\partial \dot{\theta}} \cdot \dot{\theta} - \bar{\mathcal{L}}\left(\theta, \dot{\theta}, \theta'\right) \tag{11.226}$$

where the generalized momentum is given by

$$\frac{\partial \bar{\mathcal{L}}}{\partial \dot{\theta}} = \rho J\left(x\right)\dot{\theta} \tag{11.227}$$

The Total Energy density function reduces to

$$\bar{\mathcal{E}}\left(\theta, \dot{\theta}, \theta'\right) = \frac{1}{2}\rho J\left(x\right)\dot{\theta}^2 + \frac{1}{2}GJ\left(x\right)\left\{\frac{\partial \theta\left(x,t\right)}{\partial x}\right\}^2 \tag{11.228}$$

The variation of the Total Energy density function is given by

$$\delta\bar{\mathcal{E}}\left(\theta, \dot{\theta}, \theta'\right) = \frac{\partial \bar{\mathcal{E}}}{\partial \theta}\delta\theta + \frac{\partial \bar{\mathcal{E}}}{\partial \dot{\theta}}\delta\dot{\theta} + \frac{\partial \bar{\mathcal{E}}}{\partial \theta'}\delta\theta' \tag{11.229}$$

Then, the variation of the Lagrangian density function can be represented as

$$\delta\bar{\mathcal{L}}\left(\theta, \dot{\theta}, \theta'\right) = \frac{\partial \bar{\mathcal{L}}}{\partial \dot{\theta}}\delta\dot{\theta} + \dot{\theta}\delta\left(\frac{\partial \bar{\mathcal{L}}}{\partial \dot{\theta}}\right) - \delta\bar{\mathcal{E}}\left(\theta, \dot{\theta}, \theta'\right) \tag{11.230}$$

In view of Equation (11.228) and Equation (11.230), we have the relation

$$\frac{\partial \bar{\mathcal{L}}}{\partial \dot{\theta}} = \frac{\partial \bar{\mathcal{E}}}{\partial \dot{\theta}} \tag{11.231}$$

Hence, Equation (11.230) can be rewritten as

$$\delta\bar{\mathcal{L}}\left(\theta,\dot{\theta},\theta'\right)=\dot{\theta}\delta\left(\frac{\partial\bar{\mathcal{E}}}{\partial\dot{\theta}}\right)-\frac{\partial\bar{\mathcal{E}}}{\partial\theta}\delta\theta-\frac{\partial\bar{\mathcal{E}}}{\partial\theta'}\delta\theta' \tag{11.232}$$

The first term on the right-hand side above can be written as

$$\dot{\theta}\delta\left(\frac{\partial\bar{\mathcal{E}}}{\partial\dot{\theta}}\right)=\frac{d}{dt}\left(\frac{\partial\bar{\mathcal{E}}}{\partial\dot{\theta}}\delta\theta\right)-\frac{d}{dt}\left(\frac{\partial\bar{\mathcal{E}}}{\partial\dot{\theta}}\right)\delta\theta \tag{11.233}$$

since $\frac{\partial\bar{\mathcal{E}}}{\partial\dot{\theta}}=\rho J\left(x\right)\dot{\theta}$. Therefore, Equation (11.232) is given by

$$\begin{aligned}\delta\bar{\mathcal{L}}\left(\theta,\dot{\theta},\theta'\right) &=\left\{\frac{\partial}{\partial x}\left(\frac{\partial\bar{\mathcal{E}}}{\partial\theta'}\right)-\frac{\partial\bar{\mathcal{E}}}{\partial\theta}-\frac{d}{dt}\left(\frac{\partial\bar{\mathcal{E}}}{\partial\dot{\theta}}\right)\right\}\delta\theta\\ &\quad\frac{d}{dt}\left(\frac{\partial\bar{\mathcal{E}}}{\partial\dot{\theta}}\delta\theta\right)-\frac{\partial}{\partial x}\left(\frac{\partial\bar{\mathcal{E}}}{\partial\theta'}\delta\theta\right)\end{aligned} \tag{11.234}$$

Therefore, Hamilton's law of varying action yields

$$\begin{aligned}&\int_{t_0}^{t_1}\left[\int_L\left\{\left(\frac{\partial}{\partial x}\left(\frac{\partial\bar{\mathcal{E}}}{\partial\theta'}\right)-\frac{\partial\bar{\mathcal{E}}}{\partial\theta}-\frac{d}{dt}\left(\frac{\partial\bar{\mathcal{E}}}{\partial\dot{\theta}}\right)+\bar{m}_T\left(x\right)\right)\delta\theta\right\}dx\right]dt\\ &+\int_{t_0}^{t_1}\left(\left.\frac{\partial\bar{\mathcal{E}}}{\partial\theta'}\delta\theta\right|_0^L\right)dt=0\end{aligned} \tag{11.235}$$

Therefore, the governing equations are obtained as

$$\frac{d}{dt}\left(\frac{\partial\bar{\mathcal{E}}}{\partial\dot{\theta}}\right)=\frac{\partial}{\partial x}\left(\frac{\partial\bar{\mathcal{E}}}{\partial\theta'}\right)-\frac{\partial\bar{\mathcal{E}}}{\partial\theta}+\bar{m}_T\left(x\right) \tag{11.236}$$

subject to the boundary conditions

$$\left.\frac{\partial\bar{\mathcal{E}}}{\partial\theta'}\delta\theta\right|_0^L=0 \tag{11.237}$$

Eventually, substituting the Total Energy density function in Equation (11.228) into Equation (11.236) and Equation (11.237), we obtain the same governing equation of motion as given by Equation (10.113) and the same boundary condition as given by Equation (10.114).

Space-Discrete Finite Element Formulation

Henceforth, we shall discuss the development of the semi-discrete second-order equation of motion via the space-discrete Total Energy finite element formulation. The trial functions may be approximated by Equation (10.116). Thus, we obtain the space-discrete kinetic energy given by Equation (10.117) and the space-discrete potential energy given by Equation (10.119).

Consequently, the space-discrete Total Energy can be constructed as

$$\mathcal{E}_e^h(\mathbf{q}_e, \dot{\mathbf{q}}_e) = \frac{1}{2}\dot{\mathbf{q}}_e \cdot \mathbf{M}^e \dot{\mathbf{q}}_e + \frac{1}{2}\mathbf{q}_e \cdot \mathbf{K}^e \mathbf{q}_e - \mathbf{q}_e \cdot \mathbf{F}_{\text{ext}}^e \tag{11.238}$$

where $\mathbf{K}^e$ and $\mathbf{F}_{\text{ext}}^e$ can be found in Equation (10.120), and $\mathbf{M}^e$ is given in Equation (10.118). Substituting the space-discrete Total Energy into the space-discrete Total Energy differential operator given in Equation (11.203), we have the following semi-discrete finite element equation of motion:

$$\mathfrak{E}_D\left(\mathcal{E}^h\right) = \mathbf{M}^e \ddot{\mathbf{q}}_e + \mathbf{K}^e \mathbf{q}_e - \mathbf{F}_{\text{ext}}^e = \mathbf{0} \tag{11.239}$$

which is the semi-discrete second-order finite element equation of motion.

Example 11.8 Euler-Bernoulli Beam

In a similar manner as we illustrated the Euler-Bernoulli beam problem with the Lagrangian descriptive function, we next consider the same dynamical behavior of the Euler-Bernoulli beam with the descriptive scalar function, namely, the Total Energy.

Governing Equations: Total Energy Density Based Representations

The total rotation of the Euler-Bernoulli beam is defined as in Equation (11.83). Hamilton's law of varying action gives

$$\delta \int_{t_0}^{t_1} \left\{ \int_L \frac{\partial \bar{\mathcal{E}}(\boldsymbol{u}, \dot{\mathbf{u}}, \mathbf{u}')}{\partial \dot{\mathbf{u}}} \cdot \dot{\mathbf{u}}(x,t)\, dx - \mathcal{E}(\mathbf{u}, \dot{\boldsymbol{u}}) \right\} dt - \int_L \frac{\partial \bar{\mathcal{E}}}{\partial \dot{\mathbf{u}}} \cdot \delta \dot{\mathbf{u}}\, dx \Bigg|_{t_0}^{t_1} = 0, \tag{11.240}$$

$$\dot{\mathbf{u}}(x,t) = \left(\dot{w}(x,t), \dot{\phi}(x,t)\right), \quad \boldsymbol{u}'(x,t) = \left(w'(x,t), \phi'(x,t)\right)$$

where the autonomous Total Energy is given by

$$\mathcal{E}(\mathbf{u}, \dot{\mathbf{u}}) = \int_l \bar{\mathcal{E}}\left(\mathbf{u}, \dot{\boldsymbol{u}}, \mathbf{u}'\right) dx - \int_l \omega(x)\, \mathbf{a} \cdot \mathbf{u}\, dx, \quad \mathbf{a} = (1, 0) \tag{11.241}$$

and the Total Energy density function is defined as in Equation (11.86). Hamilton's law of varying action in Equation (11.240) can be rewritten as

$$\delta \int_{t_0}^{t_1} \left\{ \int_L \frac{\partial \bar{\mathcal{E}}(\boldsymbol{u}, \dot{\mathbf{u}}, \mathbf{u}')}{\partial \dot{\mathbf{u}}} \cdot \dot{\mathbf{u}}(x,t)\, dx - \mathcal{E}(\mathbf{u}, \dot{\boldsymbol{u}}) \right\} dt - \int_L \frac{\partial \bar{\mathcal{E}}}{\partial \dot{\mathbf{u}}} \cdot \delta \dot{\mathbf{u}}\, dx \Bigg|_{t_0}^{t_1} =$$

$$\int_{t_0}^{t_1} \left[\int_L \left\{ \begin{pmatrix} \dfrac{\partial}{\partial x}\left(\dfrac{d}{dt}\left(\dfrac{\partial \bar{\mathcal{E}}}{\partial \dot{w}'}\right)\right) - \dfrac{d}{dt}\left(\dfrac{\partial \bar{\mathcal{E}}}{\partial \dot{w}}\right) - \dfrac{\partial \bar{\mathcal{E}}}{\partial w} \\ -\dfrac{\partial^2}{\partial x^2}\left(\dfrac{\partial \bar{\mathcal{E}}}{\partial w''}\right) + \omega(x) \end{pmatrix} \delta w \right\} dx \right] dt$$

$$-\int_{t_0}^{t_1}\left(\frac{\partial\bar{\mathcal{E}}}{\partial w''}\delta w'\Bigg|_0^L\right)dt+\int_{t_0}^{t_1}\left(\left\{\frac{\partial}{\partial x}\left(\frac{\partial\bar{\mathcal{E}}}{\partial w''}\right)-\frac{d}{dt}\left(\frac{\partial\bar{\mathcal{E}}}{\partial\dot{w}'}\right)\right\}\delta w\Bigg|_0^L\right)dt=0 \tag{11.242}$$

where both governing equations and natural boundary conditions are included. Consequently, we have derived the governing equation of motion with the traction boundary conditions via Hamilton's law of varying action involving the Total Energy. Since the variation of the dependent variable, the transverse deflection, is arbitrary, the governing equation for the Euler-Bernoulli beam in terms of the Total Energy density function is obtained as

$$\frac{\partial}{\partial x}\left(\frac{d}{dt}\left(\frac{\partial\bar{\mathcal{E}}}{\partial\dot{w}'}\right)\right)-\frac{d}{dt}\left(\frac{\partial\bar{\mathcal{E}}}{\partial\dot{w}}\right)-\frac{\partial\bar{\mathcal{E}}}{\partial w}-\frac{\partial^2}{\partial x^2}\left(\frac{\partial\bar{\mathcal{E}}}{\partial w''}\right)+\omega\left(x\right)=0 \tag{11.243}$$

subject to the boundary conditions

$$\frac{\partial\bar{\mathcal{E}}}{\partial w''}\delta w'\Bigg|_0^L=0,\quad \left\{\frac{\partial}{\partial x}\left(\frac{\partial\bar{\mathcal{E}}}{\partial w''}\right)-\frac{d}{dt}\left(\frac{\partial\bar{\mathcal{E}}}{\partial\dot{w}'}\right)\right\}\delta w\Bigg|_0^L=0 \tag{11.244}$$

Subsequently, substituting the Total Energy density function in Equation (11.243) into Equation (11.244) and Equation (11.246), we obtain the same differential equations of motion as given in Equation (10.137) and the same boundary conditions as given in Equation (10.138).

Space-Discrete Finite Element Formulation

Next, we discuss the development of second-order semi-discrete differential equations of motion, via the space-discrete Total Energy finite element formulation. The trial displacement function may be approximated using the Hermite polynomial interpolation functions as given in Equation (10.140). In favor of matrix and vector notation, we have the space-discrete kinetic energy given by Equation (10.142) and the space-discrete potential energy given by Equation (10.144). We now construct the space-discrete autonomous Total Energy as follows

$$\mathcal{E}_e^h\left(\mathbf{q}_e,\dot{\mathbf{q}}_e\right)=\frac{1}{2}\dot{\mathbf{q}}_e\left(t\right)\cdot\mathbf{M}^e\dot{\mathbf{q}}_e\left(t\right)+\frac{1}{2}\mathbf{q}_e\left(t\right)\cdot\mathbf{K}^e\mathbf{q}_e\left(t\right)-\mathbf{q}_e\left(t\right)\cdot\mathbf{F}_{\text{ext}}^e \tag{11.245}$$

where $\mathbf{M}^e$, $\mathbf{K}^e$ and $\mathbf{F}_{\text{ext}}^e$ can be found in Section 9.2.6. Substituting the space-discrete autonomous Total Energy given in Equation (11.245) into the space-discrete Total Energy differential operators in Equation (11.203), readily yields the semi-discrete equations of motion as

$$\mathfrak{E}_D\left(\mathcal{E}_e^h\right)=\mathbf{M}^e\ddot{\mathbf{q}}_e+\mathbf{K}^e\mathbf{q}_e-\mathbf{F}_{\text{ext}}^e=\mathbf{0} \tag{11.246}$$

Example 11.9 Timoshenko Beam

In the Total Energy framework via Hamilton's law of varying action in the variational calculus setting, next we shall discuss the Timoshenko beam problem.

Governing Equations: Total Energy Density Based Representations

The Total Energy of the Timoshenko beam is given by Equation (11.95), and the Total Energy density function is given by Equation (11.96). Hamilton's law of varying action in the context of the Total Energy gives

$$\delta \int_{t_0}^{t_1} \left\{ \int_L \left(\frac{\partial \bar{\mathcal{E}}\left(\mathbf{u}, \dot{\mathbf{u}}, \mathbf{u}'\right)}{\partial \dot{\mathbf{u}}} \cdot \dot{\mathbf{u}} \right) dx - \mathcal{E}\left(\mathbf{u}, \dot{\mathbf{u}}\right) \right\} dt - \int_L \frac{\partial \bar{\mathcal{E}}}{\partial \dot{\mathbf{u}}} \cdot \delta \mathbf{u} \, dx \Bigg|_{t_0}^{t_1} = 0 \tag{11.247}$$

where the continuous Total Energy of the Euler-Bernoulli beam can be obtained as

$$\mathcal{E}\left(\mathbf{u}, \dot{\mathbf{u}}\right) = \int_\Omega \bar{\mathcal{E}}\left(\mathbf{u}, \dot{\mathbf{u}}, \mathbf{u}'\right) dx dy - \int_\Omega \bar{\omega}\left(x\right) \mathbf{a} \cdot \mathbf{u} \, dx dy, \quad \mathbf{a} = (1, 0) \tag{11.248}$$

Employing integration by parts and the chain rule, Hamilton's law of varying action yields

$$\int_{t_0}^{t_1} \left[\int_L \left\{ \left\{ \frac{\partial}{\partial x} \left(\frac{\partial \bar{\mathcal{E}}}{\partial \mathbf{u}'} \right) - \frac{d}{dt} \left(\frac{\partial \bar{\mathcal{E}}}{\partial \dot{\mathbf{u}}} \right) - \frac{\partial \bar{\mathcal{E}}}{\partial \mathbf{u}} + \bar{\omega}\left(x\right) \mathbf{a} \right\} \cdot \delta \mathbf{u} \right\} dx \right] dt$$
$$- \int_{t_0}^{t_1} \left(\frac{\partial \bar{\mathcal{E}}}{\partial \mathbf{u}'} \cdot \delta \mathbf{u} \Bigg|_0^L \right) dt = 0 \tag{11.249}$$

Note that Equation (11.249) includes both governing equations and natural boundary conditions. The scalar representation of the governing equations in the context of the Total Energy are given by

$$\frac{\partial}{\partial x} \left(\frac{\partial \bar{\mathcal{E}}}{\partial \mathbf{u}'} \right) - \frac{d}{dt} \left(\frac{\partial \bar{\mathcal{E}}}{\partial \dot{\mathbf{u}}} \right) - \frac{\partial \bar{\mathcal{E}}}{\partial \mathbf{u}} + \bar{\omega}\left(x\right) \mathbf{a} = 0 \tag{11.250}$$

subject to the natural boundary conditions

$$\frac{\partial \bar{\mathcal{E}}}{\partial \mathbf{u}'} \cdot \delta \mathbf{u} \Bigg|_0^L = 0 \tag{11.251}$$

Therefore, the governing equations in Equation (11.250) can be rewritten as

$$\frac{d}{dt} \left(\frac{\partial \bar{\mathcal{E}}}{\partial \dot{w}} \right) = \frac{\partial}{\partial x} \left(\frac{\partial \bar{\mathcal{E}}}{\partial w'} \right) - \frac{\partial \bar{\mathcal{E}}}{\partial w} + \bar{\omega}\left(x\right),$$
$$\frac{d}{dt} \left(\frac{\partial \bar{\mathcal{E}}}{\partial \dot{\theta}_y} \right) = \frac{\partial}{\partial x} \left(\frac{\partial \bar{\mathcal{E}}}{\partial \theta'_y} \right) - \frac{\partial \bar{\mathcal{E}}}{\partial \theta_y} \tag{11.252}$$

subject to the boundary conditions

$$\frac{\partial \bar{\mathcal{E}}}{\partial w'} \delta w \Bigg|_0^L = 0, \quad \frac{\partial \bar{\mathcal{E}}}{\partial \theta'_y} \delta \theta_y \Bigg|_0^L = 0 \tag{11.253}$$

Eventually, upon substitution of the Total Energy density function in Equation (11.96) into Equation (11.252) and Equation (11.253), we obtain the same differential equations of motion as given in Equation (10.163) and the same boundary conditions as given in Equation (10.164).

Space-Discrete Finite Element Formulation

Next, we shall derive the semi-discretized equations of motion via the space-discrete Total Energy finite element formulation. The admissible trial functions are given by Equation (10.166). The space-discrete Total Energy may be constructed as Equation (11.107). Substituting the space-discrete Total Energy into the space-discrete Total Energy differential operator given in Equation (11.203), simply and naturally leads to the semi-discrete finite element equations of motion within the Total Energy framework:

$$\mathfrak{E}_D\left(\mathcal{E}^h\right) = \mathbf{M}^e \ddot{\mathbf{q}}_e + \mathbf{K}^e \mathbf{q}_e - \mathbf{F}^e_{\text{ext}} = \mathbf{0} \tag{11.254}$$

where details about $\mathbf{M}^e$, $\mathbf{K}^e$ and $\mathbf{F}^e_{\text{ext}}$ can be found in Section 9.2.7.

Example 11.10 Reissner-Mindlin Plate

In a previous chapter, we have discussed the Reissner-Mindlin plate problem within the Lagrangian and Hamiltonian mechanics frameworks. Similarly, we shall discuss the same problem in the Total Energy framework.

Governing Equations: Total Energy Density Based Representations

The kinetic and potential energies for the plate were given previously. Consequently, we find that Hamilton's law of varying action in the context of the Total Energy yields

$$\delta \int_{t_0}^{t_1} \left\{ \int_\Omega \left(\frac{\partial \bar{\mathcal{E}}}{\partial \dot{\mathbf{u}}} \cdot \dot{\mathbf{u}} \right) dxdy - \mathcal{E}(\mathbf{u}, \dot{\mathbf{u}}) \right\} dt - \int_\Omega \frac{\partial \bar{\mathcal{E}}}{\partial \dot{\mathbf{u}}} \cdot \delta \mathbf{u}\, dxdy \Bigg|_{t_0}^{t_1} = 0,$$

$$\mathbf{u} = (w, \theta_x, \theta_y), \quad \dot{\mathbf{u}} = (\dot{w}, \dot{\theta}_x, \dot{\theta}_y), \quad \mathbf{u}_{,x} = \left(\frac{\partial w}{\partial x}, \frac{\partial \theta_x}{\partial x}, \frac{\partial \theta_y}{\partial x} \right), \quad \mathbf{u}_{,y} = \left(\frac{\partial w}{\partial y}, \frac{\partial \theta_x}{\partial y}, \frac{\partial \theta_y}{\partial y} \right) \tag{11.255}$$

where the continuous Total Energy of the Reissner-Mindlin plate is given by Equation (11.110) and the Total Energy density function can be defined as Equation (11.111). The variation of the autonomous Total Energy is given by

$$\begin{aligned} \delta \mathcal{E}(\mathbf{u}, \dot{\mathbf{u}}) &= \int_\Omega \delta \bar{\mathcal{E}}\left(\mathbf{u}, \dot{u}, \mathbf{u}_{,x}, \mathbf{u}_{,y}\right) dxdy - \int_\Omega \omega(x, y)\, \mathbf{a} \cdot \mathbf{u}\, dxdy \\ &= \int_\Omega \frac{\partial \bar{\mathcal{E}}}{\partial \mathbf{u}} \cdot \delta \mathbf{u} + \frac{\partial \bar{\mathcal{E}}}{\partial \dot{\mathbf{u}}} \cdot \delta \dot{\mathbf{u}} + \frac{\partial \bar{\mathcal{E}}}{\partial \mathbf{u}_{,x}} \cdot \delta \mathbf{u}_{,x} + \frac{\partial \bar{\mathcal{E}}}{\partial \mathbf{u}_{,y}} \cdot \delta \mathbf{u}_{,y} dxdy \\ &- \int_\Omega p(x, y)\, \mathbf{a} \cdot \delta \mathbf{u}\, dxdy \end{aligned} \tag{11.256}$$

and, integration by parts gives

$$\int_{t_0}^{t_1} \left(\int_\Omega \frac{\partial \bar{\mathcal{E}}}{\partial \dot{\mathbf{u}}} \cdot \delta \dot{\mathbf{u}} dxdy \right) dt = \int_{t_0}^{t_1} \left(- \int_\Omega \frac{d}{dt} \left(\frac{\partial \bar{\mathcal{E}}}{\partial \dot{\mathbf{u}}} \right) \cdot \delta \mathbf{u}\, dxdy \right) dt \tag{11.257}$$

$$+\int_{\Omega} \frac{\partial \bar{\mathcal{E}}}{\partial \dot{\mathbf{u}}} \cdot \delta \mathbf{u}\, dxdy \Bigg|_{t_0}^{t_1}$$

Thus, Hamilton's law of varying action leads to

$$\int_{t_0}^{t_1} \left\{ -\int_{\Omega} \left(\begin{array}{c} \dfrac{d}{dt}\left(\dfrac{\partial \bar{\mathcal{E}}}{\partial \dot{\mathbf{u}}}\right) \cdot \delta \mathbf{u} + \dfrac{\partial \bar{\mathcal{E}}}{\partial \mathbf{u}} \cdot \delta \mathbf{u} \\ + \dfrac{\partial \bar{\mathcal{E}}}{\partial \mathbf{u}_{,x}} \cdot \delta \mathbf{u}_{,x} + \dfrac{\partial \bar{\mathcal{E}}}{\partial \mathbf{u}_{,y}} \cdot \delta \mathbf{u}_{,y} - \omega(x,y)\mathbf{a} \cdot \delta \mathbf{u} \end{array} \right) dxdy \right\} dt = 0 \qquad (11.258)$$

By means of Green's theorem, we have the following:

$$\begin{aligned} &\int_{\Omega} \left(\frac{\partial \bar{\mathcal{E}}}{\partial \mathbf{u}_{,x}} \cdot \delta \mathbf{u}_{,x} + \frac{\partial \bar{\mathcal{E}}}{\partial \mathbf{u}_{,y}} \cdot \delta \mathbf{u}_{,y} \right) dxdy \\ &= -\int_{\Omega} \left\{ \frac{\partial}{\partial x}\left(\frac{\partial \bar{\mathcal{E}}}{\partial \mathbf{u}_{,x}}\right) + \frac{\partial}{\partial y}\left(\frac{\partial \bar{\mathcal{E}}}{\partial \mathbf{u}_{,y}}\right) \right\} \cdot \delta \mathbf{u}\, dxdy \\ &\quad + \int_{\partial\Omega} \left\{ \frac{\partial \bar{\mathcal{E}}}{\partial \mathbf{u}_{,x}} \cdot \delta \mathbf{u}_x n_x + \frac{\partial \bar{\mathcal{E}}}{\partial \mathbf{u}_{,y}} \cdot \delta \mathbf{u}_y n_y \right\} d\Gamma = 0, \quad n_x = \frac{\partial x}{\partial n},\ n_y = \frac{\partial y}{\partial n} \end{aligned} \qquad (11.259)$$

where n_x, n_y denote the directional cosines respectively and n stands for the coordinate of the outward unit vector normal to the boundary surface.

$$\begin{aligned} &-\int_{t_0}^{t_1} \left\{ \int_{\Omega} \left\{ \begin{array}{c} \dfrac{d}{dt}\left(\dfrac{\partial \bar{\mathcal{E}}}{\partial \dot{\mathbf{u}}}\right) + \dfrac{\partial \bar{\mathcal{E}}}{\partial \mathbf{u}} - \dfrac{\partial}{\partial x}\left(\dfrac{\partial \bar{\mathcal{E}}}{\partial \mathbf{u}_{,x}}\right) \\ -\dfrac{\partial}{\partial y}\left(\dfrac{\partial \bar{\mathcal{E}}}{\partial \mathbf{u}_{,y}}\right) - \omega(x,y)\mathbf{a} \end{array} \right\} \cdot \delta \mathbf{u}\, dxdy \right\} dt \\ &+ \int_{\partial\Omega} \left\{ \frac{\partial \bar{\mathcal{E}}}{\partial \mathbf{u}_{,x}} \cdot \delta \mathbf{u}_x n_x + \frac{\partial \bar{\mathcal{E}}}{\partial \mathbf{u}_{,y}} \cdot \delta \mathbf{u}_y n_y \right\} d\Gamma = 0 \end{aligned} \qquad (11.260)$$

Consequently, we have the governing equations for the Reissner-Mindlin plate as given by Equation (11.117).

Space-Discrete Finite Element Formulation

Next, we show that the space-discrete Total Energy formulation facilitates the derivation of the semi-discrete finite element equations of motion in contrast to the principle of virtual work. Instead of substituting the admissible functions into the Bubnov-Galerkin weighted residual form in Equation (11.260), we first construct the space-discrete autonomous Total Energy. The admissible trial functions are given by Equation (10.194). Introducing new notations for convenience, $\mathbf{q}_e(t) = \begin{bmatrix} \mathbf{w}^e(t) & \boldsymbol{\theta}_x^e(t) & \boldsymbol{\theta}_y^e(t) \end{bmatrix}^T$, the trial functions are rewritten as Equation (10.195). The space-discrete Total Energy is constructed as Equation (11.122). Substituting the space-discrete Total Energy into the space-discrete Total Energy differential operator given by Equation (11.203), simply and naturally leads to the following semi-discrete finite element equation of motion:

$$\mathfrak{E}_D\left(\mathcal{E}_e^h\right) = \mathbf{M}^e\ddot{\mathbf{q}} + \mathbf{K}^e\mathbf{q}_e - \mathbf{F}_{\text{ext}}^e = \mathbf{0} \tag{11.261}$$

More details about $\mathbf{M}^e$, $\mathbf{K}^e$ and $\mathbf{F}_{\text{ext}}^e$ can be found in Section 9.2.9.

EXERCISES

11.1. Circular-Axial Bar Undergoing Rotational and Longitudinal Vibration Let us consider the dynamical behavior of the same axial bar in Exercise 9.1 of Chapter 9.

(a) Establish the Total Energy density function and write down the principle of balance of mechanical energy (the rate of change of the Total Energy) in terms of the Total Energy density function. Derive the governing equation in terms of the Total Energy density function and the natural boundary condition from the principle of balance of mechanical energy.

(b) Establish the space-discrete Total Energy for the element. Obtain the finite element equations of motion for the circular-axial bar undergoing rotational and longitudinal vibration by use of the Total Energy differential operator.

(c) Write down Hamilton's law of varying action in terms of the Total Energy density function. Derive the Total Energy density equations of motion in terms of the Total Energy density function and the natural boundary condition from Hamilton's law of varying action. Repeat (b). Do the same with Total Energy replaced by the Lagrangian and the Hamiltonian.

(d) Establish the Lagrangian density function and write down the principle of balance of mechanical energy (the rate of Jacobi's integral) in terms of the Lagrangian density function. Derive the governing equation in terms of the Lagrangian density function and the natural boundary condition from the principle of balance of mechanical energy. Repeat (b) with Total Energy replace by Lagrangian.

(e) Establish the Hamiltonian density function and write down the principle of balance of mechanical energy (the rate of change of the Hamiltonian) in terms of the Hamiltonian density function. Derive Hamilton's equations of motion in terms of the Hamiltonian density function and the natural boundary condition from the principle of balance of mechanical energy. Repeat (b) with Total Energy replaced by Hamiltonian.

11.2. Euler-Bernoulli Beam under Axial and Transverse Loads Let us consider the dynamical behavior of the same Euler-Bernoulli beam in Exercise 9.2 of Chapter 9.

(a) Repeat the same as in Exercise 11.1.

11.3. Timoshenko Beam on an Elastic Foundation Let us consider the dynamical behavior of the same Timoshenko beam in Exercise 9.3 of Chapter 9.

(a) Repeat the same as in Exercise 11.1.

11.4. Kirchhoff-Love Plate under In-plane and Transverse Loads Let us consider the dynamical behavior of the same Kirchhoff-Love plate in Exercise 9.4 of Chapter 9.

(a) Repeat the same as in Exercise 11.1.

11.5. Reissner-Mindlin Plate under In-plane and Transverse Loads Let us consider the dynamical behavior of the same Reissner-Mindlin plate in Exercise 9.5 of Chapter 9.

(a) Repeat the same as in Exercise 11.1.

CHAPTER TWELVE

EQUIVALENCE OF EQUATIONS

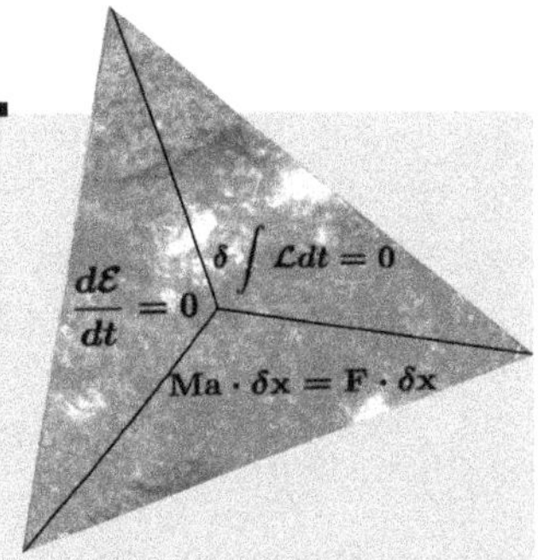

In the previous chapters, we have shown via the three independent fundamental principles, namely, the principle of virtual work, Hamilton's principle, or equivalently, Hamilton's law of varying action, and the principle of balance of mechanical energy or equivalently, the theorem of power expended, that we can obtain the strong form of the governing equations of motion for N-body systems, and the strong and weak forms of the equations of motion together with the boundary conditions for continuous – body systems. Furthermore, we have demonstrated that the strong and/or weak forms can be represented via vector formalism or via several scalar frameworks such as the Lagrangian, the Hamiltonian and the Total Energy. Within each framework for continuous-body systems, the spatial discretization naturally leads to the semi-discrete finite element equations of motion.

In this section, for continuous-body systems, we shall investigate the equivalence between the strong forms, and between the weak forms from the various approaches. That is, we will establish the equivalences in : 1) the principle of virtual work in dynamics, 2) Hamilton's principle or, equivalently, Hamilton's law of varying action, and 3) the principle of balance of mechanical energy. The equivalences of the semi-discretized finite element equations of motion and the finite element formulations are also shown.

12.1 EQUIVALENCE IN THE PRINCIPLE OF VIRTUAL WORK IN DYNAMICS

Note that Cauchy's equation motion is a vector representation formalism for describing the governing equation of motion, and the virtual displacement is also a vector. Also, recall that the equivalent scalar formalism of representation of the equation of motion via the scalar descriptive functions, namely, the Lagrangian, the Hamiltonian, and the Total Energy can also be formulated. We next discuss the various equivalences in the sense of the principle of virtual work in dynamics.

As discussed in Chapter 9, the Bubnov-Galerkin weighted residual form is given by the inner product of Cauchy's equation of motion and the virtual displacement as follows:

$$\int_{\mathcal{B}} (\mathrm{DIV}\,\mathbf{P}(\mathbf{X},t) + \rho_0 \mathbf{B}(\mathbf{X},t) - \rho_0 \ddot{\boldsymbol{\varphi}}(\mathbf{X},t)) \cdot \delta\boldsymbol{\varphi}(\mathbf{X},t)\, dV = 0 \qquad (12.1)$$

Advances in Computational Dynamics of Particles, Materials and Structures, First Edition. Jason Har and Kumar K. Tamma.

By use of Gauss's divergence theorem and integration by parts, the Bubnov-Galerkin weighted residual form gives the weak form as

$$\int_{\mathcal{B}} \rho_0 \ddot{\boldsymbol{\varphi}} \cdot \delta\boldsymbol{\varphi}\, dV + \int_{\mathcal{B}} \mathbf{P} : \delta\mathbf{F}\, dV = \int_{\mathcal{B}} \rho_0 \mathbf{B}(\mathbf{X}, t) \cdot \delta\boldsymbol{\varphi}\, dV + \int_{\partial\mathcal{B}_\sigma} \bar{\mathbf{T}} \cdot \delta\boldsymbol{\varphi}\, dA \tag{12.2}$$

Since the natural boundary condition is imposed on the Bubnov-Galerkin weighted residual form after performing integration by parts, the weak form above leads to

$$\int_{\mathcal{B}} (\mathrm{DIV}\,\mathbf{P} + \rho_0\mathbf{B} - \rho_0\ddot{\boldsymbol{\varphi}}) \cdot \delta\boldsymbol{\varphi}\, dV + \int_{\partial\mathcal{B}_\sigma} (\bar{\mathbf{T}} - \mathbf{P}\cdot\hat{\mathbf{N}}) \cdot \delta\boldsymbol{\varphi}\, dV = 0 \tag{12.3}$$

which is the modified Bubnov-Galerkin weighted residual form. For convenience, we are confined to the case that the external forces are dead loads. For such cases, the external potential energy can be defined. Then, the principle of virtual work in elastodynamics is written as

$$\begin{aligned} &\int_{\mathcal{B}} \rho_0 \mathbf{B}(\mathbf{X}) \cdot \delta\boldsymbol{\varphi}\, dV + \int_{\partial\mathcal{B}_\sigma} \bar{\mathbf{T}}(\mathbf{X}, \hat{\mathbf{N}}) \cdot \delta\boldsymbol{\varphi}\, dA - \int_{\mathcal{B}} \mathbf{P} : \delta\mathbf{F}\, dV \\ &- \int_{\mathcal{B}} \rho_0 \ddot{\boldsymbol{\varphi}} \cdot \delta\boldsymbol{\varphi}\, dV = 0 \end{aligned} \tag{12.4}$$

By introducing the total potential and kinetic energy for the continuous-body dynamical system, the above equation can be rewritten as

$$\delta\mathcal{K}(\dot{\boldsymbol{\varphi}}) - \delta\mathcal{U}(\boldsymbol{\varphi}) - \frac{d}{dt}\int_{\mathcal{B}} \rho_0 \dot{\boldsymbol{\varphi}} \cdot \delta\boldsymbol{\varphi}\, dV = 0 \tag{12.5}$$

which is a restatement of the principle of virtual work in terms of the kinetic and total potential energies.

Next, we wish to introduce the scalar formalism that is equivalent to the vector formalism via the descriptive scalar function, i.e., the autonomous Lagrangian $\mathcal{L}(\boldsymbol{\varphi}, \dot{\boldsymbol{\varphi}}) : TC \to \mathbb{R}$ in the Lagrangian mechanics framework. Then we have

$$\delta\mathcal{L}(\boldsymbol{\varphi}, \dot{\boldsymbol{\varphi}}) - \frac{d}{dt}\int_{\mathcal{B}} \rho_0 \dot{\boldsymbol{\varphi}} \cdot \delta\boldsymbol{\varphi}\, dV = 0 \tag{12.6}$$

where the autonomous Lagrangian is defined as

$$\begin{aligned} \mathcal{L}(\boldsymbol{\varphi}, \dot{\boldsymbol{\varphi}}) &= \mathcal{K}(\dot{\boldsymbol{\varphi}}) - \mathcal{U}(\boldsymbol{\varphi}) \\ &= \frac{1}{2}\int_{\mathcal{B}} \rho_0 \dot{\boldsymbol{\varphi}} \cdot \dot{\boldsymbol{\varphi}} dV - \int_{\mathcal{B}} \rho_0 \Psi(\mathbf{F})\, dV \\ &\quad + \int_{\partial\mathcal{B}_\sigma} \bar{\mathbf{T}}(\mathbf{X}, \hat{\mathbf{N}}) \cdot \boldsymbol{\varphi}(\mathbf{X}, t)\, dA + \int_{\mathcal{B}} \rho_0 \mathbf{B}(\mathbf{X}) \cdot \boldsymbol{\varphi}(\mathbf{X}, t)\, dV \\ &= \int_{\mathcal{B}} \bar{\mathcal{L}}(\boldsymbol{\varphi}, \dot{\boldsymbol{\varphi}}, \mathbf{F})\, dV + \int_{\partial\mathcal{B}_\sigma} \bar{\mathbf{T}}(\mathbf{X}, \hat{\mathbf{N}}) \cdot \boldsymbol{\varphi}(\mathbf{X}, t)\, dA \end{aligned} \tag{12.7}$$

where the Lagrangian density function $\bar{\mathcal{L}}(\boldsymbol{\varphi}, \dot{\boldsymbol{\varphi}}, \mathbf{F})$ is given by

$$\bar{\mathcal{L}}(\boldsymbol{\varphi}, \dot{\boldsymbol{\varphi}}, \mathbf{F}) = \frac{1}{2}\rho_0 \dot{\boldsymbol{\varphi}}(\mathbf{X}, t) \cdot \dot{\boldsymbol{\varphi}}(\mathbf{X}, t) - \rho_0 \Psi(\mathbf{F}) + \rho_0 \mathbf{B} \cdot \boldsymbol{\varphi}(\mathbf{X}, t) \tag{12.8}$$

In the Lagrangian framework, Equation (12.6) can be expressed in terms of the Lagrangian density function as

$$\begin{aligned}&\int_{\mathcal{B}}\left[\frac{\partial\bar{\mathcal{L}}}{\partial\boldsymbol{\varphi}}-\left(\mathrm{DIV}\,\frac{\partial\bar{\mathcal{L}}}{\partial\mathbf{F}}\right)-\frac{d}{dt}\left(\frac{\partial\bar{\mathcal{L}}}{\partial\dot{\boldsymbol{\varphi}}}\right)\right]\cdot\delta\boldsymbol{\varphi}\,dV\\&+\int_{\partial\mathcal{B}_\sigma}\left(\bar{\mathbf{T}}(\mathbf{X},\hat{\mathbf{N}})+\frac{\partial\bar{\mathcal{L}}}{\partial\mathbf{F}}\cdot\hat{\mathbf{N}}\right)\cdot\delta\boldsymbol{\varphi}\,dA=0\end{aligned} \tag{12.9}$$

As the virtual displacement is arbitrary, we have the governing equation of motion in the body as

$$\frac{\partial\bar{\mathcal{L}}}{\partial\boldsymbol{\varphi}}-\frac{d}{dt}\left(\frac{\partial\bar{\mathcal{L}}}{\partial\dot{\boldsymbol{\varphi}}}\right)-\left(\mathrm{DIV}\,\frac{\partial\bar{\mathcal{L}}}{\partial\mathbf{F}}\right)=\mathbf{0},\ \text{ for }(\mathbf{X},t)\in\mathcal{B}\times\mathcal{T} \tag{12.10}$$

and the traction boundary condition on the prescribed traction boundary surface as

$$\bar{\mathbf{T}}(\mathbf{X},\hat{\mathbf{N}})=-\frac{\partial\bar{\mathcal{L}}}{\partial\mathbf{F}}\cdot\hat{\mathbf{N}},\ \text{ for }(\mathbf{X},t)\in\partial\mathcal{B}_\sigma\times\mathcal{T} \tag{12.11}$$

Alternatively, in the Hamiltonian mechanics framework, Equation (12.5) can be represented as

$$\delta\left\{\int_{\mathcal{B}}\boldsymbol{\phi}\cdot\dot{\boldsymbol{\varphi}}\,dV-\mathcal{H}(\boldsymbol{\phi},\boldsymbol{\varphi})\right\}-\frac{d}{dt}\int_{\mathcal{B}}\boldsymbol{\phi}\cdot\delta\boldsymbol{\varphi}\,dV=0 \tag{12.12}$$

where the descriptive scalar function, i.e. the autonomous Hamiltonian $\mathcal{H}(\boldsymbol{\phi},\boldsymbol{\varphi}):T^*\mathcal{C}\to\mathbb{R}$, is defined as

$$\begin{aligned}\mathcal{H}(\boldsymbol{\phi},\boldsymbol{\varphi})&=\mathcal{K}(\boldsymbol{\phi})+\mathcal{U}(\boldsymbol{\varphi})\\&=\int_{\mathcal{B}}\bar{\mathcal{H}}(\boldsymbol{\phi},\boldsymbol{\varphi},\mathbf{F})\,dV-\int_{\partial\mathcal{B}_\sigma}\bar{\mathbf{T}}(\mathbf{X},\hat{\mathbf{N}})\cdot\boldsymbol{\varphi}(\mathbf{X},t)\,dA\end{aligned} \tag{12.13}$$

where the Hamiltonian density function is given by

$$\bar{\mathcal{H}}(\boldsymbol{\phi},\boldsymbol{\varphi},\mathbf{F})=\frac{1}{2\rho_0}\boldsymbol{\phi}(\mathbf{X},t)\cdot\boldsymbol{\phi}(\mathbf{X},t)+\rho_0\Psi(\mathbf{F})-\rho_0\mathbf{B}(\mathbf{X})\cdot\boldsymbol{\varphi}(\mathbf{X},t) \tag{12.14}$$

Consequently, we have

$$\begin{aligned}&\int_{\partial\mathcal{B}_\sigma}\left[\left(\bar{\mathbf{T}}(\mathbf{X},\hat{\mathbf{N}})-\frac{\partial\bar{\mathcal{H}}}{\partial\mathbf{F}}\cdot\hat{\mathbf{N}}\right)\cdot\delta\boldsymbol{\varphi}\right]dA\\&+\int_{\mathcal{B}}\left[\left(\dot{\boldsymbol{\varphi}}-\frac{\partial\bar{\mathcal{H}}}{\partial\boldsymbol{\phi}}\right)\cdot\delta\boldsymbol{\phi}-\left(\dot{\boldsymbol{\phi}}+\frac{\partial\bar{\mathcal{H}}}{\partial\boldsymbol{\varphi}}-\mathrm{DIV}\,\frac{\partial\bar{\mathcal{H}}}{\partial\mathbf{F}}\right)\cdot\delta\boldsymbol{\varphi}\right]dV=0\end{aligned} \tag{12.15}$$

Since the two variations, $\delta\boldsymbol{\varphi}$ and $\delta\boldsymbol{\varphi}$, are arbitrary, we have the following Hamiltonian density equations of motion, which hold inside the body

$$\dot{\boldsymbol{\phi}}+\frac{\partial\bar{\mathcal{H}}}{\partial\boldsymbol{\varphi}}=\mathrm{DIV}\,\frac{\partial\bar{\mathcal{H}}}{\partial\mathbf{F}},\quad\dot{\boldsymbol{\varphi}}=\frac{\partial\bar{\mathcal{H}}}{\partial\boldsymbol{\phi}},\quad\text{for }(\mathbf{X},t)\in\mathcal{B}\times\mathcal{T} \tag{12.16}$$

and the natural boundary condition on the prescribed traction boundary surface $\partial\mathcal{B}_\sigma$ as

$$\bar{\mathbf{T}}(\mathbf{X},\hat{\mathbf{N}}) = \frac{\partial\bar{\mathcal{H}}}{\partial\mathbf{F}}\cdot\hat{\mathbf{N}} \quad \text{for } (\mathbf{X},t)\in\partial\mathcal{B}_\sigma\times\mathcal{T} \tag{12.17}$$

In addition, in the Total Energy framework Equation (12.5) can be written as

$$\delta\left\{\int_{\mathcal{B}}\rho_0\dot{\boldsymbol{\varphi}}\cdot\dot{\boldsymbol{\varphi}}\,dV - \mathcal{E}(\boldsymbol{\varphi},\dot{\boldsymbol{\varphi}})\right\} - \frac{d}{dt}\int_{\mathcal{B}}\rho_0\dot{\boldsymbol{\varphi}}\cdot\delta\boldsymbol{\varphi}\,dV = 0 \tag{12.18}$$

where the descriptive scalar function, i.e. the autonomous Total Energy $\mathcal{E}(\boldsymbol{\varphi},\dot{\boldsymbol{\varphi}}) : T\mathcal{C}\to\mathbb{R}$, for dead load cases is defined as

$$\begin{aligned}\mathcal{E}(\boldsymbol{\varphi},\dot{\boldsymbol{\varphi}}) &= \mathcal{K}(\dot{\boldsymbol{\phi}}) + \mathcal{U}(\boldsymbol{\varphi})\\ &= \int_{\mathcal{B}}\bar{\mathcal{E}}(\boldsymbol{\varphi},\dot{\boldsymbol{\varphi}},\mathbf{F})\,dV - \int_{\partial\mathcal{B}_\sigma}\bar{\mathbf{T}}(\mathbf{X},\hat{\mathbf{N}})\cdot\boldsymbol{\varphi}(\mathbf{X},t)\,dA\end{aligned} \tag{12.19}$$

where the Total Energy density function $\bar{\mathcal{E}}(\boldsymbol{\varphi},\dot{\boldsymbol{\varphi}},\mathbf{F})$ is defined as

$$\bar{\mathcal{E}}(\boldsymbol{\varphi},\dot{\boldsymbol{\varphi}},\mathbf{F}) = \frac{1}{2}\rho_0\dot{\boldsymbol{\varphi}}\cdot\dot{\boldsymbol{\varphi}} + \rho_0\Psi(\mathbf{F}) - \rho_0\mathbf{B}(\mathbf{X})\cdot\boldsymbol{\varphi} \tag{12.20}$$

Consequently, we have

$$\begin{aligned}&\int_{\mathcal{B}}\left[\mathrm{DIV}\frac{\partial\bar{\mathcal{E}}}{\partial\mathbf{F}} - \frac{\partial\bar{\mathcal{E}}}{\partial\boldsymbol{\varphi}} - \frac{d}{dt}\left(\frac{\partial\bar{\mathcal{E}}}{\partial\dot{\boldsymbol{\varphi}}}\right)\right]\cdot\delta\boldsymbol{\varphi}\,dV\\ &+\int_{\partial\mathcal{B}_\sigma}\left[\left(\bar{\mathbf{T}}(\mathbf{X},\hat{\mathbf{N}}) - \frac{\partial\bar{\mathcal{E}}}{\partial\mathbf{F}}\cdot\hat{\mathbf{N}}\right)\cdot\delta\boldsymbol{\varphi}\right]dA = 0\end{aligned} \tag{12.21}$$

Since the virtual displacement $\delta\boldsymbol{\varphi}$ is arbitrary, we have the Total Energy density representation of the equation of motion, which holds inside the body

$$\frac{d}{dt}\left(\frac{\partial\bar{\mathcal{E}}}{\partial\dot{\boldsymbol{\varphi}}}\right) + \frac{\partial\bar{\mathcal{E}}}{\partial\boldsymbol{\varphi}} - \left(\mathrm{DIV}\frac{\partial\bar{\mathcal{E}}}{\partial\mathbf{F}}\right) = \mathbf{0},\ \text{for } (\mathbf{X},t)\in\mathcal{B}\times\mathcal{T} \tag{12.22}$$

and the natural boundary condition on the prescribed traction boundary surface $\partial\mathcal{B}_\sigma$ as

$$\bar{\mathbf{T}}(\mathbf{X},\hat{\mathbf{N}}) = \frac{\partial\bar{\mathcal{E}}}{\partial\mathbf{F}}\cdot\hat{\mathbf{N}} \quad \text{for } (\mathbf{X},t)\in\partial\mathcal{B}_\sigma\times\mathcal{T} \tag{12.23}$$

12.2 EQUIVALENCE IN HAMILTON'S PRINCIPLE OR HAMILTON'S LAW OF VARYING ACTION

As discussed in Chapter 10, in elastodynamics, Hamilton's principle or Hamilton's law of varying action can be expressed in several forms of representation. In this section, we shall investigate the equivalence (see Figure 12.1; we prefer and advocate Hamilton's law of varying action due to the inconsistency with Hamilton's principle) between the various expressions given in several forms.

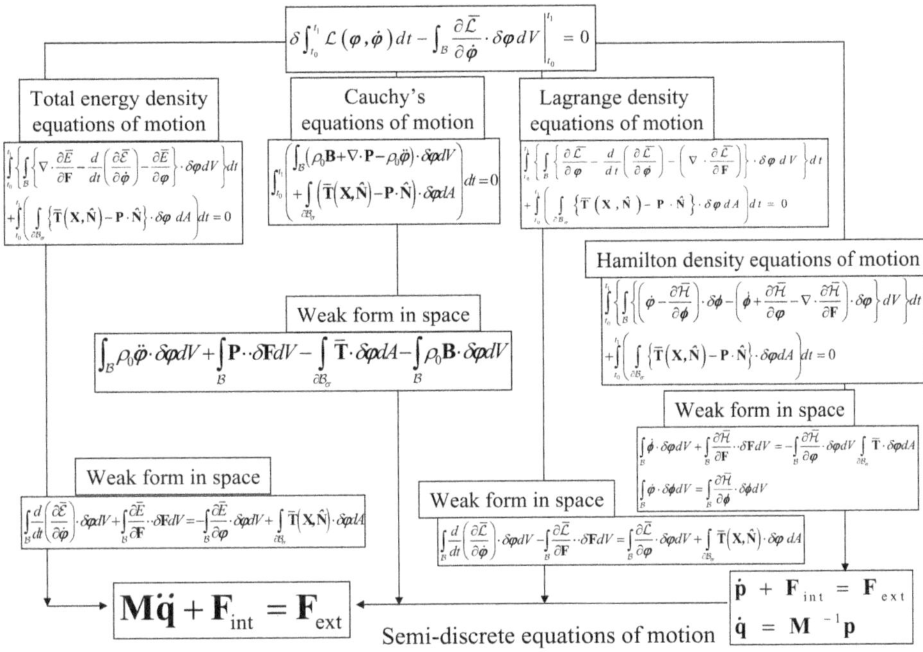

Figure 12.1. Lagrangian/Hamiltonian/Total Energy framework for continuous-body problems

As a standard expression, Hamilton's principle is defined as

$$\mathrm{HP}_{\mathcal{L}} : \delta \int_{t_0}^{t_1} \mathcal{L}(\boldsymbol{\varphi}, \dot{\boldsymbol{\varphi}})\, dt = 0, \qquad \delta\boldsymbol{\varphi}(\mathbf{X}, t_0) = \delta\boldsymbol{\varphi}(\mathbf{X}, t_1) = \mathbf{0} \tag{12.24}$$

and alternatively, Hamilton's law of varying action is given by

$$\mathrm{HLVA}_{\mathcal{L}} : \ \delta \int_{t_0}^{t_1} \mathcal{L}(\boldsymbol{\varphi}, \dot{\boldsymbol{\varphi}})\, dt - \int_{\mathcal{B}} \rho_0 \dot{\boldsymbol{\varphi}} \cdot \delta\boldsymbol{\varphi}\, dV \Big|_{t_0}^{t_1} = 0 \tag{12.25}$$

For dead load cases, the autonomous Lagrangian $\mathcal{L}(\boldsymbol{\varphi}, \dot{\boldsymbol{\varphi}}) : T\mathcal{C} \to \mathbb{R}$ above is defined as

$$\mathcal{L}(\boldsymbol{\varphi}, \dot{\boldsymbol{\varphi}}) = \frac{1}{2}\int_{\mathcal{B}} \rho_0 \dot{\boldsymbol{\varphi}} \cdot \dot{\boldsymbol{\varphi}}\, dV - \int_{\mathcal{B}} \rho_0 \Psi(\mathbf{F})\, dV + \int_{\partial\mathcal{B}_\sigma} \bar{\mathbf{T}}(\mathbf{X}, \hat{\mathbf{N}}) \cdot \boldsymbol{\varphi}(\mathbf{X}, t)\, dA + \int_{\mathcal{B}} \rho_0 \mathbf{B}(\mathbf{X}) \cdot \boldsymbol{\varphi}(\mathbf{X}, t)\, dV \tag{12.26}$$

Substituting the Lagrangian above into Equation (12.24) or Equation (12.25) gives rise to the following

$$\int_{t_0}^{t_1} \left[\int_{\mathcal{B}} (\mathrm{DIV}\,\mathbf{P} + \rho_0 \mathbf{B} - \rho_0 \ddot{\boldsymbol{\varphi}}) \cdot \delta\boldsymbol{\varphi}\, dV + \int_{\partial\mathcal{B}_\sigma} (\bar{\mathbf{T}} - \mathbf{P} \cdot \hat{\mathbf{N}}) \cdot \delta\boldsymbol{\varphi}\, dV \right] dt = 0 \tag{12.27}$$

As the virtual displacement is arbitrary, we obtain the governing equation in the continuous-body system in the vector formalism as

$$\text{DIV}\,\mathbf{P} + \rho_0\mathbf{B} - \rho_0\ddot{\boldsymbol{\varphi}} = \mathbf{0}, \ \text{for } (\mathbf{X}, t) \in \mathcal{B} \times \mathcal{T} \tag{12.28}$$

and the traction boundary condition on the prescribed traction boundary surface as

$$\bar{\mathbf{T}}(\mathbf{X}, \hat{\mathbf{N}}) = \mathbf{P} \cdot \hat{\mathbf{N}}, \ \text{for } (\mathbf{X}, t) \in \partial\mathcal{B}_\sigma \times \mathcal{T} \tag{12.29}$$

On the other hand, in the scalar formalism via the Lagrangian framework, Equation (12.27) can be expressed in terms of the Lagrangian density function as

$$\begin{aligned} &\int_{t_0}^{t_1} \left\{ \int_{\mathcal{B}} \left[\frac{\partial \bar{\mathcal{L}}}{\partial \boldsymbol{\varphi}} - \left(\text{DIV} \frac{\partial \bar{\mathcal{L}}}{\partial \mathbf{F}} \right) - \frac{d}{dt} \left(\frac{\partial \bar{\mathcal{L}}}{\partial \dot{\boldsymbol{\varphi}}} \right) \right] \cdot \delta\boldsymbol{\varphi} \, dV \right\} dt \\ &+ \int_{t_0}^{t_1} \left[\int_{\partial\mathcal{B}_\sigma} \left(\bar{\mathbf{T}}(\mathbf{X}, \hat{\mathbf{N}}) + \frac{\partial \bar{\mathcal{L}}}{\partial \mathbf{F}} \cdot \hat{\mathbf{N}} \right) \cdot \delta\boldsymbol{\varphi} \, dA \right] dt = 0 \end{aligned} \tag{12.30}$$

As the virtual displacement is arbitrary, we have the scalar form of the governing equation in the continuous-body system as

$$\frac{\partial \bar{\mathcal{L}}}{\partial \boldsymbol{\varphi}} - \frac{d}{dt} \left(\frac{\partial \bar{\mathcal{L}}}{\partial \dot{\boldsymbol{\varphi}}} \right) - \left(\text{DIV} \frac{\partial \bar{\mathcal{L}}}{\partial \mathbf{F}} \right) = \mathbf{0}, \ \text{for } (\mathbf{X}, t) \in \mathcal{B} \times \mathcal{T} \tag{12.31}$$

and the traction boundary condition on the prescribed traction boundary surface as

$$\bar{\mathbf{T}}(\mathbf{X}, \hat{\mathbf{N}}) = -\frac{\partial \bar{\mathcal{L}}}{\partial \mathbf{F}} \cdot \hat{\mathbf{N}}, \ \text{for } (\mathbf{X}, t) \in \partial\mathcal{B}_\sigma \times \mathcal{T} \tag{12.32}$$

In Hamiltonian mechanics framework, the modified Hamilton's principle reads as

$$\begin{aligned} &\text{HP}_{\mathcal{H}} : \delta \int_{t_0}^{t_1} \left(\int_{\mathcal{B}} \boldsymbol{\phi} \cdot \dot{\boldsymbol{\varphi}} dV - \mathcal{H}(\boldsymbol{\phi}, \boldsymbol{\varphi}) \right) dt = 0, \\ &\delta\boldsymbol{\varphi}(\mathbf{X}, t_0) = \delta\boldsymbol{\varphi}(\mathbf{X}, t_1) = \mathbf{0} \end{aligned} \tag{12.33}$$

and Hamilton's law of varying action is given by

$$\text{HLVA}_{\mathcal{H}} : \ \delta \int_{t_0}^{t_1} \left\{ \int_{\mathcal{B}} \boldsymbol{\phi} \cdot \dot{\boldsymbol{\varphi}} \, dV - \mathcal{H}(\boldsymbol{\phi}, \boldsymbol{\varphi}) \right\} dt - \left[\int_{\mathcal{B}} \boldsymbol{\phi} \cdot \delta\boldsymbol{\varphi} \, dV \right] \Bigg|_{t_0}^{t_1} = 0 \tag{12.34}$$

In the above equations, for dead load cases the autonomous Hamiltonian $\mathcal{H}(\boldsymbol{\phi}, \boldsymbol{\varphi}) : T^*\mathcal{C} \to \mathbb{R}$ is defined as

$$\mathcal{H}(\boldsymbol{\phi}, \boldsymbol{\varphi}) = \int_{\mathcal{B}} \bar{\mathcal{H}}(\boldsymbol{\phi}, \boldsymbol{\varphi}, \mathbf{F}) \, dV - \int_{\partial\mathcal{B}_\sigma} \bar{\mathbf{T}}(\mathbf{X}, \hat{\mathbf{N}}) \cdot \boldsymbol{\varphi}(\mathbf{X}, t) \, dA \tag{12.35}$$

Therefore, Hamilton's law of varying action can be written as

$$\begin{aligned}&\int_{t_0}^{t_1}\left\{\int_{\partial\mathcal{B}_\sigma}\left[\left(\bar{\mathbf{T}}(\mathbf{X},\hat{\mathbf{N}})-\frac{\partial\bar{\mathcal{H}}}{\partial\mathbf{F}}\cdot\hat{\mathbf{N}}\right)\cdot\delta\boldsymbol{\varphi}\right]dA\right\}dt\\&+\int_{t_0}^{t_1}\left\{\int_{\mathcal{B}}\left[\left(\dot{\boldsymbol{\varphi}}-\frac{\partial\bar{\mathcal{H}}}{\partial\boldsymbol{\phi}}\right)\cdot\delta\boldsymbol{\phi}-\left(\dot{\boldsymbol{\phi}}+\frac{\partial\bar{\mathcal{H}}}{\partial\boldsymbol{\varphi}}-\mathrm{DIV}\,\frac{\partial\bar{\mathcal{H}}}{\partial\mathbf{F}}\right)\cdot\delta\boldsymbol{\varphi}\right]dV\right\}dt=0\end{aligned}\tag{12.36}$$

Since the two variations, $\delta\boldsymbol{\varphi}$ and $\delta\boldsymbol{\varphi}$, are arbitrary, we have the Hamiltonian density equations of motion, which hold inside the body

$$\dot{\boldsymbol{\phi}}+\frac{\partial\bar{\mathcal{H}}}{\partial\boldsymbol{\varphi}}=\mathrm{DIV}\,\frac{\partial\bar{\mathcal{H}}}{\partial\mathbf{F}},\quad \dot{\boldsymbol{\varphi}}=\frac{\partial\bar{\mathcal{H}}}{\partial\boldsymbol{\phi}},\quad \text{for } (\mathbf{X},t)\in\mathcal{B}\times\mathcal{T}\tag{12.37}$$

and the natural boundary condition on the prescribed traction boundary surface $\partial\mathcal{B}_\sigma$ as

$$\bar{\mathbf{T}}(\mathbf{X},\hat{\mathbf{N}})=\frac{\partial\bar{\mathcal{H}}}{\partial\mathbf{F}}\cdot\hat{\mathbf{N}}\quad\text{for } (\mathbf{X},t)\in\partial\mathcal{B}_\sigma\times\mathcal{T}\tag{12.38}$$

Lastly, in the Total Energy framework, Hamilton's principle can be expressed as

$$\begin{aligned}&\mathrm{HP}_{\mathcal{E}}:\delta\int_{t_0}^{t_1}\left(\int_{\mathcal{B}}\frac{\partial\bar{\mathcal{E}}(\boldsymbol{\varphi},\dot{\boldsymbol{\varphi}},\mathbf{F})}{\partial\dot{\boldsymbol{\varphi}}}\cdot\dot{\boldsymbol{\varphi}}dV-\mathcal{E}(\boldsymbol{\varphi},\dot{\boldsymbol{\varphi}})\right)dt=0,\\&\delta\boldsymbol{\varphi}(\mathbf{X},t_0)=\delta\boldsymbol{\varphi}(\mathbf{X},t_1)=\mathbf{0}\end{aligned}\tag{12.39}$$

and Hamilton's law of varying action is given by

$$\begin{aligned}&\mathrm{HLVA}_{\mathcal{E}}:\delta\int_{t_0}^{t_1}\left(\int_{\mathcal{B}}\frac{\partial\bar{\mathcal{E}}(\boldsymbol{\varphi},\dot{\boldsymbol{\varphi}},\mathbf{F})}{\partial\dot{\boldsymbol{\varphi}}}\cdot\dot{\boldsymbol{\varphi}}\,dV-\mathcal{E}(\boldsymbol{\varphi},\dot{\boldsymbol{\varphi}})\right)dt\\&-\int_{\mathcal{B}}\frac{\partial\bar{\mathcal{E}}}{\partial\dot{\boldsymbol{\varphi}}}\cdot\delta\boldsymbol{\varphi}dV\bigg|_{t_0}^{t_1}=0\end{aligned}\tag{12.40}$$

In the above equations, for dead load cases the autonomous Total Energy $\mathcal{E}(\boldsymbol{\varphi},\dot{\boldsymbol{\varphi}}):T\mathcal{C}\to\mathbb{R}$ is defined as

$$\mathcal{E}(\boldsymbol{\varphi},\dot{\boldsymbol{\varphi}})=\int_{\mathcal{B}}\bar{\mathcal{E}}(\boldsymbol{\varphi},\dot{\boldsymbol{\varphi}},\mathbf{F})\,dV-\int_{\partial\mathcal{B}_\sigma}\bar{\mathbf{T}}(\mathbf{X},\hat{\mathbf{N}})\cdot\boldsymbol{\varphi}(\mathbf{X},t)\,dA\tag{12.41}$$

Therefore, Hamilton's law of varying action can be written as

$$\begin{aligned}&\int_{t_0}^{t_1}\left\{\int_{\mathcal{B}}\left[\mathrm{DIV}\,\frac{\partial\bar{\mathcal{E}}}{\partial\mathbf{F}}-\frac{\partial\bar{\mathcal{E}}}{\partial\boldsymbol{\varphi}}-\frac{d}{dt}\left(\frac{\partial\bar{\mathcal{E}}}{\partial\dot{\boldsymbol{\varphi}}}\right)\right]\cdot\delta\boldsymbol{\varphi}\,dV\right\}dt\\&+\int_{t_0}^{t_1}\left\{\int_{\partial\mathcal{B}_\sigma}\left[\left(\bar{\mathbf{T}}(\mathbf{X},\hat{\mathbf{N}})-\frac{\partial\bar{\mathcal{E}}}{\partial\mathbf{F}}\cdot\hat{\mathbf{N}}\right)\cdot\delta\boldsymbol{\varphi}\right]dA\right\}dt=0\end{aligned}\tag{12.42}$$

Since the virtual displacement $\delta\boldsymbol{\varphi}$ is arbitrary, we have the Total Energy density representation of the equation of motion, which holds inside the body given by

$$\frac{d}{dt}\left(\frac{\partial\bar{\mathcal{E}}}{\partial\dot{\boldsymbol{\varphi}}}\right)+\frac{\partial\bar{\mathcal{E}}}{\partial\boldsymbol{\varphi}}-\left(\mathrm{DIV}\,\frac{\partial\bar{\mathcal{E}}}{\partial\mathbf{F}}\right)=\mathbf{0},\ \text{for } (\mathbf{X},t)\in\mathcal{B}\times\mathcal{T} \tag{12.43}$$

and the natural boundary condition on the prescribed traction boundary surface $\partial\mathcal{B}_\sigma$

$$\bar{\mathbf{T}}(\mathbf{X},\hat{\mathbf{N}})=\frac{\partial\bar{\mathcal{E}}}{\partial\mathbf{F}}\cdot\hat{\mathbf{N}}\quad\text{for}\quad(\mathbf{X},t)\in\partial\mathcal{B}_\sigma\times\mathcal{T} \tag{12.44}$$

12.3 EQUIVALENCE IN THE PRINCIPLE OF BALANCE OF MECHANICAL ENERGY

Next, our attention will be confined to the case that the Total Energy for the continuous-body dynamical system is conserved. Therefore, the external forces are dead loads. Then, the rate of change of the Total Energy can be written as

$$\begin{aligned}\frac{\partial\mathcal{E}(\boldsymbol{\varphi},\dot{\boldsymbol{\varphi}})}{\partial t}&=\int_{\mathcal{B}}\left[(\rho_0\ddot{\boldsymbol{\varphi}}(\mathbf{X},t)-\mathrm{DIV}\,\mathbf{P}-\rho_0\mathbf{B}(\mathbf{X}))\cdot\dot{\boldsymbol{\varphi}}\right]dV\\&\quad-\int_{\partial\mathcal{B}_\sigma}\left[\left(\bar{\mathbf{T}}(\mathbf{X},\hat{\mathbf{N}})-\mathbf{P}\cdot\hat{\mathbf{N}}\right)\cdot\dot{\boldsymbol{\varphi}}(\mathbf{X},t)\right]dA=0\end{aligned} \tag{12.45}$$

Note that in the body $\mathcal{B}$ or on the surface $\mathcal{B}_\sigma$, $\dot{\boldsymbol{\varphi}}(\mathbf{X},t)$ is not arbitrary, but is only a dependent variable as a field function. For any value of the dependent variable, when the terms enclosed in the parenthesis (simple brackets) vanish, the total energy is conserved. As a consequence, we obtain the vector formalism of the governing equation of motion in the body

$$\mathrm{DIV}\,\mathbf{P}+\rho_0\mathbf{B}-\rho_0\ddot{\boldsymbol{\varphi}}=\mathbf{0},\ \text{for } (\mathbf{X},t)\in\mathcal{B}\times\mathcal{T} \tag{12.46}$$

and the traction boundary condition on the prescribed traction boundary surface as

$$\bar{\mathbf{T}}(\mathbf{X},\hat{\mathbf{N}})=\mathbf{P}\cdot\hat{\mathbf{N}},\ \text{for } (\mathbf{X},t)\in\partial\mathcal{B}_\sigma\times\mathcal{T} \tag{12.47}$$

In terms of the scalar formalism such as the Total Energy density function, the rate of change of the Total Energy can be rewritten as

$$\begin{aligned}\frac{d\mathcal{E}}{dt}&=\int_{\mathcal{B}}\left(\frac{\partial\bar{\mathcal{E}}(\boldsymbol{\varphi},\dot{\boldsymbol{\varphi}},\mathbf{F})}{\partial\boldsymbol{\varphi}}-\mathrm{DIV}\,\frac{\partial\bar{\mathcal{E}}}{\partial\mathbf{F}}+\frac{d}{dt}\left(\frac{\partial\bar{\mathcal{E}}}{\partial\dot{\boldsymbol{\varphi}}}\right)\right)\cdot\dot{\boldsymbol{\varphi}}(\mathbf{X},t)\,dV\\&\quad-\int_{\partial\mathcal{B}_\sigma}\left(\bar{\mathbf{T}}-\frac{\partial\bar{\mathcal{E}}}{\partial\mathbf{F}}\cdot\hat{\mathbf{N}}\right)\cdot\dot{\boldsymbol{\varphi}}(\mathbf{X},t)\,dA=0\end{aligned} \tag{12.48}$$

Consequently, we have the Total Energy density representation of the equation of motion, which holds inside the body

$$\frac{d}{dt}\left(\frac{\partial\bar{\mathcal{E}}}{\partial\dot{\boldsymbol{\varphi}}}\right)+\frac{\partial\bar{\mathcal{E}}}{\partial\boldsymbol{\varphi}}-\left(\mathrm{DIV}\,\frac{\partial\bar{\mathcal{E}}}{\partial\mathbf{F}}\right)=\mathbf{0},\ \text{for } (\mathbf{X},t)\in\mathcal{B}\times\mathcal{T} \tag{12.49}$$

and the natural boundary condition on the prescribed traction boundary surface $\partial\mathcal{B}_\sigma$ as

$$\bar{\mathbf{T}}(\mathbf{X}, \hat{\mathbf{N}}) = \frac{\partial \bar{\mathcal{E}}}{\partial \mathbf{F}} \cdot \hat{\mathbf{N}} \quad \text{for} \ \ (\mathbf{X}, t) \in \partial\mathcal{B}_\sigma \times \mathcal{T} \tag{12.50}$$

In terms of the Lagrangian density function, the rate of change of Jacobi's integral can be written as

$$\begin{aligned} &\int_{\mathcal{B}} \left(\frac{d}{dt}\left(\frac{\partial \bar{\mathcal{L}}}{\partial \dot{\boldsymbol{\varphi}}}\right) + \mathrm{DIV}\,\frac{\partial \bar{\mathcal{L}}}{\partial \mathbf{F}} - \frac{\partial \bar{\mathcal{L}}\,(\boldsymbol{\varphi}, \dot{\boldsymbol{\varphi}}, \mathbf{F})}{\partial \boldsymbol{\varphi}} \right) \cdot \dot{\boldsymbol{\varphi}}\, dV \\ &- \int_{\partial\mathcal{B}_\sigma} \left(\bar{\mathbf{T}}(\mathbf{X}, \hat{\mathbf{N}}) + \frac{\partial \bar{\mathcal{L}}\,(\boldsymbol{\varphi}, \dot{\boldsymbol{\varphi}}, \mathbf{F})}{\partial \mathbf{F}} \cdot \hat{\mathbf{N}} \right) \cdot \dot{\boldsymbol{\varphi}}\,(\mathbf{X}, t)\, dA = 0 \end{aligned} \tag{12.51}$$

Hence, we have the Lagrangian density equation of motion, which holds inside the body

$$\frac{\partial \bar{\mathcal{L}}}{\partial \boldsymbol{\varphi}} - \frac{d}{dt}\left(\frac{\partial \bar{\mathcal{L}}}{\partial \dot{\boldsymbol{\varphi}}}\right) - \left(\mathrm{DIV}\,\frac{\partial \bar{\mathcal{L}}}{\partial \mathbf{F}}\right) = \mathbf{0}, \ \ \text{for} \ (\mathbf{X}, t) \in \mathcal{B} \times \mathcal{T} \tag{12.52}$$

and the traction boundary condition on the prescribed traction boundary surface as

$$\bar{\mathbf{T}}(\mathbf{X}, \hat{\mathbf{N}}) = -\frac{\partial \bar{\mathcal{L}}}{\partial \mathbf{F}} \cdot \hat{\mathbf{N}}, \ \ \text{for} \ (\mathbf{X}, t) \in \partial\mathcal{B}_\sigma \times \mathcal{T} \tag{12.53}$$

In terms of the Hamiltonian density function, the rate of change of the Hamiltonian can be written as

$$\begin{aligned} \frac{d\mathcal{H}}{dt} = &\int_{\mathcal{B}} \left[\left(\dot{\boldsymbol{\phi}} + \frac{\partial \bar{\mathcal{H}}}{\partial \boldsymbol{\varphi}} - \mathrm{DIV}\,\frac{\partial \bar{\mathcal{H}}}{\partial \mathbf{F}} \right) \cdot \frac{d\boldsymbol{\varphi}}{dt} + \left(\frac{\partial \bar{\mathcal{H}}}{\partial \boldsymbol{\phi}} - \dot{\boldsymbol{\varphi}} \right) \cdot \frac{d\boldsymbol{\phi}}{dt} \right] dV \\ &- \int_{\partial\mathcal{B}_\sigma} \left(\bar{\mathbf{T}}(\mathbf{X}, \hat{\mathbf{N}}) - \frac{\partial \bar{\mathcal{H}}}{\partial \mathbf{F}} \cdot \hat{\mathbf{N}} \right) \cdot \dot{\boldsymbol{\varphi}}\,(\mathbf{X}, t)\, dA = 0 \end{aligned} \tag{12.54}$$

For any values of the two dependent variations, we have the Hamiltonian density equations of motion, which hold inside the body

$$\dot{\boldsymbol{\phi}} + \frac{\partial \bar{\mathcal{H}}}{\partial \boldsymbol{\varphi}} = \mathrm{DIV}\,\frac{\partial \bar{\mathcal{H}}}{\partial \mathbf{F}}, \quad \dot{\boldsymbol{\varphi}} = \frac{\partial \bar{\mathcal{H}}}{\partial \boldsymbol{\phi}}, \qquad \text{for} \ (\mathbf{X}, t) \in \mathcal{B} \times \mathcal{T} \tag{12.55}$$

and the natural boundary condition on the prescribed traction boundary surface $\partial\mathcal{B}_\sigma$

$$\bar{\mathbf{T}}(\mathbf{X}, \hat{\mathbf{N}}) = \frac{\partial \bar{\mathcal{H}}}{\partial \mathbf{F}} \cdot \hat{\mathbf{N}} \quad \text{for} \ (\mathbf{X}, t) \in \partial\mathcal{B}_\sigma \times \mathcal{T} \tag{12.56}$$

12.4 EQUIVALENCE OF STRONG AND WEAK FORMS FOR INITIAL BOUNDARY-VALUE PROBLEMS

Our attention is again confined to the case when the total energy for the continuous-body dynamical system is conserved.

Strong Forms In continuum elastodynamics, the local form of balance of linear momentum is given by Cauchy's equation of motion, which is the vector formalism of the governing equation of motion for the initial boundary-value problem as follows:

$$\rho_0 \ddot{\boldsymbol{\varphi}}(\mathbf{X}, t) = \mathrm{DIV}\,\mathbf{P} + \rho_0 \mathbf{B}(\mathbf{X}), \quad \text{in } \mathcal{B} \tag{12.57}$$

subject to the natural boundary condition

$$\mathbf{T}(\mathbf{X}, \hat{\mathbf{N}}) = \mathbf{P} \cdot \hat{\mathbf{N}}(\mathbf{X}, t) \quad \text{for } \mathbf{X} \in \partial\mathcal{B} \tag{12.58}$$

In terms of the scalar Lagrangian density function, Equation (12.57) can be represented as

$$\frac{\partial \bar{\mathcal{L}}(\boldsymbol{\varphi}, \dot{\boldsymbol{\varphi}}, \mathbf{F})}{\partial \boldsymbol{\varphi}} - \frac{d}{dt}\left(\frac{\partial \bar{\mathcal{L}}(\boldsymbol{\varphi}, \dot{\boldsymbol{\varphi}}, \mathbf{F})}{\partial \dot{\boldsymbol{\varphi}}}\right) - \mathrm{DIV}\,\frac{\partial \bar{\mathcal{L}}(\boldsymbol{\varphi}, \dot{\boldsymbol{\varphi}}, \mathbf{F})}{\partial \mathbf{F}} = \mathbf{0} \tag{12.59}$$

where the Lagrangian density function is given by

$$\bar{\mathcal{L}}(\boldsymbol{\varphi}, \dot{\boldsymbol{\varphi}}, \mathbf{F}) = \frac{1}{2}\rho_0 \dot{\boldsymbol{\varphi}} \cdot \dot{\boldsymbol{\varphi}} - \rho_0 \Psi(\mathbf{F}) + \rho_0 \mathbf{B}(\mathbf{X}) \cdot \boldsymbol{\varphi}(\mathbf{X}, t) \tag{12.60}$$

And in terms of the scalar Hamiltonian density function, we have

$$\dot{\boldsymbol{\varphi}} - \frac{\partial \bar{\mathcal{H}}(\boldsymbol{\phi}, \boldsymbol{\varphi}, \mathbf{F})}{\partial \boldsymbol{\phi}} = \mathbf{0}, \quad \dot{\boldsymbol{\phi}} + \frac{\partial \bar{\mathcal{H}}(\boldsymbol{\phi}, \boldsymbol{\varphi}, \mathbf{F})}{\partial \boldsymbol{\varphi}} - \mathrm{DIV}\,\frac{\partial \bar{\mathcal{H}}(\boldsymbol{\phi}, \boldsymbol{\varphi}, \mathbf{F})}{\partial \mathbf{F}} = \mathbf{0} \tag{12.61}$$

where the Hamiltonian density function is given by

$$\bar{\mathcal{H}}(\boldsymbol{\phi}, \boldsymbol{\varphi}, \mathbf{F}) = \frac{1}{2\rho_0}\boldsymbol{\phi} \cdot \boldsymbol{\phi} + \rho_0 \Psi(\mathbf{F}) - \rho_0 \mathbf{B}(\mathbf{X}) \cdot \boldsymbol{\varphi}(\mathbf{X}, t) \tag{12.62}$$

Lastly, in terms of the scalar Total Energy density function, we have

$$\frac{d}{dt}\left(\frac{\partial \bar{\mathcal{E}}(\boldsymbol{\varphi}, \dot{\boldsymbol{\varphi}}, \mathbf{F})}{\partial \dot{\boldsymbol{\varphi}}}\right) + \frac{\partial \bar{\mathcal{E}}(\boldsymbol{\varphi}, \dot{\boldsymbol{\varphi}}, \mathbf{F})}{\partial \boldsymbol{\varphi}} - \mathrm{DIV}\,\frac{\partial \bar{\mathcal{E}}(\boldsymbol{\varphi}, \dot{\boldsymbol{\varphi}}, \mathbf{F})}{\partial \mathbf{F}} = \mathbf{0} \tag{12.63}$$

where the Total Energy density function is given by

$$\bar{\mathcal{E}}(\boldsymbol{\varphi}, \dot{\boldsymbol{\varphi}}, \mathbf{F}) = \frac{1}{2}\rho_0 \dot{\boldsymbol{\varphi}} \cdot \dot{\boldsymbol{\varphi}} + \rho_0 \Psi(\mathbf{F}) - \rho_0 \mathbf{B}(\mathbf{X}) \cdot \boldsymbol{\varphi}(\mathbf{X}, t) \tag{12.64}$$

Note that the following relations are true between these frameworks

$$\begin{aligned}
&\frac{\partial \bar{\mathcal{E}}}{\partial \dot{\boldsymbol{\varphi}}} = \frac{\partial \bar{\mathcal{L}}}{\partial \dot{\boldsymbol{\varphi}}} = \boldsymbol{\phi} = \rho_0 \dot{\boldsymbol{\varphi}}, \\
&\frac{\partial \bar{\mathcal{E}}}{\partial \boldsymbol{\varphi}} = -\frac{\partial \bar{\mathcal{L}}}{\partial \boldsymbol{\varphi}} = \frac{\partial \bar{\mathcal{H}}}{\partial \boldsymbol{\varphi}} = -\rho_0 \mathbf{B}(\mathbf{X}), \\
&\frac{\partial \bar{\mathcal{E}}}{\partial \mathbf{F}} = -\frac{\partial \bar{\mathcal{L}}}{\partial \mathbf{F}} = \frac{\partial \bar{\mathcal{H}}}{\partial \mathbf{F}} = \mathbf{P}
\end{aligned} \tag{12.65}$$

Therefore, in view of Equation (12.65), we can arrive at the conclusion that the differential governing equations of motion in Equations (12.57), (12.59), (12.61) and (12.63) are equivalent to each other as shown in Figure 12.2.

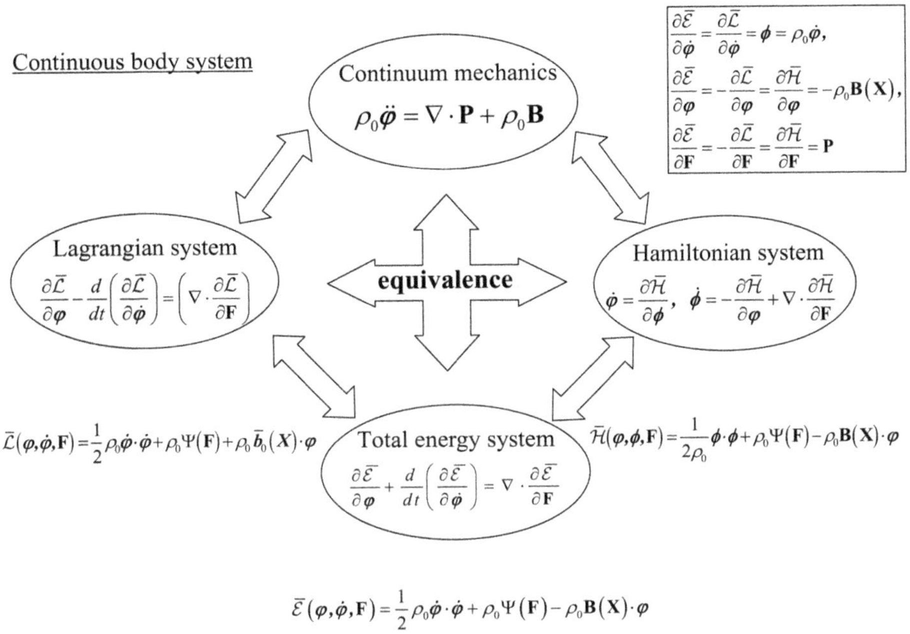

Figure 12.2. Equivalence of governing equations of motion for continuous-body systems

Weak Forms in Variational Calculus Setting Next, we would like to investigate the equivalence of the weak forms for the initial boundary-value problem. As mentioned previously, the weighted residual statement produces the weak form, which is the principle of virtual work in dynamics. It is very important to recall that Cauchy's law (the natural boundary condition) should be imposed to derive the weak form (a variational form for the initial boundary-value problem). As a consequence, the weak form contains Cauchy's equation of motion and Cauchy's law, while the weighted residual statement contains only Cauchy's equation of motion (on the other hand, note that Hamilton's law of varying action contains both Cauchy's equation of motion and Cauchy's law just as the weak form does; and in addition it also contains the weighted residual form in time).

The principle of virtual work in dynamics is given by

$$\int_{\mathcal{B}} \rho_0\ddot{\boldsymbol{\varphi}}\cdot\delta\boldsymbol{\varphi}\,dV + \int_{\mathcal{B}} \mathbf{P} : \delta\mathbf{F}\,dV = \int_{\mathcal{B}} \rho_0\mathbf{B}\cdot\delta\boldsymbol{\varphi}\,dV + \int_{\partial\mathcal{B}_\sigma} \bar{\mathbf{T}}\cdot\delta\boldsymbol{\varphi}\,dA \tag{12.66}$$

But, in terms of the Lagrangian density function, we have

$$\int_{\mathcal{B}} \frac{d}{dt}\left(\frac{\partial\bar{\mathcal{L}}}{\partial\dot{\boldsymbol{\varphi}}}\right)\cdot\delta\boldsymbol{\varphi}\,dV - \int_{\mathcal{B}} \frac{\partial\bar{\mathcal{L}}}{\partial\mathbf{F}} : \delta\mathbf{F}\,dV = \int_{\mathcal{B}} \frac{\partial\bar{\mathcal{L}}}{\partial\boldsymbol{\varphi}}\cdot\delta\boldsymbol{\varphi}\,dV + \int_{\partial\mathcal{B}_\sigma} \bar{\mathbf{T}}\cdot\delta\boldsymbol{\varphi}\,dA \tag{12.67}$$

And in terms of the Hamiltonian density function, we have

$$\begin{gathered}\int_{\mathcal{B}} \dot{\boldsymbol{\phi}}\cdot\delta\boldsymbol{\varphi}\,dV + \int_{\mathcal{B}} \frac{\partial\bar{\mathcal{H}}}{\partial\mathbf{F}} : \delta\mathbf{F}\,dV = -\int_{\mathcal{B}} \frac{\partial\bar{\mathcal{H}}}{\partial\boldsymbol{\varphi}}\cdot\delta\boldsymbol{\varphi}\,dV \int_{\partial\mathcal{B}_\sigma} \bar{\mathbf{T}}\cdot\delta\boldsymbol{\varphi}\,dA \\ \int_{\mathcal{B}} \dot{\boldsymbol{\varphi}}\cdot\delta\boldsymbol{\phi}\,dV = \int_{\mathcal{B}} \frac{\partial\bar{\mathcal{H}}}{\partial\boldsymbol{\phi}}\cdot\delta\boldsymbol{\phi}\,dV\end{gathered} \tag{12.68}$$

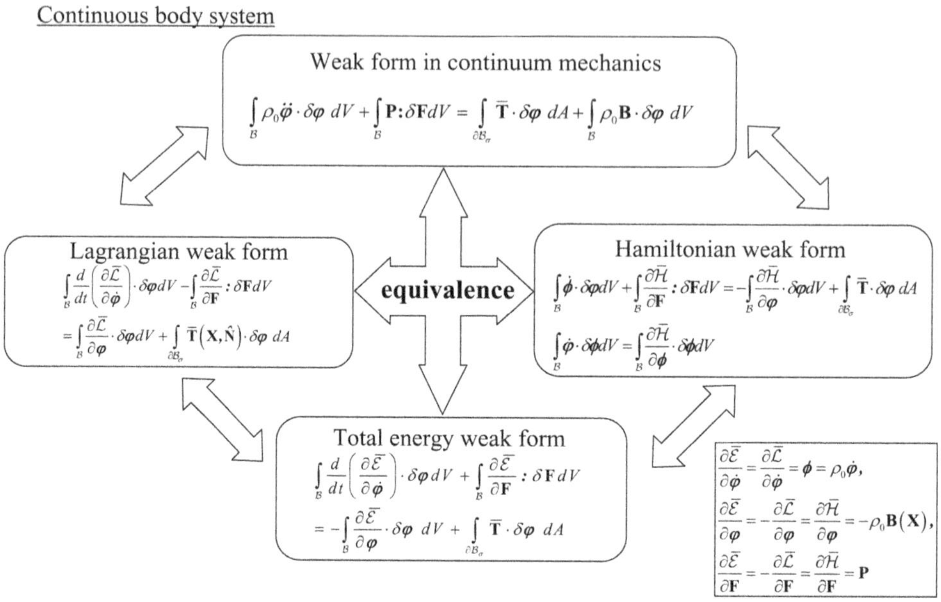

Figure 12.3. Equivalence of weak forms for initial boundary-value problems for continuous-body systems in the variational calculus setting

Lastly, in terms of the Total Energy density function, we have

$$\int_{\mathcal{B}} \frac{d}{dt}\left(\frac{\partial \bar{\mathcal{E}}}{\partial \dot{\boldsymbol{\varphi}}}\right) \cdot \delta\boldsymbol{\varphi}\, dV + \int_{\mathcal{B}} \frac{\partial \bar{\mathcal{E}}}{\partial \mathbf{F}} : \delta\mathbf{F}\, dV = -\int_{\mathcal{B}} \frac{\partial \bar{\mathcal{E}}}{\partial \boldsymbol{\varphi}} \cdot \delta\boldsymbol{\varphi}\, dV + \int_{\partial\mathcal{B}_\sigma} \bar{\mathbf{T}} \cdot \delta\boldsymbol{\varphi}\, dA \tag{12.69}$$

Likewise, in view of Equation (12.65), it is easy to conclude that the weak forms in Equations (12.66) through (12.69) are equivalent to each other as shown in Figure 12.3.

Weak Forms in Differential Calculus Setting In addition, we wish to investigate the equivalence of the weak forms in the differential calculus setting. The weighted residual statement, where the weighting function is velocity, yields the principle of balance of mechanical energy, or equivalently, the theorem of power expended; which is effectively a weak form. Cauchy's law (the natural boundary condition) is also imposed into the weighted residual statement to derive the theorem of power expended, or equivalently, the principle of balance of mechanical energy. Consequently, it contains Cauchy's equation of motion and Cauchy's law, while the weighted residual statement contains only Cauchy's equation of motion.

The principle of balance of mechanical energy, or equivalently, the theorem of power expended as a weak form in the vector formalism is given by

$$\int_{\mathcal{B}} \rho_0 \ddot{\boldsymbol{\varphi}} \cdot \dot{\boldsymbol{\varphi}}\, dV + \int_{\mathcal{B}} \mathbf{P} : \dot{\mathbf{F}}\, dV = \int_{\mathcal{B}} \rho_0 \mathbf{B} \cdot \dot{\boldsymbol{\varphi}}\, dV + \int_{\partial\mathcal{B}_\sigma} \bar{\mathbf{T}} \cdot \dot{\boldsymbol{\varphi}}\, dA \tag{12.70}$$

But, in terms of the scalar Lagrangian density function, we have

$$\int_{\mathcal{B}} \frac{d}{dt}\left(\frac{\partial \bar{\mathcal{L}}}{\partial \dot{\boldsymbol{\varphi}}}\right) \cdot \dot{\boldsymbol{\varphi}}\, dV - \int_{\mathcal{B}} \frac{\partial \bar{\mathcal{L}}}{\partial \mathbf{F}} : \dot{\mathbf{F}}\, dV = \int_{\mathcal{B}} \frac{\partial \bar{\mathcal{L}}}{\partial \boldsymbol{\varphi}} \cdot \dot{\boldsymbol{\varphi}}\, dV + \int_{\partial\mathcal{B}_\sigma} \bar{\mathbf{T}} \cdot \dot{\boldsymbol{\varphi}}\, dA \tag{12.71}$$

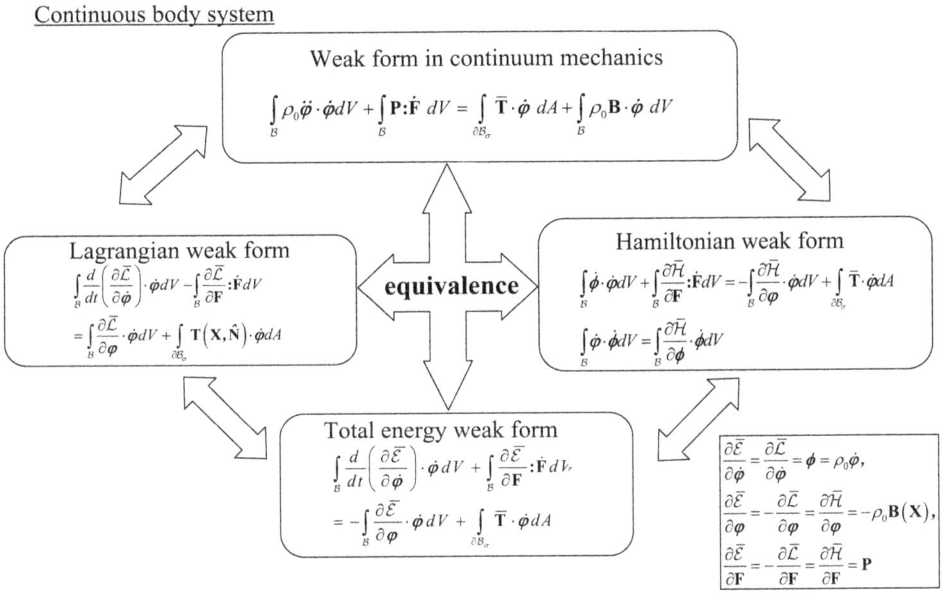

Figure 12.4. Equivalence of weak forms for initial boundary-value problems for continuous-body systems in the differential calculus setting

And in terms of the scalar Hamiltonian density function, we have

$$\begin{aligned} \int_{\mathcal{B}} \dot{\boldsymbol{\phi}} \cdot \boldsymbol{\varphi}\, dV + \int_{\mathcal{B}} \frac{\partial \bar{\mathcal{H}}}{\partial \mathbf{F}} : \dot{\mathbf{F}}\, dV &= -\int_{\mathcal{B}} \frac{\partial \bar{\mathcal{H}}}{\partial \boldsymbol{\varphi}} \cdot \boldsymbol{\varphi}\, dV \int_{\partial \mathcal{B}_\sigma} \bar{\mathbf{T}} \cdot \boldsymbol{\varphi}\, dA \\ \int_{\mathcal{B}} \dot{\boldsymbol{\varphi}} \cdot \dot{\boldsymbol{\phi}}\, dV &= \int_{\mathcal{B}} \frac{\partial \bar{\mathcal{H}}}{\partial \boldsymbol{\phi}} \cdot \dot{\boldsymbol{\phi}}\, dV \end{aligned} \tag{12.72}$$

Lastly, in terms of the scalar Total Energy density function, we have

$$\int_{\mathcal{B}} \frac{d}{dt}\left(\frac{\partial \bar{\mathcal{E}}}{\partial \dot{\boldsymbol{\varphi}}}\right) \cdot \dot{\boldsymbol{\varphi}}\, dV + \int_{\mathcal{B}} \frac{\partial \bar{\mathcal{E}}}{\partial \mathbf{F}} : \dot{\mathbf{F}}\, dV = -\int_{\mathcal{B}} \frac{\partial \bar{\mathcal{E}}}{\partial \boldsymbol{\varphi}} \cdot \dot{\boldsymbol{\varphi}}\, dV + \int_{\partial \mathcal{B}_\sigma} \bar{\mathbf{T}} \cdot \dot{\boldsymbol{\varphi}}\, dA \tag{12.73}$$

Likewise, in view of Equation (12.65), it is easy to conclude that the weak forms in Equations (12.70) through (12.73) are equivalent to each other as shown in Figure 12.4.

12.5 EQUIVALENCE OF THE SEMI-DISCRETE FINITE ELEMENT EQUATIONS OF MOTION

After an infinite dimensional continuous-body dynamical system is projected onto a finite dimensional continuous-body dynamical system via the modified Bubnov-Galerkin weighted residual method, or the space-discrete Lagrangian/Hamiltonian/Total Energy finite element formulations, we readily obtain the semi-discrete finite element equations of motion, i.e.,

$$\mathbf{M}\ddot{\mathbf{q}} + \mathbf{F}_{\text{int}} - \mathbf{F}_{\text{ext}} = \mathbf{0} \tag{12.74}$$

where the consistent mass matrix is denoted by $\mathbf{M}$, the internal force vector by $\mathbf{F}_{\text{int}}$, the external force vector by $\mathbf{F}_{\text{ext}}$, and the nodal acceleration vector by $\ddot{\mathbf{q}}$. In terms of the space-discrete Lagrangian, the above equation can be expressed as

$$\frac{\partial \mathcal{L}^h(\mathbf{q}, \dot{\mathbf{q}})}{\partial \mathbf{q}} - \frac{d}{dt}\left(\frac{\partial \mathcal{L}^h(\mathbf{q}, \dot{\mathbf{q}})}{\partial \dot{\mathbf{q}}}\right) = \mathbf{0} \tag{12.75}$$

where the space-discrete Lagrangian is defined as

$$\mathcal{L}^h(\mathbf{q}, \dot{\mathbf{q}}) = \frac{1}{2}\dot{\mathbf{q}} \cdot \mathbf{M}\dot{\mathbf{q}} - \int_{\mathcal{B}} \rho_0 \Psi\left(\tilde{\mathbf{F}}^h\right) dV + \mathbf{q} \cdot \mathbf{F}_{\text{ext}} \tag{12.76}$$

Next, the space-discrete Hamiltonian equation of motion gives

$$\dot{\mathbf{p}} + \frac{\partial \mathcal{H}^h(\mathbf{p}, \mathbf{q})}{\partial \mathbf{q}} = \mathbf{0}, \quad \dot{\mathbf{q}} - \frac{\partial \mathcal{H}^h(\mathbf{p}, \mathbf{q})}{\partial \mathbf{p}} = \mathbf{0} \tag{12.77}$$

where the space-discrete Hamiltonian is defined as

$$\mathcal{H}^h(\mathbf{p}, \mathbf{q}) = \frac{1}{2}\mathbf{p} \cdot \mathbf{M}^{-1}\mathbf{p} + \int_{\mathcal{B}} \rho_0 \Psi\left(\tilde{\mathbf{F}}^h\right) dV - \mathbf{q} \cdot \mathbf{F}_{\text{ext}} \tag{12.78}$$

Lastly, we have the space-discrete Total Energy representation of the equation of motion as shown below,

$$\frac{\partial \mathcal{E}^h(\mathbf{q}, \dot{\mathbf{q}})}{\partial \mathbf{q}} + \frac{d}{dt}\left(\frac{\partial \mathcal{E}^h(\mathbf{q}, \dot{\mathbf{q}})}{\partial \dot{\mathbf{q}}}\right) = \mathbf{0} \tag{12.79}$$

where the space-discrete Total Energy is defined as

$$\mathcal{E}^h(\mathbf{q}, \dot{\mathbf{q}}) = \frac{1}{2}\dot{\mathbf{q}} \cdot \mathbf{M}\dot{\mathbf{q}} + \int_{\mathcal{B}} \rho_0 \Psi\left(\tilde{\mathbf{F}}^h\right) dV - \mathbf{q} \cdot \mathbf{F}_{\text{ext}} \tag{12.80}$$

Recall that the following relations are true

$$\begin{aligned} \mathbf{p} &= \frac{\partial \mathcal{E}^h(\mathbf{q}, \dot{\mathbf{q}})}{\partial \dot{\mathbf{q}}} = \frac{\partial \mathcal{L}^h(\mathbf{q}, \dot{\mathbf{q}})}{\partial \dot{\mathbf{q}}} = \mathbf{M}\dot{\mathbf{q}} \\ \frac{\partial \mathcal{E}^h(\mathbf{q}, \dot{\mathbf{q}})}{\partial \mathbf{q}} &= \frac{\partial \mathcal{H}^h(\mathbf{p}, \mathbf{q})}{\partial \mathbf{q}} = \frac{-\partial \mathcal{L}^h(\mathbf{q}, \dot{\mathbf{q}})}{\partial \mathbf{q}} = \mathbf{F}_{\text{int}} - \mathbf{F}_{\text{ext}} \end{aligned} \tag{12.81}$$

In view of Equation (12.81), we observe that Equations (12.74), (12.75), (12.77) and (12.79) are equivalent to each other as shown in Figure 12.5.

12.6 EQUIVALENCE OF FINITE ELEMENT FORMULATIONS

In the preceding chapters, we have demonstrated that the semi-discrete finite element equations of motion can be obtained by either the variational calculus setting or the differential calculus setting.

Variational Calculus Setting Employing the Cauchy's equation of motion as the starting point in the vector formalism for the space discretization, the modified Bubnov-Galerkin weighted residual method requires a weak form to be discretized in space. The spatial discretization procedure results in

Continuous-body problem

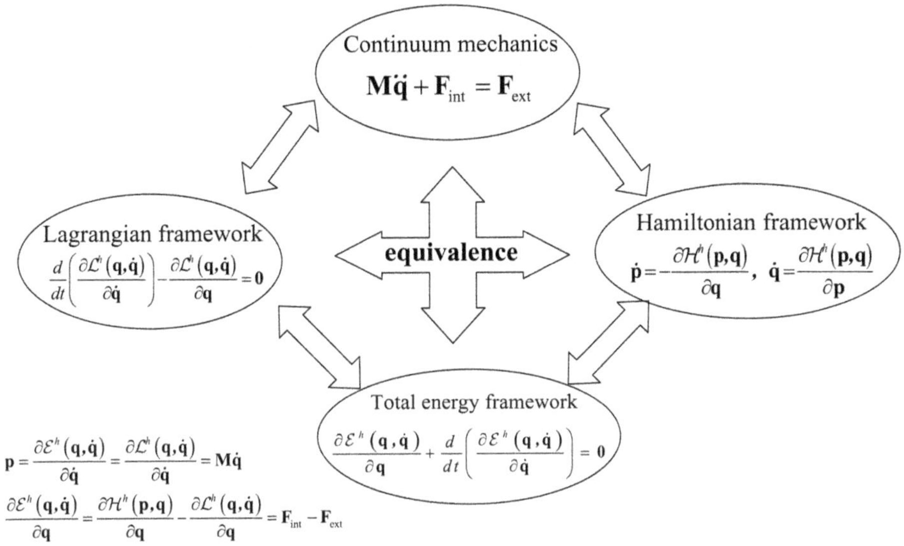

Figure 12.5. Equivalence of the semi-discrete governing equations of motion for continuous-body systems

the following form

$$\delta \mathbf{q} \cdot (\mathbf{M}\ddot{\mathbf{q}} + \mathbf{F}_{\text{int}} - \mathbf{F}_{\text{ext}}) = 0 \tag{12.82}$$

After discretizing the weak form in space leading to the semi-discretized equation of motion, and taking the integral over a time step, it yields the time weighted residual form. Such a form can be readily employed for the time discretization as described in the later chapters. However, note that in contrast, Hamilton's law of varying action is itself a statement of the weak form in space, and in addition, the time weighted residual representation is inherently built-in within the formulation itself. Hamilton's law of varying action is presented below.

Hamilton's law of varying action via the space-discrete Lagrangian finite element formulation in the Lagrangian framework, naturally yields the following representation:

$$\int_{t_0}^{t_1} \delta \mathbf{q} \cdot \left(\frac{\partial \mathcal{L}^h(\mathbf{q}, \dot{\mathbf{q}})}{\partial \mathbf{q}} - \frac{d}{dt}\left(\frac{\partial \mathcal{L}^h(\mathbf{q}, \dot{\mathbf{q}})}{\partial \dot{\mathbf{q}}} \right) \right) dt = 0 \tag{12.83}$$

Likewise, the Hamiltonian framework in the context of Hamilton's law of varying action yields

$$\int_{t_0}^{t_1} \left[\left(\dot{\mathbf{q}} - \frac{\partial \mathcal{H}^h(\mathbf{p}, \mathbf{q})}{\partial \mathbf{p}} \right) \cdot \delta \mathbf{p} - \left(\dot{\mathbf{p}} + \frac{\partial \mathcal{H}^h(\mathbf{p}, \mathbf{q})}{\partial \mathbf{q}} \right) \cdot \delta \mathbf{q} \right] dt = 0 \tag{12.84}$$

Finally, the Total Energy framework in the context of Hamilton's law of varying action yields

$$\int_{t_0}^{t_1} \left\{ \left[\frac{\partial \mathcal{E}^h(\mathbf{q}, \dot{\mathbf{q}})}{\partial \mathbf{q}} + \frac{d}{dt}\left(\frac{\partial \mathcal{E}^h(\mathbf{q}, \dot{\mathbf{q}})}{\partial \dot{\mathbf{q}}} \right) \right] \cdot \delta \mathbf{q} \right\} dt = 0 \tag{12.85}$$

We observe that Equation (12.82) is equivalent to the integrands in Equations (12.83) through (12.85). Note that the spatially discretized form of Hamilton's law of varying action in Equation (12.83) or (12.84) or (12.85) naturally takes the weighted residual statement over the time domain, which is useful in developing time integration schemes as described in the later chapters.

Differential Calculus Setting After the spatial discretization via the principle of balance of mechanical energy, or equivalently, the theorem of power expended, in the vector formalism we arrive at the following form:

$$\dot{\mathbf{q}} \cdot (\mathbf{M}\ddot{\mathbf{q}} + \mathbf{F}_{\text{int}} - \mathbf{F}_{\text{ext}}) = 0 \tag{12.86}$$

Next, taking the integral over a time step yields the time weighted residual form which can be employed for the time discretization of the equation of motion.

On the other hand, for the scalar formalism representations via the principle of balance of mechanical energy, or equivalently, the theorem of power expended, for example, the space-discrete Lagrangian finite element formulation in the Lagrangian framework after taking the integral over a time step, yields

$$\int_{t_0}^{t_1} \dot{\mathbf{q}} \cdot \left(\frac{\partial \mathcal{L}^h(\mathbf{q}, \dot{\mathbf{q}})}{\partial \mathbf{q}} - \frac{d}{dt}\left(\frac{\partial \mathcal{L}^h(\mathbf{q}, \dot{\mathbf{q}})}{\partial \dot{\mathbf{q}}} \right) \right) dt = 0 \tag{12.87}$$

Likewise, the Hamiltonian frameworks yields

$$\int_{t_0}^{t_1} \left[\left(\dot{\mathbf{q}} - \frac{\partial \mathcal{H}^h(\mathbf{p}, \mathbf{q})}{\partial \mathbf{p}} \right) \cdot \dot{\mathbf{p}} - \left(\dot{\mathbf{p}} + \frac{\partial \mathcal{H}^h(\mathbf{p}, \mathbf{q})}{\partial \mathbf{q}} \right) \cdot \dot{\mathbf{q}} \right] dt = 0 \tag{12.88}$$

Finally, the Total Energy framework yields

$$\int_{t_0}^{t_1} \left\{ \left[\frac{\partial \mathcal{E}^h(\mathbf{q}, \dot{\mathbf{q}})}{\partial \mathbf{q}} + \frac{d}{dt}\left(\frac{\partial \mathcal{E}^h(\mathbf{q}, \dot{\mathbf{q}})}{\partial \dot{\mathbf{q}}} \right) \right] \cdot \dot{\mathbf{q}} \right\} dt = 0 \tag{12.89}$$

We observe that Equation (12.86) is equivalent to the integrands in Equations (12.87) through (12.89). Note that only after taking the time integral over a time step of the spatially discretized form of the principle of balance of mechanical energy, or equivalently, the theorem of power expended in Equation (12.87) or (12.88) or (12.89), it then yields the weighted residual statement over the time domain which is useful for developing time integration schemes as discussed in later chapters.

12.7 CONSERVATION LAWS

Due to the symmetry in physics, conservation laws are considered to be critical for assessing the dynamical behavior of the continuous body in motion. The symmetries of descriptive scalar functions such as the Lagrangian, Hamiltonian, and Total Energy for infinite dimensional continuous-body systems with an infinite number of degrees of freedom can be readily shown (we defer to this as an exercise at the end of this chapter). Considering finite dimensional continuous-body dynamical systems after discretizing the space, one can show the invariance properties such as translational, rotational, and time invariance properties for the descriptive scalar functions mentioned above. That is, we can readily prove the conservation of linear momentum, angular momentum, and energy for these descriptive scalar functions (in the sense of Noether). One-parameter Lie group theory (Bluman and Kumei 1989) can be employed to investigate the invariance properties of the descriptive scalar functions for continuous-body systems.

EXERCISES

12.1. Consider the non-dead load case where external potential energy cannot be defined. Discuss the equivalence in the principle of virtual work in dynamics.

12.2. For the same non-dead load case, discuss the equivalence in Hamilton's principle or Hamilton's law of varying action.

12.3. Consider nonconservative continuous-body dynamical systems. Discuss the equivalence in the principle of mechanical energy.

12.4. For the axial bar problem we have discussed in Chapter 9, discuss the equivalence in Hamilton's law of varying action along with the three descriptive scalar functions.

12.5. For the dynamical behavior of the Euler-Bernoulli beam we have discussed in Chapter 9, discuss the equivalence in Hamilton's law of varying action along with the three descriptive scalar functions.

12.6. For the dynamical behavior of the Timoshenko beam we have discussed in Chapter 9, discuss the equivalence in Hamilton's law of varying action along with the three descriptive scalar functions.

12.7. For the dynamical behavior of the Kirchhoff-Love thin plate we have discussed in Chapter 9, discuss the equivalence in Hamilton's law of varying action along with the three descriptive scalar functions.

12.8. For the dynamical behavior of the Reissner-Mindlin plate we have discussed in Chapter 9, discuss the equivalence in Hamilton's law of varying action along with the three descriptive scalar functions.

12.9. If the continuous Lagrangian in an infinite dimensional system is translationally invariant under the Euclidean group, then the linear momentum is conserved. If the continuous Lagrangian is rotationally invariant, then the angular momentum is conserved. Lastly, the total energy is conserved, if the continuous Lagrangian is invariant in time for holonomic-scleronomic systems. Use Lagrangian density equation of motion to examine the invariance properties (symmetries) of the descriptive scalar function, namely, the continuous Lagrangian. (Hint: Use an infinitesimal continuous parameter in the Lie group (Bluman and Kumei 1989)).

12.10. Repeat Exercise 12.9 for the descriptive scalar function, namely, the continuous Hamiltonian.

12.11. Repeat Exercise 12.9 for the descriptive scalar function, namely, the continuous Total Energy.

12.12. If the space-discrete Lagrangian in an infinite dimensional system is translationally invariant under the Euclidean group, then the linear momentum is conserved. If the space-discrete Lagrangian is rotationally invariant, then the angular momentum is conserved. Lastly, the total energy is conserved, if the space-discrete Lagrangian is invariant in time for holonomic-scleronomic systems. Use space-discrete Lagrange's equation of motion to examine the invariance properties (symmetries) of the descriptive scalar function, namely, the space-discrete Lagrangian.

12.13. Repeat Exercise 12.12 for the descriptive scalar function, namely, the space-discrete Hamiltonian.

12.14. Repeat Exercise 12.12 for the descriptive scalar function, namely, the space-discrete Total Energy.

PART III

THE TIME DIMENSION

CHAPTER THIRTEEN

TIME DISCRETIZATION OF EQUATIONS OF MOTION: OVERVIEW AND CONVENTIONAL PRACTICES

Kumar K. Tamma, Xiangmin Zhou, Masao Shimada, Jason Har and Andrew Hoitink

Of primary interest and the focus in Chapter 13 is an overview and highlights which mostly encompasses conventional practices including the theoretical and numerical aspects as related to the design and development of time integration algorithms for the equations of motion for holonomic-scleronomic systems. Such systems are routinely encountered in a wide class of problems in computational dynamics of particles, materials, and continuous-body systems such as unconstrained elastodynamics problems as well as constrained dynamics applications and typical of multi-body systems dynamics.

13.1 INTRODUCTION

Previously, we have described the strict limitations for establishing equivalences of the strong and/or weak forms amongst the various frameworks (Newtonian, Lagrangian, Hamiltonian, and the Total Energy). These limitations are that the kinetic energy is quadratic in generalized velocities and the potential energy is not velocity dependent for N-body systems, and for continuous-body dynamics applications we additionally restricted attention to the case of holonomic-scleronomic systems.

However, for the time integration developments that are highlighted here for both N-body and continuous-body dynamics applications with/without constraints, we only consider the class of applications where the kinetic energy is quadratic in generalized velocities and is not dependent upon generalized coordinates, and the potential energy is velocity independent. It is to be noted that, with the exception of the Hamiltonian framework which yields a system of first-order (two-field form with the dependent variables as canonical coordinates and canonical momentum) in time semi-discretized equations of motion, all the other frameworks naturally give rise to a semi-discretized system of second-order (single-field form with the dependent variable as generalized coordinates) in time differential equations. Historically within the computational mechanics community, most time integration methods have been derived in the traditional sense, for the single-field form of representation of the equations of motion. In general, the design of computational algorithms for the integration of the equation of motion can be based upon either the two-field (first-order) form or the single-field (second-order) form of representation

Advances in Computational Dynamics of Particles, Materials and Structures, First Edition. Jason Har and Kumar K. Tamma.

for conducting the process of time integration. Consequently, the former representation does not naturally permit a spurious root leading to only numerically non-dissipative algorithm designs (a drawback of the two-field form for certain applications when significant high frequency participation of modes is present which is a result of space discretization). However, unless accelerations are not treated as variables in which case a spurious root does not exist leading to numerically non-dissipative algorithm designs, in general, the latter single-field form naturally involves a spurious root that is associated with the underlying algorithmic amplification matrix (such as algorithm designs with controllable numerical dissipation, which can be employed to control the amount of participation of the high frequency modes resulting from the numerical space discretization). For solving real world practical applications, this feature plays an important role. In certain real world applications, especially, involving flexible bodies and high stiffness, numerically non-dissipative algorithms such as symplectic-momentum conserving or energy-momentum conserving may not be strictly (practically) capable of resolving the various subtle issues leading to incompletion of the analysis after a certain time point. However, the single-field form has the flexibility to not only mimic the two-field form representations such as symplectic-momentum and/or energy-momentum features, but in addition, the single-field form also has the added flexibility to embed controllable numerical dissipation (helps to foster completion of the total analysis, and/or at least the total time duration window for the analysis) to these parent symplectic-momentum and/or energy-momentum based representations such that, when the controllable numerical dissipation is turned off, one recovers the original structure preserving representations.

In this chapter, covering both two-field and single-field form of representations, we particularly focus attention upon the time integration as related to the class of LMS methods involving a single system of equations and a single solve within each single time step as these are the most practical and economical; and they are commonly employed in commercial software settings. It is assumed that the reader is aware that linear multi-step methods can be cast as equivalent single step methods. For the implicit linear dynamic algorithm designs with unconditional stability and second-order time accuracy, we are not interested in solving more than a single system of equations and a single solve within each time step (since we would like to keep the computational complexity to a minimum or optimal for the purposes of practical considerations). Caution should be exercised in noting that there exist approaches in the open literature with the objective to claim a special algorithmic property, but they involve more than a single solve of the implicit equation system leading to high computational expense (this defeats the original purpose of meeting the desired theoretical capability; see barrier theorem in (Zhou and Tamma 2004a) for other details). For clarity, the following is useful to note. Explicit methods are those methods where a solver (direct or iterative) is not needed (since the mass matrix is diagonalized), and second-order time accuracy is preferred. On the other hand, implicit methods require the solution of a system of simultaneous equations, and in principle for linear dynamic situations, no more than a single system of equations within a time step should be solved (from the viewpoint of practical considerations of real-world applications), and second-order time accuracy is preferred. Caution should be exercised in that some implicit methods also claim high-order of time accuracy; but it is equally important to note that this can be only possible with an extra cost of solving more than a single system of equations within a given time step (see barrier theorem; (Zhou and Tamma 2004a)) and/or may entail loss of unconditional stability. In general most explicit methods other than some exceptions (Zhou et al. 2004; Zhou et al. 2005) are conditionally stable, and do not need nonlinear iterations within a time step for nonlinear transient situations; furthermore, the result of an explicit algorithm can be trusted to be reasonably accurate (for the given mesh size under consideration), and typical time step studies as with implicit methods may not be warranted. This is unlike implicit methods which are unconditionally stable mostly for linear systems, but require a solver within a given time step; furthermore, the accuracy of the result is not guaranteed at any arbitrary time step value and time step studies need to be conducted. However, for nonlinear transient situations, the implicit methods necessitate some type of nonlinear iterations (such as the Newton-Raphson type) and unconditional stability does not necessarily hold (see exceptions in later parts of this and the next chapter). It is quite customary in commercial software to employ explicit methods for wave propagation type applications where all the modes participate in the solution, unlike the class of inertial or vibrations type applications where mostly the low frequency modes are dominant and implicit methods are preferred.

Traditionally, most of the classical time integration methods including linear multistep methods, which are fairly popular, have been motivated from Taylor's series expansions for the approximation of the variables such as displacement, velocity, and acceleration appearing in the space discretized equations of motion. Even though several of these methods have been developed to a certain degree of maturity and from various differing viewpoints, the geometric nature which the ordinary or partial differential systems possess has been ignored to an extent in certain cases (especially in the early years of time integration developments) in constructing time integration algorithms. Consequently, the time integration for highly nonlinear differential systems may become unstable and often blows up (most often due to convergence of nonlinear Newton type iterations) for certain values of the time steps (often large time steps).

To somewhat improve upon some of these concerns, geometric mechanics stemming from Lagrangian mechanics which has been pioneered by Lie (1842–1899) appears to play a role to a certain extent, as evident from more recent developments over the past two decades. In particular, it serves to foster certain physical attributes to be conserved/preserved and/or bounded in the algorithm design. It has been believed that the preservation of the geometric nature of the differential system can give rise to ease of long-time computations, especially for nonlinear differential systems (see subsequent details in this and the next chapter regarding the validity of such claims). Furthermore, preserving the structure appears to provide more physically accurate time discretized algorithms (McLachlan and Quispel 2001) that are useful for a certain class of applications. For stiff systems and real world applications, there is no guarantee that a solution can be obtained at any given large time step value (even for long term simulations) with structure preserving algorithms.

Understanding the symmetry in physics is very important in developing and designing time integration schemes that inherit particular desirable algorithmic attributes useful for certain classes of problems. Noether (Noether 1918) first discovered that the Lagrangian has symmetry. Conservation of quantities such as energy, linear, and angular momentum results from the symmetry of the Lagrangian or the Hamiltonian. Hence, structure preserving algorithms have attracted some degree of attention, for example, the symplectic-momentum and energy-momentum conserving algorithms. Besides, variational time integrators derived from the discretization of Hamilton's principle and which have followed the underlying theoretical aspects described by (Veselov 1988) have also been developed as useful methods for a particular class of applications preserving both symplectic structure and symmetry, where the energy is bounded for a fixed time step. Besides the Lagrangian and the Hamiltonian descriptive functions, alternatively, the Total Energy framework is a computationally convenient, attractive, and a viable setting that is very natural for the computational mechanics developments (for both space and time discretization). We strongly advocate and place heavy emphasis on employing the Total Energy framework as a starting point since it provides improved physical insight and easy to grasp fundamentals; and the descriptive scalar function, namely, the Total Energy, also has symmetries. It is our contention that the Total Energy framework is a self-contained framework in that it particularly provides for the particular class of holonomic-scleronomic systems, yet another viewpoint for easily deriving the strong forms with good physical insight. Furthermore, it naturally inherits certain added advantages for subsequently deriving the space/time discretized equations of motion in a unified manner with improved physical insight, and provides new avenues that permit such discretizations to be easily conducted, and with certain computationally attractive features (see details in the next chapter).

The organization of this chapter on overview and conventional practices that exist in time integration of the equation of motion is as follows. It is divided into three subsections, namely:

(i) Overview of classical developments from historical perspectives in chronological order that have been developed and described over the past fifty years or so from various viewpoints (we first overview and highlight a wide variety of linear multistep methods and equivalent single step methods encompassing both numerically non-dissipative algorithms, and algorithms with controllable numerical dissipation).

(ii) Variational time integrators that have been described in the literature stemming from Discrete Euler-Lagrange equations.

(iii) Energy-momentum conserving/dissipating algorithm designs for finite dimensional systems following the original methods of development (classical practices) through enforcing energy constraints.

In Chapter 14, we present recent advances and the state-of-the-art, and draw special attention to and highlight the more recent advances and developments directly emanating from the Total Energy representation of the equations of motion as a starting point. This is in conjunction with a generalized time weighted residual approach which provides new perspectives and a unified viewpoint for developing algorithms by design including optimal algorithm designs; and it also covers (recovers) most of the developments presented in this chapter that have been previously derived from various other classical viewpoints as described in (i), (ii), and (iii) above from conventional practices.

In this chapter, we first focus attention upon linear algorithmic designs which are an important first step in the design of computational algorithms for the integration of the equations of motion. Consequently, the generalized time weighted residual approach that is highlighted in the next chapter not only recovers a wide variety of most of these time integration methods described independently following traditional and/or past practices that have been developed over the past fifty years or so, but it also provides additional avenues towards a wide class of other new and optimal algorithm designs; the developments in the sense of the single-field form of representation of the equations of motion need to be carefully understood. We also briefly highlight the two-field form of representation in the design of computational algorithms for integrating the equations of motion. Subsequently, we particularly draw attention to the important fact that firstly, the parent linear dynamic algorithm designs are no doubt of fundamental importance. The question that follows next is that instead of following ad hoc approaches for extending the applicability to nonlinear transient situations, how does one appropriately provide the necessary theoretical basis for proper extensions of the parent linear dynamic algorithm designs to nonlinear dynamics applications? In this regard, a novel normalized time weighted residual approach is particularly stressed in conjunction with the Total Energy framework to help provide the proper theoretical basis and foundation for such extensions to nonlinear dynamic situations. These issues are presented in Chapter 14 to follow. *A point of clarification is in order via the generalized time weighted residual approach and the Total Energy framework that is described in the next chapter for developing algorithms by design. Some of the developments leading to particular conclusions regarding algorithmic attributes (physical properties) leading to the conclusions of structure preservation such as symplectic-momentum conservation and/or energy-momentum conservation that are described in the next chapter, are in the context of either relating the resulting formulations and algorithm designs developed through the approaches presented in the next chapter to those independently developed and designed employing the past classical and/or traditional practices as outlined in (i), (ii), and (iii) above in this chapter, and/or through demonstration via proofs*. One final comment that is noteworthy is the fact that, most of the original classical developments in symplectic-momentum and energy-momentum conserving algorithmic designs are in the sense that the unknowns are the displacement and the velocity as in the two-field form representations that are described in this chapter. These are unlike the single-field form of representations that are additionally presented in Chapter 14 with unknowns, namely, displacement, velocity, as well as acceleration which also readily yield equivalent representations to the two-field form, but involve a spurious root to accommodate both numerically non-dissipative and numerically (controllable) dissipative algorithms and representations. The single-field form provides an additional mechanism with flexibility (in contrast to the two-field form) to enable practical applications to be more easily simulated; especially structural systems with stiff and/or flexible structural components.

Guidelines and Typical Criteria for the Selection of a Time Integration Algorithm for Time Dependent Applications In summary, some viable guidelines and typical criteria one should carefully investigate in evaluating the preferred choice of one time integration algorithm over another algorithm can be based upon the following considerations (and, the following is in the sense of implicit algorithms and implementation aspects for general situations and for practical computational dynamics applications):

(i) *Does the time integration algorithm under consideration involve the solution of only a single system of equations with a single solve within a single time step?*

(ii) *Does the time integration method inherit unconditional stability (for linear transient situations)?*

(iii) *Do the convergence rates for all the unknowns that are being solved attain the necessary expected theoretical maximum of second-order time accuracy, or is one sacrificing this very important attribute in any one of the unknowns either due to lack of proper numerical implementation and/or drawback of the particular algorithm under consideration including violation of order preserving property (for computational dynamics, first-order time accurate integration algorithms are not favorably looked upon as they may not be practical for real world applications)?*

(iv) *Does the algorithm inherit controllable numerical dissipation so as to resolve unwanted high frequency participation of modes due to the spatially discretized numerical mesh (this is an important and useful feature in certain applications, especially, involving flexible bodies and high stiffness since numerically non-dissipative algorithms such as symplectic-momentum conserving or energy-momentum conserving may not be capable of resolving the various subtle issues)?*

(v) *Is the algorithmic numerical dissipation and numerical dispersion the minimum in contrast to the other existing time integration algorithms?*

(vi) *It is highly recommended that the time integration algorithm overshoot behavior in the unknowns be preferably of zero-order in displacement and velocity for both numerically non-dissipative algorithm designs and/or for time integration algorithms with features such as controllable numerical dissipation (or else, if it is higher or excessive in these variables it may cause increased error, convergence and/or other non-physical problems down the line for any given set of arbitrary initial conditions)?, and lastly*

(vii) *How does one extend the parent linear transient time integration algorithm properly to nonlinear transient situations, and can one provide a theoretical basis (see details in the next chapter) for enabling such extensions to nonlinear transient applications or is the implementation being conducted in an ad hoc manner (these latter ad hoc practices are, in general, the basis of poor performance of time integration algorithms and their lack of robustness; also there are certain misconceptions in the literature regarding time integration algorithms for linear versus nonlinear dynamics applications)? In particular, in the sense of applications to computational dynamics, are the algorithm numerical attributes such as structure preservation and/or energy-momentum type conservation properties inherited within the parent numerically non-dissipative structure preserving and/or energy-momentum conserving algorithm design, and are the controllable numerically dissipative algorithm designs the direct extensions of these underlying parent numerically non-dissipative algorithm designs?*

Other very important and subtle issues as related to time level consistency of an algorithm (for example, every time integration algorithm satisfies the discrete equation of motion only at a particular time point, and this time level aspect is not well known for the general class of LMS methods to-date (Hoitink et al. 2008*)), evaluation of the nonlinear operator with parameters inside or outside the operator, the adverse effect of accelerations from the previous time step upon the computation of the unknowns in the current time step, etc., play a critical role and all relevant factors should be very clearly understood. Caution should be exercised in noting that misconceptions exist in the literature such as* *for problems with smooth versus non-smooth solutions that certain structure preserving algorithms are preferred over others; that time discontinuous algorithms have certain features over time continuous algorithms, when in fact, time continuous algorithms also indeed inherit the same identical attributes (Tamma et al. 2002), etc. Finally, the most desirable attribute as the next generation computational toolkit is that which has the flexibility and which can provide a wide variety of choices to the analyst in a unified setting which also includes optimal algorithms and designs as well to tackle a wide variety of applications in computational science and engineering.*

All these are described in the next chapter in a unified setting dealing with a generalized single step single solve (GSSSS) framework of algorithms and designs. *In contrast to understanding the traditional practices as outlined in this chapter, in general, the reader is strongly encouraged to become familiar with several very important concepts described in the next chapter which are an outcome of the theoretical foundations related to three principal aspects: a) the Unified theory underlying computational algorithms (Zhou and Tamma 2004a; Tamma et al. 2002; Tamma et al. 2000) which explains the design spaces of time integration algorithms, the associated classifications, and the generalized barrier theorem wherein Dahlquist's theorem is a particular case, b) the so-called design spaces, measures and metrics associated with time integration algorithms (Zhou et al.* 2005*) which helps one to understand how to place future new developments in time integration algorithms in the proper context and how to evaluate the new design in contrast to existing designs, and c) the concepts underlying the notion of algorithms by design (Zhou and Tamma* 2006*) whose objectives are to foster the design of new time integration algorithms with a wish list of a priori desired algorithmic attributes*.

In contrast to implicit algorithms, it is worth noting that explicit algorithms can be readily extracted from the parent implicit counterparts; however, these algorithms are conditionally stable, and they do not require a solver, and no nonlinear iterations are needed within each time step (the reader is, however, encouraged to be aware of some of the exceptions in the literature (Zhou et al. 2004; Zhou et al. 2005) which are superior to the traditional practices and which do not suffer from drawbacks of conditional consistency issues and the like).

A brief overview of conventional practices and summary of some common single-step and linear multi-step methods are first highlighted for first-order systems. Following this, the majority of the presentation is devoted to second-order systems which are of primary interest in this chapter. In the next chapter to follow, new and more recent advances and the general theoretical design considerations including particular time integration algorithm designs meeting several of the criteria that are mentioned previously for time dependent applications are described in detail.

13.2 SINGLE-STEP METHODS FOR FIRST-ORDER ORDINARY DIFFERENTIAL EQUATIONS

Let us consider an initial value problem with the ordinary differential equation of the form,

$$\dot{y} = f(y, t) \tag{13.1}$$

where the dependent variable $y(t) : \mathcal{T} \rightarrow \mathbb{R}$ denotes a scalar-valued function of the real variable $t \in \mathcal{T} = [0, T]$ and the independent variable for the differential equation is the time t. The first-order ordinary differential equation given in Equation (13.1) is subjected to the initial condition, i.e., $y_0 = y(t_0)$ at the time $t = t_0 = 0$. Let us consider that the time interval $[t_0, t_N = T]$ can be assumed to be a disjoint union

$$\mathcal{T} = \bigcup_{n=0}^{N-1} \left[t_n, t_{n+1}\right] \tag{13.2}$$

with a typical time step, $\Delta t = t_{n+1} - t_n$, where the time at the nth time-step is denoted by t_n. The time step here is assumed to be constant and n represents a whole number.

Then, the first-order time derivative at $t = t_n$ is denoted by

$$\dot{y}_n = f(y_n, t_n) \tag{13.3}$$

where y_n denotes the approximation to the solution of the independent variable at the time $t = t_n$.

13.2.1 Forward Euler Method

The forward Euler method can be obtained by a Taylor series of order 1 around the point t_n (Henrici 1962). The approximation to the solution y_{n+1} at the time $t = t_{n+1}$ is expanded in the neighborhood of the approximation to the solution y_n at the time $t = t_n$ as follows:

$$y_{n+1} = y_n + \Delta t \dot{y}_n + \frac{\Delta t^2}{2!}\ddot{y}_n + \frac{\Delta t^3}{3!}\dddot{y}_n + \frac{\Delta t^4}{4!}\ddddot{y}_n + \ldots \tag{13.4}$$

The above can be rewritten as

$$y_{n+1} = y_n + \Delta t \dot{y}_n + \mathcal{O}(\Delta t^2) \tag{13.5}$$

where $\mathcal{O}(\Delta t^2)$ is referred to as the local truncation error. Then, the first-order time derivative at $t = t_n$ can be rewritten as

$$\dot{y}_n = \frac{y_{n+1} - y_n}{\Delta t} + E_{trunc}(y, \Delta t), \; E_{trunc}(y, \Delta t) = \mathcal{O}(\Delta t^1) \tag{13.6}$$

where $E_{trunc}(y, \Delta t)$ is called the global truncation error for the method. The local truncation error has a power higher than the global truncation error. Hence, the forward Euler method is given by

$$\dot{y}_n = \frac{y_{n+1} - y_n}{\Delta t} \tag{13.7}$$

As the method is first-order time accurate, it is called a first order method (Gear 1971; Mathews and Fink 2004). Therefore, the approximation to the solution of the independent variable at the time $t = t_{n+1}$ is updated using the forward Euler method as follows:

$$y_{n+1} = y_n + \Delta t \dot{y}_n = y_n + \Delta t f(y_n, t_n) \tag{13.8}$$

where an error is induced at every time-step due to the truncation of the Taylor series. The forward Euler method is a one-step explicit method, which prescribes the numerical integration scheme for updating the approximation to the solution at the time $t = t_{n+1}$ in terms of the approximation to the solution at the previous step time $t = t_n$.

13.2.2 Backward Euler Method

The backward Euler method can also be obtained by a Taylor series of order 1 around the point t_n. The approximation to the solution y_{n-1} at the time $t = t_{n-1}$ is expanded in the neighborhood of the approximation to the solution y_n at the time $t = t_n$ as follows:

$$y_{n-1} = y_n - \Delta t \dot{y}_n + \frac{\Delta t^2}{2!}\ddot{y}_n - \frac{\Delta t^3}{3!}\dddot{y}_n + \frac{\Delta t^4}{4!}\ddddot{y}_n - \ldots \tag{13.9}$$

The above can be rewritten as

$$y_{n-1} = y_n - \Delta t \dot{y}_n + \mathcal{O}(\Delta t^2) \tag{13.10}$$

Then, the first-order time derivative at $t = t_n$ can be rewritten as

$$\dot{y}_n = \frac{y_n - y_{n-1}}{\Delta t} + E_{trunc}(y, \Delta t) \tag{13.11}$$

where the truncation error $E_{trunc}(y, \Delta t) = \mathcal{O}(\Delta t^1)$. Hence, the backward Euler method is given by

$$\dot{y}_{n+1} = \frac{y_{n+1} - y_n}{\Delta t} \tag{13.12}$$

It is also called a first-order method. Therefore, the approximation to the solution of the independent variable at the time $t = t_{n+1}$ is updated using the backward Euler method as follows:

$$y_{n+1} = y_n + \Delta t \dot{y}_{n+1} = y_n + \Delta t f\left(y_{n+1}, t_{n+1}\right) \tag{13.13}$$

where an error is also induced at every time-step due to the truncation of the Taylor series. The backward Euler method is a one-step implicit method, which prescribes the numerical integration scheme for updating the approximation to the solution at the time $t = t_{n+1}$ in terms of the approximation to the solution at the current step time $t = t_{n+1}$.

13.2.3 Generalized Trapezoidal Method

By use of the convex combination (Simo and Hughes 1998), we have

$$y_{n+1} - y_n = \Delta t \dot{y}_{n+\alpha} \tag{13.14}$$

where $0 \leq \alpha = \frac{\tau}{\Delta t} \leq 1$ and $\tau = t - t_n$. A one-parameter α-family of the variable $\dot{y}$ can be written as

$$\dot{y}_{n+\alpha} = \dot{y}_n + \alpha(\dot{y}_{n+1} - \dot{y}_n) \tag{13.15}$$

Therefore, the approximation to the solution of the independent variable at the time $t = t_{n+1}$ is updated using the central difference method as follows:

$$y_{n+1} = y_n + \Delta t \left\{\dot{y}_n + \alpha(\dot{y}_{n+1} - \dot{y}_n)\right\} \tag{13.16}$$

which is called the generalized trapezoidal family for the first-order ordinary differential equation. Hence, when $\alpha = 0$, it produces the forward Euler method; when $\alpha = 1$, it produces the backward Euler method; and when $\alpha = \frac{1}{2}$, it produces the trapezoidal rule, i.e.,

$$y_{n+1} = y_n + \frac{1}{2}\Delta t(\dot{y}_n + \dot{y}_{n+1}) \tag{13.17}$$

which is often called the Crank-Nicolson method.

13.2.4 Midpoint Method

A one-parameter α-family of the variable y can be written as

$$y_{n+\alpha} = \alpha y_{n+1} + (1 - \alpha)\, y_n \tag{13.18}$$

A one-parameter β-family of the time variable, t, can be written as

$$t_{n+\beta} = \beta t_{n+1} + (1-\beta)\, t_n \tag{13.19}$$

Then the midpoint of the variable y is defined at $\alpha = \frac{1}{2}$, that is,

$$y_{n+\frac{1}{2}} = \frac{y_{n+1} + y_n}{2} \tag{13.20}$$

On the other hand, the midpoint of the time variable can be defined at $\beta = \frac{1}{2}$, that is,

$$t_{n+\frac{1}{2}} = \frac{t_{n+1} + t_n}{2} \tag{13.21}$$

We observe that the midpoint of the y-variable is different from the y-value at the midpoint of the time variable for the general case of nonlinear situations, that is,

$$y_{n+\frac{1}{2}} \neq y\left(t = t_{n+\frac{1}{2}}\right) \tag{13.22}$$

Thus, there are several types of midpoint methods. For example, the explicit midpoint method (Gear 1971) can be written as

$$\begin{aligned} y_{n+1} &= y_n + \Delta t \dot{y}_{n+\frac{1}{2}} = y_n + \Delta t f\left(y_{n+\frac{1}{2}}, t_{n+\frac{1}{2}}\right), \\ y_{n+\frac{1}{2}} &= y_n + \frac{\Delta t}{2} f\left(y_n, t_n\right) \end{aligned} \tag{13.23}$$

and the implicit midpoint method (Leimkuhler and Reich 2004) can be given by

$$\begin{aligned} y_{n+1} &= y_n + \Delta t \dot{y}_{n+\frac{1}{2}} = y_n + \Delta t f\left(y_{n+\frac{1}{2}}, t_{n+\frac{1}{2}}\right), \\ t_{n+\frac{1}{2}} &= \frac{t_{n+1} + t_n}{2} \end{aligned} \tag{13.24}$$

For linear transient problems, the midpoint method above leads to the equivalent implicit trapezoidal method (Crank-Nicolson Method)

$$y_{n+1} = y_n + \Delta t \frac{1}{2}\left(\dot{y}_{n+1} + \dot{y}_{n+1}\right) = y_n + \Delta t \frac{1}{2}\left\{f\left(y_n, t_n\right) + f\left(y_{n+1}, t_{n+1}\right)\right\} \tag{13.25}$$

13.2.5 Runge-Kutta Method

The second-order Runge-Kutta method is a one-step explicit method, and the explicit midpoint method was introduced earlier. The approximation to the solution at the time $t = t_{n+1}$ for the first-order linear, ordinary, differential equation given in Equation (13.1) can be written as

$$y_{n+1} = y_n + \int_{t_n}^{t_{n+1}} f\left(y, t\right) dt \tag{13.26}$$

By means of the explicit midpoint method, the above equation can be rewritten as

$$y_{n+1} = y_n + \Delta t f\left(y_1, t_{n+\frac{1}{2}}\right) \tag{13.27}$$

where y_1, the midpoint of the variable y, can be approximated by the forward Euler method as follows:

$$y_1 = y_n + \frac{\Delta t}{2} f\left(y_n, t_n\right) \tag{13.28}$$

Thus, the second-order Runge-Kutta method is given by

$$\begin{aligned} y_{n+1} &= y_n + k_2, \\ k_1 &= \Delta t f\left(y_n, t_n\right), \\ k_2 &= \Delta t f\left(y_n + \frac{k_1}{2}, t_{n+\frac{1}{2}}\right) \end{aligned} \tag{13.29}$$

The second-order Runge-Kutta method has the third-order local truncation error, that is, $\mathcal{O}(\Delta t^3)$, and the truncation error, that is, $E_{trunc}(y, \Delta t^2) = \mathcal{O}(\Delta t^2)$.

The fourth-order Runge-Kutta method is a one-step explicit method, which is much more accurate than the one-step methods we have discussed so far, but requires additional computations.

$$\begin{aligned} y_{n+1} &= y_n + \frac{1}{6}(k_1 + 2k_2 + 2k_3 + k_4), \\ k_1 &= \Delta t f\left(y_n, t_n\right), \\ k_2 &= \Delta t f\left(y_n + \frac{k_1}{2}, t_{n+\frac{1}{2}}\right), \\ k_3 &= \Delta t f\left(y_n + \frac{k_2}{2}, t_{n+\frac{1}{2}}\right), \\ k_4 &= \Delta t f\left(y_n + k_3, t_{n+\frac{1}{2}}\right) \end{aligned} \tag{13.30}$$

The fourth-order Runge-Kutta method has the local truncation error, that is, $\mathcal{O}(\Delta t^5)$, and the truncation error, that is, $E_{trunc}(y, \Delta t) = \mathcal{O}(\Delta t^4)$.

13.2.6 Generalized Trapezoidal Family for a Vector-Valued Function

Let us consider an initial value problem with the ordinary differential equation of the form,

$$\mathbf{C}\dot{\mathbf{u}} + \mathbf{K}\mathbf{u} = \mathbf{F} \tag{13.31}$$

where the dependent variable $\mathbf{u} : T \to \mathbb{R}^n$ denotes a vector-valued function of the real variable $t \in \mathcal{T} = [0, T]$ and the independent variable for the differential equation is the time t. Owing to the convex combination, the time domain can be discretized in the following manner

$$\begin{aligned} \mathbf{C}\frac{\mathbf{u}_{n+1} - \mathbf{u}_n}{\Delta t} + \mathbf{K}\mathbf{u}_{n+\alpha} &= \mathbf{F}_{n+\alpha}, \\ \mathbf{u}_{n+\alpha} &= \mathbf{u}_n + \alpha(\mathbf{u}_{n+1} - \mathbf{u}_n), \\ \mathbf{F}_{n+\alpha} &= \mathbf{F}_n + \alpha(\mathbf{F}_{n+1} - \mathbf{F}_n) \end{aligned} \tag{13.32}$$

where $0 \leq \alpha = \frac{\tau}{\Delta t} \leq 1$ and $\tau = t - t_n$. Therefore, the approximation to the solution at the time $t = t_{n+1}$ is given by

$$\mathbf{u}_{n+1} = (\mathbf{C} + \alpha \Delta t \mathbf{K})^{-1}\left[(\mathbf{C} - (1-\alpha)\Delta t \mathbf{K})\mathbf{u}_n - \Delta t \mathbf{F}_{n+\alpha}\right] \tag{13.33}$$

which is the one-step time integrator for the first-order differential equation given in Equation (13.31). When $\alpha = 0$, the one-step time integrator produces the forward Euler method; when $\alpha = 1$, it produces the backward Euler method; and when $\alpha = \frac{1}{2}$, it produces the trapezoidal rule.

13.3 LINEAR MULTISTEP METHODS

So far, we have discussed one-step methods for first-order ordinary differential equations. Henceforth, we shall consider linear multistep methods, which use several previous steps. The equivalent single step representations can also be formulated.

13.3.1 Central Difference Method

First-Order Ordinary Differential Equations One form of the central difference explicit method for first-order ordinary differential equations prescribes the numerical integration scheme for updating the approximation to the solution at the time $t = t_{n+1}$ in terms of the approximation to the solution at the previous step time $t = t_{n-1}$. The central difference method is also derived from the Taylor series. By means of Equations (13.4) and (13.9), the first-order time derivative at $t = t_n = n\Delta t$ can be written as

$$\dot{y}_n = \frac{y_{n+1} - y_{n-1}}{2\Delta t} - \frac{\Delta t^2}{3}\dddot{y}_n + \ldots \tag{13.34}$$

Note that the approximation to the solution y_{n+1} at the time $t = t_{n+1}$ can be written as

$$y_{n+1} = y_{n-1} + 2\Delta t \dot{y}_n + \mathcal{O}(\Delta t^3) \tag{13.35}$$

where the local truncation error is $\mathcal{O}(\Delta t^3)$. Then, the first-order time derivative at $t = t_n$ can be rewritten as

$$\dot{y}_n = \frac{y_{n+1} - y_{n-1}}{2\Delta t} + E_{trunc}(y, \Delta t^2) \tag{13.36}$$

where the truncation error $E_{trunc}(y, \Delta t) = \mathcal{O}(\Delta t^2)$. Hence, the central difference method for the first-order ordinary differential equation is given by

$$\dot{y}_n = \frac{y_{n+1} - y_{n-1}}{2\Delta t} \tag{13.37}$$

The central difference method is a second-order time accurate method in contrast to the Euler method. Therefore, the approximation to the solution of the independent variable at the time $t = t_{n+1}$ is updated using the central difference method as follows:

$$y_{n+1} = y_{n-1} + 2\Delta t \dot{y}_n = y_{n-1} + 2\Delta t f(y_n, t_n) \tag{13.38}$$

where an error is also induced at every time step due to the truncation of the Taylor series.

Second-Order Ordinary Differential Equations Let us consider a system of second-order ordinary differential equations in the form

$$\ddot{y} = f(y, \dot{y}, t) \tag{13.39}$$

A form of the central difference method for second-order ordinary differential equations prescribes the numerical integration scheme for updating the approximation to the solution at the time $t = t_{n+1}$ in terms of the approximations to the solutions at the previous step times $t = t_{n-1}$ and $t = t_n$.

By means of Equations (13.4) and (13.9), the second-order time derivative at $t = t_n = n\Delta t$ can be written as

$$\ddot{y}_n = \frac{y_{n+1} - 2y_n + y_{n-1}}{\Delta t^2} - \frac{2\Delta t^2}{4!} y_n^{(4)} - \frac{2\Delta t^4}{6!} y_n^{(6)} - \dots \tag{13.40}$$

where $y_n^{(i)}$ denotes the derivative of order i at $t = t_n$ with respect to the time t. Note that the approximation to the solution y_{n+1} at the time $t = t_{n+1}$ can be written as

$$y_{n+1} = -y_{n-1} + 2y_n + \Delta t^2 \ddot{y}_n + \mathcal{O}(\Delta t^4) \tag{13.41}$$

where the local truncation error is $\mathcal{O}(\Delta t^4)$. Then, the second-order time derivative at $t = t_n$ can be rewritten as

$$\ddot{y}_n = \frac{y_{n+1} - 2y_n + y_{n-1}}{\Delta t^2} + E_{trunc}(y, \Delta t^2) \tag{13.42}$$

where the global truncation error $E_{trunc}(y, \Delta t) = \mathcal{O}(\Delta t^2)$. Hence, the central difference method for the second-order ordinary differential equation is given by

$$\ddot{y}_n = \frac{y_{n+1} - 2y_n + y_{n-1}}{\Delta t^2} \tag{13.43}$$

Hence, the central difference method for the second-order ordinary differential equation is also a second-order time accurate method. Therefore, the approximation to the solution of the independent variable at the time $t = t_{n+1}$ is updated using the central difference method as follows:

$$y_{n+1} = -y_{n-1} + 2y_n + \Delta t^2 \ddot{y}_n = -y_{n-1} + 2y_n + \Delta t^2 f\,(y_n, \dot{y}_n, t_n) \tag{13.44}$$

where an error is also induced at every time-step due to the truncation of the Taylor series.

13.3.2 Linear Multistep Methods for First-Order Ordinary Differential Equations

A k-step linear multistep method (Gear 1971; Henrici 1962; Hughes 1987) for Equation (13.1) is defined as

$$\begin{aligned} &\alpha_0 y_{n+1} + \alpha_1 y_n + \dots + \alpha_k y_{n+1-k} \\ &= \Delta t\,\{\beta_0 f\,(y_{n+1}, t_{n+1}) + \beta_1 f\,(y_n, t_n) + \dots + \beta_k f\,(y_{n+1-k}, t_{n+1-k})\} \end{aligned} \tag{13.45}$$

where α_i and β_i are parameters. The above equation can be rewritten as

$$\sum_{i=0}^{k} \{\alpha_i y_{n+1-i} + \Delta t \beta_i f(y_{n+1-i}, t_{n+1-i})\} = 0 \tag{13.46}$$

Note that the relation of the terms is linear. Note that an explicit method is obtained when $\beta_0 = 0$, otherwise the k-step linear multistep method gives an implicit method.

13.3.3 Linear Multistep Methods for Second-Order Ordinary Differential Equations

Let us consider a system of second-order ordinary differential equations in the form

$$\ddot{y} = f(y, \dot{y}, t) = g_0 y + g_1 \dot{y} + h(t) \tag{13.47}$$

where g_0 and g_1 are constants. Note that the above differential equation is linear. Then, a k-step linear multistep method (Geradin 1974) for the second-order linear differential equation is defined as

$$\sum_{i=0}^{k} \left\{ \alpha_i y_{n+1-i} + \Delta t \beta_i g_1 y_{n+1-i} + \Delta t^2 \gamma_i \left[g_0 y_{n+1-i} + h(t_{n+1-i}) \right] \right\} = 0 \tag{13.48}$$

where α_i and βi are parameters. Note that an explicit method is obtained when $\beta_0 = 0$, otherwise the k-step linear multistep method gives an implicit method.

13.4 SECOND-ORDER SYSTEMS AND SINGLE STEP AND/OR EQUIVALENT LMS METHODS: BRIEF OVERVIEW OF CLASSICAL METHODS FROM HISTORICAL PERSPECTIVES AND CHRONOLOGICAL DEVELOPMENTS

For the class of single step methods and/or equivalent LMS methods in the single-field form of representation for second-order systems, various methods of time integration following traditional practices and/or the original approaches of algorithm development are first briefly summarized. We mostly present these in the context of linear dynamic systems, historical perspectives, and chronological order. Of primary interest are second-order time accurate algorithms since they are the most practical and are customarily employed in most commercial software. In contrast, first-order time accurate algorithms are strictly not customarily employed as they are not practical for large scale dynamical applications, and they are also not looked upon favorably. For the class of LMS methods in the single-field form of representation for second-order systems, the three basic building blocks which serve as the backbone of this class of LMS methods are the well known Newmark average acceleration method (Newmark 1959), the commonly employed numerically non-dissipative classical version of the Midpoint Rule method or strictly the Midpoint Rule with end-point acceleration (MPR-EPA), and a new version that is presented here and termed as the Midpoint Rule with mid-point acceleration (MPR-MPA); the well known Velocity based scheme (Tamma and Namburu 1990) in the two-field form is equivalent to this last approach in the single-field form of representation (with the exception of Newmark method, the latter two approaches are energy, linear and angular momentum conserving for linear dynamical situations). The foundations and the fundamental theoretical considerations underlying LMS methods which comprise of these basic building blocks are not well known in the traditional literature, and the reader is strongly encouraged to understand the subtle details presented in the next Chapter 14, as well as have a good grasp of the big picture. The above three numerically non-dissipative algorithms are distinctly different from each other as described in the next chapter to follow, and also serve as the backbone for LMS methods in the sense of the single-field form of representation. Furthermore, controllable numerical dissipative algorithms and the corresponding designs are those which emanate from these three basic building blocks as a starting point. Particular attention is drawn in the next chapter, to a generalized unified framework and representations which, in a nutshell, encompasses most of these algorithms that have been developed over the past fifty years or so from classical and/or traditional practices. Furthermore, the generalized framework and representations that are described in the next chapter additionally includes more recent advances leading to new and optimal algorithms and designs under the umbrella of algorithms by design (and, the well-known Generalized Single Step Single Solve GSSSS framework is described from new perspectives, and emanates from the Total Energy representation of the equations of motion and a generalized time weighted residual approach with which the reader is strongly advised to become familiar). The associated terminology and issues regarding the classification based upon displacement and velocity overshoot behavior, and in particular, the various algorithmic parameters as well (Tamma et al. 2000, 2001; Zhou and Tamma 2004a,c, 2006) need to be clearly understood. The optimality of an algorithm is in the sense of the least amount of algorithmic numerical dissipation, dispersion, and overshooting behavior for any given arbitrary set of initial starting conditions (again, such issues are related to which algorithm is preferable for any arbitrary set of initial conditions are not well known and the reader is strongly encouraged to carefully understand the subtle details). Subsequently, in the various sections of the next chapter, firstly, parent linear dynamic algorithmic designs in both the single-field and the two-field forms are overviewed directly emanating from the Total Energy framework and a generalized time weighted residual approach. Finally, how to appropriately provide the necessary theoretical basis for general extensions of the parent linear dynamic algorithms to nonlinear dynamic applications are also particularly highlighted (note that nonlinear dynamics algorithms are extensions of parent linear dynamic counterparts, and this is not well understood to-date). The overview in the next Chapter 14 via the GSSSS framework and setting encompasses most of the developments to-date in the sense of the classical developments in time integration, as well as algorithm designs arising from variational integrators and energy-momentum conserving/dissipating algorithms; in particular, the GSSSS framework also additionally provides new avenues and optimal algorithm designs as well for practical applications. In this chapter, we first provide an overview of conventional practices so as to place the developments to follow in the next chapter in the proper context.

13.4.1 The Houbolt Method [1950]

The Houbolt method (Houbolt 1950) is one of the earliest which has been employed in structural dynamics computations. It uses the equation of motion at time t_{n+1} and two backward difference formulae, obtained by considering a cubic polynomial passing through four successive time levels, for approximating the velocity and acceleration at time level t_{n+1}:

$$\mathbf{M}\ddot{\mathbf{u}}_{n+1} + \mathbf{C}\dot{\mathbf{u}}_{n+1} + \mathbf{K}\mathbf{u}_{n+1} = \mathbf{f}_{n+1} \tag{13.49}$$

$$\dot{\mathbf{u}}_{n+1} = \frac{1}{6\Delta t}(11\mathbf{u}_{n+1} - 18\mathbf{u}_n + 9\mathbf{u}_{n-1} - 2\mathbf{u}_{n-2}) \tag{13.50}$$

$$\ddot{\mathbf{u}}_{n+1} = \frac{1}{\Delta t^2}(2\mathbf{u}_{n+1} - 5\mathbf{u}_n + 4\mathbf{u}_{n-1} - \mathbf{u}_{n-2}) \tag{13.51}$$

where $\mathbf{M}$, $\mathbf{C}$, and $\mathbf{K}$ denote the mass, damping, and stiffness matrices, respectively; $\mathbf{u}$, $\dot{\mathbf{u}}$, and $\ddot{\mathbf{u}}$ denote the displacement, velocity, and acceleration vectors, respectively; and $\mathbf{f}$ denotes the external time-dependent force vector. Subscripts $n+1$, n, $n-1$, and $n-2$ refer to time levels t_{n+1}, t_n, t_{n-1}, and t_{n-2}, respectively. Substitution of Equations (13.50)–(13.51) into the equation of motion (Equation (13.49)) leads to a linear three-step method, from which the dynamic response is given by:

$$\left[\frac{2}{\Delta t^2}\mathbf{M} + \frac{11}{6\Delta t}\mathbf{C} + \mathbf{K}\right]\mathbf{u}_{n+1} = \mathbf{f}_{n+1} + \left[\frac{5}{\Delta t^2}\mathbf{M} + \frac{3}{\Delta t}\mathbf{C}\right]\mathbf{u}_n - \left[\frac{4}{\Delta t^2}\mathbf{M} + \frac{3}{2\Delta t}\mathbf{C}\right]\mathbf{u}_{n-1} + \left[\frac{1}{\Delta t^2}\mathbf{M} + \frac{1}{3\Delta t}\mathbf{C}\right]\mathbf{u}_{n-2} \tag{13.52}$$

While the initial conditions give the displacement vector at $t = 0$, other direct integration algorithms must be used for finding the displacement vector $\mathbf{u}$ at the previous time levels $t = t_{n-1}$ and $t = t_{n-2}$. Equations (13.50)–(13.51) are then employed to calculate the values of the velocity and acceleration vectors at time level $t = t_{n+1}$. The Houbolt algorithm is implicit, unconditionally stable, second-order time accurate and not self-starting, since it requires a special starting procedure for the determination of the displacement vector at the previous time levels (t_{n-2}, t_{n-1} and t_n). The Houbolt algorithm possesses strong damping in the high frequency regime (the spectral radius at infinity, i.e., $\rho_\infty \to 0$ as $\omega\Delta t \to \infty$). Its algorithmic damping and relative period error are higher than most of the widely used methods. This is seen as an important disadvantage of the Houbolt algorithm, since its excessive algorithmic damping affects the low frequency modes too strongly. Note that the Houbolt method can be obtained by selecting

$$\begin{aligned} & k = 3 \\ & \alpha_0 = 11,\ \alpha_1 = -18,\ \alpha_2 = 9,\ \alpha_3 = -2 \\ & \beta_0 = 6,\ \beta_1 = \beta_2 = \beta_3 = 0 \\ & \gamma_0 = 2,\ \gamma_1 = -5,\ \gamma_2 = 4,\ \gamma_3 = -1 \\ & \eta_0 = 1,\ \eta_1 = \eta_2 = \eta_3 = 0 \end{aligned} \tag{13.53}$$

in the following approximated equations (Geradin et al. 1983) for the second-order equation of motion,

$$\sum_{i=0}^{k} \alpha_i \mathbf{u}_{n+1-i} = \Delta t \sum_{i=0}^{k} \beta_i \dot{\mathbf{u}}_{n+1-i} \tag{13.54}$$

$$\sum_{i=0}^{k} \gamma_i \mathbf{u}_{n+1-i} = \Delta t^2 \sum_{i=0}^{k} \eta_i \ddot{\mathbf{u}}_{n+1-i} \tag{13.55}$$

with Equation (13.49).

13.4.2 Classical Midpoint Rule Method

In the classical version of that which is commonly referred to as the midpoint method, the semidiscrete second-order finite element equations of motion at the time $t = t_{n+\frac{1}{2}}$ is given by

$$\mathbf{M}\ddot{\mathbf{u}}_{n+\frac{1}{2}} + \mathbf{C}\dot{\mathbf{u}}_{n+\frac{1}{2}} + \mathbf{K}\mathbf{u}_{n+\frac{1}{2}} = \mathbf{f}_{n+\frac{1}{2}} \tag{13.56}$$

with velocities and displacements updated as follows:

$$\begin{aligned} \mathbf{u}_{n+1} &= \mathbf{u}_n + \Delta t \dot{\mathbf{u}}_{n+\frac{1}{2}}, \\ \dot{\mathbf{u}}_{n+1} &= \dot{\mathbf{u}}_n + \Delta t \ddot{\mathbf{u}}_{n+\frac{1}{2}} \end{aligned} \tag{13.57}$$

where displacement, velocities, displacements and external forces at the time $t = t_{n+\frac{1}{2}}$ are approximated by

$$\begin{aligned} \mathbf{u}_{n+\frac{1}{2}} &= \frac{1}{2}(\mathbf{u}_n + \mathbf{u}_{n+1}), \\ \dot{\mathbf{u}}_{n+\frac{1}{2}} &= \frac{1}{2}(\dot{\mathbf{u}}_n + \dot{\mathbf{u}}_{n+1}), \\ \ddot{\mathbf{u}}_{n+\frac{1}{2}} &= \frac{1}{2}(\ddot{\mathbf{u}}_n + \ddot{\mathbf{u}}_{n+1}), \\ \mathbf{f}_{n+\frac{1}{2}} &= \frac{\mathbf{f}_n + \mathbf{f}_{n+1}}{2} \end{aligned} \tag{13.58}$$

Consequently, by means of Equations (13.57) and (13.58), the classical midpoint rule method for the semidiscrete second-order ordinary differential equation yields the recursion relation

$$\begin{aligned}&\left[\frac{1}{2}\mathbf{M}+\frac{\Delta t}{4}\mathbf{C}+\frac{\Delta t^2}{8}\mathbf{K}\right]\ddot{\mathbf{u}}_{n+1}\\&=-\left[\frac{1}{2}\mathbf{M}+\frac{\Delta t}{4}\mathbf{C}+\frac{\Delta t^2}{8}\mathbf{K}\right]\ddot{\mathbf{u}}_n\\&-\left[\mathbf{C}+\frac{\Delta t}{2}\mathbf{K}\right]\dot{\mathbf{u}}_n-\mathbf{K}\mathbf{u}_n+\frac{\mathbf{f}_n+\mathbf{f}_{n+1}}{2}\end{aligned}\tag{13.59}$$

with the following updates:

$$\begin{aligned}\mathbf{u}_{n+1}&=\mathbf{u}_n+\Delta t\dot{\mathbf{u}}_n+\frac{\Delta t^2}{4}(\ddot{\mathbf{u}}_n+\ddot{\mathbf{u}}_{n+1}),\\\dot{\mathbf{u}}_{n+1}&=\dot{\mathbf{u}}_n+\frac{\Delta t}{2}(\ddot{\mathbf{u}}_n+\ddot{\mathbf{u}}_{n+1})\end{aligned}$$

It is important to note that the unknowns are $\mathbf{u}_{n+1}$, $\dot{\mathbf{u}}_{n+1}$ and $\ddot{\mathbf{u}}_{n+1}$ in the time integration scheme. In particular, note that the unknown acceleration is evaluated at time level $\ddot{\mathbf{u}}_{n+1}$, and hence we will strictly refer to this method as the classical Midpoint Rule method with end-point (that is, at the end of the time step) acceleration (MPR-EPA). Also note that the acceleration from the previous time step plays a role in the current computations. It is important to distinguish and understand some very important and subtle details of this classical version of the method from the relatively new representation of the Midpoint Rule with mid-point acceleration (MPR-MPA) that is described subsequently; and this and various other issues are also particularly stressed in the next chapter as well.

In comparison, when we follow the single-field form of the GSSSS unified framework encompassing the entire class of LMS methods that is described subsequently in the next chapter, the time integrator for the classical version of the Midpoint Rule method (MPR-EPA) is naturally represented in the form (and appears as)

$$\begin{aligned}&\left[\frac{1}{2}\mathbf{M}+\frac{\Delta t}{4}\mathbf{C}+\frac{\Delta t^2}{8}\mathbf{K}\right]\Delta\ddot{\mathbf{u}}\\&=-\mathbf{M}\ddot{\mathbf{u}}_n-\mathbf{C}\left[\dot{\mathbf{u}}_n+\frac{\Delta t}{2}\ddot{\mathbf{u}}_n\right]-\mathbf{K}\left[\mathbf{u}_n+\frac{\Delta t}{2}\dot{\mathbf{u}}_n+\frac{\Delta t^2}{4}\ddot{\mathbf{u}}_n\right]+\frac{\mathbf{f}_n+\mathbf{f}_{n+1}}{2}\end{aligned}\tag{13.60}$$

with the following updates:

$$\begin{aligned}\mathbf{u}_{n+1}&=\mathbf{u}_n+\dot{\mathbf{u}}_n\Delta t+\frac{\Delta t^2}{2}\ddot{\mathbf{u}}_n+\frac{\Delta t^2}{4}\Delta\ddot{\mathbf{u}}\\\dot{\mathbf{u}}_{n+1}&=\dot{\mathbf{u}}_n+\ddot{\mathbf{u}}_n\Delta t+\frac{\Delta t}{2}\Delta\ddot{\mathbf{u}}\\\ddot{\mathbf{u}}_{n+1}&=\ddot{\mathbf{u}}_n+\Delta\ddot{\mathbf{u}}\end{aligned}$$

At this juncture, some very important and noteworthy remarks are in order. Note that the recursion relation given in Equation (13.60) is identical to Equation (13.59). In the context of the single-field GSSSS unified framework encompassing the entire class of LMS methods that is described subsequently, the classical Midpoint Rule method (MPR-EPA) is strictly that algorithm which is common to both the

U- and V-family of algorithms and the acceleration that is computed from the algorithm is at time level t_{n+1} and it is not at time level $t_{n+1/2}$; and the (MPR-EPA) is one of the three principal building blocks within this class (the other two are the Newmark average acceleration method and the new version of the Midpoint Rule method with mid-point acceleration (MPR-MPA) where the acceleration is strictly at time level $t_{n+1/2}$ and is highlighted subsequently; the well known velocity based scheme (Tamma and Namburu 1990) in the sense of the two-field form is equivalent to this last approach in the single-field form of representation (with the exception of the Newmark method, the latter two approaches are energy, linear and angular momentum conserving for linear dynamical situations). For other related efforts regarding the velocity based scheme, see (Tamma and Namburu 1988, 1990, 1997).

13.4.3 The Newmark Family of Algorithms [1959]

Perhaps the most widely used family of direct time integration methods for solving the semi-discretized equations of motion are the Newmark family of algorithms (Newmark 1959). The finite difference approximations employed consider the acceleration in the time interval $[t_n, t_{n+1}]$ as equal to a constant average value (average acceleration) given by:

$$\ddot{\mathbf{u}}_{avg} = (1-\gamma)\,\ddot{\mathbf{u}}_n + \gamma\ddot{\mathbf{u}}_{n+1} \tag{13.61}$$

where γ represents a free parameter. By using the dynamic equilibrium equations at time level t_{n+1}, and upon integrating Equation (13.61) between t_n and t_{n+1}, the following finite difference equations are obtained for the Newmark family of algorithms:

$$\mathbf{M}\ddot{\mathbf{u}}_{n+1} + \mathbf{C}\dot{\mathbf{u}}_{n+1} + \mathbf{K}\mathbf{u}_{n+1} = \mathbf{f}_{n+1} \tag{13.62}$$

$$\mathbf{u}_{n+1} = \mathbf{u}_n + \Delta t\dot{\mathbf{u}}_n + \tfrac{1}{2}\Delta t^2\left[(1-2\beta)\,\ddot{\mathbf{u}}_n + 2\beta\ddot{\mathbf{u}}_{n+1}\right] \tag{13.63}$$

$$\dot{\mathbf{u}}_{n+1} = \dot{\mathbf{u}}_n + \Delta t\left[(1-\gamma)\,\ddot{\mathbf{u}}_n + \gamma\ddot{\mathbf{u}}_{n+1}\right] \tag{13.64}$$

where β represents a free parameter, and $\mathbf{u}_n$, $\mathbf{u}_{n+1}$ are approximations of the displacement vector at time levels t_n and t_{n+1}, respectively. From the finite difference equations above, the unknowns $\mathbf{u}_{n+1}$, $\dot{\mathbf{u}}_{n+1}$ and $\ddot{\mathbf{u}}_{n+1}$ are determined, assuming that $\mathbf{u}_n$, $\dot{\mathbf{u}}_n$ and $\ddot{\mathbf{u}}_n$ are given or can be calculated. The modal decomposition of Equation (13.62)–Equation (13.64) leads to

$$\ddot{u}_{n+1} + 2\xi\omega\dot{u}_{n+1} + \omega^2 u_{n+1} = f_{n+1} \tag{13.65}$$

$$u_{n+1} = u_n + \Delta t\dot{u}_n + \tfrac{1}{2}\Delta t^2\left[(1-2\beta)\,\ddot{u}_n + 2\beta\ddot{u}_{n+1}\right] \tag{13.66}$$

$$\dot{u}_{n+1} = \dot{u}_n + \Delta t\left[(1-\gamma)\,\ddot{u}_n + \gamma\ddot{u}_{n+1}\right] \tag{13.67}$$

where ξ and ω represent the modal damping ratio and the natural frequency, respectively.

There are several possible implementation aspects of the family of Newmark algorithms. One of the most widely used is the *a-form* (for linear dynamic systems), in which the following predictors are defined:

$$\widetilde{\mathbf{u}}_{n+1} = \mathbf{u}_n + \Delta t\dot{\mathbf{u}}_n + \frac{1}{2}\Delta t^2\,(1-2\beta)\,\ddot{\mathbf{u}}_n \tag{13.68}$$

$$\widetilde{\dot{\mathbf{u}}}_{n+1} = \dot{\mathbf{u}}_n + (1-\gamma)\,\Delta t\ddot{\mathbf{u}}_n \tag{13.69}$$

Using Equations (13.68)–(13.69), Equations (13.63) and (13.64) become:

$$\mathbf{u}_{n+1} = \widetilde{\mathbf{u}}_{n+1} + \beta \Delta t^2 \ddot{\mathbf{u}}_{n+1} \tag{13.70}$$

$$\dot{\mathbf{u}}_{n+1} = \widetilde{\dot{\mathbf{u}}}_{n+1} + \gamma \Delta t \ddot{\mathbf{u}}_{n+1} \tag{13.71}$$

To start the implementation process, the acceleration at time $t = 0$ is obtained from

$$\ddot{\mathbf{u}}_0 = \mathbf{M}^{-1} \left(\mathbf{f}_0 - \mathbf{C}\dot{\mathbf{u}}_0 - \mathbf{K}\mathbf{u}_0 \right) \tag{13.72}$$

Next, the recursion relation for the determination of the acceleration vector is obtained by substituting Equations (13.70)–(13.71) into Equation (13.62):

$$\left(\mathbf{M} + \gamma \Delta t \mathbf{C} + \beta \Delta t^2 \mathbf{K} \right) \ddot{\mathbf{u}}_{n+1} = \mathbf{f}_{n+1} - \mathbf{C}\widetilde{\dot{\mathbf{u}}}_{n+1} - \mathbf{K}\widetilde{\mathbf{u}}_{n+1} \tag{13.73}$$

Finally, Equations (13.63) and (13.64) are used for obtaining the velocity and displacement vectors, respectively. The parameters β and γ are free parameters, which determine the stability and accuracy properties of the algorithm under consideration. The unconditional stability conditions for the Newmark family of algorithms are:

$$2\beta \geq \gamma \geq \frac{1}{2} \tag{13.74}$$

For conditional stability, the Newmark parameters must satisfy:

$$\gamma < \frac{1}{2}, \quad \beta < \frac{1}{2}\gamma \quad \text{and} \quad \Omega \leq \Omega_{crit} \tag{13.75}$$

where $\Omega = \omega \Delta t$, and

$$\Omega_{crit} = \frac{\xi\left(\gamma - \frac{1}{2}\right) + \left[\frac{1}{2}\gamma - \beta + \xi^2\left(\gamma - \frac{1}{2}\right)^2\right]^{\frac{1}{2}}}{\left(\frac{1}{2}\gamma - \beta\right)} \tag{13.76}$$

The stability conditions must be satisfied for each mode in the system. Consequently, the highest natural frequency must satisfy Equation (13.76) in order for the algorithm to be conditionally stable. It can be shown that, for values of the parameter $\gamma < \frac{1}{2}$, the general form of the Newmark algorithm introduces negative damping, leading ultimately to an unbounded, oscillatory response. For $\gamma > \frac{1}{2}$, a positive damping is introduced, ultimately annihilating the transient response. For this reason, the discussion will be restricted to the case $\gamma = \frac{1}{2}$. See also Hughes (Hughes 1987) for instance for further discussions on the Newmark family of algorithms.

As special cases, the Newmark family contains several well-known and widely used direct time integration methods. These are presented next.

The Central Difference Method For $\beta = 0$ and $\gamma = \frac{1}{2}$, the finite difference equations for the Newmark algorithm (Equations (13.63) and (13.64)) yield the classical first and second central difference approximations, respectively. The central difference method is an explicit, conditionally stable, and second-order accurate algorithm in time. The mass and damping matrices are diagonalized employing special techniques. For the undamped case ($\xi = 0$) the stability condition becomes $\Omega_{crit} = 2$. When the time step restriction is not too severe, the central difference algorithm is generally regarded as an economical procedure for the direct integration of the equations of motion and is thus widely used. Alternatively, other forms of explicit methods exist which have certain computationally attractive features (Tamma and Namburu 1988) such as direct self-starting features.

The Average Acceleration Method For $\beta = \frac{1}{4}$ and $\gamma = \frac{1}{2}$, the Newmark algorithm yields the implicit, unconditionally stable and second-order accurate average acceleration algorithm. The method has the smallest error among all second-order accurate direct integration methods (see (Dahlquist 1963)). The spectral radius equals unity for any value of the time step (assuming no physical damping) and the algorithmic damping ratio is zero. While its unconditional stability and low error make it attractive for computer implementation, the value of its spectral radius and algorithmic damping error indicates that the Average Acceleration Method has the disadvantage of not damping the spurious oscillations due to the higher frequency modes. For general nonlinear dynamic applications, it has various other disadvantages (Hoitink et al. 2008; Masuri et al. 2009a,b,c; Zhou and Tamma 2004c, 2006). Under the GSSSS framework described subsequently, the implicit average acceleration method belongs to the U-family of algorithms.

The Linear Acceleration Method For $\beta = \frac{1}{6}$ and $\gamma = \frac{1}{2}$, the Newmark family of algorithms yield the linear acceleration algorithm, which is implicit and conditionally stable ($\Omega_{crit} = 3.46$ for the undamped case).

The Fox-Goodwin Method For $\beta = \frac{1}{12}$ and $\gamma = \frac{1}{2}$, the Newmark family algorithms yield the Fox-Goodwin algorithm, which is implicit, conditionally stable ($\Omega_{crit} = \sqrt{6}$ for the undamped case) and second-order accurate. For the case of undamped systems (i.e., $\mathbf{C} = \mathbf{0}$), the Fox-Goodwin algorithm is fourth-order time accurate, but is only conditionally stable.

13.4.4 The Wilson-θ Method [1968]

The Wilson-θ algorithm (Wilson 1968) is essentially an extension of the average acceleration method, in which a linear variation of the acceleration in the time interval $(t_n, t_n + \theta\Delta t)$ is assumed, according to:

$$\ddot{\mathbf{u}}_{n+\tau} = \ddot{\mathbf{u}}_n + \frac{\tau}{\theta\Delta t}\left(\ddot{\mathbf{u}}_{n+\theta} - \ddot{\mathbf{u}}_n\right) \tag{13.77}$$

where θ is a free parameter which is used to control the stability and accuracy of the algorithm and $0 \leq \tau \leq \theta\Delta t$. Integrating the acceleration twice as a function of τ in the specified interval, we obtain:

$$\dot{\mathbf{u}}_{n+\tau} = \dot{\mathbf{u}}_n + \ddot{\mathbf{u}}_n\tau + \frac{\tau^2}{2\theta\Delta t}\left(\ddot{\mathbf{u}}_{n+\theta} - \ddot{\mathbf{u}}_n\right) \tag{13.78}$$

$$\mathbf{u}_{n+\tau} = \mathbf{u}_n + \dot{\mathbf{u}}_n\tau + \frac{1}{2}\ddot{\mathbf{u}}_n\tau^2 + \frac{1}{6\theta\Delta t}\tau^3\left(\ddot{\mathbf{u}}_{n+\theta} - \ddot{\mathbf{u}}_n\right) \tag{13.79}$$

where $\mathbf{u}_{n+\tau} = \mathbf{u}(t_n + \tau)$ and $\mathbf{u}_{n+\theta} = \mathbf{u}(t_n + \theta\Delta t)$. Taking $\tau = \theta\Delta t$ in Equations (13.78)–(13.79) and employing the equation of motion at time level $t + \theta\Delta t$, we have:

$$\ddot{\mathbf{u}}_{n+\theta} = \frac{6}{\theta^2\Delta t^2}\left(\mathbf{u}_{n+\theta} - \mathbf{u}_n\right) - \frac{6}{\theta}\dot{\mathbf{u}}_n - 2\ddot{\mathbf{u}}_n \tag{13.80}$$

$$\dot{\mathbf{u}}_{n+\theta} = \frac{3}{\theta\Delta t}\left(\mathbf{u}_{n+\theta} - \mathbf{u}_n\right) - 2\dot{\mathbf{u}}_n - \frac{\theta\Delta t}{2}\ddot{\mathbf{u}}_n \tag{13.81}$$

and

$$\mathbf{M}\ddot{\mathbf{u}}_{n+\theta} + \mathbf{C}\dot{\mathbf{u}}_{n+\theta} + \mathbf{K}\mathbf{u}_{n+\theta} = \mathbf{f}_{n+\theta} \tag{13.82}$$

where

$$\mathbf{f}_{n+\theta} = \mathbf{f}_n + \theta\left(\mathbf{f}_{n+1} - \mathbf{f}_n\right) \tag{13.83}$$

represents the linear approximation of the load vector. Substituting Equations (13.80)–(13.81) into the equation of motion (Equation (13.82)), allows for the solution of $\mathbf{u}_{n+\theta}$:

$$\left[\frac{6}{\theta^2\Delta t^2}\mathbf{M} + \frac{3}{\theta\Delta t}\mathbf{C} + \mathbf{K}\right]\mathbf{u}_{n+\theta} = \mathbf{f}_n + \theta\left[\mathbf{f}_{n+1} - \mathbf{f}_n\right] + \mathbf{M}\left[\frac{6}{\theta^2\Delta t^2}\mathbf{u}_n + \frac{6}{\theta\Delta t}\dot{\mathbf{u}}_n + 2\ddot{\mathbf{u}}_n\right] + \mathbf{C}\left[\frac{3}{\theta\Delta t}\mathbf{u}_n + 2\dot{\mathbf{u}}_n + \frac{\theta\Delta t}{2}\ddot{\mathbf{u}}_n\right] \tag{13.84}$$

Furthermore, substituting $\mathbf{u}_{n+\theta}$ into the first Equation (13.81) yields $\ddot{\mathbf{u}}_{n+\theta}$, which, when used in Equations (13.78)–(13.79) evaluated at $\tau = \Delta t$, allows the calculation of $\ddot{\mathbf{u}}_{n+1}$, $\dot{\mathbf{u}}_{n+1}$ and $\mathbf{u}_{n+1}$:

$$\ddot{\mathbf{u}}_{n+1} = \frac{6}{\theta^3\Delta t^2}\left(\mathbf{u}_{n+\theta} - \mathbf{u}_n\right) - \frac{6}{\theta^2\Delta t}\dot{\mathbf{u}}_n + \left(1 - \frac{3}{\theta}\right)\ddot{\mathbf{u}}_n \tag{13.85}$$

$$\dot{\mathbf{u}}_{n+1} = \dot{\mathbf{u}}_n + \frac{\Delta t}{2}\left(\ddot{\mathbf{u}}_{n+1} + \ddot{\mathbf{u}}_n\right) \tag{13.86}$$

$$\mathbf{u}_{n+1} = \mathbf{u}_n + \Delta t\dot{\mathbf{u}}_n + \frac{\Delta t^2}{6}\left(\ddot{\mathbf{u}}_{n+1} + 2\ddot{\mathbf{u}}_n\right) \tag{13.87}$$

Equation (13.84) together with Equations (13.85)–(13.87) allow the computation of the dynamic response at time level t_{n+1}. The Wilson-θ method is an implicit algorithm. Unlike the Houbolt algorithm, it has the advantage of not requiring any special starting procedures. Since it requires the evaluation of the acceleration vector at $t = 0$, it is strictly not a direct self-starting algorithm. The Wilson-θ algorithm reduces to the linear acceleration method for $\theta = 1$, but it is unconditionally stable only for values of the parameter $\theta \geq 1.37$, with an usually employed value of $\theta = 1.4$.

The order of time accuracy is two, with reasonably good numerical dissipation properties, controllable by the values of parameter θ. Values of $\theta > 2$ reduce the numerical dissipation of the algorithm and increase the relative period error. A disadvantage is the high overshooting behavior (Zhou and Tamma 2004c), and the Wilson-θ method is a member of the U-family of algorithms in the context of the GSSSS framework described later.

13.4.5 The Park Method [1975]

The Park method (Park 1975a) is a linear three-step method which is implicit, unconditionally stable, and second-order time accurate. Similar to the Houbolt method, it is based on the equation of motion at time t_{n+1} and two backward difference formulae, obtained by considering a cubic polynomial passing through four successive time levels, for approximating the velocity and acceleration at time level t_{n+1}. The Park method is defined by

$$\mathbf{M}\ddot{\mathbf{u}}_{n+1} + \mathbf{C}\dot{\mathbf{u}}_{n+1} + \mathbf{K}\mathbf{u}_{n+1} = \mathbf{f}_{n+1} \tag{13.88}$$

$$\dot{\mathbf{u}}_{n+1} = \frac{1}{6\Delta t}\left(10\mathbf{u}_{n+1} - 15\mathbf{u}_n + 6\mathbf{u}_{n-1} - \mathbf{u}_{n-2}\right) \tag{13.89}$$

$$\ddot{\mathbf{u}}_{n+1} = \frac{1}{6\Delta t}\left(10\dot{\mathbf{u}}_{n+1} - 15\dot{\mathbf{u}}_n + 6\dot{\mathbf{u}}_{n-1} - \dot{\mathbf{u}}_{n-2}\right) \tag{13.90}$$

Note that it is stiffly stable (Gear 1967). As described in (Park 1975b), the method can be developed by combining Gear's two-step and three-step stiffly stable algorithms. For Gear's two-step and three-step stiffly stable algorithms, see (Gear 1971). Therefore, it may be used to integrate the equation of motion in a stiff system with a fairly large time step size. It shows strong dissipative characteristics in the high-frequency region similar to the Houbolt method, but it possesses improved characteristics for the algorithmic damping ratio and relative period error, compared to the Houbolt method. The second-order equation of motion in linear dynamic systems can be cast into the following first-order system by defining $\mathbf{y}(t) := (\dot{\mathbf{u}}(t), \mathbf{u}(t))^T$:

$$\dot{\mathbf{y}} = \mathbf{G}\mathbf{y} + \mathbf{H}(t) \tag{13.91}$$

where $\dot{\mathbf{y}}(t) = (\ddot{\mathbf{u}}(t), \dot{\mathbf{u}}(t))^T$,

$$\mathbf{G} = \begin{bmatrix} -\mathbf{M}^{-1}\mathbf{C} & -\mathbf{M}^{-1}\mathbf{K} \\ \mathbf{I} & \mathbf{0} \end{bmatrix} \text{ and } \mathbf{H}(t) = \begin{Bmatrix} \mathbf{M}^{-1}\mathbf{f}(t) \\ \mathbf{0} \end{Bmatrix} \tag{13.92}$$

For the first-order equation of motion in linear dynamic systems, i.e., Equation (13.91), the k-step linear-multistep (LMS) method is generally given as

$$\sum_{i=0}^{k} \alpha_i \mathbf{y}_{n+1-i} = \Delta t \sum_{i=0}^{k} \beta_i \dot{\mathbf{y}}_{n+1-i} \tag{13.93}$$

where $\alpha_i \in \mathbb{R}$ and $\beta_i \in \mathbb{R}$ are algorithmic parameters; see Geradin (Geradin et al. 1983). If $\beta_i = 0$, an LMS method defined by Equation (13.93) is explicit. Park's method is obtained by selecting

$$\begin{aligned} &k = 3 \\ &\alpha_0 = 10,\ \alpha_1 = -15,\ \alpha_2 = 6,\ \alpha_3 = -1 \\ &\beta_0 = 6,\ \beta_1 = \beta_2 = \beta_3 = 0 \end{aligned} \tag{13.94}$$

with Equation (13.88).

13.4.6 The Hilber-Hughes-Taylor-α Method [1977]

The Hilber-Hughes-Taylor-α algorithm (HHT-α) (Hilber et al. 1977) has been developed with the motivation, that the resulting algorithm is unconditionally stable, possesses high frequency numerical dissipation that can be controlled by a free parameter rather than by the size of the time step such that it does not affect the lower modes too strongly. The finite difference approximations of the Newmark algorithm are retained in the time interval $[t_n, t_{n+1}]$ (fundamentally, one of the principal building blocks in LMS methods, namely, the Newmark method is used as the basis and is the starting point):

$$\mathbf{u}_{n+1} = \mathbf{u}_n + \Delta t \dot{\mathbf{u}}_n + \frac{1}{2}\Delta t^2 \left[(1-2\beta)\,\ddot{\mathbf{u}}_n + 2\beta\ddot{\mathbf{u}}_{n+1}\right] \tag{13.95}$$

$$\dot{\mathbf{u}}_{n+1} = \dot{\mathbf{u}}_n + \Delta t \left[(1-\gamma)\,\ddot{\mathbf{u}}_n + \gamma\ddot{\mathbf{u}}_{n+1}\right] \tag{13.96}$$

while the discretized equation of motion is modified as:

$$\mathbf{M}\ddot{\mathbf{u}}_{n+1} + (1+\alpha)\,\mathbf{C}\dot{\mathbf{u}}_{n+1} - \alpha\mathbf{C}\dot{\mathbf{u}}_n + (1+\alpha)\,\mathbf{K}\mathbf{u}_{n+1} - \alpha\mathbf{K}\mathbf{u}_n = \mathbf{f}_{t_{n+\alpha}} \tag{13.97}$$

where α is the algorithmic free parameter, and $t_{n+1+\alpha} = (1+\alpha)\,t_{n+1} - \alpha t_n = t_{n+1} + \alpha\Delta t$.

Substitution of Equations (13.95)–(13.96) in the modified equation of motion, Equation (13.97), yields the following expression for the computation of the acceleration vector $\ddot{\mathbf{u}}_{n+1}$:

$$\begin{aligned}\left[\mathbf{M} + \Delta t\gamma\,(1+\alpha)\,\mathbf{C} + \Delta t^2\beta\,(1+\alpha)\,\mathbf{K}\right]\ddot{\mathbf{u}}_{n+1} \\ = -\,(1+\alpha)\,\mathbf{C}\left[\dot{\mathbf{u}}_n + \Delta t\,(1-\gamma)\,\ddot{\mathbf{u}}_n\right] - \alpha\mathbf{C}\dot{\mathbf{u}}_n \\ +\,(1+\alpha)\,\mathbf{K}\left[\mathbf{u}_n + \Delta t\dot{\mathbf{u}}_n + \Delta t^2\left(\frac{1}{2}-\beta\right)\ddot{\mathbf{u}}_n\right] - \alpha\mathbf{K}\mathbf{u}_n + \mathbf{f}\,(t_{n+1}+\alpha\Delta t)\end{aligned} \tag{13.98}$$

Employing Equations (13.95)–(13.96) and (13.98), the unknowns, namely, $\mathbf{u}_{n+1}$, $\dot{\mathbf{u}}_{n+1}$ and $\ddot{\mathbf{u}}_{n+1}$ at the end of time level t_{n+1} can be computed. Clearly, if the parameter $\alpha = 0$, the algorithm reduces to the implicit Newmark ($\gamma = 1/2$, $\beta = 1/2$) method. In addition, the H-H-T-α method is implicit and second-order time accurate with and without the presence of physical damping. The conditions for unconditional stability are:

$$-\frac{1}{3} \le \alpha \le 0, \quad \beta = \frac{(1-\alpha)^2}{4} \quad \text{and} \quad \gamma = \frac{1-2\alpha}{2} \tag{13.99}$$

The parameter α also governs the numerical dissipation of the algorithm. Larger negative values of α will increase the amount of numerical dissipation (algorithmic damping ratio) and the relative period error, while small negative values will have the opposite effect (see (Hughes 1987)). In both cases, it is to be noticed that the amount of numerical dissipation and relative period error is less than that exhibited by the Houbolt or Wilson-θ methods. It is not an optimal algorithm within the class of LMS methods, yet, it has served as a prelude to foster other algorithm designs with significantly improved optimal features (Zhou and Tamma 2004c). It is a U0-V1 algorithm (zero-order overshoot in displacement and first-order overshoot in velocity) and is a member of the U-family of algorithms in the context of the GSSSS framework described later.

13.4.7 The SSpj Family of Algorithms [1977]

The SSpj family of time stepping algorithms, introduced by Zienkiewicz et al. (see (Zienkiewicz et al. 1984)), are formulated based upon a weighted residual approach in time. Consider the displacement vector Taylor series approximation polynomial of degree p, given by:

$$\mathbf{u} = \mathbf{u}_n + \dot{\mathbf{u}}_n\tau + \ddot{\mathbf{u}}_n\frac{1}{2}\tau^2 + \ldots \boldsymbol{\alpha}_n^{(p)}\tau^p\frac{1}{p!} = \sum_{q=0}^{p-1}\mathbf{u}_n^{(q)}\frac{\tau^q}{q!} + \boldsymbol{\alpha}_n^{(p)}\tau^p\frac{1}{p!} \tag{13.100}$$

where $0 \le \tau \le \Delta t$, $\mathbf{u}_n^{(q)}$ represents the q-th time derivative of the $\mathbf{u}$ vector at $t = t_n$, and $\boldsymbol{\alpha}_n^{(p)}$ is an unknown vector that is subsequently used to determine the displacement and its $p-1$ derivatives at $t = t_{n+1}$. Taking $\tau = \Delta t$ in Equation (13.100), the displacement vector and its time derivative at $t = t_{n+1}$ are expressed as:

$$\mathbf{u}_{n+1} = \sum_{q=0}^{p-1}\mathbf{u}_n^{(q)}\frac{\Delta t^q}{q!} + \boldsymbol{\alpha}_n^{(p)}\frac{\Delta t^p}{p!} = \mathbf{U}_{n+1} + \boldsymbol{\alpha}_n^{(p)}\frac{\Delta t^p}{p!} \tag{13.101}$$

$$\dot{\mathbf{u}}_{n+1} = \sum_{q=1}^{p-1}\mathbf{u}_n^{(q)}\frac{\Delta t^{q-1}}{(q-1)!} + \boldsymbol{\alpha}_n^{(p)}\frac{\Delta t^{(p-1)}}{(p-1)!} = \dot{\mathbf{U}}_{n+1} + \boldsymbol{\alpha}_n^{(p)}\frac{\Delta t^{(p-1)}}{(p-1)!} \tag{13.102}$$

where $\mathbf{U}_{n+1}$ and $\dot{\mathbf{U}}_{n+1}$ are defined by

$$\mathbf{U}_{n+1} := \sum_{q=0}^{p-1} \mathbf{u}_n^{(q)} \frac{\Delta t^q}{q!} \tag{13.103}$$

$$\dot{\mathbf{U}}_{n+1} := \sum_{q=1}^{p-1} \mathbf{u}_n^{(q)} \frac{\Delta t^{q-1}}{(q-1)!} \tag{13.104}$$

as the extrapolated (predictor) values of the displacement and velocity vectors, respectively.

If $W(\tau)$ is a scalar weighting time field, then the time weighted residual equation is given as:

$$\int_0^{\Delta t} W(\tau)\left(\mathbf{M}\ddot{\mathbf{u}} + \mathbf{C}\dot{\mathbf{u}} + \mathbf{K}\mathbf{u} - \mathbf{f}\right) d\tau = \mathbf{0} \tag{13.105}$$

Dividing Equation (13.105) by $\int_0^{\Delta t} W(\tau)\, d\tau$, the unknown vector $\boldsymbol{\alpha}_n^{(p)}$ is found employing:

$$\begin{aligned} &\left[\frac{\Delta t^{p-2}}{(p-2)!}\theta_{p-2}\mathbf{M} + \frac{\Delta t^{p-1}}{(p-1)!}\theta_{p-1}\mathbf{C} + \frac{\Delta t^p}{p!}\theta_p\mathbf{K}\right]\boldsymbol{\alpha}_n^{(p)} \\ &= \bar{\mathbf{f}} - \mathbf{M}\ddot{\bar{\mathbf{U}}}_{n+1} - \mathbf{C}\dot{\bar{\mathbf{U}}}_{n+1} - \mathbf{K}\bar{\mathbf{U}}_{n+1} \end{aligned} \tag{13.106}$$

where the forcing vector $\bar{\mathbf{f}}$ and the dimensionless parameters θ_p are defined as:

$$\bar{\mathbf{f}} = \frac{\int_0^{\Delta t} W\mathbf{f}\, d\tau}{\int_0^{\Delta t} W\, d\tau} \tag{13.107}$$

$$\theta_p = \frac{1}{\Delta t^p}\frac{\int_0^{\Delta t} W\tau^p\, d\tau}{\int_0^{\Delta t} W\, d\tau} \tag{13.108}$$

In Equation (13.106) the vectors $\ddot{\bar{\mathbf{U}}}$, $\dot{\bar{\mathbf{U}}}$ and $\bar{\mathbf{U}}$ represent the mean predictors of the displacement, velocity and acceleration in the interval under consideration and are defined as:

$$\bar{\mathbf{U}}_{n+1} = \sum_{q=0}^{p-1} \mathbf{u}_n^{(q)} \frac{\Delta t^q}{q!}\theta_q \tag{13.109}$$

$$\dot{\bar{\mathbf{U}}}_{n+1} = \sum_{q=1}^{p-1} \mathbf{u}_n^{(q)} \frac{\Delta t^{q-1}}{(q-1)!}\theta_{q-1} \tag{13.110}$$

$$\ddot{\bar{\mathbf{U}}}_{n+1} = \sum_{q=2}^{p-1} \mathbf{u}_n^{(q)} \frac{\Delta t^{q-2}}{(q-2)!}\theta_{q-2} \tag{13.111}$$

Equations (13.101)–(13.102), (13.106) and (13.109)–(13.111) completely define the SSpj family of algorithms. It is assumed that, at the start, the values of the displacement vector and its first $p-1$ time derivatives are given and/or can be calculated and values of the parameters θ_q are specified in a manner that ensures stability and accuracy. With polynomials of order p having been chosen to represent the displacement vector $\mathbf{u}$, the error in the interval $[0, \Delta t]$ is of order up to Δt^p which ensures algorithmic consistency and allows up to p-th order accuracy. The lowest order of p applicable to structural dynamic problems is two. Next, some particular forms of the general SSpj family of algorithms are further presented; see also Zienkiewicz et al. (Zienkiewicz et al. 2005).

The SS22 Algorithm Taking $p = 2$ in Equation (13.100) and denoting $\boldsymbol{\alpha}_n^{(2)} = \boldsymbol{\alpha}$, a quadratic approximation for the displacement vector results as:

$$\mathbf{u} = \mathbf{u}_n + \dot{\mathbf{u}}_n \tau + \frac{\boldsymbol{\alpha} \tau^2}{2} \tag{13.112}$$

The mean predictors $\overline{\mathbf{U}}_{n+1}$ and $\dot{\overline{\mathbf{U}}}$ are defined as:

$$\overline{\mathbf{U}}_{n+1} = \mathbf{u}_n + \dot{\mathbf{u}}_n \Delta t \theta_1 \tag{13.113}$$

$$\dot{\overline{\mathbf{U}}}_{n+1} = \dot{\mathbf{u}}_n \tag{13.114}$$

The unknown vector $\boldsymbol{\alpha}$ can then be determined from:

$$\mathbf{M}\boldsymbol{\alpha} + \mathbf{C}\left(\dot{\overline{\mathbf{U}}}_{n+1} + \boldsymbol{\alpha}\theta_1 \Delta t\right) + \mathbf{K}\left(\overline{\mathbf{U}}_{n+1} + \frac{\boldsymbol{\alpha}\theta_2 \Delta t^2}{2}\right) + \bar{\mathbf{f}} = 0 \tag{13.115}$$

Finally, the displacement and velocity vectors at t_{n+1} are evaluated by Equation (13.112) with $\tau = \Delta t$:

$$\mathbf{u}_{n+1} = \mathbf{u}_n + \dot{\mathbf{u}}_n \Delta t + \frac{\boldsymbol{\alpha} \Delta t^2}{2} \tag{13.116}$$

$$\dot{\mathbf{u}}_{n+1} = \dot{\mathbf{u}}_n + \boldsymbol{\alpha} \Delta t \tag{13.117}$$

Equations (13.113)–(13.117) are used for the computation of the SS22 algorithm. An alternative form of the algorithm, which allows direct calculation of the displacement vector at time level t_{n+1} can be obtained by substitution of $\boldsymbol{\alpha}$ from Equation (13.115) into Equation (13.116). Thus,

$$\left(\mathbf{M} + \theta_1 \Delta t \mathbf{C} + \theta_2 \tfrac{\Delta t^2}{2} \mathbf{K}\right) \mathbf{u}_{n+1} = \left[\mathbf{M} + \theta_1 \Delta t \mathbf{C} + (\theta_2 - 1) \tfrac{\Delta t^2}{2} \mathbf{K}\right] \mathbf{u}_n + \tag{13.118}$$

$$\Delta t \left[\mathbf{M} + (\theta_1 - 1/2)\, \Delta t \mathbf{C} + (\theta_2 - \beta_1) \tfrac{\Delta t^2}{2} \mathbf{K}\right] \dot{\mathbf{u}}_n + \tfrac{\Delta t}{2} \bar{\mathbf{f}} \tag{13.119}$$

The stability conditions, as given by Wood (see (Wood 1984)), identify that the algorithm is unconditionally stable and second-order time accurate for

$$\theta_2 \geq \theta_1 = 0.5 \tag{13.120}$$

If $\theta_2 \geq \theta_1 \geq 0.5$, the SS22 algorithm is unconditionally stable but only first-order time accurate. For $\theta_1 = \frac{1}{2}$ and $\theta_2 = \frac{1}{6}$, the algorithm is conditionally stable and fourth-order time accurate for undamped systems and second-order time accurate for damped systems. The stability condition in both cases is:

$$\Omega \leq \Omega_{crit} = 2.45 \tag{13.121}$$

It can be shown that the SS22 algorithm has stability properties that are spectrally identical to the Newmark family of algorithms. An important advantage of the SS22 algorithm is the fact that it has direct self-starting features, since it does not involve any acceleration calculations at $t = 0$.

The SS32 Algorithm Following the general SSpj algorithm procedure, the implementation steps for the SS32 algorithm are as follows. Find the mean predictors according to:

$$\overline{\mathbf{U}}_{n+1} = \mathbf{u}_n + \dot{\mathbf{u}}_n \theta_1 \Delta t + \ddot{\mathbf{u}}_n \theta_2 \frac{\Delta t^2}{2} \tag{13.122}$$

$$\dot{\overline{\mathbf{U}}}_{n+1} = \dot{\mathbf{u}}_n + \ddot{\mathbf{u}}_n \theta_1 \Delta t \tag{13.123}$$

$$\ddot{\overline{\mathbf{U}}}_{n+1} = \ddot{\mathbf{u}}_n \tag{13.124}$$

Compute the vector $\boldsymbol{\alpha} = \boldsymbol{\alpha}_n^{(3)}$ from:

$$\left[\theta_1 \Delta t \mathbf{M} + \theta_2 \frac{\Delta t^2}{2} \mathbf{C} + \theta_3 \frac{\Delta t^3}{6} \mathbf{K}\right] \boldsymbol{\alpha} = \bar{\mathbf{f}} - \mathbf{M}\ddot{\overline{\mathbf{U}}}_{n+1} - \mathbf{C}\dot{\overline{\mathbf{U}}}_{n+1} - \mathbf{K}\overline{\mathbf{U}}_{n+1} \tag{13.125}$$

Evaluate the displacement, velocity and acceleration vectors at the next time step from:

$$\mathbf{u}_{n+1} = \mathbf{u}_n + \dot{\mathbf{u}}_n \Delta t + \ddot{\mathbf{u}}_n \frac{\Delta t^2}{2} + \frac{\boldsymbol{\alpha} \Delta t^3}{6} \tag{13.126}$$

$$\dot{\mathbf{u}}_{n+1} = \dot{\mathbf{u}}_n + \ddot{\mathbf{u}} \Delta t + \boldsymbol{\alpha} \frac{\Delta t^2}{2} \tag{13.127}$$

$$\ddot{\mathbf{u}}_{n+1} = \ddot{\mathbf{u}}_n + \boldsymbol{\alpha} \Delta t \tag{13.128}$$

Equations (13.122)–(13.128) are used for the computation of the SS32 algorithm.

Substitution of Equation (13.125) into Equations (13.126)–(13.128) yields an alternative expression for the displacement vector:

$$\begin{aligned}
&\left(\theta_1 \mathbf{M} + \theta_1 \frac{\Delta t}{2} \mathbf{C} + \theta_3 \frac{\Delta t^3}{6} \mathbf{K}\right) \mathbf{u}_{n+1} \\
&= \left[\theta_1 \mathbf{M} + \theta_2 \frac{\Delta t}{2} \mathbf{C} + (\theta_3 - 1) \frac{\Delta t^2}{6} \mathbf{K}\right] \mathbf{u}_n \\
&\quad + \Delta t \left[\theta_1 \mathbf{M} + \left(\theta_2 - \frac{1}{3}\right) \frac{\Delta t}{2} \mathbf{C} + (\theta_3 - \theta_1) \frac{\Delta t^2}{6} \mathbf{K}\right] \dot{\mathbf{u}}_n \\
&\quad + \frac{\Delta t^2}{2} \left[\left(\theta_1 - \frac{1}{3}\right) \mathbf{M} + \Delta t \left(\frac{\theta_2}{2} - \frac{\theta_1}{3}\right) \mathbf{C} + \left(\frac{\theta_3}{2} - \frac{\theta_2}{2}\right) \mathbf{K}\right] \ddot{\mathbf{u}}_n + \frac{\Delta t^2}{6} \mathbf{f}^*
\end{aligned} \tag{13.129}$$

The stability conditions for the SS32 algorithm, as described by Wood (see (Wood 1984)) are highlighted next. The conditions for attaining unconditional stability and second-order time accuracy are:

$$\theta_2 - \theta_1 \geq \tfrac{1}{6} \tag{13.130}$$

$$0 \leq \theta_3 - \tfrac{3}{2}\theta_2 + \tfrac{1}{4} \leq 3\left(\theta_1 - \tfrac{1}{2}\right)\left(\theta_2 - \theta_1 - \tfrac{1}{6}\right) \tag{13.131}$$

$$12\left(\theta_1 - \tfrac{1}{2}\right)^2 \geq 1 \tag{13.132}$$

For $\theta_2 - \theta_1 = -\frac{1}{6}$, the SS32 algorithm is fourth-order time accurate and only conditionally stable for undamped systems. The unconditional stability conditions are:

$$\theta_1 > \frac{1}{2} \tag{13.133}$$

$$\theta_3 - \frac{3}{2}\theta_2 + \frac{1}{4} \leq \frac{1}{2} - \theta_1 \tag{13.134}$$

$$\Omega < \Omega_{crit} = 6\left|\frac{\theta_1 - \frac{1}{2}}{\theta_3 - \frac{3}{2}\theta_2 + \frac{1}{4}}\right| \tag{13.135}$$

A comment on the SSpj and the so-called GNpj (Zienkiewicz et al. 2005) is in order. In the early stages of the developments, a key issue concerning collocation was not recognized; and hence the original formulations above have these two distinctly different representative forms in the traditional literature and in books. This drawback was identified later (Tamma et al. 2000; Tamma et al. 2001; Zhou and Tamma 2004c; Zhou and Tamma 2004a; Zhou and Tamma 2006), and is discussed under the umbrella of the well-known GSSSS framework described in the next chapter. In the GSSSS framework, a single unified representation suffices wherein both representations are an integral part of the overall theoretical framework which circumvents such issues (Tamma et al. 2000; Tamma et al. 2001; Zhou and Tamma 2004c; Zhou and Tamma 2004a; Zhou and Tamma 2006).

13.4.8 The Wood-Bosak-Zienkiewicz Method [1981]

The Wood-Bosak-Zienkiewicz (WBZ) algorithm (Wood et al. 1980) is based upon a similar idea as the HHT-α method with a controllable algorithmic dissipative parameter. It retains the Newmark finite difference formulas (see Equations (13.95)–(13.96)), while the discretized equation of motion is modified as:

$$(1 - \alpha_B)\,\mathbf{M}\ddot{\mathbf{u}}_{n+1} + \alpha_B\mathbf{M}\ddot{\mathbf{u}}_n + \mathbf{C}\dot{\mathbf{u}}_{n+1} + \mathbf{K}\mathbf{u}_{n+1} = \mathbf{f}_{n+1} \tag{13.136}$$

where α_B is the algorithmic parameter. Substitution of Equations (13.95)–(13.96) into the modified equation of motion, Equation (13.136), yields the expression for the computation of the $\ddot{\mathbf{u}}_{n+1}$ vector in the form:

$$\begin{aligned}\left[(1 - \alpha_B)\,\mathbf{M} + \Delta t\gamma\mathbf{C} + \Delta t^2\beta\mathbf{K}\right]\ddot{\mathbf{u}}_{n+1} &= \mathbf{f}_{n+1} - \alpha_B\mathbf{M}\ddot{\mathbf{u}}_n \\ -\mathbf{C}\left[\dot{\mathbf{u}}_n + \Delta t\,(1-\gamma)\,\ddot{\mathbf{u}}_n\right] - \mathbf{K}&\left[\mathbf{u}_n + \Delta t\dot{\mathbf{u}} + \Delta t^2\left(\frac{1}{2} - \beta\right)\ddot{\mathbf{u}}_n\right]\end{aligned} \tag{13.137}$$

The values of the acceleration vector determined by Equation (13.137), are then used for the calculation of the displacement and velocity vectors from Equations (13.95)–(13.96). For $\alpha_B = 0$, the WBZ algorithm reduces to the Newmark ($\gamma = 1/4$ and $\beta = 1/2$) implicit algorithm. The algorithm is second-order time accurate and unconditionally stable for:

$$\alpha_B \leq 0, \quad \gamma = \frac{1}{2} - \alpha_B \quad \text{and} \quad \beta = \frac{1}{4}(1 - \alpha_B)^2 \tag{13.138}$$

The WBZ has properties that are somewhat similar to the HHT-α algorithm. For $\alpha_B = \alpha$, both methods possess analogous algorithmic damping properties; and it is also not an optimal algorithm within the LMS methods (Zhou and Tamma 2004c). The WBZ is also a U0-V1 algorithm (zero-order overshoot in displacement and first-order overshoot in velocity) and is a member of the U-family of algorithms in the context of the GSSSS framework described later.

13.4.9 Velocity Based Scheme [1988]

Original Representation with Unknowns: Displacement and Velocity (Two-Field Form) Emanating from the Lax-Wendroff/Taylor Galerkin formulation with the equation of motion represented in the conservation form, the original γ_s-family of direct self-starting methods (Tamma and Namburu 1988, 1990, 1997) was first developed and described in the context of the two-field form. In the context of the single-field form of the GSSSS unified framework encompassing the entire class of LMS methods that is described in the next chapter, the classical Midpoint Rule method (MPR-EPA) is strictly that algorithm which is common to both the so-called U- and V-family of algorithms and the acceleration that is computed from the algorithm is at time level t_{n+1} and it is not at time level $t_{n+1/2}$; and the classical (MPR-EPA) is one of the three principal building blocks within this class (the other two are the Newmark average acceleration method and the new version of the Midpoint Rule method with midpoint acceleration (MPR-MPA) where the acceleration is strictly at time level $t_{n+1/2}$ and this algorithm is highlighted subsequently; the well known Velocity based scheme (Tamma and Namburu 1990) in the sense of the two-field form is equivalent to this last approach in the single-field form of representation (with the exception of the Newmark method, the latter two approaches are energy, linear and angular momentum conserving for linear dynamical situations)). For other related efforts regarding the velocity based scheme, see (Tamma and Namburu 1988, 1990, 1997). Notably, in the context of the general single-field form of representation, the family of MPR-EPA through MPR-MPA based representations is a member of the V-family of algorithms in the context of the GSSSS unified framework that is described in detail in the next chapter. This is in sharp contrast to most of the classical (commonly known) algorithms that have been described earlier in this section which mainly pertain to the U-family of algorithms. The original γ_s-family of direct self-starting methods with the computed unknowns as velocity and displacement only in the two-field form is given by:

$$\begin{aligned}&\left[\mathbf{M}+\Delta t\gamma_1\mathbf{C}+\Delta t^2\gamma_1\gamma_2\mathbf{K}\right]\mathbf{v}_{n+1}\\&=\left[\mathbf{M}-\Delta t\left(1-\gamma_1\right)\mathbf{C}-\Delta t^2\gamma_1\left(1-\gamma_2\right)\mathbf{K}\right]\mathbf{v}_n\\&\quad-\Delta t\mathbf{K}\mathbf{u}_n+\Delta t\left(1-\gamma_1\right)\mathbf{f}_n+\Delta t\gamma_1\mathbf{f}_{n+1}\end{aligned}\tag{13.139}$$

$$\mathbf{u}_{n+1}=\mathbf{u}_n+\Delta t\left[\gamma_3\mathbf{v}_{n+1}+\left(1-\gamma_3\right)\mathbf{v}_n\right]\tag{13.140}$$

where $\mathbf{v}_{n+1}$ and $\mathbf{u}_{n+1}$ are the velocity and displacement vectors at time t_{n+1}, respectively; and $\mathbf{v}_n$ and $\mathbf{u}_n$ are the velocity and displacement vectors at time t_n, respectively ($n=0,1,2,\ldots$).

For given initial conditions of $\mathbf{v}_n$ and $\mathbf{u}_n$, the γ_s-family of representations shown in Equation (13.139) directly yield the representative velocities at the current time interval $n+1$. Next, these velocities are substituted in Equation (13.140) to directly obtain the displacement field at time level $n+1$. The evaluation of the accelerations is not involved in the computational process (alternatively, one can describe equivalent representations under the umbrella of the GSSSS unified framework with unknowns as displacement, velocity and acceleration as highlighted next, and also described in the next chapter). The parameters γ_s, $s=1,2,3$ govern the stability and accuracy of the family of representations. For $\gamma_1=1/2$, $\gamma_2=0$, and $\gamma_3=1/2$, the scheme is explicit, conditionally stable and second-order time accurate.

Remark 13.4.1 For conservative systems and in the absence of physical damping, the original implicit velocity based scheme described above in the context of the two-field form for linear dynamic situations is energy, linear and angular momentum conserving. The original work due to Simo and Tarnow (Simo and Tarnow 1992) in the two-field form of representation for nonlinear dynamics applications is identical to the above representation of the implicit algorithm for linear dynamic situations.

A New Equivalent Representation (MPR-MPA) with Unknowns: Displacement, Velocity, and Acceleration (Single-Field Form) Alternatively, one form of representation in the single-field form that is new and equivalent to the previous, and which is implicit, second-order time accurate, and unconditionally stable in terms of displacement, velocity, and acceleration can be rewritten as shown below. It is an outcome of and directly emanates from the GSSSS unified framework encompassing the class of LMS methods that is described in the next chapter; it pertains to the so-called V-family of algorithms, and it is one of the three basic building blocks of the GSSSS unified framework encompassing the class of LMS methods.

Given $\mathbf{u}_n$, $\dot{\mathbf{u}}_n$, and $\ddot{\mathbf{u}}_n$, find $\mathbf{u}_{n+1}$, $\dot{\mathbf{u}}_{n+1}$, and $\ddot{\mathbf{u}}_{n+1}$ from

$$\left[\mathbf{M}+\frac{1}{2}\mathbf{C}\Delta t+\frac{1}{4}\mathbf{K}\Delta t^2\right]\Delta\ddot{\mathbf{u}}$$

$$=-\mathbf{M}\ddot{\mathbf{u}}_n-\mathbf{C}\left[\dot{\mathbf{u}}_n+\frac{\Delta t}{2}\ddot{\mathbf{u}}_n\right]-\mathbf{K}\left[\mathbf{u}_n+\frac{\Delta t}{2}\dot{\mathbf{u}}_n+\frac{\Delta t^2}{4}\ddot{\mathbf{u}}_n\right]+\frac{\mathbf{f}_n+\mathbf{f}_{n+1}}{2}$$

with the following updates:

$$\mathbf{u}_{n+1}=\mathbf{u}_n+\dot{\mathbf{u}}_n\Delta t+\frac{\Delta t^2}{2}\ddot{\mathbf{u}}_n+\frac{\Delta t^2}{2}\Delta\ddot{\mathbf{u}}$$

$$\dot{\mathbf{u}}_{n+1}=\dot{\mathbf{u}}_n+\ddot{\mathbf{u}}_n\Delta t+\frac{\Delta t}{2}\Delta\ddot{\mathbf{u}}$$

$$\ddot{\mathbf{u}}_{n+1}=\ddot{\mathbf{u}}_n+\Delta\ddot{\mathbf{u}}$$

It is extremely important to understand the fundamental and subtle differences between the new representation above and the classical version of the Midpoint Rule method with end-point acceleration (MPR-EPA) which was presented previously. In sharp contrast to the classical version of the Midpoint Rule, the above new representation has the analogy of the Midpoint Rule method with mid-point acceleration (MPR-MPA)(note that in the new representation of the Midpoint Rule method with mid-point acceleration (MPR-MPA), the acceleration is strictly at time level $t_{n+1/2}$, and it is not at the time level t_{n+1}). Furthermore, also note that the accelerations from the previous time step do not play a role at all in the computation of the accelerations at the current time level (and the reader should be able to figure this out). A note of caution is in order. The reader is strongly encouraged to carefully understand the GSSSS unified framework encompassing the entire class of LMS methods that is described in detail in the next chapter, including, in particular, the subtle issues regarding time level consistency to understand the fundamental concepts and time level relevant issues for displacement, velocity and acceleration (see also, (Tamma et al. 2000; Tamma et al. 2001; Tamma et al. 2011; Zhou and Tamma 2004c; Zhou and Tamma 2004a; Zhou and Tamma 2006; Hoitink et al. 2008; Masuri et al. 2009b; Masuri et al. 2009c; Masuri et al. 2009a)). Amongst all numerically non-dissipative algorithms, this algorithm is very robust and is highly recommended; it is our selected and preferred choice within the entire class of LMS methods. Alternatively, note that the optimal U0-V0 (zero-order displacement and zero-order velocity overshoot behavior) algorithm that is highlighted in a subsequent section to follow (and described in detail in the next chapter under the GSSSS unified framework) is the recommended algorithm and preferred choice amongst all algorithms with controllable numerical dissipation encompassing the class of LMS methods with a single solve representation. These are our recommendations for time integration algorithms within the class of LMS methods.

13.4.10 The Three Parameters Optimal Schemes (χ-Schemes) [1988] (and the identical Generalized-α Method)

Recently, Leontiev (Leontiev 2007) cites the original version of the "Three Parameters Optimal Schemes" due to Shao (Shao 1987) and Shao and Cai (Shao and Cai 1988) and was made aware to the present authors, where it was noted that the algorithm that is commonly referred to as the generalized-α (Chung and Hulbert 1993) is indeed identical to this original version that was named as the three parameters optimal schemes. In this section, we first summarize the original version of the Three Parameters Optimal Schemes which appeared in 1988 (Shao and Cai 1988) and recently pointed out by Leontiev (Leontiev 2007); and then we summarize the identical Generalized-α which subsequently appeared in 1993 (Chung and Hulbert 1993). In a later chapter to follow, and under the umbrella of the well known GSSSS unified framework which encompasses the class of LMS methods with a single system and a single solve representation, regardless of how the various original methods of time integration have been previously developed for both numerically non-dissipative and numerically dissipative (controllable numerical dissipation) algorithms, all these representations are shown in a unified manner and as a subset of this very general framework. Additionally, the generalized framework also includes new avenues and features for improved and optimal algorithm designs that are very robust (see U0-family and V0-family of algorithms), within the class of LMS methods in the sense of the single-field form of representation.

The Three Parameters Optimal Schemes (χ-Schemes) [1988] The basic idea due to the original contributions by Shao (Shao 1987) and Shao and Cai (Shao and Cai 1988) is the fundamental concept of combining the HHT-methods and the Bossak-methods; and therein it provides extensions to this combined framework in developing the optimal χ-schemes. Subsequently, the algorithmic properties are compared with those of the others (for example, Wilson-θ, methods, Improved Wilson-θ, etc.; and it provides reference citations and draws comparisons with Newmark methods, HHT-methods, Bossak-methods, and the single-step Houbolt form, etc.). Consequently, this representation is a member of the U-family of algorithms in the context of the GSSSS framework described in the next chapter.

Shao and Cai (Shao and Cai 1988) state that this new family of methods which was originally developed in 1988 can be obtained by extending the combined framework of the HHT-methods (α-methods) and WBZ (Wood et al. 1980) methods. Their starting point is the following basic formula:

$$\begin{aligned}(1-\alpha)\,\mathbf{M}\ddot{\mathbf{u}}_{n+1} + \alpha\mathbf{M}\ddot{\mathbf{u}}_n + (1-\eta)\,\mathbf{C}\dot{\mathbf{u}}_{n+1} + \eta\mathbf{C}\dot{\mathbf{u}}_n \\ + (1-\delta)\,\mathbf{K}\mathbf{u}_{n+1} + \delta\mathbf{K}\mathbf{u}_n = (1-\xi)\,\mathbf{f}_{n+1} + \xi\mathbf{f}_n \end{aligned} \tag{13.141}$$

$$\mathbf{u}_{n+1} = \mathbf{u}_n + \Delta t\dot{\mathbf{u}}_n + \left(\frac{1}{2} - \beta\right)\Delta t^2\ddot{\mathbf{u}}_n + \beta\Delta t^2\ddot{\mathbf{u}}_{n+1} \tag{13.142}$$

$$\dot{\mathbf{u}}_{n+1} = \dot{\mathbf{u}}_n + (1-\gamma)\,\Delta t\ddot{\mathbf{u}}_n + \gamma\Delta t\ddot{\mathbf{u}}_{n+1} \tag{13.143}$$

Using Equations (13.142) and (13.143), one can readily obtain $\ddot{\mathbf{u}}_{n+1}$ and $\dot{\mathbf{u}}_{n+1}$ in terms of the free parameters. Of the six parameters, for Newmark methods ($\alpha = \delta = \eta = \xi = 0$), HHT-methods ($\alpha = 0$), and Bossak-methods ($\delta = \eta = \xi = 0$); in fact, they prove that among the six parameters, only three can be chosen freely to obtain the optimal schemes with second-order time accuracy (for example, α, δ, β). Second-order time accurate unconditionally stable conditions (for linear dynamics) are

$$\alpha \le \delta \le \frac{1}{2} \tag{13.144}$$

$$\beta \ge \frac{1}{2}\gamma = \frac{1}{2}\left(\frac{1}{2} - \alpha + \delta\right) \tag{13.145}$$

$$\eta = \xi = \delta \tag{13.146}$$

$$\xi \ge 0 \tag{13.147}$$

Relation to ρ_∞ (is the value of the spectral radius at infinity, which is a controllable parameter)

$$\alpha = \frac{2\rho_\infty - 1}{\rho_\infty + 1} \tag{13.148}$$

$$\beta = \frac{1}{(1+\rho_\infty)^2} \tag{13.149}$$

$$\gamma = \frac{3-\rho_\infty}{2\,(1+\rho_\infty)} \tag{13.150}$$

$$\eta = \xi = \delta = \frac{\rho_\infty}{\rho_\infty + 1} \tag{13.151}$$

The Optimal χ-Schemes (Controlled by value of ρ, as $\Omega \to \infty$)

$$\alpha = -\frac{(2\rho - 1)}{(\rho + 1)}, \quad \beta = -\frac{1}{(1+\rho)^2} \tag{13.152}$$

$$\gamma = -\frac{(3-\rho)}{2\,(1+\rho)}, \quad \eta = \xi = \delta = -\frac{\rho}{1+\rho} \tag{13.153}$$

HHT-α Methods (Defined by value of ρ, as $\Omega \to \infty$)

$$\alpha = 0, \quad \beta = \frac{1}{(1+\rho)^2} \tag{13.154}$$

$$\gamma = \frac{3-\rho}{2\,(1+\rho)}, \quad \eta = \xi = \delta = \frac{1-\rho}{1+\rho} \tag{13.155}$$

Bossak Methods (Defined by value of ρ, as $\Omega \to \infty$)

$$\alpha = \frac{\rho - 1}{\rho + 1}, \quad \beta = -\frac{1}{(1+\rho)^2} \tag{13.156}$$

$$\gamma = -\frac{(3-\rho)}{2\,(1+\rho)}, \quad \eta = \xi = \delta = 0 \tag{13.157}$$

The authors note that the value of ρ_∞ between 0.7 and 0.8 yields good results for their test case under consideration. In general, it is a parameter that is arbitrarily selected. For $\alpha = \alpha_m$ and $\delta = \alpha_f$, the identical relationship between the original optimal χ-schemes and the generalized-α method (as described next) can be readily established.

The Generalized-α Method [1993] As in the χ-schemes, this method which appeared subsequently, also represents the same generalization by also combining the HHT-α and the WBZ algorithms (see (Chung and Hulbert 1993)) as described previously by Shao (Shao 1987) and Shao and Cai (Shao and Cai 1988); and it is the same identical member of the U-family of algorithms (not the V-family) in the context of the GSSSS framework described later (in the sense of the GSSSS framework, both these algorithms have the same DNA [discrete numerically assigned] markers which represent conditions that are unique to a particular algorithm and these algorithms are hence identical; that is, to be unique, no two algorithms can have the same DNA and relevant details are described in the next chapter).

The basic starting point is also the same and is as follows (compare Equation (13.141) and the following equation, and equate parameters):

$$\mathbf{M}\ddot{\mathbf{u}}_{n+1-\alpha_m} + \mathbf{C}\dot{\mathbf{u}}_{n+1-\alpha_f} + \mathbf{K}\mathbf{u}_{n+1-\alpha_f} = \mathbf{f}_{n+1-\alpha_f} \tag{13.158}$$

where:

$$\mathbf{u}_{n+1-\alpha_f} = \left(1-\alpha_f\right)\mathbf{u}_{n+1} + \alpha_f \mathbf{u}_n \tag{13.159}$$

$$\dot{\mathbf{u}}_{n+1-\alpha_f} = \left(1-\alpha_f\right)\dot{\mathbf{u}}_{n+1} + \alpha_f \dot{\mathbf{u}}_n \tag{13.160}$$

$$\ddot{\mathbf{u}}_{n+1-\alpha_m} = \left(1-\alpha_m\right)\ddot{\mathbf{u}}_{n+1} + \alpha_m \ddot{\mathbf{u}}_n \tag{13.161}$$

$$t_{n+1-\alpha_f} = \left(1-\alpha_f\right)t_{n+1} + \alpha_f t_n \tag{13.162}$$

With appropriate values for the parameters β and γ, the algorithm reduces to the HHT-α method and the WBZ method for $\alpha_m = 0$ and $\alpha_f = 0$, respectively. For $\alpha_f = \alpha_m = 0$ the method reduces to the Newmark family of algorithms. The method is second-order time accurate provided that:

$$\gamma = \frac{1}{2} - \alpha_m + \alpha_f \tag{13.163}$$

The conditions of unconditional stability are given by:

$$\alpha_m \leq \alpha_f \leq \frac{1}{2} \quad \text{and} \quad \beta \geq \frac{1}{4} + \frac{1}{2}\left(\alpha_f - \alpha_m\right) \tag{13.164}$$

The high-frequency dissipation of the algorithm is maximized for values of the parameter β given by:

$$\beta = \frac{1}{4}\left(1 - \alpha_m + \alpha_f\right)^2 \tag{13.165}$$

The method expresses the parameters α_m and α_f as functions of the amount of numerical dissipation desired, as denoted by ρ_∞. The expressions are derived from the conditions of minimizing the low-frequency dissipation and maximizing the high-frequency dissipation:

$$\alpha_m = \frac{2\rho_\infty - 1}{\rho_\infty + 1} \tag{13.166}$$

$$\alpha_f = \frac{\rho_\infty}{\rho_\infty + 1} \tag{13.167}$$

The amount of high-frequency dissipation can thus be varied from the no dissipation case ($\rho_\infty = 1$, classical version of the Midpoint Rule (MPR-EPA)), to the so-called asymptotic annihilation case ($\rho_\infty = 0$, WBZ method), in which the high-frequency response is annihilated after one time step. In the context of the GSSSS framework, both the above representations are identical to each other and inherit U0-V1 (zero-order displacement overshoot and first-order velocity overshoot) behavior.

13.4.11 Optimal U0-V0 Algorithm: The Optimal Algorithm With Controllable Numerical Dissipation Within the Class of LMS Methods in the Sense of the Single-Field Form [2004]

We next explicitly highlight in general, the optimal algorithm for any given set of arbitrary initial starting conditions (this is important to note as overshoot plays a significant role, especially in nonlinear dynamics applications involving Newton-type iterations, and several of the previously described algorithms are not optimal for the general case of arbitrary initial conditions) with controllable numerical dissipation within the class of LMS methods. This algorithm is our selected and preferred choice of the optimal and controllable numerical dissipative algorithm in the class of LMS methods from the viewpoint of applicability to practical applications and is highly recommended. The representation is in the single-field form (see details of the GSSSS framework of algorithms in the next chapter where all these developments that have been described in this chapter are presented in a unified manner (Tamma et al. 2000, 2001; Zhou and Tamma 2004a,c, 2006)). The optimality of the algorithm is in the sense of the least amount of numerical dissipation, dispersion, and algorithmic overshoot behavior for the same computational complexity.

Given $\mathbf{u}_n$, $\dot{\mathbf{u}}_n$, and $\ddot{\mathbf{u}}_n$, find $\mathbf{u}_{n+1}$, $\dot{\mathbf{u}}_{n+1}$, and $\ddot{\mathbf{u}}_{n+1}$ from

$$\begin{aligned}
&\left[\frac{3-\bar{\rho}_\infty}{2(1+\bar{\rho}_\infty)}\mathbf{M} + \frac{1}{(1+\bar{\rho}_\infty)^2}\mathbf{C}\Delta t + \frac{1}{2(1+\bar{\rho}_\infty)^2}\mathbf{K}\Delta t^2\right]\Delta\ddot{\mathbf{u}} \\
&= -\mathbf{M}\ddot{\mathbf{u}}_n - \mathbf{C}\left[\dot{\mathbf{u}}_n + \frac{\Delta t}{1+\bar{\rho}_\infty}\ddot{\mathbf{u}}_n\right] - \mathbf{K}\left[\mathbf{u}_n + \frac{\Delta t}{1+\bar{\rho}_\infty}\dot{\mathbf{u}}_n + \frac{\Delta t^2}{2(1+\bar{\rho}_\infty)}\ddot{\mathbf{u}}_n\right] \\
&\quad + \mathbf{f}_n + \frac{1}{1+\bar{\rho}_\infty}(\mathbf{f}_{n+1} - \mathbf{f}_n)
\end{aligned}$$

with the following updates:

$$\begin{aligned}
\mathbf{u}_{n+1} &= \mathbf{u}_n + \dot{\mathbf{u}}_n\Delta t + \frac{\Delta t^2}{2}\ddot{\mathbf{u}}_n + \frac{\Delta t^2}{2(1+\bar{\rho}_\infty)}\Delta\ddot{\mathbf{u}} \\
\dot{\mathbf{u}}_{n+1} &= \dot{\mathbf{u}}_n + \ddot{\mathbf{u}}_n\Delta t + \frac{\Delta t}{1+\bar{\rho}_\infty}\Delta\ddot{\mathbf{u}} \\
\ddot{\mathbf{u}}_{n+1} &= \ddot{\mathbf{u}}_n + \Delta\ddot{\mathbf{u}}
\end{aligned}$$

where $\Delta\ddot{\mathbf{u}} := \ddot{\mathbf{u}}_{n+1} - \ddot{\mathbf{u}}_n$. Note that $\bar{\rho}_\infty = \rho_{1\infty} = \rho_{3\infty}$ and $\rho_{2\infty} = 1$; see Table 14.1.

Remark 13.4.2 A note of caution is in order regarding the formulations that have been described thus far which highlighted the various developments for linear dynamics systems. Most often, for nonlinear dynamics applications, typical traditional practices simply adopt a Newton-type iterative procedure (after linearization) of the various parent linear dynamic algorithm constructions described in this section (after incorporating the nonlinear material property and/or nonlinear description of strain). Such ad-hoc practices are readily prone to implementation and other related errors (especially, in handling nonlinear dynamics applications of the same algorithm by different analysts) as evident from a variety of differing and inconsistent results that have been published in the literature. In contrast, in the next chapter and under the well known GSSSS framework (which readily encompasses all the developments in this section in a nutshell), we particularly provide the necessary theoretical basis on how to properly provide the proper extensions of the parent linear dynamic algorithms described in this and various other sections to nonlinear dynamic situations which are not prone to such implementation and other related errors. The proper approaches for providing extensions to nonlinear dynamics applications, and the various consequences leading to algorithm designs that are symplectic-momentum based or energy-momentum based respectively, with and without controllable numerical dissipation is extremely important to understand (see details in the next chapter).

Thus far, most of the time integration schemes discussed previously have been derived in the sense of the single-field form of representation of the equation of motion. They seem to be the most practical and are dominant in commercial software; and we envision that future efforts for the next generation software simulation tools for time integration will be based on the GSSSS unified framework that is described in detail in the next chapter, which, in a nutshell, encompasses most of the developments that are described in this chapter (see also Tamma et al. 2011, Tamma et al. 2012).

13.5 SYMPLECTIC-MOMENTUM CONSERVATION AND VARIATIONAL TIME INTEGRATORS

In contrast to the previous developments which are mostly based upon the single-field form of representation of the semi-discretized equations of motion and without consideration of algorithm structure preservation constraints and/or relevant concepts, algorithmic designs such as structure-preserving algorithms in the sense of symplectic algorithms and those termed as variational integrators appear more recently in the literature. The objective of these latter developments that are described in this section, is to foster the simulation of long term dynamics with structure preservation for a class of nonlinear dynamic problems. In this section, the general focus is on applications to nonlinear dynamic situations and the fundamental concepts emanate from the so-called Discrete Euler-Lagrange equations that follow the original ideas due to Veselov (Veselov 1988) and also subsequently described in (Marsden and West 2001; Marsden et al. 2000). Although the developments start with Hamilton's principle and involves the descriptive function, namely, the Lagrangian, the consequent algorithm designs have been mostly developed in the sense of the two-field form. Some brief highlights of these developments and the resulting algorithms follow next. For symplectic algorithms, while the linear and angular momenta are indeed conserved, the total energy is conserved in the sense that it is bounded for a fixed time step (this is in contrast to the energy-momentum conserving algorithms which strictly (exactly) conserve the total energy). Furthermore, depending upon the application problem at hand, the implicit form of the symplectic-momentum algorithm appears to suffer from certain drawbacks as related to convergence issues via Newton type nonlinear iterations for a given (large) time step as in energy-momentum type schemes that are subsequently described (in particular for stiff systems); alternatively, they appear to be not as robust for fairly large time steps. Other relevant and interesting efforts appear in (Gonzalez et al. 2010). In the next chapter and under the umbrella of the GSSSS family of algorithms emanating from the Total Energy framework and a generalized time weighted residual approach, we particularly describe the theoretical basis for algorithm designs in the two-field and the single-field form that are symplectic-momentum based; they cover most of the algorithm developments described from an alternative formulation such as that presented in this section via the Discrete Euler-Lagrange equations, and in addition for the single-field form also include features to incorporate controllable numerical dissipation in the parent symplectic-momentum conserving algorithm designs.

13.5.1 Discrete Euler-Lagrange Equations

For an autonomous Lagrangian system, the Lagrangian $\mathcal{L}(\mathbf{q}, \dot{\mathbf{q}}) : TQ \rightarrow \mathbb{R}$ is the difference between the kinetic energy $\mathcal{K}(\dot{\mathbf{q}}) : T_q Q \rightarrow \mathbb{R}$ and the potential energy $\mathcal{U}(\mathbf{q}) : Q \rightarrow \mathbb{R}$ as shown below

$$\mathcal{L}(\mathbf{q}, \dot{\mathbf{q}}) = \mathcal{K}(\dot{\mathbf{q}}) - \mathcal{U}(\mathbf{q}) = \frac{1}{2}\dot{\mathbf{q}} \cdot \mathbf{M}\dot{\mathbf{q}} - \mathcal{U}(\mathbf{q}) \tag{13.168}$$

Here, it is important to note that in developing variational time integrators, the mass has been assumed to be constant. Hence, the variational time integrators described in this section are limited to the Lagrangian system in which the mass associated with the kinetic energy is constant.

Hamilton's principle was used in the developments, and it states that the Euler-Lagrange equations of motion are derived from extremizing the action shown below.

$$S = \int_{t_0}^{t_N} \mathcal{L}(\mathbf{q}, \dot{\mathbf{q}})\, dt \tag{13.169}$$

It appears that Veselov (Veselov 1988) first considered discrete variational mechanics to derive variational time integrators in the context of integrable systems (Marsden and West 2001; Marsden et al. 2000), based on continuous variational mechanics. Let us consider that a time interval $[t_0, t_N]$ can be assumed to be a disjoint union $\bigcup_{n=0}^{N-1} \left[t_n, t_{n+1}\right]$ with a typical time step, $\Delta t = t_{n+1} - t_n$. Then, the discrete generalized coordinates and canonical momenta can be denoted by $\mathbf{q}_n = \mathbf{q}\left(t = t_n\right), \mathbf{p}_n = \mathbf{p}\left(t = t_n\right)$. By the discrete action map, the action in Equation (13.169) may be approximated as

$$S_d = \sum_{n=0}^{N-1} \mathcal{L}_d^{\alpha}\left(\mathbf{q}_n, \mathbf{q}_{n+1}\right) \tag{13.170}$$

where $\mathcal{L}_d^{\alpha}\left(\mathbf{q}_n, \mathbf{q}_{n+1}\right)$ is called the discrete Lagrangian $\mathcal{L}_d^{\alpha}\left(\mathbf{q}_n, \mathbf{q}_{n+1}\right) : Q \times Q \rightarrow \mathbb{R}$ in the time interval $\left[\mathbf{q}_n, \mathbf{q}_{n+1}\right]$. The discrete Lagrangian (Kane et al. 2000) can be expressed as

$$\mathcal{L}_d^{\alpha}\left(\mathbf{q}_n, \mathbf{q}_{n+1}\right) = \Delta t \left\{\frac{1}{2}\left(\frac{\mathbf{q}_{n+1} - \mathbf{q}_n}{\Delta t}\right) \cdot \mathbf{M}\left(\frac{\mathbf{q}_{n+1} - \mathbf{q}_n}{\Delta t}\right) - \mathcal{U}\left(\left(1-\alpha\right)\mathbf{q}_n + \alpha\mathbf{q}_{n+1}\right)\right\} \tag{13.171}$$

where $\alpha \in [0, 1]$ is a real value. In the case that the mass is not constant, the above equation is not valid anymore. As stated in the continuous Lagrangian system, extremizing the discrete action map in Equation (13.170) leads to

$$\delta S_d = \sum_{n=0}^{N-1} \left\{D_1\mathcal{L}_d^{\alpha}\left(\mathbf{q}_n, \mathbf{q}_{n+1}\right) \cdot \delta\mathbf{q}_n + D_2\mathcal{L}_d^{\alpha}\left(\mathbf{q}_n, \mathbf{q}_{n+1}\right) \cdot \delta\mathbf{q}_{n+1}\right\} = 0 \tag{13.172}$$

Since Hamilton's principle in N-body dynamical systems requires the first variations of the displacements at both end points to vanish (a somewhat controversial aspect as evident from the literature) in order for the discrete action to be stationary, i.e., $\delta\mathbf{q}_0 = \delta\mathbf{q}_f = \mathbf{0}$, we obtain

$$\delta S_d = \sum_{n=1}^{N-1} \left\{D_2\mathcal{L}_d^{\alpha}\left(\mathbf{q}_{n-1}, \mathbf{q}_n\right) + D_1\mathcal{L}_d^{\alpha}\left(\mathbf{q}_n, \mathbf{q}_{n+1}\right)\right\} \cdot \delta\mathbf{q}_n = 0 \tag{13.173}$$

Within the time interval $[t_0, t_N]$, $\delta\mathbf{q}_n$ is arbitrary. Hence, the above equation reduces to

$$D_2\mathcal{L}_d^{\alpha}\left(\mathbf{q}_{n-1}, \mathbf{q}_n\right) + D_1\mathcal{L}_d^{\alpha}\left(\mathbf{q}_n, \mathbf{q}_{n+1}\right) = 0, \quad for\ n = 1 \rightarrow N-1 \tag{13.174}$$

where the first term may be given by

$$D_2\mathcal{L}_d^{\alpha}\left(\mathbf{q}_{n-1}, \mathbf{q}_n\right) = \mathbf{M}\left(\frac{\mathbf{q}_n - \mathbf{q}_{n-1}}{\Delta t}\right) - \alpha\Delta t\nabla\mathcal{U}\left(\left(1-\alpha\right)\mathbf{q}_{n-1} + \alpha\mathbf{q}_n\right) \tag{13.175}$$

and the second term can be expressed as

$$D_1\mathcal{L}_d^{\alpha}\left(\mathbf{q}_n, \mathbf{q}_{n+1}\right) = -\mathbf{M}\left(\frac{\mathbf{q}_{n+1} - \mathbf{q}_n}{\Delta t}\right) - \left(1-\alpha\right)\Delta t\nabla\mathcal{U}\left(\left(1-\alpha\right)\mathbf{q}_n + \alpha\mathbf{q}_{n+1}\right) \tag{13.176}$$

Substituting Equation (13.175) and (13.176) into (13.174), we have

$$\mathbf{M}\left(\frac{\mathbf{q}_{n+1} - 2\mathbf{q}_n + \mathbf{q}_{n-1}}{\Delta t^2}\right) = -\left(1-\alpha\right)\nabla\mathcal{U}\left(\mathbf{q}_{n+\alpha}\right) - \alpha\nabla\mathcal{U}\left(\mathbf{q}_{n-1+\alpha}\right) \tag{13.177}$$

where $\mathbf{q}_{n+\alpha} = (1-\alpha)\,\mathbf{q}_n + \alpha\mathbf{q}_{n+1}$, $\mathbf{q}_{n-1+\alpha} = (1-\alpha)\,\mathbf{q}_{n-1} + \alpha\mathbf{q}_n$. We observe that for $\alpha = 0$, Equation (13.177) results in the explicit central difference time integration algorithm as shown below,

$$\mathbf{M}\left(\frac{\mathbf{q}_{n+1} - 2\mathbf{q}_n + \mathbf{q}_{n-1}}{\Delta t^2}\right) = -\nabla\mathcal{U}(\mathbf{q}_n) \tag{13.178}$$

Further, for $\alpha = \frac{1}{2}$, we have the generalized Midpoint Rule time integration algorithm,

$$\mathbf{M}\left(\frac{\mathbf{q}_{n+1} - 2\mathbf{q}_n + \mathbf{q}_{n-1}}{\Delta t^2}\right) = -\frac{1}{2}\left\{\nabla\mathcal{U}\left(\mathbf{q}_{n+\frac{1}{2}}\right) + \nabla\mathcal{U}\left(\mathbf{q}_{n-\frac{1}{2}}\right)\right\} \tag{13.179}$$

In order for Equation (13.177) to be second-order time accurate for any α, the discrete Lagrangian is alternatively defined as

$$\begin{aligned}\mathcal{L}_d^{sym,\alpha}(\mathbf{q}_n, \mathbf{q}_{n+1}) &= \frac{\Delta t}{2}\mathcal{L}\left(\frac{\mathbf{q}_{n+1} - \mathbf{q}_n}{\Delta t}, (1-\alpha)\,\mathbf{q}_n + \alpha\mathbf{q}_{n+1}\right)\\ &+ \frac{\Delta t}{2}\mathcal{L}\left(\frac{\mathbf{q}_{n+1} - \mathbf{q}_n}{\Delta t}, \alpha\mathbf{q}_n + (1-\alpha)\,\mathbf{q}_{n+1}\right)\end{aligned} \tag{13.180}$$

Consequently, instead of Equation (13.174) we have

$$D_2\mathcal{L}_d^{sym,\alpha}\left(\mathbf{q}_{n-1}, q_n\right) + D_1\mathcal{L}_d^{sym,\alpha}(\mathbf{q}_n, \mathbf{q}_{n+1}) = 0 \tag{13.181}$$

where the first term may be given by

$$D_2\mathcal{L}_d^{sym,\alpha}\left(\mathbf{q}_{n-1}, q_n\right) = \mathbf{M}\left(\frac{\mathbf{q}_n - \mathbf{q}_{n-1}}{\Delta t}\right) + \frac{1}{2}\Delta t\mathbf{M}\left(\alpha\mathbf{a}_{n-1+\alpha} + (1-\alpha)\,\mathbf{a}_{n-\alpha}\right) \tag{13.182}$$

where

$$\begin{aligned}\mathbf{a}_{n-1+\alpha} &= -\mathbf{M}^{-1}\nabla\mathcal{U}((1-\alpha)\,\mathbf{q}_{n-1} + \alpha\mathbf{q}_n),\\ \mathbf{a}_{n-\alpha} &= -\mathbf{M}^{-1}\nabla\mathcal{U}(\alpha\mathbf{q}_{n-1} + (1-\alpha)\,\mathbf{q}_n)\end{aligned} \tag{13.183}$$

And, the second term can be expressed as

$$D_1\mathcal{L}_d^{sym,\alpha}(\mathbf{q}_n, \mathbf{q}_{n+1}) = -\mathbf{M}\left(\frac{\mathbf{q}_{n+1} - q_n}{\Delta t}\right) + \frac{1}{2}\Delta t\mathbf{M}\left(\alpha\mathbf{a}_{n+1-\alpha} + (1-\alpha)\,\mathbf{a}_{n+\alpha}\right) \tag{13.184}$$

where

$$\begin{aligned}\mathbf{a}_{n+1-\alpha} &= -\mathbf{M}^{-1}\nabla\mathcal{U}((1-\alpha)\,\mathbf{q}_n + \alpha\mathbf{q}_{n+1}),\\ \mathbf{a}_{n+\alpha} &= -\mathbf{M}^{-1}\nabla\mathcal{U}(\alpha\mathbf{q}_n + (1-\alpha)\,\mathbf{q}_{n+1})\end{aligned} \tag{13.185}$$

Substituting Equation (13.182) and (13.184) into (13.181), we have

$$\left(\frac{\mathbf{q}_{n+1} - 2\mathbf{q}_n + \mathbf{q}_{n-1}}{\Delta t^2}\right) = \frac{1}{2}\left(\alpha\mathbf{a}_{n-1+\alpha} + (1-\alpha)\,\mathbf{a}_{n-\alpha} + \alpha\mathbf{a}_{n+1-\alpha} + (1-\alpha)\,\mathbf{a}_{n+\alpha}\right) \tag{13.186}$$

Equation (13.186) is called a second-order accurate variational time integrator for any α. When $\alpha = 0$, Equation (13.186) becomes equivalent to Equation (13.178). Further, when $\alpha = \frac{1}{2}$, Equation (13.186) becomes equivalent to Equation (13.179).

13.5.2 Discrete Legendre Transformation

In time-continuous Lagrangian mechanics, the autonomous Lagrangian flow map $F_{\mathcal{L}} : TQ \to TQ$ is defined in coordinate form as

$$F_{\mathcal{L}} : (\mathbf{q}, \dot{\mathbf{q}}) \to (\mathbf{q}, \dot{\mathbf{q}}) \tag{13.187}$$

The Lagrangian flow map can be written as $F_{\mathcal{L}}^t : (\mathbf{q}, \dot{\mathbf{q}}) \to (\mathbf{q}, \dot{\mathbf{q}})$ at the specific time. Likewise, in time-continuous Hamiltonian mechanics, the autonomous Hamiltonian flow map $F_{\mathcal{H}} : T^*Q \to T^*Q$ is defined in coordinate form as

$$F_{\mathcal{H}} : (\mathbf{q}, \mathbf{p}) \to (\mathbf{q}, \mathbf{p}) \tag{13.188}$$

Hence, the Hamiltonian flow map can be written as $F_{\mathcal{H}}^t : (\mathbf{q}, \mathbf{p}) \to (\mathbf{q}, \mathbf{p})$ at the specific time. The Legendre transformation relates Lagrangian mechanics to Hamiltonian mechanics. In other words, it maps the velocity phase space, which is the domain of the discrete Lagrangian, to the phase space, which is the domain of the discrete Hamiltonian. In discrete mechanics (Marsden and West 2001), the transformation is referred to as a fiber derivative of the Lagrangian as follows,

$$\mathbb{F}\mathcal{L} : (\mathbf{q}, \dot{\mathbf{q}}) \to (\mathbf{q}, \mathbf{p}) = \left(\mathbf{q}, \frac{\partial \mathcal{L}(\mathbf{q}, \dot{\mathbf{q}})}{\partial \dot{\mathbf{q}}} \right) \tag{13.189}$$

Conversely, we have a fiber derivative of the Hamiltonian as below,

$$\mathbb{F}\mathcal{H} : (\mathbf{q}, \mathbf{p}) \to (\mathbf{q}, \dot{\mathbf{q}}) = \left(\mathbf{q}, \frac{\partial \mathcal{H}(\mathbf{q}, \mathbf{p})}{\partial \mathbf{p}} \right) \tag{13.190}$$

Within this framework, we describe the so-called discrete Legendre transformation. In discrete variational mechanics, the discrete Legendre transformation or discrete fiber derivative $\mathbb{F}^+\mathcal{L}_d^{sym,\alpha}, \mathbb{F}^-\mathcal{L}_d^{sym,\alpha} : Q \times Q \to T^*Q$ can be given by

$$\begin{aligned} \mathbb{F}^+\mathcal{L}_d^{sym,\alpha}(\mathbf{q}_{n-1}, \mathbf{q}_n) \cdot \delta\mathbf{q}_n &= D_2\mathcal{L}_d^{sym,\alpha}(\mathbf{q}_{n-1}, \mathbf{q}_n) \cdot \delta\mathbf{q}_n \\ \mathbb{F}^-\mathcal{L}_d^{sym,\alpha}(\mathbf{q}_n, \mathbf{q}_{n+1}) \cdot \delta\mathbf{q}_n &= -D_1\mathcal{L}_d^{sym,\alpha}(\mathbf{q}_n, \mathbf{q}_{n+1}) \cdot \delta\mathbf{q}_n \end{aligned} \tag{13.191}$$

It can be written in coordinate form as

$$\begin{aligned} \mathbb{F}^+\mathcal{L}_d^{sym,\alpha} &: (\mathbf{q}_{n-1}, \mathbf{q}_n) \to (\mathbf{q}_n, \mathbf{p}_n) = \left(\mathbf{q}_n, D_2\mathcal{L}_d^{sym,\alpha}(\mathbf{q}_{n-1}, \mathbf{q}_n)\right) \\ \mathbb{F}^-\mathcal{L}_d^{sym,\alpha} &: (\mathbf{q}_n, \mathbf{q}_{n+1}) \to (\mathbf{q}_n, \mathbf{p}_n) = \left(\mathbf{q}_n, -D_1\mathcal{L}_d^{sym,\alpha}(\mathbf{q}_n, \mathbf{q}_{n+1})\right) \end{aligned} \tag{13.192}$$

By virtue of Equation (13.174), we have

$$\mathbb{F}^+\mathcal{L}_d^{sym,\alpha}(\mathbf{q}_{n-1}, \mathbf{q}_n) = \mathbb{F}^-\mathcal{L}_d^{sym,\alpha}(\mathbf{q}_n, \mathbf{q}_{n+1}) \tag{13.193}$$

The Discrete Euler-Lagrange (DEL) equations of motion have been derived from the discretization of Hamilton's principle. The DEL equations shown in Equation (13.181) can be written as

$$D_{DEL}\mathcal{L}_d^{\alpha}\{(\mathbf{q}_{n-1}, \mathbf{q}_n), (\mathbf{q}_n, \mathbf{q}_{n+1})\} = 0 \tag{13.194}$$

which defines the discrete Lagrangian map $F_{\mathcal{L}_d^{\alpha}} : Q \times Q \to Q \times Q$ as a one-step time integrator (Marsden and West 2001). The discrete Lagrangian map is often interpreted as $F_{\mathcal{L}_d}(\mathbf{q}_{n-1}, \mathbf{q}_n) = (\mathbf{q}_n, \mathbf{q}_{n+1})$ or $(\mathbf{q}_{n-1}, \mathbf{q}_n) \to (\mathbf{q}_n, \mathbf{q}_{n+1})$ for all $n = 1, \cdots, N-1$.

Variational Algorithm on T^*Q We examine the discrete momentum or the position momentum. Equation (13.181) can be written as

$$D_2\mathcal{L}_d^{sym,\alpha}\left(\mathbf{q}_{n-1}, q_n\right) = -D_1\mathcal{L}_d^{sym,\alpha}\left(\mathbf{q}_n, \mathbf{q}_{n+1}\right) \tag{13.195}$$

From the definition of the discrete Legendre transformation above, we obtain the discrete momentum as

$$\mathbf{p}_n = D_2\mathcal{L}_d^{sym,\alpha}\left(\mathbf{q}_{n-1}, \mathbf{q}_n\right) = -D_1\mathcal{L}_d^{sym,\alpha}\left(\mathbf{q}_n, \mathbf{q}_{n+1}\right) \tag{13.196}$$

In view of the above equation, we obtain

$$\mathbf{p}_{n+1} = D_2\mathcal{L}_d^{sym,\alpha}\left(q_n, \mathbf{q}_{n+1}\right) = -D_1\mathcal{L}_d^{sym,\alpha}\left(\mathbf{q}_{n+1}, \mathbf{q}_{n+2}\right) \tag{13.197}$$

From both the equations, we derive

$$\mathbf{p}_n = \mathbf{M}\left(\frac{\mathbf{q}_{n+1} - \mathbf{q}_n}{\Delta t}\right) - \frac{1}{2}\Delta t\mathbf{M}\left(\alpha\mathbf{a}_{n+1-\alpha} + (1-\alpha)\,\boldsymbol{a}_{n+\alpha}\right) \tag{13.198}$$

and

$$\mathbf{p}_{n+1} = \mathbf{M}\left(\frac{\mathbf{q}_{n+1} - \mathbf{q}_n}{\Delta t}\right) + \frac{1}{2}\Delta t\mathbf{M}\left(\alpha\mathbf{a}_{n+\alpha} + (1-\alpha)\,\mathbf{a}_{n+1-\alpha}\right) \tag{13.199}$$

Subtracting Equation (13.198) from Equation (13.199) gives

$$\mathbf{p}_{n+1} = \mathbf{p}_n + \frac{1}{2}\Delta t\mathbf{M}\left(\mathbf{a}_{n+\alpha} + \mathbf{a}_{n+1-\alpha}\right) \tag{13.200}$$

We also rewrite Equation (13.198) as

$$\mathbf{q}_{n+1} = \mathbf{q}_n + \Delta t\mathbf{M}^{-1}\mathbf{p}_n + \frac{1}{2}\Delta t^2\left(\alpha\mathbf{a}_{n+1-\alpha} + (1-\alpha)\,\mathbf{a}_{n+\alpha}\right) \tag{13.201}$$

It is important to note that both Equation (13.200) and Equation (13.201) represent the discrete Hamiltonian map $\tilde{F}_{L_d} : T^*Q \to T^*Q$ or $(\mathbf{q}_n, \mathbf{p}_n) \to (\mathbf{q}_{n+1}, \mathbf{p}_{n+1})$ as shown in Figure 13.1. The discrete Hamiltonian map means a mapping from the discrete phase space T^*Q at the previous time step to the discrete phase space T^*Q at the current time step.

Consequently, this algorithm is second-order time accurate for any α and preserves the canonical symplectic form and the standard momentum map Kane et al. (2000). In the case when $\alpha = \frac{1}{2}$, Equation (13.201) and Equation (13.200) can be recast in the form of a symplectic midpoint algorithm (Simo et al. 1992)

$$\begin{aligned}\frac{\mathbf{q}_{n+1} - \mathbf{q}_n}{\Delta t} &= \mathbf{M}^{-1}\left(\frac{\mathbf{p}_{n+1} + \mathbf{p}_n}{2}\right)\\ \frac{\mathbf{p}_{n+1} - \mathbf{p}_n}{\Delta t} &= -\nabla\mathcal{U}\left(\frac{\mathbf{q}_{n+1} + \mathbf{q}_n}{2}\right)\end{aligned} \tag{13.202}$$

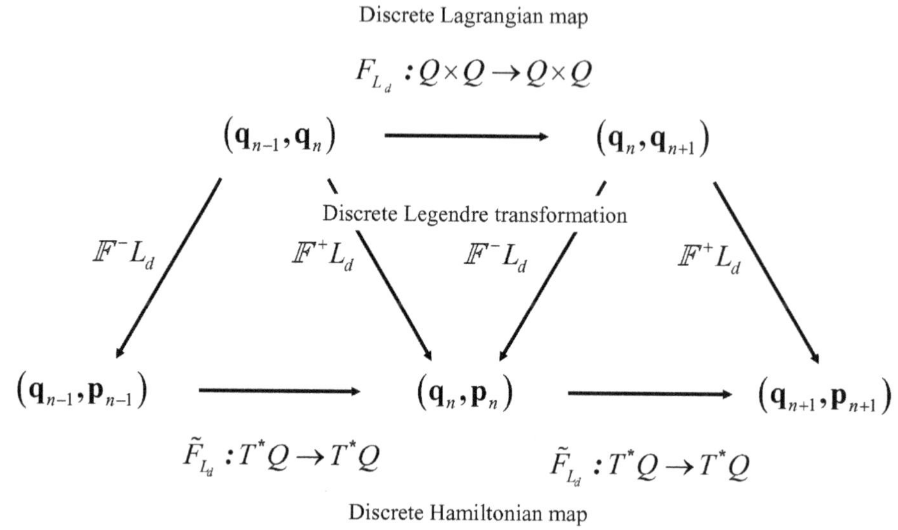

Figure 13.1. Flow diagram. (Marsden and West 2001)

Variational Algorithm on *TQ* A mapping $(\mathbf{q}_n, \dot{\mathbf{q}}_{n+1}) \to (\mathbf{q}_{n+1}, \dot{\mathbf{q}}_{n+1})$ yields

$$\begin{aligned} \mathbf{q}_{n+1} &= \mathbf{q}_n + \Delta t \dot{\mathbf{q}}_n + \frac{1}{2}\Delta t^2 [(1-\alpha)\,\mathbf{a}_{n+\alpha} + \alpha \mathbf{a}_{n+1-\alpha}] \\ \dot{\mathbf{q}}_{n+1} &= \dot{\mathbf{q}}_n + \Delta t \left(\frac{\mathbf{a}_{n+\alpha} + \mathbf{a}_{n+1-\alpha}}{2} \right) \end{aligned} \tag{13.203}$$

This algorithm is also second-order time accurate for any α and preserves the standard Lagrangian symplectic form and momentum map. Especially, for $\alpha = 1$, we obtain the shake algorithm (the Verlet algorithm is the unconstrained version of the shake algorithm). The Moser-Veselov discrete Lagrangian for the rigid body is constructed using either $\alpha = 0$ or $\alpha = 1$ (see (Marsden et al. 1999, 2000)).

13.5.3 Symplecticness of Variational Time Integrators

Considering the canonical differential one-form on the phase space and the Lagrangian differential one-form on the tangent bundle (Marsden and West 2001), the Lagrangian differential two-form can be derived from the exterior derivative of the one form as follows,

$$\omega_{\mathcal{L}} = -d\theta_{\mathcal{L}} = -d\frac{\partial \mathcal{L}}{\partial \dot{\mathbf{q}}} \wedge d\mathbf{q} = d\mathbf{q} \wedge d\frac{\partial \mathcal{L}}{\partial \dot{\mathbf{q}}} \tag{13.204}$$

From the following derivation

$$d\frac{\partial \mathcal{L}}{\partial \dot{\mathbf{q}}} = \frac{\partial^2 \mathcal{L}}{\partial \mathbf{q}\partial \dot{\mathbf{q}}} d\mathbf{q} + \frac{\partial^2 \mathcal{L}}{\partial \dot{\mathbf{q}}\partial \dot{\mathbf{q}}} d\dot{\mathbf{q}} \tag{13.205}$$

Equation (13.204) can be rewritten as

$$\omega_{\mathcal{L}} = -d\theta_{\mathcal{L}} = \frac{\partial^2 \mathcal{L}}{\partial \mathbf{q}\partial \dot{\mathbf{q}}} d\mathbf{q} \wedge d\mathbf{q} + \frac{\partial^2 \mathcal{L}}{\partial \dot{\mathbf{q}}\partial \dot{\mathbf{q}}} d\mathbf{q} \wedge d\dot{\mathbf{q}} \tag{13.206}$$

Equation (13.206) can be also expressed as

$$\omega_{\mathcal{L}} = \frac{\partial^2 \mathcal{L}}{\partial q^i \partial \dot{q}^j} dq^i \wedge dq^j + \frac{\partial^2 \mathcal{L}}{\partial \dot{q}^i \partial \dot{q}^j} dq^i \wedge d\dot{q}^j \tag{13.207}$$

It is well known that the Lagrangian flow $F_{\mathcal{L}} : TQ \to TQ$ preserves the symplectic structure, i.e., $(F_{\mathcal{L}})^* \omega_{\mathcal{L}} = \omega_{\mathcal{L}}$. Likewise, we shall show that the discrete Lagrangian flow map $F_{\mathcal{L}_d^\alpha} : Q \times Q \to Q \times Q$ also preserves the Lagrangian symplectic structure. From the discrete Euler-Lagrange's equations shown in Equation (13.181), we may define discrete Lagrangian one-forms as follows,

$$\begin{aligned}
\theta^+_{\mathcal{L}_d} &= D_2 \mathcal{L}_d^\alpha (\mathbf{q}_{n-1}, \mathbf{q}_n) \cdot d\mathbf{q}_n = \frac{\partial \mathcal{L}_d^\alpha (\mathbf{q}_{n-1}, \mathbf{q}_n)}{\partial \mathbf{q}_n} \cdot d\mathbf{q}_n \\
\theta^-_{\mathcal{L}_d} &= -D_1 \mathcal{L}_d^\alpha (\mathbf{q}_n, \mathbf{q}_{n+1}) \cdot d\mathbf{q}_n = -\frac{\partial \mathcal{L}_d^\alpha (\mathbf{q}_n, \mathbf{q}_{n+1})}{\partial \mathbf{q}_n} \cdot d\mathbf{q}_n
\end{aligned} \tag{13.208}$$

Taking the exterior derivative of the one-forms, we obtain discrete Lagrangian two-forms as below,

$$\begin{aligned}
\omega^+_{\mathcal{L}_d^\alpha} &= -d\theta^+_{\mathcal{L}_d^\alpha} = -d \frac{\partial \mathcal{L}_d^\alpha (\mathbf{q}_{n-1}, \mathbf{q}_n)}{\partial \mathbf{q}_n} \wedge d\mathbf{q}_n \\
\omega^-_{\mathcal{L}_d^\alpha} &= -d\theta^-_{\mathcal{L}_d^\alpha} = d \frac{\partial \mathcal{L}_d^\alpha (\mathbf{q}_n, \mathbf{q}_{n+1})}{\partial \mathbf{q}_n} \wedge d\mathbf{q}
\end{aligned} \tag{13.209}$$

Using the following relations

$$\begin{aligned}
d \frac{\partial \mathcal{L}_d^\alpha (\mathbf{q}_{n-1}, \mathbf{q}_n)}{\partial \mathbf{q}_n} &= \frac{\partial^2 \mathcal{L}_d^\alpha (\mathbf{q}_{n-1}, \mathbf{q}_n)}{\partial \mathbf{q}_n \cdot \partial \mathbf{q}_n} d\mathbf{q}_n + \frac{\partial^2 \mathcal{L}_d^\alpha (\mathbf{q}_{n-1}, \mathbf{q}_n)}{\partial \mathbf{q}_{n-1} \cdot \partial \mathbf{q}_n} d\mathbf{q}_{n-1} \\
d \frac{\partial \mathcal{L}_d^\alpha (\mathbf{q}_n, \mathbf{q}_{n+1})}{\partial \mathbf{q}_n} &= \frac{\partial^2 \mathcal{L}_d^\alpha (\mathbf{q}_n, \mathbf{q}_{n+1})}{\partial \mathbf{q}_n \cdot \partial \mathbf{q}_n} d\mathbf{q}_n + \frac{\partial^2 \mathcal{L}_d^\alpha (\mathbf{q}_n, \mathbf{q}_{n+1})}{\partial \mathbf{q}_{n+1} \cdot \partial \mathbf{q}_n} d\mathbf{q}_{n+1}
\end{aligned} \tag{13.210}$$

Equation (13.209) reduces to the following

$$\begin{aligned}
\omega^+_{\mathcal{L}_d^\alpha} &= -d\theta^+_{\mathcal{L}_d^\alpha} = -\frac{\partial^2 \mathcal{L}_d^\alpha (\mathbf{q}_{n-1}, \mathbf{q}_n)}{\partial \mathbf{q}_{n-1} \cdot \partial \mathbf{q}_n} d\mathbf{q}_{n-1} \wedge d\mathbf{q}_n = \frac{\partial^2 \mathcal{L}_d^\alpha (\mathbf{q}_{n-1}, \mathbf{q}_n)}{\partial \mathbf{q}_{n-1} \cdot \partial \mathbf{q}_n} d\mathbf{q}_n \wedge d\mathbf{q}_{n-1} \\
\omega^-_{\mathcal{L}_d^\alpha} &= -d\theta^-_{\mathcal{L}_d^\alpha} = \frac{\partial^2 \mathcal{L}_d^\alpha (\mathbf{q}_n, \mathbf{q}_{n+1})}{\partial \mathbf{q}_{n+1} \cdot \partial \mathbf{q}_n} d\mathbf{q}_{n+1} \wedge d\mathbf{q}_n = \frac{\partial^2 L_d^\alpha (\mathbf{q}_{n-1}, \mathbf{q}_n)}{\partial \mathbf{q}_{n-1} \cdot \partial \mathbf{q}_n} d\mathbf{q}_n \wedge d\mathbf{q}_{n-1}
\end{aligned} \tag{13.211}$$

Equation (13.211) represents the discrete Lagrangian symplectic structure. We observe from Equation (13.211) that $\omega^+_{\mathcal{L}_d^\alpha} = \omega^-_{\mathcal{L}_d^\alpha}$. Further, we note that $\theta^+_{\mathcal{L}_d^\alpha} - \theta^-_{\mathcal{L}_d^\alpha} = d\mathcal{L}_d^\alpha (\mathbf{q}_n, \mathbf{q}_{n+1})$. Therefore, we have $d^2 \mathcal{L}_d^\alpha (\mathbf{q}_n, \mathbf{q}_{n+1}) = 0$. In conclusion, the discrete Lagrangian symplectic form is given by

$$\omega_{\mathcal{L}_d^\alpha} = \frac{\partial^2 \mathcal{L}_d^\alpha (\mathbf{q}_n, \mathbf{q}_{n+1})}{\partial \mathbf{q}_n \cdot \partial \mathbf{q}_{n+1}} d\mathbf{q}_{n+1} \wedge d\mathbf{q}_n \tag{13.212}$$

And the discrete Lagrangian map $F_{\mathcal{L}_d^\alpha} : Q \times Q \to Q \times Q$ preserves the symplectic structure, i.e., $\left(F_{\mathrm{L}_d^\alpha}\right)^* \omega_{\mathcal{L}_d^\alpha} = \omega_{\mathcal{L}_d^\alpha}$. Therefore, the variational time integrators derived from the discrete Lagrangian map preserve the symplectic structure in the discrete Lagrangian flow map.

13.5.4 Discrete Noether's Theorem

Just as the continuous Lagrangian has translational and rotational symmetry, we shall show that the discrete Lagrangian has translational and rotational symmetry so that linear and angular momenta are conserved. Firstly, as in the continuous Lagrangian, the infinitesimal translational invariance of the discrete Lagrangian is true when

$$\mathcal{L}_d^{\alpha}\left(\mathbf{q}_n, \mathbf{q}_{n+1}\right) = \mathcal{L}_d^{\alpha}\left(\mathbf{q}_n', \mathbf{q}_{n+1}'\right) \tag{13.213}$$

where the new positions infinitesimally transformed are respectively given by

$$\mathbf{q}_n' = \mathbf{q}_n + \varepsilon\boldsymbol{\eta}, \quad \mathbf{q}_{n+1}' = \mathbf{q}_{n+1} + \varepsilon\boldsymbol{\eta} \tag{13.214}$$

where ε is an infinitesimal continuous parameter under the Lie group action and $\boldsymbol{\eta} \in \mathbb{R}^n$ is an arbitrary vector fixed. By the one-parameter (ε) Lie group action of infinitesimal translational transformation, the discrete Lagrangian can be expanded as

$$\phi\left(\varepsilon\right) = \mathcal{L}_d^{\alpha}\left(\mathbf{q}_n', \mathbf{q}_{n+1}'\right) = \mathcal{L}_d^{\alpha}\left(\mathbf{q}_n, \mathbf{q}_{n+1}\right) + \varepsilon\left(\left.\frac{\partial \mathcal{L}_d^{\alpha}\left(\mathbf{q}_n', \mathbf{q}_{n+1}'\right)}{\partial\varepsilon}\right|_{\varepsilon=0}\right) + O\left(\varepsilon^2\right) \tag{13.215}$$

where the infinitesimal transformation of the discrete Lagrangian is given by

$$\mathcal{L}_d^{\alpha}\left(\mathbf{q}_n, \mathbf{q}_{n+1}\right) + \varepsilon\left(\left.\frac{\partial \mathcal{L}_d^{\alpha}\left(\mathbf{q}_n', \mathbf{q}_{n+1}'\right)}{\partial\varepsilon}\right|_{\varepsilon=0}\right) \tag{13.216}$$

In order for $\phi\left(\varepsilon\right)$ to have an extremal point in the vicinity of $\varepsilon = 0$, the infinitesimal defined as below should be null.

As a consequence, the infinitesimal translational invariance of the discrete Lagrangian holds if and only if the infinitesimal of the discrete Lagrangian vanishes, i.e.,

$$\left.\frac{\partial \mathcal{L}_d^{\alpha}\left(\mathbf{q}_n', \mathbf{q}_{n+1}'\right)}{\partial\varepsilon}\right|_{\varepsilon=0} = 0 \tag{13.217}$$

The above equation can be written as

$$\left.\frac{\partial \mathcal{L}_d^{\alpha}\left(\mathbf{q}_n', \mathbf{q}_{n+1}'\right)}{\partial\varepsilon}\right|_{\varepsilon=0} = \left\{D_1\mathcal{L}_d^{\alpha}\left(\mathbf{q}_n, \mathbf{q}_{n+1}\right) + D_2\mathcal{L}_d^{\alpha}\left(\mathbf{q}_n, \mathbf{q}_{n+1}\right)\right\} \cdot \boldsymbol{\eta} = 0 \tag{13.218}$$

Owing to Equation (13.192), Equation (13.218) leads to the relation

$$\mathbf{p}_n = \mathbf{p}_{n+1} \tag{13.219}$$

Thus, we conclude that the translational symmetry of the discrete Lagrangian leads to conservation of the position momentum in the discrete Lagrangian flow.

Analogous to the previous discussion, if the discrete Lagrangian has a rotational symmetry, then the corresponding conserved quantity, that is angular momentum, is conserved. The infinitesimal rotational invariance of the discrete Lagrangian is true when

$$\mathcal{L}_d^{\alpha}\left(\mathbf{q}_n, \mathbf{q}_{n+1}\right) = \mathcal{L}_d^{\alpha}\left(\mathbf{q}_n', \mathbf{q}_{n+1}'\right) \tag{13.220}$$

Let us introduce the rotational transformation matrix $Q \in SO(3)$ denoting the special orthogonal rotation group, which is a set of the rotational matrices whose determinant equals to $+1$. Consider the one-parameter Lie group of infinitesimal rotational transformation. Then, the new positions infinitesimally rotated by the rotational transformation matrices are denoted by,

$$\mathbf{q}'_n = \exp\left(\varepsilon\hat{\boldsymbol{\xi}}\right)\mathbf{q}_n, \quad \mathbf{q}'_{n+1} = \exp\left(\varepsilon\hat{\boldsymbol{\xi}}\right)\mathbf{q}_{n+1}, \quad \forall\hat{\boldsymbol{\xi}} \in so(3) \tag{13.221}$$

We note that the rotational transformation matrix is skew-symmetric and there exists an arbitrary vector $\boldsymbol{\xi} \in \mathbb{R}^3$ with the relation

$$\hat{\boldsymbol{\xi}}\mathbf{q}_n = \boldsymbol{\xi} \times \mathbf{q}_n, \quad \hat{\boldsymbol{\xi}}\mathbf{q}_{n+1} = \boldsymbol{\xi} \times \mathbf{q}_{n+1}, \quad \forall\hat{\boldsymbol{\xi}} \in so(3) \ \ and\ \forall\boldsymbol{\xi} \in \mathbb{R}^3 \tag{13.222}$$

Then, the discrete Lagrangian can be expanded as

$$\begin{aligned}\mathcal{L}_d^\alpha\left(\mathbf{q}'_n, \mathbf{q}'_{n+1}\right) &= \mathcal{L}_d^\alpha\left(\mathbf{q}_n, \mathbf{q}_{n+1}\right)\\ &+ \varepsilon\left(\left.\frac{\partial\mathcal{L}_d^\alpha\left(\exp\left(\varepsilon\hat{\boldsymbol{\xi}}\right)\mathbf{q}_n, \exp\left(\varepsilon\hat{\boldsymbol{\xi}}\right)\mathbf{q}_{n+1}\right)}{\partial\varepsilon}\right|_{\varepsilon=0}\right) + O\left(\varepsilon^2\right)\end{aligned} \tag{13.223}$$

where the infinitesimal transformation of the discrete Lagrangian is given by

$$\mathcal{L}_d^\alpha\left(\mathbf{q}_n, \mathbf{q}_{n+1}\right) + \varepsilon\left(\left.\frac{\partial\mathcal{L}_d^\alpha\left(\exp\left(\varepsilon\hat{\boldsymbol{\xi}}\right)\mathbf{q}_n, \exp\left(\varepsilon\hat{\boldsymbol{\xi}}\right)\mathbf{q}_{n+1}\right)}{\partial\varepsilon}\right|_{\varepsilon=0}\right) \tag{13.224}$$

As a consequence, the infinitesimal rotational invariance of the discrete Lagrangian holds if and only if the infinitesimal of the discrete Lagrangian vanishes, i.e.,

$$\left.\frac{\partial\mathcal{L}_d^\alpha\left(\exp\left(\varepsilon\hat{\boldsymbol{\xi}}\right)\mathbf{q}_n, \exp\left(\varepsilon\hat{\boldsymbol{\xi}}\right)\mathbf{q}_{n+1}\right)}{\partial\varepsilon}\right|_{\varepsilon=0} = 0 \tag{13.225}$$

The above equation can be rewritten as

$$D_1\mathcal{L}_d^\alpha\left(\mathbf{q}_n, \mathbf{q}_{n+1}\right)\cdot\hat{\boldsymbol{\xi}}\mathbf{q}_n + D_2 L_d^\alpha\left(\mathbf{q}_n, \mathbf{q}_{n+1}\right)\cdot\hat{\boldsymbol{\xi}}\mathbf{q}_{n+1} = 0 \tag{13.226}$$

Substituting Equation (13.222) into Equation (13.226) reduces to

$$\boldsymbol{\xi}\cdot\left\{\mathbf{q}_n \times D_1\mathcal{L}_d^\alpha\left(\mathbf{q}_n, \mathbf{q}_{n+1}\right) + \mathbf{q}_{n+1} \times D_2\mathcal{L}_d^\alpha\left(\mathbf{q}_n, \mathbf{q}_{n+1}\right)\right\} = 0 \tag{13.227}$$

Since $\boldsymbol{\xi} \in \mathbb{R}^3$ is an arbitrary vector, we easily obtain

$$\mathbf{q}_n \times D_1\mathcal{L}_d^\alpha\left(q_n, \mathbf{q}_{n+1}\right) + \mathbf{q}_{n+1} \times D_2\mathcal{L}_d^\alpha\left(\mathbf{q}_n, \mathbf{q}_{n+1}\right) = 0 \tag{13.228}$$

Once again, the above equation can be rewritten as

$$\mathbf{q}_n \times \mathbf{p}_n = \mathbf{q}_{n+1} \times \mathbf{p}_{n+1} \tag{13.229}$$

In conclusion, the rotational invariance of the discrete Lagrangian leads to conservation of the angular momentum in the discrete Lagrangian flow.

13.6 ENERGY-MOMENTUM CONSERVATION AND TIME INTEGRATION ALGORITHMS

Besides the traditional and/or classical practices and original methods of algorithmic developments in time integration described earlier emanating from the single-field form of the semi-discretized equation of motion, and in contrast to variational time integrators emanating from the DEL that were described previously, for a class of nonlinear elastodynamics applications, energy-momentum conserving algorithms seem to be yet another alternative. Here, in principle, it is desirable to exactly conserve physical quantities such as energy, linear and angular momentum. Under certain conditions (see details subsequently), they appear to guarantee conservation of the three invariants and circumvent some of deficiencies in traditional time integration algorithms for a certain class of problems. However, although they claim unconditional stability (in the sense of energy stability), in certain situations and for arbitrary large time steps, the associated Newton type nonlinear iterative procedures may pose convergence difficulties as in symplectic-momentum conserving algorithms; such issues need to be clearly understood. Yet another concern is the unsymmetric nature of the resulting equation system that arises in the general case. These and other related aspects appear to be especially rampant for stiff systems and when high frequency participation of the modes is present. In the next chapter and under the umbrella of the GSSSS family of algorithms emanating from the Total Energy framework and a generalized time weighted residual approach (both two-field and single-field forms), we particularly describe the theoretical basis for algorithm designs that are energy-momentum based conserving/dissipating which cover most of the classical developments that are described in this section. In particular, via the GSSSS framework in both two-field and single-field form, the resulting developments cover most of the algorithm representations that are described from an alternative formulation such as that presented in this section which appear in the sense of the two-field form; and in addition also include features to incorporate controllable numerical dissipation in the parent energy-momentum conserving algorithm designs (such as in the single-field form). When the controllable numerical dissipation is turned off, it reverts to the original energy-momentum conserving representation. It is to be noted that the single-field form in (Kuhl and Crisfield 1999) is strictly not representative of the original two-field form energy-momentum and algorithm in (Simo et al. 1992). This is because it is identical to that termed as V0 (1,1,1) and not V0 (1,1,0) within the GSSSS framework described in the next chapter to follow, wherein, the latter single-field form is exactly representative (MPR-MPA) unlike the former (MPR-EPA).

13.6.1 Energy-Momentum Conserving Scheme for Discrete Systems

Greenspan (Greenspan 1973; LaBudde and Greenspan 1974) appears to be one of the first to stress the importance that conserved quantities play a crucial role in developing time stepping algorithms. The conserved quantities include the linear and angular momenta, and the total energy. The development of time stepping algorithms conserving exactly the linear and angular momenta, and the total energy was proposed initially for particle dynamics. Subsequent developments and more extensive discussion on exact energy-momentum conserving algorithms for autonomous nonlinear Hamiltonian systems are then followed by (Simo et al. 1992). Owing to the Hamiltonian version of Noether's theorem (Noether 1918), it has been well known that autonomous nonlinear or linear Hamiltonian systems possess symmetries (as discussed in this chapter), namely, the resulting three invariance properties. We first highlight applicability to discrete systems, and then briefly describe the corresponding developments for continuum dynamics applications as well.

It has been well known that the trapezoidal method preserves the total energy of the system only for linear dynamic problems. For nonlinear dynamic problems, however, the method fails to conserve the total energy of the system (see Table 13.1). In addition, the angular momentum is not preserved for both linear dynamic and nonlinear dynamic problems when the trapezoidal method (Newmark) is applied.

Table 13.1. Some representative time integration schemes and characterization

Methods	Linear Momentum	Angular Momentum Linear	Angular Momentum Nonlinear	Total Energy Linear	Total Energy Nonlinear	Symplectic $\rho = 1$
Trapezoidal Rule	✓	✓	X	✓	X	X
Central Difference	✓	✓	✓	✓	X	✓
Classical Midpoint Rule	✓	✓	✓	✓	X	✓
Classical Methods	✓	X	X	✓	X	X

On the other hand, the explicit central difference or the implicit midpoint method can conserve the angular momentum of an autonomous nonlinear Hamiltonian system, but they do not strictly (exactly) preserve the total energy of the system (it is bounded when fixed time steps are selected). In the following, we briefly highlight one of the exact energy momentum conserving algorithms initially proposed for N-body problems (Simo et al. 1992).

13.6.2 Energy-Momentum Conserving Algorithms for N-Body Systems

The Energy-Momentum Conserving Algorithm with a Distance-Dependent Internal Potential Energy The exact energy-momentum conserving algorithm in a nonlinear dynamical system of N particles when the internal potential energy is given as a function of the distance between the particles only, has been proposed by Greenspan (Greenspan 1995) and Simo and Gonzalez (Simo and Gonzalez 1993, 1994). Introduce the inter-particle distance, i.e., $r^{AB} := \|\mathbf{q}^{AB}\| := \|\mathbf{q}^B - \mathbf{q}^A\| = \sqrt{(\mathbf{q}^B - \mathbf{q}^A)\cdot(\mathbf{q}^B - \mathbf{q}^A)}$, for the internal potential energy, i.e., $\mathcal{U}^{\text{int}}_{AB}(\mathbf{q}^A, \mathbf{q}^B) \equiv \mathcal{V}^{\text{int}}_{AB}(r^{AB})$. Assume the total potential energy is given only by the internal potential energy. We assume the system free from constraints is holonomic-scleronomic. Note the generalized momentum is given as $\mathbf{p}_A = m_A\dot{\mathbf{q}}^A$ for $A = 1, 2, \cdots, N$. Then, the following time integration scheme for a conservative system is obtained where the total linear momentum, total angular momentum, and total energy are exactly conserved in $P[t_n, t_{n+1}]$:

Algorithm 1: ***Energy-Momentum Conserving Algorithm with a Distance-Dependent Internal Potential Energy***

Given $\mathbf{q}_n$ and $\mathbf{v}_n$, find $\mathbf{q}_{n+1}$ and $\mathbf{v}_{n+1}$ from

$$\mathbf{q}^A_{n+1} = \mathbf{q}^A_n + \Delta t m_A^{-1}\frac{\mathbf{p}_{A\,n+1} + \mathbf{p}_{A\,n}}{2} \tag{13.230a}$$

$$\mathbf{p}_{A\,n+1} = \mathbf{p}_{A\,n} + \Delta t \sum_{B\neq A}^{N} \frac{\mathcal{V}^{\text{int}}_{AB\,n+1} - \mathcal{V}^{\text{int}}_{AB\,n}}{r^{AB}_{n+1} - r^{AB}_n}\,\frac{\mathbf{q}^{AB}_{n+1} + \mathbf{q}^{AB}_n}{r^{AB}_{n+1} + r^{AB}_n} \tag{13.230b}$$

for $A = 1, 2, \ldots, N$, where m_A denotes the mass of Ath particle, and $\mathcal{V}^{\text{int}}_{AB\,n+1} := \mathcal{V}^{\text{int}}(r^{AB}_{n+1})$ and $\mathcal{V}^{\text{int}}_{AB\,n} := \mathcal{V}^{\text{int}}(r^{AB}_n)$.

Remark 13.6.1

1. Algorithm 1 is an implicit, single-step, second-order time accurate time integration scheme which exactly conserves the total linear/angular momenta and the total energy within the time step.

2. Assuming the algorithm is well-defined, employ

$$\left.\frac{d\mathcal{V}^{\text{int}}_{AB}}{dr^{AB}}\right|_{r^{AB}_{n+1/2}} \quad \text{instead of} \quad \left.\frac{d\mathcal{V}^{\text{int}}_{AB}}{dr^{AB}}\right|_{r^{AB}_{n+\gamma_0}} = \frac{\mathcal{V}^{\text{int}}_{AB\,n+1} - \mathcal{V}^{\text{int}}_{AB\,n}}{r^{AB}_{n+1} - r^{AB}_{n}} \tag{13.231}$$

for the limit case ($r^{AB}_{n+1} - r^{AB}_{n} \to 0$).

3. Algorithm 1 is identical to the so-called assumed distance method proposed by Betsch and Steinmann (Betsch and Steinmann 2000) when the internal potential energy is given as a polynomial function of degree two or less; see (Graham et al. 2002) for details. The assumed distance method is:

$$\begin{aligned} m_A \frac{\mathbf{v}^A_{n+1} - \mathbf{v}^A_n}{\Delta t} - \sum_{B\neq A}^{N} \left.\frac{d\mathcal{V}^{\text{int}}_{AB}}{dr^{AB}}\right|_{r^{AB}_{n+1/2}} \frac{\mathbf{q}^{AB}_{n+1} + \mathbf{q}^{AB}_n}{r^{AB}_{n+1} + r^{AB}_n} &= \mathbf{0} \\ \frac{\mathbf{q}^A_{n+1} - \mathbf{q}^A_n}{\Delta t} &= \frac{\mathbf{v}^A_{n+1} + \mathbf{v}^A_n}{2} \end{aligned} \tag{13.232}$$

Simo-Tarnow-Wong Framework Here, we summarize the original formalism towards constructing an exact energy-momentum conserving algorithm for a N-body system, described in Simo et al. (Simo et al. 1992). In contrast to the Lagrangian, the Hamiltonian is used leading to a first-order representation of the second-order dynamic system of equations. Assume the autonomous Hamiltonian $\mathcal{H} : T^*Q \to \mathbb{R}$ to be completely separable; namely, $\mathcal{H}(\mathbf{q}, \mathbf{p}) = \mathcal{K}(\mathbf{p}) + \mathcal{U}(\mathbf{q})$, where the kinetic energy and the potential energy of the system are $\mathcal{K} : T^*Q \to \mathbb{R}$ and $\mathcal{U} : Q \to \mathbb{R}$, respectively. Recall the Hamiltonian is defined on the cotangent bundle T^*Q ($\equiv P$, phase space). Assume the configuration manifold Q of a finite dimensional system is a linear space. Consider the case the kinetic energy is a positive definite quadratic form in the canonical momenta, i.e.,

$$\mathcal{K}(\mathbf{p}) = \frac{1}{2}\mathbf{p} \cdot \mathbf{M}^{-1}\mathbf{p} \geq 0 \tag{13.233}$$

where the symmetric mass matrix $\mathbf{M}$ is *constant*, and a set of the generalized momenta, $\mathbf{p} = (\mathbf{p}_1, \mathbf{p}_2, \ldots, \mathbf{p}_N) \in T^*_qQ \equiv \mathbb{R}^{3N}$ is defined with the Lagrangian $\mathcal{L}$ by

$$\mathbf{p}_A = \frac{\partial \mathcal{L}}{\partial \dot{\mathbf{q}}^A} \tag{13.234}$$

for $A = 1, 2, \ldots, N$. Note that the Lagrangian is defined on the tangent bundle TQ ($\equiv S$, velocity phase space), and the transformation from the (regular) Lagrangian to the Hamiltonian is achieved via the Legendre transformation: $(\mathbf{q}, \dot{\mathbf{q}}) \in TQ \mapsto (\mathbf{q}, \mathbf{p}) \in T^*Q;\ \mathcal{L}(\mathbf{q}, \dot{\mathbf{q}}) \mapsto \mathcal{H}(\mathbf{q}, \mathbf{p}) = \sum_{A=1}^{N} \mathbf{p}_i \cdot \boldsymbol{\psi}^A - \mathcal{L}(\mathbf{q}, \dot{\mathbf{q}})$, where $\dot{\mathbf{q}} = \boldsymbol{\psi}(\mathbf{q}, \mathbf{p}) \in T_qQ$ obtained from Equation (13.234). Note the canonical coordinates $\mathbf{q} \in Q$, the canonical momenta $\mathbf{p} \in T^*_qQ$, and the canonical variables $(\mathbf{q}, \mathbf{p}) \in T^*Q$. In this special case where the kinetic energy is given in the form shown in Equation (13.233) and the potential energy depends only on a set of generalized coordinates $\mathbf{q} = (\mathbf{q}^1, \mathbf{q}^2, \ldots, \mathbf{q}^N) \in Q \equiv \mathbb{R}^{3N}$ explicitly, the Hamiltonian is identical to the total energy of the system. The equation of motion of the system is described by the following Hamilton's canonical equations:

$$\dot{\mathbf{q}} = \frac{\partial \mathcal{H}(\mathbf{q}, \mathbf{p})}{\partial \mathbf{p}} = \mathbf{M}^{-1}\mathbf{p} \tag{13.235a}$$

$$\dot{\mathbf{p}} = -\frac{\partial \mathcal{H}(\mathbf{q}, \mathbf{p})}{\partial \mathbf{q}} = -\nabla \mathcal{U}(\mathbf{q}) \tag{13.235b}$$

The algorithms derived in this framework can be used for the motion of finite N-dimensional structural dynamical systems as well as the motion of small oscillation about an equilibrium position in a system of N particles. It is important to note the generalized coordinates are supposed to be taken as a set of (nodal) displacement vectors: $\mathbf{q} = \mathbf{u} := \mathbf{x} - \bar{\mathbf{x}}$ where $\mathbf{x}$ and $\bar{\mathbf{x}}$ denote a set of (nodal) position vector and a set of equilibrium (nodal) position vectors. We usually take $\bar{\mathbf{x}} \equiv \mathbf{x}(t_0)$ for simplicity. Hence, in a linear dynamical system, linearization of the system of equations (13.235), assuming $\mathbf{M}$ is constant, leads to

$$\dot{\mathbf{q}} = \frac{\partial \mathcal{H}(\mathbf{q}, \mathbf{p})}{\partial \mathbf{p}} = \mathbf{M}^{-1}\mathbf{p} \tag{13.236a}$$

$$\dot{\mathbf{p}} = -\frac{\partial \mathcal{H}(\mathbf{q}, \mathbf{p})}{\partial \mathbf{q}} = -\mathbf{K}\mathbf{q} \tag{13.236b}$$

where $\mathbf{K}$, the Hessian matrix, is given as

$$\mathbf{K} := \left.\frac{\partial^2 \mathcal{U}}{\partial \mathbf{q} \partial \mathbf{q}}\right|_{\mathbf{q}=\mathbf{0}} \tag{13.237}$$

Here, we have assumed $\mathcal{U}(\mathbf{q} = \mathbf{0}) = 0$ and $\mathcal{U}'(\mathbf{q} = \mathbf{0}) = \mathbf{0}$.

Consider a time interval $\mathcal{T} = [t_0, t_N]$ split into subintervals, i.e., $\mathcal{T} = [t_0, t_N] = \bigcup_{n=0}^{N-1} \left[t_n, t_{n+1}\right]$ $(T := t_N - t_0)$, and consider a typical time interval $\left[t_n, t_{n+1}\right] \subset \mathcal{T}$. Based on the invariance properties of the autonomous Hamiltonian system, developing a consistent algorithmic time integration scheme satisfying the conservation of the linear momentum $\mathbf{L}$, angular momentum $\mathbf{J}$, and the total energy $\mathcal{H}$ in the typical time interval as shown below is desired:

$$\mathbf{L}_n = \mathbf{L}_{n+1}, \quad \mathbf{J}_n = \mathbf{J}_{n+1}, \quad \mathcal{H}_n = \mathcal{H}_{n+1} \tag{13.238}$$

where $\mathbf{L}_n := \mathbf{L}(\mathbf{p}_n)$ and $\mathbf{L}_{n+1} := \mathbf{L}(\mathbf{p}_{n+1})$; $\mathbf{J}_n := \mathbf{J}(\mathbf{q}_n, \mathbf{p}_n)$ and $\mathbf{J}_{n+1} := \mathbf{J}(\mathbf{q}_{n+1}, \mathbf{p}_{n+1})$; $\mathcal{H}_n := \mathcal{H}$ $(\mathbf{q}_n, \mathbf{p}_n)$ and $\mathcal{H}_{n+1} := \mathcal{H}(\mathbf{q}_{n+1}, \mathbf{p}_{n+1})$.

Next, we define the convex combinations for the canonical coordinates and the canonical momenta as

$$\mathbf{q}_{n+\alpha} := (1 - \alpha)\, \mathbf{q}_n + \alpha \mathbf{q}_{n+1} \in Q \equiv \mathbb{R}^{3N} \tag{13.239}$$

$$\mathbf{p}_{n+(1-\alpha)} := \alpha \mathbf{p}_n + (1 - \alpha)\, \mathbf{p}_{n+1} \in T_q^* Q \equiv \mathbb{R}^{3N} \tag{13.240}$$

for $\alpha \in [0, 1]$, respectively. The translational invariance of the Hamiltonian leads to conservation of the linear momentum, i.e.,

$$\frac{\partial \mathcal{H}(\mathbf{q}, \mathbf{p})}{\partial \mathbf{q}} = \nabla \mathcal{U}(\mathbf{q}) = \dot{\mathbf{p}} = \mathbf{0} \tag{13.241}$$

Thus, in view of Equation (13.235), the linear momentum is automatically conserved in the absence of conservative loads. The linear momentum is conserved regardless of the type of time integration scheme in dynamics.

Assuming the kinetic energy $\mathcal{K}(\mathbf{p})$ and potential energy $\mathcal{U}(\mathbf{q})$ are $SO(3)$-frame invariant, i.e.,

$$\mathcal{K}(\mathbf{Q} \circ \mathbf{p}) := \mathcal{K}(\mathbf{Q}\mathbf{p}_1, \mathbf{Q}\mathbf{p}_2, \ldots, \mathbf{Q}\mathbf{p}_N) \equiv \mathcal{K}(\mathbf{p}) \tag{13.242}$$

$$\mathcal{U}(\mathbf{Q} \circ \mathbf{q}) := \mathcal{U}(\mathbf{Q}\mathbf{q}^1, \mathbf{Q}\mathbf{q}^2, \ldots, \mathbf{Q}\mathbf{q}^N) \equiv \mathcal{U}(\mathbf{q}) \tag{13.243}$$

$\forall \mathbf{Q} \in SO(3)$, we can show that the Hamiltonian is also $SO(3)$-frame invariant: $\mathcal{H}(\mathbf{Q} \circ \mathbf{q}, \mathbf{Q} \circ \mathbf{p}) \equiv \mathcal{H}(\mathbf{q}, \mathbf{p})$. For the satisfaction of the $SO(3)$-frame invariance of the Hamiltonian and the potential energy, the following are required, respectively:

$$\sum_{A=1}^{N} \left[\mathbf{q}^A \times \frac{\partial \mathcal{H}}{\partial \mathbf{q}^A} + \mathbf{p}_A \times \frac{\partial \mathcal{H}}{\partial \mathbf{p}_A} \right] = \mathbf{0} \tag{13.244}$$

$$\sum_{A=1}^{N} \left[\mathbf{q}^A \times \frac{\partial \mathcal{U}}{\partial \mathbf{q}^A} \right] = \mathbf{0} \tag{13.245}$$

Notice $\mathbf{p}_A$ is the covariant vector; and $\mathbf{q}^A$ are neither covariant nor contravariant vectors. Since their derivatives $d\mathbf{q}^A$ are contravariant vectors, the superscript A is used. Directly from the definition of the angular momentum $\mathbf{J} : T^*Q \to \mathbb{R}^3$; that is,

$$\mathbf{J}(\mathbf{q}, \mathbf{p}) := \sum_{A=1}^{N} \mathbf{q}^A \times \mathbf{p}_A \tag{13.246}$$

the finite increment of the angular momentum within a time step $[t_n, t_{n+1}]$ can be written as

$$\mathbf{J}_{n+1} - \mathbf{J}_n = \sum_{A=1}^{N} \left[\mathbf{q}^A_{n+\alpha} \times \left(\mathbf{p}_{A\,n+1} - \mathbf{p}_{A\,n} \right) - \mathbf{p}_{A\,n+(1-\alpha)} \times \left(\mathbf{q}^A_{n+1} - \mathbf{q}^A_n \right) \right] \tag{13.247}$$

Comparison of Equation (13.247) and the following relations obtained from Equation (13.244) and Equation (13.245) with the convex combinations $\mathbf{q}_{n+\alpha}$ and $\mathbf{p}_{n+\alpha}$, results in

$$\sum_{A=1}^{N} \mathbf{q}^A_{n+\alpha} \times \frac{\partial \mathcal{U}(\mathbf{q}_{n+\alpha})}{\partial \mathbf{q}^A_{n+\alpha}} = \mathbf{0} \tag{13.248}$$

$$\sum_{A=1}^{N} \mathbf{p}_{A\,n+(1-\alpha)} \times \frac{\partial \mathcal{K}\left(\mathbf{p}_{n+(1-\alpha)}\right)}{\partial \mathbf{p}_{A\,n+(1-\alpha)}} = \mathbf{0} \tag{13.249}$$

which yields

$$\mathbf{q}_{n+1} - \mathbf{q}_n \propto \frac{\partial \mathcal{K}\left(\mathbf{p}_{n+(1-\alpha)}\right)}{\partial \mathbf{p}_{n+(1-\alpha)}} = \mathbf{M}^{-1} \mathbf{p}_{n+(1-\alpha)} \tag{13.250}$$

$$\mathbf{p}_{n+1} - \mathbf{p}_n \propto \frac{\partial \mathcal{U}(\mathbf{q}_{n+\alpha})}{\partial \mathbf{q}_{n+\alpha}} = \nabla \mathcal{U}(\mathbf{q}_{n+\alpha}) \tag{13.251}$$

Therefore, introducing arbitrary functions $\kappa_\ell : T^*Q \to \mathbb{R}$ $(\ell = 1, 2)$, the family of algorithms which conserve the angular momentum within a time step $[t_n, t_{n+1}]$ exactly is given as follows (note $\Delta t := t_{n+1} - t_n > 0$ denotes the time step size):

Algorithm 2: ***Family of single-step angular momentum conserving algorithms***

Given $\mathbf{q}_n$ and $\mathbf{p}_n$, find $\mathbf{q}_{n+1}$ and $\mathbf{p}_{n+1}$ from

$$\mathbf{q}_{n+1} = \mathbf{q}_n + \Delta t \kappa_1 \mathbf{M}^{-1} \mathbf{p}_{n+(1-\alpha)} \tag{13.252a}$$

$$\mathbf{p}_{n+1} = \mathbf{p}_n - \Delta t \kappa_2 \nabla \mathcal{U}(\mathbf{q}_{n+\alpha}) \tag{13.252b}$$

for $\alpha \in [0, 1] \subset \mathbb{R}$ and $\kappa_\ell : T^*Q \rightarrow \mathbb{R}$ $(\ell = 1, 2)$.

Remark 13.6.2

1. Consistency condition: The arbitrary functions κ_ℓ need to satisfy $\kappa_\ell = 1 + \mathcal{O}(\Delta t)$ for a sufficiently small time step size, i.e., $\lim\limits_{\Delta t \rightarrow 0} \kappa_\ell = 1$.
2. When $\kappa_1 = \kappa_2 = 1$ and $\alpha \in [0, 1]$, the algorithms are a family of single-step *symplectic* schemes:

$$\mathbf{q}_{n+1} = \mathbf{q}_n + \Delta t \mathbf{M}^{-1} \mathbf{p}_{n+(1-\alpha)} \tag{13.253a}$$

$$\mathbf{p}_{n+1} = \mathbf{p}_n - \Delta t \nabla \mathcal{U}(\mathbf{q}_{n+\alpha}) \tag{13.253b}$$

Equations (13.253) are second-order time accurate and unconditionally stable only when $\alpha = 1/2$ (Mid-point Rule); otherwise, they are first-order time accurate and conditionally stable with a critical time step $\Delta t_{\text{crit}} = 1/\left(\mid \alpha - \frac{1}{2} \mid \Omega_{\max}\right)$ where $\Omega_{\max}$ is the maximum frequency of the linearized system.

From Equation (13.233), a finite increment of the kinetic energy within a time step $\left[t_n, t_{n+1}\right]$ can be written as

$$\begin{aligned} \mathcal{K}(\mathbf{p}_{n+1}) - \mathcal{K}(\mathbf{p}_n) &= \left[\mathbf{p}_{n+1} - \mathbf{p}_n\right] \cdot \mathbf{M}^{-1} \left[\frac{\mathbf{p}_{n+1} + \mathbf{p}_n}{2}\right] \\ &= \Delta \mathbf{p} \cdot \mathbf{M}^{-1} \mathbf{p}_{n+1/2} \end{aligned} \tag{13.254}$$

where $\Delta \mathbf{p} := \mathbf{p}_{n+1} - \mathbf{p}_n$. Substituting $\mathbf{p}_{n+1/2} = \mathbf{p}_{n+(1-\alpha)} + \left(\alpha - \frac{1}{2}\right)(\mathbf{p}_{n+1} - \mathbf{p}_n)$, the right-hand side of the above equation yields

$$\text{RHS} = \Delta \mathbf{p} \cdot \mathbf{M}^{-1} \mathbf{p}_{n+(1-\alpha)} + \left(\alpha - \tfrac{1}{2}\right) \Delta \mathbf{p} \cdot \mathbf{M}^{-1} \Delta \mathbf{p} \tag{13.255}$$

In view of Algorithm 2, we obtain

$$\begin{aligned} \mathcal{K}(\mathbf{p}_{n+1}) - \mathcal{K}(\mathbf{p}_n) &= \left(\alpha - \tfrac{1}{2}\right) \Delta \mathbf{p} \cdot \mathbf{M}^{-1} \Delta \mathbf{p} - \tfrac{\kappa_2}{\kappa_1} \nabla \mathcal{U}(\mathbf{q}_{n+\alpha}) \cdot \left[\mathbf{q}_{n+1} - \mathbf{q}_n\right] \\ &= \kappa_2^2 \Delta t^2 \left(\alpha - \tfrac{1}{2}\right) \nabla \mathcal{U}(\mathbf{q}_{n+\alpha}) \cdot \mathbf{M}^{-1} \nabla \mathcal{U}(\mathbf{q}_{n+\alpha}) \\ &\quad - \tfrac{\kappa_2}{\kappa_1} \nabla \mathcal{U}(\mathbf{q}_{n+\alpha}) \cdot \left[\mathbf{q}_{n+1} - \mathbf{q}_n\right] \end{aligned} \tag{13.256}$$

Enforcing the exact energy conservation in $[t_n, t_{n+1}]$; namely,

$$\mathcal{K}(\mathbf{p}_{n+1}) - \mathcal{K}(\mathbf{p}_n) = -\left[\mathcal{U}(\mathbf{q}_{n+1}) - \mathcal{U}(\mathbf{q}_n)\right] \tag{13.257}$$

the following relation is readily obtained:

$$\mathcal{U}(\mathbf{q}_{n+1}) - \mathcal{U}(\mathbf{q}_n) = \tfrac{\kappa_2}{\kappa_1}\nabla\mathcal{U}(\mathbf{q}_{n+\alpha}) \cdot \left[\mathbf{q}_{n+1} - \mathbf{q}_n\right] + \left(\tfrac{1}{2} - \alpha\right) ||| \nabla\mathcal{U}(\mathbf{q}_{n+\alpha}) |||^2 \kappa_2^2 \Delta t^2 \tag{13.258}$$

where $||| \nabla\mathcal{U}(\mathbf{q}_{n+\alpha}) ||| := \sqrt{\nabla\mathcal{U}(\mathbf{q}_{n+\alpha}) \cdot \mathbf{M}^{-1}\nabla\mathcal{U}(\mathbf{q}_{n+\alpha})}$. Hence, the family of single-step exact energy-momentum conserving algorithms may be summarized as follows:

Algorithm 3: ***Family of single-step energy-momentum conserving algorithms***

Given $\mathbf{q}_n$ and $\mathbf{p}_n$, find $\mathbf{q}_{n+1}$ and $\mathbf{p}_{n+1}$ from

$$\mathbf{q}_{n+1} = \mathbf{q}_n + \Delta t \kappa_1 \mathbf{M}^{-1}\mathbf{p}_{n+(1-\alpha)} \tag{13.259a}$$

$$\mathbf{p}_{n+1} = \mathbf{p}_n - \Delta t \kappa_2 \nabla\mathcal{U}(\mathbf{q}_{n+\alpha}) \tag{13.259b}$$

with

$$\begin{aligned}\mathcal{U}(\mathbf{q}_{n+1}) - \mathcal{U}(\mathbf{q}_n) &= \tfrac{\kappa_2}{\kappa_1}\nabla\mathcal{U}(\mathbf{q}_{n+\alpha}) \cdot \left[\mathbf{q}_{n+1} - \mathbf{q}_n\right] \\ &\quad + \left(\tfrac{1}{2} - \alpha\right) ||| \nabla\mathcal{U}(\mathbf{q}_{n+\alpha}) |||^2 \kappa_2^2 \Delta t^2\end{aligned} \tag{13.259c}$$

for $\alpha \in [0, 1] \subset \mathbb{R}$ and $\kappa_\ell : T^*Q \to \mathbb{R}$ $(\ell = 1, 2)$.

Remark 13.6.3

1. Suppose Equation (13.259) is solvable; then, Algorithm 3 is unconditionally stable in the sense of exact energy conservation within a time step. Recall that the total energy defines the Lyapunov function of the dynamical system.
2. Projection methods: Setting $\alpha = 1/2$ and $\kappa_1 = 1$, Algorithm 3 is reduced to the following second-order time accurate algorithms:

$$\mathbf{q}_{n+1} = \mathbf{q}_n + \Delta t \mathbf{M}^{-1}\mathbf{p}_{n+1/2} \tag{13.260a}$$

$$\mathbf{p}_{n+1} = \mathbf{p}_n - \Delta t \kappa_2 \nabla\mathcal{U}\left(\mathbf{q}_{n+1/2}\right) \tag{13.260b}$$

   and solve κ_2 from

$$\mathcal{U}(\mathbf{q}_{n+1}) - \mathcal{U}(\mathbf{q}_n) = \kappa_2 \nabla\mathcal{U}\left(\mathbf{q}_{n+1/2}\right) \cdot \left[\mathbf{q}_{n+1} - \mathbf{q}_n\right] \tag{13.260c}$$

   For linear Hamiltonian systems, we have $\kappa_2 = 1$, and the total energy is exactly conserved.
3. Collocation schemes: Setting $\kappa_1 = \kappa_2 = 1$, Algorithm 3 is reduced to

$$\mathbf{q}_{n+1} = \mathbf{q}_n + \Delta t \mathbf{M}^{-1}\mathbf{p}_{n+(1-\alpha)} \tag{13.261a}$$

$$\mathbf{p}_{n+1} = \mathbf{p}_n - \Delta t \nabla\mathcal{U}(\mathbf{q}_{n+\alpha}) \tag{13.261b}$$

with

$$\mathcal{U}(\mathbf{q}_{n+1}) - \mathcal{U}(\mathbf{q}_n) = \nabla\mathcal{U}(\mathbf{q}_{n+\alpha}) \cdot \left[\mathbf{q}_{n+1} - \mathbf{q}_n\right] + \left(\tfrac{1}{2} - \alpha\right) ||| \; \nabla\mathcal{U}(\mathbf{q}_{n+\alpha}) \; |||^2 \; \Delta t^2 \tag{13.261c}$$

Note there exists at least one $\alpha \in [0, 1]$ which satisfies Equation (13.261c). For linear Hamiltonian systems, we have $\alpha = 1/2$, and the total energy is exactly conserved.

4. Premultiplying in Equation (13.259b) by $\left[\mathbf{q}_{n+1} - \mathbf{q}_n\right]$ and then substituting it into Equation (13.259c), we can remove κ_2:

$$\mathcal{U}(\mathbf{q}_{n+1}) - \mathcal{U}(\mathbf{q}_n) = -\frac{\mathbf{p}_{n+1} - \mathbf{p}_n}{\kappa_1 \Delta t} \cdot \left[\mathbf{q}_{n+1} - \mathbf{q}_n\right] + \left(\tfrac{1}{2} - \alpha\right) ||| \; \nabla\mathcal{U}(\mathbf{q}_{n+\alpha}) \; |||^2 \; \kappa_2^2 \Delta t^2 \tag{13.262}$$

5. For *linear* Hamiltonian systems we have $\alpha = 1/2$ with $\kappa_1 = \kappa_2 = 1$, and Algorithm 3 becomes the Midpoint Rule which is second-order time accurate, unconditionally stable, symplectic, and is also energy-momentum conserving. It is important to note that the symplectic properties and energy conservation within a time step co-exist for this case.
An illustration of the single-step algorithms discussed above may be summarized as shown in Figure 13.2.

13.6.3 Energy-Momentum Conserving Algorithms for Continuum Elastodynamics

Simo-Tarnow Framework In this section, we summarize, in the sense of continuum elastodynamics, the original exact energy-momentum conserving algorithm formulated for a compressible hyperelastic material model in Simo and Tarnow (Simo and Tarnow 1992). Although this framework has been formulated for a general compressible hyperelastic material model, this version of the algorithm was originally implemented only for the St. Venant-Kirchhoff material model because of certain numerical

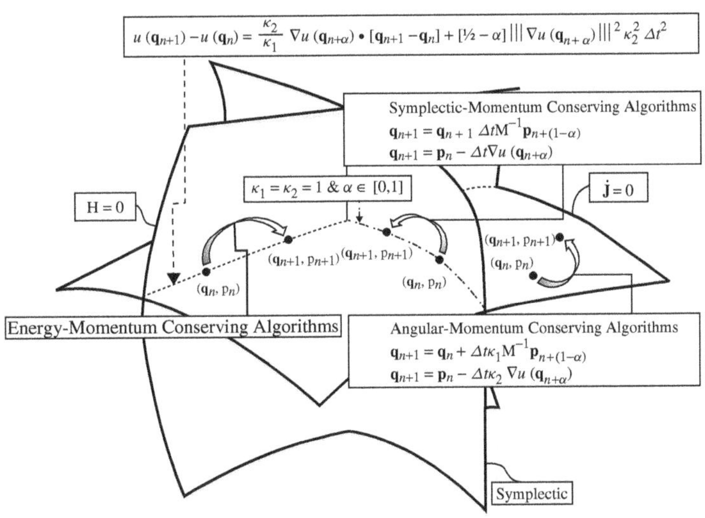

Figure 13.2. Illustration of the single-step algorithms with a parameter $\alpha \in [0, 1]$ and arbitrary functions $\kappa_\ell : T^*Q\,(= P) \rightarrow \mathbb{R},\ \ell = 1, 2$

and computational non-robustness issues that were needed to be addressed for the general case. However, some of the numerical difficulties have been to an extent, investigated by Laursen and Meng (Laursen and Meng 2001) for a general compressible hyperelastic material model.

Consider a partition $\mathcal{T} = [t_0, t_N] = \bigcup_{n=0}^{N-1} \left[t_n, t_{n+1}\right]$, and focus on a typical time interval $\left[t_n, t_{n+1}\right] \subset \mathcal{T}$. Define the convex combinations of configurations and the material velocity fields by

$$\boldsymbol{\varphi}_{n+\alpha} := \alpha \boldsymbol{\varphi}_{n+1} + (1 - \alpha) \boldsymbol{\varphi}_n \in \mathcal{C} \tag{13.263}$$

$$\mathbf{V}_{n+\alpha} := \alpha \mathbf{V}_{n+1} + (1 - \alpha) \mathbf{V}_n \tag{13.264}$$

for $\alpha \in [0, 1]$, respectively. The smooth manifold of admissible configurations $\mathcal{C}$ may be defined by

$$\mathcal{C} := \left\{\boldsymbol{\varphi} : \mathcal{B} \to \mathbb{R}^3 \mid J := \det \mathbf{F} > 0 \text{ and } \boldsymbol{\varphi}|_{\partial \mathcal{B}_\varphi} = \bar{\boldsymbol{\varphi}}\right\} \tag{13.265}$$

Introduce the test function, $\boldsymbol{\eta}$, and define the space of test functions as

$$\mathcal{V} := \left\{\boldsymbol{\eta} : \mathcal{B} \to \mathbb{R}^3 \mid \ \boldsymbol{\eta}|_{\partial \mathcal{B}_\varphi} = \mathbf{0}\right\} \tag{13.266}$$

The weak formulation of the initial boundary-value problem, is

$$\int_{\mathcal{B}} \rho_0 \dot{\mathbf{V}} \cdot \boldsymbol{\eta} dV + \int_{\mathcal{B}} \mathbf{F} \cdot \mathbf{S} : \mathrm{GRAD}\, \boldsymbol{\eta} dV = \int_{\mathcal{B}} \rho_0 \mathbf{B} \cdot \boldsymbol{\eta} dV + \int_{\partial \mathcal{B}_\sigma} \bar{\mathbf{T}} \cdot \boldsymbol{\eta} dA \tag{13.267}$$

where $\rho_0 : \mathcal{B} \to \mathbb{R}_+$ is the reference density and $\mathrm{GRAD}\, \boldsymbol{\eta} = \frac{\partial \boldsymbol{\eta}}{\partial \mathbf{X}}$. Then, the associated time discrete form is proposed as

$$\int_{\mathcal{B}} \rho_0 \frac{\mathbf{V}_{n+1} - \mathbf{V}_n}{\Delta t} \cdot \boldsymbol{\eta} dV + \int_{\mathcal{B}} \mathbf{F}\left(\boldsymbol{\varphi}_{n+\alpha}\right) \cdot \tilde{\mathbf{S}} : \mathrm{GRAD}\, \boldsymbol{\eta} dV$$

$$= \int_{\mathcal{B}} \rho_0 \mathbf{B}_{n+\alpha} \cdot \boldsymbol{\eta} dV + \int_{\partial \mathcal{B}_\sigma} \bar{\mathbf{T}}_{n+\alpha} \cdot \boldsymbol{\eta} dA, \quad \forall \boldsymbol{\eta} \in \mathcal{V} \tag{13.268a}$$

$$\int_{\mathcal{B}} \frac{\boldsymbol{\varphi}_{n+1} - \boldsymbol{\varphi}_n}{\Delta t} \cdot \boldsymbol{\eta} dV = \int_{\mathcal{B}} \mathbf{V}_{n+\alpha} \cdot \boldsymbol{\eta} dV \tag{13.268b}$$

and we assume that the initial conditions are given as

$$\int_{\mathcal{B}} \boldsymbol{\varphi}\left(\mathbf{X}, t_0\right) \cdot \boldsymbol{\eta} dV = \int_{\mathcal{B}} \boldsymbol{\varphi}_0 \cdot \boldsymbol{\eta} dV, \quad \mathbf{X} \in \mathcal{B} \tag{13.268c}$$

$$\int_{\mathcal{B}} \mathbf{V}\left(\mathbf{X}, t_0\right) \cdot \boldsymbol{\eta} dV = \int_{\mathcal{B}} \mathbf{V}_0 \cdot \boldsymbol{\eta} dV, \quad \mathbf{X} \in \mathcal{B} \tag{13.268d}$$

Notice that the algorithmic second Piola-Kirchhoff stress tensor, $\tilde{\mathbf{S}}$, has been introduced. In order to achieve the exact energy-momentum conservation within a time step $\left[t_n, t_{n+1}\right]$ under the framework of Equation (13.268) in the sense of the following

$$\mathbf{L}_n = \mathbf{L}_{n+1}, \quad \mathbf{J}_n = \mathbf{J}_{n+1} \quad (n = 0, 1, \ldots, k - 1) \tag{13.269a}$$

where $\mathbf{L}$ and $\mathbf{J}$ denote the linear and angular momenta, respectively, and

$$\mathcal{E}_n \geq \mathcal{E}_{n+1} \text{ and } \mathcal{E}_n \equiv \mathcal{E}_{n+1} \Leftrightarrow \mathcal{D}^{\text{int}} \equiv 0 \quad (n = 0, 1, \ldots, k-1) \tag{13.269b}$$

where $\mathcal{E}$ and $\mathcal{D}^{\text{int}}$ denotes the total energy and the internal dissipation function of the system, respectively, we need to find the conditions for the parameter $\alpha \in [0, 1]$ which satisfy the conditions given by Equation (13.269). Note $\mathbf{L}_n := \mathbf{L}(\mathbf{V}_n)$ and $\mathbf{L}_{n+1} := \mathbf{L}(\mathbf{V}_{n+1})$; $\mathbf{J}_n := \mathbf{J}(\boldsymbol{\varphi}_n, \mathbf{V}_n)$ and $\mathbf{J}_{n+1} := \mathbf{J}(\boldsymbol{\varphi}_{n+1}, \mathbf{V}_{n+1})$; $\mathcal{E}_n := \mathcal{E}(\boldsymbol{\varphi}_n, \mathbf{V}_n)$ and $\mathcal{E}_{n+1} := \mathcal{E}(\boldsymbol{\varphi}_{n+1}, \mathbf{V}_{n+1})$.

Theorem 1: Time discrete momentum conservation within a time step *Suppose the algorithmic 2nd Piola-Kirchhoff stress* $\tilde{\mathbf{S}}$ *is* arbitrary *and* symmetric ($\tilde{\mathbf{S}}^T = \tilde{\mathbf{S}}$). *Then, in the absence of the resultant external loading* $\mathbf{F}^{\text{ext}}_{n+\alpha} = \mathbf{0}$ *and* $\mathbf{T}^{\text{ext}}_{n+\alpha} = \mathbf{0}$, *the discrete linear momentum and the discrete angular momentum are conserved in* $[t_n, t_{n+1}]$ *for any* $\alpha \in [0, 1]$ *and for* $\alpha = 1/2$, *respectively (* $\mathbf{T}^{\text{ext}}_{n+\alpha} := \boldsymbol{\varphi}_{n+\alpha} \times \mathbf{F}^{\text{ext}}_{n+\alpha}$ *denotes the resultant momentum due to the external loading). Therefore, for the satisfaction of the conservation of time discrete linear and angular momenta, we must have* $\alpha = 1/2$:

$$\mathbf{F}^{\text{ext}}_{n+\alpha} = \mathbf{0},\ \mathbf{T}^{\text{ext}}_{n+\alpha} = \mathbf{0},\ \textit{and } \alpha = \tfrac{1}{2} \Rightarrow \mathbf{L}_n = \mathbf{L}_{n+1} \textit{ and } \mathbf{J}_n = \mathbf{J}_{n+1} \tag{13.270a}$$

$$\frac{\mathbf{L}_{n+1} - \mathbf{L}_n}{\Delta t} = \mathbf{F}^{\text{ext}}_{n+\alpha} \quad \textit{for any } \alpha \in [0, 1] \tag{13.270b}$$

$$\frac{\mathbf{J}_{n+1} - \mathbf{J}_n}{\Delta t} = \mathbf{T}^{\text{ext}}_{n+\alpha} \quad \textit{only if } \alpha = \tfrac{1}{2} \tag{13.270c}$$

Proof 1:

1. Time discrete linear momentum conservation: Setting $\boldsymbol{\eta} = \boldsymbol{\xi}$ (= const.) $\in \mathcal{B}$ such that $\text{GRAD}\,\boldsymbol{\eta} = \text{GRAD}\,\boldsymbol{\xi} = \mathbf{0}$ in Equation (13.268a) yields

$$\int_{\mathcal{B}} \rho_0 \frac{\mathbf{V}_{n+1} - \mathbf{V}_n}{\Delta t} \cdot \boldsymbol{\xi}\, dV = \underbrace{\int_{\mathcal{B}} \rho_0 \mathbf{B}_{n+\alpha} \cdot \boldsymbol{\xi}\, dV + \int_{\partial \mathcal{B}_\sigma} \bar{\mathbf{T}}_{n+\alpha} \cdot \boldsymbol{\xi}\, dA}_{\mathbf{F}^{\text{ext}}_{n+\alpha} \cdot \boldsymbol{\xi}} \tag{13.271}$$

With the definition of the linear momentum; namely, $\mathbf{L} := \int_{\mathcal{B}} \rho_0 \mathbf{V}\, dV$, we get

$$\frac{\mathbf{L}_{n+1} - \mathbf{L}_n}{\Delta t} \cdot \boldsymbol{\xi} = \mathbf{F}^{\text{ext}}_{n+\alpha} \cdot \boldsymbol{\xi} \tag{13.272}$$

Hence, the discrete linear momentum is conserved in $[t_n, t_{n+1}]$ for any $\alpha \in [0, 1]$ in the absence of the resultant external loading.

2. Time discrete angular momentum conservation: Setting $\boldsymbol{\eta} = \boldsymbol{\xi} \times \boldsymbol{\varphi}_{n+\alpha}$, where $\boldsymbol{\xi}$ (= const.) $\in \mathbb{R}^3$, such that $\text{GRAD}\,\boldsymbol{\eta} = \hat{\boldsymbol{\eta}}\mathbf{F}(\boldsymbol{\varphi}_{n+\alpha})$, Equation (13.268a) yields

$$\begin{aligned}
&\int_{\mathcal{B}} \rho_0 \frac{\mathbf{V}_{n+1} - \mathbf{V}_n}{\Delta t} \cdot [\boldsymbol{\xi} \times \boldsymbol{\varphi}_{n+\alpha}]\, dV + \int_{\mathcal{B}} \mathbf{F}(\boldsymbol{\varphi}_{n+\alpha}) \cdot \tilde{\mathbf{S}} : [\hat{\boldsymbol{\eta}}\mathbf{F}(\boldsymbol{\varphi}_{n+\alpha})]\, dV \\
&= \int_{\mathcal{B}} \rho_0 \mathbf{B}_{n+\alpha} \cdot [\boldsymbol{\xi} \times \boldsymbol{\varphi}_{n+\alpha}]\, dV + \int_{\partial \mathcal{B}_\sigma} \bar{\mathbf{T}}_{n+\alpha} \cdot [\boldsymbol{\xi} \times \boldsymbol{\varphi}_{n+\alpha}]\, dA \\
\Rightarrow &\int_{\mathcal{B}} [\boldsymbol{\varphi}_{n+\alpha} \times \rho_0 (\mathbf{V}_{n+1} - \mathbf{V}_n)] \cdot \boldsymbol{\xi}\, dV
\end{aligned}$$

$$= \Delta t \underbrace{\left(\int_{\mathcal{B}} \left[\boldsymbol{\varphi}_{n+\alpha} \times \rho_0 \boldsymbol{b}_{n+\alpha} \right] \cdot \boldsymbol{\eta} dV + \int_{\partial \mathcal{B}_\sigma} \left[\boldsymbol{\varphi}_{n+\alpha} \times \bar{\mathbf{T}}^0_{n+\alpha} \right] \cdot \boldsymbol{\eta} dA \right)}_{\mathbf{T}^{\text{ext}}_{n+\alpha} \cdot \boldsymbol{\xi}}$$

$$- \Delta t \underbrace{\int_{\mathcal{B}} \left[\mathbf{F}^T \left(\boldsymbol{\varphi}_{n+\alpha} \right) \cdot \tilde{\mathbf{S}} \cdot \mathbf{F} \left(\boldsymbol{\varphi}_{n+\alpha} \right) \right] \cdot \cdot \hat{\boldsymbol{\xi}} dV}_{0}$$

$$= \Delta t \boldsymbol{\xi} \cdot \mathbf{T}^{\text{ext}}_{n+\alpha} \tag{13.273}$$

Using the definition of the angular momentum; namely, $\mathbf{J} := \int_{\mathcal{B}} \boldsymbol{\varphi} \times \rho_0 \mathbf{V} dV$,

$$\begin{aligned} \boldsymbol{\xi} \cdot (\mathbf{J}_{n+1} - \mathbf{J}_n) &= \boldsymbol{\xi} \cdot \int_{\mathcal{B}} \left(\boldsymbol{\varphi}_{n+1} \times \rho_0 \mathbf{V}_{n+1} - \boldsymbol{\varphi}_n \times \rho_0 \mathbf{V}_n \right) dV \\ &= \boldsymbol{\xi} \cdot \int_{\mathcal{B}} \boldsymbol{\varphi}_{n+\alpha} \times \rho_0 \left(\mathbf{V}_{n+1} - \mathbf{V}_n \right) dV + \boldsymbol{\xi} \cdot \int_{\mathcal{B}} \left(\boldsymbol{\varphi}_{n+1} - \boldsymbol{\varphi}_n \right) \times \rho_0 \mathbf{V}_{n+\alpha} dV \\ &= \boldsymbol{\xi} \cdot \int_{\mathcal{B}} \boldsymbol{\varphi}_{n+\alpha} \times \rho_0 \left(\mathbf{V}_{n+1} - \mathbf{V}_n \right) dV + \boldsymbol{\xi} \cdot 2\Delta t \int_{\mathcal{B}} \rho_0 \mathbf{V}_{n+\alpha} \times \mathbf{V}_{n+1/2} dV \end{aligned} \tag{13.274}$$

Notice Equation (13.268b) has been used in the last step above. Substituting Equation (13.273) into the first term of Equation (13.274) yields

$$\boldsymbol{\xi} \cdot (\mathbf{J}_{n+1} - \mathbf{J}_n) = \Delta t \boldsymbol{\xi} \cdot \mathbf{T}^{\text{ext}}_{n+\alpha} + 2\Delta t \boldsymbol{\xi} \cdot \int_{\mathcal{B}} \rho_0 \mathbf{V}_{n+\alpha} \times \mathbf{V}_{n+1/2} dV \tag{13.275}$$

When $\alpha = 1/2$, Equation (13.275) becomes

$$\frac{\mathbf{J}_{n+1} - \mathbf{J}_n}{\Delta t} \cdot \boldsymbol{\xi} = \mathbf{T}^{\text{ext}}_{n+1/2} \cdot \boldsymbol{\xi} \tag{13.276}$$

Hence, the discrete angular momentum is conserved in $[t_n, t_{n+1}]$ only if $\alpha = 1/2$ in the absence of the resultant external momentum (torque). □

Therefore, the exact momentum conserving algorithm within a time step $\left[t_n, t_{n+1}\right]$ may be written as follows:

Algorithm 4: ***Momentum conserving algorithm***

Suppose the algorithmic 2nd Piola-Kirchhoff stress $\tilde{\mathbf{S}}$ is *arbitrary* and *symmetric* ($\tilde{\mathbf{S}}^T = \tilde{\mathbf{S}}$); then, the following algorithm conserves the linear and angular momenta within the time step $\left[t_n, t_{n+1}\right]$:

$$\begin{aligned} &\int_{\mathcal{B}} \rho_0 \frac{\mathbf{V}_{n+1} - \mathbf{V}_n}{\Delta t} \cdot \boldsymbol{\eta} dV + \int_{\mathcal{B}} \mathbf{F} \left(\boldsymbol{\varphi}_{n+1/2} \right) \cdot \tilde{\mathbf{S}} : \text{GRAD}\, \boldsymbol{\eta} dV \\ &= \int_{\mathcal{B}} \rho_0 \mathbf{B}_{n+1/2} \cdot \boldsymbol{\eta} dV + \int_{\partial \mathcal{B}_\sigma} \bar{\mathbf{T}}_{n+1/2} \cdot \boldsymbol{\eta} dA, \quad \forall \boldsymbol{\eta} \in \mathcal{V} \end{aligned} \tag{13.277a}$$

$$\int_{\mathcal{B}} \frac{\boldsymbol{\varphi}_{n+1} - \boldsymbol{\varphi}_n}{\Delta t} \cdot \boldsymbol{\eta} dV = \int_{\mathcal{B}} \mathbf{V}_{n+1/2} \cdot \boldsymbol{\eta} dV \tag{13.277b}$$

Remark 13.6.4

1. It is important to note that the discrete linear and angular momenta are conserved with an *arbitrary* and *symmetric* algorithmic 2nd Piola-Kirchhoff stress $\tilde{\mathbf{S}}$.
2. Algorithm 4 satisfies the algorithmic form (in the sense of the theorem of expended power) within the time step $[t_n, t_{n+1}]$: Note the internal force term can be equivalently written as

$$\int_{\mathcal{B}} \mathbf{F}(\varphi_{n+1/2}) \cdot \tilde{\mathbf{S}} : \mathrm{GRAD}\, \eta dV = \int_{\mathcal{B}} \tilde{\mathbf{S}} : \mathbf{F}^T(\varphi_{n+1/2}) \cdot \mathrm{GRAD}\, \eta dV \tag{13.278}$$

in the Cartesian coordinate system. Using Equation (13.278) for convenience and setting $\eta = \mathbf{V}_{n+1/2}$ in Equation (13.277a) yields

$$\underbrace{\int_{\mathcal{B}} \rho_0 (\mathbf{V}_{n+1} - \mathbf{V}_n) \cdot \mathbf{V}_{n+1/2} dV}_{\mathcal{K}_{n+1} - \mathcal{K}_n} + \int_{\mathcal{B}} \tilde{\mathbf{S}} : \underbrace{\mathbf{F}^T(\varphi_{n+1/2}) \cdot \mathrm{GRAD}(\Delta t \mathbf{V}_{n+1/2})}_{\frac{1}{2}[\mathbf{C}(\varphi_{n+1}) - \mathbf{C}(\varphi_n)]} dV$$

$$= \Delta t \int_{\mathcal{B}} \rho_0 \mathbf{B}_{n+1/2} \cdot \mathbf{V}_{n+1/2} dV + \Delta t \int_{\partial \mathcal{B}_\sigma} \bar{\mathbf{T}}_{n+1/2} \cdot \mathbf{V}_{n+1/2} dA \tag{13.279}$$

where $\mathcal{K}$ and $\mathbf{C}$ denote the kinetic energy of the system and the right Cauchy-Green strain tensor, respectively, and the right-hand side implies the change of the potential energy loading within the time step.
To study the criteria of the discrete energy balance within the time step $[t_n, t_{n+1}]$, consider the following theorem:

Theorem 2: Energy balance *In a mechanical system, the following inequality is true:*

$$\dot{\mathcal{E}} - \mathcal{P}^{\mathrm{ext}} = -\mathcal{D}^{\mathrm{int}} \leq 0, \quad \forall t \in \mathcal{T} \tag{13.280a}$$

where $\mathcal{D}^{\mathrm{int}} \geq 0$ *and*

$$\mathcal{P}^{\mathrm{ext}} = \int_{\mathcal{B}} \rho_0 \mathbf{B}(\mathbf{X}, t) \cdot \mathbf{V}(\mathbf{X}, t)\, dV + \int_{\partial \mathcal{B}_\sigma} \bar{\mathbf{T}}(\mathbf{X}, t) \cdot \mathbf{V}(\mathbf{X}, t)\, dA \tag{13.280b}$$

denote the internal dissipation function and the expended power of the external loading of the system, respectively. The total energy of the system $\mathcal{E}$ *is defined by the sum of the kinetic energy* $\mathcal{K}$ *and the internal energy* $\mathcal{U}^{\mathrm{int}}$*:*

$$\mathcal{E} := \mathcal{K} + \mathcal{U}^{\mathrm{int}} = \int_{\mathcal{B}} \tfrac{1}{2} \rho_0 \mathbf{V}(\mathbf{X}, t) \cdot \mathbf{V}(\mathbf{X}, t)\, dV + \textstyle\int_{\mathcal{B}} W dV \tag{13.280c}$$

where W *denotes the stored energy density function defined per unit reference volume.*

Proof 2: The first law of thermodynamics in a mechanical system is described as

$$\dot{\mathcal{E}} = \mathcal{P}^{\mathrm{ext}} \tag{13.281}$$

With Equation (13.280c), the left-hand side of above equation can be also written as

$$\dot{\mathcal{E}} = \dot{\mathcal{K}} + \dot{\mathcal{U}}^{\mathrm{int}} = \int_{\mathcal{B}} \left[\rho_0 \mathbf{V}(\mathbf{X}, t) \cdot \frac{\partial \mathbf{V}(\mathbf{X}, t)}{\partial t} + \frac{\partial W}{\partial t} \right] dV \tag{13.282}$$

According to the Clausius-Planck inequality, the internal dissipation function in the local form is given by

$$\mathcal{D}^{\text{int}} := \mathbf{P} : \dot{\mathbf{F}} - \frac{\partial W}{\partial t} \geq 0 \tag{13.283}$$

where $\mathbf{P}$ is the nominal stress tensor in the mechanical system. In the continuum system, the internal dissipation function may be given by

$$\begin{aligned} \mathcal{D}^{\text{int}} &:= \int_{\mathcal{B}} \mathbf{P} : \dot{\mathbf{F}} dV - \int_{\mathcal{B}} \frac{\partial W}{\partial t} dV \\ &= \mathcal{P}^{\text{int}} - \dot{\mathcal{U}}^{\text{int}} \geq 0 \end{aligned} \tag{13.284}$$

where $\mathcal{P}^{\text{int}} := \int_{\mathcal{B}} \mathbf{P} : \dot{\mathbf{F}}^T dV$ is called the total stress power of the system. Recalling the following relation which is true in the mechanical system:

$$\dot{\mathcal{U}}^{\text{int}} = \int_{\mathcal{B}} \frac{\partial W}{\partial t} dV = \int_{\mathcal{B}} \mathbf{P} : \dot{\mathbf{F}} dV \tag{13.285}$$

inequality (13.280a) can be readily derived from Equations (13.281), (13.282) and (13.284). □

In view of Equation (13.279), the discrete form of Theorem 2 in the absence of the external loading (therefore, $\mathcal{P}^{\text{ext}} = 0$) may be written as

$$\mathcal{U}_{n+1}^{\text{int}} - \mathcal{U}_n^{\text{int}} = \int_{\mathcal{B}} \tilde{\mathbf{S}} : \tfrac{1}{2} \left[\mathbf{C}(\boldsymbol{\varphi}_{n+1}) - \mathbf{C}(\boldsymbol{\varphi}_n) \right] dV - \Delta\mathcal{D}^{\text{int}} \quad \text{with} \quad \Delta\mathcal{D}^{\text{int}} \geq 0 \tag{13.286}$$

For the exact energy conservation within time step ($\Delta\mathcal{D}^{\text{int}} \equiv 0$), the following relation must hold:

$$\mathcal{U}_{n+1}^{\text{int}} - \mathcal{U}_n^{\text{int}} = \int_{\mathcal{B}} \tilde{\mathbf{S}} : \tfrac{1}{2} \left[\mathbf{C}(\boldsymbol{\varphi}_{n+1}) - \mathbf{C}(\boldsymbol{\varphi}_n) \right] dV \tag{13.287}$$

To observe the expression of the algorithmic 2nd Piola-Kirchhoff stress $\tilde{\mathbf{S}}$ for the exact energy-momentum conserving algorithm for smooth nonlinear elastodynamics with the restriction of Equation (13.287), define the *algorithmic* right Cauchy-Green strain tensor as

$$\mathbf{C}_{n+\beta} := \beta \mathbf{C}(\boldsymbol{\varphi}_n) + [1 - \beta] \mathbf{C}(\boldsymbol{\varphi}_{n+1}) \tag{13.288}$$

for $\beta \in [0, 1]$. Note $\mathbf{C}(\boldsymbol{\varphi}_n) \equiv \mathbf{C}_n$ and $\mathbf{C}(\boldsymbol{\varphi}_{n+1}) \equiv \mathbf{C}_{n+1}$; however, $\mathbf{C}(\boldsymbol{\varphi}_{n+\beta}) \neq \mathbf{C}_{n+\beta}$ for $\beta \in (0, 1)$ in general. It is very important to use $\mathbf{C}_{n+\beta}$ instead of $\mathbf{C}(\boldsymbol{\varphi}_{n+\beta})$ to avoid non-physical couplings in the calculation which may cause numerical instabilities; see Gonzalez and Simo (Gonzalez and Simo 1996). Note that the algorithmic right Cauchy-Green strain tensor remains symmetric and positive definite for all $\beta \in [0, 1]$. By the definition, the algorithmic Lagrange-Green strain tensor may be correspondingly written as

$$\mathbf{E}_{n+\beta} := \tfrac{1}{2} \left(\mathbf{C}_{n+\beta} - \mathbf{I} \right) \tag{13.289}$$

For the exact energy conservation within the time step $[t_n, t_{n+1}]$, the mean value theorem must be used; namely, there exist some $\beta_0 \in (0, 1)$ such that

$$\begin{aligned} \hat{W}(\mathbf{C}_{n+1}) - \hat{W}(\mathbf{C}_n) &= \frac{\partial \hat{W}(\mathbf{C}_{n+\beta_0})}{\partial \mathbf{C}_{n+\beta_0}} : [\mathbf{C}_{n+1} - \mathbf{C}_n] \\ &= \nabla \hat{W}(\mathbf{C}_{n+\beta_0}) : [\mathbf{C}_{n+1} - \mathbf{C}_n] \end{aligned} \tag{13.290}$$

A comparison with the absence of external loading and $\Delta\mathcal{D}^{\text{int}} \equiv 0$, i.e.,

$$\int_{\mathcal{B}} [\hat{W}(\mathbf{C}_{n+1}) - \hat{W}(\mathbf{C}_n)]\, dV = \tfrac{1}{2}\textstyle\int_{\mathcal{B}} \tilde{\mathbf{S}} : [\mathbf{C}_{n+1} - \mathbf{C}_n]\, dV \tag{13.291}$$

yields

$$\tilde{\mathbf{S}} = 2\nabla \hat{W}(\mathbf{C}_{n+\beta_0}) \quad \text{for} \quad \beta_0 \in (0, 1) \tag{13.292}$$

Using Equation (13.292) with Algorithm 4 gives only first-order time accuracy unless $\beta_0 = 1/2$. To resolve this limitation, Simo and Tarnow (Simo and Tarnow 1992) proposed the following algorithmic 2nd Piola-Kirchhoff stress which also satisfies the Equation (13.291):

$$\tilde{\mathbf{S}} = \nabla \hat{W}(\mathbf{C}_{n+\beta_0}) + \nabla \hat{W}(\mathbf{C}_{n+(1-\beta_0)}) \quad \text{for} \quad \beta_0 \in (0, 1) \tag{13.293}$$

Using Equation (13.293) with Algorithm 4 gives second-order time accuracy for any $\beta_0 = (0, 1)$. Hence, with the algorithmic 2nd Piola-Kirchhoff stress, the exact energy-momentum conserving algorithm within time step $[t_n, t_{n+1}]$ may be given as follows:

Algorithm 5: ***Energy-momentum conserving algorithm, Simo-Tarnow framework***

$$\begin{aligned} &\int_{\mathcal{B}} \rho_0 \frac{\mathbf{V}_{n+1} - \mathbf{V}_n}{\Delta t} \cdot \eta\, dV + \int_{\mathcal{B}} \mathbf{F}(\varphi_{n+1/2}) \cdot \tilde{\mathbf{S}} : \text{GRAD}\, \eta\, dV \\ &= \int_{\mathcal{B}} \rho_0 \mathbf{B}_{n+1/2} \cdot \eta\, dV + \int_{\partial\mathcal{B}_\sigma} \bar{\mathbf{T}}_{n+1/2} \cdot \eta\, dA, \quad \forall \eta \in \mathcal{V} \end{aligned} \tag{13.294a}$$

$$\int_{\mathcal{B}} \frac{\varphi_{n+1} - \varphi_n}{\Delta t} \cdot \eta\, dV = \int_{\mathcal{B}} \mathbf{V}_{n+1/2} \cdot \eta\, dV \tag{13.294b}$$

where the algorithmic 2nd Piola-Kirchhoff stress $\tilde{\mathbf{S}}$ is given by

$$\text{Option 1:} \quad \tilde{\mathbf{S}} = \tilde{\mathbf{S}}_1 := 2\nabla \hat{W}(\mathbf{C}_{n+\beta_0}) \tag{13.294c}$$

$$\text{Option 2:} \quad \tilde{\mathbf{S}} = \tilde{\mathbf{S}}_2 := \nabla \hat{W}(\mathbf{C}_{n+\beta_0}) + \nabla \hat{W}(\mathbf{C}_{n+(1-\beta_0)}) \tag{13.294d}$$

with $\beta_0 \in (0, 1)$ such that

$$\hat{W}(\mathbf{C}_{n+1}) - \hat{W}(\mathbf{C}_n) = \tilde{\mathbf{S}} : \tfrac{1}{2}[\mathbf{C}_{n+1} - \mathbf{C}_n] \tag{13.294e}$$

Remark 13.6.5

1. Accuracy in time - The algorithmic 2nd Piola-Kirchhoff stress $\tilde{\mathbf{S}}$: Both Option 1, Equation (13.294c), and Option 2, Equation (13.294d), satisfy Equation (13.294e). Although second-order time accuracy is achieved for any $\beta_0 \in (0, 1)$ in Option 2, it is given only when $\beta_0 = 1/2$ in Option 1; otherwise, we have only first-order time accuracy for $\beta_0 \neq 1/2$ in Option 1.
2. For St. Venant-Kirchhoff type materials, we always have $\beta_0 = 1/2$.
3. Algorithm 5 is unconditionally stable in the sense of exact energy conservation within a time step. Recall the total energy defines the Lyapunov function of the dynamical system.
4. During the implementation of the fully discretized form of Algorithm 5 via finite-element space discretization, an iterative method, such as the Newton-Raphson method, needs to be used during every time step to find β_0 which satisfies Equation (13.294e). Dealing with β_0 calculated at each quadrature point and $\boldsymbol{\varphi}_{n+1}$ consistently when computing the tangent stiffness matrix is not guaranteed in Algorithm 5, which may cause convergence issues. For this reason, Algorithm 5 has been implemented only for the St. Venant-Kirchhoff materials in the paper (Simo and Tarnow 1992). For the solution procedure to address this issue to an extent, see Laursen and Meng (Laursen and Meng 2001).
5. During the implementation of the fully discretized form of Algorithm 5 via finite-element space discretization by means of the Newton-Raphson method, the tangent stiffness matrix becomes *asymmetric* in general; however, for a sufficient small Δt, it tends to become close to a symmetric matrix form.
   Even after the Galerkin finite element space discretization (through the Galerkin projection of infinite dimensional state space onto a finite dimensional state subspace $C^h \subset C$), all the algorithmic properties such as discrete total linear/angular momenta conservation and discrete total energy conservation are still preserved under the same algorithmic conditions.

13.6.4 Extension of Energy-Momentum Conserving Algorithms for a General Hyperelastic Material Model

Simo-Gonzalez Framework A further improvement of the Simo-Tarnow general framework of the exact energy-momentum conserving algorithm for compressible elastodynamics has been proposed by Simo and Gonzalez (Simo and Gonzalez 1994) and Gonzalez (Gonzalez 2000) by applying the so-called *discrete derivative* for the evaluation of the algorithmic stress. For details about the discrete derivative, see Gonzalez (Gonzalez 1996). In this framework, an algorithmic stress field $\tilde{\mathbf{S}}\left(\boldsymbol{\varphi}_n, \boldsymbol{\varphi}_{n+1}\right)$ is defined as

$$\tilde{\mathbf{S}}\left(\boldsymbol{\varphi}_n, \boldsymbol{\varphi}_{n+1}\right) := 2\mathrm{d}\hat{W}\left(\mathbf{C}_n, \mathbf{C}_{n+1}\right) \tag{13.295}$$

where $\hat{W}(\mathbf{C}) \equiv \bar{W}(\mathbf{E})$ is the stored energy potential, and "d" denotes a discrete derivative (Gonzalez 1996). With the proposition, Equation (13.295) may be written as

$$\tilde{\mathbf{S}}\left(\boldsymbol{\varphi}_n, \boldsymbol{\varphi}_{n+1}\right) = 2D\hat{W}\left(\mathbf{C}_{n+1/2}\right) + 2\frac{\hat{W}\left(\mathbf{C}_{n+1}\right) - \hat{W}\left(\mathbf{C}_n\right) - D\hat{W}\left(\mathbf{C}_{n+1/2}\right) : \Delta\mathbf{C}}{\|\Delta\mathbf{C}\|^2}\Delta\mathbf{C} \tag{13.296}$$

where $\Delta\mathbf{C} := \mathbf{C}_{n+1} - \mathbf{C}_n$ and $\|\Delta\mathbf{C}\| := \sqrt{\Delta\mathbf{C} : \Delta\mathbf{C}}$. In terms of the second Piola-Kirchhoff tensor $\mathbf{E}$, Equation (13.296) can be written as

$$\tilde{\mathbf{S}}\left(\boldsymbol{\varphi}_n, \boldsymbol{\varphi}_{n+1}\right) = D\hat{W}\left(\mathbf{E}_{n+1/2}\right) + \frac{\hat{W}\left(\mathbf{E}_{n+1}\right) - \hat{W}\left(\mathbf{E}_n\right) - 2D\hat{W}\left(\mathbf{E}_{n+1/2}\right) : \Delta\mathbf{E}}{2\|\Delta\mathbf{E}\|^2}\Delta\mathbf{E} \tag{13.297}$$

where $\Delta\mathbf{E} := \mathbf{E}_{n+1} - \mathbf{E}_n$ and $\|\Delta\mathbf{E}\| := \sqrt{\Delta\mathbf{E} : \Delta\mathbf{E}}$. For St. Venant-Kirchhoff type materials, the stored energy potential is given by

$$\bar{W} = \int \mathbf{S} : d\mathbf{E} = \int \mathbb{C} : \mathbf{E} : d\mathbf{E} = \frac{1}{2}\mathbf{E} : \mathbb{C} : \mathbf{E} \tag{13.298}$$

where $\mathbb{C}$ is the constant fourth-order tensor of elastic moduli which is symmetric and positive-definite. Therefore, the algorithmic stress field for this special material model can be reduced as follows:

$$\begin{aligned}\tilde{\mathbf{S}}\left(\varphi_n, \varphi_{n+1}\right) &= \mathbb{C} : \mathbf{E}_{n+1/2} \\ &+ \frac{(\mathbf{E}_{n+1} - \mathbf{E}_n) : \mathbb{C} : \mathbf{E}_{n+1/2} - 2\left(\mathbf{E}_{n+1} - \mathbf{E}_n\right) : \mathbb{C} : \mathbf{E}_{n+1/2}}{2\left(\mathbf{E}_{n+1} - \mathbf{E}_n\right) : \left(\mathbf{E}_{n+1} - \mathbf{E}_n\right)}\left[\mathbf{E}_{n+1} - \mathbf{E}_n\right] \\ &= \mathbb{C} : \mathbf{E}_{n+1/2}\end{aligned} \tag{13.299}$$

Hence, using the algorithmic 2nd Piola-Kirchhoff stress $\tilde{\mathbf{S}}$, the conserving scheme proposed by Simo and Gonzalez (Simo and Gonzalez 1994) may be summarized as follows:

Algorithm 6: ***Simo-Gonzalez Framework***

$$\begin{aligned}&\int_{\mathcal{B}} \rho_0 \frac{\mathbf{V}_{n+1} - \mathbf{V}_n}{\Delta t} \cdot \eta dV + \int_{\mathcal{B}} \tilde{\mathbf{S}} : \mathbf{F}^T\left(\varphi_{n+1/2}\right) \mathrm{GRAD}\, \eta dV \\ &= \int_{\mathcal{B}} \rho_0 \mathbf{B}_{n+1/2} \cdot \eta dV + \int_{\partial\mathcal{B}_\sigma} \bar{\mathbf{T}}_{n+1/2} \cdot \eta dA, \quad \forall \eta \in \mathcal{V}\end{aligned} \tag{13.300a}$$

$$\int_{\mathcal{B}} \frac{\varphi_{n+1} - \varphi_n}{\Delta t} \cdot \eta dV = \int_{\mathcal{B}} \mathbf{V}_{n+1/2} \cdot \eta dV \tag{13.300b}$$

where the algorithmic 2nd Piola-Kirchhoff stress $\tilde{\mathbf{S}}$ is given by

$$\begin{aligned}\tilde{\mathbf{S}}\left(\varphi_n, \varphi_{n+1}\right) &:= 2\mathrm{d}\hat{W}\left(\mathbf{C}_n, \mathbf{C}_{n+1}\right) \\ &= 2D\hat{W}\left(\mathbf{C}_{n+1/2}\right) + 2\frac{\hat{W}\left(\mathbf{C}_{n+1}\right) - \hat{W}\left(\mathbf{C}_n\right) - D\hat{W}\left(\mathbf{C}_{n+1/2}\right) : \Delta\mathbf{C}}{\|\Delta\mathbf{C}\|^2}\Delta\mathbf{C}\end{aligned} \tag{13.300c}$$

where $\Delta\mathbf{C} := \mathbf{C}_{n+1} - \mathbf{C}_n$ and $\|\Delta\mathbf{C}\| := \sqrt{\Delta\mathbf{C} : \Delta\mathbf{C}}$.

Remark 13.6.6

1. The algorithmic 2nd Piola-Kirchhoff stress $\tilde{\mathbf{S}}$ satisfies the directionality condition:

$$\hat{W}\left(\mathbf{C}_{n+1}\right) - \hat{W}\left(\mathbf{C}_n\right) = \tilde{\mathbf{S}} : \tfrac{1}{2}\left[\mathbf{C}_{n+1} - \mathbf{C}_n\right] \tag{13.301}$$

2. The algorithmic 2nd Piola-Kirchhoff stress $\tilde{\mathbf{S}}$ satisfies the consistency condition, and

$$\tilde{\mathbf{S}}\left(\varphi_n, \varphi_{n+1}\right) = 2\mathrm{d}\hat{W}\left(\mathbf{C}_n, \mathbf{C}_{n+1}\right) = \mathbf{S}\left(\varphi_{n+1/2}\right) + \mathcal{O}\left(\Delta t^2\right) \tag{13.302}$$

Therefore, Algorithm 6 is second-order time accurate.

3. Since the algorithmic 2nd Piola-Kirchhoff stress $\tilde{\mathbf{S}}$ given by Equation (13.300c) is symmetric, the linear and angular momenta are conserved within the time step $\left[t_n, t_{n+1}\right]$; see Algorithm 4.
4. When $\Delta\mathbf{C} = \mathbf{0}$, we have $\tilde{\mathbf{S}} = 2D\hat{W}\left(\mathbf{C}_{n+1/2}\right)$. Note that there is no singularity issue for any initial conditions due to Equation (13.300c).
5. The extra iteration needed for computing the parameter β_0 in the implementation of Algorithm 5 is not required for Algorithm 6.

CHAPTER FOURTEEN

TIME DISCRETIZATION OF EQUATIONS OF MOTION: RECENT ADVANCES

Kumar K. Tamma, Masao Shimada, Xiangmin Zhou, Andrew Hoitink, Siti U. Masuri and Jason Har

In Chapter 14 which deals with the *time discretization of the equations of motion and advances in algorithms by design for time integration*, we mostly focus attention upon recent advances directly emanating from the discrete form of the Total Energy representation of the equation of motion as a starting point (an outcome of the Principle of Balance of Energy) which naturally reflects the semi-discretized equation system. In conjunction with the Total Energy framework and a generalized time weighted residual approach, we describe the various underlying concepts leading to developing time integration algorithms by design. *In contrast to understanding the traditional practices as outlined in the previous chapter, in general, the reader is strongly encouraged to become familiar with several very important concepts that are particularly described in this chapter which are an outcome of the theoretical foundations related to three principal aspects: a) the Unified theory underlying computational algorithms (Zhou and Tamma 2004a; Tamma et al. 2002; Tamma et al. 2000) for transient systems which explains the concept of design spaces of general time integration algorithms, the associated classifications, and the very important generalized barrier theorem wherein Dahlquist's theorem is a particular case, b) the so-called design spaces, measures and metrics associated with time integration algorithms (Zhou et al. 2005) which helps one to understand how to place future new developments in time integration algorithms in the proper context and how to evaluate future new algorithm designs in contrast to existing algorithm designs, and c) the subsequent concepts underlying the notion of algorithms by design (Zhou and Tamma 2006) whose objectives are to foster the design of new time integration algorithms with a wish list of a priori desired algorithmic attributes for the application under consideration.*

14.1 INTRODUCTION

An overview of the big picture dealing with recent advances emanating from the Total Energy framework and a generalized time weighted residual approach with particular focus upon the class of LMS methods is highlighted next. In the previous Chapter 13, we have mostly described conventional practices of time discretization of the equations of motion. Here in Chapter 14, we describe recent advances emanating

Advances in Computational Dynamics of Particles, Materials and Structures, First Edition. Jason Har and Kumar K. Tamma.

from new and different perspectives in both single-field and two-field forms of representations. They not only encompass most of the classical developments described earlier in Chapter 13, but also additionally provide new insights, avenues, and optimal algorithms and designs that resolve a wide variety of issues that have plagued the community at large; and therein, they are ideally suited to solve practical computational science and engineering problems. It is important to note that for the particular class of LMS methods which are of focus here, both the two-field form and the single-field form of representation are highlighted and overviewed under the umbrella of the well known Generalized Single System Single Solve (GSSSS) framework. Of noteworthy importance are also issues related to time level consistency of algorithms within the GSSSS unified framework. Of noteworthy importance are also issues related to time level consistency of algorithms within the GSSSS unified framework. That is, for every time integration algorithm a collocation time point exists, at which the discrete equations of motion are exactly satisfied; and it is essential that the various terms in the semi-discrete equation of motion be carefully evaluated at this particular time level so that it paves the way for proper implementation and also provides the necessary and theoretically expected second-order time accuracy for all the algorithm unknowns (Hoitink et al. 2008). Failure to recognize these subtle, yet very important issues will cause misinterpretation and/or improper implementation leading to poor analysis results and conclusions (the literature is filled with a wide variety of misconceptions). The GSSSS unified framework encompasses the class of Linear Multi Step (LMS) methods, and the related designs of the underlying time stepping algorithms within this framework are of second-order time accuracy (first-order time accurate algorithms are not of interest in the commercial setting as they are not practical and are not viewed favorably), and involve the solution of a single system of equations with only a single solve within each single time step. They are the most practical and popular in commercial and research software due to their ability to simulate large scale practical applications. Although one can formulate time integration algorithms from various differing viewpoints, a point worth mentioning is that the overall developments outlined in the sections to follow emanate from the discrete Total Energy framework, and highlight and cover the following:

1. Firstly, restricting attention to the general class of LMS methods, the present concepts readily enable the design of most of the past developments in time integration described in the previous chapter that have been presented from various different viewpoints (regardless of how the original algorithms have been formulated in developing the time stepping schemes). Both the single-field and two-field form are an integral part and subset of the present unified framework and design process. In a nutshell, the unified framework of the GSSSS family of algorithms not only permits one to recover the various developments in time integration within the class of LMS methods over the past fifty years or so (although the original representation of the algorithms may have been derived from various different viewpoints), but it also additionally provides new perspectives and new avenues to include optimal time stepping algorithm designs (the optimality is in the sense of the least amount of numerical dissipation, dispersion, and overshooting behavior). The focus is upon the initial design of the parent linear dynamic algorithms and the associated designs which are a necessary first step (and includes both numerically non-dissipative and numerically dissipative algorithm designs with controllable numerical dissipation), and also forms the fundamental basis for subsequent extensions to nonlinear dynamics applications properly,
2. Secondly, via this unified GSSSS framework encompassing LMS methods as the basic foundation, a sound theoretical basis via a new normalized time weighted residual approach is particularly described next on how to properly provide in general, extensions of the parent linear dynamic algorithms to nonlinear dynamics applications; this is important to understand (there are certain misconceptions in general, in the literature regarding time integration algorithms for linear versus nonlinear dynamics applications). *In general, there exist three distinctly different options, namely: (i) the classical time weighted residual approach which fails to properly preserve the underlying physics for nonlinear dynamics applications, (ii) the new normalized time weighted residual approach which is ideally suitable for extensions of the parent linear dynamic algorithms to nonlinear*

dynamics applications such that, as one option, it naturally leads to symplectic-momentum based representations for particular types of approximation, and (iii) alternatively, for other particular types of approximation, as another option, it naturally leads to energy-momentum based representations. Besides the common numerically non-dissipative time integration algorithms and the various others with controllable numerical dissipation, it is important to understand this unified framework which encompasses both symplectic-momentum based and energy-momentum based time stepping schemes and the associated frameworks (in particular, the virgin symplectic-momentum or energy-momentum conserving features of the basic time integration algorithm are also readily recovered in this framework when the controllable numerical dissipation that can be embedded in them is turned off). A note of clarification is in order. Although the developments emanate from the Total Energy framework which is a result of the principle of balance of mechanical energy, since a generalized time weighed residual approach is employed, the consequences leading to the conclusions of structure preservation such as symplectic-momentum conservation and/or energy-momentum conservation can only be drawn either by comparison and contrast with algorithms that inherit such features and/or through demonstrations via proofs. Alternatively, directly emanating from the discrete total energy and approaches such as the mean value theorem, one can also readily derive algorithms inheriting energy-momentum conservation (but without controllable numerical dissipation features). Equivalently, emanating from the discrete Hamilton's principle or Hamilton's law of varying action, one can readily derive symplectic-momentum conserving algorithms. Finally in the Newtonian mechanics setting, since no scalar representations exist, the generalized time weighted residual approach serves well for the development of a wide variety of numerically non-dissipative and also algorithms with controllable numerical dissipation. It is also to be noted that most of the original classical developments in symplectic-momentum and energy-momentum conserving algorithmic designs are in the sense that the unknowns are the displacement and the velocity as in the two-field form representations. These are unlike the single-field form of representations with unknowns, namely, displacement, velocity, as well as acceleration which, it must be noted, also yield equivalent representations to the two-field form; but involve a spurious root to accommodate both numerically non-dissipative and numerically (controllable) dissipative algorithms and representations. Consequently, the single-field form provides an additional mechanism to enable practical applications to be more easily simulated; especially structural systems with stiff and/or flexible structural components due to the fact that controllable numerical dissipation can be readily introduced to the parent symplectic-momentum and/or energy-momentum algorithm designs, and when such features are turned off, they readily recover the parent structure preserving algorithm designs,

3. Thirdly, applicability to N-body systems are highlighted, and
4. Finally, applicability to conservative/non-conservative mechanical systems with holonomic-scleronomic constraints are also highlighted.

14.2 TIME DISCRETIZATION AND THE TOTAL ENERGY FRAMEWORK: LINEAR DYNAMIC ALGORITHMS AND DESIGNS – GENERALIZED SINGLE STEP SINGLE SOLVE [GSSSS] UNIFIED FRAMEWORK ENCOMPASSING LMS METHODS

For simplicity in understanding the basic concepts and for clarity, let us restrict attention to the general class of LMS methods involving a single system of equations and a single solve within each single time step. As mentioned previously, there exist two schools of thought for deriving time stepping algorithms for integrating the equations of motion in computational dynamics. One is the so-called two-field form (where the unknowns are displacement and velocity) resulting in a first-order system of equations, and the other is the so-called single-field form (where the unknowns are displacement, velocity and acceleration) resulting in a second-order system of equations. In most commercial software, and from historical perspectives, time stepping algorithms derived from the single-field form of representation which contains two principal roots and a spurious root have been fairly prominent. Furthermore, the representation in the single-field form also readily permits the development of both numerically non-dissipative time integration algorithms and numerically dissipative time integration algorithms (with controllable numerical dissipation) in the design process. This is unlike the two-field form of representation which does not have a spurious root and hence, in general, it mostly yields numerically non-dissipative algorithms in the design of time stepping algorithms (under certain conditions such as the case when accelerations are not treated as unknowns, then via the single-field form, time stepping algorithms equivalent to that obtained via the two-field form may be constructed). It is to be noted that the single-field form in (Kuhl and Crisfield 1999) is strictly not representative of the original two-field form energy-momentum algorithm in (Simo et al. 1992). This is because it is identical to that termed as V0 (1,1,1) and not V0 (1,1,0) within the GSSSS framework described in the next chapter to follow, wherein, the latter single-field form is exactly representative (MPR-MPA) unlike the former (MPR-EPA).

A brief background of the important and subtle issues, underlying foundations, and terminology/classification of the well known Generalized Single System Single Solve (GSSSS) unified theoretical and computational framework which encompasses LMS methods is overviewed next.

Recently, a novel concept and the underlying theory which describes new avenues to design time integration algorithms, firstly, for linear dynamic problems via a unified framework, and which also unifies all of the existing LMS algorithms, including providing new designs of time integration algorithms involving a single solve within a given time step, and with second-order time accuracy was introduced in a series of papers by Zhou et al. (Tamma et al. 2000, 2001; Zhou and Tamma 2004a,c, 2006); a notion called *algorithms by design*. Key algorithmic parameters termed as the algorithmic DNA (Discrete Numerically Assigned) markers uniquely define the underlying relations which must hold in order to characterize and differentiate one given algorithm with another (that is, no two time stepping algorithms can have the same DNA). This is essential for establishing the necessary metrics and measures and for assessing new developments in time integration and contributions in the future; and it also helps to place in the proper context existing algorithms and the corresponding designs. The same characteristics hold for both single-field and two-field forms of representation. Furthermore, as an illustration of the significance of the present efforts, consider those algorithms designed from the single-field of representation and the unified GSSSS framework. From the viewpoint of implementation, especially in large scale research and commercial software, all that is subsequently required in the GSSSS framework to switch from one time integration algorithm design to another algorithm design, are simply the values of the two principal roots and the spurious root at infinity (in essence, it simply requires three parameters which are the only variables). This is truly remarkable as it provides to the analyst, a general purpose simulation software module that can be readily implemented with a wide variety of algorithms and optimal designs as well at their fingertips. In this regard, under the umbrella of the family of the generalized single solve single step (GSSSS) time integrators, the unified framework enables one to uniquely design the underlying algorithms. Consequently, the underlying design enables one to not only recover most of the existing time integrators in the literature (numerically non-dissipative and numerically dissipative with controllable numerical dissipation) in the sense of LMS methods, but also additionally yields new time integration algorithmic designs with optimal algorithmic features. As highlighted in the previous chapter, the following is worth noting. *For the class of LMS methods in the single-field form of representation for second-order systems, the three basic building blocks which serve as the backbone of this class of LMS methods are the well known Newmark average acceleration method (Newmark 1959), the commonly employed numerically non-dissipative classical version of the Midpoint rule method or strictly we refer to this as the mid point rule with end-point acceleration (MPR-EPA), and a new version that is presented here and termed as the Midpoint Rule with mid-point acceleration (MPR-MPA); the well known Velocity based scheme (Tamma and Namburu 1990) in the two-field form is equivalent to this latter approach in the single-field form of representation (and, with the exception of the Newmark method, the latter two approaches are energy, linear and angular momentum conserving for linear dynamical situations). These three basic time integration algorithms are distinctly different from each other (see details in a subsequent section) and serve as the backbone for the class of LMS methods. Furthermore, it is important to note that emanating from these three basic building blocks as a starting point are the various other algorithms which can be designed involving features with controllable numerical dissipation such as those described in the previous chapter (and they can be readily recovered here via the present framework); and the framework additionally includes various other new time integration algorithm designs with optimal algorithmic attributes as well which are a natural outcome and an added bonus. The optimality is in the sense of the least amount of algorithmic numerical dissipation, dispersion and overshoot behavior for any given set of arbitrary initial conditions. Symplectic-momentum based and energy-momentum based algorithmic designs can be constructed as an integral part of the framework through the theoretical basis (see normalized time weighted residual approach) that is highlighted in this chapter; and such designs are in the sense of the single-field form of representation and involve two principal roots and a spurious root.* Similar characteristics also hold for the two-field form as also briefly highlighted (however, it does not involve the spurious root). Various other efforts related to time integration also appear in the literature (Aharoni and Bar-Yoseph 1992; Borri and Atluri 1988; Borri and Bottasso 1991; Borri et al. 1990, 1991; Simkins 1981).

The big picture regarding the class of LMS methods is as follows. Consider for example, the general representation of the single-field form involving a single system of equations and a single solve within each time step. In the sense of LMS methods in this form, there exist three roots participating in the algorithm, namely, the two principal roots $\rho_{1\infty}$, $\rho_{2\infty}$ and the spurious root $\rho_{3\infty}$. It has been shown (Zhou and Tamma 2004c, 2006) that the second-order time accurate, unconditionally stable LMS framework is basically comprised of two distinct algorithmic structures, termed as constrained U (displacement overshooting behavior aspects) and constrained V (velocity overshooting behavior aspects) family of algorithms for linear structural dynamic systems. This classification (characterization) is based upon the time integration algorithm overshoot behavior, and it is important to understand the essential details. Algorithm overshoot plays a critical role, especially, for any given set of arbitrary initial conditions in which some time integration algorithms perform better than others in the sense that certain overshoot characteristics may inherently cause increased errors and/or convergence problems such as with Newton-type iterations for nonlinear dynamics applications. *In this regard, one particular time integration algorithm and representation termed as the optimal U0-V0 (zero-order displacement and zero-order velocity overshoot) algorithm is the most optimal and is the recommended choice amongst the class of algorithms with controllable numerical dissipation. Alternatively, amongst the time integration algorithms that are numerically non-dissipative within the class of LMS methods, for any given set of arbitrary initial conditions, the so-called (MPR-MPA) is the most robust and is the preferred choice and is also highly recommended; it also inherits U0-V0 (zero-order displacement and zero-order velocity overshoot) characteristics which are ideal.* Most of the developments following past practices for both numerically non-dissipative and numerically dissipative algorithm designs with controllable numerical dissipation, pertain only to the constrained U-family of algorithms within the LMS class of methods for finite element computations. The underlying reason is that there did not exist any theory explaining such a classification; and only recently we have provided a new theory explaining the fundamental concepts underlying the design of algorithms encompassing the class of LMS methods in a general purpose manner. With these considerations in mind, the constrained V-family of algorithms are, hence, relatively new to the community at large. The various recent advances appear in the series of papers described earlier, and are also highlighted here from yet another new perspective; namely, via the Total Energy framework as a starting point. Also as reported in the literature (Zhou and Tamma 2006), only those with either zero-order displacement overshooting behavior in U (U0 family) or zero-order velocity overshooting behavior in V (V0 family) are competitive (computationally attractive with improved accuracy) amongst the class of LMS methods, and hence only these are highlighted and considered here. This is because for a class of nonlinear dynamics applications, the others pose quite severe and problematic issues as related to convergence of nonlinear iterations in the step by step time marching process which may be partly due to the overshooting behavior and the like, and are hence undesirable.

The U0 and V0 families of algorithms can be viewed as a mirror reflection of each other, wherein, the U0 families have no displacement overshoot, while the V0 families have no velocity overshoot. Some insight into the U0 family and the V0 family of algorithms discussed above follows next. The discussion firstly focuses upon linear dynamical systems. The U0 family of algorithms considered here contain all the algorithms with zero-order displacement overshooting behavior, while the V0 family of algorithms contain all the algorithms with zero-order velocity overshooting behavior. These two families of algorithms have been derived and described in detail in (Zhou and Tamma 2006). However, for the purpose of easy reference in this chapter, the U0 family and the V0 family are summarized next. For illustration purposes, the developments are described: 1) Starting from the traditional paradigm which is the semi-discretized equations of motion with given initial conditions, and 2) Thenceforth, for the remainder of the sections in this chapter, the traditional paradigm is altogether discarded, and an alternative perspective is presented which provides an improved physical insight directly starting from the discrete form of the Total Energy representation of the equation of motion. This is in the sense of the Total Energy framework with the descriptive scalar function, namely, the Total Energy.

14.2.1 GSSSS Framework Encompassing LMS Methods: Nonconservative Systems and Linear Dynamics Algorithms and Designs in Two-field/Single-field Form Via the Semi-Discretized Equations of Motion

We first purposely highlight the developments starting from the traditional practices which employ the standard form of the semi-discretized equations of motion and the time weighted residual approach. Both two-field form and single-field form are described. Thenceforth, in the remainder of the chapter we primarily focus attention upon the discrete Total Energy framework as a starting point as it is equivalent to the former. Furthermore, the latter scalar formalism additionally provides improved physical insight, computationally convenient and attractive features, and new and different perspectives.

Two-field Form Algorithm Designs Consider the semi-discretized equation of motion for a linear structural dynamic problem in the two-field form:

$$\left.\begin{array}{ll}\mathbf{M}\dot{\mathbf{v}}(t)+\mathbf{C}\mathbf{v}(t)+\mathbf{K}\mathbf{u}(t) & =\mathbf{f}(t)\\ \mathbf{v}(t)=\dot{\mathbf{u}}(t) & \end{array}\right\}\ \forall t\in\mathcal{T}=[t_0,T]$$
$$\text{with}\ \ \mathbf{u}_0=\mathbf{u}(t_0),\quad \mathbf{v}_0=\mathbf{v}(t_0) \tag{14.1}$$

where $\mathbf{M}$, $\mathbf{C}$ and $\mathbf{K}$ denote the mass matrix, the damping matrix and the stiffness matrix, respectively; $\mathbf{u}(t):\mathcal{T}\rightarrow\mathbb{R}^{n_{\text{dof}}}$ and $\dot{\mathbf{u}}(t)=d\mathbf{u}/dt:\mathcal{T}\rightarrow\mathbb{R}^{n_{\text{dof}}}$ denotes the nodal displacement vector and nodal velocity vector, respectively; and $\mathbf{f}(t):\mathcal{T}\rightarrow\mathbb{R}^{n_{\text{dof}}}$ denotes the discrete time-dependent external force. The second equation in Equation (14.1), i.e., $\mathbf{v}(t)=\dot{\mathbf{u}}(t):\mathcal{T}\rightarrow\mathbb{R}^{n_{\text{dof}}}$, is regarded as the kinematic constraint or the second-order condition. Representing Equation (14.1) as a time weighted residual representation with an arbitrary weighting time field, $\mathbf{W}(t)$, leads to

$$\int_{t_n}^{t_{n+1}}\mathbf{W}^T\left(\mathbf{M}\dot{\mathbf{v}}+\mathbf{C}\mathbf{v}+\mathbf{K}\mathbf{u}-\mathbf{f}\right)dt=\mathbf{0} \tag{14.2}$$

Assuming the weighting time field as a degenerated *scalar* polynomial function, $\mathbf{W}$, as

$$W=1+w_1\Gamma+w_2\Gamma^2 \tag{14.3}$$

where $\Gamma(t)=\frac{\tau(t)}{\Delta t}$ and $\tau(t)\in[0,\Delta t]$ ($\Delta t=t_{n+1}-t_n$ and $\tau=t-t_n$), the corresponding approximations dictated for the time dependent variables $\mathbf{u}$, $\dot{\mathbf{u}}$, and $\ddot{\mathbf{u}}$ by an asymptotic series expansion are given as

$$\mathbf{u} = \mathbf{u}_n + \Lambda_1 \mathbf{v}_n \tau + \Lambda_2 \frac{\mathbf{v}_{n+1} - \mathbf{v}_n}{\Delta t} \tau^2 \tag{14.4}$$

$$\mathbf{v} = \mathbf{v}_n + \Lambda_3 \frac{\mathbf{v}_{n+1} - \mathbf{v}_n}{\Delta t} \tau \tag{14.5}$$

$$\dot{\mathbf{v}} = \frac{\mathbf{v}_{n+1} - \mathbf{v}_n}{\Delta t} \tag{14.6}$$

respectively, and the linear approximation for the external time-dependent force vector by the Taylor series expansion (in general, this is a special case of an asymptotic series expansion and is sufficient for the cases of interest here; but the general case is a function evaluation at a particular time point and not necessarily a linear interpolation) is given as

$$\mathbf{f} = \mathbf{f}_n + \dot{\mathbf{f}}_n \tau = \mathbf{f}_n + \frac{\mathbf{f}_{n+1} - \mathbf{f}_n}{\Delta t} \tau \tag{14.7}$$

Note that the kinematic constraints at time t_n and t_{n+1} are assumed $\mathbf{u}_n = \mathbf{v}_n$ and $\mathbf{u}_{n+1} = \mathbf{v}_{n+1}$. Therefore, Equation (14.2) becomes

$$\int_{t_n}^{t_{n+1}} W \left(\mathbf{M}\dot{\mathbf{v}} + \mathbf{C}\mathbf{v} + \mathbf{K}\mathbf{u} - \mathbf{f} \right) dt = \mathbf{0} \tag{14.8}$$

and it leads to the following family of time integration schemes with the approximations as shown in Equations (14.3)–(14.7) via the classical time weighted residual methodology (see subsection 14.2.1):

$$\begin{aligned} &\left[\frac{\mathbf{M}}{\Delta t} + \mathbf{C} W_1 \Lambda_3 + \Delta t \mathbf{K} W_2 \Lambda_2 \right] (\mathbf{v}_{n+1} - \mathbf{v}_n) \\ &= -\mathbf{C}\mathbf{v}_n - \mathbf{K} \left[\mathbf{u}_n + \Delta t W_1 \Lambda_1 \mathbf{v}_n \right] + (1 - W_1)\mathbf{f}_n + W_1 \mathbf{f}_{n+1} \end{aligned} \tag{14.9}$$

with the associated updates as

$$\mathbf{u}_{n+1} = \mathbf{u}_n + \Delta t \lambda_1 \mathbf{v}_n + \Delta t \lambda_2 (\mathbf{v}_{n+1} - \mathbf{v}_n) \tag{14.10}$$

$$\mathbf{v}_{n+1} = \mathbf{v}_n + (\mathbf{v}_{n+1} - \mathbf{v}_n) \tag{14.11}$$

where we define

$$W_i := \frac{\int_{t_n}^{t_{n+1}} \left(\frac{\tau}{\Delta t} \right)^i W dt}{\int_{t_n}^{t_{n+1}} W dt} = \frac{\sum_{j=0}^{3} \frac{w_j}{1+i+j}}{\sum_{j=0}^{3} \frac{w_j}{1+j}} \tag{14.12}$$

Denoting $\Delta \mathbf{v} := \mathbf{v}_{n+1} - \mathbf{v}_n$, the v-form representation of the algorithms can be summarized as follows:

Algorithm 7

Given $\mathbf{u}_n$ and $\mathbf{v}_n$, find $\mathbf{u}_{n+1}$ and $\mathbf{v}_{n+1}$ from

$$\begin{aligned} &\left[\frac{\mathbf{M}}{\Delta t} + \mathbf{C} W_1 \Lambda_3 + \Delta t \mathbf{K} W_2 \Lambda_2 \right] \Delta \mathbf{v} \\ &= -\mathbf{C}\mathbf{v}_n - \mathbf{K} \left[\mathbf{u}_n + \Delta t W_1 \Lambda_1 \mathbf{v}_n \right] + (1 - W_1)\mathbf{f}_n + W_1 \mathbf{f}_{n+1} \end{aligned} \tag{14.13}$$

with the associated updates

$$\mathbf{u}_{n+1} = \mathbf{u}_n + \Delta t \lambda_1 \mathbf{v}_n + \Delta t \lambda_2 \Delta \mathbf{v} \tag{14.14}$$

$$\mathbf{v}_{n+1} = \mathbf{v}_n + \Delta \mathbf{v} \tag{14.15}$$

Remark 14.2.1

1. By enforcing that no bifurcation occurs for the principal roots at infinity, i.e., $\rho_{1\infty} = \rho_{2\infty}$, we can show

$$W_1\Lambda_1 = \frac{3 + \rho_{1\infty} + \rho_{2\infty} - \rho_{1\infty}\rho_{2\infty}}{2(1 + \rho_{1\infty})(1 + \rho_{2\infty})} \quad \text{and} \quad W_2\Lambda_2 = \frac{1}{(1 + \rho_{1\infty})(1 + \rho_{2\infty})} \tag{14.16}$$

   assuming $\Lambda_3 = 1$, $\lambda_1 = 1$ and $\lambda_2 = 1/2$. Notice there does not exist the spurious root in this two-field form framework.

2. The only member which is unconditionally stable and second-order time accurate in the framework of Algorithm 7 is obtained when we select

$$\begin{aligned} W_1\Lambda_1 &= \tfrac{1}{2}\,, \ \lambda_1 = 1 \\ W_2\Lambda_2 &= \tfrac{1}{4}\,, \ \lambda_2 = \tfrac{1}{2} \\ W_1\Lambda_3 &= \tfrac{1}{2} \end{aligned}$$

   which leads to the midpoint/trapezoidal rule (they are the same for the two-field form and linear dynamic systems unlike the single-field form); only upon selecting $\rho_{1\infty} = \rho_{2\infty} = 1$ in Equation (14.16) leads to an unconditionally stable, second-order time accurate algorithm.

3. The kinematic constraint $\mathbf{v} = \dot{\mathbf{u}}$ is discretized within the time step by Equation (14.14) which yields

$$\lambda_1\mathbf{v}_n + \lambda_2\Delta\mathbf{v} = \frac{\mathbf{u}_{n+1} - \mathbf{u}_n}{\Delta t} \tag{14.17}$$

Single-field Form Algorithm Designs Next, we consider the algorithm designs for the single-field form. Consider the semi-discretized system of equations of linear structural dynamic problems by space discretization of the single-field form as

$$\begin{gathered} \mathbf{M}\ddot{\mathbf{u}}(t) + \mathbf{C}\dot{\mathbf{u}}(t) + \mathbf{K}\mathbf{u}(t) = \mathbf{f}(t) \\ \mathbf{u}(0) = \mathbf{u}_0 \quad , \quad \dot{\mathbf{u}}(0) = \dot{\mathbf{u}}_0 \end{gathered} \tag{14.18}$$

In particular, the generalized framework of time integration operators for the single-field form of representation for the second-order time accurate GSSSS algorithms are obtained from assuming the weighting time field in Equation (14.2) as a degenerated scalar polynomial function as (see (Tamma et al. 2001) for other details),

$$W = 1 + w_1\Gamma + w_2\Gamma^2 + w_3\Gamma^3 \tag{14.19}$$

where $\Gamma = \frac{\tau}{\Delta t}$ and $\tau = t - t_n \in [0, \Delta t]$. The corresponding approximations dictated for the dependent variables $\mathbf{u}$, $\dot{\mathbf{u}}$, and $\ddot{\mathbf{u}}$ using an asymptotic series expansion is given as,

$$\mathbf{u} = \mathbf{u}_n + \Lambda_1\dot{\mathbf{u}}_n\tau + \Lambda_2\ddot{\mathbf{u}}_n\tau^2 + \Lambda_3\frac{\ddot{\mathbf{u}}_{n+1} - \ddot{\mathbf{u}}_n}{\Delta t}\tau^3 \tag{14.20}$$

$$\dot{\mathbf{u}} = \dot{\mathbf{u}}_n + \Lambda_4\ddot{\mathbf{u}}_n\tau + \Lambda_5\frac{\ddot{\mathbf{u}}_{n+1} - \ddot{\mathbf{u}}_n}{\Delta t}\tau^2 \tag{14.21}$$

$$\ddot{\mathbf{u}} = \ddot{\mathbf{u}}_n + \Lambda_6\frac{\ddot{\mathbf{u}}_{n+1} - \ddot{\mathbf{u}}_n}{\Delta t}\tau \tag{14.22}$$

and using a Taylor series expansion (in general, this is a special case of an asymptotic series expansion and is sufficient for the cases of interest here; but the general case is a function evaluation at a particular time point and not necessarily a linear interpolation) to approximate the load vector to first-order as,

$$\mathbf{f} = \mathbf{f}_n + \dot{\mathbf{f}}_n \tau = \mathbf{f}_n + \frac{\mathbf{f}_{n+1} - \mathbf{f}_n}{\Delta t}\tau \tag{14.23}$$

Substituting Equations (14.19)–(14.23) into the time weighted residual equation

$$\int_{t_n}^{t_{n+1}} W\left(\mathbf{M}\ddot{\mathbf{u}} + \mathbf{C}\dot{\mathbf{u}} + \mathbf{K}\mathbf{u} - \mathbf{f}\right) dt = \mathbf{0} \tag{14.24}$$

the resulting GSSSS family of algorithms are thus represented as

$$\begin{aligned} &\mathbf{M}\left[\ddot{\mathbf{u}}_n + \Lambda_6 W_1\left(\ddot{\mathbf{u}}_{n+1} - \ddot{\mathbf{u}}_n\right)\right] + \mathbf{C}\left[\dot{\mathbf{u}}_n + \Lambda_4 W_1 \ddot{\mathbf{u}}_n \Delta t + \Lambda_5 W_2\left(\ddot{\mathbf{u}}_{n+1} - \ddot{\mathbf{u}}_n\right)\Delta t\right] \\ &+ \mathbf{K}\left[\mathbf{u}_n + \Lambda_1 W_1 \dot{\mathbf{u}}_n \Delta t + \Lambda_2 W_2 \ddot{\mathbf{u}}_n \Delta t^2 + \Lambda_3 W_3\left(\ddot{\mathbf{u}}_{n+1} - \ddot{\mathbf{u}}_n\right)\Delta t^2\right] \\ &= (1 - W_1)\mathbf{f}_n + W_1 \mathbf{f}_{n+1} \end{aligned} \tag{14.25}$$

The associated design of the updates at the next time level is chosen as

$$\mathbf{u}_{n+1} = \mathbf{u}_n + \lambda_1 \dot{\mathbf{u}}_n \Delta t + \lambda_2 \ddot{\mathbf{u}}_n \Delta t^2 + \lambda_3\left(\ddot{\mathbf{u}}_{n+1} - \ddot{\mathbf{u}}_n\right)\Delta t^2 \tag{14.26}$$

$$\dot{\mathbf{u}}_{n+1} = \dot{\mathbf{u}}_n + \lambda_4 \ddot{\mathbf{u}}_n \Delta t + \lambda_5\left(\ddot{\mathbf{u}}_{n+1} - \ddot{\mathbf{u}}_n\right)\Delta t \tag{14.27}$$

Denoting $\Delta\ddot{\mathbf{u}} = \ddot{\mathbf{u}}_{n+1} - \ddot{\mathbf{u}}_n$ leads to the GSSSS a-form representations for a wide variety of single step time integration algorithms; they can be described via a unified framework encompassing the class of LMS methods (both the U family and V family of algorithms) as follows:

Algorithm 8

Given $\mathbf{u}_n$, $\dot{\mathbf{u}}_n$, and $\ddot{\mathbf{u}}_n$, find $\mathbf{u}_{n+1}$, $\dot{\mathbf{u}}_{n+1}$, and $\ddot{\mathbf{u}}_{n+1}$ from

$$\begin{aligned} &\left(\Lambda_6 W_1 \mathbf{M} + \Lambda_5 W_2 \mathbf{C}\Delta t + \Lambda_3 W_3 \mathbf{K}\Delta t^2\right)\Delta\ddot{\mathbf{u}} \\ &= -\mathbf{M}\ddot{\mathbf{u}}_n - \mathbf{C}\left(\dot{\mathbf{u}}_n + \Lambda_4 W_1 \ddot{\mathbf{u}}_n \Delta t\right) - \mathbf{K}\left(\mathbf{u}_n + \Lambda_1 W_1 \dot{\mathbf{u}}_n \Delta t + \Lambda_2 W_2 \ddot{\mathbf{u}}_n \Delta t^2\right) \\ &\quad + (1 - W_1)\mathbf{f}_n + W_1 \mathbf{f}_{n+1} \end{aligned} \tag{14.28}$$

with the associated updates

$$\mathbf{u}_{n+1} = \mathbf{u}_n + \lambda_1 \dot{\mathbf{u}}_n \Delta t + \lambda_2 \ddot{\mathbf{u}}_n \Delta t^2 + \lambda_3 \Delta\ddot{\mathbf{u}}\Delta t^2 \tag{14.29a}$$

$$\dot{\mathbf{u}}_{n+1} = \dot{\mathbf{u}}_n + \lambda_4 \ddot{\mathbf{u}}_n \Delta t + \lambda_5 \Delta\ddot{\mathbf{u}}\Delta t \tag{14.29b}$$

$$\ddot{\mathbf{u}}_{n+1} = \ddot{\mathbf{u}}_n + \Delta\ddot{\mathbf{u}} \tag{14.29c}$$

Likewise, the associated d-form and v-form representations can be readily constructed (Masuri et al. 2009a,b,c; Zhou and Tamma 2004c, 2006). As related to the above unified framework and its general structure, for clarity and conciseness, this section simply highlights the parent linear dynamic algorithms that pertain to the distinctly different U0 family and the V0 family of algorithms following our previous efforts in (Zhou and Tamma 2004c, 2006). As mentioned earlier, these parent linear dynamic algorithms can be readily used as the basis for subsequent extensions to nonlinear dynamic cases, provided careful attention is paid to the manner in which the proper extensions are addressed. How to appropriately provide such extensions is described subsequently with three different schools of thought that employ distinctly different theoretical basis for designing computational algorithms for time integration of nonlinear dynamic situations. The noteworthy point here is that the so-called *normalized time weighted residual approach* is the key to appropriately providing proper extensions to nonlinear dynamics situations. It respectively leads to symplectic-momentum based and energy-momentum based representations depending upon how the approximations for the various variables are designed as described subsequently. *The different schools of thought include (see brief highlights in the section on nonlinear dynamics applications): (i) the classical approach which fails to provide proper extensions to nonlinear dynamic situations, (ii) the displacement based approach which leads to that termed as symplectic-momentum based algorithm designs (described in detail subsequently), and (iii) the hybrid displacement and strain based approach (described in detail subsequently) which leads to that termed as energy-momentum based algorithm designs*. They respectively yield different implementation aspects associated with the nonlinear dynamic terms and different physical interpretation for computational analysis. The pros/cons are also briefly highlighted. Within the framework of the particular class of LMS methods involving a single system of equations and a single solve representation within each single time step, the GSSSS unified framework readily encompasses both numerically non-dissipative and numerically controllable dissipative methods. In this framework, by optimal, we mean that algorithm which has the least amount of numerical dissipation, period error and overshoot behavior. Obviously, all the algorithms have the same computational complexity (that is, in the linear dynamic sense only a single system of equations is solved within each time step) and the implicit algorithms are all unconditionally stable and second-order time accurate in all the variables in the sense of linear dynamic situations.

The subclass of the U0 family of algorithms within the unified GSSSS framework is given as follows:

Algorithm 9: ***U0 Family of Algorithms***

Given $\mathbf{u}_n$, $\dot{\mathbf{u}}_n$, and $\ddot{\mathbf{u}}_n$, find $\mathbf{u}_{n+1}$, $\dot{\mathbf{u}}_{n+1}$, and $\ddot{\mathbf{u}}_{n+1}$ from

$$\begin{aligned}&\left(\Lambda_6 W_1\mathbf{M} + \Lambda_5 W_2\mathbf{C}\Delta t + \Lambda_3 W_3\mathbf{K}\Delta t^2\right)\Delta\ddot{\mathbf{u}}\\ &= -\mathbf{M}\ddot{\mathbf{u}}_n - \mathbf{C}\left(\dot{\mathbf{u}}_n + \Lambda_4 W_1\ddot{\mathbf{u}}_n\Delta t\right)\\ &\quad - \mathbf{K}\left(\mathbf{u}_n + \Lambda_1 W_1\dot{\mathbf{u}}_n\Delta t + \Lambda_2 W_2\ddot{\mathbf{u}}_n\Delta t^2\right)\\ &\quad + (1 - W_1)\mathbf{f}_n + W_1\mathbf{f}_{n+1}\end{aligned} \tag{14.30a}$$

$$\mathbf{u}_{n+1} = \mathbf{u}_n + \lambda_1\dot{\mathbf{u}}_n\Delta t + \lambda_2\ddot{\mathbf{u}}_n\Delta t^2 + \lambda_3\Delta\ddot{\mathbf{u}}\Delta t^2 \tag{14.30b}$$

$$\dot{\mathbf{u}}_{n+1} = \dot{\mathbf{u}}_n + \lambda_4\ddot{\mathbf{u}}_n\Delta t + \lambda_5\Delta\ddot{\mathbf{u}}\Delta t \tag{14.30c}$$

$$\ddot{\mathbf{u}}_{n+1} = \ddot{\mathbf{u}}_n + \Delta\ddot{\mathbf{u}} \tag{14.30d}$$

where

$$W_1\Lambda_1 = \frac{1}{1+\rho_{3\infty}},\ \lambda_1 = 1$$

$$W_2\Lambda_2 = \frac{1}{2(1+\rho_{3\infty})},\ \lambda_2 = \frac{1}{2}$$

$$W_3\Lambda_3 = \frac{1}{(1+\rho_{1\infty})(1+\rho_{2\infty})(1+\rho_{3\infty})}, \lambda_3 = \frac{1}{(1+\rho_{1\infty})(1+\rho_{2\infty})}$$
$$W_1\Lambda_4 = \frac{1}{1+\rho_{3\infty}}, \lambda_4 = 1$$
$$W_2\Lambda_5 = \frac{3+\rho_{1\infty}+\rho_{2\infty}-\rho_{1\infty}\rho_{2\infty}}{2(1+\rho_{1\infty})(1+\rho_{2\infty})(1+\rho_{3\infty})}, \lambda_5 = \frac{3+\rho_{1\infty}+\rho_{2\infty}-\rho_{1\infty}\rho_{2\infty}}{2(1+\rho_{1\infty})(1+\rho_{2\infty})}$$
$$W_1\Lambda_6 = \frac{2+\rho_{1\infty}+\rho_{2\infty}+\rho_{3\infty}-\rho_{1\infty}\rho_{2\infty}\rho_{3\infty}}{(1+\rho_{1\infty})(1+\rho_{2\infty})(1+\rho_{3\infty})}$$

The weighting time field is suggested to be,

$$W = 1 - \frac{15(1-2\rho_{3\infty})}{1-4\rho_{3\infty}}\frac{\tau}{\Delta t} + \frac{15(3-4\rho_{3\infty})}{1-4\rho_{3\infty}}\left(\frac{\tau}{\Delta t}\right)^2 - \frac{35(1-\rho_{3\infty})}{1-4\rho_{3\infty}}\left(\frac{\tau}{\Delta t}\right)^3; \quad \tau \in [0, \Delta t]$$

and

$$W_i = \frac{\sum_{j=0}^{3}\frac{w_j}{1+i+j}}{\sum_{j=0}^{3}\frac{w_j}{1+j}}; \; i = 1, 2, 3$$

$\rho_{1\infty}$, $\rho_{2\infty}$, and $\rho_{3\infty}$ are the first or minimum principal root, the second or maximum principal root, and the spurious root at the high-frequency limit, respectively, and they satisfy the following relation:

$$0 \le \rho_{3\infty} \le \rho_{1\infty} \le \rho_{2\infty} \le 1$$

Additionally, the subclass related to the V0 family of algorithms within the unified GSSSS framework is given as follows.

Algorithm 10: ***V0 Family of Algorithms***

Given $\mathbf{u}_n$, $\dot{\mathbf{u}}_n$, and $\ddot{\mathbf{u}}_n$, find $\mathbf{u}_{n+1}$, $\dot{\mathbf{u}}_{n+1}$, and $\ddot{\mathbf{u}}_{n+1}$ from

$$\begin{aligned}&\left(\Lambda_6 W_1\mathbf{M} + \Lambda_5 W_2\mathbf{C}\Delta t + \Lambda_3 W_3\mathbf{K}\Delta t^2\right)\Delta\ddot{\mathbf{u}}\\ &= -\mathbf{M}\ddot{\mathbf{u}}_n - \mathbf{C}\left(\dot{\mathbf{u}}_n + \Lambda_4 W_1\ddot{\mathbf{u}}_n\Delta t\right)\\ &\quad - \mathbf{K}\left(\mathbf{u}_n + \Lambda_1 W_1\dot{\mathbf{u}}_n\Delta t + \Lambda_2 W_2\ddot{\mathbf{u}}_n\Delta t^2\right)\\ &\quad + (1-W_1)\mathbf{f}_n + W_1\mathbf{f}_{n+1}\end{aligned} \tag{14.31a}$$

$$\mathbf{u}_{n+1} = \mathbf{u}_n + \lambda_1\dot{\mathbf{u}}_n\Delta t + \lambda_2\ddot{\mathbf{u}}_n\Delta t^2 + \lambda_3\Delta\ddot{\mathbf{u}}\Delta t^2 \tag{14.31b}$$

$$\dot{\mathbf{u}}_{n+1} = \dot{\mathbf{u}}_n + \lambda_4\ddot{\mathbf{u}}_n\Delta t + \lambda_5\Delta\ddot{\mathbf{u}}\Delta t \tag{14.31c}$$

$$\ddot{\mathbf{u}}_{n+1} = \ddot{\mathbf{u}}_n + \Delta\ddot{\mathbf{u}} \tag{14.31d}$$

where

$$W_1\Lambda_1 = \frac{3+\rho_{1\infty}+\rho_{2\infty}-\rho_{1\infty}\rho_{2\infty}}{2(1+\rho_{1\infty})(1+\rho_{2\infty})},\ \lambda_1 = 1$$

$$W_2\Lambda_2 = \frac{1}{(1+\rho_{1\infty})(1+\rho_{2\infty})},\ \lambda_2 = \frac{1}{2}$$

$$W_3\Lambda_3 = \frac{1}{(1+\rho_{1\infty})(1+\rho_{2\infty})(1+\rho_{3\infty})},\ \lambda_3 = \frac{1}{2(1+\rho_{3\infty})}$$

$$W_1\Lambda_4 = \frac{3+\rho_{1\infty}+\rho_{2\infty}-\rho_{1\infty}\rho_{2\infty}}{2(1+\rho_{1\infty})(1+\rho_{2\infty})},\ \lambda_4 = 1$$

$$W_2\Lambda_5 = \frac{2}{(1+\rho_{1\infty})(1+\rho_{2\infty})(1+\rho_{3\infty})},\ \lambda_5 = \frac{1}{1+\rho_{3\infty}}$$

$$W_1\Lambda_6 = \frac{2+\rho_{1\infty}+\rho_{2\infty}+\rho_{3\infty}-\rho_{1\infty}\rho_{2\infty}\rho_{3\infty}}{(1+\rho_{1\infty})(1+\rho_{2\infty})(1+\rho_{3\infty})}$$

The weighting time field is suggested to be,

$$\begin{aligned} W =& 1 - \frac{30(3-4\rho_{1\infty}-4\rho_{2\infty}+6\rho_{1\infty}\rho_{2\infty})}{9-11\rho_{1\infty}-11\rho_{2\infty}+19\rho_{1\infty}\rho_{2\infty}}\frac{\tau}{\Delta t} \\ &+ \frac{15(25-37\rho_{1\infty}-37\rho_{2\infty}+53\rho_{1\infty}\rho_{2\infty})}{2(9-11\rho_{1\infty}-11\rho_{2\infty}+19\rho_{1\infty}\rho_{2\infty})}\left(\frac{\tau}{\Delta t}\right)^2 \\ &- \frac{35(3-5\rho_{1\infty}-5\rho_{2\infty}+7\rho_{1\infty}\rho_{2\infty})}{9-11\rho_{1\infty}-11\rho_{2\infty}+19\rho_{1\infty}\rho_{2\infty}}\left(\frac{\tau}{\Delta t}\right)^3; \\ &\tau \in [0, \Delta t] \end{aligned}$$

and

$$W_i = \frac{\sum_{j=0}^{3}\frac{w_j}{1+i+j}}{\sum_{j=0}^{3}\frac{w_j}{1+j}}\ ;\ i = 1, 2, 3$$

$\rho_{1\infty}$, $\rho_{2\infty}$, and $\rho_{3\infty}$ are the first or minimum principal root, the second or maximum principal root, and the spurious root at the high-frequency limit, respectively, and they satisfy the following relation:

$$0 \le \rho_{3\infty} \le \rho_{1\infty} \le \rho_{2\infty} \le 1$$

Remark 14.2.2

1. The various algorithms starting from the Houbolt, and thenceforth the Newmark method, the Classical Midpoint Rule and several others including controllable numerically dissipative algorithms that have been developed over the past fifty years or so including other new developments that are more promising are inherently contained in this unified GSSSS framework which encompasses the class of LMS methods involving a single system and single solve. The respective designs of the original methods of development as well as new and optimal designs of algorithms within

this class of LMS methods can be readily obtained as highlighted in Table 14.1, and involve the relations shown in Algorithms 9 and 10 by proper selection of the algorithmic parameters. Within this unified GSSSS framework, all that is necessary to switch from one algorithm design to another algorithm design, are simply three parameters, namely, the values of the two principal roots and the spurious root at infinity (see also Figures 14.1 and 14.2).

Table 14.1. Commonly known and optimal algorithms within the class of LMS methods

Algorithms Family($\rho_{1\infty}$,$\rho_{2\infty}$,$\rho_{3\infty}$)	Common name	Conditions
U0(1,1,0)	Newmark	-
U0/V0(1,1,1)	Classical Midpoint Rule [MPR-EPA]	-
U0($\rho_{1\infty}$,$\rho_{2\infty}$,$\rho_{3\infty}$)	Three-parameter optimal schemes (the generalized-α is identical)	$\rho_{1\infty} = \rho_{2\infty} = \rho_{3\infty} = \rho_\infty, 0 \le \rho_\infty \le 1$
U0($\rho_{1\infty}$,$\rho_{2\infty}$,0)	WBZ	$\rho_{1\infty} = \rho_{2\infty} = \rho_\infty, 0 \le \rho_\infty \le 1$
U0($\rho_{1\infty}$,$\rho_{1\infty}$,$\rho_{3\infty}$)	HHT-α	$\rho_{1\infty} = \rho_{2\infty}, \rho_{3\infty} = \frac{1-\rho_{1\infty}\rho_{2\infty}}{\rho_{1\infty}+\rho_{2\infty}+2\rho_{1\infty}\rho_{2\infty}}$
U0/V0($\rho_{1\infty}$,1,$\rho_{3\infty}$)	U0-V0 Optimal[a]	$\rho_{1\infty} = \rho_{3\infty} = \rho_\infty, 0 \le \rho_\infty \le 1$
V0(1,1,0)	New Version Midpoint Rule [MPR-MPA]	-

[a]The U0-V0 optimal algorithm which pertains to both U0 and V0 family of algorithms is in general, the optimal algorithm with controllable numerical dissipation within the class of LMS methods in the sense of possessing the least amount of numerical dissipation, dispersion, and overshoot behavior, and is highly recommended; alternatively, the [MPR-MPA] pertains only to the V0 family (the V0 family of algorithms are not shown explicitly in this table and are relatively new; and they have zero-order overshoot in velocity and can be viewed as simply mirror reflections of the U0 family of algorithms which, in contrast, have zero-order overshoot in the displacement) and is the recommended numerically non-dissipative optimal algorithm and not the classical [MPR-EPA].

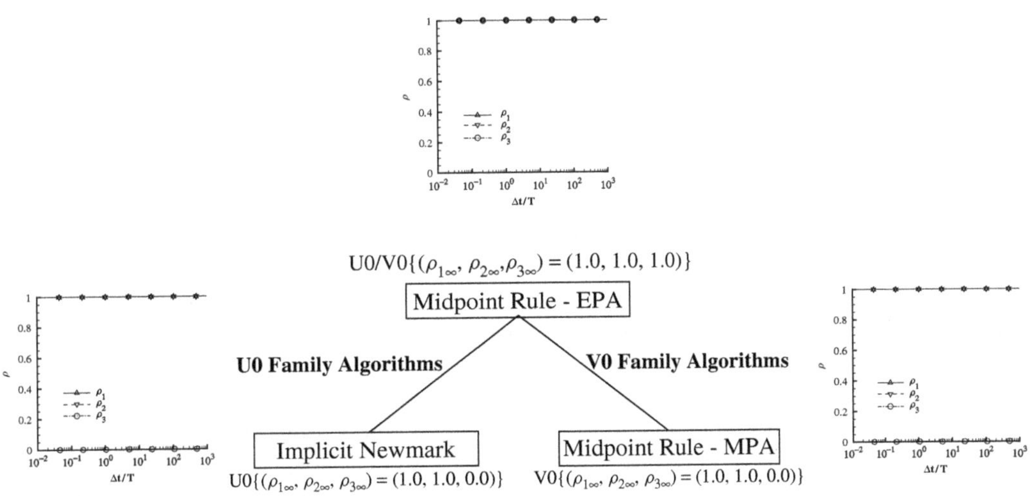

Figure 14.1. GSSSS Framework: The three basic building blocks or the backbone (Newmark, Classical Midpoint Rule [MPR-EPA] and the new version Midpoint Rule [MPR-MPA] are primitives) underlying LMS methods (numerically controllable dissipative schemes use these building blocks as a starting point)

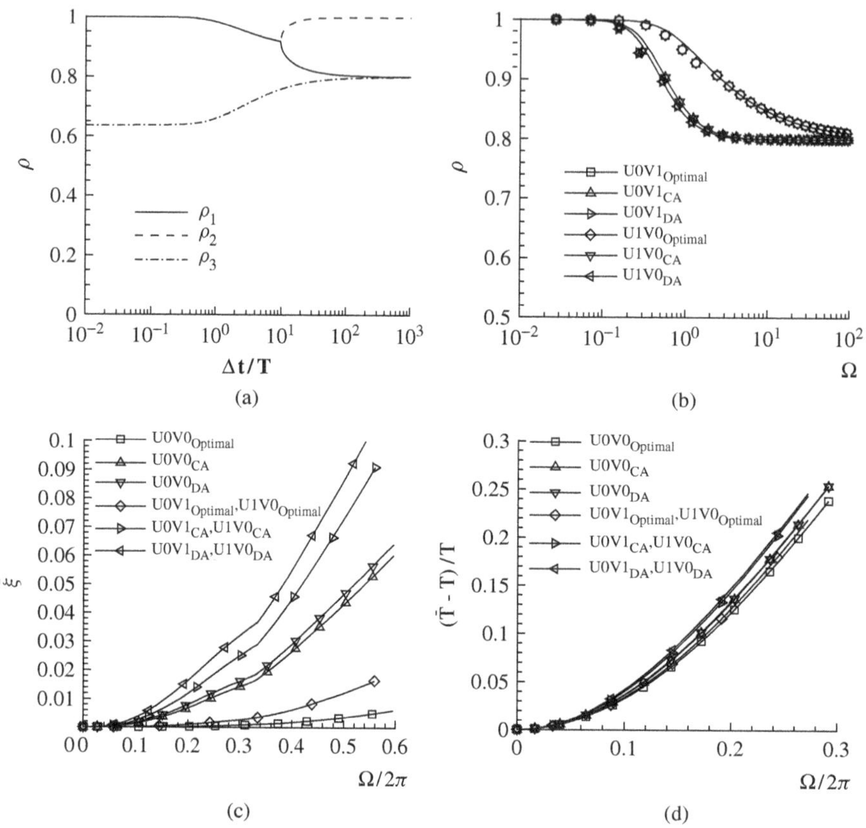

Figure 14.2. (a) The principal roots and the spurious root for U0-V0 optimal algorithm U0 – V0($\rho_{1\infty}$, $\rho_{2\infty}$, $\rho_{3\infty}$) = (0.8, 1.0, 0.8), (b) The principal roots for U0-V1 and U1-V0 algorithms ($\rho_\infty = 0.8$; and the spurious root is not shown), (c) Numerical dissipation for U0-V0, U0-V1 and U1-V0 algorithms, and (d) Numerical dispersion for U0-V0, U0-V1 and U1-V0 algorithms

14.2.2 GSSSS Framework Encompassing LMS Methods: Conservative Systems and Linear Dynamics Algorithms and Designs in Two-field/Single-field Form Via the Discrete Total Energy Framework

In contrast to the previous subsection, we provide a different viewpoint in this section via the Total Energy framework for conservative systems as an illustration. It accomplishes the same objectives, but from different perspectives and easy to grasp physical interpretation. *Henceforth, all relevant developments in time integration are described via the discrete form of the Total Energy representation of the equation of motion and framework as a starting point. It provides new and different perspectives and is very natural for time integration developments (as well as space as described in a previous chapter) in contrast to traditional practices and employing the semi-discretized equations of motion as a starting point as described in the previous chapter*. Earlier in this book, the various symmetries of the respective descriptive scalar functions were described starting from the continuous to the discrete case in space. The equivalences of the strong and weak forms amongst the different frameworks were established for holonomic-scleronomic systems and with particular imposed limitations. The developments to follow are confined to the case where the mass matrix is assumed constant. However, of the various frameworks (Newtonian, Lagrangian, Hamiltonian, and Total Energy) we described previously, in this section we particularly focus upon and advocate the Total Energy framework as a viable alternative for conducting the general designs of time integration methods for linear and nonlinear dynamics applications. Via the Total Energy framework, the relation between the energy conservation law and the governing equation in terms of the equation of motion that is continuous in space and time is established via Noether's first theorem (Noether 1918). The underlying reasons for the selection of this framework are that: i) The Total Energy framework provides a unified framework and is self-contained in that it can independently be employed for deriving the strong forms, and consequently the associated weak forms in space and/or time (and in a natural setting with improved and easy physical interpretation) without resorting to other avenues for arriving at the same such as employing the principle of virtual work in dynamics or Hamilton's principle respectively, ii) It readily inherits certain advantages such as ensuring the proof of satisfaction of conservation laws for the space discretized system, and consequently enables the design of the fully discretized system satisfying conservation laws quite naturally via a measurable built-in descriptive scalar function, namely, the Total Energy, and iii) Furthermore, it has inherent computationally attractive features and enables improved physical interpretation.

With attention focused upon the Total Energy framework and the single-field form and two-field form of representation, we herein first provide a somewhat different perspective in the sense that we altogether discard the developments starting from the continuous representations of the equations of motion to derive the semi-discretized equation. Instead, we describe yet another route directly starting from the fundamental supposition of energy conservation with regards to the space discretized system, and therein: 1) We first describe an approach to establish a direct relation between the energy conservation law and the semi-discretized equation of motion (in general and for complex structural systems, this is not trivial via traditional practices emanating from the Newtonian mechanics framework; but it is easier to establish such proofs of satisfaction via descriptive scalar functions such as the Total Energy and the like as highlighted previously and utilizing Noether's theorem (Noether 1918)); and subsequently it is also easier to ensure the satisfaction of the other symmetries, namely, such as those related to linear and angular momentum conservation for the space discretized system, and 2) Furthermore, we readily explain the general design of computational algorithms that are numerically non-dissipative and with controllable numerical dissipation first for linear dynamic systems; subsequently, via three different schools of thought we provide the theoretical foundations on how to appropriately ensure proper extensions for nonlinear dynamic applications. Both numerically non-dissipative schemes such as symplectic-momentum conserving and energy-momentum conserving algorithms, and algorithms by design with optimal controllable numerical dissipation features which are based upon the aforementioned symplectic-momentum and/or energy-momentum based representations, and in the sense of and within the scope of the particular class of LMS methods are described. They involve a single system of equations with a single solve within a single time step representation to foster the solution of practical and real-world problems. The focus is upon LMS methods simply for illustration as they are the most practical and popular in commercial codes. We demonstrate conserving (numerically non-dissipative) and controllable numerically dissipative algorithms by design for linear dynamic systems, which is a necessary first step and basis that is required for consequent extensions of the parent linear dynamics algorithms to general nonlinear dynamic applications as described in the sections to follow.

We now turn attention to discretization in time. Once the relations between the conservation laws and the semi-discretized equation of motion are established, one is then ready to design computational algorithms in general including those which satisfy all three conservation laws. For algorithmic designs in computational dynamics, the stability of algorithms only depends upon a conservative system without any external forces. This is an added motivation in the selection of the Total Energy framework in this chapter. In this regard, it is important to point out that the Total Energy does not include the potential energy from external forces, that is $E^h \in \mathbb{R}^+$, (previously, we did include these aspects in the definition of the total energy where $E \in \mathbb{R}$) to simply demonstrate the applicability of the space discretization approach to general situations encountered in finite element computations.

The design of time integration algorithms for conducting the time discretization process needs to ensure satisfaction of two properties: (i) convergence (in the sense of the Lax equivalence theorem), and (ii) discrete energy-momentum conservation. And, there exist two principal approaches for designing energy-momentum conserving algorithms: (a) First, starting with the established discrete Total Energy representation of the equation of motion that satisfies energy conservation and its relation to the semi-discretized equation of motion which likewise satisfies the conservation properties, we design a general structure and representation of all the possible convergent time discretized algorithms (attention is confined here for illustration to the class of LMS methods), and then we determine the particular types of algorithms that inherit features such as, for example, energy-momentum conservation within this designed family of algorithms, and (b) Secondly, starting with the Total Energy representation of the equation of motion which inherently guarantees the conservation properties, one can design those algorithms that also inherit these attributes, and then study the convergence aspects of the resulting algorithmic designs. In this section, we only focus upon the first approach to investigate all possible energy-momentum conserving algorithms within the particular class of linear multi-step (LMS) methods, for illustration of the basic ideas (the second approach is described elsewhere). Since a time integration algorithm must be convergent (satisfying the requirements of stability and consistency) in the first place, we first derive the general families of time integration algorithms which are convergent within this particular class of LMS methods since they are the most popular. Then, we identify the specific designs of algorithms within the general families of convergent algorithms which satisfy the energy, linear momentum, and angular momentum conservation laws. It should be noted that for the design of computational time integration algorithms encompassing the class of LMS methods that are described here, the generalized time weighted residual approach together with the Total Energy framework yields algorithm designs that are symplectic-momentum conserving and/or energy-momentum conserving which can only be guaranteed either through comparison and/or through subsequent proofs.

As discussed previously, the nodal velocity phase space, TQ, can be defined after discretizing the space. Since it is customary to define the total energy of a dynamical system in continuum mechanics as the summation of the kinetic energy and the internal potential energy in the absence of external forces, the space-discretized completely separable autonomous total energy, $\mathcal{E}^h(\mathbf{q}, \dot{\mathbf{q}}) : TQ \to \mathbb{R}$, may be written as

$$\mathcal{E}^h(\mathbf{q}, \dot{\mathbf{q}}) = \mathcal{K}^h(\dot{\mathbf{q}}) + \mathcal{U}^h_{\text{int}}(\mathbf{q}) \tag{14.32}$$

where the space-discrete kinetic energy, $\mathcal{K}^h(\dot{\mathbf{q}}) : T_q Q \to \mathbb{R}$, and the space-discrete internal potential energy, $\mathcal{U}^h_{\text{int}}(\mathbf{q}) : Q \to \mathbb{R}$, are given by

$$\mathcal{K}^h(\dot{\mathbf{q}}) = \frac{1}{2}\dot{\mathbf{q}} \cdot \mathbf{M}\dot{\mathbf{q}} > 0 \quad \text{and} \quad \mathcal{U}^h_{\text{int}}(\mathbf{q}) = \mathbf{q} \cdot \mathbf{f}_0^{\text{int}} \tag{14.33}$$

respectively. The **constant** mass matrix $\mathbf{M}$, i.e., $M_{AB} := \int_{\mathcal{B}} \rho_0 N_A(\mathbf{X}) N_B(\mathbf{X}) dV$, has the symmetry properties ($\mathbf{M} = \mathbf{M}^T$) and positive definiteness ($\boldsymbol{\xi} \cdot \mathbf{M}\boldsymbol{\xi} > 0$ for $\boldsymbol{\xi} \neq \mathbf{0}$). It is important to note that we assume the mass matrix is constant.

Conservation of Total Mechanical Energy and the Equation of Motion The discrete total energy conservation of a mechanical system implies that the discrete total energy of the dynamical system is time invariant. Therefore, we have the following theorem, which we simply highlight and proofs are presented elsewhere.

Theorem 3: *For an autonomous non-dissipative dynamic system with Holonomic-Scleronomic constraints only involving potential forces, if the discrete total energy described by the following*

$$\frac{d}{dt}\left(\frac{\partial \mathcal{E}^h(\mathbf{q}, \dot{\mathbf{q}})}{\partial \dot{\mathbf{q}}}\right) + \frac{\partial \mathcal{E}^h(\mathbf{q}, \dot{\mathbf{q}})}{\partial \mathbf{q}} = \mathbf{0} \quad \forall t \in \mathcal{T} \subset \mathbb{R} \tag{14.34}$$

holds, then the dynamic system conserves the system discrete Total Energy, $\frac{d\mathcal{E}^h}{dt} = 0$.

According to Theorem 3, we have the following corollary:

Corollary 1 *For an autonomous non-dissipative dynamics system involving only potential forces, the satisfaction of the above semi-discrete equation of motion implies that the dynamic system conserves system discrete total energy.*

14.2.3 Total Energy Framework and Semi-Discretized Equation of Motion for Conservative Dynamic Systems

In the first approach discussed earlier, Theorem 3 is utilized to establish the relation between the discrete total energy conservation and the underlying finite element semi-discretized equation of motion. In addition, the discrete total linear momentum and angular momentum conservations for the finite element semi-discretized equation of motion can also be subsequently established. Since the finite element semi-discretized equation possesses the properties of discrete total energy, linear and angular momentum conservation, we would expect that within all the possible convergent time integration algorithms in the LMS framework, there exist time integration algorithms that can preserve the corresponding discrete properties. Focusing on the linear dynamical situations, we first construct all the possible unconditionally stable and second-order accurate time integration algorithms with single-step-single-solve representation within the LMS framework and then determine which algorithms within this framework conserve the discrete properties. Since the governing partial differential equation and consequently the corresponding semi-discretized equation satisfy the conservation laws for the conservative dynamical systems, the next step is to describe the design of conserving time integrators for the fully discretized system. Recall that we define $\mathbf{q}$ as a set of *nodal displacement vectors*; namely, $\mathbf{u} \equiv \mathbf{q} = (\mathbf{q}^1, \mathbf{q}^2, \ldots, \mathbf{q}^{n_{\text{node}}}) \in Q \equiv \mathbb{R}^{n_{\text{dof}}}$. The discrete internal potential energy, $\mathcal{U}^h_{\text{int}}(\mathbf{q}) : Q \to \mathbb{R}$, for the linear dynamical systems may be defined as

$$\mathcal{U}^h_{\text{int}}(\mathbf{q}) = \frac{1}{2}\sum_{A=1}^{n_{\text{node}}} \mathbf{q}^A \cdot \sum_{B=1}^{n_{\text{node}}} k_{AB}\mathbf{q}^B = \frac{1}{2}\mathbf{q} \cdot \mathbf{K}\mathbf{q} \geq 0 \tag{14.35}$$

where $\mathbf{K}$ is the stiffness matrix with properties of symmetry ($\mathbf{K} = \mathbf{K}^T$) and semi-positive definiteness ($\boldsymbol{\xi} \cdot \mathbf{K}\boldsymbol{\xi} \geq 0$ for $\boldsymbol{\xi} \neq \mathbf{0}$). Therefore, the completely separable autonomous discrete total energy $\mathcal{E}^h(\mathbf{q}, \dot{\mathbf{q}}) : TQ \to \mathbb{R}$ is given by

$$\mathcal{E}^h(\mathbf{q}, \dot{\mathbf{q}}) = \frac{1}{2}\dot{\mathbf{q}} \cdot \mathbf{M}\dot{\mathbf{q}} + \frac{1}{2}\mathbf{q} \cdot \mathbf{K}\mathbf{q} > 0 \tag{14.36}$$

From Theorem 3 and Corollary 1, it can be shown that the autonomous total energy representation of the equation of motion is equivalent to the semi-discretized equation of motion, i.e.,

$$\frac{d}{dt}\left(\frac{\partial \mathcal{E}^h(\mathbf{q}, \dot{\mathbf{q}})}{\partial \dot{\mathbf{q}}}\right) + \frac{\partial \mathcal{E}^h(\mathbf{q}, \dot{\mathbf{q}})}{\partial \mathbf{q}} = \mathbf{0} \Leftrightarrow \mathbf{M}\ddot{\mathbf{q}} + \mathbf{K}\mathbf{q} = \mathbf{0} \quad \forall t \in \mathcal{T} \tag{14.37}$$

Therefore, from the previous discussions, the corresponding dynamical system should conserve the discrete total linear and angular momenta, and the total energy.

Two-field Form Algorithm Designs: Time Discretization and the Generalized Single-Step-Single-Solve Framework, and Linear Conservative Dynamics Systems The semi-discretized total energy representation of the equation of motion with given initial conditions for the linear dynamical systems,

$$\frac{d}{dt}\left(\frac{\partial \mathcal{E}^h(\mathbf{q}, \mathbf{v})}{\partial \mathbf{v}}\right) + \frac{\partial \mathcal{E}^h(\mathbf{q}, \mathbf{v})}{\partial \mathbf{q}} = \mathbf{0} \tag{14.38}$$

with the kinematic constraint $\mathbf{v} = \dot{\mathbf{q}}$ and given initial conditions, $\mathbf{q}(t_0) = \mathbf{q}_0$ and $\mathbf{v}(t_0) = \mathbf{v}_0$, leads to

$$\mathbf{M}\dot{\mathbf{v}} + \mathbf{K}\mathbf{q} = \mathbf{0} \tag{14.39}$$

Employing the classical time weighted residual methodology, we readily have

$$\int_{t_n}^{t_{n+1}} W \left[\frac{d}{dt} \left(\frac{\partial \mathcal{E}^h (\mathbf{q}, \mathbf{v})}{\partial \mathbf{v}} \right) + \frac{\partial \mathcal{E}^h (\mathbf{q}, \mathbf{v})}{\partial \mathbf{q}} \right] dt = \mathbf{0} \tag{14.40}$$

or

$$\int_{t_n}^{t_{n+1}} W \left[\mathbf{M}\dot{\mathbf{v}} + \mathbf{K}\mathbf{q} \right] dt = \mathbf{0} \tag{14.41}$$

with the appropriate approximations for the time dependent variables; see the derivation of Algorithm 7. In this linear dynamical conservative system in the two-field form, we readily obtain the following single step single solve framework of algorithms:

Algorithm 11

Given $\mathbf{q}_n$ and $\mathbf{v}_n$, find $\mathbf{q}_{n+1}$ and $\mathbf{v}_{n+1}$ from

$$\left[\frac{\mathbf{M}}{\Delta t} + \Delta t \mathbf{K} W_2 \Lambda_2 \right] \Delta \mathbf{v} = -\mathbf{K} \left[\mathbf{q}_n + \Delta t W_1 \Lambda_1 \mathbf{v}_n \right] \tag{14.42}$$

with the associated updates

$$\mathbf{q}_{n+1} = \mathbf{q}_n + \Delta t \lambda_1 \mathbf{v}_n + \Delta t \lambda_2 \Delta \mathbf{v} \tag{14.43}$$

$$\mathbf{v}_{n+1} = \mathbf{v}_n + \Delta \mathbf{v} \tag{14.44}$$

Remark 14.2.3

1. By enforcing that no bifurcation occurs for the principal roots at infinity, i.e., $\rho_{1\infty} = \rho_{2\infty}$, we can show

$$W_1 \Lambda_1 = \frac{3 + \rho_{1\infty} + \rho_{2\infty} - \rho_{1\infty}\rho_{2\infty}}{2(1 + \rho_{1\infty})(1 + \rho_{2\infty})} \quad \text{and} \quad W_2 \Lambda_2 = \frac{1}{(1 + \rho_{1\infty})(1 + \rho_{2\infty})} \tag{14.45}$$

   assuming $\Lambda_3 = 1$, $\lambda_1 = 1$ and $\lambda_2 = 1/2$.
2. The discrete total angular momentum and total energy within a time step are exactly conserved only if

$$W_1 \Lambda_1 = \tfrac{1}{2}\,,\ \lambda_1 = 1$$
$$W_2 \Lambda_2 = \tfrac{1}{4}\,,\ \lambda_2 = \tfrac{1}{2}$$
$$W_1 \Lambda_3 = \tfrac{1}{2}$$

The discrete total linear momentum within a time step is exactly conserved for any conditions of the parameters. Therefore, the only member which is the exact energy-momentum conserving time integration scheme in Algorithm 11 is the midpoint/trapezoidal rule. The proof is straightforward; hence, it is omitted.

Single-field Form Algorithm Designs: Time Discretization and the Generalized Single-Step-Single-Solve Framework, and Linear Conservative Dynamics Systems The semi-discretized total energy representation of the equation of motion with given initial conditions for the linear dynamical systems is given as

$$\frac{d}{dt}\left(\frac{\partial \mathcal{E}^h(\mathbf{q},\dot{\mathbf{q}})}{\partial \dot{\mathbf{q}}}\right)+\frac{\partial \mathcal{E}^h(\mathbf{q},\dot{\mathbf{q}})}{\partial \mathbf{q}}=\mathbf{0} \quad \text{or} \quad \mathbf{M}\ddot{\mathbf{q}}+\mathbf{K}\mathbf{q}=\mathbf{0} \;\; \forall t\in\mathcal{T}\subset\mathbb{R} \\ \mathbf{q}(t_0)=\mathbf{q}_0 \quad \text{and} \quad \dot{\mathbf{q}}(t_0)=\dot{\mathbf{q}}_0 \tag{14.46}$$

We adopt the approach in (Zhou and Tamma 2004c, 2006) as described previously for designing the general class of time integration operators with the single-step-single-solve representation, unconditional stability and second-order time accuracy. For a time interval $\mathcal{T}=[t_0,t_N]$ split into subintervals, i.e., $\mathcal{T}=[t_0,t_N]=\bigcup_{n=0}^{N-1}\left[t_n,t_{n+1}\right]$, the classical generalized time weighted residual representation with an arbitrary time weighting function, $W(t):\mathcal{T}\to\mathbb{R}$, leads to

$$\int_{t_n}^{t_{n+1}} W\left[\frac{d}{dt}\left(\frac{\partial \mathcal{E}^h(\mathbf{q},\dot{\mathbf{q}})}{\partial \dot{\mathbf{q}}}\right)+\frac{\partial \mathcal{E}^h(\mathbf{q},\dot{\mathbf{q}})}{\partial \mathbf{q}}\right]dt=\mathbf{0} \quad \text{or} \quad \int_{t_n}^{t_{n+1}} W\left[\mathbf{M}\ddot{\mathbf{q}}+\mathbf{K}\mathbf{q}\right]dt=\mathbf{0} \tag{14.47}$$

where the time weighting function is given by

$$W=w_0+w_1\Gamma+w_2\Gamma^2+w_3\Gamma^3 \quad \text{with} \quad \Gamma:=\frac{\tau}{\Delta t}=\frac{t-t_n}{t_{n+1}-t_n} \tag{14.48}$$

Next, we approximate the time dependent nodal variables $\mathbf{q}$, $\dot{\mathbf{q}}$, and $\ddot{\mathbf{q}}$ by means of an asymptotic series expansions in time as follows:

$$\mathbf{q}=\mathbf{q}_n+\Lambda_1\dot{\mathbf{q}}_n\tau+\Lambda_2\ddot{\mathbf{q}}_n\tau^2+\Lambda_3\frac{\ddot{\mathbf{q}}_{n+1}-\ddot{\mathbf{q}}_n}{\Delta t}\tau^3 \tag{14.49}$$

$$\dot{\mathbf{q}}=\dot{\mathbf{q}}_n+\Lambda_4\ddot{\mathbf{q}}_n\tau+\Lambda_5\frac{\ddot{\mathbf{q}}_{n+1}-\ddot{\mathbf{q}}_n}{\Delta t}\tau^2 \tag{14.50}$$

$$\ddot{\mathbf{q}}=\ddot{\mathbf{q}}_n+\Lambda_6\frac{\ddot{\mathbf{q}}_{n+1}-\ddot{\mathbf{q}}_n}{\Delta t}\tau \tag{14.51}$$

Note that $\ddot{\mathbf{q}}$ is approximated linearly. The time weighting function given by Equation (14.48) is a scalar approximation of the weighting time field, and we usually assume $w_0\equiv 1$ (generally, $w_0\neq 0$); see (Tamma et al. 2000). Following (Zhou and Tamma 2004c, 2006), we obtain the following representation of the generalized family of algorithms:

$$\mathbf{M}\tilde{\mathbf{a}}+\mathbf{K}\tilde{\mathbf{q}}=\mathbf{0} \tag{14.52}$$

where the algorithmic unknowns are given by

$$\tilde{\mathbf{a}}=\tilde{\ddot{\mathbf{q}}}=\ddot{\mathbf{q}}_n+W_1\Lambda_6(\ddot{\mathbf{q}}_{n+1}-\ddot{\mathbf{q}}_n)=:\ddot{\mathbf{q}}_{n+W_1\Lambda_6} \tag{14.53}$$

$$\tilde{\mathbf{q}}=\mathbf{q}_n+W_1\Lambda_1\dot{\mathbf{q}}_n\Delta t+W_2\Lambda_2\ddot{\mathbf{q}}_n\Delta t^2+W_3\Lambda_3(\ddot{\mathbf{q}}_{n+1}-\ddot{\mathbf{q}}_n)\Delta t^2 \tag{14.54}$$

and the updates are given by

$$\mathbf{q}_{n+1} = \mathbf{q}_n + \lambda_1 \dot{\mathbf{q}}_n \Delta t + \lambda_2 \ddot{\mathbf{q}}_n \Delta t^2 + \lambda_3 \Delta \mathbf{a} \Delta t^2 \tag{14.55a}$$

$$\dot{\mathbf{q}}_{n+1} = \dot{\mathbf{q}}_n + \lambda_4 \ddot{\mathbf{q}}_n \Delta t + \lambda_5 \Delta \mathbf{a} \Delta t \tag{14.55b}$$

$$\ddot{\mathbf{q}}_{n+1} = \ddot{\mathbf{q}}_n + \Delta \mathbf{a} \tag{14.55c}$$

where $\Delta \mathbf{a} := \ddot{\mathbf{q}}_{n+1} - \ddot{\mathbf{q}}_n$. As mentioned earlier, the general class of LMS methods can be characterized by two distinct families algorithmic structures, namely, constrained U (displacement overshooting aspects) and constrained V (velocity overshooting aspects).

The U0 family of algorithms and the V0 family of algorithms within the general representation in the framework of the linear multi-step methods for linear dynamical systems can be summarized as follows (see Table 14.2):

Algorithm 12: ***U0 Family of Algorithms for Conservative Linear Dynamical Systems***

Given $\mathbf{q}_n$, $\dot{\mathbf{q}}_n$, and $\ddot{\mathbf{q}}_n$, find $\mathbf{q}_{n+1}$, $\dot{\mathbf{q}}_{n+1}$, and $\ddot{\mathbf{q}}_{n+1}$ from

$$\left(W_1 \Lambda_6 \mathbf{M} + W_3 \Lambda_3 \mathbf{K} \Delta t^2\right) \Delta \mathbf{a} = -\mathbf{M} \ddot{\mathbf{q}}_n - \mathbf{K}\left(\mathbf{q}_n + W_1 \Lambda_1 \dot{\mathbf{q}}_n \Delta t + W_2 \Lambda_2 \ddot{\mathbf{q}}_n \Delta t^2\right) \tag{14.56a}$$

$$\mathbf{q}_{n+1} = \mathbf{q}_n + \lambda_1 \dot{\mathbf{q}}_n \Delta t + \lambda_2 \ddot{\mathbf{q}}_n \Delta t^2 + \lambda_3 \Delta \mathbf{a} \Delta t^2 \tag{14.56b}$$

$$\dot{\mathbf{q}}_{n+1} = \dot{\mathbf{q}}_n + \lambda_4 \ddot{\mathbf{q}}_n \Delta t + \lambda_5 \Delta \mathbf{a} \Delta t \tag{14.56c}$$

$$\ddot{\mathbf{q}}_{n+1} = \ddot{\mathbf{q}}_n + \Delta \mathbf{a} \tag{14.56d}$$

where the respective relations for W, Λ and λ were described previously.

Algorithm 13: ***V0 Family of Algorithms for Conservative Linear Dynamical Systems***

Given $\mathbf{q}_n$, $\dot{\mathbf{q}}_n$, and $\ddot{\mathbf{q}}_n$, find $\mathbf{q}_{n+1}$, $\dot{\mathbf{q}}_{n+1}$, and $\ddot{\mathbf{q}}_{n+1}$ from

$$\left(W_1 \Lambda_6 \mathbf{M} + W_3 \Lambda_3 \mathbf{K} \Delta t^2\right) \Delta \mathbf{a} = -\mathbf{M} \ddot{\mathbf{q}}_n - \mathbf{K}\left(\mathbf{q}_n + W_1 \Lambda_1 \dot{\mathbf{q}}_n \Delta t + W_2 \Lambda_2 \ddot{\mathbf{q}}_n \Delta t^2\right) \tag{14.57a}$$

$$\mathbf{q}_{n+1} = \mathbf{q}_n + \lambda_1 \dot{\mathbf{q}}_n \Delta t + \lambda_2 \ddot{\mathbf{q}}_n \Delta t^2 + \lambda_3 \Delta \mathbf{a} \Delta t^2 \tag{14.57b}$$

$$\dot{\mathbf{q}}_{n+1} = \dot{\mathbf{q}}_n + \lambda_4 \ddot{\mathbf{q}}_n \Delta t + \lambda_5 \Delta \mathbf{a} \Delta t \tag{14.57c}$$

$$\ddot{\mathbf{q}}_{n+1} = \ddot{\mathbf{q}}_n + \Delta \mathbf{a} \tag{14.57d}$$

where the respective relations for W, Λ and λ were described previously.

Discrete Total Linear Momentum Conservation The total linear momentum of the semi-discretized system is given as

$$\begin{aligned}
\mathbf{L}^h_{3\times 1} &= \int_{\mathcal{B}} \rho_0 \dot{\boldsymbol{\varphi}}^h dV = \sum_{B=1}^{n_{\text{node}}} \int_{\mathcal{B}} \rho_0 N_B \dot{\mathbf{q}}^B dV = \sum_{A=1}^{n_{\text{node}}} \sum_{B=1}^{n_{\text{node}}} \int_{\mathcal{B}} \rho_0 N_A N_B dV \dot{\mathbf{q}}^B \\
&= \sum_{A=1}^{n_{\text{node}}} \sum_{B=1}^{n_{\text{node}}} M_{AB} \dot{\mathbf{q}}^B
\end{aligned}$$

Therefore, the exact conservation of the discrete total linear momentum in $[t_n, t_{n+1}]$ in the sense of $\mathbf{L}^h_{n+1} = \mathbf{L}^h_n$, where $\mathbf{L}^h_{n+1} := \mathbf{L}^h(\dot{\mathbf{q}}_{n+1})$ and $\mathbf{L}^h_n := \mathbf{L}^h(\dot{\mathbf{q}}_n)$ can be written as

$$\frac{\mathbf{L}^h_{n+1} - \mathbf{L}^h_n}{\Delta t} = \sum_{A=1}^{n_{\text{node}}} \sum_{B=1}^{n_{\text{node}}} M_{AB} \frac{\dot{\mathbf{q}}^B_{n+1} - \dot{\mathbf{q}}^B_n}{\Delta t} = \mathbf{0} \tag{14.58}$$

To study the condition of the exact total linear momentum conservation within a time step for Algorithm 12/Algorithm 13, consider the following; that is, Algorithm 12/Algorithm 13 can be cast into

$$\begin{aligned} &\sum_{B=1}^{n_{\text{node}}} M_{AB} \tilde{\mathbf{a}}^B + \sum_{B=1}^{n_{\text{node}}} K_{AB} \tilde{\mathbf{q}}^B = \mathbf{0} \\ \Leftrightarrow &\sum_{B=1}^{n_{\text{node}}} M_{AB} \ddot{\mathbf{q}}^B(t_{n+W_1}) + \sum_{B=1}^{n_{\text{node}}} K_{AB} \mathbf{q}^B(t_{n+W_1}) = \mathbf{0} \end{aligned} \tag{14.59}$$

where $\ddot{\mathbf{q}}(t_{n+W_1}) = \tilde{\mathbf{a}} = \ddot{\mathbf{q}}_{n+W_1\Lambda_6}$ and $\mathbf{q}(t_{n+W_1}) = \tilde{\mathbf{q}} = \mathbf{q}_{n+W_3} + \Delta t(W_1 - W_3)\dot{\mathbf{q}}_n$; see Hoitink et al. (Hoitink et al. 2008) for discussions about the time level consistency of the discrete equation of motion. By taking the summation for $A = 1, 2, \ldots, n_{\text{node}}$ for Equation (14.59), we obtain

$$\sum_{A=1}^{n_{\text{node}}} \sum_{B=1}^{n_{\text{node}}} M_{AB} \tilde{\mathbf{a}}^B + \sum_{A=1}^{n_{\text{node}}} \sum_{B=1}^{n_{\text{node}}} K_{AB} \tilde{\mathbf{q}}^B = \mathbf{0} \tag{14.60}$$

In the above, note that the total internal force vanishes, i.e.,

$$\sum_{A=1}^{n_{\text{node}}} \sum_{B=1}^{n_{\text{node}}} K_{AB} \tilde{\mathbf{q}}^B = \mathbf{0} \tag{14.61}$$

Therefore, Equation (14.60) yields

$$\begin{aligned} \mathbf{0} &= \sum_{A=1}^{n_{\text{node}}} \sum_{B=1}^{n_{\text{node}}} M_{AB} \tilde{\mathbf{a}}^B = \sum_{A=1}^{n_{\text{node}}} \sum_{B=1}^{n_{\text{node}}} M_{AB} \ddot{\mathbf{q}}^B(t_{n+W_1}) \\ &= \sum_{A=1}^{n_{\text{node}}} \sum_{B=1}^{n_{\text{node}}} M_{AB} \frac{\dot{\mathbf{q}}^B_{n+1} - \dot{\mathbf{q}}^B_n}{\Delta t} \end{aligned} \tag{14.62}$$

where we have used the relation $\dot{\mathbf{q}}_{n+1} = \dot{\mathbf{q}}_n + \Delta t \ddot{\mathbf{q}}(t_{n+W_1})$. Hence, in view of Equations (14.58) and (14.62), we conclude that the discrete total linear momentum is exactly conserved for any $\{\rho_{1\infty}, \rho_{2\infty}, \rho_{3\infty}\}$; that is,

$$\mathbf{L}^h_{n+1} = \mathbf{L}^h_n \quad \text{for any} \quad \{\rho_{1\infty}, \rho_{2\infty}, \rho_{3\infty}\} \tag{14.63}$$

for Algorithm 12/Algorithm 13 for the Neumann problem. Note the relation $0 \le \rho_{3\infty} \le \rho_{1\infty} \le \rho_{2\infty} \le 1$.

Remark 14.2.4

1. Algorithm 12 and Algorithm 13 conserve the discrete total linear momentum within a time step regardless of the choice of $\rho_{1\infty}$, $\rho_{2\infty}$, and $\rho_{3\infty}$.
2. The numerically non-dissipative / dissipative algorithms within Algorithm 12 and Algorithm 13 do not affect the conservation of the discrete total linear momentum within a time step.

Discrete Total Angular Momentum Conservation The total angular momentum of the semi-discretized system is given as

$$\begin{aligned}
\mathbf{J}^h_{3\times 1} &= \int_{\mathcal{B}} \boldsymbol{\varphi}^h \times \rho_0 \dot{\boldsymbol{\varphi}}^h dV = \int_{\mathcal{B}} \left[\mathbf{X} + \sum_{A=1}^{n_{\text{node}}} N_A \mathbf{q}^A \right] \times \rho_0 \sum_{B=1}^{n_{\text{node}}} N_B \dot{\mathbf{q}}^B dV \\
&= \sum_{A=1}^{n_{\text{node}}} \sum_{B=1}^{n_{\text{node}}} \int_{\mathcal{B}} \mathbf{X} \times \rho_0 N_A N_B dV \dot{\mathbf{q}}^B + \sum_{A=1}^{n_{\text{node}}} \mathbf{q}^A \times \sum_{B=1}^{n_{\text{node}}} \int_{\mathcal{B}} \rho_0 N_A N_B dV \dot{\mathbf{q}}^B \\
&= \mathbf{X} \times \mathbf{L}^h + \sum_{A=1}^{n_{\text{node}}} \mathbf{q}^A \times \sum_{B=1}^{n_{\text{node}}} M_{AB} \dot{\mathbf{q}}^B
\end{aligned} \tag{14.64}$$

The discrete total angular momentum conservation in $[t_n, t_{n+1}]$ is defined by $\mathbf{J}^h_{n+1} = \mathbf{J}^h_n$ where $\mathbf{J}^h_{n+1} := \mathbf{J}^h(\mathbf{q}_{n+1}, \dot{\mathbf{q}}_{n+1})$ and $\mathbf{J}^h_n := \mathbf{J}^h(\mathbf{q}_n, \dot{\mathbf{q}}_n)$. In view of Equation (14.64) and the discrete total linear momentum conservation as discussed previously, the discrete total angular momentum difference in $[t_n, t_{n+1}]$ yields

$$\begin{aligned}
\mathbf{J}^h_{n+1} - \mathbf{J}^h_n &= \sum_{A=1}^{n_{\text{node}}} \mathbf{q}^A_{n+1} \times \sum_{B=1}^{n_{\text{node}}} M_{AB} \dot{\mathbf{q}}^B_{n+1} - \sum_{A=1}^{n_{\text{node}}} \mathbf{q}^A_n \times \sum_{B=1}^{n_{\text{node}}} M_{AB} \dot{\mathbf{q}}^B_n \\
&= \sum_{A=1}^{n_{\text{node}}} \mathbf{q}^A_{n+1/2} \times \sum_{B=1}^{n_{\text{node}}} M_{AB} (\dot{\mathbf{q}}^B_{n+1} - \dot{\mathbf{q}}^B_n) \\
&\quad + \sum_{A=1}^{n_{\text{node}}} (\dot{\mathbf{q}}^A_{n+1} - \dot{\mathbf{q}}^A_n) \times \sum_{B=1}^{n_{\text{node}}} M_{AB} \mathbf{q}^B_{n+1/2}
\end{aligned} \tag{14.65}$$

for any $\{\rho_{1\infty}, \rho_{2\infty}, \rho_{3\infty}\}$. Employing Algorithm 12 and Algorithm 13, we obtain the following identity after some cumbersome manipulations:

$$\begin{aligned}
&\mathbf{J}^h_{n+1} - \mathbf{J}^h_n = \\
&\frac{\Delta t^2}{2} (W_2 \Lambda_2 - W_3 \Lambda_3 + W_1 \Lambda_1 (W_2 \Lambda_5 - W_1 \Lambda_4)) \\
&\sum_{A=1}^{n_{\text{node}}} \left(\ddot{\mathbf{q}}^A_n \times \sum_{B=1}^{n_{\text{node}}} M_{AB} \dot{\mathbf{q}}^B_n - \ddot{\mathbf{q}}^A_{n+1} \times \sum_{B=1}^{n_{\text{node}}} M_{AB} \dot{\mathbf{q}}^B_{n+1} \right) \\
&+ \frac{W_3 \Lambda_3 \Delta t^2}{2} \sum_{A=1}^{n_{\text{node}}} \left(\ddot{\mathbf{q}}^A_{n+W_1} \times \sum_{B=1}^{n_{\text{node}}} M_{AB} \dot{\mathbf{q}}^B_{n+W_1} - \ddot{\mathbf{q}}^A_{n+1-W_1} \times \sum_{B=1}^{n_{\text{node}}} M_{AB} \dot{\mathbf{q}}^B_{n+1-W_1} \right)
\end{aligned} \tag{14.66}$$

In view of Equation (14.66), for the satisfaction of the discrete angular momentum conservation in $[t_n, t_{n+1}]$ for the Neumann problem, we conclude

$$\mathbf{J}^h_{n+1} = \mathbf{J}^h_n \;\; if \;\; \rho_{1\infty} = \rho_{2\infty} = \rho_{3\infty} = 1 \tag{14.67}$$

for Algorithm 12, and

$$\mathbf{J}^h_{n+1} = \mathbf{J}^h_n \;\; if \;\; \rho_{2\infty} = \rho_{1\infty} = 1 \quad \text{and} \quad \rho_{3\infty} \in [0, 1] \tag{14.68}$$

for Algorithm 13.

Remark 14.2.5

1. In Algorithm 12, there exists only one algorithm that conserves the discrete total angular momentum within a time step, and it is the Classical Midpoint Rule with end-point acceleration [MPR-EPA] which corresponds to the spectral condition $\rho_{1\infty} = \rho_{2\infty} = \rho_{3\infty} = 1$.
2. In Algorithm 13, there exists a family of algorithms ($\rho_{1\infty} = \rho_{2\infty} = 1$ and $\rho_{3\infty} \in [0, 1]$) that conserve the discrete total angular momentum within a time step. The case of $\rho_{1\infty} = \rho_{2\infty} = \rho_{3\infty} = 1$ corresponds to the Classical Midpoint Rule method with end point acceleration [MPR-EPA], and the case of $\rho_{1\infty} = \rho_{2\infty} = 1$ and $\rho_{3\infty} = 0$ corresponds to the New Version of the Midpoint Rule with mid-point acceleration [MPR-MPA]; the well known Velocity based scheme (Tamma and Namburu 1990) in the two-field form is equivalent to this last approach in the single-field form of representation (with the exception of the Newmark method, the latter two approaches are energy, linear and angular momentum conserving for linear dynamical situations).

Discrete Total Energy Conservation The autonomous total energy of the semi-discrete system for the Neumann problem in a *linear* dynamical system is defined as

$$\mathcal{E}^h(\mathbf{q}, \dot{\mathbf{q}}) = \mathcal{K}^h(\dot{\mathbf{q}}) + \mathcal{U}^h_{\text{int}}(\mathbf{q}) = \frac{1}{2}\dot{\mathbf{q}} \cdot \mathbf{M}\dot{\mathbf{q}} + \frac{1}{2}\mathbf{q} \cdot \mathbf{K}\mathbf{q} > 0 \tag{14.69}$$

where the discrete internal potential energy $\mathcal{U}^h_{\text{int}}(\mathbf{q}) = \frac{1}{2}\mathbf{q} \cdot \mathbf{K}\mathbf{q}$ is often called the discrete strain energy. As a necessary condition for an algorithm to be consistent, the modal decomposition method should be applicable to the algorithm for the linear case (Zhou and Tamma 2006). This necessary condition implies that the energy of the system does not transfer between different modes. As a consequence, for a dynamic system if the energy of each mode is conserved, the total energy of the system is conserved. However, for Algorithm 12 and Algorithm 13, the modal decomposition method is indeed applicable. Therefore, we only need to study the energy property of a particular mode. And the energy of given mode is defined as,

$$\begin{aligned}\mathcal{E}^h &= \tfrac{1}{2}m\dot{q}^2 + \tfrac{1}{2}kq^2 = \tfrac{1}{2}\begin{pmatrix} q & \dot{q}\end{pmatrix}\begin{bmatrix} k & 0 \\ 0 & m\end{bmatrix}\begin{pmatrix} q \\ \dot{q}\end{pmatrix} \\ &= \frac{1}{2}\begin{pmatrix} q & \dot{q} & \ddot{q}\end{pmatrix}\begin{bmatrix} k & 0 & 0 \\ 0 & m & 0 \\ 0 & 0 & 0\end{bmatrix}\begin{pmatrix} q \\ \dot{q} \\ \ddot{q}\end{pmatrix} = \frac{1}{2}\mathbf{d} \cdot \begin{bmatrix} k & 0 & 0 \\ 0 & m & 0 \\ 0 & 0 & 0\end{bmatrix}\mathbf{d}\end{aligned} \tag{14.70}$$

where we have defined $\mathbf{d} := (q, \dot{q}, \ddot{q})^T$. It is trivial to show that the nodal displacement and velocity for energy conserving dynamic systems are bounded. Therefore, we have,

$$\mid q \mid_{\max} = \sqrt{\frac{\mathcal{E}^h}{k}} \quad \text{and} \quad \mid \dot{q} \mid_{\max} = \sqrt{\frac{\mathcal{E}^h}{m}} \tag{14.71}$$

However, when the dynamical problem involves both the initial nodal displacement and initial velocity conditions, an algorithm exhibiting overshooting behavior will cause the displacement or the velocity to be unbounded. For example, when selecting $\rho_{2\infty} \neq 1$ in Algorithm 12, one has, $\dot{q}_1 = \infty$ while $\Delta t \to \infty$; when selecting $\rho_{2\infty} \neq 1$ in Algorithm 13, one has, $q_1 = \infty$ while $\Delta t \to \infty$ (Zhou and Tamma 2004c). Therefore, the energy of the dynamic system will be unbounded, such that

$$\lim_{\Delta t \to \infty} \mathcal{E}_1 = \infty \tag{14.72}$$

Therefore, a necessary condition for the total energy of the system to be bounded for Algorithm 12 and Algorithm 13 is that $\rho_{2\infty} = 1$.

Following (Zhou and Tamma 2004c), and selecting $\rho_{2\infty} \neq 1$ for Algorithm 12 and Algorithm 13, we have

$$\mathbf{X}^{-1}\mathbf{d}_{n+1} = \mathbf{\Phi}\mathbf{X}^{-1}\mathbf{d}_n \tag{14.73}$$

where $\mathbf{X}$ is the unified right eigenvector matrix, $\mathbf{\Phi}$ is the eigenvalue matrix given by

$$\mathbf{\Phi} = \begin{bmatrix} \phi_1 & 0 & 0 \\ 0 & \phi_2 & 0 \\ 0 & 0 & \phi_3 \end{bmatrix} \tag{14.74}$$

where $|\phi_1| \leq 1$, $|\phi_2| \leq 1$, and $|\phi_3| \leq 1$. Therefore, we have,

$$\mathcal{E}^h_{n+1} = \frac{1}{2}\mathbf{d}_{n+1} \cdot \begin{bmatrix} k & 0 & 0 \\ 0 & m & 0 \\ 0 & 0 & 0 \end{bmatrix} \mathbf{d}_{n+1} = \frac{1}{2}\mathbf{d}_n \cdot \begin{bmatrix} k\phi_1^2 & 0 & 0 \\ 0 & m\phi_2^2 & 0 \\ 0 & 0 & 0 \end{bmatrix} \mathbf{d}_n \tag{14.75}$$

Hence, for the satisfaction of the exact discrete total energy conservation within a time step, we must have $\rho_{2\infty} = \rho_{1\infty} = 1$; that is, we conclude

$$\Delta\mathcal{E}^h = \Delta\mathcal{K}^h + \Delta\mathcal{U}^h_{\text{int}} = 0 \;\; if \;\; \rho_{2\infty} = \rho_{1\infty} = 1 \quad \text{and} \quad \rho_{3\infty} \in [0, 1] \tag{14.76}$$

for Algorithm 12 and Algorithm 13.

Remark 14.2.6

1. When selecting $\rho_{1\infty} = \rho_{2\infty} = 1$ and $\rho_{3\infty} \in [0, 1]$ in Algorithm 12 and Algorithm 13, the corresponding time integration schemes conserve the discrete total energy within a time step. The summary of the spectral conditions for the discrete momenta and total energy conservation is given in Table 14.2.
2. When the spurious root is not a function of time step (spurious root is constant over the spectrum), the corresponding algorithms in Algorithm 12 and Algorithm 13 conserve the discrete total energy energy within a time step.
3. The well-known implicit Newmark algorithm conserves the discrete total energy within a time step for a linear dynamical system (angular momentum is not conserved), but does not conserve energy for a nonlinear dynamical system.

Illustrative Example

Example 14.1 Linear Mass-Spring System

Consider a simple linear mass-spring system problem shown in Figure 14.3. The point masses are selected as $m_1 = 1$, $m_2 = 1$, $m_3 = 1$, and $m_4 = 1$. The four springs are linear springs with an unstressed length of 1 and stiffness of 1000. The dynamic system is subjected to an initial velocity of {0,0,2} at one of the masses. The time step is selected as 0.01 second. The results of the L_2 norm of the linear momentum, the L_2 norm of the angular momentum, and the total energy for selected numerically non-dissipative algorithms within the U0 and V0 algorithms are shown in Figure 14.4. These results also confirm the conservative properties of the respective algorithms within the LMS framework with the single-field form of representation. Also note that $\rho_{3\infty}$ does not play a role for the V0-family of algorithms as long as $\rho_{1\infty} = \rho_{2\infty} = 1$ in order to achieve conserving attributes.

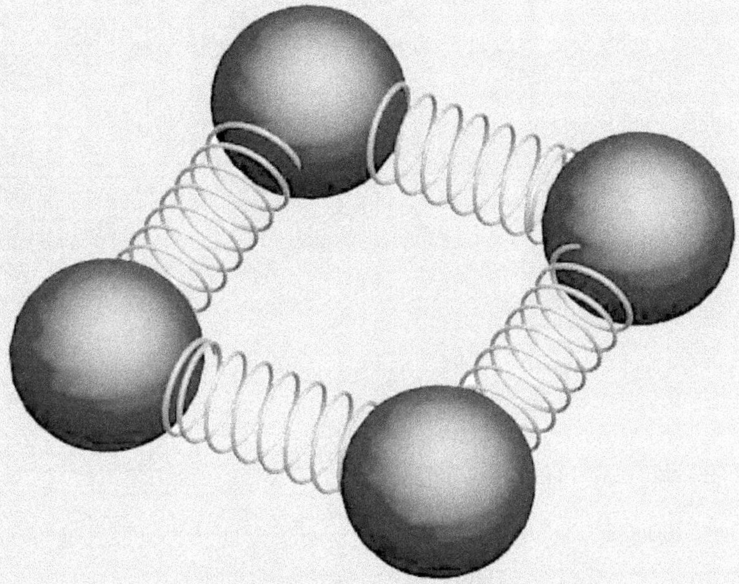

Figure 14.3. The configuration of a mass-spring dynamic system

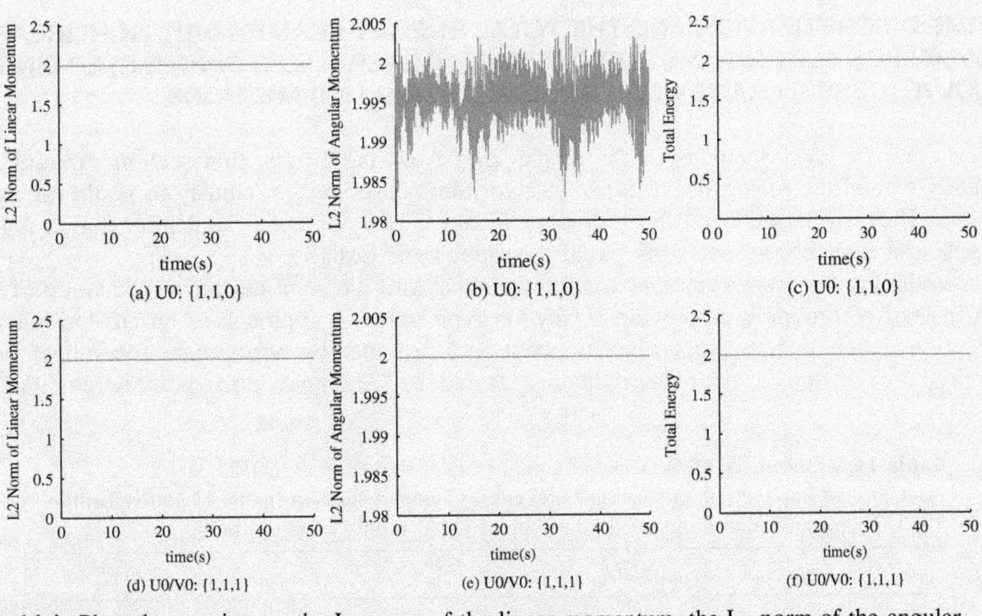

Figure 14.4. Plotted row wise are the L_2 norm of the linear momentum, the L_2 norm of the angular momentum, and the total energy of selected U0 and V0 Algorithms: Newmark average acceleration method (U0: $\{\rho_{1\infty}, \rho_{2\infty}, \rho_{3\infty}\} = \{1,1,0\}$), Classical Version of Midpoint Rule method with end-point acceleration [MPR-EPA] (U0/V0: $\{\rho_{1\infty}, \rho_{2\infty}, \rho_{3\infty}\} = \{1, 1, 1\}$), arbitrarily selected V0 method (V0: $\{\rho_{1\infty}, \rho_{2\infty}, \rho_{3\infty}\} = \{1, 1, 0.21\}$), and the New Version of the Midpoint Rule method with mid-point acceleration [MPR-MPA](V0: $\{\rho_{1\infty}, \rho_{2\infty}, \rho_{3\infty}\} = \{1, 1, 0\}$)

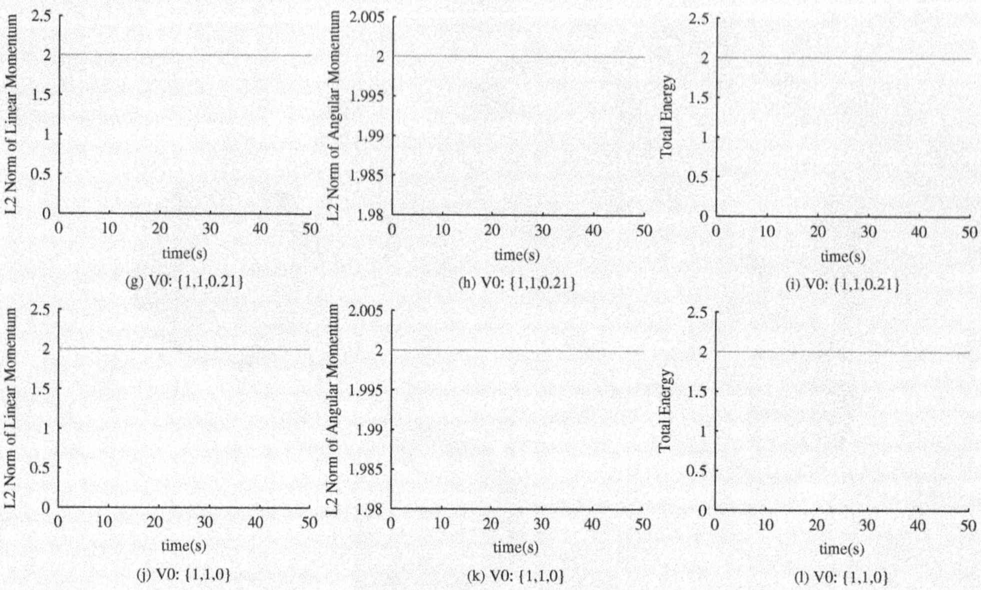

Figure 14.4. *(continued)*

14.3 TIME DISCRETIZATION AND THE TOTAL ENERGY FRAMEWORK: NONLINEAR DYNAMICS ALGORITHMS AND DESIGNS - GENERALIZED SINGLE STEP SINGLE SOLVE [GSSSS] FRAMEWORK ENCOMPASSING LMS METHODS

Within the class of LMS methods in the single- and two-field forms, this section explains how to appropriately extend the parent linear dynamic algorithms described previously to nonlinear dynamics applications. They are based upon the following features: i) Symplectic-momentum based algorithms and designs, and ii) Energy-momentum based algorithms and designs.

After introducing the approximations for the variables into the semi-discretized equation of motion, and then instead of routinely employing a Newton type iterative approach in an ad hoc manner, the fundamental question is how does one go about and provide the appropriate theoretical basis for enabling proper extensions to nonlinear dynamical situations? The noteworthy point here is that a novel

Table 14.2. Summary of the spectral conditions for the discrete total linear momentum/total angular momentum/total energy conservation in $[t_n, t_{n+1}]$ for Algorithm 12 (U0 family of algorithms) and Algorithm 13 (V0 family of algorithms)

	$\mathbf{L}^h_{n+1} = \mathbf{L}^h_n$	$\mathbf{J}^h_{n+1} = \mathbf{J}^h_n$	$\mathcal{E}^h_{n+1} = \mathcal{E}^h_n$
Algorithm 12	Any $\{\rho_{1\infty}, \rho_{2\infty}, \rho_{3\infty}\}$	$\rho_{1\infty} = \rho_{2\infty} = \rho_{3\infty} = 1$	$\rho_{1\infty} = \rho_{2\infty} = 1$, $\rho_{3\infty} \in [0, 1]$
Algorithm 13	Any $\{\rho_{1\infty}, \rho_{2\infty}, \rho_{3\infty}\}$	$\rho_{1\infty} = \rho_{2\infty} = 1$, $\rho_{3\infty} \in [0, 1]$	$\rho_{1\infty} = \rho_{2\infty} = 1$, $\rho_{3\infty} \in [0, 1]$

normalized time weighted residual approach that is presented here is the key to appropriately provide the proper extensions to nonlinear dynamics situations. It leads to symplectic-momentum based and energy-momentum based representations depending upon how the approximations for the variables are introduced. One can also embed controllable numerical dissipation into these parent algorithm designs. The brief highlights follow next.

14.3.1 Classical/Normalized Time Weighted Residual Methodology

The method of time weighted residuals is a method of discretizing the semi-discretized equation of motion in time. We approximate each of the dependent variables, by an asymptotic series expansion with the truncation terms up to a certain order, and then we substitute the approximated solutions into the original differential equation of motion to form the residual $\mathbf{R}$. Of course, if the approximations are exact, we have $\mathbf{R} = \mathbf{0}$. Introducing the weighting time field function $\mathbf{W}(t)$, the time weighted integrals of the residual within a time step from time t_n to t_{n+1} are given as

$$\int_{t_n}^{t_{n+1}} \mathbf{W}^T \mathbf{R} dt = 0 \tag{14.77}$$

by enforcing to zero. Note that the weighting time field function is arbitrary. In the following, we briefly summarize the classical time weighted residual methodology as well as the so-called normalized time weighted residual methodology for discretizing the equations of motion in the time-domain for nonlinear dynamical systems.

Classical Time Weighted Residual Methodology Consider the semi-discretized equation of motion for structural dynamics of the form

$$\mathbf{M}\ddot{\mathbf{q}}(t) + \mathbf{C}\dot{\mathbf{q}}(t) + \mathbf{f}^{\text{int}}(t) = \mathbf{f}(t) \quad \forall t \in \mathcal{T} \tag{14.78}$$

in the single-field form representation, with the initial conditions $\mathbf{q}_0 = \mathbf{q}(t_0)$ and $\dot{\mathbf{q}}_0 = \dot{\mathbf{q}}(t_0)$. Note $\mathbf{f}^{\text{int}}(t) = \mathbf{K}\mathbf{q}(t)$ and $\mathbf{f}(t)$ denote the (discrete) internal force vector in linear systems and the time-dependent external force vector, respectively. Assume that the mass matrix $\mathbf{M}$ and damping matrix $\mathbf{C}$ are constant. Consider the time dependent solutions evaluated numerically as

$$\mathbf{f}^{\text{int}}(t) = \mathbf{f}^{\text{int}}_{\text{approx}} + \mathbf{e}^p \tag{14.79}$$

$$\dot{\mathbf{q}}(t) = \dot{\mathbf{q}}_{\text{approx}} + \mathbf{e}^v \tag{14.80}$$

$$\ddot{\mathbf{q}}(t) = \ddot{\mathbf{q}}_{\text{approx}} + \mathbf{e}^a \tag{14.81}$$

$$\mathbf{f}(t) = \mathbf{f}_{\text{approx}} + \mathbf{e}^f \tag{14.82}$$

where $\mathbf{e}^p$, $\mathbf{e}^v$, $\mathbf{e}^a$, and $\mathbf{e}^f$ are error vectors for the internal force, (nodal) velocity, acceleration, and external force. Assume

$$\mathbf{f}^{\text{int}}(t) \cong \mathbf{f}^{\text{int}}_{\text{approx}} \tag{14.83}$$

$$\dot{\mathbf{q}}(t) \cong \dot{\mathbf{q}}_{\text{approx}} \tag{14.84}$$

$$\ddot{\mathbf{q}}(t) \cong \ddot{\mathbf{q}}_{\text{approx}} \tag{14.85}$$

$$\mathbf{f}(t) \cong \mathbf{f}_{\text{approx}} \tag{14.86}$$

the residual vector is formed as

$$\mathbf{R} := \mathbf{M}\ddot{\mathbf{q}}_{\text{approx}} + \mathbf{C}\dot{\mathbf{q}}_{\text{approx}} + \mathbf{f}^{\text{int}}_{\text{approx}} - \mathbf{f}_{\text{approx}} \tag{14.87}$$

Therefore, the time weighted integral of the residual from time t_n to t_{n+1} is given as

$$\int_{t_n}^{t_{n+1}} \mathbf{W}^T \left[\mathbf{M}\ddot{\mathbf{q}}_{\text{approx}} + \mathbf{C}\dot{\mathbf{q}}_{\text{approx}} + \mathbf{f}^{\text{int}}_{\text{approx}} - \mathbf{f}_{\text{approx}}\right] dt = 0 \tag{14.88}$$

by enforcing to zero. Transition from the theoretical weighting time field function, $\mathbf{W}$, to an approximated scalar function, W, may be done by introducing the damping coefficient $\xi := c/(2\sqrt{mk})$, the undamped natural frequency $\omega := \sqrt{k/m}$, and the damped natural frequency $\omega_d := \omega\sqrt{1-\xi^2}$, where c, m, and k denote the (viscous) damping coefficient, mass, and stiffness constant, respectively, in the mean sense as

$$\bar{\xi} := \frac{1}{n_{\text{dof}}} \sum_{i=1}^{n_{\text{dof}}} \xi_i \tag{14.89}$$

$$\bar{\omega} := \frac{1}{n_{\text{dof}}} \sum_{i=1}^{n_{\text{dof}}} \omega_i \tag{14.90}$$

$$\bar{\omega}_d := \frac{1}{n_{\text{dof}}} \sum_{i=1}^{n_{\text{dof}}} \omega_{d_i} \tag{14.91}$$

where n_{dof} denotes the number of degrees of freedom. Consider the theoretical weighting time field function approximated as

$$\mathbf{W} \approx W\delta\mathbf{q}_{n+1} \tag{14.92}$$

where $\delta\mathbf{q}_{n+1}$ is an arbitrary parameter, and

$$W := \frac{1}{\bar{\omega}_d} \exp\left[\bar{\xi}\bar{\omega}\tau\right] \sin(\bar{\omega}_d \tau) \tag{14.93}$$

We assume $t - t_n =: \tau \geq 0$. Approximating Equation (14.93) by an asymptotic series expansion about time t_n yields

$$W = w_0 + w_1\Gamma + w_2\Gamma^2 + w_3\Gamma^3 + \cdots \tag{14.94}$$

where $\Gamma := (t - t_n)/\Delta t$, and w_j $(j = 0, 1, 2, 3, \cdots)$ are constant coefficients. Note it is natural to set $w_0 = 1$. For the detailed discussions on the theoretical and approximated weighting time field functions, refer to Tamma et al. . (Tamma et al. 2000). Using the approximated scalar weighting time field function, Equation (14.88) becomes

$$\int_{t_n}^{t_{n+1}} W \left[\mathbf{M}\ddot{\mathbf{q}}_{\text{approx}} + \mathbf{C}\dot{\mathbf{q}}_{\text{approx}} + \mathbf{f}^{\text{int}}_{\text{approx}} - \mathbf{f}_{\text{approx}}\right] dt = \mathbf{0} \tag{14.95}$$

Dividing Equation (14.95) by $\int_{t_n}^{t_{n+1}} W dt$, we have

$$\mathbf{M}\frac{\int_{t_n}^{t_{n+1}} W\ddot{\mathbf{q}}_{\text{approx}} dt}{\int_{t_n}^{t_{n+1}} W dt} + \mathbf{C}\frac{\int_{t_n}^{t_{n+1}} W\dot{\mathbf{q}}_{\text{approx}} dt}{\int_{t_n}^{t_{n+1}} W dt} + \frac{\int_{t_n}^{t_{n+1}} W\mathbf{f}^{\text{int}}_{\text{approx}} dt}{\int_{t_n}^{t_{n+1}} W dt} = \frac{\int_{t_n}^{t_{n+1}} W\mathbf{f}_{\text{approx}} dt}{\int_{t_n}^{t_{n+1}} W dt} \tag{14.96}$$

Normalized Time Weighted Residual Methodology In a nonlinear dynamical system, the semi-discretized equation of motion is of the form

$$\mathbf{M}\ddot{\mathbf{q}}(t) + \mathbf{C}\dot{\mathbf{q}}(t) + \mathbf{f}^{\text{int}}(\mathbf{q}(t)) = \mathbf{f}(t) \tag{14.97}$$

in the single-field form representation, with the initial conditions $\mathbf{q}_0 = \mathbf{q}(t_0)$ and $\dot{\mathbf{q}}_0 = \dot{\mathbf{q}}(t_0)$. The internal force is a nonlinear function in the displacement vector, and it is defined as $\mathbf{f}^{\text{int}}(\mathbf{q}(t)) := \nabla\mathcal{U}_{\text{int}}(\mathbf{q}(t))$. Approximating $\mathbf{q}$, $\dot{\mathbf{q}}$, $\ddot{\mathbf{q}}$, and $\mathbf{f}$ in time as shown in Equations (14.83)–(14.86), respectively, the residual vector for the nonlinear system may be formed as

$$\mathbf{R} := \mathbf{M}\ddot{\mathbf{q}}_{\text{approx}} + \mathbf{C}\dot{\mathbf{q}}_{\text{approx}} + \mathbf{f}^{\text{int}}_{\text{approx}} - \mathbf{f}_{\text{approx}} \tag{14.98a}$$

with

$$\mathbf{f}^{\text{int}} \cong \mathbf{f}^{\text{int}}_{\text{approx}} = \nabla\mathcal{U}_{\text{int}}(\mathbf{q}_{\text{approx}}) \tag{14.98b}$$

To establish the time weighted integral of the residual in the time domain $[t_n, t_{n+1}]$ for Equation (14.98), using the scalar W, employ the time weighted integral for each of the time dependent functions and normalize by dividing each by $\int_{t_n}^{t_{n+1}} W dt$, respectively; that is,

$$\begin{aligned} &\mathbf{M}\frac{\int_{t_n}^{t_{n+1}} W\ddot{\mathbf{q}}_{\text{approx}}dt}{\int_{t_n}^{t_{n+1}} W dt} + \mathbf{C}\frac{\int_{t_n}^{t_{n+1}} W\dot{\mathbf{q}}_{\text{approx}}dt}{\int_{t_n}^{t_{n+1}} W dt} \\ &+ \nabla\mathcal{U}\left(\frac{\int_{t_n}^{t_{n+1}} W\mathbf{q}_{\text{approx}}dt}{\int_{t_n}^{t_{n+1}} W dt}\right) - \frac{\int_{t_n}^{t_{n+1}} W\mathbf{f}_{\text{approx}}dt}{\int_{t_n}^{t_{n+1}} W dt} = \mathbf{0} \end{aligned} \tag{14.99}$$

by enforcing to zero.

When the nonlinear internal function is, for example, expressed as a function

$$\mathbf{f}^{\text{int}}(\mathbf{q}(t)) \equiv \mathbf{f}^{\text{int}}(\mathbf{q}(t), \mathbf{g}(\mathbf{q}(t))) \tag{14.100}$$

where $\mathbf{g}$ can be a scalar, vector, or tensor, we may have an alternative choice to approximate the internal force by approximating $\mathbf{q}$ and $\mathbf{g}$ independently in time as

$$\mathbf{q}(t) = \mathbf{q}_{\text{approx}} + \mathbf{e}^q \quad \text{and} \quad \mathbf{g}(\mathbf{q}(t))) \equiv \mathbf{g}(t) = \mathbf{g}_{\text{approx}} + \mathbf{e}^g \tag{14.101}$$

where $\mathbf{e}^q$ and $\mathbf{e}^g$ denote the error vectors of the (nodal) displacement vector and function $\mathbf{g}$. Assuming

$$\mathbf{q}(t) \cong \mathbf{q}_{\text{approx}} \quad \text{and} \quad \mathbf{g}(\mathbf{q}(t)) \equiv \mathbf{g}(t) \cong \mathbf{g}_{\text{approx}} \tag{14.102}$$

the residual vector may be formed as

$$\mathbf{R} := \mathbf{M}\ddot{\mathbf{q}}_{\text{approx}} + \mathbf{C}\dot{\mathbf{q}}_{\text{approx}} + \mathbf{f}^{\text{int}}_{\text{approx}} - \mathbf{f}_{\text{approx}} \tag{14.103a}$$

with

$$\mathbf{f}^{\text{int}} \cong \mathbf{f}^{\text{int}}_{\text{approx}} = \nabla\mathcal{U}_{\text{int}}(\mathbf{q}_{\text{approx}}, \mathbf{g}_{\text{approx}}) \tag{14.103b}$$

In this case, the time weighted integral of the residual in $[t_n, t_{n+1}]$, using the scalar W, leads to

$$\mathbf{M}\frac{\int_{t_n}^{t_{n+1}} W\ddot{\mathbf{q}}_{\text{approx}}dt}{\int_{t_n}^{t_{n+1}} W dt} + \mathbf{C}\frac{\int_{t_n}^{t_{n+1}} W\dot{\mathbf{q}}_{\text{approx}}dt}{\int_{t_n}^{t_{n+1}} W dt}$$
$$+ \nabla\mathcal{U}(\frac{\int_{t_n}^{t_{n+1}} W\mathbf{q}_{\text{approx}}dt}{\int_{t_n}^{t_{n+1}} W dt}, \frac{\int_{t_n}^{t_{n+1}} W\mathbf{g}_{\text{approx}}dt}{\int_{t_n}^{t_{n+1}} W dt}) - \frac{\int_{t_n}^{t_{n+1}} W\mathbf{f}_{\text{approx}}dt}{\int_{t_n}^{t_{n+1}} W dt} = \mathbf{0} \tag{14.104}$$

by enforcing to zero.

Remark 14.3.1

1. In linear N-body dynamical systems, the internal force vector is given by $\mathbf{f}^{\text{int}} = \mathbf{Kq}$ where the stiffness matrix $\mathbf{K}$ is constant. Therefore, the algorithmic internal force term in Equation (14.96) readily leads to
$$\frac{\int_{t_n}^{t_{n+1}} W\mathbf{f}^{\text{int}}_{\text{approx}}dt}{\int_{t_n}^{t_{n+1}} W dt} = \mathbf{K}\frac{\int_{t_n}^{t_{n+1}} W\mathbf{q}_{\text{approx}}dt}{\int_{t_n}^{t_{n+1}} W dt} \tag{14.105}$$
where $\mathbf{q}_{\text{approx}}$ denotes the approximated displacement vector. In fact, the classical time weighted residual methodology is identical to the normalized time weighted residual methodology for linear dynamic systems.
2. For nonlinear dynamical systems, there exist significant differences between the classical and normalized time weighted residual methodologies. The subtle, yet important differences need to be clearly understood.

14.3.2 Time Discretization and Total Energy Framework: Brief Highlights of the Two-/Single-Field Form of LMS Methods, and Conserving Algorithms and Designs - General Hyperelastic Material Models

Here, we briefly provide the theoretical basis and highlight the three distinctly different schools of thought mentioned earlier for enabling the parent linear dynamic algorithms to be appropriately extended to nonlinear elastodynamics applications within the class of LMS methods. Both the two-field form and the single-field form of representation are highlighted. The three distinctly different options, namely, (i) the classical time weighted residual approach fails to properly preserve the underlying physics for nonlinear dynamics applications, (ii) the new normalized time weighted residual approach is ideally suitable for extensions of the parent linear dynamic algorithms to nonlinear dynamics applications, and for particular types of approximations it naturally leads to symplectic-momentum based representations, and (iii) alternatively, for other particular approximations, it naturally leads to energy-momentum based representations. Of the three distinctly different approaches, only the latter two are suitable as proper extensions to nonlinear dynamics situations; and they respectively lead to symplectic-momentum and energy-momentum based algorithm designs. Of the three approaches, the classical time weighted residual approach is not the proper way for nonlinear dynamics applications as highlighted in the subsequent sections.

Two-field Form Algorithm Designs

Approach 1: Classical Framework - Classical Time Weighted Residual Approach Consider the semi-discretized representation for conservative systems via the Total Energy framework and the corresponding equations of motion with given initial conditions:

$$\frac{d}{dt}\left(\frac{\partial \mathcal{E}^h(\mathbf{q},\mathbf{v})}{\partial \mathbf{v}^A}\right) + \frac{\partial \mathcal{E}^h(\mathbf{q},\mathbf{v})}{\partial \mathbf{q}^A} = \mathbf{0} \ \ \forall t \in \mathcal{T} \qquad (14.106)$$
$$\mathbf{q}(t_0) = \mathbf{q}_0 \quad \text{and} \quad \mathbf{v}(t_0) = \mathbf{v}_0$$

with the kinematic constraint $\dot{\mathbf{q}} = \mathbf{v} \in T_q Q$ for $A = 1, 2, \ldots, n_{\text{node}}$, and we assume the autonomous discrete total energy, $\mathcal{E}^h(\mathbf{q},\mathbf{v}) : TQ \to \mathbb{R}$, is given by the summation of the discrete kinetic energy, $\mathcal{K}^h(\mathbf{v}) : T_q Q \to \mathbb{R}$, and the discrete total potential energy, $\mathcal{U}^h(\mathbf{q}) : Q \to \mathbb{R}$; that is,

$$\mathcal{E}^h(\mathbf{q},\mathbf{v}) = \mathcal{K}^h(\mathbf{v}) + \mathcal{U}^h(\mathbf{q}) = \frac{1}{2}\mathbf{v}\cdot\mathbf{M}\mathbf{v} + \mathcal{U}^h(\mathbf{q}) \qquad (14.107)$$

where we assume the mass matrix is constant. Assume the discrete total potential energy is defined as the summation of the discrete internal potential energy and the discrete external potential energy as

$$\mathcal{U}^h(\mathbf{q}) = \mathcal{U}^h_{\text{int}}(\mathbf{q}) + \mathcal{U}^h_{\text{ext}}(\mathbf{q}) \qquad (14.108)$$

In this framework, we directly employ the classical time weighted residual methodology for the nonlinear dynamic system with the time weighting function $W \in \mathbb{R}$ approximated as

$$W = w_0 + w_1\Gamma + w_2\Gamma^2; \quad \Gamma := \frac{\tau}{\Delta t} = \frac{t - t_n}{t_{n+1} - t_n} \qquad (14.109)$$

And, assume $w_0 \equiv 1$ for the time discretization of Equation (14.106). For a time interval $\mathcal{T} = [t_0, t_N]$ split into subintervals of interest, i.e., $\mathcal{T} = [t_0, t_N] = \bigcup_{n=0}^{N-1} \left[t_n, t_{n+1}\right]$, assuming $\mathbf{q}(t_n) = \mathbf{q}_n$ and $\mathbf{v}(t_n) = \mathbf{v}_n$ for $n = 1, 2, \cdots, N$, we obtain

$$\int_{t_n}^{t_{n+1}} W\left[\frac{d}{dt}\left(\frac{\partial \mathcal{E}^h(\mathbf{q},\mathbf{v})}{\partial \mathbf{v}}\right) + \frac{\partial \mathcal{E}^h(\mathbf{q},\mathbf{v})}{\partial \mathbf{q}}\right] dt = \mathbf{0} \qquad (14.110)$$

or

$$\int_{t_n}^{t_{n+1}} W\left[\mathbf{M}\dot{\mathbf{v}} + \mathbf{F}^{\text{int}^h} - \mathbf{F}^{\text{ext}^h}\right] dt = \mathbf{0} \qquad (14.111)$$

where the discrete internal and external forces are defined by

$$\mathbf{F}^{\text{int}^h}_A := \frac{\partial \mathcal{U}_{\text{int}}}{\partial \mathbf{q}^A} = \int_{\mathcal{B}} \mathbf{F}^h \cdot \mathbf{S}^h \,\text{GRAD}\, N_A dV \qquad (14.112)$$

$$\mathbf{F}^{\text{ext}^h}_A := -\frac{\partial \mathcal{U}_{\text{ext}}}{\partial \mathbf{q}^A} = \int_{\mathcal{B}} N_A \rho_0 \mathbf{B} dV + \int_{\partial \mathcal{B}_\sigma} N_A \bar{\mathbf{T}} dA \qquad (14.113)$$

respectively. We approximate $\mathbf{F}^{\text{int}^h}$, $\mathbf{v}$ and $\dot{\mathbf{v}}$ by means of the asymptotic series expansions in time and approximate the external nodal forces $\mathbf{F}^{\text{ext}^h}$ by means of the Taylor series expansion (in general, this is a special case of an asymptotic series expansion and is sufficient for the cases of interest here; but the general case is a function evaluation at a particular time point and not necessarily a linear interpolation) in time as follows ($\tau := t - t_n$):

$$\mathbf{F}^{\text{int}^h} = \mathbf{F}^{\text{int}^h}_n + \Lambda_1 \dot{\mathbf{F}}^{\text{int}^h}_n \tau + \Lambda_2 \frac{\dot{\mathbf{F}}^{\text{int}^h}_{n+1} - \dot{\mathbf{F}}^{\text{int}^h}_n}{\Delta t}\tau^2 \qquad (14.114)$$

$$\mathbf{v} = \mathbf{v}_n + \Lambda_3 \frac{\mathbf{v}_{n+1} - \mathbf{v}_n}{\Delta t}\tau \tag{14.115}$$

$$\dot{\mathbf{v}} = \frac{\mathbf{v}_{n+1} - \mathbf{v}_n}{\Delta t} \tag{14.116}$$

$$\mathbf{F}^{\text{ext}^h} = \mathbf{F}_n^{\text{ext}^h} + \frac{\mathbf{F}_{n+1}^{\text{ext}^h} - \mathbf{F}_n^{\text{ext}^h}}{\Delta t}\tau \tag{14.117}$$

It should be noted that the discrete internal nodal force vector is assumed to be a function depending on time explicitly: $\mathbf{F}^{\text{int}^h}(t) : \mathcal{T} \rightarrow \mathbb{R}^{n_{\text{dof}}}$. To design the update for the nodal displacement vector, introduce $\lambda_i \in \mathbb{R}$ $(i = 1, 2)$ for

$$\mathbf{q} = \mathbf{q}_n + \lambda_1 \mathbf{v}_n \tau + \lambda_2 \frac{\mathbf{v}_{n+1} - \mathbf{v}_n}{\Delta t}\tau^2 \tag{14.118}$$

and set $\tau = \Delta t$; hence, we have

$$\mathbf{q}_{n+1} = \mathbf{q}_n + \Delta t \lambda_1 \mathbf{v}_n + \Delta t \lambda_2 (\mathbf{v}_{n+1} - \mathbf{v}_n) \tag{14.119}$$

Define

$$W_i := \frac{\int_{t_n}^{t_{n+1}} \left(\frac{\tau}{\Delta t}\right)^i W dt}{\int_{t_n}^{t_{n+1}} W dt} \quad for \;\; i = 1, 2 \tag{14.120}$$

for convenience, and following the same procedure as shown before for linear dynamic systems, we obtain the following classical representation in two-field form for nonlinear elastodynamics:

Algorithm 14: ***Approach 1***

Given $\mathbf{q}_n$ and $\mathbf{v}_n$, find $\mathbf{q}_{n+1}$ and $\mathbf{v}_{n+1}$ from

$$\sum_{B=1}^{n_{\text{node}}} M_{AB} \frac{\mathbf{v}_{n+1}^B - \mathbf{v}_n^B}{\Delta t} + \tilde{\mathbf{F}}_A^{\text{int}^h} = \tilde{\mathbf{F}}_A^{\text{ext}^h} \quad (A = 1, 2, \ldots, n_{\text{node}}) \tag{14.121a}$$

$$\tilde{\mathbf{F}}^{\text{int}^h} = \mathbf{F}_n^{\text{int}^h} + \frac{W_2 \Lambda_2}{\lambda_2}(\mathbf{F}_{n+1}^{\text{int}^h} - \mathbf{F}_n^{\text{int}^h}) + \Delta t \left(W_1 \Lambda_1 - \frac{\lambda_1}{\lambda_2} W_2 \Lambda_2 \right) \dot{\mathbf{F}}_n^{\text{int}^h} \tag{14.121b}$$

$$\tilde{\mathbf{F}}^{\text{ext}^h} = (1 - W_1)\mathbf{F}_n^{\text{ext}^h} + W_1 \mathbf{F}_{n+1}^{\text{ext}^h} \tag{14.121c}$$

with the following updates:

$$\mathbf{q}_{n+1} = \mathbf{q}_n + \Delta t \lambda_1 \mathbf{v}_n + \Delta t \lambda_2 \Delta \mathbf{v} \tag{14.121d}$$

$$\mathbf{v}_{n+1} = \mathbf{v}_n + \Delta \mathbf{v} \tag{14.121e}$$

where $\Delta \mathbf{v} := \mathbf{v}_{n+1} - \mathbf{v}_n$.

Remark 14.3.2

1. Algorithm 14 may be the simplest extension via the classical time weighted residual methodology for nonlinear dynamical conservative systems. Here, we have considered the family of algorithms derived for linear dynamic conservative systems shown in a previous section as the parent algorithms.

2. For second-order time accuracy, we must have

$$\lambda_1 = 1, \quad \lambda_2 = \tfrac{1}{2}, \quad \Lambda_1 = \Lambda_3 = 1, \quad \text{and} \quad W_1 = \tfrac{1}{2} \tag{14.122}$$

In the absence of the external force, the condition is reduced to

$$\lambda_1 = 1, \quad \lambda_2 = \tfrac{1}{2}, \quad \text{and} \quad W_1\Lambda_1 = W_1\Lambda_3 = \tfrac{1}{2} \tag{14.123}$$

3. The discrete algorithmic internal force vector can be written as

$$\tilde{\mathbf{F}}^{\text{int}^h} = \mathbf{F}_n^{\text{int}^h} + \frac{W_2\Lambda_2}{\lambda_2}(\mathbf{F}_{n+1}^{\text{int}^h} - \mathbf{F}_n^{\text{int}^h}) + \Delta t(W_1\Lambda_1 - \tfrac{\lambda_1}{\lambda_2}W_2\Lambda_2)\dot{\mathbf{F}}_n^{\text{int}^h} \tag{14.124}$$

by substituting the update. When $\Lambda_1 = \lambda_1 = 1$, $\Lambda_2 = \lambda_2 = 1/2$, $\Lambda_3 = 1$ and $W_1 = W_2 = 1/2$, Equation (14.124) yields

$$\begin{aligned} \tilde{\mathbf{F}}_A^{\text{int}^h} &= \frac{\mathbf{F}_{A\,n+1}^{\text{int}^h} + \mathbf{F}_{A\,n}^{\text{int}^h}}{2} \\ &= \int_{\mathcal{B}} \frac{1}{2}\left[\mathbf{F}^h(q_{n+1}) \cdot \mathbf{S}_{n+1}^h \,\text{GRAD}N_A + \mathbf{F}^h(q_n) \cdot \mathbf{S}_n^h \,\text{GRAD}N_A\right] dV \end{aligned} \tag{14.125}$$

where $\mathbf{S}_{n+1}^h = \mathbf{S}^h(\mathbf{C}_{n+1}^h) = \mathbf{S}^h(\mathbf{C}^h(\mathbf{q}_{n+1}))$ and $\mathbf{S}_n^h = \mathbf{S}^h(\mathbf{C}_n^h) = \mathbf{S}^h(\mathbf{C}^h(\mathbf{q}_n))$, and Algorithm 14 becomes

$$\begin{aligned} \mathbf{M}\frac{\mathbf{v}_{n+1} - \mathbf{v}_n}{\Delta t} + \frac{\mathbf{F}_{n+1}^{\text{int}^h} + \mathbf{F}_n^{\text{int}^h}}{2} &= \frac{\mathbf{F}_{n+1}^{\text{ext}^h} + \mathbf{F}_n^{\text{ext}^h}}{2} \\ \frac{\mathbf{q}_{n+1} - \mathbf{q}_n}{\Delta t} &= \frac{\mathbf{v}_{n+1} + \mathbf{v}_n}{2} \end{aligned} \tag{14.126}$$

4. The algorithmic external nodal forces are given by

$$\begin{aligned} \tilde{\mathbf{F}}_A^{\text{ext}^h} &= (1 - W_1)\mathbf{F}_{A\,n}^{\text{ext}^h} + W_1\mathbf{F}_{A\,n+1}^{\text{ext}^h} \\ &= \int_{\mathcal{B}} N_A\rho_0\mathbf{B}_{n+W_1}\,dV + \int_{\partial\mathcal{B}_\sigma} N_A\bar{\mathbf{T}}_{n+W_1}\,dA \end{aligned} \tag{14.127}$$

where $\mathbf{B}_{n+W_1} := (1 - W_1)\mathbf{B}_n + W_1\mathbf{B}_{n+1}$ and $\bar{\mathbf{T}}_{n+W_1} := (1 - W_1)\bar{\mathbf{T}}_n + W_1\bar{\mathbf{T}}_{n+1}$.

5. In general, no time integration scheme in this classical framework exactly preserves symplecticness or conserves discrete total energy in $[t_n, t_{n+1}]$ for general nonlinear cases and this is the drawback with this approach; however, the discrete total linear momentum within the time step is exactly conserved. In general, the discrete total angular momentum within the time step also cannot be conserved.

Approach 2: Symplectic-Momentum Framework–Normalized Time Weighted Residual Approach Consider the semi-discretized representation for conservative systems via the Total Energy framework and the corresponding equations of motion with given initial conditions as shown in Equation (14.106). In view of Equation (14.107), Equation (14.108), Equation (14.112) and Equation (14.113), it can be written as

$$\frac{d}{dt}\left(\frac{\partial\mathcal{E}^h(\mathbf{q},\mathbf{v})}{\partial\mathbf{v}^A}\right)+\frac{\partial\mathcal{E}^h(\mathbf{q},\mathbf{v})}{\partial\mathbf{q}^A}=\mathbf{0}$$
$$\Rightarrow\sum_{B=1}^{n_{\text{node}}}M_{AB}\dot{\mathbf{v}}^B+\int_{\mathcal{B}}\mathbf{F}^h\cdot\mathbf{S}^h\,\text{GRAD}\,N_A dV=\underbrace{\int_{\mathcal{B}}N_A\rho_0\mathbf{B}dV+\int_{\partial\mathcal{B}_\sigma}N_A\bar{\mathbf{T}}dA}_{\mathbf{F}_A^{\text{ext}^h}:\mathcal{T}\to\mathbb{R}^3} \tag{14.128}$$

Assume the discrete deformation gradient tensor and the discrete second Piola-Kirchhoff stresses tensor as functions of the nodal displacement vectors explicitly. Employing the *normalized* time weighted residual methodology on $\mathbf{q}(t)$ and $\ddot{\mathbf{q}}(t)$ yields

$$\sum_{B=1}^{n_{\text{node}}}M_{AB}\frac{\int_{t_n}^{t_{n+1}}W\dot{\mathbf{v}}^B dt}{\int_{t_n}^{t_{n+1}}W dt}+\int_{\mathcal{B}}\mathbf{F}^h\left(\frac{\int_{t_n}^{t_{n+1}}W\mathbf{q}dt}{\int_{t_n}^{t_{n+1}}W dt}\right)\cdot\mathbf{S}^h\left(C^h\left(\frac{\int_{t_n}^{t_{n+1}}W\mathbf{q}dt}{\int_{t_n}^{t_{n+1}}W dt}\right)\right)\text{GRAD}\,N_A dV$$
$$=\frac{\int_{t_n}^{t_{n+1}}W\mathbf{F}_A^{\text{ext}^h}dt}{\int_{t_n}^{t_{n+1}}W dt} \tag{14.129}$$

where the scalar time weighting function is approximated by

$$W=1+w_1\Gamma+w_2\Gamma^2 \tag{14.130}$$

with the approximations of $\mathbf{q}$, $\mathbf{v}$, and $\dot{\mathbf{v}}$ by means of the asymptotic series expansions, and the approximations of the set of the external nodal forces, $\mathbf{F}^{\text{ext}^h}$, by means of the Taylor series expansion (in general, this is a special case of an asymptotic series expansion and is sufficient for the cases of interest here; but the general case is a function evaluation at a particular time point and not necessarily a linear interpolation) in time as follows:

$$\mathbf{q}=\mathbf{q}_n+\Lambda_1\mathbf{v}_n\tau+\Lambda_2\frac{\mathbf{v}_{n+1}-\mathbf{v}_n}{\Delta t}\tau^2 \tag{14.131}$$

$$\mathbf{v}=\mathbf{v}_n+\Lambda_3\frac{\mathbf{v}_{n+1}-\mathbf{v}_n}{\Delta t}\tau \tag{14.132}$$

$$\dot{\mathbf{v}}=\frac{\mathbf{v}_{n+1}-\mathbf{v}_n}{\Delta t} \tag{14.133}$$

$$\mathbf{F}^{\text{ext}^h}=\mathbf{F}_n^{\text{ext}^h}+\frac{\mathbf{F}_{n+1}^{\text{ext}^h}-\mathbf{F}_n^{\text{ext}^h}}{\Delta t}\tau \tag{14.134}$$

The algorithmic nodal unknowns are defined as

$$\tilde{\mathbf{q}}:=\frac{\int_{t_n}^{t_{n+1}}W\mathbf{q}dt}{\int_{t_n}^{t_{n+1}}W dt}=\mathbf{q}_n+\Delta t W_1\Lambda_1\mathbf{v}_n+\Delta t W_2\Lambda_2\Delta\mathbf{v} \tag{14.135}$$

$$\tilde{\mathbf{v}}:=\frac{\int_{t_n}^{t_{n+1}}W\mathbf{v}dt}{\int_{t_n}^{t_{n+1}}W dt}=\mathbf{v}_n+W_1\Lambda_3\Delta\mathbf{v}=:\mathbf{v}_{n+W_1\Lambda_3} \tag{14.136}$$

$$\tilde{\dot{\mathbf{v}}}:=\frac{\Delta\mathbf{v}}{\Delta t} \tag{14.137}$$

$$\tilde{\mathbf{F}}^{\text{ext}^h} := \frac{\int_{t_n}^{t_{n+1}} W \mathbf{F}^{\text{ext}^h} dt}{\int_{t_n}^{t_{n+1}} W dt} = \mathbf{F}_n^{\text{ext}^h} + W_1(\mathbf{F}_{n+1}^{\text{ext}^h} - \mathbf{F}_n^{\text{ext}^h}) =: \mathbf{F}_{n+W_1}^{\text{ext}^h} \tag{14.138}$$

where $\Delta\mathbf{v} := \mathbf{v}_{n+1} - \mathbf{v}_n$. Therefore, the algorithmic deformation gradient and the algorithmic right Cauchy-Green deformation tensor may be defined as

$$\tilde{\mathbf{F}}^h := \mathbf{F}^h(\tilde{\mathbf{q}}) \quad \text{and} \quad \tilde{\mathbf{C}}^h := \mathbf{F}^{h^T}(\tilde{\mathbf{q}}) \cdot \mathbf{F}^h(\tilde{\mathbf{q}}) \tag{14.139}$$

respectively. For the semi-discrete system, we define the generalized coordinates as a set of nodal displacement vectors via $\boldsymbol{\varphi}^h(\mathbf{X}, t) \cong \mathbf{X} + \sum_{A=1}^{n_{\text{node}}} N_A(\mathbf{X})\mathbf{q}^A(t)$, and the algorithmic deformation gradient may be given as

$$\tilde{\mathbf{F}}^h = \mathbf{F}^h(\tilde{\mathbf{q}}) = \mathbf{I} + \sum_{A=1}^{n_{\text{node}}} \tilde{\mathbf{q}}^A \otimes \text{GRAD}\, N_A \tag{14.140}$$

where $\mathbf{I}$ denotes the identity tensor. The algorithmic right Cauchy-Green deformation tensor is given from Equation (14.139), using the above $\tilde{\mathbf{F}}^h$. Note that we can alternatively define the generalized coordinates as a set of nodal position vectors via $\boldsymbol{\varphi}^h(\mathbf{X}, t) \cong \sum_{A=1}^{n_{\text{node}}} N_A(\mathbf{X})\mathbf{q}^A(t)$, and the algorithmic deformation gradient and the algorithmic right Cauchy-Green deformation tensor are defined as

$$\tilde{\mathbf{F}}^h = \mathbf{F}^h(\tilde{\mathbf{q}}) = \sum_{A=1}^{n_{\text{node}}} \tilde{\mathbf{q}}^A \otimes \text{GRAD}\, N_A \tag{14.141}$$

$$\tilde{\mathbf{C}}^h = \sum_{A=1}^{n_{\text{node}}} \sum_{B=1}^{n_{\text{node}}} \tilde{\mathbf{q}}^A \cdot \tilde{\mathbf{q}}^B \ \text{GRAD}\, N_A \otimes \text{GRAD}\, N_B \tag{14.142}$$

respectively. As can be seen, the position-based discretization is sometimes more convenient for the evaluations of algorithmic quantities, but we follow the displacement-based discretization in view of the consistency of notations from the discussions for linear dynamic systems. Summarizing, the following generalized family of algorithms are obtained:

Algorithm 15: ***Approach 2***

Given $\mathbf{q}_n$ and $\mathbf{v}_n$, find $\mathbf{q}_{n+1}$ and $\mathbf{v}_{n+1}$ from

$$\sum_{B=1}^{n_{\text{node}}} M_{AB} \frac{\mathbf{v}_{n+1}^B - \mathbf{v}_n^B}{\Delta t} + \int_{\mathcal{B}} \mathbf{F}^h(\tilde{\mathbf{q}}) \cdot \mathbf{S}^h\left(\mathbf{C}^h(\tilde{\mathbf{q}})\right) \text{GRAD} N_A dV = \tilde{\mathbf{F}}_A^{\text{ext}^h} \tag{14.143a}$$

$$\tilde{\mathbf{q}} = \mathbf{q}_n + \Delta t W_1 \Lambda_1 \mathbf{v}_n + \Delta t W_2 \Lambda_2 \Delta\mathbf{v} \tag{14.143b}$$

$$\tilde{\mathbf{v}} = \mathbf{v}_n + W_1 \Lambda_3 \Delta\mathbf{v} \tag{14.143c}$$

$$\tilde{\mathbf{F}}^{\text{ext}^h} = (1 - W_1)\mathbf{F}_n^{\text{ext}^h} + W_1 \mathbf{F}_{n+1}^{\text{ext}^h} \tag{14.143d}$$

with the following updates:

$$\mathbf{q}_{n+1} = \mathbf{q}_n + \Delta t \lambda_1 \mathbf{v}_n + \Delta t \lambda_2 \Delta\mathbf{v} \tag{14.143e}$$

$$\mathbf{v}_{n+1} = \mathbf{v}_n + \Delta\mathbf{v} \tag{14.143f}$$

Remark 14.3.3

1. When $\Lambda_1 = \lambda_1 = 1$, $\Lambda_2 = \lambda_2 = 1/2$, $\Lambda_3 = 1$ and $W_1 = W_2 = 1/2$, Algorithm 15 becomes

$$\begin{aligned}
&\sum_{B=1}^{n_{\text{node}}} M_{AB} \frac{\mathbf{v}_{n+1}^B - \mathbf{v}_n^B}{\Delta t} + \int_{\mathcal{B}} \mathbf{F}^h(\mathbf{q}_{n+1/2}) \cdot \mathbf{S}^h\left(\mathbf{C}^h(\mathbf{q}_{n+1/2})\right) \text{GRAD} N_A dV \\
&= \int_{\mathcal{B}} N_A \rho_0 \mathbf{B}_{n+1/2} dV + \int_{\partial \mathcal{B}_\sigma} N_A \bar{\mathbf{T}}_{n+1/2} dA \qquad (14.144) \\
&\frac{\mathbf{q}_{n+1} - \mathbf{q}_n}{\Delta t} = \mathbf{v}_{n+1/2}
\end{aligned}$$

which is identical to the symplectic midpoint rule. This is the only member which is second-order time accurate in this framework. Unlike the classical approach, this representation is applicable to nonlinear dynamic situations and yields symplectic-momentum based representations.
2. In general, no member in this framework exactly conserves the discrete total energy in $[t_n, t_{n+1}]$. The discrete total angular momentum is exactly conserved only when $\tilde{\mathbf{q}} = \mathbf{q}_{n+1/2}$ while the discrete total linear momentum is exactly conserved for any conditions within the time step.

Approach 3: Energy-Momentum Framework - Normalized Time Weighted Residual Approach
Consider the semi-discretized representation for conservative systems via the Total Energy framework and the corresponding equations of motion with given initial conditions as shown in Equation (14.106), which yields

$$\begin{aligned}
&\frac{d}{dt}\left(\frac{\partial \mathcal{E}^h(\mathbf{q}, \mathbf{v})}{\partial \mathbf{v}^A}\right) + \frac{\partial \mathcal{E}^h(\mathbf{q}, \mathbf{v})}{\partial \mathbf{q}^A} = \mathbf{0} \\
&\Rightarrow \sum_{B=1}^{n_{\text{node}}} M_{AB} \dot{\mathbf{v}}^B + \int_{\mathcal{B}} \mathbf{F}^h \cdot \mathbf{S}^h \,\text{GRAD}\, N_A dV = \underbrace{\int_{\mathcal{B}} N_A \rho_0 \mathbf{B} dV + \int_{\partial \mathcal{B}_\sigma} N_A \bar{\mathbf{T}} dA}_{\mathbf{F}_A^{\text{ext}^h}: \mathcal{T} \to \mathbb{R}^3} \qquad (14.145)
\end{aligned}$$

where M_{AB} is assumed constant. In this framework, we assume the discrete deformation gradient and the discrete second Piola-Kirchhoff stress tensor as the mappings, $\mathbf{q}(t) \longmapsto \mathbf{F}^h$ and $\mathbf{C}^h(t) \longmapsto \mathbf{S}^h$, respectively. Notice the discrete second Piola-Kirchhoff stress tensor is a function of the discrete right Cauchy-Green deformation tensor which is assumed to be a function of time directly. Employing the *normalized* time weighted residual methodology, yields

$$\begin{aligned}
&\sum_{B=1}^{n_{\text{node}}} M_{AB} \frac{\int_{t_n}^{t_{n+1}} W \dot{\mathbf{v}}^B dt}{\int_{t_n}^{t_{n+1}} W dt} \\
&+ \int_{\mathcal{B}} \mathbf{F}^h \left(\frac{\int_{t_n}^{t_{n+1}} W q dt}{\int_{t_n}^{t_{n+1}} W dt}\right) \cdot \mathbf{S}^h \left(\frac{\int_{t_n}^{t_{n+1}} W C^h dt}{\int_{t_n}^{t_{n+1}} W dt}\right) \text{GRAD}\, N_A dV \qquad (14.146) \\
&= \frac{\int_{t_n}^{t_{n+1}} W \mathbf{F}_A^{\text{ext}^h} dt}{\int_{t_n}^{t_{n+1}} W dt}
\end{aligned}$$

where the scalar time weighting function is approximated by

$$W = 1 + w_1 \Gamma + w_2 \Gamma^2 \qquad (14.147)$$

with the approximations of $\mathbf{q}$, $\mathbf{v}$, and $\dot{\mathbf{v}}$ by means of the asymptotic series expansions in time, and the external nodal forces, $\mathbf{F}^{ext^h}$, are approximated by means of the Taylor series expansions (in general, this is a special case of an asymptotic series expansion and is sufficient for the cases of interest here; but the general case is a function evaluation at a particular time point and not necessarily a linear interpolation) in time as follows:

$$\mathbf{q} = \mathbf{q}_n + \Lambda_1 \mathbf{v}_n \tau + \Lambda_2 \frac{\mathbf{v}_{n+1} - \mathbf{v}_n}{\Delta t} \tau^2 \tag{14.148}$$

$$\mathbf{v} = \mathbf{v}_n + \Lambda_3 \frac{\mathbf{v}_{n+1} - \mathbf{v}_n}{\Delta t} \tau \tag{14.149}$$

$$\dot{\mathbf{v}} = \frac{\mathbf{v}_{n+1} - \mathbf{v}_n}{\Delta t} \tag{14.150}$$

$$\mathbf{F}^{ext^h} = \mathbf{F}_n^{ext^h} + \frac{\mathbf{F}_{n+1}^{ext^h} - \mathbf{F}_n^{ext^h}}{\Delta t} \tau \tag{14.151}$$

Therefore, the algorithmic unknowns are defined as

$$\tilde{\mathbf{q}} := \frac{\int_{t_n}^{t_{n+1}} W \mathbf{q} dt}{\int_{t_n}^{t_{n+1}} W dt} = \mathbf{q}_n + \Delta t W_1 \Lambda_1 \mathbf{v}_n + \Delta t W_2 \Lambda_2 \Delta \mathbf{v} \tag{14.152}$$

$$\tilde{\mathbf{v}} := \frac{\int_{t_n}^{t_{n+1}} W \mathbf{v} dt}{\int_{t_n}^{t_{n+1}} W dt} = \mathbf{v}_n + W_1 \Lambda_3 \Delta \mathbf{v} \tag{14.153}$$

$$\tilde{\dot{\mathbf{v}}} := \frac{\int_{t_n}^{t_{n+1}} W \frac{d\mathbf{v}}{dt} dt}{\int_{t_n}^{t_{n+1}} W dt} = \frac{\Delta \mathbf{v}}{\Delta t} \tag{14.154}$$

$$\tilde{\mathbf{F}}^{ext^h} := \frac{\int_{t_n}^{t_{n+1}} W \mathbf{F}^{ext^h} dt}{\int_{t_n}^{t_{n+1}} W dt} = (1 - W_1) \mathbf{F}_n^{ext^h} + W_1 \mathbf{F}_{n+1}^{ext^h} \tag{14.155}$$

The algorithmic discrete right Cauchy-Green deformation tensor is defined by

$$\tilde{\mathbf{C}}^h := \frac{\int_{t_n}^{t_{n+1}} W \mathbf{C}^h dt}{\int_{t_n}^{t_{n+1}} W dt} = (1 - W_1 \Lambda_4) \mathbf{C}_n^h + W_1 \Lambda_4 \mathbf{C}_{n+1}^h \tag{14.156}$$

from the following approximation by the linear asymptotic series expansions in time:

$$\mathbf{C}^h = \mathbf{C}_n^h + \Lambda_4 \frac{\mathbf{C}_{n+1}^h - \mathbf{C}_n^h}{\Delta t} \tau \tag{14.157}$$

Following a similar procedure as described in the previous sections (Approaches 1 and 2), we readily obtain the following as options of the algorithmic representation.

Algorithm 16: ***Approach 3-α***

Given $\mathbf{q}_n$ and $\mathbf{v}_n$, find $\mathbf{q}_{n+1}$ and $\mathbf{v}_{n+1}$ from

$$\sum_{B=1}^{n_{\text{node}}} M_{AB}\frac{\mathbf{v}_{n+1}^B - \mathbf{v}_n^B}{\Delta t} + \int_{\mathcal{B}} \mathbf{F}^h(\tilde{\mathbf{q}}) \cdot \mathbf{S}^h(\tilde{\mathbf{C}}^h)\text{GRAD}N_A dV = \tilde{\mathbf{F}}_A^{\text{ext}^h} \tag{14.158a}$$

$$\tilde{\mathbf{q}} = \mathbf{q}_n + \Delta t W_1\Lambda_1\mathbf{v}_n + \Delta t W_2\Lambda_2\Delta\mathbf{v} \tag{14.158b}$$

$$\tilde{\mathbf{C}}^h = (1 - W_1\Lambda_4)\mathbf{C}_n^h + W_1\Lambda_4\mathbf{C}_{n+1}^h \tag{14.158c}$$

$$\tilde{\mathbf{F}}^{\text{ext}^h} = (1 - W_1)\mathbf{F}_n^{\text{ext}^h} + W_1\mathbf{F}_{n+1}^{\text{ext}^h} \tag{14.158d}$$

with the following updates:

$$\mathbf{q}_{n+1} = \mathbf{q}_n + \Delta t\lambda_1\mathbf{v}_n + \Delta t\lambda_2\Delta\mathbf{v} \tag{14.158e}$$

$$\mathbf{v}_{n+1} = \mathbf{v}_n + \Delta\mathbf{v} \tag{14.158f}$$

Remark 14.3.4 (Algorithm 16)

1. For second-order time accuracy, we must have

$$\lambda_1 = 1, \quad \lambda_2 = \tfrac{1}{2}, \quad \Lambda_1 = \Lambda_3 = \Lambda_4 = 1, \quad \text{and} \quad W_1 = \tfrac{1}{2} \tag{14.159}$$

In the absence of the external force, the condition is reduced to

$$\lambda_1 = 1, \quad \lambda_2 = \tfrac{1}{2}, \quad \text{and} \quad W_1\Lambda_1 = W_1\Lambda_3 = W_1\Lambda_4 = \tfrac{1}{2} \tag{14.160}$$

2. **Discrete energy conservation within a time step**: Consider the case of $\sum_{B=1}^{n_{\text{node}}} C_{AB} = 0$, and define

$$\begin{aligned}\Theta := &\sum_{A=1}^{n_{\text{node}}}\sum_{B=1}^{n_{\text{node}}} M_{AB}\frac{\mathbf{v}_{n+1}^B - \mathbf{v}_n^B}{\Delta t} \cdot \left[\lambda_1\mathbf{v}_n^A + \lambda_2\Delta\mathbf{v}^A\right] \\ &+ \sum_{B=1}^{n_{\text{node}}}\left[\tilde{\mathbf{F}}_A^{\text{int}^h} - \tilde{\mathbf{F}}_A^{\text{ext}^h}\right] \cdot \frac{\mathbf{q}_{n+1}^A - \mathbf{q}_n^A}{\Delta t} = 0\end{aligned} \tag{14.161}$$

We observe $\Theta = (\mathcal{E}_{n+1}^h - \mathcal{E}_n^h)/\Delta t = 0$ (exact energy conservation in $[t_n, t_{n+1}]$) only for St. Venant-Kirchhoff material model when selecting $\Lambda_1 = \lambda_1 = 1$, $\Lambda_2 = \lambda_2 = 1/2$, $\Lambda_4 = 1$ and $W_1 = W_2 = 1/2$, assuming the external force and the boundary conditions do not depend on time explicitly; therefore,

$$\begin{aligned}0 &= \Theta \\ &= \sum_{A=1}^{n_{\text{node}}}\sum_{B=1}^{n_{\text{node}}} M_{AB}\frac{\mathbf{v}_{n+1}^B - \mathbf{v}_n^B}{\Delta t} \cdot \mathbf{v}_{n+1/2}^A + \int_{\mathcal{B}} \mathbf{S}^h(\mathbf{C}_{n+1/2}^h) : \frac{\mathbf{C}_{n+1}^h - \mathbf{C}_n^h}{\Delta t} dV \\ &\quad - \sum_{A=1}^{n_{\text{node}}}\int_{\mathcal{B}} N_A\rho_0\mathbf{B}_{n+W_1}dV \cdot \frac{\mathbf{q}_{n+1}^A - \mathbf{q}_n^A}{\Delta t}\end{aligned}$$

$$
\begin{aligned}
&-\sum_{A=1}^{n_{\text{node}}} \int_{\partial\mathcal{B}_\sigma} N_A \bar{\mathbf{T}}_{n+W_1} dA \cdot \frac{\mathbf{q}_{n+1}^A - \mathbf{q}_n^A}{\Delta t} \\
&\equiv \frac{\mathcal{E}_{n+1}^h - \mathcal{E}_n^h}{\Delta t} \quad \text{only for St. Venant-Kirchhoff material model}
\end{aligned} \tag{14.162}
$$

where the internal potential difference term yields

$$
\begin{aligned}
\int_{\mathcal{B}} \mathbf{S}^h(\mathbf{C}_{n+1/2}^h) : \frac{\mathbf{C}_{n+1}^h - \mathbf{C}_n^h}{\Delta t} dV &= \int_{\mathcal{B}} \mathbb{C} : \mathbf{E}_{n+1/2}^h : \frac{2(\mathbf{E}_{n+1}^h - \mathbf{E}_n^h)}{\Delta t} dV \\
&= \int_{\mathcal{B}} \frac{\bar{W}_{n+1}^h - \bar{W}_n^h}{\Delta t} dV
\end{aligned} \tag{14.163}
$$

for St. Venant-Kirchhoff material model (notice $\tilde{\mathbf{E}}^h = \frac{1}{2}[\tilde{\mathbf{C}}^h - \mathbf{I}]$, and we have $\mathbf{E}_{n+1/2}^h = \frac{1}{2}[\mathbf{C}_{n+1/2}^h - \mathbf{I}]$ if $W_1\Lambda_4 = 1/2$). For $W_1 = W_2 = 1/2$, we have $w_1 = -5$ and $w_2 = 5$ (constants) during simulation in $\mathcal{T} = [t_0, t_N]$; Algorithm 16 naturally becomes the exact energy-momentum conserving algorithm for this simplest nonlinear material model, and inherits second-order time accuracy.

To extend Algorithm 16 for a general hyperelastic material model with the second-order time accuracy keeping W_i $(i = 1, 2)$ constant in $\mathcal{T} = [t_0, t_N]$, we compare the developments to the energy-momentum conserving algorithms proposed by Simo and Tarnow (Simo and Tarnow 1992) and Simo and Gonzalez (Simo and Gonzalez 1994)/Gonzalez (Gonzalez 2000); it shows the algorithmic discrete second Piola-Kirchhoff stress tensor in Algorithm 16, i.e., $\mathbf{S}^h(\tilde{\mathbf{C}}^h)$, needs to be replaced with $\mathbf{S}^h(\mathbf{C}_{n+\beta_0}^h) + \mathbf{S}^h(\mathbf{C}_{n+(1-\beta_0)}^h)$ or $\tilde{\mathbf{S}}_{\text{DD}}^h$ (evaluation by the discrete derivative), respectively. We briefly show the resultant algorithms as an extension of Algorithm 16 for the general hyperelastic material model as follows (Algorithm 17 and Algorithm 18):

Algorithm 17: ***Approach 3-β***

Given $\mathbf{q}_n$ and $\mathbf{v}_n$, find $\mathbf{q}_{n+1}$ and $\mathbf{v}_{n+1}$ from

$$
\sum_{B=1}^{n_{\text{node}}} M_{AB} \frac{\mathbf{v}_{n+1}^B - \mathbf{v}_n^B}{\Delta t} + \int_{\mathcal{B}} \mathbf{F}^h(\tilde{\mathbf{q}}) \cdot \left[\mathbf{S}^h(\mathbf{C}_{n+\beta_0}^h) + \mathbf{S}^h(\mathbf{C}_{n+(1-\beta_0)}^h)\right] \text{GRAD}\, N_A dV = \tilde{\mathbf{F}}_A^{\text{ext}^h} \tag{14.164a}
$$

where $\beta_0 \in (0, 1)$ is obtained from

$$
\hat{W}^h(\mathbf{C}_{n+1}^h) - \hat{W}^h(\mathbf{C}_n^h) = [\mathbf{C}_{n+1}^h - \mathbf{C}_n^h] : \frac{1}{2}\left[\mathbf{S}^h(\mathbf{C}_{n+\beta_0}^h) + \mathbf{S}^h(\mathbf{C}_{n+(1-\beta_0)}^h)\right] \tag{14.164b}
$$

and the algorithmic nodal displacement and external force vectors are given by

$$
\tilde{\mathbf{q}} = \mathbf{q}_n + \Delta t W_1 \Lambda_1 \mathbf{v}_n + \Delta t W_2 \Lambda_2 \Delta \mathbf{v} \tag{14.164c}
$$

$$
\tilde{\mathbf{F}}^{\text{ext}^h} = (1 - W_1)\mathbf{F}_n^{\text{ext}^h} + W_1 \mathbf{F}_{n+1}^{\text{ext}^h} \tag{14.164d}
$$

respectively. The following are the updates:

$$\mathbf{q}_{n+1} = \mathbf{q}_n + \Delta t \lambda_1 \mathbf{v}_n + \Delta t \lambda_2 \Delta \mathbf{v} \tag{14.164e}$$

$$\mathbf{v}_{n+1} = \mathbf{v}_n + \Delta \mathbf{v} \tag{14.164f}$$

Remark 14.3.5

1. Algorithm 17 recovers the second-order time accurate energy-momentum conserving algorithm proposed by Simo and Tarnow (Simo and Tarnow 1992) for $\Lambda_1 = \lambda_1 = 1$, $\Lambda_2 = \lambda_2 = 1/2$, $\Lambda_4 = 1$ and $W_1 = W_2 = 1/2$ with the time-independent external force and boundary conditions.

Algorithm 18: ***Approach 3-γ***

Given $\mathbf{q}_n$ and $\mathbf{v}_n$, find $\mathbf{q}_{n+1}$ and $\mathbf{v}_{n+1}$ from

$$\sum_{B=1}^{n_{\text{node}}} M_{AB} \frac{\mathbf{v}_{n+1}^B - \mathbf{v}_n^B}{\Delta t} + \int_B \mathbf{F}^h(\tilde{\mathbf{q}}) \cdot \tilde{\mathbf{S}}_{\text{DD}}^h \, \text{GRAD} N_A dV = \tilde{\mathbf{F}}_A^{\text{ext}^h} \tag{14.165a}$$

where the algorithmic second Piola-Kirchhoff stress tensor $\tilde{\mathbf{S}}^h$ is defined via the discrete derivative as

$$\tilde{\mathbf{S}}_{\text{DD}}^h := 2D\hat{W}^h(\tilde{\mathbf{C}}^h) + 2\frac{\hat{W}^h(\mathbf{C}_{n+1}^h) - \hat{W}^h(\mathbf{C}_n^h) - D\hat{W}^h(\tilde{\mathbf{C}}^h) : \Delta \mathbf{C}^h}{||\Delta \mathbf{C}^h||^2} \Delta \mathbf{C}^h \tag{14.165b}$$

where $\Delta \mathbf{C}^h := \mathbf{C}_{n+1}^h - \mathbf{C}_n^h$ and $||\Delta \mathbf{C}^h||^2 := \Delta \mathbf{C}^h : \Delta \mathbf{C}^h$; and the algorithmic nodal displacement vector, discrete right Cauchy-Green tensor and external force vector are given by

$$\tilde{\mathbf{q}} = \mathbf{q}_n + \Delta t W_1 \Lambda_1 \mathbf{v}_n + \Delta t W_2 \Lambda_2 \Delta \mathbf{v} \tag{14.165c}$$

$$\tilde{\mathbf{C}}^h = (1 - W_1 \Lambda_4) \mathbf{C}_n^h + W_1 \Lambda_4 \mathbf{C}_{n+1}^h \tag{14.165d}$$

$$\tilde{\mathbf{F}}^{\text{ext}^h} = (1 - W_1) \mathbf{F}_n^{\text{ext}^h} + W_1 \mathbf{F}_{n+1}^{\text{ext}^h} \tag{14.165e}$$

respectively. The following are the updates:

$$\mathbf{q}_{n+1} = \mathbf{q}_n + \Delta t \lambda_1 \mathbf{v}_n + \Delta t \lambda_2 \Delta \mathbf{v} \tag{14.165f}$$

$$\mathbf{v}_{n+1} = \mathbf{v}_n + \Delta \mathbf{v} \tag{14.165g}$$

Remark 14.3.6

1. Algorithm 18 recovers the energy-momentum conserving algorithm and the underlying features described by Simo and Gonzalez (Simo and Gonzalez 1994) or Gonzalez (Gonzalez 2000) for $\Lambda_1 = \lambda_1 = 1$, $\Lambda_2 = \lambda_2 = 1/2$, $\Lambda_4 = 1$ and $W_1 = W_2 = 1/2$ with the time-independent external force and boundary conditions.

Energy Dissipative and Momentum Conserving Algorithm for General Hyperelastic Material Models Armero and Romeo (Armero and Romero 2001a) proposed the so-called energy dissipative and momentum conserving (EDMC) algorithm from the original energy-momentum conserving algorithm design. It can be obtained by the selection of the algorithmic parameters as $\Lambda_1 = \lambda_1 = 1$, $\Lambda_2 = \lambda_2 = 1/2$, $\Lambda_4 = 1$ and $W_1 = W_2 = 1/2$ in Algorithm 18, for the general hyperelastic material models. That is,

$$\sum_{B=1}^{n_{\text{node}}} M_{AB}\dot{\mathbf{v}}^B + \tilde{\mathbf{F}}^{\text{int}^h}_{\text{con }A} = \mathbf{F}^{\text{ext}^h}_{A\ n+1/2} \tag{14.166}$$

$$\frac{\mathbf{q}^A_{n+1} - \mathbf{q}^A_n}{\Delta t} = \mathbf{v}^A_{n+1/2} \tag{14.167}$$

where the constant mass matrix is defined as $M_{AB} := \int_{\mathcal{B}} \rho_0 N_A N_B dV$ and the algorithmic discrete internal and external forces for the conserving schemes are given by

$$\tilde{\mathbf{F}}^{\text{int}^h}_{\text{con }A} := \int_{\mathcal{B}} \mathbf{F}^h \cdot \tilde{\mathbf{S}}^h \,\text{GRAD}\, N_A dV \tag{14.168}$$

$$\mathbf{F}^{\text{ext}^h}_{A\ n+1/2} := \int_{\mathcal{B}} N_A \rho_0 \mathbf{B}_{n+1/2} dV + \int_{\partial\mathcal{B}_\sigma} N_A \bar{\mathbf{T}}_{n+1/2} dA \tag{14.169}$$

respectively, with the algorithmic discrete 2nd Piola-Kirchhoff stress $\tilde{\mathbf{S}}^h$ defined by

$$\tilde{\mathbf{S}}^h := 2D\hat{W}^h\left(\mathbf{C}^h_{n+1/2}\right) + 2\frac{\hat{W}^h\left(\mathbf{C}^h_{n+1}\right) - \hat{W}^h\left(\mathbf{C}^h_n\right) - D\hat{W}^h\left(\mathbf{C}^h_{n+1/2}\right) : \Delta\mathbf{C}^h}{||\Delta\mathbf{C}^h||^2}\Delta\mathbf{C}^h \tag{14.170}$$

Note $\mathbf{v}^A(t) = \dot{\mathbf{q}}^A(t)$ denotes the nodal velocity vector.

Algorithmic Dissipation - Internal Force Introduce the algorithmic dissipation internal force function, $\tilde{\mathbf{F}}^{\text{int}^h}_{\text{diss}}$, in Equation (14.166) as

$$\sum_{B=1}^{n_{\text{node}}} M_{AB}\frac{\mathbf{v}^B_{n+1} - \mathbf{v}^B_n}{\Delta t} + \tilde{\mathbf{F}}^{\text{int}^h}_{\text{con }A} + \tilde{\mathbf{F}}^{\text{int}^h}_{\text{diss }A} = \mathbf{F}^{\text{ext}^h}_{A\ n+1/2} \tag{14.171}$$

Equation (14.171) is first-order time approximation if $\tilde{\mathbf{F}}^{\text{int}^h}_{\text{diss }A} = \mathcal{O}(\Delta t)$, and it is second-order time approximation if $\tilde{\mathbf{F}}^{\text{int}^h}_{\text{diss }A} = \mathcal{O}(\Delta t^2)$. To define the algorithmic dissipation internal force function for the first-order time accurate energy dissipative and momentum conserving (EDMC-1) algorithm, consider the discrete algorithmic 2nd Piola-Kirchhoff stress $\tilde{\mathbf{S}}^h$ defined by Equation (14.170):

$$\begin{aligned}\tilde{\mathbf{S}}^h &:= 2D\hat{W}^h\left(\mathbf{C}^h_{n+1/2}\right) + 2\frac{\hat{W}^h\left(\mathbf{C}^h_{n+1}\right) - \hat{W}^h\left(\mathbf{C}^h_n\right) - D\hat{W}^h\left(\mathbf{C}^h_{n+1/2}\right) : \Delta\mathbf{C}^h}{||\Delta\mathbf{C}^h||^2}\Delta\mathbf{C}^h \\ &= 2D\hat{W}^h\left(\mathbf{C}^h_{n+1/2}\right) - 2\frac{D\hat{W}^h\left(\mathbf{C}^h_{n+1/2}\right) : \Delta\mathbf{C}^h}{||\Delta\mathbf{C}^h||}\frac{\Delta\mathbf{C}^h}{||\Delta\mathbf{C}^h||} + 2\Psi\frac{\Delta\mathbf{C}^h}{||\Delta\mathbf{C}^h||}\end{aligned} \tag{14.172}$$

where

$$\Psi := \frac{\hat{W}^h\left(\mathbf{C}^h_{n+1}\right) - \hat{W}^h\left(\mathbf{C}^h_n\right)}{||\Delta\mathbf{C}^h||} \tag{14.173}$$

Modify Equation (14.173) by introducing a new algorithmic parameter $\chi_1 \in \mathbb{R}$ as

$$\Psi \approx \frac{(1+2\chi_1)\hat{W}^h\left(\mathbf{C}^h_{n+1}\right) - (1-2\chi_1)\hat{W}^h\left(\mathbf{C}^h_n\right) - 4\chi_1\hat{W}^h\left(\mathbf{C}^h_{n+1/2}\right)}{||\Delta\mathbf{C}^h||}$$
$$= \frac{\hat{W}^h\left(\mathbf{C}^h_{n+1}\right) - \hat{W}^h\left(\mathbf{C}^h_n\right) + \mathcal{D}_\mathrm{W}}{||\Delta\mathbf{C}^h||} \tag{14.174}$$

where the algorithmic dissipation function $\mathcal{D}_\mathrm{W} \in \mathbb{R}_+$ is defined by

$$\mathcal{D}_\mathrm{W} := 4\chi_1\left[\frac{\hat{W}^h\left(\mathbf{C}^h_{n+1}\right) + \hat{W}^h\left(\mathbf{C}^h_n\right)}{2} - \hat{W}^h\left(\mathbf{C}^h_{n+1/2}\right)\right] \geq 0 \tag{14.175}$$

for $\chi_1 \geq 0$. Note Equation (14.175) satisfies the consistency condition, i.e.,

$$\frac{\mathcal{D}_\mathrm{W}}{||\Delta\mathbf{C}^h||} \to 0 \quad as \quad ||\Delta\mathbf{C}^h|| \to 0 \tag{14.176}$$

Therefore, Equation (14.172) can be written as

$$\tilde{\mathbf{S}}^h \approx \tilde{\mathbf{S}}^h_\mathrm{con} + \tilde{\mathbf{S}}^h_\mathrm{diss} \tag{14.177}$$

where the algorithmic discrete 2nd Piola-Kirchhoff stresses for the conserving scheme and the dissipative scheme are defined by

$$\tilde{\mathbf{S}}^h_\mathrm{con} := 2D\hat{W}^h\left(\mathbf{C}^h_{n+1/2}\right) \tag{14.178}$$

$$+\,2\frac{\hat{W}^h\left(\mathbf{C}^h_{n+1}\right) - \hat{W}^h\left(\mathbf{C}^h_n\right) - D\hat{W}^h\left(\mathbf{C}^h_{n+1/2}\right) : \Delta\mathbf{C}^h}{||\Delta\mathbf{C}^h||^2}\Delta\mathbf{C}^h \tag{14.179}$$

$$\tilde{\mathbf{S}}^h_\mathrm{diss} := 2\frac{\mathcal{D}_\mathrm{W}}{||\Delta\mathbf{C}^h||^2}\Delta\mathbf{C}^h \tag{14.180}$$

respectively. Hence, the algorithmic dissipation internal force in Equation (14.171) can be defined as

$$\tilde{\mathbf{F}}^{\mathrm{int}^h}_{\mathrm{diss}\,A} := \int_\mathcal{B} \mathbf{F}^h\left(\varphi_{n+1/2}\right) \cdot \tilde{\mathbf{S}}^h_\mathrm{diss}\mathrm{GRAD}N_A dV \tag{14.181}$$

Algorithmic Dissipation - Velocity In order to introduce the algorithmic dissipation, the internal force is not enough to show the dissipation in the total energy of the system. The Taylor series expansion of the nodal displacement vector at time t_{n+1} about t_n yields

$$\mathbf{q}^A(t_{n+1}) = \mathbf{q}^A(t_n) + \Delta t\dot{\mathbf{q}}^A(t_n) + \frac{\Delta t^2}{2}\ddot{\mathbf{q}}^A(t_n) + \mathcal{O}(\Delta t^3)$$
$$= \mathbf{q}^A(t_n) + \Delta t\mathbf{v}^A(t_n) + \frac{\Delta t^2}{2}\dot{\mathbf{v}}^A(t_n) + \mathcal{O}(\Delta t^3)$$
$$= \mathbf{q}^A(t_n) + \frac{\Delta t}{2}[\mathbf{v}^A(t_{n+1}) + \mathbf{v}^A(t_n)] + \mathcal{O}(\Delta t^3) \tag{14.182}$$

Assuming $\mathbf{q}^A(t_{n+1}) = \mathbf{q}_{n+1}^A$, $\mathbf{q}^A(t_n) = \mathbf{q}_n^A$, $\mathbf{v}^A(t_{n+1}) = \mathbf{v}_{n+1}^A$ and $\mathbf{v}^A(t_n) = \mathbf{v}_n^A$ and introducing the algorithmic dissipative nodal velocity function $\tilde{\mathbf{v}}_{\text{diss}}^A$, we have

$$\begin{aligned}\mathbf{q}_{n+1}^A &= \mathbf{q}_n^A + \frac{\Delta t}{2}[\mathbf{v}_{n+1}^A + \mathbf{v}_n^A] + \Delta t \tilde{\mathbf{v}}_{\text{diss}}^A + \mathcal{O}(\Delta t^3) \\ &\cong \mathbf{q}_n^A + \frac{\Delta t}{2}[\mathbf{v}_{n+1}^A + \mathbf{v}_n^A] + \Delta t \tilde{\mathbf{v}}_{\text{diss}}^A\end{aligned} \tag{14.183}$$

That is,

$$\frac{\mathbf{q}_{n+1}^A - \mathbf{q}_n^A}{\Delta t} = \mathbf{v}_{n+1/2}^A + \tilde{\mathbf{v}}_{\text{diss}}^A \tag{14.184}$$

As can be seen from Equation (14.183), the approximation is first-order in time if $\tilde{\mathbf{v}}_{\text{diss}}^A = \mathcal{O}(\Delta t)$, and it is second-oder in time if $\tilde{\mathbf{v}}_{\text{diss}}^A = \mathcal{O}(\Delta t^2)$. Armero and Romeo (Armero and Romero 2001a) suggested that the expression for the algorithmic dissipative nodal velocity function is obtained by solving

$$\sum_{B=1}^{n_{\text{node}}} M_{AB} \tilde{\mathbf{v}}_{\text{diss}}^B = \int_{\mathcal{B}} N_A \hat{\mathcal{D}}_{\mathcal{K}} \frac{\mathbf{V}_{n+1}^h + \mathbf{V}_n^h}{||\mathbf{V}_{n+1}^h||^2 - ||\mathbf{V}_n^h||^2} dV \tag{14.185}$$

with the numerical dissipation function given by

$$\mathcal{D}_{\mathcal{K}} = \int_{\mathcal{B}} \hat{\mathcal{D}}_{\mathcal{K}} dV = \frac{\chi_2}{2} \int_{\mathcal{B}} \rho_0 \left[||\mathbf{V}_{n+1}^h|| - ||\mathbf{V}_n^h||\right]^2 dV \geq 0 \tag{14.186}$$

for a new algorithmic scalar parameter $\chi_2 \geq 0$. Notice we calculate $\mathbf{V}_{n+1}^h$ and $\mathbf{V}_n^h$ from

$$\mathbf{V}_{n+1}^h \cong \sum_{A=1}^{n_{\text{node}}} N_A(\mathbf{X}) \dot{\mathbf{q}}_{n+1}^A = \sum_{A=1}^{n_{\text{node}}} N_A(\mathbf{X}) \mathbf{v}_{n+1}^A \tag{14.187}$$

$$\mathbf{V}_n^h \cong \sum_{A=1}^{n_{\text{node}}} N_A(\mathbf{X}) \dot{\mathbf{q}}_n^A = \sum_{A=1}^{n_{\text{node}}} N_A(\mathbf{X}) \mathbf{v}_n^A \tag{14.188}$$

respectively. Substituting Equation (14.186) into Equation (14.185) yields

$$\sum_{B=1}^{n_{\text{node}}} M_{AB} \tilde{\mathbf{v}}_{\text{diss}}^B = \chi_2 \int_{\mathcal{B}} \rho_0 N_A \frac{||\mathbf{V}_{n+1}^h|| - ||\mathbf{V}_n^h||}{||\mathbf{V}_{n+1}^h|| + ||\mathbf{V}_n^h||} \mathbf{V}_{n+1/2}^h dV \tag{14.189}$$

In a lumped system, we may write Equation (14.189) as

$$\sum_{B=1}^{n_{\text{node}}} M_{AB} \tilde{\mathbf{v}}_{\text{diss}}^B \approx \sum_{B=1}^{n_{\text{node}}} \delta_{AB} m_B \frac{\hat{\mathcal{D}}_{\mathcal{K}}^B}{||\mathbf{v}_{n+1}^B||^2 - ||\mathbf{v}_n^B||^2} \mathbf{v}_{n+1/2}^B \tag{14.190}$$

where δ_{AB} is Kronecker's delta and the dissipation function is given as

$$\mathcal{D}_{\mathcal{K}} = \sum_{A=1}^{n_{\text{node}}} \mathcal{D}_{\mathcal{K}}^A = \frac{\chi_2}{2} \sum_{A=1}^{n_{\text{node}}} m_A \left[||\mathbf{v}_{n+1}^A|| - ||\mathbf{v}_n^A||\right]^2 \geq 0 \tag{14.191}$$

for $\chi_2 \geq 0$. Note Equation (14.191) satisfies the consistency condition, i.e.,

$$\frac{\mathcal{D}_{\mathcal{K}}^A}{||\mathbf{v}_{n+1}^A|| - ||\mathbf{v}_n^A||} \rightarrow 0 \quad as \quad ||\mathbf{v}_{n+1}^A|| \rightarrow ||\mathbf{v}_n^A|| \tag{14.192}$$

Hence, in this case, the algorithmic dissipative nodal velocity function can be written as

$$\tilde{\mathbf{v}}_{\text{diss}}^A = \chi_2 \frac{||\mathbf{v}_{n+1}^A|| - ||\mathbf{v}_n^A||}{||\mathbf{v}_{n+1}^A|| + ||\mathbf{v}_n^A||} \mathbf{v}_{n+1/2}^A \tag{14.193}$$

Algorithm 19: ***Energy Dissipative Momentum Conserving Algorithm (EDMC-1)***

$$\sum_{B=1}^{n_{\text{node}}} M_{AB} \frac{\mathbf{v}_{n+1}^B - \mathbf{v}_n^B}{\Delta t} + \int_{\mathcal{B}} \mathbf{F}^h \cdot \tilde{\mathbf{S}}^h \,\text{GRAD}\, N_A \, dV$$

$$= \int_{\mathcal{B}} N_A \rho_0 \mathbf{B}_{n+1/2} dV + \int_{\partial \mathcal{B}_\sigma} N_A \bar{\mathbf{T}}_{n+1/2} dA \tag{14.194a}$$

$$\frac{\mathbf{q}_{n+1}^A - \mathbf{q}_n^A}{\Delta t} = \mathbf{v}_{n+1/2}^A + \tilde{\mathbf{v}}_{\text{diss}}^A \tag{14.194b}$$

where the algorithmic discrete 2nd Piola-Kirchhoff stress is given by

$$\tilde{\mathbf{S}}^h = \tilde{\mathbf{S}}_{\text{con}}^h + \tilde{\mathbf{S}}_{\text{diss}}^h \tag{14.194c}$$

$$\tilde{\mathbf{S}}_{\text{con}}^h := 2D\hat{W}^h \left(\mathbf{C}_{n+1/2}^h\right) + 2\frac{\hat{W}^h \left(\mathbf{C}_{n+1}^h\right) - \hat{W}^h \left(\mathbf{C}_n^h\right) - D\hat{W}^h \left(\mathbf{C}_{n+1/2}^h\right) : \Delta \mathbf{C}^h}{||\Delta \mathbf{C}^h||^2} \Delta \mathbf{C}^h \tag{14.194d}$$

$$\tilde{\mathbf{S}}_{\text{diss}}^h := 2\frac{\mathcal{D}_W}{||\Delta \mathbf{C}^h||^2} \Delta \mathbf{C}^h \tag{14.194e}$$

and the algorithmic dissipation nodal velocity is given by

$$\tilde{\mathbf{v}}_{\text{diss}}^A = \frac{\mathcal{D}_{\mathcal{K}}}{||\mathbf{v}_{n+1}^A||^2 - ||\mathbf{v}_n^A||^2} \mathbf{v}_{n+1/2}^A \tag{14.194f}$$

Remark 14.3.7

1. When $\chi_1 = \chi_2 = 0$, Algorithm 19 recovers the energy-momentum conserving algorithm for the general hyperleastic material models.
2. Algorithm 19 exactly conserves the discrete total linear and angular momenta within the time step, while the discrete total energy is dissipative.
3. Algorithm 19 was proved to be unconditionally stable in nonlinear dynamic systems. Although the exact energy-momentum conserving algorithm remains unconditionally stable in nonlinear dynamic systems, it sometimes causes divergence of the Newton-Raphson iterative method because

of the participation of the high-frequency modes due to the finite element space discretization. Algorithm 19 tends to avoid this issue that is caused by the presence of high-frequency modes.

4. For the second-order time accurate EDMC (EDMC-2) algorithm design, see Armero and Romero (Armero and Romero 2001b); it, however, requires the solution of two systems of equations and is not a single system single solve procedure (for such a higher computational expense, the typical time accuracy of computational algorithms that should be theoretically attained is upto a maximum of four with unconditional stability (see generalized barrier theorem; (Zhou and Tamma 2004a)).

Single-field Form Algorithm Designs

Approach 1: Classical Approach - Classical Time Weighted Residual Approach In this subsection, we turn our attention to the single-field form, and we derive the single-field form versions of Algorithm 14. Again, consider the semi-discretized representation for conservative systems via the Total Energy framework and the corresponding equations of motion with given initial conditions:

$$\frac{d}{dt}\left(\frac{\partial \mathcal{E}^h(\mathbf{q},\dot{\mathbf{q}})}{\partial \dot{\mathbf{q}}^A}\right)+\frac{\partial \mathcal{E}^h(\mathbf{q},\dot{\mathbf{q}})}{\partial \mathbf{q}^A}=\mathbf{0}\ \ \forall t\in\mathcal{T} \qquad \mathbf{q}(t_0)=\mathbf{q}_0 \ \text{ and } \ \dot{\mathbf{q}}(t_0)=\dot{\mathbf{q}}_0 \tag{14.195}$$

for $A=1,2,\ldots,n_{\text{node}}$. Employing the classical time weighted residual methodology, yields

$$\int_{t_n}^{t_{n+1}} W\left[\frac{d}{dt}\left(\frac{\partial \mathcal{E}^h(\mathbf{q},\dot{\mathbf{q}})}{\partial \dot{\mathbf{q}}}\right)+\frac{\partial \mathcal{E}^h(\mathbf{q},\dot{\mathbf{q}})}{\partial \mathbf{q}}\right]dt=\mathbf{0} \quad or \quad \int_{t_n}^{t_{n+1}} W\left[\mathbf{M}\frac{d\dot{\mathbf{q}}}{dt}+\mathbf{F}^{\text{int}^h}-\mathbf{F}^{\text{ext}^h}\right]dt=\mathbf{0} \tag{14.196}$$

with the time weighting function $W\in\mathbb{R}$ given by

$$W=w_0+w_1\Gamma+w_2\Gamma^2+w_3\Gamma^3 \ \ (w_0=1) \tag{14.197}$$

and correspondingly approximate $\mathbf{F}^{\text{int}^h}$, $\dot{\mathbf{q}}$ and $\ddot{\mathbf{q}}$ by means of the asymptotic series expansions while we approximate the external nodal forces $\mathbf{F}^{\text{ext}^h}$ by means of the Taylor series expansion (in general, this is a special case of an asymptotic series expansion and is sufficient for the cases of interest here; but the general case is a function evaluation at a particular time point and not necessarily a linear interpolation).

$$\mathbf{F}^{\text{int}^h}=\mathbf{F}_n^{\text{int}^h}+\Lambda_1\dot{\mathbf{F}}_n^{\text{int}^h}\tau+\Lambda_2\ddot{\mathbf{F}}_n^{\text{int}^h}\tau^2+\Lambda_3\frac{\ddot{\mathbf{F}}_{n+1}^{\text{int}^h}-\ddot{\mathbf{F}}_n^{\text{int}^h}}{\Delta t}\tau^3 \tag{14.198}$$

$$\dot{\mathbf{q}}=\dot{\mathbf{q}}_n+\Lambda_4\ddot{\mathbf{q}}_n\tau+\Lambda_5\frac{\ddot{\mathbf{q}}_{n+1}-\ddot{\mathbf{q}}_n}{\Delta t}\tau^2 \tag{14.199}$$

$$\ddot{\mathbf{q}}=\ddot{\mathbf{q}}_n+\Lambda_6\frac{\ddot{\mathbf{q}}_{n+1}-\ddot{\mathbf{q}}_n}{\Delta t}\tau \tag{14.200}$$

$$\mathbf{F}^{\text{ext}^h}=\mathbf{F}_n^{\text{ext}^h}+\frac{\mathbf{F}_{n+1}^{\text{ext}^h}-\mathbf{F}_n^{\text{ext}^h}}{\Delta t}\tau \tag{14.201}$$

To design the update for the nodal displacement and velocity vectors, introduce $\lambda_i\in\mathbb{R}$ $(i=1,2,3,4,5)$ for

$$\mathbf{q}=\mathbf{q}_n+\lambda_1\dot{\mathbf{q}}_n\tau+\lambda_2\ddot{\mathbf{q}}_n\tau^2+\lambda_3\frac{\ddot{\mathbf{q}}_{n+1}-\ddot{\mathbf{q}}_n}{\Delta t}\tau^3 \tag{14.202}$$

$$\dot{\mathbf{q}} = \dot{\mathbf{q}}_n + \lambda_4 \ddot{\mathbf{q}}_n \tau + \lambda_5 \frac{\ddot{\mathbf{q}}_{n+1} - \ddot{\mathbf{q}}_n}{\Delta t} \tau^2 \tag{14.203}$$

and set $\tau = \Delta t$; hence, we have

$$\mathbf{q}_{n+1} = \mathbf{q}_n + \lambda_1 \dot{\mathbf{q}}_n \Delta t + \lambda_2 \ddot{\mathbf{q}}_n \Delta t^2 + \lambda_3 \Delta (\ddot{\mathbf{q}}_{n+1} - \ddot{\mathbf{q}}_n) \Delta t^2 \tag{14.204}$$

$$\dot{\mathbf{q}}_{n+1} = \dot{\mathbf{q}}_n + \lambda_4 \ddot{\mathbf{q}}_n \Delta t + \lambda_5 (\ddot{\mathbf{q}}_{n+1} - \ddot{\mathbf{q}}_n) \Delta t \tag{14.205}$$

Define

$$W_i := \frac{\int_{t_n}^{t_{n+1}} \left(\frac{\tau}{\Delta t}\right)^i W dt}{\int_{t_n}^{t_{n+1}} W dt} \quad for \ \ i = 1, 2, 3 \tag{14.206}$$

for convenience, and as a result, the following generalized single-step single-solve family of algorithms are obtained:

Algorithm 20: ***Approach 1***

Given $\mathbf{q}_n$, $\dot{\mathbf{q}}_n$, and $\ddot{\mathbf{q}}_n$, find $\mathbf{q}_{n+1}$, $\dot{\mathbf{q}}_{n+1}$, and $\ddot{\mathbf{q}}_{n+1}$ from

$$\sum_{B=1}^{n_{\text{node}}} M_{AB} \tilde{\mathbf{a}}^B + \tilde{\mathbf{F}}_A^{\text{int}^h} = \tilde{\mathbf{F}}_A^{\text{ext}^h} \tag{14.207a}$$

$$\tilde{\mathbf{a}} = \ddot{\mathbf{q}}_n + W_1 \Lambda_6 \Delta \mathbf{a} \tag{14.207b}$$

$$\begin{aligned} \tilde{\mathbf{F}}^{\text{int}^h} &= \mathbf{F}_n^{\text{int}^h} + \frac{W_3 \Lambda_3}{\lambda_3} \Delta \mathbf{F}^{\text{int}^h} \\ &+ \left(W_1 \Lambda_1 - \frac{W_3 \Lambda_3 \lambda_1}{\lambda_3} \right) \dot{\mathbf{F}}_n^{\text{int}^h} \Delta t + \left(W_2 \Lambda_2 - \frac{W_3 \Lambda_3 \lambda_2}{\lambda_3} \right) \ddot{\mathbf{F}}_n^{\text{int}^h} \Delta t^2 \end{aligned} \tag{14.207c}$$

$$\tilde{\mathbf{F}}^{\text{ext}^h} = (1 - W_1) \mathbf{F}_n^{\text{ext}^h} + W_1 \mathbf{F}_{n+1}^{\text{ext}^h} \tag{14.207d}$$

with the following updates:

$$\mathbf{q}_{n+1} = \mathbf{q}_n + \lambda_1 \dot{\mathbf{q}}_n \Delta t + \lambda_2 \ddot{\mathbf{q}}_n \Delta t^2 + \lambda_3 \Delta \mathbf{a} \Delta t^2 \tag{14.207e}$$

$$\dot{\mathbf{q}}_{n+1} = \dot{\mathbf{q}}_n + \lambda_4 \ddot{\mathbf{q}}_n \Delta t + \lambda_5 \Delta \mathbf{a} \Delta t \tag{14.207f}$$

$$\ddot{\mathbf{q}}_{n+1} = \ddot{\mathbf{q}}_n + \Delta \mathbf{a} \tag{14.207g}$$

where $\Delta \mathbf{a} := \ddot{\mathbf{q}}_{n+1} - \ddot{\mathbf{q}}_n$.

Remark 14.3.8

1. Algorithm 20 is an extension of GSSSS family of algorithms originally derived for linear dynamic conservative systems to nonlinear dynamic conservative systems. The DNA parameters $\{W, \Lambda, \lambda\}$ are given as shown in Algorithm 9 and Algorithm 10 for the U0 family of algorithms and the V0 family of algorithms, respectively. This approach fails to provide proper extensions for nonlinear dynamic situations in contrast to the latter two approaches that are described next (Hoitink et al. 2008; Masuri et al. 2009a,b,c).

2. The algorithmic external nodal forces are given by

$$\begin{aligned}\tilde{\mathbf{F}}^{\text{ext}^h} &= (1 - W_1)\mathbf{F}_n^{\text{ext}^h} + W_1\mathbf{F}_{n+1}^{\text{ext}^h} =: \mathbf{F}_{n+W_1}^{\text{ext}^h} \\ &= \int_{\mathcal{B}} N_A \rho_0 \mathbf{B}_{n+W_1} dV + \int_{\partial\mathcal{B}_\sigma} N_A \bar{\mathbf{T}}_{n+W_1} dA \end{aligned} \tag{14.208}$$

where $\mathbf{B}_{n+W_1} := (1 - W_1)\mathbf{B}_n + W_1\mathbf{B}_{n+1}$ and $\bar{\mathbf{T}}_{n+W_1} := (1 - W_1)\bar{\mathbf{T}}_n + W_1\bar{\mathbf{T}}_{n+1}$.

3. In general, symplecticness, discrete total energy, and discrete total angular momentum are not exactly preserved/conserved within the time step $[t_n, t_{n+1}]$ within the framework of Algorithm 20. Note the discrete total linear momentum within the time step is exactly conserved for any conditions.

Approach 2: Symplectic-Momentum Framework - Normalized Time Weighted Residual Approach For the derivation of the single-field form versions of Algorithm 15 (Approach 2), consider Equation (14.195) which yields

$$\begin{aligned}&\frac{d}{dt}\left(\frac{\partial \mathcal{E}^h(\mathbf{q}, \dot{\mathbf{q}})}{\partial \dot{\mathbf{q}}^A}\right) + \frac{\partial \mathcal{E}^h(\mathbf{q}, \dot{\mathbf{q}})}{\partial \mathbf{q}^A} = \mathbf{0} \\ &\Rightarrow \sum_{B=1}^{n_{\text{node}}} M_{AB}\ddot{\mathbf{q}}^B + \int_{\mathcal{B}} \mathbf{F}^h \cdot \mathbf{S}^h \text{ GRAD } N_A dV = \underbrace{\int_{\mathcal{B}} N_A \rho_0 \mathbf{B} dV + \int_{\partial\mathcal{B}_\sigma} N_A \bar{\mathbf{T}} dA}_{\mathbf{F}_A^{\text{ext}^h} : \mathcal{T} \to \mathbb{R}^3}\end{aligned} \tag{14.209}$$

Following the same concept for the time discretization of the discrete internal force vector in the two-field form algorithm design, assume the discrete deformation gradient and the discrete second Piola-Kirchhoff stress tensor as functions of $\mathbf{q}$. Employing the *normalized* time weighted residual methodology, to $\ddot{\mathbf{q}}$, $\mathbf{q}$ and $\mathbf{F}^{\text{ext}^h}$, we have

$$\begin{aligned}&\sum_{B=1}^{n_{\text{node}}} M_{AB} \frac{\int_{t_n}^{t_{n+1}} W \ddot{\mathbf{q}}^B dt}{\int_{t_n}^{t_{n+1}} W dt} \\ &+ \int_{\mathcal{B}} \mathbf{F}^h \left(\frac{\int_{t_n}^{t_{n+1}} W \mathbf{q} dt}{\int_{t_n}^{t_{n+1}} W dt}\right) \cdot \mathbf{S}^h \left(C^h \left(\frac{\int_{t_n}^{t_{n+1}} W \mathbf{q} dt}{\int_{t_n}^{t_{n+1}} W dt}\right)\right) \text{GRAD } N_A dV \\ &= \frac{\int_{t_n}^{t_{n+1}} W \mathbf{F}_A^{\text{ext}^h} dt}{\int_{t_n}^{t_{n+1}} W dt}\end{aligned} \tag{14.210}$$

where the scalar time weighting function is approximated by

$$W = 1 + w_1\Gamma + w_2\Gamma^2 + w_3\Gamma^3 \tag{14.211}$$

with the corresponding approximations of $\mathbf{q}$, $\dot{\mathbf{q}}$, and $\ddot{\mathbf{q}}$ by means of the asymptotic series expansions, and the approximations of the set of the external nodal forces, $\mathbf{F}^{\text{ext}^h}$, by means of the Taylor series expansion (in general, this is a special case of an asymptotic series expansion and is sufficient for the cases of interest here; but the general case is a function evaluation at a particular time point and not necessarily a linear interpolation) in time as follows:

$$\mathbf{q} = \mathbf{q}_n + \Lambda_1 \dot{\mathbf{q}}_n \tau + \Lambda_2 \ddot{\mathbf{q}}_n \tau^2 + \Lambda_3 \frac{\ddot{\mathbf{q}}_{n+1} - \ddot{\mathbf{q}}_n}{\Delta t} \tau^3 \tag{14.212}$$

$$\dot{\mathbf{q}} = \dot{\mathbf{q}}_n + \Lambda_4 \ddot{\mathbf{q}}_n \tau + \Lambda_5 \frac{\ddot{\mathbf{q}}_{n+1} - \ddot{\mathbf{q}}_n}{\Delta t} \tau^2 \tag{14.213}$$

$$\ddot{\mathbf{q}} = \ddot{\mathbf{q}}_n + \Lambda_6 \frac{\ddot{\mathbf{q}}_{n+1} - \ddot{\mathbf{q}}_n}{\Delta t} \tau \tag{14.214}$$

$$\mathbf{F}^{\text{ext}^h} = \mathbf{F}_n^{\text{ext}^h} + \frac{\mathbf{F}_{n+1}^{\text{ext}^h} - \mathbf{F}_n^{\text{ext}^h}}{\Delta t} \tau \tag{14.215}$$

Therefore, the algorithmic nodal unknowns are defined as follows:

$$\tilde{\mathbf{q}} := \frac{\int_{t_n}^{t_{n+1}} W \mathbf{q} dt}{\int_{t_n}^{t_{n+1}} W dt} = \mathbf{q}_n + W_1 \Lambda_1 \dot{\mathbf{q}}_n \Delta t + W_2 \Lambda_2 \ddot{\mathbf{q}}_n \Delta t^2 + W_3 \Lambda_3 \left(\ddot{\mathbf{q}}_{n+1} - \ddot{\mathbf{q}}_n\right) \Delta t^2 \tag{14.216}$$

$$\tilde{\mathbf{v}} := \frac{\int_{t_n}^{t_{n+1}} W \dot{\mathbf{q}} dt}{\int_{t_n}^{t_{n+1}} W dt} = \dot{\mathbf{q}}_n + W_1 \Lambda_4 \ddot{\mathbf{q}}_n \Delta t + W_2 \Lambda_5 \left(\ddot{\mathbf{q}}_{n+1} - \ddot{\mathbf{q}}_n\right) \Delta t \tag{14.217}$$

$$\tilde{\mathbf{a}} := \frac{\int_{t_n}^{t_{n+1}} W \ddot{\mathbf{q}} dt}{\int_{t_n}^{t_{n+1}} W dt} = \ddot{\mathbf{q}}_n + W_1 \Lambda_6 \left(\ddot{\mathbf{q}}_{n+1} - \ddot{\mathbf{q}}_n\right) =: \ddot{\mathbf{q}}_{n+W_1 \Lambda_6} \tag{14.218}$$

$$\tilde{\mathbf{F}}^{\text{ext}^h} := \frac{\int_{t_n}^{t_{n+1}} W \mathbf{F}^{\text{ext}^h} dt}{\int_{t_n}^{t_{n+1}} W dt} = \mathbf{F}_n^{\text{ext}^h} + W_1 \left(\mathbf{F}_{n+1}^{\text{ext}^h} - \mathbf{F}_n^{\text{ext}^h}\right) =: \mathbf{F}_{n+W_1}^{\text{ext}^h} \tag{14.219}$$

The algorithmic deformation gradient and the algorithmic right Cauchy-Green deformation tensor can be obtained as shown in Equation (14.139). Summarizing, the following generalized family of algorithms are obtained:

Algorithm 21: ***Approach 2***

Given $\mathbf{q}_n$, $\dot{\mathbf{q}}_n$, and $\ddot{\mathbf{q}}_n$, find $\mathbf{q}_{n+1}$, $\dot{\mathbf{q}}_{n+1}$, and $\ddot{\mathbf{q}}_{n+1}$ from

$$\sum_{B=1}^{n_{\text{node}}} M_{AB} \tilde{\mathbf{a}}^B + \int_B \mathbf{F}^h(\tilde{\mathbf{q}}) \cdot \mathbf{S}(\mathbf{C}^h(\tilde{\mathbf{q}})) \text{GRAD}\, N_A dV = \tilde{\mathbf{F}}_A^{\text{ext}^h} \tag{14.220a}$$

$$\tilde{\mathbf{a}} = \ddot{\mathbf{q}}_n + W_1 \Lambda_6 \Delta \mathbf{a} \tag{14.220b}$$

$$\tilde{\mathbf{q}} = \mathbf{q}_n + W_1 \Lambda_1 \dot{\mathbf{q}}_n \Delta t + W_2 \Lambda_2 \ddot{\mathbf{q}}_n \Delta t^2 + W_3 \Lambda_3 \Delta \mathbf{a} \Delta t^2 \tag{14.220c}$$

$$\tilde{\mathbf{F}}^{\text{ext}^h} = (1 - W_1) \mathbf{F}_n^{\text{ext}^h} + W_1 \mathbf{F}_{n+1}^{\text{ext}^h} \tag{14.220d}$$

with the following updates:

$$\mathbf{q}_{n+1} = \mathbf{q}_n + \lambda_1 \dot{\mathbf{q}}_n \Delta t + \lambda_2 \ddot{\mathbf{q}}_n \Delta t^2 + \lambda_3 \Delta \mathbf{a} \Delta t^2 \tag{14.220e}$$

$$\dot{\mathbf{q}}_{n+1} = \dot{\mathbf{q}}_n + \lambda_4 \ddot{\mathbf{q}}_n \Delta t + \lambda_5 \Delta \mathbf{a} \Delta t \tag{14.220f}$$

$$\ddot{\mathbf{q}}_{n+1} = \ddot{\mathbf{q}}_n + \Delta \mathbf{a} \tag{14.220g}$$

Remark 14.3.9

1. Algorithm 21 is an extension of GSSSS family of parent linear algorithms originally developed for linear dynamic conservative systems to nonlinear dynamic conservative systems. The DNA parameters $\{W, \Lambda, \lambda\}$ are given as shown in Algorithm 9 and Algorithm 10 for the U0 family of algorithms and the V0 family of algorithms, respectively. The representations yield symplectic-momentum based algorithm designs (and, when controllable numerical dissipation is turned off, they readily yield a family of symplectic-momentum conserving algorithms in the strict sense of the single-field form of representation of the equation of motion (Hoitink et al. 2008; Masuri et al. 2009b,c)).
2. **Symplecticness**: Algorithm 21 yields a family of symplectic algorithms in the sense of the LMS single-field form (three root scheme) for the conditions $U0(\{\rho_{1\infty}, \rho_{2\infty}, \rho_{3\infty}\} = \{1, 1, 1\})$ or $V0(\{\rho_{1\infty}, \rho_{2\infty}, \rho_{3\infty}\} = \{1, 1, \rho_{3\infty}\})$ where $\rho_{3\infty} \in [0, 1]$.
3. The discrete total angular momentum in $[t_n, t_{n+1}]$ is exactly conserved under the same conditions of the symplectic members; that is,

$$U0(\{\rho_{1\infty}, \rho_{2\infty}, \rho_{3\infty}\} = \{1, 1, 1\}) \quad \text{or} \quad V0(\{\rho_{1\infty}, \rho_{2\infty}, \rho_{3\infty}\} = \{1, 1, \rho_{3\infty}\}) \tag{14.221}$$

   where $\rho_{3\infty} \in [0, 1]$. The discrete total linear momentum within the time step is exactly conserved for any conditions. The discrete total energy is not conserved exactly, but is bounded.

The algorithmic right Cauchy-Green deformation tensor with algorithmic nodal *position* vectors $\tilde{\mathbf{x}} = \mathbf{x}_{n+W_1}$ in Algorithm 21 is evaluated as

$$\begin{aligned}
\tilde{\mathbf{C}}^h &= \mathbf{C}^h(\tilde{\mathbf{x}}) := \mathbf{F}^{h^T}(\mathbf{x}_{n+W_1})\mathbf{F}^h(\mathbf{x}_{n+W_1}) = \mathbf{F}^{h^T}_{n+W_1}\mathbf{F}^h_{n+W_1} \\
&= \left[(1-W_1)\mathbf{F}^{h^T}_n + W_1\mathbf{F}^{h^T}_{n+1}\right]\left[(1-W_1)\mathbf{F}^h_n + W_1\mathbf{F}^h_{n+1}\right] \\
&= (1-W_1)^2\mathbf{F}^{h^T}_n\mathbf{F}^h_n + W_1^2\mathbf{F}^{h^T}_{n+1}\mathbf{F}^h_{n+1} \\
&\quad + W_1(1-W_1)(\mathbf{F}^{h^T}_n\mathbf{F}^h_{n+1} + \mathbf{F}^{h^T}_{n+1}\mathbf{F}^h_n) \\
&= (1-W_1)^2\mathbf{C}^h_n + W_1^2\mathbf{C}^h_{n+1} + W_1(1-W_1)\bar{\mathbf{C}}^h
\end{aligned} \tag{14.222}$$

where $\bar{\mathbf{C}}^h := \mathbf{F}^{h^T}_n\mathbf{F}^h_{n+1} + \mathbf{F}^{h^T}_{n+1}\mathbf{F}^h_n$. As can be seen, evaluating the right Cauchy-Green deformation tensor using $\tilde{\mathbf{x}} = \mathbf{x}_{n+W_1}$ introduces the artificial deformation tensor $\bar{\mathbf{C}}^h$. This situation is also obviously true for the case of evaluation in terms of the nodal displacement. To avoid this issue, we evaluate the algorithmic right Cauchy-Green deformation tensor independently from the position vectors; therefore, the algorithmic second Piola-Kirchhoff stress tensor is directly evaluated as $\tilde{\mathbf{S}}^h = \mathbf{S}^h(\tilde{\mathbf{C}}^h)$ instead of $\tilde{\mathbf{S}}^h = \mathbf{S}^h(\mathbf{C}^h(\tilde{\mathbf{q}}))$.

Approach 3: Energy-Momentum Framework – Normalized Time Weighted Residual Approach
Consider an algorithmic extension of Approach 3 discussed for the two-field form. Consider the semi-discretized equations of motion with given initial conditions, i.e., Equation (14.195) which leads to

$$\begin{aligned}
&\frac{d}{dt}\left(\frac{\partial \mathcal{E}^h(\mathbf{q}, \dot{\mathbf{q}})}{\partial \dot{\mathbf{q}}^A}\right) + \frac{\partial \mathcal{E}^h(\mathbf{q}, \dot{\mathbf{q}})}{\partial \mathbf{q}^A} = \mathbf{0} \\
&\Rightarrow \sum_{B=1}^{n_{\text{node}}} M_{AB}\ddot{\mathbf{q}}^B + \int_{\mathcal{B}} \mathbf{F}^h \cdot \mathbf{S}^h \, \text{GRAD}\, N_A dV = \underbrace{\int_{\mathcal{B}} N_A \rho_0 \mathbf{B} dV + \int_{\partial\mathcal{B}_\sigma} N_A \bar{\mathbf{T}} dA}_{\mathbf{F}^{\text{ext}^h}_A : \mathcal{T} \to \mathbb{R}^3}
\end{aligned} \tag{14.223}$$

Employ the normalized time weighted residual methodology for $\ddot{\mathbf{q}}$, $\mathbf{q}$, $\mathbf{C}^h$ and $\mathbf{F}^{\text{ext}^h}$ in Equation (14.223) as

$$\begin{aligned}&\int_B \mathbf{F}^h\left(\frac{\int_{t_n}^{t_{n+1}} W\mathbf{q}dt}{\int_{t_n}^{t_{n+1}} Wdt}\right)\cdot\mathbf{S}^h\left(\frac{\int_{t_n}^{t_{n+1}} W\mathbf{C}^h dt}{\int_{t_n}^{t_{n+1}} Wdt}\right)\text{GRAD}\, N_A dV\\&+\sum_{B=1}^{n_{\text{node}}} M_{AB}\frac{\int_{t_n}^{t_{n+1}} W\ddot{\mathbf{q}}^B dt}{\int_{t_n}^{t_{n+1}} Wdt}=\frac{\int_{t_n}^{t_{n+1}} W\mathbf{F}_A^{\text{ext}^h} dt}{\int_{t_n}^{t_{n+1}} Wdt}\end{aligned} \tag{14.224}$$

where the scalar time weighting function is approximated by

$$W = 1 + w_1\Gamma + w_2\Gamma^2 + w_3\Gamma^3 \tag{14.225}$$

with the corresponding approximations of $\mathbf{q}$, $\dot{\mathbf{q}}$, $\ddot{\mathbf{q}}$ and the discrete right Cauchy-Green deformation tensor, $\mathbf{C}^h$, by means of the asymptotic series expansions in time, and the approximations of the set of the external nodal forces, $\mathbf{F}^{\text{ext}^h}$, by means of a Taylor series expansion (in general, this is a special case of an asymptotic series expansion and is sufficient for the cases of interest here; but the general case is a function evaluation at a particular time point and not necessarily a linear interpolation) in time as follows:

$$\mathbf{q} = \mathbf{q}_n + \Lambda_1\dot{\mathbf{q}}_n\tau + \Lambda_2\ddot{\mathbf{q}}_n\tau^2 + \Lambda_3\frac{\ddot{\mathbf{q}}_{n+1}-\ddot{\mathbf{q}}_n}{\Delta t}\tau^3 \tag{14.226}$$

$$\dot{\mathbf{q}} = \dot{\mathbf{q}}_n + \Lambda_4\ddot{\mathbf{q}}_n\tau + \Lambda_5\frac{\ddot{\mathbf{q}}_{n+1}-\ddot{\mathbf{q}}_n}{\Delta t}\tau^2 \tag{14.227}$$

$$\ddot{\mathbf{q}} = \ddot{\mathbf{q}}_n + \Lambda_6\frac{\ddot{\mathbf{q}}_{n+1}-\ddot{\mathbf{q}}_n}{\Delta t}\tau \tag{14.228}$$

$$\mathbf{C}^h = \mathbf{C}_n^h + \Lambda_7\frac{\mathbf{C}_{n+1}^h-\mathbf{C}_n^h}{\Delta t}\tau \tag{14.229}$$

$$\mathbf{F}^{\text{ext}^h} = \mathbf{F}_n^{\text{ext}^h} + \frac{\mathbf{F}_{n+1}^{\text{ext}^h}-\mathbf{F}_n^{\text{ext}^h}}{\Delta t}\tau \tag{14.230}$$

Therefore, the algorithmic nodal unknowns are defined as follows:

$$\begin{aligned}\tilde{\mathbf{q}} := \frac{\int_{t_n}^{t_{n+1}} W\mathbf{q}dt}{\int_{t_n}^{t_{n+1}} Wdt} &= \mathbf{q}_n + W_1\Lambda_1\dot{\mathbf{q}}_n\Delta t + W_2\Lambda_2\ddot{\mathbf{q}}_n\Delta t^2\\&+W_3\Lambda_3\left(\ddot{\mathbf{q}}_{n+1}-\ddot{\mathbf{q}}_n\right)\Delta t^2\end{aligned} \tag{14.231}$$

$$\tilde{\mathbf{v}} := \frac{\int_{t_n}^{t_{n+1}} W\dot{\mathbf{q}}dt}{\int_{t_n}^{t_{n+1}} Wdt} = \dot{\mathbf{q}}_n + W_1\Lambda_4\ddot{\mathbf{q}}_n\Delta t + W_2\Lambda_5\left(\ddot{\mathbf{q}}_{n+1}-\ddot{\mathbf{q}}_n\right)\Delta t \tag{14.232}$$

$$\tilde{\mathbf{a}} := \frac{\int_{t_n}^{t_{n+1}} W\ddot{\mathbf{q}}dt}{\int_{t_n}^{t_{n+1}} Wdt} = \ddot{\mathbf{q}}_n + W_1\Lambda_6\left(\ddot{\mathbf{q}}_{n+1}-\ddot{\mathbf{q}}_n\right) \tag{14.233}$$

$$\tilde{\mathbf{C}}^h := \frac{\int_{t_n}^{t_{n+1}} W\mathbf{C}^h dt}{\int_{t_n}^{t_{n+1}} Wdt} = \mathbf{C}_n^h + W_1\Lambda_7(\mathbf{C}_{n+1}^h-\mathbf{C}_n^h) \tag{14.234}$$

$$\tilde{\mathbf{F}}^{\text{ext}^h} := \frac{\int_{t_n}^{t_{n+1}} W\mathbf{F}^{\text{ext}^h} dt}{\int_{t_n}^{t_{n+1}} W dt} = \mathbf{F}_n^{\text{ext}^h} + W_1 \left(\mathbf{F}_{n+1}^{\text{ext}^h} - \mathbf{F}_n^{\text{ext}^h} \right) \tag{14.235}$$

Following a similar procedure as described in the previous sections (Approaches 1 and 2), we readily obtain the following Algorithm 22. Algorithm 22 yields the energy-momentum conserving algorithm designs with the conditions for $U0(\{\rho_{1\infty}, \rho_{2\infty}, \rho_{3\infty}\} = \{1, 1, 1\})$ or $V0(\{\rho_{1\infty}, \rho_{2\infty}, \rho_{3\infty}\} = \{1, 1, \rho_{3\infty}\})$ only for the St. Venant-Kirchhoff material models (with $\Lambda_7 = 1$). To remove this limitation for the general case, we have modified (Algorithm 22) so that it recovers the second-order time accurate energy-momentum conserving algorithm by Simo and Tarnow (Simo and Tarnow 1992) (see Algorithm 23) and by Simo and Gonzalez (Simo and Gonzalez 1994) or Gonzales (Gonzalez 2000) (see Algorithm 24) in the sense of the single-field form (two principal roots and spurious root) algorithm designs.

Algorithm 22: ***Approach 3-α***

Given $\mathbf{q}_n$, $\dot{\mathbf{q}}_n$, and $\ddot{\mathbf{q}}_n$, find $\mathbf{q}_{n+1}$, $\dot{\mathbf{q}}_{n+1}$, and $\ddot{\mathbf{q}}_{n+1}$ from

$$\sum_{B=1}^{n_{\text{node}}} M_{AB}\tilde{\mathbf{a}}^B + \int_{\mathcal{B}} \mathbf{F}^h(\tilde{\mathbf{q}}) \cdot \mathbf{S}^h(\tilde{\mathbf{C}}^h)\,\text{GRAD}\,N_A dV = \tilde{\mathbf{F}}_A^{\text{ext}^h} \tag{14.236a}$$

$$\tilde{\mathbf{a}} = \ddot{\mathbf{q}}_n + W_1\Lambda_6\Delta\mathbf{a} \tag{14.236b}$$

$$\tilde{\mathbf{q}} = \mathbf{q}_n + W_1\Lambda_1\dot{\mathbf{q}}_n\Delta t + W_2\Lambda_2\ddot{\mathbf{q}}_n\Delta t^2 + W_3\Lambda_3\Delta\mathbf{a}\Delta t^2 \tag{14.236c}$$

$$\tilde{\mathbf{C}}^h = (1 - W_1\Lambda_7)\mathbf{C}_n^h + W_1\Lambda_7\mathbf{C}_{n+1}^h \tag{14.236d}$$

$$\tilde{\mathbf{F}}^{\text{ext}^h} = (1 - W_1)\mathbf{F}_n^{\text{ext}^h} + W_1\mathbf{F}_{n+1}^{\text{ext}^h} \tag{14.236e}$$

with the following updates:

$$\mathbf{q}_{n+1} = \mathbf{q}_n + \lambda_1\dot{\mathbf{q}}_n\Delta t + \lambda_2\ddot{\mathbf{q}}_n\Delta t^2 + \lambda_3\Delta\mathbf{a}\Delta t^2 \tag{14.236f}$$

$$\dot{\mathbf{q}}_{n+1} = \dot{\mathbf{q}}_n + \lambda_4\ddot{\mathbf{q}}_n\Delta t + \lambda_5\Delta\mathbf{a}\Delta t \tag{14.236g}$$

$$\ddot{\mathbf{q}}_{n+1} = \ddot{\mathbf{q}}_n + \Delta\mathbf{a} \tag{14.236h}$$

Remark 14.3.10

1. Algorithm 22 is an extension of GSSSS family of parent linear algorithms originally derived for linear dynamic conservative systems to nonlinear dynamic conservative systems; therefore, the algorithmic DNA parameters $\{W, \Lambda, \lambda\}$ are given as shown in Algorithm 9 and Algorithm 10 for the U0 family of algorithms and the V0 family of algorithms, respectively.
2. The algorithmic right Cauchy-Green deformation tensors at t_n and t_{n+1} are evaluated by $\mathbf{C}_n^h = \mathbf{C}^h(\mathbf{q}_n)$ and $\mathbf{C}_{n+1}^h = \mathbf{C}^h(\mathbf{q}_{n+1})$, respectively, but $\tilde{\mathbf{C}}^h = \mathbf{C}_{n+W_1\Lambda_7}^h \neq \mathbf{C}^h(\mathbf{q}_{n+W_1\Lambda_7})$. Note $\tilde{\mathbf{C}}^h$ keeps the symmetric properties and positive definiteness for any $W_1\Lambda_7 \in \mathbb{R}$.
3. Algorithm 22 conserves the discrete total energy in $[t_n, t_{n+1}]$ in the sense of $\mathcal{E}_{n+1}^h = \mathcal{E}_n^h$ only for the St. Venant-Kirchhoff material models for $U0(\{\rho_{1\infty}, \rho_{2\infty}, \rho_{3\infty}\} = \{1, 1, 1\})$ or $V0(\{\rho_{1\infty}, \rho_{2\infty}, \rho_{3\infty}\} = \{1, 1, \rho_{3\infty}\})$ with $\Lambda_7 = 1$ (note $\Lambda_7 = 1$ is required to guarantee the second-order time accuracy of the algorithm in general).

From a comparison with the second-order time accurate two field form energy-momentum conserving algorithm by Simo and Tarnow (Simo and Tarnow 1992), the equivalent single-field form of Algorithm 22 may be rewritten as follows:

Algorithm 23: ***Approach 3-β***

Given $\mathbf{q}_n$, $\dot{\mathbf{q}}_n$, and $\ddot{\mathbf{q}}_n$, find $\mathbf{q}_{n+1}$, $\dot{\mathbf{q}}_{n+1}$, and $\ddot{\mathbf{q}}_{n+1}$ from

$$\sum_{B=1}^{n_{\text{node}}} M_{AB}\tilde{\mathbf{a}}^B + \int_B \mathbf{F}^h(\tilde{\mathbf{q}}) \cdot \tilde{\mathbf{S}}^h \,\text{GRAD}\, N_A dV = \tilde{\mathbf{F}}_A^{\text{ext}^h} \tag{14.237a}$$

$$\tilde{\mathbf{a}} = \ddot{\mathbf{q}}_n + W_1\Lambda_6\Delta\mathbf{a} \tag{14.237b}$$

$$\tilde{\mathbf{q}} = \mathbf{q}_n + W_1\Lambda_1\dot{\mathbf{q}}_n\Delta t + W_2\Lambda_2\ddot{\mathbf{q}}_n\Delta t^2 + W_3\Lambda_3\Delta\mathbf{a}\Delta t^2 \tag{14.237c}$$

$$\tilde{\mathbf{F}}^{\text{ext}^h} = (1 - W_1)\mathbf{F}_n^{\text{ext}^h} + W_1\mathbf{F}_{n+1}^{\text{ext}^h} \tag{14.237d}$$

And, the algorithmic second Piola-Kirchhoff stress tensor $\tilde{\mathbf{S}}^h$ defined by

$$\hat{W}^h(\mathbf{C}_{n+1}^h) - \hat{W}^h(\mathbf{C}_n^h) = \tilde{\mathbf{S}}^h : \tfrac{1}{2}[\mathbf{C}_{n+1}^h - \mathbf{C}_n^h] \tag{14.237e}$$

where

$$\tilde{\mathbf{S}}^h = D\hat{W}^h(\mathbf{C}_{n+\beta_0}^h) + D\hat{W}^h(\mathbf{C}_{n+(1-\beta_0)}^h) \;\; for \;\; \beta_0 \in (0, 1) \tag{14.237f}$$

for $U0(\{\rho_{1\infty}, \rho_{2\infty}, \rho_{3\infty}\} = \{1, 1, 1\})$ or $V0(\{\rho_{1\infty}, \rho_{2\infty}, \rho_{3\infty}\} = \{1, 1, \rho_{3\infty}\})$; otherwise,

$$\tilde{\mathbf{S}}^h = 2D\hat{W}^h(\tilde{\mathbf{C}}^h) \quad \text{with} \quad \tilde{\mathbf{C}}^h = (1 - W_1)\mathbf{C}_n^h + W_1\mathbf{C}_{n+1}^h \;\; (\Lambda_7 \equiv 1) \tag{14.237g}$$

with the following updates:

$$\mathbf{q}_{n+1} = \mathbf{q}_n + \lambda_1\dot{\mathbf{q}}_n\Delta t + \lambda_2\ddot{\mathbf{q}}_n\Delta t^2 + \lambda_3\Delta\mathbf{a}\Delta t^2 \tag{14.237h}$$

$$\dot{\mathbf{q}}_{n+1} = \dot{\mathbf{q}}_n + \lambda_4\ddot{\mathbf{q}}_n\Delta t + \lambda_5\Delta\mathbf{a}\Delta t \tag{14.237i}$$

$$\ddot{\mathbf{q}}_{n+1} = \ddot{\mathbf{q}}_n + \Delta\mathbf{a} \tag{14.237j}$$

Remark 14.3.11

1. Algorithm 23 is an extension of GSSSS family of parent linear algorithms originally derived for linear dynamic conservative systems to nonlinear dynamic conservative systems; therefore, the algorithmic DNA parameters $\{W, \Lambda, \lambda\}$ are given as shown in Algorithm 9 and Algorithm 10 for the U0 family of algorithms and the V0 family of algorithms, respectively.
2. Algorithm 23 recovers the energy-momentum conserving algorithm and the underlying features proposed by Simo and Tarnow (Simo and Tarnow 1992) for $U0(\{\rho_{1\infty}, \rho_{2\infty}, \rho_{3\infty}\} = \{1, 1, 1\})$ or $V0(\{\rho_{1\infty}, \rho_{2\infty}, \rho_{3\infty}\} = \{1, 1, \rho_{3\infty}\})$ in the sense of the single-field form (with two principal roots and a spurious root) algorithm design. The algorithmic right Cauchy-Green deformation tensor at t_n and t_{n+1} is evaluated by $\mathbf{C}_n^h = \mathbf{C}^h(\mathbf{q}_n)$ and $\mathbf{C}_{n+1}^h = \mathbf{C}^h(\mathbf{q}_{n+1})$, respectively, and a scalar parameter $\beta_0 \in (0, 1)$ is evaluated iteratively from the mean value theorem; that is,

$$\hat{W}^h(\mathbf{C}_{n+1}^h) - \hat{W}^h(\mathbf{C}_n^h) = \tilde{\mathbf{S}}^h(\mathbf{C}_{n+\beta_0}^h) : \tfrac{1}{2}[\mathbf{C}_{n+1}^h - \mathbf{C}_n^h] \tag{14.238}$$

where $\mathbf{C}^h_{n+\beta_0} := (1-\beta_0)\mathbf{C}^h_n + \beta_0\mathbf{C}^h_{n+1}$. The mean value theorem enables the algorithmic second Piola-Kirchhoff stress tensor $\tilde{\mathbf{S}}^h$ to be defined as

$$\tilde{\mathbf{S}}^h = 2D\hat{W}^h(\mathbf{C}^h_{n+\beta_0}) \quad for \quad \beta_0 \in (0,1) \tag{14.239}$$

Unfortunately, Equation (14.239) gives second-order time accuracy for the algorithm only when $\beta_0 = 1/2$; otherwise, it yields only first-order time accuracy. However, Simo and Tarnow (Simo and Tarnow 1992) proved that

$$\tilde{\mathbf{S}}^h = D\hat{W}^h(\mathbf{C}^h_{n+\beta_0}) + D\hat{W}^h(\mathbf{C}^h_{n+(1-\beta_0)}) \quad for \quad \beta_0 \in (0,1) \tag{14.240}$$

where $\mathbf{C}^h_{n+(1-\beta_0)} := \beta_0\mathbf{C}^h_n + (1-\beta_0)\mathbf{C}^h_{n+1}$ also satisfies the mean value theorem, and it guarantees second-order time accuracy of the algorithm.

From a comparison with the second-order time accurate two-field energy-momentum conserving algorithm by Simo and Gonzalez (Simo and Gonzalez 1994) or Gonzalez (Gonzalez 2000), the equivalent single-field form of Algorithm 22 may be rewritten as follows for general hyperelastic material models:

Algorithm 24: *Approach 3-γ*

Given $\mathbf{q}_n$, $\dot{\mathbf{q}}_n$, and $\ddot{\mathbf{q}}_n$, find $\mathbf{q}_{n+1}$, $\dot{\mathbf{q}}_{n+1}$, and $\ddot{\mathbf{q}}_{n+1}$ from

$$\sum_{B=1}^{n_{\text{node}}} M_{AB}\tilde{\mathbf{a}}^B + \int_{\mathcal{B}} \mathbf{F}^h(\tilde{\mathbf{q}}) \cdot \tilde{\mathbf{S}}^h \,\text{GRAD}\, N_A dV = \tilde{\mathbf{F}}^{\text{ext}^h}_A \tag{14.241a}$$

$$\tilde{\mathbf{a}} = \ddot{\mathbf{q}}_n + W_1\Lambda_6\Delta\mathbf{a} \tag{14.241b}$$

$$\tilde{\mathbf{q}} = \mathbf{q}_n + W_1\Lambda_1\dot{\mathbf{q}}_n\Delta t + W_2\Lambda_2\ddot{\mathbf{q}}_n\Delta t^2 + W_3\Lambda_3\Delta\mathbf{a}\Delta t^2 \tag{14.241c}$$

$$\tilde{\mathbf{C}}^h = (1-W_1)\mathbf{C}^h_n + W_1\mathbf{C}^h_{n+1} \quad (\Lambda_7 \equiv 1) \tag{14.241d}$$

$$\tilde{\mathbf{F}}^{\text{ext}^h} = (1-W_1)\mathbf{F}^{\text{ext}^h}_n + W_1\mathbf{F}^{\text{ext}^h}_{n+1} \tag{14.241e}$$

And, the algorithmic second Piola-Kirchhoff stress tensor $\tilde{\mathbf{S}}^h$ which in turn is defined as

$$\tilde{\mathbf{S}}^h := 2D\hat{W}(\tilde{\mathbf{C}}^h) + 2\chi\frac{\hat{W}(\mathbf{C}^h_{n+1}) - \hat{W}(\mathbf{C}^h_n) - D\hat{W}(\tilde{\mathbf{C}}^h) : \Delta\mathbf{C}^h}{||\Delta\mathbf{C}^h||^2}\Delta\mathbf{C}^h \tag{14.241f}$$

where χ is defined by

$$\chi := 1 - \text{sgn}^2(W_1 - \tfrac{1}{2}) \tag{14.241g}$$

with the following updates:

$$\mathbf{q}_{n+1} = \mathbf{q}_n + \lambda_1\dot{\mathbf{q}}_n\Delta t + \lambda_2\ddot{\mathbf{q}}_n\Delta t^2 + \lambda_3\Delta\mathbf{a}\Delta t^2 \tag{14.241h}$$

$$\dot{\mathbf{q}}_{n+1} = \dot{\mathbf{q}}_n + \lambda_4\ddot{\mathbf{q}}_n\Delta t + \lambda_5\Delta\mathbf{a}\Delta t \tag{14.241i}$$

$$\ddot{\mathbf{q}}_{n+1} = \ddot{\mathbf{q}}_n + \Delta\mathbf{a} \tag{14.241j}$$

Remark 14.3.12

1. Algorithm 24 is an extension of GSSSS family of parent linear algorithms originally derived for linear dynamic conservative systems to nonlinear dynamic conservative systems; therefore, the algorithmic DNA parameters $\{W, \Lambda, \lambda\}$ are given as shown in Algorithm 9 and Algorithm 10 for the U0 family of algorithms and the V0 family of algorithms, respectively.
2. Algorithm 24 recovers the energy-momentum conserving algorithm and the underlying features described by Simo and Gonzalez (Simo and Gonzalez 1994) or Gonzales (Gonzalez 2000) for $U0(\{\rho_{1\infty}, \rho_{2\infty}, \rho_{3\infty}\} = \{1, 1, 1\})$ or $V0(\{\rho_{1\infty}, \rho_{2\infty}, \rho_{3\infty}\} = \{1, 1, \rho_{3\infty}\})$ in the sense of the single-field form (with two principal roots and a spurious root) algorithm design. The algorithmic right Cauchy-Green deformation tensor at t_n and t_{n+1} is evaluated by $\mathbf{C}_n^h = \mathbf{C}^h(\mathbf{q}_n)$ and $\mathbf{C}_{n+1}^h = \mathbf{C}^h(\mathbf{q}_{n+1})$, respectively, and we define $\Delta\mathbf{C}^h := \mathbf{C}_{n+1}^h - \mathbf{C}_n^h$ and $||\Delta\mathbf{C}^h||^2 := \Delta\mathbf{C}^h : \Delta\mathbf{C}^h$. The function χ defined with the sign function is given as

$$\begin{aligned}\chi &:= 1 - \text{sgn}^2(W_1 - \tfrac{1}{2}) = 4H_{1/2}(W_1 - \tfrac{1}{2})\left[1 - H_{1/2}(W_1 - \tfrac{1}{2})\right] \\ &= \begin{cases} 1 & for\ \ W_1 = \frac{1}{2} \\ 0 & for\ \ W_1 \neq \frac{1}{2} \end{cases}\end{aligned} \tag{14.242}$$

where $H_{1/2}$ is the Heaviside step function. Therefore, the algorithmic second Piola-Kirchhoff stress tensor $\tilde{\mathbf{S}}^h$ is given as

$$\tilde{\mathbf{S}}^h = 2D\hat{W}(\tilde{\mathbf{C}}^h) + 2\frac{\hat{W}(\mathbf{C}_{n+1}^h) - \hat{W}(\mathbf{C}_n^h) - D\hat{W}(\tilde{\mathbf{C}}^h) : \Delta\mathbf{C}^h}{||\Delta\mathbf{C}^h||^2}\Delta\mathbf{C}^h \ \ for\ \ W_1 = \tfrac{1}{2} \tag{14.243}$$

and

$$\tilde{\mathbf{S}}^h = 2D\hat{W}(\tilde{\mathbf{C}}^h) \ \ for\ \ W_1 \neq \tfrac{1}{2} \tag{14.244}$$

3. Since $\bar{\mathbf{S}}^h := 2\frac{\hat{W}(\mathbf{C}_{n+1}^h) - \hat{W}(\mathbf{C}_n^h) - D\hat{W}(\tilde{\mathbf{C}}^h):\Delta\mathbf{C}^h}{||\Delta\mathbf{C}^h||^2}\Delta\mathbf{C}^h = \mathcal{O}(\Delta t^2)$ and the time level of the evaluation of the Lagrange strain tensor is at $t \approx t_{n+1/2}$, it is appropriate to recover the energy-momentum conserving algorithm under the conditions of U0(1, 1, 1) or V0(1, 1, $\rho_{3\infty}$) in the displacement-strain hybrid GSSSS nonlinear framework.

14.3.3 Numerical Implementation Aspects of Classical Framework in Single-field Form: Internal Force Based Numerically Non-Dissipative and Dissipative Algorithms and Designs

For illustration of the basic ideas presented in the previous section of the single-field form algorithms by design, we consider the classical framework for the truss element problem in the three dimensional Euclidean space as a simple example of a St. Venant Kirchhoff hyperelastic material model (the two-field form is not presented here). For this material model, define the generalized coordinates as a set of nodal displacement vectors as $\mathbf{q}(t) = \mathbf{u}(t) : \mathcal{T} \rightarrow \mathcal{Q}$, and define $\dot{\mathbf{q}}(t) = \dot{\mathbf{u}}(t) : \mathcal{T} \rightarrow T_q\mathcal{Q}$. The total energy in the semi-discrete system is defined as $\mathcal{E}^h := \mathcal{K}^h + \mathcal{U}_{\text{int}}^h + \mathcal{U}_{\text{ext}}^h$ where the semi-discrete kinetic energy and internal and external potential energy are given as

$$\mathcal{K}^h = \tfrac{1}{2}\dot{\mathbf{q}} \cdot \mathbf{M}\dot{\mathbf{q}} = \tfrac{1}{2}\dot{\mathbf{u}} \cdot \mathbf{M}\dot{\mathbf{u}} \tag{14.245}$$

$$\mathcal{U}_{\text{int}}^h = \tfrac{1}{2}A_0 l_0 E \varepsilon^2 \tag{14.246}$$

$$\mathcal{U}_{\text{ext}}^h = \mathbf{q} \cdot \mathbf{F}^{\text{ext}^h} = \mathbf{u} \cdot \mathbf{F}^{\text{ext}^h} \tag{14.247}$$

respectively, where the constant mass matrix is defined as $\mathbf{M} := \frac{1}{2} A_0 l_0 \rho_0 \mathbf{I}_6$; A_0, l_0 and ρ_0 denote the cross-sectional area, the length and the density of the truss in the reference configuration; E denotes Young's modulus; and the scalar discrete Lagrange-Green (Green) strain is defined as

$$\varepsilon = \varepsilon(\mathbf{q}) = \frac{1}{2}\left[\left(\frac{l(\mathbf{q})}{l_0}\right)^2 - 1\right] = \mathbf{b}\cdot\mathbf{q} + \frac{2}{l_0^2}\mathbf{q}\cdot\mathbf{A}\mathbf{q} = \mathbf{b}\cdot\mathbf{u} + \frac{2}{l_0^2}\mathbf{u}\cdot\mathbf{A}\mathbf{u} \tag{14.248}$$

where l denotes the element length in the current configuration, and the constant vector $\mathbf{b}$ and constant matrix $\mathbf{A}$ are defined as

$$\mathbf{b} := \tfrac{4}{l_0^2}\mathbf{A}\boldsymbol{\varphi}_0 \quad \text{and} \quad \mathbf{A} := \tfrac{1}{4}\begin{bmatrix} \mathbf{I}_3 & -\mathbf{I}_3 \\ -\mathbf{I}_3 & \mathbf{I}_3 \end{bmatrix} \tag{14.249}$$

respectively. For the general discussion including the damping force, consider the following balance of the total mechanical energy as

$$\frac{d\mathcal{E}^h}{dt} = \mathbf{F}^{\mathrm{nc}^h}\cdot\dot{\mathbf{q}} \tag{14.250}$$

where $\mathbf{F}^{\mathrm{nc}^h}$ is the nodal nonconservative force. Define the nonconservative force as $\mathbf{F}^{\mathrm{nc}^h} = -\mathbf{C}\dot{\mathbf{q}} = -\mathbf{C}\dot{\mathbf{u}}$ where $\mathbf{C}$ is the constant damping matrix; therefore, Equation (14.250) leads to

$$\left(\frac{d}{dt}\left(\frac{\partial\mathcal{K}^h}{\partial\dot{\mathbf{q}}}\right) + \frac{\partial\mathcal{U}^h_{\mathrm{int}}}{\partial\mathbf{q}} + \frac{\partial\mathcal{U}^h_{\mathrm{ext}}}{\partial\mathbf{q}} + \mathbf{C}\dot{\mathbf{q}}\right)\cdot\dot{\mathbf{q}} = \mathbf{0} \tag{14.251}$$

Since $\dot{\mathbf{q}}$ is linearly independent, we have

$$\frac{d}{dt}\left(\frac{\partial\mathcal{K}^h}{\partial\dot{\mathbf{q}}}\right) + \frac{\partial\mathcal{U}^h_{\mathrm{int}}}{\partial\mathbf{q}} + \frac{\partial\mathcal{U}^h_{\mathrm{ext}}}{\partial\mathbf{q}} + \mathbf{C}\dot{\mathbf{q}} = \mathbf{0} \tag{14.252}$$

which yields

$$\mathbf{M}\ddot{\mathbf{u}} + \mathbf{C}\dot{\mathbf{u}} + \mathbf{F}^{\mathrm{int}^h} = \mathbf{F}^{\mathrm{ext}^h} \tag{14.253}$$

where the discrete internal force is given as

$$\mathbf{F}^{\mathrm{int}^h} = A_0 l_0 E\varepsilon\left[\mathbf{b} + \frac{4}{l_0^2}\mathbf{A}\mathbf{u}\right] \tag{14.254}$$

In view of Equation (14.248) and Equation (14.254), the discrete internal force may also be written as

$$\mathbf{F}^{\mathrm{int}^h} = \mathbf{K}_1\mathbf{u} + \mathbf{K}_2\varepsilon + \mathbf{K}_3\mathbf{u}\varepsilon \tag{14.255}$$

$$= \mathbf{K}_1\mathbf{u} + (\mathbf{b}\cdot\mathbf{u})\mathbf{K}_2 + \frac{2}{l_0^2}(\mathbf{u}\cdot\mathbf{A}\mathbf{u})\mathbf{K}_2 + (\mathbf{b}\cdot\mathbf{u})\mathbf{K}_3\mathbf{u} \tag{14.256}$$

$$+ \frac{2}{l_0^2}(\mathbf{u}\cdot\mathbf{A}\mathbf{u})\mathbf{K}_3\mathbf{u} \tag{14.257}$$

where

$$\mathbf{K}_1 := \mathbf{0}, \quad \mathbf{K}_2 := A_0 l_0 E\mathbf{b}, \quad \text{and} \quad \mathbf{K}_3 := \frac{4A_0E}{l_0}\mathbf{A} \tag{14.258}$$

Computational and Implementation Aspects In the implementation of the GSSSS framework of algorithms encompassing LMS methods, one can have predictor multi-corrector incremental a, v, or d-form representations. That is, the unknowns to be solved are related to incremental quantities of the acceleration, velocity or displacement. All forms yield the same results. Next, simply for illustration, the a-form representation is described (likewise, the v- and d-forms can be derived). Employing the classical time weighted residual methodology to the constant terms in the equation of motion described above, one can readily derive the a-, v-, and d-form representations as highlighted next. We employ the Newton-Raphson method to solve the nonlinear structural dynamic equation iteratively within each time step.

Approximation 1 ***Classical Time Weighted Residual Approach in a-form Representation***
The effective structural dynamic equations employing the classical time weighted residual approach is given by

$$\mathbf{M}\widetilde{\mathbf{a}} + \mathbf{C}\widetilde{\mathbf{v}} + \widetilde{\mathbf{F}}^{\text{int}^h} = \widetilde{\mathbf{F}}^{\text{ext}^h} \tag{14.259}$$

where

$$\widetilde{\mathbf{a}} = \ddot{\mathbf{u}}_n + \Lambda_6 W_1(\ddot{\mathbf{u}}_{n+1} - \ddot{\mathbf{u}}_n) \tag{14.260}$$

$$\widetilde{\mathbf{v}} = \dot{\mathbf{u}}_n + \Lambda_4 W_1 \Delta t \ddot{\mathbf{u}}_n + \Lambda_5 W_2 \Delta t(\ddot{\mathbf{u}}_{n+1} - \ddot{\mathbf{u}}_n) \tag{14.261}$$

$$\begin{aligned}\widetilde{\mathbf{F}}^{\text{int}^h} &= \mathbf{F}_n^{\text{int}^h} + \frac{W_3\Lambda_3}{\lambda_3}(\mathbf{F}_{n+1}^{\text{int}^h} - \mathbf{F}_n^{\text{int}^h}) + (W_1\Lambda_1 - \frac{W_3\Lambda_3\lambda_1}{\lambda_3})\Delta t \dot{\mathbf{F}}_n^{\text{int}^h} + \\ &(W_2\Lambda_2 - \frac{W_3\Lambda_3\lambda_2}{\lambda_3})\Delta t^2 \ddot{\mathbf{F}}_n^{\text{int}^h}\end{aligned} \tag{14.262}$$

$$\widetilde{\mathbf{F}}^{\text{ext}^h} = (1 - W_1)\mathbf{F}_n^{\text{ext}^h} + W_1\mathbf{F}_{n+1}^{\text{ext}^h} \tag{14.263}$$

Employ Newton-Raphson iterative method to solve the nonlinear effective structural dynamic equation above:

At the beginning of n+1 time level, we predict (k = 1) as follows (see also Table 4.3).

$$\widetilde{\mathbf{a}}_{n+1}^k = \Xi_{Pa}^{(1)}\mathbf{u}_n + \Xi_{Pa}^{(2)}\dot{\mathbf{u}}_n + \Xi_{Pa}^{(3)}\ddot{\mathbf{u}}_n \tag{14.264}$$

$$\widetilde{\mathbf{v}}_{n+1}^k = \Xi_{Pv}^{(1)}\mathbf{u}_n + \Xi_{Pv}^{(2)}\dot{\mathbf{u}}_n + \Xi_{Pv}^{(3)}\ddot{\mathbf{u}}_n \tag{14.265}$$

$$\mathbf{u}_{n+1}^k = \Xi_{Pu}^{(1)}\mathbf{u}_n + \Xi_{Pu}^{(2)}\dot{\mathbf{u}}_n + \Xi_{Pu}^{(3)}\ddot{\mathbf{u}}_n \tag{14.266}$$

$$\widetilde{\mathbf{F}}_{n+1}^{\text{int}^h\,k} = \Xi_{Pp}^{(1)}\mathbf{F}_n^{\text{int}^h} + \Xi_{Pp}^{(2)}\dot{\mathbf{F}}_n^{\text{int}^h} + \Xi_{Pp}^{(3)}\ddot{\mathbf{F}}_n^{\text{int}^h} + \Xi_{Pp}^{(4)}(\mathbf{F}_{n+1}^{\text{int}^h\,k} - \mathbf{F}_n^{\text{int}^h}) \tag{14.267}$$

$$\widetilde{\mathbf{F}}_{n+1}^{\text{ext}^h\,k} = (1 - W_1)\mathbf{F}_n^{\text{ext}^h} + W_1\mathbf{F}_{n+1}^{\text{ext}^h} \tag{14.268}$$

Start nonlinear iteration. Solve for $\Delta\ddot{\mathbf{u}}_{n+1}^{k+1}$ from

$$\begin{aligned}&[\Xi_{Ca}\mathbf{M} + \Xi_{Cv}\mathbf{C} + \Xi_{Cd}^t\mathbf{K}_{n+1}^k]\Delta\ddot{\mathbf{u}}_{n+1}^{k+1} = \\ &-(\mathbf{M}\widetilde{\mathbf{a}}_{n+1}^k + \mathbf{C}\widetilde{\mathbf{v}}_{n+1}^k + \widetilde{\mathbf{F}}_{n+1}^{\text{int}^h\,k} - \widetilde{\mathbf{F}}_{n+1}^{\text{ext}^h\,k})\end{aligned} \tag{14.269}$$

where

$${}^t\mathbf{K}_{n+1}^k = \mathbf{K}_1 + \mathbf{K}_2(\mathbf{b} + \frac{4}{l_0^2}\mathbf{A}\mathbf{u}_{n+1}^k) + \mathbf{K}_3[\mathbf{u}_{n+1}^k(\mathbf{b} + \frac{4}{l_0^2}\mathbf{A}\mathbf{u}_{n+1}^k) + \epsilon_{n+1}^k] \tag{14.270}$$

Then correct the variables as follows.

$$\widetilde{\mathbf{a}}_{n+1}^{k+1} = \widetilde{\mathbf{a}}_{n+1}^{k} + \Xi_{Ca}\Delta\delta_{n+1}^{k+1} \tag{14.271}$$

$$\widetilde{\mathbf{v}}_{n+1}^{k+1} = \widetilde{\mathbf{v}}_{n+1}^{k} + \Xi_{Cv}\Delta\delta_{n+1}^{k+1} \tag{14.272}$$

$$\mathbf{u}_{n+1}^{k+1} = \mathbf{u}_{n+1}^{k} + \Xi_{Cu}\Delta\delta_{n+1}^{k+1} \tag{14.273}$$

$$\widetilde{\mathbf{F}}_{n+1}^{\text{int}^h\, k+1} = \Xi_{Cp}^{(1)}\mathbf{F}_n^{\text{int}^h} + \Xi_{Cp}^{(2)}\dot{\mathbf{F}}_n^{\text{int}^h} + \Xi_{Cp}^{(3)}\ddot{\mathbf{F}}_n^{\text{int}^h} + \Xi_{Cp}^{(4)}(\mathbf{F}_{n+1}^{\text{int}^h\, k+1} - \mathbf{F}_n^{\text{int}^h}) \tag{14.274}$$

until the solution converges

$$||\Delta\delta_{n+1}^{k+1}|| < tolerance \tag{14.275}$$

Once the solution converges, update the unknowns at the end of the time step as follows.

$$\Delta\ddot{\mathbf{u}} = (\widetilde{\mathbf{a}}_{n+1}^{k+1} - \ddot{\mathbf{u}}_n)/(\Lambda_6 W_1) \tag{14.276}$$

$$\mathbf{u}_{n+1} = \mathbf{u}_n + \lambda_1\dot{\mathbf{u}}_n\Delta t + \lambda_2\ddot{\mathbf{u}}_n\Delta t^2 + \lambda_3\Delta\ddot{\mathbf{u}}\Delta t^2 \tag{14.277}$$

$$\dot{\mathbf{u}}_{n+1} = \dot{\mathbf{u}}_n + \lambda_4\ddot{\mathbf{u}}_n\Delta t + \lambda_5\Delta\ddot{\mathbf{u}}\Delta t \tag{14.278}$$

$$\ddot{\mathbf{u}}_{n+1} = \ddot{\mathbf{u}}_n + \Delta\ddot{\mathbf{u}} \tag{14.279}$$

Likewise, one can readily derive the v-form and d-form representations (not shown here).

Illustrative Example

Example 14.2 Four Mass-Spring System

We again consider a four mass-spring system constructed of four springs connecting four masses as depicted earlier in Figure 14.3. However, the problem consists of truss elements characterizing the four springs that are connected to four masses. This model problem has potential energy for the continuum truss elements as a quadratic function of Green strain, and is representative of the St. Venant Kirchhoff material model. In this example, each spring has initial length of 1m and stiffness of $10^2 N/m$ while each mass weighs 1kg. The initial position of the four nodes are [0,0,0;1,0,0;1,1,0;0,1,0]m. The initial condition is as follows simply for purposes of illustration.

i) given initial displacement problem, $\mathbf{u}_0 = [0,0,0,0,0,0,0,0,0,0,0,1]$

Classical Time Weighted Residual Approach We study the conservation of the physical quantities including energy, linear and angular momentum to evaluate the performance of the classical time weighted residual approach when analyzed with the numerically non-dissipative algorithms which include U0/V0 ($\{\rho_{1\infty}, \rho_{2\infty}, \rho_{3\infty}\} = \{1.00, 1.00, 1.00\}$) (Classical Version Midpoint Rule [MPR-EPA]), V0 ($\{\rho_{1\infty}, \rho_{2\infty}, \rho_{3\infty}\} = \{1.00, 1.00, 0.00\}$) (New Version Midpoint Rule [MPR-MPA]), and U0 ($\{\rho_{1\infty}, \rho_{2\infty}, \rho_{3\infty}\} = \{1.00, 1.00, 0.00\}$) (Newmark Method) using a time step of $\Delta t = 1$ s and the results are shown in Figures 14.5–14.7. The results clearly show the inability of the classical time weighted residual approach to accurately and properly and handle the physics of the nonlinear dynamics situations. For satisfactory performance and proper extensions to nonlinear dynamics situations, the reader is encouraged to notice the next two sections dealing with symplectic-momentum and energy-momentum based frameworks via the normalized time weighted residual approach.

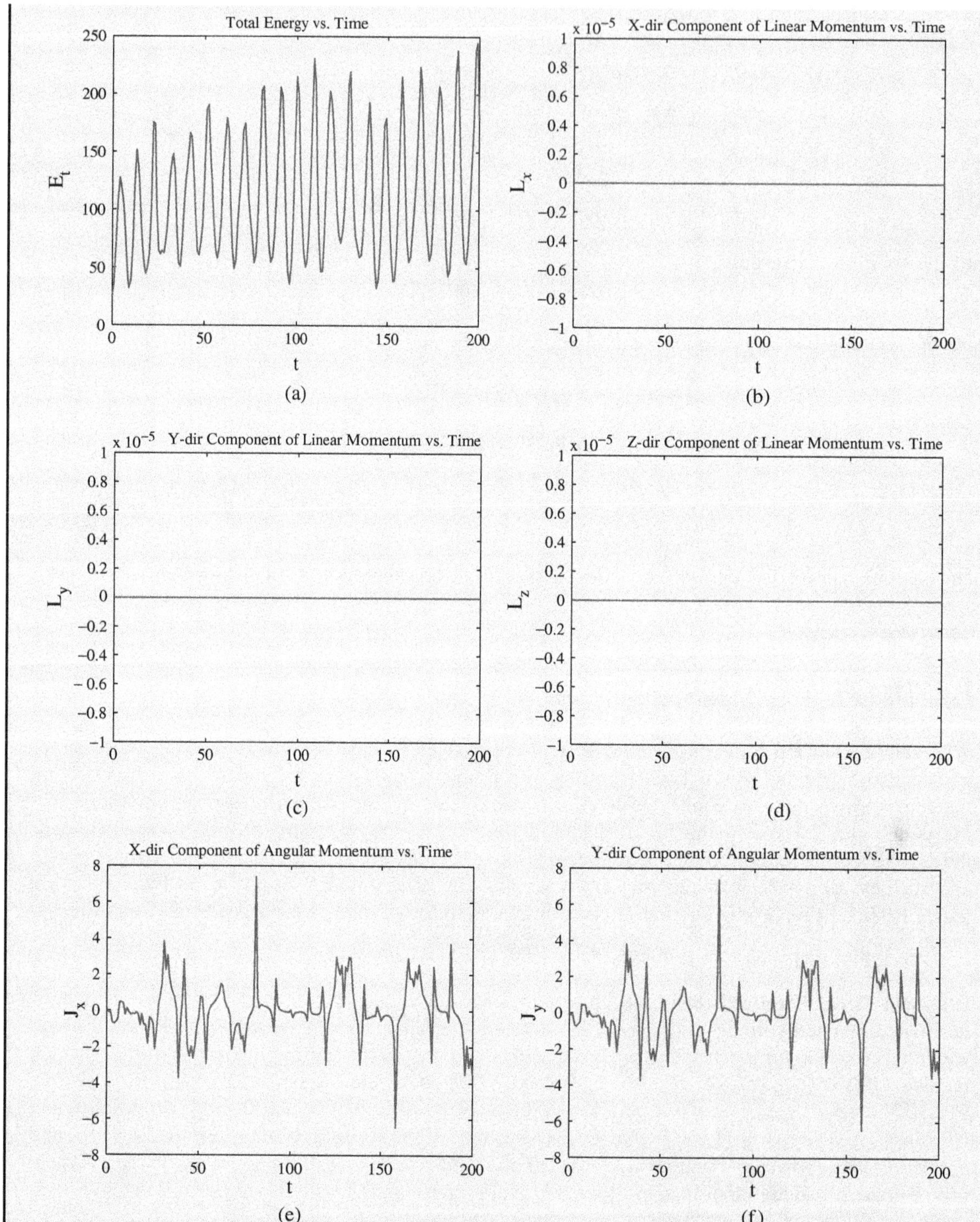

Figure 14.5. Plots of energy, components of linear momentum, and components of angular momentum for the four spring-mass problem with given initial displacement simulated using U0/V0 ($\{\rho_{1\infty}, \rho_{2\infty}, \rho_{3\infty}\} = \{1, 1, 1\}$) (Classical Midpoint Rule-[MPR-EPA]) employing the classical time weighted residual approach

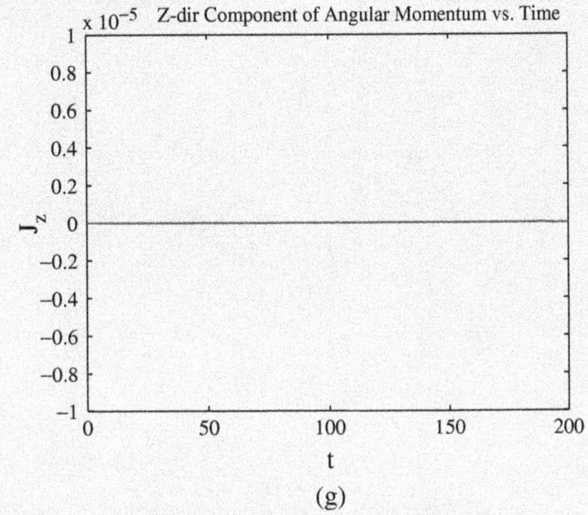

Figure 14.5. *(continued)*

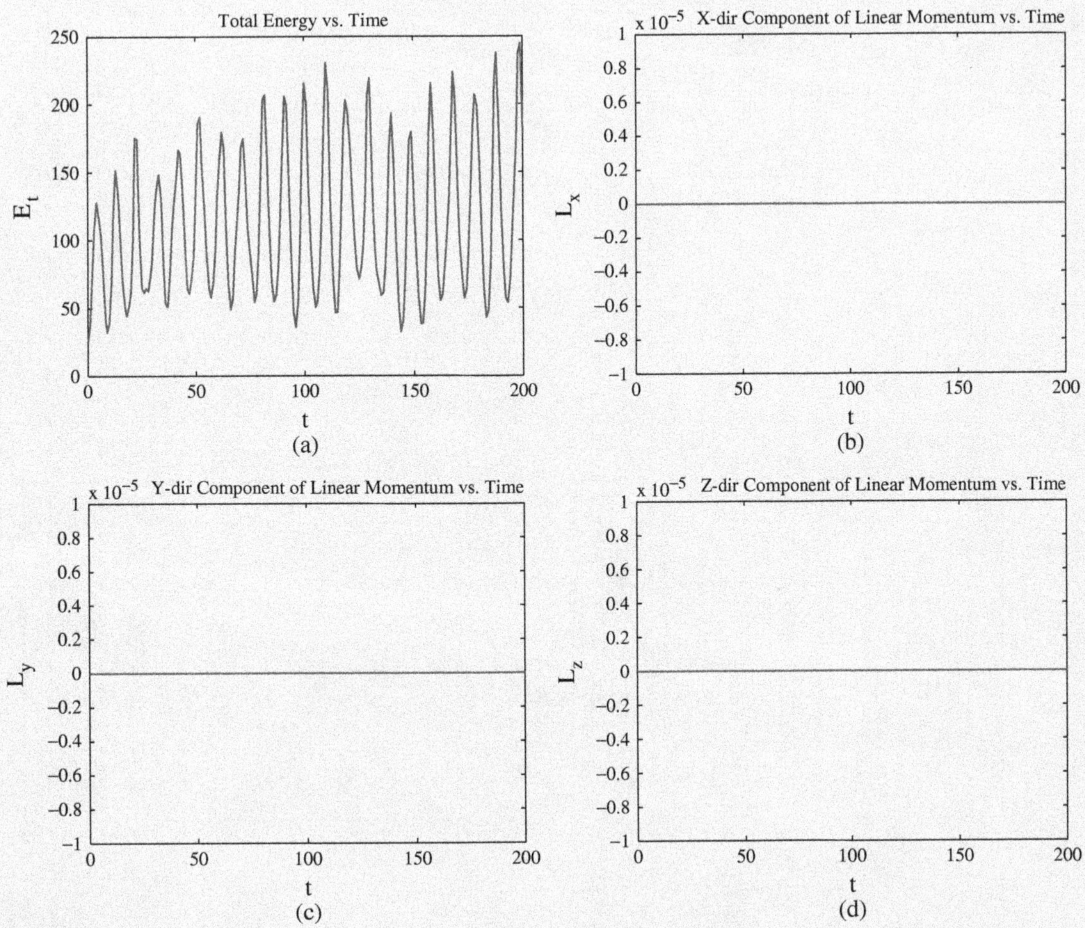

Figure 14.6. Plots of energy, components of linear momentum, and components of angular momentum for the four spring-mass problem with given initial displacement simulated using V0 ($\{\rho_{1\infty}, \rho_{2\infty}, \rho_{3\infty}\} = \{1, 1, 0\}$) (New Version Midpoint Rule-[MPR-MPA]) employing the classical time weighted residual approach

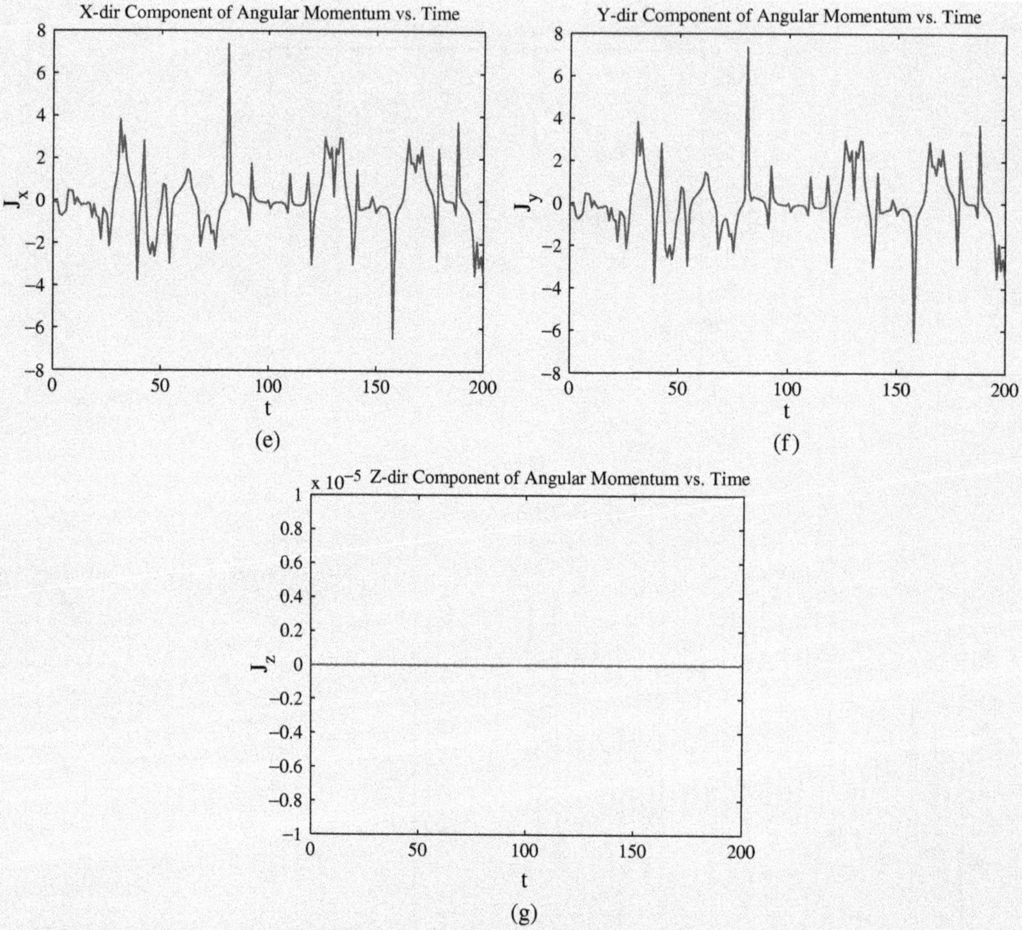

Figure 14.6. *(continued)*

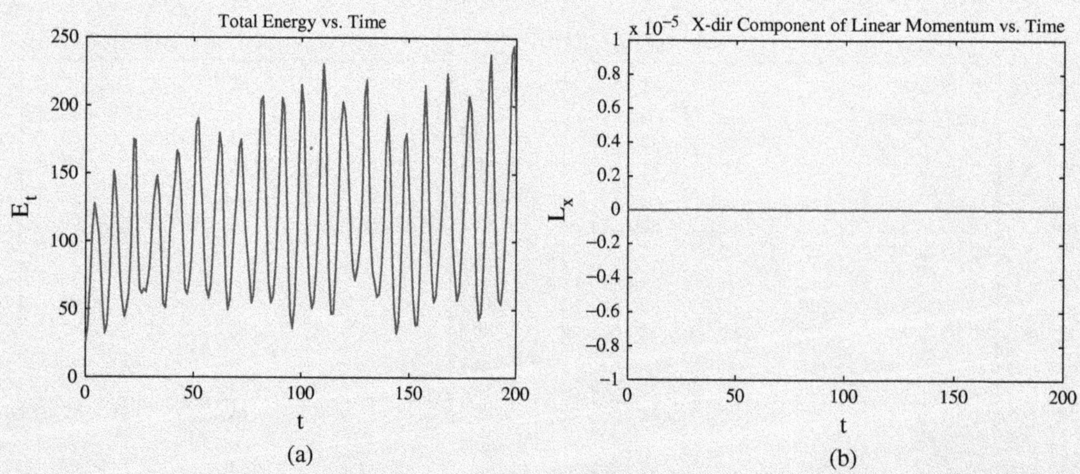

Figure 14.7. Plots of energy, components of linear momentum, and components of angular momentum for the four spring-mass problem with given initial displacement simulated using U0 ($\{\rho_{1\infty}, \rho_{2\infty}, \rho_{3\infty}\} = \{1, 1, 0\}$) (Newmark Method) employing the classical time weighted residual approach

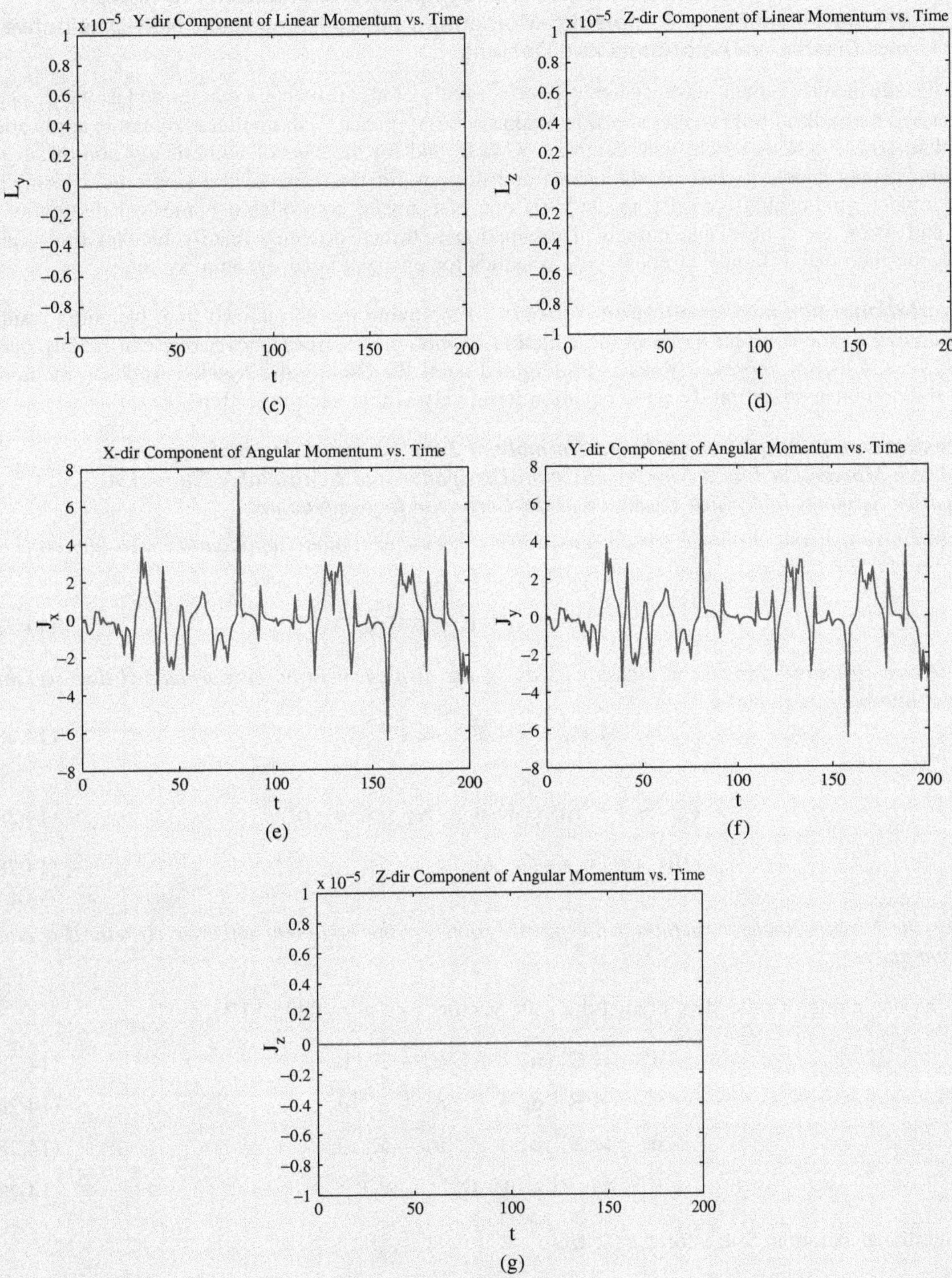

Figure 14.7. *(continued)*

14.3.4 Numerical Implementation Aspects of Symplectic-Momentum Framework in Single-field Form: Symplectic-Momentum Based Numerically Non-Dissipative and Dissipative Algorithms and Designs

Next, we advance the parent linear dynamic GSSSS family of algorithms via that termed as the displacement based normalized time weighted residual approach (Approach 2) to nonlinear dynamic applications such that: (i) for case of no physical damping ($\mathbf{C} = \mathbf{0}$) and for the case of numerically non-dissipative algorithms, they enable a class of algorithms and designs (in the sense of the single-field form) that are symplectic-momentum conserving, and (ii) one can embed controllable numerical dissipation as well, and when the controllable numerical dissipation is turned off, they readily recover the original symplectic-momentum family of conserving schemes for unconstrained dynamic systems.

Computational and Implementation Aspects Employing the normalized time weighted residual methodology to the constant terms in the equation of motion described above, one can readily derive the a-, v-, and d-form representations as highlighted next. We employ the Newton-Raphson method to solve the nonlinear structural dynamic equation iteratively within each time step.

Approximation 2 ***Displacement Based Normalized Time Weighted Residual Approach: Symplectic-Momentum Based Numerically Non-Dissipative and Controllable Numerical Dissipative Schemes in Unified Predictor Multi-Corrector Representation***

For structural dynamic problems for an illustrative truss element model application with (general) discrete internal force representation of the form

$$\mathbf{F}^{\text{int}^h} = \mathbf{K}_1\mathbf{u} + \mathbf{K}_2\epsilon + \mathbf{K}_3\mathbf{u}\epsilon \tag{14.280}$$

the effective structural dynamic equation employing the displacement based normalized time weighted residual approach is given by

$$\mathbf{M}\widetilde{\mathbf{a}} + \mathbf{C}\widetilde{\mathbf{v}} + \widetilde{\mathbf{F}}^{\text{int}^h} = \widetilde{\mathbf{F}}^{\text{ext}^h} \tag{14.281}$$

where

$$\widetilde{\mathbf{F}}^{\text{int}^h} = \mathbf{F}^{\text{int}^h}(\widetilde{\mathbf{u}}) = \mathbf{K}_1\widetilde{\mathbf{u}} + (\mathbf{K}_2 + \mathbf{K}_3\widetilde{\mathbf{u}})\epsilon(\widetilde{\mathbf{u}}) \tag{14.282}$$

$$\epsilon(\widetilde{\mathbf{u}}) = \mathbf{b}\cdot\widetilde{\mathbf{u}} + \frac{2}{l_0^2}\widetilde{\mathbf{u}}\cdot\mathbf{A}\widetilde{\mathbf{u}} \tag{14.283}$$

Employ the Newton-Raphson method to iteratively solve for the nonlinear effective structural dynamic equation above:

At the beginning of time step, predict the state vectors (see also Table 14.3),

$$\widetilde{\mathbf{a}}_{n+1}^k = \Xi_{Pa}^{(1)}\mathbf{u}_n + \Xi_{Pa}^{(2)}\dot{\mathbf{u}}_n + \Xi_{Pa}^{(3)}\ddot{\mathbf{u}}_n \tag{14.284}$$

$$\widetilde{\mathbf{v}}_{n+1}^k = \Xi_{Pv}^{(1)}\mathbf{u}_n + \Xi_{Pv}^{(2)}\dot{\mathbf{u}}_n + \Xi_{Pv}^{(3)}\ddot{\mathbf{u}}_n \tag{14.285}$$

$$\widetilde{\mathbf{u}}_{n+1}^k = \Xi_{Pd}^{(1)}\mathbf{u}_n + \Xi_{Pd}^{(2)}\dot{\mathbf{u}}_n + \Xi_{Pd}^{(3)}\ddot{\mathbf{u}}_n \tag{14.286}$$

$$\widetilde{\mathbf{F}}_{n+1}^{\text{ext}^h\,k} = (1 - W_1)\mathbf{F}_n^{\text{ext}^h} + W_1\mathbf{F}_{n+1}^{\text{ext}^h} \tag{14.287}$$

Start nonlinear iteration. Solve for $\Delta\boldsymbol{\delta}_{n+1}^{k+1}$ from

$$\begin{aligned}&[\Xi_{Ca}\mathbf{M} + \Xi_{Cv}\mathbf{C} + \Xi_{Cd}\mathbf{K}_{n+1}^k]\Delta\boldsymbol{\delta}_{n+1}^{k+1} = \\ &-\left[\mathbf{M}\widetilde{\mathbf{a}}_{n+1}^k + \mathbf{C}\widetilde{\mathbf{v}}_{n+1}^k + \mathbf{K}_1\widetilde{\mathbf{u}}_{n+1}^k + \mathbf{K}_2\widetilde{\epsilon}_{n+1}^k + \mathbf{K}_3\widetilde{\mathbf{u}}\epsilon(\widetilde{\mathbf{u}}_{n+1}^k) - \widetilde{\mathbf{F}}_{n+1}^{\text{ext}^h\,k}\right]\end{aligned} \tag{14.288}$$

$$ {}^{t}\mathbf{K}^{k}_{n+1} = \mathbf{K_1} + \mathbf{K_2}(\mathbf{b} + \frac{4}{l_0^2}\mathbf{A}\tilde{\mathbf{u}}^{k}_{n+1}) + \mathbf{K_3}[\tilde{\mathbf{u}}^{k}_{n+1}(\mathbf{b} + \frac{4}{l_0^2}\mathbf{A}\tilde{\mathbf{u}}^{k}_{n+1}) + \epsilon(\tilde{\mathbf{u}}^{k}_{n+1})] \tag{14.289} $$

Then correct the unknowns as follows.

$$ \tilde{\mathbf{a}}^{k+1}_{n+1} = \tilde{\mathbf{a}}^{k}_{n+1} + \Xi_{Ca}\Delta\boldsymbol{\delta}^{k+1}_{n+1} \tag{14.290} $$

$$ \tilde{\mathbf{v}}^{k+1}_{n+1} = \tilde{\mathbf{v}}^{k}_{n+1} + \Xi_{Cv}\Delta\boldsymbol{\delta}^{k+1}_{n+1} \tag{14.291} $$

$$ \tilde{\mathbf{u}}^{k+1}_{n+1} = \tilde{\mathbf{u}}^{k}_{n+1} + \Xi_{Cd}\Delta\boldsymbol{\delta}^{k+1}_{n+1} \tag{14.292} $$

until the solution converges

$$ ||\Delta\boldsymbol{\delta}^{k+1}_{n+1}|| = ||\boldsymbol{\delta}^{k+1}_{n+1} - \boldsymbol{\delta}^{k}_{n+1}|| < tolerance \tag{14.293} $$

Table 14.3. Predictor multi-corrector coefficients for the incremental a-, v- and d-form representations

	a-form	v-form	d-form
$\Xi^{(1)}_{Pd}$	1	1	1
$\Xi^{(2)}_{Pd}$	$\lambda_1\Delta t$	$\lambda_1\Delta t$	0
$\Xi^{(3)}_{Pd}$	$\lambda_2\Delta t^2$	$(\lambda_2 - \frac{\lambda_3\lambda_4}{\lambda_5})\Delta t^2$	0
$\Xi^{(1)}_{Pv}$	0	0	0
$\Xi^{(2)}_{Pv}$	1	1	$1 - \frac{\Lambda_5 W_2 \lambda 1}{\lambda_3}$
$\Xi^{(3)}_{Pv}$	$\Lambda_4 W_1\Delta t$	$(\Lambda_4 W_1 - \frac{\Lambda_5 W_2\lambda_4}{\lambda_5})\Delta t$	$(\Lambda_4 W_1 - \frac{\Lambda_5 W_2\lambda_2}{\lambda_3})\Delta t$
$\Xi^{(1)}_{Pa}$	0	0	0
$\Xi^{(2)}_{Pa}$	0	0	$-\frac{\Lambda_6 W_1\lambda_1}{\lambda_3\Delta t}$
$\Xi^{(3)}_{Pa}$	1	$1 - \frac{\Lambda_6 W_1\lambda_4}{\lambda_5}$	$1 - \frac{\Lambda_6 W_1\lambda_2}{\lambda_3}$
$\Xi^{(1)}_{Pp}$	1	1	1
$\Xi^{(2)}_{Pp}$	$(\Lambda_1 W_1 - \frac{\Lambda_3 W_3\lambda_1}{\lambda_3})\Delta t$	$(\Lambda_1 W_1 - \frac{\Lambda_3 W_3\lambda_1}{\lambda_3})\Delta t$	$(\Lambda_1 W_1 - \frac{\Lambda_3 W_3\lambda_1}{\lambda_3})\Delta t$
$\Xi^{(3)}_{Pp}$	$(\Lambda_2 W_2 - \frac{\Lambda_3 W_3\lambda_2}{\lambda_3})\Delta t^2$	$(\Lambda_2 W_2 - \frac{\Lambda_3 W_3\lambda_2}{\lambda_3})\Delta t^2$	$(\Lambda_2 W_2 - \frac{\Lambda_3 W_3\lambda_2}{\lambda_3})\Delta t^2$
$\Xi^{(4)}_{Pp}$	$\frac{\Lambda_3 W_3}{\lambda_3}$	$\frac{\Lambda_3 W_3}{\lambda_3}$	$\frac{\Lambda_3 W_3}{\lambda_3}$
Ξ_{Cd}	$\Lambda_3 W_3\Delta t^2$	$\frac{\Lambda_3 W_3}{\lambda_5}\Delta t$	$\frac{\Lambda_3 W_3}{\lambda_3}$
Ξ_{Cv}	$\Lambda_5 W_2\Delta t$	$\frac{\Lambda_5 W_2}{\lambda_5}$	$\frac{\Lambda_5 W_2}{\lambda_3\Delta t}$
Ξ_{Ca}	$\Lambda_6 W_1$	$\frac{\Lambda_6 W_1}{\lambda_5\Delta t}$	$\frac{\Lambda_6 W_1}{\lambda_3\Delta t^2}$
$\Xi^{(1)}_{Cp}$	1	1	1
$\Xi^{(2)}_{Cp}$	$(\Lambda_1 W_1 - \frac{\Lambda_3 W_3\lambda_1}{\lambda_3})\Delta t$	$(\Lambda_1 W_1 - \frac{\Lambda_3 W_3\lambda_1}{\lambda_3})\Delta t$	$(\Lambda_1 W_1 - \frac{\Lambda_3 W_3\lambda_1}{\lambda_3})\Delta t$
$\Xi^{(3)}_{Cp}$	$(\Lambda_2 W_2 - \frac{\Lambda_3 W_3\lambda_2}{\lambda_3})\Delta t^2$	$(\Lambda_2 W_2 - \frac{\Lambda_3 W_3\lambda_2}{\lambda_3})\Delta t^2$	$(\Lambda_2 W_2 - \frac{\Lambda_3 W_3\lambda_2}{\lambda_3})\Delta t^2$
$\Xi^{(4)}_{Cp}$	$\frac{\Lambda_3 W_3}{\lambda_3}$	$\frac{\Lambda_3 W_3}{\lambda_3}$	$\frac{\Lambda_3 W_3}{\lambda_3}$

Once the solution converges, update the unknowns at the end of the time step as follows.

$$\ddot{\mathbf{u}}_{n+1} = \ddot{\mathbf{u}}_n + (\tilde{\mathbf{a}}_{n+1}^{k+1} - \ddot{\mathbf{u}}_n)/(\Lambda_6 W_1) \tag{14.294}$$

$$\dot{\mathbf{u}}_{n+1} = \dot{\mathbf{u}}_n + \lambda_4 \ddot{\mathbf{u}}_n \Delta t + \lambda_5 (\ddot{\mathbf{u}}_{n+1} - \ddot{\mathbf{u}}_n)\Delta t \tag{14.295}$$

$$\mathbf{u}_{n+1} = \mathbf{u}_n + \lambda_1 \dot{\mathbf{u}}_n \Delta t + \lambda_2 \ddot{\mathbf{u}}_n \Delta t^2 + \lambda_3 (\ddot{\mathbf{u}}_{n+1} - \ddot{\mathbf{u}}_n)\Delta t^2 \tag{14.296}$$

The predictor-corrector coefficients Ξ above in the corresponding a, v, and d-form are listed in Table 14.3. We show and explain in (Hoitink et al. 2008; Masuri et al. 2009b,c), that a different representation for the update of the unknowns are preferable to help foster angular momentum conservation, thereby avoiding errors due to acceleration computations.

Remark 14.3.13 It is to be noted that with the displacement based normalized time weighted residual approach within the GSSSS framework, U0/V0 ($\{\rho_{1\infty}, \rho_{2\infty}, \rho_{3\infty}\} = \{1, 1, 1\}$) and V0 ($\{\rho_{1\infty}, \rho_{2\infty}, \rho_{3\infty}\} = \{1, 1, \rho_{3\infty}\}$) all yield symplectic-momentum conservation (numerically non-dissipative family of algorithms) in the sense of the single-field form.

Illustrative Example

Example 14.3 Four Mass-Spring System St. Venant Material Model

We again consider a four mass-spring system constructed of four springs connecting four masses as depicted earlier in Figure 14.3. However, the problem consists of truss elements characterizing the four springs that are connected to four masses. This model problem has potential energy for the continuum truss elements as a quadratic function of Green strain, and is representative of the St. Venant Kirchhoff material model. In this example, each spring has initial length of 1m and stiffness of $10^2 N/m$ while each mass weighs 1kg. The initial position of the four nodes are [0,0,0;1,0,0;1,1,0;0,1,0]m. The initial conditions are as follows, and selected examples to follow subsequently in this section use either i) or ii) as shown below simply for purposes of illustration.

i) given initial displacement problem, $\mathbf{u}_0$ = [0,0,0,0,0,0,0,0,0,0,0,1]
ii) given initial displacement and velocity problem, $\mathbf{u}_0$ = [0,0,1,0,0,1,1,0,0,0,0,1], and $\mathbf{v}_0$ = [0,0,0, 0,0,0,0,0,0,0,0,1]

Numerically Non-Dissipative Displacement Based Normalized Time Weighted Residual Approach The nonlinear four mass-spring system described above is first analyzed here with given initial displacement only (see Figures 14.8–14.10). To evaluate the performance of the displacement based normalized time weighted residual approach, we study the conservation of the physical quantities including energy, linear and angular momentum to evaluate the performance of the displacement based normalized time weighted residual approach. The backbone of LMS methods, namely, the numerically non-dissipative algorithms which include U0/V0 ($\{\rho_{1\infty}, \rho_{2\infty}, \rho_{3\infty}\} = \{1.00, 1.00, 1.00\}$) (Classical Version Midpoint Rule-[MPR-EPA]), V0 ($\{\rho_{1\infty}, \rho_{2\infty}, \rho_{3\infty}\} = \{1.00, 1.00, 0.00\}$) (New Version Midpoint Rule-[MPR-MPA]), and U0 ($\{\rho_{1\infty}, \rho_{2\infty}, \rho_{3\infty}\} = \{1.00, 1.00, 0.00\}$) (Newmark Method) are employed to solve the nonlinear structural dynamic problem.

We see that the Newmark method fails to adequately capture the underlying physics. This is in contrast to both the Classical Midpoint Rule-[MPR-EPA] (common to both U- and V-family of

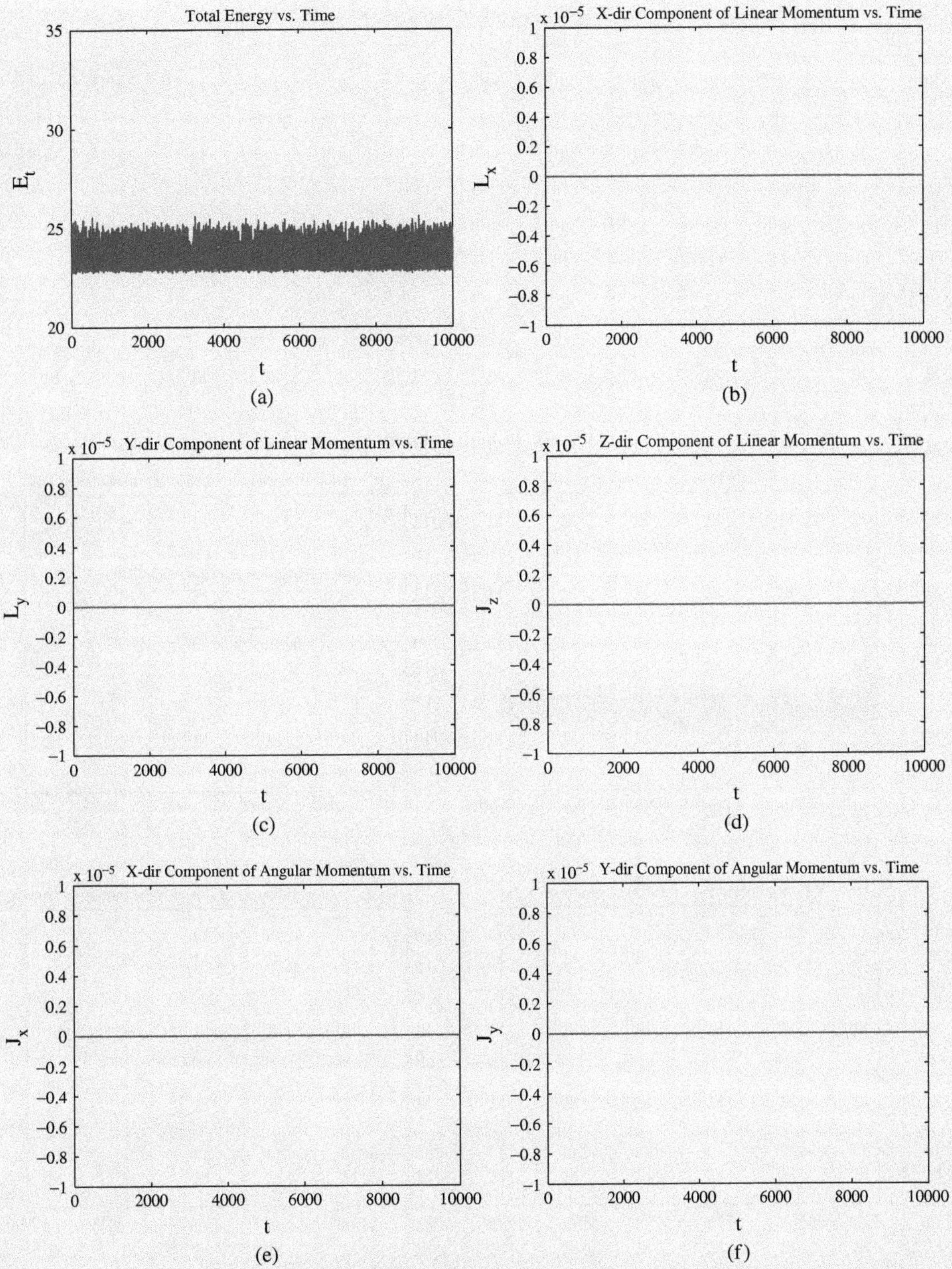

Figure 14.8. Plots of energy, components of linear momentum, and components of angular momentum for the four spring-mass problem with given initial displacement simulated using U0/V0 ($\{\rho_{1\infty}, \rho_{2\infty}, \rho_{3\infty}\} = \{1, 1, 1\}$) (Classical Midpoint Rule-[MPR-EPA]) employing the displacement based normalized time weighted residual approach

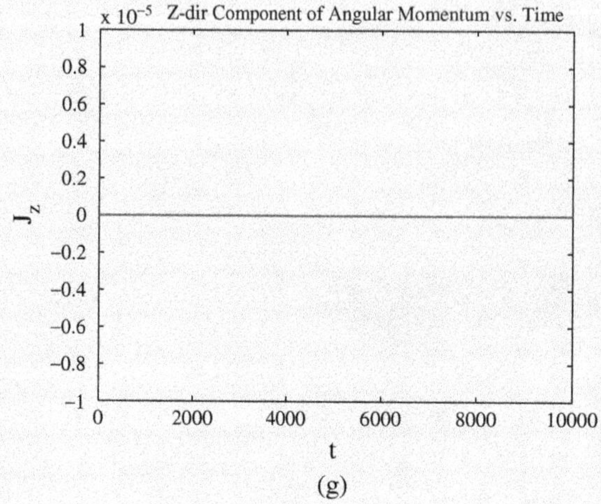

Figure 14.8. *(continued)*

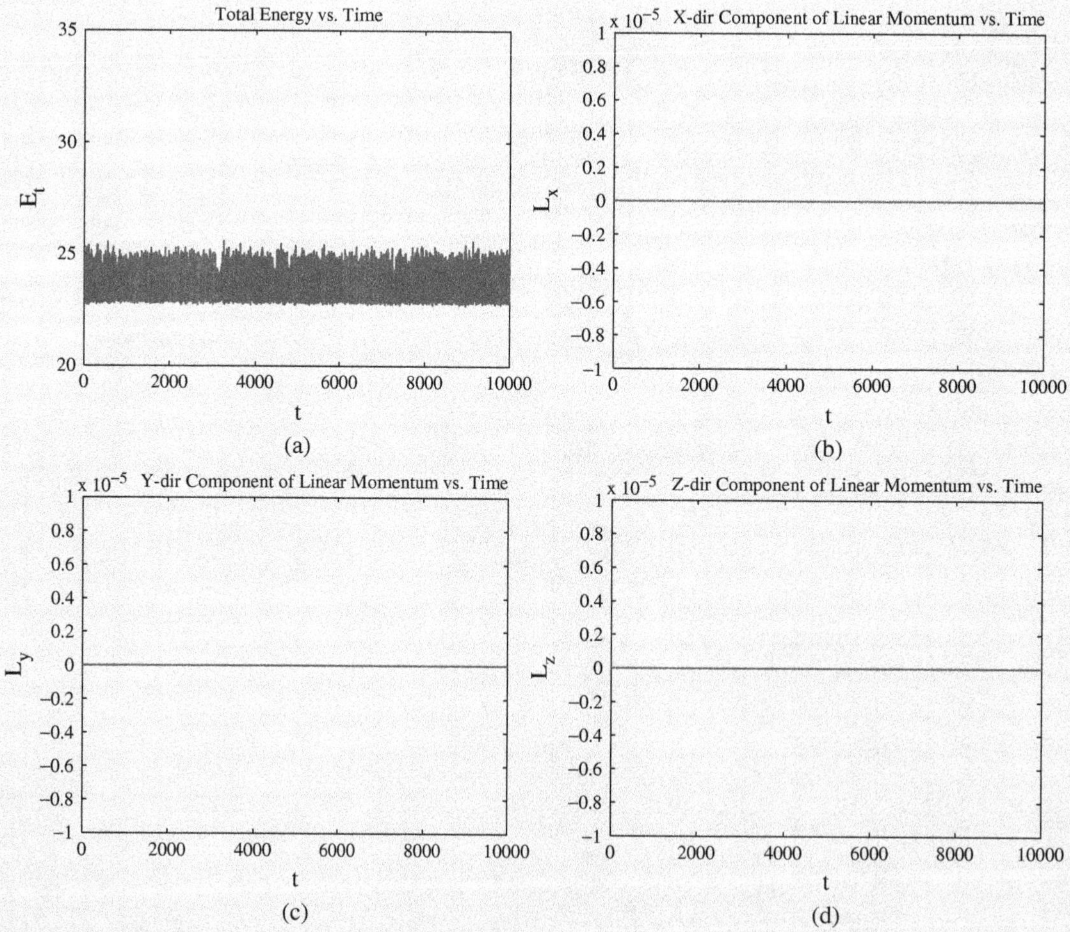

Figure 14.9. Plots of energy, components of linear momentum, and components of angular momentum for the four spring-mass problem with given initial displacement simulated using V0 ($\{\rho_{1\infty}, \rho_{2\infty}, \rho_{3\infty}\} = \{1, 1, 0\}$) (New Version Midpoint Rule-[MPR-MPA]) employing the displacement based normalized time weighted residual approach

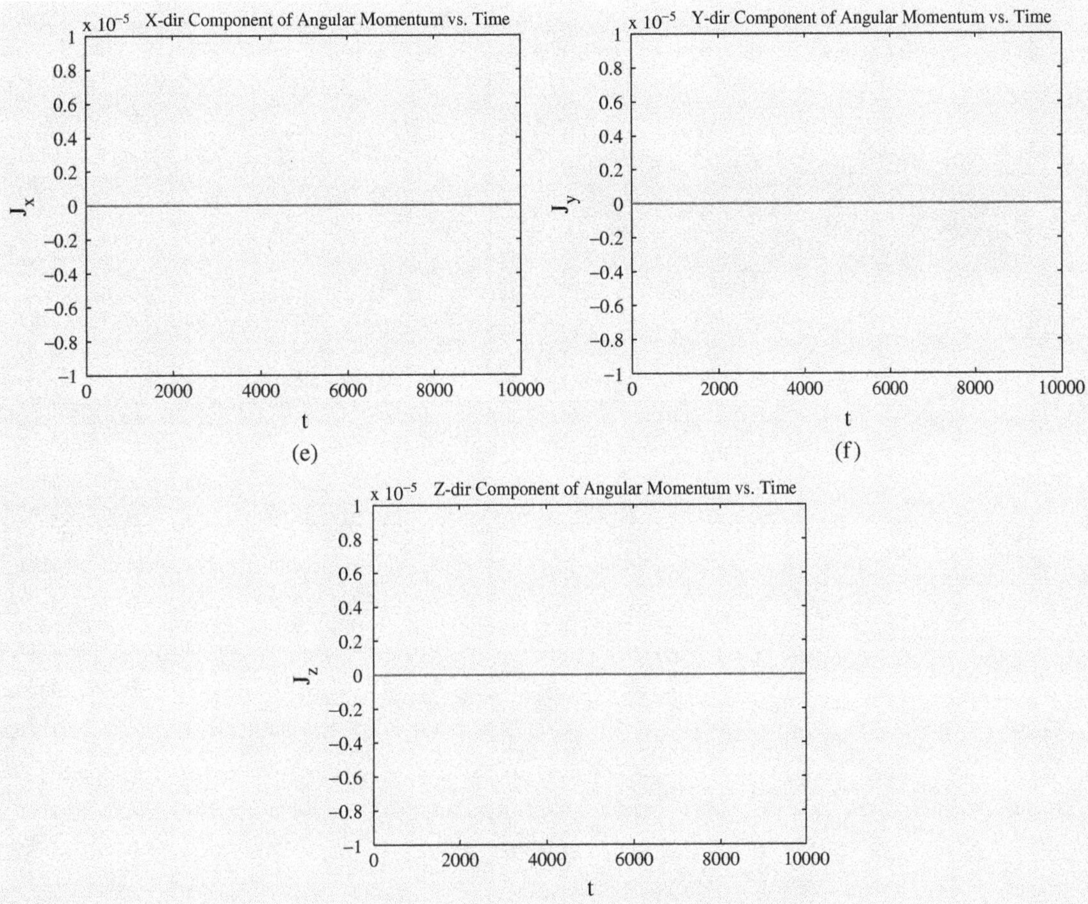

(e) (f) (g)

Figure 14.9. *(continued)*

algorithms) and the New Version Midpoint Rule-[MPR-MPA] within the V-family of algorithms and in the sense of the single-field form which inherit symplectic-momentum conserving features via the present developments.

Numerically Dissipative Displacement Based Normalized Time Weighted Residual Approach We next consider the same nonlinear four mass-spring system with both given initial displacement and velocity as an illustration. We simulate the problem using the following optimal dissipative time integrator within the class of LMS methods as the basic parent algorithm, implemented via both the classical approach and the displacement based normalized time weighted residual approach (note the improved performance via the normalized time weighted residual approach):

i) U0/V0 ($\{\rho_{1\infty}, \rho_{2\infty}, \rho_{3\infty}\} = \{0.8, 1.0, 0.8\}$)

The results (see Figure 14.11) indicate that when the controllable numerical dissipation is turned off, that is when U0 ($\{\rho_{1\infty}, \rho_{2\infty}, \rho_{3\infty}\} = \{1.0, 1.0, 1.0\}$) or V0 ($\{\rho_{1\infty}, \rho_{2\infty}, \rho_{3\infty}\} = \{1.0, 1.0, \rho_{3\infty}\}$), then the underlying physics is preserved via the symplectic-momentum conserving schemes.

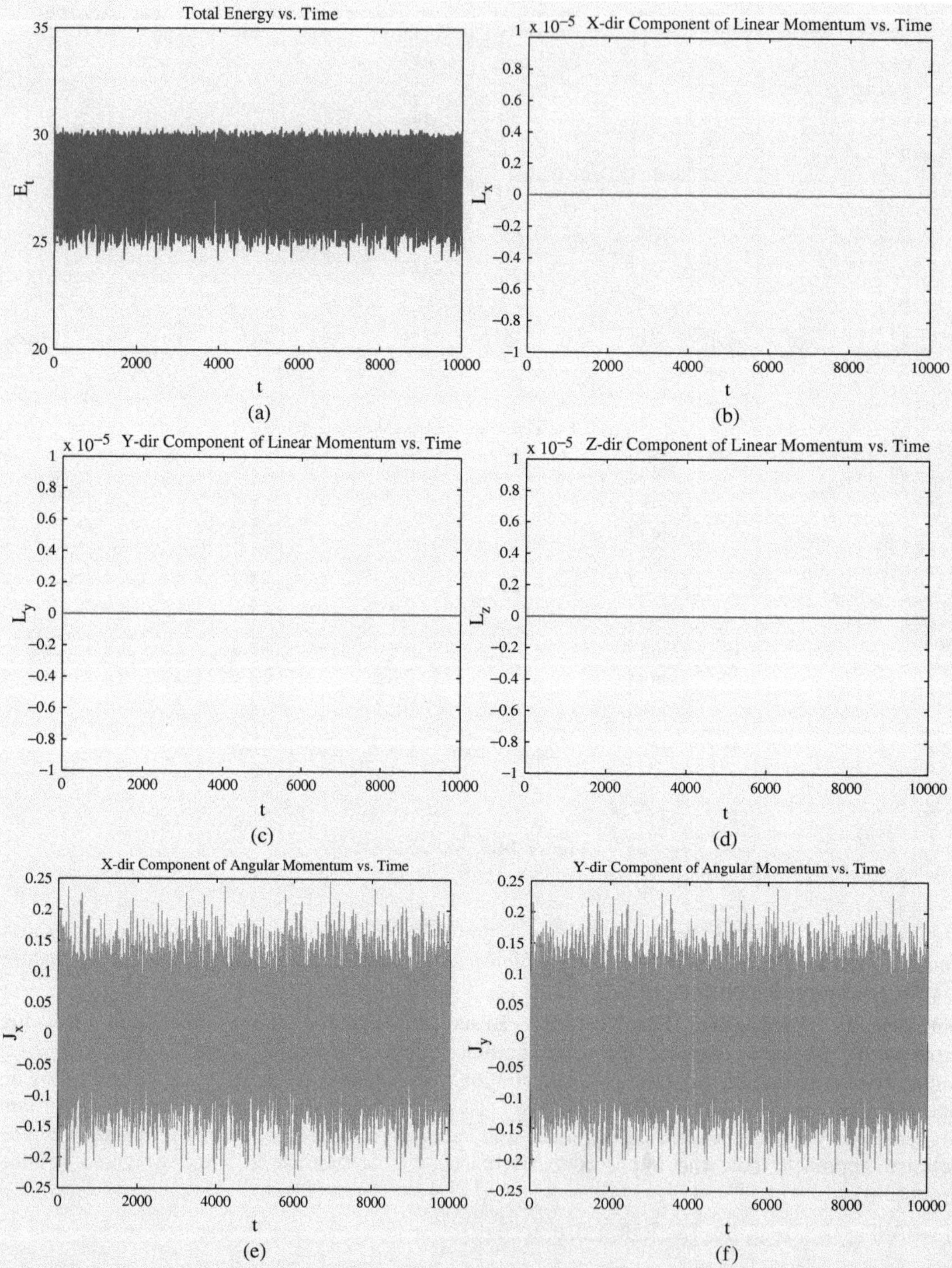

Figure 14.10. Plots of energy, components of linear momentum, and components of angular momentum for the four spring-mass problem with given initial displacement simulated using U0 ($\{\rho_{1\infty}, \rho_{2\infty}, \rho_{3\infty}\} = \{1, 1, 0\}$) (Newmark Method) employing the displacement based normalized time weighted residual approach

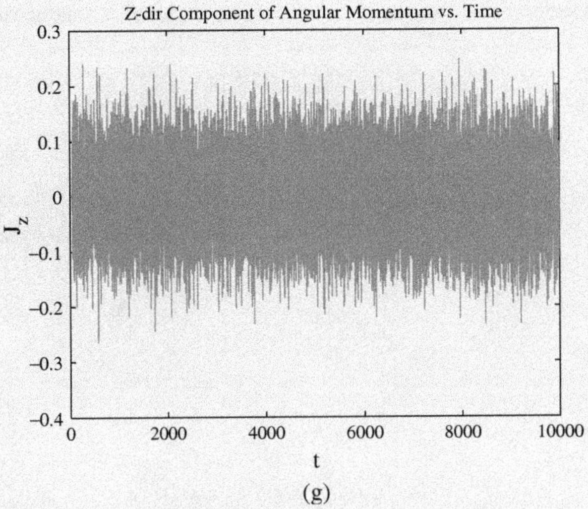

Figure 14.10. *(continued)*

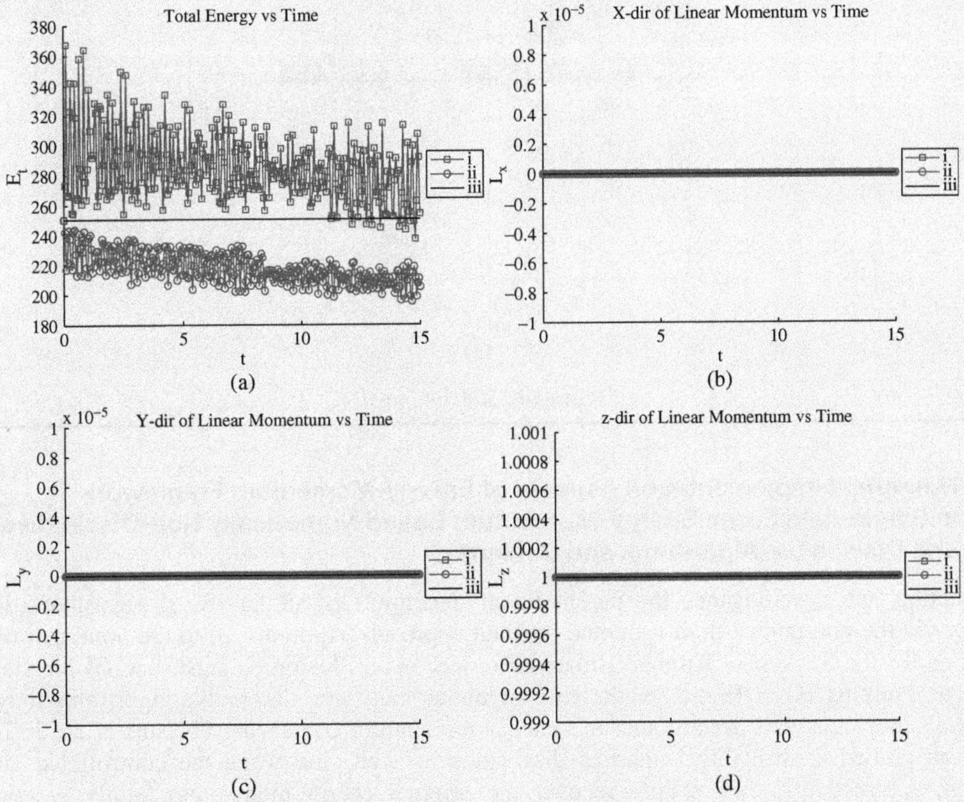

Figure 14.11. Plots of energy, components of linear momentum, and components of angular momentum for the four mass-spring problem with given initial displacement and velocity simulated using (i) U0/V0 ($\{\rho_{1\infty}, \rho_{2\infty}, \rho_{3\infty}\} = \{0.8, 1.0, 0.8\}$) implemented via the the classical time weighted residual approach with controllable numerical dissipation turned on, (ii) U0/V0 ($\{\rho_{1\infty}, \rho_{2\infty}, \rho_{3\infty}\} = \{0.8, 1.0, 0.8\}$) implemented via the displacement based normalized time weighted residual approach with controllable numerical dissipation turned on, and (iii) Exact solution

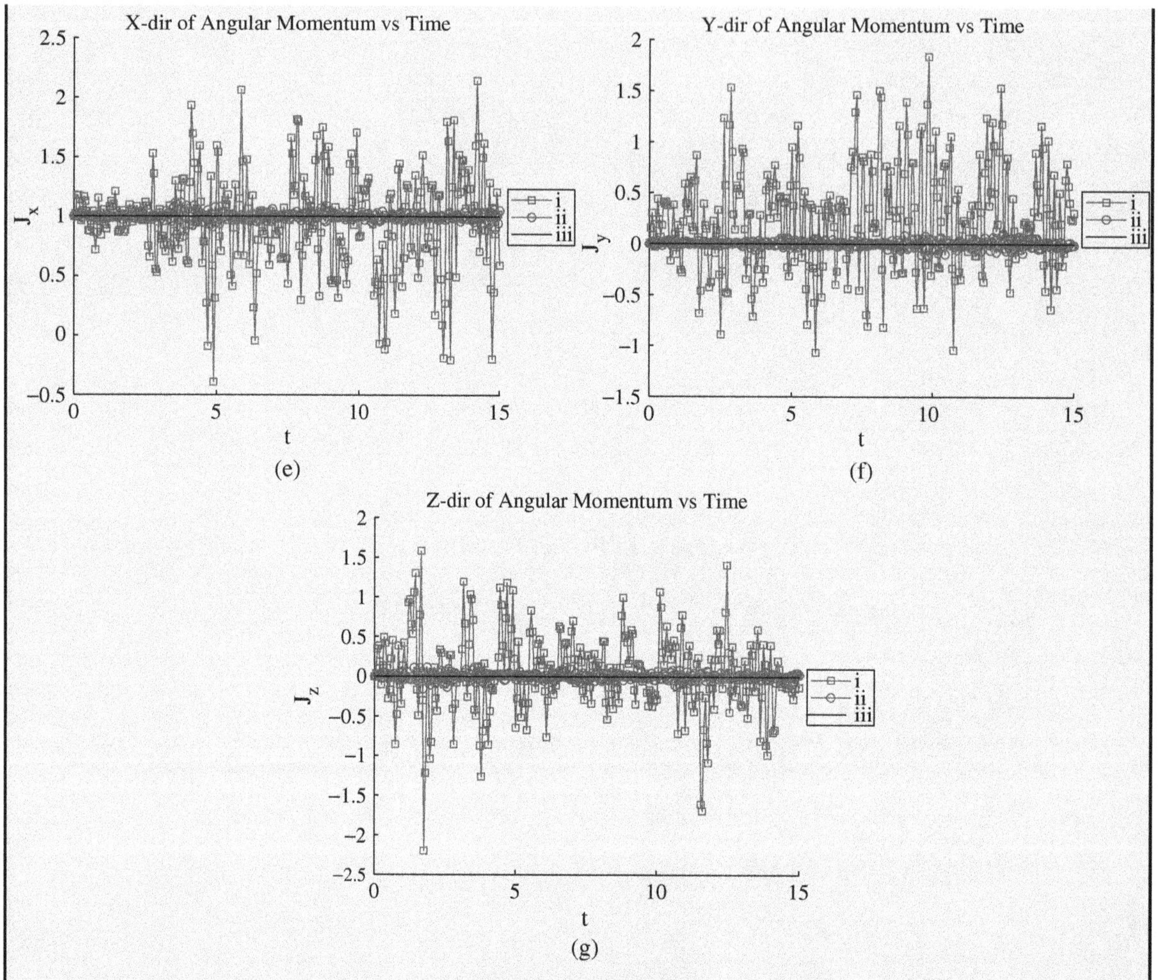

Figure 14.11. (*continued*)

14.3.5 Numerical Implementation Aspects of Energy-Momentum Framework in Single-field Form: Energy-Momentum Based Numerically Non-Dissipative and Dissipative Algorithms and Designs

In this section, we next advance the parent linear dynamic GSSSS family of algorithms described previously via the normalized time weighted residual approach (Approach 3) to the nonlinear dynamics applications for the St. Venant Kirchhoff material model as an illustration such that: (i) for the case of no physical damping ($\mathbf{C} = \mathbf{0}$) and for the case of numerically non-dissipative algorithms they enable a family of algorithms and designs that are energy-momentum conserving (Masuri et al. 2009a), and (ii) one can embed controllable numerical dissipation as well, and when the controllable numerical dissipation is turned off, they readily recover the original energy-momentum family of conserving schemes for unconstrained conservative dynamic systems.

Again for illustration of the basic ideas, we simply consider the St. Venant-Kirchhoff material model (see various details described earlier for general hyperelastic models) and a simple truss element model given by

$$\mathbf{M}\ddot{\mathbf{u}}(t) + \mathbf{C}\dot{\mathbf{u}}(t) + \mathbf{K}_1\mathbf{u}(t) + \mathbf{K}_2\epsilon(t) + \mathbf{K}_3\mathbf{u}(t)\epsilon(t) = \mathbf{F}^{\text{ext}^h}(t) \tag{14.297}$$

We next employ the normalized time weighted residual methodology described earlier in conjunction with the hybrid displacement-strain based approach, and consequently approximate the non-constant variables $\ddot{\mathbf{u}}(t)$, $\dot{\mathbf{u}}(t)$, $\mathbf{u}(t)$, $\epsilon(t)$ and $\mathbf{f}(t)$ in Equation (14.297) as follows:

$$\ddot{\mathbf{u}} = \ddot{\mathbf{u}}_n + \Lambda_6 \frac{\Delta\ddot{\mathbf{u}}}{\Delta t}\tau \tag{14.298}$$

$$\dot{\mathbf{u}} = \dot{\mathbf{u}}_n + \Lambda_4 \ddot{\mathbf{u}}_n \tau + \Lambda_5 \frac{\Delta\ddot{\mathbf{u}}}{\Delta t}\tau^2 \tag{14.299}$$

$$\mathbf{u} = \mathbf{u}_n + \Lambda_1 \dot{\mathbf{u}}_n \tau + \Lambda_2 \ddot{\mathbf{u}}_n \tau^2 + \Lambda_3 \frac{\Delta\ddot{\mathbf{u}}}{\Delta t}\tau^3 \tag{14.300}$$

$$\epsilon = \epsilon_n + \Lambda_1 \dot{\epsilon}_n \tau + \Lambda_2 \ddot{\epsilon}_n \tau^2 + \Lambda_3 \frac{\Delta\ddot{\epsilon}}{\Delta t}\tau^3 \tag{14.301}$$

$$\mathbf{F}^{\text{ext}^h} = \mathbf{F}_n^{\text{ext}^h} + \frac{\mathbf{F}_{n+1}^{\text{ext}^h} - \mathbf{F}_n^{\text{ext}^h}}{\Delta t}\tau \tag{14.302}$$

where

$$\Delta\ddot{\mathbf{u}} = \ddot{\mathbf{u}}_{n+1} - \ddot{\mathbf{u}}_n \quad \text{and} \quad \Delta\ddot{\epsilon} = \ddot{\epsilon}_{n+1} - \ddot{\epsilon}_n \tag{14.303}$$

With the weighting time field approximated as the degenerated scalar polynomial function given by

$$W = 1 + w_1\Gamma + w_2\Gamma^2 + w_3\Gamma^3 \tag{14.304}$$

we obtain the effective nonlinear structural dynamic equation employing the hybrid displacement-strain based normalized time weighted residual approach

$$\mathbf{M}\widetilde{\mathbf{a}} + \mathbf{C}\widetilde{\mathbf{v}} + \widetilde{\mathbf{F}}^{\text{int}^h} = \widetilde{\mathbf{F}}^{\text{ext}^h} \tag{14.305}$$

where

$$\widetilde{\mathbf{a}} = \ddot{\mathbf{u}}_n + \Lambda_6 W_1 \Delta\ddot{\mathbf{u}} \tag{14.306}$$

$$\widetilde{\mathbf{v}} = \dot{\mathbf{u}}_n + \Lambda_4 W_1 \Delta t \ddot{\mathbf{u}}_n + \Lambda_5 W_2 \Delta t \Delta\ddot{\mathbf{u}} \tag{14.307}$$

$$\widetilde{\mathbf{F}}^{\text{int}^h} = \mathbf{K}_1\widetilde{\mathbf{u}} + \mathbf{K}_2\widetilde{\epsilon} + \mathbf{K}_3\widetilde{\mathbf{u}}\widetilde{\epsilon} \tag{14.308}$$

$$\widetilde{\mathbf{u}} = \mathbf{u}_n + \Lambda_1 W_1 \Delta t \dot{\mathbf{u}}_n + \Lambda_2 W_2 \Delta t^2 \ddot{\mathbf{u}}_n + \Lambda_3 W_3 \Delta t^2 \Delta\ddot{\mathbf{u}} \tag{14.309}$$

$$\widetilde{\epsilon} = \epsilon_n + \Lambda_1 W_1 \Delta t \dot{\epsilon}_n + \Lambda_2 W_2 \Delta t^2 \ddot{\epsilon}_n + \Lambda_3 W_3 \Delta t^2 \Delta\ddot{\epsilon} \tag{14.310}$$

$$\widetilde{\mathbf{F}}^{\text{ext}^h} = (1 - W_1)\mathbf{F}_n^{\text{ext}^h} + W_1\mathbf{F}_{n+1}^{\text{ext}^h} \tag{14.311}$$

$\Delta\ddot{\epsilon}$ in Equation (14.310) can be obtained from the update equation on the strain as follows. Following the update on displacement (as described in (Zhou and Tamma 2004c, 2006)) as

$$\mathbf{u}_{n+1} = \mathbf{u}_n + \lambda_1 \Delta t \dot{\mathbf{u}}_n + \lambda_2 \Delta t^2 \ddot{\mathbf{u}}_n + \lambda_3 \Delta\ddot{\mathbf{u}} \Delta t^2 \tag{14.312}$$

we have the update on the strain at the end of each time step as follows,

$$\epsilon_{n+1} = \epsilon_n + \lambda_1 \dot{\epsilon}_n \Delta t + \lambda_2 \ddot{\epsilon}_n \Delta t^2 + \lambda_3 \Delta\ddot{\epsilon} \Delta t^2 \tag{14.313}$$

Solving for $\Delta\ddot{\epsilon}$ from Equation (14.313) and substituting it into Equation (14.310), the algorithmic strain becomes

$$\widetilde{\epsilon} = \epsilon_n + \frac{W_3\Lambda_3}{\lambda_3}\Delta\epsilon + (W_1\Lambda_1 - \frac{W_3\Lambda_3\lambda_1}{\lambda_3})\Delta t\dot{\epsilon}_n + (W_2\Lambda_2 - \frac{W_3\Lambda_3\lambda_2}{\lambda_3})\Delta t^2\ddot{\epsilon}_n \tag{14.314}$$

where

$$\Delta\epsilon = \epsilon_{n+1} - \epsilon_n \tag{14.315}$$

We note that we have approximated the scalar ϵ in time in a similar manner as for the nodal displacement vector although the discrete strain tensor has been approximated linearly in time in subsection 14.2.

Computational and Implementation Aspects We again employ the Newton-Raphson method to solve the nonlinear structural dynamic equation system iteratively within each time step. Here we summarize the computational procedure in a unified and concise framework to follow under various forms, namely, the a, v, and d-form representations.

Approximation 3 ***Hybrid Displacement-Strain Based Normalized Time Weighted Residual Approach: Energy-Momentum Based Numerically Non-Dissipative and Controllable Numerical Dissipative Schemes in Unified Predictor Multi-Corrector Representation***

For structural dynamic problems of a truss element model with general representation of the internal force of the form

$$\mathbf{F}^{\text{int}^h} = \mathbf{K}_1\mathbf{u} + \mathbf{K}_2\epsilon + \mathbf{K}_3\mathbf{u}\epsilon \tag{14.316}$$

the effective structural dynamic equation employing the hybrid displacement-strain based normalized time weighted residual approach is given by

$$\mathbf{M}\widetilde{\mathbf{a}} + \mathbf{C}\widetilde{\mathbf{v}} + \widetilde{\mathbf{F}}^{\text{int}^h} = \widetilde{\mathbf{F}}^{\text{ext}^h} \tag{14.317}$$

where

$$\widetilde{\mathbf{F}}^{\text{int}^h} = \mathbf{F}^{\text{int}^h}(\widetilde{\mathbf{u}}) = \mathbf{K}_1\widetilde{\mathbf{u}} + (\mathbf{K}_2 + \mathbf{K}_3\widetilde{\mathbf{u}})\epsilon(\widetilde{\mathbf{u}}) \tag{14.318}$$

$$\epsilon(\widetilde{\mathbf{u}}) = \mathbf{b}\cdot\widetilde{\mathbf{u}} + \frac{2}{l_0^2}\widetilde{\mathbf{u}}\cdot\mathbf{A}\widetilde{\mathbf{u}} \tag{14.319}$$

Employ the Newton-Raphson method to iteratively solve for the nonlinear effective structural dynamic equation above:

At the beginning of time step, predict the state vectors

$$\widetilde{\mathbf{a}}_{n+1}^{k} = \Xi_{Pa}^{(1)}\mathbf{u}_n + \Xi_{Pa}^{(2)}\dot{\mathbf{u}}_n + \Xi_{Pa}^{(3)}\ddot{\mathbf{u}}_n \tag{14.320}$$

$$\widetilde{\mathbf{v}}_{n+1}^{k} = \Xi_{Pv}^{(1)}\mathbf{u}_n + \Xi_{Pv}^{(2)}\dot{\mathbf{u}}_n + \Xi_{Pv}^{(3)}\ddot{\mathbf{u}}_n \tag{14.321}$$

$$\widetilde{\mathbf{u}}_{n+1}^{k} = \Xi_{Pd}^{(1)}\mathbf{u}_n + \Xi_{Pd}^{(2)}\dot{\mathbf{u}}_n + \Xi_{Pd}^{(3)}\ddot{\mathbf{u}}_n \tag{14.322}$$

$$\widetilde{\epsilon}_{n+1}^{k} = \Xi_{P\epsilon}^{(1)}\epsilon_n + \Xi_{P\epsilon}^{(2)}\dot{\epsilon}_n + \Xi_{P\epsilon}^{(3)}\ddot{\epsilon}_n + \Xi_{P\epsilon}^{(4)}\Delta\epsilon \tag{14.323}$$

Start nonlinear iteration. Solve for $\Delta\boldsymbol{\delta}_{n+1}^{k+1}$ from

$$[\Xi_{Ca}\mathbf{M} + \Xi_{Cv}\mathbf{C} + \Xi_{Cd}\mathbf{K}_{n+1}^{k}]\Delta\boldsymbol{\delta}_{n+1}^{k+1} = -\left[\mathbf{M}\widetilde{\mathbf{a}}_{n+1}^{k} + \mathbf{C}\widetilde{\mathbf{v}}_{n+1}^{k} + \mathbf{K}_1\widetilde{\mathbf{u}}_{n+1}^{k} + \mathbf{K}_2\widetilde{\epsilon}_{n+1}^{k} + \mathbf{K}_3\widetilde{\mathbf{u}}_{n+1}^{k}\widetilde{\epsilon}_{n+1}^{k} - \widetilde{\mathbf{F}}_{n+1}^{\text{ext}^h\,k}\right] \tag{14.324}$$

$${}^{t}\mathbf{K}_{n+1}^{k} = \mathbf{K}_1 + \mathbf{K}_2(\mathbf{b} + \frac{4}{l_0^2}\mathbf{A}\mathbf{u}_{n+1}^{k}) + \mathbf{K}_3[\widetilde{\mathbf{u}}_{n+1}^{k}\cdot(\mathbf{b} + \frac{4}{l_0^2}\mathbf{A}\mathbf{u}_{n+1}^{k}) + \widetilde{\epsilon}_{n+1}^{k}] \tag{14.325}$$

Then correct the unkowns as follows.

$$\widetilde{\mathbf{a}}_{n+1}^{k+1} = \widetilde{\mathbf{a}}_{n+1}^{k} + \Xi_{Ca}\Delta\boldsymbol{\delta}_{n+1}^{k+1} \tag{14.326}$$

$$\widetilde{\mathbf{v}}_{n+1}^{k+1} = \widetilde{\mathbf{v}}_{n+1}^{k} + \Xi_{Cv}\Delta\boldsymbol{\delta}_{n+1}^{k+1} \tag{14.327}$$

$$\widetilde{\mathbf{u}}_{n+1}^{k+1} = \widetilde{\mathbf{u}}_{n+1}^{k} + \Xi_{Cd}\Delta\boldsymbol{\delta}_{n+1}^{k+1} \tag{14.328}$$

$$\widetilde{\epsilon}_{n+1}^{k+1} = \Xi_{C\epsilon}^{(1)}\epsilon_n + \Xi_{C\epsilon}^{(2)}\dot{\epsilon}_n + \Xi_{C\epsilon}^{(3)}\ddot{\epsilon}_n + \Xi_{C\epsilon}^{(4)}\Delta\epsilon \tag{14.329}$$

until the solution converges

$$||\Delta\boldsymbol{\delta}_{n+1}^{k+1}|| = ||\boldsymbol{\delta}_{n+1}^{k+1} - \boldsymbol{\delta}_{n+1}^{k}|| < tolerance \tag{14.330}$$

Once the solution converges, update the unknowns at the end of the time step as follows.

$$\ddot{\mathbf{u}}_{n+1} = \ddot{\mathbf{u}}_n + (\widetilde{\mathbf{a}}_{n+1}^{k+1} - \ddot{\mathbf{u}}_n)/(\Lambda_6 W_1) \tag{14.331}$$

$$\dot{\mathbf{u}}_{n+1} = \dot{\mathbf{u}}_n + \lambda_4\ddot{\mathbf{u}}_n\Delta t + \lambda_5(\ddot{\mathbf{u}}_{n+1} - \ddot{\mathbf{u}}_n)\Delta t \tag{14.332}$$

$$\mathbf{u}_{n+1} = \mathbf{u}_n + \lambda_1\dot{\mathbf{u}}_n\Delta t + \lambda_2\ddot{\mathbf{u}}_n\Delta t^2 + \lambda_3(\ddot{\mathbf{u}}_{n+1} - \ddot{\mathbf{u}}_n)\Delta t^2 \tag{14.333}$$

$$\epsilon_{n+1} = \epsilon(\mathbf{u}_{n+1}) \tag{14.334}$$

The predictor-corrector coefficients Ξ above in the corresponding a, v, and d-form were previously listed in Table 14.3.

Remark 14.3.14 It is to be noted that with the hybrid displacement-strain based normalized time weighted residual approach within the GSSSS framework, U0/V0 ($\{\rho_{1\infty}, \rho_{2\infty}, \rho_{3\infty}\} = \{1, 1, 1\}$) and V0 ($\{\rho_{1\infty}, \rho_{2\infty}, \rho_{3\infty}\} = \{1, 1, \rho_{3\infty}\}$) all yield energy-momentum conservation.

Illustrative Example

Example 14.4 Four Mass-Spring System with Numerically Non-Dissipative Schemes

Numerically Non-Dissipative Hybrid Displacement-Strain Based Normalized Time Weighted Residual Approach

To demonstrate the effectiveness of the numerically non-dissipative hybrid displacement-strain based normalized time weighted residual approach we have just discussed, the previous four mass (depicted earlier in Figure 14.3) example with given initial displacement is simulated using the three basic building blocks within LMS methods, namely, the numerically non-dissipative time integrators including U0/V0 ($\{\rho_{1\infty}, \rho_{2\infty}, \rho_{3\infty}\}$ $= \{1.00, 1.00, 1.00\}$) (Classical Version Midpoint Rule-[MPR-EPA]), V0 ($\{\rho_{1\infty}, \rho_{2\infty}, \rho_{3\infty}\} =$ $\{1.00, 1.00, 0.00\}$) (New Version Midpoint Rule-[MPR-MPA]), and U0 ($\{\rho_{1\infty}, \rho_{2\infty}, \rho_{3\infty}\} =$

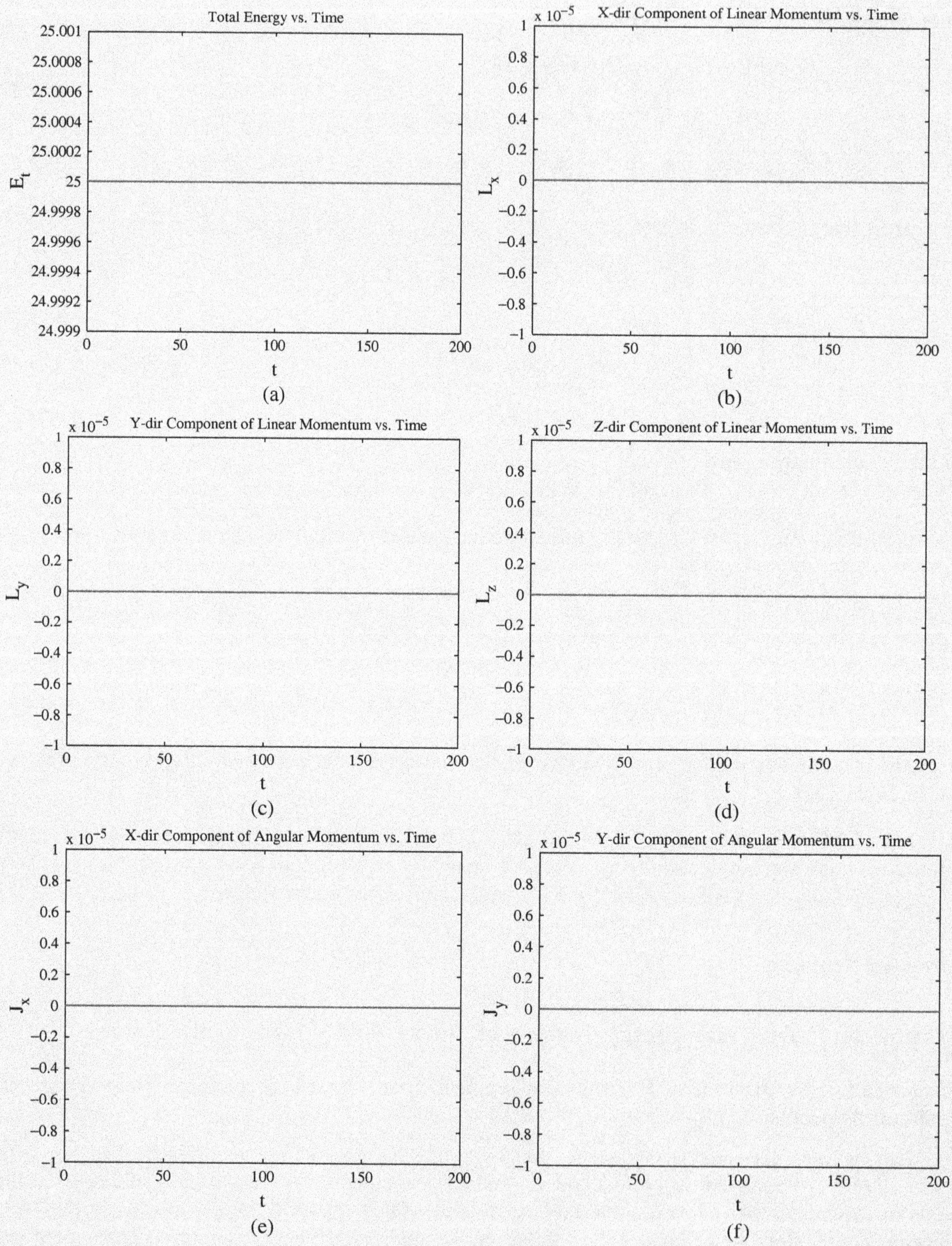

Figure 14.12. Plots of energy, components of linear momentum, and components of angular momentum for the four spring-mass problem with given initial displacement simulated using U0/V0 ($\{\rho_{1\infty}, \rho_{2\infty}, \rho_{3\infty}\} = \{1, 1, 1\}$) which is the Classical Version Midpoint Rule-[MPR-EPA] employing the hybrid displacement-strain based normalized time weighted residual approach

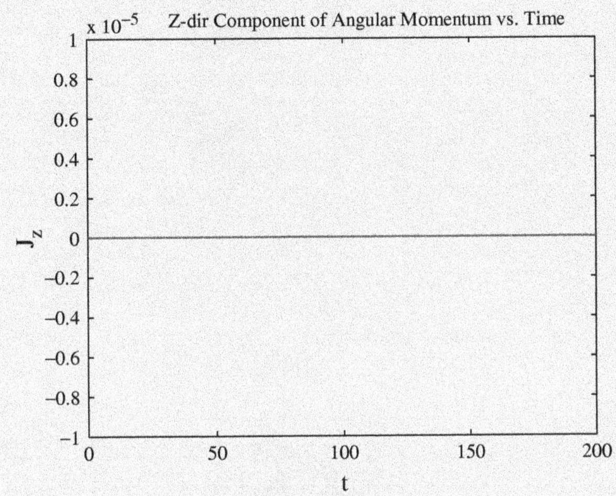

(g)

Figure 14.12. *(continued)*

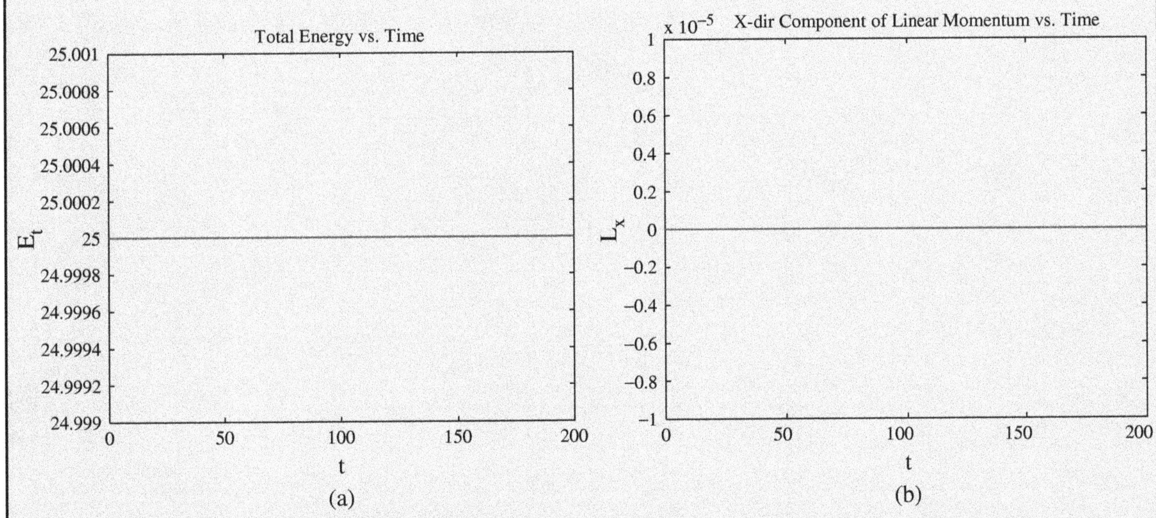

Figure 14.13. Plots of energy, components of linear momentum, and components of angular momentum for the four spring-mass problem with given initial displacement simulated using V0 ($\{\rho_{1\infty}, \rho_{2\infty}, \rho_{3\infty}\} = \{1, 1, 0\}$) which is the New Version Midpoint Rule-[MPR-MPA] employing the hybrid displacement-strain based normalized time weighted residual approach

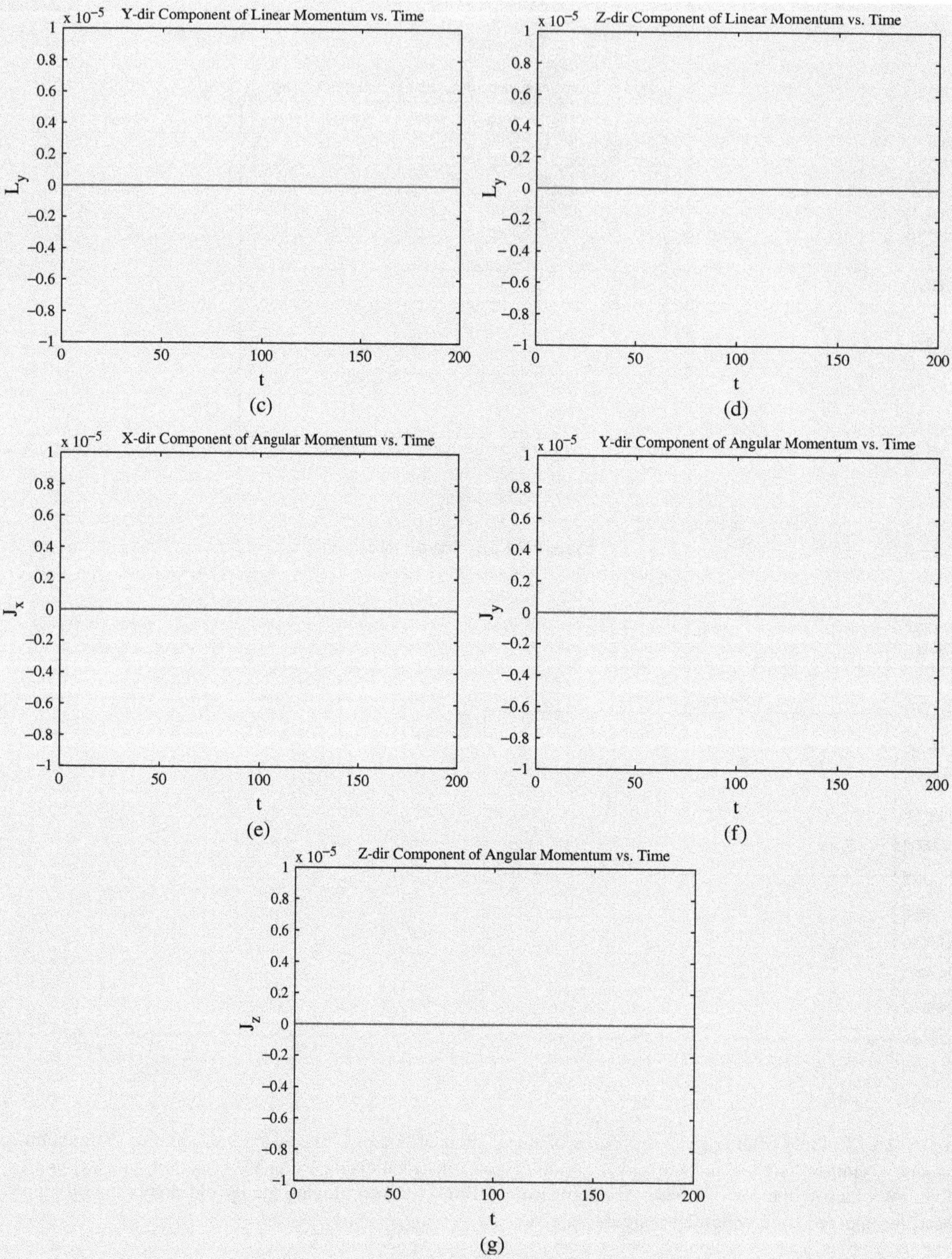

Figure 14.13. *(continued)*

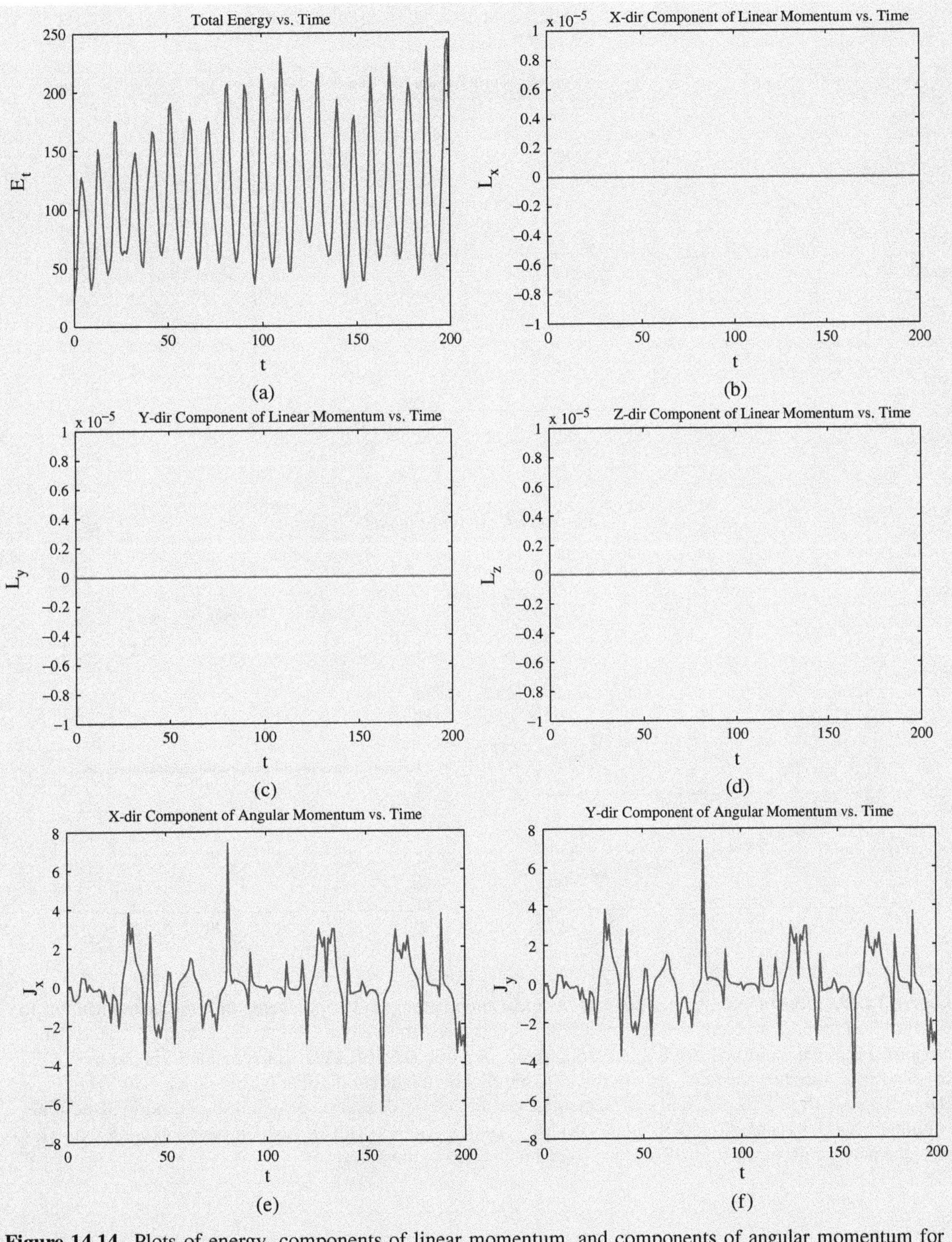

Figure 14.14. Plots of energy, components of linear momentum, and components of angular momentum for the four spring-mass problem with given initial displacement simulated using U0 ($\{\rho_{1\infty}, \rho_{2\infty}, \rho_{3\infty}\} = \{1, 1, 0\}$) which is the Newmark method employing the hybrid displacement-strain based normalized time weighted residual approach

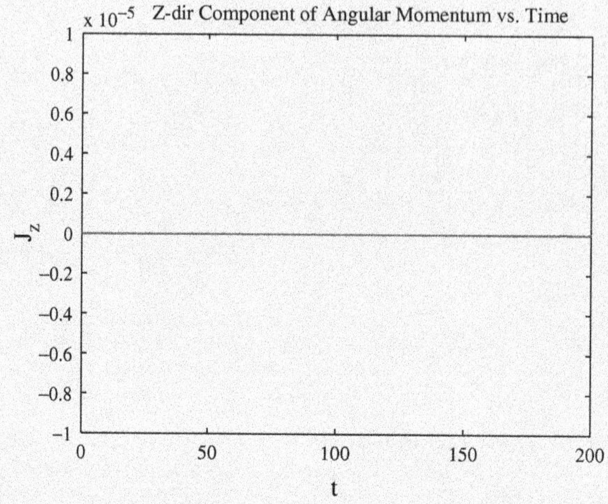

(g)

Figure 14.14. *(continued)*

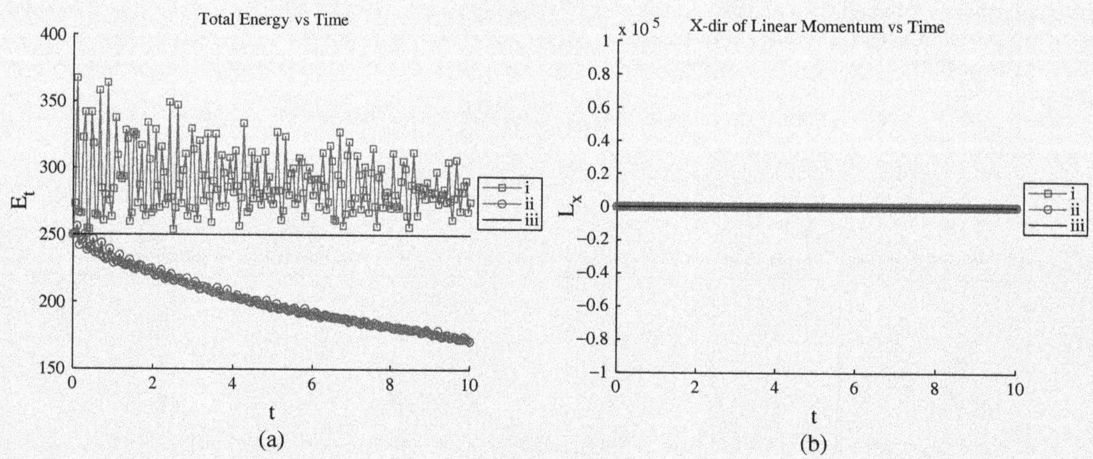

Figure 14.15. Plots of energy, components of linear momentum, and components of angular momentum for the four spring-mass problem with given initial displacement and velocity simulated using: (i) Controllable numerical dissipation turned on−U0/V0 ($\{\rho_{1\infty}, \rho_{2\infty}, \rho_{3\infty}\} = \{0.8, 1.0, 0.8\}$) implemented via the the classical time weighted residual approach, (ii) Controllable numerical dissipation turned on−U0/V0 ($\{\rho_{1\infty}, \rho_{2\infty}, \rho_{3\infty}\} = \{0.8, 1.0, 0.8\}$) implemented via the hybrid displacement-strain based normalized time weighted residual approach, and (iii) U0/V0 ($\{\rho_{1\infty}, \rho_{2\infty}, \rho_{3\infty}\} = \{1.0, 1.0, 1.0\}$) implemented via the hybrid displacement-strain based normalized time weighted residual approach

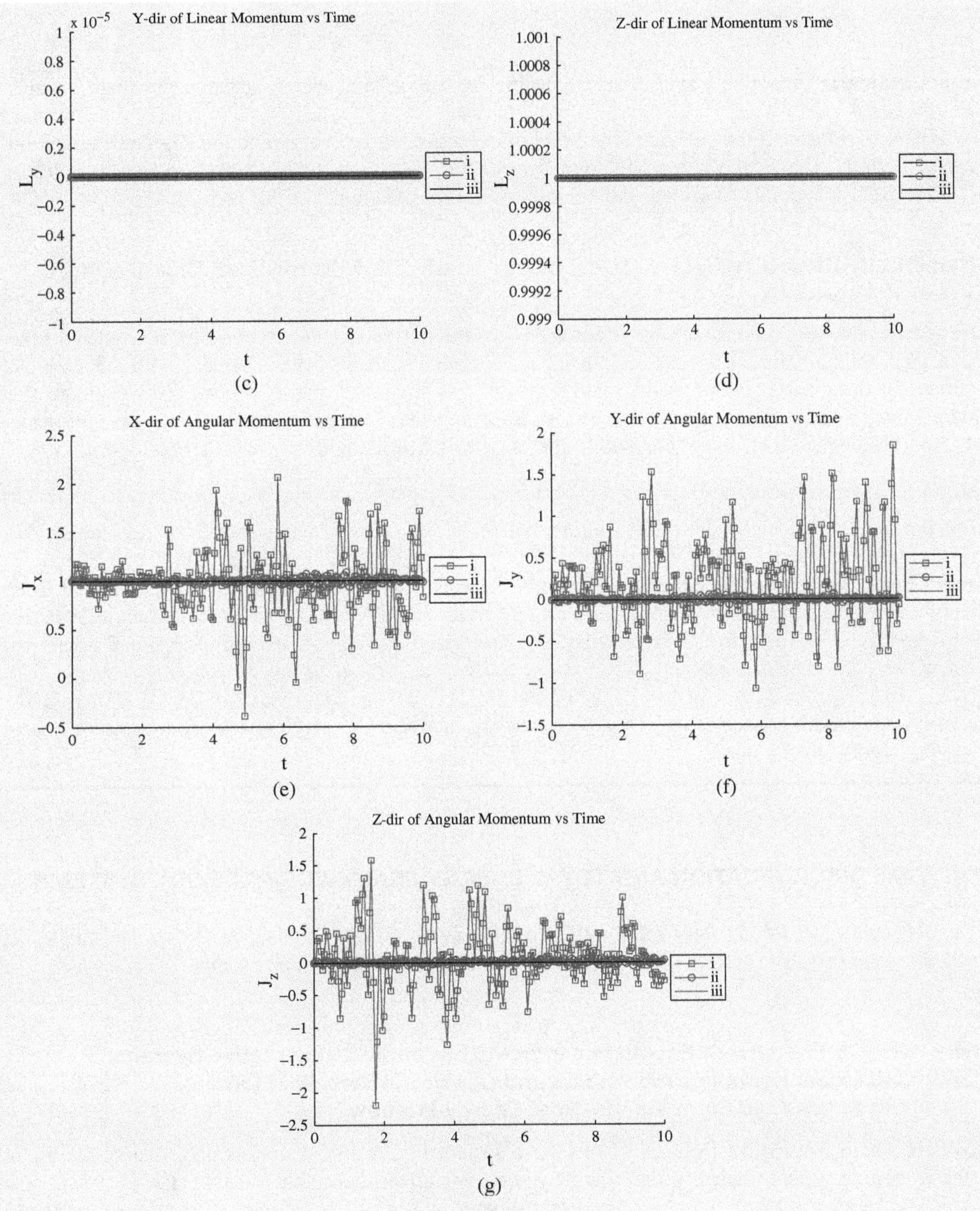

Figure 14.15. *(continued)*

{1.00, 1.00, 0.00}) (Newmark Method) as the basic parent algorithms (see Figures 14.12–14.14). For the problem under investigation, the motion is simulated using a time step of $\Delta t = 1$ s with the aforementioned numerically non-dissipative time integrators and the results are shown in Figures 14.12–14.14.

It is to be noted that the Newmark method fails to conserve the energy and the angular momentum unlike both the Classical Version Midpoint Rule-[MPR-EPA] and the New Version Midpoint Rule-[MPR-MPA] which inherit energy-momentum conserving features via the present developments.

Numerically Dissipative Hybrid Displacement - Strain Based Normalized Time Weighted Residual Approach

In this section, we demonstrate the effectiveness of the hybrid displacement-strain normalized time weighted residual approach when employed to numerically dissipative schemes (with controllable numerical dissipation) which is the main focus of this section. For this purpose, we simulate the same four mass-spring problem with given initial displacement and velocity using the following optimal dissipative time integrator within the class of LMS methods as the selected algorithm.

i) U0/V0 ($\{\rho_{1\infty}, \rho_{2\infty}, \rho_{3\infty}\} = \{0.8, 1.0, 0.8\}$)

For the simulation, with controllable numerical dissipation turned on, we study and compare the resulting energy and momentum resulting from the implementation via the hybrid displacement-strain based normalized time weighted residual approach to those generated by the classical time weighted residual counterparts (see Figure 14.15). In the same Figure 14.15, we also demonstrate the performance of the hybrid displacement-strain based normalized time weighted residual approach when the controllable numerical dissipative features are turned off by setting, for example, U0 ($\{\rho_{1\infty}, \rho_{2\infty}, \rho_{3\infty}\} = \{1.0, 1.0, 1.0\}$) or V0 ($\{\rho_{1\infty} = \rho_{2\infty} = 1$, and any value of $\rho_{3\infty}\}$) which yields energy-momentum conservation. That is, it readily recovers the original energy-momentum conserving algorithmic features.

14.4 TIME DISCRETIZATION AND TOTAL ENERGY FRAMEWORK: N-BODY SYSTEMS

In this subsection, we briefly show the single-step implicit algorithm designs in both two- and single-field forms for a conservative system of N particles in holonomic-scleronomic systems.

14.4.1 GSSSS Framework Encompassing LMS Methods: Conservative Systems and Linear Dynamics Algorithms and Designs in Two-field form and Single-field Form Via the Total Energy Framework

Two-field Form Algorithm Designs Consider a system free from constraints and assume the motion in the system can be completely described by the generalized coordinates $\mathbf{q}(t) : \mathcal{T} = [t_0, t_f] \subset \mathbb{R} \rightarrow Q \equiv \mathbb{R}^{n_{\mathrm{dof}}}$. Here, we define the generalized coordinates as a set of position vectors of each particle in $\mathbb{R}^3$ and assume the number of degrees of freedom is given by $n_{\mathrm{dof}} = 3N$. The principle of conservation of Total Energy together with the kinematic constraint $\mathbf{v} = \dot{\mathbf{q}} \in T_q Q$ yields

$$\frac{d\mathcal{E}(\mathbf{q}, \mathbf{v})}{dt} = 0 \ \ \forall t \in \mathcal{T} \qquad (14.335)$$

where the autonomous total energy of the system is assumed to consist of the kinetic energy $\mathcal{K}(\mathbf{v}) : T_q Q \equiv \mathbb{R}^{n_{\text{dof}}} \to \mathbb{R}$ and the total potential energy $\mathcal{U}(\mathbf{q}) : Q \to \mathbb{R}$ given by

$$\mathcal{K}(\mathbf{v}) = \frac{1}{2}\sum_{A=1}^{N} m_A \mathbf{v}^A \cdot \mathbf{v}^A = \frac{1}{2}\mathbf{v} \cdot \mathbf{M}\mathbf{v} > 0 \tag{14.336}$$

$$\mathcal{U}(\mathbf{q}) = \mathcal{U}^{\text{int}}(\mathbf{q}) + \mathcal{U}^{\text{ext}}(\mathbf{q}) \tag{14.337}$$

respectively. We assume the mass matrix, $\mathbf{M} = \text{diag}(\mathbf{m}_A)$ with $\mathbf{m}_A = m_A \mathbf{I}_3$ for $A = 1, 2, \ldots, N$, is **constant** and has the properties of symmetry ($\mathbf{M} = \mathbf{M}^T$) and positive definiteness ($\boldsymbol{\xi} \cdot \mathbf{M}\boldsymbol{\xi} > 0$ for $\boldsymbol{\xi} \in \mathbb{R}^{n_{\text{dof}}}$ and $\boldsymbol{\xi} \neq \mathbf{0}$). The total potential energy defined in Equation (14.334) is given as the summation of internal potential energy $\mathcal{U}^{\text{int}} : Q \to \mathbb{R}$ and external potential energy $\mathcal{U}^{\text{ext}} : Q \to \mathbb{R}$, and they are given by

$$\mathcal{U}^{\text{int}}(\mathbf{q}) = \sum_{A=1}^{N}\sum_{B>A}^{N} \mathcal{U}_{AB}^{\text{int}}(\mathbf{q}^A, \mathbf{q}^B) \text{ and } \mathcal{U}^{\text{ext}}(\mathbf{q}) = \sum_{A=1}^{N} \mathcal{U}_A^{\text{ext}}(\mathbf{q}^A) \tag{14.338}$$

respectively. As shown before, Equation (14.335) leads to the so-called Total Energy representation of the equation of motion:

$$\frac{d}{dt}\left(\frac{\partial \mathcal{E}(\mathbf{q}, \mathbf{v})}{\partial \mathbf{v}}\right) + \frac{\partial \mathcal{E}(\mathbf{q}, \mathbf{v})}{\partial \mathbf{q}} = \mathbf{0} \tag{14.339}$$

with $\mathbf{v} = \dot{\mathbf{q}}$, which yields

$$\begin{cases} \mathbf{M}\dot{\mathbf{v}} + \mathbf{f}^{\text{int}}(\mathbf{q}) = \mathbf{f}^{\text{c}}(\mathbf{q}) \\ \mathbf{v} = \dot{\mathbf{q}} \end{cases} \tag{14.340}$$

where $\mathbf{f}^{\text{int}}$ and $\mathbf{f}^{\text{c}}$ denote the internal force due to the interactions among particles and the conservative force defined by

$$\mathbf{f}_A^{\text{int}} := \sum_{B \neq A}^{N} \frac{\partial \mathcal{U}_{AB}^{\text{int}}(\mathbf{q}^A, \mathbf{q}^B)}{\partial \mathbf{q}^A} \text{ and } \mathbf{f}_A^{\text{c}} := -\frac{\partial \mathcal{U}_A^{\text{ext}}(\mathbf{q}^A)}{\partial \mathbf{q}^A} \tag{14.341}$$

for $A = 1, 2, \ldots, N$, respectively. Here, we consider a small motion about equilibrium points. To study this special situation of the motion, we linearize the equation of motion about equilibrium points $\bar{\mathbf{q}}$. Since we assume the mass matrix is constant, we only consider the potential term here. The Taylor series expansion of the total potential energy about $\bar{\mathbf{q}}$ yields

$$\mathcal{U}(\mathbf{q}) = \mathcal{U}(\bar{\mathbf{q}}) + [\mathbf{q} - \bar{\mathbf{q}}] \cdot \left.\frac{\partial \mathcal{U}}{\partial \mathbf{q}}\right|_{\bar{\mathbf{q}}} + \frac{1}{2}[\mathbf{q} - \bar{\mathbf{q}}] \cdot \left.\frac{\partial^2 \mathcal{U}}{\partial \mathbf{q}\partial \mathbf{q}}\right|_{\bar{\mathbf{q}}} [\mathbf{q} - \bar{\mathbf{q}}] + \cdots \tag{14.342}$$

Without loss of generality, set $\mathcal{U}(\bar{\mathbf{q}}) = 0$. Also, we assume

$$\left.\frac{\partial \mathcal{U}}{\partial \mathbf{q}}\right|_{\bar{\mathbf{q}}} = \mathbf{0} \tag{14.343}$$

Ignoring the higher-order terms in Equation (14.342), the following quadratic total potential energy is obtained:

$$\mathcal{U}(\mathbf{q}) \cong \frac{1}{2}[\mathbf{q} - \bar{\mathbf{q}}] \cdot \left.\frac{\partial^2 \mathcal{U}}{\partial \mathbf{q}\partial \mathbf{q}}\right|_{\bar{\mathbf{q}}} [\mathbf{q} - \bar{\mathbf{q}}] \geq 0 \tag{14.344}$$

Defining $\mathbf{d} := \mathbf{q} - \bar{\mathbf{q}}$ the linearized equation of motion yields

$$\begin{cases} \mathbf{M}\dot{\mathbf{v}} + \mathbf{K}\mathbf{d} = \mathbf{0} \\ \mathbf{v} = \dot{\mathbf{d}} \end{cases} \tag{14.345}$$

where $\mathbf{K}$ denotes the constant symmetric stiffness matrix, defined by

$$\mathbf{K} := \left. \frac{\partial^2 \mathcal{U}}{\partial \mathbf{d} \partial \mathbf{d}} \right|_{\mathbf{d}=\mathbf{0}} \tag{14.346}$$

If $\bar{\mathbf{q}} = \mathbf{q}(t_0)$, $\mathbf{d} = \mathbf{u}$ (the displacement vector). It is important to notice that the system of equations of motion, Equation (14.345), is equivalent to the linearized system of equations of motion in two-field form as described in subsection 14.1.2. Therefore, we can exactly follow the same procedures of the GSSSS framework of algorithm designs to obtain a family of single-step algorithms similar to Algorithm 11 as follows.

Algorithm 25

Given $\mathbf{d}_n$ and $\mathbf{v}_n$, find $\mathbf{d}_{n+1}$ and $\mathbf{v}_{n+1}$ from

$$\left[\frac{\mathbf{M}}{\Delta t} + \Delta t \mathbf{K} W_2 \Lambda_2 \right] \Delta \mathbf{v} = -\mathbf{K} \left[\mathbf{d}_n + \Delta t W_1 \Lambda_1 \mathbf{v}_n \right] \tag{14.347}$$

with the associated updates

$$\mathbf{d}_{n+1} = \mathbf{d}_n + \Delta t \lambda_1 \mathbf{v}_n + \Delta t \lambda_2 \Delta \mathbf{v} \tag{14.348}$$

$$\mathbf{v}_{n+1} = \mathbf{v}_n + \Delta \mathbf{v} \tag{14.349}$$

Remark 14.4.1

1. By enforcing that no bifurcation occurs for the principal roots at infinity, i.e., $\rho_{1\infty} = \rho_{2\infty}$,

$$W_1 \Lambda_1 = \frac{3 + \rho_{1\infty} + \rho_{2\infty} - \rho_{1\infty}\rho_{2\infty}}{2(1 + \rho_{1\infty})(1 + \rho_{2\infty})} \quad \text{and} \quad W_2 \Lambda_2 = \frac{1}{(1 + \rho_{1\infty})(1 + \rho_{2\infty})} \tag{14.350}$$

   with $\lambda_1 = 1$ and $\lambda_2 = 1/2$.
2. The total angular momentum and total energy within a time step are exactly conserved only if $\rho_{1\infty} = \rho_{2\infty} = 1$. The discrete total linear momentum within a time step is exactly conserved for any conditions of the parameters.

Single-field Form Algorithm Designs In the single-field-form representation, the total energy may be written as

$$\mathcal{E}(\mathbf{q}, \dot{\mathbf{q}}) = \mathcal{K}(\dot{\mathbf{q}}) + \mathcal{U}(\mathbf{q}) = \frac{1}{2} \dot{\mathbf{q}} \cdot \mathbf{M} \dot{\mathbf{q}} + \mathcal{U}(\mathbf{q}) \tag{14.351}$$

where the total potential energy is given by Equation (14.337) and Equation (14.338). Linearization about the equilibrium points $\bar{\mathbf{q}}$ shows that the total energy is given as

$$\mathcal{E}(\mathbf{d}, \dot{\mathbf{d}}) = \mathcal{K}(\dot{\mathbf{d}}) + \mathcal{U}(\mathbf{d}) \cong \frac{1}{2} \dot{\mathbf{d}} \cdot \mathbf{M} \dot{\mathbf{d}} + \frac{1}{2} \mathbf{d} \cdot \mathbf{K} \mathbf{d} \tag{14.352}$$

Therefore, the principle of balance of Total Energy in the linear dynamic system, that is,

$$\frac{d\mathcal{E}(\mathbf{d}, \dot{\mathbf{d}})}{dt} = 0 \ \ \forall t \in \mathcal{T} \tag{14.353a}$$

leads to

$$\frac{d}{dt}\left(\frac{\partial\mathcal{E}(\mathbf{d}, \dot{\mathbf{d}})}{\partial\dot{\mathbf{d}}}\right) + \frac{\partial\mathcal{E}(\mathbf{d}, \dot{\mathbf{d}})}{\partial\mathbf{d}} = \mathbf{0} \tag{14.353b}$$

and it yields

$$\mathbf{M}\ddot{\mathbf{d}} + \mathbf{K}\mathbf{d} = \mathbf{0} \tag{14.353c}$$

Again, note that Equation (14.355) is equivalent to the equations of motion in single-field form in subsection 14.1.2. Therefore, following the similar procedures, the U0 family and the V0 family of algorithms for the conservative, linear dynamical system of N particles can be readily obtained as ($\Delta\ddot{\mathbf{d}} := \ddot{\mathbf{d}}_{n+1} - \ddot{\mathbf{d}}_n$):

Algorithm 26: ***U0 Family of Algorithms for Conservative Linear Dynamical Systems***

Given $\mathbf{d}_n$, $\dot{\mathbf{d}}_n$, and $\ddot{\mathbf{d}}_n$, find $\mathbf{d}_{n+1}$, $\dot{\mathbf{d}}_{n+1}$, and $\ddot{\mathbf{d}}_{n+1}$ from

$$\left(W_1\Lambda_6\mathbf{M} + W_3\Lambda_3\mathbf{K}\Delta t^2\right)\Delta\ddot{\mathbf{d}} = -\mathbf{M}\ddot{\mathbf{d}}_n - \mathbf{K}\left(\mathbf{d}_n + W_1\Lambda_1\dot{\mathbf{d}}_n\Delta t + W_2\Lambda_2\ddot{\mathbf{d}}_n\Delta t^2\right) \tag{14.354a}$$

$$\mathbf{d}_{n+1} = \mathbf{d}_n + \lambda_1\dot{\mathbf{d}}_n\Delta t + \lambda_2\ddot{\mathbf{d}}_n\Delta t^2 + \lambda_3\Delta\ddot{\mathbf{d}}\Delta t^2 \tag{14.354b}$$

$$\dot{\mathbf{d}}_{n+1} = \dot{\mathbf{d}}_n + \lambda_4\ddot{\mathbf{d}}_n\Delta t + \lambda_5\Delta\ddot{\mathbf{d}}\Delta t \tag{14.354c}$$

$$\ddot{\mathbf{d}}_{n+1} = \ddot{\mathbf{d}}_n + \Delta\ddot{\mathbf{d}} \tag{14.354d}$$

where the respective relations for W, Λ and λ were described previously.

Algorithm 27: ***V0 Family of Algorithms for Conservative Linear Dynamical Systems***

Given $\mathbf{d}_n$, $\dot{\mathbf{d}}_n$, and $\ddot{\mathbf{d}}_n$, find $\mathbf{d}_{n+1}$, $\dot{\mathbf{d}}_{n+1}$, and $\ddot{\mathbf{d}}_{n+1}$ from

$$\left(W_1\Lambda_6\mathbf{M} + W_3\Lambda_3\mathbf{K}\Delta t^2\right)\Delta\ddot{\mathbf{d}} = -\mathbf{M}\ddot{\mathbf{d}}_n - \mathbf{K}\left(\mathbf{d}_n + W_1\Lambda_1\dot{\mathbf{d}}_n\Delta t + W_2\Lambda_2\ddot{\mathbf{d}}_n\Delta t^2\right) \tag{14.355a}$$

$$\mathbf{d}_{n+1} = \mathbf{d}_n + \lambda_1\dot{\mathbf{d}}_n\Delta t + \lambda_2\ddot{\mathbf{d}}_n\Delta t^2 + \lambda_3\Delta\ddot{\mathbf{d}}\Delta t^2 \tag{14.355b}$$

$$\dot{\mathbf{d}}_{n+1} = \dot{\mathbf{d}}_n + \lambda_4\ddot{\mathbf{d}}_n\Delta t + \lambda_5\Delta\ddot{\mathbf{d}}\Delta t \tag{14.355c}$$

$$\ddot{\mathbf{d}}_{n+1} = \ddot{\mathbf{d}}_n + \Delta\ddot{\mathbf{d}} \tag{14.355d}$$

where the respective relations for W, Λ and λ were described previously.

14.4.2 GSSSS Framework Encompassing LMS Methods: Conservative Systems and Nonlinear Dynamics Algorithms and Designs in Two-field Form and Single-field Form Via the Total Energy Framework

Here, we briefly show the algorithm designs via the normalized time weighted residual methodology for a nonlinear dynamic system when the internal potential energy is given as a function of the inter-particle distances among particles in a conservative system of N particles. With the inter-particle distance defined by $r^{AB} = ||\mathbf{q}^{AB}|| = ||\mathbf{q}^B - \mathbf{q}^A|| = \sqrt{(\mathbf{q}^B - \mathbf{q}^A) \cdot (\mathbf{q}^B - \mathbf{q}^A)}$, the internal potential energy for A^{th} and B^{th} particles may be written as $\mathcal{U}^{\text{int}}_{AB}(\mathbf{q}^A, \mathbf{q}^B) = \mathcal{V}_{AB}(r^{AB})$. For simplicity, we confine our attention to the case with no external potential; therefore, the total potential energy may be written as

$$\mathcal{U} = \mathcal{V} = \sum_{A=1}^{N} \sum_{B>A}^{N} \mathcal{V}_{AB}(r^{AB}) \tag{14.356}$$

In this particular nonlinear dynamic system, the internal force due to the interactions on the Ath particle is given by

$$\mathbf{f}_A^{\text{int}} = -\sum_{B \neq A}^{N} \frac{d\mathcal{V}_{AB}(r^{AB})}{dr^{AB}} \frac{\mathbf{q}^{AB}}{r^{AB}} = -\sum_{B \neq A}^{N} \mathcal{V}'_{AB}(r^{AB}) \frac{\mathbf{q}^{AB}}{r^{AB}} \tag{14.357}$$

Note that the summation of the internal force vanishes due to Newton's third law; that is, $\sum_{A=1}^{N} \mathbf{f}_A^{\text{int}} = \mathbf{0}$. Following the three possible approaches towards discretizing the internal and conservative forces in time via the normalized time weighted residual methodology, we have the resulting options as shown. Option 1: Discretization based on $\mathbf{f}_A^{\text{int}}(t) : \mathcal{T} \rightarrow \mathbb{R}^3$ and $\mathbf{f}_A^{\text{c}}(t) : \mathcal{T} \rightarrow \mathbb{R}^3$, Option 2: Discretization based on $\mathbf{q}^A(t) : \mathcal{T} \rightarrow \mathbb{R}^3$ only, and Option 3: Discretization based on $\mathbf{q}^A(t) : \mathcal{T} \rightarrow \mathbb{R}^3$ and $r^{AB}(t) : \mathcal{T} \rightarrow \mathbb{R}$. In the following, we focus upon the third approach and show the procedures for the time discretization of the equation of motion in the two- and single-field forms, respectively.

Two-field Form Algorithm Designs With the autonomous total energy defined by

$$\mathcal{E} = \mathcal{K} + \mathcal{V} = \frac{1}{2} \sum_{A=1}^{N} m_A \mathbf{v}^A \cdot \mathbf{v}^A + \sum_{A=1}^{N} \sum_{B>A}^{N} \mathcal{V}_{AB}(r^{AB}) \tag{14.358}$$

the principle of conservation of Total Energy leads to the following system of equations of motion:

$$\frac{d\mathcal{E}}{dt} = 0 \Longrightarrow \begin{cases} m_A \dot{\mathbf{v}}^A + \mathbf{f}_A^{\text{int}} = \mathbf{0} \\ \mathbf{v}^A = \dot{\mathbf{q}}^A \end{cases} \tag{14.359}$$

for $A = 1, 2, \cdots, N$. The internal force on the Ath particle is given by Equation (14.357). With the scalar time weighting function given by

$$W = w_0 + w_1 \Gamma + w_2 \Gamma^2 \quad (w_0 = 1) \tag{14.360}$$

where $\Gamma := \tau / \Delta t$ and $\tau := t - t_n$, and with the following approximations for $\mathbf{q}$, $\dot{\mathbf{v}}$ and $r^{AB} \in \mathbb{R}$ by the asymptotic series expansions in time with the algorithmic parameters $\Lambda_i \in \mathbb{R}$ and $\Theta_i \in \mathbb{R} (i = 1, 2)$ as

$$\mathbf{q} = \mathbf{q}_n + \Lambda_1 \dot{\mathbf{q}}_n \tau + \Lambda_2 \frac{\dot{\mathbf{q}}_{n+1} - \dot{\mathbf{q}}_n}{\Delta t} \tau^2 \tag{14.361}$$

$$= \mathbf{q}_n + \Lambda_1 \mathbf{v}_n \tau + \Lambda_2 \frac{\mathbf{v}_{n+1} - \mathbf{v}_n}{\Delta t} \tau^2 \tag{14.362}$$

$$\dot{\mathbf{v}} = \frac{\mathbf{v}_{n+1} - \mathbf{v}_n}{\Delta t} \tag{14.363}$$

and

$$r^{AB} = r_n^{AB} + \Theta_1 \dot{r}_n^{AB}\tau + \Theta_2 \frac{\dot{r}_{n+1}^{AB} - \dot{r}_n^{AB}}{\Delta t}\tau^2 \tag{14.364}$$

respectively, the normalized time weighted integral of the residual from time t_n to t_{n+1} may be formulated as

$$m_A \tilde{\dot{\mathbf{v}}}^A - \sum_{B\neq A}^{N} \mathcal{V}_{AB}(\tilde{r}^{AB}) \frac{\tilde{\mathbf{q}}^{AB}}{\tilde{r}^{AB}} = \mathbf{0} \tag{14.365}$$

where the algorithmic variables are

$$\begin{aligned}
\tilde{\mathbf{q}}^{AB} &:= \frac{\int_{t_n}^{t_{n+1}} W\mathbf{q}^{AB}dt}{\int_{t_n}^{t_{n+1}} Wdt} = \frac{\int_{t_n}^{t_{n+1}} W\mathbf{q}^{B}dt}{\int_{t_n}^{t_{n+1}} Wdt} - \frac{\int_{t_n}^{t_{n+1}} W\mathbf{q}^{A}dt}{\int_{t_n}^{t_{n+1}} Wdt} \\
&= [\mathbf{q}_n^B + \Delta t W_1\Lambda_1\mathbf{v}_n^B + \Delta t W_2\Lambda_2(\mathbf{v}_{n+1}^B - \mathbf{v}_n^B)] \\
&\quad -[\mathbf{q}_n^A + \Delta t W_1\Lambda_1\mathbf{v}_n^A + \Delta t W_2\Lambda_2(\mathbf{v}_{n+1}^A - \mathbf{v}_n^A)] \\
&= \mathbf{q}_n^{AB} + \Delta t W_1\Lambda_1\mathbf{v}_n^{AB} + \Delta t W_2\Lambda_2(\mathbf{v}_{n+1}^{AB} - \mathbf{v}_n^{AB})
\end{aligned} \tag{14.366}$$

$$\tilde{\dot{\mathbf{v}}}^A := \frac{\int_{t_n}^{t_{n+1}} W\dot{\mathbf{v}}^A dt}{\int_{t_n}^{t_{n+1}} Wdt} = \frac{\mathbf{v}_{n+1}^A - \mathbf{v}_n^A}{\Delta t} \tag{14.367}$$

where $\mathbf{v}_{n+1}^{AB} := \mathbf{v}_{n+1}^B - \mathbf{v}_{n+1}^A$ and $\mathbf{v}_n^{AB} := \mathbf{v}_n^B - \mathbf{v}_n^A$, and the algorithmic inter-particle distance between A^{th} and B^{th} particles is

$$\tilde{r}^{AB} := \frac{\int_{t_n}^{t_{n+1}} Wr^{AB}dt}{\int_{t_n}^{t_{n+1}} Wdt} = r_n^{AB} + \Delta t W_1\Theta_1\dot{r}_n^{AB} + \Delta t W_2\Theta_2(\dot{r}_{n+1}^{AB} - \dot{r}_n^{AB}) \tag{14.368}$$

Notice we assume $\mathbf{v}_n = \dot{\mathbf{q}}_n$ and $\mathbf{v}_{n+1} = \dot{\mathbf{q}}_{n+1}$ at time t_n and t_{n+1}, respectively. And we define

$$W_i := \frac{\int_{t_n}^{t_{n+1}} \left(\frac{\tau}{\Delta t}\right)^i Wdt}{\int_{t_n}^{t_{n+1}} Wdt} \quad \text{for } i = 1, 2 \tag{14.369}$$

To design the updates, approximate $\mathbf{q}(t) : \mathcal{T} \to \mathcal{Q} \equiv \mathbb{R}^{n_{\text{dof}}}$ by the asymptotic series expansions in time with new scalar algorithmic parameters λ_1 and λ_2, and set $\tau = \Delta t$. Then, we obtain

$$\mathbf{q}_{n+1} = \mathbf{q}_n + \lambda_1\mathbf{v}_n\Delta t + \lambda_2(\mathbf{v}_{n+1} - \mathbf{v}_n)\Delta t \tag{14.370}$$

Also, approximate r^{AB} by the asymptotic series expansions in time with new scalar algorithmic parameters θ_1 and θ_2, and set $\tau = \Delta t$. With Equation (14.364), we obtain

$$\tilde{r}^{AB} = r_n^{AB} + \frac{W_2\Theta_2}{\theta_2}(r_{n+1}^{AB} - r_n^{AB}) + \Delta t\left(W_1\Theta_1 - \frac{\theta_1}{\theta_2}W_2\Theta_2\right)\dot{r}_n^{AB} \tag{14.371}$$

Summarizing, the following single-step family of time integration schemes are readily obtained:

Algorithm 28

Given $\mathbf{q}_n$ and $\mathbf{v}_n$, find $\mathbf{q}_{n+1}$ and $\mathbf{v}_{n+1}$ from

$$m_A \frac{\mathbf{v}_{n+1}^A - \mathbf{v}_n^A}{\Delta t} - \sum_{B \neq A}^{N} \mathcal{V}_{AB}'(\tilde{r}^{AB}) \frac{\tilde{\mathbf{q}}^{AB}}{\tilde{r}^{AB}} = \mathbf{0} \tag{14.372a}$$

$$\tilde{\mathbf{q}}^{AB} = \mathbf{q}_n^{AB} + \Delta t W_1 \Lambda_1 \mathbf{v}_n^{AB} + \Delta t W_2 \Lambda_2 (\mathbf{v}_{n+1}^{AB} - \mathbf{v}_n^{AB}) \tag{14.372b}$$

$$\tilde{r}^{AB} = r_n^{AB} + \frac{W_2 \Theta_2}{\theta_2}(r_{n+1}^{AB} - r_n^{AB}) + \Delta t \left(W_1 \Theta_1 - \frac{\theta_1}{\theta_2} W_2 \Theta_2 \right) \dot{r}_n^{AB} \tag{14.372c}$$

with the following updates:

$$\mathbf{q}_{n+1}^A = \mathbf{q}_n^A + \Delta t \lambda_1 \mathbf{v}_n^A + \Delta t \lambda_2 \Delta \mathbf{v}^A \tag{14.372d}$$

$$\mathbf{v}_{n+1}^A = \mathbf{v}_n^A + \Delta \mathbf{v}^A \tag{14.372e}$$

where $\Delta \mathbf{v}^A := \mathbf{v}_{n+1}^A - \mathbf{v}_n^A$.

Remark 14.4.2

1. Second-order time accuracy and unconditional stability are obtained only if selecting $\rho_{1\infty} = \rho_{2\infty} = 1$ in Equation (14.350) and $\Lambda_1 = \lambda_1 = 1$, $\Lambda_2 = \lambda_1 = 2 = 1/2$, $\Theta_1 = \theta_1 = 1$, $\Theta_2 = \theta_2 = 1/2$. Under these conditions, Algorithm 28 is cast into

$$m_A \frac{\mathbf{v}_{n+1}^A - \mathbf{v}_n^A}{\Delta t} - \sum_{B \neq A}^{N} \mathcal{V}_{AB}'(r_{n+1/2}^{AB}) \frac{\mathbf{q}_n^{AB} + \mathbf{q}_{n+1}^{AB}}{r_n^{AB} + r_{n+1}^{AB}} = \mathbf{0}$$

$$\mathbf{q}_{n+1}^A = \mathbf{q}_n^A + \Delta t \mathbf{v}_n^A + \frac{\Delta t}{2} \Delta \mathbf{v}^A$$

$$\mathbf{v}_{n+1}^A = \mathbf{v}_n^A + \Delta \mathbf{v}^A$$

which is equivalent to the assumed distance method proposed by Betsch and Steinmann (Betsch and Steinmann 2000). It is the exact energy-momentum conserving algorithm when the internal potential energy is given as a polynomial function of degree two or less; see (Graham et al. 2002).
2. For any conditions, the total linear momentum is exactly conserved.
3. In general, no member in the framework of Algorithm 28 is symplectic. Symplectic midpoint rule is a member of the framework via option 2 when we discretize the internal force only by the position vector, i.e., the algorithmic inter-particle distance between the Ath and B^{th} particles is given by $\tilde{r}^{AB} = ||\tilde{\mathbf{q}}^{AB}||$.

No member in Algorithm 28 is the exact energy-momentum conserving algorithm for a general nonlinear dynamic system. By modifying Algorithm 28 it can recover the exact energy-momentum conserving algorithm by selecting the spectral conditions $\rho_{1\infty} = \rho_{2\infty} = 1$ as the basic time integration scheme with second-order time accuracy and unconditional stability within the family of algorithms; consequently the following algorithm may be proposed:

Algorithm 29

Given $\mathbf{q}_n$ and $\mathbf{v}_n$, find $\mathbf{q}_{n+1}$ and $\mathbf{v}_{n+1}$ from

$$m_A \frac{\mathbf{v}_{n+1}^A - \mathbf{v}_n^A}{\Delta t} - \sum_{B \neq A}^{N} \widetilde{\mathcal{V}'_{AB}} \frac{\tilde{\mathbf{q}}^{AB}}{\tilde{r}^{AB}} = \mathbf{0} \tag{14.373a}$$

$$\tilde{\mathbf{q}}^{AB} = \mathbf{q}_n^{AB} + \Delta t W_1 \Lambda_1 \mathbf{v}_n^{AB} + \Delta t W_2 \Lambda_2 (\mathbf{v}_{n+1}^{AB} - \mathbf{v}_n^{AB}) \tag{14.373b}$$

$$\tilde{r}^{AB} = r_n^{AB} + \frac{W_2 \Theta_2}{\theta_2}(r_{n+1}^{AB} - r_n^{AB}) + \Delta t \left(W_1 \Theta_1 - \frac{\theta_1}{\theta_2} W_2 \Theta_2 \right) \dot{r}_n^{AB} \tag{14.373c}$$

with the following updates:

$$\mathbf{q}_{n+1}^A = \mathbf{q}_n^A + \Delta t \lambda_1 \mathbf{v}_n^A + \Delta t \lambda_2 \Delta \mathbf{v}^A \tag{14.373d}$$

$$\mathbf{v}_{n+1}^A = \mathbf{v}_n^A + \Delta \mathbf{v}^A \tag{14.373e}$$

where $\Delta \mathbf{v}^A := \mathbf{v}_{n+1}^A - \mathbf{v}_n^A$.

Remark 14.4.3

1. The algorithmic quantity of the derivative of the internal potential energy between the Ath and B^{th} particles with respect to the inter-particle distance r_{AB} is denoted by $\widetilde{\mathcal{V}'_{AB}}$. It is defined to be

$$\widetilde{\mathcal{V}_{AB}}' = \frac{\mathcal{V}_{AB}(r_{n+1}^{AB}) - \mathcal{V}_{AB}(r_n^{AB})}{r_{n+1}^{AB} - r_n^{AB}} \tag{14.374}$$

when selecting $\rho_{1\infty} = \rho_{2\infty} = 1$ with $\Lambda_1 = \lambda_1 = 1$, $\Lambda_2 = \lambda_1 = 2 = 1/2$, $\Theta_1 = \theta_1 = 1$, $\Theta_2 = \theta_2 = 1/2$; otherwise,

$$\widetilde{\mathcal{V}'_{AB}} = \mathcal{V}'_{AB}(\tilde{r}^{AB}) \tag{14.375}$$

2. Selecting $\rho_{1\infty} = \rho_{2\infty} = 1$ with $\Lambda_1 = \lambda_1 = 1$, $\Lambda_2 = \lambda_1 = 2 = 1/2$, $\Theta_1 = \theta_1 = 1$, $\Theta_2 = \theta_2 = 1/2$, Algorithm 29 recovers the exact energy-momentum conserving algorithm proposed by Greenspan (Greenspan 1995) and Simo and Gonzalez (Simo and Gonzalez 1993, 1994).
3. Algorithm 29 recovers the exact energy-momentum conserving algorithm for a special case when the potential energy is given as a function of the inter-particle distance of the particles in the mechanical system free from constraints. Notice that the exact energy conserving algorithm proposed by Simo et al. (Simo et al. 1992), i.e., Algorithm 3, for a nonlinear conservative system with a completely separable total energy given by $\mathcal{E}(\mathbf{q}, \mathbf{v}) = \mathcal{T}(\mathbf{v}) + \mathcal{U}(\mathbf{q})$ where the system kinetic energy $\mathcal{T}(\mathbf{v}) = (1/2)\mathbf{v} \cdot \mathbf{M}\mathbf{v}$ with a constant $\mathbf{M}$ and the system potential energy $\mathcal{U}(\mathbf{q})$ which is not a function of the inter-particle distance is not discussed here.

Single-field Form Algorithm Designs With the autonomous total energy defined by

$$\mathcal{E} = \mathcal{K} + \mathcal{V} = \frac{1}{2} \sum_{A=1}^{N} m_A \dot{\mathbf{q}}^A \cdot \dot{\mathbf{q}}^A + \sum_{A=1}^{N} \sum_{B>A}^{N} \mathcal{V}_{AB}(r^{AB}) \tag{14.376}$$

the principle of conservation of Total Energy leads to the following system of equations of motion:

$$\frac{d\mathcal{E}}{dt} = 0 \Longrightarrow m_A \ddot{\mathbf{q}}^A + \mathbf{f}_A^{\text{int}} = \mathbf{0} \tag{14.377}$$

for $A = 1, 2, \cdots, N$. The internal force on the Ath particle is given by Equation (14.357). With the scalar time weighting function given by

$$W = w_0 + w_1\Gamma + w_2\Gamma^2 + w_3\Gamma^3 \quad (w_0 = 1) \tag{14.378}$$

where $\Gamma := \tau/\Delta t$ and $\tau := t - t_n$, and the following approximations for $\mathbf{q}$, $\ddot{\mathbf{q}}$, and $r^{AB} \in \mathbb{R}$ by the asymptotic series expansions in time with the algorithmic parameters $\Lambda_i \in \mathbb{R}$ and $\Theta_i \in \mathbb{R}$ $(i = 1, 2, 3)$ as

$$\mathbf{q} = \mathbf{q}_n + \Lambda_1 \dot{\mathbf{q}}_n \tau + \Lambda_2 \ddot{\mathbf{q}}_n \tau^2 + \Lambda_3 \frac{\ddot{\mathbf{q}}_{n+1} - \ddot{\mathbf{q}}_n}{\Delta t} \tau^3 \tag{14.379}$$

$$\ddot{\mathbf{q}} = \ddot{\mathbf{q}}_n + \Lambda_6 \frac{\ddot{\mathbf{q}}_{n+1} - \ddot{\mathbf{q}}_n}{\Delta t} \tau \tag{14.380}$$

and

$$r^{AB} = r_n^{AB} + \Theta_1 \dot{r}_n^{AB} \tau + \Theta_2 \ddot{r}_n^{AB} \tau^2 + \Theta_3 \frac{\ddot{r}_{n+1}^{AB} - \ddot{r}_n^{AB}}{\Delta t} \tau^3 \tag{14.381}$$

respectively, the normalized time weighted integral of the residual from time t_n to t_{n+1} may be formulated as

$$m_A \tilde{\mathbf{a}}^A + \sum_{B \neq A}^{N} -\mathcal{V}_{AB}(\tilde{r}^{AB}) \frac{\tilde{\mathbf{q}}^{AB}}{\tilde{r}^{AB}} = \mathbf{0} \tag{14.382}$$

where the algorithmic variables are defined as

$$\tilde{\mathbf{q}}^{AB} := \frac{\int_{t_n}^{t_{n+1}} W \mathbf{q}^{AB} dt}{\int_{t_n}^{t_{n+1}} W dt} = \frac{\int_{t_n}^{t_{n+1}} W \mathbf{q}^{B} dt}{\int_{t_n}^{t_{n+1}} W dt} - \frac{\int_{t_n}^{t_{n+1}} W \mathbf{q}^{A} dt}{\int_{t_n}^{t_{n+1}} W dt}$$

$$= \mathbf{q}_n^{AB} + W_1 \Lambda_1 \dot{\mathbf{q}}_n^{AB} \Delta t \tag{14.383}$$

$$+ W_2 \Lambda_2 \ddot{\mathbf{q}}_n^{AB} \Delta t^2 + W_3 \Lambda_3 \left(\ddot{\mathbf{q}}_{n+1}^{AB} - \ddot{\mathbf{q}}_n^{AB}\right) \Delta t^2 \tag{14.384}$$

$$\tilde{\mathbf{a}}^A := \frac{\int_{t_n}^{t_{n+1}} W \ddot{\mathbf{q}}^A dt}{\int_{t_n}^{t_{n+1}} W dt} = \ddot{\mathbf{q}}_n^A + W_1 \Lambda_6 \left(\ddot{\mathbf{q}}_{n+1}^A - \ddot{\mathbf{q}}_n^A\right) \tag{14.385}$$

where $\ddot{\mathbf{q}}_{n+1}^{AB} := \ddot{\mathbf{q}}_{n+1}^{B} - \ddot{\mathbf{q}}_{n+1}^{A}$ and $\ddot{\mathbf{q}}_n^{AB} := \ddot{\mathbf{q}}_n^{B} - \ddot{\mathbf{q}}_n^{A}$, and the algorithmic inter-particle distance between A^{th} and B^{th} particles is defined as

$$\tilde{r}^{AB} := \frac{\int_{t_n}^{t_{n+1}} W r^{AB} dt}{\int_{t_n}^{t_{n+1}} W dt} = r_n^{AB} + W_1 \Theta_1 \dot{r}_n^{AB} \Delta t + W_2 \Theta_2 \ddot{r}_n^{AB} \Delta t^2 + W_3 \Theta_3 \left(\ddot{r}_{n+1}^{AB} - \ddot{r}_n^{AB}\right) \Delta t^2 \tag{14.386}$$

To design the updates, approximate $\mathbf{q}(t) : \mathcal{T} \rightarrow \mathcal{Q}$ and $\dot{\mathbf{q}}(t) : \mathcal{T} \rightarrow T_q\mathcal{Q}$ by the asymptotic series expansions in time with new scalar algorithmic parameters λ_i $(i = 1, 2, 3, 4, 5)$, and set $\tau = \Delta t$. Then, we obtain

$$\mathbf{q}_{n+1}^A = \mathbf{q}_n^A + \lambda_1 \dot{\mathbf{q}}_n^A \Delta t + \lambda_2 \ddot{\mathbf{q}}_n^A \Delta t^2 + \lambda_3 (\ddot{\mathbf{q}}_{n+1}^A - \ddot{\mathbf{q}}_n^A)\Delta t^2 \tag{14.387}$$

$$\dot{\mathbf{q}}_{n+1}^A = \dot{\mathbf{q}}_n^A + \lambda_4 \ddot{\mathbf{q}}_n^A \Delta t + \lambda_5 (\ddot{\mathbf{q}}_{n+1}^A - \ddot{\mathbf{q}}_n^A)\Delta t \tag{14.388}$$

For the approximate r^{AB} by the asymptotic series expansions in time with new scalar algorithmic parameters θ_i $(i = 1, 2, 3)$ and set $\tau = \Delta t$. Then, we have

$$r_{n+1}^{AB} = r_n^{AB} + \theta_1 \dot{r}_n^{AB} \Delta t + \theta_2 \ddot{r}_n^{AB} \Delta t^2 + \theta_3 (\ddot{r}_{n+1}^{AB} - \ddot{r}_n^{AB})\Delta t^2 \tag{14.389}$$

Substituting Equation (14.389) into Equation (14.386), we obtain

$$\begin{aligned} \tilde{r}^{AB} &= r_n^{AB} + \frac{W_3\Theta_3}{\theta_3}(r_{n+1}^{AB} - r_n^{AB}) + \left(W_1\Theta_1 - \frac{W_3\Theta_3\theta_1}{\theta_3} \right) \dot{r}_n^{AB} \Delta t \\ &+ \left(W_2\Theta_2 - \frac{W_3\Theta_3\theta_2}{\theta_3} \right) \ddot{r}_n^{AB} \Delta t^2 \end{aligned} \tag{14.390}$$

Summarizing, the following family of single-step time integration schemes are readily obtained:

Algorithm 30

Given $\mathbf{q}_n$, $\dot{\mathbf{q}}_n$, and $\ddot{\mathbf{q}}_n$, find $\mathbf{q}_{n+1}$, $\dot{\mathbf{q}}_{n+1}$, and $\ddot{\mathbf{q}}_{n+1}$ from

$$m_A \tilde{\mathbf{a}}^A - \sum_{B \neq A}^{N} \mathcal{V}_{AB}(\tilde{r}^{AB}) \frac{\tilde{\mathbf{q}}^{AB}}{\tilde{r}^{AB}} = \mathbf{0} \tag{14.391a}$$

$$\tilde{\mathbf{a}}^A = \ddot{\mathbf{q}}_n^A + W_1\Lambda_6 \left(\ddot{\mathbf{q}}_{n+1}^A - \ddot{\mathbf{q}}_n^A\right) \tag{14.391b}$$

$$\tilde{\mathbf{q}}^{AB} = \mathbf{q}_n^{AB} + W_1\Lambda_1 \dot{\mathbf{q}}_n^{AB} \Delta t + W_2\Lambda_2 \ddot{\mathbf{q}}_n^{AB} \Delta t^2 + W_3\Lambda_3 \Delta \mathbf{a}^A \Delta t^2 \tag{14.391c}$$

$$\begin{aligned} \tilde{r}^{AB} &= r_n^{AB} + \frac{W_3\Theta_3}{\theta_3}(r_{n+1}^{AB} - r_n^{AB}) + \left(W_1\Theta_1 - \frac{W_3\Theta_3\theta_1}{\theta_3} \right) \dot{r}_n^{AB} \Delta t \\ &+ \left(W_2\Theta_2 - \frac{W_3\Theta_3\theta_2}{\theta_3} \right) \ddot{r}_n^{AB} \Delta t^2 \end{aligned} \tag{14.391d}$$

with the following updates:

$$\mathbf{q}_{n+1}^A = \mathbf{q}_n^A + \lambda_1 \dot{\mathbf{q}}_n^A \Delta t + \lambda_2 \ddot{\mathbf{q}}_n^A \Delta t^2 + \lambda_3 \Delta \mathbf{a}^A \Delta t^2 \tag{14.391e}$$

$$\dot{\mathbf{q}}_{n+1}^A = \dot{\mathbf{q}}_n^A + \lambda_4 \ddot{\mathbf{q}}_n^A \Delta t + \lambda_5 \Delta \mathbf{a}^A \Delta t \tag{14.391f}$$

$$\ddot{\mathbf{q}}_{n+1}^A = \ddot{\mathbf{q}}_n^A + \Delta \mathbf{a}^A \tag{14.391g}$$

where $\Delta \mathbf{a}^A := \ddot{\mathbf{q}}_{n+1}^A - \ddot{\mathbf{q}}_n^A$.

Remark 14.4.4

1. The DNA parameters $\{W, \Lambda, \lambda\}$ are given as shown in Algorithm 9 and Algorithm 10 for the U0 family of algorithms and the V0 family of algorithms, respectively. Assume $\Lambda_i = \Theta_i$ and $\lambda_i = \theta_i$ $(i = 1, 2, 3)$ for simplicity.

2. When selecting

$$\mathrm{U0}(\{\rho_{1\infty}, \rho_{2\infty}, \rho_{3\infty}\} = \{1, 1, 1\}) \text{ or } \mathrm{V0}(\{\rho_{1\infty}, \rho_{2\infty}, \rho_{3\infty}\} = \{1, 1, \rho_{3\infty}\})$$

where $\rho_{3\infty} \in [0, 1]$, assuming $\Lambda_i = \Theta_i$ and $\lambda_i = \theta_i$ $(i = 1, 2, 3)$, Algorithm 30 becomes the single-field form version of the assumed distance method. Notice there exists a family of algorithms in the framework of the V0 family of algorithms due to arbitrary conditions of $\rho_{3\infty} \in [0, 1]$.

Algorithm 30 cannot recover the exact energy-momentum conserving algorithm naturally in a general nonlinear dynamic system. To circumvent this, following a similar procedure, we propose the following family of algorithms in the sense of the single-field form:

Algorithm 31

Given $\mathbf{q}_n$, $\dot{\mathbf{q}}_n$, and $\ddot{\mathbf{q}}_n$, find $\mathbf{q}_{n+1}$, $\dot{\mathbf{q}}_{n+1}$, and $\ddot{\mathbf{q}}_{n+1}$ from

$$m_A \tilde{\mathbf{a}}^A - \sum_{B \neq A}^{N} \widetilde{\mathcal{V}'_{AB}} \frac{\tilde{\mathbf{q}}^{AB}}{\tilde{r}^{AB}} = \mathbf{0} \tag{14.392a}$$

$$\tilde{\mathbf{a}}^A = \ddot{\mathbf{q}}_n^A + W_1 \Lambda_6 (\ddot{\mathbf{q}}_{n+1}^A - \ddot{\mathbf{q}}_n^A) \tag{14.392b}$$

$$\tilde{\mathbf{q}}^{AB} = \mathbf{q}_n^{AB} + W_1 \Lambda_1 \dot{\mathbf{q}}_n^{AB} \Delta t + W_2 \Lambda_2 \ddot{\mathbf{q}}_n^{AB} \Delta t^2 + W_3 \Lambda_3 \Delta \mathbf{a}^A \Delta t^2 \tag{14.392c}$$

$$\begin{aligned} \tilde{r}^{AB} &= r_n^{AB} + \frac{W_3 \Theta_3}{\theta_3}(r_{n+1}^{AB} - r_n^{AB}) + \left(W_1 \Theta_1 - \frac{W_3 \Theta_3 \theta_1}{\theta_3} \right) \dot{r}_n^{AB} \Delta t \\ &\quad + \left(W_2 \Theta_2 - \frac{W_3 \Theta_3 \theta_2}{\theta_3} \right) \ddot{r}_n^{AB} \Delta t^2 \end{aligned} \tag{14.392d}$$

with the following updates:

$$\mathbf{q}_{n+1}^A = \mathbf{q}_n^A + \lambda_1 \dot{\mathbf{q}}_n^A \Delta t + \lambda_2 \ddot{\mathbf{q}}_n^A \Delta t^2 + \lambda_3 \Delta \mathbf{a}^A \Delta t^2 \tag{14.392e}$$

$$\dot{\mathbf{q}}_{n+1}^A = \dot{\mathbf{q}}_n^A + \lambda_4 \ddot{\mathbf{q}}_n^A \Delta t + \lambda_5 \Delta \mathbf{a}^A \Delta t \tag{14.392f}$$

$$\ddot{\mathbf{q}}_{n+1}^A = \ddot{\mathbf{q}}_n^A + \Delta \mathbf{a}^A \tag{14.392g}$$

where $\Delta \mathbf{a}^A := \ddot{\mathbf{q}}_{n+1}^A - \ddot{\mathbf{q}}_n^A$.

Remark 14.4.5

1. The DNA parameters $\{W, \Lambda, \lambda\}$ are given as shown in Algorithm 9 and Algorithm 10 for the U0 family of algorithms and the V0 family of algorithms, respectively. Algorithm 31 is the modified version of Algorithm 30. Assume $\Lambda_i = \Theta_i$ and $\lambda_i = \theta_i$ $(i = 1, 2, 3)$ for simplicity.
2. The algorithmic quantity of the derivative of the internal potential energy between the Ath and B^{th} particles with respect to the inter-particle distance r_{AB}, $\widetilde{\mathcal{V}'_{AB}}$, is given as

$$\widetilde{\mathcal{V}'_{AB}} = \frac{\mathcal{V}_{AB}(r_{n+1}^{AB}) - \mathcal{V}_{AB}(r_n^{AB})}{r_{n+1}^{AB} - r_n^{AB}} \tag{14.393}$$

when $W_1 = 1/2$; otherwise,

$$\widetilde{\mathcal{V}_{AB}} = \mathcal{V}_{AB}(\tilde{r}^{AB}) \tag{14.394}$$

3. When selecting

$$\text{U0}(\{\rho_{1\infty}, \rho_{2\infty}, \rho_{3\infty}\} = \{1, 1, 1\}) \text{ or } \text{V0}(\{\rho_{1\infty}, \rho_{2\infty}, \rho_{3\infty}\} = \{1, 1, \rho_{3\infty}\})$$

where $\rho_{3\infty} \in [0, 1]$, assuming $\Lambda_i = \Theta_i$ and $\lambda_i = \theta_i$ $(i = 1, 2, 3)$, Algorithm 31 becomes the single-field form version of the exact energy-momentum conserving algorithm. Notice there exists a family of algorithms in the framework of the V0 family of algorithms due to arbitrary conditions of $\rho_{3\infty} \in [0, 1]$.

Illustrative Examples The numerical results for the algorithms in the single-field form design are briefly shown next for the Kepler's problem and the Lennard-Jones (5, 3) Potential 2-body problem.

Example 14.5 Kepler's Problem

Consider the famous Kepler problem in $\mathbb{R}^3$ space $(N = 1)$. The generalized coordinate is defined as the position of a particle of mass m, moving in the space, and the generalized velocity is defined as its velocity:

$$\mathbf{q}(t) \equiv \mathbf{x}(t) : \mathcal{T} \to \mathcal{Q} \equiv \mathbb{R}^3 \tag{14.395}$$

$$\dot{\mathbf{q}}(t) \equiv \dot{\mathbf{x}}(t) : \mathcal{T} \to T_q\mathcal{Q} \equiv \mathbb{R}^3 \tag{14.396}$$

The total energy is defined as the summation of the kinetic energy and the potential energy given by

$$\mathcal{K}(\dot{\mathbf{q}}) = \frac{1}{2}\dot{\mathbf{q}} \cdot \mathbf{M}\dot{\mathbf{q}} \quad \text{and} \quad \mathcal{V}(r) = -\frac{k}{r} \tag{14.397}$$

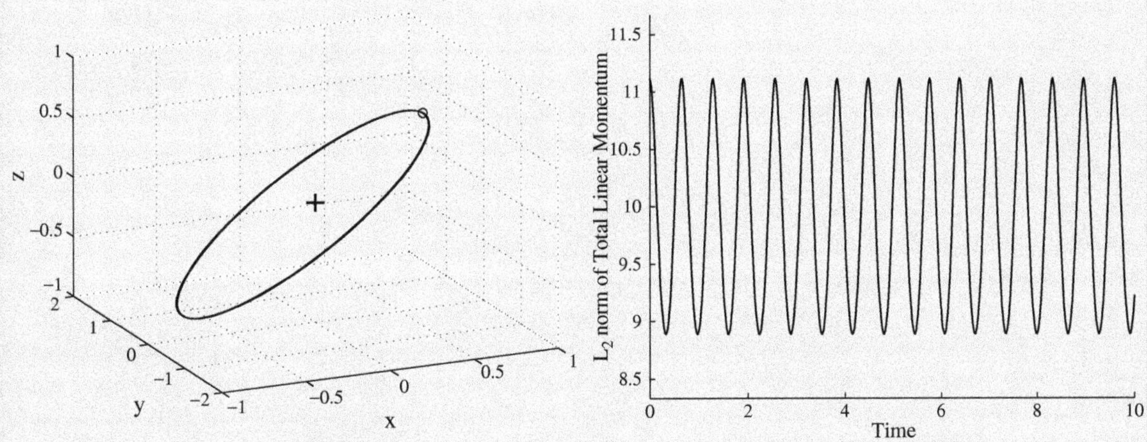

(a) Configuration (b) Linear Momentum

Figure 14.16. Simulation results for the Kepler problem with $\Delta t = 0.01$ by Algorithm 30: U0(1.0, 1.0, 1.0)/V0(1.0, 1.0, $\rho_{3\infty}$) with $\rho_{3\infty} \in [0, 1]$

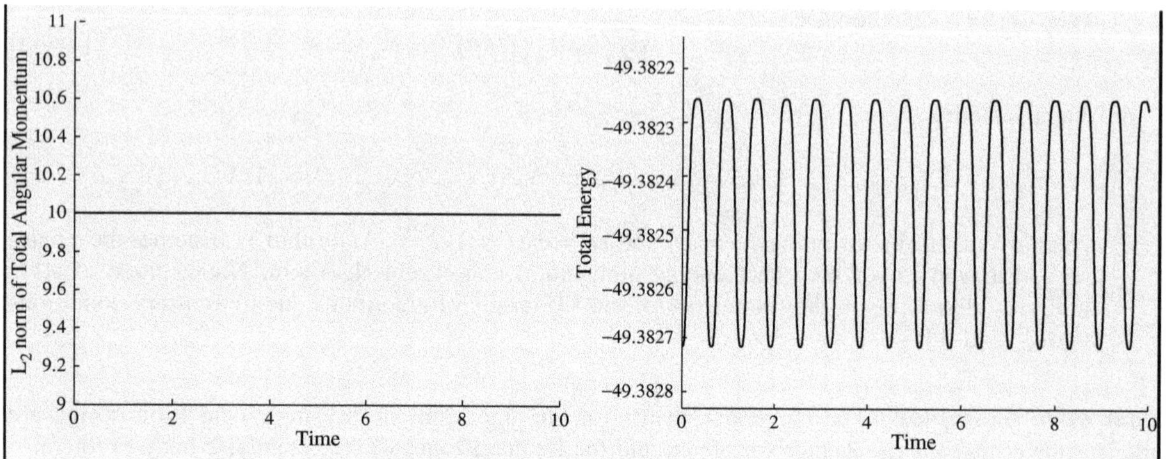

(c) Angular Momentum (d) Total Energy

Figure 14.16. *(continued)*

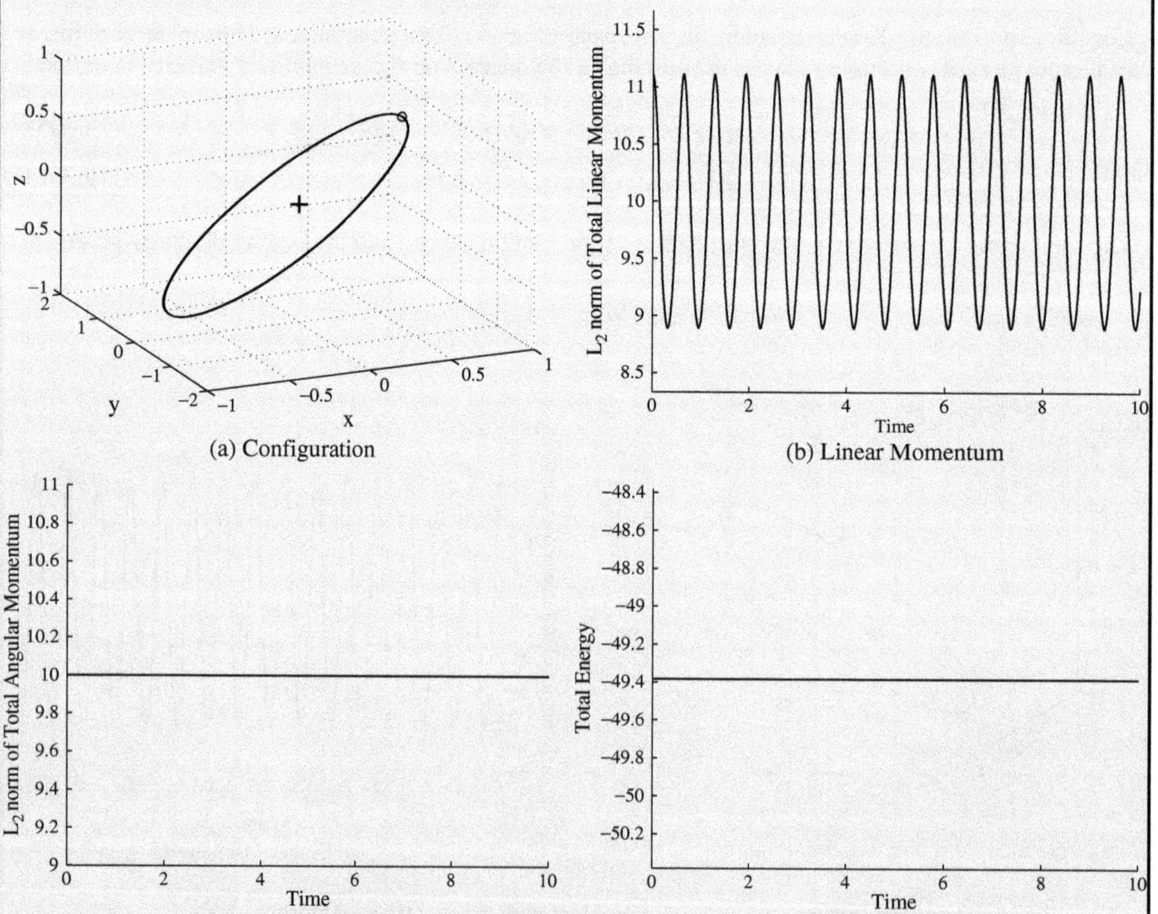

(a) Configuration (b) Linear Momentum

(c) Angular Momentum (d) Total Energy

Figure 14.17. Simulation results for the Kepler problem with $\Delta t = 0.01$ by Algorithm 31: U0(1.0, 1.0, 1.0)/V0(1.0, 1.0, $\rho_{3\infty}$) with $\rho_{3\infty} \in [0, 1]$

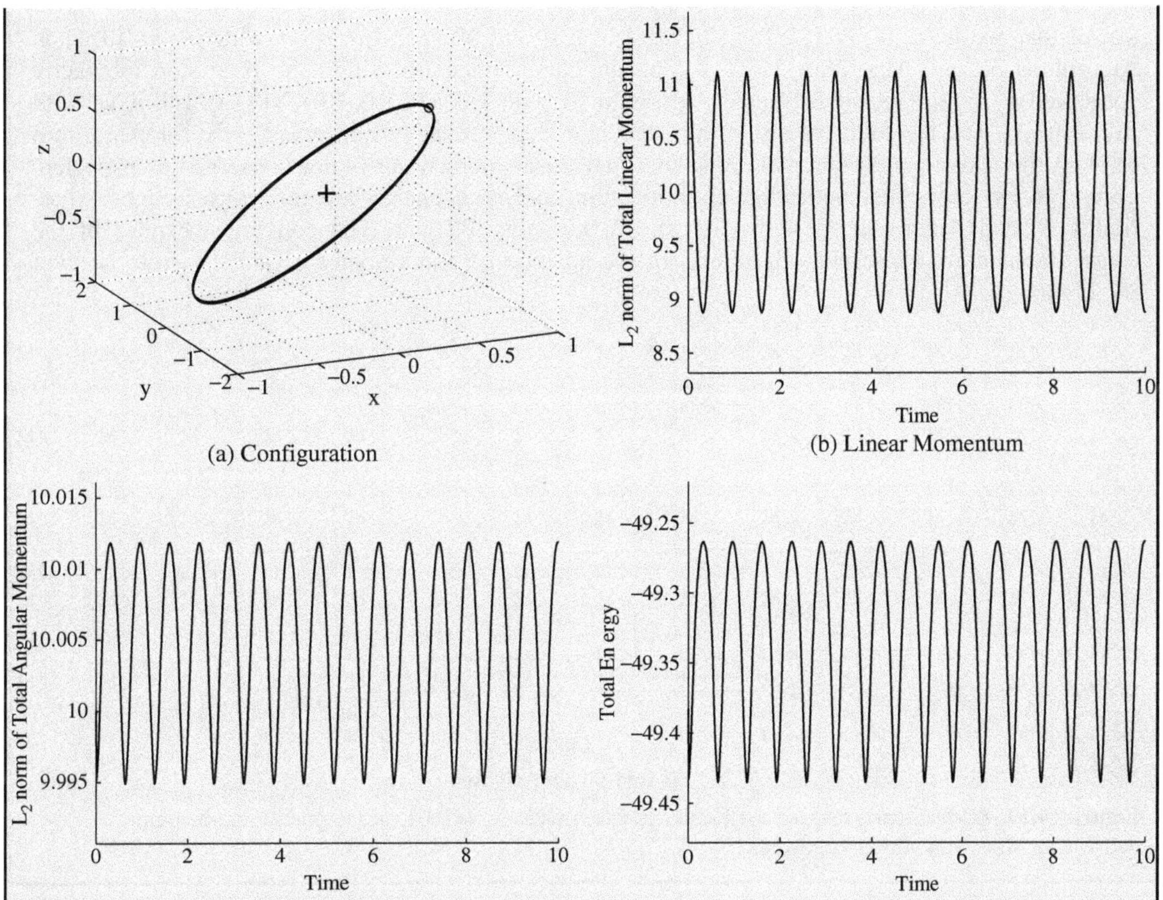

(a) Configuration

(b) Linear Momentum

(c) Angular Momentum

(d) Total Energy

Figure 14.18. Simulation results for the Kepler problem with $\Delta t = 0.01$ by Algorithms 30/31: U0(1.0, 1.0, 0.0)

respectively, and the constant mass matrix is defined by $\mathbf{M} = m\mathbf{I}_3$, $r = ||\mathbf{q}|| := \sqrt{\mathbf{q} \cdot \mathbf{q}}$ denotes the distance from the origin, and k is a constant. Note the potential energy satisfies $\lim_{r \to 0} \mathcal{V}(r) = 0$. Therefore, the Total Energy representation of the equation of motion

$$\frac{d}{dt}\left(\frac{\partial \mathcal{E}}{\partial \dot{\mathbf{q}}}\right) + \frac{\partial \mathcal{E}}{\partial \mathbf{q}} = \mathbf{0} \tag{14.398}$$

leads to

$$\mathbf{M}\ddot{\mathbf{q}} + \frac{k}{r^2}\frac{\mathbf{q}}{r} = \mathbf{0} \tag{14.399}$$

The problem description, convergence rates, and the numerical results via Algorithms 30 and Algorithm 31 with the particular conditions of $(\rho_{1\infty}, \rho_{2\infty}, \rho_{3\infty})$ within the extension of the U0 family and V0 family frameworks for the nonlinear dynamic problem are shown in Figures 14.16–14.19. The initial conditions are $\mathbf{q}_0 = \mathbf{q}(t_0 = 0) = (\frac{0.9}{\sqrt{2}}, 0, \frac{0.9}{\sqrt{2}})$ and $\dot{\mathbf{q}}_0 = \dot{\mathbf{q}}(t_0 = 0) = (0, -\frac{100}{9}, 0)$; $m = 1$ and $k = 100$; and time step size $\Delta t = 0.01$, final time $T = t_N = 10.00$ and the tolerance is 10^{-8}.

Recall that we have $W_1 = 1/2$ for U0(1.0, 1.0, 1.0) and V0(1.0, 1.0, $\rho_{3\infty}$) with $\rho_{3\infty} \in [0, 1]$ in Algorithms 30 and Algorithm 31. The total angular momentum within the time step is exactly conserved by both Algorithm 30 and Algorithm 31 with U0(1.0, 1.0, 1.0)/V0(1.0, 1.0, $\rho_{3\infty}$) with $\rho_{3\infty} \in [0, 1]$, but the total energy within the time step is exactly conserved only for Algorithm 31 with the above conditions. Note that the total linear momentum is not conserved as expected. Figure 14.18 shows that both angular momentum and total energy are not conserved by Algorithms 30 and Algorithm 31 with U0(1.0, 1.0, 0.0), i.e., $W_1 = 1$. The convergence rates of the energy-momentum conserving algorithm for the position q^1 and velocity $\dot{q}^1$ are shown to be two; see Figure 14.19.

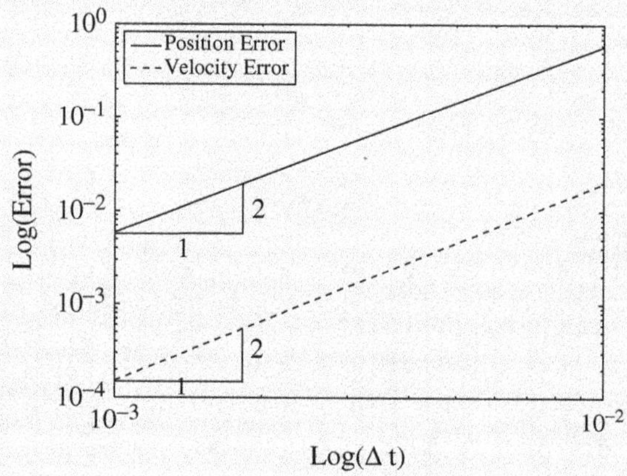

(a) Convergence Rate

Figure 14.19. Convergence rate for the Kepler problem with $\Delta t = 0.01$ by the energy-momentum conserving algorithm within Algorithm 29

Example 14.6 Lennard-Jones (5,3) Potential 2-Body Problem

Consider a system of two particles as shown in Figure 14.20. This model problem in $\mathbb{R}^2$ space has been given in (Betsch and Steinmann 2000). Define the generalized coordinate and velocity as

$$\mathbf{q}(t) = (\mathbf{q}^1(t), \mathbf{q}^2(t)) \equiv \mathbf{x} : \mathcal{T} \rightarrow Q \equiv \mathbb{R}^6 \tag{14.400}$$

$$\dot{\mathbf{q}}(t) = (\dot{\mathbf{q}}^1(t), \dot{\mathbf{q}}^2(t)) \equiv \dot{\mathbf{x}} : \mathcal{T} \rightarrow T_q Q \equiv \mathbb{R}^6 \tag{14.401}$$

respectively. The total energy of the system is defined by the kinetic energy and the internal potential energy given by

$$\mathcal{K}(\dot{\mathbf{q}}) = \frac{1}{2}\dot{\mathbf{q}} \cdot \mathbf{M}\dot{\mathbf{q}} \quad \text{and} \quad \mathcal{V}^{\text{int}}(r) = A\left[\left(\frac{\sigma}{r}\right)^5 - \left(\frac{\sigma}{r}\right)^3\right] \tag{14.402}$$

respectively, where the constant mass matrix is defined by $\mathbf{M} = \text{diag}(\mathbf{m}_1, \mathbf{m}_2)$ for $\mathbf{m}_i = m_i\mathbf{I}_3$ $(i = 1, 2)$; $r = r_{12} = r_{21} := ||\mathbf{q}^1 - \mathbf{q}^2||$ is the inter-particle distance, and A and σ are constant parameters. Therefore, the Total Energy representation of the equation of motion given by

$$\frac{d}{dt}\left(\frac{\partial \mathcal{E}}{\partial \dot{\mathbf{q}}}\right) + \frac{\partial \mathcal{E}}{\partial \mathbf{q}} = \mathbf{0} \tag{14.403}$$

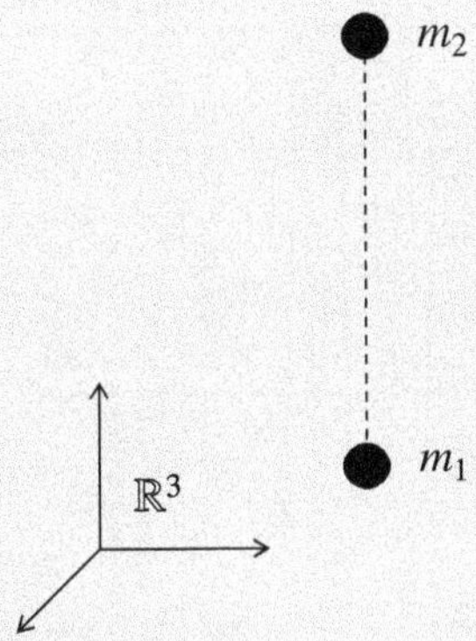

Figure 14.20. The problem description of the 2-body Problem

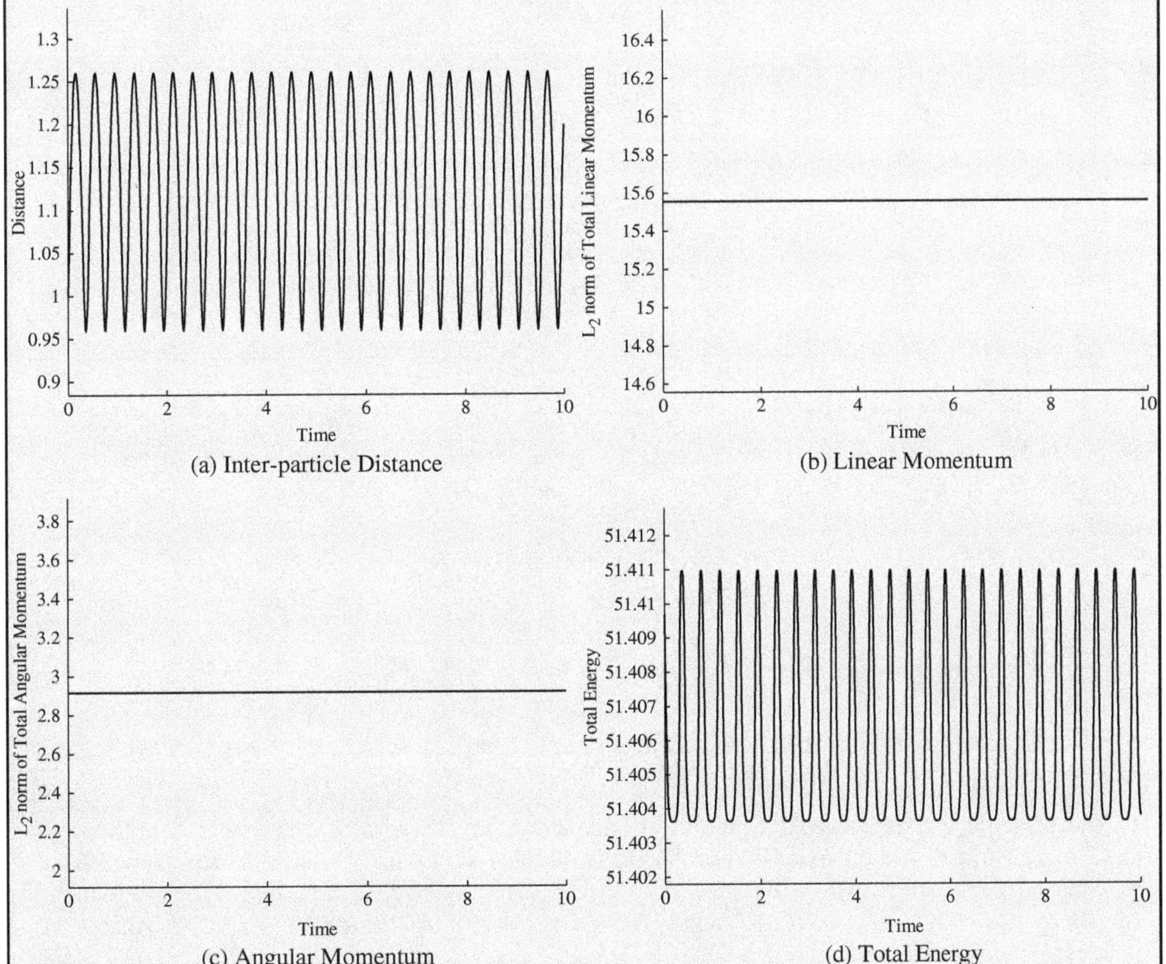

Figure 14.21. Simulation results for the Lennard-Jones (5, 3) Potential 2-body problem with $\Delta t = 0.01$ by Algorithm 30: U0(1.0, 1.0, 1.0)/V0(1.0, 1.0, $\rho_{3\infty}$) with $\rho_{3\infty} \in [0, 1]$

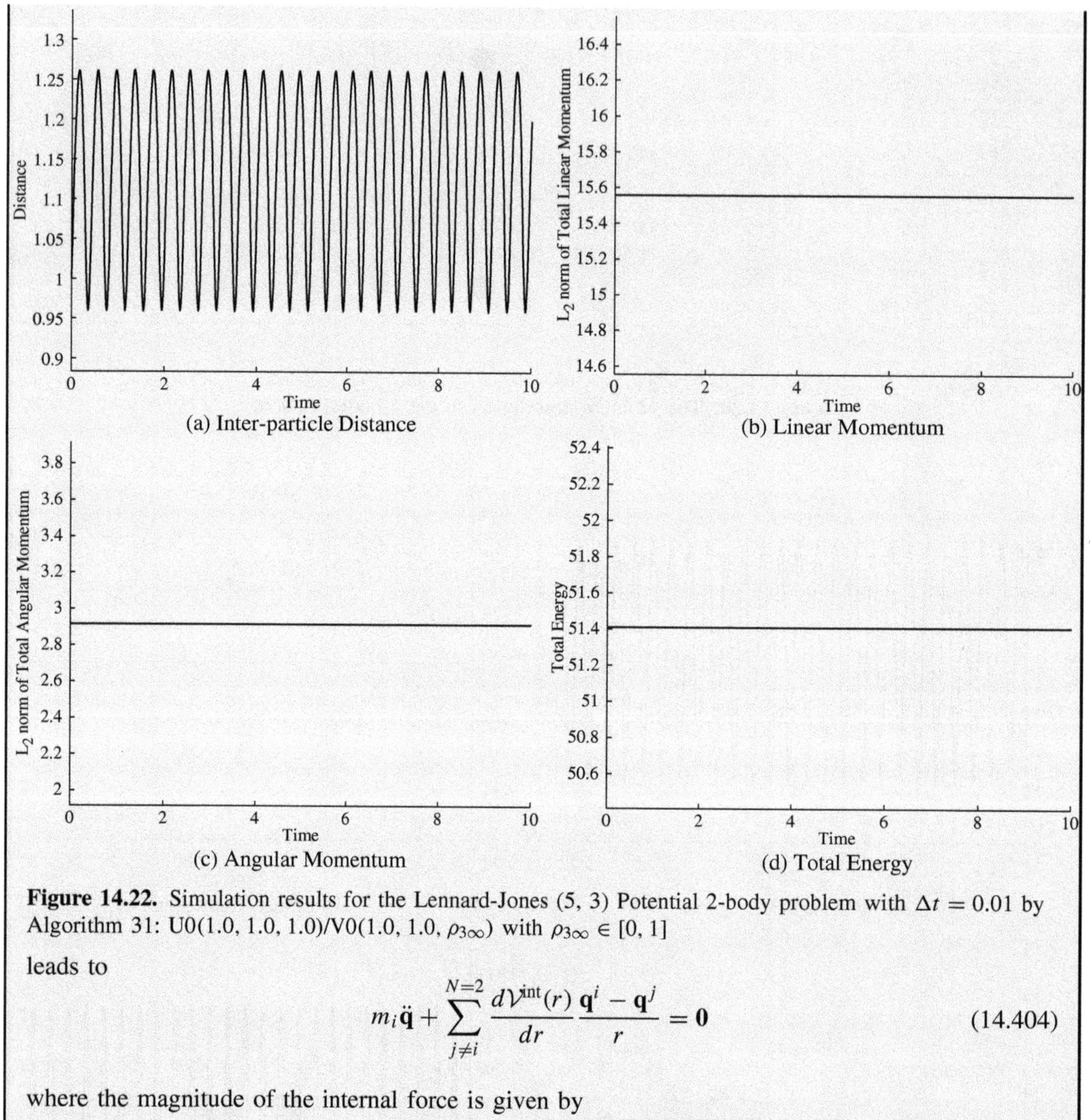

Figure 14.22. Simulation results for the Lennard-Jones (5, 3) Potential 2-body problem with $\Delta t = 0.01$ by Algorithm 31: U0(1.0, 1.0, 1.0)/V0(1.0, 1.0, $\rho_{3\infty}$) with $\rho_{3\infty} \in [0, 1]$

leads to

$$m_i \ddot{\mathbf{q}} + \sum_{j \neq i}^{N=2} \frac{d\mathcal{V}^{\text{int}}(r)}{dr} \frac{\mathbf{q}^i - \mathbf{q}^j}{r} = \mathbf{0} \tag{14.404}$$

where the magnitude of the internal force is given by

$$\frac{d\mathcal{V}^{\text{int}}(r)}{dr} = \frac{A}{\sigma}\left[3\left(\frac{\sigma}{r}\right)^4 - 5\left(\frac{\sigma}{r}\right)^6\right] \tag{14.405}$$

The numerical results via Algorithms 30 and Algorithm 31 with the particular conditions of $(\rho_{1\infty}, \rho_{2\infty}, \rho_{3\infty})$ within the extension of the U0 family and V0 family frameworks for the nonlinear dynamic problem are shown in Figures 14.21–14.23. The initial conditions are $\mathbf{q}_0 = \mathbf{q}(t_0 = 0) = (\mathbf{q}^1(0), \mathbf{q}^2(0)) = (0, -\frac{1}{2}, 0, 0, \frac{1}{2}, 0)$ and $\dot{\mathbf{q}}_0 = \dot{\mathbf{q}}(t_0 = 0) = (\dot{\mathbf{q}}^1(0), \dot{\mathbf{q}}^2(0)) = (5, 1, 2, 10, 3, -1)$; masses of particles 1 and 2 are $m_1 = m_2 = 1$ kg, $A = 100$; and time step size $\Delta t = 0.01$, final time $T = t_N = 10.00$ and the tolerance is 10^{-8}. Select $\sigma = \sqrt{\frac{3}{5}}$ such that $\left.\frac{d\mathcal{V}^{\text{int}}(r)}{dr}\right|_{r(t_0)=1} = 0$.

The total linear and angular momenta within the time step are exactly conserved by both Algorithms 30 and Algorithm 31 with U0(1.0, 1.0, 1.0)/V0(1.0, 1.0, $\rho_{3\infty}$) with $\rho_{3\infty} \in [0, 1]$, but the total energy within the time step is exactly conserved only by Algorithm 31 with the above conditions. Algorithm 31 is designed to recover the exact energy-momentum conserving algorithm when $W_1 = 1/2$. Note that total linear and angular momenta, and total energy are not conserved by Algorithms 30 and Algorithm 31 with U0(1.0, 1.0, 0.0).

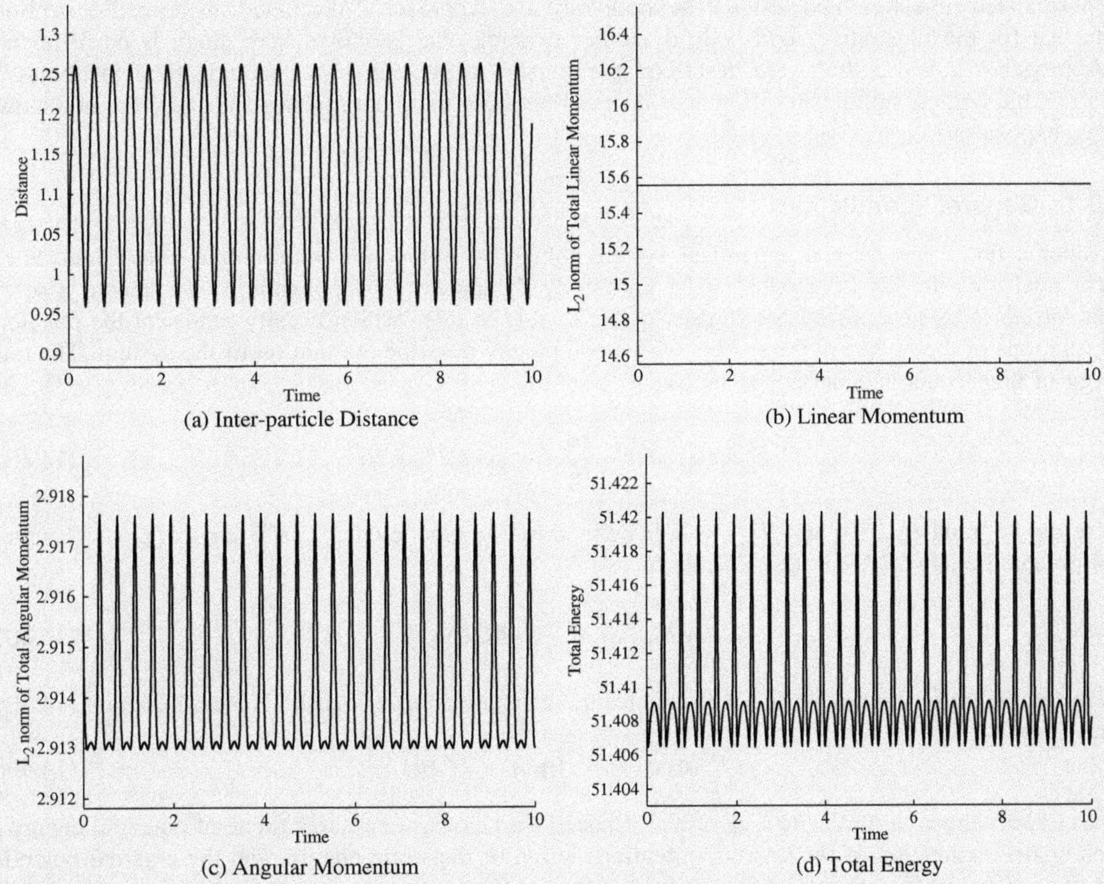

Figure 14.23. Simulation results for the Lennard-Jones (5, 3) Potential 2-body problem with $\Delta t = 0.01$ by Algorithms 30/31: U0(1.0, 1.0, 0.0)

14.5 TIME DISCRETIZATION AND TOTAL ENERGY FRAMEWORK: NONCONSERVATIVE/CONSERVATIVE MECHANICAL SYSTEMS WITH HOLONOMIC-SCLERONOMIC CONSTRAINTS

In this section, we highlight further developments for applications involving nonconservative/conservative mechanical systems (by appropriate selection of terms) with holonomic-scleronomic constraints, and typical of those termed in the literature as multi-body dynamics problems. Simply for illustration, we describe the applications for the single-field form of representation of the equation of motion via the GSSSS framework described previously. Starting from the principle of balance of Total Energy (total mechanical energy), we first describe the generic mathematical model equations valid for flexible and/or rigid bodies (by appropriate selection of terms) which lead to the semi-discrete differential algebraic equations (DAE) in the so-called Index 3 (which naturally arise and are the most practical from the viewpoint of finite element commercial software developments) which are directly employed instead of resorting to index reduction techniques. Next, simply for illustration of the basic concepts to applications involving multi-body dynamics with rigid bodies, we describe the time discretization approach via the GSSSS unified framework. Purposely, the formulations are shown via the normalized time weighted residual methodology and Approach 2 discussed in the earlier sections (note that for the illustrative rigid body dynamics example that is shown here, there is no difference in Approaches 2 and 3 that were described previously; in general, if we include flexible bodies or rigid/flexible bodies, significant differences do exist between these approaches and these are not shown in this exposition).

14.5.1 General Formulations

Consider a finite dimensional dynamical system subject to holonomic-scleronomic constraints. For a mechanical motion with constraints as in multibody dynamics type applications, we denote a set of nodal variables by generalized coordinates $\mathbf{q}(t) : \mathcal{T} \rightarrow \mathcal{Q} \equiv \mathbb{R}^{n_{\text{dof}}}$ which usually represent the positions and rotations of the bodies in the system which completely describe the motion of the system. The total energy of the system is generally given as

$$\mathcal{E}^h(\mathbf{q}, \dot{\mathbf{q}}) = \frac{\partial \mathcal{L}^h(\mathbf{q}, \dot{\mathbf{q}})}{\partial \dot{\mathbf{q}}} \cdot \dot{\mathbf{q}} - \mathcal{L}^h(\mathbf{q}, \dot{\mathbf{q}}) \tag{14.406}$$

where the Lagrangian $\mathcal{L}^h(\mathbf{q}, \dot{\mathbf{q}}) : T\mathcal{Q} \rightarrow \mathbb{R}$ is defined by the kinetic energy, $\mathcal{K}^h(\mathbf{q}, \dot{\mathbf{q}}) : T\mathcal{Q} \rightarrow \mathbb{R}$, minus total potential energy, $\mathcal{U}^h(\mathbf{q}) : \mathcal{Q} \rightarrow \mathbb{R}$, with

$$\mathcal{K}^h(\mathbf{q}, \dot{\mathbf{q}}) = \frac{1}{2}\dot{\mathbf{q}} \cdot \mathbf{M}(\mathbf{q})\dot{\mathbf{q}} \tag{14.407}$$

where symmetric positive-definite $\mathbf{M}(\mathbf{q})$ denotes the generalized mass matrix. Hence, Equation (14.406) gives

$$\mathcal{E}^h(\mathbf{q}, \dot{\mathbf{q}}) = \mathcal{K}^h(\mathbf{q}, \dot{\mathbf{q}}) + \mathcal{U}^h(\mathbf{q}) \tag{14.408}$$

due to Euler's theorem, $(\partial \mathcal{K}^h / \partial \dot{\mathbf{q}}) \cdot \dot{\mathbf{q}} = 2\mathcal{K}^h$. In dead load cases, we assume the total potential energy is given as the summation of the internal potential energy, or the strain energy, and the external potential energy as $\mathcal{U}^h(\mathbf{q}) = \mathcal{U}^h_{\text{int}}(\mathbf{q}) + \mathcal{U}^h_{\text{ext}}(\mathbf{q})$ where $\mathcal{U}^h_{\text{ext}}(\mathbf{q}) = \mathbf{q} \cdot \mathbf{Q}$ with the constant external force $\mathbf{Q}$, and the power input is defined by

$$\mathcal{P}^h_{\text{input}} = \dot{\mathbf{q}} \cdot \mathbf{Q}^{\text{nc}} \tag{14.409}$$

where $\mathbf{Q}^{\text{nc}}$ denotes the nonconservative force. In non dead load cases, the total potential energy is defined only by the internal potential energy (strain energy), i.e., $\mathcal{U}^h(\mathbf{q}) = \mathcal{U}^h_{\text{int}}(\mathbf{q})$, and the power input is defined by

$$\mathcal{P}^h_{\text{input}} = \dot{\mathbf{q}} \cdot \mathbf{Q}^{\text{nc}} + \dot{\mathbf{q}} \cdot \mathbf{Q}(t) \tag{14.410}$$

where $\mathbf{Q}(t) : \mathcal{T} \rightarrow \mathbf{R}^{n_{\text{dof}}}$ denotes the time-dependent external force. In the following discussion, consider the nondead load case, and we assume the nonconservative force is given as the dissipative force, i.e.,

$$\mathbf{Q}^{\text{diss}}(\dot{\mathbf{q}}) = -\mathbf{C}\dot{\mathbf{q}} \tag{14.411}$$

where $\mathbf{C}$ is the damping matrix. Therefore, the principle of balance of mechanical energy yields

$$\frac{d\mathcal{E}^h(\mathbf{q}, \dot{\mathbf{q}})}{dt} = \mathcal{P}^h_{\text{input}} \quad \forall t \in \mathcal{T} = [t_0, t_N] \tag{14.412}$$

is subject to a holonomic-scleronomic constraint equation of the form

$$\mathbf{0} = \mathbf{\Phi}(\mathbf{q}) \in \mathbb{R}^k \tag{14.413}$$

where k denotes the number of constraint conditions. Note $k \leq n_{\text{dof}}$. Assume the constraint equation is linearly independent. That is, it satisfies

$$\text{rank}\mathbf{G}(\mathbf{q}) = k \tag{14.414}$$

with the constraint Jacobian matrix defined by $\mathbf{G}(\mathbf{q}) := \nabla\mathbf{\Phi} = \partial\mathbf{\Phi}/\partial\mathbf{q}$. The constraint on the velocity level is obtained by differentiating Equation (14.413) once with respect to time. That is,

$$\mathbf{0} = \dot{\mathbf{\Phi}} = \nabla\mathbf{\Phi} \cdot \dot{\mathbf{q}} \tag{14.415}$$

Similarly, the constraint on the acceleration level is obtained by differentiating Equation (14.415) once with respect to time. That is,

$$\mathbf{0} = \ddot{\mathbf{\Phi}} = \nabla\mathbf{\Phi} \cdot \ddot{\mathbf{q}} + \left(\nabla^2\mathbf{\Phi} \cdot \dot{\mathbf{q}}\right) \cdot \dot{\mathbf{q}} \tag{14.416}$$

Here, we introduce a modified total energy function or an augmented total energy, $\tilde{\mathcal{E}}^h(\mathbf{q}, \dot{\mathbf{q}}, \boldsymbol{\lambda}) : TQ \times \mathbb{R}^k \to \mathbb{R}$, defined by

$$\tilde{\mathcal{E}}^h(\mathbf{q}, \dot{\mathbf{q}}, \boldsymbol{\lambda}) = \mathcal{E}^h(\mathbf{q}, \dot{\mathbf{q}}) + \mathcal{U}'^h(\mathbf{q}, \boldsymbol{\lambda}) \tag{14.417}$$

In Equation (14.417), the Lagrange multiplier, $\boldsymbol{\lambda}(t) : \mathcal{T} \to \mathbb{R}^k$, has been introduced to define

$$\mathcal{U}'^h(\mathbf{q}, \boldsymbol{\lambda}) = -\boldsymbol{\lambda} \cdot \mathbf{\Phi}(\mathbf{q}) \tag{14.418}$$

Instead of Equation (14.412) and Equation (14.413), we can equivalently deal with the following:

$$\frac{d\tilde{\mathcal{E}}^h(\mathbf{q}, \dot{\mathbf{q}}, \boldsymbol{\lambda})}{dt} = \mathcal{P}^h_{\text{input}} \quad \forall t \in \mathcal{T} = [t_0, t_N] \tag{14.419}$$

Equation (14.419) yields

$$\begin{aligned}
0 &= \frac{d\tilde{\mathcal{E}}^h(\mathbf{q}, \dot{\mathbf{q}}, \boldsymbol{\lambda})}{dt} - \mathcal{P}^h_{\text{input}} \\
&= \frac{\partial\tilde{\mathcal{E}}^h}{\partial\dot{\mathbf{q}}} \cdot \ddot{\mathbf{q}} + \frac{\partial\tilde{\mathcal{E}}^h}{\partial\mathbf{q}} \cdot \dot{\mathbf{q}} + \frac{\partial\tilde{\mathcal{E}}^h}{\partial\boldsymbol{\lambda}} \cdot \dot{\boldsymbol{\lambda}} - \mathbf{Q}^{\text{diss}} \cdot \dot{\mathbf{q}} - \mathbf{Q}(t) \cdot \dot{\mathbf{q}} \\
&= \frac{\partial\mathcal{K}^h}{\partial\dot{\mathbf{q}}} \cdot \ddot{\mathbf{q}} + \left[\frac{\partial\mathcal{E}^h}{\partial\mathbf{q}} + \frac{\partial\mathcal{U}'^h}{\partial\mathbf{q}} - \mathbf{Q}^{\text{diss}} - \mathbf{Q}(t)\right] \cdot \dot{\mathbf{q}} + \frac{\partial\mathcal{U}'^h}{\partial\boldsymbol{\lambda}} \cdot \dot{\boldsymbol{\lambda}} \\
&= \frac{\partial\mathcal{K}^h}{\partial\dot{\mathbf{q}}} \cdot \ddot{\mathbf{q}} + \left[\frac{\partial\mathcal{E}^h}{\partial\mathbf{q}} - \boldsymbol{\lambda} \cdot \nabla\mathbf{\Phi}(\mathbf{q}) - \mathbf{Q}^{\text{diss}} - \mathbf{Q}(t)\right] \cdot \dot{\mathbf{q}} - \mathbf{\Phi}(\mathbf{q}) \cdot \dot{\boldsymbol{\lambda}}
\end{aligned} \tag{14.420}$$

Here, we have the following relation:

$$\begin{aligned}\frac{\partial \mathcal{K}^h(\mathbf{q},\dot{\mathbf{q}})}{\partial \dot{\mathbf{q}}}\cdot\ddot{\mathbf{q}} &= \mathbf{M}(\mathbf{q})\dot{\mathbf{q}}\cdot\ddot{\mathbf{q}} = \mathbf{M}(\mathbf{q})\ddot{\mathbf{q}}\cdot\dot{\mathbf{q}}\\ &= \left[\frac{d(\mathbf{M}(\mathbf{q})\dot{\mathbf{q}})}{dt} - \dot{\mathbf{M}}(\mathbf{q})\dot{\mathbf{q}}\right]\cdot\dot{\mathbf{q}}\\ &= \left[\frac{d}{dt}\left(\frac{\partial \mathcal{K}^h(\mathbf{q},\dot{\mathbf{q}})}{\partial \dot{\mathbf{q}}}\right) - 2\frac{\partial}{\partial \mathbf{q}}\left(\frac{1}{2}\dot{\mathbf{q}}\cdot\mathbf{M}(\mathbf{q})\dot{\mathbf{q}}\right)\right]\cdot\dot{\mathbf{q}}\\ &= \left[\frac{d}{dt}\left(\frac{\partial \mathcal{E}^h(\mathbf{q},\dot{\mathbf{q}})}{\partial \dot{\mathbf{q}}}\right) - 2\frac{\partial \mathcal{K}^h(\mathbf{q},\dot{\mathbf{q}})}{\partial \mathbf{q}}\right]\cdot\dot{\mathbf{q}}\end{aligned} \tag{14.421}$$

due to $\mathbf{M}^T(\mathbf{q}) = \mathbf{M}(\mathbf{q})$. Notice that

$$\mathbf{M}\ddot{\mathbf{q}} = \frac{d}{dt}\left(\frac{\partial \mathcal{E}^h(\mathbf{q},\dot{\mathbf{q}})}{\partial \dot{\mathbf{q}}}\right) \tag{14.422}$$

if $\mathbf{M}$ is constant. Therefore, Equation (14.420) can be written as

$$\begin{aligned}\left[\frac{d}{dt}\left(\frac{\partial \mathcal{E}^h(\mathbf{q},\dot{\mathbf{q}})}{\partial \dot{\mathbf{q}}}\right) - 2\frac{\partial \mathcal{K}^h(\mathbf{q},\dot{\mathbf{q}})}{\partial \mathbf{q}} + \frac{\partial \mathcal{E}^h(\mathbf{q},\dot{\mathbf{q}})}{\partial \mathbf{q}} - \boldsymbol{\lambda}\cdot\nabla\boldsymbol{\Phi}(\mathbf{q}) - \mathbf{Q}^{\text{diss}}(\dot{\mathbf{q}}) - \mathbf{Q}(t)\right]\cdot\dot{\mathbf{q}}&\\ = \boldsymbol{\Phi}(\mathbf{q})\cdot\dot{\boldsymbol{\lambda}}&\end{aligned} \tag{14.423}$$

which leads to

$$\frac{d}{dt}\left(\frac{\partial \mathcal{E}^h(\mathbf{q},\dot{\mathbf{q}})}{\partial \dot{\mathbf{q}}}\right) + \frac{\partial \mathcal{E}^h(\mathbf{q},\dot{\mathbf{q}})}{\partial \mathbf{q}} - 2\frac{\partial \mathcal{K}^h(\mathbf{q},\dot{\mathbf{q}})}{\partial \mathbf{q}} - \boldsymbol{\lambda}\cdot\nabla\boldsymbol{\Phi}(\mathbf{q}) - \mathbf{Q}^{\text{diss}}(\dot{\mathbf{q}}) - \mathbf{Q}(t) = \mathbf{0} \tag{14.424}$$

with $\boldsymbol{\Phi} = \mathbf{0}$ because $\dot{\mathbf{q}}(t)$ are linearly independent. Hence, the following system of equations are obtained from Equation (14.419):

$$\begin{aligned}\frac{d}{dt}\left(\frac{\partial \mathcal{E}^h(\mathbf{q},\dot{\mathbf{q}})}{\partial \dot{\mathbf{q}}}\right) + \frac{\partial \mathcal{E}^h(\mathbf{q},\dot{\mathbf{q}})}{\partial \mathbf{q}} - 2\frac{\partial \mathcal{K}^h(\mathbf{q},\dot{\mathbf{q}})}{\partial \mathbf{q}} - \mathbf{Q}^{\text{diss}}(\dot{\mathbf{q}}) &= \boldsymbol{\lambda}\cdot\nabla\boldsymbol{\Phi}(\mathbf{q}) + \mathbf{Q}(t)\\ \boldsymbol{\Phi}(\mathbf{q}) &= \mathbf{0}\end{aligned} \tag{14.425}$$

or

$$\begin{aligned}\mathbf{M}(\mathbf{q})\ddot{\mathbf{q}} + \bar{\mathbf{Q}}(\mathbf{q},\dot{\mathbf{q}},t) - \boldsymbol{\lambda}(t)\cdot\nabla\boldsymbol{\Phi}(\mathbf{q}) &= \mathbf{0}\\ \boldsymbol{\Phi}(\mathbf{q}) &= \mathbf{0}\end{aligned} \tag{14.426}$$

where $\bar{\mathbf{Q}}(\mathbf{q},\dot{\mathbf{q}},t) : TQ \times \mathcal{T} \to \mathbb{R}^{n_{\text{dof}}}$ denotes the sum of the internal and external force vectors, i.e.,

$$\bar{\mathbf{Q}}(\mathbf{q},\dot{\mathbf{q}},t) := \frac{\partial \mathcal{E}^h(\mathbf{q},\dot{\mathbf{q}})}{\partial \mathbf{q}} + \mathbf{C}\dot{\mathbf{q}} - \mathbf{Q}(t) \tag{14.427}$$

If $\mathbf{M}$ is constant, we have

$$\frac{d}{dt}\left(\frac{\partial \mathcal{E}^h(\mathbf{q},\dot{\mathbf{q}})}{\partial \dot{\mathbf{q}}}\right)+\frac{\partial \mathcal{E}^h(\mathbf{q},\dot{\mathbf{q}})}{\partial \mathbf{q}}-\mathbf{Q}^{\text{diss}}(\dot{\mathbf{q}})=\boldsymbol{\lambda}\cdot\nabla\boldsymbol{\Phi}(\mathbf{q})+\mathbf{Q}(t) \qquad \boldsymbol{\Phi}(\mathbf{q})=\mathbf{0} \tag{14.428}$$

or

$$\mathbf{M}\ddot{\mathbf{q}}+\mathbf{C}\dot{\mathbf{q}}+\nabla\mathcal{U}^h(\mathbf{q})-\boldsymbol{\lambda}\cdot\nabla\boldsymbol{\Phi}(\mathbf{q})=\mathbf{Q}(t) \qquad \boldsymbol{\Phi}(\mathbf{q})=\mathbf{0} \tag{14.429}$$

From Equation (14.429), define the residual vectors as

$${}^{q}\mathbf{R}(\mathbf{q},\dot{\mathbf{q}},\ddot{\mathbf{q}},\boldsymbol{\lambda},t):=\mathbf{M}\ddot{\mathbf{q}}+\mathbf{C}\dot{\mathbf{q}}+\nabla\mathcal{U}^h(\mathbf{q})-\boldsymbol{\lambda}\cdot\nabla\boldsymbol{\Phi}(\mathbf{q})-\mathbf{Q}(t)\equiv\mathbf{0} \tag{14.430}$$

$${}^{\Phi}\mathbf{R}(\mathbf{q}):=\boldsymbol{\Phi}(\mathbf{q})\equiv\mathbf{0} \tag{14.431}$$

with the given initial data, $\mathbf{q}(t_0)=\mathbf{q}_0$ and $\dot{\mathbf{q}}(t_0)=\dot{\mathbf{q}}_0$.

To directly solve the Index 3 DAE system, we now turn attention to the time discretization via the GSSSS unified framework and Approach 2 for reasons that were mentioned earlier. Consider a time interval $\mathcal{T}=[t_0,t_N]$ split into subintervals, i.e., $\mathcal{T}=[t_0,t_N]=\bigcup_{n=0}^{N-1}\left[t_n,t_{n+1}\right]$. Employing the *normalized* time weighted residual methodology for Equation (14.429) about $\ddot{\mathbf{q}}(t)$, $\ddot{\mathbf{q}}(t)$, $\mathbf{q}(t)$, $\boldsymbol{\lambda}(t)$, and $\mathbf{Q}(t)$, we have

$$\mathbf{M}\frac{\int_{t_n}^{t_{n+1}}W\ddot{\mathbf{q}}dt}{\int_{t_n}^{t_{n+1}}Wdt}+\mathbf{C}\frac{\int_{t_n}^{t_{n+1}}W\dot{\mathbf{q}}dt}{\int_{t_n}^{t_{n+1}}Wdt}+\nabla\mathcal{U}^h\left(\frac{\int_{t_n}^{t_{n+1}}W\mathbf{q}dt}{\int_{t_n}^{t_{n+1}}Wdt}\right)-\mathbf{G}^T\left(\frac{\int_{t_n}^{t_{n+1}}W\mathbf{q}dt}{\int_{t_n}^{t_{n+1}}Wdt}\right)\tilde{\boldsymbol{\lambda}}=\frac{\int_{t_n}^{t_{n+1}}W\mathbf{Q}dt}{\int_{t_n}^{t_{n+1}}Wdt} \tag{14.432}$$

where the scalar time weighting function is approximated by

$$W=1+w_1\Gamma+w_2\Gamma^2+w_3\Gamma^3 \tag{14.433}$$

with the corresponding approximations of $\mathbf{q}$, $\dot{\mathbf{q}}$, and $\ddot{\mathbf{q}}$ by an asymptotic series expansion in time with the algorithmic parameters $\Lambda_i\in\mathbb{R}$ $(i=1,2,3,4,5,6)$ as

$$\mathbf{q}=\mathbf{q}_n+\Lambda_1\dot{\mathbf{q}}_n\tau+\Lambda_2\ddot{\mathbf{q}}_n\tau^2+\Lambda_3\frac{\ddot{\mathbf{q}}_{n+1}-\ddot{\mathbf{q}}_n}{\Delta t}\tau^3 \tag{14.434}$$

$$\dot{\mathbf{q}}=\dot{\mathbf{q}}_n+\Lambda_4\ddot{\mathbf{q}}_n\tau+\Lambda_5\frac{\ddot{\mathbf{q}}_{n+1}-\ddot{\mathbf{q}}_n}{\Delta t}\tau^2 \tag{14.435}$$

$$\ddot{\mathbf{q}}=\ddot{\mathbf{q}}_n+\Lambda_6\frac{\ddot{\mathbf{q}}_{n+1}-\ddot{\mathbf{q}}_n}{\Delta t}\tau \tag{14.436}$$

and with the approximations of $\boldsymbol{\lambda}$ and $\mathbf{Q}$ by the Taylor series expansion in time as

$$\mathbf{Q}=\mathbf{Q}_n+\tau\frac{\mathbf{Q}_{n+1}-\mathbf{Q}_n}{\Delta t} \tag{14.437}$$

where $\tau := t - t_n$ and $\Delta t := t_{n+1} - t_n > 0$. Therefore,

$$\mathbf{M}\tilde{\mathbf{a}} + \mathbf{C}\tilde{\mathbf{v}} + \nabla\mathcal{U}^h(\tilde{\mathbf{q}}) - \mathbf{G}^T(\tilde{\mathbf{q}})\tilde{\boldsymbol{\lambda}} = \tilde{\mathbf{Q}} \tag{14.438}$$

with the following algorithmic unknowns which are defined as follows, notice that $\tilde{\boldsymbol{\lambda}}$ is directly approximated in time at $t = t_{n+W_1}$, based on the theory of the time level consistency in the discrete equation of motion:

$$\tilde{\mathbf{q}} := \frac{\int_{t_n}^{t_{n+1}} W\mathbf{q}dt}{\int_{t_n}^{t_{n+1}} Wdt} \tag{14.439}$$

$$= \mathbf{q}_n + W_1\Lambda_1\dot{\mathbf{q}}_n\Delta t + W_2\Lambda_2\ddot{\mathbf{q}}_n\Delta t^2 + W_3\Lambda_3\left(\ddot{\mathbf{q}}_{n+1} - \ddot{\mathbf{q}}_n\right)\Delta t^2$$

$$\tilde{\mathbf{v}} := \frac{\int_{t_n}^{t_{n+1}} W\dot{\mathbf{q}}dt}{\int_{t_n}^{t_{n+1}} Wdt} = \dot{\mathbf{q}}_n + W_1\Lambda_4\ddot{\mathbf{q}}_n\Delta t + W_2\Lambda_5\left(\ddot{\mathbf{q}}_{n+1} - \ddot{\mathbf{q}}_n\right)\Delta t \tag{14.440}$$

$$\tilde{\mathbf{a}} := \frac{\int_{t_n}^{t_{n+1}} W\ddot{\mathbf{q}}dt}{\int_{t_n}^{t_{n+1}} Wdt} = \ddot{\mathbf{q}}_n + W_1\Lambda_6\left(\ddot{\mathbf{q}}_{n+1} - \ddot{\mathbf{q}}_n\right) \tag{14.441}$$

$$\tilde{\boldsymbol{\lambda}} := \boldsymbol{\lambda}(t_{n+W_1}) \tag{14.442}$$

$$\tilde{\mathbf{Q}} := \frac{\int_{t_n}^{t_{n+1}} W\mathbf{Q}dt}{\int_{t_n}^{t_{n+1}} Wdt} = \mathbf{Q}_n + W_1\left(\mathbf{Q}_{n+1} - \mathbf{Q}_n\right) \tag{14.443}$$

The associated design for the updates at the next time level are chosen as

$$\mathbf{q}_{n+1} = \mathbf{q}_n + \lambda_1\dot{\mathbf{q}}_n\Delta t + \lambda_2\ddot{\mathbf{q}}_n\Delta t^2 + \lambda_3\left(\ddot{\mathbf{q}}_{n+1} - \ddot{\mathbf{q}}_n\right)\Delta t^2 \tag{14.444}$$

$$\dot{\mathbf{q}}_{n+1} = \dot{\mathbf{q}}_n + \lambda_4\ddot{\mathbf{q}}_n\Delta t + \lambda_5\left(\ddot{\mathbf{q}}_{n+1} - \ddot{\mathbf{q}}_n\right)\Delta t \tag{14.445}$$

with the algorithmic parameters $\lambda_i \in \mathbb{R}$ $(i = 1, 2, 3, 4, 5)$. For the constraint equation, we have

$$\boldsymbol{\Phi}(\mathbf{q}_n) = \boldsymbol{\Phi}(\mathbf{q}_{n+1}) = \mathbf{0} \tag{14.446}$$

Algorithm 32

Given $\mathbf{q}_n$, $\dot{\mathbf{q}}_n$, and $\ddot{\mathbf{q}}_n$, find $\mathbf{q}_{n+1}$, $\dot{\mathbf{q}}_{n+1}$, and $\ddot{\mathbf{q}}_{n+1}$ from

$$\mathbf{M}\tilde{\mathbf{a}} + \mathbf{C}\tilde{\mathbf{v}} + \nabla\mathcal{U}^h(\tilde{\mathbf{q}}) - \mathbf{G}^T(\tilde{\mathbf{q}})\tilde{\boldsymbol{\lambda}} = \tilde{\mathbf{Q}} \tag{14.447a}$$

$$\boldsymbol{\Phi}(\mathbf{q}_n) = \boldsymbol{\Phi}(\mathbf{q}_{n+1}) = \mathbf{0} \tag{14.447b}$$

where the algorithmic unknowns are given in Equation (14.432)–Equation (14.436). The associated updates are:

$$\mathbf{q}_{n+1} = \mathbf{q}_n + \lambda_1\dot{\mathbf{q}}_n\Delta t + \lambda_2\ddot{\mathbf{q}}_n\Delta t^2 + \lambda_3\Delta\mathbf{a}\Delta t^2 \tag{14.447c}$$

$$\dot{\mathbf{q}}_{n+1} = \dot{\mathbf{q}}_n + \lambda_4\ddot{\mathbf{q}}_n\Delta t + \lambda_5\Delta\mathbf{a}\Delta t \tag{14.447d}$$

$$\ddot{\mathbf{q}}_{n+1} = \ddot{\mathbf{q}}_n + \Delta\mathbf{a} \tag{14.447e}$$

Computational and Implementation Aspects Employ the Newton-Raphson method to iteratively solve the coupled nonlinear effective multibody dynamics equation above:

At the beginning of time step, predict the state vectors (see also Table 14.3),

$$\tilde{\mathbf{a}}_{n+1}^{k} = \Xi_{Pa}^{(1)}\mathbf{q}_n + \Xi_{Pa}^{(2)}\dot{\mathbf{q}}_n + \Xi_{Pa}^{(3)}\ddot{\mathbf{q}}_n \tag{14.448}$$

$$\tilde{\mathbf{v}}_{n+1}^{k} = \Xi_{Pv}^{(1)}\mathbf{q}_n + \Xi_{Pv}^{(2)}\dot{\mathbf{q}}_n + \Xi_{Pv}^{(3)}\ddot{\mathbf{q}}_n \tag{14.449}$$

$$\tilde{\mathbf{q}}_{n+1}^{k} = \Xi_{Pd}^{(1)}\mathbf{q}_n + \Xi_{Pd}^{(2)}\dot{\mathbf{q}}_n + \Xi_{Pd}^{(3)}\ddot{\mathbf{q}}_n \tag{14.450}$$

$$\mathbf{q}_{n+1}^{k} = \Xi_{Pu}^{(1)}\mathbf{q}_n + \Xi_{Pu}^{(2)}\dot{\mathbf{q}}_n + \Xi_{Pu}^{(3)}\ddot{\mathbf{q}}_n \tag{14.451}$$

$$\tilde{\mathbf{Q}}_{n+1}^{k} = (1 - W_1)\mathbf{Q}_n + W_1\mathbf{Q}_{n+1} \quad \text{or} \quad \tilde{\mathbf{Q}}_{n+1}^{k} = \mathbf{Q}(t_{n+W_1}) \tag{14.452}$$

$$\tilde{\boldsymbol{\lambda}}_{n+1}^{k} = \tilde{\boldsymbol{\lambda}}_{n}^{k} \tag{14.453}$$

Start nonlinear iteration. Solve for $\Delta\boldsymbol{\delta}_{n+1}^{k+1}$ and $\Delta\boldsymbol{\lambda}_{n+1}^{k+1}$ from the following:

$$\begin{bmatrix} {}^{q}\mathbf{J}_q & {}^{q}\mathbf{J}_\lambda \\ {}^{\Phi}\mathbf{J}_q & \mathbf{0} \end{bmatrix} \begin{bmatrix} \Delta\boldsymbol{\delta}_{n+1}^{k+1} \\ \Delta\boldsymbol{\lambda}_{n+1}^{k+1} \end{bmatrix} = -\begin{bmatrix} {}^{q}\mathbf{R}(\tilde{\mathbf{q}}_{n+1}^{k}, \tilde{\mathbf{v}}_{n+1}^{k}, \tilde{\mathbf{a}}_{n+1}^{k}, \tilde{\boldsymbol{\lambda}}_{n+1}^{k}) \\ {}^{\Phi}\mathbf{R}(\mathbf{q}_{n+1}^{k}) \end{bmatrix} \tag{14.454}$$

where

$${}^{q}\mathbf{J}_q = \frac{\partial\, {}^{q}\mathbf{R}(\tilde{\mathbf{q}}_{n+1}^{k}, \tilde{\mathbf{v}}_{n+1}^{k}, \tilde{\mathbf{a}}_{n+1}^{k}, \tilde{\boldsymbol{\lambda}}_{n+1}^{k})}{\partial \boldsymbol{q}_{n+1}^{k}} \tag{14.455}$$

$${}^{q}\mathbf{J}_\lambda = \frac{\partial\, {}^{q}\mathbf{R}(\tilde{\mathbf{q}}_{n+1}^{k}, \tilde{\mathbf{v}}_{n+1}^{k}, \tilde{\mathbf{a}}_{n+1}^{k}, \tilde{\boldsymbol{\lambda}}_{n+1}^{k})}{\partial \boldsymbol{\lambda}_{n+1}^{k}} \tag{14.456}$$

$${}^{\Phi}\mathbf{J}_q = \frac{\partial\, {}^{\Phi}\mathbf{R}(\mathbf{q}_{n+1}^{k})}{\partial \boldsymbol{q}_{n+1}^{k}} \tag{14.457}$$

Then correct the unknowns as follows

$$\tilde{\mathbf{a}}_{n+1}^{k+1} = \tilde{\mathbf{a}}_{n+1}^{k} + \Xi_{Ca}\Delta\boldsymbol{\delta}_{n+1}^{k+1} \tag{14.458}$$

$$\tilde{\mathbf{v}}_{n+1}^{k+1} = \tilde{\mathbf{v}}_{n+1}^{k} + \Xi_{Cv}\Delta\boldsymbol{\delta}_{n+1}^{k+1} \tag{14.459}$$

$$\tilde{\mathbf{q}}_{n+1}^{k+1} = \tilde{\mathbf{q}}_{n+1}^{k} + \Xi_{Cd}\Delta\boldsymbol{\delta}_{n+1}^{k+1} \tag{14.460}$$

$$\mathbf{q}_{n+1}^{k+1} = \mathbf{q}_{n+1}^{k} + \Xi_{Cu}\Delta\boldsymbol{\delta}_{n+1}^{k+1} \tag{14.461}$$

$$\boldsymbol{\lambda}_{n+1}^{k+1} = \boldsymbol{\lambda}_{n+1}^{k} + \Delta\boldsymbol{\lambda}_{n+1}^{k+1} \tag{14.462}$$

until the solution converges on both residuals as follows:

$$\max\left|{}^{q}\mathbf{R}(\tilde{\mathbf{q}}_{n+1}^{k+1}, \tilde{\mathbf{v}}_{n+1}^{k+1}, \tilde{\mathbf{a}}_{n+1}^{k+1}, \tilde{\boldsymbol{\lambda}}_{n+1}^{k+1})\right| < \text{tolerance} \tag{14.463}$$

$$\max\left|{}^{\Phi}\mathbf{R}(\mathbf{q}_{n+1}^{k+1})\right| < \text{tolerance} \tag{14.464}$$

Once the solution converges, update the unknowns at the end of the time step as follows.

$$\ddot{\mathbf{q}}_{n+1} = \ddot{\mathbf{q}}_n + (\tilde{\mathbf{a}}_{n+1}^{k+1} - \ddot{\mathbf{q}}_n)/(\Lambda_6 W_1) \tag{14.465}$$

$$\dot{\mathbf{q}}_{n+1} = \dot{\mathbf{q}}_n + \lambda_4 \ddot{\mathbf{q}}_n \Delta t + \lambda_5 (\ddot{\mathbf{q}}_{n+1} - \ddot{\mathbf{q}}_n)\Delta t \tag{14.466}$$

$$\mathbf{q}_{n+1} = \mathbf{q}_n + \lambda_1 \dot{\mathbf{q}}_n \Delta t + \lambda_2 \ddot{\mathbf{q}}_n \Delta t^2 + \lambda_3 (\ddot{\mathbf{q}}_{n+1} - \ddot{\mathbf{q}}_n)\Delta t^2 \tag{14.467}$$

The numerical results via Algorithm 32 in the single-field form are briefly shown for application to the double pendulum problem.

Illustrative Example

Example 14.7 Double Pendulum Problem

Consider the double pendulum problem as shown in Figure 14.24. The generalized coordinates and the generalized velocities are described by

$$\mathbf{q}(t) = \begin{pmatrix} x_1 & y_1 & \theta_1 & x_2 & y_2 & \theta_2 \end{pmatrix}^T \tag{14.468}$$

$$\dot{\mathbf{q}}(t) = \begin{pmatrix} \dot{x}_1 & \dot{y}_1 & \dot{\theta}_1 & \dot{x}_2 & \dot{y}_2 & \dot{\theta}_2 \end{pmatrix}^T \tag{14.469}$$

The total energy of the system is defined as the summation of the kinetic energy and the external potential energy as

$$\mathcal{K} = \frac{1}{2}\dot{\mathbf{q}} \cdot \mathbf{M}\dot{\mathbf{q}} = \frac{1}{2}m_1(\dot{x}_1^2 + \dot{y}_1^2) + \frac{1}{2}m_2(\dot{x}_2^2 + \dot{y}_2^2) \tag{14.470}$$

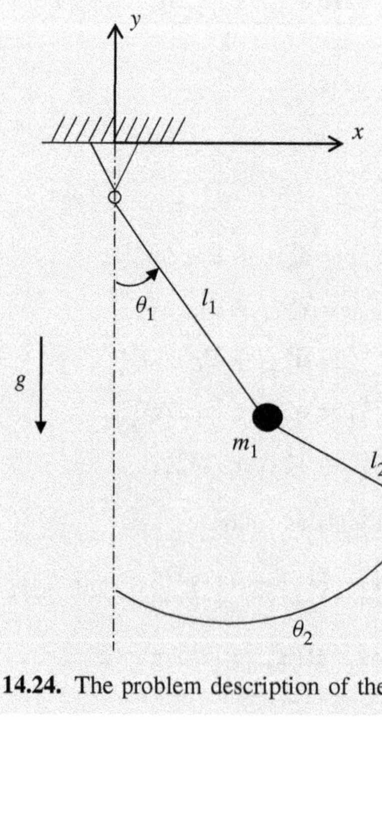

Figure 14.24. The problem description of the double pendulum

$$\mathcal{U} = m_1 g y_1 + m_2 g y_2 \tag{14.471}$$

where the constant mass matrix is given by

$$\mathbf{M} = \begin{bmatrix} m_1 & 0 & 0 & 0 & 0 & 0 \\ 0 & m_1 & 0 & 0 & 0 & 0 \\ 0 & 0 & 0 & 0 & 0 & 0 \\ 0 & 0 & 0 & m_2 & 0 & 0 \\ 0 & 0 & 0 & 0 & m_2 & 0 \\ 0 & 0 & 0 & 0 & 0 & 0 \end{bmatrix} \tag{14.472}$$

For this double pendulum problem, we do not have internal potential energy, i.e., $\mathcal{U}_{\text{int}} = 0$; however, we have the external potential energy defined by Equation (14.471), and the generalized conservative external force vector is defined by

$$\mathbf{Q}^{c} := -\nabla \mathcal{U} = \begin{pmatrix} 0 & -m_1 g & 0 & 0 & -m_2 g & 0 \end{pmatrix}^T \tag{14.473}$$

The constraints imposed on the motion of the system are

$$\mathbf{\Phi} = \begin{pmatrix} x_1 - \ell_1 \sin\theta_1 \\ y_1 + \ell_1 \cos\theta_1 \\ x_2 - x_1 - \ell_2 \sin\theta_2 \\ y_2 - y_1 + \ell_2 \cos\theta_2 \end{pmatrix} \tag{14.474}$$

Therefore, we have

$$\mathbf{G} := \nabla \mathbf{\Phi} = \begin{bmatrix} 1 & 0 & -\ell_1 \cos\theta_1 & 0 & 0 & 0 \\ 0 & 1 & -\ell_1 \sin\theta_1 & 0 & 0 & 0 \\ -1 & 0 & 0 & 1 & 0 & -\ell_2 \cos\theta_2 \\ 0 & -1 & 0 & 0 & 1 & -\ell_2 \sin\theta_2 \end{bmatrix} \tag{14.475}$$

Hence, the Total Energy representation of the equation of motion for this system,

$$\begin{gathered} \frac{d}{dt}\left(\frac{\partial \mathcal{E}(\mathbf{q}, \dot{\mathbf{q}})}{\partial \dot{\mathbf{q}}}\right) + \frac{\partial \mathcal{E}(\mathbf{q}, \dot{\mathbf{q}})}{\partial \mathbf{q}} = \boldsymbol{\lambda} \cdot \nabla \mathbf{\Phi} \\ \mathbf{\Phi} = \mathbf{0} \end{gathered} \tag{14.476}$$

leads to

$$\begin{gathered} \mathbf{M}\ddot{\mathbf{q}} = \mathbf{Q}^{\text{con}} + \boldsymbol{\lambda} \cdot \nabla \mathbf{\Phi} \\ \mathbf{\Phi} = \mathbf{0} \end{gathered} \tag{14.477}$$

For demonstration, this example is analyzed using the three numerically non-dissipative algorithms (Newmark, Classical Version Midpoint Rule [MPR-EPA], and the New Version Midpoint Rule [MPR-MPA]), as well as the controllable numerically dissipative U0/V0 optimal algorithm with $\rho_\infty = 0.25$. The initial conditions and geometric properties are shown below.

$$m_1 = m_2 = 1, \quad l_1 = l_2 = 1$$
$$\mathbf{q}(0) = \begin{pmatrix} 0 & -1 & 0 & 1 & -1 & \frac{\pi}{2} \end{pmatrix}^T$$

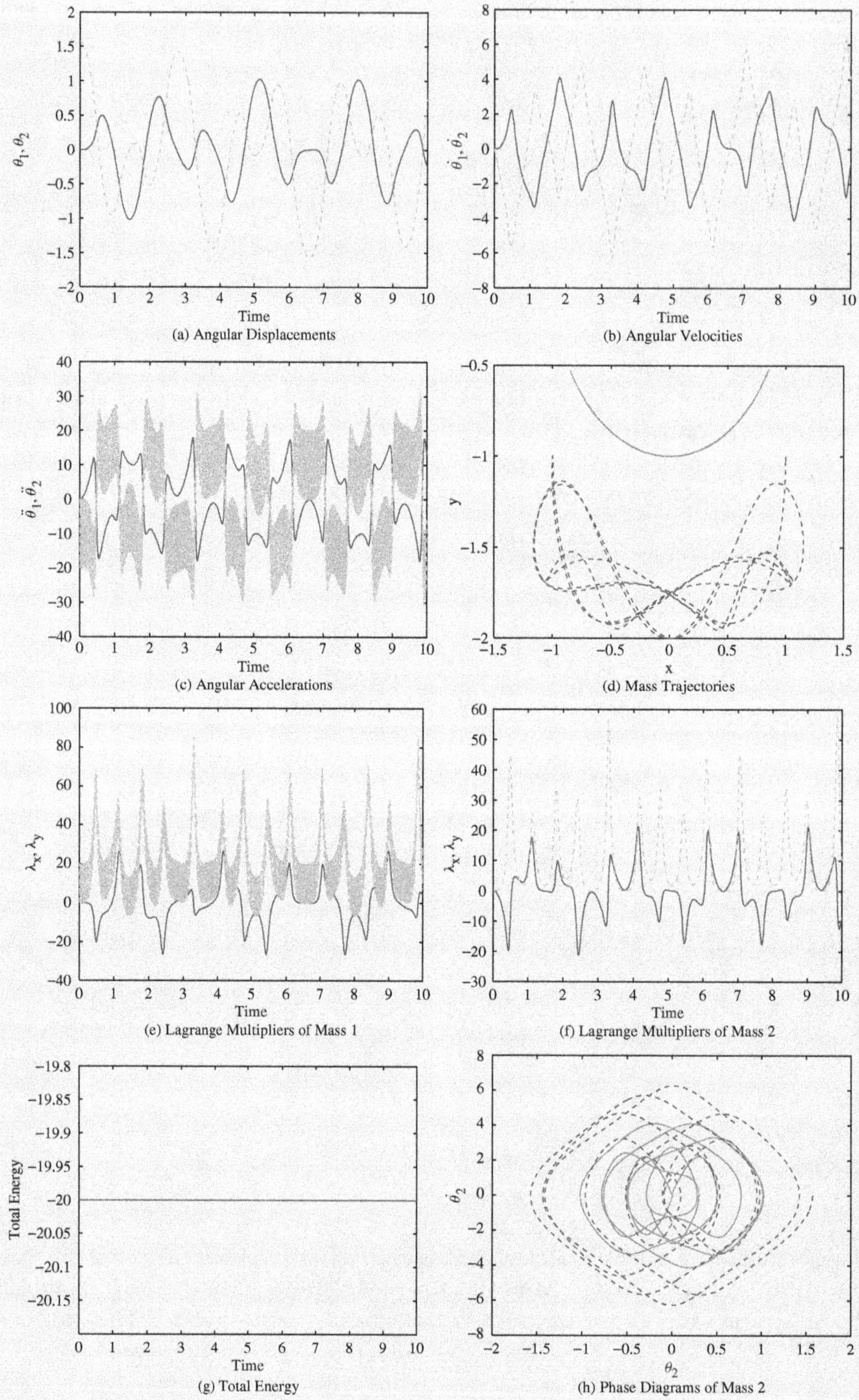

Figure 14.25. Simulation results for the double pendulum problem with $\Delta t = 0.01$ by Algorithm 32: U0/V0(1.00, 1.00, 1.00) (Classical Version Midpoint Rule [MPR-EPA])

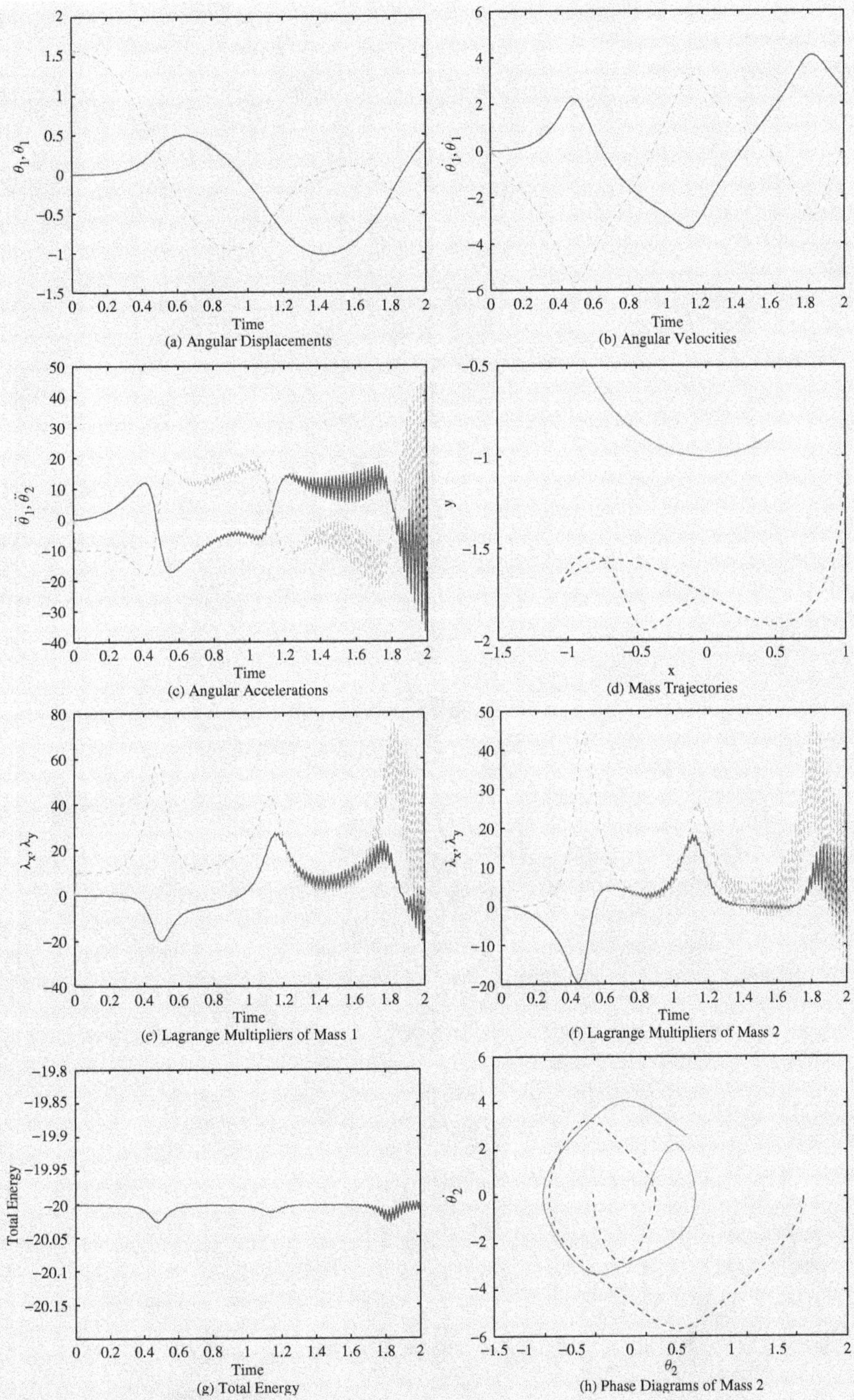

Figure 14.26. Simulation results for the double pendulum problem with $\Delta t = 0.01$ by Algorithm 32: U0(1.00, 1.00, 0.00) (Implicit Newmark Algorithm)

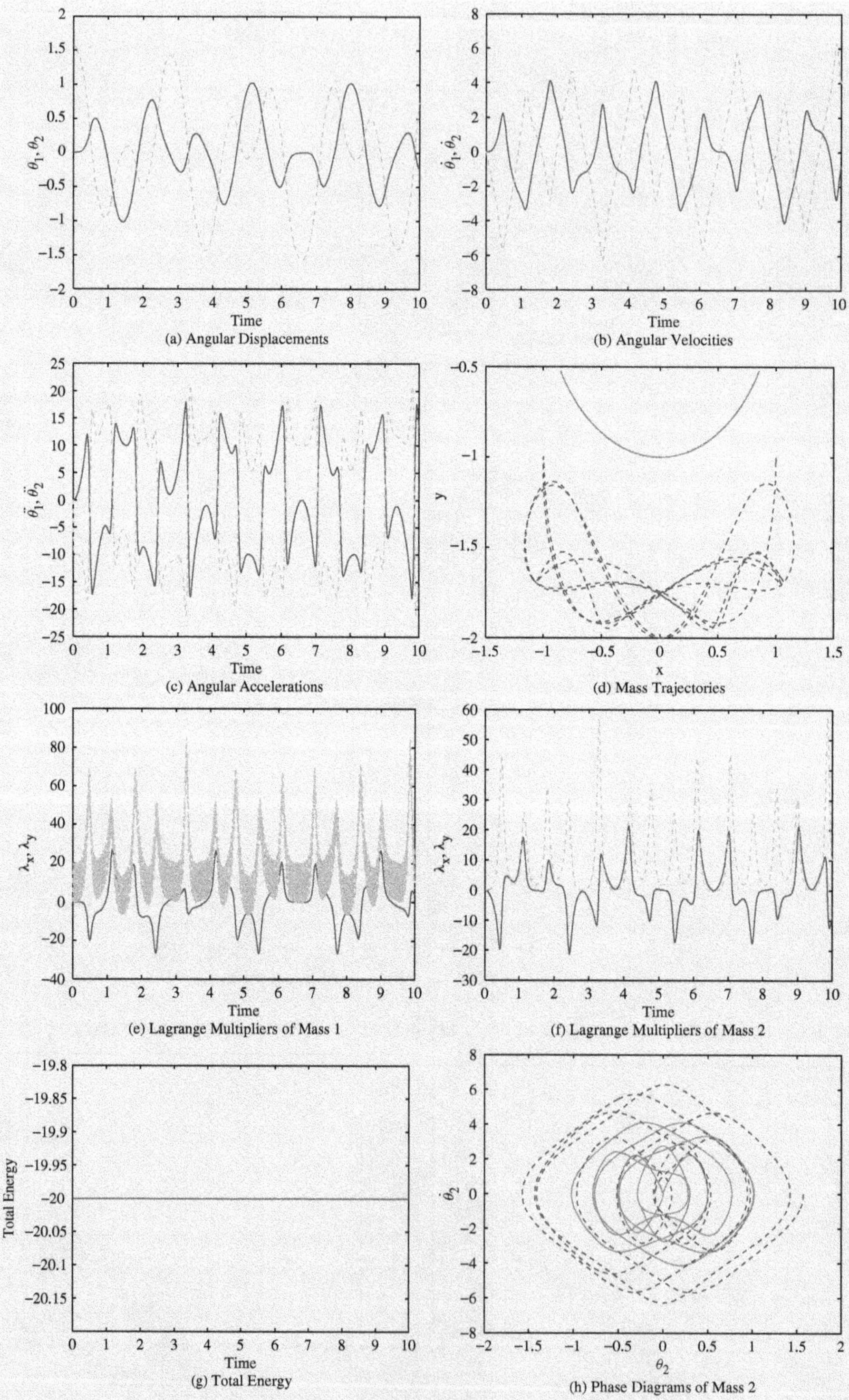

Figure 14.27. Simulation results for the double pendulum problem with $\Delta t = 0.01$ by Algorithm 32: V0(1.00, 1.00, 0.00) (New Version Midpoint Rule [MPR-MPA]

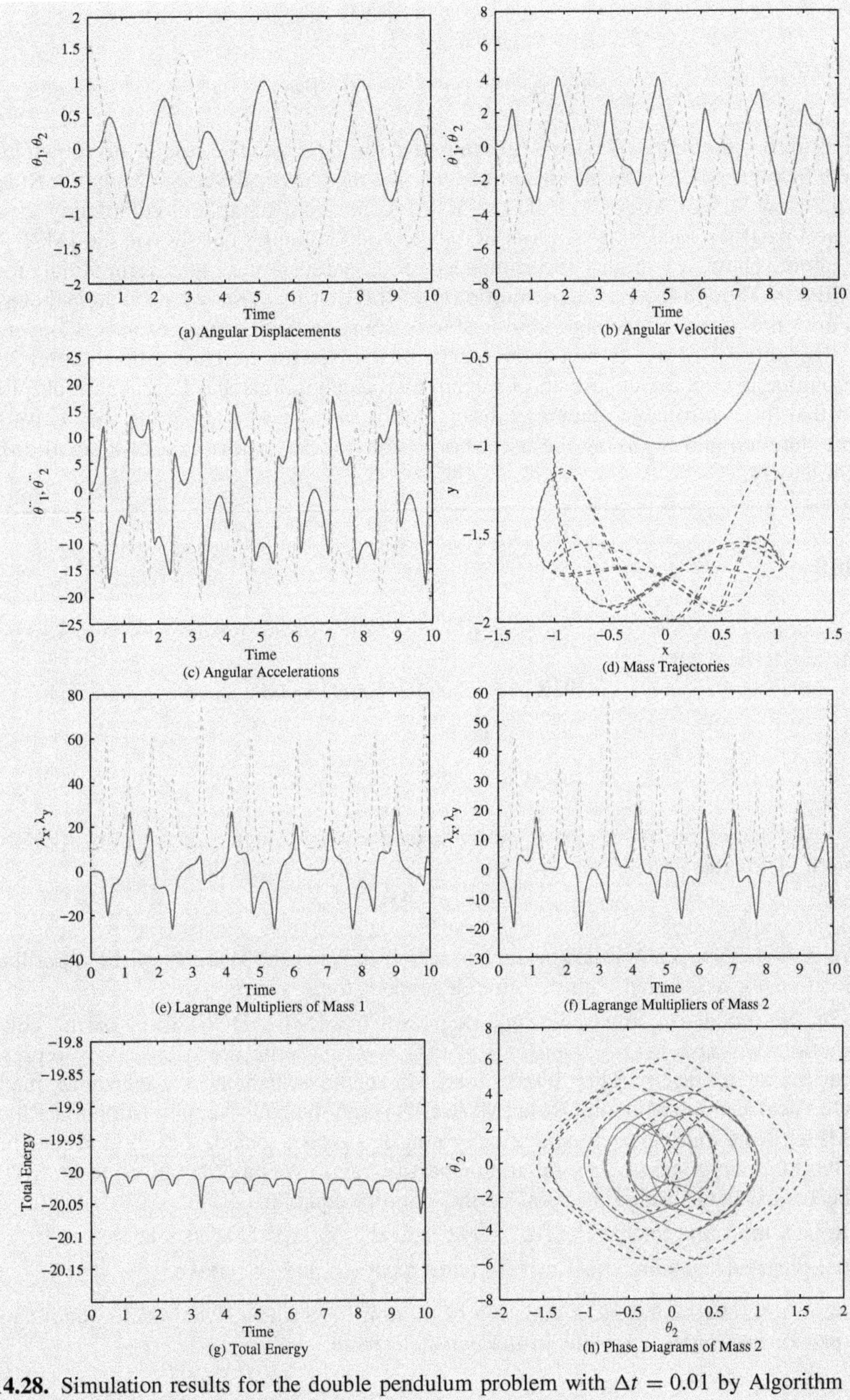

Figure 14.28. Simulation results for the double pendulum problem with $\Delta t = 0.01$ by Algorithm 32: U0(0.25, 1.00, 0.25) (Optimal U0/V0 Algorithm)

$$\dot{\mathbf{q}}(t) = \begin{pmatrix}0 & 0 & 0 & 0 & 0 & 0\end{pmatrix}^T$$

$$\ddot{\mathbf{q}}(t) = \begin{pmatrix}0 & 0 & 0 & 0 & 0 & 0\end{pmatrix}^T$$

Each algorithm was employed using a timestep of $\Delta t = 0.01\text{sec}$. The results are shown in Figures 14.25–14.28 below. For this test case, it can be seen that the Classical Version Midpoint Rule [MPR-EPA] and the New Version Midpoint Rule [MPR-MPA] are both capable of completing the solution upto (and well past) the total time duration of 10 seconds (although not shown, the [MPR-MPA] is in general, more robust, it is highly recommended, and enables a very long term simulation unlike the [MPR-EPA]). Though the Lagrange multiplier of the first mass shows oscillatory behavior, the oscillation does not grow and the algorithm is able to conserve total energy exactly. Conversely, the Newmark algorithm failed due to non-convergence of the Newton iterations shortly after 3 seconds of calculation due to continuous growth of the angular accelerations and Lagrange multipliers. It is also shown that the controllable numerical dissipation is capable of completing the analysis till the desired time duration and resolving the oscillatory behavior, but at the cost of a small drop in the total system energy.

EXERCISES

14.1. The semidiscrete second-order equation of motion for a linear dynamical system can be written in the single-field form as

$$\mathbf{M}\ddot{\mathbf{u}}(t) + \mathbf{C}\dot{\mathbf{u}}(t) + \mathbf{K}\mathbf{u}(t) = \mathbf{f}(t),$$

$$\mathbf{u}(t_0) = \mathbf{u}_0,$$

$$\dot{\mathbf{u}}(t_0) = \dot{\mathbf{u}}_0$$

The modal reduction of the above problem to the single degree of freedom (SDOF) model problem yields the form

$$\mathbf{y}_{n+1} = \mathbf{A}\mathbf{y}_n + \mathbf{L}_n$$

where $\mathbf{A}$ is the 3×3 amplification matrix which governs the stability of the algorithm. $\mathbf{L}_n$ is the load vector associated with the time-dependent force vector.

(a) For the modal decomposed single-degree-of-freedom (SDOF) form of the equation of motion, we have $\ddot{u}(t) + 2\xi\omega\dot{u}(t) + \omega^2 u(t) = f(t)$. Write down the above representation for the three basic building blocks for LMS methods, namely, the Newmark method, the Classical Version Midpoint Rule [MPR-EPA], and the New Version Midpoint Rule [MPR-MPA]. Note that these are numerically non-dissipative schemes. Study the time accuracy, stability, numerical dissipation and dispersion, and overshoot behavior. Use MATLAB as necessary to determine the roots of the stability equation.

(b) Repeat the same for the HHT-α, WBZ, and the optimal U0-V0 schemes.

(c) Set physical damping equal to zero and repeat (**a**) and (**b**) above.

14.2. The semi-discrete second-order equation of motion for a linear dynamical system can be cast as a pair of first-order equations in the two-field form

$$\mathbf{M}\dot{\mathbf{v}}(t) + \mathbf{C}\mathbf{v}(t) + \mathbf{K}\mathbf{u}(t) = f(t),$$

$$\mathbf{v}(t) = \dot{\mathbf{u}}(t),$$
$$\mathbf{u}(t_0) = \mathbf{u}_0,$$
$$\mathbf{v}(t_0) = \mathbf{v}_0$$

(a) For the modal decomposed single-degree-of-freedom (SDOF) form of the equation of motion, we have $\dot{v}(t) + 2\xi\omega v(t) + \omega^2 u(t) = f(t)$ with $\dot{u}(t) = v(t)$. Repeat (**a**) in Exercise 14.1 above. Note that **A** is the 2×2 amplification matrix.

(b) Comment on the issues regarding (**b**) in Exercise 14.1 above, and explain the fundamental aspects for your reasons. Hint: There is no spurious root for this representation.

(c) Set physical damping equal to zero and repeat (**a**) and (**b**) above.

14.3. The Generalized Single-Step Single-Solve (GSSSS) framework encompasses most of the methods developed over the past fifty years or so under the umbrella of LMS methods with a single solve. Because of the practical and powerful nature of this unified framework and representation, the objective is to make sure that the reader is familiar with the overall derivation and the underlying physics of the various formulations. First we will consider the single-field form representation of the equation of motion. That is,

$$\mathbf{M}\ddot{\mathbf{u}}(t) + \mathbf{C}\dot{\mathbf{u}}(t) + \mathbf{K}\mathbf{u}(t) = \mathbf{f}(t),$$
$$\mathbf{u}(t_0) = \mathbf{u}_0,$$
$$\dot{\mathbf{u}}(t_0) = \dot{\mathbf{u}}_0$$

(a) Assume the approximations in Equations (14.20-14.23) and perform the time weighted residual form as in Equation 14.24, and derive Algorithm 8; note that this is the a-form.

(b) Likewise derive the v-form and d-form and show that all representations are identical.

(c) Algorithms 9 and 10 are the same representations as in Algorithm 8, but are specifically classified as U0-family and V0-family. All that is required to switch from one algorithm design to another are simply the values of $\rho_{1\infty}$, $\rho_{2\infty}$, and $\rho_{3\infty}$ (only three parameters). Plugging in these values as shown in Table 14.1, derive the various algorithms in Table 14.1. Compare the results of each via the GSSSS framework to the original algorithm design and convince yourself that they are identical.

(d) Repeat the same as the above via the GSSSS framework for the two-field form. That is, starting from Equation (14.8) with approximations in Equations (14.4)–(14.7), derive Algorithm 7. Prove Remark 14.2.1.

(e) Starting from the discrete Total Energy representation for the two-field form, derive Algorithm 11 and prove Remark 14.2.3.

(f) Starting from the discrete Total Energy representation for the single-field form, derive Algorithms 12 and 13, and prove Remarks 14.2.4–14.2.6.

14.4. Write computer programs for the Generalized Single-Step Single-Solve (GSSSS) family of algorithms in the single-field form for linear/nonlinear dynamic systems to solve the following problems with time step sizes $\Delta t = 0.01, 0.1, 1, 10$. Also, check the accuracy (convergence rate) of the algorithms with various $(\rho_{1\infty}, \rho_{2\infty}, \rho_{3\infty})$. Note that you may wish to initially select the commonly encountered algorithms in Table 14.1. Likewise, you may wish to repeat the same for the two-field form GSSSS unified framework.

(a) **Linear Systems with Damping Neglected** Consider the equation of motion,

$$\mathbf{M}\ddot{\mathbf{q}} + \mathbf{K}\mathbf{q} = \mathbf{F}$$

where the mass matrix, stiffness matrix, and the external load vector are given by

$$\mathbf{M} = \begin{bmatrix} 2 & 0 \\ 0 & 1 \end{bmatrix}, \quad \mathbf{K} = \begin{bmatrix} 6 & -2 \\ -2 & 4 \end{bmatrix}, \quad \text{and } \mathbf{F} = \begin{pmatrix} 0 \\ 10 \end{pmatrix}$$

respectively. Suppose the vectors of the initial displacement and velocity are null. After solving the associated eigenvalue problem and finding the modal solution, the exact solution is obtained as

$$\mathbf{q}(t) = \begin{bmatrix} \sqrt{\frac{1}{3}} & \frac{1}{2}\sqrt{\frac{2}{3}} \\ \sqrt{\frac{1}{3}} & -\sqrt{\frac{2}{3}} \end{bmatrix} \begin{pmatrix} \frac{5}{\sqrt{3}}(1 - \cos(\sqrt{2}t)) \\ 2\sqrt{\frac{2}{3}}(-1 + \cos(\sqrt{5}t)) \end{pmatrix}$$

(b) **Linear SDOF System with Damping Included** Consider the equation of motion in a single-degree-of-freedom system,

$$\ddot{q} + 2\xi\omega\dot{q} + \omega^2 q = 0$$

where $\xi = 0.25$ and $\omega = 10$. The initial conditions are given by $q(t_0) = q_0 = 0$ and $\dot{q}(t_0) = \dot{q}_0 = 1$. The exact solution $q(t)$ is given as

$$q(t) = e^{-\xi\omega t}\left[q_0 \cos(\omega_{\mathrm{d}} t) + \frac{\dot{q}_0 + u_0\xi\omega}{\omega_{\mathrm{d}}} \sin(\omega_{\mathrm{d}} t)\right]$$

with the damped natural frequency of the system defined as $\omega_{\mathrm{d}} := \omega\sqrt{1 - \xi^2}$. Therefore,

$$q(t) = e^{-\frac{\sqrt{10}}{4}t}\left[\frac{4}{\sqrt{150}} \sin \frac{\sqrt{150}}{4}\right]$$

(c) **Linear Systems with Damping Included** Consider the equation of motion,

$$\mathbf{M}\ddot{\mathbf{q}} + \mathbf{C}\dot{\mathbf{q}} + \mathbf{K}\mathbf{q} = \mathbf{0}$$

where the mass matrix, damping matrix, and stiffness matrix are given by

$$\mathbf{M} = \begin{bmatrix} 2 & 0 \\ 0 & 1 \end{bmatrix}, \quad \mathbf{C} = \begin{bmatrix} 0.4 & -0.05 \\ -0.05 & 0.2 \end{bmatrix}, \quad \text{and } \mathbf{K} = \begin{bmatrix} 3 & -1 \\ -1 & 1 \end{bmatrix}$$

respectively. The initial conditions are given by $\mathbf{q}(t_0) = \mathbf{q}_0 = (1, 2)^T$ and $\dot{\mathbf{q}}(t_0) = \dot{\mathbf{q}}_0 = (0, 0)^T$. Denoting the modal matrix of the eigenvalue problem associated with the undamped system by $\mathbf{X}$, the transformed damping matrix becomes

$$\mathbf{X}^T\mathbf{C}\mathbf{X} = \begin{bmatrix} 0.233 & 0.01178 \\ 0.01178 & 0.167 \end{bmatrix}$$

The exact solutions are found to be $\mathbf{q}(t) = (u^1(t), u^2(t))^T$ with

$$q^1(t) = e^{-0.08334t}(1.00028\cos 0.70221t + 0.11076\sin 0.70221t) + e^{-0.11667t}(-0.00028\cos 1.490933t + 0.00395\sin 1.40933t)$$

$$q^2(t) = e^{-0.08334t}(1.99983\cos 0.70221t + 0.24528\sin 0.70221t) + e^{-0.11667t}(-0.00019\cos 1.490933t - 0.00395\sin 1.40933t)$$

(d) **Nonlinear Damped Driven Quadratic Oscillator** The equation of motion for a damped driven quadratic oscillator is represented by

$$\ddot{q} + \alpha\dot{q} + k_1 q(1 + k_2 q^2) = f_1 \sin\Omega t + f_2 \sin\Omega t$$

where k_1 and $k_2 > 0$ are the stiffness constants. The right-hand side of equation above can be written as $f_0 \sin(\Omega t + \gamma)$ where $f_0 = \sqrt{f_1^2 + f_2^2}$ denotes the amplitude, and γ denotes the phase difference which can be found from $\tan\gamma = f_2/f_1$. The damping is linear in $\dot{u}$ with a constant parameter α. Let $\alpha = 10$, $k_1 = 100$, $k_2 = 10$, and $f_1 = f_2 = 10$; and the initial conditions are given by $q(t_0) = q_0 = 0$ and $\dot{q}(t_0) = \dot{q}_0 = 1$.

(e) **Nonlinear Hardening Spring** The equation of motion for a cubic hardening spring is represented by

$$\ddot{q} + k_1 q(1 + k_2 q^2) = 0$$

where k_1 and $k_2 > 0$ are the stiffness constants. Let $k_1 = 100$ and $k_2 = 10$; and the initial conditions are given by $q(t_0) = q_0 = 0$ and $\dot{q}(t_0) = \dot{q}_0 = 1$.

(f) **Nonlinear Softening Spring** The equation of motion for a softening spring is represented by

$$\ddot{q} + k\tanh q = 0$$

where k denotes the stiffness constants. Let $k = 100$; and the initial conditions are given by $q(t_0) = q_0 = 0$ and $\dot{q}(t_0) = \dot{q}_0 = 1$.

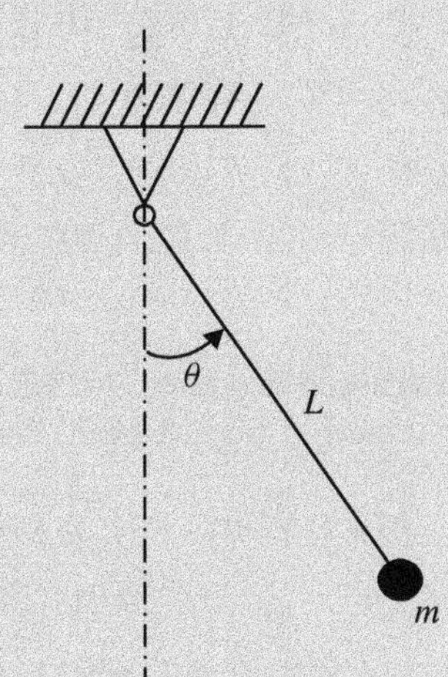

Figure 14.29. The problem description of simple pendulum problem

(g) **Nonlinear Simple Mathematical Pendulum** The equation of motion of a simple mathematical pendulum is represented by

$$\ddot{q} + \frac{g}{L}\sin q = 0$$

where L denotes the length of the pendulum, and g denotes the gravitational acceleration; see Figure 14.29. Notice $q = \theta$. Let $L = 10$ m; and the initial conditions are given by $q(t_0) = q_0 = \pi/4$ rad and $\dot{q}(t_0) = \dot{q}_0 = 0$ rad.

(h) **Nonlinear N-Body Problem** Consider a dynamical system of N atoms of argon in Euclidean three space. The internal potential energy due to the interactions among the atoms is approximated by the Lennard-Jones potential energy function represented by

$$\mathcal{V}(r) = 4\epsilon\left[\left(\frac{\sigma}{r}\right)^{12} - \left(\frac{\sigma}{r}\right)^{6}\right]$$

where $r := \left|\left|\mathbf{q}^j - \mathbf{q}^i\right|\right|$ is the inter-particle distance. Assume the system is free from constraints. The equation of motion is given by

$$\sum_{i=1}^{N} m_i\ddot{\mathbf{q}}^i + \sum_{j\neq i}^{N}\frac{d\mathcal{V}}{dr}\frac{\mathbf{q}^i - \mathbf{q}^j}{r} = \mathbf{0}$$

where

$$\frac{d\mathcal{V}}{dr} = -24\frac{\epsilon}{\sigma}\left[2\left(\frac{\sigma}{r}\right)^{13} - \left(\frac{\sigma}{r}\right)^{7}\right]$$

The internal potential energy becomes minimum at $r = \bar{r} = 2^(1/6)\sigma$: $\mathcal{V}(\bar{r}) = -\epsilon$ and $\mathcal{V}'(\bar{r}) = 0$. Consider $N = 4$ and each atom has a mass of $m_i = 6.69 \times 10^{-26}$ kg ($i = 1, 2, 3, 4$). Assume $\sigma = 3.405 \times 10^{-10}$ m and $\epsilon = 1.654 \times 10^{-21}$ J. The initial conditions are given by

$$\mathbf{q}^1(0) = (\mathring{A}, \mathring{A}, 0), \dot{\mathbf{q}}^1(0) = (0, 0, 0),$$
$$\mathbf{q}^2(0) = (\mathring{A}, \mathring{A}, 0), \dot{\mathbf{q}}^2(0) = (0, 0, 0),$$
$$\mathbf{q}^3(0) = (\mathring{A}, \mathring{A}, 0), \dot{\mathbf{q}}^3(0) = (0, 0, 0),$$
$$\mathbf{q}^4(0) = (\mathring{A}, \mathring{A}, 0), \dot{\mathbf{q}}^4(0) = (0, 0, 0)$$

where $\mathring{A} = 1.0 \times 10^{-10}\,m$.

(i) **Nonlinear Four Mass Constrained System** The equations of motion for the four mass constrained system which is described in Figure 14.30 can be written as

$$\mathbf{M}\ddot{\mathbf{q}} + \nabla\mathcal{U}(\mathbf{q}) + \mathbf{G}^T\boldsymbol{\mu} = 0$$
$$\boldsymbol{\phi}(\mathbf{q}) = 0$$
$$\mathbf{q}(0) = \mathbf{q}_0;\ \ \dot{\mathbf{q}}(0) = \dot{\mathbf{q}}_0$$

where, $\mathbf{q} \in \mathbb{R}^{12} = [u_1\ u_2\ u_3\ u_4]^T$ is the vector of the degrees-of-freedom and u_i $(i = 1, 2, 3, 4)$ is the position vector of ith mass in three dimension, $\mathbf{M} \in \mathbb{R}^{12\times 12}$ is the mass matrix, $\mathbf{p} \in \mathbb{R}^{12}$ is the vector of nonlinear internal force, $\boldsymbol{\mu} \in \mathbb{R}^2$ is the vector of Lagrange multipliers to enforce the constraints, $\phi \in \mathbb{R}^2$, and $\mathbf{G} = \frac{\partial \boldsymbol{\phi}}{\partial \mathbf{q}}$ is the Jacobian of constraints. The parameters describing the problem are given as $m_1 = 1.0$, $m_2 = 3.0$, $m_3 = 2.3$, and $m_4 = 1.7$; $L_1 = 1.0$ and $L_2 = 1.0$; and $K_1 = 100$ and $K_2 = 1,000$. The diagonal mass matrix is given by,

$$M_{ii} = \text{diag}(\mathbf{M}) = (m_1\mathbb{I}, m_2\mathbb{I}, m_3\mathbb{I}, m_4\mathbb{I})$$

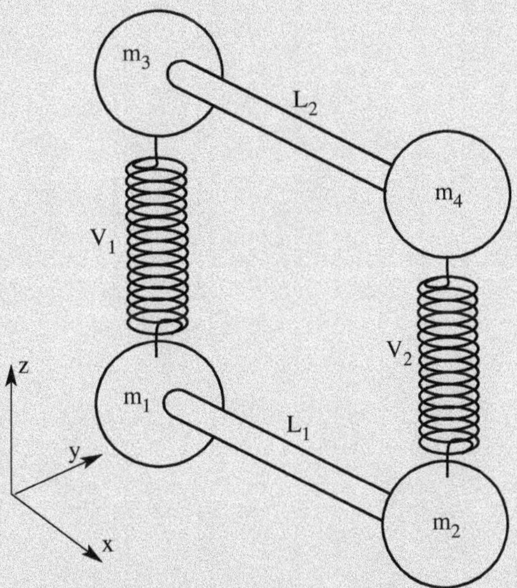

Figure 14.30. Four mass system: Problem description

where, m_i $(i = 1, 2, 3, 4)$ is the mass of each of the four particles and $\mathbb{I} \in \mathbb{R}^{3\times 3}$ is an identity matrix. The vector of non-linear internal forces is given by,

$$\nabla\mathcal{U}(\mathbf{q}) = K_1((u_1 - u_3)^2 - 1)\begin{bmatrix} u_1 - u_3 \\ 0 \\ u_3 - u_1 \\ 0 \end{bmatrix} + K_2((u_2 - u_4)^2 - 1)\begin{bmatrix} u_2 - u_4 \\ 0 \\ u_4 - u_2 \\ 0 \end{bmatrix}$$

The vector of constraints is defined as

$$\boldsymbol{\phi}(\mathbf{q}) = \begin{bmatrix} \sqrt{(u_1 - u_2)^2} - L_1 \\ \sqrt{(u_3 - u_4)^2} - L_2 \end{bmatrix}$$

and consistent initial conditions are taken to be

$$\mathbf{q}_0 = (0,\ 0,\ 0,\ 1,\ 0,\ 0,\ 0,\ 1,\ 0,\ 1,\ 1,\ 0,)^T$$
$$\dot{\mathbf{q}}_0 = (0,\ 0,\ 0,\ 0,\ 0,\ 0,\ 0,\ 0,\ 0,\ 0,\ 0,\ 1.1765)^T$$

REFERENCES

Abraham R, Marsden JE and Ratiu TS 1988 *Manifolds, Tensor Analysis, and Applications*. Springer-Verlag, New York.

Adams RA 1975 *Sobolev Spaces*. Academic Press Inc., New York.

Aharoni D and Bar-Yoseph P 1992 Mixed Finite Element Formulations in the Time Domain for Solution of Dynamic Problems. *Computational Mechanics* **9**(5), 359–374.

Ardema MD 2005 *Analytical Dynamics: Theory and Applications*. Kluwer Academic/Plenum Publishers, New York.

Argyris JH and Scharpf DW 1969 Finite Elements in Time and Space. *The Aeronautical Journal of the Royal Aeronautical Society* **73**, 1041–1044.

Armero F and Romero I 2001a On the Formulation of High-frequency Dissipative Time-stepping Algorithms for Nonlinear Dynamics. Part I: Low-order Methods for Two Model Problems and Nonlinear Elastodynamics. *Computer Methods in Applied Mechanics and Engineering* **190**(20–21), 2603–2649.

Armero F and Romero I 2001b On the Formulation of High-frequency Dissipative Time-stepping Algorithms for Nonlinear Dynamics. Part II: Second-order Methods. *Computer Methods in Applied Mechanics and Engineering* **190**(51–52), 6783–6824.

Arnold VI 1989 *Mathematical Methods of Classical Mechanics*. Springer-Verlag, New York.

Atluri SN 1984 On Constitutive Relations at Finite Strain: Hypo-Elasticity and Elasto-Plasticity with Isotropic or Kinematic Hardening. *Computer Methods in Applied Mechanics and Engineering* **43**(2), 137–171.

Atluri SN and Cazzani A 1995 Rotations in Computational Solid Mechanics. *Archives of Computational Methods in Engineering* **2**(1), 49–138.

Bailey CD 1975a A New Look at Hamilton's Principle. *Foundation of Physics* **5**, 433–451.

Bailey CD 1975b Application of Hamilton's Law of Varying Action. *AIAA Journal* **13**(9), 1154–1157.

Bailey CD 1982 Hamilton's Principle and the Calculus of Variations. *Acta Mechanica* **44**(1–2), 49–57.

Bailey CD 2004 Hamilton and the Law of Varying Action Revisited. *Foundations of Physics* **34**(9), 1385–1406.

Advances in Computational Dynamics of Particles, Materials and Structures, First Edition. Jason Har and Kumar K. Tamma.

Balmer H and Witmer E 1964 Theoretical Experimental Correlation of Large Dynamic and Permanent Deformation of Impulsively Loaded Simple Structures. Technical Report Rep. FDQ-TDR-64-108, Air Force flight Dynamic Laboratory, Wright-Patterson AFB, Ohio, USA.

Barber JR 2010 *Elasticity* 3rd edn. Springer, New York.

Baruh H 1999 *Analytical Dynamics*. McGraw-Hill, New York.

Basar Y and Weichert D 2000 *Nonlinear Continuum Mechanics of Solids: Fundamental Mathematical and Physical Concepts*. Springer, New York.

Bathe KJ 1982 *Finite Element Procedures in Engineering Analysis*. Prentice-Hall, Englewood Cliffs, New Jersey.

Bathe KJ 1996 *Finite Element Procedures in Engineering Analysis*. Prentice-Hall, Englewood Cliffs, New Jersey.

Bathe KJ and Bolourchi S 1980 A Geometric and Material Nonlinear Plate and Shell Element. *Computers & Structures* **11**(1–2), 23–48.

Bathe KJ, Ramm E and Wilson EL 1975 Finite Element Formulations for Large Deformation Dynamic Analysis. *International Journal for Numerical Methods in Engineering* **9**(2), 353–386.

Belytschko T and Hughes TJR 1983 *Computational Methods in Transient Analysis*. North Holland.

Belytschko T, Lin JI and Tsay C 1984 Explicit Algorithms for The Nonlinear Dynamics of Shells. *Computer Methods in Applied Mechanics and Engineering* **42**(2), 225–251.

Belytschko T, Liu WK and Moran B 2000 *Nonlinear Finite Elements for Continua and Structures*. John Wiley & Sons.

Betsch P and Steinmann P 2000 Conservation Properties of a Time FE Method. Part I: Time-Stepping Schemes for N-body Problems. *International Journal for Numerical Methods in Engineering* **49**, 599–638.

Bluman GW and Kumei S 1989 *Symmetries and Differential Equations*. Springer-Verlag.

Bonet J and Wood RD 2008 *Nonlinear Continuum Mechanics for Finite Element Analysis*. Cambridge University Press.

Borri M and Atluri SN 1988 *Time Finite Element Method for the Constrained Dynamics of a Rigid Body* Springer chapter Computational Mechanics '88: -Theory and Applications, p. 41.

Borri M and Bottasso C 1991 Variational Approaches for Dynamics and Time-Finite-Elements: Numerical Studies. *Computational Mechanics* **7**(1), 49–76.

Borri M, Mello F and Atluri SN 1990 Time Finite Element Methods for Large Rotational Dynamics of Multibody Systems. *Computers & Structures* **37**(2), 231–240.

Borri M, Mello F and Atluri SN 1991 Primal and Mixed Forms of Hamilton's Principle for Constrained Rigid Body Systems:-Numerical Studies. *Computational Mechanics* **21**, 205–220.

Bowen RM and Wang CC 1976 *Introduction to Vectors and Tensors*. Plenum Press.

Bubnov IG 1913 Report on the Works of Prof. Timoshenko Which were Awarded the Zhuranskii Prize. *Symposium of the Institute of Communication Engineers, No. 81, All Union Special Planning Office*.

Byers N 1996 *The Life and Times of Emmy Noether. Contributions of Emmy Noether to Particle Physics* Plenum Press New York chapter History of Original Ideas and Basic Discoveries in Particle Physics, pp. 945–976.

Byers N 1999 E. Noether's Discovery of the Deep Connection between Symmetries and Conservation Laws *The Heritage of Emmy Noether, Israel Mathematical Conference Proceedings, Vol. 12*, pp. 67–82, Gelbart Research Institute for Mathematical Sciences, and Emmy Noether Research Institute of Mathematics Ramat-Gan, Israel.

Calin O and Chang DC 2005 *Geometric Mechanics on Riemannian Manifolds: Applications to Partial Differential Equations*. Birkhauser, Boston.

Capecchi D and Drago A 2005 On Lagrange's History of Mechanics. *Meccanica* **40**(1), 19–33.

Carey G and Oden JT 1983 *Finite Elements - A Second Course*. Prentice-Hall, Englewood Cliffs, N.J.

Casey J 1985 Approximate Kinematical Relations in Plasticity. *International Journal of Solids and Structures* **21**(7), 671–682.

Chadwick P 1976 *Continuum Mechanics: Concise Theory and Problems*. John Wiley & Sons, New York.

Chow TL 1995 *Classical Mechanics*. John Wiley & Sons, New York.

Chung J and Hulbert G 1993 A Time Integration Method for Structural Dynamics With Improved Numerical Dissipation: The Generalized α-Method. *Journal of Applied Mechanics* **30**, 371–375.

Chung TJ 2007 *General Continuum Mechanics*. Cambridge University Press, Cambridge.

Ciarlet PG 1978 *The Finite Element Method for Elliptic Problems*. North-Holland Publishing Co., New York.

Ciarlet PG 1988 *Mathematical Elasticity. Volume I: Three-Dimensional Elasticity* vol. I. North-Holland Publishing Co., Amsterdam.

Coleman BD and Noll W 1963 The Thermodynamics of Elastic Materials with Heat Conduction and Viscosity. *Archive for Rational Mechanics and Analysis* **13**(1), 167–178.

Cook RD, Malkus DS, Plesha ME and Witt RJ 2002 *Concepts and Applications of Finite Element Analysis*. Wiley, New York.

Courant R 1974 *Introduction to Calculus and Analysis* vol. 2. Interscience Publishers, New York.

Crandall SH 1956 *Engineering Analysis*. McGraw-Hill, New York.

Crandall SH, Dahl NC and Lardner TJ 1978 *An Introduction to the Mechanics of Solids*. McGraw-Hill, New York.

Dahlquist G 1963 A Special Stability Problem for Linear Multistep Methods. *BIT* **3**, 27–43.

D'Alembert J 1758 *Traité de Dynamique* 2nd edn. Book Company David Street and vis-a-vis the Gate of the Mathurins, Paris.

de Castro A 2000 Exploring a Rheonomic System. *Eur. J. Phys.* **21**(1), 23–26.

Debnath L 2005 *Introduction to Hilbert Spaces with Applications* 3rd edn. Elsevier, Inc., Burlington, MA.

Degas R 1955 *A History of Mechanics*. Central Book Co, New York.

Dienes JK 1979 On the Analysis of Rotation and Stress Rate in Deforming Bodies. *Acta Mechanica* **32**(4), 217–232.

Drew TB 1961 *Handbook of Vector and Polyadic Analysis*. Reinhold Pub. Corp., New York.

(ed. Gear CW) 1971 *Numerical Initial Value Problems in Ordinary Differential Equations*. Prentice-Hall, Englewood Cliffs, NJ.

(ed. Gear JM) 2004 *Mechanics of Materials*. Brooks/Cole-Thomas Learning, Belmont, CA.

Edwards CH 1994 *Advanced Calculus of Several Variables*. Dover Publications, New York.

Fasano A and Marmi S 2006 *Analytical Mechanics: an Introduction*. Oxford University Press, Oxford.

Feynman RP 2006 *The Feynman Lectures on Physics*. Pearson/Addison-Wesley.

Finlayson BA 1972 *The Method of Weighted Residuals and Variational Principles*. Academic Press.

Fish J and Belytschko T 2007 *A First Course in Finite Elements*. John Wiley & Sons.

Flory RJ 1961 Thermodynamic Relations for Highly Elastic Materials. *Transactions of the Faraday Society* **57**, 829–838.

Fosdick RL and Serrin J 1979 On the Impossibility of Linear Cauchy and Piola-Kirchhoff Constitutive Theories for Stress in Solids. *Journal of Elasticity* **9**(1), 83–89.

Fried I 1969 Finite Element Analysis of Time-Dependent Phenomena. *AIAA J* **7**(6), 1170–1173.

Fried I and Malkus DD 1976 Finite Element Mass Matrix Lumping by Numerical Integration Without Convergence Rate Loss. *International Journal of Solids and Structures* **11**(6), 461–466.

Fung YC 1965 *Foundation of Solid Mechanics*. Prentice Hall.

Galerkin BG 1915 Rods and Plates. Series in Some Problems of Elastic Equilibrium of Rods and Plates. *Vestn. Inzh, Tech*.

Gallagher RH 1975 *Finite Element Analysis: Fundamentals*. Englewood Cliffs, N.J.

Gear CW 1967 Numerical Integration of Stiff Ordinary Differential Equations. *Math. Comp*. **21**, 146–156.

Gelfand I and Fomin S 1963 *Calculus of Variations*. Prentice-Hall, Englewood Cliffs, NJ.

Geradin M 1974 A Classification and Discussion of Integration Operator for Transient Structural Response *AIAA 12th Aerospace Science Meeting*, Washington, D. C.

Geradin M, Hogge M and Idelshon S 1983 *Computational Methods for Transient Analysis* Elsevier Amsterdam, North Holland chapter Implicit Finite Element Methods, pp. 417–471.

Ginsberg JH and Genin J 1984 *Dynamics* 2 edn. John Wiley & Sons, New York.

Goldstein H 2002 *Classical Mechanics*. Addison Wesley, San Francisco.

Gonzalez M, Schmidt B and Ortiz M 2010 Energy-Stepping Integrators in Lagrangian Mechanics. *International Journal for Numerical Methods in Engineering* **82**(2), 205–241.

Gonzalez O 1996 Time Integration and Discrete Hamiltonian Systems. *Journal of Nonlinear Science* **6**, 449–467.

Gonzalez O 2000 Exact Energy and Momentum Conserving Algorithms for General Models in Nonlinear Elasticity. *Computer Methods in Applied Mechanics and Engineering* **190**, 1763–1783.

Gonzalez O and Simo JC 1996 On the Stability of Symplectic and Energy-Momentum Algorithms for Non-linear Hamiltonian Systems with Symmetry. *Computer Methods in Applied Mechanics and Engineering* **134**, 197–222.

Gosz MR 2006 *Finite Element Method: Applications in Solids, Structures, and Heat Transfer*. Taylor & Francis, Boca Raton, Florida.

Graham E, Jelenic G and Crisfield MA 2002 A Note on the Equivalence of Two Recent Time-Integration Schemes for N-Body Problems. *Communications in Numerical Methods in Engineering* **18**, 615–620.

Green AE and Naghdi PM 1965 A General Theory of an Elastic-Plastic Continuum. *Archive for Rational Mechanics and Analysis* **18**(4), 251–281.

Greenberg MD 1998 *Advanced Engineering Mathematics* 2nd edn. Prentice Hall, Upper Saddle River, N.J.

Greenberg MD 2001 *Differential Equations and Linear Algebra*. Prentice Hall, Upper Saddle River, N.J.

Greenspan D 1973 *Discrete Models*. Addison-Wesley, Reading, Mass.

Greenspan D 1995 Completely Conservative, Covariant Numerical Methodology. *Computers & Mathematics with Applications* **29**, 37–43.

Greenwood DT 1977 *Classical Dynamics*. Prentice-Hall, Reading, Mass.

Gron O and Hervik S 2007 *Einstein's General Theory of Relativity*. Springer, New York.

Grossman SI 1986 *Multivariable Calculus, Linear Algebra, and Differential Equations*. Academic Press, Orlando.

Gurtin M 1964a Variational Principles for Linear Elastodynamics. *Archive for Rational Mechanics and Analysis* **16**(3), 34–50.

Gurtin M 1964b Variational Principles for Linear Initial-Value Problems. *Quarterly of Applied Mathematics* **22**(3), 252–256.

Gurtin M 1981 *An Introduction to Continuum Mechanics*. Academic Press.

Gurtin ME 1972 *The Linear Theory of Elasticit* vol. VIa/2 of *Mechanics of Solids II* Springer-Verlag Berlin pp. 1–295.

Halmos PR 1958 *Finite-Dimensional Vector Spaces*. Van Nostrand, Princeton, N.J.

Hamel F 1948 *Theoretische Mechanik*. Springer, Berlin.

Hamilton WR 1834a On a General Method in Dynamics. *Philosophical Transactions of the Royal Society of London* **124**, 247–308.

Hamilton WR 1834b On the Application to Dynamics of a General Mathematical Method Previously Applied to Optics. *Report of the British Association for the Advancement of Science* pp. 513–518.

Hamilton WR 1835 Second Essay on a General Method in Dynamics. *Philosophical Transactions of the Royal Society of London* **125**, 95–144.

Hand LH and Finch JD 1998 *Analytical Mechanics*. Cambridge University Press, Cambridge.

Har J 2007 A Unified Stress Update Algorithm for Explicit Transient Shell Dynamics with Combined Isotropic-Kinematic Hardening in Eulerian Rate-Type Phenomenological Finite Elasto-Plasticity models. *Computer Methods in Applied Mechanics and Engineering* **196**(33–34), 3248–3275.

Har J and Tamma KK 2009a Finite Element Formulation and Discretization of Hamilton's Law of Varying Action: Total Energy Mechanics Framework. *Acta Mechanica* **206**(1–2), 105–132.

Har J and Tamma KK 2009b Theorem of Expended Power and Finite Element Formulations: Hamiltonian Mechanics Framework. *AIAA Journal* **47**(11), 2757–2769.

Har J and Tamma KK 2010 On the Numerical Discretization in Space and Time: Part I - Hamilton's Law of Varying Action Involving Lagrangian/Hamiltonian/Total Energy Frameworks. *International Journal of Computational Methods in Engineering Science* **11**, 264–279.

Harris PJF 2008 *Carbon Nanotubes*. McGraw-Hill Companies, AccessScience.

Haupt P 2000 *Continuum Mechanics and Theory of Materials*. Springer, New York.

Henrici PK 1962 *Discrete Variable Methods in Ordinary Differential Equation*. John Wiley & Sons, New York.

Hibbitt H, Marcal P and Rice J 1970 A Finite Element Formulation for Problems of Large Strain and Large Displacements. *International Journal of Solids and Structures* **6**(8), 1069–1086.

Hilber HM, Hughes TJR and Taylor RL 1977 Improved Numerical Dissipation for Time Integration Algorithms in Structural Dynamics. *Earthquake Engineering and Structural Dynamics* **5**(3), 283–292.

Hjelmstad KD 2005 *Fundamentals of Structural Mechanics* 2nd edn. Springer, New York.

Hoitink A, Masuri S, Zhou X and Tamma KK 2008 Algorithms by Design: Part I - On the Hidden Point Collocation Within LMS Methods and Implications for Nonlinear Dynamics Applications. *International Journal for Computational Methods in Engineering Science and Mechanics* **8**, 383–407.

Holzapfel GA 2000 *Nonlinear Solid Mechanics: A Continuum Approach for Engineering*. John Wiley & Sons, New York.

Houbolt JC 1950 A Recurrence Matrix Solution for the Dynamic Response of Elastic Aircraft. *Journal of Aeronautical Science* **17**, 540–550.

Huebner KH 1982 *The Finite Element Method for Engineers*. John Wiley & Sons, New York.

Hughes TJR 1987 *The Finite Element Method, Linear Static and Dynamic Finite Element Analysis*. Prentice-Hall, Englewood Cliffs, New Jersey.

Hurley JF 1981 *Multivariable Calculus*. Saunders College Pub., Philadelphia.

Iijima S, Brabec C, Maiti A and Bernholc J 1996 Structural flexibility of carbon nanotubes. *The Journal of Chemical Physics* **104**(5), 2089–2092.

Irgens F 2008 *Continuum Mechanics*. Springer, Berlin.

Irons BM 1966 Engineering Applications of Numerical Integration in Stiffness Method. *AIAA Journal* **4**(11), 2035–2037.

Irons BM and Ahmad S 1980 *Techniques of Finite Elements*. Ellis Horwood Limited, Chichester, England.

Jaumann G 1911 Geschlossenes System Physikalischer und Chemischer Differentialgesefze. *Sitz Ser. Akad. Wein, (IIa)* **120**, 385–530.

Johnson C 1987 *Numerical Solution of Partial Differential Equations by the Finite Element Method*. Cambridge Univ. Press., Cambridge, UK.

Jose JV and Saletan EJ 1998 *Classical Dynamics: A Contemporary Approach*. Cambridge Univ. Press., Cambridge, UK.

Kane C, Marsden JE, Ortiz M and West M 2000 Variational Integrators and the Newmark Algorithm for Conservative and Dissipative Mechanical Systems. *International Journal for Numerical Methods in Engineering* **49**, 1295–1325.

Key SW and Beisinger ZE 1971 The Transient Dynamic Analysis of Thin Shells by the Finite Element Method *Proceedings of the 3rd Conference on Matrix Methods in Structural Mechanics, Wright-Patterson Air Force Base*.

Khan AS and Huang S 1995 *Continuum Theory of Plasticity*. John Wiley & Sons.

Kirchhoff G 1877 *Vorlesungen uber mathematische Physik* 2nd edn. Teubner, Leipzig.

Kouhia R and Tuomala M 1993 Static and Dynamic Analysis of Space Frames Using Simple Timoshenko Type Elements. *International Journal for Numerical Methods in Engineering* **36**(7), 1189–1221.

Kreyszig E 1978 *An Introductory Functional Analysis with Applications*. John Wiley & Sons, New York.

Kreyszig E 2006 *Advanced Engineering Mathematics* 9th edn. John Wiley & Sons, Hoboken, NJ.

Kuhl D and Crisfiled MA 1999 Energy-Conserving and Decaying Algorithms in Non-Linear Structural Dynamics, *Int. J. Num. Methods in Engr.,* Vol 45, 569–599.

LaBudde RA and Greenspan D 1974 Discrete Mechanics - A General Treatment. *Journal of Computational Physics* **15**(2), 134–167.

Lagrange JL 1788 *Mecanique Analytique*. Vvc Desaint, Paris.

Lanczos C 1970 *The Variational Principles of Mechanics*. University of Toronto Press, Toronto, CA.

Langhaar HL 1962 *Energy Methods in Applied Mechanics*. John Wiley & Sons, New York, USA.

Laursen TA and Meng XN 2001 A New Solution Procedure for Application of Energy-Conserving Algorithms to General Constitutive Models in Nonlinear Elastodynamics. *Computer Methods in Applied Mechanics and Engineering* **190**, 6309–6322.

Lee E and Liu D 1967 Finite Strain Elastic-Plastic Theory Particularly for Plane Wave Analysis. *Journal of Applied Physics* **38**(1), 19–27.

Leimkuhler B and Reich S 2004 *Simulating Hamiltonian Dynamics*. Cambridge University Press, New York, USA.

Leontiev VA 2007 Extension of LMS formulations for L-Stable Optimal Integration Methods with U0-V0 Overshoot Properties in Structural Dynamics: The Level-Symmetric (LS) Integration Methods. *International Journal for Numerical Methods in Engineering* **71**, 1598–1632.

Lindsay RB and Margenau H 1936 *Foundations of Physics*. Ox Bow Press, Woodbridge, Conn.

Love AEH 1927 *Mathematical Theory of Elasticity* 4th edn. Cambridge University Press, Cambridge.

Lubarda VA 2002 *Elastoplasticity Theory*. CRC Press, Boca Raton, FL.

Mach E 1906 *Space and Geometry in the Light of Physiological, Psychological and Physical Inquiry*. The Open Court Publishing Company, Chicago.

Mach E 1907 *The Science of Mechanics* 2nd edn. The Open Court Publishing Company, Chicago.

Maddox IJ 1988 *Elements of Functional Analysis* 2nd edn. Cambridge University Press, Cambridge, Great Britain.

Malvern LE 1969 *Introduction to the Mechanics of a Continuous Medium*. Prentice-Hall, New Jersey.

Marsden JE and Hoffman MJ 1993 *Elementary Classical Analysis*. W.H. Freeman, New York.

Marsden JE and Hughes TJR 1983 *Mathematical Foundations of Elasticity*. Prentice-Hall Inc., Englewood Cliffs, New Jersey.

Marsden JE and Tromba AJ 2003 *Vector Calculus*. W.H. Freeman, New York.

Marsden JE and West M 2001 Discrete Mechanics and Variational Integrators. *Acta Numerica* **10**, 357–514.

Marsden JE, Pekarsky S and Shkoller S 1999 Discrete Euler-Poincare and Lie-Poisson equations: equation and reduction. *Nonlinearlity* **12**, 1647–1662.

Marsden JE, Pekarsky S and Shkoller S 2000 Symmetry reduction of discrete Lagrangian mechanics on Lie groups. *Journal of Geometry and Physics* **36**, 140–151.

Masuri S, Hoitink A, Zhou X and Tamma KK 2009a Algorithms by Design: A New Normalized Time-Weighted Residual Methodology and Design Leading to a Family of Energy–Momentum Conserving Algorithms for Non-linear Structural Dynamics. *International Journal of Numerical Methods in Engineering* **79**, 1094–1146.

Masuri S, Hoitink A, Zhou X and Tamma KK 2009b Algorithms by Design: Part II - A Novel Normalized Time Weighted Residual Methodology and Design of a Family of Symplectic-Momentum Conserving Algorithms for Nonlinear Structural Dynamics. *International Journal for Computational Methods in Engineering Science and Mechanics* **10**, 27–56.

Masuri S, Hoitink A, Zhou X and Tamma KK 2009c Algorithms by Design: Part III - A Novel Normalized Time Weighted Residual Methodology and Design of Optimal Symplectic-Momentum Based Controllable Numerical Dissipative Algorithms for Nonlinear Structural Dynamics. *International Journal for Computational Methods in Engineering Science and Mechanics* **10**, 57–90.

Mathews JH and Fink KK 2004 *Numerical Methods Using Matlab*. Prentice-Hall Inc., Upper Saddle River, N.J.

McComb WD 1999 *Dynamics and Relativity*. Oxford University Press, New York.

McLachlan RI and Quispel G 2001 *Six Lectures on the Geometric Integration of ODEs* Cambridge University Press Cambridge, UK pp. 155–210.

McMeeking RM and Rice JR 1975 Finite-Element Formulations for Problems of Large Elastic-Plastic Deformation. *International Journal of Solids and Structures* **11**(5), 601–616.

McNamara JF 1974 Solution Schemes for Problems of Nonlinear Structural Dynamics. *Transactions of ASME, Journal of Pressure Vessel Technology* **96**, 147–155.

Meirovitch L 1967 *Analytical Methods in Vibrations*. The MacMillan Co., New York.

Meirovitch L 2004 *Methods of Analytical Dynamics*. Dover, New York.

Mikhlin SG 1964 *Variational Methods in Mathematical Physics*. Macmillan, New York.

Mindlin RD 1951 Influence of Rotary Inertia and Shear on Flexural Motions of Elastic Plates. *Journal of Applied Mechanics* **18**, 31–38.

Mondkar DP and Powell GH 1977 Finite Element Analysis of Non-linear Static and Dynamic Response. *International Journal for Numerical Methods in Engineering* **11**(3), 499–520.

Mooney M 1940 A Theory of Large Elastic Deformation. *Journal of Applied Physics* **11**(9), 582–592.

Moore EN 1983 *Theoretical Mechanics*. John Wiley & Sons, New York.

Murnaghan FD 1951 *Finite Deformation of an Elastic Solids*. John Wiley & Sons, New York.

Nakane M and Fraser CG 2002 The Early History of Hamilton-Jacobi Dynamics 1834-1837. *Centaurus* **44**(3–4), 161–227.

Narasimhan MN 1993 *Principles of Continuum Mechanics*. John Wiley & Sons, New York.

Newmark NM 1959 A Method of Computation for Structural Dynamics. *Journal for American Society of Civil Engineers* **1**, 67–94.

Noether E 1918 Invariante Variationsprobleme *Nachr. d. König. Gesellsch. d. Wiss. zu Göttingen*, pp. 235–257, Math-Phys. Klasse.

Oden JT 1972 *Finite Elements of Nonlinear Continua*. McGraw-Hill, New York.

Oden JT 1979 *Applied Functional Analysis: A First Course for Students of Mechanics and Engineering Science*. Prentice-Hall, Englewood Cliffs, NJ.

Oden JT and Reddy JN 1976 *An Introduction to the Mathematical Theory of Finite Elements*. John Wiley & Sons, New York.

Ogden RW 1972 Large Deformation Isotropic Elasticity-on the Correlation of Theory and Experiment for Incompressible Rubberlike Solids. *Proceedings of the Royal Society of London* **A 326**, 565–584.

Ogden RW 1984 *Non-Linear Elastic Deformations*. Ellis Horwood Limited, Chichester.

Osgood WF 1937 *Mechanics*. The Macmillan Company, New York.

Papastavridis JG 1998 *Tensor Calculus and Analytical Dynamics*. CRC Press, Boca Raton, Fl.

Park KC 1975a An Improved Stiffly Stable Method for Direct Integration of Nonlinear Structural Dynamic Equations. *Journal of Applied Mechanics* **42**, 464–470.

Park KC 1975b *Finite Element Analysis of Transient Nonlinear Behavior* ASME New York chapter Evaluating Time Integration Methods for Nonlinear Dynamic Analysis, pp. 35–58.

Park KC 2009 The d'Alembert-Lagrange Principal Equations and Applications to Floating Flexible Systems. *International Journal for Numerical Methods in Engineering* **77**, 1072–1099.

Pars L 1965 *A Treatise on Analytical Dynamics*. John Wiley & Sons, New York.

Pian THH 1995 State-of-the-Art Development of Hybrid/Mixed Finite Element Method. *Finite Elements in Analysis and Design* **21**, 5–20.

Pian THH and Tong P 1969 Basis of Finite Element Methods for Solid Continua. *International Journal for Numerical Methods in Engineering* **1**, 3–281.

Pian THH and Tong P 1972 Finite Element Methods in Continuum Mechanics. *Advances in Applied Mechanics* **12**, 1–58.

Pian THH and Wu CC 2006 *Hybrid and Incompatible Finite Element Methods*. Boca Raton: Chapman & Hall/CRC, Boca Raton.

Prager W 1961 *Introduction to Mechanics of Continua*. Ginn and Co., New York.

Rao JS 1954 *Dynamics of Plates*. Narosa Pub. House, New Delhi, India.

Rayleigh L 1870 On the theory of resonance. *Trans. Roy. Soc*. **161**, 77–118.

Reddy JN 1986 *Applied Functional Analysis and Variational Methods in Engineering*. McGraw-Hill, New York.

Reddy JN 2002 *Energy Principles and Variational Methods in Applied Mechanics* 2nd edn. John Wiley & Sons, New York.

Reddy JN 2006 *An Introduction to the Finite Element Method*. McGraw-Hill, New York.

Reismann H and Lee YC 1969 Forced Motion of Rectangular Plates. *Deuelopments in Thcoreticol and Applied Mechanics* **4**, 3–18.

Reismann H and Pawlik PS 1980 *Elasticity, Theory and Applications*. John Wiley & Sons, New York.

Reissner E 1945 The Effect of Transverse Shear Deformation on the Bending of Elastic Plates. *Journal of Applied Mechanics* **12**, 69–77.

Reissner E 1950 On a Variational Theorem in Elasticity. *Journal of Mathematics and Physics* **29**(2), 90–95.

Rektorys K 1980 *Variational Methods in Mathematics, Science and Engineering* 2 edn. D. Reidel, Dordrecht, Holland, Boston.

Richards TH 1977 *Energy Methods in Stress Analysis: with an Introduction to Finite Element Techniques*. Ellis-Horwood Limited, Chichester.

Ritz W 1909 Über eine neue Methode zur Lösung gewisser Variationsproblem der mathematicshen Physik. *Journal für die reine und angewandte Mathematik* **135**, 1–61.

Rivlin RS 1948 Some Applications of Elasticity Theory to Rubber Engineering. *Proc. 2nd Tech. Conf., Cambridge*.

Rock T and Hinton E 1974 Free Vibration and Transient Response of Thick and Thin Plates Using the Finite Element Method. *Earthquake Engineering and Structural Dynamics* **3**, 51–63.

Rosenberg R 1977 *Analytical Dynamics of Discrete Systems*. Plennum.

Saxe K 2002 *Beginning Functional Analysis*. Springer, New York.

Sha D, Zhou X and Tamma KK 2003 Time Discretized Operators. Part 2: Towards the Theoretical Design of a New Generation of a Generalized Family of Unconditionally Stable Implicit and Explicit Representations of Arbitrary Order for Computational Dynamics. *Computer Methods in Applied Mechanics and Engineering* **192**, 291–329.

Shao HP 1987 *The Studying on the Direct Time Integration Algorithms for Structural Dynamics Response* Master's thesis Zheng Jiang University.

Shao HP and Cai CW 1988 The Direct Integration Three-Parameters Optimal Schemes for Structural Dynamics (in English) *Proceeding of the International Conference: Machine Dynamics and Engineering Applications*, pp. C16–C20. Xi'an Jiaotong University Press (see also Chinese Journal of Applied Mechanics (in Chinese), Volume 5, No. 4, Dec. 1988).

Shibberu Y 1992 *Discrete-Time Hamiltonian Dynamics* PhD Thesis University of Texas at Arlington.

Shibberu Y 1994 Time-Discretization of Hamiltonian Dynamics. *Computers and Mathematics with Applications* **28**, 123–145.

Simkins TE 1981 Finite Elements for Initial Value Problems in Dynamics. *AIAA Journal* **19**(10), 1357–1362.

Simmons JG 1994 *A Brief on Tensor Analysis* 2nd edn. Springer-Verlag, New York.

Simo J and Tarnow N 1992 The Discrete Energy-Momentum Method. Conserving Algorithms for Nonlinear Elastodynamics. *Journal of Applied Mathematics and Physics* **43**, 757–792.

Simo JC 1992 Algorithm for Static and Dynamic Multiplicative Plasticity That Preserve the Classical return Mapping Schemes of the Infinitesimal Theory. *Computer Methods in Applied Mechanics and Engineering* **99**, 61–112.

Simo JC and Gonzalez O 1993 Assessment of Energy-Momentum and Symplectic Schemes for Stiff Dynamical Systems. *In: ASME Winter Ann. Meeting*.

Simo JC and Gonzalez O 1994 Recent Results on the Numerical Integration of Infinite-Dimensional Hamiltonian Systems. *In: Recent Developments in Finite Element Analysis, CIMNE* pp. 255–271.

Simo JC and Hughes TJR 1998 *Computational Inelasticity*. Springer, New York.

Simo JC and Laursen TA 1992 An Augmented Lagrangian Treatment of Contact Problem Friction Problems. *Computers & Structures* **42**, 97–116.

Simo JC and Pister KS 1984 Remarks on Rate Constitutive Equations for Finite Deformation Problems: Computational Implications. *Computer Methods in Applied Mechanics and Engineering* **46**, 201–215.

Simo JC and Taylor RL 1991 Quasi-Incompressible Finite Elasticity in Principle Stretches, Continuum Basis and Numerical Algorithms. *International Journal for Numerical Methods in Engineering* **85**, 273.

Simo JC, Marsden JE and Krishnaprasad PS 1988 The Hamiltonian Structure of Nonlinear Elasticity: The Material and Convective Representations of Solids, Rods, and Plates. *Archive for Rational Mechanics and Analysis* **104**(2), 125–183.

Simo JC, Tarnow N and Wong KK 1992 Exact Energy-Momentum Conserving Algorithms and Symplectic Schemes for Nonlinear Dynamics. *Computer Methods in Applied Mechanics and Engineering* **100**(1), 63–116.

Smith DR and Smith CVJ 1974 When is Hamilton's principle an extreme principle. *AIAA Journal* **12**, 1573–1576.

Sokolnikoff IS 1951 *Tensor Analysis: Theory and Applications*. John Wiley & Sons, New York.

Spencer AJ 2004 *Continuum Mechanics*. Dover, Mineola, N.Y.

Strang G 1988 *Linear Algebra and Its Applications*. Harcourt, Brace, Jovanovich, San Diego.

Strang G and Fix GJ 1973 *An Analysis of the Finite Element Method*. Prentice-Hall, Englewood Cliffs, N.J.

Suetake Y, Iura M and Atluri SN 2003 Variational Formulation and Symmetric Tangent Operator for Shells with Finite Rotation Field. *CMES:Computer Modeling in Engineering and Sciences* **4**, 329–336.

Synge JL 1960 *Encyclopedia of Physics* (Flügge, S. Ed.) vol. III Springer Berlin/Gettingen/Heidelberg chapter Classical Dynamics, pp. 1–225.

Taig IC 1961 Structural Analysis by the Matrix Displacement Method. Technical Report No. S017, Engl. Electric Aviation.

Tamma KK 2012 *iINTEGRATORS:* Isochronous Integrators and the Next Generation Simulation Toolkit for First/Second Order Transient Systems, *10th World Congress on Computational Mechanics (WCCM – 10), July 8–13, 2012, Sao Paulo, Brazil (Semi-Plenary Lecture).*

Tamma KK and Namburu RR 1988 A New Finite Element Based Lax-Wendroff/Taylor-Galerkin Methodology for Computational Dynamics. *Computer Methods in Applied Mechanics and Engineering* **71**, 137–150.

Tamma KK and Namburu RR 1990 Applicability and Evaluation of An Implicit Self-Starting Unconditionally Stable Methodology for Dynamics of Structures. *Computers & Structures* **34**, 835–842.

Tamma KK and Namburu RR 1997 Computational Approaches With Applications to Non-classical and Classical Thermomechanical Problems. *Applied Mechanics Reviews* **50**, 514.

Tamma KK, Har J, Zhou X, Shimada M and Hoitink A 2011 An Overview and Recent Advances in Vector and Scalar Formalisms: Space/Time Discretizations in Computational Dynamics: A Unified Approach. *Arch Comput Methods Eng*, Springer, **18**(2), 119–283.

Tamma KK, Sha D and Zhou X 2003 Time Discretized Operators. Part 1: Towards the Theoretical Design of a New Generation of a Generalized Family of Unconditionally Stable Implicit and Explicit Representations of Arbitrary Order for Computational Dynamics. *Computer Methods in Applied Mechanics and Engineering* **192**, 257–290.

Tamma KK, Zhou X and Kanapady R 2002 The Time Dimension and a Unified Mathematical Framework for First-Order Parabolic Systems. *Numerical Heat Transfer: Part B: Fundamentals* **50**, 239–262.

Tamma KK, Zhou X and Sha D 2000 The Time Dimension: A Theory Towards the Evolution, Classification, Characterization and Design of Computational Algorithms for Transient/Dynamic Applications. *Archives of Computational Methods in Engineering* **7**(2), 67–290.

Tamma KK, Zhou X and Sha D 2001 A Theory of Development and Design of Generalized Integration Operators for Computational Structural Dynamics. *International Journal of Numerical Methods in Engineering* **50**, 1619–1664.

Taylor RL 1972 On Completeness of Shape Functions for Finite Element Analysis. *International Journal of Numerical Methods in Engineering* **4**, 17–22.

Teodorescu PP 2007 *Mechanical Systems, Classical Models*. Springer, Dordrecht.

Timoshenko S and Goodier JN 1951 *Theory of Elasticity* 2nd edn. McGraw-Hill, New York.

Timoshenko S and Woinowsky-Krieger S 1959 *Theory of Plates and Shells* 2nd edn. McGraw-Hill, New York.

Timoshenko S, Young DH and Weaver W 1974 *Vibration Problems in Engineering* 4th edn. John Wiley & Sons, New York.

Troutman JL 1983 *Variational Calculus with Elementary Convexity*. Springer-Verlag, New York.

Troutman JL 1996 *Variational Calculus and Optimal Control: Optimization with Elementary Convexity* 2nd edn. Springer-Verlag, New York.

Truesdell C 1955 Hypoelasticity. *Journal of Rational Mechanics and Analysis* **4**, 83–133.

Truesdell C 1991 *A First Course in Rational Continuum Mechanics* second edition edn. Academic Press, Boston.

Truesdell C and Noll W 2004 *The Non-Linear Field Theory of Mechanics* 3rd edn. Springer-Verlag, Berlin.

Truesdell C and Toupin RA 1960 *The Classical Field Theories* Handbuch der Physik, III/1. Springer-Verlag, Berlin.

Ugural AC 2008 *Mechanics of Materials*. John Wiley & Sons, Hoboken, N.J.

Ugural AC and Fenster SK 2003 *Advanced Strength and Applied Elasticity*. Prentice Hall, Upper Saddle River, N.J.

Veselov AP 1988 Integrable Discrete-Time Systems and Difference Operators. *Functional Analysis and Its Applications* **22**, 83–93. Translated from Funktsional'nyi Analiz i Ego Prilozheniya, vol. 22, pp. 1-13, 1988.

Washizu K 1982 *Variational Methods in Elasticity and Plasticity* 3 edn. Pergamon Press, Oxford.

Weeks G 1972 Temporal Operators for Nonlinear Structural Dynamics Problems. *Journal of Engineering Mechanics Division, ASCE* **98**(5), 1087–1104.

Wempner G and Talaslidis D 2003 *Mechanics of Solids and Shells: Theories and Approximations*. CRC Press, Boca Raton.

Wendlandta JM and Marsden JE 1997 Mechanical Integrators Derived from a Discrete Variational Principle. *Physica D: Nonlinear Phenomena* **106**, 223–246.

Whittaker ET 1942 *The Calculus of Observations; a Treatise on Numerical Mathematics*. Blackie & Son Limited, London.

Wilson EL 1968 *A Computer Program for Dynamic Stress Analysis of Underground Structures*. SESM, University of California, Berkeley.

Wood WL 1984 A Unified Set of Single-Step Algorithms, Part 2. *International Journal for Numerical Methods in Engineering* **20**, 2303–2309.

Wood WL 1990 *Practical Time Stepping Schemes*. Clarendon Press, Oxford.

Wood WL, Bossak M and Zienkiewicz OC 1980 An Alpha Modification of Newmark's Method. *International Journal for Numerical Methods in Engineering* **15**, 1562–1566.

Wu RWH and Witmer EA 1971 Finite-Element Analysis of Large Elastic-Plastic Transient Deformations of Simple Structures. *AIAA Journal* **9**(9), 1719–1724.

Yang TY and Saigal S 1984 A Simple Element for Static and Dynamic Response of Beams with Material and Geometric Nonlinearities. *International Journal for Numerical Methods in Engineering* **20**(5), 851–867.

Yourgrau W and Mandelstam S 1968 *Variational Methods in Elasticity and Plasticity*. W. B. Saunders Co., Philadelphia.

Zhou X and Tamma KK 2004a A New Unified Theory Underlying Time Dependent Linear First-Order Systems: A Prelude to Algorithms by Design. *International Journal for Numerical Methods in Engineering* **60**, 1699–1740.

Zhou X and Tamma KK 2004c Design, Analysis, and Synthesis of Generalized Single Step Single Solve and Optimal Algorithms for Structural Dynamics. *International Journal for Numerical Methods in Engineering* **59**, 597–668.

Zhou X and Tamma KK 2006 Algorithms by Design with Illustrations to Solid and Structural Mechanics/Dynamics. *International Journal for Numerical Methods in Engineering* **66**, 1738–1790.

Zhou X, Sha D and Tamma KK 2004 A Novel Nonlinearly Explicit Second-Order Accurate L-Stable Methodology for Finite Deformation Hypoelastic/Hypoelasto-Plastic Structural Dynamics Problems with Total/Updated Lagrangian Formulations. *International Journal for Numerical Methods in Engineering* **59**, 795–823.

Zhou X, Sha D and Tamma KK 2008 A Robust Consistent Configuration Framework and Formulation for 3D Finite Strain Dynamic Impact Problems. *Computer Methods in Applied Mechanics and Engineering* **197**(51-52), 4571–4590.

Zhou X, Tamma KK and Sha D 2001 Linear Multi-Step and Optimal Dissipative Single-Step Algorithms for Structural Dynamics In *Computational Fluid and Solid Mechanics* (ed. Bathe KJ), vol. I, pp. 571–574. Elsevier. First MIT Conference on Computational Fluid and Solid Mechanics.

Zhou X, Tamma KK and Sha D 2005 Design Spaces, Measures and Metrics for Evaluating Quality of Time Operators and Consequences Leading to Improved Algorithms By Design - Illustration to Structural Dynamics. *International Journal for Numerical Methods in Engineering* **64**, 1841–1870.

Zhou X, Tamma KK, Kanapady R and Sha D 2000 The Time Dimension: New and Recent Advances and a Unified Framework Towards Design of Time Discretized Operators for Structural Dynamics *European Congress on Computational Methods in Applied Science and Engineering (ECCOMAS) 2000*, pp. 11–14, Barcelona, Spain.

Ziegler E 1991 *Mechanics of Solids and Fluids*. Springer-Verlag, New York.

Zienkiewicz OC and Cheung YK 1964 The Finite Element Method for Analysis of Elastic Isotropic and Orthotropic Slabs. *Proc. Inst. Civ. Eng*. **28**, 471–488.

Zienkiewicz OC and Taylor RL 2006 *The Finite Element Method for Solid and Structural Mechanics* 6th edn. Elsevier Butterworth-Heinemann, Oxford.

Zienkiewicz OC, Taylor RL and Zhu JZ 2005 *The Finite Element Method: Its Basis and Fundamentals* 6th edn. Elsevier Butterworth-Heinemann, Oxford.

Zienkiewicz OC, Wood WL, Hine NW and Taylor RL 1984 A Unified Set of Single-Step Algorithms, Part 1: General Formulations and Applications. *International Journal for Numerical Methods in Engineering* **20**, 1529–1552.

Zill DG and Cullen MR 2006 *Advanced Engineering Mathematics* 3rd edn. Jones and Bartlett Publishers, Sudbury, Mass.

INDEX

A

Almansi strain tensor, 185
Area coordinate system, 310
Augmented Lagrangian, 134
Augmented Total Energy, 159
Autonomous Hamiltonian systems, 99
Autonomous Lagrangian system, 87
Axial bar element, 301

B

Backward Euler method, 501
Banach space, 24
Bilinear quadratic functional, 238
Bubnov-Galerkin weighted residual form, 223
Bubnov-Galerkin weighted residual method, 244

C

Calculus of variations, 125
  variational principle
    Brachistochrone, 128
    Hamilton's principle, 129
    shortest distance, 127
Canonical coordinates, 92
Cartesian metric tensor, 49
Cartesian product, 16, 18
Cartesian tensor, 49
Cauchy deformation tensor, 186
Cauchy sequence, 23
Cauchy stress tensor, 194
Cauchy traction vector, 194
Cauchy's equation of equilibrium, 202
Cauchy's equation of motion, 202
Cauchy's Equations of Motion, 432
Cauchy's first law of motion, *see* Cauchy's equation of motion, 202
Cauchy's fundamental theorem, *see* Cauchy's law, 201
Cauchy's law, 201
Cauchy's second equation of motion, 203
Cauchy's second law of motion, 203
Cayley-Hamilton equation, 48
Central difference method, 505
Christoffel symbols, 51
Clasius-Duhem inequality, 205
Classical Midpoint rule method
  Second-order differential system, 509
Classical time weighted residual approach, 608, 609
Classical time weighted residual methodology, 579
Clausius-Planck inequality, 205
Configuration space
  continuous-body dynamical system, 176
  continuous-body systems, 268
  N-body systems, 72
Conservative force, 60
Conservative system, 145

Advances in Computational Dynamics of Particles, Materials and Structures, First Edition. Jason Har and Kumar K. Tamma.

Constraints, 61
holonomic constraints
rheonomic, 62
scleronomic, 63, 145
Continuity equation, 200
Contravariant vector with covariant base vectors, 178
Convected coordinate system, 177
Convected Riemannian metric tensor, 192
Convective coordinate system, 179
Coordinate transformation, 61
Cotangent bundle, 92
Cotter-Rivlin stress rate, 219
Current body, 177
Current configuration, 176

D

D'Alembert's principle, 69
Dead loads, 221
Deformation gradient tensor, 181
determinant of, 182
inverse of, 182
transpose of, 182
transpose of the inverse of, 183
Deformed configuration, 176
Descriptive scalar function, 78
Direct variational methods, 237
Bubnov-Galerkin weighted residual method, 244
modified Bubnov-Galerkin weighted-residual method, 247
Rayleigh-Ritz method, 237
Directional derivative, *see* Gâteaux derivative, 222
Discrete Euler-Lagrange equations, 527
Discrete Hamiltonian map, 531
Discrete Legendre transformation, 530
Discrete Noether's theorem, 534
Discrete Total Energy framework
Discrete total angular momentum conservation, 574
Discrete Total Energy conservation, 575
Discrete total linear momentum conservation, 572
Semidiscretized equation of motion, 569
Disjoint subsets, 177
Displacements
Eulerian Description, 181
Lagrangian Description, 179
Divergence theorem, *see* Gauss's theorem, 198
Dot cancellation, 77

E

Elastic
Cauchy elastic material, 206
hyperelastic material, 207
hypoelastic material, 213
Elastic axial bar, 271, 378, 407, 438, 464
Elastodynamics, 175
Energy dissipative and momentum conserving algorithm, 593
Energy-momentum conserving scheme
Continuous-body dynamical systems, 543
N-body dynamical systems, 537
Energy-momentum framework, 536, 588, 601, 622
Entropy production inequality, *see* Clasius-Duhem inequality, 205
Euler-Bernoulli beam, 276, 387, 413, 442, 469
Euler-Bernoulli beam element, 329
Euler-Lagrange equations of motion, 125
Eulerian description, 179
Eulerian rate of deformation tensor, 186
Eulerian stress measure
Cauchy stress tensor, 194
Kirchhoff stress tensor, 195
External power, *see* Power input, 203

F

Finger deformation tensor, 186
Finite Element Library, 301
First law of thermodynamics, 204
First Piola-Kirchhoff stress tensor, 196
First Piola-Kirchhoff traction vector, 196
Force function, 77
Forward Euler method, 501
Fréchet derivative, 28, 238
Fréchet differentiable, 110, 238
Function Spaces, 34
Banach space, 36
Hilbert space, 38
Lebesgue space, 36
metric space, 36
normed space, 35
Sobolev space, 37
Functions, 17
domain of, 17
Fréchet differentiable, 28
range of, 17
scalar-valued, 28
tensor-valued, 49
vector-valued, 30
multivariables, 31
one real, 30

G

Gâteaux derivative, 30, 221, 222
Gâteaux operator, 222
Gâteaux variation, 111, 126
Gauss's theorem, 33, 198

Generalized coordinates, 72
Generalized Single Step Single Solve, 508, 553
Generalized-α method, 524
Green deformation tensor, 184
Green's theorem, 32
Green-Lagrange strain tensor, 184
Green-McInnis-Naghdi stress rate, *see* Green-Naghdi stress rate, 220
Green-Naghdi stress rate, 220

H

Hamel's central principle, 123
Hamilton's equations of motion, 96
Hamilton's integral, 124
Hamilton's law of varying, 133
Hamilton's law of varying action, 124, 165, 367
  Hamilton's equations of motion, 137
  Lagrange's equations of motion, 136
  Newton's equations of motion, 135
Hamilton's Principle, 122, 129, 165, 368, 530
  admissible function, 130
Hamilton's Principle/Hamilton's Law of Varying Action, 158
Hamiltonian density equations of motion, 400, 454
Hamiltonian density function, 401
Hamiltonian differential operator, 405, 457
Hamiltonian dynamical system, 138, 154
Hamiltonian mechanics, 91
Hellinger-Reissner variational principle, 235
Helmholtz free energy, 205
Hessian matrix, 30
Hexahedral element, 325
Hilber-Hughes-Taylor-α algorithm, 515
Holonomic constraints, 62
Houbolt method, 508
Hu-Washizu variational principle, 233
Hydrostatic stress, 195
Hyperelastic material, 207
  compressible Neo-Hookean, 211
  incompressible Mooney-Rivlin, 213
  incompressible Neo-Hookean, 212
  St. Venant-Kirchhoff, 211

I

Initial configuration, 176
Internal dissipation, 205
Intrinsic coordinate system, 177
Isochoric materials, 200
Isothermal process, 206
Isotropic Elastic Material, 208

J

Jacobian, 179
Jacobian matrix, 179, 182
Jaumann stress rate, 218

K

Kepler's Problem, 643
Kinematically admissible displacement, 224
Kinetic energy in generalized coordinates, 84
Kirchhoff stress tensor, 195
Kirchhoff's equation of energy, 204
Kirchhoff-Love plate element, 336
Kirchhoff-Love thin plate, 285

L

Lagrange multiplier method, 86, 145, 152
Lagrange's equations of motion, 76
Lagrangian density equations of motion, 371, 449
Lagrangian density function, 371
Lagrangian description, 179
Lagrangian differential operator, 376, 452
Lagrangian dynamical system, 146, 152
Lagrangian flow, 137
Lagrangian form of D'Alembert's principle, 69, 122, 133, 136, 156, 163
Lagrangian mechanics, 60
Lagrangian strain tensor, 184
Lagrangian stress measure
  Second Piola-Kirchhoff stress tensor, 197
Left (spatial) stretch tensor, 183
Left Cauchy-Green tensor, 186
Legendre transformation, 93, 205, 230
Lie derivative, 192
Lie group, 172, 491
Linear multistep methods, 505
  Central difference method, 505
  Houbolt method, 508
  Park method, 514
Local form of balance of energy, *see* Kirchhoff's equation of energy, 204

M

Material acceleration, 180
Material coordinates, 177
Material frame-indifference, 214
Material point, 177
Material strain tensor, 184
Material velocity, 180
Material velocity gradient tensor, 183
Matrix
  adjoint of, 27

Matrix (*continued*)
cofactor of, 26
Cramer's rule, 25, 26
determinant, 25
identity matrix, 27
inverse of, 27
matrix and vector notation, 25
minor of, 26
rectangular matrix, 24
square matrix, 24
Mean value theorem, 33
Mechanical power, *see* Power input, 203
Metric tensor
reference configuration, 184
spatial configuration, 185
Midpoint method
First-order differential system, 502
Midpoint Rule [MPR-EPA], 509
Midpoint Rule [MPR-MPA], 521
Modified Bubnov-Galerkin weighted residual form, 269
Modified Bubnov-Galerkin weighted-residual method, 247
Modified Hamilton's principle, 131
Multibody dynamics, 650

N

Nanson's formula, 195
Natural coordinate system, 315
Newmark family of algorithms, 511
Newton's equations of motion, 58
Newton's laws of motion, 57
Newtonian dynamical system, 136, 141, 151
Newtonian path, 58
Noether's theorem, 171
Normalized time weighted residual approach
Numerically dissipative displacement, 622
Numerically dissipative hybrid displacement, 632
Numerically non-dissipative hybrid displacement-strain, 625
Normalized time weighted residual methodology, 579, 581
Numerical integration procedure, 305

O

Objective stress rates, 218
Cotter-Rivlin stress rate, 219
Green-Naghdi stress rate, 220
Jaumann stress rate, 218
Oldroyd stress rate, 218
Truesdell stress rate, 219
Objectivity, *see* Material frame-indifference, 214
Cauchy stress tensor, 216
co-rotational stress rates, 217
deformation gradient tensor, 216
Green deformation tensor, 216
rate of deformation tensor, 216
Oldroyd stress rate, 218
Optimal algorithm in LMS, 525
Orthogonal tensor, 46

P

Park method, 514
Pendulum
double planar pendulum, 75, 82
double spherical pendulum, 76
planar pendulum, 64, 75, 79, 87, 98, 132
spherical pendulum, 66, 76, 81, 98
Perfect elastic material, 206
Perfectly elastic material, 204
Phase space, 91
Piola deformation tensor, 185
Piola transformation, 226, 257
Polar decomposition theorem, 185
Power input, 203
Principal axes, 194
Principal function, 121, 124
Principal stresses, 194
Principle balance of angular momentum, 196, 202
Principle of balance of energy, 203
Principle of balance of linear momentum, 200
Principle of balance of mechanical energy, 150
N-body dynamical system, 166
N-body dynamical system, 149
Principle of complementary virtual work, 230
Principle of conservation of mass, 200
Principle of conservation of total energy, 145
Principle of entropy inequality, 205
Principle of minimum potential energy, 230
Principle of stationary complementary energy, 232
Principle of stationary potential energy, 228
Principle of virtual displacements, 225
Principle of virtual forces, 231
Principle of virtual power, 228
Principle of virtual work, 225, 226
continuous-body dynamical system, 223
Principle of virtual work in dynamics, 115
Principle of virtual work in statics, 111
Pseudo-force, 197
Pseudo-traction, 197
Pull-back operation, 191
Push-forward operation, 191

Q

Quadrilateral element, 314

R

Rate of deformation tensor, 187
Rate of entropy production, 205
Rate of rotation tensor, 188
Rayleigh-Ritz method, 237
Reference body, 176
Reference configuration, 176
Reissner-Mindlin plate, 294, 395, 420, 446, 472
Reissner-Mindlin plate element, 339
Reversible process, 206
Reynolds transport theorem, 198
Riemannian material metric, 176
Riemannian metrics, 177
Riemannian spatial metric, 176
Right (material) stretch tensor, 183
Right Cauchy-Green tensor, 184
Ritz coefficients, 239
Rotated rate of deformation tensor, 187
Rotating bar, 274, 384, 410, 440, 467
Rotation tensor, 183, 187
Runge-Kutta method, 503

S

Second Piola-Kirchhoff stress tensor, 197
Set, 15
 intersection, 16
 proper subset, 16
 union, 16
Single-field form, 571, 597, 639
 Time integration design, 560
Single-Step Methods, 500
 Backward Euler method, 501
 Forward Euler method, 501
 Midpoint method, 502
 Runge-Kutta method, 503
 Trapezoidal method, 502
Space-discrete Hamiltonian finite element formulation, 403, 456
Space-discrete Lagrangian finite element formulation, 374, 451
Space-discrete Total Energy finite element formulation, 436, 462
Spatial configuration, 176
Spatial coordinates, 177
Spatial point, 177
Spatial velocity gradient tensor, 183
Specific internal energy function, 205
Spectral decomposition
 of the Cauchy deformation tensor, 190
 of the deformation gradient tensor, 190
 of the Green deformation tensor, 189
 of the left stretch tensor, 190
 of the rotation tensor, 190
Spin tensor, 188
Spring-mass discrete system, 148
SSpj family of time stepping algorithms, 516
Symplectic manifold, 100
Symplectic structures, 100
Symplectic transformation, 102
Symplectic two-form, 100
Symplectic-momentum framework, 527, 585, 599, 614

T

Tangent bundle, 74
Taylor's series expansion theorem, 28, 30, 109, 125
Tensor
 change of basis, 46
 contravariant, 42
 covariant, 42
 determinant, 45
 deviatoric, 44
 dot product, 42
 dyad, 41
 dyadic, 41
 identity tensor, 42
 inverse of, 43
 metric tensor, 40
 mixed, 42
 principal invariants, 47
 scalar product, 43
 second-order tensor, 41
 spherical, 44
 trace, 45
 transpose of, 44
Tetrahedral element, 320
Theorem of power expended, 430
Three Parameters Optimal Schemes, 522
Time discretization, 632
Timoshenko beam, 280, 391, 417, 444, 470
Timoshenko beam element, 333
Total Energy density equations of motion, 435
Total Energy density function, 434, 459
Total Energy density representation of the equation of motion, 459
Total Energy differential operator, 437, 463
Total Energy equation of motion, 147, 157
Total Energy:N-body dynamical system, 145
Total Lagrangian formulation, 344
Triangular element, 307
Truesdell stress rate, 219
Truncation error
 Global truncation error, 501
 Local truncation error, 501

Two-field form, 569, 582, 636
  Time integration design, 558
Two-field variational principle, 131
Two-point tensor, 182, 185

U

U0 Family of Algorithms, 562, 572
Undeformed configuration, 176
Updated Lagrangian formulation, 347

V

U0-V0 Algorithms and Design, 565, 566, 572
U0 Family of Algorithms, 562, 572
U0-V0 Optimal Algorithms and Designs, 525, 565
V0 Family of Algorithms, 563, 572
Variational time integrators, 527, 585, 599, 614
Vector
  components of, 18
  contravariant, 39
  covariant, 39
  covector, 39
  dual space, 38
  linear dependence of, 19
  linear independence of, 19
Vector space, 17
  n-space, 19
  n-vectors, 19
  base vectors, 18
  basis of, 18
  coordinates, 18
  Euclidean n-space, 19, 23, 40
  Euclidean metric, 20
  inner product space, 20
  metric spaces, 22
  normed spaces, 23
  pre-Hilbert space, 21
Velocity based scheme, 521
Velocity strain tensor, 187
Virtual displacement
  continuous-body dynamical system, 221
  N-body dynamical system, 110
Virtual work, 220
  N-body dynamical systems, 108
Vorticity tensor, 188

W

Wilson-θ method, 513
Wood-Bosak-Zienkiewicz algorithm, 520